Mathematical Foundations of Infin
Statistical Models

In nonparametric and high-dimensional statistical models, the classical Gauss-Fisher-Le Cam theory of the optimality of maximum likelihood estimators and Bayesian posterior inference does not apply, and new foundations and ideas have been developed in the past several decades. This book gives a coherent account of the statistical theory in infinite-dimensional parameter spaces. The mathematical foundations include self-contained 'mini-courses' on the theory of Gaussian and empirical processes, on approximation and wavelet theory, and on the basic theory of function spaces. The theory of statistical inference in such models - hypothesis testing, estimation and confidence sets – is then presented within the minimax paradigm of decision theory. This includes the basic theory of convolution kernel and projection estimation, but also Bayesian nonparametrics and nonparametric maximum likelihood estimation. In a final chapter the theory of adaptive inference in nonparametric models is developed, including Lepski's method, wavelet thresholding, and adaptive inference for self-similar functions.

Winner of the 2017 PROSE Award for Mathematics

EVARIST GINÉ (1944–2015) was Head of the Department of Mathematics at the University of Connecticut. Giné was a distinguished mathematician who worked on mathematical statistics and probability in infinite dimensions. He was the author of two books and more than 100 articles.

RICHARD NICKL is Professor of Mathematical Statistics in the Statistical Laboratory within the Department of Pure Mathematics and Mathematical Statistics at the University of Cambridge.

CAMBRIDGE SERIES IN STATISTICAL AND PROBABILISTIC MATHEMATICS

This series of high-quality upper-division textbooks and expository monographs covers all aspects of stochastic applicable mathematics. The topics range from pure and applied statistics to probability theory, operations research, optimization and mathematical programming. The books contain clear presentations of new developments in the field and of the state of the art in classical methods. While emphasising rigorous treatment of theoretical methods, the books also contain applications and discussions of new techniques made possible by advances in computational practice.

A complete list of books in the series can be found at www.cambridge.org/statistics. Recent titles include the following:

14. *Statistical Analysis of Stochastic Processes in Time*, by J. K. Lindsey
15. *Measure Theory and Filtering*, by Lakhdar Aggoun and Robert Elliott
16. *Essentials of Statistical Inference*, by G. A. Young and R. L. Smith
17. *Elements of Distribution Theory*, by Thomas A. Severini
18. *Statistical Mechanics of Disordered Systems*, by Anton Bovier
19. *The Coordinate-Free Approach to Linear Models*, by Michael J. Wichura
20. *Random Graph Dynamics*, by Rick Durrett
21. *Networks*, by Peter Whittle
22. *Saddlepoint Approximations with Applications*, by Ronald W. Butler
23. *Applied Asymptotics*, by A. R. Brazzale, A. C. Davison and N. Reid
24. *Random Networks for Communication*, by Massimo Franceschetti and Ronald Meester
25. *Design of Comparative Experiments*, by R. A. Bailey
26. *Symmetry Studies*, by Marlos A. G. Viana
27. *Model Selection and Model Averaging*, by Gerda Claeskens and Nils Lid Hjort
28. *Bayesian Nonparametrics*, edited by Nils Lid Hjort et al.
29. *From Finite Sample to Asymptotic Methods in Statistics*, by Pranab K. Sen, Julio M. Singer and Antonio C. Pedrosa de Lima
30. *Brownian Motion*, by Peter Mörters and Yuval Peres
31. *Probability* (Fourth Edition), by Rick Durrett
32. *Stochastic Processes*, by Richard F. Bass
33. *Regression for Categorical Data*, by Gerhard Tutz
34. *Exercises in Probability* (Second Edition), by Loïc Chaumont and Marc Yor
35. *Statistical Principles for the Design of Experiments*, by R. Mead, S. G. Gilmour and A. Mead
36. *Quantum Stochastics*, by Mou-Hsiung Chang
37. *Nonparametric Estimation under Shape Constraints*, by Piet Groeneboom and Geurt Jongbloed
38. *Large Sample Covariance Matrices*, by Jianfeng Yao, Zhidong Bai and Shurong Zheng

Mathematical Foundations of Infinite-Dimensional Statistical Models

Evarist Giné

Richard Nickl
University of Cambridge

CAMBRIDGE
UNIVERSITY PRESS

CAMBRIDGE
UNIVERSITY PRESS

University Printing House, Cambridge CB2 8BS, United Kingdom

One Liberty Plaza, 20th Floor, New York, NY 10006, USA

477 Williamstown Road, Port Melbourne, VIC 3207, Australia

314–321, 3rd Floor, Plot 3, Splendor Forum, Jasola District Centre, New Delhi – 110025, India

79 Anson Road, #06–04/06, Singapore 079906

Cambridge University Press is part of the University of Cambridge.

It furthers the University's mission by disseminating knowledge in the pursuit of education, learning, and research at the highest international levels of excellence.

www.cambridge.org
Information on this title: www.cambridge.org/9781108994132
DOI: 10.1017/9781009022811

First published 2016
Revised edition 2021

Printed in the United Kingdom by TJ Books Limited, Padstow Cornwall

A catalogue record for this publication is available from the British Library.

ISBN 978-1-108-99413-2 Paperback

A la meva esposa Rosalind

Dem Andenken meiner Mutter Reingard, 1940–2010

Contents

Preface

The classical theory of statistics was developed for parametric models with *finite-dimensional parameter spaces*, building on fundamental ideas of C. F. Gauss, P. S. Laplace, R. A. Fisher and L. Le Cam, among others. It has been successful in providing modern science with a paradigm for making statistical inferences, in particular, in the 'frequentist large sample size' scenario. A comprehensive account of the mathematical foundations of this classical theory is given in the monograph by A. van der Vaart, *Asymptotic Statistics* (Cambridge University Press, 1998).

The last three decades have seen the development of statistical models that are infinite (or 'high') dimensional. The principal target of statistical inference in these models is a function or an infinite vector f that itself is not modelled further parametrically. Hence, these models are often called, in some abuse of terminology, *nonparametric models*, although f itself clearly also is a parameter. In view of modern computational techniques, such models are tractable and in fact attractive in statistical practice. Moreover, a mathematical theory of such nonparametric models has emerged, originally driven by the Russian school in the early 1980s and since then followed by a phase of very high international activity.

This book is an attempt to describe some elements of the mathematical theory of statistical inference in such *nonparametric*, or infinite-dimensional, models. We will first establish the main probabilistic foundations: the theory of Gaussian and empirical processes, with an emphasis on the 'nonasymptotic concentration of measure' perspective on these areas, including the pathbreaking work by M. Talagrand and M. Ledoux on concentration inequalities for product measures. Moreover, since a thorough understanding of infinite-dimensional models requires a solid background in functional analysis and approximation theory, some of the most relevant results from these areas, particularly the theory of wavelets and of Besov spaces, will be developed from first principles in this book.

After these foundations have been laid, we turn to the statistical core of the book. Comparing nonparametric models in a very informal way with classical parametric models, one may think of them as models in which the number of parameters that one estimates from the observations is *growing proportionally to sample size n* and has to be carefully selected by the statistician, ideally in a data-driven way. In practice, nonparametric modelling is often driven by the honesty of admitting that the traditional assumption that n is large compared to the number of unknown parameters is too strong. From a mathematical point of view, the frequentist theory that validates statistical inferences in such models undergoes a radical shift: leaving the world of finite-dimensional statistical models behind implies that the likelihood function no longer provides 'automatically optimal' statistical methods ('maximum likelihood estimators') and that extreme care has to be exercised when constructing inference procedures. In particular, the Gauss–Fisher–Le Cam efficiency theory based on the Fisher information typically yields nothing informative about what optimal procedures are in nonparametric statistics, and a new theoretical framework is required. We will show how the minimax paradigm can serve as a benchmark by which a theory of optimality in nonparametric models can be developed. From this paradigm arises the 'adaptation' problem, whose solution has been perhaps one of the major achievements of the theory of nonparametric statistics and which will be presented here for nonparametric function estimation problems. Finally, likelihood-based procedures can be relevant in nonparametric models as well, particularly after some regularisation step that can be incorporated by adopting a 'Bayesian' approach or by imposing qualitative a priori shape constraints. How such approaches can be analysed mathematically also will be shown here.

Our presentation of the main statistical materials focusses on function estimation problems, such as density estimation or signal in white-noise models. Many other nonparametric models have similar features but are formally different. Our aim is to present a unified statistical theory for a canonical family of infinite-dimensional models, and this comes at the expense of the breadth of topics that could be covered. However, the mathematical mechanisms described here also can serve as guiding principles for many nonparametric problems not covered in this book.

Throughout this book, we assume familiarity with material from real and functional analysis, measure and probability theory on the level of a US graduate course on the subject. We refer to the monographs by G. Folland, *Real Analysis* (Wiley, 1999), and R. Dudley, *Real Analysis and Probability* (Cambridge University Press, 2002), for relevant background. Apart from this, the monograph is self-contained, with a few exceptions and 'starred sections' indicated in the text.

This book would not have been possible without the many colleagues and friends from whom we learnt, either in person or through their writings. Among them, we would like to thank P. Bickel, L. Birgé, S. Boucheron, L. Brown, T. Cai, I. Castillo, V. Chernozhukov, P. Dawid, L. Devroye, D. Donoho, R. Dudley, L. Dümbgen, U. Einmahl, X. Fernique, S. Ghosal, A. Goldenshluger, Y. Golubev, M. Hoffmann, I. Ibragimov, Y. Ingster, A. Iouditski, I. Johnstone, G. Kerkyacharian, R. Khasminskii, V. Koltchinskii, R. Latala, M. Ledoux, O. Lepski, M. Low, G. Lugosi, W. Madych, E. Mammen, D. Mason, P. Massart, M. Nussbaum, D. Picard, B. Pötscher, M. Reiß, P. Rigollet, Y. Ritov, R. Samworth, V. Spokoiny, M. Talagrand, A. Tsybakov, S. van de Geer, A. van der Vaart, H. van Zanten, J. Wellner, H. Zhou and J. Zinn.

We are grateful to A. Carpentier, I. Castillo, U. Einmahl, D. Gauthier, D. Heydecker, K. Ray, J. Söhl and B. Szabò for proofreading parts of the manuscript and providing helpful corrections.

Moreover, we are indebted to Diana Gillooly of Cambridge University Press for her support, patience and understanding in the process of this book project since 2011.

R.N. would also like to thank his friends N. Berestycki, C. Damböck, R. Dawid and M. Neuber for uniquely stimulating friendships that have played a large role in the intellectual development that led to this book (and beyond).

Outline and Reading Guide

In principle, all the chapters of this book can be read independently. In particular, the chapters on Gaussian and empirical processes, as well as the one on function spaces and approximation theory, are mostly self-contained. A reader interested primarily in the 'statistical chapters' (5 through 8) may choose to read those first and then turn to the mathematical foundations laid out in Chapters 2 through 4 later, when required. A short outline of the contents of each chapter is given in the following paragraphs:

Chapter 1 introduces the kinds of statistical models studied in this book. In particular, we will discuss why many common 'regular' regression models with normally distributed error terms can be mathematically accommodated within one Gaussian function estimation problem known as the *Gaussian white noise model.*

Chapters 2 and 3 lay the probabilistic foundations of much of the statistical theory that follows: one chapter on Gaussian processes and one on empirical processes. The Gaussian theory is mostly classical, presented with a focus on statistically relevant materials, such as the isoperimetric inequality for Gaussian measures and its consequences on concentration, as well as a study of suprema of Gaussian processes. The theory for empirical measures reflects the striking recent developments around the concentration-of-measure phenomenon. Effectively, here, the classical role of the central limit theorem in statistics is replaced by nonasymptotic concentration properties of product measures, as revealed in fundamental work by Talagrand, Ledoux, Massart and others. This is complemented by a treatment of abstract empirical process theory, including metric entropy methods, Vapnik-Červonenkis classes and uniform central limit theorems.

Chapter 4 develops from first principles some key aspects of approximation theory and its functional analytic foundations. In particular, we give an account of wavelet theory and of Besov spaces, with a focus on results that are relevant in subsequent chapters.

Chapter 5 introduces basic linear estimation techniques that are commonly used in nonparametric statistics, based on convolution kernels and finite-dimensional projection operators. Tools from Chapters 3 and 4 are used to derive a variety of probabilistic results about these estimators that will be useful in what follows.

Chapter 6 introduces a theoretical paradigm – the *minimax paradigm* – that can be used to objectively measure the performance of statistical methods in nonparametric models. The basic information-theoretic ideas behind it are developed, and it is shown how statistical inference procedures – estimators, tests and confidence sets – can be analysed and compared from a minimax point of view. For a variety of common nonparametric models, concrete constructions of minimax optimal procedures are given using the results from previous chapters.

Chapter 7 shows how the likelihood function can still serve as a successful guiding principle in certain nonparametric problems if a priori information is used carefully. This can be done by imposing certain qualitative constraints on the statistical model or by formally adopting a Bayesian approach which then can be analysed from a frequentist point of view. The key role of the Hellinger distance in this theory (as pointed out in work by Le Cam, Birgé, van de Geer, van der Vaart and others) is described in some detail.

Chapter 8 presents the solution to the nonparametric adaptation problem that arises from the minimax paradigm and gives a theory of statistical inference for 'fully automatic' statistical procedures that perform well over maximal collections of nonparametric statistical models. Surprising differences are shown to arise when considering the existence of adaptive estimation procedures in contrast to the existence of associated adaptive confidence sets. A resolution of this discrepancy can be obtained by considering certain nonparametric models of 'self-similar' functions, which are discussed in some detail and for which a unified theory of optimal statistical inference can be developed.

Each chapter is organised in several sections, and historical notes complementing each section can be found at the end of each chapter – these are by no means exhaustive and only indicate our understanding of the literature.

At the end of each section, exercises are provided: these, likewise, complement the main results of the text and often indicate interesting applications or extensions of the materials presented.

Postscript

It is a terrible tragedy that Evarist Giné passed away shortly after we completed the manuscript. His passion for mathematics was exceeded only by his love for his wife, Rosalind; his daughters, Núria and Roser; and his grandchildren, Liam and Mireia. He mentioned to me in September 2014, when I last met him in Cambridge (MA), that perhaps he wanted to dedicate this book to all of them, but in an e-mail to me in January 2015, he mentioned explicitly that he wanted it to be for Rosalind. I have honoured his decision; however, I know that with this last work he wanted to thank all of them for having been his wonderful family – who continue his infectious passion into new generations.

I am myself deeply grateful to my father, Harald, for all his support and inspiration throughout my life in all domains. I dedicate this book to the memory of my mother, Reingard, in loving gratitude for all her courage and everything she has done for me. And of course, insofar as this book relates to the future, it is for Ana and our son, Julian, with love and affection.

Postscript (2020)

In this paperback edition a large number of (mostly minor) corrections have been incorporated. I would like to thank the various readers and students, specifically Kweku Abraham, who pointed them out to me.

1

Nonparametric Statistical Models

In this chapter we introduce and motivate the statistical models that will be considered in this book. Some of the materials depend on basic facts developed in subsequent chapters – mostly the basic Gaussian process and Hilbert space theory. This will be hinted at when necessary.

Very generally speaking, a *statistical model* for a random observation Y is a family

$$\{P_f : f \in \mathcal{F}\}$$

of probability distributions P_f, each of which is a candidate for having generated the observation Y. The parameter f belongs to the *parameter space* $\mathcal{F}$. The problem of *statistical inference* on f, broadly speaking, can be divided into three intimately connected problems of using the observation Y to

(a) *Estimate* the parameter f by an estimator $T(Y)$,
(b) *Test hypotheses* on f based on test functions $\Psi(Y)$ and/or
(c) *Construct confidence sets* $C(Y)$ that contain f with high probability.

To interpret inferential results of these kinds, we will typically need to specify a distance, or loss function on $\mathcal{F}$, and for a given model, different loss functions may or may not lead to very different conclusions.

The statistical models we will introduce in this chapter are, on the one hand, conceptually closely related to each other in that the parameter space $\mathcal{F}$ is infinite or high dimensional and the loss functions relevant to the analysis of the performance of statistical procedures are similar. On the other hand, these models are naturally divided by the different probabilistic frameworks in which they occur – which will be either a *Gaussian noise model* or an *independent sampling model*. These frameworks are asymptotically related in a fundamental way (see the discussion after Theorem 1.2.1). However, the most effective probabilistic techniques available are based on a direct, nonasymptotic analysis of the Gaussian or product probability measures that arise in the relevant sampling context and hence require a separate treatment.

Thus, while many of the statistical intuitions are common to both the sampling and the Gaussian noise models and in fact inform each other, the probabilistic foundations of these models will be laid out independently.

1.1 Statistical Sampling Models

Let X be a random experiment with associated sample space $\mathcal{X}$. We take the mathematical point of view of probability theory and model X as a random variable, that is, as a measurable mapping defined on some underlying probability space that takes values in the measurable space $(\mathcal{X},\mathcal{A})$, where $\mathcal{A}$ is a σ-field of subsets of $\mathcal{X}$. The law of X is described by the probability measure P on $\mathcal{A}$. We may typically think of $\mathcal{X}$ equal to $\mathbb{R}^d$ or a measurable subset thereof, equipped with its Borel σ-field $\mathcal{A}$.

The perhaps most basic problem of statistics is the following: consider repeated outcomes of the experiment X, that is, a random sample of independent and identically distributed (i.i.d.) copies $X_1,\dots,X_n$ from X. The joint distribution of the X_i equals the product probability measure $P^n = \otimes_{i=1}^n P$ on $(\mathcal{X}^n,\mathcal{A}^n)$. The goal is to recover P from the n observations. 'Recovering P' can mean many things. Classical statistics has been concerned mostly with models where P is explicitly parameterised by a finite-dimensional parameter, such as the mean and variance of the normal distribution, or the 'parameters' of the usual families of statistical distributions (gamma, beta, exponential, Poisson, etc.). Recovering P then simply means to use the observations to make inferences on the unknown parameter, and the fact that this parameter is finite dimensional is crucial for this traditional paradigm of statistical inference, in particular, for the famous likelihood principle of R. A. Fisher. In this book, we will follow the often more realistic assumption that no such parametric assumptions are made on P. For most sample spaces $\mathcal{X}$ of interest, this will naturally lead to models that are infinite dimensional, and we will investigate how the theory of statistical inference needs to be developed in this situation.

1.1.1 Nonparametric Models for Probability Measures

In its most elementary form, without imposing any parameterisations on P, we can simply consider the problem of making inferences on the unknown probability measure P based on the sample. Natural loss functions arise from the usual metrics on the space of probability measures on $\mathcal{X}$, such as the total variation metric

$$\|P-Q\|_{TV} = \sup_{A\in\mathcal{A}} |P(A)-Q(A)|$$

or weaker metrics that generate the topology of weak convergence of probability measures on $\mathcal{X}$. For instance, if $\mathcal{X}$ itself is endowed with a metric d, we could take the bounded Lipschitz metric

$$\beta_{(\mathcal{X},d)}(P,Q) = \sup_{f\in BL(1)} \left| \int_{\mathcal{X}} f(dP-dQ) \right|$$

for weak convergence of probability measures, where

$$BL(M) = \left\{ f:\mathcal{X}\to\mathbb{R},\ \sup_{x\in\mathcal{X}}|f(x)| + \sup_{x\neq y}\frac{|f(x)-f(y)|}{d(x,y)} \le M \right\}, \quad 0<M<\infty.$$

If $\mathcal{X}$ has some geometric structure, we can consider more intuitive loss functions. For example, if $\mathcal{X}=\mathbb{R}$, we can consider the cumulative distribution function

$$F(x) = P(X\le x), \quad x\in\mathbb{R},$$

or, if X takes values in $\mathbb{R}^d$, its multivariate analogue. A natural distance function on distribution functions is simply the supremum-norm metric ('Kolmogorov distance')

$$\|F_P - F_Q\|_\infty = \sup_{x\in\mathbb{R}} |F_P(x) - F_Q(x)|.$$

Since the indicators $\{1_{(-\infty,x]} : x \in \mathbb{R}\}$ generate the Borel σ-field of $\mathbb{R}$, we see that, on $\mathbb{R}$, the statistical parameter P is characterised entirely by the functional parameter F, and vice versa. The parameter space is thus the infinite-dimensional space of all cumulative distribution functions on $\mathbb{R}$.

Often we will know that P has some more structure, such as that P possesses a probability-density function $f : \mathbb{R} \to [0,\infty)$, which itself may have further properties that will be seen to influence the complexity of the statistical problem at hand. For probability-density functions, a natural loss function is the L^1-distance

$$\|f_P - f_Q\|_1 = \int_{\mathbb{R}} |f_P(x) - f_Q(x)|dx$$

and in some situations also other L^p-type and related loss functions. Although in some sense a subset of the other, the class of probability densities is more complex than the class of probability-distribution functions, as it is not described by monotonicity constraints and does not consist of functions bounded in absolute value by 1. In a heuristic way, we can anticipate that estimating a probability density is harder than estimating the distribution function, just as the preceding total variation metric is stronger than any metric for weak convergence of probability measures (on nontrivial sample spaces $\mathcal{X}$). In all these situations, we will see that the theory of statistical inference on the parameter f significantly departs from the usual finite-dimensional setting.

Instead of P, a particular functional $\Phi(P)$ may be the parameter of statistical interest, such as the moments of P or the quantile function F^{-1} of the distribution function F – examples for this situation are abundant. The nonparametric theory is naturally compatible with such functional estimation problems because it provides the direct plug-in estimate $\Phi(T)$ based on an estimator T for P. Proving closeness of T to P in some strong loss function then gives access to 'many' continuous functionals Φ for which $\Phi(T)$ will be close to $\Phi(P)$, as we shall see later in this book.

1.1.2 Indirect Observations

A common problem in statistical sampling models is that some systematic measurement errors are present. A classical problem of this kind is the statistical regression problem, which will be introduced in the next section. Another problem, which is more closely related to the sampling model from earlier, is where one considers observations in $\mathbb{R}^d$ of the form

$$Y_i = X_i + \varepsilon_i, \quad i = 1,\ldots,n, \tag{1.1}$$

where the X_i are i.i.d. with common law P_X, and the ε_i are random 'error' variables that are independent of the X_i and have law P_ε. The law P_ε is assumed to be known to the observer – the nature of this assumption is best understood by considering examples: the attempt is to model situations in which a scientist, for reasons of cost, complexity or lack of precision of the involved measurement device, is forced to observe Y_i instead of the realisations X_i of interest. The observer may, however, have very concrete knowledge of the source of the error, which could, for example, consist of light emissions of the Milky Way interfering with cosmic rays from deeper space, an erratic optical device through which images are observed (e.g., a space telescope which cannot be repaired except at very high cost) or transmissions of signals through a very busy communication channel. Such situations of implicit measurements are encountered frequently in the applied sciences and are often called *inverse problems*, as one wishes to 'undo' the errors inflicted on the signal in which one is interested. The model (1.1) gives a simple way to model the main aspects of such statistical inverse problems. It is also known as the *deconvolution model* because the law of the Y_i equals

$$P_Y = P_X * P_\varepsilon,$$

the convolution of the two probability measures P_X, P_ε, and one wishes to 'deconvolve' P_ε.

As earlier, we will be interested in inference on the underlying distribution P_X of the signal X when the statistical model for P_X is infinite dimensional. The loss functions in this problem are thus typically the same as in the preceding subsection.

1.2 Gaussian Models

The randomness in the preceding sampling model was encoded in a general product measure P^n describing the joint law of the observations. Another paradigm of statistical modelling deals with situations in which the randomness in the model is described by a Gaussian (normal) distribution. This paradigm naturally encompasses a variety of nonparametric models, where the infinite-dimensional character of the problem does not necessarily derive from the probabilistic angle but from a functional relationship that one wishes to model.

1.2.1 Basic Ideas of Regression

Perhaps the most natural occurrence of a statistical model in the sciences is the one in which observations, modelled here as numerical values or vectors, say, (Y_i, x_i), arise according to a functional relationship

$$Y_i = f(x_i) + \varepsilon_i, \quad i = 1, \dots, n, \tag{1.2}$$

where n is the number of observations (sample size), f is some function of the x_i and the ε_i are random noise. By 'random noise', we may mean here either a probabilistic model for certain measurement errors that we believe to be intrinsic to our method of making observations, or some innate stochastic nature of the way the Y_i are generated from the $f(x_i)$. In either case, we will model the ε_i as random variables in the sense of axiomatic probability theory – the question of the genuine physical origin of this random noise will not concern us here. It is sometimes natural to assume also that the x_i are realisations of random variables X_i – we can either take this into account explicitly in our analysis or make statements conditional on the observed values $X_i = x_i$.

The function f often will be unknown to the observer of observations (Y_i, x_i), and the goal is to recover f from the (Y_i, x_i). This may be of interest for various reasons, for instance, for predicting new values Y_{n+1} from $f(x_{n+1})$ or to gain quantitative and qualitative understanding of the functional relationship $Y_i = f(x_i)$ under consideration.

In the preceding context, a statistical model in the broad sense is an a priori specification of both a parameter space for the functions f that possibly could have generated (1.2) and a family of probability measures that describes the possible distributions of the random variables ε_i. By 'a priori', we mean here that this is done independently of (e.g., before) the observational process, reflecting the situation of an experimentalist.

A systematic use and study of such models was undertaken in the early nineteenth century by Carl Friedrich Gauss, who was mostly interested in predicting astronomical observations. When the model is translated into the preceding formalisation, Gauss effectively assumed that the x_i are vectors $(x_{i1},\ldots,x_{ip})^T$ and thought of f as a linear function in that vector, more precisely,

$$f(x_i) = x_{i1}\theta_i + \ldots x_{ip}\theta_p, \quad i = 1,\ldots,n,$$

for some real-valued parameters $\theta_j, j = 1,\ldots,p$. The parameter space for f is thus the Euclidean space $\mathbb{R}^p$ expressed through all such linear mappings. In Gauss's time, the assumption of linearity was almost a computational necessity.

Moreover, Gauss modelled the random noise ε_i as independent and identically distributed samples from a normal distribution $N(0,\sigma^2)$ with some variance σ^2. His motivation behind this assumption was twofold. First, it is reasonable to assume that $E(\varepsilon_i) = 0$ for every i. If this expectation were nonzero, then there would be some deterministic, or 'systematic', measurement error $e_i = E(\varepsilon_i)$ of the measurement device, and this could always be accommodated in the functional model by adding a constant $x_{10} = \cdots = x_{n0} = 1$ to the preceding linear relationship. The second assumption that ε_i has a normal distribution is deeper. If we think of each measurement error ε_i as the sum of many 'very small', or infinitesimal, independent measurement errors $\varepsilon_{ik}, k = 1,2,\ldots$, then, by the central limit theorem, $\varepsilon_i = \sum_k \varepsilon_{ik}$ should be approximately normally distributed, regardless of the actual distribution of the ε_{ik}. By the same reasoning, it is typically natural to assume that the ε_i are also independent among themselves. This leads to what is now called the *standard Gaussian linear model*

$$Y_i = f(x_i) + \varepsilon_i \equiv \sum_{j=1}^{p} x_{ij}\theta_j + \varepsilon_i, \qquad \varepsilon_i \sim^{i.i.d.} N(0,\sigma^2), \quad i = 1,\ldots,n, \tag{1.3}$$

which bears this name both because Gauss studied it and, since the $N(0,\sigma^2)$ distribution is often called the *Gaussian distribution*, because Gauss first made systematic use of it. The unknown parameter (θ,σ^2) varies in the $(p+1)$-dimensional parameter space

$$\Theta \times \Sigma = \mathbb{R}^p \times (0,\infty).$$

This model constitutes perhaps *the* classical example of a *finite-dimensional model*, which has been studied extensively and for which a fairly complete theory is available. For instance, when p is smaller than n, the least-squares estimator of Gauss finds the value $\hat{\theta} \in \mathbb{R}^p$ which solves the optimisation problem

$$\min_{\theta \in \mathbb{R}^p} \sum_{i=1}^{n} \left(Y_i - \sum_{j=1}^{p} x_{ij}\theta_j \right)^2$$

and hence minimises the Euclidean distance of the vector $Y = (Y_1,\ldots,Y_n)^T$ to the p-dimensional subspace spanned by the p vectors $(x_{1j},\ldots,x_{nj})^T, j = 1,\ldots,p$.

1.2.2 Some Nonparametric Gaussian Models

We now give a variety of models that generalise Gauss's ideas to infinite-dimensional situations. In particular, we will introduce the Gaussian white noise model, which serves as a generic surrogate for a large class of nonparametric models, including even non-Gaussian ones, through the theory of equivalence of experiments (discussed in the next section).

Nonparametric Gaussian Regression

Gauss's model and its theory basically consist of two crucial assumptions: one is that the ε_i are normally distributed, and the other is that the function f is linear. The former assumption was argued to be in some sense natural, at least in a measurement-error model (see also the remarks after Theorem 1.2.1 for further justification). The latter assumption is in principle quite arbitrary, particularly in times when computational power does not constrain us as much any longer as it did in Gauss's time. A nonparametric approach therefore attempts to assume as little structure of f as possible. For instance, by the *nonparametric regression model with fixed, equally spaced design on* $[0,1]$, we shall understand here the model

$$Y_i = f(x_i) + \varepsilon_i, \qquad x_i = \frac{i}{n}, \qquad \varepsilon_i \sim^{i.i.d.} N(0,\sigma^2), \quad i = 1,\dots,n. \tag{1.4}$$

where f is any function defined on $[0,1]$. We are thus sampling the unknown function f at an equally spaced grid of $[0,1]$ that, as $n \to \infty$, grows dense in the interval $[0,1]$ as $n \to \infty$.

The model immediately generalises to bounded intervals $[a,b]$, to 'approximately' equally spaced designs $\{x_i : i = 1,\dots,n\} \subset [a,b]$ and to multivariate situations, where the x_i are equally spaced points in some hypercube. We note that the assumption that the x_i are equally spaced is important for the theory that will follow – this is natural as we cannot hope to make inference on f in regions that contain no or too few observations x_i.

Other generalisations include the *random design regression model*, in which the x_i are viewed as i.i.d. copies of a random variable X. One can then either proceed to argue conditionally on the realisations $X_i = x_i$, or one takes this randomness explicitly into account by making probability statements under the law of X and ε simultaneously. For reasonable design distributions, this will lead to results that are comparable to the fixed-design model – one way of seeing this is through the equivalence theory for statistical experiments (see after Theorem 1.2.1).

A priori it may not bc rcasonablc to assume that f has any specific properties other than that it is a continuous or perhaps a differentiable function of its argument. Even if we assumed that f has infinitely many continuous derivatives the set of all such f would be infinite dimensional and could never be fully captured by a p-dimensional parameter space. We thus have to expect that the theory of statistical inference in this nonparametric model will be different from the one in Gauss's classical linear model.

The Gaussian White Noise Model

For the mathematical development in this book we shall work with a mathematical idealisation of the regression model (1.4) in continuous time, known as the *Gaussian white noise model*, and with its infinite sequence space analogue. While perhaps at first appearing more complicated than the discrete model, once constructed, it allows for a clean and intuitive mathematical exposition that mirrors all the main ideas and challenges of the discrete case with no severe loss of generality.

Consider the following stochastic differential equation:

$$dY(t) \equiv dY_f^{(n)}(t) = f(t)dt + \frac{\sigma}{\sqrt{n}}dW(t), \qquad t \in [0,1], \quad n \in \mathbb{N}, \tag{1.5}$$

where $f \in L^2 \equiv L^2([0,1])$ is a square integrable function on $[0,1]$, $\sigma > 0$ is a dispersion parameter and dW is a *standard Gaussian white noise process*. When we observe a realisation of (1.5), we shall say that we observe the function or signal f in Gaussian white noise, at the noise level, or a signal-to-noise ratio $\sigma/\sqrt{n}$. We typically think of n large, serving as a proxy for sample size, and of $\sigma > 0$ a fixed known value. If σ is unknown, one can usually replace it by a consistent estimate in the models we shall encounter in this book.

The exact meaning of dW needs further explanation. Heuristically, we may think of dW as a weak derivative of a standard Brownian motion $\{W(t) : t \in [0,1]\}$, whose existence requires a suitable notion of stochastic derivative that we do not want to develop here explicitly. Instead, we take a 'stochastic process' approach to define this stochastic differential equation, which for statistical purposes is perfectly satisfactory. Let us thus agree that 'observing the trajectory (1.5)' will simply mean that we observe a realisation of the Gaussian process defined by the application

$$g \mapsto \int_0^1 g(t)dY^{(n)}(t) \equiv \mathbb{Y}_f^{(n)}(g) \sim N\left(\langle f,g\rangle, \frac{\|g\|_2^2}{n}\right), \tag{1.6}$$

where g is any element of the Hilbert space $L^2([0,1])$ with inner product $\langle \cdot,\cdot \rangle$ and norm $\|\cdot\|_2$. Even more explicitly, we observe all the $N(\langle f,g\rangle, \|g\|_2^2/n)$ variables, as g runs through $L^2([0,1])$. The randomness in the equation (1.5) comes entirely from the additive term dW, so after translating by $\langle f,g\rangle$ and scaling by $1/\sqrt{n}$, this means that dW is defined through the Gaussian process obtained from the action

$$g \mapsto \int_0^1 g(t)dW(t) \equiv \mathbb{W}(g) \sim N(0, \|g\|_2^2), \quad g \in L^2([0,1]). \tag{1.7}$$

Note that this process has a diagonal covariance in the sense that for any *finite* set of orthonormal vectors $\{e_k\} \subset L^2$ we have that the family $\{\mathbb{W}(e_k)\}$ is a multivariate standard normal variable, and as a consequence of the Kolmogorov consistency theorem (Proposition 2.1.10), $\mathbb{W}$ and $\mathbb{Y}^{(n)}$ indeed define Gaussian processes on L^2.

The fact that the model (1.5) can be interpreted as a Gaussian process indexed by L^2 means that the natural sample space $\mathcal{Y}$ in which dY from (1.5) takes values is the 'path' space $\mathbb{R}^{L^2([0,1])}$. This space may be awkward to work with in practice. In Section 6.1.1 we shall show that we can find more tractable choices for $\mathcal{Y}$ where dY concentrates with probability 1.

Gaussian Sequence Space Model

Again, to observe the stochastic process $\{\mathbb{Y}_f^{(n)}(g) : g \in L^2\}$ just means that we observe $\mathbb{Y}_f^{(n)}(g)$ for all $g \in L^2$ simultaneously. In particular, we may pick any orthonormal basis $\{e_k : k \in \mathbb{Z}\}$ of L^2, giving rise to an observation in the *Gaussian sequence space model*

$$Y_k \equiv Y_{f,k}^{(n)} = \langle f, e_k\rangle + \frac{\sigma}{\sqrt{n}}g_k, \qquad k \in \mathbb{Z}, \quad n \in \mathbb{N}, \tag{1.8}$$

where the g_k are i.i.d. of law $\mathbb{W}(e_k) \sim N(0, \|e_k\|_2^2) = N(0,1)$. Here we observe all the basis coefficients of the unknown function f with additive Gaussian noise of variance σ^2/n. Note that since the $\{e_k : k \in \mathbb{Z}\}$ realise a sequence space isometry between L^2 and the sequence space ℓ_2 of all square-summable infinite sequences through the mapping $f \mapsto \langle f, e_k \rangle$, the law of $\{Y_{f,k}^{(n)} : k \in \mathbb{Z}\}$ completely characterises the finite-dimensional distributions, and thus the law, of the process $\mathbb{Y}_f^{(n)}$. Hence, models (1.5) and (1.8) are observationally equivalent to each other, and we can prefer to work in either one of them (see also Theorem 1.2.1).

We note that the random sequence $Y = (Y_k : k \in \mathbb{Z})$ itself does not take values in ℓ_2, but we can view it as a random variable in the 'path' space $\mathbb{R}^{\ell_2}$. A more tractable, separable sample space on which $(Y_k : k \in \mathbb{Z})$ can be realised is discussed in Section 6.1.1.

A special case of the Gaussian sequence model is obtained when the space is restricted to n coefficients

$$Y_k = \theta_k + \frac{\sigma}{\sqrt{n}} g_k, \quad k = 1, \dots, n, \tag{1.9}$$

where the θ_k are equal to the $\langle f, e_k \rangle$. This is known as the *normal means model*. While itself a finite-dimensional model, it cannot be compared to the standard Gaussian linear model from the preceding section as its dimension increases as fast as n. In fact, for most parameter spaces that we will encounter in this book, the difference between model (1.9) and model (1.8) is negligible, as follows, for instance, from inspection of the proof of Theorem 1.2.1.

Multivariate Gaussian Models

To define a Gaussian white noise model for functions of several variables on $[0,1]^d$ through the preceding construction is straightforward. We simply take, for $f \in L^2([0,1]^d)$,

$$dY(t) = f(t)dt + \frac{\sigma}{\sqrt{n}} dW(t), \qquad t \in [0,1]^d, \quad n \in \mathbb{N}, \quad \sigma > 0, \tag{1.10}$$

where dW is defined through the action

$$g \mapsto \int_{[0,1]^d} g(t) dW(t) \equiv \mathbb{W}(g) \sim N(0, \|g\|_2^2) \tag{1.11}$$

on elements g of $L^2([0,1]^d)$, which corresponds to multivariate stochastic integrals with respect to independent Brownian motions $W_1(t_1), \dots, W_d(t_d)$. Likewise, we can reduce to a sequence space model by taking an orthonormal basis $\{e_k : k \in \mathbb{Z}^d\}$ of $L^2([0,1]^d)$.

1.2.3 Equivalence of Statistical Experiments

It is time to build a bridge between the preceding abstract models and the statistically more intuitive nonparametric fixed-design regression model (1.4). Some experience with the preceding models reveals that a statistical inference procedure in any of these models constructively suggests a procedure in the others with comparable statistical properties. Using a suitable notion of distance between statistical experiments, this intuition can be turned into a theorem, as we show in this subsection. We present results for Gaussian regression models; the general approach, however, can be developed much further to show that even highly non-Gaussian models can be, in a certain sense, asymptotically equivalent to the standard Gaussian white noise model (1.5). This gives a general justification for a rigorous study of the Gaussian white noise model in itself. Some of the proofs in this subsection require material from subsequent chapters, but the main ideas can be grasped without difficulty.

The Le Cam Distance of Statistical Experiments

We employ a general notion of distance between statistical experiments $\mathcal{E}^{(i)}, i = 1,2$, due to Le Cam. Each experiment $\mathcal{E}^{(i)}$ consists of a sample space $\mathcal{Y}_i$ and a probability measure $P_f^{(i)}$ defined on it, indexed by a common parameter $f \in \mathcal{F}$. Let $\mathcal{T}$ be a measurable space of 'decision rules', and let

$$L : \mathcal{F} \times \mathcal{T} \to [0,\infty)$$

be a 'loss function' measuring the performance of a decision procedure $T^{(i)}(Y^{(i)}) \in \mathcal{T}$ based on observations $Y^{(i)}$ in experiment i. For instance, $T^{(i)}(Y^{(i)})$ could be an estimator for f so that $\mathcal{T} = \mathcal{F}$ and $L(f,T) = d(f,T)$, where d is some metric on $\mathcal{F}$, but other scenarios are possible. The risk under $P_f^{(i)}$ for this loss is the $P_f^{(i)}$-expectation of $L(f,T^{(i)}(Y^{(i)}))$, denoted by $R^{(i)}(f,T^{(i)},L)$. Define also

$$|L| = \sup\{L(f,T) : f \in \mathcal{F}, T \in \mathcal{T}\}.$$

The *Le Cam distance* between two experiments is defined as

$$\Delta_{\mathcal{F}}(\mathcal{E}^{(1)},\mathcal{E}^{(2)}) \equiv \max\left[\sup_{T^{(2)}}\inf_{T^{(1)}}\sup_{f,L:|L|=1}\left|R^{(1)}(f,T^{(1)},L) - R^{(2)}(f,T^{(2)},L)\right|,\right. \tag{1.12}$$

$$\left.\sup_{T^{(1)}}\inf_{T^{(2)}}\sup_{f,L:|L|=1}\left|R^{(1)}(f,T^{(1)},L) - R^{(2)}(f,T^{(2)},L)\right|\right].$$

If this quantity equals zero, this means that any decision procedure $T^{(1)}$ in experiment $\mathcal{E}^{(1)}$ can be translated into a decision procedure $T^{(2)}$ in experiment $\mathcal{E}^{(2)}$, and vice versa, and that the statistical performance of these procedures in terms of the associated risk $R^{(i)}$ will be the same for any bounded loss function L. If the distance is not zero but small, then, likewise, the performance of the corresponding procedures in both experiments will differ by at most their Le Cam distance.

Some useful observations on the Le Cam distance are the following: if both experiments have a common sample space $\mathcal{Y}^{(1)} = \mathcal{Y}^{(2)} = \mathcal{Y}$ equal to a complete separable metric space, and if the probability measures $P_f^{(1)}, P_f^{(2)}$ have a common dominating measure μ on $\mathcal{Y}$, then

$$\Delta_{\mathcal{F}}(\mathcal{E}^{(1)},\mathcal{E}^{(2)}) \le \sup_{f\in\mathcal{F}}\int_{\mathcal{Y}}\left|\frac{dP_f^{(1)}}{d\mu} - \frac{dP_f^{(2)}}{d\mu}\right| d\mu \equiv \|P^{(1)} - P^{(2)}\|_{1,\mu,\mathcal{F}}. \tag{1.13}$$

This follows from the fact that in this case we can always use the decision rule $T^{(2)}(Y)$ in experiment $\mathcal{E}^{(1)}$ and vice versa and from

$$|R^{(1)}(f,T,L) - R^{(2)}(f,T,L)| \le \int_{\mathcal{Y}} |L(f,T(Y))||dP_f^{(1)}(Y) - dP_f^{(2)}(Y)| \le |L|\|P^{(1)} - P^{(2)}\|_{1,\mu,\mathcal{F}}.$$

The situation in which the two experiments are not defined on the sample space needs some more thought. Suppose, in the simplest case, that we can find a bi-measurable isomorphism B of $\mathcal{Y}^{(1)}$ with $\mathcal{Y}^{(2)}$, independent of f, such that

$$P_f^{(2)} = P_f^{(1)} \circ B^{-1}, \qquad P_f^{(1)} = P_f^{(2)} \circ B \quad \forall f \in \mathcal{F}.$$

Then, given observations $Y^{(2)}$ in $\mathcal{Y}^{(2)}$, we can use the decision rule $T^{(2)}(Y^{(2)}) \equiv T^{(1)}(B^{-1}(Y^{(2)}))$ in $\mathcal{E}^{(2)}$, and vice versa, and the risks $R^{(i)}$ in both experiments coincide by the image measure theorem. We can conclude in this case that

$$\Delta_{\mathcal{F}}(\mathcal{E}^{(1)}, \mathcal{E}^{(2)}) = \Delta_{\mathcal{F}}(\mathcal{E}^{(1)}, B^{-1}(\mathcal{E}^{(2)})) = 0. \tag{1.14}$$

In the absence of such a bijection, the theory of sufficient statistics can come to our aid to bound the Le Cam distance. Let again $\mathcal{Y}^{(i)}, i = 1,2$, be two sample spaces that we assume to be complete separable metric spaces. Let $\mathcal{E}^{(1)}$ be the experiment giving rise to observations $Y^{(1)}$ of law $P_f^{(1)}$ on $\mathcal{Y}^{(1)}$, and suppose that there exists a mapping $S : \mathcal{Y}^{(1)} \to \mathcal{Y}^{(2)}$ independent of f such that

$$Y^{(2)} = S(Y^{(1)}), \qquad Y^{(2)} \sim P_f^{(2)} \quad \text{on } \mathcal{Y}^{(2)}.$$

Assume, moreover, that $S(Y^{(1)})$ is a sufficient statistic for $Y^{(1)}$; that is, the conditional distribution of $Y^{(1)}$ given that we have observed $S(Y^{(1)})$ is independent of $f \in \mathcal{F}$. Then

$$\Delta_{\mathcal{F}}(\mathcal{E}^{(1)}, \mathcal{E}^{(2)}) = 0. \tag{1.15}$$

The proof of this result, which is an application of the *sufficiency principle* from statistics, is left as Exercise 1.1.

Asymptotic Equivalence for Nonparametric Gaussian Regression Models

We can now give the main result of this subsection. We shall show that the experiments

$$Y_i = f(x_i) + \varepsilon_i, \qquad x_i = \frac{i}{n}, \qquad \varepsilon_i \sim^{i.i.d.} N(0, \sigma^2), \quad i = 1, \dots, n, \tag{1.16}$$

and

$$dY(t) = f(t)dt + \frac{\sigma}{\sqrt{n}} dW(t), \qquad t \in [0,1], \quad n \in \mathbb{N}, \tag{1.17}$$

are asymptotically ($n \to \infty$) equivalent in the sense of Le Cam distance. In the course of the proofs, we shall show that any of these models is also asymptotically equivalent to the sequence space model (1.8). Further models that can be shown to be equivalent to (1.17) are discussed after the proof of the following theorem.

We define classes

$$\mathcal{F}(\alpha, M) = \left\{ f : [0,1] \to \mathbb{R}, \sup_{x \in [0,1]} |f(x)| + \sup_{x \neq y} \frac{|f(x) - f(y)|}{|x-y|^\alpha} \leq M \right\},$$
$$0 < \alpha \leq 1, \quad 0 < M < \infty,$$

of α-Hölderian functions. Moreover, for $(x_i)_{i=1}^n$ the design points of the fixed-design regression model (1.16) and for f any bounded function defined on $[0,1]$, let $\pi_n(f)$ be the unique function that interpolates f at the x_i and that is piecewise constant on each interval $(x_{i_1}, x_i] \subset [0,1]$.

Theorem 1.2.1 *Let $(\mathcal{E}_n^{(i)} : n \in \mathbb{N}), i = 1,2,3$, equal the sequence of statistical experiments given by $i = 1$ the fixed-design nonparametric regression model (1.16); $i = 2$, the standard Gaussian white noise model (1.17); and $i = 3$, the Gaussian sequence space model (1.8),* *respectively. Then, for $\mathcal{F}$ any family of bounded functions on $[0,1]$, for $\pi_n(f)$ as earlier and for any $n \in \mathbb{N}$,*

$$\Delta_{\mathcal{F}}(\mathcal{E}_n^{(2)}, \mathcal{E}_n^{(3)}) = 0, \quad \Delta_{\mathcal{F}}(\mathcal{E}_n^{(1)}, \mathcal{E}_n^{(2)}) \leq \sqrt{\frac{n\sigma^2}{2}} \sup_{f \in \mathcal{F}} \|f - \pi_n(f)\|_2. \tag{1.18}$$

In particular, if $\mathcal{F} = \mathcal{F}(\alpha, M)$ for any $\alpha > 1/2, M > 0$, then all these experiments are asymptotically equivalent in the sense that their Le Cam distance satisfies, as $n \to \infty$,

$$\Delta_{\mathcal{F}}(\mathcal{E}_n^{(i)}, \mathcal{E}_n^{(j)}) \to 0, \quad i,j \in \{1,2,3\}. \tag{1.19}$$

Proof In the proof we shall say that two experiments are equivalent if their Le Cam distance is exactly equal to zero. The first claim in (1.18) immediately follows from (1.14) and the isometry between $L^2([0,1])$ and ℓ_2 used in the definition of the sequence space model (1.8).

Define $\mathcal{V}_n$ to equal the n-dimensional space of functions $f : [0,1] \to \mathbb{R}$ that are piecewise constant on the intervals

$$I_{in} = (x_{i-1}, x_i] = \left(\frac{i-1}{n}, \frac{i}{n}\right], \quad i = 1, \dots, n.$$

The indicator functions $\phi_{in} = 1_{I_{in}}$ of these intervals have disjoint support, and they form an orthonormal basis of $\mathcal{V}_n$ for the inner product

$$\langle f, g \rangle_n = \sum_{j=1}^{n} f(x_j) g(x_j),$$

noting that $\sum_{j=1}^{n} \phi_{in}^2(x_j) = 1$ for every i. Given bounded $f : [0,1] \to \mathbb{R}$, let $\pi_n(f)$ be the $\langle \cdot, \cdot \rangle_n$-projection of f onto $\mathcal{V}_n$. Since

$$\langle f, \phi_{in} \rangle_n = \sum_{j=1}^{n} f(x_j) \phi_{in}(x_j) = f(x_i) \; \forall i,$$

we see

$$\pi_n(f)(t) = \sum_{i=1}^{n} f(x_i) \phi_{in}(t), \quad t \in [0,1],$$

so this projection interpolates f at the design points x_i, that is, $\pi_n(f)(x_j) = f(x_j)$ for all j. Note that the functions $\{\sqrt{n}\phi_{in} : i = 1, \dots, n\}$ also form a basis of $\mathcal{V}_n$ in the standard $L^2([0,1])$ inner product $\langle \cdot, \cdot \rangle$. This simultaneous orthogonality property will be useful in what follows.

Observing $Y_i = f(x_i) + \varepsilon_i$ in $\mathbb{R}^n$ from model (1.16) with bounded f is, by (1.14), equivalent to observations in the n-dimensional functional space $\mathcal{V}_n$ given by

$$\sum_{i=1}^{n} Y_i \phi_{in}(t) = \sum_{i=1}^{n} f(x_i) \phi_{in}(t) + \sum_{i=1}^{n} \varepsilon_i \phi_{in}(t), \quad t \in [0,1]. \tag{1.20}$$

We immediately recognise that $\sum_{i=1}^{n} f(x_i) \phi_{in}$ is the interpolation $\pi_n(f)$ of f at the x_i. Moreover, the error process is a scaled white noise process restricted to the space $\mathcal{V}_n$: indeed, its $L^2([0,1])$ action on $h \in \mathcal{V}_n$ is given by

$$\int_0^1 \sum_{i=1}^{n} \varepsilon_i \phi_{in}(t) h(t) dt = \frac{1}{\sqrt{n}} \sum_{i=1}^{n} \varepsilon_i \langle h, \sqrt{n}\phi_{in} \rangle \sim N\left(0, \frac{\sigma^2}{n} \sum_{i=1}^{n} \langle h, \sqrt{n}\phi_{in} \rangle^2\right) = N\left(0, \frac{\sigma^2}{n} \|h\|_2^2\right)$$

using Parseval's identity and that the $\sqrt{n}\phi_{in}$ form an $L^2([0,1])$ orthonormal basis of $\mathcal{V}_n$. If Π_n is the $L^2([0,1])$ projector onto $\mathcal{V}_n$ spanned by the $\{\sqrt{n}\phi_{in}\}$, then one shows, by the same arguments, that this process can be realised as a version of the Gaussian process defined on L^2 by the action $h \mapsto \mathbb{W}(\Pi_n(h))$, where $\mathbb{W}$ is as in (1.7). In other words, it equals the L^2-projection of the standard white noise process dW onto the finite-dimensional space $\mathcal{V}_n$, justifying the notation

$$\frac{\sigma}{\sqrt{n}}dW_n(t) \equiv \sum_{i=1}^{n} \varepsilon_i \phi_{in}(t)dt.$$

To summarise, (1.16) is equivalent to model (1.20), which itself can be rewritten as

$$d\tilde{Y}(t) \equiv \pi_n(f)(t) + \frac{\sigma}{\sqrt{n}}dW_n(t), \quad t \in [0,1]. \tag{1.21}$$

Next, consider the model

$$d\bar{Y}(t) = \pi_n(f)(t) + \frac{\sigma}{\sqrt{n}}dW(t), \quad t \in [0,1], \tag{1.22}$$

which is the standard white noise model (1.17) but with f replaced by its interpolation $\pi_n(f)$ at the design points x_i. Since $\pi_n(f) \in \mathcal{V}_n$, we have $\Pi_n(\pi_n(f)) = \pi_n(f)$, and since $dW_n = \Pi_n(dW) \in \mathcal{V}_n$, the statistics

$$d\tilde{Y} = \Pi_n(d\bar{Y}) = \left\{ \int_0^1 h(t)d\tilde{Y}(t) : h \in \mathcal{V}_n \right\}$$

are sufficient for $d\bar{Y}$, so by (1.15) the models (1.21) and (1.22) are equivalent. [To use (1.15) rigorously, we interpret $d\tilde{Y}, d\bar{Y}$ as tight random variables in a large enough, separable Banach space (see Section 6.1.1).]

To prove the second claim in (1.18), we relate (1.22) to (1.17), that is, to

$$dY(t) = f(t) + \frac{\sigma}{\sqrt{n}}dW(t), \quad t \in [0,1].$$

Both experiments have the same sample space, which in view of Section 6.1.1 we can take to be, for instance, the space of continuous functions on $[0,1]$, and the standard white noise $\mathbb{W}$ gives a common dominating measure P_0^Y on that space for the corresponding probability measures $P_f^Y, P_{\pi_n(f)}^Y$. In view of (1.13) and using Proposition 6.1.7a) combined with (6.16), we see that the Le Cam distance is bounded by

$$\sup_{f\in\mathcal{F}} \|P_f^Y - P_{\pi_n(f)}^Y\|_{1,\mu,\mathcal{F}}^2 \leq \frac{n}{\sigma^2} \sup_{f\in\mathcal{F}} \|f - \pi_n(f)\|_2^2, \tag{1.23}$$

which gives (1.18). Finally, for (1.19), uniformly in $f \in \mathcal{F}(\alpha, M)$,

$$\|f - \pi_n(f)\|_2^2 = \sum_{i=1}^{n} \int_{(i-1)/n}^{i/n} (f(x) - f(x_i))^2 dx \leq M^2 \sum_{i=1}^{n} \int_{(i-1)/n}^{i/n} |x - x_i|^{2\alpha} dx$$

$$\leq M^2 n^{-2\alpha} \sum_{i=1}^{n} \int_{(i-1)/n}^{i/n} dx = O(n^{-2\alpha}),$$

so for $\alpha > 1/2$, the quantity in (1.23) converges to zero, completing the proof. ■

In the preceding theorem the Hölder classes $\mathcal{F}(\alpha,M)$ could be replaced by balls in the larger Besov-Sobolev spaces $B^{\alpha}_{2\infty}$ (defined in Chapter 4) whenever $\alpha > 1/2$. The condition on α, however, cannot be relaxed, as we discuss in the notes.

The theory of asymptotic equivalence can be taken much further, to include results like the one preceding for random design regression experiments in possibly multivariate settings and with possibly non-Gaussian noise ε. The theory also extends to non-Gaussian settings that are not of regression type: one can show that nonparametric models for probability or spectral densities, or ergodic diffusions, are asymptotically equivalent to a suitable Gaussian white noise model. We discuss relevant references in the notes.

Asymptotic equivalence theory, which is a subject in its own, justifies that the Gaussian white noise model is, in the sense of the Le Cam distance, a canonical limit experiment in which one can develop some main theoretical ideas of nonparametric statistics. For Gaussian regression problems, the closeness of the experiments involved is in fact of a nonasymptotic nature, as shown by Theorem 1.2.1, and in this book we thus shall concentrate on the white noise model as the natural continuous surrogate for the standard fixed-design regression model. For other, non-Gaussian models, such as density estimation, asymptotic equivalence theory is, however, often overly simplistic in its account of the probabilistic structure of the problem at hand, and for the purposes of this book, we hence prefer to stay within the product-measure setting of Section 1.1, such that a nonasymptotic analysis is possible.

Exercises

1.1 Prove (1.15). [*Hint*: Use the fact that the proof of the standard sufficiency reduction principle extends to complete separable metric spaces (see Le Cam 1986).]

1.3 Notes

The modern understanding of statistical inference as consisting of the three related branches of estimation, testing and confidence statements probably goes back, in its most fundamental form, to the work of Fisher (1922; 1925a, b), who considered mostly parametric (finite-dimensional) statistical models. The need to investigate nonparametric statistical models was realised not much later, roughly at the same time at which the axiomatic approach to probability theory was put forward by Kolmogorov (1933). Classic papers on fully nonparametric sampling models for the cumulative distribution function are, for instance, Glivenko (1933), Cantelli (1933), Kolmogorov (1933a), and Smirnov (1939). More recent developments will be reviewed in later chapters of this book.

The linear regression model with normally distributed errors was initiated by Gauss (1809), who used it successfully in the context of observational astronomy. Gauss most likely was the first to use the least-squares algorithm, although Legendre and even some others can claim priority as well. The history is reviewed, for example, in Plackett (1972) and Stigler (1981).

Nonparametric regression models were apparently not studied systematically before the 1960s; see Nadaraya (1964) and Watson (1964). The Gaussian white noise model and its sequence space analogue were systematically developed in the 1970s and later by the Russian school – we refer to the seminal monograph by Ibragimov and Khasminskii (1981). The asymptotic equivalence theory for statistical experiments was developed by Le Cam; we refer to his fundamental book Le Cam (1986) and also to Le Cam and Yang (1990). Landmark contributions in nonparametric asymptotic equivalence theory are the papers Brown and Low (1996) and Nussbaum (1996), who treated univariate regression models with fixed design and density estimation, respectively. The necessity of the assumption $\alpha \geq 1/2$ is the subject of the paper by Brown and Zhang (1998). Asymptotic equivalence for random design regression is somewhat more involved: the univariate case is considered in Brown et al. (2002), and the general, multivariate random design regression case is considered in Reiß (2008). Further important results include asymptotic equivalence for nonparametric regression with non-Gaussian error distributions in Grama and Nussbaum (2002), asymptotic equivalence for spectral density estimation in Golubev, Nussbaum and Zhou (2010), and asymptotic equivalence for ergodic diffusions in Dalalyan and Reiß (2006).

2

Gaussian Processes

This chapter develops some classical theory and fundamental tools for Gaussian random processes. We start with the basic definitions of Gaussian processes indexed by abstract parameter spaces and, by way of introduction to the subject, derive some elementary yet powerful properties. We present the isoperimetric and log-Sobolev inequalities for Gaussian measures in $\mathbb{R}^n$ and apply them to establish concentration properties for the supremum of a Gaussian process about its median and mean, which are some of the deepest and most useful results on Gaussian processes. Then we introduce Dudley's metric entropy bounds for moments of suprema of (sub-) Gaussian processes as well as for their a.s. modulus of continuity. The chapter also contains a thorough discussion of convexity and comparison properties of Gaussian measures and of reproducing kernel Hilbert spaces and ends with an exposition of the limit theory for suprema of stationary Gaussian processes.

2.1 Definitions, Separability, 0-1 Law, Concentration

We start with some preliminaries about stochastic processes, mainly to fix notation and terminology. Then these concepts are specialised to Gaussian processes, and some first properties of Gaussian processes are developed. The fundamental observation is that a Gaussian process X indexed by a set T induces an intrinsic distance d_X on T ($d_X(s,t)$ is the L^2-distance between $X(s)$ and $X(t)$), and all the probabilistic information about X is contained in the metric or pseudo-metric space (T,d). This is tested on some of the first properties, such as the 0-1 law and the existence of separable versions of X. One of the main properties of Gaussian processes, namely, their concentration about the mean, is introduced; this subject will be treated in the next section, but a first result on it, which is not sharp but that has been chosen for its simplicity, is given in this section.

2.1.1 Stochastic Processes: Preliminaries and Definitions

Let $(\Omega, \Sigma, \Pr)$ be a probability space, and let T be a set. A stochastic process X indexed by T and defined on the probability space $(\Omega, \Sigma, \Pr)$ is a function $X : T \times \Omega \mapsto \mathbb{R}$, $(t,\omega) \mapsto X(t,\omega)$ such that, for each $t \in T$, $X(t,\cdot)$ is a random variable. Then, for any finite set $F \subset T$, the maps $\Omega \mapsto \mathbb{R}^F$ given by $\omega \mapsto \{X(t,\omega) : t \in F\}$ are also measurable, and their probability laws $\mu_F = \Pr \circ \{X(t,\cdot) : t \in F\}^{-1}$ are the *finite-dimensional distributions* (or finite-dimensional marginal distributions or finite-dimensional marginals) of X. If $F \subset G \subset T$ and G is finite and π_{GF} is the natural projection from $\mathbb{R}^G$ onto $\mathbb{R}^F$, then, obviously, the *consistency conditions* $\mu_F = \mu_G \circ \pi_{GF}^{-1}$ are satisfied ($\pi_{GF}(\{X(t) : t \in G\}) = \{X(t) : t \in F\}$). Conversely, the Kolmogorov consistency theorem shows that any collection of Borel probability measures μ_F on $\mathbb{R}^F$, indexed by the finite subsets $F \subset T$ and satisfying the consistency conditions, is the collection of finite-dimensional distributions of a stochastic process X indexed by T. In other words, a consistent family of probability measures μ_F, $F \subset T$, F finite, defines a unique probability measure μ on the cylindrical σ-algebra $\mathcal{C}$ of $\mathbb{R}^T$ such that $\mu_F = \mu \circ \pi_{TF}^{-1}$. (The cylindrical σ-algebra $\mathcal{C}$ is the σ-algebra generated by the cylindrical sets with finite-dimensional base, $\pi_{TF}^{-1}(A)$, $A \in \mathcal{B}(\mathbb{R}^F)$, $F \subset T$, F finite.) Then the map $X : T \times \mathbb{R}^T \mapsto \mathbb{R}$, $(t,x) \mapsto x(t)$, is a process defined on the probability space $(\mathbb{R}^T, \mathcal{C}, \mu)$. If μ is the probability measure on $(\mathbb{R}^T, \mathcal{C})$ defined by the finite-dimensional distributions of a process X, then we say that μ is the *probability law of* X (which can be thought of as a 'random variable' taking values on the measurable space $(\mathbb{R}^T, \mathcal{C})$). See almost any probability textbook, for example, Dudley (2002).

Definition 2.1.1 Two processes X and Y of index set T are said to be a *version* of each other if both have the same finite-dimensional distributions $\mathcal{L}(X(t_1), \ldots, X(t_n)) = \mathcal{L}(Y(t_1), \ldots, Y(t_n))$ for all $n \in \mathbb{N}$ and $t_i \in T$ or, what is the same, if both have the same probability law on $(\mathbb{R}^T, \mathcal{C})$. They are said to be a *strict version* or a *modification* of each other if $\Pr\{X(t) = Y(t)\} = 1$ for all t.

It is convenient to recall the definition of pseudo-distance and pseudo-metric space. A pseudo-distance d on T is a nonnegative symmetric function of two variables $s,t \in T$ that satisfies the triangle inequality but for which $d(s,t) = 0$ does not necessarily imply $s = t$. A pseudo-metric space (T,d) is a set T equipped with a pseudo-distance d. Clearly, a pseudo-metric space becomes a metric space by taking the quotient with respect to the equivalence relation $s \simeq t$ iff $d(s,t) = 0$. For instance, the space $\mathcal{L}^p$ of functions is a pseudo-metric space for the L^p (pseudo-)norm, and the space of equivalence classes, L^p, is a metric space for the same norm. One only seldom needs to distinguish between the two.

If the index set T of a process X is a metric or pseudo-metric space (T,d), we say that X is *continuous in probability* if $X(t_n) \to X(t)$ in probability whenever $d(t_n,t) \to 0$. In this case, if T_0 is a d-dense subset of T, the law of the process on $(\mathbb{R}^T, \mathcal{C})$ is determined by the finite-dimensional distributions $\mathcal{L}(X(t_1), \ldots, X(t_n))$ for all $n \in \mathbb{N}$ and $t_i \in T_0$.

Here are two more definitions of interest.

Definition 2.1.2 A process $X(t)$, $t \in T$, (T,d) a metric or pseudo-metric space, is separable if there exists $T_0 \subset T$, T_0 countable, and $\Omega_0 \subset \Omega$ with $\Pr(\Omega_0) = 1$ such that for all $\omega \in \Omega_0$, $t \in T$ and $\varepsilon > 0$,

$$X(t,\omega) \in \overline{\{X(s,\omega) : s \in T_0 \cap B_d(t,\varepsilon)\}},$$

where $B_d(t,\varepsilon)$ is the open d-ball about t of radius ε. X is *measurable* if the map $(\Omega \times T, \Sigma \otimes \mathcal{T}) \to \mathbb{R}$ given by $(\omega,t) \longrightarrow X(\omega,t)$ is jointly measurable, where $\mathcal{T}$ is the σ-algebra generated by the d-balls of T.

By definition, if $X(t)$, $t \in T$, is separable, then there are points from T_0 in any neighborhood of t, $t \in T$; hence (T,d) is separable; that is, (T,d) possesses a countable dense subset. Note that if X is separable, then $\sup_{t\in T} X(t) = \sup_{s\in T_0} X(s)$ a.s., and the latter, being a countable supremum, is measurable; that is, suprema over uncountable sets are measurable. The same holds for $|X(t)|$.

Often we require the *sample paths* $t \mapsto X(t,\omega)$ to have certain properties for almost every ω, notably, to be bounded or bounded and uniformly continuous ω a.s.

Definition 2.1.3 A process $X(t)$, $t \in T$, is sample bounded if it has a version $\tilde{X}$ whose sample paths $t \mapsto \tilde{X}(t,\omega)$ are almost all uniformly bounded, that is, $\sup_{t\in T}|\tilde{X}(t)| < \infty$ a.s. If (T,d) is a metric or pseudo-metric space, then X is sample continuous (more properly, sample bounded and uniformly continuous) if it has a version $\tilde{X}(t)$ whose sample paths are almost all bounded and uniformly d-continuous.

Note that if X is sample continuous, then the finite-dimensional distributions of X are the marginals of a probability measure μ defined on the cylindrical σ-algebra $\mathcal{C} \cap C_u(T,d)$ of $C_u(T,d)$, the space of bounded uniformly continuous functions on (T,d), $\mathcal{L}(X(t_1),\ldots,X(t_k)) = \mu \circ (\delta_{t_1},\ldots,\delta_{t_k})^{-1}$, $t_i \in T$, $k < \infty$ (here and in what follows, δ_t is unit mass at t). The vector space $C_u(T,d)$, equipped with the supremum norm $\|f\|_\infty = \sup_{t\in T}|f(t)|$, is a Banach space, that is, a complete normed space for which the vector space operations are continuous. The Banach space $C_u(T,d)$ is separable if (and only if) (T,d) is totally bounded, and in this case, $C_u(T,d)$ is isometric to $C(\bar{T},d)$, where $(\bar{T},d)$ is the completion of (T,d), which is compact. Then, assuming (T,d) totally bounded, we have $\|f\|_\infty = \sup_{t\in T_0}|f(t)|$, where T_0 is any countable dense subset of T; in particular, the closed balls of $C_u(T,d)$ are measurable for the cylindrical σ-algebra: $\{f : \|f - f_0\|_\infty \le r\} = \cap_{t\in T_0}\{f : |f(t) - f_0(t)| \le r\}$. This implies that the open sets are also measurable because, by separability of $C_u(T,d)$, every open set in this space is the union of a countable number of closed balls. This proves that the Borel and the cylindrical σ-algebras of $C_u(T,d)$ coincide if (T,d) is totally bounded. Hence, in this case, the finite-dimensional distributions of X are the marginal measures of a Borel probability measure μ on $C_u(T,d)$. Since $C_u(T,d)$ is separable and complete (for the supremum norm), the probability law μ of X is tight in view of the following basic result that we shall use frequently in this book (see Exercise 2.1.6 for its proof). Recall that a probability measure μ is tight if for all $\varepsilon > 0$ there is K compact such that $\mu(K^c) < \varepsilon$.

Proposition 2.1.4 (Oxtoby-Ulam) *If μ is a Borel probability measure on a complete separable metric space, then μ is tight.*

In general, given a Banach space B, a *B-valued random variable* X is a Borel measurable map from a probability space into B. Thus, the preceding considerations prove the following proposition. It is convenient to introduce an important Banach space: given a set T, $\ell_\infty(T) \subset \mathbb{R}^T$ will denote the set of bounded functions $x : T \mapsto \mathbb{R}$. Note that this is a Banach space if we equip it with the supremum norm $\|x\|_T = \sup_{t\in T}|x(t)|$ and that the inclusion of $C_u(T)$ into $\ell_\infty(T)$ is isometric. Observe that $\ell_\infty(T)$ is separable for the supremum norm if and only if T is finite.

Proposition 2.1.5 *If (T,d) is a totally bounded metric or pseudo-metric space and $X(t)$, $t \in T$, is a sample continuous process, then X has a version which is a $C_u(T,d)$-valued random variable, and its probability law is a tight Borel measure with support contained in $C_u(T,d)$ and hence a tight Borel probability measure on $\ell_\infty(T)$.*

Example 2.1.6 (Banach space–valued random variables as sample continuous processes.) Let B be a separable Banach space, let B^* be its dual space and let B_1^* denote the closed-unit ball of B_1^* about the origin. Then there exists a countable set $D \subset B_1^*$ such that $\|x\| = \sup_{f \in D} f(x)$ for all $x \in B$: if $\{x_i\} \subset B$ is a countable dense subset of B and $f_i \in B_1^*$ are such that $f_i(x_i) = \|x_i\|$ (note that f_i exists by the Hahn-Banach theorem), then $D = \{f_i\}$ is such a set. The inclusion $B \mapsto C_u(D, \|\cdot\|)$, where $\|\cdot\|$ is the norm on B_1^*, is an isometric imbedding, and every B-valued random variable X defines a process $f \mapsto f(X)$, $f \in D$, with all its sample paths bounded and uniformly continuous. Hence, any results proved for sample bounded and uniformly continuous processes indexed by totally bounded metric spaces do apply to Banach space–valued random variables for B separable.

If $X(t)$, $t \in T$, is a sample bounded process, then its probability law is defined on the cylindrical σ-algebra of $\ell_\infty(T)$, $\Sigma = \mathcal{C} \cap \ell_\infty(T)$. Since $\ell_\infty(T)$ is a metric space for the supremum norm, it also has another natural σ-algebra, the Borel σ-algebra. We conclude with the interesting fact that if the law of the bounded process X extends to a tight Borel measure on $\ell_\infty(T)$, then X is sample continuous with respect to a metric d for which (T,d) is totally bounded.

Proposition 2.1.7 *Let $X(t)$, $t \in T$, be a sample bounded stochastic process. Then the finite-dimensional probability laws of X are those of a tight Borel probability measure on $\ell_\infty(T)$ if and only if there exists on T a pseudo-distance d for which (T,d) is totally bounded and such that X has a version with almost all its sample paths uniformly continuous for d.*

Proof Let us assume that the probability law of X is a tight Borel measure μ on $\ell_\infty(T)$; let K_n, $n \in \mathbb{N}$, be an increasing sequence of compact sets in $\ell_\infty(T)$ such that $\mu\left(\cup_{n=1}^\infty K_n\right) = 1$; and set $K = \cup_{n=1}^\infty K_n$. Define a pseudo-metric d as

$$d(s,t) = \sum_{n=1}^{\infty} 2^{-n}\left(1 \wedge d_n(s,t)\right),$$

where

$$d_n(s,t) = \sup\left\{|f(t) - f(s)| : f \in K_n\right\}.$$

To prove that (T,d) is totally bounded, given $\varepsilon > 0$, let m be such that $\sum_{n=m+1}^\infty 2^{-n} < \varepsilon/4$. Since the set $\cup_{n=1}^m K_n$ is compact, it is totally bounded, and therefore, it contains a finite subset $\{f_1, \ldots, f_r\}$ which is $\varepsilon/4$ dense in $\cup_{n=1}^m K_n$ for the supremum norm; that is, for each $f \in \cup_{n=1}^m K_n$, there is $i \le r$ such that $\|f - f_i\|_\infty \le \varepsilon/4$. Since $\cup_{n=1}^m K_n$ is a bounded subset of $\ell_\infty(T)$ (as it is compact), it follows that the subset $A = \{(f_1(t), \ldots, f_r(t)) : t \in T\}$ of $\mathbb{R}^r$ is bounded, hence precompact, hence totally bounded, and therefore there exists a finite set $T_\varepsilon = \{t_i : 1 \le i \le N\}$ such that for each $t \in T$ there is $i = i(t) \le N$ such that $\max_{1 \le s \le r} |f_s(t) - f_s(t_i)| \le \varepsilon/4$. It follows that T_ε is ε dense in T for the pseudo-metric d: for $n \le m$, $t \in T$ and $t_i = t_{i(t)}$, we have

$$d_n(t,t_i) = \sup_{f \in K_n} |f(t) - f(t_i)| \le \max_{s \le r} |f_s(t) - f_s(t_i)| + \varepsilon/2 \le \frac{3\varepsilon}{4}$$

and therefore

$$d(t,t_i) \le \frac{\varepsilon}{4} + \sum_{n=1}^{m} 2^{-n} d_n(t,t_i) \le \varepsilon,$$

proving that (T,d) is totally bounded.

Next, since $\mu(K) = 1$, the identity map of $(\ell_\infty(T), \mathcal{B}, \mu)$ is a version of X with almost all its trajectories in K. Thus, to prove that X has a version with almost all its sample paths bounded and uniformly d-continuous, it suffices to show that the functions from K have these properties. If $f \in K_n$, then $|f(s) - f(t)| \le d_n(s,t) \le 2^n d(s,t)$ for all $s,t \in T$ with $d(s,t) < 2^{-n}$, proving that f is uniformly continuous, and f is bounded because K_n is bounded.

Conversely, let $X(t)$, $t \in T$, be a process with a version whose sample paths are almost all in $C_u(T,d)$ for a distance or pseudo-distance d on T for which (T,d) is totally bounded, and let us continue denoting X such a version (recall the notation $C_u(T,d)$ as the space of bounded uniformly continuous functions on (T,d)). Then X is a random variable taking values in $C_u(T,d)$, and its marginal laws correspond to a Borel probability measure on $C_u(T,d)$ (see the argument following Definition 2.1.3). But since (T,d) is precompact, $C_u(T,d)$ is separable, and the law of X is in fact a tight Borel measure by the Oxtoby-Ulam theorem (Proposition 2.1.4). But a tight Borel probability measure on $C_u(T,d)$ is a tight Borel measure on $\ell_\infty(T)$ because the inclusion of $C_u(T,d)$ into ℓ_∞ is continuous. ■

2.1.2 Gaussian Processes: Introduction and First Properties

We now look at Gaussian processes. Recall that a finite-dimensional random vector or a multivariate random variable $Z = (Z_1, \ldots, Z_n)$, $n \in \mathbb{N}$, is an n-dimensional Gaussian vector, or a multivariate normal random vector, or its coordinates are jointly normal, if the random variables $\langle a, Z\rangle = \sum_{i=1}^n a_i Z_i$, $a = (a_1, \ldots, a_n) \in \mathbb{R}^n$, are normal variables, that is, variables with laws $N(m(a), \sigma^2(a))$, $\sigma(a) \ge 0$, $m \in \mathbb{R}$. If $m = m(a) = 0$ for all $a \in \mathbb{R}^n$, we say that the Gaussian vector is *centred*.

Definition 2.1.8 A stochastic process $X(t)$, $t \in T$, is a Gaussian process if for all $n \in \mathbb{N}$, $a_i \in \mathbb{R}$ and $t_i \in T$, the random variable $\sum_{i=1}^n a_i X(t_i)$ is normal or, equivalently, if all the finite-dimensional marginals of X are multivariate normal. X is a centred Gaussian process if all these random variables are normal with mean zero.

Definition 2.1.9 A covariance Φ on T is a map $\Phi : T \times T \to \mathbb{R}$ such that for all $n \in \mathbb{N}$ and $t_1, \ldots, t_n \in T$, the matrix $(\Phi(t_i,t_j))_{i,j=1}^n$ is symmetric and nonnegative definite (i.e., $\Phi(t_i,t_j) = \Phi(t_j,t_i)$ and $\sum_{i,j} a_i a_j \Phi(t_i,t_j) \ge 0$ for all a_i).

The following is a consequence of the Kolmogorov consistency theorem.

Proposition 2.1.10 *Given a covariance Φ on T and a function f on T, there is a Gaussian process $X(t)$ such that $E(X(t)) = f(t)$ and $E[(X(t) - f(t))(X(s) - f(s))] = \Phi(s,t)$ for all $s,t \in T$. Φ is called the covariance of the process and f its expectation, and we say that X is a centred Gaussian process if and only if $f \equiv 0$.*

Proof If $F \subset T$ is finite, take $\mu_F = N\left((f(t) : t \in F), \Phi|_{F\times F}\right)$. It is easy to see that the set $\{\mu_F : F \subset T,\ F \text{ finite}\}$ is a consistent system of marginals. Hence, by the Kolmogorov consistency theorem, there is a probability on $(\mathbb{R}^T, \mathcal{C})$, hence a process, with $\{\mu_F\}$ as its set of finite-dimensional marginals. ■

Example 2.1.11 A basic example of a Gaussian process is the *isonormal* or *white noise process* on a separable Hilbert space H, where $\{X(h) : h \in H\}$ has a covariance diagonal for the inner product $\langle \cdot, \cdot \rangle$ of H: $EX(h) = 0$ and $EX(h)X(g) = \langle h, g \rangle_H$ for all $g, h \in H$. The existence of this process does not even require the Kolmogorov consistency theorem but only the existence of an infinite sequence of random variables (i.e., the existence of an infinite product probability space): if $\{g_i\}$ is a sequence of independent $N(0,1)$ random variables and $\{\psi_i\}$ is an orthonormal basis of H, the process defined by linear and continuous extension of $\tilde{X}(\psi_i) = g_i$ (i.e., by $\tilde{X}(\sum a_i \psi_i) = \sum a_i g_i$ whenever $\sum a_i^2 < \infty$) is clearly a version of X. Note for further use that if $V \subset L^2(\Omega, \Sigma, \Pr)$ is the closed linear span of the sequence $\{g_i\}$, then the map $\tilde{X} : H \mapsto V$ is an isometry.

From now on, all our Gaussian processes will be *centred*, even if sometimes we omit mentioning it. If X is a centred Gaussian process on T, the L^2-pseudo-distance between $X(t)$ and $X(s)$ defines a pseudo-distance d_X on T

$$d_X^2(s,t) := E\,(X(t) - X(s))^2 = \Phi(t,t) + \Phi(s,s) - 2\Phi(s,t)$$

that we call the *intrinsic distance* of the process. With this pseudo-metric, T is isometric to the subspace $\{X(t) : t \in T\}$ of $L^2(\Omega, \Sigma, \Pr)$. Clearly, a centred Gaussian process X is continuous in probability for the pseudo-distance d_X; in particular, its probability law in $(\mathbb{R}^T, \mathcal{C})$ is determined by the finite-dimensional marginals based on subsets of any d_X-dense subset T_0 of T.

It is important to note that the probability law of a centred Gaussian process X is completely determined by its intrinsic distance d_X (or by the covariance Φ). Thus, all the probabilistic information about a centred Gaussian process is contained in the metric (or pseudo-metric) space (T, d_X). This is a very distinctive feature of Gaussian processes.

Here is a first, albeit trivial, example of the exact translation of a property of the metric space (T, d_X) into a probabilistic property of X, actually, necessarily of a version of X.

Proposition 2.1.12 *For a Gaussian process X indexed by T, the following are equivalent:*

1. *The pseudo-metric space (T, d_X) is separable, and*
2. *X, as a process on (T, d_X), has a separable, measurable (strict) version.*

Proof If point 2 holds, let $\bar{X}$ be a separable and measurable version of X (in particular, $d_{\bar{X}} = d_X$), and let T_0 be a countable set as in the definition of separability. Then, as mentioned earlier, the very definition of separability implies that $T_0 \cap B_{d_X}(t, \varepsilon) \neq \emptyset$ for all $t \in T$ and $\varepsilon > 0$. Thus, T_0 is dense in (T, d_X), and therefore, (T, d_X) is separable.

Assume now that (T, d_X) is separable, and let T_0 be a countable d_X-dense subset of T. Also assume, as we may by taking equivalence classes, that $d_X(s,t) \neq 0$ for all $s, t \in T_0$, $s \neq t$. If $T_0 = \{s_i : i \in \mathbb{N}\}$, define, for each n, the following partition of T:

$$C_n(s_m) = B\left(s_m, 2^{-n}\right) \setminus \bigcup_{k<m} B\left(s_k, 2^{-n}\right), \quad m \in \mathbb{N}.$$

For each $t \in T$, let $s_n(t)$ be the only $s \in T_0$ such that $t \in C_n(s)$, and define $X_n(t) = X(s_n(t))$. Now $X_n(t, \omega)$ is jointly measurable because $X_n^{-1}(A) = \bigcup_{i \in \mathbb{N}} [C_n(s_i) \times \{\omega : X(s_i, \omega) \in A\}]$. Since, for any $t \in T$, $\Pr\{|X_n(t) - X(t)| > 1/n\} \leq n^2\ E\ (X(s_n(t)) - X(t))^2 \leq n^2/2^{2n}$, it follows by Borel-Cantelli that $X_n(t) \to X(t)$ a.s.

Define $\bar{X}(t, \omega) = \limsup_n X_n(t, \omega)$, which, for each t, is ∞ at most on a set of measure 0. Then the process $\bar{X}(t, \omega)$ is measurable because it is a lim sup of measurable functions. Also, for each t, $\bar{X}(t) = X(t)$ on a set of measure 1; that is, $\bar{X}$ is a strict version of X. Next we show that $\bar{X}$ is separable. Given $r \in \mathbb{N}$, there exists n_r large enough that $d_X(s_r, s_l) > 1/2^{n_r}$ for all $l < r$; hence, for $n \geq n_r$, $X_n(s_r) = X(s_r)$. This shows that $\bar{X}(s) = X(s)$ for all $s \in T_0$. Then, for all $\omega \in \Omega$,

$$\bar{X}(t,\omega) = \limsup X_n(t,\omega) = \limsup X(s_n(t),\omega) = \limsup \bar{X}(s_n(t),\omega),$$

proving that $\bar{X}$ is separable. ■

Just as with normal random variables, Gaussian processes also satisfy the Gaussian stability property, namely, that if two Gaussian processes with index set T are independent, then their sum is a Gaussian process with covariance the sum of covariances (and mean the sum of means); in particular, if X and Y are independent and equally distributed Gaussian processes (meaning that they have the same finite-dimensional marginal distributions or, what is the same, the same law on the cylindrical σ-algebra $\mathcal{C}$ of $\mathbb{R}^T$), then the process $\alpha X + \beta Y$ has the same law as $(\alpha^2 + \beta^2)^{1/2} X$. This property has many consequences, and here is a nice instance of its use.

Theorem 2.1.13 (0-1 law) *Let $F \subset \mathbb{R}^T$ be a $\mathcal{C}$-measurable linear subspace, and let X be a (centred) Gaussian process indexed by T. Then*

$$\Pr\{X \in F\} = 0 \text{ or } 1.$$

Proof Let X_1 and X_2 be independent copies of X. Define sets

$$A_n = \{X_1 + nX_2 \in F\} \quad \text{and} \quad B_n = \{X_2 \notin F\} \cap A_n, \quad n \in \mathbb{N}.$$

Since $X_1 + nX_2$ is a version of $\sqrt{1+n^2}X$ and F is a vector space, we have

$$\begin{aligned}
\Pr\{B_n\} &= \Pr\{A_n\} - \Pr[A_n \cap \{X_2 \in F\}] \\
&= \Pr\{X \in F\} - \Pr\{X_1 + n\,X_2 \in F,\, X_2 \in F\} \\
&= \Pr\{X \in F\} - \Pr\{X_1 \in F, X_2 \in F\} \\
&= \Pr\{X \in F\} - [\Pr\{X \in F\}]^2.
\end{aligned}$$

Clearly, $B_n \cap B_m = \emptyset$ if $n \neq m$; hence, since by the preceding equalities $\Pr\{B_n\}$ does not depend on n, it follows that $\Pr\{B_n\} = 0$ for all n. But then, again by the same inequalities, $\Pr\{X \in F\}$ can only be 0 or 1. ■

Corollary 2.1.14 *Let X be a centred Gaussian process on T and $\|\cdot\|$ be a $\mathcal{C}$-measurable pseudo-norm on $\mathbb{R}^T$. Then*

$$P\{\|X\| < \infty\} = 0 \text{ or } 1.$$

Proof The set $\{x \in \mathbb{R}^T : \|x\| < \infty\} = \cup_n \{x \in \mathbb{R}^T : \|x\| < n\}$ is a measurable vector space, and the 0-1 law yields the result. ■

Example 2.1.15 If X is Gaussian, separable and centred, then there exists $T_0 \subset T$, T_0 countable, such that $\sup_{t\in T} |X(t)| = \sup_{t\in T_0} |X(t)|$ a.s, but $\|x\|_{T_0} := \sup_{t\in T_0} |x(t)|$ is a measurable pseudo-norm, and hence it is finite with probability 0 or 1.

Example 2.1.16 *The B-valued Gaussian variables where B is a separable Banach space constitute a very general and important class of Gaussian processes, and we define them now. Given a separable Banach space B, a B-valued random variable X is centred Gaussian if $f(X)$ is a mean zero normal variable for every $f \in B^*$, the topological dual of B. By linearity, this is equivalent to the statement that $f_1(X), \ldots, f_n(X)$ are jointly centred normal for every $n \in \mathbb{N}$ and $f_i \in B^*$. In particular, if X is a B-valued centred Gaussian random variable, then the map $X : B^* \mapsto \mathcal{L}^2(\Omega, \Sigma, \Pr)$, defined by $X(f) = f(X)$, is a centred Gaussian process. If $B = E$ has dimension d, X is centred Gaussian iff the coordinates of X in a basis of E are jointly normal with mean zero (hence, the same is true for the coordinates of X in any basis).*

Now we turn to a very useful property of Gaussian processes X, namely, that *the supremum norm of a Gaussian process concentrates about its mean, as well as about its median, with very high probability*, in fact as if it were a real normal variable with variance the largest variance of the individual variables $X(t)$. This result is a consequence of an even deeper result, the isoperimetric inequality for Gaussian measures, although it has simpler direct proofs, particularly if one is allowed some latitude and does not aim at the best result. Here is one such proof that uses the stability property in an elegant and simple way.

We should recall that a function $f : V \mapsto \mathbb{R}$, where V is a metric space, is Lipschitz with Lipschitz constant $c = \|f\|_{\text{Lip}}$ if $c := \sup_{x \neq y} |f(x) - f(y)|/d(x,y) < \infty$. Rademacher's theorem asserts that if $f : \mathbb{R}^n \mapsto \mathbb{R}$ is Lipschitz, then it is a.e. differentiable and the essential supremum of the norm of its derivative is bounded by its Lipschitz constant $\|f\|_{\text{Lip}}$. We remark that although we will use this result in the theorem that follows, it is not needed for its application to a concentration of maxima of jointly normal variables because one can compute by hand the derivative of the Lipschitz function $x \mapsto \max_{i \le d} |x_i|$, $x \in \mathbb{R}^d$.

Theorem 2.1.17 *Let $(B, \|\cdot\|_B)$ be a finite-dimensional Banach space, and let X be an B-valued centred Gaussian random variable. Let $f : B \mapsto \mathbb{R}$ be a Lipschitz function. Let $\Psi : \mathbb{R} \mapsto \mathbb{R}$ be a nonnegative, convex, measurable function. Then the following inequality holds:*

$$E[\Psi(f(X) - Ef(X))] \le E\left[\Psi\left(\frac{\pi}{2}\langle f'(X), Y\rangle\right)\right], \tag{2.1}$$

where Y is an independent copy of X (X and Y have the same probability law and are independent), and $\langle\cdot,\cdot\rangle$ denotes the duality action of B^* on B.

Proof Since the range of X is a full subspace, we may assume without loss of generality that B equals the range of X (i.e., the support of the law of X is B). This has the effect that the law of X and Lebesgue measure on B are mutually absolutely continuous (as the density of X is strictly positive on its supporting subspace). For $\theta \in [0, 2\pi)$, define $X(\theta) = X\sin\theta + Y\cos\theta$. Then $X'(\theta) = X\cos\theta - Y\sin\theta$, and notice that $X(\theta)$ and $X'(\theta)$ are (normal and) independent: it suffices to check covariances, and if $f, g \in B^*$, we have

$$E[f(X(\theta))g(X'(\theta))] = E(f(X)g(X))\sin\theta\cos\theta - E(f(Y)g(Y))\sin\theta\cos\theta = 0.$$

In other words, the joint probability laws of X and Y and of $X(\theta)$ and $X'(\theta)$ coincide.

Since for any increasing sequence θ_i

$$\begin{aligned}\sum |f(X(\theta_i)) - f(X(\theta_{i-1}))| &\le \|f\|_{\mathrm{Lip}} \sum \|X(\theta_i) - X(\theta_{i-1})\| \\ &\le \|f\|_{\mathrm{Lip}}(\|X\| + \|Y\|) \sum |\theta_i - \theta_{i-1}|,\end{aligned}$$

it follows that the function $\theta \mapsto f(X(\theta))$ is absolutely continuous, and therefore, we have

$$f(X) - f(Y) = f(X(\pi/2)) - f(X(0)) = \int_0^{\pi/2} \frac{d}{d\theta} f(X(\theta))\, d\theta.$$

Using convexity of Ψ, Fubini's theorem and the preceding, we obtain

$$\begin{aligned}&E\Psi(f(X) - Ef(X)) = E\Psi(f(X) - Ef(Y)) \le E\Psi(f(X) - f(Y)) \\ &\quad = E\Psi\left(\int_0^{\pi/2} \frac{d}{d\theta} f(X(\theta)) d\theta\right) \le \frac{2}{\pi} E \int_0^{\pi/2} \Psi\left(\frac{\pi}{2}\frac{d}{d\theta} f(X(\theta))\right) d\theta \\ &\quad = \frac{2}{\pi}\int_0^{\pi/2} E\Psi\left(\frac{\pi}{2}\frac{d}{d\theta} f(X(\theta))\right) d\theta.\end{aligned}$$

Now f is m a.e. differentiable with a bounded derivative by Rademacher's theorem, where m is Lebesgue measure on B, and since $\mathcal{L}(X(\theta))$ is absolutely continuous with respect to Lebesgue measure for every $\theta \in [0, \pi/2)$ ($X(\theta)$ has the same support as X), f' exists a.s. relative to the law of $X(\theta)$. Since $X'(\theta)$ exists for each θ, it follows from the chain rule that given θ, $df(X(\theta))/d\theta = \langle f'(X(\theta)), X'(\theta)\rangle$ a.s. Then, since $\mathcal{L}(X, Y) = \mathcal{L}(X(\theta), X'(\theta))$, we have

$$E\Psi\left(\frac{\pi}{2}\frac{d}{d\theta} f(X(\theta))\right) = E\Psi\left(\frac{\pi}{2}\langle f'(X), Y\rangle\right),$$

which, combined with the preceding string of inequalities, proves the theorem. ■

Remark 2.1.18 It turns out, as we will see in the next section, that Lipschitz functions are the natural tool for extracting concentration results from isoperimetric inequalities, on the one hand, and on the other, as we will see now, the supremum norm of a vector in $\mathbb{R}^n$ is a Lipschitz function, so concentration inequalities for Lipschitz functions include as particular cases concentration inequalities for the supremum norm and for other norms as well.

Example 2.1.19 (Concentration for the maximum of a finite number of jointly normal variables) To estimate the distribution of $\max_{i\le n} |g_i|$ for a finite sequence $g_1, \dots, g_n$ of jointly normal variables using the preceding theorem, we take $B = \ell^n_\infty$, which is $\mathbb{R}^n$ with the norm $f(x) = \max_{i\le n} |x_i|$, where $x = (x_1, \dots, x_n)$, which we take as our function f, and we take $X = (g_1, \dots, g_n)$. f is obviously Lipschitz, so the previous theorem will apply to it. We also have that for each $1 \le i \le n$, $f(x) = x_i$ on the set $\{x : x_i > |x_j|, 1 \le j \le n, j \ne i\}$ and $f(x) = -x_i$ on $\{x : -x_i > |x_j|, 1 \le j \le n, j \ne i\}$. It follows that m a.s. the gradient of f has all but one coordinate equal to zero, and this coordinate is 1 or -1. If $g_i \ne \pm g_j$ for $i \ne j$, which we can assume without loss of generality (by deleting repeated coordinates without changing the maximum), then this also holds a.s. for the law of X. Let $\sigma_i^2 = Eg_i^2$ and $\sigma^2 = \max_{i\le n} \sigma_i^2$. For almost every $X = x$ fixed, $\langle f'(x), Y\rangle$ is $\pm g_i$ for some i, that is, in law, the same as $\sigma_i g$, g standard normal. Therefore, if we assume that the function Ψ is as in the preceding theorem and that, moreover, it is even and nondecreasing on $[0,\infty)$, then, letting E_Y denote integration with respect to the variable Y only, the preceding observation implies that, X a.s.,

$$E_Y\Psi\left(\frac{\pi}{2}\langle f'(x),Y\rangle\right)\leq E\Psi\left(\frac{\pi}{2}\sigma g\right).$$

We conclude that for any $n\in\mathbb{N}$, if $g_1,\ldots,g_n$ are jointly normal random variables and if $\sigma^2=\max_{i\leq n}Eg_i^2$, then for any nonnegative, even, convex function Ψ nondecreasing on $[0,\infty)$,

$$E\Psi\left(\max_{i\leq n}|g_i|-E\max_{i\leq n}|g_i|\right)\leq E\Psi\left(\frac{\pi}{2}\sigma g\right), \tag{2.2}$$

where g denotes a standard normal random variable.

Now $Ee^{t|g|}\leq E(e^{tg}+e^{-tg})=2e^{t^2/2}$. Thus, if $\Psi_\lambda(x)=e^{\lambda|x|}$, we have

$$E\Psi_\lambda\left(\frac{\pi}{2}\sigma g\right)\leq 2e^{\lambda^2\pi^2\sigma^2/8}$$

and, by (2.2) and Chebyshev's inequality,

$$\Pr\left\{\left|\max_{i\leq n}|g_i|-E\max_{i\leq n}|g_i|\right|>u\right\}\leq 2e^{-\lambda u+\lambda^2\pi^2\sigma^2/8},\quad u\geq 0.$$

With $\lambda u/2=\lambda^2\pi^2\sigma^2/8$, that is, $\lambda=4u/(\pi^2\sigma^2)$, this inequality gives the following approximate concentration inequality about its mean for the maximum of any finite number of normal random variables:

$$\Pr\left\{\left|\max_{i\leq n}|g_i|-E\max_{i\leq n}|g_i|\right|>u\right\}\leq 2e^{-\frac{4}{\pi^2}\frac{u^2}{2\sigma^2}},\quad u\geq 0. \tag{2.3}$$

The last inequality and the one in the next theorem are suboptimal: the factor $4/\pi^2$ in the exponent is superfluous, as we will see in two of the sections that follow. We can translate (2.2) and (2.3) into a concentration inequality for the supremum norm of a separable Gaussian process (and draw as well some consequences).

Theorem 2.1.20 *Let $\{X(t),\ t\in T\}$ be a separable centred Gaussian process such that*

$$\Pr\{\sup_{t\in T}|X(t)|<\infty\}>0.$$

Let Ψ be an even, convex, measurable function, nondecreasing on $[0,\infty)$. Let g be $N(0,1)$. Then,

a. $\sigma=\sigma(X):=\sup_{t\in T}\left(EX^2(t)\right)^{1/2}<\infty$ and $E\sup_{t\in T}|X(t)|<\infty$ and
b. The following inequalities hold:

$$E\Psi\left(\sup_{t\in T}|X(t)|-E\sup_{t\in T}|X(t)|\right)\leq E\Psi\left(\frac{\pi}{2}\sigma g\right)$$

and

$$\Pr\left\{\left|\sup_{t\in T}|X(t)|-E\sup_{t\in T}|X(t)|\right|>u\right\}\leq 2e^{-(Ku^2/2\sigma^2)},$$

where $K=\frac{4}{\pi^2}$.

(As mentioned earlier, the optimal constant K in this theorem will be shown to be 1.)

Proof By assumption and the 0-1 law (Theorem 2.1.13; see the example following Corollary 2.1.14), $\sup_{t\in T}|X(t)| < \infty$ a.s. Let $0 < z_{1/2} < 1$ be such that $\Pr\{|g| > z_{1/2}\} = 1/2$, and let $M < \infty$ be such that $\Pr\left\{\sup_{t\in T}|X(t)| > M\right\} < 1/2$. Then, for each t,

$$1/2 > \Pr\{|X(t)| > M\} = \Pr\{|g| > M/(EX(t)^2)^{1/2}\},$$

which implies that $\sigma = \sup_{t\in T}(EX^2(t))^{1/2} \le M/z_{1/2} < \infty$.

Let $T_0 = \{t_i\}_{i=1}^{\infty}$ be a countable set such that $\sup_{t\in T}|X(t)| = \sup_{t\in T_0}|X(t)|$. For every $n \in \mathbb{N}$, we have, by inequality (2.3),

$$\Pr\left\{\left|\max_{i\le n}|X(t_i)| - E\max_{i\le n}|X(t_i)|\right| > \sigma u\right\} \le 2e^{-2u^2/\pi^2}.$$

Since $\sup_{t\in T}|X(t)| < \infty$ a.s., this variable has a finite median m, and also for all n,

$$\Pr\left\{\max_{i\le n}|X(t_i)| \le m\right\} \ge \frac{1}{2}.$$

If u_0 is such that $2e^{-2u_0^2/\pi^2} < 1/2$, these two inequalities imply that for all $n \in \mathbb{N}$, the intersection of the two sets $\left\{x : \left|E\max_{i\le n}|X(t_i)| - x\right| \le \sigma u_0\right\}$ and $\{x : x \le m\}$ is not empty and hence that $E\max_{i\le n}|X(t_i)| \le m + \sigma u_0 < \infty$. a) is proved.

We have $\sup_{t\in T}|X(t)| = \lim_{n\to\infty}\max_{i\le n}|X(t_i)|$ a.s. and, by monotone convergence, also in $L^1(\Pr)$. Hence, the first inequality in (b) follows by inequality (2.2), continuity of Ψ and Fatou's lemma. The second inequality follows from the first by Chebyshev's inequality in the same way as (2.3) follows from (2.2). ■

Exercises

In Exercises 2.1 to 2.4 we write $\|X\|$ for $\sup_{t\in T}|X(t)|$, and X denotes a separable, centred Gaussian process such that $\Pr\left\{\sup_{t\in T}|X(t)| < \infty\right\} > 0$. Also, for any random variable ξ, $\|\xi\|_p$ will denote its L^p-norm.

2.1.1 Prove that there exists $\alpha > 0$ such that $Ee^{\alpha\|X\|^2} < \infty$.

2.1.2 Use results from this section to show that for all $p \ge 1$,

$$(E\|X\|^p)^{1/p} \le K\sqrt{p}E\|X\|$$

for a universal constant $K < \infty$. *Hint*: Integrating the exponential inequality in Theorem 2.1.20 with respect to $pt^{p-1}dt$ yields $\|\,\|X\| - E\|X\|\,\|_p \le c\sigma\|g\|_p$, where g is standard normal and c a universal constant. Check that $\|g\|_p$ is of the order of $\sqrt{p}$.

2.1.3 Prove that the median m of $\|X\|$, satisfies $KE\|X\| \le m \le 2E\|X\|$ for a universal constant $K > 0$. *Hint*: The second inequality is obvious, and the first is contained in the proof of Theorem 2.1.20.

2.1.4 Prove that if X_n are separable, centred Gaussian processes such that $\Pr\{\|X_n(t)\| < \infty\} > 0$, then $\|X_n\| \to 0$ in pr. iff $\|X_n\| \to 0$ in L^p for some $p \ge 1$ iff $\|X_n\| \to 0$ in L^p for all $p \ge 1$. *Hint*: L^p convergences for different p are equivalent by Exercise 2.1.2, and the equivalence extends to convergence in probability by Exercise 2.1.2 and the *Paley-Zygmund argument* as follows: for any $0 < \tau < 1$,

$$E\|X\| \le \tau E\|X\| + E(\|X\|I_{\|X\|>\tau E\|X\|}) \le \tau E\|X\| + \left(E\|X\|^2\right)^{1/2}(\Pr\{\|X\| > \tau E\|X\|\})^{1/2},$$

so

$$\Pr\{\|X\| > \tau E\|X\|\} \ge \left[\frac{(1-\tau)E\|X\|}{\left(E\|X\|^2\right)^{1/2}}\right]^2 \ge K(1-\tau)^2$$

for a universal constant K. Thus, if $\|X_n\| \to 0$ in probability, then $E\|X_n\| \to 0$.

2.1.5 Let B be a separable Banach space, and let X be a B-valued Gaussian (centred) random variable. Show that the previous theorems apply to $\|X\|$ where now $\|\cdot\|$ is the Banach space norm. *Hint*: Use Example 2.1.6.

2.1.6 Prove Proposition 2.1.4. *Hint*: Recall that a subset of a complete separable metric space S is compact if and only if it is closed and totally bounded. Given $\varepsilon > 0$, by separability, for each n there exists a finite collection $\{F_{n,k}\}_{k=1}^{k_n}$ of closed sets of diameter not exceeding n^{-1} and such that $\mu\left(\cup_{k=1}^{k_n} F_{n,k}\right)^c < \varepsilon/2^n$. The set $K = \cap_{n=1}^{\infty} \cup_{k=1}^{k_n} F_{n,k}$ is compact and satisfies $\mu(K^c) < \varepsilon$.

2.2 Isoperimetric Inequalities with Applications to Concentration

The Gaussian isoperimetric inequality, in its simplest form, identifies the half-spaces as the sets of $\mathbb{R}^n$ with the smallest Gaussian perimeter among those with a fixed Gaussian measure, where the Gaussian measure in question is the standard one, that is, the probability law of n independent standard normal random variables, and where the Gaussian perimeter of a set is taken as the limit of the measure of the difference of an ε-enlargement of the set and the set itself divided by ε. The proof of this theorem was obtained originally by translating the isoperimetric inequality on the sphere to the Gaussian setting by means of Poincaré's lemma, which states that the limiting distribution of the orthogonal projection onto a Euclidean space of fixed dimension n of the uniform distribution on the sphere of $\mathbb{R}^{m+1}$ with radius $\sqrt{m}$ is the standard Gaussian measure of $\mathbb{R}^n$. The isoperimetric inequality on the sphere is a deep result that goes back to P. Lévy and E. Schmidt, ca. 1950 (although the equivalent isoperimetric problem on the plane goes back to the Greeks–recall, for instance, 'Dido's problem'). The Gaussian isoperimetric inequality does imply best possible concentration inequalities for Lipschitz functions on $\mathbb{R}^n$ and for functions on $\mathbb{R}^{\mathbb{N}}$ that are Lipschitz 'in the direction of ℓ_2', although concentration inequalities have easier proofs, as seen in the preceding section and as will be seen again in further sections. The Gaussian isoperimetric inequality in general Banach spaces requires the notion of reproducing kernel Hilbert space and will be developed in a further section as well. This section contains proofs as short as we could find of the isoperimetric inequalities on the sphere and for the standard Gaussian measure on $\mathbb{R}^n$, $n \leq \infty$, with applications to obtain the best possible concentration inequality with respect to the standard Gaussian measure for Lipschitz functions f about their medians and for the supremum norm of a separable Gaussian process X when $\sup_{t\in T}|X(t)| < \infty$ a.s.

2.2.1 The Isoperimetric Inequality on the Sphere

Let $S^n = \left\{x \in \mathbb{R}^{n+1} : \|x\|^2 = \sum_{i=1}^{n+1} x_i^2 = 1\right\}$, where $x = (x_1,\dots,x_{n+1})$; let p be an arbitrary point in S^n that we take to be the north pole, $p = (0,\dots,0,1)$; and let μ be the uniform probability distribution on S^n (equal to the normalized volume element – surface area for S^2 – equal also to the normalized Haar measure of the rotation group). Let d be the geodesic distance on S^n, defined, for any two points, as the length of the shortest segment of the great circle joining them.

A closed *cap* centred at a point $x \in S^n$ is a geodesic closed ball around x, that is, a set of the form $C(x,\rho) := \{y : d(x,y) \leq \rho\}$. Here ρ is the radius of the cap, and clearly, the μ-measure of a cap is a continuous function of its radius, varying between 0 and 1. Often we will not specify the centre or the radius of $C = C(x, \rho)$, particularly if the centre is the north pole.

The isoperimetric inequality on the sphere states that the caps are the sets of shortest perimeter among all the measurable sets of a given surface area. What we will need is an equivalent formulation, in terms of neighbourhoods of sets, defined as follows: the closed ε neighbourhood of a set A is defined as $A_\varepsilon = \{x : d(x,A) \le \varepsilon\}$, with the distance between a point and a set being defined, as usual, by $d(x,A) = \inf\{d(x,y) : y \in A\}$. The question is: among all measurable subsets of the sphere with surface area equal to the surface area of A, find sets B for which the surface areas of their neighbourhoods B_ε, $0 < \varepsilon < 1$, are smallest. The following theorem shows that an answer are the caps (they are in fact *the* answer, but uniqueness will not be considered: we are only interested in the value of $\inf \mu(A_\varepsilon)$, $\varepsilon > 0$).

Theorem 2.2.1 *Let $A \neq \emptyset$ be a measurable subset of S^n, and let C be a cap such that $\mu(C) = \mu(A)$. Then, for all $\varepsilon > 0$,*

$$\mu(C_\varepsilon) \le \mu(A_\varepsilon). \tag{2.4}$$

The proof is relatively long, and some prior digression may help. The idea is to construct transformations $A \mapsto A^*$ on measurable subsets of the sphere that preserve area, that is, $\mu(A) = \mu(A^*)$, and decrease perimeter, a condition implied by $\mu((A^*)_\varepsilon) \le \mu(A_\varepsilon) = \mu((A_\varepsilon)^*)$, $\varepsilon > 0$, because the perimeter of A is the limit as $\varepsilon \to 0$ of $\mu(A_\varepsilon \setminus A)/\varepsilon$. Then iterating transformations that satisfy these two properties should eventually produce the solution, in our case a cap. Or, more directly, one may obtain a cap using a more synthetic compactness argument instead of iteration. In the sense that A^* concentrates the same area as A on a smaller perimeter, A^* is closer to the solution of the problem than A is. A^* is called a *symmetrisation* of A.

Proof If $\mu(A) = 0$, then C consists of a single point, and (2.4) holds. Next, we observe that by regularity of the measure μ, it suffices to prove the theorem for A compact. By regularity, there exist A^m compact, $A^m \subset A$, A^m increasing and such that $\mu(A^m) \nearrow \mu(A)$. Let C^m be caps with the same centre as C and with $\mu(C^m) = \mu(A^m)$. Since the measure of a cap is a continuous one-to-one function of its geodesic radius, we also have $\mu(C^m_\varepsilon) \nearrow \mu(C_\varepsilon)$, and if the theorem holds for compact sets, then

$$\mu(A_\varepsilon) \ge \lim \mu(A^m_\varepsilon) \ge \lim \mu(C^m_\varepsilon) = \mu(C_\varepsilon),$$

and the theorem holds in general. Thus, we will assume that A is compact and that $\mu(A) \neq 0$. We divide the proof into several parts.

Part 1: Construction and main properties of the symmetrisation operation. Given an n-dimensional subspace $H \subset \mathbb{R}^{n+1}$ that does not contain the point p, let $\sigma = \sigma_H$ be the reflection about H; that is, if $x = u + v$ with $u \in H$ and v orthogonal to H, then $\sigma(x) = u - v$. Clearly, σ is an isometry (so it preserves μ-measure), and it is involutive; that is, $\sigma^2 = \sigma$. It also satisfies a property that, together with the preceding two, is crucial for the symmetrisation operation to work, namely, that if x and y are on the same half-space with respect to H, then

$$d(x,y) \le d(x, \sigma(y)). \tag{2.5}$$

To see this, observe that the geodesic distance is an increasing function of the Euclidean distance, so it suffices to prove (2.5) for the Euclidean distance. Changing orthogonal coordinates if necessary, we may and do assume that $H = \{x : x_{n+1} = 0\}$, so if x and y are in the same hemisphere, then $\mathrm{sign}(x_{n+1}) = \mathrm{sign}(y_{n+1})$, which implies that the $(n+1)$th coordinate of $x - y$ is dominated in absolute value by the $(n+1)$th coordinate of $x - \sigma(y)$, whereas the first n coordinates of these two vectors coincide. Hence, $\sum_{i=1}^{n+1}(x_i - y_i)^2 \le \sum_{i=1}^{n+1}(x_i - \sigma(y)_i)^2$.

H divides S^n into two open hemispheres, and we denote by S_+ the open hemisphere that contains p, S_- the other hemisphere, and $S_0 = S^n \cap H$. The symmetrisation of A with respect to $\sigma = \sigma_H$, $s_H(A) = A^*$ is defined as

$$s_H(A) = A^* := [A \cap (S_+ \cup S_0)] \cup \{a \in A \cap S_- : \sigma(a) \in A\} \cup \{\sigma(a) : a \in A \cap S_-, \sigma(a) \notin A\}. \tag{2.6}$$

Note that A^* is obtained from A by reflecting towards the northern hemisphere every $a \in A \cap S_-$ for which $\sigma(A)$ is not already in A. It is easy to see (Exercise 2.2.1) that if A is compact, then so is A^* and that if C is a cap with centre at p or at any other point in the northern hemisphere, then $C^* = C$. Next, observe that the three sets in the definition are disjoint and that, σ being an isometry, the measure of the third set equals $\mu\{a \in A \cap S_- : \sigma(a) \notin A\}$, which implies that

$$\mu(A^*) = \mu(A), \quad A \in \mathcal{B}(S^{n+1}). \tag{2.7}$$

This is one of the two properties of the symmetrisation operation that we need.

We now show that the ε-neighbourhoods of A^* are less massive than those of A (thus making A^* ‘closer’ to being a cap than A is), actually, we prove more, namely, that for all $A \in \mathcal{B}(S^n)$ and $\varepsilon > 0$, then

$$(A^*)_\varepsilon \subseteq (A_\varepsilon)^*, \quad \text{hence} \quad \mu((A^*)_\varepsilon) \le \mu((A_\varepsilon)^*) = \mu(A_\varepsilon). \tag{2.8}$$

To see this, let $x \in (A^*)_\varepsilon$ and let $y \in A^*$ be such that $d(x,y) \le \varepsilon$ (such a $y \in A^*$ exists by compactness). Then, using (2.5) and that σ is an involutive isometry, we obtain, when x and y lay on different half-spaces,

$$d(\sigma(x),y) = d(x,\sigma(y)) \le d(\sigma(x),\sigma(y)) = d(x,y) \le \varepsilon.$$

Thus, since $y \in A^*$ implies that either $y \in A$ or $\sigma(y) \in A$, in either case we have that both $x \in A_\varepsilon$ and $\sigma(x) \in A_\varepsilon$; hence, $x \in (A_\varepsilon)^*$. If x and y are in S_-, then y and $\sigma(y)$ are both in A, and therefore, by the last identity earlier, $x \in A_\varepsilon$ and $\sigma(x) \in A_\varepsilon$; hence, $x \in (A_\varepsilon)^*$ in this case as well. If x and y are in S_+, then either y or $\sigma(y)$ is in A; hence, either x or $\sigma(x)$ is in A_ε, which together with $x \in S_+$ implies that $x \in (A_\varepsilon)^*$. The cases where x and/or y are in S_0 are similar, even easier, and they are omitted. The inclusion in (2.8) is proved, and the inequality there follows from the inclusion and from (2.7).

Part 2: Preparation for the compactness argument. Let $(\mathcal{K}, h)$ denote the set of nonempty compact subsets of S^n equipped with the Hausdorff distance, defined as $h(A,B) = \inf\{\varepsilon : A \subseteq B_\varepsilon,\ B \subseteq A_\varepsilon\}$, $A, B \in \mathcal{K}$. $(\mathcal{K}, h)$ is a compact metric space (Exercise 2.2.2). Given a compact nonempty set $A \subseteq S^n$, let $\mathcal{A}$ be the minimal closed subset of $\mathcal{K}$ that contains A and is preserved by s_H for all n-dimensional subspaces H of $\mathbb{R}^{n+1}$ that do not contain the north pole p (meaning that if $A \in \mathcal{K}$, then $s_H(A) \in \mathcal{K}$ for all H with $p \notin H$). $\mathcal{A}$ exists and is nonempty because $\mathcal{K}$ is a closed $\{s_H\}$-invariant collection of sets that contains A. Also note that since $(\mathcal{K},h)$ is compact and $\mathcal{A}$ closed, $\mathcal{A}$ is compact. We have

Claim: If $B \in \mathcal{A}$, then (a) $\mu(B) = \mu(A)$, and (b) for all $\varepsilon > 0$, $\mu(B_\varepsilon) \le \mu(A_\varepsilon)$.
Proof of the claim. It suffices to show that the collection of closed sets $\mathcal{F}$ satisfying a) and b) is preserved by s_H for all H not containing p and is a closed subset of $\mathcal{K}$ because then $\mathcal{A} \subseteq \mathcal{F}$ follows by minimality of $\mathcal{A}$. That $s_H(\mathcal{F} \Rightarrow \subseteq \mathcal{F}$ follows from (2.7) and (2.8). Let now $B^n \in \mathcal{F}$ and $h(B^n,B) \to 0$. Let $\varepsilon > 0$ be fixed. Given $\delta > 0$, there exists n_δ such that $B \subseteq B^n_\delta$ for all $n \ge n_\delta$; hence, $B_\varepsilon \subseteq B^n_{\delta+\varepsilon}$ and $\mu(B_\varepsilon) \le \mu(B^n_{\delta+\varepsilon}) \le \mu(A_{\delta+\varepsilon})$. Letting $\delta \searrow 0$ shows that B satisfies condition (b). Letting $\varepsilon \searrow 0$ in condition (b) for B shows that $\mu(B) \le \mu(A)$. Using that for all n large enough we also have $B^n \subseteq B_\delta$, we get that $\mu(A) = \mu(B^n) \le \mu(B_\delta)$ and, letting $\delta \searrow 0$, that $\mu(A) \le \mu(B)$, proving condition (a). The claim is proved.

Part 3: Completion of the proof of Theorem 2.2.1. Clearly, because of the claim about $\mathcal{A}$, it suffices to show that if C is the cap centred at p such that $\mu(A) = \mu(C)$, then $C \in \mathcal{A}$.

Define $f(B) = \mu(B \cap C)$, $B \in \mathcal{A}$. We show first that f is upper semicontinuous on $\mathcal{A}$. If $h(B^n,B) \to 0$, then, given $\delta > 0$, for all n large enough, $B^n \subseteq B_\delta$, which, as is easy to see, implies that $B^n \cap C \subseteq (B \cap C_\delta)_\delta$. Hence, $\limsup_n \mu(B^n \cap C) \le \mu((B \cap C_\delta)_\delta)$, but because B and C are closed, if $\delta_n \searrow 0$, then $\cap_n (B \cap C_{\delta_n})_{\delta_n} = B \cap C$, thus obtaining $\limsup_n \mu(B^n \cap C) \le \mu(B \cap C)$.

Since f is upper semicontinuous on $\mathcal{A}$ and $\mathcal{A}$ is compact, f attains its maximum at some $B \in \mathcal{A}$. The theorem will be proved if we show that $C \subseteq B$. Assume that $C \not\subseteq B$. Then, since $\mu(C) = \mu(A) = \mu(B)$ and both C and B are closed, we have that both $B \setminus C$ and $C \setminus B$ have positive μ-measure. Thus, the Lebesgue density theorem, which holds on S^n (see Exercise 2.2.3 for definitions and a sketch of the proof), implies that there exist points of density $x \in B \setminus C$ and $y \in C \setminus B$. Let H be the subspace of dimension n orthogonal to the vector $x - y$, and let us keep the shorthand notation σ for the reflection with respect to H, D^* for $s_H(D)$, S_+, S_- for the two hemispheres determined by H, and S_0 for $S^n \cap H$. Then $\sigma(y) = x$. Since $y \in C$ and $x \notin C$, we have both, that p is not in H (the reflection of a point in C with respect to a hyperplane through p is necessarily in C) and that y is closer to p than x is; that is, $d(y,p) \le d(x,p) = d(\sigma_H(y),p)$. Then it follows from this last obsesrvation and (2.5) that $y \in S_+$ and $x \in S_-$.

Let $x \in (B \cap C)^*$. Then, if $x \in B \cap C \cap (S_+ \cup S_0)$ or if $x \in B \cap C \cap S_-$ and $\sigma(x) \in B \cap C$, we obviously have $x \in B^* \cap C$. Now, if $z \in C \cap S_-$, then $\sigma(z) \in C$ (as $\sigma(z)$ is closer to p than z is); hence, if $x = \sigma(z)$ with $z \in B \cap C \cap S_-$ and $\sigma(z) \notin B \cap C$, then $\sigma(z)$ is not in B and therefore $x \in B^* \cap C$. We conclude that $(B \cap C)^* \subseteq B^* \cap C$ and, in particular, that

$$\mu(B \cap C) = \mu((B \cap C)^*) \le \mu(B^* \cap C). \tag{2.9}$$

By definition of density point, for $\delta > 0$ small enough, $C(x,\delta) \subset S_-$, $\sigma(C(x,\delta)) = C(y,\delta) \subset S_+$, $\mu((B \setminus C) \cap C(x,\delta)) \ge 2\mu(C(x,\delta))/3$, and $\mu((C \setminus B) \cap C(y,\delta)) \ge 2\mu(C(y,\delta))/3$. Then the set

$$D = ((B \setminus C) \cap C(x,\delta)) \cap \sigma((C \setminus B) \cap C(y,\delta))$$

satisfies

$$\mu(D) \ge \mu(C(x,\delta))/3 > 0, \quad D \subset (B \setminus C) \cap S_- \quad \text{and} \quad \sigma(D) \subset C \setminus B. \tag{2.10}$$

The inclusions in (2.10) imply that $\sigma(D) \subset B^* \cap C$ and $\sigma(D) \cap (B \cap C)^* = \emptyset$ (as $z \in (B \cap C)^*$ implies either $z \in B \cap C$ or $\sigma(z) \in B \cap C$). This together with (2.9) and $\mu(D) > 0$ proves

$$\mu(B^* \cap C) \geq \mu((B \cap C)^* \cup \sigma(D)) = \mu((B \cap C)^*) + \mu(D) > \mu((B \cap C)^*),$$

which, because $B^* \in \mathcal{A}$, contradicts the fact that f attains it maximum at B. ■

2.2.2 The Gaussian Isoperimetric Inequality for the Standard Gaussian Measure on $\mathbb{R}^{\mathbb{N}}$

In this subsection we translate the isoperimetric inequality on the sphere to an isoperimetric inequality for the probability law γ_n of n independent $N(0,1)$ random variables by means of Poincaré's lemma, which states that this measure can be obtained as the limit of the projection of the uniform distribution on $\sqrt{m}S^{n+m}$ onto $\mathbb{R}^n$ when $m \to \infty$. We also let $n \to \infty$.

In what follows, g_i, $i \in \mathbb{N}$, is a sequence of independent $N(0,1)$ random variables, and as mentioned earlier, $\gamma_n = \mathcal{L}(g_1, \ldots, g_n)$. We call γ_n the *standard Gaussian measure on* $\mathbb{R}^n$. We also set $\gamma = \mathcal{L}(\{g_i\}_{i=1}^{\infty})$, the law of the process $i \mapsto g_i$, $i \in \mathbb{N}$, a probability measure on the cylindrical σ-algebra $\mathcal{C}$ of $\mathbb{R}^{\mathbb{N}}$, which we also refer to as the *standard Gaussian measure on* $\mathbb{R}^{\mathbb{N}}$.

Here is the Gaussian isoperimetric problem: for a measurable subset A of $\mathbb{R}^n$, and $\varepsilon > 0$, define its Euclidean neighbourhoods A_ε as $A_\varepsilon := \{x \in \mathbb{R}^n : d(x,A) \leq \varepsilon\} = A + \varepsilon O_n$, where d denotes Euclidean distance and O_n is the closed d-unit ball centred at $0 \in \mathbb{R}^n$. The problem is this: given a Borel set A, find among the Borel sets $B \subset \mathbb{R}^n$ with the same γ_n-measure as A those for which the γ_n-measure of the neighbourhood B_ε is smallest, for all $0 < \varepsilon < 1/2$. The solution will be shown to be the affine half-space ($\{x : \langle x,u \rangle \leq \lambda\}$, u any unit vector, $\lambda \in \mathbb{R}$) of the same measure as A. Note that $\gamma_n\{x : \langle x,u \rangle \leq \lambda\} = \gamma_1\{x \leq \lambda\}$.

Prior to stating and proving the main results, we describe the relationship between the uniform distribution on the sphere of increasing radius and dimension and the standard Gaussian measure on $\mathbb{R}^n$.

Lemma 2.2.2 (Poincaré's lemma) *Let μ_{n+m} be the uniform distribution on $\sqrt{m}S^{n+m}$, the sphere of $\mathbb{R}^{n+m+1}$ of radius $\sqrt{m}$ and centred at the origin. Let π_m be the orthogonal projection $\mathbb{R}^{n+m+1} \mapsto \mathbb{R}^n = \{x \in \mathbb{R}^{n+m+1} : x_i = 0, n < i \leq n+m+1\}$, and let $\tilde{\pi}_m$ be the restriction of π_m to $\sqrt{m}S^{n+m}$. Let $\nu_m = \mu_{n+m} \circ \tilde{\pi}_m^{-1}$ be the projection onto $\mathbb{R}^n$ of μ_{n+m}. Then ν_m has a density f_m such that if ϕ_n is the density of γ_n, $\lim_{m\to\infty} f_m(x) = \phi_n(x)$ for all $x \in \mathbb{R}^n$. Therefore,*

$$\gamma_n(A) = \lim_{m\to\infty} \mu_{n+m}(\tilde{\pi}_m^{-1}(A)) \tag{2.11}$$

for all Borel sets A of $\mathbb{R}^n$.

Proof Set $G_n := (g_1, \ldots, g_n)$ and $G_{n+m+1} := (g_1, \ldots, g_{n+m+1})$. The rotational invariance of the standard Gaussian law on Euclidean space implies that μ_{n+m} is the law of the vector $\sqrt{m}G_{n+m+1}/|G_{n+m+1}|^{1/2}$. Hence, ν_m is the law of $\sqrt{m}G_n/|G_{n+m+1}|^{1/2}$. This allows for computations with normal densities that we only sketch. For any measurable set A of $\mathbb{R}^n$,

$$\nu_m(A) = \frac{1}{(2\pi)^{(n+m+1)/2}} \int_{\mathbb{R}^{m+1}} \int_{\tilde{A}(y)} e^{-(|z|^2+|y|^2)/2} dz\, dy$$

where $z \in \mathbb{R}^n$ and $y \in \mathbb{R}^{m+1}$, and $\tilde{A} = \left\{z \in \mathbb{R}^n : \sqrt{m/(|z|^2+|y|^2)}\, z \in A\right\}$. Make the change of variables $z \mapsto x$, $x = \sqrt{m/(|z|^2+|y|^2)}\, z$ or $z = |y|x/\sqrt{m-|x|^2}$, $|x| \le \sqrt{m}$. Its Jacobian is $\partial(z)/\partial(x) = m|y|^n/(m-|x|^2)^{1+n/2}$, thus obtaining

$$\begin{aligned}\nu_m(A) &= \frac{1}{(2\pi)^{(n+m+1)/2}} \int_A I(|x|^2 < m) \frac{m}{(m-|x|^2)^{n/2+1}} \int_{\mathbb{R}^{m+1}} |y|^n \exp\left(-\frac{1}{2}\frac{m|y|^2}{m-|x|^2}\right) dy\, dx \\ &= \frac{E(|G_{m+1}|^n)}{m^{n/2}} \frac{1}{(2\pi)^{n/2}} \int_A \left(1-\frac{|x|^2}{m}\right)^{(m-1)/2} I(|x|^2 < m)\, dx.\end{aligned}$$

Hence, the density of ν_m is $f_m(x) = C_{n,m}(2\pi)^{-n/2}(1-|x|^2/m)^{(m-1)/2} I(|x|^2 < m)$, $x \in \mathbb{R}^n$. Clearly, $(2\pi)^{-n/2}(1-|x|^2/m)^{(m-1)/2} I(|x|^2 < m) \to (2\pi)^{-n/2} e^{-|x|^2/2}$ for all x as $m \to \infty$. Moreover, since for $0 \le a < m$ and $m \ge 2$ we have $1 - a/m \le e^{-a/2(m-1)}$, it follows that $(1-|x|^2/m)^{(m-1)/2} I(|x|^2 < m)$ is dominated by the integrable function $e^{-|x|^2/4}$. Thus, by the dominated convergence theorem, $f_m(x)/C_{n,m} \to (2\pi)^{-n/2} e^{-|x|^2/2}$ in L^1, which implies that $C_{n,m}^{-1} \to 1$, proving the lemma. (Alternatively, just show that $C_{n,m} = E(|G_{m+1}|^n)/m^{n/2} \to 1$ as $m \to \infty$ by taking limits on well-known expressions for the moments of chi-square random variables.) Now the limit (2.11) for any Borel set follows by dominated convergence. ■

Theorem 2.2.3 *For $n < \infty$, let γ_n be the standard Gaussian measure of $\mathbb{R}^n$, let A be a measurable subset of $\mathbb{R}^n$, and let H be a half-space $H = \{x \in \mathbb{R}^n : \langle x, u\rangle \le a\}$, u a unit vector, such that $\gamma_n(H) = \gamma_n(A)$ and hence with $a := \Phi^{-1}(\gamma_n(A))$, where Φ denotes the standard normal distribution function. Then, for all $\varepsilon > 0$,*

$$\gamma_n(H + \varepsilon O_n) \le \gamma_n(A + \varepsilon O_n), \tag{2.12}$$

which, by the definition of a, is equivalent to

$$\gamma_n(A + \varepsilon O_n) \ge \Phi(\Phi^{-1}(\gamma_n(A)) + \varepsilon). \tag{2.13}$$

Proof First, we check the behaviour of distances under $\tilde{\pi}_m$. If d_{n+m} denotes the geodesic distance of $\sqrt{m}S^{n+m}$, it is clear that the projection $\tilde{\pi}_m$ is a contraction from the sphere onto $\mathbb{R}^n$; that is, $|\tilde{\pi}_m(x) - \tilde{\pi}_m(y)| \le d_{n+m}(x,y)$ for any $x, y \in \sqrt{m}S^{n+m}$. Moreover, if in the half-space $H_b := \{x \in \mathbb{R}^n : \langle x, u\rangle \le b\}$, we have $-\sqrt{m} < b < \sqrt{m}$; then its pre-image $\tilde{\pi}^{-1}(H_b)$ is a nonempty cap, and for $0 < \varepsilon < \sqrt{m} - b$, we have $(\tilde{\pi}^{-1}(H_b))_\varepsilon = \tilde{\pi}^{-1}(H_b + \tau(b,\varepsilon)O_n) = \tilde{\pi}^{-1}(H_{b+\tau(b,\varepsilon)})$, where

$$b + \tau = \sqrt{m}\cos\left(\cos^{-1}\frac{b}{\sqrt{m}} \pm \frac{\varepsilon}{\sqrt{m}}\right),$$

which, taking limits in the addition formula for the cosine, immediately gives $\lim_{m\to\infty} \tau(b,\varepsilon) = \varepsilon$.

Let now $b < a = \Phi^{-1}(\gamma_n(A))$ so that $H_b = \{x : \langle x, u\rangle \le b\} \subset H$. Then, by Poincaré's lemma,

$$\lim_m \mu_{n+m}(\tilde{\pi}_m^{-1}(A)) = \gamma_n(A) > \gamma_n(H_b) = \lim_m \mu_{n+m}(\tilde{\pi}_m^{-1}(H_b)),$$

so for all m large enough, we have both $b \in (-\sqrt{m}, \sqrt{m})$, such that $\tilde{\pi}_m^{-1}(H_b)$ is a nonempty cap in the sphere, and $\mu_{n+m}(\tilde{\pi}_m^{-1}(A)) \ge \mu_{n+m}(\tilde{\pi}_m^{-1}(H_b))$. Then the isoperimetric inequality for μ_{n+m} (Theorem 2.2.1) yields that for each $\varepsilon > 0$, $b + \varepsilon < \sqrt{m}$, for all m large enough,

$$\mu_{n+m}\left((\tilde{\pi}_m^{-1}(A))_\varepsilon\right) \ge \mu_{n+m}\left((\tilde{\pi}_m^{-1}(H_b))_\varepsilon\right) = \mu_{n+m}\left(\tilde{\pi}_m^{-1}(H_{b+\tau(b,\varepsilon)})\right),$$

so by Poincaré's lemma again,

$$\gamma_n(A+\varepsilon O_n) \geq \limsup_m \mu_{n+m}\left((\tilde{\pi}_m^{-1}(A))_\varepsilon\right) \geq \limsup_m \mu_{n+m}\left((\tilde{\pi}_m^{-1}(H_{b+\tau(b,\varepsilon)})\right) = \gamma_n(H_{b+\varepsilon}).$$

Since this holds for all $b < a$, it also holds with b replaced by a. ■

Theorem 2.2.3 extends to infinite dimensions, as will be shown in Theorem 2.6.12. An extension to the standard Gaussian measure on $\mathbb{R}^{\mathbb{N}}$, that is, for the law γ of a sequence of independent standard normal random variables, can be obtained directly. Before stating the theorem, it is convenient to make some topological and measure-theoretic considerations. The distance $\rho(x,y) = \sum_{k=1}^{\infty} \min(|x_k - y_k|, 1)/2^k$ metrises the product topology of $\mathbb{R}^{\mathbb{N}}$, and $(\mathbb{R}^{\mathbb{N}}, \rho)$ is a separable and complete metric space, as is easy to see. That is, $\mathbb{R}^{\mathbb{N}}$ is a Polish space (a topological space that admits a metric for which it is separable and complete). Then the cylindrical σ-algebra $\mathcal{C}$ coincides with the Borel σ-algebra of $\mathbb{R}^{\mathbb{N}}$, and any finite cylindrical (hence Borel) measure is tight (Radon). The product space $\mathbb{R}^{\mathbb{N}} \times \ell_2$ is also Polish, and for each $t \in \mathbb{R}$, the map $f_t : \mathbb{R}^{\mathbb{N}} \times \ell_2 \mapsto \mathbb{R}^{\mathbb{N}}$, $f_t(x,y) = x + ty$ is continuous. Then the image of f_t is universally measurable, that is, measurable for any Radon measure, in particular, in our case, measurable for any finite measure on the cylindrical σ-algebra $\mathcal{C}$ of $\mathbb{R}^{\mathbb{N}}$. See, for example, theorem 13.2.6 in section 13.2 in Dudley (2002).

Theorem 2.2.4 *Let A be a Borel set of $\mathbb{R}^{\mathbb{N}}$ (i.e., $A \in \mathcal{C}$), and let γ be the probability law of $(g_i : i \in \mathbb{N})$, g_i independent standard normal. Let O denote the unit ball about zero of $\ell_2 \subset \mathbb{R}^{\mathbb{N}}$, $O = \{x \in \mathbb{R}^{\mathbb{N}} : \sum_i x_i^2 \leq 1\}$. Then, for all $\varepsilon > 0$,*

$$\gamma(A+\varepsilon O) \geq \Phi(\Phi^{-1}(\gamma(A)) + \varepsilon). \tag{2.14}$$

The proof is indicated in Exercises 2.2.5 through 2.2.7.

2.2.3 Application to Gaussian Concentration

We would like to translate the isoperimetric inequality in Theorem 2.2.4 into a concentration inequality for functions of $\{g_i\}_{i=1}^n$ about their medians, that is, into a bound for $\gamma\{|f(x) - M| > \varepsilon\}$ for all $\varepsilon > 0$. The following definition describes the functions for which such a translation is almost obvious.

Definition 2.2.5 A function $f : \mathbb{R}^{\mathbb{N}} \mapsto \mathbb{R}$ is Lipschitz in the direction of ℓ_2, or ℓ_2-Lipschitz for short, if it is measurable and if

$$\|f\|_{\mathrm{Lip2}} := \sup\left\{\frac{|f(x) - f(y)|}{|x-y|} : x,y \in \mathbb{R}^{\mathbb{N}}, x \neq y, x - y \in \ell_2\right\} < \infty,$$

where $|x - y|$ is the ℓ_2 norm of $x - y$.

For a measurable function f on $\mathbb{R}^{\mathbb{N}}$, we denote by M_f the median of f with respect to the Gaussian measure γ, defined as $M_f = \inf\{t : \gamma\{x : f(x) \leq t\} > 1/2\}$. Then $\gamma(f \leq M_f) \geq 1/2$ and $\gamma(f \geq M_f) \geq 1/2$, and M is the largest number satisfying these two inequalities.

Theorem 2.2.6 *If f is an ℓ_2-Lipschitz function on $\mathbb{R}^{\mathbb{N}}$, and if M_f is its median with respect to γ, then*

$$\gamma\{x : f(x) \geq M_f + \varepsilon\} \leq (1 - \Phi(\varepsilon/\|f\|_{\text{Lip2}})),$$
$$\gamma\{x : f(x) \leq M_f - \varepsilon\} \leq (1 - \Phi(\varepsilon/\|f\|_{\text{Lip2}})), \tag{2.15}$$

in particular

$$\gamma\{x : |f(x) - M_f| \geq \varepsilon\} \leq 2(1 - \Phi(\varepsilon/\|f\|_{\text{Lip2}})) \leq e^{-\varepsilon^2/2\|f\|_{\text{Lip2}}}, \tag{2.16}$$

for all $\varepsilon > 0$.

Proof Let $A^+ = \{x \in \mathbb{R}^{\mathbb{N}} : f(x) \geq M_f\}$ and $A^- = \{x \in \mathbb{R}^{\mathbb{N}} : f(x) \leq M_f\}$. Then $\gamma(A^+) \geq 1/2$, $\gamma(A^-) \geq 1/2$. Moreover, if $x \in A^+ + \varepsilon O$, then there exists $h \in O$ such that $x - \varepsilon h \in A^+$; hence, $f(x - \varepsilon h) \geq M_f$ and $f(x) + \varepsilon\|f\|_{\text{Lip2}} \geq f(x - \varepsilon h) \geq M_f$; that is, $A^+ + \varepsilon O \subset \{x : f(x) \geq M_f - \varepsilon\|f\|_{\text{Lip2}}\}$. Then the Gaussian isoperimetric inequality (2.14) for $A = A^+$ gives (recall $\Phi^{-1}(1/2) = 0$)

$$\gamma\{f < M_f - \varepsilon\|f\|_{\text{Lip2}}\} \leq 1 - \gamma(A^+ + \varepsilon O) \leq 1 - \Phi(\varepsilon),$$

which is the second inequality in (2.15). Likewise, $A^- + \varepsilon O \subset \{x : f(x) \leq M_f + \varepsilon\|f\|_{\text{Lip2}}\}$, and the isoperimetric inequality applied to A^+ gives the first inequality in (2.15). Finally, (2.16) follows by combination of the previous two inequalities and a known bound for the tail probabilities of a normal variable (Exercise 2.2.8). ■

Let now $X(t)$, $t \in T$, be a separable centred Gaussian process such that $\Pr\{\sup_{t \in T} |X(t)| < \infty\} > 0$. Then $\sup_{t \in T} |X(t)| = \sup_{t \in T_0} |X(t)| < \infty$ a.s., where $T_0 = \{t_k\}_{k=1}^{\infty}$ is a countable subset of T (see Example 2.1.15). Ortho-normalizing (in $L^2(\Pr)$), the jointly normal sequence $\{X(t_k)\}$ yields $X(t_k) = \sum_{i=1}^k a_{ki} g_i$, where g_i are independent standard normal variables, and $\sum_{i=1}^k a_{ki}^2 = EX^2(t_k)$. Then the probability law of the process $X(t_k)$, $k \in \mathbb{N}$, coincides with the law of the random variable defined on the probability space $(\mathbb{R}^{\mathbb{N}}, \mathcal{C}, \gamma)$, $\tilde{X} : \mathbb{R}^{\mathbb{N}} \mapsto \mathbb{R}$, $\tilde{X}(t_k, x) = \sum_{i=1}^k a_{ki} x_i$. This is so because the coordinates of $\mathbb{R}^{\mathbb{N}}$, considered as random variables on the probability space $(\mathbb{R}^{\mathbb{N}}, \mathcal{C}, \gamma)$, are i.i.d. $N(0,1)$. Now define a function $f : \mathbb{R}^{\mathbb{N}} \mapsto \mathbb{R}$ by

$$f(x) = \sup_k \left| \sum_{i=1}^k a_{ki} x_i \right|.$$

The probability law of f under γ is the same as the law of $\sup_{t \in T_0} |X(t)|$, which, in turn, is the same as the law of $\sup_{t \in T} |X(t)|$. Moreover, if $h \in O$, the unit ball of ℓ_2, by Cauchy-Schwarz,

$$|f(x+h) - f(x)|^2 = \sup_k \left| \sum_{i=1}^k a_{ki} h_i \right|^2 \leq \sup_k \left[\sum_{i=1}^k a_{ki}^2 \sum_{i=1}^k h_i^2 \right] \leq \sup_k \sum_{i=1}^k a_{ki}^2 = \sup_k EX^2(t_k).$$

Therefore,

$$\|f\|_{\text{Lip2}} \leq \sigma^2(X), \quad \text{where } \sigma^2 = \sigma^2(X) := \sup_{t \in T} EX^2(t).$$

Recall from an argument at the beginning of the proof of Theorem 2.1.20 that for the processes X we are considering here, $\sigma^2 < \infty$ and the median $M < \infty$. Then Theorem 2.2.6 applies to the function f and gives the following concentration inequality:

Theorem 2.2.7 (The Borell-Sudakov-Tsirelson concentration inequality for Gaussian processes) *Let $X(t)$, $t \in T$, be a centred separable Gaussian process such that $\Pr\{\sup_{t\in T}|X(t)| < \infty\} > 0$, and let M be the median of $\sup_{t\in T}|X(t)|$ and σ^2 the supremum of the variances $EX^2(t)$. Then, for all $u > 0$,*

$$\Pr\left\{\sup_{t\in T}|X(t)| > M+u\right\} \le 1-\Phi(u/\sigma), \quad \Pr\left\{\sup_{t\in T}|X(t)| < M-u\right\} \le 1-\Phi(u/\sigma), \tag{2.17}$$

and hence,

$$\Pr\left\{\left|\sup_{t\in T}|X(t)| - M\right| > u\right\} \le 2(1-\Phi(u/\sigma)) \le e^{-u^2/2\sigma^2}. \tag{2.18}$$

Inequality (2.18) is also true with the median M of $\sup_{t\in T}|X(t)|$ replaced by the expectation $E\left(\sup_{t\in T}|X(t)|\right)$, as we will see in Section 2.5 as a consequence of the Gaussian logarithmic Sobolev inequality (other proofs are possible; see Section 2.1 for a simple proof of a weaker version). But such a result, in its sharpest form, does not seem to be obtainable from (2.18). However, notice that if we integrate in (2.18) and let g be a $N(0,1)$ random variable, we obtain

$$\left|E\sup_{t\in T}|X(t)| - M\right| \le E\left|\sup_{t\in T}|X(t)| - M\right| \le \sigma E|g| = \sqrt{2/\pi}\,\sigma, \tag{2.19}$$

an inequality which is interesting in its own right and which gives, by combining with the same (2.18),

$$\Pr\left\{\left|\sup_{t\in T}|X(t)| - E\sup_{t\in T}|X(t)|\right| > u+\sqrt{2/\pi}\sigma\right\} \le e^{-u^2/2\sigma^2}, \tag{2.20}$$

which is of the right order for large values of u.

Theorem 2.2.7, or even (2.20), expresses the remarkable fact that the supremum of a Gaussian process $X(t)$, centred at its mean or at its median, has tail probabilities not worse than those of a normal variable with the largest of the variances $EX^2(t)$, $t \in T$. In particular, if we knew the size of $E\sup_{t\in T}|X(t)|$, we would have a very exact knowledge of the distribution of $\sup_{t\in T}|X(t)|$. This will be the object of the next two sections.

We complete this section with simple applications of Theorem 2.2.7 to integrability and moments of the supremum of a Gaussian processes.

Corollary 2.2.8 *Let $X(t)$, $t \in T$, be a Gaussian process as in Theorem 2.2.7. Let M and σ also be as in this theorem, and write $\|X\| := \sup_{t\in T}|X(t)|$ to ease notation. Then there exists $K < \infty$ such that with the same hypothesis and notation as in the preceding corollary, for all $p \ge 1$,*

$$(E\|X\|^p)^{1/p} \le 2E\|X\| + (E|g|^p)^{1/p}\sigma \le K\sqrt{p}E\|X\|$$

for some absolute constant K.

Proof Just integrate inequality (2.18) with respect to $pt^{p-1}dt$ and then use that $M \le 2E\|X\|$ (by Chebyshev) and that $\sigma \le \sqrt{\pi/2}\sup_{t\in T}E|X(t)|$. See Exercise 2.1.2. ■

Corollary 2.2.9 *Let $X(t)$, $t \in T$, be a Gaussian process as in Theorem 2.2.7, and let $\|X\|$, M and σ be as in Corollary 2.2.8. Then*

$$\lim_{u\to\infty} \frac{1}{u^2} \log \Pr\{\|X\| > u\} = -\frac{1}{2\sigma^2}$$

and

$$Ee^{\lambda\|X\|^2} < \infty \text{ if and only if } \lambda < \frac{1}{2\sigma^2}.$$

Proof The first limit follows from the facts that the first inequality in (2.17) can be rewritten as

$$\frac{1}{(u-M)^2} \log \Pr\{\|X\| > u\} \le -\frac{1}{\sigma^2}$$

and that $\Pr\{\|X\| > u\} \ge \Pr\{|X(t)| > u\}$ for all $t \in T$ (as, for a $N(0,1)$ variable g, we do have $u^{-2} \log \Pr\{|g| > u/a\} \to -1/2a^2$, e.g., by l'Hôpital's rule). For the second statement, just apply the first limit to $Ee^{\lambda\|X\|} = 1 + \int_0^\infty \int_0^{\lambda\|X\|^2} e^v dv\, d\mathcal{L}(\|X\|)(u) = 1 + \int_0^\infty e^v \Pr\{\|X\| > \sqrt{v/\lambda}\}dv$. ■

Exercises

2.2.1 Prove that if A is closed, so is $s_H(A)$ for any subspace H of dimension n. *Hint*: Conveniently enlarge some of the components in the definition of $s_H(A)$ to make them compact and still keep the same union.

2.2.2 Prove that $(\mathcal{K}, h)$, the space of nonempty compact subsets of S^n with the Hausdorff distance, is a compact metric space. *Hint*: Show that the map $\mathcal{K} \mapsto C(S^n)$, $A \mapsto d(\cdot, A)$, is an isometry between $(\mathcal{K}, h)$ and its image in $(C(S^n), \|\cdot\|_\infty)$ and that this image is compact in $C(S^n)$ (note that $x \mapsto d(x,A)$ is bounded and Lipschitz or see Beers (1993)).

2.2.3 Show that the Lebesgue density theorem holds in S^n for the uniform measure; that is, show that if $\mu(E) > 0$, then μ-almost all points of E satisfy $\lim_{\rho\to 0} [\mu(E \cap C(x,\rho))]/[\mu(C(x,\rho))] = 1$. *Hint*: First adapt the usual proof of the Vitali covering theorem to the sphere, using that $L_n < \infty$ such that any cap of radius 2ρ can be covered by L_n caps of radius ρ. Then use the Vitali covering theorem to show that if for each $0 < \alpha < 1$, A_α is the set of those points in E for which $\liminf_{\rho\to 0} [\mu(E \cap C(x,\rho))]/[\mu(C(x,\rho))] < \alpha < 1$, then $\mu(A_\alpha) = 0$ as follows: if G is an open set containing A_α with $\mu(G) < \mu(A_\alpha)/\alpha$, let $\mathcal{V}$ be the set of caps $C(x,\rho)$ that satisfy $[\mu(E \cap C(x,\rho))]/[\mu(C(x,\rho))] < \alpha$ and are contained in G; get a Vitali subcover and show that its total measure, which is at most $\mu(G)$, is larger than or equal than $\mu(A_\alpha)/\alpha$, a contradiction. Or refer to Mattila (1995).

2.2.4 Prove that for $n \ge 2$, if $\mu(A) \ge 1/2$, then $\mu(A_\varepsilon) \ge 1 - (\pi/8)^{1/2} e^{-(n-1)\varepsilon^2/2}$, where μ is the uniform probability measure on S^n.

2.2.5 Let $\pi_n : \mathbb{R}^{\mathbb{N}} \mapsto \mathbb{R}^n$ be the projection $\pi_n(x) = \pi_n(x_k : k \in \mathbb{N}) = (x_1, \dots, x_n)$. Then show that (a) $\gamma_n = \gamma \circ \pi_n^{-1}$, (b) if $K \subset \mathbb{R}^{\mathbb{N}}$ is compact, then $K = \cap_{n=1}^\infty \pi_n^{-1}(\pi_n(K))$, and (c) $K + tO$, where O is the closed unit ball of ℓ_2, is compact if K is.

2.2.6 Use Theorem 2.2.3 and Exercise 2.2.5 to prove Theorem 2.2.4 in the particular case where A is a compact set.

2.2.7 Since $\mathbb{R}^{\mathbb{N}}$ is Polish, it follows that γ is tight (Proposition 2.1.4). Use this and Exercise 2.2.2 to prove Theorem 2.2.4 for any $A \in \mathcal{C}$.

In the remaining exercises, the process X is as in Theorem 2.2.7.

2.2.8 Let Φ be the $N(0,1)$ distribution function. Then, for all $u \geq 0$, show that $2(1-\Phi(u)) \leq e^{-u^2/2}$. *Hint*: Use the well-known bound $\int_u^\infty e^{-t^2/2}dt \leq u^{-1}e^{-u^2/2}$ for $u \geq \sqrt{2/\pi}$ and differentiation for $0 \leq u \leq \sqrt{2/\pi}$.

2.2.9 Prove the analogue of Theorem 2.2.7 for $\sup_{t\in T} X(t)$ and its median.

2.2.10 Show that $\Pr\{\sup_{t\in T}|X(t)| > u\} \leq 2\Pr\{\sup_{t\in T} X(t) > u\}$.

2.2.11 Show that: (a) The random variable $\sup_{t\in T}|X(t)|$ has a unique median, meaning that M is the only number for which both $\Pr\{\sup_{t\in T}|X(t)| \geq M\}$ and $\Pr\{\sup_{t\in T}|X(t)| \leq M\}$ are larger than or equal to $1/2$. In particular, the distribution function of $\sup_{t\in T}|X(t)|$ is continuous at M. *Hint*: The second equation in (2.15) implies that no number below the largest median of f for the measure γ can be a median; now apply this to the appropriate f. (b) Use the same reasoning to conclude that if $M_a = \inf\{u : \Pr\{\sup_{t\in T}|X(t)| \leq u\} > a\}$, $0 < a < 1$, then the distribution function of $\sup_{t\in T}|X(t)|$ is continuous at M_a.

2.2.12 Let B be a Banach space whose norm $\|\cdot\|$ satisfies the following: there exists a countable subset D of the unit ball of its (topological) dual space B^* such that $\|x\| = \sup_{f\in D}|f(x)|$ for all $x \in B$. For instance, this is true for separable Banach spaces as well as for ℓ_∞. Define a Gaussian random variable X with values in B as a map from some probability space $(\Omega, \Sigma, \Pr)$ into B such that $f(X)$ is a centred normal random variable for every $f \in B^*$. Prove that if $\|X\|$ is finite almost surely, if M is a median of $\|X\|$ and $\sigma^2 = \sup_{f\in D} Ef^2(X)$, then

$$\Pr\{|\|X\| - M| > u\} \leq 2(1-\Phi(u/\sigma)) \leq e^{-u^2/2\sigma^2}. \tag{2.21}$$

2.2.13 Let B be a Banach space as in Exercise 2.2.12, and let X be a centred Gaussian B-valued random variable. Use Exercise 2.2.12 to show that the distribution function $F_{\|X\|}$ of $\|X\|$ is continuous at M_a for all $0 < a < 1$, where M_a is as defined in Exercise 2.2.11 with the supremum of the process replaced by $\|X\|$.

2.3 The Metric Entropy Bound for Suprema of Sub-Gaussian Processes

In this section we define sub-Gaussian processes and obtain the celebrated Dudley's entropy bound for their supremum norm. We are careful about the constants, as they are of some consequence in statistical estimations, at the expense of making the 'chaining argument' (proof of Theorem 2.3.6) slightly more complicated than it could be. Combined with concentration inequalities, these bounds yield good estimates of the distribution of the supremum of a Gaussian process. They also constitute sufficient conditions for sample boundedness and sample continuity of Gaussian and sub-Gaussian processes and provide moduli of continuity for their sample paths which are effectively sharp in light of Sudakov's inequality derived in the next section.

A square integrable random variable ξ is said to be *sub-Gaussian* with parameter $\sigma > 0$ if for all $\lambda \in \mathbb{R}$,

$$Ee^{\lambda\xi} \leq e^{\lambda^2\sigma^2/2}.$$

Developing the two exponentials, dividing by $\lambda > 0$ and by $\lambda < 0$ and letting $\lambda \to 0$ in each case yield $E\xi = 0$; that is, sub-Gaussian random variables are automatically centred. Then, if in the two developments once the expectation term is cancelled, we divide by λ^2 and let $\lambda \to 0$, we obtain $E\xi^2 \leq \sigma^2$.

Aside from normal variables, perhaps the main examples of sub-Gaussian variables are the linear combinations of independent *Rademacher* (or symmetric *Bernoulli*) random variables $\xi = \sum_{i=1}^{n} a_i\varepsilon_i$, where ε_i are independent identically distributed and $\Pr\{\varepsilon_i = 1\} = \Pr\{\varepsilon_i = -1\} = 1/2$. To see that these variables are sub-Gaussian, just note that by Taylor expansion, if ε is a Rademacher variable,

$$Ee^{\lambda\varepsilon} = (e^{\lambda} + e^{-\lambda})/2 \le e^{\lambda^2/2}, \quad \lambda \in \mathbb{R},$$

so that, by independence,

$$Ee^{\lambda\sum a_i\varepsilon_i} \le e^{\lambda^2\sum a_i^2/2}.$$

Both for Gaussian and for linear combinations of independent Rademacher variables, $\sigma^2 = E\xi^2$.

The distributions of sub-Gaussian variables have *sub-Gaussian tails*: Chebyshev's inequality in exponential form, namely,

$$\Pr\{\xi \ge t\} = \Pr\left\{e^{\lambda\xi} \ge e^{\lambda t}\right\} \le e^{\lambda^2\sigma^2/2-\lambda t}, \quad t > 0,\ \lambda > 0,$$

with $\lambda = t/\sigma^2$ and applied as well to $-\xi$, gives that if ξ is sub-Gaussian for σ^2, then

$$\begin{aligned} &\Pr\{\xi \ge t\} \le e^{-t^2/2\sigma^2} \quad \text{and} \quad \Pr\{\xi \le -t\} \le e^{-t^2/2\sigma^2}, \quad \text{hence,} \\ &\Pr\{|\xi| \ge t\} \le 2e^{-t^2/2\sigma^2}, \quad t > 0. \end{aligned} \tag{2.22}$$

The last inequality in (2.22) in the case of linear combinations of independent Rademacher variables is called *Hoeffding's inequality*. Of course, we can be more precise about the tail probabilities of normal variables: simple calculus gives that for all $t > 0$,

$$\frac{t}{t^2+1}e^{-t^2/2} \le \int_t^{\infty} e^{-u^2/2}du \le \min\left(t^{-1}, \sqrt{\pi/2}\right)e^{-t^2/2}, \tag{2.23}$$

(see Exercise 2.2.8).

Back to the inequalities (2.22), we notice that if they hold for ξ, then ξ/c enjoys square exponential integrability for some $0 < c < \infty$: if $c^2 > 2\sigma^2$, then

$$Ee^{\xi^2/c^2} - 1 = \int_0^{\infty} 2te^{t^2}\Pr\{|\xi| > ct\}\,dt \le \frac{2}{c^2/2\sigma^2 - 1} < \infty. \tag{2.24}$$

The collection of random variables ξ on $(\Omega, \Sigma, \Pr)$ that satisfy this integrability property constitutes a vector space, denoted by $L^{\psi_2}(\Omega, \Sigma, \Pr)$, and the functional

$$\|\xi\|_{\psi_2} = \inf\{c > 0 : E\psi_2(|\xi|/c) \le 1\},$$

where $\psi_2(x) := e^{x^2} - 1$ (a convex function which is zero at zero) is a pseudo-norm on it for which L^{ψ_2}, with identification of a.s. equal functions, is a Banach space (Exercise 2.3.5). With this definition, inequality (2.24) shows that

$$\Pr\{|\xi| \ge t\} \le 2e^{-t^2/2\sigma^2} \quad \text{for all } t > 0 \quad \text{implies} \quad \|\xi\|_{\psi_2} \le \sqrt{6}\sigma. \tag{2.25}$$

To complete the set of relationships developed so far, suppose that $\xi \in L^{\psi_2}$ and $E\xi = 0$, and let us show that ξ is sub-Gaussian. We have

$$Ee^{\lambda\xi} - 1 \le E\sum_{k=2}^{\infty}|\lambda^k\xi^k|/k \le \frac{\lambda^2}{2}E\left(\xi^2 e^{|\lambda\xi|}\right).$$

Now we estimate the exponent $|\lambda\xi|$ on the region $|\xi| > 2\lambda\|\xi\|_{\psi_2}^2$ and on its complement to obtain, after multiplying and dividing by $\|\xi\|_{\psi_2}^2$ and using that $a < e^{a/2}$ for all $a > 0$,

$$\frac{\lambda^2}{2}E\left(\xi^2 e^{|\lambda\xi|}\right) \le \frac{\lambda^2\|\xi\|_{\psi_2}^2}{2}e^{2\lambda^2\|\xi\|_{\psi_2}^2}E\left(\frac{\xi^2}{\|\xi\|_{\psi_2}^2}e^{\xi^2/2\|\xi\|_{\psi_2}^2}\right)$$
$$\le \lambda^2\|\xi\|_{\psi_2}^2 e^{2\lambda^2\|\xi\|_{\psi_2}^2}Ee^{\xi^2/\|\xi\|_{\psi_2}^2}/2 \le \lambda^2\|\xi\|_{\psi_2}^2 e^{2\lambda^2\|\xi\|_{\psi_2}^2}.$$

Using $1+a \le e^a$, the last two bounds give

$$Ee^{\lambda\xi} \le e^{3\lambda^2\|\xi\|_{\psi_2}^2}, \tag{2.26}$$

showing that ξ is sub-Gaussian with $\sigma \le \sqrt{6}\|\xi\|_{\psi_2}$. If ξ is symmetric, just developing the exponential gives the better inequality $Ee^{\lambda\xi} \le e^{\lambda^2\|\xi\|_{\psi_2}^2/2}$.

We collect these facts:

Lemma 2.3.1 *If ξ is sub-Gaussian for a constant $\sigma > 0$, then it satisfies the sub-Gaussian tail inequalities (2.22), and therefore, $\xi \in L^{\psi_2}$, with $\|\xi\|_{\psi_2} \le \sqrt{6}\sigma$. Conversely, if ξ is in L^{ψ_2} and is centred, then it is sub-Gaussian for the constant $\sigma \le \sqrt{6}\|\xi\|_{\psi_2}$, and in particular, it also satisfies the inequalities (2.22) for $\sigma = \sqrt{6}\|\xi\|_{\psi_2}$.*

In other words, ignoring constants, for ξ centred, the conditions (a) $\xi \in L^{\psi_2}$ and (b) ξ satisfies the sub-Gaussian tail inequalities (2.22) for some σ_1 and (c) ξ is sub-Gaussian for some σ_2 are all equivalent.

Lemma 2.3.1 extends to random variables whose tail probabilities are bounded by a constant times the sub-Gaussian probabilities in (2.22) as follows.

Lemma 2.3.2 *Assume that*

$$\Pr\{|\xi| \ge t\} \le 2Ce^{-t^2/2\sigma^2}, \quad t > 0, \tag{2.27}$$

for some $C \ge 1$ and $\sigma > 0$, a condition implied by the Laplace transform condition

$$Ee^{\lambda\xi} \le Ce^{\lambda^2\sigma^2/2}, \quad \lambda \in \mathbb{R}. \tag{2.28}$$

Then ξ also satisfies

$$\|\xi\|_{\psi_2} \le \sqrt{2(2C+1)}\sigma. \tag{2.29}$$

Moreover, if in addition $E\xi = 0$, then also

$$Ee^{\lambda\xi} \le e^{3\lambda^2(2(2C+1))\sigma^2}, \quad \lambda \in \mathbb{R}, \tag{2.30}$$

that is, ξ is sub-Gaussian with constant $\tilde{\sigma}^2 = 12(2C+1)\sigma^2$.

Proof The proof of inequality (2.22) shows that (2.28) implies (2.27). The preceding proof showing that (2.22) implies (2.25), with only formal changes, proves that (2.27) implies (2.29). Finally, inequality (2.30) follows from (2.29) and (2.26). ■

This lemma is useful in that showing that a variable ξ is sub-Gaussian reduces to proving the tail probability bounds (2.27) for some $C > 1$, which may be easier than proving them for $C = 1$.

Lemma 2.3.1 (or, more precisely, the inequalities that make it possible) has many important consequences on the size of maxima of sub-Gaussian stochastic processes. The simplest examples of such processes are finite collections of sub-Gaussian variables. The following lemma contains a maximal inequality for variables in $\xi_i \in L^{\psi_2}$ not necessarily centred, and it applies by Lemma 2.3.1 to finite collections of sub-Gaussian variables.

Lemma 2.3.3 *Let $\xi_i \in L^{\psi_2}$, $i = 1, \ldots, N$, $2 \le N < \infty$. Then*

$$\left\| \max_{i \le N} |\xi_i| \right\|_{\psi_2} \le 4\sqrt{\log N} \max_{i \le N} \|\xi_i\|_{\psi_2}, \tag{2.31}$$

and, in particular, there exist $K_p < \infty$, $1 \le p < \infty$, such that

$$\left\| \max_{i \le N} |\xi_i| \right\|_{L^p} \le K_p \sqrt{\log N} \max_{i \le N} \|\xi_i\|_{\psi_2}. \tag{2.32}$$

Proof To prove (a), we may assume that $\max \|\xi_i\|_{\psi_2} = 1$. Then the definition of the ψ_2 norm together with the exponential Chebyshev's inequality gives

$$\begin{aligned} E \max_{i \le N} e^{\xi_i^2/(16 \log N)} &= \int_0^\infty \Pr\left\{ \max_{i \le N} e^{\xi_i^2/(16 \log N)} \ge t \right\} dt \\ &\le e^{1/8} + \sum_{i=1}^N \int_{e^{1/8}}^\infty \Pr\left\{ e^{\xi_i^2/(16 \log N)} \ge t \right\} dt \\ &\le e^{1/8} + 2N \int_{e^{1/8}}^\infty e^{-8(\log N)(\log t)} dt = e^{1/8} + 2N \int_{e^{1/8}}^\infty t^{-8 \log N} dt \\ &= e^{1/8} \left(1 + \frac{2}{8(\log N) - 1} \right) < 2, \end{aligned}$$

proving (2.31). For part (b), use that $\|\zeta\|_{L^{2k}} \le (k)^{1/2k} \|\zeta\|_{L^{\psi_2}}$ for any random variable $\zeta \in L^{\psi_2}$ (as observed earlier) and part (a) to obtain inequality (2.32). ■

It is convenient to have sensible values of K_p at hand, particularly for $p = 1$. The method to obtain the following bound is quite simple and general: let Φ be a nonnegative, strictly increasing, convex function on a finite or infinite interval I, and let ξ_i, $1 \le i \le N$, be random variables taking values in I and such that $E\Phi(\xi_i) < \infty$. We then have, by Jensen's inequality and the properties of Φ,

$$\begin{aligned} \Phi\left(E \max_{i \le N} \xi_i \right) &\le E\Phi\left(\max_{i \le N} \xi_i \right) = E \max_{i \le N} \Phi(\xi_i) \\ &\le \sum_{i=1}^N E\Phi(\xi_i) \le N \max_{i \le N} E\Phi(\xi_i), \end{aligned} \tag{2.33}$$

and, inverting Φ,

$$E \max_{i \le N} \xi_i \le \Phi^{-1}\left(N \max_{i \le N} E\Phi(\xi_i) \right). \tag{2.34}$$

Lemma 2.3.4 *For any $N \geq 1$, if ξ_i, $i \leq N$, are sub-Gaussian random variables admitting constants σ_i, then*

$$E \max_{i \leq N} \xi_i \leq \sqrt{2 \log N} \max_{i \leq N} \sigma_i, \quad E \max_{i \leq N} |\xi_i| \leq \sqrt{2 \log 2N} \max_{i \leq N} \sigma_i. \tag{2.35}$$

Proof We take $\Phi(x) = e^{\lambda x}$ in (2.34). Since ξ_i is sub-Gaussian, we have $E\Phi(\xi_i) \leq e^{\lambda^2 \sigma_i^2/2}$, and (2.34) gives

$$E \max_{i \leq N} \xi_i \leq \frac{\log N}{\lambda} + \frac{1}{2} \lambda \max_{i \leq N} \sigma_i^2.$$

The first inequality in the lemma follows by minimizing in λ in this inequality (*i.e.*, by taking $\lambda = (2 \log N)^{1/2} / \max_{i \leq N} \sigma_i$). The second inequality follows by applying the first to the collection of $2N$ random variables $\eta_i = \xi_i$, $\eta_{n+i} = -\xi_i$, $1 \leq i \leq N$. ■

We now consider more general sub-Gaussian processes.

Definition 2.3.5 A centred stochastic process $X(t)$, $t \in T$, is sub-Gaussian with respect to a distance or pseudo-distance d on T if its increments satisfy the sub-Gaussian inequality, that is, if

$$E e^{\lambda(X(t) - X(s))} \leq e^{\lambda^2 d^2(s,t)/2} \ \lambda \in \mathbb{R}, \quad s, t \in T. \tag{2.36}$$

If instead of condition (2.36) the centred process X satisfies

$$E e^{\lambda(X(t) - X(s))} \leq C e^{\lambda^2 d^2(s,t)/2} \quad \text{or} \quad \Pr\{|X(t) - X(s)| \geq u\} \leq 2C e^{-u^2/2d^2(s,t)},$$

for all $\lambda \in \mathbb{R}$, $u > 0$ and $s, t \in T$ and some $C > 1$, then, by Lemma 2.3.2, X is sub-Gaussian for the distance $\tilde{d}(s,t) := \sqrt{12(2C+1)}d$. Then all the results that follow for sub-Gaussian processes apply as well to processes X satisfying this condition, and the effects on the results themselves of the dilation of the distance d can be easily quantified.

Gaussian processes, that is, processes $X(t)$ such that for every finite set of indices $t_1, \ldots, t_k$, $k < \infty$, the vectors $(X(t_i) : 1 \leq i \leq k)$ are multivariate normal and are sub-Gaussian with respect to the L^2-distance $d_X(s,t) = \|X(t) - X(s)\|_{L^2}$. *Randomized empirical processes* constitute another important class of examples. Let $(S, \mathcal{S}, P)$ be a probability space, and let $X_i : S^{\mathbb{N}} \mapsto S$, $i \in \mathbb{N}$, be the coordinate functions (which are i.i.d. with law P). Given a collection $\mathcal{F}$ of measurable functions on $(S, \mathcal{S})$, the empirical measures indexed by $\mathcal{F}$ and based on $\{X_i\}$ are defined as

$$\left\{ P_n(f) := \frac{1}{n} \sum_{i=1}^{n} f(X_i) : f \in \mathcal{F} \right\}, \quad n \in \mathbb{N},$$

and a related process that has turned out to be an excellent tool in the study of empirical measures is the randomized empirical process, defined for each $n \in \mathbb{N}$ as

$$\left\{ \frac{1}{\sqrt{n}} \sum_{i=1}^{n} \varepsilon_i f(X_i) : f \in \mathcal{F} \right\},$$

where $\{\varepsilon_i\}$ is a sequence of independent Rademacher variables, independent of the variables X_i. Since linear combinations of independent Rademacher variables are sub-Gaussian with respect to their variance, it follows that randomized empirical processes are *sub-Gaussian with respect to $d(f,g) = \|f - g\|_{L^2(P_n)}$ conditionally on the variables X_i.*

Here are two useful observations about sub-Gaussian processes: if X is a Gaussian process with respect to d, then the definition immediately implies that

$$E(X(t)-X(s))^2 \leq d^2(s,t)$$

(as observed earlier, just after the definition of sub-Gaussian variables). Moreover, since for any s,t, $(X(t)-X(s))/d(s,t)$ is a sub-Gaussian variable with variance not exceeding 1, Lemma 2.3.4 implies that if F is a finite subset of $T\times T$ of cardinality N, then

$$E\max_{(s,t)\in F}|X(t)-X(s)| \leq \sqrt{2\log 2N}\max_{(s,t)\in F} d(s,t). \tag{2.37}$$

Inequalities analogous to those in Lemma 2.3.3 for these maxima hold as well.

Given a sub-Gaussian process $X(t)$, $t\in T$, it is of great interest to determine the (stochastic) size of $\sup_{t\in T}|X(t)|$ or of $\sup_{s,t\in T, d_X(s,t)\leq\delta}|X(t)-X(s)|$ or whether X has a version with bounded sample paths or with uniformly d_X-continuous sample paths (or perhaps continuous in another metric). For Gaussian processes, these questions should and have been answered exclusively in terms of the properties of the metric space (T,d), and for sub-Gaussian processes, properties of this metric space do provide good control of these quantities and good sufficient conditions for sample boundedness and continuity. Although there are much more refined analyses (see the notes at the end of the section), we will develop only the very neat and useful chaining method based on Dudley's metric entropy. The reason for not presenting this subject in more generality is that it is not needed in this book.

The following theorem indicates a way to control $\sup_{t\in T}|X(t)|$ based on a combination of the bound in Lemma 2.3.4 with the size of the (pseudo-) metric space (T,d), measured in terms of the size of the most economical coverings. Given a metric or pseudo-metric space (T,d), for any $\varepsilon>0$, its *covering number* $N(T,d,\varepsilon)$ is defined as the smallest number of closed d-balls of radius ε needed to cover T, formally, if $B(t,\varepsilon):=\{s\in T: d(s,t)\leq\varepsilon\}$,

$$N(T,d,\varepsilon):=\min\left\{n:\ \text{there exist } t_1,\dots,t_n\in T \text{ such that } T\subseteq \cup_{i=1}^n B(t_i,\varepsilon)\right\},$$

where we take the minimum of the empty set to be infinite. The *packing numbers*

$$D(T,d,\varepsilon):=\max\left\{n:\ \text{there exist } t_1,\dots,t_n\in T \text{ such that } \min_{1\leq i,j\leq n} d(t_i,t_j)>\varepsilon\right\}$$

are sometimes useful and are equivalent to the covering numbers: it is easy to see (and we will use it without explicit mention) that, for all $\varepsilon>0$,

$$N(T,d,\varepsilon)\leq D(T,d,\varepsilon)\leq N(T,d,\varepsilon/2).$$

The logarithm of the covering number of (T,d) is known as its *metric entropy*.

Theorem 2.3.6 *Let (T,d) be a pseudo-metric space, and let $X(t)$, $t\in T$, be a stochastic process sub-Gaussian with respect to the pseudo-distance d, that is, one whose increments satisfy condition (2.36). Then, for all finite subsets $S\subseteq T$ and points $t_0\in T$, the following inequalities hold:*

$$E\max_{t\in S}|X(t)| \leq E|X(t_0)| + 4\sqrt{2}\int_0^{D/2}\sqrt{\log 2N(T,d,\varepsilon)}\,d\varepsilon, \tag{2.38}$$

where D is the diameter of (T,d)*, and*

$$E \max_{\substack{s,t\in S\\ d(s,t)\le\delta}} |X(t)-X(s)| \le (16\sqrt{2}+2)\int_0^{\delta}\sqrt{\log 2N(T,d,\varepsilon)}\,d\varepsilon, \tag{2.39}$$

for all $\delta > 0$*, where the integrals are taken to be 0 if* $D=0$*.*

Proof If the d-diameter D of T is zero, or if $\int_0^{D/2}\sqrt{\log N(T,d,\varepsilon)}\,d\varepsilon=\infty$, there is nothing to prove. Thus, we assume that $D>0$ and that the entropy integral is finite, in which case (T,d) is totally bounded and, in particular, $D<\infty$. By taking $(X(t)-X(t_0))/((1+\delta)D)$ instead of $X(t)$ and $d/((1+\delta)D)$ for any small δ instead of d, we may assume that $X(t_0)=0$ and $1/2<D<1$. Given $S\subset T$ finite, since $d(s,t)=0$ implies $X(t)=X(s)$ a.s., we can identify points of S at d-distance zero from each other; that is, we can assume that d is a proper distance on S. We also can assume that S has cardinality at least 2. Since S is finite, there is $k_1\in\mathbb{N}$ such that for each $t\in T$, the ball $B(t,2^{-k_1})$ contains at most one point from S. Set $T_{k_1}=S$, which has cardinality at most $N(T,d,2^{-k_1})$, set $T_0=\{t_0\}$ and for $1\le k<k_1$, let T_k be a set of centres of $N(T,d,2^{-k})$ d-balls of radius 2^{-k} covering T. For each $s\in S$, we construct a chain $(\pi_{k_1}(s),\pi_{k_1-1}(s),\dots,\pi_0(s))$ with links $\pi_k(s)\in T_k$, $0\le k\le k_1$, as follows: $\pi_{k_1}(s)=s$ and, given $\pi_k(s)$, $k_1\ge k>0$, $\pi_{k-1}(s)$ is taken to be a point in T_{k-1}, for which the ball $B(\pi_{k-1}(s),2^{-(k-1)})$ contains $\pi_k(s)$, this being done in such a way that $\pi_{k-1}(s)$ depends only on $\pi_k(s)$ in the sense that if $\pi_k(s)=\pi_k(t)$, then $\pi_{k-1}(s)=\pi_{k-1}(t)$. Note that $\pi_0(s)=t_0$ for all s. In particular, for each $0\le k\le k_1$, the number of 'subchains' $(\pi_k(s),\pi_{k-1}(s),\dots,\pi_0(s))$, $s\in S$, is exactly $\mathrm{Card}\{\pi_k(s):s\in S\}\le N(T,d,2^{-k})$. In particular, for $k=1,\dots,k_1$,

$$\mathrm{Card}\{(X(\pi_k(s))-X(\pi_{k-1}(s))):s\in S\}=\mathrm{Card}\{\pi_k(s):s\in S\}\le N(T,d,2^{-k}).$$

Moreover, since $\pi_k(s)\in B(\pi_{k-1}(s),2^{-(k-1)})$,

$$\left[E(X(\pi_k(s))-X(\pi_{k-1}(s)))^2\right]^{1/2}\le d(\pi_k(s),\pi_{k-1}(s))\le 2^{-k+1},\quad k=1,\dots,k_1.$$

Hence, by inequality (2.37),

$$E\max_{s\in S}|X(\pi_k(s))-X(\pi_{k-1}(s))|\le 2^{-k+1}\sqrt{2\log 2N(T,d,2^{-k})},\quad k=1,\dots,k_1.$$

(Note that $N(T,d,2^{-k})\ge 2$ for $k\ge 1$ because $D>1/2$, so this inequality holds even if $\mathrm{Card}(\pi_k(s):s\in S\}=1$.) Therefore, noting that $X(\pi_0(s))=X(t_0)=0$ and $X(\pi_{k_1}(s))=X(s)$, we have

$$\begin{aligned}
E\max_{s\in S}|X(s)| &\le \sum_{k=1}^{k_1} E\max_{s\in S}|X(\pi_k(s))-X(\pi_{k-1}(s))|\\
&\le \sum_{k=1}^{\infty} 2^{-k+1}\sqrt{2\log 2N(T,d,2^{-k})}\\
&\le 4\int_0^{1/2}\sqrt{2\log 2N(T,d,\varepsilon)}\,d\varepsilon.
\end{aligned}$$

Replacing $X(t)$ by $(X(t)-X(t_0))/D$ and d by $d/(1+\delta)D$ and letting $\delta\to 0$, we obtain inequality (2.38).

Given $\delta < \text{diam}(T)$, let $k(\delta) = \min\{k \in \mathbb{N} : 2^{-k} \le \delta\}$. Define

$$U = \{(x,y) \in T_{k(\delta)} \times T_{k(\delta)} : \exists\, u,v \in S, d(u,v) \le \delta, \pi_{k(\delta)}(u) = x, \pi_{k(\delta)}(v) = y\},$$

and given $(x,y) \in U$, fix $u_{x,y},\ v_{x,y} \in S$, such that $\pi_{k(\delta)}(u_{x,y}) = x$, $\pi_{k(\delta)}(v_{x,y}) = y$, $d(u_{x,y}, v_{x,y}) \le \delta$. For $s,t \in S$ such that $d(s,t) \le \delta$, obviously, $(x,y) := (\pi_{k(\delta)}(s), \pi_{k(\delta)}(t)) \in U$, and we can write

$$\begin{aligned}|X(t) - X(s)| &\le |X(t) - X(\pi_{k(\delta)}(t))| + |X(\pi_{k(\delta)}(t)) - X(v_{x,y})| + |X(v_{x,y}) - X(u_{x,y})| + |X(u_{x,y}) \\ &\quad - X(\pi_{k(\delta)}(s))| + |X(\pi_{k(\delta)}(s)) - X(s)| \\ &\le \sup_{(x,y)\in U} |X(u_{x,y}) - X(v_{x,y})| + 4\max_{r\in S} |X(r) - X(\pi_{k(\delta)}(r))|.\end{aligned}$$

Since $\text{Card}(U) \le (N(T,d,2^{-k(\delta)}))^2$ and, for $(x,y) \in U$, $d(u_{x,y}, v_{x,y}) \le \delta$, inequality (2.37) gives

$$E \sup_{(x,y)\in U} |X(u_{x,y}) - X(v_{x,y})| \le \delta\sqrt{2\log 2N^2(T,d,2^{-k(\delta)})}.$$

Next, the proof of (2.38) gives

$$E\max_{r\in S} |X(r) - X(\pi_{k(\delta)}(r))| \le \sum_{k>k(\delta)} 2^{-k+1}\sqrt{2\log 2N(T,d,2^{-k})}.$$

We then conclude from the last three inequalities that

$$\begin{aligned}E \max_{\substack{s,t\in S \\ d(s,t)\le\delta}} |X(t) - X(s)| &\le 2\delta\sqrt{\log\sqrt{2}N(T,d,2^{-k(\delta)})} + 4\sum_{k>k(\delta)} 2^{-k+1}\sqrt{2\log 2N(T,d,2^{-k})} \\ &\le (16\sqrt{2}+2)\int_0^\delta \sqrt{\log 2N(T,d,\varepsilon)}\, d\varepsilon. \quad \blacksquare\end{aligned}$$

Theorem 2.3.6 implies the existence of versions of $X(t)$ whose sample paths are bounded and uniformly continuous for d, actually, that this holds for all the separable versions of X, and they do exist (recall Proposition 2.1.12 complemented by Exercise 2.3.6, and note that the entropy condition obviously implies that (T,d) is a separable pseudo-metric space). For the next theorem, recall the definition of sample bounded and sample continuous processes (Definition 2.1.3).

Theorem 2.3.7 *Let (T,d) be a metric or pseudo-metric space, and let $X(t)$, $t \in T$, be a sub-Gaussian process relative to d. Assume that*

$$\int_0^\infty \sqrt{\log N(T,d,\varepsilon)}\, d\varepsilon < \infty. \tag{2.40}$$

Then

(a) $X(t)$, $t \in T$, is sample d-continuous (in particular, X admits a separable version), and
(b) Any separable version of $X(t)$, $t \in T$, that we keep denoting by $X(t)$ has almost all its sample paths bounded and uniformly d-continuous, and satisfies the inequalities

$$E\sup_{t\in T} |X(t)| \le E|X(t_0)| + 4\sqrt{2}\int_0^{D/2} \sqrt{\log 2N(T,d,\varepsilon)}\, d\varepsilon, \tag{2.41}$$

where $t_0 \in T$, D is the diameter of (T,d) and

$$E \sup_{\substack{s,t\in T\\ d(s,t)\le\delta}} |X(t)-X(s)| \le (16\sqrt{2}+2)\int_0^\delta \sqrt{\log 2N(T,d,\varepsilon)}\, d\varepsilon, \tag{2.42}$$

for all $\delta > 0$.

Proof The entropy condition implies that (T,d) is totally bounded, in particular, separable. Then, if T_0 is a countable dense set and $T_n \nearrow T_0$, T_n finite, the monotone convergence theorem together with inequality (2.38) implies that both this inequality holds for $\sup_{t\in T_0}|X(t)|$ and this random variable is almost surely finite. Likewise, monotone convergence also proves inequality (2.39) for T_0 and, in particular, that for any sequence $\delta_n \searrow 0$,

$$E \sup_{\substack{s,t\in T_0\\ d(s,t)\le\delta_n}} |X(t)-X(s)| \searrow 0.$$

This implies not only that these random variables are finite a.s. but also that $\sup_{s,t\in T_0, d(s,t)\le\delta_n}|X(t)-X(s)| \searrow 0$ a.s. Hence, there exists a set $\Omega_0 \subseteq \Omega$ with $\Pr(\Omega_0)=1$ such that the restriction $X|_{T_0}$ of X to T_0 has bounded and d-uniformly continuous sample paths $t \mapsto X(t,\omega)$, $t\in T_0$, for all $\omega\in\Omega_0$. If we extend each of these paths to T by continuity, we obtain a separable version $\tilde{X}$ of the process X with almost all its sample paths bounded and d-uniformly continuous and such that the inequalities (2.38) and (2.39) hold for $\sup_{t\in T}|\tilde{X}(t)|$ and $\sup_{s,t\in T, d(s,t)\le\delta}|\tilde{X}(t)-\tilde{X}(s)|$, respectively (as these suprema equal the corresponding suprema over T_0 for all $\omega\in\Omega_0$). This proves part (a) and the inequalities in part (b) for the version just constructed. Now, if $\bar{X}$ is any separable version of X and T_0 is the countable set from Definition 2.1, we can apply to them the same reasoning as earlier and conclude part (b). ■

The chaining argument also can be adapted to obtain a metric entropy bound on the *modulus of continuity* of a sample continuous Gaussian or sub-Gaussian process.

Theorem 2.3.8 (Dudley's theorem) *If $X(t)$, $t\in T$, is a Gaussian process for a pseudo-metric d such that (T,d) has positive d-diameter and satisfies the metric entropy condition (2.40), then, for any separable version of X (still denoted by X), we have, with the convention $0/0=0$,*

$$E\left[\sup_{s,t\in T}\frac{|X(t)-X(s)|}{\int_0^{d(s,t)}\sqrt{\log N(T,d,\varepsilon)}\,d\varepsilon}\right] < \infty. \tag{2.43}$$

Proof The main part of the proof consists of showing that

$$\sup_{s,t\in T}\frac{|X(t)-X(s)|}{\int_0^{d(s,t)}\sqrt{\log N(T,d,\varepsilon)}\,d\varepsilon} < \infty \ a.s. \tag{2.44}$$

Once this is proved, (2.43) will follow from general properties of Gaussian processes. The proof of (2.44) (which in fact applies also to sub-Gaussian processes) consists of a delicate chaining argument. Set $H(\varepsilon) = \log N(T,d,\varepsilon)$. Instead of discretising at $\varepsilon = 2^{-k}$ as in the proof of Theorem 2.3.6, we define $\varepsilon_1 = 1$ and, inductively, $\delta_k \searrow 0$ and $\varepsilon_k \searrow 0$ as

$$\delta_k = 2\inf\{\varepsilon : H(\varepsilon) \le 2H(\varepsilon_k)\}, \quad \varepsilon_{k+1} = \min(\varepsilon_k/3, \delta_k), \quad k \in \mathbb{N}.$$

Then, since $\varepsilon_{k+1} \le \varepsilon_k/3$, we have $\varepsilon_k \le 3(\varepsilon_k - \varepsilon_{k+1})/2$; also, if $\varepsilon_{k+1} = \delta_k$, then $H(\varepsilon_{k+1}) \le H(2\delta_k/3) \le 2H(\varepsilon_k)$, so $\int_{\varepsilon_{k+1}}^{\varepsilon_k} H^{1/2}(x)dx \le 2\varepsilon_k H^{1/2}(\varepsilon_k)$, whereas if $\varepsilon_{k+1} = \varepsilon_k/3$, then $\int_{\varepsilon_{k+1}}^{\varepsilon_k} H^{1/2}(x)dx \le 2\varepsilon_{k+1} H^{1/2}(\varepsilon_{k+1})$. This gives, for each n,

$$\frac{2}{3}\sum_{k=n}^{\infty} \varepsilon_k H^{1/2}(\varepsilon_k) \le \sum_{k=n}^{\infty} (\varepsilon_k - \varepsilon_{k+1}) H^{1/2}(\varepsilon_k) \le \int_0^{\varepsilon_n} H^{1/2}(x)dx \le 4\sum_{k=n}^{\infty} \varepsilon_k H^{1/2}(\varepsilon_k), \tag{2.45}$$

and the sums converge because, by (2.40), so does the integral. We also have, for each k,

$$H(\varepsilon_{k+2}) \ge H(\varepsilon_{k+1}/3) \ge H(\delta_k/3) \ge 2H(\varepsilon_k). \tag{2.46}$$

Finally, $\{\delta_k\}$ relates to $\{\varepsilon_k\}$ as follows: by definition, if $\tau < \delta_k/2$, then $H(\tau) > 2H(\varepsilon_k) \ge H(\varepsilon_k)$ so that $\delta_k \le 2\varepsilon_k$, which gives

$$\varepsilon_{k+1} \le \delta_k \le 6\varepsilon_{k+1}. \tag{2.47}$$

For each k, let T_k be a set of cardinality $N(\delta_k) = N(T,d,\delta_k)$ and δ_k-dense in T for d, and let $G_k = \{(s,t) : s \in T_{k-1}, t \in T_k\}$. Then Card($T_k$)$= e^{H(\delta_k)} \le e^{2H(\varepsilon_k)}$ by definition of δ_k, and Card(G_k)$\le e^{4H(\varepsilon_k)}$. Then the sub-Gaussian tail bound (2.22) combined with the bound on the cardinality of G_k gives

$$\sum_k \Pr\left\{\max_{s\in T_{k-1}, t\in T_k} \frac{|X(t)-X(s)|}{d(s,t)} \ge 3H^{1/2}(\varepsilon_k)\right\} \le 2\sum_k e^{4H(\varepsilon_k)-9H(\varepsilon_k)/2} \le 2\sum_k e^{-H^{1/2}(\varepsilon_k)/2},$$

which is finite because by (2.46) this last series is dominated by the sum of two convergent geometric series. Hence, by the Borel-Cantelli lemma, there exists $n_0(\omega) < \infty$ a.s. such that

$$\frac{|X(t,\omega)-X(s,\omega)|}{d(s,t)} \le 3H^{1/2}(\varepsilon_n), \quad \text{for all } (s,t) \in G_n \text{ and } n \ge n_0(\omega). \tag{2.48}$$

Next, given $n \in \mathbb{N}$, and $t \in T$, let $\pi_n(t) \in T_n$ be such that $d(t,\pi_n(t)) < \delta_n$. The metric entropy being finite, any separable version of X has almost all its sample paths continuous by Theorem 2.3.7; hence, there is a set of measure one Ω_1 such that if $\omega \in \Omega_1$, both $n_0(\omega) < \infty$ and $X(\pi_k(t),\omega)$ converges to $X(t,\omega)$ for all $t \in T$ (actually, there is no need to invoke this theorem because it is easy to see that $\{X(\pi_k(t),\omega))\}$ is a Cauchy sequence for all ω such that $n_0(\omega) < \infty$ by (2.48) and finiteness of the entropy integral. Hence, we can take a version of X such that, for these ω, $X(t,\omega) = \lim X(\pi_n(t),\omega))$.

Then, if $n \ge n_0(\omega)$ and $\varepsilon_{n-1} < d(s,t) \le \varepsilon_n$, $s,t \in T$, the preceding two observations and the fact that $d(\pi_k(s), \pi_k(t)) \le d(s,t) + 2\delta_k$, give

$$\begin{aligned}
|X(t,\omega) - X(s,\omega)| &\le |X(\pi_n(t),\omega) - X(\pi_n(s),\omega)| + \sum_{k=n}^{\infty} |X(\pi_k(t),\omega) - X(\pi_k(t),\omega)| \\
&\quad + \sum_{k=n}^{\infty} |X(\pi_k(s),\omega) - X(\pi_k(s),\omega)| \\
&\le 3(d(s,t) + 2\delta_n)H^{1/2}(\varepsilon_n) + 12\sum_{k=n}^{\infty} \delta_k H^{1/2}(\varepsilon_{k+1}) \\
&\le 39 d(s,t) H^{1/2}(d(s,t)) + 108 \int_0^{d(s,t)} H^{1/2}(x)dx \\
&\le 147 \int_0^{d(s,t)} H^{1/2}(x)dx,
\end{aligned}$$

where, besides (2.48) and the convergence of $X(\pi_k(t))$, we have used (2.47) and (2.45). Thus, the modulus $\int_0^{d(s,t)} H^{1/2}(x)dx$ for $X(t,\omega)$ is valid for $d(s,t) \le \varepsilon_{n_0(\omega)}$, and hence, by total boundedness of T, it is valid for all $d(s,t)$ and for all $\omega \in \Omega_1$. This proves (2.44)

Next, we show how (2.44) implies (2.43). Set

$$U = \{u = (u_1, u_2) : u_1, u_2 \in T, d(u_1, u_2) \neq 0\},$$

and define on U the pseudo-metric $D(u,v) = d(u_1, v_1) + d(u_2, v_2)$. Then (U, D) is a separable metric or pseudo-metric space because (T, d) is separable by Proposition 2.1.12. Consider the Gaussian process

$$Y(u) = \frac{X(u_2) - X(u_1)}{J(d(u_1, u_2))}, \quad u \in U,$$

where $J(x) = \int_0^x \sqrt{\log N(T, d, \varepsilon)}d\varepsilon$, and note that $J(x) > 0$ for all $x > 0$ (as the diameter of T is not zero). This is a Gaussian process on U with bounded sample paths (by (2.44)). It also has continuous paths for D because

$$\begin{aligned}
|Y(u) - Y(u^0)| &\le \frac{|X(u_2) - X(u_1) - (X(u_2^0) - X(u_1^0))|}{J((d(u_1^0, u_2^0))} \\
&\quad + \left(\sup_{s,t \in T} |X(t) - X(s)| \right) \left| \frac{1}{J((d(u_1, u_2))} - \frac{1}{J((d(u_1^0, u_2^0))} \right|
\end{aligned}$$

tends to zero as $u \to u^0$ in the D-distance because of (a) the sample continuity of X, (b) the first part of the theorem, (c) the continuity of $J(x)$ and (d) $J(d(u_1^0, u_2^0)) > 0$. In particular, Y is a separable Gaussian process on (U, D) with bounded sample paths; hence,

$$E \sup_{u \in U} |Y(u)| < \infty$$

by part (a) of Theorem 2.1.20, proving (2.43). ■

In fact, Theorem 2.1.20 yields more than just first-moment integrability in (2.43) once (2.44) is proved, namely, square exponential integrability.

Exercises

2.3.1 The main ingredient in the basic estimates of Theorem 2.3.6 is clearly the first maximal inequality in (2.37) (hence, Lemma 2.3.4). Replace this inequality with the maximal inequality (2.31) for the ψ_2-norm from Lemma 2.3.3 in the proof of Theorem 2.3.6 to obtain that if $X(t)$, $t \in T$, is a sub-Gaussian process for a pseudo-distance d for which (T,d) satisfies the entropy condition (2.40), then the following inequalities hold for any separable version of X:

$$\left\| \sup_{t\in T} |X(t)| \right\|_{\psi_2} \le \|X(t_0)\|_{\psi_2} + 16\sqrt{6}\int_0^D \sqrt{\log N(T,d,\varepsilon)}\,d\varepsilon,$$

where $t_0 \in T$ is arbitrary and D is the d-diameter of T, and

$$\left\| \sup_{\substack{s,t\in T\\ d(s,t)\le\delta}} |X(t)-X(s)| \right\|_{\psi_2} \le 128\sqrt{3}\int_0^\delta \sqrt{\log N(T,d,\varepsilon)}\,d\varepsilon,$$

for any $\delta > 0$ In particular, these inequalities also hold for the L^p-norms of these random variables, $p < \infty$, possibly with different constants.

2.3.2 Brownian motion on $[0,1]$ is defined as a centred Gaussian process $X(t)$ with continuous sample paths and such that $X(0) = 0$ a.s., $E(X(s) - X(t))^2 = |t - s|$, $s,t \in [0,1]$. Prove the existence of Brownian motion, and show that $\sup_{s,t\in[0,1]} |X(t) - X(s)|/\sqrt{|t-s||\log|t-s||} < \infty$ almost surely.

2.3.3 For real random variables X_i, give an upper bound for $E\sup_{t\in\mathbb{R}} \left|1/\sqrt{n}\sum_{i=1}^n \varepsilon_i I(X_i \le t)\right|$, $n \in \mathbb{N}$; in particular, prove that $E\sup_{t\in\mathbb{R}} \left|1/n\sum_{i=1}^n \varepsilon_i I(X_i \le t)\right| \to 0$ (Glivenko-Cantelli theorem). *Hint*: Conditionally on $\{X_i\}$, take $d^2(s,t) = 1/n\sum_{i=1}^n (I(X_i \le t) - I(X_i \le s)|)^2$, and notice that if $X_{(i)}$, $i = 1,\dots,n$, are the order statistics, $d(s,t) = 0$ if (and only if) both s and t belong to one of the sets $(-\infty, X_{(1)}]$, $(X_{(n)}, \infty)$ or $(X_{(i)}, X_{(i+1)}]$, $i = 1,\dots,n-1$. Note also that $d(s,t) \le 1$ for all s,t. Deduce that $N(\mathbb{R},d,\varepsilon) \le n+1$ for all $\varepsilon > 0$ and that $D \le 1$. The bound follows from this estimate and the entropy integral bound.

2.3.4 (Alternate proof of inequality (2.39) with a slightly larger constant.) Define $V = \{(s,t) \in T\times T : d(s,t) \le \delta\}$ and on V the process $Y(u) = X(t_u) - X(s_u)$, where $u = (s_u,t_u) \in V$. Take on V the pseudo-distance $\rho(u,v) := \|Y(u) - Y(v)\|_{\psi_2}$. One has that $Y(v)$ is sub-Gaussian for $\sqrt{6}\rho$ on V, that $2\max_{u\in V}\|Y(u)\|_{\psi_2} \le 2\sqrt{6}\delta$ and that $\rho(u,v) \le \sqrt{6}(d(t_u,t_v) + d(s_u,s_v))$, all by Lemma 2.3.1. Thus, one can apply inequality (2.38) to Y for ρ, using that the first of the preceding two inequalities gives a bound for the ρ-diameter of V and that the second implies $N(V,\rho,4\sqrt{6}\varepsilon) \le N^2(T,d,\varepsilon)$.

2.3.5 Use the fact that the function $e^{x^2} - 1$ is convex and zero at zero to show that $\|\cdot\|_{\psi_2}$ is a (pseudo-)norm on the space L^{ψ_2} of all the random variables $\xi : \Omega \mapsto \mathbb{R}$ such that $Ee^{\lambda\xi^2} < \infty$ for some $\lambda > 0$ (with identification of a.s. equal functions). Show that the resulting normed space is complete.

2.3.6 Show that Proposition 2.1.12 holds true for sub-Gaussian processes.

2.3.7 Show that a separable stochastic process $X(t)$, $t \in T$, is sample continuous on (T,d) iff there exists a Borel probability measure on $C_u(T,d)$, the Banach space of bounded and uniformly continuous functions on (T,d), whose finite-dimensional marginals $\mu \circ (\delta_{t_1},\dots,\delta_{t_n})^{-1}$ are the marginals $\mathcal{L}(X(t_1),\dots,X(t_n))$, for all $t_i \in T$, $i \le n$, $n \in \mathbb{N}$.

2.3.8 Prove the following inequality, which is a qualitative improvement on (2.31) as it does not assume a finite number of variables: there exists a universal constant $K < \infty$ such that

$$\left\| \sup_k \frac{|\xi_k|}{\psi_2^{-1}(k)} \right\|_{\psi_2} \le K \sup_k \|\xi_k\|_{\psi_2}$$

with $\|\xi_k\|_{\psi_2}$ replaced by $\|\xi_k\|_{L^2}$ if the variables ξ_i are normal. *Hint*: Assume $\|\xi_k\|_{\psi_2} \le 1$. Then using a union bound,

$$\Pr\left\{\exp\left[\sup_{k\ge 9}\left(\frac{|\xi_k|}{\sqrt{6\log k}}\right)^2\right] > t\right\} \le \sum_{k=9}^{\infty}\Pr\left\{e^{|\xi_k|^2} > e^{6(\log k)(\log t)}\right\},$$

and then apply inequality (2.22) together with the fact that for $t \ge 3/2$ and $k \ge 9$, $\log(kt) \le 3(\log k)(\log t)$. Use the resulting bound to show that

$$E\exp\left[\sup_{k\ge 9}\left(|\xi_k|/(\sqrt{6\log k})\right)^2\right] < 2.$$

2.3.9 Let X_i, $i \le n$, be separable centred Gaussian processes such that $E\|X_i\|_\infty < \infty$ (where $\|\cdot\|_\infty$ denotes the supremum norm), and let σ_i^2 and M_i be, respectively, their sup of second moments and median. Prove that

$$E\max_{i\le n}\|X_i\|_\infty \le \max_{i\le n} E\|X_i\|_\infty + (8\sqrt{\log n} + \sqrt{2/\pi})\max_{i\le n}\sigma_i.$$

Hint: By Theorem 2.2.7 and Lemma 2.3.2, the variables $|\|X_i\| - M_i|$ have ψ_2-norm bounded by $2\sigma_i$, and the result then follows from Lemma 2.3.3 and inequality (2.19).

2.3.10 Show that there exists $K < \infty$ such that if $Y(t)$, $t \in T$, is a centred Gaussian process such that $d_Y^2(s,t) = E(Y(t) - Y(s))^2 \le d^2(s,t)$ and (T,d) is totally bounded, then

$$E\sup_{d(s,t)<\delta}|Y(t) - Y(s)| \le K\left[\sup_{t\in T} E\sup_{s\in T:d(s,t)<\delta}|Y(t) - Y(s)| + \delta(\log N(T,d,\delta))^{1/2}\right].$$

Hint: Let U be the set of centres of $N(T,d,\delta)$ d-balls of radius δ covering T. Apply the result in Exercise 2.3.9 to the processes $Y_u = Y - Y(u)$, $u \in U$, and inequality (2.35) to $\max_{u,v\in U:d(u,v)<3\delta}|Y(u) - Y(v)|$.

2.4 Anderson's Lemma, Comparison and Sudakov's Lower Bound

In this section we deal with the general question of how comparison of the distributions of the supremum of two Gaussian processes follows from comparison of their covariances or of their induced metric structures. Perhaps the most important results of this kind are Anderson's inequality regarding the probability, relative to a centred Gaussian measure on $\mathbb{R}^n$, of a convex symmetric set and its translates, and Slepian's lemma that allows comparing the distributions of the suprema of $X(t)$ and $Y(t)$ if the covariance of one of the processes dominates the other. Anderson's lemma is related to the fact that centred Gaussian measures on $\mathbb{R}^n$ are log-concave.

These results have several important consequences, and we will examine two particularly interesting ones, the Khatri-Sidak inequality and Sudakov's inequality, that compare, for a jointly normal variable $(g_1,\ldots,g_n)$, the distribution of $\max_{1\le i\le n}|g_i|$ with the maximum of related independent normal random variables. Sudakov's inequality shows that Dudley's entropy bound is effectively sharp and, in this sense, complements it.

2.4.1 *Anderson's Lemma*

A set C in a vector space is convex and symmetric if $\sum_{i=1}^n \lambda_i x_i \in C$ whenever $x_i \in C$ and $\lambda_i \in \mathbb{R}$ satisfy $\sum_{i=1}^n |\lambda_i| = 1$, $n < \infty$. Example: Balls centred at the origin in Banach spaces,

$\{x : \|x\| \le c\}$. Anderson's lemma states that for a centred Gaussian measure μ on $\mathbb{R}^n$, if C is a measurable, convex, symmetric set, then

$$\mu(C+x) \le \mu(C),$$

for all $x \in \mathbb{R}^n$. Suppose now that $X = Y + Z$, where Y and Z are two independent centred Gaussian random vectors in $\mathbb{R}^n$, which holds if and only if the difference of covariances $C_X - C_Y$ is nonnegative definite. Then

$$\Pr\{X \in C\} = \int \Pr\{Y \in C - z\} d\mathcal{L}(Z)(z) \le \Pr\{Y \in C\}.$$

This inequality is stronger than $E\|Y+Z\|^p \ge E\|Y\|^p$ for all $p \ge 1$, which follows from it and also from Jensen's inequality. Both Anderson's inequality and its corollary on comparison of Gaussian probabilities are quite useful. The modern proof of Anderson's lemma uses the Brunn-Minkowski inequality, or inequalities similar to it, expressing the log-concavity of the function $A \mapsto m(A)$, where m is Lebesgue measure and, as a consequence (of a slightly stronger inequality) of $A \mapsto \mu(A)$, μ-Gaussian and centred.

We start with the Brunn-Minkowski inequality for Lebesgue measure in $\mathbb{R}$. Given two sets A and B in a vector space, their Minkowski addition is $A+B = \{x+y : x \in A, y \in B\}$, and λA is defined as $\lambda A = \{\lambda x : x \in A\}$. In this subsection, m will stand for Lebesgue measure on $\mathbb{R}^n$ for any n.

Lemma 2.4.1 *Let A and B be Borel measurable sets in $\mathbb{R}$. Then*

$$m(A+B) \ge m(A) + m(B).$$

Proof Note that $A+B$ is Lebesgue measurable as it is the image by a continuous function of the Borel set $A \times B$, hence, analytic. Regularity of m by compact sets reduces the problem to A and B compact. Since m is invariant by translations, neither side of the inequality changes if we translate the sets A and/or B; hence, by taking $A + \{-\sup A\}$ and $B + \{-\inf B\}$ instead of A and B, we can assume $A \subset \{x \le 0\}$, $B \subseteq \{x \ge 0\}$ and $A \cap B = \{0\}$. But then $m(A+B) \ge m(A \cup B) = m(A) + m(B)$. ■

Theorem 2.4.2 (Précopa-Leindler theorem) *Let f, g, φ be Lebesgue measurable functions on $\mathbb{R}^n$ taking values in $[0,\infty]$ and satisfying, for some $0 < \lambda < 1$ and all $u, v \in \mathbb{R}^n$,*

$$\varphi(\lambda u + (1-\lambda)v) \ge f^\lambda(u) g^{1-\lambda}(v). \tag{2.49}$$

Then

$$\int \varphi\, dm \ge \left(\int f\, dm\right)^\lambda \left(\int g\, dm\right)^{1-\lambda}. \tag{2.50}$$

Proof The proof is by induction on the dimension n. Assume that $n = 1$. We can divide both sides of inequality (2.49) by $\|f\|_\infty^\lambda \|g\|_\infty^{1-\lambda}$; that is, we can assume without loss of generality that $\|f\|_\infty = \|g\|_\infty = 1$. Then, for $0 \le t < 1$, the sets $\{x : f(x) \ge t\}$ and $\{x : g(x) \ge t\}$ are not empty, and we have

$$\lambda\{f \ge t\} + (1-\lambda)\{g \ge t\} \subseteq \{\varphi \ge t\},$$

since, by (2.49), if $f(u) \ge t$ and $g(v) \ge t$, then $\varphi(\lambda u + (1-\lambda)v) \ge t$. But then, by Lemma 2.4.1,

$$m\{\varphi \ge t\} \ge \lambda m\{f \ge t\} + (1-\lambda) m\{g \ge t\}.$$

Integrating with respect to t and using the concavity of the logarithm, we obtain

$$\int \varphi\, dm \geq \lambda \int f\, dm + (1-\lambda)\int g\, dm \geq \left(\int f\, dm\right)^{\lambda}\left(\int g\, dm\right)^{1-\lambda},$$

proving the theorem for $n = 1$. Assume now that the result holds for $n-1$, and let φ, f, g, λ be as in the statement of the theorem. Fix a coordinate, say, $x_n = x$, and consider $\varphi_x : \mathbb{R}^{n-1} \mapsto [0,\infty]$, defined by $\varphi_x(t) = \varphi(t,x)$, and likewise define f_x and g_x. Then, for x_1, x_2 such that $x = \lambda x_1 + (1-\lambda)x_2$ and for any $u, v \in \mathbb{R}^{n-1}$,

$$\varphi_x(\lambda u + (1-\lambda)v) = \varphi(\lambda(u,x_1) + (1-\lambda)(v,x_2)) \geq f^{\lambda}(u,x_1)g^{1-\lambda}(v,x_2) = f_{x_1}^{\lambda}(u)g_{x_2}^{1-\lambda}(v).$$

Hence, induction gives

$$\int_{\mathbb{R}^{n-1}} \varphi_x\, dm \geq \left(\int_{\mathbb{R}^{n-1}} f_{x_1}\, dm\right)^{\lambda}\left(\int_{\mathbb{R}^{n-1}} g_{x_2}\, dm\right)^{1-\lambda},$$

and (2.50) now follows by application of the very same result in dimension one. ∎

We sketch in Exercise 2.4.1 how to obtain the Brunn-Minkowski inequality from Theorem 2.4.2. Of course, we are primarily interested in using this theorem to prove that centred Gaussian measures are logarithmically concave.

Theorem 2.4.3 (Log-concavity of Gaussian measures in $\mathbb{R}^n$) *Let μ be a centred Gaussian measure on $\mathbb{R}^n$. Then, for any Borel sets A, B in $\mathbb{R}^n$ and $0 \leq \lambda \leq 1$, we have*

$$\mu\,(\lambda A + (1-\lambda)B) \geq (\mu(A))^{\lambda}(\mu(B))^{1-\lambda}. \tag{2.51}$$

Proof Let μ be a centred Gaussian measure on $\mathbb{R}^n$. Then μ is supported by a subspace $V \subset \mathbb{R}^n$, and the density of the restriction of μ to V with respect to Lebesgue measure on V is $\phi(x) = ce^{-|\Gamma x|^2/2}$, where $\Gamma : V \mapsto V$ is the positive square root of the inverse of the restriction to V of the covariance of μ and is a strictly positive definite operator. It is easy to see, for example, by diagonalising Γ, that the function $x \mapsto \log\phi(x) = -|\Gamma x|^2$, $x \in V$, is concave and therefore that

$$\phi(\lambda u + (1-\lambda)v) \geq \phi^{\lambda}(u)\phi^{1-\lambda}(v), \quad u, v \in V. \tag{2.52}$$

Now, if A and B are Borel sets of $\mathbb{R}^n$, we define, on V,

$$\varphi = \phi I_{\lambda(A\cap V)+(1-\lambda)(B\cap V)},\; f = \phi I_{A\cap V},\; g = \phi I_{B\cap V}.$$

Note that the set $\lambda(A \cap V) + (1-\lambda)(B \cap V)$ is the image by a continuous function of a Borel set on $V \times V$; hence, it is measurable for the completion of any Borel measure on V (e.g., Dudley (2002), section 13.2)). Inequality (2.52) shows that these functions satisfy the hypothesis (2.49) with $\mathbb{R}^n$ replaced by V. Hence, Theorem 2.4.2 applies to give

$$\int_{\lambda(A\cap V)+(1-\lambda)(B\cap V)} \phi\, dm \geq \left(\int_{A\cap V} \phi\, dm\right)^{\lambda}\left(\int_{B\cap V} \phi\, dm\right)^{1-\lambda},$$

where m is Lebesgue measure on V. This inequality implies the theorem because

$$\begin{aligned}\mu(\lambda A + (1-\lambda)B) &= \mu[(\lambda A + (1-\lambda)B) \cap V] \\ &\geq \mu(\lambda(A\cap V) + (1-\lambda)(B\cap V)) = \int_{\lambda(A\cap V)+(1-\lambda)(B\cap V)} \phi\, dm,\end{aligned}$$

and $\mu(A) = \int_{A\cap V} \phi\, dm$ and likewise for $\mu(B)$. ∎

An immediate consequence of this theorem is Anderson's inequality for *any* centred Gaussian measure on $\mathbb{R}^n$.

Theorem 2.4.4 (Anderson's lemma) *Let $X = (g_1, \ldots, g_n)$ be a centred jointly normal vector in $\mathbb{R}^n$, and let C be a measurable convex symmetric set of $\mathbb{R}^n$. Then, for all $x \in \mathbb{R}^n$,*

$$\Pr\{X + x \in C\} \le \Pr\{X \in C\}. \tag{2.53}$$

Proof Let $\mu = \mathcal{L}(X)$. Let $A = C + x$, $B = C - x$ and $\lambda = 1/2$ in (2.51), and note that by symmetry of μ and symmetry of C, $\mu(A) = \mu(B)$, so we obtain $\mu(C) \ge \mu(C + x)$, which is (2.53). ■

The assumption of measurability for C in the statement of the preceding theorem is superfluous because the boundary of a convex set C has μ-measure zero (whereas obviously its closure and its interior are measurable), but in applications, C is usually open or closed and hence measurable.

Theorem 2.4.4 extends to infinite dimensions, both for B-valued random variables, B separable (next theorem) and processes (Exercise 2.4.3).

Theorem 2.4.5 *Let B be a separable Banach space, let X be a B-valued centred Gaussian random variable and let C be a closed, convex, symmetric subset of B. Then, for all $x \in B$,*

$$\Pr\{X + x \in C\} \le \Pr\{X \in C\}.$$

In particular, $\Pr(\|X\| \le \varepsilon) > 0$, *for all* $\varepsilon > 0$.

Proof By the Hahn-Banach separation theorem in locally convex topological spaces, there exists a set $D_C \subset B^*$ such that $C = \cap_{f \in D_C}\{|f| \le 1\}$. Then $C^c = \cup_{f \in D_C}\{|f| > 1\}$. Since C^c is separable, its topology has a countable base, and therefore, this covering admits a countable subcovering; that is, there exists a countable subset $T_C \subset D_C$ such that $C^c = \cup_{f \in T_C}\{|f| > 1\}$ or $C = \cap_{f \in T_C}\{|f| \le 1\}$. Then, if $T_n \nearrow T_C$, T_n finite, we have

$$\begin{aligned}\Pr\{X \in C\} &= \Pr\{\sup_{f \in T_C} |f(X)| \le 1\} = \lim_{n \to \infty} \Pr\left\{\max_{f \in T_n} |f(X)| \le 1\right\} \\ &\ge \lim_{n \to \infty} \Pr\left\{\max_{f \in T_n} |f(X + x)| \le 1\right\} = \Pr\{X + x \in C\},\end{aligned}$$

where the inequality follows from Theorem 2.4.4 applied to the Gaussian vector $(f(X) : f \in T_n)$ and the convex set $\{x \in \mathbb{R}^{\mathrm{Card}(T_n)} : |x_i| \le 1, i = 1, \ldots, \mathrm{Card}(T_n)\}$. For the last claim, apply the first part to closed balls $C_i = \{x : \|x - x_i\| \le \varepsilon\}$ for x_i a countable dense subset of B. ■

Anderson's lemma applies to the comparison of the probabilities that $X = Y + Z$ and Y fall in convex symmetric sets C, where Y and Z are independent centred Gaussian $\mathbb{R}^n$-valued random vectors (Exercise 2.4.2), and gives

$$\Pr\{X \in C\} \le \Pr\{Y \in C\}.$$

Here is another application of Anderson's lemma, in the version of Exercise 2.4.5, to comparison of Gaussian processes, concretely, to proving the simplest yet useful instance of the famous Gaussian correlation conjecture, known as the *Khatri-Sidak inequality*. The

Gaussian correlation conjecture itself states that for symmetric convex sets A, B, if X and Y are arbitrary centred Gaussian vectors, $\Pr\{X \in A, Y \in B\} \geq \Pr\{X \in A\}\Pr\{Y \in B\}$; that is, the independent case gives the smallest probability of the intersection of two symmetric convex sets.

Corollary 2.4.6 (Khatri-Sidak inequality) *Let $n \geq 2$, and let $g_1, \ldots, g_n$ be jointly normal centred random variables. Then, for all $x \geq 0$,*

$$\Pr\{\max_{1\leq i\leq n} |g_i| \leq x\} \geq \Pr\{|g_1| \leq x\}\Pr\{\max_{2\leq i\leq n} |g_i| \leq x\},$$

and hence, iterating,

$$\Pr\{\max_{1\leq i\leq n} |g_i| \leq x\} \geq \prod_{i=1}^{n} \Pr\{|g_i| \leq x\}.$$

Proof Note that $\Pr\{\max_{2\leq i\leq n} |g_i| \leq x\} = \lim_{t\to\infty} \Pr\{\max_{2\leq i\leq n} |g_i| \leq x, |g_1| \leq t\}$. Hence, it suffices to show that for any convex symmetric subset A of $\mathbb{R}^{n-1}$, the function

$$f(t)/g(t) := \Pr\{|g_1| \leq t, (g_2, \ldots, g_n) \in A\}/\Pr\{|g_1| \leq t\}$$

is monotone decreasing. Let ϕ_1 denote the density of g_1, and set $X = (g_2, \ldots, g_n)$. Since

$$\begin{aligned}\Pr\{X \in A \,|\, |g_1| \leq t\} &= \int_{-t}^{t} \Pr\{X \in A | g_1 = u\} d\mathcal{L}(g_1 \,|\, |g_1| \leq t)(u) \\ &= \int_{-t}^{t} \Pr\{X \in A | g_1 = u\}\phi_1(u)du/\Pr\{|g_1| \leq t\},\end{aligned}$$

we have (using symmetry of the different laws) that

$$f(t) = \int_{-t}^{t} \Pr\{X \in A | g_1 = u\}\phi_1(u)du, \;\; f'(t) = 2\Pr\{X \in A | g_1 = t\}\phi_1(t)$$

and that, by Exercises 2.4.5 and 2.4.6,

$$\Pr\{X \in A \,|\, |g_1| \leq t\} \geq \Pr\{X \in A | g_1 = t\}.$$

These two observations give

$$\begin{aligned} g^2(t)(f/g)'(t) &= 2\phi_1(t)\Pr\{X \in A | g_1 = t\}\Pr\{|g_1| \leq t\} - 2\Pr\{|g_1| \leq t, (g_2, \ldots, g_n) \in A\}\phi_1(t) \\ &= 2\phi_1(t)\Pr\{|g_1| \leq t\}\,[\Pr\{X \in A | g_1 = t\} - \Pr\{X \in A \,|\, |g_1| \leq t\}] \leq 0. \end{aligned}$$

Thus, the function f/g is monotone decreasing, proving the corollary. ∎

2.4.2 Slepian's Lemma and Sudakov's Minorisation

Before proving the basic comparison result, it is convenient to consider a useful identity regarding derivatives of the multidimensional normal density. Let $f(C, x) = ((2\pi)^n \det C)^{-1/2} e^{-xC^{-1}x^T/2}$ be the $N(0, C)$ density in $\mathbb{R}^n$, where $C = (C_{ij})$ is an $n \times n$ symmetric strictly positive definite matrix $x = (x_1, \ldots, x_n)$ and x^T is the transpose of x. Consider f as a function of the real variables C_{ij}, $1 \leq i \leq j \leq n$, and x_i, $1 \leq i \leq n$. Then

$$\frac{\partial f(C,x)}{\partial C_{ij}} = \frac{\partial^2 f(C,x)}{\partial x_i \partial x_j} = \frac{\partial^2 f(C,x)}{\partial x_j \partial x_i}, \quad 1 \leq i < j \leq n. \tag{2.54}$$

To see this, just note that by the inversion formula for characteristic functions,

$$f(C,x) = \frac{1}{(2\pi)^n}\int_{\mathbb{R}^n} e^{-ixu^T} e^{-uCu^T/2}\,du$$

and that differentiation under the integral sign is justified by dominated convergence, so the three partial derivatives in (2.54) are all equal to $-x_ix_jf(C,x)$.

We can now prove the following comparison result:

Theorem 2.4.7 *Let $X = (X_1,\dots,X_n)$ and $Y = (Y_1,\dots,Y_n)$ be centred normal vectors in $\mathbb{R}^n$ such that $EX_i^2 = EY_j^2 = 1$, $1 \le i,j \le n$. Set, for each $1 \le i < j \le n$, $C_{ij}^1 = E(X_iX_j)$, $C_{ij}^0 = E(Y_iY_j)$ and $\rho_{ij} = \max\{|C_{ij}^0|, |C_{ij}^1|\}$. Then, for any $\lambda_i \in \mathbb{R}$,*

$$\Pr\bigcap_{i=1}^n\{X_i \le \lambda_i\} - \Pr\bigcap_{i=1}^n\{Y_i \le \lambda_i\} \le \frac{1}{2\pi}\sum_{1\le i<j\le n}(C_{ij}^1 - C_{ij}^0)^+\frac{1}{(1-\rho_{ij}^2)^{1/2}}\exp\left(-\frac{(\lambda_i^2+\lambda_j^2)/2}{1+\rho_{ij}}\right). \tag{2.55}$$

Moreover, if $\mu_i \le \lambda_i$ and $v = \min\{|\lambda_i|, |\mu_i| : i = 1,\dots n\}$, then

$$\left|\Pr\bigcap_{i=1}^n\{\mu_i \le X_i \le \lambda_i\} - \Pr\bigcap_{i=1}^n\{\mu_i \le Y_i \le \lambda_i\}\right| \le \frac{2}{\pi}\sum_{1\le i<j\le n}|C_{ij}^1 - C_{ij}^0|\frac{1}{(1-\rho_{ij}^2)^{1/2}} \times \exp\left(-\frac{v^2}{1+\rho_{ij}}\right). \tag{2.56}$$

Proof We may assume that the covariances of X and Y are invertible (so that both X and Y have densities): just take, if necessary, $X_\varepsilon = (1-\varepsilon^2)^{1/2}X + \varepsilon G$, $Y_\varepsilon = (1-\varepsilon^2)^{1/2}Y + \varepsilon G$ instead, where G is a standard normal random vector on $\mathbb{R}^n$ independent of X and Y. Then the result for X_ε and Y_ε implies the result for X and Y by letting $\varepsilon \to 0$. Moreover, since both the hypotheses and the conclusions of the theorem involve the probability laws of X and Y but not their joint law, we may also assume that X and Y are independent.

Under these two assumptions, define $X(t) = t^{1/2}X + (1-t)^{1/2}Y$. Then $X(0) = Y$, $X(1) = X$ and $C^t := \mathrm{Cov}(X(t)) = tC^1 + (1-t)C^0$. This curve in $\mathbb{R}^{n(n-1)/2}$ has a neighbourhood consisting only of (symmetric) strictly positive definite matrices. Let f_t denote the density of $X(t)$, and define

$$F(t) = \int_{-\infty}^{\lambda_1}\cdots\int_{-\infty}^{\lambda_n} f_t(x)dx, \tag{2.57}$$

which can be easily seen to be in $C([0,1])$. Then the left-hand side of (2.55) is precisely

$$F(1) - F(0) = \int_0^1 F'(t)\,dt.$$

We can still differentiate under the integral sign in (2.57), and since by (2.54)

$$\frac{df_t}{dt} = \sum_{1\le i<j\le n}\frac{\partial f_t}{\partial C_{ij}}\frac{dC_{ij}}{dt} = \sum_{1\le i<j\le n}(C_{ij}^1 - C_{ij}^0)\frac{\partial^2 f_t}{\partial x_i\partial x_j},$$

we obtain

$$F'(t) = \sum_{1\le i<j\le n}(C_{ij}^1 - C_{ij}^0)\int_{-\infty}^{\lambda_1}\cdots\int_{-\infty}^{\lambda_n}\frac{\partial^2 f_t}{\partial x_i\partial x_j}dx.$$

Integrating $\partial f_t/(\partial x_i \partial x_j)$ with respect to x_i and x_j, we obtain $f_t(x')$, where $x'_k = x_k$ if $k \neq i,j$, $x'_i = \lambda_i$, $x'_j = \lambda_j$. Moreover, we can bound the integrals with respect to the other coordinates, $\int_{-\infty}^{\lambda_k}$, by integrals over $\mathbb{R}$ and obtain

$$\int_{-\infty}^{\lambda_1} \cdots \int_{-\infty}^{\lambda_n} \frac{\partial^2 f_t}{\partial x_i \partial x_j} dx \leq \int_{\mathbb{R}^{n-2}} f_t(x_1, \ldots, x_{i-1}, \lambda_i, x_{i+1}, \ldots, x_{j-1}, \lambda_j, x_{j+1}, \ldots, x_n) dx.$$

This last integral is just the evaluation at the point (λ_i, λ_j) of the joint density of $X_i(t)$ and $X_j(t)$, that is, the density of the centred normal probability law in $\mathbb{R}^2$ with covariance

$$\begin{pmatrix} 1 & C_{ij}^t \\ C_{ij}^t & 1 \end{pmatrix},$$

$$\frac{1}{2\pi(1-(C_{ij}^t)^2)^{1/2}} \exp\left(-\frac{\lambda_i^2 - 2C_{ij}^t \lambda_i \lambda_j + \lambda_j^2)}{2(1-(C_{ij}^t)^2)} \right).$$

Replacing C_{ij}^t with its absolute value and noting that the minimum of the function of u, $(a^2 - 2uab + b^2)/(1-u)$ on $[0,\infty)$, is attained at $u = 0$, to obtain $(\lambda_i^2 - 2C_{ij}^t \lambda_i \lambda_j + \lambda_j^2)/(2(1-(C_{ij}^t)^2)) \geq (\lambda_i^2 + \lambda_j^2)/2(1+|C_{ij}^t|)$, and then using that $\rho_{ij} \geq |C_{ij}^t|$, we see that the quantity in the last display is dominated by

$$\frac{1}{2\pi(1-\rho_{ij}^2)^{1/2}} \exp\left(-\frac{(\lambda_i^2 + \lambda_j^2)/2}{1+\rho_{ij}} \right).$$

This shows that

$$F'(t) \leq \frac{1}{2\pi} \sum_{1 \leq i < j \leq n} (C_{ij}^1 - C_{ij}^0)^+ \frac{1}{(1-\rho_{ij}^2)^{1/2}} \exp\left(-\frac{(\lambda_i^2 + \lambda_j^2)/2}{1+\rho_{ij}} \right)$$

and that this is a bound for its integral over $[0,1]$ as well, that is, for $F(1) - F(0)$, proving (2.55).

To prove (2.56), we define

$$\tilde{F}(t) = \int_{\mu_1}^{\lambda_1} \cdots \int_{\mu_n}^{\lambda_n} f_t(x) dx$$

and proceed as before to obtain, as a result of the double integration $\int_{\mu_i}^{\lambda_i} \int_{\mu_j}^{\lambda_j} (\partial^2 f_t)/(\partial x_i \partial x_j)$, the sum of four functions of $n-2$ variables, two of them obtained from f_t by, respectively, setting $(x_i, x_j) = (\lambda_i, \lambda_j)$ and $(x_i, x_j) = (\mu_i, \mu_j)$ and the other two from $-f_t$ by, respectively, setting $(x_i, x_j) = (\lambda_i, \mu_j)$ and $(x_i, x_j) = (\mu_i, \lambda_j)$. Then, on integrating over $\mathbb{R}^{n-2}$ as earlier (instead of between μ_k and λ_k for each $k \neq i,j$), we obtain

$$|\tilde{F}'(t)| \leq \frac{4}{2\pi} \sum_{1 \leq i < j \leq n} |C_{ij}^1 - C_{ij}^0| \frac{1}{(1-\rho_{ij}^2)^{1/2}} \exp\left(-\frac{v^2}{1+\rho_{ij}} \right),$$

which yields inequality (2.56) by integrating between 0 and 1. ■

In this section we need a little less, in fact, only the following consequence of Theorem 2.4.7:

Theorem 2.4.8 (Slepian's lemma) *Let $X = (X_1,\ldots,X_n)$ and $Y = (Y_1,\ldots,Y_n)$ be centred jointly normal vectors in $\mathbb{R}^n$ such that*

$$E(X_iX_j) \le E(Y_iY_j) \quad and \quad EX_i^2 = EY_i^2 \quad for\ 1 \le i,j \le n. \tag{2.58}$$

Then, for all $\lambda_i \in \mathbb{R}$, $i \le n$,

$$\Pr\left(\bigcup_{i=1}^{n}\{Y_i > \lambda_i\}\right) \le \Pr\left(\bigcup_{i=1}^{n}\{X_i > \lambda_i\}\right), \tag{2.59}$$

and therefore,

$$E\max_{i\le n} Y_i \le E\max_{i\le n} X_i. \tag{2.60}$$

Proof Under assumptions (2.58), the right-hand side of (2.55) is less than or equal to zero, so (2.59) follows from Theorem 2.4.7. Inequality (2.60) follows from (2.58) by integration by parts ($E|\xi| = \int_0^\infty \Pr\{|\xi| > \lambda\}d\lambda$). ■

Remark 2.4.9 Sometimes one wishes to compare expected values of the maximum of the absolute values, and to this end, the following may be useful: for X_i symmetric, for any $i_0 \in \{1,\ldots,n\}$,

$$E\max_{i\le n} X_i \le E\max_{i\le n}|X_i| \le E|X_{i_0}| + E\max_{i,j}|X_i - X_j| \le E|X_{i_0}| + 2E\max_{i\le n} X_i,$$

where the last inequality follows because

$$E\max_{i,j}|X_i - X_j| = E\max_{i,j}(X_i - X_j) = E\max_i X_i + E\max_j(-X_j) = 2E\max_i X_i.$$

It is also worth mentioning that for any real random variable with mean zero, $E\max_i(X_i + Z) = EZ + E\max_i X_i = E\max_i X_i$.

The following corollary of Slepian's lemma is sometimes easier to apply than Theorem 2.4.8 because it does not require $EX_i^2 = EY_i^2$, $i \le n$.

Corollary 2.4.10 *Let $X = (X_1,\ldots,X_n)$ and $Y = (Y_1,\ldots,Y_n)$ be two centred, jointly normal vectors in $\mathbb{R}^n$, and assume that*

$$E(Y_i - Y_j)^2 \le E(X_i - X_j)^2, \quad i,j \in \{1,\ldots,n\}.$$

Then

$$E\max_{i\le n} Y_i \le 2E\max_{i\le n} X_i.$$

Proof Replacing X_i by $X_i - X_1$ and Y_i by $Y_i - Y_1$, we may assume that $X_1 = Y_1 = 0$ (see the preceding remark), which in particular implies that $EY_i^2 \le EX_i^2$. Set $\sigma_X^2 = \max_{i\le n} EX_i^2$, and let $\bar{X}$ and $\bar{Y}$ be Gaussian vectors whose coordinates are defined by

$$\bar{X}_i = X_i + (\sigma_X^2 + EY_i^2 - EX_i^2)^{1/2}g, \quad \bar{Y}_i = Y_i + \sigma_X g, \quad i = 1,\ldots,n,$$

where g is standard normal and independent of X and Y. Then

$$E\bar{X}_i^2 = E\bar{Y}_i^2 = EY_i^2 + \sigma_X^2$$

and

$$E(\bar{Y}_i - \bar{Y}_j)^2 = E(Y_i - Y_j)^2 \le E(X_i - X_j)^2 \le E(\bar{X}_i - \bar{X}_j)^2.$$

Therefore, we also have $E(\bar{X}_i\bar{X}_j) \le E(\bar{Y}_i\bar{Y}_j)$ and can apply Slepian's lemma to $\bar{X}$ and $\bar{Y}$ to obtain $E\max_i \bar{Y}_i \le E\max_i \bar{X}_i$. We clearly have $E\max_i \bar{Y}_i = E\max_i Y_i$. Moreover, since $EY_i^2 \le EX_i^2$, we have

$$E\max_i \bar{X}_i \le E\max_i X_i + \sigma_X Eg^+$$

and, using Remark 2.4.9,

$$\begin{aligned}\sigma_X = \max(EX_i^2)^{1/2} &= \sqrt{\frac{\pi}{2}}\max E|X_i| \\ &\le 2\sqrt{\frac{\pi}{2}} E\max_i X_i = \frac{1}{Eg^+} E\max_i X_i.\end{aligned}$$

Thus, $E\max_i \bar{X}_i \le 2E\max_i X_i$, and the result follows. ■

In fact, the constant 2 in this corollary is suboptimal: considerably more work gives a constant of 1, which is best (see the notes at the end of this chapter).

Finally, we will apply the comparison results to obtain a lower bound for $E\sup_t X(t)$ in terms of the metric entropy of the space (T, d_X), where X is a Gaussian process and $d_X^2(s,t) = E(X(t) - X(s))^2$. This result will be based on a comparison between a Gaussian vector X and an appropriate vector of independent normal variables. Then the entropy lower bound will follow from the following evaluation of the maximum of a finite number of independent normal variables.

Lemma 2.4.11 *Let g_i, $i \in \mathbb{N}$, be independent standard normal random variables. Then*

a. $\displaystyle\lim_{n\to\infty} \frac{E\max_{i\le n}|g_i|}{\sqrt{2\log n}} = 1$, *and*
b. There exists $K < \infty$ such that, for all $n > 1$,

$$K^{-1}(\log n)^{1/2} \le E\max_{i\le n} g_i \le E\max_{i\le n}|g_i| \le K(\log n)^{1/2}.$$

Proof The right-hand side of the inequality in part (b) is contained in Lemma 2.3.4. Since $E\max(g_1, g_2) > 0$ (in fact equal to $1/\sqrt{\pi}$), it suffices to prove the left-hand inequality for all $n \ge n_0$, for some n_0 (large enough). Since, by Remark 2.4.9, $E\max_{i\le n}|g_i| < \sqrt{2/\pi} + 2E\max_{i\le n} g_i$, it follows that, for n large enough,

$$E\max_{i\le n} g_i \ge 3^{-1} E\max_{i\le n}|g_i|,$$

and part (b) is therefore a consequence of part (a). To prove part (a), first note that if in the estimate

$$\begin{aligned}E\max_{i\le n}|g_i| &= \int_0^\infty \Pr\left\{\max_{i\le n}|g_i| > t\right\} dt \le \delta + n\int_\delta^\infty \Pr\{|g| > t\}dt \\ &= \delta + n\sqrt{\frac{2}{\pi}}\int_\delta^\infty \int_t^\infty e^{-u^2/2}du dt = \delta + n\sqrt{\frac{2}{\pi}}\int_\delta^\infty e^{-u^2/2}\int_\delta^u dt du\end{aligned}$$

$$\leq \delta + n\sqrt{\frac{2}{\pi}}e^{-\delta^2/2} - n\sqrt{\frac{2}{\pi}}\frac{\delta^2}{\delta^2+1}e^{-\delta^2/2}$$
$$= \delta + n\sqrt{\frac{2}{\pi}}\frac{1}{\delta^2+1}e^{-\delta^2/2}$$

we take $\delta = \sqrt{2\log n}$, we obtain $E\max_{i\leq n}|g_i| \leq \sqrt{2\log n} + \sqrt{2/\pi}/(1+2\log n)$, giving

$$\limsup_{n\to\infty}\frac{E\max_{i\leq n}|g_i|}{\sqrt{2\log n}} \leq 1.$$

In the opposite direction, since by (2.23), for $t \leq \sqrt{(2-\delta)\log n}$, for $0 < \delta < 2$,

$$\Pr\{|g| > t\} \geq \sqrt{\frac{2}{\pi}}\frac{t}{t^2+1}e^{-t^2/2} \geq \sqrt{\frac{2}{\pi}}\frac{\sqrt{(2-\delta)\log n}}{(2-\delta)\log n+1}n^{-(2-\delta)/2} = \frac{c(n,\delta)}{n},$$

with $\lim_{n\to\infty} c(n,\delta) = \infty$, for all $\delta > 0$, and

$$\Pr\{\max_{i\leq n}|g_i| > t\} = 1 - (1 - P\{|g| > t\})^n \geq 1 - e^{-nP\{|g|>t\}} \geq 1 - e^{-c(n,\delta)},$$

we have

$$E\max_{i\leq n}|g_i| > \int_0^{\sqrt{(2-\delta)\log n}} \left(1 - e^{-c(n,\delta)}\right)dt = \left(1 - e^{-c(n,\delta)}\right)\sqrt{(2-\delta)\log n},$$

which yields

$$\liminf_n \frac{E\max_{i\leq n}|g_i|}{\sqrt{(2-\delta)\log n}} \geq 1, \quad \text{for all } 0 < \delta < 2.$$

Letting $\delta \to 0$ completes the proof. ■

Recall that given a metric or pseudo-metric space (T,d), $N(T,d,\varepsilon)$ denotes the ε-covering number of (T,d) and that the packing numbers $D(T,d,\varepsilon)$ are comparable to the covering numbers, concretely, $N(T,d,\varepsilon) \leq D(T,d,\varepsilon)$ (see immediately preceding Theorem 2.3.6).

Theorem 2.4.12 (Sudakov's lower bound) *There exists $K < \infty$ such that if $X(t)$, $t \in T$, is a centred Gaussian process and $d_X(s,t) = (E(X(t) - X(s))^2)^{1/2}$ denotes the associated pseudo-metric on T, then, for all $\varepsilon > 0$,*

$$\varepsilon\sqrt{\log N(T,d_X,\varepsilon)} \leq K \sup_{S\subset T,\ S \text{ finite}} E\max_{t\in S} X(t).$$

Proof Let N be any finite number not exceeding $N(T,d_X,\varepsilon)$ (which may or may not be finite). Then, since $D(T,d_X,\varepsilon) \geq N$, there exist N points in T, say, $S = \{t_1,\ldots,t_N\}$, such that $d_X(t_i,t_j) \geq \varepsilon$, for $1 \leq i \neq j \leq N$. Let g_i, $i \leq N$, be i.i.d. standard normal random variables, and set $X'(t_i) = \varepsilon g_i/2$, $i \leq N$. Then $E(X'(t_i) - X'(t_j))^2 = \varepsilon^2/2 \leq d_X^2(t_i,t_j)$, $i \neq j$, and Corollary 2.4.10 implies that

$$E\max_{s\in S} X'(s) \leq 2E\max_{s\in S} X(s).$$

The theorem now follows from part (b) of Lemma 2.4.11. ■

Corollary 2.4.13 (Sudakov's theorem) *Let $X(t)$, $t \in T$, be a centred Gaussian process, and let d_X be the associated pseudo-distance. If $\limsup_{\varepsilon\downarrow 0}\varepsilon\sqrt{\log N(T,d_X,\varepsilon)} = \infty$, then $\sup_{t\in T}|X(t)| = \infty$ a.s., so X is not sample bounded.*

Proof Under the hypothesis of the corollary, by Sudakov's lower bound, there exists an increasing sequence of finite subsets $S_n \subset T$ such that $E\sup_{t\in S_n}|X(t)| \nearrow \infty$, and by monotone convergence, this gives $E\sup_{t\in\cup_{n=1}^{\infty}S_n}|X(t)| = \infty$. The process X restricted to $\cup_n S_n$ is separable, and applying Theorem 2.1.20 to it implies that

$$\Pr\left\{\sup_{t\in\cup_{n=1}^{\infty}S_n}|X(t)| < \infty\right\} = 0. \quad \blacksquare$$

This corollary shows that if a centred Gaussian process X is sample bounded, then the covering numbers $N(T,d_X,\varepsilon)$ are all finite; that is, (T,d_X) is not only separable but totally bounded, and in particular, X is separable by Theorem 2.1.12.

If X is sample continuous, the preceding theorem admits a stronger version:

Corollary 2.4.14 *Let $X(t)$, $t\in T$, be a sample continuous centred Gaussian process. Then*

$$\lim_{\varepsilon\to 0}\varepsilon\sqrt{\log N(T,d_X,\varepsilon)} = 0.$$

Proof If X is sample continuous, let X itself denote a process with the same law (hence the same d_X) with bounded, uniformly continuous sample paths. Thus, in particular, X is separable because it is sample bounded. Then $E\sup_{t\in T}|X(t)| < \infty$ by Theorem 2.1.20, and since $\sup_{d_X(s,t)<\delta}|X(t)-X(s)| \le 2\sup_{t\in T}|X(t)|$, continuity of X and dominated convergence give

$$\eta(\delta) := E\sup_{d_X(s,t)<\delta}|X(t)-X(s)| \to 0, \quad \text{as } \delta\to 0.$$

Given $\delta>0$, since by Theorem 2.4.12 (T,d_X) is totally bounded, there exists a finite set A in T which is δ-dense in T (i.e., for all $t\in T$ there is $s\in A$ such that $d_X(s,t)<\delta$). Then we can partition T into Card(A) sets, each within the sphere of radius δ about a point s in A. Call these sets T_s, $s\in A$. For each $s\in A$, consider the process $Y_t = X_t - X_s$, $t\in T_s$. We have $d_Y = d_X$ on T_s, and by the preceding theorem, T_s has an ε-dense subset, say, B_s, whose cardinality satisfies

$$\varepsilon\sqrt{\log \mathrm{Card}(B_s)} \le K\eta(\delta).$$

Now the set $\bigcup_{s\in A}B_s$ is ε-dense in T, and by definition of $N(T,d_X,\varepsilon)$ and the preceding inequality we have

$$\begin{aligned}\varepsilon\sqrt{\log N(T,d_X,\varepsilon)} &\le \varepsilon\sqrt{\log \mathrm{Card}(\cup_{s\in A}B_s)} \le \varepsilon\sqrt{\log[\mathrm{Card}(A)\times \max_s \mathrm{Card}(B_s)]}\\ &\le \varepsilon\sqrt{\log\mathrm{Card}(A) + \frac{K^2\eta^2(\delta)}{\varepsilon^2}} \le \varepsilon\sqrt{\log\mathrm{Card}(A)} + K\eta(\delta).\end{aligned}$$

Thus, for all $\delta>0$,

$$\limsup_{\varepsilon\to 0}\varepsilon\sqrt{\log N(T,d_X,\varepsilon)} \le K\eta(\delta),$$

proving the corollary because $\eta(\delta)\to 0$, as $\delta\to 0$. $\blacksquare$

The lower bound for $E\sup_{t\in T}|X(t)|$ in Theorem 2.4.12 should be compared with the upper bound in Theorem 2.3.6 for X a centred Gaussian process with $X(t_0)=0$ a.s. for some $t_0\in T$. Note that if $\log N(T,d_X,1/\tau)$ is bounded above and below by a constant times a regularly varying (at infinity) function of τ, then both bounds combine to give that there exists $K<\infty$ such that

$$\frac{1}{K}\sigma_X\sqrt{\log N(T,d_X,\sigma_X)}\le E\sup_{t\in T}|X(t)|\le K\sigma_X\sqrt{\log N(T,d_X,\sigma_X)}. \tag{2.61}$$

Exercises

2.4.1 Let A and B be two compact subsets of $\mathbb{R}^n$. Take $\varphi=I_{\lambda A+(1-\lambda)B}$, $f=I_A, g=I_B$ in Theorem 2.4.2 to obtain

$$\mathrm{Vol}(\lambda A+(1-\lambda)B)\ge(\mathrm{Vol}(A))^{\lambda}(\mathrm{Vol}(B))^{1-\lambda},$$

where Vol indicates volume in $\mathbb{R}^n$. Now prove that, as a consequence,

$$(\mathrm{Vol}(A+B))^{1/n}\ge(\mathrm{Vol}(A))^{1/n}+(\mathrm{Vol}(B))^{1/n},$$

which is *Brunn-Minkowski's inequality*. *Hint*: Apply the first inequality to $\tilde{A}=(\mathrm{Vol}(A))^{-1/n}A$, $\tilde{B}=(\mathrm{Vol}(B))^{-1/n}B$ and $\tilde{\lambda}=(\mathrm{Vol}(A))^{1/n}/((\mathrm{Vol}(A))^{1/n}+(\mathrm{Vol}(B))^{1/n})$.

2.4.2 Let X and Y be two centred Gaussian random vectors in $\mathbb{R}^n$ such that $E\langle x,Y\rangle^2\le E\langle x,X\rangle^2$, $x\in\mathbb{R}^n$. Show that C_X-C_Y, the difference between the covariance of X and the covariance of Y, is positive definite. Conclude that if Z is a centred random vector with covariance C_X-C_Y and independent of Y, then X and $Y+Z$ have the same probability law. Prove, using Fubini and Anderson's lemma, that $\Pr\{X\in C\}\le\Pr\{Y\in C\}$ for any convex symmetric set $C\subset\mathbb{R}^n$.

2.4.3 Let $X(t)$, $t\in T$, be a centred Gaussian process. Let $C\subset\mathbb{R}^T$ be a convex symmetric set such that $C=\cap_{k=1}^{\infty}\{x\in\mathbb{R}^T:|f_k(x)|\le1\}$, where $f_k(x)=\sum_{i=1}^{r_k}a_{ki}x(t_{ki})$ for some $r_k<\infty$ and collections of r_k points $t_{k,i}\in T$ and r_k coefficients $a_{ki}\in\mathbb{R}$. Then, for all $x\in\mathbb{R}^T$, $\Pr\{X+x\in C\}\le\Pr\{X\in C\}$. (In this sense, there is no need for B to be separable, or even a Banach space, in Theorem 2.4.5, but only that the definition of the convex set involve only a countable set of measurable linear functionals.)

2.4.4 Let X be a centred Gaussian B-valued random variable and B a separable Banach space. Show that the distribution function $F_{\|X\|}(t)=\Pr\{\|X\|\le t\}$ is continuous for all $t>0$. Deduce that if $\Pr(X=0)=0$ then $F_{\|X\|}$ is uniformly continuous on $[0,\infty)$. *Hint*: For any $t>0$, B is a countable union of translates of balls of radius t by separability; hence, at least one of them has strictly positive measure for the probability law of X. Thus, by Theorem 2.4.5, the same is true for the ball centred at the origin. This allows use of Exercise 2.2.13.

2.4.5 Here is a formal improvement of Theorems 2.4.4 and 2.4.5: prove that in the notation of each of these theorems, the function $\lambda\mapsto\Pr\{X+\lambda x\in C\}$, $0\le\lambda\le1$, is monotone decreasing. *Hint*: Use these theorems after applying the log-concavity inequality to $A=C-\lambda x$ and $B=C$.

2.4.6 Let $X=(g_0,g_1,\dots,g_n)$ be a centred jointly normal random vector in $\mathbb{R}^{n+1}$, set $Y=(g_1,\dots,g_n)$ and let C be a measurable convex symmetric set of $\mathbb{R}^n$. Use the preceding exercise to show that $\Pr\{Y\in C|g_0=x\}\le\Pr\{Y\in C||g_0|\le x\}$. *Hint*: There exists Z centred normal in $\mathbb{R}^n$, independent of g_0, and a vector $a=(a_1,\dots,a_n)$ such that $Y=ag_0+Z$. Then $\Pr\{Y\in C|g_0=x\}=\Pr\{Z\in C-ax\}$. Also, $\Pr\{Y\in C||g_0|\le x\}=\int_{-x}^{x}\Pr\{Y\in C|g_0=t\}d\mathcal{L}(g_0||g_0|\le x)(t)$.

2.4.7 Complete the details of the proof of identity (2.54) by showing that differentiation under the integral sign in the Fourier inversion formula for f is justified.

2.4.8 Likewise for the differentiation of F in (2.57).

2.4.9 Prove the following fact: If X and Y are centred Gaussian processes such that $d_X(s,t)\ge d_Y(s,t)$, and if X is sample continuous, then so is Y. *Hint*: Use Exercise 2.3.10 and Sudakov's inequality.

2.4.10 Extend Corollary 2.4.10 to separable centred Gaussian processes: if X and Y are two separable centred Gaussian processes on T, and if $E(Y(t)-Y(s))^2\le E(X(t)-X(s))^2$ for all $s,t\in T$, then

$E\sup_{t\in T} Y(t) \le 2E\sup_{t\in T} X(t)$ (meaning, in particular, that if the second quantity is finite, then so is the first).

2.4.11 If the processes X and Y in Exercise 2.4.10 have zero in their range for every ω (meaning that for each ω there is t_ω such that $X(\omega, t_\omega) = 0$, and likewise for Y), then the inequality between their intrinsic distances also implies that $E\sup_{t\in T}|Y(t)| \le 4E\sup_{t\in T}|X(t)|$ (the constant is not best possible, neither in this exercise nor in Exercise 2.4.10). *Hint*: By nonnegativity, $\sup_t X^+(t) = \sup_t X(t)$ and $\sup_t X^-(t) = \sup_t(-X(t))$, and likewise for Y. Now apply the comparison theorem (Exercise 2.4.10) to both $\{X, Y\}$ and $\{-X, -Y\}$.

2.4.12 (Comparison of moduli of continuity.) Let X and Y be as in Exercise 2.4.11, and set $d_X^2(s,t) = E(X(t) - X(s))^2$, $s,t \in T$, and likewise for Y. Exercise 2.3.10 'localizes' the increments by reducing the estimation of $E\sup_{s,t:d_Y(s,t)\le\delta}|Y(t) - Y(s)|$ to that of $\sup_t E\sup_{s:d_Y(s,t)\le\delta}|Y(t) - Y(s)|$ (plus a metric entropy term), but then, since $d_Y = d_{Y_t}$, where $Y_t(s) = Y(s) - Y(t)$, this localization allows for comparison. Concretely, prove that there exists $K < \infty$ such that if X and Y are as in Exercise 2.4.11, then for all $\delta > 0$,

$$E\sup_{d_X(s,t)\le\delta}|Y(t) - Y(s)| \le K\left[\sup_{t\in T} E\sup_{s:d_X(s,t)\le\delta}|X(t) - X(s)| + \delta(\log N(T,d,\delta))^{1/2}\right].$$

2.5 The Log-Sobolev Inequality and Further Concentration

In this section we present another approach to concentration via log-Sobolev inequalities and the *Herbst method*. This gives, by way of solving a differential inequality for the Laplace transform of Lipschitz functions of a Gaussian process, sharp concentration inequalities about its mean. The method is of particular interest as it is amenable to generalisation to non-Gaussian situations.

2.5.1 Some Properties of Entropy: Variational Definition and Tensorisation

In this subsection, μ is a probability measure on some measurable space $(S, \mathcal{S})$, and f is a measurable, real nonnegative function on this space. Convention: $0\log 0 := \lim_{x\to 0+} x \log x = 0$.

Definition 2.5.1 The entropy of $f \ge 0$ with respect to a probability measure μ is defined as

$$\mathrm{Ent}_\mu f = \int f\log f\,d\mu - \left(\int f\,d\mu\right)\left(\log\int f\,d\mu\right)$$

if $\int f\log(1+f)d\mu < \infty$ and as ∞ otherwise.

Note that since $x\log x$ (extended by continuity at 0) is a convex function on $\mathbb{R}_+ \cup \{0\}$, the entropy functional is nonnegative. It is easy to see from the definition that the functional Ent_μ is homogeneous of degree 1, that is, $\mathrm{Ent}_\mu(\lambda f) = \lambda \mathrm{Ent}_\mu f$.

Recall the following Young's inequality: for $x \in \mathbb{R}$ and $y \ge 0$,

$$xy \le y\log y - y + e^x. \tag{2.62}$$

To prove it, observe that for each $x \in \mathbb{R}$ the function $z_x(y) = xy - y\log y + y - e^x$, $y > 0$, has an absolute maximum equal to 0 at $y = e^x$. This inequality yields the following useful expression for the entropy:

Lemma 2.5.2

$$\mathrm{Ent}_\mu f = \sup\left\{\int fg\,d\mu : \int e^g d\mu \le 1,\ g \text{ measurable}\right\}.$$

Proof By homogeneity, we can assume that $\int f d\mu = 1$. By Young's inequality (2.62) for $y = f \ge 0$ and $x = g$,

$$\int fg\,d\mu \le \int f \log f\,d\mu - 1 + \int e^g d\mu \le \int f \log f\,d\mu = \mathrm{Ent}_\mu f,$$

which gives that the preceding sup is dominated by the entropy. To see that it equals the entropy, take $g = \log f$. ■

Lemma 2.5.2 yields an inequality about the behavior of Ent for product measures, 'tensorisation of entropy', which is basic for the proof of concentration inequalities. Given a product measure $P = \mu_1 \times \cdots \times \mu_n$ and a function f of n variables, we denote $\mathrm{Ent}_{\mu_i} f$ the function of $n-1$ variables obtained by computing the entropy with respect to μ_i of the function of one variable $f_i(x) = f(x_1, \ldots, x_{i-1}, x, x_{i+1}, \ldots, x_n)$. With this notation, we have the following:

Proposition 2.5.3 *Let $P = \mu_1 \times \cdots \times \mu_n$, and let $f \ge 0$ on a product space. Then*

$$\mathrm{Ent}_P f \le \sum_{i=1}^n \int \left(\mathrm{Ent}_{\mu_i} f\right) dP.$$

Proof Given g on the product space such that $\int e^g dP \le 1$, set, for any $x = (x_1, \ldots, x_n)$,

$$g_1(x) = \log \frac{e^{g(x)}}{\int e^{g(x)} d\mu_1(x_1)},$$

$$g_i(x) = \log \frac{\int e^{g(x)} d\mu_1(x_1) \cdots d\mu_{i-1}(x_{i-1})}{\int e^{g(x)} d\mu_1(x_1) \cdots d\mu_i(x_i)}, \quad i = 2, \ldots, n.$$

Thus, for each g_i, the integral with respect to μ_i of the numerator is the denominator. Note then that

$$g \le g - \log \int e^g dP = \sum_{i=1}^n g_i \quad \text{and} \quad \int e^{g_i} d\mu_i = 1.$$

Then, by Lemma 2.5.2,

$$\int fg\,dP \le \sum_{i=1}^n \int fg_i\,dP = \sum_{i=1}^n \int \int fg_i\,d\mu_i\,dP \le \sum_{i=1}^n \int (\mathrm{Ent}_{\mu_i} f)\,dP. \quad ■$$

We still need another variational definition of entropy and its consequence for exponential functions.

Let ξ be a convex function on a finite or infinite interval (e.g., $\xi(u) = u \log u$ on $[0, \infty)$), differentiable on its interior, and let the range of f be contained in it. Then, assuming the existence of the integrals involved,

$$\int \xi(f)\,d\mu - \xi\left(\int f\,d\mu\right) = \inf_t \int \left[\xi(f) - \xi(t) + (t - f)\xi'(t)\right] d\mu.$$

To see this, note that the integral on the right-hand side for $t = \int f d\mu$ is just the left-hand side. Now note that the convex function $y = \xi(x)$ at $\int f d\mu$ is larger than or equal to the value at $\int f d\mu$ of the tangent line to the graph of this function at $(t, \xi(t))$, which gives

$$\xi\left(\int f d\mu\right) \geq \xi(t) + \left(\int f d\mu - t\right)\xi'(t),$$

proving the claim. Applied to entropy, this gives the following identity:

Lemma 2.5.4

$$\mathrm{Ent}_\mu f = \inf_{t \geq 0} \int [f \log f - (\log t + 1) f + t] d\mu.$$

In the case of exponential functions e^f, this lemma gives the following:

Lemma 2.5.5 *Setting*

$$\phi(u) := e^{-u} + u - 1, \quad u \in \mathbb{R},$$

we have

$$\mathrm{Ent}_\mu e^f = \inf_{u \in \mathbb{R}} \int \phi(f - u) e^f d\mu. \tag{2.63}$$

Proof The last lemma and the change of variables $u = \log t$ give

$$\mathrm{Ent}_\mu e^f = \inf_{t \geq 0} \int [f e^f - (\log t + 1) e^f + t] d\mu = \inf_{u \in \mathbb{R}} \int \phi(f - u) e^f d\mu. \quad \blacksquare$$

2.5.2 A First Instance of the Herbst (or Entropy) Method: Concentration of the Norm of a Gaussian Variable about Its Expectation

An application of Theorem 2.4.2 (the Prékopa-Leindler theorem) yields the logarithmic Sobolev inequality for Gaussian measures, which then can be integrated to provide an upper bound for the Laplace transform of the norm of the associated Gaussian vector. Given a smooth function of several variables $f(x_1, \ldots, x_n)$, we let $f' = (\partial f/\partial x_1, \ldots, \partial f/\partial x_n)$ denote its gradient and let $|f'|$ denote its Euclidean norm.

Theorem 2.5.6 *Let γ be the canonical Gaussian measure on $\mathbb{R}^n$, and let $f : \mathbb{R}^n \mapsto \mathbb{R}$ be a twice continuously differentiable function such that f^2 and $|f'|^2$ are γ-integrable. Then*

$$\mathrm{Ent}_\gamma(f^2) \leq 2 \int |f'|^2 d\gamma. \tag{2.64}$$

Proof By Proposition 2.5.3 on the tensorisation of entropy, it suffices to prove this theorem for $n = 1$. Also, by standard approximation arguments, we can assume that $f^2 = e^g$, where g is twice continuously differentiable and has compact support. For completeness sake, an approximation argument is sketched in Exercise 2.5.1. Set $V(x) = (x^2 + \ln 2\pi)/2$ so that the standard normal density becomes $e^{-V(x)}$. For $0 < \lambda < 1$, define

$$g_\lambda(z) = \sup_{u, v \in \mathbb{R} : \lambda u + (1-\lambda) v = z} \left[g(u) - \big(\lambda V(u) + (1-\lambda) V(v) - V(\lambda u + (1-\lambda) v)\big)\right],$$

so that taking $\phi(z) = e^{g_\lambda(z) - V(z)}$, we have the relation

$$\phi(\lambda u + (1-\lambda) v) \geq e^{\lambda(g(u)/\lambda - V(u))} e^{-(1-\lambda) V(v)},$$

that is, the functions $e^{g/\lambda - V}$, e^{-V} and ϕ satisfy the log-concavity relation (2.49) from Theorem 2.4.2, respectively, as the functions f, g and ϕ there (the different use of the notation g in this proof and in Theorem 2.4.2 should not lead to confusion). Hence, the conclusion of this theorem gives

$$\int e^{g_\lambda} d\gamma \geq \left(\int e^{g/\lambda} d\gamma\right)^\lambda. \tag{2.65}$$

This is relevant to entropy because letting $H(\lambda) = \left(\int e^{g/\lambda} d\gamma\right)^\lambda$, we obtain, by logarithmic differentiation,

$$H'(1) = -\mathrm{Ent}_\gamma(e^g).$$

Hence, by Taylor expansion about $\lambda = 1$,

$$\left(\int e^{g/\lambda} d\gamma\right)^\lambda = \int e^g d\gamma + (1-\lambda)\mathrm{Ent}_\gamma(e^g) + O\left((1-\lambda)^2\right), \tag{2.66}$$

as $\lambda \to 1$. To finish the proof, we must find an upper bound for the left-hand side of (2.65).

We observe that with the change of variables $h = z - v$ and $\eta = (1-\lambda)/\lambda$ and using the definition of V,

$$g_\lambda(z) = \sup_{u,v \in \mathbb{R}: \lambda u + (1-\lambda)v = z} \left[g(u) - \frac{\lambda(1-\lambda)}{2}(u-v)^2\right] = \sup_{h \in \mathbb{R}} \left[g(z+\eta h) - \frac{\eta h^2}{2}\right].$$

Since g'' is continuous and has compact support, we have that

$$g(z+\eta h) \leq g(z) + g'(z)\eta h + C\eta^2 h^2$$

for some constant C independent of z. This gives

$$g_\lambda(z) \leq g(z) + \eta \sup_h \left[g'(z)h - (1/2 - C\eta)h^2\right] = g(z) + \frac{\eta}{2}(g'(z))^2 + O(\eta^2)$$

uniformly over all $z \in \mathbb{R}$ and for η small enough because g' is bounded (we must have $1/2 - C\eta > 0$ and observe that for a real and b positive, the maximum of $ah - bh^2$ is attained at $h = a/2b$). Applying Taylor's theorem again, we obtain

$$e^{g_\lambda(z)} \leq e^{g(z)} + \frac{\eta}{2}(g'(z))^2 e^{g(z)} + O(\eta^2)$$

uniformly in z (recall that g has compact support). Integrating with respect to γ and using (2.65) and (2.66), we obtain (letting $\lambda \to 1$)

$$\mathrm{Ent}_\gamma(e^g) \leq \frac{1}{2}\int (g')^2 e^g d\gamma,$$

which, since $2ff' = g'e^g$ and hence $4(f')^2 = (g')^2 e^g$, gives the log-Sobolev inequality for f. ■

This proof works as well in $\mathbb{R}^n$ for any n, so tensorisation of entropy is in fact redundant here. Notice the independence on the dimension of the preceding log-Sobolev inequality. Such independence may be seen either as a consequence of the tensorisation property of entropy or as a consequence of the the log-concavity inequality from Theorem 2.4.2 (which, as just mentioned, could have been used in any dimensions).

Let now $f^2 = e^{\lambda F}$, where $F : \mathbb{R}^n \mapsto \mathbb{R}$ is twice continuously differentiable with $|F'| \le 1$. Then

$$E|f'|^2 = \frac{\lambda^2}{4} E\left(|F'|^2 e^{\lambda F}\right) \le \frac{\lambda^2}{4} E\left(e^{\lambda F}\right). \tag{2.67}$$

With this bound, the Gaussian log-Sobolev inequality becomes a solvable *differential inequality* for the Laplace transform of the random variable $F(X)$, where X is a random variable with probability law γ. This is the first, simplest instance of the Herbst method or entropy method for obtaining exponential inequalities. Applied in this case it gives optimal concentration of the supremum norm of a Gaussian process about its mean as follows: recall that $\|F\|_{\rm Lip} = \sup_{x \neq y} |F(x) - F(y)|/|x - y|$.

Theorem 2.5.7 *Let F be a Lipschitz function on $\mathbb{R}^n$, with $\|F\|_{\rm Lip} \le 1$, and let $X = (g_1, \ldots, g_n)$ with g_i independent standard normal random variables. Then, for all $\lambda \in \mathbb{R}$,*

$$E\left(e^{\lambda F(X)}\right) \le e^{\lambda E(F(X)) + \lambda^2/2}. \tag{2.68}$$

As a consequence,

$$\Pr\{F(X) \ge E(F(X)) + t\} \le e^{-t^2/2}, \quad \Pr\{F(X) \le E(F(X)) - t\} \le e^{-t^2/2}. \tag{2.69}$$

Proof Assume first that F is twice continuously differentiable with $|F'| \le 1$, and set $H(\lambda) = E\left(e^{\lambda F(X)}\right) = \int e^{\lambda F} d\gamma$. Then the log-Sobolev inequality (2.64) applied to $f^2 = e^{\lambda F}$ together with inequality (2.67) gives

$$\lambda H'(\lambda) - H(\lambda) \log H(\lambda) = \mathrm{Ent}_\gamma\left(e^{\lambda F}\right) \le \frac{\lambda^2}{2} H(\lambda).$$

This inequality simplifies if we write it in terms of $K(\lambda) := \lambda^{-1} \log H(\lambda)$, becoming

$$K'(\lambda) \le \frac{1}{2}.$$

Now, using l'Hôpital's rule, we see that $K(\lambda)$ satisfies the initial condition $K(0) = H'(0)/H(0) = E(F(X))$, and therefore,

$$K(\lambda) = K(0) + \int_0^\lambda K'(t) dt \le E(F(X)) + \frac{\lambda}{2}.$$

Hence, $H(\lambda) = e^{\lambda K(\lambda)} \le e^{\lambda E(F(X)) + \lambda^2/2}$ for all $\lambda \in \mathbb{R}$, which is inequality (2.68) in the case of differentiable F. The case of F Lipschitz follows by convolution of F with a smooth approximate identity and a standard approximation argument (Exercise 2.5.2). For the first inequality in (2.69), we observe that, by Chebyshev's inequality, for $\lambda \ge 0$,

$$\Pr\{F(X) \ge E(F(X)) + t\} \le \frac{H(\lambda)}{e^{\lambda(E(F(X)) + t)}} \le e^{-\lambda t + \lambda^2/2}$$

and then take $\lambda = t$. The second inequality in (2.69) follows by applying the preceding one to the Lipschitz function $-F$. ■

The same arguments used to derive Theorem 2.2.7 from Theorem 2.2.6 show how the preceding theorem applies to the supremum of a separable Gaussian process or to the norm of a Gaussian random vector. With the notation of processes we thus obtain the following version of the Borell-Sudakov-Tsirelson concentration inequality, now around the mean. Recall from the 0-1 law and the integrability properties of Gaussian processes (e.g., Theorem 2.1.20) that if the sup norm of a Gaussian process X is finite with positive probability, then $E\|X\|_\infty < \infty$ and $\sigma^2 = \sup_{t\in T} EX^2(t) < \infty$.

Theorem 2.5.8 *Let $X(t)$, $t \in T$, be a separable centred Gaussian process whose supremum norm is finite with positive probability. Let σ^2 be the supremum of the variances $EX^2(t)$, and set $\|X\|_\infty := \sup_{t\in T} |X(t)|$. Then,*

$$\Pr\{\|X\|_\infty \geq E\|X\|_\infty + u\} \leq e^{-u^2/2\sigma^2}, \quad \Pr\{\|X\|_\infty \leq E\|X\|_\infty - u\} \leq e^{-u^2/2\sigma^2}. \tag{2.70}$$

Note that inequality (2.70) gives $\Pr\{|\|X\|_\infty - E\|X\|_\infty| \geq u\} \leq 2e^{-u^2/2\sigma^2}$, whereas the Borell-Sudakov-Tsirelson concentration inequality about the median $\Pr\{|\|X\|_\infty - M| \geq u\} \leq e^{-u^2/2\sigma^2}$ is better by a factor of 2 (see (2.18)). Compare also with (2.20).

Exercises

2.5.1 (The approximation argument in Theorem 2.5.6). Let γ be the standard normal measure on $\mathbb{R}$, and let $f : \mathbb{R} \mapsto \mathbb{R}$ satisfy $f \in L^2(\gamma)$ and $f' \in L^2(\gamma)$. Take $1 > \varepsilon_n \downarrow 0$, and choose $L_n \uparrow \infty$ such that

$$\int_{(-L_n,L_n)^c} \left(\frac{1}{2}(f^2+1)^{1-\varepsilon_n}(\log(f^2+1))^2 + 2\varepsilon_n^{-\varepsilon_n}(f')^2\right) d\gamma < \varepsilon_n,$$

which exists by the hypotheses on f. Let w_n be an even twice continuously differentiable function such that w_n is $1-\varepsilon_n$ on $[-L_n, L_n]$ and zero on $[-L_n-2, L_n+2]^c$; it is decreasing on $[0,\infty)$ and $|w_n'| \leq 1$. Define

$$g_n = w_n \log(f^2+\varepsilon_n), \quad h_n^2 = e^{g_n} = (f^2+\varepsilon_n)^{w_n}.$$

Then (a) g_n is twice continuously differentiable and has bounded support, so, by the proof of Theorem 2.5.6 (without the approximation argument), we have

$$\mathrm{Ent}_\gamma(h_n^2) \leq 2\int (h_n')^2 d\gamma.$$

(b) $h_n^2 \leq f^2+1$, so, by dominated convergence, $\int h_n^2 d\gamma \to \int f^2 d\gamma$, and we also have, by bounded convergence, that $\int_{h_n \leq 1} h_n^2 \log h_n^2\, d\gamma \to \int_{f^2\leq 1} f^2 \log f^2\, d\gamma$. Then, by Fatou's lemma applied to $\int_{h_n>1} h_n^2 \log h_n^2\, d\gamma$ and by part (a), we obtain

$$\mathrm{Ent}_\gamma(f^2) \leq \liminf_n \mathrm{Ent}_\gamma(h_n^2) \leq 2\liminf_n \int (h_n')^2 d\gamma.$$

(c) Next, $2h_n'/h_n = w_n' \log(f^2+\varepsilon_n) + 2w_n f f'/(f^2+\varepsilon_n)$, and recall that $w_n' = 0$ on $[-L_n, L_n]$ and on $[-L_n-2, L_n+2]^c$. Hence, $|h_n'| \leq |f'|$ on $[-L_n, L_n]$ and $\int_{[-L_n,L_n]^c} (h_n')^2 d\gamma \leq \varepsilon_n$, concluding that $\int (h_n')^2 d\gamma \leq \int (f')^2 d\gamma + \varepsilon_n$, which gives

$$\mathrm{Ent}_\gamma(f^2) \leq 2\int (f')^2 d\gamma$$

by part (b).

2.5.2 Show that if Theorem 2.5.7 holds for F infinitely differentiable with $\|F'\|_\infty \leq 1$, it also holds for F Lipschitz with $\|F\|_{\mathrm{Lip}} \leq 1$. *Hint*: Apply the result for smooth functions to the convolution of F Lipschitz with, for example, the density of the $N(0,\varepsilon^2 I)$ distribution, and let $\varepsilon \to 0$. Since we will be using this type of approximation more than once, just note the following two easy estimates: if $\phi(u)$ is the $N(0,I)$ density in $\mathbb{R}^d$ and for $F:\mathbb{R}^d \mapsto \mathbb{R}$ with $\|F\|_{\mathrm{Lip}} < \infty$, then

$$F_\varepsilon(x) = F * \phi_\varepsilon(x) := \varepsilon^{-d} \int_{\mathbb{R}^d} \phi((x-y)/\varepsilon)F(y)dy = \int_{\mathbb{R}^d} \phi(u)F(x-\varepsilon u)du$$

is infinitely differentiable, $\|F_\varepsilon - F\|_\infty \le \varepsilon \int_{\mathbb{R}^d} \phi(u)|u|du \to 0$ and $|F_\varepsilon(x_1) - F_\varepsilon(x_2)| \le \sup_v |F(x_1 - v) - F(x_2 - v)|$ such that $\|F_\varepsilon\|_{Lip} \le \|F\|_{Lip}$.

2.6 Reproducing Kernel Hilbert Spaces

In the first subsection the reproducing kernel Hilbert spaces of Gaussian processes and of Gaussian random variables taking values in separable Banach spaces are defined, and their very basic properties are given. In the next subsection, several applications are developed, particularly to isoperimetry, equivalence and singularity, and small ball estimation.

2.6.1 Definition and Basic Properties

Let $X(t)$, $t \in T$, be a centred Gaussian process, and let $C(s,t) = E(X(s)X(t))$, $s,t \in T$, be its covariance. Let F be the linear span of the collection of (square integrable) random variables $\{X(t) : t \in T\}$ and $\bar{F}$ its closure in $L^2(\Omega, \Sigma, \Pr)$. This space is isometric to a Hilbert space H of functions on T, which is called the *reproducing Hilbert space of* X (or, more properly, of the covariance function C) as follows: define

$$\phi : F \mapsto \mathbb{R}^T, \quad \phi\left(\sum_{i=1}^k a_i X(t_i)\right) = \sum_{i=1}^k a_i C(t_i, \cdot) \tag{2.71}$$

for $k < \infty$, $a_i \in \mathbb{R}$ and $t_i \in T$, $i = 1, \dots, k$, and define on $\phi(F)$ the inner product induced by the L^2 inner product via ϕ, that is,

$$\left\langle \sum_{i=1}^k a_i C(t_i, \cdot), \sum_{i=1}^{\ell} b_i C(s_i, \cdot) \right\rangle_H = \sum_{i \le k, j \le \ell} a_i b_j C(t_i, s_j), \tag{2.72}$$

in particular, $\langle C(t,\cdot), C(s,\cdot)\rangle = C(s,t)$ for all $s,t \in T$. The reproducing kernel Hilbert space $H = H_X$ of X is defined as the completion of $\phi(F)$ by this inner product. In order to see that H_X can be identified to a space of functions (or of classes of functions, identifying those at distance zero from each other), note that the definitions (2.71) of ϕ and (2.72) of the inner product $\langle \cdot, \cdot \rangle_H$ can be restated as

$$\phi\left(\sum_{i=1}^k a_i X(t_i)\right)(t) = E\left(\left(\sum_{i=1}^k a_i X(t_i)\right) X(t)\right), \quad t \in T,$$

and

$$\left\langle \phi\left(\sum_{i=1}^k a_i X(t_i)\right), \ \phi\left(\sum_{j=1}^{\ell} b_j X(s_j)\right)\right\rangle_H = E\left[\sum_{i=1}^k a_i X(t_i) \sum_{j=1}^{\ell} b_j X(s_j)\right].$$

That is,

$$\phi(h)(t) = E(hX(t)), \quad \langle \phi(h_1), \phi(h_2)\rangle_H = \langle h_1, h_2 \rangle_{L^2}, \quad h, h_1, h_2 \in F.$$

Thus, the function $\phi(h) = E(hX)$ is a linear isometry between $(F, \|\cdot\|_{L^2})$ and $(\phi(F), \|\cdot\|_H)$ (where $\|v\|_H^2 = \langle v, v\rangle_H$). It follows that the completion of $\phi(F)$ for the induced inner product $\langle\cdot,\cdot\rangle_H$ is simply the collection of functions

$$\{\phi(h) : \phi(h)(t) := E(hX(t)),\ t \in T, h \in \bar{F}\} = \{E(hX) : h \in \bar{F}\},$$

with inner product $\langle E(h_1X), E(h_2X)\rangle_H = E(h_1h_2)$. In short:

Definition 2.6.1 The reproducing kernel Hilbert space (RKHS) of a centred Gaussian process $X(t)$, $t \in T$, (or of its probability law, or of its covariance) is the Hilbert space of functions

$$t \mapsto (E(hX))(t) := E(hX(t)), \quad t \in T,$$

where h ranges over the closure $\bar{F}$ in $L^2(\Omega, \Sigma, \Pr)$ of the linear span F of the range of the process $\{X(t) : t \in T\} \subset L^2(\Omega, \Sigma, \Pr)$, with inner product $\langle E(h_1X), E(h_2X)\rangle_H = E(h_1h_2)$, $h_i \in \bar{F}$.

Example 2.6.2 The RKHS of an orthogaussian sequence $\{g_n\}_{n=1}^\infty$. The closure in L^2 of the linear span of $X(n) = g_n$, $n \in \mathbb{N}$, g_n independent $N(0,1)$, is $\bar{F} = \left\{\sum_{i=1}^\infty a_i g_i : \sum a_i^2 < \infty\right\}$, and if $h = \sum_{i=1}^\infty a_i g_i \in \bar{F}$, then $(E(hX))(n) = E\left(\left(\sum_{i=1}^\infty a_i g_i\right) g_n\right) = a_n$. Hence, $E(hX) = \{a_n\}_{n=1}^\infty \in \ell_2 \subset \mathbb{R}^{\mathbb{N}}$. That is, the RKHS of the standard Gaussian measure on $\mathbb{R}^{\mathbb{N}}$ is ℓ_2 (as a subset of $\mathbb{R}^{\mathbb{N}}$).

Often one is interested in sample continuous Gaussian processes, as in Section 2.3, or more generally in processes whose sample paths belong to a Banach space of functions or even more generally in Banach space–valued random variables (defined in Section 2.1). If X is a centred Gaussian B-valued random variable, it certainly can be viewed as a Gaussian process on the unit ball of the topological dual B^* of B, and then the preceding definition applies. However, more can be said, for instance, that the RKHS is not only a collection of continuous linear functionals on B^*, that is, a subset of B^{**}, but also a subset of B. Thus, there is something to gain from adapting the preceding definition to this situation. In what follows, for $x \in B$, $\|x\|$ will denote its B-norm, and the same symbol will be used for the B^*-norm; that is, $\|f\| = \sup_{\|x\|\le 1} |f(x)|$ will denote the B^*-norm of f.

Before proceeding, it is convenient to recall some definitions and basic properties of Banach-valued random variables and their expectations. Let $(\Omega, \Sigma, \Pr)$ be a probability space, and let B be a separable Banach space, equipped with its Borel σ-algebra $\mathcal{B}$. Let X be a B-valued random variable, that is, a $\Sigma - \mathcal{B}$ measurable function $X : \Omega \mapsto B$. X is simple if it is finitely valued, and the expected value of a finitely valued random variable $X = \sum_{i=1}^n x_i I_{A_i}$, $x_i \in B$, $A_i \in \Sigma$, is naturally defined as $EX = \sum_{i=1}^n \Pr(A_i)x_i$. X is *Bochner integrable* or *strongly integrable* if there exists a sequence of simple functions X_n such that $E\|X_n - X\| \to 0$. Then

$$\|EX_n - EX_m\| \le E\|X_n - X_m\| \le E\|X_n - X\| + E\|X_m - X\|$$

by convexity and the triangle inequality; hence, the sequence $\{EX_n\}$ is Cauchy. The expected value of X is then defined as $EX = \lim_{n\to\infty} EX_n$. It is immediate that EX is well defined. Since $E|\|X_n\| - \|X\|| \le E\|X_n - X\| \to 0$, we obtain $E\|X\| = \lim_n E\|X_n\| < \infty$. Also, if $f \in B^*$, then $E|f(X_n) - f(X)| \le \|f\|E\|X_n - X\| \to 0$; hence, $Ef(X) = \lim_n Ef(X_n)$ exists and is finite. Moreover, $f(EX_n) = Ef(X_n)$ and $|f(EX_n) - f(EX)| \le \|f\|\|EX_n - EX\| \to 0$. These observations imply that $f(EX) = Ef(X)$.

A random variable X is *Pettis or weakly integrable* with integral x if $f(X) \in L^1(\Pr)$ for all $f \in B^*$ and there exists $x \in B$ such that $Ef(X) = f(x)$, $f \in B^*$. The preceding paragraph shows that if X is Bochner integrable, then it is Pettis integrable and both integrals coincide.

If $E\|X\| < \infty$, then the map $\nu : A \mapsto \nu(A) := E(\|X\| I_{X\in A})$ is a finite measure on the Borel sets of B and hence, by the Oxtoby-Ulam theorem (Proposition 2.1.4), a tight Borel measure. Given $0 < \varepsilon_n \to 0$, let K_n be a compact set such that $\nu(K_n^c) < \varepsilon_n/2$, let $A_{n,1}, \ldots, A_{n,k_n}$ be a finite partition of K_n consisting of sets of diameter at most $\varepsilon_n/2$, pick up a point $x_{n,k} \in A_{n,k}$ for each k and define the simple function $X_n = \sum_{k=1}^{k_n} x_{n,k} I_{(X\in A_{n,k})}$. Then $E\|X_n - X\| \le \varepsilon_n/2 + \nu(K_n^c) < \varepsilon_n \to 0$, showing that X is Bochner integrable. Summarizing:

Lemma 2.6.3 *Let B be a separable Banach space, and let X be a B-valued random variable. Then X is Bochner integrable if and only if $E\|X\| < \infty$. Moreover, if X is Bochner integrable, then X is also Pettis integrable and both integrals coincide.*

Recall that as defined in Section 2.1, a B-valued random variable X is centred Gaussian if $f(X)$ is a mean zero normal random variable for every $f \in B^*$. Then, as shown in Example 2.1.6, $\|X\| = \sup_{f\in D} f(X)$, where D is a countable subset of B_1^*, the unit ball of the dual of B. Thus, the process $f \mapsto f(X)$, $f \in D$, is Gaussian, separable and $\sup_{f\in D}|f(X)| = \|X\| < \infty$ a.s. Hence, the Borell-Sudakov-Tsirelson theorem (Theorem 2.2.7) applies, and so do its corollaries, so we have not only $E\|X\| < \infty$ but also $Ee^{\lambda\|X\|^2} < \infty$ for all $\lambda < 1/2\sigma^2$, for $\sigma = \sup_{f\in D}(Ef^2(X))^{1/2} \le E\|X\|$. In particular, then X is Bochner and hence also Pettis integrable.

Recall that in the case of a separable Banach space, the process $f \mapsto f(X)$, $f \in B_1^*$, defined by a centred Gaussian B-valued random variable is also a separable Gaussian process, and therefore, so is the process defined over all of B^*. To see this, note that by the Banach-Alaoglu theorem (see Exercise 2.6.2), the unit ball B_1^* is compact for the weak* topology, which is metrisable. Recall that the weak* topology of B^* is the topology of pointwise convergence over B, $f_n \to_{w*} f$ iff $f_n(x) \to f(x)$, for all $x \in B$. Then, if $\tilde{D}$ is countable and weak-*dense in B_1^*, $f \in B_1^*$, there are $h_k \in \tilde{D}$ such that $h_k(X(\omega)) \to f(X(\omega))$ for all ω, proving the separability of the process defined on B_1^*. Since $B^* = \cup_n(nB_1^*)$, separability of the whole process $\{f(X) : f \in B^*\}$ follows as well: $B_0^* := \cup_n(n\tilde{D})$ is countable and weak* sequentially dense in B^*.

Now we construct the RKHS of a centred Gaussian B-valued random variable X. Let $F = \{f(X) : f \in B^*\} \subset L^2(\Omega, \Sigma, \Pr)$, which is a vector space, and let $\bar{F}$ be its completion in L^2. For every $h \in \bar{F}$, define $\phi(h) = E(hX)$. Then hX is measurable, and $E\|hX\| \le (Eh^2)^{1/2}(E\|X\|^2)^{1/2} < \infty$ because $h \in L^2(\Pr)$ and $\|X\|$ enjoys very strong integrability, in particular being square integrable. Hence, by Lemma 2.6.3, hX is Bochner integrable and its integral satisfies $f(E(hX)) = E(hf(X))$ for all $f \in B^*$.

Definition 2.6.4 Let B be a separable Banach space, and let X be a B-valued centred Gaussian variable. The reproducing kernel Hilbert space $H = H_X$ of X (or of its probability law or of its covariance) is the vector space $H = \{E(hX) : h \in \bar{F}\} \subset B$ with inner product $\langle E(h_1X), E(h_2X)\rangle_H := E(h_1h_2)$.

This definition is very similar to Definition 2.6.1 because the space F in the process case is also the linear span of $f(X)$, where $f \in (\mathbb{R}^T)^*$ (this is immediate from Exercise 2.6.1). The only difference is that $\mathbb{R}^T$ is not a Banach space. We could unify both definitions by considering Fréchet topological vector spaces, but on the one hand, this requires some technicalities about duality, and on the other hand, this level of generality is not necessary for the applications developed in this book.

Remark 2.6.5 In both definitions we do have $H = \{E(kX) : k \in L^2(\Omega, \Sigma, \Pr)\}$, the reason being that if $k \in L^2$ and $\pi_{\bar{F}}(k)$ is its orthogonal projection onto $\bar{F}$, then $E(kX) = E(\pi_{\bar{F}}(k)X)$. In the case of $\mathbb{R}^T$, this follows just by definition, and in the case of a separable Banach space B, it follows because it is obviously true that $E(kf(X)) = E(\pi_{\bar{F}}(k)f(X))$ for all $f \in B^*$, as $f(X) \in \bar{F}$, and the Bochner integral coincides with the Pettis integral.

The following lemma is helpful in the construction of RKHSs:

Lemma 2.6.6 *The map* $\varphi : B^* \mapsto H$ *defined as* $\varphi(f) = \phi(f(X)) = E(f(X)X)$ *is weak* sequentially continuous. Consequently, if* B_0^* *is sequentially dense in* B^* *for the weak* topology,* H *is the completion of* $\varphi(B_0^*)$ *for the norm of* H, $\|\cdot\|_H$.

Proof If $f_n \to_{w*} f$, then $f_n(X) \to f(X)$ a.s. and hence also in L^2 (e.g., Exercise 2.1.4). Then, as $n \to \infty$,

$$\|\varphi(f_n) - \varphi(f)\|_{\mathbb{H}} = \sqrt{E(f_n(X) - f(X))^2} \to 0. \quad \blacksquare$$

Example 2.6.7 (The RKHS of Brownian motion.) Brownian motion on $[0,1]$ is a centred sample continuous Gaussian process W whose covariance is $E(W(s)W(t)) = s \wedge t$, $s,t \in [0,1]$ (see Exercise 2.3.2). It also can be equivalently thought of as a B-valued random variable, where $B = C([0,1])$, the space of continuous function on $[0,1]$, endowed with the supremum norm. Then B^* is the space of finite signed measures on $[0,1]$, and the subspace B_0^* of finite linear combinations of point masses $\sum_{i=1}^n a_i \delta_{t_i}$ is weak* sequentially dense in B^* (Exercise 2.6.3). Then, assuming t_i increasing and letting $t_0 = 0$, $t_{k+1} = 1$, we have

$$\varphi\left(\sum_{i=1}^k a_i \delta_{t_i}\right)(t) = \sum_{i=1}^k a_i (t_i \wedge t) = \sum_{r=0}^{k-1}\left(\sum_{i=1}^r a_i t_i + t\sum_{i=r+1}^k a_i\right) I_{(t_r,t_{r+1}]}(t) + \sum_{i=1}^k a_i t_i I_{(t_k,t_{k+1}]},$$

a piecewise linear continuous function which is zero at zero and has constant slope $\sum_{i=r+1}^k a_i$ on the interval (t_r, t_{r+1}), $r = 0, \ldots, k$. Hence, $\varphi(B_0^*)$ is the set of all piecewise linear continuous functions on $[0,1]$ which are zero at zero. Then

$$\begin{aligned}
\left\|\varphi\left(\sum_{i=1}^k a_i \delta_{t_i}\right)\right\|_H^2 &= E\left(\sum_{i=1}^k a_i W(t_i)\right)^2 \\
&= E\left(\left(\sum_{i=1}^k a_i\right) W(t_1) + \left(\sum_{i=2}^k a_i\right)(W(t_2) - W(t_1)) + \cdots + a_k(W(t_k) - W(t_{k-1}))\right)^2 \\
&= \sum_{r=0}^{k-1}\left(\sum_{i=r+1}^k a_i\right)^2 (t_{r+1} - t_r).
\end{aligned}$$

That is, writing $F(t) = \varphi\left(\sum_{i=1}^k a_i \delta_{t_i}\right)$, we have $\|F\|_H^2 = \int_0^1 (F'(t))^2 dt$, where F' is a step function. Since step functions are dense in L^2, the closure of $\varphi(B_0^*)$ for the $\|\cdot\|_H$ norm is the set of absolutely continuous functions on $[0,1]$ which are zero at zero and whose derivative is in $L^2([0,1])$. By Lemma 2.6.6, this set of functions is the RKHS of Brownian motion, that is,

$$H = \{f : f(0) = 0,\ f \text{ is absolutely continuous},\ f' \in L^2([0,1])\},\ \|f\|_H^2 = \int_0^1 (f'(x))^2 dx. \tag{2.73}$$

Not surprisingly, one obtains the same RKHS if Brownian motion is considered as a stochastic process (as opposed to a $C([0,1])$-valued random variable) and Definition 2.6.1 is applied. (The fact that both RKHS definitions produce the same object is not a coincidence and can in fact be deduced from a general proposition, which can be inferred by modifying the statement of Exercise 2.6.5).

Here is another characterization of H:

Proposition 2.6.8 *Let X be a centred Gaussian B-valued random variable, let H be its RKHS and let $C_X(f,g) = E(f(X)g(X))$, $f,g \in B^*$, be its covariance. Then*

$$H = \left\{x \in B : \sup_{f \in B^*} f^2(x)/C_X(f,f) < \infty\right\},$$

and $\|x\|_H^2$ equals this supremum.

The proof is sketched in Exercise 2.6.4. The following topological properties of $H \subseteq B$ are important:

Proposition 2.6.9 *Let X be a centred B-valued Gaussian variable and B a separable Banach space. Then H is a separable Hilbert space and a measurable subset of B. The imbedding of H into B is continuous, and in fact, the unit ball $O_H = \{h \in H : \|h\|_H \le 1\}$ is a compact subset of B.*

Proof By the argument just preceding Definition 2.6.4, there exists a countable set B_0^* which is weak* sequentially dense in B^*, and by Lemma 2.6.6, H is the completion of $\varphi(B_0^*)$ for the norm of H. Applying the Gram-Schmidt ortho-normalisation to $\varphi(B_0^*)$ (or to a maximal linearly independent subset of it) produces a countable ortho-normal basis of H, in particular showing that H is separable.

Next, observe that, by definition, if $h \in H$, then $h = E(kX)$ for some $k \in \overline{F} \subset L^2$, and we obtain

$$\|h\| = \sup_{f \in B_1^*} |E(kf(X))| \le (Ek^2)^{1/2} \sup_{f \in B_1^*} (E(f^2(X)))^{1/2} = \sigma \|h\|_H,$$

where $\sigma^2 := \sup_{f \in B_1^*} E(f^2(X))$, showing that the imbedding of H into B is continuous. We should recall that here and elsewhere in this section, given a centred Gaussian B-valued random variable X, $\overline{F}$ stands for the closure in L^2 of $F = \{f(X) : f \in B^*\}$, as in the definition of H.

In fact, let K be a convex symmetric compact subset of B such that $\Pr\{X \in K\} > 0$, which exists by tightness of X, and consider the semi-norm induced by K, $\|x\|_K = \inf\{\lambda : x \in \lambda K\}$. By Exercise 2.6.7, $E\|X\|_K^2 < \infty$, which implies that for any $h \in H$, if $k \in \overline{F}$ defines h,

$$\|h\|_K = \|E(kX)\|_K \le E(|k|\|X\|_K) \le (Ek^2)^{1/2}(E\|X\|_K^2)^{1/2} = \|h\|_H(E\|X\|_K^2)^{1/2}.$$

Hence the unit ball O_H of H is contained in the compact set $(E\|X\|_K)^{1/2}K$. To see that O_H is closed, let x belong to its boundary for the topology of B, and let $h_n \in O_H$, $\|h_n - x\| \to 0$. Let $k_n \in \bar{F}$ be such that $h_n = E(k_n X)$, and note that $\|k_n\|_2 = \|h_n\|_H \le 1$. The unit ball of L^2 is compact for its weak* topology, so there exists a subsequence k_{n_ℓ} converging in the weak* topology of L^2 to some k in the unit ball of L^2. Then, for all $f \in B^*$, we have

$$f(x) = \lim_{n\to\infty} f(h_n) = \lim_{\ell\to\infty} E(k_{n_\ell} f(X)) = E(kf(X)) = f(E(kX)),$$

showing that $x = E(kX)$. Since $\|k\|_2 \le 1$, Remark 2.6.5 shows that $x \in O_H$. Thus, O_H is closed for the topology of B and hence compact in this topology. Moreover, since H is the countable union of the closed subsets nO_H of B, it is a Borel set of B. ■

The following result, *the Karhunen-Loève expansion of X*, develops the random variable X as a series of independent Gaussian vectors of the form hg, with $h \in H$ and g standard normal. If we equip $\bar{F} \subset L^2(\Omega, \Sigma, \Pr)$ with the L^2-norm, then, by definition, the map ϕ : $\bar{F} \mapsto H$ given by $\phi(k) = E(kX)$ is a linear isometry. In particular, since H is separable by Proposition 2.6.9, it follows that $\bar{F}$ is a separable closed subspace of L^2.

Theorem 2.6.10 *Let X be a centred B-valued Gaussian variable and B a separable Banach space, and let H be its RKHS. Let k_j, $j \in \mathbb{N}$, be a complete ortho-normal system of $\bar{F}$. Then the series $\sum_{j=1}^{\infty} E(k_j X)k_j$ converges a.s. to X in the norm of B (and the series reduces to a finite sum if* $\dim(H) < \infty$*).*

Alternatively, we may state the theorem in terms of $h_j \in H$: let h_j, $j \in \mathbb{N}$, be a complete ortho-normal system of H, and let $k_j = \phi^{-1}(h_j)$. Then $\sum_{j=1}^{\infty} h_j k_j$ converges a.s. to X in the norm of B. Note that by definition of $\bar{F}$, the k_j are i.i.d. $N(0,1)$ random variables.

Proof Note that for each $f \in B^*$, the random series $\sum_{j=1}^{\infty} f(E(k_j X))k_j = \sum_{j=1}^{\infty} E(k_j f(X))k_j$ is just the L^2-expansion of $f(X)$ in the ortho-normal basis $\{k_j\}$ (as $f(X) \in F \subset \bar{F}$), and therefore, this series converges a.s. to $f(X)$ by Lévy's theorem on convergence equivalence for series of independent random variables (or, somewhat less directly, by the three-series theorem). We infer that if $V \subset B^*$ is countable then for almost every ω, the series $\sum_{j=1}^{\infty} f(E(k_j X)k_j(\omega)$ converges to $f(X(\omega))$ for all $f \in V$. We may take $V = B_0^*$ as the countable weak-*dense subset of B^* constructed before Definition 2.6.4 (and already used in the proof of Proposition 2.6.9). Since this set is weak* dense, it determines points of B; that is, if $f(x) = f(y)$ for all $f \in B_0^*$, then $x = y$. Therefore, if we show that the sequence of partial sums $\{S_m(\omega) := \sum_{j=1}^{m} E(k_j X)k_j(\omega) : m \in \mathbb{N}\}$ is relatively compact in the norm of B for almost every ω, it will follow that its subsequential limits $s(\omega)$ will all satisfy $f(s(\omega)) = f(X(\omega))$ on a set of probability 1 and hence that $S_m(\omega) \to X(\omega)$ a.s.

Next, we show that $\{S_m(\omega)\}$ is relatively compact ω a.s. For each $n \in \mathbb{N}$, there exists a compact convex symmetric set $K_n \subset B$ such that $\Pr\{X \in K_n^c\} < 1/n$, and we can assume that K_n is increasing with n. Fix n and momentarily set $K = K_n$ to ease notation. Let v_i, $i \in \mathbb{N}$, be an enumeration of a countable set $D_K \subset B^*$ such that $\|x\|_K = \sup_{f \in D_K} |f(x)|$ for all $x \in B$ (see the proof of Proposition 2.6.9 for the definition of $\|x\|_K$ and Exercise 2.6.7 for the existence of D_K). For $M < \infty$, set $\|x\|_M = \max_{n\le M} |v_n(x)| \le \|x\|_K$, and note that

$$\|S_N\|_M = \max_{n\le M}\left|\sum_{j=1}^N E(k_j v_n(X))k_j\right| \to \max_{n\le M}|v_n(X)| = \|X\|_M \text{ a.s.}$$

Then, by Lévy's maximal inequality (Exercise 2.6.8) applied for the semi-norms $\|x\|_M$, we have

$$\begin{aligned}\Pr\left\{\sup_{m\in\mathbb{N}}\|S_m\|_M > 1\right\} &= \lim_{N\to\infty}\Pr\left\{\max_{m\le N}\|S_m\|_M > 1\right\}\\ &\le 2\lim_{N\to\infty}\Pr\{\|S_N\|_M > 1\}\\ &= 2\Pr\{\|X\|_M > 1\} \le 2\Pr\{\|X\|_K > 1\},\end{aligned}$$

for each $M < \infty$ (note that $\Pr\{\|X\|_M = 1\} = 0$ because $\|X\|_M$ has a density except perhaps at 0; see Exercise 2.4.4). Since $\|S\|_M \nearrow \|S\|_K$, we obtain from these inequalities applied to $K = K_n$ that

$$\Pr\left\{\sup_{m\in\mathbb{N}}\|S_m\|_{K_n} > 1\right\} \le 2\Pr\{\|X\|_K > 1\} < 2/n$$

or

$$\Pr\left(\bigcap_n\left\{\sup_{m\in\mathbb{N}}\|S_m\|_{K_n} > 1\right\}\right) = 0,$$

proving that for almost every ω there is a compact set $K_{n(\omega)}$ that contains $S_m(\omega)$ for all m. ■

Using Ito-Nisio's theorem, which is a Banach space version of Lévy's theorem on convergence equivalence for sums of independent random variables, one can produce a slicker proof of the preceding theorem, but the present proof is more elementary.

Corollary 2.6.11 *With the same notation as in the preceding theorem, we have* $\Pr\{X \in \bar{H}\} = 1$, *where* $\bar{H}$ *is the closure of* H *in* B. *If* H *is finite dimensional, then* $\Pr\{X \in H\} = 1$, *whereas if* H *is infinite dimensional,* $\Pr\{X \in H\} = 0$.

Proof Recall that by Proposition 2.6.9, H is a Borel set of B. Now the result follows immediately from the preceding theorem and the fact that $\left\|\sum_{i=1}^n k_i E(k_i X)\right\|_H^2 = \sum_{i=1}^n k_i^2$ tends almost surely to infinity in the infinite-dimensional case (recall that the variables k_i are i.i.d. standard normal). ■

2.6.2 Some Applications of RKHS: Isoperimetric Inequality, Equivalence and Singularity, Small Ball Estimates

In this section we collect three of the most interesting results on Gaussian processes where the RKHS plays an important role. They will be used at different instances in this book.

The General Form of the Gaussian Isoperimetric Theorem

Theorem 2.6.12 *Let O_H be the unit ball centred at zero of the RKHS H of X, where X is a centred Gaussian B-valued random variable, B a separable Banach space. Let μ be the probability law of X, a probability measure on the Borel σ-algebra $\mathcal{B}_B$ of B. Then, for every set $A \in \mathcal{B}_B$ and every $\varepsilon > 0$,*

$$\mu(A + \varepsilon O_H) \geq \Phi\big(\Phi^{-1}(\mu(A)) + \varepsilon\big), \tag{2.74}$$

where Φ is the standard normal distribution function.

Proof Let $h_j, j \in \mathbb{N}$ in the infinite-dimensional case and $j \leq n$ in the finite-dimensional case, be an ortho-normal basis of H. By Theorem 2.6.10, we may as well assume (by restricting the probability space if necessary) that $X = \sum_j \phi^{-1}(h_j)h_j$, with convergence in the norm of B. Let $A \in \mathcal{B}_B$, define $\tilde{A} = \{x \in \mathbb{R}^{\mathbb{N}} : \sum_j x_j h_j \in A\}$ and set $\rho = \Phi^{-1}(\mu(A))$. Let γ denote the probability law of a sequence of independent standard normal random variables, say, of $\{\phi^{-1}(h_j)\}$, which is a Borel probability measure on $\mathbb{R}^{\mathbb{N}}$ in infinite dimensions or $\mathbb{R}^n$ in finite dimensions. Then

$$\gamma(\tilde{A}) = \gamma\{x \in \mathbb{R}^{\mathbb{N}} : \sum_j x_j h_j \in A\} = \Pr\{X \in A\} = \mu(A) = \Phi(\rho)$$

(replace $\mathbb{N}$ by n in finite dimensions). Hence, by the isoperimetric inequality for γ, Theorem 2.2.4, we have

$$\gamma(\tilde{A} + \varepsilon O) \geq \Phi(\rho + \varepsilon),$$

where O is the unit ball of ℓ_2 (or of $\mathbb{R}^n$). Now, if $v \in O$, then, by definition, $\sum v_j h_j \in O_H$, and this series converges in H and then also in B (as the imbedding $H \mapsto B$ is continuous); therefore, if $x \in \tilde{A} + \varepsilon O$, then $\sum_j x_j h_j$ converges in B and is an element of the set $A + \varepsilon O_H$. Therefore,

$$\mu(A + \varepsilon O_H) = \Pr\{X \in A + \varepsilon O_H\} \geq \gamma\{\tilde{A} + \varepsilon O_H\} \geq \Phi(\rho + \varepsilon). \quad \blacksquare$$

This result is often more readily applicable than Theorem 2.2.4.

Equivalence and Singularity of Translates of Gaussian Measures

The following theorem about equivalence of translates of Gaussian measures is known as the *Cameron-Martin theorem*, and it states that a Gaussian measure and its translates by vectors in the RKHS of μ are mutually absolutely continuous. It is also true that μ and any translate of μ by a vector not in its RKHS are mutually singular.

Recall that given a centred Gaussian random variable X, since $\bar{F} \subset L^2(\Omega, \Sigma, \Pr)$ is in the closure in L^2 of functions of X ($f(X)$, $f \in B^*$), any function $k \in \bar{F}$ equals a.s. a function of X; that is, $k(\omega) = \tilde{k}(X(\omega))$ a.s. for some measurable function $\tilde{k} : B \mapsto \mathbb{R}$. We drop the tilde from $\tilde{k}$; that is, we identify it with k. Recall also that $\phi(k) = E(kX) \in H$.

Theorem 2.6.13 (Cameron-Martin formula) *Let B be a separable Banach space, let μ be a centred Gaussian Borel measure on B, let H be its RKHS and let $h \in H$. Then the probability measure $\tau_h\mu$ defined as $\tau_h\mu(A) = \mu(A - h) = \mu\{x : x + h \in A\}$, is absolutely continuous with respect to μ, and*

$$\frac{d\tau_h\mu}{d\mu}(x) = e^{(\phi^{-1}h)(x) - \|h\|_H^2/2}.$$

Moreover, if $v \notin H$, then $\tau_v\mu$ and μ are mutually singular.

Proof Let X be a random vector with law μ. Recall that $\phi^{-1}h = (\phi^{-1}h)(X)$ (with the previously mentioned abuse of notation) is a normal random variable as it is the L^2-limit of $f_k(X)$ for some sequence $f_k \in B^*$. Moreover, $E(f(X)\phi^{-1}h) = f(E(X\phi^{-1}h)) = f(h)$ by the definitions, and $E(\phi^{-1}h)^2 = \|h\|_H^2$, because ϕ is an isometry. This shows that for all $f \in B^*$, the random vector of $\mathbb{R}^2$ $(f(X), \phi^{-1}h)$ has covariance matrix

$$A := \begin{pmatrix} C_\mu(f,f) & f(h) \\ f(h) & \|h\|_H^2 \end{pmatrix}.$$

Consider the Borel probability measure on B

$$d\pi(x) = e^{(\phi^{-1}h)(x) - \frac{\|h\|_H^2}{2}} d\mu(x).$$

The theorem will be proved if we show that $\pi = \tau_h\mu$, and for this it suffices to show that both measures have the same characteristic functional (Exercise 2.6.10). Let $f \in B^*$. Then the usual computation for the characteristic function of a multivariate normal variable gives

$$\begin{aligned}
\hat{\pi}(f) &= \int e^{if(x)} e^{(\phi^{-1}h)(x) - \|h\|_H^2/2} d\mu(x) = e^{-\|h\|_H^2/2} E\left(e^{if(X) + \phi^{-1}(h)}\right) \\
&= e^{-\|h\|_H^2/2} E \exp\left((i,1)\begin{pmatrix} f(X) \\ \phi^{-1}h \end{pmatrix}\right) = e^{-\|h\|_H^2/2} e^{\frac{1}{2}(i,1)A(i,1)^t} \\
&= e^{-C_\mu(f,f)/2 + if(h)} = e^{if(h)}\hat{\mu}(f) = (\tau_h\mu)\hat{}\,(f).
\end{aligned}$$

To prove the converse, if $\dim(H) < \infty$, then $\mu(H) = 1$, which together with the fact that if $v \notin H$, then $(H - v) \cap H = \emptyset$ gives $\tau_v\mu(H) = 0$. If $\dim(H) = \infty$, we assume that $\tau_v\mu$ and μ are not mutually singular, and we would like to conclude $v \in H$. Since F is dense in $\bar{F}$, there exist $f_i \in B^*$, $i \in \mathbb{N}$, such that the sequence $\{f_i(X)\}$ is an ortho-normal basis of $\bar{F}$. Hence, $\{h_i := \phi(f_i(X))\}$ is an ortho-normal basis of H, and we have by Theorem 2.6.10 that the series $\sum_i f_i(X)h_i$ converges a.s. to X in the sense of convergence in B. Expressed in terms of μ, if

$$C = \{x \in B : \sum_i f_i(x)h_i \text{ converges in } B \text{ to } x\},$$

then, $\mu(C) = 1$. But then, since $\tau_v\mu$ is not singular with μ, we have that $\mu(C - v) \neq 0$ and hence $C \cap (C - v) \neq \emptyset$, or there exists $x \in C$ such that $x - v \in C$ as well. Subtracting the series for x and $x - v$, we obtain that the series $\sum_i f_i(v)h_i$ converges in B to v. Since $f_i(X)$ are independent standard normal, Lévy's theorem on convergence of series of independent terms (or, alternatively, the three-series theorem) shows that if $u = (u_i) \in \ell_2$, then the series of real random variables $\sum_i u_i f_i(X)$ converges a.s., and the preceding argument applied to $C_u = \{x \in B : \sum_i u_i f_i(x) \text{ converges}\}$, $u \in \ell_2$, shows that the series $\sum_i u_i f_i(v)$ converges for all $u \in \ell_2$. This implies that the vector $(f_i(v) : i \in \mathbb{N})$ is in ℓ_2; that is, $\sum_i f_i^2(v) < \infty$. Therefore, the series $\sum_i f_i(v)h_i$ converges in the sense of H (H is a Hilbert space, and h_i is an ortho-normal basis of H). Since the embedding $H \subset B$ is continuous, it follows that the sum of the series $\sum_i f_i(v)h_i$ in the sense of H must be the same as its sum in the sense of B, that is, v, which shows $v \in H$ by completeness of H. ■

Remark 2.6.14 Since, by the Cameron-Martin formula, the Radon-Nikodym derivative of $\tau_h\mu$ with respect to μ is everywhere positive, it follows that μ and $\tau_h\mu$ are mutually absolutely continuous for any $h \in H$. As a consequence, if $h, \ell \in H$, we have both

$$\frac{d\tau_h\mu}{d\mu}\frac{d\mu}{d\tau_h\mu} = 1 \quad \text{and} \quad \frac{d\tau_h\mu}{d\tau_\ell\mu} = \frac{d\tau_h\mu}{d\mu}\frac{d\mu}{d\tau_\ell\mu}, \qquad \mu \text{ a.s.},$$

by the chain rule for Radon-Nikodym derivatives.

Example 2.6.15 Let B be a separable Hilbert space and X a B-valued centred Gaussian random variable. Without loss of generality, we may take $X = \sum_{i=1}^{\infty} \lambda_i g_i e_i$, where $\{g_i\}$ is a sequence of independent $N(0,1)$ random variables, $\{e_i\}$ is a complete ortho-normal system in B and $\lambda_i \geq 0$ satisfy $\sum_{i=1}^{\infty} \lambda_i^2 < \infty$ (Exercise 2.6.14). It follows from Remark 2.6.5 that since $E(kX) = 0$ if $k \in L^2(\Omega, \sigma, \Pr)$ is orthogonal to the sequence $\{g_i\}$, the RKHS of X is

$$\begin{aligned} H &= \left\{E\left(\left(\sum k_i g_i\right)X\right) : \sum k_i^2 < \infty\right\} \\ &= \left\{\sum \lambda_i k_i e_i : \sum k_i^2 < \infty\right\} = \left\{h \in B : \sum h_i^2/\lambda_i^2 < \infty\right\}, \end{aligned} \tag{2.75}$$

where $h_i = \langle h, e_i\rangle$, with inner product $\langle h, \bar{h}\rangle_H = \sum h_i \bar{h}_i/\lambda_i^2$. Given $h = \sum h_i e_i \in H$, clearly $\phi^{-1}(h)(\omega) = \sum_i (h_i/\lambda_i) g_i(\omega)$ or, as a function of X, $\phi^{-1}(h)(X(\omega)) = \langle \sum (h_i/\lambda_i^2) e_i, X(\omega)\rangle$. Then the Cameron-Martin formula for the probability law μ of X can formally be written as

$$\frac{d\tau_h\mu}{d\mu}(x) = e^{\langle h,x\rangle_H - \sum h_i^2/2\lambda_i^2}. \tag{2.76}$$

In a coordinate-free formulation when B is the range of X, if Q is the covariance operator of X ($Q : B \mapsto B$ is defined as $\langle Q(v), w\rangle = E(\langle X, v\rangle\langle X, w\rangle)$ for all $w \in B$), (2.75) and (2.76) become

$$H = \{h = Q^{-1/2}(v) : v \in B\}, \qquad \frac{d\tau_h\mu}{d\mu} = e^{\langle Q^{-1}h,x\rangle - \langle Q^{-1}h,h\rangle/2}.$$

The case $B = \mathbb{R}^n$ and $X = (g_1, \ldots, g_n)$ corresponds to $\lambda_1 = \cdots = \lambda_n = 1$ (or $Q = I$), and we have:

$$H = \mathbb{R}^n, \qquad \frac{d\tau_h\mu}{d\mu} = e^{\langle h,x\rangle - \|h\|^2/2}, \quad x, h \in \mathbb{R}^n.$$

The Cameron-Martin formula in $\mathbb{R}^n$ can, of course, be derived much more easily. In $\mathbb{R}$, if $\mu = N(0, \sigma^2)$ and $\tau_h\mu = N(h, \sigma^2)$, $h \in \mathbb{R}$, $\sigma \neq 0$, then, for all $x \in \mathbb{R}$,

$$\frac{d\tau_h\mu}{d\mu}(x) = \frac{e^{-(x-h)^2/2\sigma^2}}{e^{-x^2/2\sigma^2}} = e^{hx/\sigma^2 - \frac{1}{2}h^2/\sigma^2},$$

and then, since the density of a (finite) product of measures is the product of their densities, we obtain that if $\mu_n = \prod_{i=1}^n N(0, \sigma_i^2)$ and $\tau_h\mu_n = \prod_{i=1}^n N(h_i, \sigma_i^2)$, then, for all $x \in \mathbb{R}^n$,

$$\varphi_n(x) := \frac{d\tau_h\mu_n}{d\mu_n}(x) = \exp\left(\sum_{i=1}^n h_i x_i/\sigma_i^2 - \frac{1}{2}\sum_{i=1}^n h_i^2/\sigma_i^2\right). \tag{2.77}$$

Example 2.6.15 shows how Theorem 2.6.13 extends this differentiation formula to countably infinite products of normal random variables when they are tight Borel measures, and the question arises as to whether there is a further extension. The setting is $\Omega = \mathbb{R}^{\mathbb{N}}$, $\Sigma = \mathcal{C}$

the cylindrical σ-algebra of $\mathbb{R}^{\mathbb{N}}$ (see Section 2.1), and $\mu = \prod_{i=1}^{\infty} N(0,\sigma_i^2)$ for a sequence of strictly positive numbers σ_i; that is, μ is the probability law in $\mathcal{C}$ of the sequence $\{\sigma_i g_i\}$, g_i i.i.d. standard normal. It can be argued as in Example 2.6.2 that the RKHS of μ (or of the process $\{\sigma_i g_i\}$) is the set of all sequences $\{h = (h_i : i \in \mathbb{N})\}$ such that $\sum_i h_i^2/\sigma_i^2 < \infty$, which we denote by $\ell_2(\{\sigma_i^{-2}\})$, the L^2-space for the measure $\sum_{n=1}^{\infty} \sigma_n^{-2}\delta_n$ on $\mathbb{N}$. Since $\mathbb{R}^{\mathbb{N}}$ is not a separable Banach space, Theorem 2.6.13 does not apply to μ or its translates.

Proposition 2.6.16 *Let $\{g_i\}_{i=1}^{\infty}$ be a sequence of independent standard normal random variables, let $\sigma_i > 0$, $i \in \mathbb{N}$, be arbitrary positive numbers and let $\mu = \prod_{i=1}^{\infty} N(0,\sigma_i^2)$ be the probability law of the sequence $\{\sigma_i g_i\}$, defined on the cylindrical σ-algebra $\mathcal{C}$ of $\mathbb{R}^{\mathbb{N}}$. If $h \in \ell_2(\{\sigma_i^{-2}\})$ and $x = (x_i : i \in \mathbb{N}) \in \mathbb{R}^{\mathbb{N}}$, then the measures $\tau_h\mu$ and μ are mutually absolutely continuous, and*

$$\frac{d\tau_h\mu}{d\mu}(x) = \varphi(x) := \exp\left(\sum_{i=1}^{\infty} h_i x_i/\sigma_i^2 - \frac{1}{2}\sum_{i=1}^{\infty} h_i^2/\sigma_i^2\right), \quad \mu \text{ a.s.} \tag{2.78}$$

If $h \in \mathbb{R}^{\mathbb{N}} \setminus \ell_2(\{\sigma_i^{-2}\})$, then μ and $\tau_h\mu$ are mutually singular.

Proof For any $1 \le n < m \le \infty$ and $\lambda > 0$,

$$\int_{\mathbb{R}^{\mathbb{N}}} e^{\lambda\sum_{i=n+1}^{m} h_i x_i/\sigma^2} d\mu(x) = Ee^{\lambda\sum_{i=n+1}^{m} h_i g_i/\sigma_i} = e^{\lambda^2\sum_{i=n+1}^{m} h_i^2/2\sigma_i^2}.$$

In particular, assuming that $h \in \ell_2(\{\sigma_i^{-2}\})$, we have (a) $\int_{\mathbb{R}^{\mathbb{N}}} e^{\lambda\sum_{i=1}^{\infty} h_i x_i/2\sigma_i^2} d\mu(x) = e^{-\lambda^2\sum_{i=1}^{\infty} h_i^2/2\sigma_i^2} < \infty$, which shows that $\varphi(x)$ is μ a.s. finite, and (b)

$$E\left[e^{\sum_{i=1}^{m} h_i g_i/2\sigma_i} - e^{\sum_{i=1}^{n} h_i g_i/2\sigma_i}\right]^2 = e^{\sum_{i=1}^{n} h_i^2/\sigma_i^2} + e^{\sum_{i=1}^{m} h_i^2/\sigma_i^2} - 2e^{\sum_{i=1}^{n} h_i^2/\sigma_i^2} e^{\sum_{i=n+1}^{m} h_i^2/2\sigma^2}$$

converges to zero as $n \to \infty$, which proves that if we set, with some abuse of notation, $\varphi_n(x) = \varphi_n(x_1,\dots,x_n)$, for $x \in \mathbb{R}^{\mathbb{N}}$, where φ_n is as in (2.77), then

$$\sqrt{\varphi_n} \to \sqrt{\varphi} \quad \text{in} \quad L^2(\mathbb{R}^{\mathbb{N}},\mathcal{C},\mu).$$

Therefore, if E is a finite-dimensional cylinder, say,

$$E = \{x \in \mathbb{R}^{\mathbb{N}} : (x_1,\dots,x_n) \in E_{1,\dots,n}, x_j \in \mathbb{R} \text{ for } j \ge n\},$$

where $E_{1,\dots,n}$ is a Borel set of $\mathbb{R}^n$, then, using also (2.77), we have, for all $m > n$,

$$\begin{aligned}\tau_h\mu(E) &= \int_{E_{1,\dots,n}} \left(\sqrt{\varphi_n(x_1,\dots,x_n)}\right)^2 d\mu_n(x_1,\dots,x_n)\\ &= \int_E \left(\sqrt{\varphi_n(x)}\right)^2 d\mu(x) = \int_E \left(\sqrt{\varphi_m(x)}\right)^2 d\mu(x) = \int_E \left(\sqrt{\varphi(x)}\right)^2 d\mu(x).\end{aligned}$$

The algebra $\mathcal{A}$ of finite-dimensional cylinders generates the cylindrical σ-algebra, and therefore, given any $E \in \mathcal{C}$, we can find $E_n \in \mathcal{A}$, decreasing and containing E, such that both $\tau_h\mu(E_n) \searrow \tau_h\mu(E)$ and $\mu(E_n) \searrow \mu(E)$ and the preceding identities then give, by dominated convergence, that

$$\tau_h\mu(E) = \lim_n \tau_h\mu(E_n) = \lim_n \int_{E_n} \varphi d\mu = \int_E \varphi d\mu,$$

proving (2.78).

For the converse, it suffices to prove that for every $\varepsilon > 0$ there exists $B \in \mathcal{C}$ such that $\mu(B) < \varepsilon$ and $\tau_h\mu(B^c) < \varepsilon$ (if B_n corresponds to $\varepsilon = 2^{-n}$, then $B = (B_n \ i.o.)$ satisfies $\mu(B) = 0$ and $\tau_h\mu(B^c) = 0$). If $\sum_{i=1}^{\infty} h_i^2/\sigma_i^2 = \infty$, then

$$\int_{\mathbb{R}^n} \sqrt{\varphi_n} d\mu_n = e^{-\sum_{i=1}^n h_i^2/8\sigma_i^2} \to 0, \quad \text{as } n \to \infty.$$

Then, given $\varepsilon > 0$, let n be such that $\int_{\mathbb{R}^n} \sqrt{\varphi_n} d\mu_n < \varepsilon$, let $B_{1,\dots,n} = \{x \in \mathbb{R}^n : \varphi_n(x) > 1\}$ and $B = B_{1,\dots,n} \times \mathbb{R} \times \cdots \in \mathcal{C}$. We have

$$\mu(B) = \int_{B_{1,\dots,n}} d\mu_n \le \int_{B_{1,\dots,n}} \sqrt{\varphi_n} d\mu_n < \varepsilon$$

and

$$\tau_h\mu(B^c) = \int_{\mathbb{R}^n \setminus B_{1,\dots,n}} \varphi_n d\mu \le \int_{\mathbb{R}^n \setminus B_{1,\dots,n}} \sqrt{\varphi_n} d\mu_n < \varepsilon. \quad \blacksquare$$

We conclude with an interesting consequence of Theorem 2.6.13 that adds to the relevance of the RKHS. Recall that the support of a Borel measure on a metric space is the set of points x such that $\mu\{y : d(x,y) < \varepsilon\} > 0$, for all $\varepsilon > 0$.

Corollary 2.6.17 *Let μ be a centred Gaussian measure on a separable Banach space B, and let H be its RKHS. Then the support of μ is $\bar{H}$, the closure in B of H.*

Proof Let X be a Gaussian random variable with law μ. First, by Anderson's lemma, (Theorem 2.4.5), $\Pr(\|X\| < \varepsilon) > 0$ for all $\varepsilon > 0$. Next, by Theorem 2.6.13, the law of $X + h$, $h \in H$, has a positive density λ_h with respect to the law of X, and therefore, $\Pr\{\|X + h\| < \varepsilon\} = \int_{\|x\|<\varepsilon} \lambda_h(x) d\mu(x) > 0$, for all $h \in H$ and $\varepsilon > 0$. Then, since any ball around $x \in \bar{H}$ contains a ball around a point in H, the same is true for all $x \in \bar{H}$. Hence, the support of μ contains $\bar{H}$.

For the reverse inclusion, just note that as recorded in Corollary 2.6.11, the Karhunen-Loève expansion of X (Theorem 2.6.10) implies that $\Pr\{X \in \bar{H}\} = 1$. Thus, the support of μ is contained in $\bar{H}$. $\blacksquare$

Application to Estimation of the Probabilities of Small Balls

Theorem 2.6.13 together with Anderson's lemma (Theorem 2.4.5) provides a quite exact two-sided relationship between the Gaussian probability of small balls centred at the origin and that of small balls centred at other points in the support of the measure.

Here is a useful first consequence of Theorem 2.6.13 regarding balls shifted by $h \in H$, in the opposite direction to Anderson's lemma (already somewhat implicit in the preceding proof):

Corollary 2.6.18 *Let $C \subset B$ be a symmetric Borel set, where B is a separable Banach space, and let X be a centred Gaussian B-valued random variable. Then, for every $h \in H$,*

$$\Pr\{X - h \in C\} \ge e^{-\|h\|_H^2/2} \Pr\{X \in C\}.$$

Proof By Theorem 2.6.13 and by symmetry of C, and since $(e^a + e^{-a})/2 \geq 1$, for all $a \in \mathbb{R}$,

$$\begin{aligned}2\Pr\{X - h \in C\} &= \Pr\{X - h \in C\} + \Pr\{X + h \in C\} \\ &= E\left[\left(e^{\phi^{-1}(h)} + e^{\phi^{-1}(-h)}\right) e^{-\|h\|_H^2/2} I_C(X)\right] \\ &\geq 2e^{-\|h\|_H^2/2} E(I_C(X)). \quad \blacksquare\end{aligned}$$

Given a centred Gaussian B-valued random variable X with law μ, define its *concentration function* $\phi_x(\varepsilon)$, for x in the support of μ and hence $x \in \bar{H}$ and $\varepsilon \geq 0$, as

$$\phi_x(\varepsilon) = \inf_{h \in H, \|h - x\| \leq \varepsilon} \left[\frac{1}{2}\|h\|_H^2 - \log \Pr\{\|X\| < \varepsilon\}\right], \tag{2.79}$$

and note that $\phi_0(\varepsilon) = -\log \Pr\{\|X\| < \varepsilon\}$; that is, $\Pr\{\|X\| < \varepsilon\} = e^{-\phi_0(\varepsilon)}$. We will prove the following proposition:

Proposition 2.6.19 *Let X be a centred Gaussian B-valued random variable, where B is a separable Banach space. Let $x \in \mathrm{supp}(\mathcal{L}(X)) = \bar{H}$ (see Corollary 2.6.17) and $\varepsilon > 0$. Then*

$$\phi_x(\varepsilon) \leq -\log \Pr\{\|X - x\| < \varepsilon\} \leq \phi_x(\varepsilon/2). \tag{2.80}$$

Proof If $\|h - x\| \leq \varepsilon/2$, then $\|X - x\| \leq \varepsilon/2 + \|X - h\|$, so if moreover $h \in H$, by Corollary 2.6.18,

$$\Pr\{\|X - x\| < \varepsilon\} \geq \Pr\{\|X - h\| < \varepsilon/2\} \geq e^{-\|h\|_H^2/2} \Pr\{\|X\| < \varepsilon/2\},$$

which yields the upper estimate in (2.80).

For the lower estimate, suppose that given $x \in \bar{H}$ and $\varepsilon > 0$ there exists h_ε such that

$$\phi^{-1}(h_\varepsilon) \geq 0 \text{ a.s.} \quad \text{on the event} \quad \{\|X + h_\varepsilon - x\| < \varepsilon\}. \tag{2.81}$$

Then, applying first the Cameron-Martin theorem (Theorem 2.6.13) and then Anderson's lemma (Theorem 2.4.5) and finally the continuity of the distribution of $\|X\|$ outside 0 (Exercise 2.4.4), we have

$$\begin{aligned}\Pr\{\|X - x\| < \varepsilon\} &= \Pr\{\|(X - h_\varepsilon) - x + h_\varepsilon\| < \varepsilon\} = E\left(e^{-\phi^{-1}(h_\varepsilon) - \|h_\varepsilon\|_H^2/2} I_{\|X - x + h_\varepsilon\| < \varepsilon}\right) \\ &\leq e^{-\|h_\varepsilon\|_H^2/2} EI_{\|X - x + h_\varepsilon\| < \varepsilon} \leq e^{-\|h_\varepsilon\|_H^2/2} EI_{\|X\| \leq \varepsilon} = e^{-\|h_\varepsilon\|_H^2/2} EI_{\|X\| < \varepsilon},\end{aligned}$$

proving the proposition modulo the statement in (2.81). The proof of (2.81) is carried out in Lemmas 2.6.20 and 2.6.21. $\blacksquare$

Lemma 2.6.20 *With the notation of Proposition 2.6.19, given $x \in \bar{H}$ and $\varepsilon > 0$, set $B_{\varepsilon,x} = \{h \in H : \|h - x\| \leq \varepsilon\}$. Then there exists $h_\varepsilon \in B_{\varepsilon,x}$ such that $\|h_\varepsilon\|_H = \inf\{\|h\|_H : h \in B_{\varepsilon,x}\}$, and h_ε satisfies $\langle h, h_\varepsilon\rangle_H \geq \|h_\varepsilon\|_H^2$, for all $h \in B_{\varepsilon,x}$.*

Proof The set $B_{\varepsilon,x}$ is closed in H: if $h_n \to h$ in H, then $h_n \to h$ in B by continuity of the embedding of H into B (Proposition 2.6.9), so if, moreover, $h_n \in B_{\varepsilon,x}$, then $\|h - x\| \leq \varepsilon$, that is, $h \in B_{\varepsilon,x}$. If the infimum of $\|h\|_H$ in $B_{\varepsilon,x}$ is zero, then there exist $h_n \in B_{\varepsilon,x}$ such that $\|h_n\| \to 0$ in H, and since $B_{\varepsilon,x}$ is closed, it contains 0. Suppose now that the infimum is $c > 0$. If $c_n \searrow c$, $c_n > c$, then we have $\{h \in H : \|h\|_H \leq c_n\} \cap \bar{B}_{\varepsilon,x} \neq \emptyset$, where $\bar{B}_{\varepsilon,x}$ is the closure in B of $\bar{B}_{\varepsilon,x}$, but these sets are compact by Proposition 2.6.9, and hence their intersection is not empty. Now $\bar{B}_{\varepsilon,x} \subset \{y \in B : \|y - x\| \le \varepsilon\}$, so the not-empty intersection is precisely $\{h \in H : \|h\| \le c\} \cap \bar{B}_{\varepsilon,x} = \{h : \|h\| = c, \|h - x\| \le \varepsilon\} \subset B_{\varepsilon,x}$. Thus, the infimum of $\|\cdot\|_H$ on $B_{\varepsilon,x}$ is attained at some $h_\varepsilon \in B_{\varepsilon,x}$.

Next, $B_{\varepsilon,x}$ is convex, and therefore, for any $h \in B_{\varepsilon,x}$ and $0 \le \lambda \le 1$, we have $\|\lambda h + (1-\lambda)h_\varepsilon\|_H^2 \ge \|h_\varepsilon\|_H^2$, which gives, after developing and dividing by λ,

$$\lambda \langle h, h\rangle_H + 2(1-\lambda)\langle h, h_\varepsilon\rangle_H \ge 2(1-\lambda)\|h_\varepsilon\|_H^2, \quad 0 < \lambda \le 1.$$

Letting $\lambda \to 0$, $\lambda > 0$, we obtain $\langle h, h_\varepsilon\rangle_H \ge \|h_\varepsilon\|_H^2$. ■

Lemma 2.6.21 *With the notation of Proposition 2.6.19 and Lemma 2.6.20, $\phi^{-1}(h_\varepsilon) \ge 0$ almost surely on the event $\{\|X + h_\varepsilon - x\| < \varepsilon\}$.*

Proof Let h_i, $i \in \mathbb{N}$, be an ortho-normal basis of H. Then, by Theorem 2.6.10, $X = \sum_{i=1}^{\infty} \phi^{-1}(h_i)h_i$ a.s., with convergence in B. The partial sums $X_m = \sum_{i=1}^{m} \phi^{-1}(h_i)h_i$ are in H, and

$$\langle X_m, h_\varepsilon\rangle_H = \sum_{i=1}^{m} \phi^{-1}(h_i)\langle h_i, h_\varepsilon\rangle_H = \phi^{-1}\left(\sum_{i=1}^{m} \langle h_i, h_\varepsilon\rangle_H h_i\right).$$

The right-hand side converges to $\phi^{-1}(h_\varepsilon)$ in L^2 (recall that ϕ^{-1} is an isometry onto $\bar{F} \subset L^2$); hence, so does the left-hand side. Therefore, there is a subsequence $m_k \to \infty$ such that $\langle X_{m_k}, h_\varepsilon\rangle_H \to \phi^{-1}(h_\varepsilon)$. Now, on the event $\|X + h_\varepsilon - x\| < \varepsilon$, eventually a.s. $\|X_m + h_\varepsilon - x\| < \varepsilon$; that is, $X_m + h_\varepsilon \in B_{\varepsilon,x}$, for all $m \ge m_0(\omega)$, $m_0(\omega) < \infty$ a.s., and therefore, Lemma 2.6.20 implies that eventually a.s. $\langle X_m + h_\varepsilon, h_\varepsilon\rangle_H \ge \|h_\varepsilon\|_H^2$, implying $\langle X_m, h_\varepsilon\rangle_H \ge 0$ eventually a.s. This, together with $\langle X_{m_k}, h_\varepsilon\rangle_H \to \phi^{-1}(h_\varepsilon)$ a.s., implies that on the event $\|X + h_\varepsilon - x\| < \varepsilon$, we have $\phi^{-1}(h_\varepsilon) \ge 0$ a.s. ■

2.6.3 An Example: RKHS and Lower Bounds for Small Ball Probabilities of Integrated Brownian Motion

In this subsection we obtain the reproducing kernel Hilbert space of Gaussian processes defined as iterated integrals of Brownian motion and of their 'released' versions and estimate as well their small ball concentration functions. These results will be put to use later in Bayesian density estimation.

The Reproducing Kernel Hilbert Space of Integrated Brownian Motion Released at Zero

Let W be Brownian motion on $[0,1]$. See Exercise 2.3.2 for its definition and smoothness properties. In particular, it follows from this exercise that the sample paths of W are all in $C^\alpha([0,1])$, the space of Hölder continuous functions of order α on $[0,1]$ for all $0 < \alpha < 1/2$.

Let $(I_{0+}f)(t) = \int_0^t f(x)dx$ denote the primitive of f which is zero at zero for any continuous function f on $[0,1]$, and let $(I_{0+}^k f)(t) = \int_0^t (I_{0+}^{k-1} f)(x)\,dx$ denote its iterations for $k \in \mathbb{N}$. We consider now the successive integrals of Brownian motion

$$I_{0+}^0 W = W, \quad (I_{0+}^k W)(t) = \int_0^t (I_{0+}^{k-1} W)(s)ds, \quad k \geq 1.$$

This is a string of stochastic processes with sample paths of increasing smoothness: for each k, the sample paths of $I_{0+}^k(W)$ are almost all in $C^{k+\alpha}([0,1])$, the space of functions on $C([0,1])$ with k derivatives, the kth being Hölder continuous with index α, for any $\alpha < 1/2$. They would constitute nice prior models for smooth functions if the sample paths and their derivatives were not zero at zero. This can be easily remedied, for example, by considering instead the 'released' processes

$$W^k(t) = \sum_{j=0}^{k} t^j g_j/j + (I_{0+}^k W)(t), \quad t \in [0,1], \quad k \geq 0, \tag{2.82}$$

where g_i, $i \geq 0$, are i.i.d. standard normal variables independent of W (W^0 is the released Brownian motion, $W^0(t) = g_0 + W(t)$). Next, we obtain the RKHS of W^k as a $C([0,1])$-valued random variable.

Recall that $C^k([0,1])$, the space of functions on $[0,1]$ with k continuous derivatives, has norm $\|f\|_{C^k} = \sum_{j=0}^k \|f^{(j)}\|_\infty$ and hence that the map $I_{0+}^k : C([0,1]) \mapsto C^k([0,1])$ is continuous and one to one. Then Example 2.6.7 in Section 2.6.1 and Exercise 2.6.5 immediately give the following lemma.

Lemma 2.6.22 *The RKHS $H_{I^k W}$ of $I_{0+}^k W$ as a $C([0,1])$-valued random variable is*

$$H_{I^k W} = I_{0+}^k H_W = \{I_{0+}^k f : f \in H_W\}$$
$$= \{f : [0,1] \mapsto \mathbb{R} : \ f^{(j)}(0) = 0, j = 0,1,\ldots,k, \ f^{(k)} \text{ abs. cont}, \ f^{(k+1)} \in L^2([0,1])\}$$

with inner product

$$\langle f, g \rangle_{H_{I^k W}} = \int_0^1 f^{(k+1)}(s) g^{(k+1)}(s) ds.$$

Note that, again by Exercise 2.6.5, since the inclusion $C^j \subset C$ is continuous, the RKHS of $I_{0+}^k W$ as a $C^j([0,1])$-valued random variable, $j \leq k$, is also $H_{I^k W}$.

Let $Z_k = \sum_{j=0}^k t^j g_j/j$. Then, for any points s, s_1, s_2 in $[0,1]$, we have

$$\delta_s(Z_k) = \sum_{j=0}^k s^j g_j/j, \quad E(\delta_s(Z_k)Z_k(t)) = \sum_{j=0}^k (s_j^j/j)t^j/j, \quad E(\delta_{s_1}(Z_k)\delta_{s_2}(Z_k)) = \sum_{j=0}^k s_1^j s_2^j/(j)^2.$$

This implies the following lemma by the definitions and elementary properties in Section 2.6.1:

Lemma 2.6.23 *The RKHS H_{Z_k} of the process $Z_k = \sum_{j=0}^k t^j g_j/j$ as a $C([0,1])$-valued random variable, hence as a $C^\ell([0,1])$-valued random variable for all ℓ, is the set of polynomials of degree k or lower with the inner product*

$$\langle P, Q \rangle_{H_{Z_k}} = \sum_{j=0}^k P^{(j)}(0) Q^{(j)}(0).$$

We now combine these two lemmas into the following proposition:

Proposition 2.6.24 *For $k \ge 0$, the RKHS of W^k as a $C([0,1])$-valued random variable is*

$$H_{W,k} = \left\{ f : [0,1] \mapsto \mathbb{R} : \ f \text{ is } k \text{ times differentiable, } f^{(k)} \text{ is abs. cont. and } f^{(k+1)} \in L^2([0,1]) \right\}$$

with inner product

$$\langle f, g \rangle_{H_{W,k}} = \sum_{j=0}^{k} f^{(j)}(0) g^{(j)}(0) + \int_0^1 f^{(k+1)}(s) g^{(k+1)}(s)\, ds.$$

Proof To ease notation, set $Z = Z_k$ and $V = I_{0+}^k W$. As indicated earlier, it suffices to obtain the RKHS of $Z + W$ as a $C^k = C^k([0,1])$-valued random variable. The support of Z is the set of polynomials of degree k or lower, say, B^Z, and the support B^V of V is contained in the closed linear subspace of functions in C^k that vanish at zero together with their first k derivatives. Therefore, if we define $\Pi : C^k \mapsto B^Z$ on C^k by $\Pi f(t) = \sum_{j=0}^k f^{(j)}(0) t^j / j$, then Π is continuous, $\Pi(B^Z) = B^Z$ and $\Pi(B^V) = 0$. Now take $f \in (C^k)^*$, and note that by independence of Z and V, and since $\Pi(Z) = Z$ and $\Pi(W) = 0$ a.s.,

$$\phi_{Z+V}(f) = E(f(Z+V)(Z+V)) = E(f(Z)Z) + E(f(V)V) = \phi_Z(f) + \phi_V(f)$$

$$\phi_{Z+V}(f \circ \Pi) = E(f(Z)(Z+V)) = E(f(Z)Z) = \phi_Z(f)$$

$$\phi_{Z+V}(f \circ (I - \Pi)) = \phi_V(f).$$

This implies that

$$\phi_Z((C^k)^*),\ \phi_V((C^k)^*) \subset \phi_{Z+V}((C^k)^*),\ \phi_Z((C^k)^*) + \phi_V((C^k)^*) = \phi_{Z+V}((C^k)^*).$$

Then, again by independence, if $f_1, f_2 \in (C^k)^*$,

$$\begin{aligned} \langle \phi_Z(f_1), \phi_V(f_2) \rangle_{H_{Z+V}} &= \langle \phi_{Z+V}(f \circ \Pi), \phi_{Z+V}(f \circ (I - \Pi)) \rangle_{H_{Z+V}} \\ &= E(f_1 \circ (Z+V) f_2 \circ (I - \Pi)(Z+V)) = E(f_1(Z) f_2(V) = 0, \end{aligned}$$

whereas

$$\|\phi_Z(f)\|^2_{H_{Z+V}} = \|\phi_{Z+V}(f \circ \Pi)\|^2_{H_{Z+V}} = E f^2(Z) = \|\phi_Z(f)\|^2_{H_Z},$$

and similarly, $\|\phi_V(f)\|_{H_{Z+V}} = \|\phi_V(f)\|_{H_V}$. That is,

$$\|\phi_Z(f_1) + \phi_V(f_2)\|^2_{H_{Z+V}} = \|\phi_Z(f_1)\|^2_{H_Z} + \|\phi_V(f_2)\|^2_{H_V}.$$

This proves, by completion, that $H_{Z+V} = H_Z + H_V$, where the sum is orthogonal and where the norm of H_{Z+V} restricted to H_Z (resp. H_V) coincides with the norm of H_Z (resp. of H_V). The proposition follows from this fact and the preceding two lemmas. ■

Remark 2.6.25 The RKHS $H_{W,k}$ of W^k, $k \ge 0$, coincides with the Sobolev space $H_2^{k+1}([0,1])$, and the RKHS norm is equivalent to any of the usual Sobolev norms, in particular, to $\sum_{j=0}^{k+1} \|f^{(j)}\|_2$ (see Chapter 4). In the next section, in order to estimate the small ball concentration function of these processes, we will require estimates of the covering number of the unit ball of $H_{W,k}$ with respect to the supremum norm. Let us record here that if $\mathcal{F}_{W,k}$ is the unit ball of $H_{W,k}$, then there exists $C_k < \infty$ such that

$$\log N(\mathcal{F}_{W,k}, \|\cdot\|_\infty, \varepsilon) \le C_k \varepsilon^{-1/(k+1)}, \quad \varepsilon > 0 \tag{2.83}$$

(Corollary 4.3.38).

The Small Ball Concentration Functions of Brownian Motion and Its Released Iterated Integrals

We will obtain *upper bounds* for the concentration function

$$\phi_0^X(\varepsilon) = -\log \Pr\{\|X\|_\infty < \varepsilon\}, \quad 0 < \varepsilon < \tau,$$

τ small, when X is a released at zero multiple integral of W (i.e., lower bounds for their small ball probabilities). We begin with the concentration of Brownian motion, which will provide an a priori rough bound for the concentration of the released integrated processes. Then, with this rough bound and an argument relating the metric entropy of the unit ball of the RKHS of a Gaussian process to the size of its small ball probabilities, we will obtain upper bounds of the right order for ϕ_0^X in the remaining cases.

Recall from Exercise 2.6.9 that the Karhunen-Loève expansion of Brownian motion is

$$W(t) = h_0 g_0 + \sum_{n=0}^{\infty} \sum_{k=1}^{2^n} h_{n,k}(t) g_{n,k}, \tag{2.84}$$

where g_0, $g_{n,k}$ are i.i.d. standard normal variables, $h_0(t) = t$ and $h_{n,k}$, $n \geq 0$ and, for each n, $k = 1, \ldots, 2^n$, are the 'tent' functions $h_{n,k} = \int_0^t 2^{n/2}(I_{[(k-1)/2^n,(k-1/2)/2^n)} - I_{[(k-1/2)/2^n,k/2^n]})(u)du$. Since for each n the functions $h_{n,k}$, $k = 1, \ldots, 2^n$, have supports $[(k-1)/2^n, k/2^n]$, and since $\|h_{n,k}\|_\infty = 2^{-n/2-1}$, it follows that there exists $C > 0$ such that, for all $n \geq 0$ and for all $a_1, \ldots, a_{2^n}$,

$$\left\| \sum_{k=1}^{2^n} a_k h_{n,k} \right\|_\infty = \left\| \sum_{k=1}^{2^n} |a_k||h_{n,k}| \right\|_\infty = \frac{1}{2} 2^{-n/2} \max_{i \leq 2^n} |a_k| \geq \frac{1}{2} 2^{-n/2} \left(2^{-n} \sum_{k=1}^{2^n} |a_k| \right). \tag{2.85}$$

This observation and Anderson's inequality are all that is needed to prove the following:

Theorem 2.6.26 *Let W be Brownian motion on $[0,1]$. Then there exists $C \in (0,\infty)$ such that, for all $0 < \varepsilon \leq 1$,*

$$-C\varepsilon^{-2} \leq \log \Pr\left\{ \sup_{t\in[0,1]} |W(t)| < \varepsilon \right\} \leq -\frac{1}{C}\varepsilon^{-2}, \tag{2.86}$$

that is, the exact order of (supremum norm) small ball concentration function ϕ_0^W of Brownian motion is $\phi_0^W = O(\varepsilon^{-2})$ as $\varepsilon \to 0$.

Proof Let b_n, $n \geq 0$, be positive numbers. Then, on the event

$$A_b = \left\{ |g_0| < b_0,\ |g_{n,k}| < b_n,\ n \geq 0,\ 1 \leq k \leq 2^n \text{ for each } n \right\},$$

we have, using the identities in (2.85),

$$\|W\|_\infty \leq |g_0| + \sum_{n=0}^{\infty} \left\| \sum_{k=1}^{2^n} g_{n,k} h_{n,k} \right\|_\infty < b_0 + \sum_{n=0}^{\infty} 2^{-n/2-1} b_n.$$

Now, given an integer $q \geq 1$, we choose $b_n = b_n(q) = 2^{3(n-q)/4}$ if $n \leq q$ and $b_n = b_n(q) = 2^{(n-q)/4}$, for $n > q$. A simple computation with geometric series yields

$$b_0 + \sum_{n=0}^{\infty} 2^{-n/2-1} b_n \leq C_1 2^{-q/2}.$$

Using that, for g standard normal,

$$\Pr\{|g| \le t\} \ge t/3 \text{ for } 0 \le t \le 1 \text{ and } \Pr\{|g| \le t\} \ge 1 - e^{-t^2/2} \ge \exp(-2e^{-t^2/2}) \text{ for } t \ge 1,$$

we obtain

$$\begin{aligned}\log \Pr(A_b) &= \log \Pr\{|g| < b_0\} + \sum_{n=0}^{\infty} 2^n \log \Pr\{|g| < b_n\} \\ &\ge -\frac{3(\log 2)q}{4} - (\log 3)\sum_{n=0}^{q} 2^n - \frac{3(\log 2)}{4}\sum_{n=0}^{q} 2^n(q-n) - \sum_{n=q+1}^{\infty} 2e^{-2^{(n-q)/2}/2} \\ &\ge -\left(\frac{3\log 2}{4} + 2\log 3 + \sum_{r=1}^{\infty}\frac{r}{2^r} + 4\sum_{r=1}^{\infty}\exp\left((\log 2)r - 2^{r/2}/2\right)\right)2^q.\end{aligned}$$

Then there exist C_1, C_2 positive, finite, independent of q such that

$$\log \Pr\{\|W\|_\infty < C_1 2^{-q/2}\} \ge \log \Pr(A_b) \ge C_2 2^q,$$

proving the left-hand side of (2.86).

Now let $\Pr_n$ denote the conditional probability with respect to all the normal variables in the representation (2.84) of W that do not have n in their subindex (or, what is the same, integration with respect to the $g_{n,k}$, $k = 1, \ldots, 2^n$, variables only). Then Anderson's inequality (Theorem 2.4.5) and Fubini's theorem give that, for all $\varepsilon > 0$ and all $n \ge 0$,

$$\Pr\{\|W\|_\infty \le \varepsilon\} = E\left[\Pr_n\left\{\left\|h_0 g_0 + \sum_{m=0}^{\infty}\sum_{k=1}^{2^m} h_{m,k} g_{m,k}\right\|_\infty \le \varepsilon\right\}\right] \le \Pr\left\{\left\|\sum_{k=1}^{2^n} h_{n,k} g_{n,k}\right\|_\infty \le \varepsilon\right\}.$$

Then the inequality in (2.85) gives

$$\Pr\{\|W\|_\infty \le \varepsilon\} \le \Pr\left\{\sum_{k=1}^{2^n} |g_{n,k}| \le (2^{n/2-1}\varepsilon)2^n\right\}.$$

Define c by the equation $Ee^{-|g|} = e^{-2c}$, where g is standard normal, and let $\varepsilon = \varepsilon_n = 2c2^{-n/2}$. Then, since

$$\Pr\left\{\sum_{k=1}^{2^n} |g_{n,k}| \le c2^n\right\} = \Pr\left\{\exp\left(-\sum_{k=1}^{2^n} |g_{n,k}|\right) \ge e^{-c2^n}\right\} \le \frac{(Ee^{-|g|})^{2^n}}{e^{c2^n}} \le e^{-c2^n},$$

we conclude that for these values of $\varepsilon = \varepsilon_n$, for any $n \ge 0$,

$$\Pr\{\|W\|_\infty \le \varepsilon\} \le e^{-4c^3\varepsilon^{-2}},$$

and, possibly with different constants, this extends to any $0 < \varepsilon \le 1$, proving the right-hand side of inequality (2.86). ■

This proof does not give the best constants: in fact, $\lim_{\varepsilon\to 0} \log \Pr\left\{\sup_{t\in[0,1]} |W(t)| \le \varepsilon\right\} = -\pi^2/8$. However, the proof is simple, and in fact, we are only interested in the order of magnitude of ϕ_0^W, actually in its upper bound; the lower bound is given only to show that the upper bound obtained is of the right order. It is interesting to note that exactly the same method of proof yields the small ball order of concentration for the $L^p([0,1])$ norms (which is also ε^{-2}) and for the Hölder norms of order $0<\alpha<1/2$ (which is $\varepsilon^{-2/(1-2\alpha)}$). See Exercises 2.6.12 and 2.6.13.

Corollary 2.6.27 *Let $W^0 = g + W$, g a standard normal variable independent of W. Then there exists $C>0$ such that, for all $0<\varepsilon\le 1$,*

$$-C\varepsilon^{-2} \le \log \Pr\left\{\sup_{t\in[0,1]} |W^0(t)| < \varepsilon\right\} \le -\frac{1}{C}\varepsilon^{-2}, \quad 0<\varepsilon\le 1. \tag{2.87}$$

As a consequence, for each $k\in\mathbb{N}$, there exist $C_k>0$ such that

$$\log \Pr\left\{\sup_{t\in[0,1]} |W^k(t)| < \varepsilon\right\} \ge -C_k\varepsilon^{-2}. \tag{2.88}$$

Proof Just note that $-\log\Pr\{|g|<\varepsilon\}$ is of the order of $\log\varepsilon^{-1}$ for $0<\varepsilon<1/2$, much smaller than ε^{-2} as $\varepsilon\to 0$, and use it together with the preceding theorem in the obvious inequality

$$\Pr\{\|g+W\|_\infty < \varepsilon\} \ge \Pr\{\|W\|_\infty < \varepsilon/2\}\Pr\{|g|<\varepsilon/2\}$$

in order to obtain the left-hand side inequality in (2.87). For the right-hand side, just note that integrating first with respect to W with g fixed (by Fubini) and applying Anderson's inequality (Theorem 2.4.5), we have $\Pr\{\|g+W\|_\infty<\varepsilon\}\le\Pr\{\|W\|_\infty<\varepsilon\}$.

For the second inequality, note that $W^k = g + I_{0+}W^{k-1}$ and, proceeding by induction, assume that inequality (2.88) is true for W^{k-1}. Then, since $\|I_{0+}W^{k-1}\|_\infty \le \|W^{k-1}\|_\infty$, we obtain $\Pr\{\|I_{0+}W^{k-1}\|_\infty<\varepsilon\}\ge\Pr\{\|W^{k-1}\|_\infty<\varepsilon\}\ge e^{-C\varepsilon^{-2}}$, and now we can use the first argument in the first part of the proof to obtain the inequality for $g+I_{0+}W^{k-1}$. ∎

The following lemma for general Gaussian processes expresses part of the relationship between entropy and small balls (for more on this, including the reverse direction, see the notes and references at the end of this chapter).

Lemma 2.6.28 *Let X be a B-valued centred Gaussian random variable and B a separable Banach space, and let H_1 be the unit ball centred at zero of its reproducing kernel Hilbert space. Set $N(H_1,\varepsilon) := N(H_1,\|\cdot\|,\varepsilon)$, $\varepsilon>0$, where $\|\cdot\|$ denotes the norm of B. Let Φ denote the standard normal distribution function, and let $\phi_0(\varepsilon) = -\log\Pr\{\|X\|<\varepsilon\}$. Then, for all $\lambda>0$ and $\varepsilon>0$,*

$$\phi_0(2\varepsilon) \le \log N(H_1,\varepsilon/\lambda) - \log\Phi\left(\lambda + \Phi^{-1}(e^{-\phi_0(\varepsilon)})\right). \tag{2.89}$$

Proof Let T_ε be the centres of a minimal set of $\|\cdot\|$ balls of radius ε covering λH_1 (T_ε is finite because H_1 is compact in B by Proposition 2.6.9). Then $\mathrm{Card}(T_\varepsilon) = N(H_1,\varepsilon/\lambda)$. Use the notation $B_\varepsilon(h) = \{x\in B : \|x-h\|<\varepsilon\}$. Then, since the collection of balls $\{B_\varepsilon(h) : h\in T_\varepsilon\}$ covers λH_1, we have

$$\lambda H_1 + B_\varepsilon(0) \subset \bigcup_{h\in T_\varepsilon} B_{2\varepsilon}(h).$$

Therefore, for $\mu = \mathcal{L}(X)$,

$$N(\lambda H_1,\varepsilon)\max_{h\in T_\varepsilon}\mu(B_{2\varepsilon}(h)) \ge \mu(B_\varepsilon(0)+\lambda H_1).$$

Now, applying Anderson's inequality (Theorem 2.4.5) on the left-hand side and the isoperimetric inequality (Theorem 2.6.12) on the right-hand side, we obtain

$$N(\lambda H_1, \varepsilon)\mu(B_{2\varepsilon}(0)) \geq \Phi(\Phi^{-1}(\mu(B_\varepsilon(0)) + \lambda),$$

proving the lemma. ■

When there is an a priori upper estimate of $\phi(\varepsilon)$, the preceding lemma and an iteration procedure give a precise relationship between the order of the metric entropy of H_1 and the order of ϕ_0 as follows:

Theorem 2.6.29 *With the notation of the preceding lemma, if there is $\gamma > 0$ such that, for $C_1 < \infty$ and $\tau_i > 0$,*

$$\phi_0(\varepsilon) \leq C_1 \varepsilon^{-\gamma}, \quad 0 < \varepsilon \leq \tau_1,$$

and if

$$\log N(H_1, \varepsilon) \leq C_2 \varepsilon^{-\alpha}, \quad 0 < \varepsilon < \tau_2,$$

for some $0 < \alpha < 2$, then there exist $C_3 < \infty$ and $\tau_3 > 0$ such that for every $0 < \varepsilon < \tau_3$

$$\phi_0(\varepsilon) \leq C_3 \varepsilon^{-2\alpha/(2-\alpha)}.$$

Proof Set $\theta_\varepsilon = \Phi^{-1}(e^{-\phi_0(\varepsilon)})$. By the standard normal estimate (2.23),

$$\phi_0(\varepsilon) = -\log \Phi(\theta_\varepsilon) = -\log(1 - \Phi(-\theta_\varepsilon)) \geq \theta_\varepsilon^2/2.$$

Thus, if we take $\lambda = \sqrt{2\phi_0(\varepsilon)}$, then $\lambda + \theta_\varepsilon = \sqrt{2\phi_0(\varepsilon)} + \theta_\varepsilon \geq 0$, and inequality (2.89) then implies

$$\phi_0(2\varepsilon) \leq \log N(H_1, \varepsilon/\sqrt{2\phi_0(\varepsilon)}) - \log(1/2),$$

or, re-scaling and using the hypothesis on N and that $\phi_0(\varepsilon) \to \infty$ as $\varepsilon \to 0$,

$$\phi_0(\varepsilon) \leq \log 2 + \log N(H_1, \varepsilon/\sqrt{8\phi_0(\varepsilon/2)}) \leq C\varepsilon^{-\alpha}\phi_0^{\alpha/2}(\varepsilon/2),$$

for some $C < \infty$ and for all ε small enough. Setting $\psi(\varepsilon) = C\varepsilon^{-\alpha}$, we then have

$$\log \phi_0(\varepsilon) \leq \frac{\alpha}{2} \log \phi_0(\varepsilon/2) + \log \psi(\varepsilon).$$

Iterating, this gives

$$\log \phi_0(\varepsilon) \leq \left(\frac{\alpha}{2}\right)^n \log \phi_0(\varepsilon/2^n) + \sum_{j=0}^{n-1} \left(\frac{\alpha}{2}\right)^j \log \psi(\varepsilon/2^j).$$

The hypothesis on ϕ_0 gives that $\log \phi_0(2/2^n) \leq \log(C_1 2^{n\gamma}/\varepsilon^\gamma)$, a bound of the order of n, so

$$\log \phi_0(\varepsilon) \leq \sum_{j=0}^{\infty} \left(\frac{\alpha}{2}\right)^j \log \psi(\varepsilon/2^j) = \frac{2}{2-\alpha} \log \psi(\varepsilon) + \sum_{j=1}^{\infty} \log \frac{\psi(\varepsilon/2^j)}{\psi(\varepsilon)}.$$

Now, since $\log(\psi(\varepsilon/2^j)/\psi(\varepsilon)) = \log 2^{\alpha j}$, the last series is convergent, and we obtain

$$\log \phi_0(\varepsilon) \leq \frac{2}{2-\alpha} \log \psi(\varepsilon) + C' \leq C'' \varepsilon^{-\alpha},$$

for some finite constants C', C'' and all ε small enough. ■

This theorem, Corollary 2.6.27, Proposition 2.6.24 on the RKHS of W^k and the bound (2.83) on the covering numbers of Sobolev balls yield the following:

Corollary 2.6.30 (Small Ball Concentration of Integrated Brownian Motion Released at Zero) *For all $k \in \mathbb{N} \cup \{0\}$, the concentration function $\phi_0^{W^k}$ of the k times integrated Brownian motion on $[0,1]$ released at zero, W^k, satisfies the estimate*

$$\phi_0^{W^k}(\varepsilon) = -\log \Pr\{\|W^k\|_\infty < \varepsilon\} \le C_k \varepsilon^{-2/(2k+1)},$$

for some $C_k < \infty$ and all ε small enough.

Combined with Proposition 2.6.19, this corollary gives a lower bound for $\Pr\{\|W^k - x\|_\infty < \varepsilon\}$, for all $x \in \bar{H}_{W,k}$, the closure in $C([0,1])$ of the Sobolev space $H_{W,k}$.

Exercises

2.6.1 Show that the topological dual of $\mathbb{R}^T$ is the space of measures of finite support on T; that is, $f \in (\mathbb{R}^T)^*$ if and only if there exist $n < \infty$, $t_1, \dots, t_n \in T$ and $a_1, \dots a_n \in \mathbb{R}$ such that $f(x) = \sum_{i=1}^n a_i x(t_i)$, where $x = (x(t) : t \in T)$, and conversely. *Hint*: Reduce the problem to $\mathbb{R}^\mathbb{N}$, and show that f linear is continuous at zero for the metric of $\mathbb{R}^\mathbb{N}$, defined prior to Theorem 2.2.4 if and only if f is a finite linear combination of point masses at points $k \in \mathbb{N}$.

2.6.2 Let B be a Banach space. (a) Show that the unit ball B_1^* of its dual is compact for the weak* topology. *Hint*: The weak* topology of B_1^* is the restriction of the product topology of $[-1,1]^B$, which is compact by Tychonoff's theorem, and B_1^* is closed in $[-1,1]^B$. (b) If B is separable, then the weak* topology is metrisable on B_1^*. *Hint*: Let B_0 be a countable subset of B. Then, if $f_n \in B_1'$ and $f \in B^*$, we have that $f_n \to_{w*} f$ iff $f_n(x) \to f(x)$, for all $x \in B_0$ (use the uniform boundness principle and a 3-ε argument). Now use that $[-1,1]^\mathbb{N}$ is metrisable (see the discussion before Theorem 2.2.4).

2.6.3 Show that the Borel measures with finite support on $[a,b]$ are weak* sequentially dense in the space of finite signed measures on $[a,b]$ (which is the dual of $C([a,b])$). *Hint*: By the Jordan-Hahn decomposition, it suffices to consider positive finite measures μ. The discrete part of μ obviously can be approximated in the sense of convergence in distribution by a sequence of measures with finite support, and for the continuous part, approximate the distribution function by a sequence of step functions, for example, each with all the steps of size $1/n$.

2.6.4 Prove Proposition 2.6.8. *Hint*: If $x \in H$, then $x = E(hX)$ for some $h \in L^2$, and one can use Hölder's inequality to bound $f(x)$, $f \in B^*$. Conversely, if $|f(x)|^2 \le DC_X(f,f)$ for all $f \in B^*$, then the map $x \mapsto f(x)$ extends by uniform continuity to $\bar{F}$, but, $\bar{F}$ being a Hilbert space, $\bar{F} = \bar{F}'$, and hence there is $k \in \bar{F}$ such that $E(kf(X)) = f(x)$, for all $f \in B^*$, showing that $x = \phi(k) \in H$.

2.6.5 Let $A : B_1 \mapsto B_2$ be a continuous linear map between two separable Banach spaces B_1 and B_2, let X be a B_1-valued centred Gaussian variable and let H_1 be the RKHS of X. Prove that the RKHS H_2 of the B_2-valued Gaussian variable $A(X)$ is $H_2 = A(H_1)$, that $\|Ax\|_{H_2} \le \|x\|_{H_1}$ and that equality holds for at least one pre-image of Ax. In particular, if A is one to one, then $A : H_1 \mapsto H_2$ is an isometry. What does this say if A is a continuous imbedding? *Hint*: If $x \in H_1$, $x = E(k_1 X)$ for $k_1 \in \bar{F}_1 \subset L^2$ and $\|x\|_{H_1} = \|k_1\|_{L^2}$. Then $Ax = E(k_1 A(X)) \in H_2$ by Remark 2.6.5, and $\|Ax\|_{H_2} \le \|k_1\|_{L^2}$ (as k_1 may not be in $\bar{F}_2$). Conversely, if $y \in H_2$, then $y = E(k_2 A(X))$ for some $k_2 \in \bar{F}_2 \subset L^2$, $\|y\|_{H_2} = \|k_2\|_{L^2}$ and $x := E(k_2 X)$ satisfies that $Ax = y$ and $\|x\|_{H_1} \le \|k_2\|_{L^2}$.

2.6.6 State and prove the analogues of Proposition 2.6.9 and Theorem 2.6.10 for separable Gaussian processes.

2.6.7 Let B be a separable Banach space, let $K \neq \emptyset$ be a compact, convex, symmetric subset of B, let $\|\cdot\|_K$ be the induced norm, $\|x\|_K = \inf\{\lambda : x \in \lambda K\}$, and let $B_K = \{x : \|x\|_K < \infty\}$ equipped with this norm. Show that there exists a countable subset D_K of the unit ball of B^* such that $\|x\|_K = \sup_{f \in D_K} |f(x)|$ for all $x \in B$. Hence, if X is Gaussian and $\Pr\{\|X\|_K < \infty\} > 0$, then the process $X(f) = f(X)$, $f \in D_K$, is separable, and for example Theorem 2.1.20 applies to it, in particular, $E\|X\|_K^2 < \infty$. [*Hint*: Use the Hahn-Banach separation theorem for convex sets.]

2.6.8 (Lévy's maximal inequality or the reflection principle for sums of independent random vectors.) Prove that if X_i, $i \leq n$, are independent, symmetric B-valued random vectors where B is a separable Banach space, then, for all $t > 0$,

$$\Pr\left\{\max_{1 \leq i \leq n} \|S_k\| > t\right\} \leq 2\Pr\{\|S_n\| > t\},$$

where $S_k = \sum_{i=1}^k X_i$. *Hint*: See Section 3.1.3.

2.6.9 (Lévy's series representation of Brownian motion.) On $[0,1]$, let $\phi_0 \equiv 1$, $\psi_0 = I_{[0,1/2)} - I_{[1/2,1]}$ and $\psi_{n,k}(t) = 2^{n/2}\psi_0(2^n t - k + 1)$, $n \in \mathbb{N} \cup \{0\}$, and for each n, $k = 1,\ldots,2^n$. ϕ_0 and $\psi_{n,k}$ constitute the Haar complete ortho-normal system (cons) of $L^2([0,1])$ (see Chapter 4). Show that the functions $h_0(t) = \int_0^t \phi_0(u)du = t$ and $h_{n,k}(t) = \int_0^t \psi_{n,k}(x)dx$, $t \in [0,1]$, for $n \in \mathbb{N} \cup \{0\}$, and for each n, $k = 1,\ldots,2^n$, constitute a cons of H_W, the RKHS of Brownian motion (see (2.73)). Conclude that, if g_0, $g_{n,k}$ are i.i.d. $N(0,1)$, then the series $h_0 g_0 + \sum_{n=0}^\infty \sum_{k=1}^{2^n} h_{n,k} g_{n,k}$ convergences uniformly on $[0,1]$ a.s. and also in $L^p([0,1])$ a.s., for every p, and that the process it defines is Brownian motion on $[0,1]$. *Hint*: Apply the Karhunen-Loève expansion to obtain that $W = h_0\phi^{-1}(h_0) + \sum_{n=0}^\infty \sum_{k=1}^{2^n} h_{n,k}\phi^{-1}(h_{n,k})$ a.s., and note that the random variables $\phi^{-1}(h_0)$ and $\phi^{-1}(h_{n,k})$ are i.i.d. $N(0,1)$.

2.6.10 For a Borel probability measure μ on a separable Banach space, let $\hat{\mu}(f) = \int e^{if(x)}d\mu(x)$, $f \in B^*$, be its characteristic functional. Prove that $\hat{\mu} = \hat{\nu}$ implies $\mu = \nu$. *Hints*: By separability, the Borel σ-algebra of B coincides with the cylindrical σ-algebra, the algebra generated by sets of the form $\{x : (f_1(x),\ldots,f_k(x)) \in A,\ f_i \in B^*, A \in \mathcal{B}(\mathbb{R}^k), k \in \mathbb{N}\}$. Thus, it suffices to show that for all k finite and $f_1,\ldots,f_k \in B^*$, the Borel measures $\mu \circ (f_1,\ldots,f_k)^{-1}$ and $\nu \circ (f_1,\ldots,f_k)^{-1}$ of $\mathbb{R}^k$ are equal, and for this one can apply the uniqueness of characteristic functions in $\mathbb{R}^k$ (or, by Cramér-Wold, just in $\mathbb{R}$).

2.6.11 Prove that the support of a centred Gaussian measure μ on B coincides with the closure in B of the set $F = \{E(f(X)X) : f \in B^*\}$, where $\mathcal{L}(X) = \mu$.

2.6.12 In the notation of Theorem 2.6.26, show that if $\|\cdot\|_{L^p}$ denotes the L^p-norm with respect to Lebesgue measure on $[0,1]$, $1 \leq p < \infty$, then there exists $C_p \in (0,\infty)$ such that

$$-C_p\varepsilon^{-2} \leq \log\Pr\left\{\|W\|_{L_p} < \varepsilon\right\} \leq -\frac{1}{C_p}\varepsilon^{-2}.$$

Hint: Proceed as in the proof of Theorem 2.6.26, but use the fact that, for all n,

$$\left\|\sum_{k=1}^{2^n} a_k h_{n,k}\right\|_{L^p} = C'_p 2^{-n/2}\left(2^{-n}\sum_{k=1}^{2^n}|a_k|^p\right)^{1/p}$$

and that the ℓ_p-norm of $\{a_k\}_{k=1}^{2^n}$ is between its ℓ_1- and its ℓ_∞-norms instead of (2.85).

2.6.13 Let $\|x\|_{\text{Lip}(\alpha)} = \sup_{0 \leq s \neq t < 1} |x(t) - x(s)|/|t-s|^\alpha$, for $0 < \alpha < 1$, be the α-Hölder norm of x. Show that

$$-C_\alpha\varepsilon^{-2/(1-2\alpha)} \leq \log\Pr\left\{\|W\|_{\text{Lip}(\alpha)} < \varepsilon\right\} \leq -\frac{1}{C_\alpha}\varepsilon^{-2/(1-2\alpha)},$$

for $0 < \alpha < 1/2$. *Hint*: Show first that for all n,

$$\left\| \sum_{k=1}^{2^n} a_k h_{n,k} \right\|_{\mathrm{Lip}(\alpha)}$$

is comparable to $2^{-(1/2-\alpha)n} \max_{k \le 2^n} |a_k|$, use this instead of (2.85) and take $b_n = b_n(q) = 2^{(3/2-\alpha/2)(n-q)}$ for $n \le q$ and $b_n(q) = 2^{(1/4-\alpha/2)(n-q)}$ for $n > q$ in the proof of Theorem 2.6.26.

2.6.14 In a separable Hilbert space B, the covariance operator Q of a centred random variable X satisfying $E\|X\|^2 < \infty$, defined by $\langle Q(x), y\rangle = E(\langle X, x\rangle \langle X, y\rangle)$ for all $x, y \in B$, is a self-adjoint positive definite Hilbert-Schmidt operator on B. These operators admit a complete ortho-normal system $\{e_i\}$ of eigenvectors, with eigenvalues $\lambda_i^2 \ge 0$ satisfying $\sum \lambda_i^2 < \infty$ (e.g., Gelfand and Vilenkin (1967)). For such a system of eigenvectors and eigenvalues, prove that the random variable $\tilde{X} = \sum \lambda_i g_i e_i$, where $\{g_i\}$ is an i.i.d. standard normal sequence, has the same law as X.

2.6.15 Let (Ω_i, Σ_i), $i \in \mathbb{N}$, be measurable spaces, let $\Omega = \prod_{i=1}^{\infty} \Omega_i$ and let

$$\mathcal{C} = \sigma \left\{ E_{1,\dots,n} \times \prod_{i=n+1}^{\infty} \Omega_i : n \in \mathbb{N}, E_{1,\dots,n} \in \otimes_{i=1}^{n} \Sigma_i \right\}$$

be the cylindrical σ-algebra of Ω. Let μ_i and μ_i' be equivalent probability measures on (Ω_i, Σ_i) (i.e., mutually absolutely continuous measures), and let $f_i = d\mu_i'/d\mu_i$ be the Radon-Nikodym derivative of μ_i' with respect to μ_i. Let $\mu = \prod_{i=1}^{\infty} \mu_i$ and $\mu_i' = \prod \mu_i'$ be the corresponding infinite product measures. Then either μ and μ' are equivalent or mutually singular, and this happens according to whether $\prod_{i=1}^{\infty} \int \sqrt{f_i} d\mu_i > 0$ or $\prod_{i=1}^{\infty} \int \sqrt{f_i} d\mu_i = 0$. In the first case, $(d\mu'/d\mu)(\omega) = \prod_{i=1}^{\infty} f_i(\omega_i)$ μ a.s. *Hint*: Mimick the proof of Proposition 2.6.16.

2.6.16 (The reproducing kernel Hilbert space of the Brownian bridge.) If $W(t)$, $t \in [0,1]$ is Brownian motion, the Brownian bridge $B(t)$, $t \in [0,1]$, is defined as $B(t) = W(t) - tW(1)$. Notice that $B(1) = 0$. Using computations similar to those in Example 2.6.7 but now for the covariance $s \wedge t - st$ of B, show that the RKHS H_B of B is

$$H_B = \left\{ f : f(0) = f(1) = 0,\ f \text{ is absolutely continuous and } f' \in L^2([0,1]) \right\},$$

with $\|f\|_H^2 = \int_0^1 (f'(x))^2 dx$.

2.7 Asymptotics for Extremes of Stationary Gaussian Processes

This section differs from previous ones in that the Gaussian processes treated in it are not general but very particular: stationary Gaussian process on $\mathbb{N}$ (Gaussian sequences) and on $\mathbb{R}^+$. We are interested in the limiting distribution of $\max_{1 \le k \le n} |X_n|$ for sequences and of $\sup_{0 \le t \le T} |X(t)|$ for processes on the positive real line, suitably centred and normalised. Asymptotic distributional theory for the supremum of $X(t)$ (or of $|X(t)|$) has been developed primarily for two types of Gaussian processes, stationary, for which the variance of $X(t)$ is constant, and cyclostationary, for which, among other properties, the variance of $X(t)$ is periodic and its maximum on each period is attained at a single point. These results for both types of processes are of some use in nonparametric statistics, concretely in the construction of confidence bands, although it may be argued that their interest is only theoretical because the speed of convergence for these limit theorems is unfortunately very slow (e.g., for sequences, of the order of $1/\log n$). For brevity, only the more classical case of stationary processes will be developed here. The main tool for proving these results, aside from the usual probability theory, will be the comparison inequalities from Section 2.4.2.

Gaussian Sequences

We first consider a sequence of independent standard normal random variables.

Theorem 2.7.1 *Let g_i, $i \le n$, be independent standard normal random variables. For each $n \ge e^2$, set*

$$a_n = (2\log n)^{1/2}, \quad b_n = a_n - \frac{\log\log n + \log\pi}{2a_n}.$$

Then, for all $x \in \mathbb{R}$,

$$\lim_{n\to\infty} \Pr\left\{a_n(\max_{1\le i\le n}|g_i| - b_n) \le x\right\} = \exp\left(-e^{-x}\right). \tag{2.90}$$

Proof For g standard normal, let $\Phi(x) = \Pr\{|g| \le x\} = (2/\pi)^{1/2}\int_0^x e^{-t^2/2}dt$, and let $\phi(x) = (2/\pi)^{1/2}e^{-x^2/2}$, $x \ge 0$, be, respectively, the distribution function and the density of $|g|$. Then recall that, by (2.23), for $u > 0$,

$$\frac{u}{1+u^2}\phi(u) \le 1 - \Phi(u) \le \frac{1}{u}\phi(u). \tag{2.91}$$

For $x \in \mathbb{R}$ fixed, set $u_n = x/a_n + b_n$. First we see that by the definition of u_n,

$$\frac{n\phi(u_n)}{u_n} = \frac{n}{o(1) + \sqrt{\pi\log n}}\exp\left(-\log n - x + \frac{1}{2}\log(\pi\log n) + o(1)\right) \to e^{-x}, \tag{2.92}$$

as $n \to \infty$. Then, since $1 - \Phi(u_n) \ge 0$, we have, using $|a^n - b^n| \le (a \vee b)^{n-1}|a-b|$ for $a, b > 0$,

$$\begin{aligned}\left|e^{-n(1-\Phi(u_n))} - (1-(1-\Phi(u_n))^n\right| &\le n\left|e^{-(1-\Phi(u_n))} - (1-(1-\Phi(u_n))\right| \\ &\le n(1-\Phi(u_n))^2/2 = O(1/n),\end{aligned}$$

by (2.91) and (2.92). Also, the inequalities (2.91) readily give $|n(1-\Phi(u_n)) - n\phi(u_n)/u_n| = O(1/\log n)$. Using the notation $\alpha_n \simeq \beta_n$ iff $|\alpha_n - \beta_n| \to 0$, we then see that the last two limits together with the limit in (2.92) yield

$$\begin{aligned}\Pr\left\{\max_{i\le n}|g_i| \le u_n\right\} &= \Phi^n(u_n) = [1-(1-\Phi(u_n))]^n \\ &\simeq e^{-n(1-\Phi(u_n))} \simeq e^{-n\phi(u_n)/u_n} \to \exp\left(-e^{-x}\right). \quad \blacksquare\end{aligned}$$

We will obtain the limit theorem for stationary sequences from this theorem by means of the comparison theorem for Gaussian vectors (Theorem 2.4.7). The following lemma translates the bound in that theorem to a sensible condition on the covariance. In what follows, Φ is the distribution function of $|g|$ as in the preceding proof.

Definition 2.7.2 A sequence $\{\xi_n\}$ of random variables is stationary if for any natural numbers $n_1, \ldots, n_m$, $m > 0$ and for any $k \ge 0$, the joint probability law of the random variables $\xi_{n_1+k}, \ldots, \xi_{n_m+k}$ does not depend on k.

If $\{\xi_n\}$ is a centred (jointly) Gaussian sequence, then it is stationary if and only if the covariances $E(\xi_m\xi_{n+m})$ do not depend on m. In this case, the function $r(n) = E(\xi_m\xi_{n+m})$ is the *covariance function* of the sequence.

Lemma 2.7.3 *Let $\{r_j\}$ be a sequence of numbers such that* $\sup_n |r_n| < 1$ *and* $r_n \log n \to 0$ *as* $n \to \infty$, *and let* $\{u_n\}$ *be a sequence of positive constants such that* $\sup_n n(1-\Phi(u_n)) < \infty$. *Then*

$$\lim_{n\to\infty} n \sum_{j=1}^{n} |r_j| \exp\left(-\frac{u_n^2}{1+|r_j|}\right) = 0. \tag{2.93}$$

Proof Let K be such that $\sup n(1-\Phi(u_n)) \le K$, and let v_n be defined by the equation $n(1-\Phi(v_n)) = K$ (this equation has only one solution as Φ is strictly increasing). Suppose that the lemma holds for $\{v_n\}$. Then, since $v_n \le u_n$, it follows that the lemma is also true for u_n.

By (2.23),

$$\frac{n}{v_n} e^{-v_n^2/2} \to K\sqrt{\pi/2} := K', \quad \text{and therefore,} \quad \frac{v_n}{\sqrt{2\log n}} \to 1. \tag{2.94}$$

If $\delta = \sup_n |r_n|$, let α be such that $0 < \alpha < (1-\delta)/(1+\delta)$, and let us divide the sum in (2.93) into two, the first sum only up to $[n^\alpha]$ and the second between $[n^\alpha]+1$ and n. Using the two limits in (2.94) (as inequalities), the first sum is bounded by

$$\begin{aligned} n \sum_{j=1}^{[n^\alpha]} |r_j| \exp\left(-\frac{v_n^2}{1+|r_j|}\right) &\le n^{1+\alpha} e^{-v_n^2/(1+\delta)} \le K'' n^{1+\alpha} \left(\frac{v_n}{n}\right)^{2/(1+\delta)} \\ &\le K''' n^{1+\alpha-2/(1+\delta)} (\log n)^{1/(1+\delta)} \to 0. \end{aligned}$$

For the second sum, set $\delta_n = \sup_{j\ge n} |r_j|$, and note that $\delta_n \log n \le \sup_{j\ge n} |r_j| \log j \to 0$, which implies that, by the second limit in (2.94),

$$\delta_{[n^\alpha]} v_n^2 = \left(\delta_{[n^\alpha]} \log [n^\alpha]\right) \frac{v_n^2}{\log [n^\alpha]} \to 0. \tag{2.95}$$

Then, using the first limit in (2.94) and (2.95), we obtain, for some $C < \infty$,

$$\begin{aligned} n \sum_{j=[n^\alpha]+1}^{n} |r_j| \exp\left(-\frac{v_n^2}{1+|r_j|}\right) &\le n\delta_{[n^\alpha]} e^{-v_n^2} \sum_{j=[n^\alpha]+1}^{n} e^{v_n^2 |r_j|/(1+|r_j|)} \\ &\le n^2 \delta_{[n^\alpha]} e^{-v_n^2} e^{\delta_{[n^\alpha]} v_n^2} \le C \delta_{[n^\alpha]} v_n^2 e^{\delta_{[n^\alpha]} v_n^2} \to 0, \end{aligned}$$

thus completing the proof of the lemma. ■

Theorem 2.7.4 *Let $\{\xi_n\}_{n=1}^\infty$ be a stationary sequence of standard normal random variables such that its sequence of covariances $r_n = E(\xi_m \xi_{n+m})$ satisfies $r_n \log n \to 0$. Let a_n and b_n be the constants in Theorem 2.7.1. Then, for all $x \in \mathbb{R}$,*

$$\lim_{n\to\infty} \Pr\left\{a_n\left(\max_{1\le i\le n} |\xi_i| - b_n\right) \le x\right\} = \exp\left(-e^{-x}\right). \tag{2.96}$$

Proof Inequality (2.56) in Theorem 2.4.7 applied to the sequences $\{\xi_i\}$ and $\{g_i\}$, g_i i.i.d. $N(0,1)$ with $\mu_i = -u_n$ and $\lambda_i = u_n$, where $u_n = b_n + x/a_n$, gives

$$\left|\Pr\left\{\max_{1\le i\le n} |\xi_i| \le u_n\right\} - \Pr\left\{\max_{1\le i\le n} |g_i| \le u_n\right\}\right| \le \frac{2}{\pi} n \sum_{j=1}^{n} |r_j| \exp\left(-\frac{u_n^2}{1+|r_j|}\right),$$

and the right-hand side tends to zero by Lemma 2.7.3, proving the theorem. ■

Gaussian Processes Indexed by $[0,\infty)$

As for sequences, a stochastic process $\xi(t)$, $t \geq 0$, is stationary if the finite-dimensional marginal distributions of $\xi(t_1+s),\ldots,\xi(t_n+s)$ do not depend on s, for all n and for any s such that $t_i+s \geq 0$, $i=1,\ldots,m$. If $\xi(t)$, $t \geq 0$, is a centred Gaussian process, then it is stationary if and only if the covariances $E(\xi(s)\xi(s+t))$ do not depend on s. In this case, we write $r(t)=E\xi(0)\xi(t)$ for the covariance function of the process. A stationary Gaussian process is *normalised* if $\xi(0)$, and hence $\xi(t)$, for all t, is standard normal. Now let $\xi(t)$, $t \in [0,\infty)$, be a normalised stationary Gaussian process with continuous sample paths and with covariance $r(t)=E(\xi(s)\xi(s+t))$ satisfying, for some $\alpha \in (0,2]$ and $C \in (0,\infty)$,

$$r(t) < 1, \quad \text{for } t>0 \quad \text{and} \quad r(t)=1-C|t|^{\alpha}+o(|t|)^{\alpha}, \quad \text{as } t \to 0. \tag{2.97}$$

Note that since $E(\xi(t)-\xi(s))^2 = 2-2r(|t-s|)=2C|t-s|^{\alpha}+o(|t-s|^{\alpha})$, Theorem 2.3.7 implies that there exists a version of ξ with continuous sample paths (this theorem shows that any separable version of ξ has uniformly continuous sample paths on $[0,T]$ for any $T<\infty$ with respect to $d(s,t)=|t-s|^{\alpha/2}$ and hence on $[0,\infty)$ w.r.t. $d'(s,t)=|t-s|$). In this section we will prove a theorem similar to Theorem 2.7.4 for these continuous time processes. The proof, quite long, may be divided into three parts. In the first and main part a limit theorem for the high excursions of the process over a fixed finite interval is proved, in the second part the process is replaced by its absolute value, and in the third part the limit theorem for an interval increasing to infinity is finally completed.

In what follows, we set $\overline{\Phi}(u)=(2\pi)^{-1/2}\int_u^{\infty} e^{x^2/2}dx$, the tail probabilty function for the standard normal distribution. An auxiliary nonstationary Gaussian process needs to be introduced: for each $\alpha \in (0,2)$, $\zeta(t)$ will be a Gaussian process with mean $-|t|^{\alpha}$ and covariance $r_{\zeta}(s,t)=t^{\alpha}+s^{\alpha}-|t-s|^{\alpha}$. As part of the proof of the following theorem, it will be shown that the limit

$$H_{\alpha}=\lim_{a\downarrow 0}\frac{1}{a}\Pr\left\{\sup_{k\in\mathbb{N}}\zeta(ak)+\eta\leq 0\right\} \tag{2.98}$$

exists and is finite, where η is an exponential random variable with unit mean independent of ζ. Note that for $\alpha=2$, which is the case of main interest for us, $\zeta(t)=\sqrt{2}tg-t^2$, and in this case, H_{α} is directly computable and easily seen to be $H_2=1/\sqrt{\pi}$.

Theorem 2.7.5 *Let $\xi(t)$, $t\in[0,\infty)$, be a normalised centred stationary Gaussian process with continuous sample paths and with covariance $r(t)$ satisfying condition (2.97) for some $\alpha\in[0,2]$ and $C\in(0,\infty)$. Then, for any $h\in(0,\infty)$,*

$$\lim_{u\to\infty}\frac{1}{u^{2/\alpha}h\overline{\Phi}(u)}\Pr\left\{\sup_{t\in[0,h]}\xi(t)>u\right\}=C^{1/\alpha}H_{\alpha}, \tag{2.99}$$

where $H_{\alpha}\in(0,\infty)$ is given by the limit (2.98), in particular, $H_2=1/\sqrt{\pi}$.

Proof Set $q=q(u)=u^{-2/\alpha}$ and $\overline{\Phi}=\overline{\Phi}(u)$. For each u, we discretise ξ at the points aqk, $k=0,1,\ldots[h/(aq)]$, and then let $a\downarrow 0$. Since, clearly, $\overline{\Phi}(u+x/u)/\overline{\Phi}(u)\to e^{-x}$ as $u\to\infty$, we have

$$\begin{aligned}
&\limsup_{u\to\infty}\frac{q}{\overline{\Phi}}\Pr\left\{\sup_{t\in[0,h]}\xi(t)>u\right\}\\
&=\limsup_{u\to\infty}\frac{q(u+a^{\alpha/4}/u)}{\overline{\Phi}(u+a^{\alpha/4}/u)}\Pr\left\{\sup_{t\in[0,h]}\xi(t)>u+\frac{a^{\alpha/4}}{u}\right\}\\
&\le\limsup_{u\to\infty}\frac{q}{e^{-a^{\alpha}/4}\overline{\Phi}}\Pr\left\{\sup_{t\in[0,h]}\xi(t)>u+\frac{a^{\alpha/4}}{u},\ \max_{k\in\{0,\dots,[h/aq]\}}\xi(aqk)\le u\right\}\\
&+\limsup_{u\to\infty}\frac{q}{e^{-a^{\alpha}/4}\overline{\Phi}}\Pr\left\{\max_{k\in\{0,\dots,[h/aq]\}}\xi(aqk)>u\right\}
\end{aligned}\tag{2.100}$$

and

$$\liminf_{u\to\infty}\frac{q}{\overline{\Phi}}\Pr\left\{\sup_{t\in[0,h]}\xi(t)>u\right\}\ge\limsup_{a\downarrow 0}\liminf_{u\to\infty}\frac{q}{\overline{\Phi}}\Pr\left\{\max_{k\in\{0,\dots,[h/aq]\}}\xi(aqk)>u\right\},\tag{2.101}$$

and a sensible strategy consists in studying the two probabilities appearing on the right-hand side of (2.100). We start with the second (the next three claims).

Claim 1 Let $\xi_u(t) := u(\xi(qt)-u)$. Then the finite-dimensional distributions of the process $\xi_u(t)$, $t>0$, conditioned on the event $\xi_u(0)>0$ converge in law to those of the process $\zeta(C^{1/\alpha}t)+\eta$, $t>0$, as $u\to\infty$, where η and ζ are as defined for (2.98).

Proof of Claim 1 The processes $\xi_u(t)-r(qt)\xi_u(0)$ and $r(qt)\xi_u(0)$ are independent (as is easily seen by checking covariances). Hence, it suffices to show that the finite-dimensional distributions of the first process converge in law to those of $\zeta(C^{1/\alpha}t)$ $(t>0)$ and that the finite-dimensional distributions of the second, conditioned on $\xi_u(0)>0$, converge to the exponential law with unit variance. The first limit follows easily by computing means and covariances, and the second holds because, for all $x>0$, $\Pr\{\zeta_u(0)>x|\zeta_u(0)>0\}=\overline{\Phi}(u+x/u)/\overline{\Phi}(u)\to e^{-x}$ as $u\to\infty$. Claim 1 is proved.

Claim 2 For any $a>0$,

$$\lim_{n\to\infty}\lim_{u\to\infty}\frac{1}{n\overline{\Phi}}\Pr\left\{\max_{0\le k\le n}\xi(aqk)>u\right\}=\Pr\left\{\bigcap_{\ell=1}^{\infty}\{\zeta(C^{1/\alpha}a\ell)+\eta\le 0\}\right\}.$$

Proof of Claim 2 By recurrence, for any collection of events A_i,

$$\begin{aligned}
\Pr\left\{\bigcup_{i=0}^{n}A_i\right\}&=\Pr(A_n)+\Pr\left\{\bigcup_{i=0}^{n-1}(A_i\cap A_n^c)\right\}\\
&=\Pr(A_n)+\Pr(A_{n-1}\cap A_n^c)+\Pr\left\{\bigcup_{i=0}^{n-2}(A_i\cap A_{n-1}^c\cap A_n^c)\right\}\\
&=\cdots=\Pr(A_n)+\sum_{k=0}^{n-1}\Pr\left\{A_k\cap\left(\bigcap_{\ell=k+1}^{n}A_\ell^c\right)\right\}.
\end{aligned}$$

By stationarity and the definition of ξ_u, and recalling that $\xi(t)$ is $N(0,1)$ for all t, we have

$$\Pr\{\xi(aqk) > u), \xi(aq(k+1)) \leq u, \ldots, \xi(aqn) \leq u\} = \Pr\{\xi_u(0) > 0, \xi_u(a) \leq 0, \ldots,$$
$$\times\, \xi_u(a(n-k)) \leq 0\} = \overline{\Phi} \Pr\left\{\bigcap_{\ell=1}^{n-k}\{\xi_u(a\ell) \leq 0\} \,\Big|\, \xi_u(0) > 0\right\}.$$

Putting these two observations together and applying Claim 1, we obtain

$$\frac{1}{n\overline{\Phi}} \Pr\left\{\max_{0\leq k\leq n} \xi(aqk) > u\right\} = \frac{1}{n} + \frac{1}{n}\sum_{m=1}^{n} \Pr\left\{\bigcap_{\ell=1}^{m}\{\xi_u(a\ell) \leq 0\} \,\Big|\, \xi_u(0) > 0\right\}$$
$$\to \frac{1}{n} + \frac{1}{n}\sum_{m=1}^{n} \Pr\left\{\bigcap_{\ell=1}^{m}\{\zeta(C^{1/\alpha}a\ell) + \eta \leq 0\}\right\}, \quad \text{as } u \to \infty.$$

Letting $n \to \infty$ yields the claim (as the sequence of averages of a convergent sequence tends to its limit).

Claim 3:

$$\lim_{u\to\infty} \frac{q}{\overline{\Phi}} \Pr\left\{\max_{0\leq k\leq [h/(aq(u))]} \xi(aq(u)k) > u\right\} = \frac{h}{a} \Pr\left\{\bigcap_{k=1}^{\infty}\{\zeta(C^{1/\alpha}ak) + \eta \leq 0\}\right\}, \tag{2.102}$$

where $[x]$ denotes the largest integer less than or equal to x.

Proof of Claim 3 Using that the probability of a union is dominated by the sum of the probabilities, stationarity and the preceding claim, we have

$$\limsup_{u\to\infty} \frac{q}{\overline{\Phi}} \Pr\left\{\max_{0\leq k\leq [h/(aq)]} \xi(aqk) > u\right\} \leq \limsup_{n\to\infty} \limsup_{u\to\infty} \frac{h}{an\overline{\Phi}} \Pr\left\{\max_{k=0,\ldots,n} \xi(aqk) > u\right\}$$
$$= \frac{h}{a} \Pr\left\{\bigcap_{\ell=1}^{\infty}\{\zeta(C^{1/\alpha}a\ell) + \eta \leq 0\}\right\}. \tag{2.103}$$

For the limes inferior, using Bonferroni's inequality and stationarity, we have

$$\frac{q}{\overline{\Phi}} \Pr\left\{\max_{0\leq k\leq [h/(aq)]} \xi(aqk) > u\right\}$$
$$\geq \frac{q}{\overline{\Phi}} \Pr\left\{\bigcup_{i=1}^{[h/(aqn)]}\left\{\max_{0\leq k\leq [h/(aq)]} \xi(aqk) > u\right\}\right\}$$
$$\geq \frac{q}{\overline{\Phi}} \sum_{i=1}^{[h/(aqn)]} \Pr\left\{\max_{(i-1)n\leq k\leq in-1} \xi(aqk) > u\right\}$$
$$- \frac{q}{\overline{\Phi}} \sum_{1\leq i<j\leq [h/(aqn)]} \Pr\left\{\max_{(i-1)n\leq k\leq in-1} \xi(aqk) > u, \max_{(j-1)n\leq k\leq jn-1} \xi(aqk) > u\right\}$$

$$\geq \frac{q}{\overline{\Phi}}\left[\frac{h}{aqn}\right]\Pr\left\{\max_{0\leq k\leq n-1}\xi(aqk)>u\right\}$$
$$-\frac{h}{an\overline{\Phi}}\Pr\left\{\max_{0\leq k\leq n-1}\xi(aqk)>u,\ \max_{n\leq k\leq 2n-1}\xi(aqk)>u\right\}$$
$$-\frac{h}{a}\sum_{k=n}^{[h/(aq)]}\frac{\Pr\{\xi(0)>u,\xi(aqk)>u\}}{\overline{\Phi}}. \tag{2.104}$$

Now we must show that the rightmost last two terms tend to zero as $u\to\infty$ and then $n\to\infty$. By inclusion-exclusion and stationarity,

$$\frac{1}{n\overline{\Phi}}\Pr\left\{\max_{0\leq k\leq n-1}\xi(aqk)>u,\ \max_{n\leq k\leq 2n-1}\xi(aqk)>u\right\}=\frac{2}{n\overline{\Phi}}\Pr\left\{\max_{0\leq k\leq n-1}\xi(aqk)>u\right\}$$
$$-\frac{2}{2n\overline{\Phi}}\Pr\left\{\max_{0\leq k\leq 2n-1}\xi(aqk)>u\right\},$$

and both terms have the same double limit $\lim_{n\to\infty}\lim_{u\to\infty}$ by Claim 2; hence, this double limit of the first of the last two terms in (2.104) is zero. Now we consider the last term in (2.104). Since by convexity and tangent line approximation at $x=1$, $2/\sqrt{2+2x)}\geq 1+(1-x)/4, x\geq 0$, we have

$$\Pr\{\xi(0)>u,\xi(aqk)>u\}\leq\Pr\{\xi(0)+\xi(aqk)>2u\}$$
$$=\overline{\Phi}\left(\frac{2u}{\sqrt{2+2r(aqk)}}\right)\leq\overline{\Phi}\left(u+u\frac{1-r(aqk)}{4}\right).$$

Let $\varepsilon>0$ be such that $1-r(t)\geq C|t|^\alpha/2$, for $0\leq t\leq\varepsilon$, and let $\delta(\varepsilon)=\sup_{\varepsilon\leq t\leq h}r(t)$; ε exists, and $\delta(\varepsilon)<1$ by the hypothesis (2.97) on r. Then $1-r(aqk)\geq\min(1-\delta(\varepsilon),C(ak)^\alpha/(8u^2))$ (recall that $q=q(u)=u^{-2/\alpha}$). Note also that $\overline{\Phi}(u+x/u)/\overline{\Phi}(u)\leq e^{-x}$ for all $u,x>0$. The last three observations then give

$$\limsup_{u\to\infty}\sum_{k=n}^{[h/(aq)]}\frac{\Pr\{\xi(0)>u,\xi(aqk)>u\}}{\overline{\Phi}} \tag{2.105}$$
$$\leq\limsup\left(\frac{hu^{2/\alpha}}{a}e^{-(1-\delta(\varepsilon))/4}+\sum_{k=n}^{[h/(aq)]}e^{-C(ak)^\alpha/8}\right)$$
$$=\sum_{k=n}^{\infty}e^{-C(ak)^\alpha/8}\to 0\text{ as }n\to\infty.$$

Combining these two limits and the limit in Claim 2 with (2.104) gives

$$\liminf_{u\to\infty}\frac{q}{\overline{\Phi}}\Pr\left\{\max_{0\leq k\leq[h/(aq)]}\xi(aqk)>u\right\}\geq\frac{h}{a}\Pr\left\{\bigcap_{\ell=1}^{\infty}\{\zeta(C^{1/\alpha}a\ell)+\eta\leq 0\}\right\},$$

which, together with (2.103), proves Claim 3.

Having dealt with the discretisations of ξ, we finally deal with the first probability on the right-hand side of (2.100), which constitutes the link between these discretisations and the process.

Claim 4.

$$\lim_{a\downarrow 0}\limsup_{u\to\infty}\frac{q}{\Phi}\Pr\left\{\sup_{t\in[0,h]}\xi(t)>u+\frac{a^{\alpha/4}}{u},\ \max_{0\le k\le[h/aq]\}}\xi(aqk)\le u\right\}=0. \tag{2.106}$$

Proof of Claim 4 We fix a and set $u_n = u + a^{\alpha/4}(1-2^{-n\alpha/4})/u$, and note that $u = u_0 < u_n$, for all n. Stationarity and continuity of ξ allow us to write

$$\begin{aligned}
&\Pr\left\{\sup_{t\in[0,h]}\xi(t)>u+\frac{a^{\alpha/4}}{u},\ \max_{0\le k\le[h/aq]\}}\xi(aqk)\le u\right\}\\
&\le\frac{2h}{aq}\Pr\left\{\sup_{t\in[0,aq]}\xi(t)>u+\frac{a^{\alpha/4}}{u},\xi(0)\le u\right\}\\
&=\frac{2h}{aq}\Pr\left\{\bigcup_{n=0}^{\infty}\bigcup_{k=0}^{2^n-1}\{\xi(aqk2^{-n})>u+a^{\alpha/4}u^{-1}\},\xi(0)\le u\right\}\\
&\le\frac{2h}{aq}\Pr\left\{\bigcup_{n=0}^{\infty}\bigcup_{k=0}^{2^n-1}\{\xi(aqk2^{-n})>u_n\},\xi(0)\le u\right\}.
\end{aligned}$$

First, with $F_n=\bigcup_{k=0}^{2^n-1}\{\xi(aqk2^{-n})>u_n\}$, we use that $\bigcup_{n=1}^{\infty}F_n=F_0\bigcup\left(\cup_{n=1}^{\infty}(F_n\cap F_{n-1}^c)\right)$ and that $F_0\cap\{\xi(0)\le u\}=\emptyset$ to obtain that the preceding probability is dominated by

$$\frac{2h}{aq}\sum_{n=1}^{\infty}\sum_{k=1}^{2^n-1}\Pr\left\{\{\xi(aqk2^{-n})>u_n\}\bigcap\left(\bigcap_{\ell=0}^{2^{n-1}-1}\{\xi(aq\ell2^{-n+1})\le u_{n-1}\}\right)\right\}.$$

Second, notice that, by stationarity, for each n, if k is odd, the probabilities in the nth sum are all dominated by $\Pr\{\xi(aq2^{-n})>u_n,\xi(0)\le u_{n-1}\}$ (e.g., subtract $(k-1)2^{-n}$ from the first event and from $\{\xi(aq2\ell2^{-n})\le u_{n-1}\}$ for $2\ell=k-1$), and likewise, for k even, this probability is dominated by $\Pr\{\xi(aq2\cdot2^{-n})>u_n,\xi(0)\le u_{n-1}\}$. We thus obtain

$$\begin{aligned}
&\Pr\left\{\sup_{t\in[0,h]}\xi(t)>u+\frac{a^{\alpha/4}}{u},\ \max_{0\le k\le[h/aq]\}}\xi(aqk)\le u\right\}\le\frac{2h}{aq}\sum_{n=1}^{\infty}2^{n-1}\left(\Pr\{\xi(aq2^{-n})\right.\\
&\left.>u_n,\xi(0)\le u_{n-1}\}+\Pr\{\xi(aq2^{-n+1})>u_n,\xi(0)\le u_{n-1}\}\right).
\end{aligned}\tag{2.107}$$

Third, since $1-r(t)\le C|t|^{\alpha}$, for all t small enough, in particular, for $t=aq2^{-n}$ and $t=aq2^{-n+1}$, for all n, and for all u, is large enough depending on a, it follows that for all $a>0$ sufficiently small there is u_a such that, for $u\ge u_a$, on the event $\{\xi(aq2^{-n+1})>u,\xi(0)\le u_{n-1}\}$, we have

$$\begin{aligned}
&r(aq2^{-n+1})\xi(aq2^{-n+1})-\xi(0)\\
&\ge(aq2^{-n+1})u_n-u_{n-1}\\
&\ge\frac{a^{\alpha/4}(2^{\alpha/4}-1)2^{-n\alpha/4}}{u}-(1-r(aq2^{-n+1})\left(u+\frac{a^{\alpha/4}(1-2^{-n\alpha/4})}{u}\right)
\end{aligned}$$

$$\geq \frac{a^{\alpha/4}(2^{\alpha/4}-1)2^{-n\alpha/4}}{u} - \frac{2^{1+\alpha}Ca^{\alpha}2^{-n\alpha}}{u}\left(1+\frac{a^{\alpha/4}(1-2^{-n\alpha/4})}{u^2}\right)$$
$$\geq \frac{a^{\alpha/4}(2^{\alpha/4}-1)2^{-n\alpha/4-1}}{u},$$

and, similarly,

$$r(aq2^{-n})\xi(aq2^{-n}) - \xi(0) \geq \frac{a^{\alpha/4}(2^{\alpha/4}-1)2^{-n\alpha/4-1}}{u}$$

on the event $\{\xi(aq2^{-n}) > u, \xi(0) \leq u_{n-1}\}$. Then it follows from this and (2.107) that for all a small enough,

$$\limsup_{u\to\infty} \frac{q}{\overline{\Phi}} \Pr\left\{ \sup_{t\in[0,h]} \xi(t) > u + \frac{a^{\alpha/4}}{u}, \max_{0\leq k\leq [h/aq]\}} \xi(aqk) \leq u \right\}$$
$$\leq \frac{2h}{a\overline{\Phi}} \sum_{n=1}^{\infty} 2^n \Pr\left\{ r(aq2^{-n})\xi(aq2^{-n}) - \xi(0) \geq \frac{a^{\alpha/4}(2^{\alpha/4}-1)2^{-n\alpha/4-1}}{u}, \xi(aq2^{-n}) > u \right\}$$

(where we also use that $u < u_n$). Now $r(t)\xi(t) - \xi(0)$ and $\xi(t)$ are uncorrelated and hence independent, and therefore, this quantity is dominated by

$$\limsup_{u\to\infty} \frac{2h}{a} \sum_{n=1}^{\infty} 2^n \overline{\Phi}\left(\frac{a^{\alpha/4}(2^{\alpha/4}-1)2^{-n\alpha/4-1}}{u(1-r(aq2^{-n})^2)^{1/2}} \right) \leq \limsup_{u\to\infty} \frac{2h}{a} \sum_{n=1}^{\infty} 2^n \overline{\Phi}\left(\frac{a^{\alpha/4}(2^{\alpha/4}-1)2^{-n\alpha/4-1}}{u(2(1-r(aq2^{-n}))^{1/2}} \right)$$
$$\leq \frac{2h}{a} \sum_{n=1}^{\infty} 2^n \overline{\Phi}\left(\frac{a^{\alpha/4}(2^{\alpha/4}-1)2^{-n\alpha/4-1}}{2C^{1/2}a^{\alpha/2}2^{-n\alpha/2}} \right).$$

The last sum has the form $2h\sum_n (2^n/a)\overline{\Phi}\left(-C(\alpha)(2^n/a)^{\alpha/4}\right)$ for some positive constant $C(\alpha)$, so it tends to zero as $a \to 0$ (note that for all a small enough, the sum of this series is dominated by a constant times the first term). Claim 4 is proved.

Now, combining Claims 3 and 4 with inequalities (2.100) and (2.101), we obtain

$$\limsup_{u\to\infty} \frac{q}{\overline{\Phi}} \Pr\left\{ \sup_{t\in[0,h]} \xi(t) > u \right\} \leq \liminf_{a\downarrow 0} \frac{h}{a} \Pr\left\{ \bigcap_{k-1}^{\infty} \{\zeta(C^{1/\alpha}ak) + \eta \leq 0\} \right\}$$
$$= \liminf_{a\downarrow 0} \frac{hC^{1/\alpha}}{a} \Pr\left\{ \bigcap_{k=1}^{\infty} \{\zeta(ak) + \eta \leq 0\} \right\}$$

and

$$\liminf_{u\to\infty} \frac{q}{\overline{\Phi}} \Pr\left\{ \sup_{t\in[0,h]} \xi(t) > u \right\} \geq \limsup_{a\downarrow 0} \frac{hC^{1/\alpha}}{a} \Pr\left\{ \bigcap_{k=1}^{\infty} \{\zeta(ak) + \eta \leq 0\} \right\}.$$

Thus, the limits of both quantities exist (although they could be 0 or infinity) and are equal, proving inequality (2.99). Now, the right-hand side of (2.102) is finite; therefore, by Claims 3 and 4 and (2.100), $\limsup_{u\to\infty} (q/\overline{\Phi}) \Pr\left\{\sup_{t\in[0,h]} \xi(t) > u\right\} < \infty$. This implies by the limit in the last display that $\limsup_{a\downarrow 0} ((hC^{1/\alpha})/a) \Pr\left\{\bigcap_{k=1}^{\infty} \{\zeta(ak) + \eta \leq 0\}\right\} < \infty$, so H_α is finite.

Finally, using Bonferroni's inequality and stationarity, we have

$$\lim_{u\to\infty}\frac{q}{\overline{\Phi}}\Pr\left\{\sup_{t\in[0,h]}\xi(t)>u\right\}\geq\lim_{u\to\infty}\frac{q}{\overline{\Phi}}\Pr\left\{\bigcup_{k=n}^{[h/(aqn)]}\{\xi(aqnk)>u\}\right\}$$

$$\geq\liminf_{u\to\infty}\frac{h}{an\overline{\Phi}}\left(\overline{\Phi}-\sum_{k=n}^{[h/(aqn)]}\Pr\{\xi(aqk)>u,\xi(0)>u\}\right),$$

and by (2.105), this limit inferior equals $h/(2an)$ for n large enough, hence positive. ■

It is surprising that the proof of the very precise Theorem 2.7.5 uses only elementary probability, albeit in an intricate and delicate way (Pickands' 'double sum' method).

Corollary 2.7.6 *Under the assumptions of Theorem 2.7.5,*

$$\lim_{u\to\infty}\frac{1}{u^{2/\alpha}h\overline{\Phi}(u)}\Pr\left\{\sup_{t\in[0,h]}|\xi(t)|>u\right\}=2C^{1/\alpha}H_\alpha, \tag{2.108}$$

and, for all $a>0$,

$$\lim_{u\to\infty}\frac{1}{u^{2/\alpha}h\overline{\Phi}(u)}\Pr\left\{\max_{0\leq k\leq[h/(aq)]}|\xi(aqk)|>u\right\}=\frac{2}{a}\Pr\left\{\bigcap_{k=1}^{\infty}\{\zeta(C^{1/\alpha}ak)+\eta\leq 0\}\right\}, \tag{2.109}$$

where $q=q(u)=u^{-2/\alpha}$.

Proof For the second limit, by inclusion-exclusion,

$$\Pr\left\{\max_{0\leq k\leq[h/(aq)]}|\xi(aqk)|>u\right\}=\Pr\left\{\max_{0\leq k\leq[h/(aq)]}\xi(aqk)>u\right\}+\Pr\left\{\max_{0\leq k\leq[h/(aq)]}(-\xi(aqk))>u\right\}$$

$$-\Pr\left\{\max_{0\leq k\leq[h/(aq)]}\xi(aqk)>u,\ \max_{0\leq k\leq[h/(aq)]}(-\xi(aqk))>u\right\}.$$

The first two terms are equal by symmetry and have been dealt with in Claim 3 of the preceding proof. Thus, to prove (2.109), it suffices to show that $q/\overline{\Phi}$ times the third term tends to zero. This term is bounded by $\Pr\left\{\max_{0\leq k,\ell\leq[h/(aq)]}\xi(aqk)-\xi(aq\ell)>2u\right\}$. We can decompose the maximum in this probability into two parts: one over all $k,\ell\leq[h/(aq)]$ such that $|aq(k-\ell)|\leq\varepsilon$, $0<\varepsilon<C^{-1/\alpha}$, chosen so that $1-r(t)\leq 2C|t|^\alpha$, for all $|t|\leq\varepsilon$, and the other over the remaining k,ℓ. We have, for the first part

$$\frac{q}{\overline{\Phi}}\Pr\left\{\max_{0\leq k,\ell\leq[h/(aq)],|k-\ell|\leq\varepsilon/(aq)}\xi(aqk)-\xi(aq\ell)>2u\right\}\leq\frac{q}{\overline{\Phi}}\frac{h\varepsilon}{(aq)^2}e^{-4u^2/(8C\varepsilon^\alpha)}\to 0,$$

as $u\to\infty$, since in this case $E(\xi(aqk)-\xi(aq\ell)^2=2(1-r(aq|k-\ell)\leq 4C\varepsilon^\alpha<4$. For the other part, noting that $|r(t)|\leq\delta(\varepsilon)<1$ for $|t|>\varepsilon$, we have

$$\frac{q}{\overline{\Phi}}\Pr\left\{\max_{0\leq k,\ell\leq[h/(aq)],|k-\ell|>\varepsilon/(aq)}\xi(aqk)-\xi(aq\ell)>2u\right\}\leq\frac{q}{\overline{\Phi}}\frac{h^2}{(aq)^2}e^{-4u^2/(4(1+\delta(\varepsilon))}\to 0$$

because in this case $2(1-r(aq|k-\ell))\leq 2(1+\delta(\varepsilon))<4$. This and Claim 3 in the preceding proof proves (2.109).

To prove (2.108), we can use the Borell-Sudakov-Tsirelson inequality (e.g., in its form (2.20)) after decomposing $\sup_{0\le s,t\le h}(\xi(t)-\xi(s))$ into the same two parts as in the proof of (2.109) and noting that $E\sup_{s,t\in[0,h]}|\xi(t)-\xi(s)|=M<\infty$ by Theorem 2.1.20 (a). Let ε be as earlier, so $\sup_{|s-t|\le\varepsilon,0\le s,t\le h}E(\xi(t)-\xi(s))^2\le 2C\varepsilon^\alpha$. We then have

$$\frac{q}{\overline{\Phi}}\Pr\left\{\sup_{|s-t|\le\varepsilon,0\le s,t\le h}\xi(t)-\xi(s)>2u\right\}\le\frac{q}{\overline{\Phi}}\exp\left(-\frac{(2u-2\sqrt{2/\pi}C\varepsilon^\alpha-M)^2}{4C\varepsilon^\alpha}\right)\to 0,$$

as $u\to\infty$, since $C\varepsilon^\alpha<1$. Similarly, since $|r(t)<\delta(\varepsilon)<1$ for $|t|\le\varepsilon$,

$$\frac{q}{\overline{\Phi}}\Pr\left\{\sup_{|s-t|>\varepsilon,0\le s,t\le h}\xi(t)-\xi(s)>2u\right\}\le\frac{q}{\overline{\Phi}}\exp\left(-\frac{(2u-2(1+\delta(\varepsilon))-M)^2}{4(1+\delta(\varepsilon))}\right)\to 0.$$

Then, as before,

$$\lim_{u\to\infty}\frac{q}{h\overline{\Phi}(u)}\Pr\left\{\sup_{t\in[0,h]}|\xi(t)|>u\right\}=\lim_{u\to\infty}\frac{q}{h\overline{\Phi}(u)}\Pr\left\{\sup_{t\in[0,h]}\xi(t)>u\right\}$$
$$+\lim_{u\to\infty}\frac{q}{h\overline{\Phi}(u)}\Pr\left\{\sup_{t\in[0,h]}(-\xi(t))>u\right\},$$

and (2.108) follows from Theorem 2.7.5. ∎

Use of the deep Borell-Sudakov-Tsirelson inequality in the preceding proof seems excessive, but it is certainly very convenient.

In the next theorem we obtain a limit theorem for the interval $[0,h]$ increasing to infinity, and the main ingredient of the proof will be, besides Corollary 2.7.6, the comparison theorem for Gaussian processes, Theorem 2.4.7. In order to have a sensible condition on $r(t)$ as t grows, we need a continuous analogue of Lemma 2.7.3. Since its proof is similar to that lemma and only involves calculus, we omit it. First, some notation. We set $\mu=\mu(u)=2C^{1/\alpha}H_\alpha u^{2/\alpha}\bar{\Phi}(u)$, and assume that $T=T(\mu)$ satisfies the condition $T\mu\to\tau$ for some $0<\tau<\infty$. That is,

$$\mu=\mu(u)=2C^{1/\alpha}H_\alpha u^{2/\alpha}\bar{\Phi}(u)\quad\text{and}\quad T\mu\to\tau,\quad\text{as } u\to\infty. \tag{2.110}$$

Taking logarithms and using $\overline{\Phi}(u)\simeq(2\pi)^{-1/2}e^{-u^2/2}/u$, we have

$$\log T+\log(2C^{1/\alpha}H_\alpha)+\frac{2-\alpha}{\alpha}\log u-\frac{u^2}{2}\to\log\tau,$$

as $u\to\infty$. Then $u^2/(2\log T)\to 1$ or $\log u=\frac{1}{2}(\log 2+\log\log T)+o(1)$, and replacing this in the preceding display, we obtain

$$u^2=2\log T+\frac{2-\alpha}{\alpha}\log\log T-2\log\tau+2\log((2\pi)^{-1/2}2^{(2-\alpha)/(2\alpha)}2C^{1/\alpha}H_\alpha)+o(1). \tag{2.111}$$

Conversely, if $u=u_T$ satisfies equation (2.111), then $T\mu(u_T)\to\tau$, as $T\to\infty$. This equation is used in the proof of the following analogue of Lemma 2.7.3 (see Exercise 2.7.1).

Lemma 2.7.7 *Let $\alpha\in(0,2)$, and let $r(t)$ be a function on $(0,\infty)$ such that $r(0)=1$, $|r(t)|<1$, for $t>0$ and $r(t)=1-C|t|^\alpha+o(|t|^\alpha)$. Let T and μ satisfy (2.111) for some $\tau>0$, let $q(u)=u^{-2/\alpha}$ and let $a>0$. Then, if*

$$r(t)\log t\to 0,\quad \textit{as } t\to\infty, \tag{2.112}$$

we have that for every $\varepsilon > 0$,

$$\lim_{u\to\infty} \frac{T}{q} \sum_{\varepsilon \le aqk \le T} |r(aqk)| \exp\left(-\frac{u^2}{1+|r(aqk)|}\right) = 0. \tag{2.113}$$

The following proposition contains the main steps in the proof of the limit theorem for the supremum of a stationary process over increasing intervals:

Proposition 2.7.8 *Let* $\xi(t)$, $t \in (0,\infty)$, *be a normalised centred stationary Gaussian process with continuous sample paths and with covariance function* $r(t)$ *satisfying conditions (2.97) for some* $\alpha \in (0,2)$ *and (2.112). Let* $\tau > 0$, *and for each* T, *let* $u = u_T$ *be such that condition* $T\mu(u_T) \to \tau$, *as* $T \to \infty$, *where* $\mu = \mu(u_T)$, *is as in (2.110). Then*

$$\lim_{T\to\infty} \Pr\left\{ \sup_{0\le t\le T} |\xi(t)| \le u_T \right\} \to e^{-\tau}.$$

Proof We take $0 < \varepsilon < h < T$ and define $n = [T/h]$; then we divide the interval $[0,nh]$ into n adjacent intervals of length h, say, $[t_j,t_{j+1}]$, $j = 1,\dots,n$, and split each interval into two, $I_j = [t_j, t_{j+1} - \varepsilon]$ and $I_j^* = [t_{j+1} - \varepsilon, t_{j+1}]$ of lengths $1-\varepsilon$ and ε, respectively. Given $a > 0$, let $H_\alpha(a) = \frac{1}{a}\Pr\left\{\sup_{k\in\mathbb{N}} \zeta(ak) + \eta \le 0\right\}$, and define $\rho_a(\alpha) = 1 - H_\alpha(a)/H_\alpha$. Then $\rho_a(\alpha) \to 0$ as $a \to 0$ by (2.98) (actually, by Theorem 2.7.5). The proposition will follow by comparing $\Pr\left\{\sup_{0\le t\le nh} |\xi(t)| \le u\right\}$ to $\left(\Pr\left\{\sup_{0\le t\le h} |\xi(t)| \le u\right\}\right)^n$, and this comparison will take four steps as it will go through the sup of the process over $\cup I_j$, the maximum of the discretisation over $\cup I_j$, which will be compared to the discretized process independent on the different I_k using Theorem 2.4.7 (the comparison theorem for Gaussian vectors), and then we will compare back to independent copies of the original. Three of the comparisons will only use Theorem 2.7.5 in its version for absolute values (Corollary 2.7.6).

First comparison:

$$0 \le \limsup_{u\to\infty} \left(\Pr\left\{ \sup_{t\in\cup_{j=1}^n I_j} |\xi(t)| \le u \right\} - \Pr\left\{ \sup_{t\in[0,nh]} |\xi(t)| \le u \right\} \right) \le \frac{\tau\varepsilon}{h}. \tag{2.114}$$

Proof To see this, note that the difference of the two probabilities is nonnegative and is dominated by

$$\sum_{j=1}^n \Pr\left\{ \sup_{t\in I_j^*} |\xi(t)| > u \right\} = n\mu\varepsilon + o(n\mu) \to \frac{\tau\varepsilon}{h},$$

since $n\mu \simeq T\mu/h \to \tau/h$.

Second and fourth comparisons:

$$0 \le \limsup_{u\to\infty} \left(\Pr\left\{ \max_{aqk\in\cup_{j=1}^n I_j} |\xi(aqk)| \le u \right\} - \Pr\left\{ \sup_{t\in\cup_{j=1}^n I_j} |\xi(t)| \le u \right\} \right) \le \tau\rho_a \tag{2.115}$$

and

$$0 \le \limsup_{u\to\infty} \left(\prod_{j=1}^n \Pr\left\{ \max_{aqk\in I_j} |\xi(t)| \le u \right\} - \left(\Pr\left\{ \sup_{t\in[0,h]} |\xi(t)| \le u \right\} \right)^n \right| \le \tau\rho_a. \tag{2.116}$$

Proof First, we observe that for any interval I of length h, by Corollary 2.7.6 and stationarity,

$$0 \leq \Pr\left\{\max_{akq\in I}|\xi(akq)| \leq u\right\} - \Pr\left\{\sup_{0\leq t\leq h}|\xi(t)| \leq u\right\} \leq \mu h\rho_a + o(\mu). \tag{2.117}$$

Note that, in fact, if $I = [0,h]$, this statement follows directly from Corollary 2.7.6, and by stationarity, we can replace the interval I by $[0,h]$, with equality for the second term; however, regarding the first term, the number of jqk in I may be the same or one less than in $[0,h]$ (at least for u large and a small). This difference is not significant because $\Pr\{\xi(jqk) > u\} = o(\mu)$).

Now (2.117) gives

$$\begin{aligned}
0 &\leq Pr\left\{\max_{aqk\in\cup_{j=1}^n I_j}|\xi(aqk)| \leq u\right\} - \Pr\left\{\sup_{t\in\cup_{j=1}^n I_j}|\xi(t)| \leq u\right\} \\
&\leq n\max_{1\leq j\leq n}\left(Pr\left\{\max_{aqk\in I_j}|\xi(aqk)| \leq u\right\} - \Pr\left\{\sup_{t\in I_j}|\xi(t)| \leq u\right\}\right) \\
&\leq n\mu(h-\varepsilon) + o(n\mu) \to \tau(1-\varepsilon/h) \leq \tau\rho_a.
\end{aligned}$$

Let ξ_j be independent copies of ξ, and consider the process $\overline{\xi}(t) = \sum_{j=i}^n \xi_j(t)I_{I_j}(t)$. The preceding string of inequalities applied to $\overline{\xi}(t)$, $t \in \cup_{j=1}^n I_j$, proves inequality (2.116). This completes the proof of these two comparisons.

Third comparison:

$$\lim_{u\to\infty}\left(\Pr\left\{\max_{aqk\in\cup_{i=1}^n I_j}|\xi(aqk)| \leq u\right\} - \prod_{j=1}^n \Pr\left\{\max_{aqk\in I_j}|\xi(t)| \leq u\right\}\right) = 0. \tag{2.118}$$

Proof We will apply inequality (2.56) in the comparison Theorem 2.4.7 to the finite-dimensional Gaussian vectors ($\xi(aqk) : aqk \in \cup_{j=1}^n I_j$ and $\overline{\xi}(aqk) : aqk \in \cup_{j=1}^n I_j$) with $\lambda_i = u$ and $\mu_i = u$ for all i. Now note that

$$E(\xi(aqk)\xi(aq\ell)) = E(\overline{\xi}(aqk)\overline{\xi}(aq\ell)) = r(aq|k-\ell|), \quad \text{if } aqk, aq\ell \in I_j, j = 1,\ldots,n,$$

$$E(\xi(aqk)\xi(aq\ell)) = r(aq|k-\ell|), \quad \text{but} \quad E(\overline{\xi}(aqk)\overline{\xi}(aq\ell)) = 0, \quad \text{if } aqk \in I_i, aq\ell \in I_j,\ i \neq j.$$

Then inequality (2.56) gives

$$\begin{aligned}
&\left|\Pr\left\{\max_{aqk\in\cup_{i=1}^n I_j}|\xi(aqk)| \leq u\right\} - \prod_{j=1}^n \Pr\left\{\max_{aqk\in I_j}|\xi(aqk)| \leq u\right\}\right| \\
&\leq \frac{2}{\pi}\sum_{aqk\in I_i, aq\ell\in I_j, i<j}|r(aq|k-\ell|)\exp\left(-\frac{u^2}{1+|r(aq|k-\ell|}\right) \\
&\leq \frac{2}{\pi}\frac{T}{aq}\sum_{\varepsilon\leq aqk\leq T}|r(aqk)|\exp\left(-\frac{u^2}{1+|r(aqk)|}\right),
\end{aligned} \tag{2.119}$$

which tends to zero as $u \to \infty$ by Lemma 2.7.7, proving the third comparison.

The conclusion of these four comparisons is that

$$\limsup_{T\to\infty}\left|\Pr\{\sup_{t\in[0,nh]}|\xi(t)|\le u_T\}-\left(\Pr\{\sup_{t\in[0,h]}|\xi(t)|\le u_T\}\right)^n\right|\le 2\tau(\rho_a+\varepsilon/(2h)).$$

Letting $\varepsilon\to 0$ and $a\to 0$ (recall that $\rho_a\to 0$ as $a\to 0$ by (2.98)), we obtain that this limit is zero, but, by Corollary 2.7.6,

$$\left(\Pr\{\sup_{t\in[0,h]}|\xi(t)|\le u_T\}\right)^n=(1-\mu h+o(\mu))^n,$$

and since $T\mu\to\tau$ and $n/(T/h)\to 1$, this last expression has limit e^{τ}. To complete the proof of the proposition, just note that

$$0\le\Pr\left\{\sup_{t\in[0,nh]}|\xi(t)|\le u_T\right\}-\Pr\left\{\sup_{t\in T}|\xi(t)|\le u_T\right\}\le\Pr\left\{\sup_{t\in[0,h]}|\xi(t)|>u_T\right\},$$

and by Corollary 2.7.6, the last probability is eventually smaller than a constant times $u_T^{2/\alpha}\overline{\Phi}(u_T)$, which tends to zero. ■

Let now $u_T=x/a_T+b_T$, with

$$a_T=(2\log T)^{1/2},\quad b_T=a_T+\frac{(2-\alpha)/\alpha\log\log T+2\log((2\pi)^{-1/2}2^{(2+\alpha)/\alpha}C^{1/\alpha}H_\alpha)}{2a_T}. \tag{2.120}$$

Then

$$\begin{aligned}u_T^2&=2xb_T/a_T+b_T^2+o(1)\\&=-2\log e^{-x}+(2\log T)+\frac{2-\alpha}{\alpha}\log\log T+2\log((2\pi)^{-1/2}2^{(2+\alpha)/\alpha}C^{1/\alpha}H_\alpha)+o(1),\end{aligned}$$

which is just (2.111) with $\tau=e^{-x}$. Thus, Proposition 2.7.8 gives the following weak convergence result for the supremum norm of $\xi(t)$ over increasing intervals:

Theorem 2.7.9 *Let $\xi(t)$, $t\in[0,\infty)$, be a normalised, centred stationary Gaussian process with continuous sample paths and with covariance $r(t)$ satisfying $r(t)<1$, for $t>0$, $r(t)=1-C|t|^\alpha+o(|t|^\alpha)$, for some $\alpha\in(0,2]$ as $t\to 0$, and $r(t)\log t\to 0$, as $t\to\infty$. Let H_α be defined by equation (2.98) (hence $H_2=1/\sqrt{\pi}$), and let a_T and b_T be defiened by equations (2.120). Then, for all $x\in\mathbb{R}$,*

$$\lim_{T\to\infty}\Pr\left\{a_T(\sup_{t\in[0,T]}|\xi(t)|-b_T)\le x\right\}=\exp\left(-e^{-x}\right).$$

Exercises

2.7.1 Prove Lemma 2.7.7. *Hint*: Split the sum in (2.113) into two parts, $\varepsilon\le aqk\le T^\beta$ and the rest, where $\beta\in(0,(1-\delta)/(1+\delta))$, where $\delta=\sup_{t\ge\varepsilon}|r(t)|$. The sum for the smaller k is easy to handle. Using the notation $\delta(t) := \sup_{s\geq t} |r(s)\log s|$, bound the sum for the larger k by

$$\frac{T}{q}\sum_{t^\beta \leq aqk\leq T} |r(aqk)| \exp\left(-u^2\left(1-\frac{\delta(T^\beta)}{\log T^\beta}\right)\right)$$

$$\leq \left(\frac{T}{q}\right)\exp\left(-u^2\left(1-\frac{\delta(T^\beta)}{\log T^\beta}\right)\right)\frac{1}{\log T^\beta}\sum_{T^\beta\leq aqk} \leq T|r(aqk)|\log(aqk).$$

The behaviour of r at infinity shows that $(q/t)\sum_{T^\beta\leq aqk} \leq T|r(aqk)|\log(aqk)$, and the remaining factor can be handled using (2.111).

2.7.2 Prove that under the hypotheses of Theorem 2.7.9,

$$\lim_{T\to\infty} \Pr\left\{a_T(\sup_{t\in[0,T]} \xi(t) - \bar{b}_T) \leq x\right\} = \exp\left(-e^{-x}\right),$$

where $\bar{b}_T = b_T - (\log 2)/a_T$. *Hint*: Use Theorem 2.7.5 and Claim 3 from its proof directly (without the corollary).

2.7.3 Prove that $H_{1/2} = 1/\sqrt{\pi}$, where H_2 is as defined in (2.98) for $\alpha = 2$.

2.8 Notes

Section 2.1 Proposition 2.1.4 is a classical result due to Oxtoby and Ulam (1939). Proposition 2.1.7 on bounded stochastic processes with tight Borel probability laws comes from Hoffmann-Jørgensen (1991). We learned the results on Gaussian processes in this section mainly from Fernique (1975). The proof of the 0-1 law comes from Fernique (1974) (the result itself is due to Dudley and Kanter (1974)), and the simple proof of the concentration inequality given here is due to Maurey and Pisier (see Pisier (1986)). The Paley-Zygmund argument is due to Paley and Zygmund (1932). See also Salem and Zygmund (1954) and Kahane (1968).

Section 2.2 In Section 2.2.1 we follow the excellent presentation of Benyamini (1984) of the proof of the isoperimetric inequality on the sphere by Baernstein and Taylor (1976). See Figiel, Lindenstrauss and Milman (1977) for another proof. The theorem itself if due to Lévy (1951) and Schmidt (1948).

The exposition of Sections 2.2.2 and 2.2.3 is adapted from accounts in Ledoux and Talagrand (1991) and Fernique (1997), particularly from the last reference. The Gaussian isoperimetric inequality is due independently to Borell (1975) and to Sudakov and Tsirelson (1974), and so are their consequences for Lipschitz functions and supremum norms. Corollary 2.2.9 is due to Marcus and Shepp (1972).

Section 2.3 The chaining argument, the entropy-sufficient condition for sample boundedness and continuity of Gaussian processes and the entropy bound on the modulus of continuity are due to R. M. Dudley (1967, 1973). The entropy bounds on L^p norms and Orlicz norms were noticed by G. Pisier (1983), including the argument in (2.33) and (2.34). Lemma 2.3.4 is from Boucheron, Lugosi and Massart (2013). According to R. Dudley (Math. Reviews MR0431359 (55#4359), current version), Sudakov was the first to relate metric entropy to sample boundedness and continuity of Gaussian processes, although his first results were only announced in 1969. Fernique (1975) proved that finiteness of the metric entropy integral is also necessary for sample continuity of *stationary* Gaussian processes. The exposition here is based on Dudley's papers and the expositions in de la Peña and Giné (1999) and Ledoux and Talagrand (1991).

Although metric entropy provides sharp results, it does not provide general necessary and sufficient conditions for sample boundedness or continuity of Gaussian processes. M. Talagrand (1987a) obtained such conditions in terms of another characteristic of the metric space (T, d_X), the existence of a majorising measure (sufficiency had been obtained by X. Fernique, and M. Talagrand proved necessity). This was a major achievement, solving a very natural question, namely, characterising sample boundedness of a Gaussian process X only in terms of properties of the metric space (T, d_X). Later on, Talagrand (2005) provided a nice book account of his work, with simpler proofs and several applications. Here is the sufficiency part of the simplest version of his theorem, the simplest example showing how it goes beyond metric entropy and its relation to metric entropy:

Theorem 2.8.1 (Generic chaining) *Let (T, d) be a countable pseudo-metric space, and let $X(t)$, $t \in T$, be a stochastic process sub-Gaussian with respect to d. Let $N_n = 2^{2^n}$, and let $T_n \subset T$, $n \in \mathbb{N}$, be sets such that* $\mathrm{Card}(T_n) \leq N_n$ *and $\cup_n T_n = T$. Then, for any $F \subset T$, F finite, and $t_0 \in F$, setting $T_0 = \{t_0\}$, we have*

$$E \max_{t \in F} (X(t) - X(t_0)) \leq L \max_{t \in F} \sum_{n \geq 0} 2^{n/2} d(t, T_n), \tag{2.121}$$

where L is a universal constant that can be taken to be $L = (1+\sqrt{2})(1+3\log 2)/\sqrt{3\log 2} < 5.16$. The same bound holds for the supremum over T.

Proof By identifying points of F at zero distance d from each other, we can assume that d is a distance on F. For each $t \in F$, let $\pi_n : F \mapsto T_n$ be a map such that $d(t, \pi_n(t)) = d(t, T_n)$ $(= \inf_{s \in T_n} d(s,t))$. Since $F \subseteq \cup_n T_n$, for each t, there is $n(t) < \infty$ such that $\pi_n(t) = t$. We then have

$$X(t) - X(t_0) = \sum_{n=1}^{n(t)} (X(\pi_n(t)) - X(\pi_{n-1}(t)), \quad t \in F. \tag{2.122}$$

Let $S = \max_{t \in F} \sum_{n=1}^{n(t)} 2^{n/2} d(\pi_n(t), \pi_{n-1}(t))$, and note that, by the triangle inequality,

$$S \leq (1+\sqrt{2}) \max_{t \in F} \sum_{n \geq 0} 2^{n/2} d(t, T_n).$$

Now, for $u > 0$, observe that if $X(\pi_n(t)) - X(\pi_{n-1}(t)) \leq u 2^{n/2} d(\pi_n(t), \pi_{n-1}(t))$ for all (t,n), $t \in F$, $n \leq n(t)$, then, by (2.122), $\max_{t \in F} |X(t) - X(t_0)| \leq uS$, that is,

$$\Pr\left\{\max_{t \in F} (X(t) - X(t_0)) > uS\right\}$$
$$\leq \sum_{(n, \pi_n(t), \pi_{n-1}(t)) : t \in F, 1 \leq n \leq n(t)} \Pr\left\{|X(\pi_n(t)) - X(\pi_{n-1}(t)| > u 2^{n/2} d(\pi_n(t), \pi_{n-1}(t))\right\}.$$

By the sub-Gaussian inequality (2.22), these probabilities are bounded, respectively, for each n, by $e^{-u^2 2^n/2}$ independently of t, and hence, since for each n the cardinality of the set $\{(\pi_n(t), \pi_{n-1}(t)) : t \in F\}$ is dominated by $N_n N_{n-1} \leq 2^{3 \cdot 2^{n-1}}$, the preceding inequality gives

$$\Pr\left\{\max_{t \in F} (X(t) - X(t_0)) > uS\right\} \leq \sum_{n \geq 1} 2^{3 \cdot 2^{n-1}} e^{-u^2 2^{n-1}}.$$

The random variable $\max_{t \in F} (X(t) - X(t_0))$ is nonnegative, so

$$\begin{aligned} E \max_{t \in F} (X(t) - X(t_0))/S &= \int_0^\infty \Pr\left\{\max_{t \in F} (X(t) - X(t_0)) > uS\right\} du \\ &\leq \sqrt{3 \log 2} + \sum_{n \geq 1} 2^{3 \cdot 2^{n-1}} \int_{\sqrt{3 \log 2}}^\infty e^{-u^2 2^{n-1}} du, \end{aligned}$$

which gives the theorem. ∎

Note that the same bound (2.121) with twice the constant holds for $E\max_{t\in F}|X(t)-X(t_0)|$. Of course, for this theorem to be useful, a method (or methods) to find the approximating sets is required. One of the methods leads to majorising measures, and it can be proved that there is a lower bound of the same type for the expectation of $\max_{t\in F}(X(t)-X(t_0))$, X centred Gaussian.

Example 2.8.2 Let $T=\{2,3,\ldots,\infty\}$, let $\{g_n\}$ be a sequence of independent standard normal variables and set $X(n)=g_n/\sqrt{\log n}$, for $2\le n<\infty$ and $X_\infty=0$. Then $d(n,\infty)=1/\sqrt{\log n}$ and $d(n,m)\in\left(1/\sqrt{\log(m\wedge n)},2/\sqrt{\log(m\wedge n)}\right)$, for $2\le n\ne m<\infty$. Set $T_n=\{2,\ldots,N_n,\infty\}$, for $n\ge 2$. Then $d(m,T_n)=0$ if $m\le N_n$ or if $m=\infty$, and $d(m,T_n)=1/\sqrt{\log m}$ if $N_n<m<\infty$. This shows that

$$\sup_m\sum_n 2^{n/2}d(m,T_n)\le\sup_m\frac{1}{\sqrt{\log m}}\sum_{n:2^n<\log m}2^{n/2}<4.$$

Hence, taking $t_0=\infty$, Theorem 2.8.1 proves $E\sup_n|g_n|/\sqrt{\log n}<\infty$. However, $N(T,d,1/\sqrt{\log m})\ge m$ because the set $\{2,\ldots,m+1\}$ cannot be covered by less than m balls of radius $1/\sqrt{\log m}$ (no two of these points can be in the same ball); hence, $N(T,d,\varepsilon)\ge e^{1/2\varepsilon^2}$, for all $\varepsilon\le 1/\sqrt{2}$, and the entropy integral is bounded from below by a constant times $\int_0\varepsilon^{-1}d\varepsilon=\infty$. Thus, sample boundedness of the process $X(n)$ does not follow from the metric entropy bound, but it does indeed follow from the generic chaining bound.

To see the relation of this theorem to entropy, set

$$e_n(T)=\inf_{T_n\subseteq T:\ \mathrm{Card}(T_n)\le N_n}\ \sup_{t\in T}d(t,T_n),\quad n\ge 0,\tag{2.123}$$

where, consistently with T_0 consisting of one point, we take $N_0=1$. The $e_n(T)$ are called the *entropy numbers of* T. These numbers are finite for T finite, but they can be finite or infinite if T is infinite. Let us assume that T finite for the moment. For each n, let T_n be such that $e_n(T)=\sup_t d(t,T_n)$, $n\ge 1$, $T_0=\{t_0\}$. Then Theorem 2.8.1 gives

$$\begin{aligned}E\sup_{t\in T}|X(t)-X(t_0)|&\le 2L\sup_{t\in T}\sum_{n\ge 0}2^{n/2}d(t,T_n)\\&\le 2L\sum_{n\ge 0}2^{n/2}\sup_{t\in T}d(t,T_n)\\&=2L\sum_{n\ge 0}2^{n/2}e_n(T).\end{aligned}\tag{2.124}$$

It is easy to see that this inequality also holds for T arbitrary (not just finite) because $e_n(U)\le 2e_n(T)$, if $U\subset T$. It turns out that this inequality is equivalent up to constants to the entropy bound (2.41). We show only that it implies (2.41) and leave the reverse implication to the reader. If $e_n(T)<\varepsilon$, then there exists $T_n\subseteq T$ of cardinality bounded by N_n such that $\sup_t d(t,T_n)<\varepsilon$, which implies that $N(T,d,\varepsilon)\le N_n$; conversely, if $N(T,d,\varepsilon)\le N_n$, then taking T_n to be the set of centres of the balls of d-radius ε that constitute a minimum cardinality covering of T yields $e_n(T)\le\varepsilon$. We thus have

$$e_n(T)<\varepsilon\ \Rightarrow\ N(T,d,\varepsilon)\le N_n\Rightarrow e_n(T)\le\varepsilon.$$

By the right-hand-side inequality, if $e_n(T) > \varepsilon$, then $N(T,d,\varepsilon) \geq 1 + N_n$, so

$$\sqrt{\log(1+N_n)}(e_n(T) - e_{n+1}(T)) \leq \int_{e_{n+1}(T)}^{e_n(T)} \sqrt{\log N(T,d,\varepsilon)}\, d\varepsilon,$$

which, summing over $n \geq 0$, gives

$$\sqrt{\log 2}\sum_{n\geq 0} 2^{n/2}(e_n(T) - e_{n+1}(T)) \leq \int_0^{e_0(T)} \sqrt{\log N(T,d,\varepsilon)}\, d\varepsilon.$$

But $\sum_{n\geq 0} 2^{n/2}(e_n(T) - e_{n+1}(T)) \geq (1 - 1/\sqrt{2})\sum_{n\geq 0} 2^{n/2} e_n(T)$, and $e_0(T)$ is the diameter $D(T)$ of (T,d), thus giving

$$\sqrt{\log 2}(1 - 1/\sqrt{2})\sum_{n\geq 0} 2^{n/2} e_n(T) \leq \int_0^{D(T)} \sqrt{\log N(T,d,\varepsilon)}\, d\varepsilon,$$

which shows that the bound (2.124) implies the metric entropy bound (2.41). The converse is also true, so (2.124) shows that the entropy bound follows from the generic chaining bound essentially by replacing the supremum of a sum with the sum of suprema.

Section 2.4 The proof of the main theorem in Section 2.4.1 (log-concavity of the integral with respect to Lebesgue measure on $\mathbb{R}^n$ (Theorem 2.4.2)) belongs to Ball (1986), with ideas from Prékopa (1972) and Leindler (1973), and we learned it from Pisier (1989). This theorem has a simpler proof than the usual proof of Brunn-Minkowski for volumes and directly produces the log-concavity of Gaussian measures as well as its immediate consequence, Anderson's (1955) lemma, as indicated, for example, in Bogachev (1998). The present proof of the Khatri-Sidak inequality (Khatri (1967); Sidak (1967, 1968)) belongs to Li and Shao (2001). Khatri-Sidak is a first result on the well-known 'Gaussian correlation conjecture', not yet completely settled at this writing. Gardner (2002) surveys these inequalities, their proofs and their relationships, with history and references. The comparison theorem (Theorem 2.4.8) is due to Slepian (1962), and it has been a basic tool in the development of the theory of Gaussian processes. The more quantitative Theorem 2.4.7 is due to Berman (1964, 1971), and its proof here follows Leadbetter, Lindgren and Rotzén (1983). Fernique (1975) proved the comparison theorem Corollary 2.4.10 with the constant 2 replaced by the best possible constant 1 in its conclusion. Sudakov's minorisation is due to Sudakov (1969, 1973); Chevet (1970) contains the first published proof of it. The proof of Corollary 2.4.14 comes from Ledoux and Talagrand (1991).

Section 2.5 The log-Sobolev inequality for Gaussian processes was proved by L. Gross (1975), and the idea to derive from it a bound on the Laplace transform of the (sup of the) process was described by I. Herbst in an unpublished letter to Gross (see Davies and Simon (1984)). The proof of the log-Sobolev inequality for Gaussian processes given here, using the Prékopa-Leindler convexity inequality for integrals from the preceding section, belongs to Bobkov and Ledoux (2000) and extends to measures other than Gaussian. The subsection on the different definitions and tensorisation of entropy follows Ledoux (2001).

Section 2.6 Reproducing kernel Hilbert spaces have been in use in analysis for a long time, at least since the 1920s (e.g., Bergman's kernel). See Aronszajn (1950) for early developments and historical remarks. They have become a basic component in the theory of Gaussian processes after the work of Cameron and Martin (1944), Karhunen (1947) and Loève (e.g., (1978)), Gross (1967), Kallianpur (1971), Borell (1976) and Sudakov and Tsirelson (1974), among many others.

The series expansion (Theorem 2.6.10) is due to Karhunen and Loève, the isoperimetric inequality (Theorem 2.6.12) to Borell and Sudakov and Tsirelson, the formula for differentiation of translated Gaussian measures to Cameron and Martin, and the result on equivalence and singularity of product measures (Proposition 2.6.16 and Exercise 2.6.15) to Kakutani (1948). The consequence of the Cameron-Martin theorem for the Gaussian measure of translates of symmetric sets, Corollary 2.6.18, was observed by several authors, including Borell (1976), Hoffmann-Jørgensen, Dudley and Shepp (1979) and de Acosta (1983). Proposition 2.6.19 is essentially in Kuelbs, Li and Linde (1994) and exactly in the present form in van der Vaart and van Zanten (2008a). The relationship between metric entropy and small balls was discovered by Kuelbs and Li (1993) and has been developed further by several authors; see Li and Shao (2001) for a survey; Lemma 2.6.28 comes from Kuelbs and Li (1993) and Theorem 2.6.29 from Li and Linde (1999) (which contains a stronger result). The inequality in Corollary 2.6.30 is in fact two sided; see Li and Linde (1999). Of course, the distribution of the supremum norm of Brownian motion is classical (Kolmogorov and Smirnov), but the proof of the less exact Theorem 2.6.26 is taken from Stolz (1994) and Ledoux (1996); this proof also gives the small balls bound for L^p-norms (Borovkov and Mogulskii (1991)) and Hölder norms (Baldi and Roynette (1992)) (Exercises 2.6.12 and 2.6.13). The present exposition owes much to the chapter on RKHS's in Fernique's (1997) book and to articles by van der Vaart and van Zanten (2008a) and Li and Shao (2001).

Section 2.7 The limit theorem for Gaussian stationary sequences is due to Berman (1964), and the theorem for processes is due to Pickands (1969) with complements by Berman (1971, 1971a) and others. The proof of the limit theorem for a fixed interval is taken from Albin and Choi (2010), as their proof is simpler than previous ones. The monographs of Leadbetter, Lindgren and Rootzén (1983) and Piterbarg (1996) cover different aspects of the extremal theory of Gaussian processes, the first focussing on stationary processes and the second on cyclostationary processes and fields, and we refer to them for further reading. A drawback on the limit theorems in this section is that the speed of convergence is very slow, of the order of $1/\log n$ (or $\log T$) (Hall (1979)). The convergence can be improved to $n^{-\delta}$ ($T^{-\delta}$) for some unspecified δ by changing the exponent of the limit to $e^{-x-x^2/(4\log(T/2\pi))}$ and replacing a_T and b_T by $(2\log(T/2\pi))^{1/2}$ at least for $\alpha = 2$; see Piterbarg (1996, p. 32), or Rootzén (1983) and Katz and Rootzén (1997) for similar rates in Proposition 2.7.8.

Next, we describe a result of Piterbarg and Seleznjev (1994) which will be used later; see also Konstant and Piterbarg (1993). A centred Gaussian process $Y(t)$, $t \in \mathbb{R}$, is cyclostationary if its covariance function $t \mapsto r(t,t+v) := E(Y(t)Y(t+v)$ is periodic in t for every $v \in \mathbb{R}$ with period independent of v. Such processes appear, for example, as approximations of density estimators based on wavelets (see Section 2.5). A situation in which an extremal theory has been developed as in the stationary case is for cyclostationary processes whose variance $\sigma^2(t) = EY^2(t)$ attains its absolute maximum on each of its periodicity intervals at a single point or at a finite number of points. Here is the theorem, which in this form can be found in Giné and Nickl (2010):

Theorem 2.8.3 *Let $X(t)$, $t \in \mathbb{R}$, be a cyclostationary, centred Gaussian process with period 1, variance $\sigma_X(t)$ and covariance $r_X(s,t)$. Assume that*

1. *$X(t)$ is mean square differentiable and a.s. continuous,*
2. *$r_X(s,t) = \sigma_X(s)\sigma_X(t)$ only at $s = t$,*

3. $\sup_{t\in[0,1]} \sigma_X^2(t) = \sigma_X^2(t_0) = 1$ *for a unique* $t_0 \in (0,1)$, $\sigma_X^2(t)$ *is twice continuously differentiable at* t_0, $\sigma_X'(t_0) = 0$, $\sigma_X''(t_0) < 0$ *and* $E(X'(t_0))^2 > 0$, *and*
4. $(\log v)\sup_{s,t:|s-t|\geq v} |r_X(s,t)| \to 0$, *as* $v \to \infty$.

Then, for all $x \in \mathbb{R}$,

$$\lim_{T\to\infty} \Pr\left\{a_T\left(\sup_{t\in[0,T]} |X(t)| - b_T\right) \leq x\right\} = \exp\left(-e^{-x}\right),$$

where

$$a_T = \sqrt{2\log T} \quad \text{and} \quad b_T = a_T - \frac{\log\log T + \log(\pi) - \log\left(1 - \frac{E(X'(t_0))^2}{\sigma_X''(t_0)}\right)}{2a_T}.$$

3

Empirical Processes

Empirical process theory has become, in the course of the last three or four decades, an invaluable tool in statistics. This chapter develops empirical process theory with an emphasis on finite sample sizes rather than on asymptotic theory (although asymptotics came first historically). Thus, just as with Gaussian processes, key elements of this chapter are concentration inequalities for the supremum of empirical processes about their means – the celebrated Talagrand inequalities (for empirical processes but also for Rademacher processes) – and upper bounds for these means (by chaining methods, already introduced in Chapter 2, here combined with randomisation and Vapnik-Červonenkis combinatorics or modified via 'bracketing' techniques). The chapter begins with an introduction to basic inequalities, among them exponential, randomisation and symmetrisation, moment and maximal inequalities for sums of independent random variables and vectors, needed in the rest of the chapter, and it ends with a succinct account of the more classical asymptotic theory, concretely uniform laws of large numbers and uniform central limit theorems.

3.1 Definitions, Overview and Some Background Inequalities

In this section we set up some notation, define empirical processes, give a brief outline of the chapter and prove several inequalities that will be useful throughout, namely, the classical exponential inequalities for sums of centred bounded independent random variables – the Hoeffding, Bennett, Prokhorov and Bernstein inequalities – and inequalities related to symmetrisation and randomisation.

3.1.1 Definitions and Overview

Let $(S,\mathcal{S},P)$ be a probability space, and let X_i, $i \in \mathbb{N}$, be the coordinate functions of the infinite product probability space $(\Omega,\Sigma,\Pr) := (S^{\mathbb{N}},\mathcal{S}^{\mathbb{N}},P^{\mathbb{N}})$, $X_i : S^{\mathbb{N}} \mapsto S$, which are independent identically distributed S-valued random variables with law P. In fact, we will always take independent variables (equally distributed or not) to be the coordinate functions on product probability spaces. The *empirical measure* corresponding to the 'observations' $X_1,\dots,X_n$, for any $n \in \mathbb{N}$, is defined as the random discrete probability measure

$$P_n := \frac{1}{n}\sum_{i=1}^{n}\delta_{X_i}, \tag{3.1}$$

where δ_x is Dirac measure at x, that is, unit mass at the point x. In other words, for each event A, $P_n(A)$ is the proportion of observations X_i, $1 \le i \le n$, that fall in A; that is,

$$P_n(A) = \frac{1}{n}\sum_{i=1}^{n} \delta_{X_i}(A) = \frac{1}{n}\sum_{i=1}^{n} I(X_i \in A), \quad A \in \mathcal{S}.$$

For any measure Q and Q-integrable function f, we will use the following operator notation for the integral of f with respect to Q:

$$Qf = Q(f) = \int_\Omega f\,dQ.$$

Let $\mathcal{F}$ be a collection of P-integrable functions $f : S \mapsto \mathbb{R}$, usually infinite. For any such class of functions $\mathcal{F}$, the empirical measure defines a stochastic process

$$f \mapsto P_n f, \quad f \in \mathcal{F}, \tag{3.2}$$

which we may call the *empirical process indexed by* $\mathcal{F}$, although we prefer to reserve the notation 'empirical process' for the centred and normalised process

$$f \mapsto \nu_n(f) := \sqrt{n}(P_n f - Pf), \quad f \in \mathcal{F}. \tag{3.3}$$

The object of empirical process theory is to study the properties of the approximation of Pf by $P_n f$, uniformly in $\mathcal{F}$, concretely, to obtain both probability estimates for the random quantities

$$\|P_n - P\|_{\mathcal{F}} := \sup_{f\in\mathcal{F}} |P_n f - Pf|$$

and probabilistic limit theorems for the processes $\{(P_n - P)(f) : f \in \mathcal{F}\}$.

This programme has a long history, starting with Bernoulli and de Moivre, who studied the approximation of the probability of an event PA by its frequency P_nA, continuing with Glivenko, Cantelli, Kolmogorov, Smirnov, Skorokhod, Kiefer, Wolfowitz, Kac, Doob, Donsker and many others, who considered the classical case of the empirical distribution function, where $\mathcal{F} = \{I(-\infty, t] : t \in \mathbb{R}\}$. The point of view taken here started with work of Vapnik and Červonenkis (1971) and Dudley (1978). This more general viewpoint of empirical processes has proven very useful in statistics. We could mention M-estimation: if the parameter of interest is $\text{argmax}_{\theta\in\Theta} P f_\theta$, then it makes sense to define $\hat\theta_n$ as $\text{argmax}_{\theta\in\Theta} P_n f_\theta$, so this estimator is implicitly based on the empirical process indexed by the class $\{f_\theta : \theta \in \Theta\}$. Or we can use the functional delta method, where functions of the empirical process are expanded about the function $\{Pf : f \in \mathcal{F}\}$, and the linear term is then the empirical process. Empirical processes are also pervasive in statistical learning theory. Or, closer to the subject of this book, very often linear function estimators are empirical processes. For instance, given a probability kernel K (e.g., a tent or bump function with finite support integrating to 1, or the standard normal density), the usual density estimator $\hat f_{n,h}(t)$ is defined by convolution of the compressed kernel K_h with the empirical measure, namely,

$$\hat f_{n,h} = \int \frac{1}{h} K\left(\frac{t-x}{h}\right) dP_n(x) = \frac{1}{nh}\sum_{i=1}^{n} K\left(\frac{t-X_i}{h}\right), \quad t \in \mathbb{R}, \tag{3.4}$$

where X_i are independent identically distributed real-value random variables. In this case,

$$\mathcal{F} = \{h^{-1}K((t - \cdot)/h) : t \in \mathbb{R},\ h > 0\}.$$

If we assume that

$$\sup_{f \in \mathcal{F}} |f(x) - Pf| < \infty, \quad \text{for all } x \in S, \tag{3.5}$$

then the maps from $\mathcal{F}$ to $\mathbb{R}$,

$$f \mapsto f(x) - Pf, \quad x \in S,$$

are bounded functionals over $\mathcal{F}$, and therefore, so is $f \mapsto (P_n - P)(f)$. That is,

$$P_n - P \in \ell_\infty(\mathcal{F}),$$

where $\ell_\infty(\mathcal{F})$ is the space of bounded real functions on $\mathcal{F}$, a Banach space if we equip it with the supremum norm $\|\cdot\|_{\mathcal{F}}$. A large literature is available on probability in separable Banach spaces, but unfortunately, $\ell_\infty(\mathcal{F})$ is only separable when the class $\mathcal{F}$ is finite (Exercise 3.1.3), and measurability problems arise because the probability law of the process $\{(P_n - P)(f) : f \in \mathcal{F}\}$ does not extend to the Borel σ-algebra of $\ell_\infty(\mathcal{F})$ even in simple situations (Exercise 3.1.4).

If we are interested only in $\|P_n - P\|_{\mathcal{F}}$ instead of in the process per se, then we still have a measurability problem: uncountable suprema of measurable functions may not be measurable. However, there are many situations where this is actually a countable supremum, as in the case of the empirical distribution function: for example, for probability measures P on $\mathbb{R}$, because of right continuity,

$$\sup_{t \in \mathbb{R}} |(P_n - P)(-\infty, t]| = \|F_n - F\|_\infty = \sup_{t \in \mathbb{Q}} |F_n(t) - F(t)| = \sup_{t \in \mathbb{Q}} |(P_n - P)(-\infty, t]|,$$

as well as in the case of the kernel density estimator if we take K to be right or left continuous. If $\mathcal{F}$ is *countable* or if there exists $\mathcal{F}_0$ countable such that

$$\|P_n - P\|_{\mathcal{F}} = \|P_n - P\|_{\mathcal{F}_0} \text{ a.s.},$$

then the measurability problem for $\|P_n - P\|_{\mathcal{F}}$ disappears (here stochastic separability is relevant; see Chapter 2). There are more subtle conditions on $\mathcal{F}$ to ensure that $\|P_n - P\|_{\mathcal{F}}$ is a random variable, but for the next few sections we will simply assume that the class $\mathcal{F}$ is countable.

Part of the theory of empirical processes does not require the variables X_i to be identically distributed. Thus, we will sometimes assume that the variables X_i are coordinates in the probability space $(S^{\mathbb{N}}, \mathcal{S}^{\mathbb{N}}, \prod_{i=1}^{\infty} P_i)$, where P_i is the law of X_i for each $i \in \mathbb{N}$. Moreover, since empirical processes are sample-bounded processes if condition (3.5) holds, it will sometimes be more natural to prove some facts in the context of sample-bounded processes.

At times, the empirical process variables X_i will need to be considered jointly with other random variables, such as, for example, random signs ε_i independent among themselves and of the X_j. In these instances, we assume that all the variables are defined as coordinates in an infinite product probability space, where the probability measure is the product of the individual laws of the variables and processes involved. Then E_X will denote conditional expectation with respect to the variables X_i only, and likewise for E_ε. Similarly, if we have an explicit product of probabilities, say, $P \times Q$, E_P will denote partial integration with respect to P, and likewise for E_Q, so if Fubini's theorem applies to F, we have $EF = E_P E_Q F = E_Q E_P F = E E_P F = E E_Q F$.

This chapter addresses three main questions about the empirical process, the first two being analogous to questions considered for Gaussian processes in Chapter 2. The first question has to do with *concentration* of $\|P_n - P\|_{\mathcal{F}}$ about its mean when $\mathcal{F}$ is uniformly bounded. The relevant question is: how concentrated is the variable $\|P_n - P\|_{\mathcal{F}}$ about its mean? Can we obtain exponential inequalities for the difference $\|P_n - P\|_{\mathcal{F}} - E\|P_n - P\|_{\mathcal{F}}$ that are as good as the classical inequalities for $\sum_{i=1}^n \xi_i$, ξ_i centred and bounded, or is there a penalty to be paid for the fact that we are now simultaneously considering infinitely many sums of independent random variables instead of only one? The remarkable and surprising answer to this question is that the classical exponential inequalities do hold for empirical processes as well, assuming that the parameters (size and variance) are properly defined. This is one of the most important and powerful results from the theory of empirical processes and goes by the name of *Talagrand's inequality*. Later in this section we remind the reader of the classical exponential inequalities for real random variables as background for the corresponding section on empirical processes.

The second question is, of course, as with Gaussian processes, do good estimates for $E\|P_n - P\|_{\mathcal{F}}$ exist? After all, it is good to know, from Talagrand's inequality, that the empirical process is 'almost' constant, but it is even better to know what this constant is or at least to have good approximations for it. We will examine two main techniques that give answers to this question, both related to metric entropy and chaining, as with Gaussian processes. One of them, called *bracketing*, uses chaining in combination with truncation and Bernstein's inequality. The other one applies to *Vapnik-Červonenkis classes of functions*, which are classes of functions that admit relatively small bounds for their $L^p(Q)$ covering numbers *uniformly in* $Q \in \mathcal{P}_d(S)$, the set of discrete probability measures on S, and consists only of randomisation by independent random signs (independent variables ε_i taking the values ± 1 with probability 1/2 each), together with conditional use of the sub-Gaussian metric entropy bound from Chapter 2. Concretely, one proves that moments (as well as tail probabilities) of $\|P_n - P\|_{\mathcal{F}}$ are comparable to the same quantities for $\left\|n^{-1}\sum_{i=1}^n \varepsilon_i f(X_i)\right\|_{\mathcal{F}}$ and then applies the sub-Gaussian metric entropy bound because this random variable is sub-Gaussian for every set of fixed values of the variables X_i. To carry out this program, it will be convenient to have randomisation inequalities comparing the supremum norms of the empirical process and its randomised counterpart, and we develop such inequalities in this section. We will also consider in this section the by now classical Lévy and Hoffmann-Jørgensen inequalities, respectively, for maxima of partial sums and for higher moments of the empirical process.

Finally, the last question about the empirical process refers to limit theorems, mainly the law of large numbers and the central limit theorem, in fact, the analogues of the classical Glivenko-Cantelli and Donsker theorems for the empirical distribution function. Formulation of the central limit theorem will require some more measurability because we will be considering convergence in law of random elements in not necessarily separable Banach spaces. Other limiting results for empirical processes, such as the law of the iterated logarithm and large deviations, will not be considered.

In the rest of this section we prove, as announced, several useful inequalities, namely, (1) exponential inequalities for sums of independent centred real random variables and the associated maximal inequalities, (2) Lévy's 'reflection principle' for sums of independent symmetric random processes (Lévy's inequality), which is useful when proving convergence a.s. but not only for this, and Hoffmann-Jørgensen's inequality, which allows one to obtain bounds for moments of sums of independent symmetric or centred processes from bounds on moments of lower order, and (3) several randomisation/symmetrisation inequalities, with the aim of comparing tail probabilities and moments for the supremum of a sum of independent centred random processes and the supremum of the sum of the same processes each multiplied by 1 or -1 randomly and independently (Rademacher randomisation); randomisation with other multipliers such as normal variables is also briefly considered.

3.1.2 Exponential and Maximal Inequalities for Sums of Independent Centred and Bounded Real Random Variables

Let $\xi_{n,i}$, $i \in \mathbb{N}$, be independent centred real random variables. If their sum $S_n = \sum_{i=1}^n \xi_{n,i}$ converges in law, the limiting distribution is the convolution of a normal and a (generalised) compound Poisson probability law. Therefore, it is somewhat reasonable to expect that the tail probabilities of S_n are in general similar to those of Poisson random variables and even, under the right circumstances, of a normal variable. At any rate, we should not be generally content with Chebyshev's inequality, which, even in the most favorable case of bounded and centred random variables, bounds the tail probabilities of S_n by

$$\Pr\{|S_n| \geq t\} \leq \frac{\sum_{i=1}^n E\xi_{n,i}^2}{t^2}, \; t > 0,$$

an inverse polynomial function in t instead of a negative exponential.

Typically one constructs exponential inequalities for S_n by estimating its moment-generating function, applying Markov's inequality to $e^{\lambda S_n}$ using the estimate obtained and then minimising in λ. There are two main such types of inequalities when the variables $\xi_{n,i}$ are bounded, which is mostly the case we consider: one that takes only the range of the variable into consideration but not its variance, which produces good results when the range of each variable is essentially between minus and plus its standard deviation, and the other that also takes the variance into account irrespective of its relationship to the range of the variable. Hoeffding's inequality belongs to the first type and the inequalities of Bernstein, Prokhorov and Bennett to the second. Not only will we use them repeatedly, but they also set the bar for what to expect regarding tail probability inequalities for empirical processes, that is, for collections of sums of independent random variables. We next derive all of them for the reader's convenience. We also derive a simple yet very useful bound on the expected value of the maximum of several random variables whose moment-generating functions are well behaved.

We begin with Hoeffding's inequality, which is based on the following lemma on moment-generating functions of bounded variables:

Lemma 3.1.1 *Let X be a centred random variable taking values in $[a,b]$ for some $-\infty < a < 0 \leq b < \infty$. Then, for all $\lambda > 0$, setting $L(\lambda) := \log Ee^{\lambda X}$, we have*

$$L(0) = L'(0) = 0, \quad L''(\lambda) \leq (b-a)^2/4 \tag{3.6}$$

and hence

$$Ee^{\lambda X} \le e^{\lambda^2(b-a)^2/8}. \tag{3.7}$$

Proof $L(0) = \log 1 = 0$, $L'(\lambda) = E(Xe^{\lambda X})/Ee^{\lambda X}$ so that $L'(0) = EX = 0$, and

$$\begin{aligned} L''(\lambda) &= \frac{E\left(X^2 e^{\lambda X}\right) Ee^{\lambda X} - \left(E\left(Xe^{\lambda X}\right)\right)^2}{\left(Ee^{\lambda X}\right)^2} \\ &= E\left(X^2 e^{\lambda X}\right)/Ee^{\lambda X} - \left(E\left(Xe^{\lambda X}\right)/Ee^{\lambda X}\right)^2 \\ &= \int_a^b x^2 dQ(x) - \left(\int_a^b x dQ(x)\right)^2, \end{aligned}$$

where $dQ(x) = e^{\lambda x} d\mathcal{L}(X)(x)/Ee^{\lambda X}$ is a probability measure with support contained in $[a,b]$. If a random variable Z takes values in $[a,b]$, then $Z(\omega)$ differs from the midpoint of $[a,b]$ by at most $(b-a)/2$ for all ω, and therefore, its variance is bounded by $\inf_c E(Z-c)^2 \le (b-a)^2/4$. We thus conclude that $L''(\lambda) \le (b-a)^2/4$ and that (3.6) is proved. Then, by Taylor expansion, $L(\lambda) \le \sup_{0\le\eta\le\lambda} |L''(\eta)|\lambda^2/2 \le \lambda^2(b-a)^2/8$, proving (3.7) ∎

Theorem 3.1.2 *Let X_i be independent centred random variables taking values, respectively, in $[a_i, b_i]$ for some $-\infty < a_i < 0 \le b_i < \infty$, $i = 1, \ldots, n$, for any $n \in \mathbb{N}$, and let $S_n = \sum_{i=1}^n X_i$. Then, for all $\lambda > 0$,*

$$Ee^{\lambda S_n} \le e^{\lambda^2 \sum_{i=1}^n (b_i - a_i)^2/8}, \tag{3.8}$$

and for all $t \ge 0$,

$$\Pr\{S_n \ge t\} \le \exp\left(-\frac{2t^2}{\sum_{i=1}^n (b_i-a_i)^2}\right), \quad \Pr\{S_n \le -t\} \le \exp\left(-\frac{2t^2}{\sum_{i=1}^n (b_i-a_i)^2}\right). \tag{3.9}$$

Proof By the preceding lemma and independence,

$$Ee^{\lambda S_n} = \prod_{i=1}^n Ee^{\lambda X_i} \le e^{\lambda^2 \sum_{i=1}^n (b_i-a_i)^2/8},$$

which is (3.8). We then have, by Markov's inequality,

$$\Pr\{S_n \ge t\} = \Pr\left\{e^{\lambda S_n} \ge e^{\lambda t}\right\} \le Ee^{\lambda S_n}/e^{\lambda t} \le \exp\left(\lambda^2 \sum_{i=1}^n (b_i-a_i)^2/8 - \lambda t\right).$$

This bound is smallest for $\lambda = 4t/\sum_{i=1}^n (b_i-a_i)^2$, which gives the first inequality in (3.9). The second inequality follows by applying the first to $-X_i$. ∎

Example 3.1.3 Let ε_i, $1 \le i \le n$, be a Rademacher sequence, that is, a sequence of independent identically distributed random variables such that $\Pr\{\varepsilon_i = 1\} = \Pr\{\varepsilon_i = -1\} = 1/2$. Then Hoeffding's inequality gives

$$\Pr\left\{\sum_{i=1}^n \varepsilon_i \ge t\right\} \le e^{-t^2/2n}, \quad t \ge 0,$$

and similarly for the lower tail. Note that this bound is very close to the $N(0,n)$ tail and that $ES_n^2 = n$, so this bound is what we would expect based on the central limit theorem for $S_n/\sqrt{n}$.

Example 3.1.4 (Varshamov-Gilbert bound) If Z is binomial with $p = 1/2$, then $Z - EZ = \sum_{i=1}^n \eta_i$, with $\eta_i = \pm 1/2$ and symmetric, so $\sum_{i=1}^n (b_i - a_i)^2 = n$, and Hoeffding's inequality yields the upper bound $e^{-2t^2/n}$ for both $\Pr\{Z - EZ \geq t\}$ and $\Pr\{Z - EZ \leq -t\}$. Here is a nice combinatorial application that will be used in a later chapter. Let $\Omega = \{-1, 1\}^n$, for $n \geq 8$. Write $\omega = (\omega_1, \ldots, \omega_n)$ for points in Ω, and define on Ω the distance $\rho(\omega, \omega') = \sum_{i=1}^n I_{\omega_i \neq \omega_i'}$ (the *Hamming distance*). It is of interest to find lower bounds on the largest number of αn-separated (for ρ) points that Ω can support, for some $0 < \alpha < 1$. It turns out that a natural construction can be combined with the preceeding left tail inequality for $Z - EZ$ to obtain the following result: *there exists in* $\Omega = \{-1, 1\}^n$ *an* $(n/8)$*-separated set for the Hamming distance whose cardinality is larger than* $3^{n/4}$. For the proof, we let $\overline{B}_\rho(\omega, r)$ denote the closed ball of radius $r \in \mathbb{N} \cup \{0\}$ about ω in Hamming distance and observe that, by definition of this distance, the cardinality of each such ball is $\sum_{i=0}^r \binom{n}{i}$. To construct a separated set, let $r = \lfloor n/8 \rfloor$, let $F_0 = \emptyset$ and take $\omega^{(0)} \in F_0^c = \Omega$; then define $F_1 = \overline{B}_\rho(\omega^{(0)}, r)$, and take $\omega^{(1)} \in F_1^c$; recursively, define $F_j = \cup_{i=0}^{j-1} \overline{B}_\rho(\omega^{(i)}, r)$ and take $\omega^{(j)} \in F_j^c$ unless $F_j = \Omega$. Let m be the smallest number for which $F_m = \Omega$. Obviously, the points $\omega^{(i)}$, $0 \leq i < m$, are r-separated: $\rho(\omega^{(i)}, \omega^{(j)}) \geq r + 1 \geq n/8$. Moreover, $\Omega = \cup_{j=0}^{m-1} (F_j^c \cap \overline{B}_\rho(\omega^{(j)}, r))$, and this is a union of m disjoint sets each contained in a ball of radius r. Thus, if we denote their cardinalities, respectively, by $n_j, j = 0, \ldots, m-1$, we have

$$2^n = \sum_{j=0}^{m-1} n_j \leq m \sum_{i=0}^{r} \binom{n}{i}$$

or $m \geq 1/\Pr\{Z \leq r\}$. Now

$$\Pr\{Z \leq r\} \leq \Pr\{Z - n/2 \leq -3n/8\} \leq \exp\left(-2(3n/8)^2/n\right) < 3^{-n/4},$$

which gives the result.

Note that if $b_i - a_i$ is much larger than the standard deviation σ_i of X_i then, although the tail probabilities prescribed by Hoeffding's inequality for S_n are of the normal type, they correspond to normal variables with the 'wrong' variance. Thus, perhaps we should experiment with bounds on $Ee^{\lambda X_i}$ that are closer to the moment-generating function of Poisson random variables. Let X be Poisson with parameter a (i.e., $a = EX = \mathrm{Var}(X)$). Then

$$Ee^{\lambda(X-a)} = e^{-a(\lambda+1)} \sum_{k=0}^{\infty} e^{\lambda k} a^k / k = e^{a(e^\lambda - 1 - \lambda)}, \quad \lambda \in \mathbb{R}. \tag{3.10}$$

Here is an estimate of this kind for the moment-generating function of a sum of bounded centred variables. In fact, the exponential bounds that depend on both the bound and the variance of the variables are all based on it.

Theorem 3.1.5 *Let X be a centred random variable such that $|X| \leq c$ a.s., for some $c < \infty$, and $EX^2 = \sigma^2$. Then*

$$Ee^{\lambda X} \leq \exp\left(\frac{\sigma^2}{c^2}(e^{\lambda c} - 1 - \lambda c)\right), \tag{3.11}$$

for all $\lambda > 0$. *As a consequence, if* X_i, $1 \le i \le n < \infty$, *are centred, independent and a.s. bounded by* $c < \infty$ *in absolute value, then setting*

$$\sigma^2 = \frac{1}{n}\sum_{i=1}^{n} EX_i^2 \tag{3.12}$$

and $S_n = \sum_{i=1}^n X_i$, *we have*

$$Ee^{\lambda S_n} \le \exp\left(\frac{n\sigma^2}{c^2}(e^{\lambda c} - 1 - \lambda c)\right), \tag{3.13}$$

for all $\lambda > 0$, *and the same inequality holds for* $-S_n$.

Proof Since $EX = 0$, expansion of the exponential gives

$$Ee^{\lambda X} = 1 + \sum_{k=2}^{\infty} \frac{\lambda^k EX^k}{k} \le \exp\left(\sum_{k=2}^{\infty} \frac{\lambda^k EX^k}{k}\right),$$

whereas, since $|EX^k| \le c^{k-2}\sigma^2$, for all $k \ge 2$, this exponent can be bounded by

$$\left|\sum_{k=2}^{\infty} \frac{\lambda^k EX^k}{k}\right| \le \lambda^2\sigma^2 \sum_{k=2}^{\infty} \frac{(\lambda c)^{k-2}}{k} = \frac{\sigma^2}{c^2}\sum_{k=2}^{\infty}\frac{(\lambda c)^k}{k} = \frac{\sigma^2}{c^2}(e^{\lambda c} - 1 - \lambda c).$$

This gives inequality (3.11). Inequality (3.13) follows from (3.11) by independence. The foregoing also applies to $Y_i = -X_i$. ■

Note that by (3.10), the bound (3.13) for the moment-generating function of a sum of independent bounded and centred random variables bounded by 1 and with variance $n\sigma^2$ is in fact the moment-generating function of a centred Poisson random variable with variance $n\sigma^2$.

It is standard procedure to derive tail probability bounds for a random variable Z based on a bound for its Laplace transform. We will obtain four such bounds, three of them giving rise, respectively, to the Bennett, Prokhorov and Bernstein classical inequalities for sums of independent random variables and one where the bound on the tail probability function is inverted. It is convenient to introduce two new functions

$$\phi(x) = e^{-x} - 1 + x, \quad \text{for } x \in \mathbb{R}, \quad \text{and} \quad h_1(x) = (1+x)\log(1+x) - x, \quad \text{for } x \ge 0. \tag{3.14}$$

Proposition 3.1.6 *Let* Z *be a random variable whose moment-generating function satisfies the bound*

$$Ee^{\lambda Z} \le \exp\big(v(e^{\lambda} - 1 - \lambda)\big), \quad \lambda > 0, \tag{3.15}$$

for some $v > 0$. *Then, for all* $t \ge 0$,

$$\Pr\{Z \ge t\} \le \exp(-vh_1(t/v)) \le \exp\left(-\frac{3t}{4}\log\left(1 + \frac{2t}{3v}\right)\right) \le \exp\left(-\frac{t^2}{2v + 2t/3}\right) \tag{3.16}$$

and

$$\Pr\left\{Z \ge \sqrt{2vx} + x/3\right\} \le e^{-x}, \quad x \ge 0. \tag{3.17}$$

Proof The bound (3.15) together with the exponential form of Markov's inequality gives

$$\Pr\{Z \geq t\} = \inf_{\lambda>0} \Pr\left\{e^{\lambda Z} \geq e^{\lambda t}\right\} \leq e^{v \inf_{\lambda>0}[\phi(-\lambda)-\lambda t/v]}.$$

It is a routine calculus computation to check that the infimum of the function $y = \phi(-\lambda) - \lambda z = e^{\lambda} - 1 - \lambda(1+z)$ (any $z > -1$) is attained at $\lambda = \log(1+z)$ and that its value is

$$\inf_{\lambda\in\mathbb{R}}\{\phi(-\lambda) - \lambda z\} = z - (1+z)\log(1+z) := -h_1(z). \tag{3.18}$$

This proves the first inequality in (3.16). It is equally routine to verify, by checking the value of the corresponding functions at $t = 0$ and then comparing derivatives, that

$$h_1(t) \geq \frac{3t}{4}\log\left(1+\frac{2t}{3}\right) \geq \frac{t^2}{2+2t/3}, \quad t > 0, \tag{3.19}$$

thus completing the proof of the three inequalities in (3.16).

To prove (3.17), we begin by observing that as can be seen by Taylor development, $(1-\lambda/3)(e^{\lambda} - \lambda - 1) \leq \lambda^2/2$, $\lambda \geq 0$. Thus, if

$$\varphi(\lambda) := \frac{v\lambda^2}{2(1-\lambda/3)}, \quad \lambda \in [0,3),$$

then inequality (3.15) yields, again by the exponential Markov's inequality,

$$\Pr\{Z \geq t\} \leq \exp\left[\inf_{0\leq\lambda<3}(\varphi(\lambda) - \lambda t)\right] = \exp\left[-\sup_{0\leq\lambda<3}(\lambda t - \varphi(\lambda))\right]. \tag{3.20}$$

Consider the function $\gamma(s) = \sup_{\lambda\in[0,3)}(\lambda s - \varphi(\lambda))$. Then, since φ and φ' are strictly increasing, $\varphi(0) = 0$ and $\varphi(x) \nearrow \infty$ as $x \nearrow 3$, it follows that the maximum is attained at the point λ_0 where the tangent line to the graph of $y = \varphi(\lambda)$ at $(\lambda_0, \varphi(\lambda_0))$ has slope s. Or, what is the same, $\gamma(s) = x$ if (and only if) the tangent line to the curve $y = \varphi(\lambda)$ through the point $(0, -x)$ has slope s. The slope of this tangent is precisely $(x + \varphi(\lambda_0))/\lambda_0$. Now any other straight line through $(0, -x)$ and $(\lambda, \varphi(\lambda))$, $0 < \lambda < 3$, has a larger slope $(x + \varphi(\lambda))/\lambda > (x + \varphi(\lambda_0))/\lambda_0$. It follows that

$$\gamma^{-1}(x) = \inf_{0<\lambda<3}\frac{\varphi(\lambda)+x}{\lambda} = \inf_{0<\lambda<3}\left(\frac{v\lambda}{2(1-\lambda/3)} + \frac{x}{\lambda}\right).$$

With the change of variables $u = (1-\lambda/3)/(v\lambda)$, we have $\gamma^{-1}(x) = \inf_{0<u<\infty}(1/2u + x(uv + 1/3))$, and the last function attains its infimum at $u = 1/\sqrt{2vx}$. This gives $\gamma^{-1}(x) = \sqrt{2vx} + x/3$. This together with inequality (3.20) yields (3.17). ■

The last two lines in the proof of Proposition 3.1.6 show that $\inf_{0<\lambda<3}(\varphi(\lambda)+x)/\lambda = \sqrt{2vx} + x/3$, and replacing 1/3 by any positive constant c and making the change of variables $u = (1-cx)/\lambda v$, it is equally easy to obtain

$$\inf_{0<\lambda<1/c}\frac{\varphi_{v,c}(\lambda)+x}{\lambda} = \sqrt{2vx} + cx, \quad \text{where} \quad \varphi_{v,c}(\lambda) = \frac{v\lambda^2}{2(1-c\lambda)}, \quad 0 < \lambda < 1/c, \tag{3.21}$$

which we record for further use. It is also worth noting that again from the proof of the preceding proposition,

$$v(e^{\lambda} - \lambda - 1) \leq \varphi_{v,1/3}(\lambda), \quad 0 < \lambda < 3. \tag{3.22}$$

Thus, $\varphi_{v,1/3}(\lambda)$ is a bound for the logarithm of the moment-generating function of sums of independent centred and bounded random variables; in fact, $\varphi_{v,c}(\lambda)$ for convenient parameters v, c are also bounds for the logarithm of the moment-generating function of random variables with the centred exponential and gamma distributions (see Exercise 3.1.1).

Now note that if S_n is as in Theorem 3.1.5, then $Z = S_n/c$ satisfies the hypothesis of Proposition 3.1.6 with $v = n\sigma^2/c^2$, where σ^2 is defined by (3.12). Thus we have the following exponential inequalities, which go by the names of *Bennett's, Prohorov's and Bernstein's* (in this order):

Theorem 3.1.7 (Inequalities of Bennett, Prokhorov and Bernstein) *Let X_i, $1 \le i \le n$, be independent centred random variables a.s. bounded by $c < \infty$ in absolute value. Set $\sigma^2 = (1/n)\sum_{i=1}^n EX_i^2$ and $S_n = \sum_{i=1}^n X_i$. Then, for all $u \ge 0$,*

$$\begin{aligned}\Pr\{S_n \ge u\} &\le \exp\left(-\frac{n\sigma^2}{c^2}h_1\left(\frac{uc}{n\sigma^2}\right)\right)\\ &\le \exp\left(-\frac{3u}{4c}\log\left(1+\frac{2uc}{3n\sigma^2}\right)\right)\\ &\le \exp\left(-\frac{u^2}{2n\sigma^2+2cu/3}\right) \end{aligned}\tag{3.23}$$

and

$$\Pr\left\{S_n \ge \sqrt{2n\sigma^2 u}+\frac{cu}{3}\right\} \le e^{-u}, \tag{3.24}$$

where h_1 is as defined in (3.14), and the same inequalities hold for $\Pr\{S_n < -u\}$.

Bennett's inequality is the sharpest, but Prokhorov's and Bernstein's inequalities are easier to interpret. Prokhorov's inequality exhibits two regimes for the tail probabilities of S_n: if $uc/n\sigma^2$ is small, then the logarithm is approximately $2uc/3n\sigma^2$, and the tail probability is only slightly larger than $e^{-u^2/2n\sigma^2}$, Gaussian-like, whereas if $uc/n\sigma^2$ is not small or moderate, then the exponent for the tail probability is of the order of $-(3u/4c)\log(2cu/3n\sigma^2)$, Poisson-like. Bernstein's inequality keeps the Gaussian-like regime for small values of $uc/n\sigma^2$ but replaces the Poisson regime by the larger, hence less precise, exponential regime. Since the limit distributions of sums of independent random variables $\sum_{i=1}^n X_{i,n}$ satisfying $\max_{i\le n}\Pr\{|X_{i,n}| > \delta\} \to 0$ for all $\delta > 0$ are convolutions of normal and generalised Poisson random variables by the central limit theorem, the order of the Bennet and Prokhorov bounds cannot be improved without imposing extra conditions on the random variables, and the order of the Bernstein bound is only off by a logarithm for larger values of u.

Proposition 3.1.6 together with (3.10) also yields the typical 'Poisson tail' for Poisson random variables (which may be obtained directly): if X is Poisson with parameter a, then

$$\Pr\{X-a \ge t\} \le \exp\left\{-\frac{3t}{4}\log\left(1+\frac{2t}{3a}\right)\right\}, \quad t \ge 0.$$

It should be noted that except for constants, this bound is two sided (see Exercise 3.1.2), and this is relevant to the comment in the preceding paragraph on the optimality of Prohorov's inequality. Exercise 3.1.2 also contains an upper bound for the 'left' tail of a Poisson variable.

It is natural to ask whether Theorem 3.1.7 extends to unbounded random variables. In fact, Bernstein's inequality does hold for random variables X_i with finite exponential moments, that is, such that $Ee^{\lambda|X_i|} < \infty$, for some $\lambda > 0$. The inequality is usually stated under a condition on the growth of moments which is equivalent to exponential integrability.

Proposition 3.1.8 (Bernstein's inequality) *Let X_i, $1 \le i \le n$, be centred independent random variables such that, for all $k \ge 2$ and all $1 \le i \le n$,*

$$E|X_i|^k \le \frac{k}{2}\sigma_i^2 c^{k-2}, \tag{3.25}$$

and set $\sigma^2 = \sum_{i=1}^n \sigma_i^2$, $S_n = \sum_{i=1}^n X_i$. Then

$$\Pr\{S_n \ge t\} \le \exp\left(-\frac{t^2}{2\sigma^2 + 2ct}\right), \quad t \ge 0. \tag{3.26}$$

Proof Assuming that $c|\lambda| < 1$, the moment-growth hypothesis implies that, for $1 \le i \le n$,

$$Ee^{\lambda X_i} \le 1 + \frac{\sigma_i^2}{2}\sum_{k=2}^{\infty} |\lambda|^k c^{k-2} = 1 + \frac{\lambda^2 \sigma_i^2}{2(1-|\lambda|c)} \le e^{\lambda^2\sigma_i^2/(2-2c|\lambda|)},$$

which, by independence and the exponential Markov inequality, implies that

$$\Pr\{S_n \ge t\} \le \frac{Ee^{\lambda S_n}}{e^{\lambda t}} \le \exp\left(\frac{\lambda^2\sigma^2}{2-2c|\lambda|} - \lambda t\right).$$

The result obtains by taking $\lambda = t/(\sigma^2 + ct)$. ■

It is worth noting that the inequalities of Hoeffding, Bennet, Bernstein and Prohorov also hold for the maximum of the partial-sums $\max_{k \le n} S_k$ by virtue of Doob's submartingale inequality (see Exercise 3.1.10).

Another important class of random variables that have a behaviour similar to exponential variables is the class of quadratic forms in independent normal variables, or Gaussian chaoses of order 2. The corresponding exponential inequality goes by the name of *Hanson-Wright's inequality*, and we prove it now in a version sharper than the original, as well as a related result on concentration of centred χ^2 random variables. Given a symmetric matrix A with eigenvalues λ_i, its *Hilbert-Schmidt norm* $\|A\|_{HS}$ is defined as $\|A\|_{HS}^2 = \sum \lambda_i^2$, and we will denote by $\|A\|$ the maximum of its eigenvalues (the maximum of the absolute values of its eigenvalues is the operator norm, which does dominate $\|A\|$). Note that if the trace of A is zero, then $\|A\| > 0$ because at least one eigenvalue of A must be positive. Recall the definition of the function $\varphi_{v,c}$ from (3.21).

Theorem 3.1.9 *Let $A = (a_{ij})_{i,j=1}^n$ be a symmetric matrix with all its diagonal terms a_{ii} equal to zero, let g_i, $i = 1, \ldots, n$, be independent standard normal variables and set*

$$X = \sum_{i,j} a_{ij} g_i g_j = 2\sum_{i<j} a_{ij} g_i g_j.$$

Alternatively, let A be a diagonal matrix with eigenvalues τ_i, and set

$$X = \sum_i \tau_i (g_i^2 - 1),$$

g_i independent $N(0,1)$, as earlier. Then both random variables satisfy

$$Ee^{\lambda X} \le e^{\|A\|_{HS}^2\lambda^2/(1-2\lambda\|A\|)} = e^{\varphi_{2\|A\|_{HS}^2, 2\|A\|}(\lambda)}, \quad \text{for } 0 < \lambda < 1/2\|A\|. \tag{3.27}$$

Consequently, for $t \ge 0$,

$$\Pr\{X > t\} \le e^{-t^2/4(\|A\|_{HS}^2 + \|A\|t)} \quad \textit{or} \quad \Pr\left\{X \ge \sqrt{4\|A\|_{HS}^2 t} + 2\|A\|t\right\} \le e^{-t}, \tag{3.28}$$

and the same inequalities hold for $-X$.

Proof If τ_i are the eigenvalues of $A = (a_{ij})$, an ortho-normal change of coordinates yields $X = \sum_{i=1}^n \tau_i \tilde{g}_i^2$, where the variables $\tilde{g}_i$ are also independent standard normal. Then, since $\sum \tau_i = 0$, we have $X = \sum_{i=1}^n \tau_i(\tilde{g}_i^2 - 1)$, showing that the first case reduces to the second. Now, for $t < 1/2$ and g standard normal,

$$Ee^{t(g^2-1)} = \frac{1}{\sqrt{2\pi}} \int_{\mathbb{R}} e^{t(x^2-1)-x^2/2} dx = e^{-t}/\sqrt{1-2t} = e^{1/2[-\log(1-2t)-2t]}.$$

By Taylor development, valid for $|t| < 1/2$,

$$\frac{1}{2}[-\log(1-2t) - 2t] = t^2\left(1 + \frac{2}{3}2t + \cdots + \frac{2}{k+2}(2t)^k + \cdots\right) \le \frac{t^2}{1-2t}.$$

Hence,

$$\log Ee^{\lambda X} \le \sum \frac{\tau_i^2\lambda^2}{1 - 2\tau_i\lambda} \le \frac{\lambda^2 \sum \tau_i^2}{1 - 2\lambda \max_i \tau_i} = \frac{2\lambda^2 \sum \tau_i^2}{2(1 - 2\lambda \max_i \tau_i)},$$

for $0 \le \lambda < 1/2\|A\|$, which gives (3.27). Now, inequality (3.21) gives the second inequality in (3.28) if one proceeds as in the derivation of (3.17) from (3.20). If we use the exponential Markov inequality in conjunction with (3.27), we obtain $\Pr\{X \ge t\} \le \exp[\lambda^2 v/2(1-c\lambda) - \lambda t]$, for $v = 2\|A\|_{HS}^2$ and $c = 2\|A\|$, and then the first inequality in (3.28) just follows by taking $\lambda = t/(v + ct)$. This proof applies as well to $-X$, which, in particular, shows that the bounds in (3.28) also hold for the lower tails of X. ■

For instance, in the case of centred chi-squared random variables $\sum_{i=1}^n (g_i^2 - 1)$, where $\|A\|_{HS}^2 = n$ and $\|A\| = 1$, Theorem 3.1.9 yields

$$\Pr\left\{\left|\sum_{i=1}^n (g_i^2 - 1)\right| \ge t\right\} \le 2e^{-t^2/4(n+t)} \quad \text{and} \quad \Pr\left\{\left|\sum_{i=1}^n (g_i^2 - 1)\right| \ge 2\left(\sqrt{nt} + t\right)\right\} \le 2e^{-t}, \tag{3.29}$$

where the factor 2 in front of e can be removed in the one-sided versions of these inequalities.

In general, if g_i, $i \le n$, are independent standard normal variables, then the first case in the preceding theorem covers all linear combinations of the products $g_i g_j$, $1 \le i \ne j \le n$ (since $\sum_{i<j} a_{ij} g_i g_j = 2^{-1} \sum_{i,j} a_{ij} g_i g_j$ if we define $a_{ji} = a_{ij}$ for $j < i$ and $a_{ii} = 0$), and the two cases of the theorem, in combination with the triangle inequality for probabilities, also produce exponential bounds for centred quadratic Gaussian homogeneous polynomials $\sum_{i\le j} a_{ij} g_i g_j - \sum_i a_{ii}$, with a_{ij}, $i \le j$, arbitrary.

Finally, we see that control of the moment-generating function of a collection of random variables translates into control of the expected value of their maximum: we have already seen an instance of this in the section on entropy bounds for the supremum of a Gaussian process (Section 2.3). The starting point is inequality (2.33) in that section: if $\Phi(x)$ is convex, non-negative and nondecreasing, then

$$\Phi\left(E\max_{i\le N} X_i\right) \le N \max_{i\le N} E\Phi(X_i), \tag{3.30}$$

which, in particular, holds for $\Phi(x) = e^{\lambda x}$.

Theorem 3.1.10 *(a) Let X_i, $i = 1,\dots,N$, be random variables such that $Ee^{\lambda X_i} \le e^{\lambda^2\sigma_i^2/2}$, for $0 \le \sigma_i < \infty$ for all $\lambda > 0$ and $i \le N$. Then*

$$E\max_{i\le N} X_i \le \sqrt{2\log N}\max_i \sigma_i. \tag{3.31}$$

(b) Let X_i be random variables such that $Ee^{\lambda X_i} \le e^{\varphi_{v_i,c}(\lambda)}$, for $0 < \lambda \le 1/c$ and $i = 1,\dots,N$, where $v_i, c > 0$ and $\varphi_{v,c}$ is defined in (3.21). In particular, by (3.22), this holds with $c = 1/3$ if $Ee^{\lambda X_i} \le \exp(v_i(e^\lambda - 1 - \lambda))$. Then

$$E\max_{i\le N} X_i \le \sqrt{2v\log N} + c\log N, \tag{3.32}$$

where $v = \max_{i\le N} v_i$.

Proof For part (a), see Lemma 2.3.4. For X_i as in part (b), note that with $\Phi(x) = e^{\lambda x}$, we have that $\max_{i\le N} E\Phi(X_i) \le e^{\varphi_{v,c}(\lambda)}$, so inequality (3.30) gives, by inverting Φ (i.e., taking logarithms and dividing by λ),

$$E\max_{i\le N} X_i \le \Phi^{-1}\left(Ne^{\varphi_{v,c}(\lambda)}\right) = \frac{\log N + \varphi_{v,c}(\lambda)}{\lambda}, \quad 0 < \lambda < 1/c.$$

Now the inequality in part (b) follows from (3.21). ■

3.1.3 The Lévy and Hoffmann-Jørgensen Inequalities

We now switch to sample bounded stochastic processes (see Chapter 2). In order to avoid too many measurability considerations so that we can concentrate on the purely probabilistic arguments, we will assume the index set to be *countable*. Let T be a countable set, and let $\ell_\infty(T)$ be the set of real bounded functions defined on T. Note that $\ell_\infty(T)$, with the supremum norm

$$\|x\|_T = \sup_{t\in T}|x(t)|,$$

is a Banach space, and this Banach space is separable if and only if T is finite (Exercise 3.1.3). A stochastic process with index set T, $X(t)$, $t \in T$, with bounded sample paths (or with almost all its sample paths bounded, i.e., sample bounded) is a random element of $\ell_\infty(T)$ but not necessarily an $\ell_\infty(T)$-valued random variable (i.e., it may not be Borel measurable) (see Exercise 3.1.4). However, if T is countable, then $\|X\|_T := \sup_{t\in T}|X(t)|$ is measurable because this is the supremum of a countable set of real random variables (i.e., real measurable functions). Similar observations have already been made earlier.

Let B be a separable Banach space. As shown in Example 2.1.6, there exists a countable subset D of the unit ball of the dual B^* such that $\|x\| = \sup_{f\in D}|f(x)|$, for all $x \in B$. Hence, if B

is separable, B-valued random variables are sample bounded stochastic processes indexed by a countable set D. As in most of Chapter 2, we will continue using the language of processes, and it then should be clear that any statement for sample bounded processes over countable sets also will be true for B-valued random variables, B separable, or more generally, B such that there exists a subset D of the unit ball of its dual space B^* such that if $x \in B$ and $\|x\|$ is its B-norm, then $\|x\| = \sup_{f\in D} f(x)$.

We will take independence of stochastic processes Y_i, $i = 1,\dots,n \le \infty$, to mean, without further mention, that these processes are defined on a product probability space, and each depends on the corresponding coordinate:

$$Y_i : \left(\prod_i \Omega_i, \otimes_i \Sigma_i, \Pr = \prod_i P_i\right) \mapsto (\ell_\infty(T), \mathcal{C}),$$

where $Y_i(\omega) = Y_i(\omega_i)$, with $\omega = (\omega_1,\dots,\omega_i,\dots)$. Here $\mathcal{C}$ is the cylindrical σ-algebra of $\ell_\infty(T)$, which, T being countable, contains the closed and open balls. Then, for instance, E_{Y_i} will denote integration with respect to P_i only, that is, conditional expectation with respect to $\{Y_j : j \neq i\}$.

Let Y_i be independent *symmetric* stochastic processes indexed by T. Recall that Y is symmetric if $\Pr\{Y \in A\} = \Pr\{-Y \in A\}$ for all A in the cylindrical σ-algebra. The first theorem in this subsection is *Lévy's inequality*, which is a sort of reflection principle for the partial-sum process $k \mapsto \sum_{i=1}^k Y_i$ and is quite useful, for example to derive a.s. convergence from convergence in probability. It will also be shown that although the statement is not quite true for nonsymmetrical variables, a weaker but still useful statement also holds (*Ottaviani's inequality*). Another very useful inequality to be proved here is *Hoffmann-Jørgensen's*: the L^p-norm of a sum of independent symmetric processes is dominated by the L^p-norm of the maximum of their norms plus a quantile of the sum. This is an excellent tool for the derivation of uniform integrability given tightness, hence, of convergence of moments given convergence in law or for bounding moments of $\left\|\sum Y_i\right\|$ in terms of bounds for lower moments. This inequality may be considered to be a generalisation to unbounded variables of one of the classical Kolmogorov inequalities used in the proof of the three-series theorem. We begin with Lévy's inequality.

Given a sequence of independent sample bounded processes Y_i, $i = 1,\dots,n$, indexed by T, we set

$$S_k = \sum_{i=1}^k Y_i, \; i = 1,\dots,n, \quad \text{and} \quad Y_n^* = \max_{1\le i\le n} \|Y_i\|_T.$$

Also, for conciseness, we introduce the following notation:

Notation We say that a process Y indexed by a set T is $SBC(T)$ if almost all its sample paths are bounded and the set T is countable.

Theorem 3.1.11 (Lévy's inequalities) *Let Y_i, $1 \le i \le n$, be independent symmetric $SBC(T)$ processes. Then, for every $t > 0$,*

$$\Pr\left\{\max_{1\le k\le n} \|S_k\|_T > t\right\} \le 2\Pr\left\{\|S_n\|_T > t\right\} \tag{3.33}$$

and

$$\Pr\left\{Y_n^* > t\right\} \le 2\Pr\{\|S_n\|_T > t\}. \tag{3.34}$$

In particular,

$$E\left(\max_{1\le k\le n}\|S_k\|_T\right)^p \le 2E\|S_n\|_T^p, \quad E(Y_n^*)^p \le 2E\|S_n\|_T^p,$$

for all $p > 0$.

Proof We drop the subindex T from the norms in most proofs if no confusion may arise. Consider the sets

$$A_k := \left\{\|S_i\| \le t, \text{ for } 1 \le i \le k-1, \|S_k\| > t\right\}, \quad k = 1,\dots,n,$$

which are disjoint and whose union is the event $\left\{\max_{1\le k\le n}\|S_k\| > t\right\}$. ($A_k$ is the event 'the random walk S_i leaves the ball of radius t for the first time at time k'.) For each $k \le n$, we define

$$S_n^{(k)} := S_k - Y_{k+1} - \cdots - Y_n$$

and note that, by symmetry and independence, the joint probability law of the n processes $(Y_1,\dots,Y_n)$ is the same as that of $(Y_1,\dots,Y_k,-Y_{k+1},\dots,-Y_n)$, so S_n and $S_n^{(k)}$ both have the same law. On the one hand, since A_k depends only on the first k processes, we obviously have

$$\Pr\left[A_k \cap \{\|S_n\| > t\}\right] = \Pr\left[A_k \cap \{\|S_n^{(k)}\| > t\}\right]$$

and, on the other,

$$A_k = \left[A_k \cap \{\|S_n\| > t\}\right] \cup \left[A_k \cap \{\|S_n^{(k)}\| > t\}\right],$$

since otherwise there would exist $\omega \in A_k$ such that $2\|S_k(\omega)\| = \|S_n(\omega) + S_n^{(k)}(\omega)\| \le 2t$, a contradiction with the definition of A_k. The last two identities imply that

$$\Pr(A_k) \le 2\Pr\left[A_k \cap \{\|S_n\| > t\}\right], \quad k = 1,\dots,n,$$

and therefore,

$$\Pr\left\{\max_{1\le k\le n}\|S_k\| > t\right\} = \sum_{k=1}^{n}\Pr A_k \le 2\sum_{k=1}^{n}\Pr\left[A_k \cap \{\|S_n\| > t\}\right]$$
$$\le 2\Pr\left\{\|S_n\| > t\right\},$$

which gives inequality (3.33). The second inequality is proved in the same way if we redefine A_k as

$$A_k := \{\|Y_i\| \le t, \text{ for } 1 \le i \le k-1, \|Y_k\| > t\}$$

and $S_n^{(k)}$ as

$$S_n^{(k)} := -Y_1 - \cdots - Y_{k-1} + Y_k - Y_{k+1} - \cdots - Y_n.$$

The statements about expected values follow from (3.33) and (3.34) using integration by parts ($\int |\xi|^p dP = p\int t^{p-1}\Pr\{|\xi| > t\}dt$). ■

If the random vectors are not symmetric, we have the following weaker inequality (*Lévy-Ottaviani's inequality*):

Proposition 3.1.12 *Let Y_i, $i \le n < \infty$, be independent SBC(T) processes. Then, for all $u, v > 0$,*

$$\Pr\left\{\max_{1\le k\le n} \|S_k\| > u+v\right\} \le \frac{1}{1-\max_{k\le n}\Pr\{\|S_n-S_k\|>v\}}\Pr\{\|S_n\|>u\} \tag{3.35}$$

and, for all $t \ge 0$,

$$\Pr\left\{\max_{1\le k\le n} \|S_k\|_T > t\right\} \le 3\max_{k\le n}\Pr\left\{\|S_k\|_T > \frac{t}{3}\right\}. \tag{3.36}$$

Proof Almost as in the preceding proof we define, for all $u, v \ge 0$ and $1 \le k \le n$,

$$A_k = \{\|S_i\| \le u+v, \text{ for } i<k, \text{ and } \|S_k\| > u+v\}.$$

These sets are disjoint, and their union is $\left\{\max_{1\le k\le n}\|S_k\| > u+v\right\}$. Therefore,

$$\begin{aligned}
\Pr\{\|S_n\|>u\} &\ge \Pr\{\|S_n\|>u, \max_{1\le k\le n}\|S_k\|>u+v\}\\
&\ge \sum_{k=1}^{n}\Pr\{A_k\cap\{\|S_n-S_k\|\le v\}\}\\
&= \sum_{k=1}^{n}\Pr\{A_k\}\Pr\{\|S_n-S_k\|\le v\}\\
&\ge \left[1-\max_{k\le n}\Pr\{\|S_n-S_k\|>v\}\right]\Pr\left\{\max_{1\le k\le n}\|S_k\|>u+v\right\},
\end{aligned}$$

proving (3.35). This is the typical form of the Lévy-Ottaviani inequality. Taking $u = t/3$ and $v = 2t/3$ in this inequality gives

$$\begin{aligned}
\Pr\left\{\max_{1\le k\le n}\|S_k\|>t\right\} &\le \frac{\Pr\{\|S_n\|>t/3\}}{1-\max_{k\le n}\Pr\{\|S_n-S_k\|>2t/3\}}\\
&\le \frac{\max_{k\le n}\Pr\{\|S_k\|>t/3\}}{1-2\max_{k\le n}\Pr\{\|S_k\|>t/3\}}.
\end{aligned}$$

This proves inequality (3.36) if $\max_{1\le k\le n}\Pr\{\|S_k\| > t/3\} < 1/3$. But (3.36) is trivially satisfied otherwise. ■

Next, we consider Hoffmann-Jørgensen's inequality. We emphasize that the main ingredient for Hoffmann-Jørgensen's inequality is a bound for the tail probabilities of $\|S_n\|_T$ in terms of the square of its own tail probabilities but at a smaller level and the tail probabilities of $\max_{1\le k\le n}\|Y_k\|$. We only derive the inequality for symmetric variables, and in the next subsection we will use symmetrisation to extend it to centred variables.

Lemma 3.1.13 *Let Y_i, $i \le n < \infty$, be independent symmetric SBC(T) processes, and set $S_k = \sum_{i=1}^k Y_i$, $Y_n^* = \max_{1\le k\le n}\|Y_k\|_T$. Then, for every $s, t, u > 0$,*

$$\Pr\{\|S_n\|_T > s+t+u\} \le \Pr\{Y_n^* > u\} + 4\Pr\{\|S_n\|>t\}\Pr\{\|S_n\|>s\}. \tag{3.37}$$

Proof We set $A_j = \{\|S_i\| \le t, \text{ for } 1 \le i \le j-1, \|S_j\| > t\}$ as in the proof of Lévy's inequality. Then $\|S_{j-1}\| \le t$ on A_j, and therefore, on A_j, $\|S_n\| \le t + \|Y_j\| + \|S_n - S_j\|$, which gives

$$\Pr\{A_j, \|S_n\|>s+t+u\} \le \Pr\{A_j, Y_n^*>u\} + \Pr\{A_j, \|S_n-S_j\|>s\}, \quad 1\le j\le n.$$

Then, since the sets A_j are disjoint and their union is $\{\max_{1\le k\le n}\|S_k\| > t\}$, and each A_j depends only on the first j variables, it follows, by addition followed by two applications of Lévy's inequality, that

$$\begin{aligned}\Pr\{\|S_n\| > s+t+u\} &\le \Pr\{Y_n^* > u\} + \sum_j \Pr(A_j)\Pr\left\{\|S_n - S_j\| > s\right\} \\ &\le \Pr\{Y_n^* > u\} + 2\Pr\{\max_{1\le k\le n}\|S_k\| > t\}\Pr\{\|S_n\| > s\} \\ &\le \Pr\{Y_n^* > u\} + 4\Pr\{\|S_n\| > t\}\Pr\{\|S_n\| > s\}. \quad \blacksquare\end{aligned}$$

Proposition 3.1.14 *Let Y_i, $i \le n < \infty$, be as in Lemma 3.1.13. Then, for all $t,p > 0$,*

$$E\|S_n\|_T^p \le \left[\frac{4^{1/(p+1)}t^{p/(p+1)} + \left(\|Y_n^*\|_p\right)^{p/(p+1)}}{1 - \left(4\Pr\{\|S_n\|_T > t\}\right)^{1/(p+1)}}\right]^{p+1}. \tag{3.38}$$

Proof If we take $s = \alpha v$, $t = \beta v$ and $u = \gamma v$ in inequality (3.37), multiply by pv^{p-1} and integrate with respect to dv, we obtain

$$\frac{1}{(\alpha+\beta+\gamma)^p}E\|S_n\|^p \le \frac{1}{\gamma^p}E(Y_n^*)^p + 4\int_0^\infty pv^{p-1}\Pr\{\|S_n\| > \beta v\}\Pr\{\|S_n\| > \alpha v\}dv.$$

Thus, letting $v' = \beta v$ and changing variables, we have

$$\begin{aligned}\frac{1}{(\alpha+\beta+\gamma)^p}E\|S_n\|^p &\le \frac{1}{\gamma^p}E(Y_n^*)^p + \frac{4}{\beta^p}\int_0^\infty pv^{p-1}\Pr\{\|S_n\| > v\}\Pr\{\|S_n\| > \alpha v/\beta\}dv \\ &\le \frac{1}{\gamma^p}E(Y_n^*)^p + \frac{4}{\beta^p}\int_{\frac{\beta t}{\alpha}}^\infty pv^{p-1}\Pr\{\|S_n\| > v\}\Pr\{\|S_n\| > \alpha v/\beta\}dv \\ &\quad + \frac{4}{\beta^p}\int_0^{\frac{\beta t}{\alpha}} pv^{p-1}dv \\ &\le \frac{1}{\gamma^p}E(Y_n^*)^p + \frac{1}{\beta^p}4E\|S_n\|^p\Pr\{\|S_n\| > t\} + \frac{1}{\alpha^p}4t^p.\end{aligned}$$

We now minimize the right-hand side of this inequality with respect to α, β and γ such that $\alpha+\beta+\gamma = 1$, α, β, $\gamma \ge 0$. It can be seen using Lagrange multipliers that for any a, b, c nonnegative,

$$\min_{x+y+z=1,x,y,z\ge 0}(x^{-p}a + y^{-p}b + z^{-p}c) = \left(a^{1/(p+1)} + b^{1/(p+1)} + c^{1/(p+1)}\right)^{p+1}.$$

It then follows that

$$E\|S_n\|^p \le \left[(E(Y_n^*)^p)^{1/(p+1)} + (4E\|S_n\|^p\Pr\{\|S_n\| > t\})^{1/(p+1)} + (4t^p)^{1/(p+1)}\right]^{p+1},$$

which is just inequality (3.38). $\blacksquare$

Inequality (3.38) can be put in a nicer way. In what follows, when writing L^p norms of quantities such as $\|S_n\|_T$, we will omit the subindex T even in statements of theorems (we have already been omitting them in proofs). Thus, we write, for example, $\|S_n\|_p$, for $\|\|S_n\|_T\|_p$.

Theorem 3.1.15 (Hoffmann-Jørgensen's inequality) *For each $p > 0$, if Y_i, $i \le n < \infty$, are independent symmetric $SBC(T)$ processes, and if $t_0 \ge 0$ is defined as*

$$t_0 = \inf\{t > 0 : \Pr\{\|S_n\|_T > t\} \le 1/8\},$$

then

$$\|S_n\|_p \le 2^{(p+2)/p}(p+1)^{(p+1)/p}\big[4^{1/p}t_0 + \|Y_n^*\|_p\big]. \tag{3.39}$$

Proof Since $1 - x^\alpha \ge \alpha(1-x)$, for $0 \le x \le 1$ and $0 \le \alpha \le 1$ (e.g., by convexity), and since (also by convexity) $(a+b)^{(p+1)/p} \le 2^{1/p}\big(a^{(p+1)/p} + b^{(p+1)/p}\big)$, inequality (3.38) yields

$$\|S_n\|_p \le \left[\frac{p+1}{1 - 4\Pr\{\|S_n\| > t\}}\right]^{(p+1)/p} 2^{1/p}\big[4^{1/p}t + \|Y_n^*\|_p\big].$$

Hence, by the definition of t_0,

$$\|S_n\|_p \le 2^{(p+2)/p}(p+1)^{(p+1)/p}\big[4^{1/p}t_0 + \|Y_n^*\|_p\big],$$

proving the theorem. ■

Note that by Markov's inequality, $t_0 \le 8^{1/p}\|S_n\|_p$, and that by Lévy's inequality, $\|Y_n^*\|_p \le 2^{1/p}\|S_n\|_p$; hence, Hoffmann-Jørgenssen's inequality (3.39) is two sided up to constants. Taking $t = 2 \cdot 4^{1/q}\|S_n\|_q$ in inequality (3.38), it becomes

$$E\|S_n\|^p \le \left[\frac{4^{1/(p+1)}(2 \cdot 4^{1/q} \cdot \|S_n\|_q)^{p/(p+1)} + (\|Y_n^*\|_p)^{p/(p+1)}}{1 - (1/2)^{q/(p+1)}}\right]^{p+1},$$

which, proceeding as in the preceding proof, gives

$$\|S_n\|_p \le \left[\frac{2(p+1)}{q}\right]^{(p+1)/p} 2^{1/p}\big[4^{1/(p+1)/q} \cdot 2\|S_n\|_q + \|Y_n^*\|_p\big]. \tag{3.40}$$

It is easy to check that for all $c > 0$,

$$\sup_{p,q: c < q \le p} \frac{q}{p}\left[\frac{2(p+1)}{q}\right]^{(p+1)/p} 2^{(1+1)/p}4^{1/(p+1)/q} < \infty,$$

so inequality (3.40) gives the following result about comparison of moments:

Theorem 3.1.16 *For every $c > 0$ there exists a constant $K_c < \infty$ such that if Y_i, $i \le n < \infty$, are independent symmetric $SBC(T)$ processes and $c \le q < p$, then*

$$\|S_n\|_p \le K_c\frac{p}{q}\big[\|S_n\|_q + \|Y_n^*\|_p\big]. \tag{3.41}$$

For $p = 2$ and $q = 1$, inequality (3.39) gives better constants than (3.41): with $t_0 = 8\|S_n\|_1$, the first inequality gives

$$\|S_n\|_2 \le 12\sqrt{3}\big[16\|S_n\|_1 + \|Y_n^*\|_2\big]. \tag{3.42}$$

Using symmetrisation, to be considered next, the inequality in the preceding theorem extends from symmetric to centred processes (with larger constants).

3.1.4 Symmetrisation, Randomisation, Contraction

If the independent processes Y_i are symmetric, then some computations simplify: compare, for instance, Lévy's inequality with the Lévy-Ottaviani inequality, or consider Hoffmann-Jørgensen's inequalities in the preceding section. Thus, it is sometimes useful to relate moments and tail probabilities of $\left\|\sum Y_i\right\|_T$ to the same parameters of *symmetrised* sums such as $\|\sum(Y_i - Y_i')\|_T$, where the sequence $\{Y_i'\}$ is an independent copy of the sequence $\{Y_i\}$, or $\left\|\sum \varepsilon_i Y_i\right\|_T$, where ε_i are random signs independent of the sequence $\{Y_i\}$. Randomisation has the added advantage of allowing us to conditionally treat our variable as a simpler one; for instance, conditionally on the variables Y_i, the randomised sum $\sum \varepsilon_i Y_i$ is a *Rademacher process*, which is sub-Gaussian, and therefore, the metric entropy bound from Chapter 2 applies to it. In this subsection we prove a few useful and simple randomisation and symmetrisation inequalities. The inequalities for moments are based on the following basic proposition, which is an instance of a *contraction principle*. Randomisation by multipliers different from random signs is also briefly considered.

Theorem 3.1.17 *For $n \in \mathbb{N}$, let Y_i, $i \le n < \infty$, be independent $SBC(T)$ processes, let a_i, $i \le n$, be real numbers and let F be a nonnegative, nondecreasing convex function on* $[0,\infty)$. *Then, if either*

a. $0 \le a_i \le 1$ *and the processes Y_i are centered (meaning $E\|Y_i\| < \infty$ and $EY_i = 0$), or*
b. $|a_i| \le 1$ *and the processes Y_i are symmetric,*

we have

$$EF\left(\left\|\sum_{i=1}^{n} a_i Y_i\right\|_T\right) \le EF\left(\left\|\sum_{i=1}^{n} Y_i\right\|_T\right). \tag{3.43}$$

Proof The proof of (a) reduces, by iteration, to proving that

$$EF(\|aY + Z\|) \le EF(\|Y + Z\|), \tag{3.44}$$

where $0 \le a \le 1$, and Y and Z are independent and centred. Since F is convex and nondecreasing, we have

$$\begin{aligned} EF(\|aY + Z\|) &= EF(\|a(Y + Z) + (1 - a)Z\|) \\ &\le EF(a\|Y + Z\| + (1 - a)\|Z\|) \\ &\le aEF(\|Y + Z\|) + (1 - a)EF(\|Z\|). \end{aligned}$$

But Y being centred, by the same properties of F, Jensen's inequality and Fubini's theorem, we also have

$$EF(\|Z\|) = EF(\|EY + Z\|) \le EF(E_Y\|Y + Z\|) \le EE_Y F(\|Y + Z\|) = EF(\|Y + Z\|),$$

and part (a) follows.

For part (b), assume first that $a_i \neq 0$ for all i. We then observe that, by symmetry, the processes $(a_1/|a_1|)Y_1, \ldots, (a_n/|a_n|)Y_n$ have the same joint probability law as the original processes $Y_1, \ldots, Y_n$. But now we can apply part (a) to the centred processes $Z_i = (a_i/|a_i|)Y_i$ and the constants $|a_i| \geq 0$ to obtain

$$EF\left(\left\|\sum a_i Y_i\right\|\right) = EF\left(\left\|\sum |a_i| Z_i\right\|\right) \leq EF\left(\left\|\sum Z_i\right\|\right) = EF\left(\left\|\sum Y_i\right\|\right),$$

which is part (b). If only $a_{i_1}, \ldots, a_{i_k}$ are different from zero, the preceding argument and the fact that, also by a), $EF\left(\left\|\sum_{j=1}^k Y_{i_j}\right\|\right) \leq EF\left(\left\|\sum_{i=1}^k Y_i\right\|\right)$ again yield (3.43). ■

Corollary 3.1.18 *If $|a_i| \leq 1$ and Y_i are independent and centred SBC(T) processes, $1 \leq i \leq n < \infty$, then, for all $p \geq 1$,*

$$E\left\|\sum_{i=1}^n a_i Y_i\right\|_T^p \leq 2^p E\left\|\sum_{i=1}^n Y_i\right\|_T^p.$$

Proof Setting $a_i^+ = \max(a, 0)$, $a_i^- = \max(-a, 0)$, we have

$$\left\|\sum a_i Y_i\right\|^p \leq 2^{p-1}\left(\left\|\sum a_i^+ Y_i\right\|^p + \left\|\sum a_i^- Y_i\right\|^p\right),$$

and we can apply Theorem 3.1.17, part (a), to the two random variables on the right-hand side. ■

Usually the preceding contraction inequalities are applied to random a_i, with the sequence $\{a_i\}$ independent of the sequence $\{Y_i\}$, in combination with Fubini's theorem. The most frequent random multipliers are Rademacher sequences or multiples thereof.

Definition 3.1.19 A sequence of random variables $\{\varepsilon_i : i \in I\}$, $I \subseteq \mathbb{N}$, is a Rademacher sequence if the variables ε_i are independent and $\Pr\{\varepsilon_i = 1\} = \Pr\{\varepsilon_i = -1\} = 1/2$ for all $i \in I$.

Here is an application of Theorem 3.1.17 to truncation.

Corollary 3.1.20 *Let $\{\varepsilon_i\}$ be a Rademacher sequence independent of a sequence $\{Z_i\}$ consisting of independent SBC(T) processes. Let $C_i \subset \ell_\infty(T)$ be such that the variable $\left\|\sum_{i=1}^n \tau_i Z_i I_{Z_i \in C_i}\right\|_T$ is measurable for all choices of $\tau_i = \pm 1$. Then, for all $p \geq 1$,*

$$E\left\|\sum_{i=1}^n \varepsilon_i Z_i I_{Z_i \in C_i}\right\|_T^p \leq E\left\|\sum_{i=1}^n \varepsilon_i Z_i\right\|_T^p.$$

In particular this holds for $C_i = \{\|x\|_T \leq M\}$ or $C_i = \{\|x\|_T \geq M\}$ for any M. If the processes Z_i are symmetric and the C_i's are symmetric, then the Rademacher variables in this inequality are superfluous.

Proof Let E_ε denote integration with respect to the Rademacher sequence only. Now, we apply Theorem 3.1.17 to $E = E_\varepsilon$, $Y_i = \varepsilon_i Z_i$ (that has mean zero conditionally on Z_i) and $a_i = I_{Z_i \in C_i}$, to obtain

$$E_\varepsilon\left\|\sum_{i=1}^n \varepsilon_i Z_i I_{Z_i \in C_i}\right\|^p \leq E_\varepsilon\left\|\sum_{i=1}^n \varepsilon_i Z_i\right\|^p, \tag{3.45}$$

and the result follows by integrating with respect to the variables Z_i. If Z_i are symmetric variables and C_i are symmetric sets, then each side of the inequality equals respectively $E\left\|\sum_{i=1}^n Z_i I_{Z_i\in C_i}\right\|^p$ and $E\left\|\sum_{i=1}^n Z_i\right\|^p$. ■

This corollary shows how randomization or symmetrization may allow truncation. More substantial is the following corollary on the behavior of moments under Rademacher randomization.

Theorem 3.1.21 *Let Y_i, $i \le n < \infty$, be independent centered $SBC(T)$ processes with supremum norms in L^p for some $p \ge 1$, and let ε_i, $i \le n$, be a Rademacher sequence independent of the sequence of processes Y_i. Then,*

$$2^{-p}E\left\|\sum_{i=1}^n \varepsilon_i Y_i\right\|_T^p \le E\left\|\sum_{i=1}^n Y_i\right\|_T^p \le 2^p E\left\|\sum_{i=1}^n \varepsilon_i(Y_i - c_i)\right\|_T^p, \tag{3.46}$$

for any functions $c_i = c_i(t)$, and

$$E\max_{k\le n}\left\|\sum_{i=1}^k Y_i\right\|_T^p \le 2^{p+1}E\left\|\sum_{i=1}^n \varepsilon_i Y_i\right\|_T^p. \tag{3.47}$$

Proof Let $\{Y_i'\}$ be a copy of the sequence $\{Y_i\}$ independent of $\{Y_i\}$ and of the Rademacher sequence, and let E' denote integration with respect to these variables only (i.e., conditional expectation given the variables Y_i). Jensen's inequality and Fubini's theorem, symmetry and independence give

$$\begin{aligned}
E\left\|\sum Y_i\right\|^p &= E\left\|\sum Y_i - E\sum Y_i'\right\|^p = E\left\|E'\left(\sum Y_i - \sum Y_i'\right)\right\|^p \\
&\le E\left\|\sum (Y_i - Y_i')\right\|^p = E\left\|\sum (Y_i - c_i - (Y_i' - c_i))\right\|^p \\
&= E\left\|\sum \varepsilon_i(Y_i - c_i - (Y_i' - c_i))\right\|^p \le 2^p E\left\|\sum \varepsilon_i(Y_i - c_i)\right\|^p,
\end{aligned}$$

which is the right-hand-side inequality in (3.46). The left-hand-side inequality in (3.46) follows from Corollary 3.1.18 by taking $E = E_Y$ and $a_i = \varepsilon_i$ and then integrating with respect to the Rademacher variables.

Inequality (3.47) will follow from Lévy's inequality once we introduce a new norm that incorporates the maximum. For this, we define on $T \times \{1, \dots, n\}$ the bounded random processes

$$Z_1(t,k) = Y_1(t), \quad \text{for } k = 1, \dots, n \text{ and } t \in T,$$

and in general, for $1 < i \le n$ and all $t \in T$,

$$Z_i(t,k) = 0, \quad \text{for } k = 1, \dots, i-1, \quad \text{and} \quad Z_i(t,k) = Y_i(t), \quad \text{for } k = i, \dots, n.$$

Then the processes $Z_i(t,k)$ with index set $T' := T \times \{1, \dots, n\}$, countable, satisfy that

$$\sum_{i=1}^n Z_i(t,1) = Y_1(t), \quad \sum_{i=1}^n Z_i(t,2) = \sum_{k=1}^2 Y_i(t), \dots, \quad \sum_{i=1}^n Z_i(t,n) = \sum_{i=1}^n Y_i(t),$$

and thus we have

$$\left\|\sum_{i=1}^{n} Z_i\right\|_{T'} = \max_{1\le k\le n}\left\|\sum_{i=1}^{k} Y_i\right\|_{T} \quad\text{and}\quad \left\|\sum_{i=1}^{n} \varepsilon_i Z_i\right\|_{T'} = \max_{1\le k\le n}\left\|\sum_{i=1}^{k} \varepsilon_i Y_i\right\|_{T}.$$

Hence, inequality (3.46) applied to these processes gives

$$E\max_k\left\|\sum_{i=1}^{k} Y_i\right\|^p \le 2^p E\max_k\left\|\sum_{i=1}^{k} \varepsilon_i Y_i\right\|^p$$

and since $\varepsilon_i Y_i$ are symmetric, we can apply to them the moment form of Lévy's inequality (3.33), which gives

$$E\max_k\left\|\sum_{i=1}^{k} \varepsilon_i Y_i\right\|^p \le 2E\left\|\sum_{i=1}^{n} \varepsilon_i Y_i\right\|^p,$$

and inequality (3.47) follows. ■

As an application of the preceding theorem, here is an extension of Hoffmann-Jørgensen's inequality with only moments (Theorem 3.1.16) to centred random variables. Recall the notation $S_k = \sum_{i=1}^{k} Y_i$ for partial sums.

Theorem 3.1.22 *If Y_i, $i \le n < \infty$, are independent centred SBC(T) processes and $1 \le q < p$, then*

$$\left\|\max_{1\le k\le n}\|S_k\|_T\right\|_p \le \left[\frac{2(p+1)}{q}\right]^{(p+1)/p} 2^{1+2/p}\left[4^{(1+1)/(p+1)/q}\|\|S_n\|_T\|_q + \|Y_n^*\|_p\right]. \tag{3.48}$$

Proof It follows from inequality (3.40), by combination with inequality (3.47) and the left-hand-side inequality in (3.46). ■

One may ask if it is also possible to symmetrise tail probabilities just as we have symmetrised moments. The results are not as clean as inequality (3.46) but still useful. In one direction, we have the following:

Proposition 3.1.23 *Let Y_i be SBC(T) independent processes, and let $|a_i| \le 1$, $1 \le i \le n < \infty$. Then, for all $t > 0$,*

$$\Pr\left\{\left\|\sum_{i=1}^{n} a_i Y_i\right\|_T > t\right\} \le 3\max_{j\le n}\Pr\left\{\|S_j\|_T > t/9\right\}.$$

The same inequality holds true if the sequence $\{a_i\}$ is replaced by a Rademacher sequence $\{\varepsilon_i\}$.

Proof We can assume that $-1 \le a_n \le a_{n-1} \le \cdots \le a_1 \le 1$. Then, taking $\sigma_1 = a_1 - a_2, \ldots,$ $\sigma_{n-1} = a_{n-1} - a_n$, $\sigma_n = a_n - (-1)$, we have that $a_j = -1 + \sum_{i=j}^{n}\sigma_i$ for $1 \le j \le n$, $\sigma_i \ge 0$ for

$1 \le i \le n$ and $\sum_{i=1}^{n} \sigma_i = a_1 + 1 \le 2$. Therefore, for any bounded functions x_i, we have

$$\left\| \sum_{j=1}^{n} a_j x_j \right\| = \left\| \sum_{j=1}^{n} \left(\sum_{i=j}^{n} \sigma_i - 1 \right) x_j \right\| = \left\| \sum_{j=1}^{n} \sigma_j \sum_{i=1}^{j} x_i - \sum_{i=1}^{n} x_i \right\|$$

$$\le \left(\sum_{j=1}^{n} \sigma_i \right) \max_{1 \le j \le n} \left\| \sum_{i=1}^{j} x_i \right\| + \left\| \sum_{i=1}^{n} x_i \right\| \le 3 \max_{i \le j \le n} \left\| \sum_{i=1}^{j} x_i \right\|.$$

Combined with the second Ottaviani inequality (3.36) in Proposition 3.1.12, this gives

$$\Pr\left\{ \left\| \sum_{i=1}^{n} a_i Y_i \right\|_T > t \right\} \le \Pr\left\{ \max_{1 \le j \le n} \|S_j\|_T > t/3 \right\} \le 3 \max_{j \le n} \Pr\left\{ \|S_j\|_T > t/9 \right\}.$$

The result for a_i random, in particular, for $a_i = \varepsilon_i$, follows by an extra integration and Fubini's theorem. ∎

See Exercise 3.1.5 for an inequality with better constants when $a_i = \varepsilon_i$ and the processes Y_i are identically distributed. In the other direction, we have the following classical symmetrisation inequality:

Proposition 3.1.24 *(a) Let $Y(t)$, $Y'(t)$, $t \in T$, be two $SBC(T)$ processes defined on the factors of $(\Omega \times \Omega', \Sigma \otimes \Sigma', \Pr = P \times P')$; that is, $Y(t, \omega, \omega') = Y(t, \omega)$ and $Y'(t, \omega, \omega') = Y'(t, \omega')$, $t \in T$, $\omega \in \Omega$, $\omega' \in \Omega'$. Then, for all $s > 0$ and $0 < u \le s$ such that $\sup_{t \in T} \Pr\{|Y'(t)| \ge u\} < 1$, we have*

$$\Pr\{\|Y\|_T > s\} \le \frac{1}{1 - \sup_{t \in T} \Pr\{|Y'(t)| \ge u\}} \Pr\{\|Y - Y'\|_T > s - u\}.$$

(b) If $\theta \ge \sup_{t \in T} E(Y'(t)^2)$, then for any $s \ge (2\theta)^{1/2}$,

$$\Pr\{\|Y\|_T > s\} \le 2 \Pr\left\{ \|Y - Y'\|_T > s - (2\theta)^{1/2} \right\}.$$

Proof If ω is such that $\|Y(\omega)\|_T > s$, then there exists $t \in T$ such that $|Y_t(\omega)| > s$ (note that $t = t(\omega)$), and then if, moreover, $|Y'_t| \le u$, we have $|Y_t(\omega) - Y'_t| > s - u$. This implies that for such ω,

$$\inf_{t \in T} P'\{|Y'_t| \le u\} \le P'\{\|Y(\omega) - Y'\|_T > s - u\}.$$

Then, integrating this inequality on the set $\{\|Y(\omega)\|_T > s\}$, we obtain

$$\begin{aligned}
\Pr\{\|Y - Y'\|_T > s - u\} &\ge E_P P'\{\|Y\|_T > s, \|Y - Y'\|_T > s - u\} \\
&\ge E_P \left(I_{\{\|Y\|_T > s\}} P'\{\|Y - Y'\|_T > s - u\} \right) \\
&\ge P\left(\{\|Y\|_T > s\} \inf_{t \in T} P'\{|Y'_t| \le u\} \right) \\
&= \inf_{t \in T} P'\{|Y'_t| \le u\} \Pr\{\|Y\|_T > s\}.
\end{aligned}$$

This proves part (a), and part (b) follows from part (a) and Chebyshev's inequality. ∎

Corollary 3.1.25 *Let Y_i, $1 \le i \le n < \infty$, be centred independent SBC(T) processes, and let $\{\varepsilon_i\}_{i=1}^n$ be a Rademacher sequence independent of the processes Y_i. Let $\sigma^2 = \sup_{t\in T} EY_1^2(t) < \infty$. Then, for all $s \ge \sqrt{2n\sigma^2}$ and for any real numbers a_i,*

$$\Pr\left\{\left\|\sum_{i=1}^n Y_i\right\|_T > s\right\} \le 4\Pr\left\{\left\|\sum_{i=1}^n \varepsilon_i(Y_i - a_i)\right\|_T > (s - \sqrt{2n\sigma^2})/2\right\}.$$

Proof Part (b) of the preceding proposition gives, for $\{Y_i'\}$ an independent copy of $\{Y_i\}$,

$$\Pr\left\{\left\|\sum_{i=1}^n Y_i\right\|_T > s\right\} \le 2\Pr\left\{\left\|\sum_{i=1}^n (Y_i - a_i - (Y_i' - a_i))\right\| > s - \sqrt{2n\sigma^2}\right\}.$$

But, by symmetry, the variables $\left\|\sum_{i=1}^n (Y_i - a_i - (Y_i' - a_i)\right\|$ and $\left\|\sum_{i=1}^n \varepsilon_i(Y_i - a_i - (Y_i' - a_i))\right\|$ are identically distributed, and so are the variables $\left\|\sum_{i=1}^n \varepsilon_i(Y_i - a_i)\right\|$ and $\left\|\sum_{i=1}^n \varepsilon_i(Y_i' - a_i)\right\|$. Then the result follows by the triangle inequality for probabilities $\Pr\{U + V > t\} \le \Pr\{U > t/2\} + \Pr\{V > t/2\}$. ■

Sometimes it is convenient to randomise $\sum Y_i$ not by Rademacher variables but by normal or even Poisson variables. We only consider the case of moments. If two-sided inequalities are to hold between the original partial-sum process and the randomised 1 process, the multipliers should not be too large in distribution, and the following condition turns out to be precisely what is needed for identically distributed processes. For real random variables ξ, we define

$$\Lambda_{2,1}(\xi) := \int_0^\infty \sqrt{\Pr\{|\xi| > t\}}\,dt. \tag{3.49}$$

Note that $\Lambda_{2,1}(\xi) < \infty$ implies that $E\xi^2 < \infty$ and that $E|\xi|^{2+\delta} < \infty$ for some $\delta > 0$ implies that $\Lambda_{2,1}(\xi) < \infty$.

Proposition 3.1.26 *Let $\{Y_i\}_{i=1}^n$ be a finite set of independent identically distributed SBC(T) processes such that $E\|Y_i\|_T < \infty$ for each $i \le n$, and let $\{\varepsilon_i\}_{i=1}^n$ and $\{\xi_i\}_{i=1}^n$ be, respectively, a Rademacher sequence and a sequence of symmetric i.i.d. real random variables independent of each other and of the sequence $\{Y_i\}$. Then, for $0 \le n_0 < n$, we have*

$$(E|\xi_1|)E\left\|\sum_{i=1}^n \varepsilon_i Y_i\right\|_T \le E\left\|\sum_{i=1}^n \xi_i Y_i\right\|_T$$
$$\le n_0 E\|Y_1\|_T E\max_{i\le n}|\xi_i| + \sqrt{n}\Lambda_{2,1}(\xi_1)\max_{n_0<k\le n} E\left\|\frac{\sum_{i=n_0+1}^k \varepsilon_i Y_i}{\sqrt{k}}\right\|_T. \tag{3.50}$$

If the variables ξ_i are centred but not necessarily symmetric, inequality (3.50) holds with the following modifications: $E|\xi_1|$ on the left is replaced by $E|\xi_1 - \xi_2|/2$, and the first summand on the right is multiplied by 2 and the second by $2\sqrt{2}$.

Proof The left-side inequality in (3.50) follows from the observation that, by symmetry, the joint distribution of the variables ξ_i coincides with the joint distribution of the variables

$\varepsilon_i|\xi_i|$, which gives

$$E\left\|\sum_{i=1}^{n}\xi_i Y_i\right\| = E\left\|\sum_{i=1}^{n}\varepsilon_i|\xi_i|Y_i\right\| \geq E\left\|\sum_{i=1}^{n}\varepsilon_i\,(E|\xi_i|)\,Y_i\right\|.$$

The following chain of inequalities, which are self-explanatory, gives the proof of the right-hand-side inequality:

$$\begin{aligned}
E\left\|\sum_{i=1}^{n}\xi_i Y_i\right\| &= E\left\|\sum_{i=1}^{n}\varepsilon_i|\xi_i|Y_i\right\| \\
&= E\left\|\sum_{i=1}^{n}\left(\int_0^\infty I_{t\leq|\xi_i|}dt\right)\varepsilon_i Y_i\right\| \\
&\leq \int_0^\infty E\left\|\sum_{i=1}^{n} I_{t\leq|\xi_i|}\varepsilon_i Y_i\right\| dt \\
&= \int_0^\infty E\left\|\sum_{i=1}^{\#\{i\leq n:\ |\xi_i|\geq t\}}\varepsilon_i Y_i\right\| dt \\
&\leq \int_0^\infty \left(\sum_{k=1}^{n}\Pr\left\{\sum_{i=1}^{n} I_{|\xi_i|\geq t}=k\right\}E\left\|\sum_{i=1}^{k}\varepsilon_i Y_i\right\|\right)dt \\
&\leq \left(\int_0^\infty \Pr\left\{\sum_{i=1}^{n} I_{|\xi_i|\geq t}>0\right\}dt\right)\max_{k\leq n_0} E\left\|\sum_{i=1}^{k}\varepsilon_i Y_i\right\| \\
&\quad + \sqrt{n}\left(\frac{1}{\sqrt{n}}\int_0^\infty \sum_{k=n_0+1}^{n}\sqrt{k}\Pr\left\{\sum_{i=1}^{n} I_{|\xi_i|\geq t}=k\right\}dt\right)\max_{n_0<k\leq n} E\left\|\frac{1}{\sqrt{k}}\sum_{i=n_0+1}^{k}\varepsilon_i Y_i\right\| \\
&\leq \left(\int_0^\infty \Pr\left\{\max_{i\leq n}|\xi_i|\geq t\right\}dt\right)n_0 E\|Y_1\| + \sqrt{n}\Lambda_{2,1}(\xi)\max_{n_0<k\leq n} E\left\|\frac{1}{\sqrt{k}}\sum_{i=n_0+1}^{k}\varepsilon_i Y_i\right\| \\
&= n_0 E\|Y_1\| E\max_{k\leq n}|\xi_i| + \sqrt{n}\Lambda_{2,1}(\xi)\max_{n_0<k\leq n} E\left\|\frac{1}{\sqrt{k}}\sum_{n_0<i\leq k}\varepsilon_i Y_i\right\|.
\end{aligned}$$

For the last inequality, note that

$$\sum_{k\geq 0}^{\infty}\sqrt{k}\Pr\left\{\sum_{i=1}^{n} I_{|\xi_i|\geq t}=k\right\} = E\left(\sum_{i=1}^{n} I_{|\xi_i|\geq t}\right)^{1/2} \leq \left(E\sum_{i=1}^{n} I_{|\xi_i|\geq t}\right)^{1/2}.$$

When ξ is not symmetric, but still centred, the theorem follows from the preceding estimates applied to $\xi_i - \xi_i'$, where $\{\xi_i'\}$ is an independent copy of $\{\xi_i\}$. ■

Exercises

3.1.1 If ξ is exponential with parameter $1/\alpha$ and $X = \xi - E\xi$, then $\log Ee^{\lambda X} \leq (\alpha^2\lambda^2)/(2(1-\alpha\lambda)) = \varphi_{\alpha^2,\alpha}(\lambda)$, for $0 < \lambda < 1/\alpha$. If Y is the sum of r independent centred exponential random variables, then $\log Ee^{\lambda Y} \le \varphi_{ra^2,\alpha}(\lambda)$. This also extends to any centred gamma random variables. *Hint*: $Ee^{\lambda X} = e^{-\lambda\alpha}/(1-\lambda\alpha)$, and by Taylor development, $-x - \log(1-x) \le x^2/(2(1-x))$.

3.1.2 (a) If X is Poisson with parameter a, then $\Pr\{X - a \le -t\} \le e^{-t^2/(2a)}$, $t > 0$. (b) Show also that for the same variable X and for $t > a+1$, there exists $c > 0$ such that $\Pr\{X - a \ge t\} \ge ce^{-2t\log(1+(t+1)/a)}$. *Hint*: (a) It suffices to prove this inequality for $t < a$; hence, it follows from (3.10) and the facts that $e^{-\lambda} + \lambda - 1 \le \lambda^2/2$ for $0 \le \lambda < 3$ and $\lambda^2 a/2 - \lambda t$ is smallest for $\lambda = t/a < 1$. (b) Use the nonasymptotic Stirling formula.

3.1.3 Prove that $\ell_\infty(T)$ equipped with the supremum norm is a Banach space and that it is separable if and only if T is finite.

3.1.4 Work out the details of the following example showing that the law of a simple sample bounded process, which is a probability measure on the cylindrical σ-algebra of $\ell_\infty(T)$, does not extend to its Borel σ-algebra. Set $X_t = I_{[0,t)}(U) = I_{(U,1]}(t)$, $t \in \mathbb{Q}$ (or $t \in \mathbb{R}$), where U is a uniform, for example, $(\Omega, \Sigma, \Pr) = ([0,1], \mathcal{B}, \mu)$, μ Lebesgue measure. Then X is a sample bounded process indexed by $T = \mathbb{Q} \cap [0,1]$. Let A be any subset of $[0,1]$, and set $F_A = \{I_{(s,1]} : s \in A\} \subset \ell_\infty(T)$. Then argue that F_A is closed (it is discrete) and hence Borel measurable but that if A is not measurable, then the pre-image $X^{-1}(F_A) = \{\omega : I_{(U(\omega),1]} \in F_A\} = \{\omega : U(\omega) \in A\}$ is not measurable. Thus, the law of the map $X : \Omega \mapsto \ell_\infty(T)$ is not a Borel measure; that is, the process X is not an $\ell_\infty(T)$-valued random variable.

3.1.5 Prove the following improvement on Proposition 3.1.23 in a particular case: if Y_i are i.i.d. (sample bounded processes over a countable index set T) and $\{\varepsilon_i\}$ is a Rademacher sequence independent of the processes Y_i, then

$$\Pr\left\{\left\|\sum_{i=1}^n \varepsilon_i Y_i\right\|_T > t\right\} \le 2\max_{k\le n}\Pr\left\{\left\|\sum_{i=1}^k Y_i\right\|_T > t/2\right\}.$$

(And then, using Montgomery-Smith's (1994) reflection principle for i.i.d. summands, there exist c_1, c_2 finite such that the right-hand side is dominated by $c_1 \Pr\left\{\left\|\sum_{i=1}^n Y_i\right\|_T > c_2 t\right\}$.)

3.1.6 Let ξ_i be centred independent identically distributed real random variables with $E|\xi_1|^p < \infty$ for some $p > 2$ and $E\xi = 0$. Show that $\sup_n E\left(\left|\sum_{i=1}^n \xi_i\right|/n^{1/2}\right)^p < \infty$. More generally, if Y_i are centred independent identically distributed sample bounded processes with countable index set T such that the sequence $\{S_n/n^{1/2}\}$ is stochastically bounded and $E\|Y_1\|^p < \infty$ for some $p \ge 2$, then $\sup_n E(\|S_n\|_T/n^{1/2})^p < \infty$.

3.1.7 (A proof of Glivenko-Cantelli's theorem.) Let P be a probability measure on $\mathbb{R}$, let X_i be i.i.d. random variables with probability law P and let F_n be, for each n, the empirical distribution function corresponding to $X_1,\dots,X_n$. Prove that $E\|F_n - F\|_\infty \le 4/\sqrt{n}$, which, combined with reverse submartingale convergence, gives $\|F_n - F\|_\infty \to 0$ a.s. (as well as in L_1). *Hint*: Use randomisation and Lévy's inequality: a simple randomisation result from the text gives $E\|F_n - F\|_\infty \le 2E\sup_{t\in\mathbb{R}}\left|n^{-1}\sum_{i=1}^n \varepsilon_i I(X_i \le t)\right|$, and then we observe that for each fixed value of $X_1,\dots,X_n$, there is a permutation $\{n_k\}$ of $\{1,\dots,n\}$, such that this last supremum equals $\max_{\ell\le n}\left|\sum_{k=1}^{\ell}\varepsilon_{n_k}\right|$. Now use Lévy's inequality conditionally on the X_i variables followed by Fubini's theorem.

3.1.8 (An exponential inequality for binomial probabilities.) Show that if ξ_i are i.i.d. random variables with $\Pr\{\xi_i = 1\} = p = 1 - \Pr\{\xi_i = 0\}$, then

$$\Pr\left\{\sum_{i=1}^n \xi_i \ge k\right\} \le \left(\frac{enp}{k}\right)^k,$$

and compare with the inequalities from Section 3.1.2 that apply. *Hint*: The left-hand side is obviously bounded by $\binom{n}{k}\Pr\left\{\sum_{i=1}^k \xi_i = k\right\} = \binom{n}{k}p^k$, and $\log k > \int_1^k \log x\, dx = \log(k/e)^k$, so $\binom{n}{k} \le (en/k)^k$.

3.1.9 (Expected value of maxima of independent random variables.) Let ξ_i, $i \le n$, be nonnegative independent random variables in $L^p, p > 0$. (a) Prove that, for all $\delta > 0$,

$$E \max_{1\le i\le n} \xi_i^p \le \delta^p + p\int_\delta^\infty t^{p-1}\sum_{i=1}^n \Pr\{\xi_i > t\}dt.$$

(b) Use that $1 - x \le e^{-x}$ and $1 - e^{-x} \ge x/(1-x)$ to show that

$$\Pr\left\{\max_i \xi_i > t\right\} \ge \frac{\sum_i \Pr\{\xi_i > t\}}{1 + \sum_i \Pr\{\xi_i > t\}}.$$

(c) Suppose now that $\delta_0 = \inf\{t : \sum_{i=1}^n \Pr\{\xi_i > t\} \le \lambda\}$ for some $\lambda > 0$. Use the preceding inequality and the monotonicity of the function $x/(1+x)$ to deduce $\Pr\{\max_i \xi_i > t\} \ge \sum \Pr\{\xi_i > t\}/(1+\lambda)$ for $t \ge \delta_0$ and $\Pr\{\max_i \xi_i > t\} \ge \lambda/(1+\lambda)$ otherwise. Conclude

$$E \max_{1\le i\le n} \xi_i^p \ge \frac{\lambda}{1+\lambda}\delta_0^p + \frac{p}{1+\lambda}\int_{\delta_0}^\infty t^{p-1}\sum_{i=1}^n \Pr\{\xi_i > t\}dt.$$

3.1.10 (The maximal form of some inequalities.) Let ξ_i, $i \in \mathbb{N}$, be independent centred random variables such that $Ee^{\lambda\xi_i} < \infty$, for all $0 < \lambda < \lambda_0$, $\lambda_0 \le \infty$, and set $S_k = \sum_{i=1}^k \xi_i$, $k \in \mathbb{N}$. Show that for $0 < \lambda < \lambda_0$, the sequence $\{(e^{\lambda S_k}, \mathcal{S}_k)\}$, where $\mathcal{S}_k = \sigma(\xi_i : i \le k)$, is a submartingale ($Ee^{\lambda\xi_k} \ge 1$), and apply Doob's maximal inequality to obtain

$$\Pr\left\{\max_{k\le n} S_k > t\right\} \le Ee^{\lambda S_n}/e^{\lambda t}, \quad t > 0.$$

Use this to replace S_n by $\max_{k\le n} S_k$ in the Hoeffding, Bennet, Bernstein and Prohorov inequalities.

3.1.11 Let g_i, $i \in \mathbb{N}$, be i.i.d. standard normal random variables. Prove that

$$\max_{1\le k\le n} \frac{1}{\sqrt{k}}\left|\sum_{i=1}^k (g_i^2 - 1)\right| = O_{pr}(\sqrt{\log\log n}).$$

Hint: Use Theorem 3.1.9, the preceding exercise and blocking. For the blocking, argue that if $r_k = \min\{r : k \le 2^r\}$, it suffices to consider $\pm\sum_{i=1}^k (g_i^2 - 1)/\sqrt{2^{r_k}}$; then set $S_k = \sum_{i=1}^k (g_i^2 - 1)$ and note that

$$\Pr\left\{\max_{k\le n} \frac{1}{\sqrt{2^{r_k}}} S_k > t\right\} \le \Pr\left\{\max_{r\le r_n}\max_{2^{r-1}<k\le 2^r} \frac{1}{\sqrt{2^{r_k}}} S_k \ge t\right\} \le \sum_{r\le r_n}\Pr\left\{\max_{r\le r_n} \frac{1}{\sqrt{2^{r_k}}} S_k \ge t\right\}.$$

Now it is easy to apply the two results just mentioned and obtain that if one takes $t = \sqrt{M\log\log n}$, for $M \ge 9$, $M(\log\log n)/n < 1$, then these probabilities tend to zero as $n \to \infty$.

3.2 Rademacher Processes

As can be inferred from the preceding section and will be corroborated in later sections, randomisation may be a useful tool in the study of empirical processes, particularly Rademacher randomisation, which consists in replacing $\sum_{i=1}^n (f(X_i) - Pf)$ by $\sum_{i=1}^n \varepsilon_i f(X_i)$, where ε_i, $i = 1, \ldots, n$, are independent Rademacher variables, independent of the variables

X_i: these two processes have comparable 'sizes', and the second is easier to estimate because, conditionally on the variables X_i, the process $f \mapsto \sum_{i=1}^n \varepsilon_i f(X_i)$ is sub-Gaussian. More exactly, it is a Rademacher process of the form

$$t \mapsto \sum_{i=1}^{n} t_i \varepsilon_i, \quad t = (t_1, \ldots, t_n) \in T \subseteq \mathbb{R}^n. \tag{3.51}$$

The object of this section is to collect some relevant results on Rademacher processes needed in later sections. Although they are less well understood than Gaussian processes, Rademacher processes share many properties with them, and for us, they may be at least as useful because Rademacher randomisation produces stochastically smaller processes than Gaussian randomisation; in particular, it produces bounded processes if the class $\mathcal{F}$ is bounded.

Since Rademacher processes are sub-Gaussian, the metric entropy moment bounds for sub-Gaussian processes given in Section 2.3 apply to these processes. However, analogues or partial analogues of the Slepian comparison theorem, of Sudakov's lower bound for the expected value of the supremum of a Gaussian process or of the Borell-Tsirelson-Sudakov's concentration inequality do require separate proofs. In this section we start with a comparison principle for Rademacher processes, then consider the concentration inequality in more generality than just for these processes and conclude with a Sudakov-type minorisation inequality. These results are both interesting per se and useful in the theory of empirical processes.

3.2.1 A Comparison Principle for Rademacher Processes

In the first theorem of this section we compare the sizes of the Rademacher processes $\sum \varepsilon_i t_i$ and $\sum \varepsilon_i \varphi_i(t_i)$, $t \in T \subset \mathbb{R}^n$, where the functions $\varphi_i : \mathbb{R} \mapsto \mathbb{R}$ are *contractions vanishing at* 0; that is, the φ_i satisfy

$$|\varphi_i(s) - \varphi_i(t)| \le |s - t|, \quad \text{for all } s, t \in \mathbb{R} \quad \text{and} \quad \varphi_i(0) = 0.$$

In particular, this result will greatly generalise inequality (3.45) from the preceding section.

Theorem 3.2.1 *Let F be a nonnegative, convex and nondecreasing function defined on $[0, \infty)$. Let $\varphi_i : \mathbb{R} \mapsto \mathbb{R}$ be contractions vanishing at 0, and let T be a bounded set of $\mathbb{R}^n$, $n < \infty$. Then*

$$EF\left(\frac{1}{2}\left\|\sum_{i=1}^{n} \varepsilon_i \varphi_i(t_i)\right\|_T\right) \le EF\left(\left\|\sum_{i=1}^{n} \varepsilon_i t_i\right\|_T\right), \tag{3.52}$$

where $t = (t_1, \ldots, t_n)$ and $\|\cdot\|_T$ denotes, as usual, supremum over all $t \in T$.

Proof Since F and φ_i are continuous and so is the map $t \mapsto \sum_{i=1}^n \varepsilon_i t_i$, the random quantities involved in inequality (3.52) are measurable; in fact, we may assume that T is a finite set of $\mathbb{R}^n$. We begin with several reductions of the problem to simpler ones.

First, we see that proving the theorem reduces to showing that

$$EG\left(\frac{1}{2}\sup_{t \in T} \sum \varepsilon_i \varphi_i(t_i)\right) \le EG\left(\sup_{t \in T} \sum \varepsilon_i t_i\right), \tag{3.53}$$

for all $G : \mathbb{R} \mapsto \mathbb{R}$ convex and nondecreasing. To prove that (3.53) implies (3.52), note first that since the two sequences $\{\varepsilon_i\}$ and $\{-\varepsilon_i\}$ are equi-distributed and since $a^+ = (-a)^-$, where $a^+ = \max(a, 0)$ and $a^- = -\min(a, 0)$, it follows that

$$EF\left(\frac{1}{2}\left\|\sum \varepsilon_i\varphi_i(t_i)\right\|_T\right) \le EF\left(\sup_{t\in T}\left(\sum \varepsilon_i\varphi_i(t_i)\right)^+\right) + EF\left(\sup_{t\in T}\left(\sum \varepsilon_i\varphi_i(t_i)\right)^-\right)$$
$$= 2EF\left(\sup_{t\in T}\left(\sum \varepsilon_i\varphi_i(t_i)\right)^+\right).$$

Now (3.52) follows from applying (3.53) to $G(\,\cdot\,) = F((\,\cdot\,)^+)$, which is convex and nondecreasing.

Next, we observe that by conditioning and iteration, in order to prove (3.53), it suffices to show that if $T \subset \mathbb{R}^2$, ε is a Rademacher variable, and φ is a contraction on $\mathbb{R}$ vanishing at 0; then

$$EG\left(\sup_{t\in T}(t_1 + \varepsilon\varphi(t_2))\right) \le EG\left(\sup_{t\in T}(t_1 + \varepsilon t_2)\right). \tag{3.54}$$

Moreover, since we can assume $T \subset \mathbb{R}^2$ finite, the maximum of $t_1 + \varphi(t_2)$ over T is attained, and so is the maximum of $t_1 - \varphi(t_2)$ over T. If $t \in T$ denotes an argument where the first maximum is obtained and s denotes an argument for the second maximum, then s and t satisfy

$$t_1 + \varphi(t_2) \ge s_1 + \varphi(s_2) \quad \text{and} \quad s_1 - \varphi(s_2) \ge t_1 + \varphi(t_2), \tag{3.55}$$

and

$$EG\left(\sup_{t\in T}(t_1 + \varepsilon\varphi(t_2))\right) = \frac{1}{2}G(t_1 + \varphi(t_2)) + \frac{1}{2}G(s_1 - \varphi(s_2)) =: I(t, s, \varphi).$$

Finally then, the theorem reduces to proving

$$I(t, s, \varphi) \le EG\left(\sup_{t\in T}(t_1 + \varepsilon t_2)\right), \tag{3.56}$$

for all contractions φ vanishing at 0 and points s and t in T satisfying conditions (3.55) for φ. We distinguish two cases.

Case 1: s_2 and t_2 have different signs. Suppose first that $t_2 \ge 0$ and $s_2 \le 0$. Then, since by definition $|\varphi(t)| \le |t|$ ($|\varphi(t) - \varphi(0)| \le |t - 0|$ and $\varphi(0) = 0$), we have $\varphi(t_2) \le t_2$ and $-\varphi(s_2) \le -s_2$, so

$$I(t, s, \varphi) \le \frac{1}{2}(G(t_1 + t_2) + G(s_1 - s_2)) \le EG\left(\sup_{t\in T}(t_1 + \varepsilon t_2)\right),$$

which is (3.56). The subcase $t_2 \le 0$ and $s_2 \ge 0$ reduces to the preceding one: since $-\varphi$ is a contraction and vanishes at 0, and since $I(t, s, \varphi) = I(s, t, -\varphi)$, the preceding inequality gives

$$I(t, s, \varphi) = I(s, t, -\varphi) \le \frac{1}{2}(G(t_1 - t_2) + G(s_1 + s_2)) \le EG\left(\sup_{t\in T}(t_1 + \varepsilon t_2)\right).$$

Case 2: s_2 and t_2 have equal signs. Let us assume that $0 \le s_2 \le t_2$. As in case 1, it suffices to show that $G(t_1 + \varphi(t_2)) + G(s_1 - \varphi(s_2) \le G(t_1 + t_2) + G(s_1 - s_2)$, or

$$G(s_1 - \varphi(s_2)) - G(s_1 - s_2) \le G(t_1 + t_2) - G(t_1 + \varphi(t_2)). \tag{3.57}$$

Set $a = s_1 - \varphi(s_2)$, $b = s_1 - s_2$, $a' = t_1 + t_2$ and $b' = t_1 + \varphi(t_2)$. Then $s_2 \geq 0$ implies $s_2 \geq |\varphi(s_2)|$ and hence $a \geq b$ and, using also (3.55), $b' \geq b$. Moreover, since φ is a contraction and $s_2 \leq t_2$, we have that $\varphi(t_2) - \varphi(s_2) \leq t_2 - s_2$, which gives $a - b = s_2 - \varphi(s_2) \leq t_2 - \varphi(t_2) = a' - b'$. Now, since G is convex, the function $G(\cdot + c) - G(\cdot)$ for $c > 0$ is nondecreasing (recall that for convex functions G, if $u < v < u + c$, then,

$$\frac{G(u+c) - G(u)}{u} \leq \frac{G(u+c) - G(v)}{u + c - v} \leq \frac{G(v+c) - G(v)}{c},$$

and similarly if $v \geq u + c$). Therefore, $G(a) - G(b) = G(b + (a-b)) - G(b) \leq G(b' + (a-b)) - G(b')$, which, since $b' + a - b \leq a'$ and G is nondecreasing, gives $G(a) - G(b) \leq G(a') - G(b')$, proving (3.57). The subcase $0 \leq t_2 \leq s_2$ reduces to the preceding situation just as in case 1, interchanging s and t and replacing φ with $-\varphi$. The subcase $s_2 \leq 0$, $t_2 \leq 0$ reduces to the preceding ones by taking $t' = (t_1, -t_2)$, $s' = (s_1, -s_2)$ and $\tilde{\varphi}(x) = \varphi(-x)$. ■

This theorem is very useful in estimation of the diameter of a class of functions with respect to the $L^2(P_n)$ pseudo-distance, where P_n is the empirical measure. Let X_i, $i = 1, \ldots, n$, be independent S-valued random variables, let $\mathcal{F}$ be a countable class of measurable functions $S \mapsto \mathbb{R}$ such that $F(x) = \sup_{f \in \mathcal{F}} |f(x)|$ is finite for all $x \in S$ and set $U = \max_{i=1}^n |F(X_i)|$ and $\sigma^2 = \sup_{f \in \mathcal{F}} \sum_{i=1}^n E f^2(X_i)/n$, which we assume to be finite. Let also ε_i, $i = 1, \ldots, n$, be independent Rademacher variables independent of the X variables, and let E_ε denote conditional expectation given the variables X_i. In order to compare $E_\varepsilon \left\| \sum_{i=1}^n \varepsilon_i f^2(X_i) \right\|_{\mathcal{F}}$ to $E_\varepsilon \left\| \sum_{i=1}^n \varepsilon_i f(X_i) \right\|_{\mathcal{F}}$, we apply the comparison principle just proved. For $X_1, \ldots, X_n$ fixed, we take in Theorem 3.2.1 $t_i = U f(X_i)$, $i = 1, \ldots, n$, $T = \{(U f(X_i) : i = 1, \ldots, n) : f \in \mathcal{F}\}$ and $\varphi_i(s) = \varphi(s) = s^2/2U^2 \wedge U^2/2$. It is clear that φ is a contraction and that $\varphi(0) = 0$. Also note that $U^2 f^2(X_i)/(2U^2) \leq U^2/2$, so $\varphi(U f(X_i)) = f^2(X_i)/2$. Then the preceding theorem gives

$$\frac{1}{4} E_\varepsilon \left\| \sum_{i=1}^n \varepsilon_i f^2(X_i) \right\|_{\mathcal{F}} \leq U E_\varepsilon \left\| \sum_{i=1}^n \varepsilon_i f(X_i) \right\|_{\mathcal{F}}.$$

Integrating with respect to the variables X_i and then applying the basic randomisation inequality (3.46), we further obtain

$$E \left\| \sum_{i=1}^n f^2(X_i) \right\|_{\mathcal{F}} \leq n\sigma^2 + E \left\| \sum_{i=1}^n (f^2(X_i) - E f^2(X_i)) \right\|_{\mathcal{F}} \leq n\sigma^2 + 2E \left\| \sum_{i=1}^n \varepsilon_i f^2(X_i) \right\|_{\mathcal{F}}$$
$$\leq n\sigma^2 + 8E \left[U \left\| \sum_{i=1}^n \varepsilon_i f(X_i) \right\| \right].$$

Summarising:

Corollary 3.2.2 *With the immediately preceding notation, we have*

$$E \left\| \sum_{i=1}^n \varepsilon_i f^2(X_i) \right\|_{\mathcal{F}} \leq 4E \left[U \left\| \sum_{i=1}^n \varepsilon_i f(X_i) \right\|_{\mathcal{F}} \right] \tag{3.58}$$

and

$$E \left\| \sum_{i=1}^n f^2(X_i) \right\|_{\mathcal{F}} \leq n\sigma^2 + 8E \left[U \left\| \sum_{i=1}^n \varepsilon_i f(X_i) \right\|_{\mathcal{F}} \right]. \tag{3.59}$$

Further randomising, if U is dominated by a constant K, then the preceding inequality yields

$$E\left\|\sum_{i=1}^{n} f^2(X_i)\right\|_{\mathcal{F}} \le n\sigma^2 + 16KE\left\|\sum_{i=1}^{n}(f(X_i) - Pf(X_i))\right\|_{\mathcal{F}}. \tag{3.60}$$

3.2.2 Convex Distance Concentration and Rademacher Processes

As seen in Theorem 2.2.6, Lipschitz functions of Gaussian processes are highly concentrated about their medians or, abusing language a little, are almost constant. In this subsection and the following section we will see that this phenomenon is not only typical of Gaussian processes but also that, in Talagrand's words, 'a random variable that depends in a "smooth" way on the influence of many independent variables (but not too much on any of them) is essentially constant' where 'essentially constant' in the present context means that it satisfies an exponential inequality of Hoeffding type. Whereas in the following section we will use log-Sobolev-type differential inequalities and Herbst's method, as in the case of Gaussian processes, in this subsection we use the direct method originally employed by Talagrand. For any $n \in \mathbb{N}$, let $(S_k, \mathcal{S}_k)$, $1 \le k \le n$, be measurable spaces, let X_k, $k = 1, \ldots, n$, be independent S_k-valued random variables and let X denote the random vector $X = (X_1, \ldots, X_n)$, taking values in the product space $S = \prod_{k=1}^{n} S_k$. Let us denote by P the probability law of X, $P = \mu_1 \times \cdots \times \mu_n$, where $\mu_k = \mathcal{L}(X_k)$. We have in mind $X_k = \varepsilon_k$, where ε_k are independent Rademacher variables, but the variables X_k may just be bounded variables. In fact, the smooth, in our case Lipschitz, functions of X will be almost constant because the probability law P of X will be shown to be highly concentrated about any set of large P measure. We begin with the distance that will measure the concentration about a set A, the *convex distance*.

The *Hamming distance* d on S is defined as

$$d(x,y) = \mathrm{Card}\{1 \le i \le n, x_i \ne y_i\},$$

the number of indices for which the coordinates of x and y do not coincide. Given a vector $a \in \mathbb{R}^n$ with nonnegative coordinates a_i, the weighted Hamming distance d_a is defined as

$$d_a(x,y) = \sum_{i=1}^{n} a_i I_{x_i \ne y_i}.$$

Denote by $|a|$, the Euclidean norm of $a \in \mathbb{R}^n$, $|a| = \left(\sum_{i=1}^{n} a_i^2\right)^{1/2}$. Then the *convex distance* on S is defined by

$$d_c(x,y) = \sup_{|a|\le 1} d_a(x,y),\ x,y \in S, \quad \text{and} \quad \text{for } A \subset S \text{ and } x \in S,$$

$$d_c(x,A) = \inf\{d_c(x,y) : y \in A\}. \tag{3.61}$$

We will need an alternative definition of the convex distance. Given $A \subseteq S$, let

$$U_A(x) = \{u = (u_i)_{i=1}^n \in \{0,1\}^n : \exists y \in A \text{ with } y_i = x_i \text{ if } u_i = 0\};$$

that is, $u \in U_A(x)$ if we may obtain from x a point y in A by changing only coordinates x_i such that $u_i = 1$ (without necessarily exhausting all the i such that $u_i = 1$). Let $V_A(x)$ denote the convex hull of $U_A(x)$ as a subset of $[0,1]^n$. Then:

Lemma 3.2.3

$$d_c(x,A) = \inf\{|v| : v \in V_A(x)\},$$

and the infimum is attained at a point in $V_A(x)$.

Proof The infimum is attained because $V_A(x)$ is compact. If $x \in A$, then both quantities above are zero. Thus, we may assume that $x \notin A$, in which case both are positive. Letting $\langle \cdot, \cdot \rangle$ denote inner product in $\mathbb{R}^n$, we have, by the definition of $U_A(x)$, the fact that the maximum of a linear functional over the convex hull V of a finite set U is attained at a point in U and the Cauchy-Schwarz inequality,

$$d_a(x,A) := \inf_{y \in A} d_a(x,y) = \min_{u \in U_A(x)} \langle a, u \rangle = \min_{v \in V_A(x)} \langle a, v \rangle \leq |a||v|. \tag{3.62}$$

This proves the inequality $d_c(x,A) \leq \inf\{|v| : v \in V_A(x)\}$.

For the converse inequality, if the infimum of $|v|$ is achieved at $\tilde{v} \in V_A(x)$, let $a = \tilde{v}/|\tilde{v}|$. Let $v \in V_A(x)$. Then, since V is convex, we have $\tilde{v} + \lambda(v - \tilde{v}) \in V_A(x)$, for all $0 \leq \lambda \leq 1$. Hence, by definition of $\tilde{v}$, $|\tilde{v} + \lambda(v - \tilde{v})| \geq |\tilde{v}|$, and developing the squares in this inequality and dividing by λ yield

$$2\langle \tilde{v}, v - \tilde{v} \rangle + \lambda |v - \tilde{v}|^2 \geq 0,$$

which, letting $\lambda \searrow 0$, implies $\langle \tilde{v}, v - \tilde{v} \rangle \geq 0$. Hence,

$$\langle a, v \rangle \geq \langle a, \tilde{v} \rangle = |\tilde{v}|^2/|\tilde{v}| = \min\{|v| : v \in V_A(x)\}.$$

But then, by (3.62),

$$d_c(x,A) \geq d_a(x,A) = \min_{v \in V_A(x)} \langle a, v \rangle \geq \min\{|v| : v \in V_A(x)\}. \quad \blacksquare$$

Here is Talagrand's concentration inequality for the convex distance. The elegant inequality (3.64) quantifies the concentration of the law of X about the set A (with respect to the convex distance) when the probability of X being in A is large.

Theorem 3.2.4 (Talagrand's inequality for the convex distance) *For any $n \in \mathbb{N}$, if $X = (X_1, \ldots, X_n)$ is a vector of independent random variables taking values in the product space $S^{(n)} = \prod_{k=1}^n S_k$, and $A \subseteq S^{(n)}$, then*

$$E\left(e^{d_c^2(x,A)/4}\right) \leq \frac{1}{\Pr(X \in A\}}; \tag{3.63}$$

hence, for all $t \geq 0$,

$$\Pr\{X \in A\} \Pr\{d_c(x,A) \geq t\} \leq e^{-t^2/4}. \tag{3.64}$$

Proof For $n = 1$, $d_c(x,A) = 1 - I_A(x)$, and we have

$$E\left(e^{d_c^2(X,A)/4}\right) \leq \mu_1(A) + e^{1/4}(1 - \mu_1(A)) \leq \mu_1(A) + 2(1 - \mu_1(A)) = 2 - \mu_1(A) \leq \frac{1}{\mu_1(A)},$$

proving (3.63) in this case.

Now we proceed by induction. Let $n \geq 1$, assume that inequality (3.63) holds for n (and for 1) and consider the case $n+1$. Set $S = \prod_{k=1}^{n+1} S_k$ and $S^{(n)} = \prod_{k=1}^n S_k$ so that $S = S^{(n)} \times S_{n+1}$.

Recall $\mu_k = \mathcal{L}(X_k)$, and set $X^{(n)} = (X_1,\dots,X_n)$, $P^{(n)} = \mu_1 \times \cdots \times \mu_n$, $X = (X_1,\dots,X_{n+1}) = (X^{(n)},X_{n+1})$ and $P = P^{(n)} \times \mu_{n+1}$. For $z \in S$, write $z = (x,s)$ with $x \in S^{(n)}$ and $s \in S_{n+1}$. For each $s \in S_{n+1}$, define

$$A_s = \{x \in S^{(n)} : (x,s) \in A\}, \quad B = \cup_{s\in S_{n+1}} A_s,$$

the section of A along s and the the projection of A on $S^{(n)}$, respectively. The following convexity inequality relates the squares of the convex distances $d_c(z,A)$, $d_c(x,A_s)$ and $d_c(x,B)$, for $z = (x,s) \in A$ and $0 \le \lambda \le 1$:

$$d_c^2(z,A) \le \lambda d_c^2(x,A_s) + (1-\lambda)d_c^2(x,B) + (1-\lambda)^2. \tag{3.65}$$

To see this, note that (a) if $t \in U_{A_s}(x)$, then $(t,0) \in U_A(z)$ (if $y \in A_s$, then $(y,s) \in A_z$) and (b) if $u \in U_B(x)$, then $(u,1) \in U_A(z)$ (if $y \in B$, then (y,s) is not necessarily in A). Thus, if $v \in V_{A_s}(x)$, then $(v,0) \in V_A(z)$, and if $v \in V_B(x)$, then $(v,1) \in V_A(z)$. Lemma 3.2.3 ensures that there are $t \in V_{A_s}(x)$ with $|t| = d_c(x,A_s)$ and $u \in V_B(x)$ with $|u| = d_c(x,B)$, and the preceding discussion implies that both $(t,0)$ and $(u,1) \in V_A(z)$ so that, by convexity of $V_A(z)$, we also have, for $0 \le \lambda \le 1$,

$$v = \lambda(t,0) + (1-\lambda)(u,1) = (\lambda t + (1-\lambda)u, 1-\lambda) \in V_A(z).$$

Then, by Lemma 3.2.3 again and convexity of $f(x) = x^2$, we have

$$\begin{aligned} d_c^2(z,A) &\le |v|^2 = |\lambda t + (1-\lambda)u|^2 + (1-\lambda)^2 \\ &\le \lambda |t|^2 + (1-\lambda)|u|^2 + (1-\lambda)^2 \\ &= \lambda d_c^2(x,A_s) + (1-\lambda)d_c^2(x,B) + (1-\lambda)^2, \end{aligned}$$

proving (3.65).

For $s \in S_{n+1}$ fixed, inequality (3.65) and Hölder's inequality with $p = 1/\lambda$ and $q = 1/(1-\lambda)$ give

$$\begin{aligned} Ee^{d_c^2((X^{(n)},s),A)/4} &\le e^{(1-\lambda)^2/4} E\left[e^{\lambda d_c^2(X^{(n)},A_s)/4 + (1-\lambda)d_c^2(X^{(n)},B)/4}\right] \\ &\le e^{(1-\lambda)^2/4}\left(Ee^{d_c^2(X^{(n)},A_s)/4}\right)^{\lambda}\left(Ee^{d_c^2(X^{(n)},B)/4}\right)^{1-\lambda}, \end{aligned}$$

and the induction applied to these two expected values yields

$$\begin{aligned} Ee^{d_c^2((X^{(n)},s),A)/4} &\le e^{(1-\lambda)^2/4}(P^{(n)}(A_s))^{-\lambda}(P^{(n)}(B))^{-(1-\lambda)} \\ &= \frac{1}{P^{(n)}(B)} e^{(1-\lambda)^2/4}\left(\frac{P^{(n)}(A_s)}{P^{(n)}(B)}\right)^{-\lambda}. \end{aligned}$$

At this point we optimize in $\lambda \in [0,1]$. Observe that by Exercise 3.2.1, for all $0 \le r \le 1$, $\inf_{0\le\lambda\le 1} r^{-\lambda} e^{(1-\lambda)^2/4} \le 2 - r$. Applying this bound for $r = P^{(n)}(A_s)/P^{(n)}(B)$ and integrating with respect to μ_{n+1} (note that $EP^{(n)}(A_{X_{n+1}}) = PA$), we obtain

$$Ee^{d_c^2(X,A)/4} \le \frac{1}{P^{(n)}(B)}\left(2 - \frac{P(A)}{P^{(n)}(B)}\right) \le \frac{1}{P(A)},$$

where for the preceding inequality we multiply and divide by $P(A)$ and use that $u(2-u) \le 1$ for all $u \in \mathbb{R}$. This proves inequality (3.63). Inequality (3.64) follows from it by Markov's inequality applied to $e^{d_c^2(X,A)/4}$. ■

As a consequence of Theorem 3.2.4, we have the following concentration result for Lipschitz functions:

Corollary 3.2.5 *Let $S = S_1 \times \cdots \times S_n$ be a product of measurable spaces, and let P be a product probability measure on it. Let $F : S \mapsto \mathbb{R}$ be a measurable function satisfying the following Lipschitz property for the distance d_a: for every $x \in S$, there is $a = a(x) \in \mathbb{R}^n$ with $|a| = 1$ such that*

$$F(x) \leq F(y) + d_a(x,y), \quad y \in S.$$

Let m_F be a median of F for P. Then, for all $t \geq 0$,

$$P\{|F - m_F| \geq t\} \leq 4e^{-t^2/4}.$$

Proof Taking $A = \{F \leq m\}$ (any m), we obviously have

$$F(x) \leq m + d_a(x,A) \leq m + d_c(x,A).$$

Therefore, Theorem 3.2.4 implies that

$$P\{F \geq m + t\} \leq P\{d_c(x,A) \geq t\} \leq \frac{1}{P(A)} e^{-t^2/4}, \quad t \geq 0;$$

that is,

$$P\{F \leq m\}P\{F \geq m + t\} \leq e^{-t^2/4}, \quad t \geq 0.$$

Now take $m = m_F - t$ and $m = m_F$ to obtain the result. ∎

Corollary 3.2.6 (Concentration inequality for Rademacher (and other) processes) *Let X_i, $1 \leq i \leq n$, be independent real random variables such that, for real numbers a_i, b_i,*

$$a_i \leq X_i \leq b_i, \quad 1 \leq i \leq n.$$

Let T be a countable subset of $\mathbb{R}^n$, and set

$$Z = \sup_{t \in T} \sum_{i=1}^{n} t_i X_i,$$

where $t = (t_1, \ldots, t_n)$. Let m_Z be a median of Z. Then, if $\tilde{\sigma} := \sup_{t \in T} \left(\sum_{i=1}^{n} t_i^2 (b_i - a_i)^2 \right)^{1/2}$ is finite, we have that, for every $r \geq 0$,

$$\Pr\{|Z - m_Z| \geq r\} \leq 4e^{-r^2/4\tilde{\sigma}^2},$$

$$|EZ - m_Z| \leq 4\sqrt{\pi}\,\tilde{\sigma} \quad \text{and} \quad \mathrm{Var}(Z) \leq 16\tilde{\sigma}^2.$$

Proof We will apply the preceding corollary for $S = \prod_{i=1}^{n} [a_i, b_i]$, P the product of the probability laws of $X_1, \ldots, X_n$ and $F(x) = \sup_{t \in T} \sum_{i=1}^{n} t_i X_i$. We can and do assume that T is finite. For $x \in S$, let $t = t(x)$ achieve the supremum in the definition of $F(x)$. Then, for any $y \in S$,

$$F(x) = \sum_{i=1}^{n} t_i x_i \leq \sum_{i=1}^{n} t_i y_i + \sum_{i=1}^{n} |t_i||x_i - y_i|$$

$$\leq F(y) + \tilde{\sigma} \sum_{i=1}^{n} \frac{|t_i|(b_i - a_i)}{\tilde{\sigma}} I_{x_i \neq y_i},$$

showing that the function $F/\tilde{\sigma}$ satisfies the Lipschitz condition in Corollary 3.2.5 for $a = a(x) = (|t_1|(b_1 - a_1)/\tilde{\sigma}, \ldots, |t_n|(b_n - a_n)/\tilde{\sigma})$. Then this corollary implies the concentration inequality for Z. The comparison between the mean and the median of Z follows from

$$|EZ - m_Z| \le E|Z - m_Z| \le 4\int_0^\infty e^{-r^2/4\tilde{\sigma}^2}\,dr$$

and the variance bound from

$$\mathrm{Var}(Z) \le E|Z - m_Z|^2 \le 8\int_0^\infty re^{-r^2/4\tilde{\sigma}^2}\,dr. \quad \blacksquare$$

This result is a striking generalisation of Hoeffding's inequality (3.9): it is *a Hoeffding inequality holding simultaneously for infinitely many sums of independent variables.* The constants, however, are a little worse.

For Rademacher variables $X_i = \varepsilon_i$, that is, for $Z = \sup_{t\in T}\sum_{i=1}^n t_i\varepsilon_i$, the preceding corollary yields

$$\Pr\{|Z - m_Z| \ge r\} \le 4e^{-r^2/16\sigma^2},$$

where $\sigma^2 = \sup_{t\in T}\sum_{i=1}^n t_i^2$. The constant 16 is not best possible. A more specialised convex distance inequality allows 16 to be replaced by 8. We record this result and sketch its proof in Exercises 3.2.2 and 3.2.3.

Theorem 3.2.7 *For $n < \infty$ and a countable set $T \subset \mathbb{R}^n$, set*

$$Z = \sup_{t\in T}\sum_{i=1}^n t_i\varepsilon_i,\ \sigma = \sup_{t\in T}\left(\sum_{i=1}^n t_i^2\right)^{1/2},$$

and let m_Z be a median of Z. Then, if $\sigma < \infty$,

$$\Pr\{|Z - m_Z| \ge r\} \le 4e^{-r^2/8\sigma^2} \tag{3.66}$$

and, consequently,

$$E|Z - m_Z| \le 4\sqrt{2\pi}\sigma \text{ and } \mathrm{Var}(Z) \le 32\sigma^2. \tag{3.67}$$

Next we consider an important consequence of this theorem (or Corollary 3.2.6) regarding integrability of Rademacher processes in analogy with Gaussian processes.

Proposition 3.2.8 (Khinchin-Kahane inequalities) *For Z as in Theorem 3.2.7, for all $p > q > 0$, there exists $C_q < \infty$ such that*

$$(E|Z|^p)^{1/p} \le C_q\sqrt{p}(E|Z|^q)^{1/q}. \tag{3.68}$$

Moreover, there are $\tau > 0$ and $c > 0$ such that $\Pr\{|Z| > c\|Z\|_2\} \ge \tau$.

Proof First, we see that integrating the exponential inequality (3.66) with respect to $pt^{p-1}dt, p \ge 2$, gives

$$\|Z\|_p \le m_Z + \left(8\sqrt{2/\pi}\right)^{1/p}\|g\|_p\sigma,$$

where g is a standard normal variable. Since $m_Z \le \sqrt{2}\|Z\|_2$, $\sigma \le \|Z\|_2$ and $\|g\|_p$ is of the order of $\sqrt{p}$, it follows that for a universal constant K,

$$\|Z\|_p \le K\sqrt{p}\|Z\|_2.$$

We now observe that for $0 < \tau < 1$, by Hölder's inequality,

$$\begin{aligned} EZ^2 &\le \tau^2 EZ^2 + E\left(Z^2 I(|Z| > \tau\|Z\|_2)\right) \le \tau^2 EZ^2 + \left(EZ^4\right)^{1/2} (\Pr\{|Z| > \tau\|Z\|_2\})^{1/2} \\ &\le \tau^2 EZ^2 + 4K^2 EZ^2 (\Pr\{|Z| > \tau\|Z\|_2\})^{1/2}, \end{aligned}$$

and hence,

$$\Pr\{|Z| > \tau\|Z\|_2\} \ge \frac{(1-\tau^2)^2}{16K^4}.$$

This is an instance of the *Paley-Zigmund argument*; see Exercise 2.1.4. Then, if $m'_Z = \sup\{m : \Pr\{|Z| \ge m\} \ge 9/(4K)^4\}$, this inequality with $\tau = 1/4$ gives $\|Z\|_2 \le 4m'_Z$. But for each $q > 0$, by Markov's inequality, $9/(4K)^4 \le \Pr\{|Z| \ge m'_Z\} \le E|Z|^q/m'^q_Z$; that is, $m'_Z \le ((4K)^4/9)^{1/q}(E|Z|^q)^{1/q}$ or $\|Z\|_2 \le 4((4K)^4/9)^{1/q}(E|Z|^q)^{1/q}$, for any $0 < q < 2$. ■

This result is only best possible up to constants. For instance, it is know that for $p > q > 1$,

$$\|Z\|_p \le \left(\frac{p-1}{q-1}\right)^{1/2} \|Z\|_q.$$

See the notes at the end of the chapter.

3.2.3 A Lower Bound for the Expected Supremum of a Rademacher Process

Sudakov's lower bound for centred bounded Gaussian processes X (Theorem 2.4.12), namely, that for all $\varepsilon > 0$, $\varepsilon\sqrt{\log N(T, d_X, \varepsilon)} \le KE\sup_{t\in T}|X(t)|$, does not extend to Rademacher processes without substantial modifications: if $T_n = \{e_1, \dots, e_n\}$, the canonical basis of $\mathbb{R}^n$, then $E\max_{t\in T_n}\left|\sum_{i=1}^n \varepsilon_i t_i\right| = 1$, whereas for $\varepsilon < 1$, $N(T_n, d_2, \varepsilon) = n$, where d_2 is Euclidean distance in $\mathbb{R}^n$, which is the distance induced by the process on T_n.

Here we present a first variation on Sudakov's inequality, just what we need in a later section.

Theorem 3.2.9 *There exists a finite constant $K > 0$ such that for every $n \in \mathbb{N}$ and $\varepsilon > 0$, if T is a bounded subset of $\mathbb{R}^n$ such that*

$$E\sup_{t\in T}\left|\sum_{i=1}^n \varepsilon_i t_i\right| \le \frac{1}{K}\frac{\varepsilon^2}{\max_{1\le i\le n}|t_i|}, \quad \textit{for all } t \in T, \tag{3.69}$$

then

$$\varepsilon\sqrt{\log N(T, d_2, \varepsilon)} \le KE\sup_{t\in T}\left|\sum_{i=1}^n \varepsilon_i t_i\right|. \tag{3.70}$$

We need two lemmas. The first gives a bound on the moment-generating function of a normal random variable truncated from below, and the second is just Proposition 3.2.8 for $p = 1$ and $q = 2$ when T consists of a single point, which in this case has a simpler proof.

Lemma 3.2.10 *Let g be a standard normal random variable, and let $h = gI_{|g|>s}$ for some $s > 0$. Then*

$$Ee^{\lambda h} \le 1 + 16\lambda^2 e^{-s^2/32} \le \exp\left(16\lambda^2 e^{-s^2/32}\right),$$

for all $0 \le \lambda \le s/4$.

Proof It suffices to prove the first inequality. Set $f(\lambda) = Ee^{\lambda h} - 1 - 16\lambda^2 e^{-s^2/32}$, $\lambda \ge 0$. Then $f(0) = f'(0) = 0$, and therefore, it suffices to prove that $f''(\lambda) \le 0$ for $0 \le \lambda \le s/4$. But, on the one hand,

$$f''(\lambda) = E\left(h^2 e^{\lambda h}\right) - 32e^{-s^2/32},$$

and on the other, we have, first by change of variables and then by observing that $\lambda \le s/4$ $(< s/2)$ and $s < |x+\lambda|$ imply that $\{x : |x+\lambda| > s\} \subseteq \{x : |x| > s/2\}$ because $s < |x+\lambda| \le |x| + s/2$ and that $|x+\lambda| > 2|x|$ on the first set,

$$\begin{aligned}
E\left(h^2 e^{\lambda h}\right) &= e^{\lambda^2/2} \int_{|x+\lambda|>s} (x+\lambda)^2 e^{-x^2/2} dx/\sqrt{2\pi} \\
&\le 4e^{\lambda^2/2} \int_{|x|>s/2} x^2 e^{-x^2/2} dx/\sqrt{2\pi} \\
&\le 16e^{\lambda^2/2} \int_{|x|>s/2} e^{-x^2/4} dx/\sqrt{2\pi} \\
&\le 32e^{\lambda^2/2 - s^2/16} \le 32e^{-s^2/32},
\end{aligned}$$

proving the lemma. ∎

A fast way to obtain inequalities of the type (3.2.2) for $q \le 4$ and a single point $t \in \mathbb{R}^n$ is as follows: first, we observe that, by symmetry,

$$E\left(\sum \varepsilon_i t_i\right)^4 = \sum t_i^4 + 6\sum_{i<j} t_i^2 t_j^2 \le 3\left(\sum t_i^2\right)^2 = 3\left[E\left(\sum t_i \varepsilon_i\right)^2\right]^2.$$

The Paley-Zygmund argument would allow us to obtain *reverse Hölder inequalities* down to quantiles, but if we just want to bound the L^2-norm by the L^1-norm, we may use the *Littlewood argument*, given in the proof of the next lemma. It does not give the best constant, but the constant obtained is small enough for our purposes.

Lemma 3.2.11 *For any n and $t_i \in \mathbb{R}$, $i = 1, \ldots, n$,*

$$\left\|\sum t_i \varepsilon_i\right\|_2 \le \sqrt{3}\left\|\sum t_i \varepsilon_i\right\|_1.$$

Proof Set $R = \left|\sum t_i \varepsilon_i\right|$. Applying Hölder's inequality for $1/p = \alpha$ and $1/q = 1 - \alpha$, for $0 < \alpha < 1$, we have

$$ER^2 = E\left(R^\alpha R^{2-\alpha}\right) \le (ER)^\alpha \left(ER^{(2-\alpha)/(1-\alpha)}\right)^{1-\alpha}.$$

Then, taking $\alpha = 2/3$ so that $(2-\alpha)/(1-\alpha) = 4$, this gives

$$ER^2 \le (ER)^{2/3}\left(ER^4\right)^{1/3},$$

which, combined with the bound $ER^4 \le 3(ER^2)^2$ obtained earlier, yields the lemma. ∎

The best constant in the preceding inequality is $\sqrt{2}$ (see the notes at the end of this chapter). We now prove Theorem 3.2.9.

Proof Set $B_2 := \{x \in \mathbb{R}^n : |x| \le 1\}$ and $R_T := E\sup_{t\in T}\left|\sum_{i=1}^n \varepsilon_i t_i\right|$. The main step in the proof is to show that if $T \subset B_2$ and condition (3.69) holds for $\varepsilon = 1$, then

$$(\log N(T, d_2, 1/2))^{1/2} \le KR_T, \tag{3.71}$$

for some universal constant K (and then, by increasing the constants, if necessary, we may take K to be the same in this inequality and in condition (3.69)).

To prove (3.71), we first note that if T is contained in the ball about zero of radius 1/2, there is nothing to prove. Thus, we can assume that there is a point in T with norm larger than 1/2, $|t| > 1/2$. Then, by Lemma 3.2.11, R_T is bounded from below; in fact, $R_T \ge 1/2\sqrt{3}$. Let now $U \subseteq T$ be a set of cardinality $N(T, d_2, 1/2)$ satisfying $d_2(u, v) \ge 1/2$ whenever $u, v \in U$, $u \ne v$ ($N(T, d_2, \varepsilon) < \infty$ for all $\varepsilon > 0$ because T is bounded). Such a set exists by the definition of covering numbers. Define now $\|G\|_U := \max_{u\in U}\left|\sum_{i=1}^n g_i u_i\right|$, where g_i are i.i.d. $N(0,1)$ random variables. This is the supremum of a Gaussian process, and Sudakov's inequality (Theorem 2.4.12) gives

$$E\|G\|_U \ge C_1(\log \mathrm{Card}(U))^{1/2},$$

for some universal constant C_1. Also, by the integrability properties of Gaussian processes, concretely Exercise 2.1.3, if m_G is the median of $\|G\|_U$, then $m_G \ge C_2 E\|G\|_U$, for another universal constant C_2. This and Sudakov's inequality give $m_G \ge (K')^{-1}(\log \mathrm{Card}(U))^{1/2}$ for $K' = 1/C_1C_2$, which we may and do assume larger than 1; that is,

$$\Pr\left\{\|G\|_U > (\log \mathrm{Card}(U))^{1/2}/K'\right\} \ge \frac{1}{2}. \tag{3.72}$$

Let now $K = (100K')^2$ and assume that $\max_i |t_i| \le 1/KR_T$ for all $t \in T$. Since R_T is bounded from below, there exists $\alpha \ge 1$ such that $(\log \mathrm{Card}(U))^{1/2} \le \alpha KR_T$. Suppose that we prove that for such α,

$$\Pr\left\{\|G\|_U > \alpha\frac{K}{2K'}R_T\right\} < \frac{1}{2}. \tag{3.73}$$

Then the two sets in (3.72) and (3.73) have a nonvoid intersection, and this implies that

$$(\log \mathrm{Card}(U))^{1/2} \le \frac{\alpha}{2}KR_T.$$

If $\alpha/2 \le 1$, inequality (3.71) is proved. If $\alpha/2 > 1$, we may repeat the argument and reduce the constant by another half, and so on, to conclude, in any case, that $(\log \mathrm{Card}(U))^{1/2} \le KR_T$, that is, (3.71). We proceed to prove (3.73). For $s > 0$ to be determined later, set $h_i = g_i I_{|g_i|>s}$ and $k_i = g_i - h_i$, $i = 1, \dots, n$. We may write

$$\Pr\left\{\|G\|_U > \frac{\alpha K}{2K'}R_T\right\} \le \Pr\left\{\max_{u\in U}\left|\sum_{i=1}^n k_i u_i\right| > \frac{\alpha K}{4K'}R_T\right\} + \Pr\left\{\max_{u\in U}\left|\sum_{i=1}^n h_i u_i\right| > \frac{\alpha K}{4K'}R_T\right\}.$$

By symmetry and the contraction principle (Theorem 3.1.17), the first probability is bounded by

$$\Pr\left\{\max_{u\in U}\left|\sum_{i=1}^n k_i u_i\right| > \frac{\alpha K}{4K'}R_T\right\} \le \frac{4K' s E_g E_\varepsilon \sup_{u\in U}\left|\sum_{i=1}^n \varepsilon_i u_i k_i/|k_i|\right|}{\alpha K E\max_{u\in U}\left|\sum_{i=1}^n \varepsilon_i u_i\right|} \le \frac{4K's}{\alpha K}.$$

Before estimating the second probability, let us note that if $\lambda \le sKR_T/4$ and $u \in U \subset B_2$, then, by condition (3.69), $\lambda u_i \le s/4$, and therefore, by Lemma 3.2.10,

$$Ee^{\lambda h_i u_i} \le \exp\left(16\lambda^2 u_i^2 e^{-s^2/32}\right),$$

which implies that

$$\begin{aligned} E\exp\left(\lambda\left|\sum_{i=1}^n h_i u_i\right|\right) &\le 2E\exp\left(\lambda\sum_{i=1}^n h_i u_i\right) \\ &\le 2\prod_{i=1}^n \exp\left(16\lambda^2\sum_{i=1}^n u_i^2 e^{-s^2/32}\right) \le 2\prod_{i=1}^n \exp\left(16\lambda^2 e^{-s^2/32}\right), \end{aligned}$$

where we use that if X is symmetric, then

$$Ee^{\lambda|X|} \le E\left(e^{\lambda X} I_{X\ge 0}\right) + E\left(e^{-\lambda X} I_{X\le 0}\right) \le Ee^{\lambda X} + Ee^{-\lambda X} = 2Ee^{\lambda X}.$$

Hence, by Markov's inequality after exponentiating,

$$\Pr\left\{\max_{u\in U}\left|\sum_{i=1}^n h_i u_i\right| > \frac{\alpha K}{4K'}R_T\right\} \le 2(\mathrm{Card}(U)\exp\left[-\lambda\frac{\alpha K}{4K'}R_T + 16\lambda^2 e^{-s^2/32}\right].$$

Collecting the two probability estimates and setting $s = \alpha K/10K'$ and $\lambda = \alpha K^2 R_T/40K'$, we obtain

$$\Pr\left\{\|G\|_U > \frac{\alpha K}{2K'}R_T\right\} \le \frac{2}{5} + 2\exp\left[\alpha^2 K^2 R_T^2\left(1 - \frac{K}{160(K')^2} + \frac{16K^2}{(40K')^2}\exp\left(-\frac{\alpha^2 K^2}{32(10K')^2}\right)\right)\right].$$

Since $K = (100K')^2$, $\alpha \ge 1$, and $R_T \ge 1/2\sqrt{3}$, this exponent is negative and large enough in absolute value to yield a bound of less the 1/2, proving (3.73).

Finally, it remains to be shown that the theorem follows from the case $\varepsilon = 1/2$. This will follow by iteration. Since $N(\alpha T, d_2, \varepsilon) = N(T, d_2, \varepsilon/\alpha)$ and $R_{\alpha T} = \alpha R_T$, both sides of inequality (3.70) are homogeneous of degree 1 in α (as functions of αT). Also, if αT satisfies (3.69) for a given ε, then T satisfies the same condition for ε/α. Hence, we can assume without loss of generality that $T \subset B_2$. Given $\varepsilon > 0$, let k be such that $2^{-k} < \varepsilon \le 2^{-k+1}$. Given a covering of T by balls of radius $2^{-\ell+1}$, we can always produce a covering of T by balls of radius $2^{-\ell}$ by combining coverings by balls of radius $2^{-\ell}$ of each of the balls of radius $2^{-\ell+1}$. Therefore,

$$N(T, d_2, \varepsilon) \le N(T, d_2, 2^{-k}) \le \prod_{\ell=1}^k \sup_{t\in T} N(T \cap B_2(t, 2^{-\ell+1}), d_2, 2^{-\ell}),$$

where $B_2(t,\varepsilon)$ denotes the Euclidean ball of radius ε and centre t. Now, if $T' = T \cap B_2(t, 2^{-\ell+1})$ and $\tilde{T} = 2^{\ell-1}T' \subseteq B_2$, then

$$R_{\tilde{T}} = 2^{\ell-1}R_{T'} \le 2^{\ell-1}R_T, \quad N(\tilde{T}, d_2, 1/2) = N(2^{\ell-1}T', d_2, 1/2) = N(T', d_2, 2^{-\ell}),$$

and, by (3.69),

$$\sup_{t\in\tilde{T}}\max_{i\le n}|t_i| = \sup_{t\in T'}\max_{i\le n} 2^{\ell-1}|t_i| \le 2^{\ell-1}\varepsilon^2/KR_T \le 2^{2\ell-2}\varepsilon^2/KR_{\tilde{T}} \le 1/KR_{\tilde{T}}.$$

Hence, (3.73) gives

$$\log N(T\cap B_2(t,2^{-\ell+1}),d_2,2^{-\ell})=\log N(\tilde{T},d_2,1/2)\le K^2R_{\tilde{T}}^2\le 2^{2\ell-2}K^2R_T^2,$$

and, adding up,

$$\log N(T,d_2,\varepsilon)\le K^2R_T^2\sum_{\ell=1}^{k}2^{2\ell-2}<2K^2R_T^2/\varepsilon^2,$$

proving the theorem. ■

Exercises

3.2.1 Prove that for all $0\le r\le 1$, $\inf_{0\le\lambda\le 1}r^{-\lambda}e^{(1-\lambda)^2/4}\le 2-r$. *Hints*: Take derivatives to show that the minimum of the function of λ in the statement of the exercise is attained at $\lambda=1+2\log r$, in particular, (a) the function is nondecreasing on $[0,1]$ for $0\le r\le e^{-1/2}$, its minimum is at $\lambda=0$ and it does not exceed $2-r$ (this follows from $e^{1/8}+1\le 2e^{1/2}$); for $e^{-1/2}\le r\le 1$, the minimum is $e^{-(\log r)(1+2\log r)}e^{(\log r)^2}$, and proving the inequality reduces to showing that $(\log r)^2+\log r+\log(2-r)\ge 0$. Since the value of this function is 0 at $r=1$, check that its derivative for $r\le 1$, whose value at $r=1$ is 0, is not positive by observing that the second derivative is nonnegative.

3.2.2 Set $S_n=\{-1,1\}^n$, $\mu_n=\frac{1}{2}\delta_{-1}+\frac{1}{2}\delta_1$ and $P_n=\mu^n$. For $x\in S_n$ and $A\subseteq S_n$, define $d_A(x)=\inf\{|x-y|: y\in \mathrm{Conv}(A)\}$, where $|\cdot|$ denotes Euclidean distance and $\mathrm{Conv}(A)$ is defined as the convex hull of A on $[-1,1]^n$. Then

$$P_n(A)\int_{S_n}e^{d_A^2/8}dP_n\le 1.$$

Hint: If $\mathrm{Card}(A)=1$, then the integral is $2^{-n}\sum_{i=1}^n\binom{n}{i}e^{i/2}=((1+e^{1/2})/2)^n\le 1/P_n(A)$. In particular, the theorem is proved for $n=1$ because the theorem is obviously true for the remaining case of $A=\{-1,1\}$. Thus, we can assume that the theorem is proved for $k\le n$, and we must prove that it also holds for $A\subset\{-1,1\}^{n+1}$, with A containing at least two points. We may also assume that these two points differ on the last coordinate and write $A=A_{-1}\times\{-1\}\cup A_{+1}\times\{+1\}$, with $A_i\ne\emptyset$ and, for example, $P_nA_{-1}\le P_nA_{+1}$. Prove that, in analogy with the convexity inequality (3.65), for all $0\le\lambda\le 1$ and $x\in S_n$,

$$d_A^2((x,-1))\le 4\lambda^2+\lambda d_{A_{+1}}^2(x)+(1-\lambda)d_{A_{-1}}^2(x),$$

and observe as well that $d_A((x,1))\le d_{A_{+1}}(x)$. Set $u_i=\int e^{d_i^2/8}dP_n$, $v_i=1/P_n(A_i)$, $i=-1,+1$. The induction hypothesis simply reads $u_i\le v_i$, and this and the preceding two estimates give

$$\int_{S_{n+1}}e^{d_A^2/8}dP_{n+1}\le\frac{1}{2}v_{+1}\left[1+e^{\lambda^2/2}\left(\frac{v_{-1}}{v_{+1}}\right)^{1-\lambda}\right].$$

Taking $\lambda=1-v_{+1}/v_{-1}$ (which approximates the minimiser $-\log(v_{+1}/v_{-1})$ and is dominated by 1), this bound becomes

$$\frac{1}{2}v_{+1}\left[1+e^{\lambda^2/2}(1-\lambda)^{1-\lambda}\right]\le\frac{1}{2}v_{+1}\frac{4}{2-\lambda}=1/P_{n+1}(A).$$

3.2.3 Check that the function $F(x)=\sup_{t\in T}\sum_{i=1}^n t_ix_i$, $x=(x_1,\dots,x_n)\in\{-1,1\}^n$, is convex and is Lipschitz with constant $\sigma=\sup_{t\in T}\left(\sum_{i=1}^n t_i\right)^{1/2}$ with respect to the Euclidean distance, and conclude from the preceding exercise that if $Z=\sup_{t\in T}\sum_{i=1}^n t_i\varepsilon_i$, then

$$\Pr\{|Z-m_Z|\ge t\}\le 4e^{-t^2/8\sigma^2},\quad t\ge 0.$$

Hint: Proceed by analogy with Corollary 3.2.5.

3.2.4 Let $\mathcal{E}$ denote the L^2-closure of a sequence ε_i of independent Rademacher variables. Show that on $\mathcal{E}$ the L^p-topologies are all equivalent for $0 \le p < \infty$, where the L^0-topology is the topology of convergence in probability. For $p \ge 1$, all the L^p-metrics are equivalent on $\mathcal{E}$.

3.3 The Entropy Method and Talagrand's Inequality

The object of this section is to prove Talagrand's inequality, which is one of the deepest results in the theory of empirical processes. This inequality may be thought of as a Bennett, Prokhorov or Bernstein inequality uniform over an infinite collection of sums of independent random variables, that is, for the supremum of an empirical process. As such, it constitutes an exponential inequality of the best possible kind (see the discussion about the optimality of Prohorov's inequality just after Theorem 3.1.7). Talagrand's inequality has several proofs, arguably the most efficient being the one based on log-Sobolev inequalities. These are bounds for

$$\mathrm{Ent}_\mu f := E_\mu(f \log f) - (E_\mu f)(\log E_\mu f)$$

in terms of f and its derivatives, which, when applied to $f(z) = e^{\lambda z}$, yield differential inequalities for the Laplace transform of μ, $F(\lambda) = Ee^{\lambda Z}$, where $\mathcal{L}(Z) = \mu$. Note that if $f(z) = e^{\lambda z}$, then $\mathrm{Ent}_\mu f = \lambda F'(\lambda) - F(\lambda)\log F(\lambda)$. These differential inequalities can be integrated in many important cases and produce bounds for the Laplace transform $F(\lambda)$ which, in turn, translate into exponential inequalities for the tail probabilities of Z. The prototype for this procedure is again Gaussian: the log-Sobolev inequality for Gaussian processes yields, via the entropy method (or Herbst's method), the Borell-Sudakov-Tsirelson inequality for the concentration about its mean of the supremum of a sample bounded Gaussian process, as shown in Section 2.5. Here we examine modified log-Sobolev inequalities for functions of independent random variables satisfying sets of conditions that allow for the inequalities to be integrated once suitably modified. The classes of functions examined here are, in order of increasing complexity, functions of bounded differences, self-bounding functions and subadditive functions, although they are not examined in this order. Talagrand's inequality follows from differential inequalities for subadditive functions.

3.3.1 The Subadditivity Property of the Empirical Process

There are several types of random variables $Z = f(X_1,\dots,X_n)$ defined on product spaces for which the entropy $\mathrm{Ent}_\mu(e^{\lambda f})$ can be bounded by functions of their Laplace transform and/or their first derivative (log-Sobolev-type inequalities) and such that these inequalities can be transformed into solvable differential inequalities for the logarithm of their Laplace transforms, in turn, implying useful exponential deviation or concentration inequalities. These log-Sobolev-type inequalities follow from tensorisation of entropy and are less 'user friendly' than the log-Sobolev inequality for Gaussian processes in the sense that usually one needs to transform them in clever ways to integrate them, particularly if one wishes to obtain best (or close to best) constants in the bounds.

We begin with a simple modified log-Sobolev inequality that will allow us to then isolate the properties of the variable Z that are relevant. Let $(S, \mathcal{S})$ be a measurable space. Let, for some $n \in \mathbb{N}$, $f : S^n \mapsto \mathbb{R}$ and $f_k : S^{n-1} \mapsto \mathbb{R}$, $k = 1, \ldots, n$, be measurable functions. Let X_i, $1 \le i \le n$, be independent S-valued random variables with laws $\mathcal{L}(X_i) = \mu_i$, let $\mu = \prod_{i=1}^n \mu_i$. Let $Z = f(X_1, \ldots, X_n)$ and let $Z_k = f_k(X_1, \ldots, X_{k-1}, X_{k+1}, \ldots, X_n)$. Then tensorisation of entropy together with the variational formula for entropy yield the following *modified log-Sobolev inequality* for the Laplace transform of Z.

Notation remark With μ, f, X_i and Z as in the preceding paragraph, abusing notation, we will write $\mathrm{Ent}_\mu(e^{\lambda Z})$ for $\mathrm{Ent}_\mu(e^{\lambda f})$. Also, we will let E_k denote integration with respect to X_k only (i.e., conditional expectation given X_i, $1 \le i \le n$, $i \ne k$). Finally, it is convenient to give names to two functions that appear often in the following inequalities:

$$\phi(\lambda) = e^{-\lambda} + \lambda - 1, \quad \nu(\lambda) = e^{-\lambda}\phi(-\lambda) = 1 - (1+\lambda)e^{-\lambda}, \quad \lambda \in \mathbb{R}. \tag{3.74}$$

These notations are in force for the rest of Section 3.3.

Proposition 3.3.1 *Assume that Z, Z_k have finite Laplace transforms for all λ (or for λ in an interval). Then, for all $\lambda \in \mathbb{R}$ (or in an interval),*

$$\begin{aligned} \mathrm{Ent}_\mu\left(e^{\lambda Z}\right) &\le \sum_{k=1}^n E\left(e^{\lambda Z_k} E_k\left[\phi(\lambda(Z - Z_k))e^{\lambda(Z-Z_k)}\right]\right) = \sum_{k=1}^n E\left[e^{\lambda Z_k}\nu(-\lambda(Z - Z_k))\right] \\ &= \sum_{k=1}^n E\left[\phi(\lambda(Z - Z_k))e^{\lambda Z}\right], \end{aligned} \tag{3.75}$$

Proof By tensorisation of entropy (Proposition 2.5.3), homogeneity of entropy and the variational formula for the entropy of exponentials (Lemma 2.5.5), we obtain

$$\begin{aligned} \mathrm{Ent}_\mu\left(e^{\lambda Z}\right) &\le E\sum_{k=1}^n \mathrm{Ent}_{\mu_k}\left(e^{\lambda Z}\right) \\ &= E\sum_{k=1}^n e^{\lambda Z_k}\mathrm{Ent}_{\mu_k}\left(e^{\lambda(Z-Z_k)}\right) \\ &\le E\sum_{k=1}^n e^{\lambda Z_k} E_k\left[\phi(\lambda(Z - Z_k))e^{\lambda(Z-Z_k)}\right] \\ &= \sum_{k=1}^n E\left[\phi(\lambda(Z - Z_k))e^{\lambda Z}\right]. \quad \blacksquare \end{aligned} \tag{3.76}$$

Now, ϕ is convex and $\phi(0) = 0$; therefore, for any $0 \le x \le 1$ and any $\lambda \in \mathbb{R}$,

$$\phi(\lambda x) \le x\phi(\lambda), \text{ or } \frac{\nu(-\lambda x)}{\nu(-\lambda)} \le x e^{\lambda(x-1)}. \tag{3.77}$$

Then, if Z, Z_k satisfy

$$0 \le Z - Z_k \le 1, \quad \text{for } 1 \le k \le n \quad \text{and} \quad \sum_k (Z - Z_k) \le Z, \tag{3.78}$$

we have $\phi(\lambda(Z - Z_k)) \le (Z - Z_k)\phi(\lambda)$, and the preceding inequality gives

$$\mathrm{Ent}_\mu\left(e^{\lambda Z}\right) \le \phi(\lambda)\sum_{k=1}^n E\left[(Z - Z_k)e^{\lambda Z}\right] \le \phi(\lambda)E\left(Ze^{\lambda Z}\right)$$

or, with $\tilde{Z} = Z - EZ$,

$$\mathrm{Ent}_\mu\left(e^{\lambda\tilde{Z}}\right) \le \phi(\lambda)E\left(\tilde{Z}e^{\lambda\tilde{Z}}\right) + \phi(\lambda)(EZ)Ee^{\lambda\tilde{Z}}, \quad \lambda \in \mathbb{R}.$$

Setting $F(\lambda) = Ee^{\lambda\tilde{Z}}$, this becomes

$$\lambda F'(\lambda) - F(\lambda)\log F(\lambda) \le \phi(\lambda)F'(\lambda) + \phi(\lambda)(EZ)F(\lambda), \quad \lambda \in \mathbb{R},$$

or, letting $L(\lambda) := \log F(\lambda)$,

$$(\lambda - \phi(\lambda))L'(\lambda) - L(\lambda) \le \phi(\lambda)EZ, \tag{3.79}$$

a relatively easy to integrate differential inequality. Random variables Z defined on a product probability space that satisfy the conditions in (3.78) are called *self-bounding*. However, the supremum of an empirical process is not self-bounding unless it is indexed by nonnegative functions f bounded by 1, whereas we are interested in centred empirical processes. It turns out these satisfy (3.78) except for the fact that $Z - Z_k$ may take negative values, but this lack of 'nonnegativeness' is a source of considerable complications. Other important classes of random variables that satisfy modified log-Sobolev inequalities leading to good exponential bounds are variables defined by functions with bounded differences (see later) and by functions that are Lipschitz separately in each coordinate.

Definition 3.3.2 A function $f : S^n \mapsto \mathbb{R}$ is subadditive if there exist n functions $f_k : S^{n-1} \mapsto \mathbb{R}$ such that, setting $x = (x_1, \ldots, x_n)$ and, for each k, $x^{(k)} = (x_1, \ldots, x_{k-1}, x_{k+1}, \ldots, x_n)$, we have both

$$f(x) - f_k(x^{(k)}) \le 1$$

and

$$\sum_{k=1}^{n} \left(f(x) - f_k(x^{(k)})\right) \le f(x),$$

for all $x \in S^n$. If X_i are independent random variables taking values in S and $Z = f(X_1, \ldots, X_n)$, $Z_k = f_k(X_1, \ldots, X_{k-1}, X_{k+1}, \ldots, X_n)$, where f is subadditive with respect to the functions $f_1, \ldots, f_k$, we say that Z is a subadditive random variable with respect to the variables Z_k.

Next, we show that suprema of empirical process indexed by uniformly bounded and centred functions are subadditive. We will need a little more in order to upset the lack of positivity of $Z - Z_k$: the modified log-Sobolev inequality to be obtained in Corollary 3.3.6 will be in part in terms of $\sum_{k=1}^n E_k Y_k^2$ for any variables Y_k such that both $Y_k \le Z - Z_k$ and $EY_k \ge 0$, and the following lemma also shows that such variables Y_k exist for suprema of empirical process and gives as well a nice bound for $\sum_{k=1}^n E_k Y_k^2$:

Lemma 3.3.3 *Let $\mathcal{F}$ be a finite set of measurable functions on $(S, \mathcal{S})$, P-centred and bounded above by 1, and let X, X_i be independent S-valued random variables. Let $Z = \max_{f\in\mathcal{F}} \sum_{i=1}^n f(X_i)$, and set $Z_k := \max_{f\in\mathcal{F}} \sum_{1\le i\le n, i\ne k} f(X_i)$, for some $n \in \mathbb{N}$. Then Z is subadditive with respect to Z_k, $k = 1, \ldots, n$; that is,*

$$Z - Z_k \le 1 \tag{3.80}$$

and

$$\sum_{k=1}^{n}(Z-Z_k)\le Z. \tag{3.81}$$

Moreover,

$$Z-E_kZ\le 1, \tag{3.82}$$

where we denote by E_k-integration with respect to the variable X_k only (conditional expectation given X_i, $i\neq k$). Finally, there exist random variables Y_k, $1\le k\le n$, such that $Y_k\le Z-Z_k$, $E_kY_k=0$ and, if $\mathcal{F}\subset L^2(P)$, also

$$\sum_{k=1}^{n}E_kY_k^2\le\sum_{k=1}^{n}\max_{f\in\mathcal{F}}Ef^2(X_k)=n\sigma^2, \tag{3.83}$$

where

$$\sigma^2:=\frac{1}{n}\sum_{k=1}^{n}\max_{f\in\mathcal{F}}Ef^2(X_k). \tag{3.84}$$

If the functions in $\mathcal{F}$ are bounded by 1 in absolute value, then $|Y_k|\le 1$ and $|Z-Z_k|\le 1$. In this case, the preceding conclusions also hold for $\mathcal{F}\cup(-\mathcal{F})$, that is, for $Z=\max_{f\in\mathcal{F}}|\sum_{i=1}^n f(X_i)|$ and $Z_k:=\max_{f\in\mathcal{F}}|\sum_{1\le i\le n,i\neq k}f(X_i)|$.

Proof For ω fixed, let $f_0^\omega\in\mathcal{F}$ be such that $Z(\omega)=\sum_{i=1}^n f_0^\omega(X_i(\omega))$. For ease of notation, we drop the superindex ω. Then, obviously, $Z-Z_k\le f_0(X_k)\le 1$. Also,

$$(n-1)Z=(n-1)\sum_{i=1}^{n}f_0(X_i)=\sum_{k=1}^{n}\sum_{i\neq k,1\le i\le n}f_0(X_i)\le\sum_{k=1}^{n}Z_k;$$

that is, $\sum_{k=1}^n(Z-Z_k)\le Z$.

Let now $f_k=f_k^\omega$ be such that $Z_k(\omega)=\sum_{i\neq k,1\le i\le n}f_k^\omega(X_i(\omega))$, and note that f_k does not depend of X_k, in particular, $E_kf_k(X_k)=0=Ef_k(X_k)$. (Note that f_k in this proof has a different meaning than f_k in Definition 3.3.2.) Then we have

$$Z-E_kZ\le Z-E_k\left(\sum_{i\neq k}f_k(X_i)+f_k(X_k)\right)=Z-\sum_{i\neq k}f_k(X_i)=Z-Z_k\le 1$$

by (3.80).

Set $Y_k=f_k(X_k)$, which is bounded in absolute value by 1 and, as just observed, is centred. Then

$$Y_k=\sum_{i=1}^{n}f_k(X_i)-Z_k\le Z-Z_k\le 1.$$

Also, $\sum_{k=1}^n E_kY_k^2=\sum_{k=1}^n E_kf_k^2(X_k)\le n\sigma^2$.

If the functions in $\mathcal{F}$ are bounded by 1 in absolute value, then obviously $|Y_k|=|E_kf(X_k)|\le 1$, and by the inequalities in the preceding display, also $|Z-Z_k|\le 1$. ∎

3.3.2 Differential Inequalities and Bounds for Laplace Transforms of Subadditive Functions and Centred Empirical Processes, $\lambda \geq 0$

In this subsection, $Z = Z(X_1,\dots,X_n)$, where X_i are independent and *not* necessarily identically distributed, denotes a *subadditive random variable*, that is, a random variable such that there exist random variables $Z_k = Z_k(X_1,\dots,X_{k-1},X_{k+1},\dots,X_n)$ for which inequalities (3.80) and (3.81) hold.

As indicated in the preceding subsection, in order to reduce the modified log-Sobolev inequality (3.76) to an integrable differential inequality, we need a decoupling inequality such as (3.77). Of course, (3.77) is not useful in the present situation because it only works for $x \in [0,1]$, and now the range of $Z - Z_k$ includes negative values. Thus, we must obtain a bound for $\nu(-\lambda x)/\nu(-\lambda)$ that is nonnegative on $(-\infty,1]$ and is still in terms of $xe^{\lambda x}$ and, at most, of x^2 and x. To this end, write, for $\alpha > 0$,

$$\nu(-\lambda x) = \frac{\nu(-\lambda x)}{xe^{\lambda x} + \alpha x^2 - x}(xe^{\lambda x} + \alpha x^2 - x),\ \lambda \geq 0, \quad x \leq 1,$$

and note that the function $xe^{\lambda x} + \alpha x^2 - x \geq 0$ in the stated range (as $x(e^{\lambda x} - 1) \geq 0$, for $\lambda \geq 0$); actually, it is strictly positive for $x \neq 0$. Also note that

$$\lim_{x\to 1} \frac{\nu(-\lambda x)}{xe^{\lambda x} + \alpha x^2 - x} = \frac{\nu(-\lambda)}{e^{\lambda} + \alpha - 1}.$$

We have the following:

Lemma 3.3.4 *For $\lambda \geq 0$, $\alpha > 0$ and $x \leq 1$,*

$$\nu(-\lambda x) \leq \frac{\nu(-\lambda)}{e^{\lambda} + \alpha - 1}(xe^{\lambda x} + \alpha x^2 - x).$$

As a consequence, if $f \leq 1$ μ a.s., $\lambda \geq 0$ and $\alpha > 0$, then

$$\mathrm{Ent}_{\mu}(e^{\lambda f}) \leq \frac{\nu(-\lambda)}{e^{\lambda} + \alpha - 1}\int \left(fe^{\lambda f} + \alpha f^2 - f\right) d\mu. \tag{3.85}$$

Proof By the preceding considerations, it suffices to prove that the function $f(x)/g(x) := \nu(-\lambda x)/(xe^{\lambda x} + \alpha x^2 - x)$ attains its absolute maximum over $(-\infty,1]$ at $x = 1$. Note that $f(0) = g(0) = f'(0) = g'(0) = 0$ and that $\lim_{x\to 0} f(x)/g(x)$ exists (and equals $\lambda^2/2(\lambda+\alpha)$). Also, $g(x) > 0$ for $x \neq 0$, $g'(x) > 0$ for $x > 0$ and $g'(x) < 0$ for $x < 0$. Now consider

$$\left(\frac{f}{g}\right)' = \frac{g'}{g}\left(\frac{f'}{g'} - \frac{f}{g}\right).$$

Since by the mean value theorem $f(x)/g(x) = f'(c)/g'(c)$ for some c between 0 and x, it follows that *if f'/g' is nondecreasing on $(-\infty,0)$ and on $(0,1]$, then $(f/g)' \geq 0$ on the same intervals*. If this is the case, the maximum of f/g over $(-\infty,1]$ is attained at $x = 1$. Now, the derivative of f'/g' has the same sign as the function $y = e^{\lambda x} - 1 - \lambda x + 2\lambda\alpha x^2$. For $\alpha > 0$, this function attains its absolute minimum at $x = 0$, and it is 0, proving the claim.

For the second part, we combine the variational definition of entropy (2.63) with the preceding inequality to obtain

$$\mathrm{Ent}_{\mu}(e^{\lambda f}) \leq \int \left(e^{-\lambda f} - 1 + \lambda f\right) e^{\lambda f} d\mu = \int \nu(-\lambda f) d\mu$$

$$\leq \frac{\nu(-\lambda)}{e^{\lambda} + \alpha - 1}\int \left(fe^{\lambda f} + \alpha f^2 - f\right) d\mu. \quad \blacksquare$$

Now we combine the bound in (3.85) with tensorisation of entropy as given in Proposition 3.3.1.

Proposition 3.3.5 *Let Z be a subadditive random variable such that $Z - E_k Z \le 1$. Then, for all $\alpha > 0$ and $\lambda \ge 0$,*

$$\mathrm{Ent}_\mu(e^{\lambda Z}) \le \frac{\nu(-\lambda)}{e^\lambda + \alpha - 1} E\left[Ze^{\lambda Z} + \sum_{k=1}^n (Z_k - E_k(Z) + \alpha \mathrm{Var}_k(Z))\, e^{\lambda Z} \right]. \tag{3.86}$$

Proof Let Ent_k denote entropy with respect to the probability law of X_k, $k = 1, \ldots, n$. Taking $f = Z - E_k Z \le 1$ in (3.85), we obtain

$$\begin{aligned}\mathrm{Ent}_k\left(e^{\lambda(Z - E_k Z)}\right) &\le \frac{\nu(-\lambda)}{e^\lambda + \alpha - 1} E_k\left[(Z - E_k Z)e^{\lambda(Z - E_k Z)} + \alpha(Z - E_k Z)^2 - (Z - E_k Z)\right] \\ &= \frac{\nu(-\lambda)}{e^\lambda + \alpha - 1}\left[\left(E_k\left(Ze^{\lambda Z}\right) - (E_k Z)(E_k e^{\lambda Z})\right) e^{-\lambda E_k Z} + \alpha E_k (Z - E_k Z)^2\right].\end{aligned}$$

Then, by Proposition 2.5.3 and using first the homogeneity of entropy and then that by Jensen, $e^{\lambda E_k Z} \le E_k e^{\lambda Z}$, we obtain

$$\begin{aligned}\mathrm{Ent}_\mu\left(e^{\lambda Z}\right) &\le E\left[\sum_{k=1}^n e^{\lambda E_k Z} \mathrm{Ent}_k\left(e^{\lambda(Z - E_k Z)}\right)\right] \\ &\le \frac{\nu(-\lambda)}{e^\lambda + \alpha - 1} E\left[\sum_{k=1}^n \left(E_k\left(Ze^{\lambda Z}\right) - (E_k Z)\left(E_k e^{\lambda Z}\right) + \alpha \mathrm{Var}_k(Z) E_k e^{\lambda Z}\right)\right].\end{aligned}$$

By the subadditivity property (3.81), $(n-1)Z \le \sum_{k=1}^n Z_k$, and hence,

$$E\sum_{k=1}^n E_k\left(Ze^{\lambda Z}\right) = E\left(nZe^{\lambda Z}\right) \le E\left(Ze^{\lambda Z}\right) + E\sum_{k=1}^n Z_k E_k e^{\lambda Z}.$$

Since neither Z_k nor $\mathrm{Var}_k(Z)$ nor $E_k Z$ depends on X_k, the last two inequalities yield (3.86) by Fubini. ■

Corollary 3.3.6 *Let Z be a subadditive random variable such that $Z - E_k Z \le 1$, let Y_k be random variables satisfying $Y_k \le Z - Z_k \le 1$, $E_k Y_k \ge 0$ and $Y_k \le a$ for some $a \in (0, 1]$ and let $\alpha = 1/(1 + a)$. Then, for all $\lambda \ge 0$,*

$$\mathrm{Ent}_\mu(e^{\lambda Z}) \le \frac{\nu(-\lambda)}{e^\lambda + \alpha - 1} E\left[Ze^{\lambda Z} + \alpha \sum_{k=1}^n E_k(Y_k^2) e^{\lambda Z} \right] \tag{3.87}$$

and, as a consequence, letting $\tilde{Z} = Z - EZ$,

$$\mathrm{Ent}_\mu(e^{\lambda \tilde{Z}}) \le \frac{\nu(-\lambda)}{e^\lambda + \alpha - 1} E\left[\tilde{Z} e^{\lambda \tilde{Z}} + \left(\alpha \sum_{k=1}^n E_k(Y_k^2) + EZ \right) e^{\lambda \tilde{Z}} \right]. \tag{3.88}$$

Proof Since $Z_k = E_k Z_k$ and $\mathrm{Var}_k(Z) \le E_k(Z - Z_k)^2$, we have

$$Z_k - E_k Z + \alpha \mathrm{Var}_k(Z) \le E_k\left[\alpha(Z - Z_k)^2 - (Z - Z_k)\right].$$

Next, the function $y(x) = (1+a)^{-1}x^2 - x$ is decreasing for $x \le x_0 = (1+a)/2$ and is symmetric about x_0. Since $Y_k \le a \le (1+a)/2$ and $Y_k \le Z - Z_k \le 1$, it follows that $y(Z - Z_k) \le y(1) = y(a) \le y(Y_k)$, and we obtain

$$Z_k - E_k Z + \alpha \mathrm{Var}_k(Z) \le E_k(\alpha Y_k^2 - Y_k) \le \alpha E_k Y_k^2.$$

Now, inequality (3.87) is a direct consequence of the preceding proposition. Inequality (3.88) follows from (3.87) by homogeneity of entropy. ■

Assume now that $EZ \ge 0$ and that, as in Lemma 3.3.3, $\alpha \sum_{k=1}^n E_k(Y_k^2)$ is bounded a.s., and let γ be any number such that $\alpha \sum_{k=1}^n E_k(Y_k^2) + EZ \le \gamma$. Set

$$f(\lambda) := \lambda - \frac{\nu(-\lambda)}{e^\lambda + \alpha - 1}, \quad h(\lambda) = \frac{\nu(-\lambda)}{e^\lambda + \alpha - 1}\gamma,$$
$$L(\lambda) := \log\left(Ee^{\lambda \tilde{Z}}\right).$$

Then it follows from the preceding corollary that the function L satisfies the differential inequality

$$f(\lambda)L'(\lambda) - L(\lambda) \le h(\lambda), \quad 0 \le \lambda < \infty, \tag{3.89}$$

and it also follows, for example, using L'Hôpital's rule, that $L(0) = L'(0) = 0$. We wish to show that this differential inequality yields a bound for L. Note that

$$f(\lambda) = \frac{e^\lambda + \lambda(\alpha - 1) - 1}{e^\lambda + \alpha - 1} = \frac{\kappa(\lambda)}{\kappa'(\lambda)},$$

where we define $\kappa(\lambda) := e^\lambda + \lambda(\alpha - 1) - 1$. Similarly (note that $\nu(-\lambda) = \lambda\kappa'(\lambda) - \kappa(\lambda)$),

$$h(\lambda) = \gamma \frac{\lambda\kappa'(\lambda) - \kappa(\lambda)}{\kappa'(\lambda)}.$$

Now, $\kappa'(\lambda) = e^\lambda + \alpha - 1 > 0$ on $[0, \infty)$, so we can multiply both terms of (3.89) by κ'/κ^2 to obtain that L satisfies the equation

$$\frac{\kappa(\lambda)L'(\lambda) - \kappa'(\lambda)L(\lambda)}{\kappa^2(\lambda)} \le \gamma \frac{\lambda\kappa'(\lambda) - \kappa(\lambda)}{\kappa^2(\lambda)}, \quad \lambda > 0,$$

that is,

$$\left(\frac{L(\lambda)}{\kappa(\lambda)}\right)' \le \gamma \left(-\frac{\lambda}{\kappa(\lambda)}\right)', \quad \lambda > 0. \tag{3.90}$$

In fact, both functions are differentiable from the right at zero because L is differentiable and $L(0) = L'(0) = 0$. Then, since $\lim_{\lambda \to 0} L(\lambda)/\kappa(\lambda) = L'(0)/\alpha = 0$ and $\lim_{\lambda \to 0} \gamma(-\lambda/\kappa(\lambda)) = -\gamma/\alpha$, integrating, we obtain

$$\log Ee^{\lambda \tilde{Z}} = L(\lambda) \le \gamma\kappa(\lambda)\left[\frac{1}{\alpha} - \frac{\lambda}{\kappa(\lambda)}\right] = \frac{\gamma}{\alpha}\left(e^\lambda - \lambda - 1\right) = \frac{\gamma}{\alpha}\phi(-\lambda). \tag{3.91}$$

Thus, under certain natural conditions, the differential inequality (3.86), more precisely, (3.88), integrates into the upper bound (3.91) for the logarithm of the Laplace transform of $Z - EZ$. This is another instance of the entropy or Herbst's method. Summarising, we have proved the following:

Theorem 3.3.7 *Let $Z = Z(X_1,\dots,X_n)$, X_i independent, be a subadditive random variable relative to $Z_k = Z_k(X_1,\dots,X_{k-1},X_{k+1},\dots,X_n)$, $k = 1,\dots,n$, such that $EZ \geq 0$ and for which there exist random variables $Y_k \leq Z - Z_k \leq 1$ such that $E_k Y_k \geq 0$. Let $\sigma^2 < \infty$ be any real number satisfying*

$$\frac{1}{n}\sum_{k=1}^{n} E_k(Y_k)^2 \leq \sigma^2,$$

and set

$$v := 2EZ + n\sigma^2. \tag{3.92}$$

Then

$$\log Ee^{\lambda(Z-EZ)} \leq v(e^{\lambda} - \lambda - 1) = v\phi(-\lambda), \quad \lambda \geq 0, \tag{3.93}$$

where $\phi(x) = e^{-x} - 1 + x$.

It is standard procedure to derive tail probability bounds for $Z - EZ$ based on a bound for its Laplace transform (see Proposition 3.1.6). By this proposition, we obtain four such bounds, three of them mimicking, respectively, the Bennett, Prokhorov and Bernstein classical inequalities for sums of independent random variables and one in which the bound on the probability tail function is inverted. Recall the notation

$$h_1(x) = (1+x)\log(1+x) - x, \quad x \geq 0. \tag{3.94}$$

Then Theorem 3.3.7 and Proposition 3.1.6 give the following:

Corollary 3.3.8 *Let Z be as in Theorem 3.3.7. Then, for all $t \geq 0$,*

$$\begin{aligned}\Pr\{Z \geq EZ + t\} &\leq \exp(-vh_1(t/v)) \leq \exp\left(-\frac{3t}{4}\log\left(1+\frac{2t}{3v}\right)\right)\\ &\leq \exp\left(-\frac{t^2}{2v+2t/3}\right)\end{aligned} \tag{3.95}$$

and

$$\Pr\left\{Z \geq EZ + \sqrt{2vx} + x/3\right\} \leq e^{-x}, \quad x \geq 0. \tag{3.96}$$

As another consequence of inequality (3.93), we have, by Taylor development, the following bound for the variance of Z:

$$\mathrm{Var}(Z) \leq 2EZ + n\sigma^2. \tag{3.97}$$

Combining Lemma 3.3.3 with the preceding theorem, we obtain one of the most useful results in the theory of empirical processes, *Bousquet's version of the upper half of Talagrand's inequality*:

Theorem 3.3.9 (Upper tail of Talagrand's inequality, Bousquet's version) *Let $(S,\mathcal{S})$ be a measurable space, and let $n \in \mathbb{N}$. Let $X_1,\dots,X_n$ be independent S-valued random variables. Let $\mathcal{F}$ be a countable set of measurable real-valued functions on S such that $\|f\|_\infty \leq U < \infty$ and $Ef(X_1) = \cdots = Ef(X_n) = 0$, for all $f \in \mathcal{F}$. Let*

$$S_j = \sup_{f\in\mathcal{F}}\sum_{k=1}^{j} f(X_k) \quad \text{or} \quad S_j = \sup_{f\in\mathcal{F}}\left|\sum_{k=1}^{j} f(X_k)\right|, \quad j = 1,\dots,n,$$

and let the parameters σ^2 and v_n be defined by

$$U^2 \geq \sigma^2 \geq \frac{1}{n}\sum_{k=1}^{n} \sup_{f\in\mathcal{F}} Ef^2(X_k), \quad \text{and} \quad v_n = 2UES_n + n\sigma^2.$$

Then, assuming also $U = 1$ we have

$$\log Ee^{\lambda(S_n - ES_n)} \leq v_n(e^\lambda - 1 - \lambda),\ \lambda \geq 0, \tag{3.98}$$

and also, for all $x \geq 0$,

$$\Pr\{S_n \geq ES_n + x\} \leq \Pr\left\{\max_{1\leq j\leq n} S_j \geq ES_n + x\right\} \leq e^{-v_n h_1(x/v_n)}. \tag{3.99}$$

Moreover, for any $U > 0$,

$$\Pr\{S_n \geq ES_n + x\} \leq \Pr\left\{\max_{1\leq j\leq n} S_j \geq ES_n + x\right\} \leq \exp\left[-\frac{3x}{4U}\log\left(1+\frac{2xU}{3v_n}\right)\right]$$
$$\leq \exp\left[-\frac{x^2}{2v_n + 2xU/3}\right], \tag{3.100}$$

and

$$\Pr\left\{S_n \geq ES_n + \sqrt{2v_n x} + Ux/3\right\} \leq \Pr\left\{\max_{1\leq j\leq n} S_j \geq ES_n + \sqrt{2v_n x} + Ux/3\right\} \leq e^{-x}, \tag{3.101}$$

for all $x \geq 0$, where h_1 is as in (3.94).

Proof We may assume, without loss of generality, that $U = 1$ (just apply the result for $U = 1$ to $U^{-1}\mathcal{F} = \{U^{-1}f : f \in \mathcal{F}\}$). By approximation, we also may assume the class $\mathcal{F}$ to be finite. With these two reductions, it follows from Lemma 3.3.3 that S_n is subadditive with respect to the variables S_n^k, defined by the same expression as S_n but with the kth term deleted from the sum, and that there exist $Y_k \leq S_n - S_n^k \leq 1$ satisfying $EY_k = 0$ and $\sum_k E_k Y_k^2 \leq n\sigma^2$. Therefore, Theorem 3.3.7 applies and gives inequality (3.98).

Also, $e^{\lambda S_k}$ is a nonnegative submartingale; thus, by Doob's submartingale maximal inequality,

$$\Pr\left\{\max_{j\leq n} e^{\lambda S_j} \geq e^{\lambda t + \lambda ES_n}\right\} \leq \frac{Ee^{\lambda S_n}}{e^{\lambda t - \lambda ES_n}} \leq e^{v_n\phi(-\lambda) - \lambda t},$$

for all $\lambda \geq 0$; in particular, the probability on the left side of inequality (3.99) is dominated by the infimum over λ of the preceding expression, and we obtain inequalities (3.99) and (3.100) for $U = 1$ as in Proposition 3.1.6 (Corollary 3.3.8). ■

It is worth recording that, by (3.97),

$$\mathrm{Var}(S_n) \leq 2UES_n + n\sigma^2. \tag{3.102}$$

Notice the similarity between inequality (3.100) and the Prohorov and Bernstein inequalities in Theorem 3.1.7: in the case of $\mathcal{F} = \{f\}$, with $\|f\|_\infty \leq c$, and $Ef(X_i) = 0$, U becomes c, and v_n becomes $n\sigma^2$, and the right-hand side of Talagrand's inequality becomes exactly the Bernstein and Prohorov inequalities. Clearly, then, Talagrand's inequality is essentially a best possible exponential bound for the empirical process. These comments also apply to Talagrand's inequality for the lower tails of the empirical process in the next subsection.

3.3.3 Differential Inequalities and Bounds for Laplace Transforms of Centred Empirical Processes, $\lambda < 0$

Whereas the Bousquet-Talagrand upper bound for the Laplace transform $Ee^{\lambda Z}$ of the supremum Z of an empirical process (or of a subadditive function) for $\lambda \geq 0$ is best possible, there exist quite good results for $\lambda < 0$, but these do not exactly reproduce the classical exponential bounds for sums of independent real random variables when specified to a single function. Because of this, and because the proof of the best-known result for $\lambda < 0$ is quite involved, we will only prove an inequality with slightly worse constants, although we will state here and will use throughout the best result at this writing. The method of proof will be the same as elsewhere in this section: starting from a (modified) log-Sobolev inequality, we will obtain an integrable differential inequality for the logarithm of Ee^{-tZ}, $t > 0$, and the problem reduces, as usual, to finding a good bound for the right-hand side of the log-Sobolev inequality. Here is the result we do prove:

Theorem 3.3.10 (Lower tail of Talagrand's inequality: Klein's version) *Under the same hypotheses and notation as in Theorem 3.3.9, and with $U = 1$, we have*

$$Ee^{-t(S_n - ES_n)} \leq \exp\left(v_n \frac{e^{4t} - 1 - 4t}{16}\right) = e^{\frac{v_n}{16}\phi(-4t)}, \quad \text{for } 0 \leq t < 1. \tag{3.103}$$

As a consequence, for all $x \geq 0$,

$$\Pr\{S_n \leq ES_n - x\} \leq \exp\left(-\frac{v_n}{16^2} h_1\left(\frac{4x}{v_n}\right)\right), \quad \textit{where } h_1(x) = (1+x)\log(1+x) - x, \tag{3.104}$$

$$\Pr\{S_n \leq ES_n - x\} \leq \exp\left(-\frac{3x}{16}\log\left(1 + \frac{8x}{3v + n}\right)\right) \leq \exp\left(-\frac{x^2}{2v_n + 8x/3}\right) \tag{3.105}$$

and

$$\Pr\left\{S_n \leq ES_n - \sqrt{2v_n x} - 4x/3\right\} \leq e^{-x}. \tag{3.106}$$

Remark 3.3.11 Here is the result proved by Klein and Rio (2005): setting

$$V_n = 2ES_n + \sup_f \sum_{k=1}^{n} Ef^2(X_k), \tag{3.107}$$

then

$$Ee^{-t(S_n - ES_n)} \leq \exp\left(V_n \frac{e^{3t} - 1 - 3t}{9}\right) = e^{\frac{V_n}{9}\phi(-3t)}, \quad \text{for } 0 \leq t < 1, \tag{3.108}$$

and that, as a consequence, for all $x \geq 0$,

$$\Pr\{S_n \leq ES_n - x\} \leq e^{-\frac{V_n}{9^2} h_1\left(\frac{3x}{V_n}\right)}, \quad \textit{where } h_1(x) = (1+x)\log(1+x) - x, \tag{3.109}$$

$$\Pr\{S_n \leq ES_n - x\} \leq \exp\left(-\frac{x}{4}\log\left(1 + \frac{2x}{V_n}\right)\right) \leq \exp\left(-\frac{x^2}{2V_n + 2x}\right) \tag{3.110}$$

and

$$\Pr\left\{S_n \leq ES_n - \sqrt{2V_n x} - x\right\} \leq e^{-x}. \tag{3.111}$$

We will denote these inequalities as the *Klein-Rio version of the lower tail of Talagrand's inequality*. There is not much difference, *in the i.i.d. case*, between the Klein-Rio inequalities and the Klein inequalities that we will now prove, as is readily seen by comparing (3.103)–(3.106) to (3.108)–(3.111): for instance, the denominators in the exponent for the Bernstein-type inequalities are, respectively, $2V_n + 2x$ and $2v_n + 8x/3$ and, in the i.i.d. case, $v_n = V_n$. But if the variables are not i.i.d., v_n may be much larger than V_n. In any case, both bounds fall somewhat short from what we would expect this denominator to be, namely, $2V_n + 2x/3$.

Proof We prove Theorem 3.3.10. We may assume that $U = 1$. We write $Z = S_n$ and $v = v_n$ for ease of notation, and let Z_k be Z with the kth summand deleted from the sum defining Z, as in Lemma 3.3.3, and let $Y_k = f_k(X_k)$ also be as in that lemma and its proof. The starting point is again a modified log-Sobolev inequality, namely, inequality (3.75) in Proposition 3.3.1, which, setting

$$F(t) = Ee^{-tZ}, \quad t \geq 0,$$

can be written as

$$tF'(t) - F(t)\log F(t) \leq \sum_{k=1}^{n} E\left[e^{-tZ_k}\nu(t(Z - Z_k))\right], \quad t \geq 0. \tag{3.112}$$

Adding and subtracting $t^2(1-t)^{-1}xe^{-tx}$, $\nu(tx) = 1 - (1+tx)e^{-tx}$ becomes

$$\nu(tx) = \frac{t^2}{1-t}xe^{-tx} + \left(1 - \left(\frac{1-t+tx}{1-t}\right)e^{-tx}\right) := q(t)xe^{-tx} + r(t,x),$$

where $q(t) \geq 0$ and $r(t,x)$ is decreasing in x for all $0 \leq t < 1$ and $x \leq 1$. Then the right-hand side of inequality (3.112) becomes

$$q(t)\sum_{k=1}^{n} E\left[(Z - Z_k)e^{-tZ}\right] + \sum_{k=1}^{n} E\left[e^{-tZ_k}r(t, Z - Z_k)\right].$$

By the properties of q and r, we can use $\sum_{k=1}^{n}(Z - Z_k) \leq Z$ on the first term and $Y_k \leq Z - Z_k \leq 1$ on the second to obtain

$$tF'(t) - F(t)\log F(t) \leq -q(t)F'(t) + \frac{1}{1-t}\sum_{k=1}^{n} E\left[e^{-tZ_k}\left(1 - t - e^{-tY_k}(1 - t + tY_k)\right)\right]. \tag{3.113}$$

Let $T_k = 1 - t - e^{-tY_k}(1 - t + tY_k)$, and note that $E\left(e^{-tZ_k}T_k\right) = E\left(e^{-tZ_k}E_kT_k\right)$ because Z_k does not depend on X_k. To further simplify the differential inequality (3.113), note that, for $0 \leq t < 1$,

$$T_k = (1-t)(1 - e^{-tY_k}) - tY_k e^{-tY_k} \leq -t^2 Y_k + Y_k \sum_{k=2}^{\infty}\left(\frac{1}{(k-1)} - (1-t)\frac{1}{k}\right)t^k$$

$$= -t^2 Y_k + Y_k^2 \sum_{k=2}^{\infty}\left(\frac{2}{(k-1)} - \frac{1}{k}\right)t^k = -t^2 Y_k + Y_k^2\left[e^t - 1 - t + \sum_{k=3}^{\infty}\left(\frac{2}{(k-1)} - \frac{2}{k}\right)t^k\right]$$

$$= -t^2 Y_k + Y_k^2\left[e^t - 1 - t + t^3 + \sum_{k=3}^{\infty}\frac{2}{k}(t^{k+1} - t^k)\right] \leq -t^2 Y_k + (e^t - 1 - t + t^3)Y_k^2.$$

Then, since $EY_k \geq 0$, we have $E_k T_k \leq (e^t - 1 - t + t^3) E_k Y_k^2$, and also since $Z - Z_k \leq 1$, we have $Ee^{-tZ_k} = Ee^{-tZ} e^{t(Z-Z_k)} \leq e^t F(t)$. This, together with the facts that $e^t - 1 - t + t^3 \geq 0$ and that $\sum_{k=1}^n EY_k^2 \leq n\sigma^2$, yields

$$\sum_{k=1}^{n} E\left(e^{-tZ_k} T_k\right) = \sum_{k=1}^{n} E\left(e^{-tZ_k} E_k T_k\right) \leq n\sigma^2 e^t (e^t - 1 - t + t^3),$$

which, plugged into (3.113), gives, after multiplying both sides by $1 - t$ and using $(t + q(t))(1 - t) = t$,

$$tF'(t) - (1 - t) f(t) \leq F(t) n\sigma^2 e^t (e^t - 1 - t + t^3), \quad 0 \leq t < 1,$$

or, setting as usual $L(t) = \log F(t)$,

$$tL'(t) - (1 - t)L(t) \leq n\sigma^2 e^t (e^t - 1 - t + t^3), \quad 0 \leq t < 1, \tag{3.114}$$

an inequality that can be integrated. To see this, just observe that for any differentiable function h, $d(e^t h(t)/t)/dt = e^t (th'(t) - (1 - t)h(t))/t^2$, which applied to (3.114) yields

$$\left(\frac{L(t)}{te^{-t}}\right)' = \frac{e^t}{t^2}(tL'(t) - (1 - t)L(t)) \leq n\sigma^2 e^{2t} \frac{e^t - 1 - t + t^3}{t^2}.$$

By l'Hôpital's rule, $\lim_{t \to 0} e^t L(t)/t = -EZ$, and we obtain

$$L(t) \leq -te^{-t} EZ + n\sigma^2 te^{-t} \int_0^t e^{2u} \frac{e^u - 1 - u + u^3}{u^2} \, du, \quad 0 \leq t < 1.$$

Now, differentiating and expanding, we see that the smallest $\alpha > 0$ for which

$$te^{-t} \int_0^t e^{2u} \frac{e^u - 1 - u + u^3}{u^2} \, du \leq \frac{e^{\alpha t} - 1 - \alpha t}{\alpha^2}$$

is $\alpha = 4$, and we obtain $L(t) \leq n\sigma^2 \phi(-4t)/16 - te^{-t} EZ, \quad 0 \leq t < 1$, or

$$\log Ee^{-t(Z-EZ)} \leq n\sigma^2 \phi(-4t)/16 + t(1 - e^{-t})EZ \leq (n\sigma^2 + 2EZ)\phi(-4t)/16, \quad 0 \leq t < 1,$$

since $t(1 - e^{-t}) \leq 2\phi(-4t)$, $t \geq 0$. This is just inequality (3.103).

To derive the probability inequalities in the theorem, we see first that inequality (3.103) yields

$$\Pr\{Z - EZ \leq -x\} \leq \inf_{t>0} \Pr\left\{e^{-t(Z-EZ)} \geq e^{tx}\right\} \leq \exp\left(\inf_{0 \leq t < 1} [v\phi(-4t)/16 - tx]\right).$$

The absolute minimum of the function in the exponent is attained at $t = 4^{-1} \log(1 + 4x/v)$ and is $-(v/16)h_1(4x/v)$ (see, e.g., (3.18)), which proves inequality (3.104) if $t = 4^{-1} \log(1 + 4x/v) < 1$, that is, for $x \leq v(e^4 - 1)/4$.

If $x \geq v(e^4 - 1)/4$, inequality (3.104) is a consequence of Bennet's inequality (3.16) for sums of independent real random variables as follows: for $f \in \mathcal{F}$, we have

$$\Pr\{Z \leq EZ - x\} \leq \Pr\left\{\sum_{i=1}^{n} f(X_i) \leq EZ - x\right\}$$
$$\leq \exp\left(-n\sigma^2 h_1\left(\frac{x - EZ}{n\sigma^2}\right)\right) \leq \exp\left(-vh_1\left(\frac{x - EZ}{v}\right)\right)$$

because the function $vh(u/v)$ is decreasing in v. By definition of v, for $x \geq v(e^4 - 1)/4$, we have $x - EZ \geq (e^4 - 3)/2EZ \geq 25.799EZ$ or $x - EZ \geq x - x/25 = 24x/25$, so

$$\Pr\{Z \leq EZ - x\} \leq \exp\left(-vh_1\left(\frac{24x}{25v}\right)\right).$$

Now (3.104) follows because, for these x, $h_1(24x/25v) \geq h_1(4x/v)/16$, as is easily seen, for example, by observing that both functions of x/v are zero at zero and checking first derivatives.

Inequalities (3.105) and (3.106) follow from (3.104) by change of variables in the relations (3.19) and in the proof of Proposition 3.1.6. ■

3.3.4 The Entropy Method for Random Variables with Bounded Differences and for Self-Bounding Random Variables

There are a few more results (or types of results) that are also very useful because they allow us to obtain exponential inequalities for complicated random variables that are not necessarily suprema of sums of independent centred random variables. We consider functions with bounded differences and self-bounding random variables.

We begin with the extension of Hoeffding's inequality to processes based on multivariate *functions with bounded differences*.

Definition 3.3.12 Let $(S_i, \mathcal{S}_i)$, $i = 1, \ldots, n$, be measurable spaces, and let $f : \prod_{i=1}^n S_i \mapsto \mathbb{R}$ be a measurable function. f has bounded differences if

$$\sup_{x_i, x_j' \in S, i,j \leq n} |f(x_1, \ldots, x_n) - f(x_1, \ldots, x_{i-1}, x_i', x_{i+1}, \ldots, x_n)| \leq c_i,$$

where, for each i, c_i is a measurable function of x_j, $j \neq i$, and there exists a finite constant c such that $\sum_{i=1}^n c_i^2 \leq c^2$ for all $(x_1, \ldots, x_n) \in S^n$. If $Z = f(X_1, \ldots, X_n)$, where X_i are S_i-valued independent random variables, we say that the random variable Z has bounded differences.

The typical example of a function with bounded differences is $f(x_1, \ldots, x_n) = \sum_{i=1}^n x_i$, with $x_i \in [a_i, b_i]$ (with or without absolute values).

Example 3.3.13 (a) Let X_i be independent B-valued random variables where B is Banach space. Then, if $\|X_i\| \leq c_i/2$, the random variable

$$S_n = \left\| \sum_{i=1}^n X_i \right\|$$

has bounded differences because changing one X_i by X_i' (both with norm dominated by c_i) changes the norm of the sum by at most c_i. In this case, $c^2 = \sum_{i=1}^n c_i^2$.

(b) Similarly, if $\mathcal{F}$ is a class of functions taking values on $[a,b]$ for some $-\infty < a < b < \infty$, then

$$Z = \|P_n - P\|_{\mathcal{F}}$$

has bounded differences, and in this case, $c^2 = (b-a)/n$. For instance, if $\mathcal{F}$ is a class of indicator functions, as is the case for the cumulative empirical distribution function, then $[a,b] = [0,1]$ and $c^2 = 1/n$.

(c) Let $f_n(x : x_1,\dots,x_n) = (1/nh)\sum_{i=1}^n K((x-x_i)/h)$, where K is an integrable real function that integrates to 1, that is, f_n is a kernel density estimator, and let $g(x_1,\dots,x_n) = \int |f_n(x;x_1\dots,x_n) - f(x)|dx$ for some $f \in L^1$. Then, if $(x_1,\dots,x_n)$ and $(x'_1,\dots,x'_n)$ differ only in one coordinate, say, $x_j \neq x'_j$, we have

$$|g(x_1,\dots,x_n) - g(x'_1,\dots,x'_n)| \le \frac{1}{nh}\int \left|K\left(\frac{x-x_j}{h}\right) - K\left(\frac{x-x'_j}{h}\right)\right| dx \le \frac{2\|K\|_{L^1}}{n};$$

that is, g is a function of bounded differences for $c^2 = 4\|K\|_{L^1}^2/n$.

Theorem 3.3.14 *If Z has bounded differences and $\sum c_i^2 \le c^2$, then, for all $\lambda \ge 0$,*

$$Ee^{\lambda(Z-EZ)} \le e^{\lambda^2 c^2/8} \tag{3.115}$$

so that, for all $t \ge 0$,

$$\Pr\{Z \ge EZ + t\} \le e^{-2t^2/c^2}, \quad \Pr\{Z \le EZ - t\} \le e^{-2t^2/c^2}. \tag{3.116}$$

Moreover,

$$\mathrm{Var}(Z) \le \frac{c^2}{4}. \tag{3.117}$$

Proof Let us first observe that if Y is a random variable with finite Laplace transform (for some $\lambda > 0$), and if we set $F_Y(\lambda) = Ee^{\lambda(Y-EY)}$, $L_Y = \log F_Y$, we have, for $\mu = \mathcal{L}(Y)$,

$$\mathrm{Ent}_\mu(e^{\lambda(Y-EY)}) = \lambda F'_Y(\lambda) - F_Y(\lambda)\log F_Y(\lambda) = F_Y(\lambda)(\lambda L'_Y(\lambda) - L_Y(\lambda))$$

and, by homogeneity of entropy,

$$\mathrm{Ent}_\mu e^{\lambda Y} = Ee^{\lambda Y}(\lambda L'_Y(\lambda) - L_Y(\lambda)). \tag{3.118}$$

Then, if $a \le Y - EY \le b$, (3.6) in Lemma 3.1.1 gives, by integration by parts,

$$\lambda L'_Y(\lambda) - L_Y(\lambda) = \int_0^\lambda t L''_Y(t)dt \le \lambda^2(b-a)^2/8,$$

Thus,

$$\mathrm{Ent}_\mu(e^{\lambda Y}) \le (Ee^{\lambda Y})\frac{\lambda^2(b-a)^2}{8}. \tag{3.119}$$

By hypothesis,

$$0 \le \sup_x Z(X_1,\dots,X_{i-1},x,X_{i+1},\dots,X_n) - \inf_x Z(X_1,\dots,X_{i-1},x,X_{i+1},\dots,X_n) \le c_i,$$

so, conditionally on $\{X_j : 1 \le j \le n, j \neq i\}$, $Z - E_iZ$ has range of length at most c_i, and we can apply this inequality to it with $\mu = \mu_i = \mathcal{L}(X_i)$. This observation in combination with tensorisation of entropy (Proposition 2.5.3) gives

$$\mathrm{Ent}_\mu(e^{\lambda Z}) \le \sum_{i=1}^n E(\mathrm{Ent}_{\mu_i}(e^{\lambda Z})) \le E\sum_{i=1}^n \frac{\lambda^2 c_i^2}{8} E_i e^{\lambda Z} \le \frac{\lambda^2 c^2}{8} Ee^{\lambda Z}.$$

Then, by (3.118), $\lambda L'_Z - L_Z \le (\lambda^2 c^2)/8$, where $L_Z = \log Ee^{\lambda(Z-EZ)}$, or

$$\left(\frac{L_Z(\lambda)}{\lambda}\right)' = \frac{\lambda L'_Z(\lambda) - L_Z(\lambda)}{\lambda^2} \le \frac{c^2}{8}.$$

Since $L_Z(\lambda)/\lambda \to 0$ as $\lambda \to 0$ (l'Hôpital), this yields

$$L_Z(\lambda)/\lambda \le \lambda c^2/8,$$

proving (3.115). Then, using the exponential Chebyshev's inequality, for all $\lambda \ge 0$,

$$\Pr\{Z \ge EZ + t\} \le e^{\lambda^2 c^2/8 - \lambda t},$$

and optimizing in λ ($\lambda = 4t/c^2$), we obtain the first inequality in (3.116). The second inequality follows because $-Z$ also has bounded differences with the same v as Z. Finally, the variance inequality follows from (3.115) and Taylor development. ∎

Back to the preceding example, we see that Theorem 3.3.14 gives Hoeffding's inequality for sums of independent bounded random variables: if X_i are independent with ranges, respectively, contained in $[a_i, b_i]$, then, for all $t \ge 0$,

$$\Pr\left\{\left|\sum_{i=1}^{n}(X_i - EX_i)\right| > t\right\} \le 2e^{-2t^2/\sum_{i=1}^{n}(b_i - a_i)^2},$$

a best-possible inequality, for example, for linear combinations of independent Rademacher variables. In fact, Theorem 3.3.14 is a very useful generalisation of Hoeffding's inequality. Here is what it yields in the preceding example. For Example 3.3.13(a) it gives the following generalisation of Hoeffding's inequality in Banach spaces: if X_i are independent and B-valued with $\|X_i\| \le c_i/2$, then

$$\Pr\{|\|S_n\| - \|ES_n\|| \ge t\} \le 2e^{-2t^2/\sum_{i=1}^{n} c_i^2}, \tag{3.120}$$

and $\mathrm{Var}(\|S_n\|) \le \sum c_i^2/4$.

For Example 3.3.13(c), on the L^1-norm of the kernel density estimator, Theorem 3.3.14 yields that for X_i independent, identically distributed,

$$\Pr\left\{\sqrt{n}\left|\int |f_n(x;X_1,\dots,X_n) - f(x)|dx - E\int |f_n(x;X_1,\dots,X_n) - f(x)|dx\right| \ge t\right\} \tag{3.121}$$
$$\le 2e^{-t^2/2\|K\|_{L^1}^2},$$

and $\mathrm{Var}\left(\int |f_n - f|\right) \le \|K\|_{L^1}^2/n$ for any $f \in L^1(\mathbb{R})$ and $h > 0$.

Finally, for Example 3.3.13(b), specialized to the empirical distribution function, Theorem 3.3.14 produces an inequality of the best kind, which should be compared to Massart's (1990) improvement of the classical Dvoretzky, Kiefer and Wolfowitz inequality:

$$\Pr\left\{\sqrt{n}\,|\|F_n - F\|_\infty - E\|F_n - F\|_\infty| \ge t\right\} \le 2e^{-2t^2}, \quad t \ge 0, \tag{3.122}$$

and $\mathrm{Var}(\|F_n - F\|_\infty) \le \frac{1}{4n}$. See also Exercise 3.3.2.

However, Theorem 3.3.14 produces bounds that are much weaker than those obtained from Theorem 3.2.4 for supremum norms $Z = \sup_{t \in T} \sum_{i=1}^{n} t_i(X_i - EX_i)$, X_i independent and with bounded ranges. We now turn to *self-bounding random variables*, that is, random variables $Z = f(X_1,\dots,X_n)$, X_i independent, that satisfy condition (3.78) for $Z_k = f_k(X_1,\dots,X_{k-1},X_{k+1},\dots,X_n)$, $k = 1,\dots,n$; that is

$$0 \le Z - Z_k \le 1, \quad \text{for } 1 \le k \le n \text{ and } \sum_k (Z - Z_k) \le Z.$$

We already observed that the logarithm of the Laplace transform of $Z - EZ$ for Z self-bounding satisfies the differential inequality (3.79), namely,

$$(\lambda - \phi(\lambda))L'(\lambda) - L(\lambda) \leq \phi(\lambda)EZ.$$

Integrating this inequality gives the following bounds:

Theorem 3.3.15 *Let Z be a self-bounding random variable. Then*

$$\log E\left(e^{\lambda(Z-EZ)}\right) \leq \phi(-\lambda)EZ, \quad \lambda \in \mathbb{R}. \tag{3.123}$$

This applies in particular to $Z = \sup_{f\in\mathcal{F}} \sum_{k=1}^n f(X_i)$, where $\mathcal{F}$ is countable and $0 \leq f(x) \leq 1$ for all $x \in S$ and $f \in \mathcal{F}$.

Proof To ease notation, set $v = EZ$. First, we note that since $\phi(\lambda)+\phi(-\lambda) = \phi'(\lambda)\phi'(-\lambda)$, the function $\psi_0(\lambda) := v\phi(-\lambda)$ solves the differential equation

$$(1-e^{-\lambda})\psi_0'(\lambda) - \psi_0(\lambda) = v\phi(\lambda).$$

Next, we show that if a function L satisfies the differential inequality (3.79), then

$$L \leq \psi_0,$$

which will prove the theorem because, as a consequence of Proposition 3.3.1, $L = \log Ee^{\lambda\tilde{Z}}$ does satisfy inequality (3.79). The function $\psi_1 := L - \psi_0$ satisfies the inequality

$$(1-e^{-\lambda})\psi_1'(\lambda) - \psi_1(\lambda) \leq 0,$$

which can be written as $(e^\lambda - 1)\psi_1'(\lambda) - e^\lambda\psi_1(\lambda) \leq 0$. This inequality has the form $fg' - f'g \leq 0$, with $f \neq 0$ for $\lambda \neq 0$, which implies that $(g/f)' \leq 0$. In our case, $g(\lambda) = \psi_1(\lambda)/(e^\lambda - 1)$, and the conclusion is that g is nonincreasing. Now $\psi_1(0) = 0$, and since $E\tilde{Z} = 0$ and $\psi_1'(\lambda) = E\left(\tilde{Z}e^{\lambda\tilde{Z}}\right) - v(1-e^{-\lambda})$, also $\psi_1'(0) = 0$. Hence, using l'Hôpital's rule, $g(0) = 0$. This implies that $g \leq 0$ on $[0,\infty)$ and $g \geq 0$ on $(-\infty, 0]$, showing that $\psi_1(\lambda) \leq 0$, for all $\lambda \in \mathbb{R}$. ■

Now (the proof of) Proposition 3.1.6 together with the easy-to-check fact that $h_1(-t) \geq \frac{t^2}{2}$ for $0 \leq t \leq 1$ (note also that $t^2/2 \geq (3t/4)\log(1+2t/3)$) gives that if Z is self-bounding, then for $t \geq 0$,

$$\begin{aligned} \Pr\{Z \geq EZ + t\} &\leq \exp(-(EZ)h_1(t/EZ)), \\ \Pr\{Z \leq EZ - t\} &\leq \exp(-(EZ)h_1(-t/EZ)) \end{aligned} \tag{3.124}$$

and, as a consequence,

$$\begin{aligned} \Pr\{Z \geq EZ + t\} &\leq \exp\left(-\frac{3t}{4}\log\left(1+\frac{2t}{3EZ}\right)\right) \leq \exp\left(-\frac{t^2}{2EZ+2t/3}\right), \\ \Pr\{Z \leq EZ - t\} &\leq \exp\left(-t^2/(2EZ)\right) \end{aligned}$$

and

$$\mathrm{Var}(Z) \leq EZ.$$

Suprema of empirical processes indexed by nonnegative, bounded functions are not the only examples of self-bounding random variables. Here we only note that conditional expectations of suprema of randomised empirical processes are self-bounding; that is, if $\mathcal{F}$ is a countable class of functions bounded by 1 in absolute value and X_i are independent identically distributed random variables, independent of a Rademacher sequence $\varepsilon_1,\ldots,\varepsilon_n$, then

$$E_\varepsilon \left\| \sum_{i=1}^n \varepsilon_i f(X_i) \right\|_{\mathcal{F}} := E\left(\sup_{f\in\mathcal{F}} \left| \sum_{i=1}^n \varepsilon_i f(X_i) \right| \Big| X_1,\ldots,X_n \right) \tag{3.125}$$

is self-bounding. The proof is not different from the proof of Lemma 3.3.3 and is omitted.

3.3.5 The Upper Tail in Talagrand's Inequality for Nonidentically Distributed Random Variables*

Consider the 'variance' parameters occurring in Talagrand's inequality for empirical processes, namely,

$$v_n = 2UES_n + \sum_{k=1}^n \sup_{f\in\mathcal{F}} Ef^2(X_k) \quad \text{and} \quad V_n = 2UES_n + \sup_{f\in\mathcal{F}} \sum_{k=1}^n Ef^2(X_k).$$

(In this section, the first occurs in Theorems 3.3.9 and 3.3.10 and the second in the Klein-Rio inequality (3.108) for the lower tails of the empirical process.) They coincide if the variables X_i are identically distributed, but not in general, and the first could be quite larger than the second. This is not much of an inconvenience in this book as we will deal mostly with independent identically distributed random variables. For these, Bousquet's version (Theorem 3.3.9) is best, and Klein's version (Theorem 3.3.10) is quite close to being best. However, in the next section we will find a situation in which we need to apply the upper-tail Talagrand inequality for nonidentically distributed summands and Theorem 3.3.9 does not apply precisely because it is in terms of v_n, not V_n.

The object of this subsection is to prove an upper-tail version of Talagrand's inequality for non-i.d. summands. The proof is quite involved.

Theorem 3.3.16 *Let X_i, $i\in\mathbb{N}$, be independent S-valued random variables, and let* $\mathbf{F}$ *be a countable class of functions $f=(f^1,\ldots,f^n): S\mapsto[-1,1]^n$ such that $Ef^k(X_k)=0$ for all $f_i\in\mathbf{F}$ and $k=1,\ldots,n$. Set*

$$T_n(f) = \sum_{k=1}^n f^k(X_k), \quad Z = \sup_{f\in\mathbf{F}} T_n(f)$$

and

$$\mathcal{V}_n = \sup_{f\in\mathbf{F}} ET_n^2(f) = \sup_{f\in\mathbf{F}} \sum_{k=1}^n E[f^k(X_k)]^2, \quad V_n = 2EZ + \mathcal{V}_n. \tag{3.126}$$

Then, for all $t\in[0,2/3]$,

$$L(t) := \log(Ee^{tZ}) \le tEZ + \frac{t^2}{2-3t} V_n, \tag{3.127}$$

and therefore, for all $x\ge 0$,

$$\Pr\left\{ Z \ge EZ + \sqrt{2V_n x} + \frac{3x}{2} \right\} \le e^{-x}. \tag{3.128}$$

Note that $\mathbf{F}$ could be $\mathbf{F} = \{(f,\dots,f) : f \in \mathcal{F}\}$, and then we would obtain $T_n(f) = \sum_{k=1}^n f(X_k)$, the empirical process indexed by $\mathcal{F}$, and in this case

$$\mathcal{V}_n = \sup_{f\in\mathcal{F}} \sum_{k=1}^n E f^2(X_k) = \sup_{f\in\mathcal{F}} \mathrm{Var}\left(\sum_{k=1}^n f(X_k)\right).$$

But the present setting allows for f changing with k. Also, taking $\mathbf{F}\cup(-\mathbf{F})$, we obtain $Z = \sup_{f\in\mathbf{F}} |T_n(f)|$, so the supremum of the absolute values of the empirical process are also included in the theorem.

We now prove the theorem. First, we observe that it suffices to prove the theorem for $\mathbf{F}$ *finite*, say, $\mathbf{F} = \{f_1,\dots,f_m\}$ for m finite, and $f_i = (f_i^1,\dots,f_i^m)$. We begin with a decomposition of entropy based on tensorisation and the variational formula. Recall that E_k denotes integration with respect to the variable X_k. We set $\mu_k = \mathcal{L}(X_k)$, $k = 1,\dots,n$, and $P = \mu_1 \times \cdots \times \mu_n$. The range of the variable t in what follows will be $[0,\infty)$.

Lemma 3.3.17 *Let $F(t) = Ee^{tZ}$, let $g(t;X_1,\dots,X_n) = e^{tZ}$ and let $g_k(t;X_1,\dots,X_n)$, $k = 1,\dots,n$, be nonnegative functions such that $E(g_k \log g_k) < \infty$ for all $t \ge 0$. Then*

$$tF'(t) - F(t)\log F(t) = \mathrm{Ent}_P(g(t)) \le \sum_{k=1}^n E[g_k \log(g_k/E_k g_k)]$$
$$+\sum_{k=1}^n E[(g-g_k)\log(g/E_k g)]. \tag{3.129}$$

Proof By tensorisation of entropy (Proposition 2.5.3),

$$\begin{aligned}\mathrm{Ent}_P(g) &\le \sum_{k=1}^n E\left(\mathrm{Ent}_{\mu_k}(g)\right) = \sum_{k=1}^n E\left(E_k(g\log g) - (E_k(g\log E_k g)\right)\\ &= \sum_{k=1}^n (E(g\log g) - (E(g\log E_k g)) = \sum_{k=1}^n E\left(g\log(g/E_k g)\right)\\ &= \sum_{k=1}^n E\left(g_k\log(g/E_k g)\right) + E\left((g-g_k)\log(g/E_k g)\right).\end{aligned}$$

Now $E_k(g/E_k g) = 1$, and therefore, by Lemma 2.5.2,

$$E\left(g_k\log(g/E_k g)\right) \le E\sup\left\{E_k(g_k h) : E_k e^h = 1\right\} = E(\mathrm{Ent}_{\mu_k}(g_k)) = E\left(g_k\log(g_k/E_k g_k)\right),$$

which combined with the preceding inequality gives the lemma. ∎

The point is now to choose functions $g_k(t;X_1,\dots,X_n)$ whose μ_k-entropy is computable and such that $g - g_k \ge 0$. The functions

$$g_k(t;X_1,\dots,X_n) = \sum_{i=1}^m \mu_k\{\tau = i\} e^{tT_n(f_i)}, \tag{3.130}$$

where τ is the first $i \le m$ such that $Z = T_n(f_\tau)$ will be shown to work. Note that g_k is a weighted average of variables $e^{tT_n(f_i)}$; hence, $g_k \le e^{tZ} = g$.

The next lemma bounds the second term at the right-hand side of inequality (3.129):

Lemma 3.3.18 *For $g = e^{tZ}$ and the functions g_k, $1 \le k \le n$, defined by (3.130), we have*

$$E\left((g - g_k)\log(g/E_k g)\right) \le tE(g - g_k).$$

Proof Let $T_n^k(f) = T_n(f) - f^k(X_k)$, and $Z_k = \sup_{f\in\mathbf{F}} T_n^k(f)$. Then Z_k is independent of X_k and $Z - Z_k \le 1$. In particular,

$$g = e^{tZ} \le e^t e^{tZ_k}.$$

Next, if τ_k is the smallest integer such that $Z_k = T_n^k(f_{\tau_k})$, then τ_k is independent of X_k, and therefore, $E_k f_{\tau_k}^k(X_k) = 0$ (recall that $Ef^k(X_k) = 0$, for all $f \in \mathbf{F}$); hence,

$$Z_k = T_n^k(f_{\tau_k}) = E_k T_n(f_{\tau_k}) \le E_k Z.$$

By conditional Jensen's inequality this gives

$$e^{tZ_k} \le e^{tE_k Z} \le E_k e^{tZ} = E_k g \text{ a.s.}$$

We conclude that $g \le e^t E_k g$ a.s., and the lemma follows. ■

Before estimating the first term on the right-hand side of (3.129), we will slightly modify its inequality. Set

$$h_k = E_k e^{tT_n^k(f_\tau)} = \sum_{i=1}^m \mu_k\{\tau = i\} e^{tT_n^k(f_i)}, \tag{3.131}$$

a strictly positive function of the random variables X_i, $i \ne k$. Then, by Young's inequality (2.62),

$$\frac{g_k}{h_k}\log\frac{g_k}{E_k g_k} \le \frac{g_k}{h_k}\log\frac{g_k}{h_k} - \frac{g_k}{h_k} + \frac{g_k}{E_k g_k},$$

which, multiplying by h_k and integrating with respect to the variable X_k only, gives

$$E_k\left(g_k \log\frac{g_k}{E_k g_k}\right) \le E_k\left(g_k\log\frac{g_k}{h_k}\right) - E_k g_k + h_k,$$

and, integrating, gives

$$E\left(g_k\log\frac{g_k}{E_k g_k}\right) \le E\left(g_k\log\frac{g_k}{h_k} - g_k + h_k\right).$$

Using this inequality and Lemma 3.3.18 in inequality (3.129), we obtain

$$tF'(t) - F(t)\log F(t) \le \sum_{k=1}^n E\left(g_k\log\frac{g_k}{h_k} + (1+t)(h_k - g_k)\right) + t\sum_{k=1}^n E(g - h_k). \tag{3.132}$$

We further estimate the last term: by convexity of the exponential function,

$$\sum_{k=1}^n E(g - h_k) = n\left(Eg - \frac{1}{n}\sum_{k=1}^n Ee^{tT_n^k(f_\tau)}\right) \le n\left(Eg - Ee^{t\sum_{k=1}^n T_n^k(f_\tau)/n}\right),$$

and note that $\sum_{k=1}^n T_n^k(f_\tau) = (n-1)T_n(f_\tau) = (n-1)Z$. Therefore, using the convexity of F, we obtain

$$\sum_{k=1}^n E(g - h_k) \le n(F(t) - F(t(n-1)/n) \le tF'(t). \tag{3.133}$$

For the first term on the right-hand side of inequality (3.132), we have the following upper bound:

Lemma 3.3.19

$$E\left(g_k \log \frac{g_k}{h_k} + (1+t)(h_k - g_k)\right) \le \frac{t^2 e^t \mathcal{V}_n}{2} F(t). \tag{3.134}$$

Proof It is convenient to introduce the function

$$s(t,x) = x\log x + (1+t)(1-x), \tag{3.135}$$

which is convex in x for all t. With this notation

$$g_k \log \frac{g_k}{h_k} + (1+t)(h_k - g_k) = h_k s(t, g_k/h_k)$$

and the convexity of s in x together with the fact that $\sum_{i=1}^m \mu_k\{\tau = i\} e^{tT_n^k(f_i)}/h_k = 1$ by definition of h_k, we obtain

$$h_k s(t, g_k/h_k) \le \sum_{i=1}^m \mu_k\{\tau = i\} e^{tT_n^k(f_i)} s(t, e^{tf_i^k(X_k)}).$$

Integrating the variable X_k, we obtain

$$E_k(h_k s(t, g_k/h_k)) \le \sum_{k=1}^n \mu_k\{\tau = i\} e^{tT_n^k(f_i)} E\left(s(t, e^{tf_i^k(X_k)})\right).$$

Now it is a calculus exercise to show that for each $t \ge 0$, the function $\eta_t(x) = s(t, e^{tx})$ satisfies the inequality $\eta_t(x) \le x\eta_t'(0) + (tx)^2/2$ for all $x \le 1$. Since $Ef^k(X_k) = 0$ for all $f \in \mathbf{F}$, this inequality implies that

$$Es(t, e^{tf_i^k(X_k)}) \le \frac{t^2}{2} E(f_i^k(X_k))^2. \tag{3.136}$$

These two estimates give (recall that $T_n^k(f) = T_n(f) - f^k(X_k) \le 1 + T_n(f)$)

$$\sum_{k=1}^n E_k(h_k s(t, g_k/h_k)) \le \frac{t^2 e^t}{2} E_k\left(\sum_{i=1}^m 1_{\tau=i} e^{tT_n(f_i)} \sum_{k=1}^n E(f_i^k(X_k))^2\right)$$

or, since $\sum_{k=1}^n (Ef_i(X_k))^2 \le \mathcal{V}_n$ and $E\sum_{i=1}^m 1_{\tau=i} e^{tT_n(f_i)} = F(t)$,

$$\sum_{k=1}^n E\left(h_k s(t, g_k/h_k)\right) \le \frac{t^2 e^t \mathcal{V}_n}{2} F(t),$$

proving the lemma. ■

Setting as usual $L(t) = \log Ee^{tZ} = \log F(t)$, the decomposition (3.132) in combination with the bounds (3.134) and (3.133) gives the differential inequality

$$t(1-t)L'(t) - L(t) \le t^2 e^t \mathcal{V}_n/2.$$

Dividing both sides by t^2 and noting that $(L/t)' = L'/t - L/t^2$, it becomes

$$\left(\frac{L}{t}\right)' - L' \le e^t \frac{\mathcal{V}_n}{2}.$$

Since $L(0) = 0$ and $L(t)/t \to L'(0) = EZ$, integrating between 0 and t, we obtain

$$\frac{L(t)}{t} - L(t) - EZ \leq \frac{e^t - 1}{2} \mathcal{V}_n$$

or

$$\frac{1-t}{t} L(t) \leq \frac{e^t - 1}{2} \mathcal{V}_n + EZ.$$

Now,

$$\frac{e^t - 1}{2} = \sum_{k=1}^{\infty} \frac{t^k}{2 \cdot k} \leq \sum_{k=1}^{\infty} \left(\frac{t}{2}\right)^k = \frac{t}{2-t},$$

so, for $0 \leq t < 2/3$,

$$L(t) - tEZ \leq \frac{t^2}{(2-t)(1-t)} \mathcal{V}_n + \frac{t^2}{1-t} EZ = \frac{t^2(\mathcal{V}_n + (2-t)EZ)}{(2-t)(1-t)} \leq \frac{t^2(\mathcal{V}_n + 2EZ)}{2-3t}.$$

This proves (3.127). To prove (3.128), one proceeds as in the last part of the proof of Proposition 3.1.6 with the function $\varphi(\lambda)$ there redefined as $\varphi(\lambda) = V_n\lambda^2/(2(1-3\lambda/2))$.

Exercises

3.3.1 If X_i, $1 \leq i \leq n$, are independent symmetric random variables, prove that $Ee^{\lambda \sum_{i=1}^n X_i / (\sum_{i=1}^n X_i^2)^{1/2}} \leq e^{\lambda^2/2}$ and hence

$$\Pr\left\{\sum_{i=1}^n X_i \Big/ \left(\sum_{i=1}^n X_i^2\right)^{1/2} > t\right\} \leq e^{-t^2/2}.$$

3.3.2 Let $\mathcal{F}$ be a countable class of measurable functions bounded by 1. Prove that

$$Ee^{\lambda(\|P_n - P\|_{\mathcal{F}} - E\|P_n - P\|_{\mathcal{F}})} \leq e^{\lambda^2/2n}$$

and hence that $\Pr\{|\|P_n - P\|_{\mathcal{F}} - E\|P_n - P\|_{\mathcal{F}})| \geq t\} \leq 2e^{-2nt^2}$. In particular, the sequence $\{\sqrt{n}\|P_n - P\|_{\mathcal{F}}\}$ is stochastically bounded if and only if the sequence of its expected values is bounded.

3.3.3 (The Dvoretzky-Kiefer-Wolfowitz inequality.) Let F be a cumulative distribution function (c.d.f.) on $\mathbb{R}$, and let F_n be the empirical c.d.f. corresponding to n i.i.d. random variables with common c.d.f. F. Massart's (1990) improvement of the bounds in the classical Dvoretzky-Kiefer-Wolfowitz (1956) inequality states that, for all $t \geq 0$,

$$\Pr\left\{\sqrt{n}\|F_n - F\|_\infty \geq t\right\} \leq 2e^{-2t^2}.$$

The proof is a real tour de force, and we do not reproduce it here (see Dudley (2014) for a detailed proof). However, use techniques from this Section and from Section 3.1.3 to prove that, for all $u \geq 4$,

$$\Pr\left\{\sqrt{n}\|F_n - F\|_\infty \geq u\right\} \leq 2e^{-2(u-4)^2}.$$

Hint: From Exercise 3.1.7, $\sqrt{n}E\|F_n - F\|_\infty \leq 4$ (actually, this bound can be improved to 1 by Massart's inequality!), and the result follows from this and inequality (3.122) applied for $t \geq 4$.

3.3.4 (a) Use Theorem 3.3.16 to show that if $S_n = \sup_{f\in\mathcal{F}} \left|\sum_{k=1}^n f(X_k)\right|$ with X_i independent and $\mathcal{F}$ countable and such that for each $f \in \mathcal{F}$, $\|f\|_\infty \le U/2$, then

$$E(S_n - ES_n)_+^p \le N_p((1+\delta)v_n)^{p/2} + E_p(3U(1+\delta^{-1}))^p,$$

for all $p > 1$ and $\delta > 0$, where $N_p = \int_0^\infty pu^{p-1}e^{-u^2/2}du = 2^{p/2}\Gamma(p/2+1)$, which is bounded by $3(p/2)^{1/2}(p/e)^{p/2}$, and $E_p = \int_0^\infty pu^{p-1}e^{-u}du = \Gamma(p+1)$, bounded by $3p^{1/2}(p/e)^p$ (by Stirling's formula). *Hint*: First obtain a Bernstein-type inequality for $S_n - ES_n$ from (3.127) like (3.100) but with 3 instead of 2/3 as coefficient of xU.

(b) Deduce from (a) and the fact that S_n is nonnegative, which implies that $(S_n - ES_n)_- \le ES_n$, that for all $p > 1$ and $\delta, \tau > 0$,

$$\|S_n\|_p \le \|(S_n - ES_n)_+\|_p + ES_n \le (1+\tau)ES_n + N_p^{1/p}(1+\delta)^{1/2}v_n^{1/2}$$
$$+ \left[\frac{N_p^{2/p}(1+\delta)}{\tau} + 3E_p^{1/p}(1+\delta^{-1})\right]U,$$

where $\|X\|_p := (E|X|^p)^{1/p}$. For instance, taking $\delta = \tau = 1$, we obtain

$$\|S_n\|_p \le 2ES_n + \left(\frac{9p}{2}\right)^{1/(2p)}\sqrt{\frac{2p}{e}v_n} + (9p)^{1/p}\frac{4}{e}pU$$
$$\le 2ES_n + 1.24\cdot 3^{1/p}\sqrt{pv_n} + 2.13\cdot 9^{1/p}pU.$$

3.3.5 Prove that the coefficient of U in the preceding inequality can be improved if the variables X_i are i.i.d. (use Theorem 3.3.9 instead of Theorem 3.3.16).

3.3.6 Show that if $\mathcal{F}$ is a countable class of functions bounded by 1 in absolute value and X_i are independent identically distributed random variables, independent of a Rademacher sequence $\varepsilon_1, \ldots, \varepsilon_n$, then for all n, the random variables $E_\varepsilon \left\|\sum_{i=1}^n \varepsilon_i f(X_i)\right\|_{\mathcal{F}}$ in (3.125) are self-bounding.

3.3.7 Show that the exponential inequality (3.128) holds for $Z^* = \max_{k\le n}\sup_{f\in\mathcal{F}}\sum_{i=1}^k f^i(X_i)$ (keeping EZ unchanged).

3.3.8 For what theorems for empirical processes in this section can the class $\mathcal{F} = \{(f, \ldots, f)\}$ be replaced by $\mathcal{F} = \{(f^1, \ldots, f^n)\}$ as in Theorem 3.3.16?

3.3.9 (Sums of independent Banach space–valued random variables.) Use the equivalence between empirical processes and norms of sums of independent random variables to translate all the exponential inequalities for empirical processes in this section into exponential inequalities for sums of independent random variables. More concretely, as an example: let $\mathbb{B}$ be a Banach space satisfying the property that there exists a countable subset D of the unit ball of its dual space such that, for all $x \in \mathbb{B}$, $\|x\| = \sup_{f\in B} f(x)$, where $\|x\|$ denotes the Banach space norm of x. (For instance, by the Hahn-Banach theorem, separable Banach spaces satisfy this property, but other Banach spaces satisfy it as well, such as ℓ_∞.) Then, if X_i are independent identically distributed B-valued random variables such that $\|X_i\| \le U$ and $EX_i = 0$ – recall Lemma 2.6.3 – and setting $\sigma^2 = \sup_{f\in D} Ef^2(X_1)$ and $v_n = n\sigma^2 + 2UE\left\|\sum_{i=1}^n X_i\right\|$, we have

$$\Pr\left\{\left\|\sum_{i=1}^n X_i\right\| \ge E\left\|\sum_{i=1}^n X_i\right\| + x\right\} \le \exp\left(-\frac{3x}{4U}\log\left(1+\frac{2xU}{3v_n}\right)\right),$$

$$\Pr\left\{\left\|\sum_{i=1}^n X_i\right\| \le E\left\|\sum_{i=1}^n X_i\right\| - x\right\} \le \exp\left(-\frac{x}{4U}\log\left(1+\frac{2xU}{v_n}\right)\right).$$

Slightly weaker inequalities hold if the variables X_i are not identically distributed.

3.4 First Applications of Talagrand's Inequality

In this section we present a few important results related to empirical processes and U-statistics. They are somewhat disconnected, perhaps their only connection being that their proofs require Talagrand's inequality in essential ways. The first is a moment inequality for empirical processes which allows us to reduce the estimation of any moments to that of the first or second moments. The second type of result is statistically important: Talagrand's inequality and symmetrisation (actually, randomisation by Rademacher variables) allow us to replace the expected value of the supremum of the empirical process appearing in Talagrand's inequality by a completely data-based surrogate, thus rendering the inequality statistically useful. The third consists of a Bernstein-type inequality for completely degenerate or canonical U-statistics of order 2.

3.4.1 Moment Inequalities

As seen in Exercise 3.3.4, Talagrand's inequality allows us to bound moments of the empirical process based on a bounded class of functions in terms of the first moment, the uniform bound on the functions and the supremum of the individual second moments $n\sigma^2 = \sum_{i=1}^n \sup_{f\in\mathcal{F}} Ef^2(X_i)$. Combining this with Hoffmann-Jørgensen's inequality (after symmetrising) yields a bound that applies to moments of the empirical processes over classes of functions whose envelope is not necessarily bounded but just satisfies integrability conditions. These inequalities are the analogues for the empirical process of very sharp moment inequalities for sums of independent random variables: there exist $C, K < \infty$ such that if ξ_i are independent centred random variables, then, for all $p \ge 2$,

$$E\left|\sum_{i=1}^n \xi_i\right|^p \le CK^p\left[p^p E\max_{i\le n}|\xi_i|^p + p^{p/2}\left(\sum_{i=1}^n E\xi_i^2\right)^{p/2}\right],$$

for instance, with $C = 16e+4$ and $K = 2e^{1/2}$. The dependence on p of the bounds as $p\to\infty$ is optimal in the sense that these inequalities for all $p \ge 2$ and for bounded variables do imply Bernstein's inequality up to multiplicative constants (both in the inequality itself and in the exponent). The constants in the following theorem may not be best possible (in large part due to the use of symmetrisation), but they are reasonable. This theorem is designed for classes with *unbounded envelope*: if a class of functions is uniformly bounded, the bound on expectations given in Exercise 3.3.4 produces better constants.

Theorem 3.4.1 *Let $\mathcal{F}$ be a countable collection of measurable functions on $(S, \mathcal{S})$, and let X_i be independent S-valued random variables such that $V_n := \sup_{f\in\mathcal{F}}\sum_{i=1}^n Ef^2(X_i) < \infty$*

and $Ef(X_i)=0$, *for all* $f\in\mathcal{F}$ *and all* i. *Set* $F=\sup_{f\in\mathcal{F}}|f|$,

$$S_n=\left\|\sum_{i=1}^n f(X_i)\right\|_{\mathcal{F}} \quad \text{and} \quad S_{n,M}=\left\|\sum_{i=1}^n (f(X_i)I_{F(X_i)\le M}-Ef(X_i)I_{F(X_i)\le M})\right\|_{\mathcal{F}},$$

where $M>0$ *is a positive constant. Then, for any* $n\in\mathbb{N}$ *and any* $p>1$,

$$\|S_n\|_p\le 2ES_{n,M_p}+\left(\frac{9p}{2}\right)^{1/(2p)}\sqrt{\frac{2p}{e}\mathcal{V}_n}+\left(\frac{4}{e}(72p)^{1/p}+16(4p)^{1/p}\right)p\left\|\max_i F(X_i)\right\|_p, \quad (3.137)$$

where $M_p^p=8E\max_i F^p(X_i)$.

Proof We decompose S_n as

$$S_n\le S_{n,M_p}+\left\|\sum_{i=1}^n (f(X_i)I_{F(X_i)>M_p}-Ef(X_i)I_{F(X_i)>M_p})\right\|_{\mathcal{F}}.$$

Then we apply the inequality in Exercise 3.3.4 to the first term of this decomposition:

$$\|S_{n,M_p}\|_p\le 2ES_{n,M_p}+\left(\frac{9p}{2}\right)^{1/(2p)}\sqrt{\frac{2p}{e}\mathcal{V}_n}+(9p)^{1/p}\frac{4}{e}pM_p.$$

To estimate the second term, we note that by Rademacher randomisation (Theorem 3.1.21),

$$E\left(\left\|\sum_{i=1}^n (f(X_i)I_{F(X_i)>M_p}-Ef(X_i)I_{F(X_i)>M_p})\right\|_{\mathcal{F}}\right)^{1/p}\le 2\|\tilde{S}_n^{(M_p)}\|_p,$$

where

$$\tilde{S}_n^{(M_p)}:=\left\|\sum_{i=1}^n \varepsilon_i f(X_i)I_{F(X_i)>M_p}\right\|_{\mathcal{F}}$$

and ε_i are independent Rademacher variables independent of the sequence $\{X_i\}$. To estimate $\|\tilde{S}_n^{(M_p)}\|_p$, we use Hoffmann-Jørgensen's inequality (Theorem 3.1.15), which gives

$$\|\tilde{S}_n^{(M_p)}\|_p\le 2^{(p+2)/p}(p+1)^{(p+1)/p}\left[4^{1/p}t_0+M_p/8^{1/p}\right],$$

where t_0 is any number such that $\Pr\left\{\|\sum_{i=1}^n \varepsilon_i f(X_i)I_{F(X_i)>M_p}\|_{\mathcal{F}}>t_0\right\}\le 1/8$. But since

$$\Pr\left\{\left\|\sum_{i=1}^n \varepsilon_i f(X_i)I_{F(X_i)>M_p}\right\|_{\mathcal{F}}>0\right\}=\Pr\left\{\max_i F(X_i)>M_p\right\}\le 1/8,$$

we can take $t_0=0$. Hence,

$$\|\tilde{S}_n^{(M_p)}\|_p\le 2\cdot 4^{1/p}(p+1)^{(p+1)/p}M_p/8^{1/p}.$$

To simplify, we may use $(p+1)^{(p+1)/p}=((p+1)/p)^{(p+1)/p}p^{1+1/p}\le 4p^{1+1/p}$. Collecting these bounds yields the theorem. ■

In concrete situations, as with metric entropy expectation bounds for *VC* classes of functions, one may have as good an estimate for $ES_{n,M}$ as for ES_n, if not better. In general, one can prove that $ES_{n,M}\le 2ES_n$ (and that if the variables $f(X_i)$ are symmetric, then $ES_{n,M}\le ES_n$): this follows by Theorem 3.1.21 (Rademacher randomisation) and Corollary 3.1.20.

Remark 3.4.2 It is also worth noting that in (3.137) the coefficient 2 for ES_{n,M_p} can be replaced by $1+\delta$ at the expense of increasing the coefficients of the other two summands from the bound for $\|S_n\|_p$. This follows readily, with the same proof as that of Theorem 3.4.1, from one of the inequalities in Exercise 3.3.4.

Inequality (3.137) simplifies a bit by using the bound $p^{1/p} \le e^{1/e}$. Also, in the i.i.d. case, one can do a little better in the last summand by using Exercise 3.3.5 instead of Exercise 3.3.6, that is, by basing the derivation of the inequality on Theorem 3.3.9 instead of Theorem 3.3.16.

3.4.2 Data-Driven Inequalities: Rademacher Complexities

Talagrand's inequality (Theorems 3.3.7 and 3.3.10) gives an essentially best-possible rate of concentration of the (supremum of the) empirical process about its expectation, whereas, in general, the available expectation bounds for empirical processes are much less precise, such as, for example, the metric entropy or the bracketing bounds for expected values. Moreover, considering for simplicity the case of i.i.d. random variables X_i, the 'parameters' ES_n and $\sigma^2 = n\max_{f\in\mathcal{F}} Ef^2(X_1)$ depend on the distribution of X_1, which is usually partially or totally unknown in statistical inference. Therefore, Talagrand's inequality would be much more useful if these quantities could be replaced by data-dependent surrogates or estimates, particularly if the constants involved were reasonable. A similar comment applies to the probability inequalities for empirical processes derived from bounded differences (Theorem 3.3.14). Of the two parameters ES_n and σ^2, the first is more complex than the second, and the second can always be bounded by U and usually by much smaller quantities, as, for instance, in density estimation. In this subsection we replace ES_n in Talagrand's and in the bounded differences inequalities by random surrogates, namely,

$$\left\|\sum_{i=1}^n \varepsilon_i f(X_i)\right\|_{\mathcal{F}} \quad \text{or} \quad E_\varepsilon \left\|\sum_{i=1}^n \varepsilon_i f(X_i)\right\|_{\mathcal{F}}.$$

These variables are sometimes called *Rademacher complexities.*

Theorem 3.4.3 *Let $\mathcal{F}$ be a countable collection of measurable functions on $(S,\mathcal{S})$ with absolute values bounded by 1/2, let X_i, $i\in\mathbb{N}$, be independent, identically distributed S-valued random variables with common probability law P, let ε_i, $i\in\mathbb{N}$, be a Rademacher sequence independent from the sequence $\{X_i\}$ and let $\sigma^2 \ge \sup_{f\in\mathcal{F}} Pf^2$. Then, for all $n\in\mathbb{N}$ and $x\ge 0$,*

$$\Pr\left\{\left\|\frac{1}{n}\sum_{i=1}^n (f(X_i)-Pf)\right\|_{\mathcal{F}} \ge 3\left\|\frac{1}{n}\sum_{i=1}^n \varepsilon_i f(X_i)\right\|_{\mathcal{F}} + 4\sqrt{\frac{2\sigma^2 x}{n}} + \frac{70}{3n}x\right\} \le 2e^{-x}. \tag{3.138}$$

Proof Set $S_n = \|\sum_{i=1}^n (f(X_i)-Pf)\|_{\mathcal{F}}$ and $\tilde{S}_n = \|\sum_{i=1}^n \varepsilon_i f(X_i)\|_{\mathcal{F}}$ (as in the preceding proof). Note that the second variable is also the supremum of an empirical process: the variables are $\tilde{X}_i = (\varepsilon_i, X_i)$, defined on $\{-1,1\}\times S$, and the functions are $\tilde{f}(\varepsilon,x) = \varepsilon f(x)$. Thus, Talagrand's inequalities apply to both S_n and $\tilde{S}_n$. Then, using

$$\sqrt{2x(n\sigma^2+2E\tilde{S}_n)} \le \sqrt{2xn\sigma^2} + 2\sqrt{xE\tilde{S}_n} \le \sqrt{2xn\sigma^2} + \frac{1}{\delta}x + \delta E\tilde{S}_n,$$

for any $\delta > 0$, the Klein-Rio (3.111) version of Talagrand's lower-tail inequality (that we use instead of Theorem 3.3.10) gives

$$e^{-x} \geq \Pr\left\{\tilde{S}_n \leq E\tilde{S}_n - \sqrt{2x(n\sigma^2 + 2E\tilde{S}_n)} - x\right\}$$
$$\geq \Pr\left\{\tilde{S}_n \leq (1-\delta)E\tilde{S}_n - \sqrt{2xn\sigma^2} - \frac{1+\delta}{\delta}x\right\}.$$

Similarly, Theorem 3.3.7 gives

$$\Pr\left\{S_n > (1+\delta)ES_n + \sqrt{2xn\sigma^2} + \frac{3+\delta}{3\delta}x\right\} \leq e^{-x}$$

(and the analogous inequality for $\tilde{S}_n$, which we will not use in this proof). Recall also that by Theorem 3.1.21, $ES_n \leq 2E\tilde{S}_n$. Then we have, on the intersection of the complement of the events in the last two inequalities for, for example, $\delta = 1/5$,

$$S_n < \frac{6}{5}ES_n + \sqrt{2xn\sigma^2} + \frac{16}{3}x \leq \frac{12}{5}E\tilde{S}_n + \sqrt{2xn\sigma^2} + \frac{16}{3}x$$
$$< \frac{12}{5}\left[\frac{5}{4}\tilde{S}_n + \frac{5}{4}\sqrt{2xn\sigma^2} + 7.5x\right] + \sqrt{2xn\sigma^2} + \frac{16}{3}x$$
$$= 3\tilde{S}_n + 4\sqrt{2xn\sigma^2} + \frac{70}{3}x;$$

that is, this inequality holds with probability at least $1 - 2e^{-x}$. ■

Different values of δ (or even different δ from each of the two inequalities used) produce different coefficients in inequality (3.138) (e.g., $\delta = 1/2$ gives the coefficients 6, 7 and 61/3).

Remark 3.4.4 Since Rademacher complexities are self-bounding (see Exercise 3.3.6) and the lower-tail exponential inequality for self-bounding variables is tighter than the Klein-Rio inequality, if one is willing to use $E_\varepsilon\tilde{S}_n$ instead of the simpler $\tilde{S}_n$ as a surrogate for ES_n in inequality (3.138), one obtains a slightly better bound. The self-bounding inequality (3.124) yields (see two lines below the inequality)

$$\Pr\left\{E_\varepsilon\tilde{S}_n \leq E\tilde{S}_n - t\right\} \leq e^{-t^2/(2E\tilde{S}_n)},$$

and, with a change of variables $t^2 = 2xE\tilde{S}_n$ and using as in the preceding proof that the arithmetic mean dominates the geometric mean,

$$\Pr\left\{E_\varepsilon\tilde{S}_n \leq (1-\delta/2)E\tilde{S}_n - x/\delta\right\} \leq e^{-x}.$$

Then replacing Klein-Rio's inequality with this inequality with $\delta = 2/7$ in the preceding proof yields

$$\Pr\left\{\left\|\frac{1}{n}\sum_{i=1}^n (f(X_i) - Pf)\right\|_{\mathcal{F}} \geq 3E_\varepsilon\left\|\frac{1}{n}\sum_{i=1}^n \varepsilon_i f(X_i)\right\|_{\mathcal{F}} + 4\sqrt{\frac{2\sigma^2 x}{n}} + \frac{12x}{n}\right\} \leq 2e^{-x}. \quad (3.139)$$

If no estimate of the variance of $f \in \mathcal{F}$ other than the supremum norm of the envelope of the class is available, then a cruder but more purely data-driven estimate can be obtained: one just uses the fact that the class $\mathcal{F}$ has bounded differences. If a class of functions $\mathcal{F}$ is bounded by 1, then when one replaces a single variable X_i in $\|\sum_{i=1}^n (f(X_i) - Pf)/n\|_{\mathcal{F}}$, the variable changes by at most $2/n$, which means that these random variables have bounded differences with constant $c^2 = 4/n$, and the same is true for the variables $\left\|\frac{1}{n}\sum_{i=1}^n \varepsilon_i f(X_i)\right\|_{\mathcal{F}}$. Then, as a consequence of the exponential inequality for functions of bounded differences (3.116), we have the following:

Theorem 3.4.5 *Let $\mathcal{F}$ be a countable collection of measurable functions on $(S, \mathcal{S})$ with absolute values bounded by 1, let X_i, $i \in \mathbb{N}$, be independent, identically distributed S-valued random variables with common probability law P and let ε_i, $i \in \mathbb{N}$, be a Rademacher sequence independent from the sequence $\{X_i\}$. Then, for all $n \in \mathbb{N}$ and $x \geq 0$,*

$$\Pr\left\{\left\|\frac{1}{n}\sum_{i=1}^n (f(X_i) - Pf)\right\|_{\mathcal{F}} \geq 2\left\|\frac{1}{n}\sum_{i=1}^n \varepsilon_i f(X_i)\right\|_{\mathcal{F}} + 3\sqrt{\frac{2x}{n}}\right\} \leq 2e^{-x}. \tag{3.140}$$

Proof With the same notation as in the preceding proof, Theorem 3.3.14 gives, after a change of variables, both

$$\Pr\left\{\tilde{S}_n \leq E\tilde{S}_n - \sqrt{\frac{2x}{n}}\right\} \leq e^{-x} \quad \text{and} \quad \Pr\left\{S_n \geq ES_n + \sqrt{\frac{2x}{n}}\right\} \leq e^{-x},$$

and we recall that $ES_n \leq 2E\tilde{S}_n$. Hence, with probability at least $1 - 2e^{-x}$,

$$S_n < ES_n + \sqrt{\frac{2x}{n}} < 2\left[\tilde{S}_n + \sqrt{\frac{2x}{n}}\right] + \sqrt{\frac{2x}{n}},$$

and the result follows. ■

Consider the class of functions $\mathcal{F}_h = \{y \mapsto K((x-y)/h) : x \in \mathbb{R}\}$, where K is in $L^1(\mathbb{R}) \cap L^\infty(\mathbb{R})$, and a probability measure $dP(x) = f(x)dx$, where f is bounded and continuous. Then the envelope of the class is $U = \|K\|_\infty$, whereas

$$\sigma^2 = \sup_x \int K^2\left(\frac{x-y}{h}\right) f(y)dy = \int K^2(u) f(x - uh)du \leq \|f\|_\infty \|K\|_{L^2}^2 h,$$

much smaller than the envelope as $h \to 0$. For the empirical process based on P and indexed by classes of functions, which will occur in density estimation later, Theorem 3.4.3 is more adequate than Theorem 3.4.5.

3.4.3 A Bernstein-Type Inequality for Canonical U-statistics of Order 2

A U-statistic is a sum of the form

$$U_n = \sum_{1 \leq i < j \leq n} h_{ij}(X_i, X_j), \tag{3.141}$$

where X_i are independent random variables taking values in a measurable space $(S, \mathcal{S})$ and with respective laws P_i and h_{ij} which are measurable functions of two variables $h_{ij} : S^2 \mapsto \mathbb{R}$

such that $E|h_{ij}(X_i,X_j)| < \infty$ for all i,j. The U-statistic is degenerate or *canonical* if for all i,j and $x,y \in S$,

$$Eh_{ij}(X_i,y) = Eh_{ij}(x,X_j) = 0. \tag{3.142}$$

If U_n is not canonical, it decomposes into a 'linear' term and a canonical U-statistic (Hoeffding decomposition). For instance, in the case $h_{ij} = h$ with $h(x,y) = h(y,x)$ and X_i identically distributed, this decomposition is as follows:

$$2(U_n - EU_n) = \sum_{i \neq j} [h(X_i,X_j) - E_X h(X,X_j) - E_X h(X_i,X) + Eh(X_i,X_j)]$$
$$+ 2(n-1)\sum_{i=1}^{n} [E_X h(X_i,X) - Eh(X_i,X_j)].$$

The second term is a sum of independent random variables, and its tail probabilities assuming that h is bounded are well understood: they have two regimes, a Gaussian tail regime and a Poisson tail regime, as made clear by Prokhorov's inequality. The first sum has more complex tail probabilities: they will be shown to have four regimes, with tail probabilities of orders $e^{-c_1 t^2}$, $e^{-c_2 t}$, $e^{-c_3 t^{2/3}}$ and $e^{-c_4 t^{1/2}}$ on different ranges of $t > 0$; these correspond, respectively, to tail probabilities such as those of Gaussian chaos (the first two) and, up to logarithmic factors, of the product of a normal and a Poisson variables and of the product of two Poisson variables. Whereas Bernstein's inequality for sums of independent random variables is in terms of two parameters, the supremum norm of the variables and the sum of their variances, for canonical U-statistics we will need two more parameters, which correspond to other norms of the matrix (h_{ij}). Here are the parameters entering in the concentration inequality to be presented later:

$$A := \max_{i,j} \|h_{ij}\|_\infty, \quad C^2 := \sum_{j=2}^{n}\sum_{i=1}^{j-1} Eh_{ij}^2(X_i,X_j), \tag{3.143}$$

$$B^2 := \max\left\{ \max_j \left\| \sum_{i=1}^{j-1} E_i h_{ij}^2(X_i,x) \right\|_\infty, \max_i \left\| \sum_{j=i+1}^{n} E_j h_{ij}^2(x,X_j) \right\|_\infty \right\}, \tag{3.144}$$

$$D = \sup\left\{ \sum_{j=2}^{n}\sum_{i=1}^{j-1} E(h_{ij}(X_i,X_j)\xi_i(X_i)\zeta_j(X_j)) : \sum_{i=1}^{n-1} E\xi_i^2(X_i) \le 1, \sum_{j=2}^{n} \zeta_j^2(X_j) \le 1 \right\}. \tag{3.145}$$

In the case of a single function h symmetric in its entries and the variables X_i identically distributed, these parameters become

$$A = \|h\|_\infty,\ C^2 = \frac{n(n-1)}{2} Eh^2(X_1,X_2), \quad B^2 = (n-1)\|E_1 h^2(X_1,x)\|_\infty, \tag{3.146}$$

$$D = \frac{n}{2}\sup\left\{E\left(h(X_1,X_2)\xi(X_1)\zeta(X_2)\right) : E\xi^2(X_1) \le 1, E\xi^2(X_2) \le 1\right\} = \frac{n}{2}\|h\|_{L^2 \mapsto L^2}, \tag{3.147}$$

where $\|h\|_{L^2 \mapsto L^2}$ is the norm of the operator of $L^2(\mathcal{L}(X_1))$ with kernel h, $f \mapsto E(h(X_1,\cdot)f(X_1))$.

Let us assume for the rest of this subsection that the U-statistic U_n is canonical. We can write U_n as

$$U_n = \sum_{j=2}^{n}\left(\sum_{i=1}^{j-1} h_{ij}(X_i,X_j)\right) =: \sum_{j=2}^{n} Y_j.$$

Note that $E_jY_j := E(Y_j|X_1,\ldots,X_{j-1}) = 0$ by (3.142), whereas Y_ℓ is $\sigma(X_1,\ldots,X_{j-1})$ measurable for $\ell < j$; hence, $\{U_k : k \geq 2\}$ is a martingale relative to the σ-algebras $\mathcal{G}_k = \sigma(X_1,\ldots,X_k)$, $k \geq 2$. This martingale can be extended to $n = 0$ and $n = 1$ by taking $U_0 = U_1 = 0$, $\mathcal{G}_0 = \{\emptyset,\Omega\}$, $\mathcal{G}_1 = \sigma(X_1)$. We will use the martingale structure of U_n to effectively reduce it to an empirical process that can be handled using Talagrand's inequality. Before describing this reduction, we need a lemma on martingales. In its proof and elsewhere in this subsection, we make free use of discrete martingale theory, as found in most graduate probability texts.

Lemma 3.4.6 *Let $(U_n,\mathcal{G}_n)$, $n \in \mathbb{N}\cup\{0\}$, be a martingale with respect to a filtration $\mathcal{G}_n$ such that $U_0 = U_1 = 0$. For each $n \geq 1$ and $k \geq 2$, define the 'angle brackets' $A_n^k = A_n^k(U)$ of the martingale U by*

$$A_n^k = \sum_{i=1}^{n} E[(U_i - U_{i-1})^k|\mathcal{G}_{i-1}]$$

(and note $A_1^k = 0$ for all k). Suppose that for $\lambda > 0$ and all $i \geq 1$, $Ee^{\lambda|U_i-U_{i-1}|} < \infty$. Then

$$\left(\mathcal{E}_n := e^{\lambda U_n - \sum_{k=2}^{\infty}\lambda^k A_n^k/k}, \mathcal{G}_n\right), \quad n \in \mathbb{N},$$

is a supermartingale. In particular, $E\mathcal{E}_n \leq E\mathcal{E}_1 = 1$, so, if $A_n^k \leq w_n^k$ for constants $w_n^k \geq 0$; then

$$Ee^{\lambda U_n} \leq e^{\sum_{k\geq 2}\lambda^k w_n^k/k}. \tag{3.148}$$

Proof Obviously,

$$\begin{aligned} E(\mathcal{E}_n|\mathcal{G}_{n-1}) &= E\left[\mathcal{E}_{n-1}e^{\lambda(U_n-U_{n-1})}e^{-\sum_{k\geq 2}\lambda^k E((U_n-U_{n-1})^k|\mathcal{G}_{n-1})}\Big|\mathcal{G}_{n-1}\right] \\ &= \mathcal{E}_{n-1}e^{-\sum_{k\geq 2}\lambda^k E((U_n-U_{n-1})^k|\mathcal{G}_{n-1})}E(e^{\lambda(U_n-U_{n-1})}|\mathcal{G}_{n-1}). \end{aligned}$$

Now, using that $\{U_n\}$ is a martingale, the dominated convergence theorem for conditional expectations and that $1 + x \leq e^x$, we have

$$\begin{aligned} E(e^{\lambda(U_n-U_{n-1})}|\mathcal{G}_{n-1}) &= 1 + E\left(\sum_{k\geq 2}\frac{\lambda^k}{k}(U_n - U_{n-1})^k\Big|\mathcal{G}_{n-1}\right) \\ &= 1 + \sum_{k\geq 2}\frac{\lambda^k}{k}E((U_n - U_{n-1})^k\Big|\mathcal{G}_{n-1}) \\ &\leq e^{\sum_{k\geq 2}\lambda^k(\lambda^k/k)E((U_n-U_{n-1})^k|\mathcal{G}_{n-1})}, \end{aligned}$$

which, plugged into the preceding identities, yields $E(\mathcal{E}_n|\mathcal{G}_{n-1}) \leq \mathcal{E}_{n-1}$, proving the supermartingale property for $\mathcal{E}_n$. Inequality (3.148) follows immediately from this. ∎

In our case, with U_n a canonical U-statistic as defined in (3.141) and (3.142), we have

$$A_n^k = \sum_{j=2}^{n} E_j\left[\sum_{i=1}^{j-1} h_{ij}(X_i,X_j)\right]^k \le V_n^k := \sum_{j=2}^{n} E_j\left|\sum_{i=1}^{j-1} h_{ij}(X_i,X_j)\right|^k, \tag{3.149}$$

for all $k \ge 2$ and $n \ge 1$, and $A_1^k = V_1^k = 0$, for all $k \ge 2$. Then, by duality (see Exercise 3.4.1),

$$\begin{aligned}(V_n^k)^{1/k} &= \sup_{\xi_j \in L^{k/(k-1)}(P):\sum_{j=2}^n E|\xi_j(X_j)|^{k/(k-1)}=1} \sum_{j=2}^{n}\sum_{i=1}^{j-1} E_j\left(h_{ij}(X_i,X_j)\xi_j(X_j)\right)\\ &= \sup_{\xi_j \in L^{k/(k-1)}(P):\sum_{j=2}^n E|\xi_j(X_j)|^{k/(k-1)}=1} \sum_{i=1}^{n-1}\sum_{j=i+1}^{n} E_j\left(h_{ij}(X_i,X_j)\xi_j(X_j)\right).\end{aligned}$$

Thus, if we define random vectors $\mathbf{X}_i$, $i = 1,\dots,n-1$, on $\mathbb{R}^n$ by

$$\mathbf{X}_i = (0,\dots,0,h_{i,i+1}(X_i,x_{i+1}),\dots,h_{i,n}(X_i,x_n))$$

and for $\xi = (\xi_2,\dots,\xi_n) \in \prod_{i=2}^n L^{k/(k-1)}(P_i)$, the function $f_\xi : S \mapsto \mathbb{R}$ defined as $f_\xi(h_2,\dots,h_n) = \sum_{j=2}^n \int h_j(x)\xi_j(x)dP(x)$, then, setting $\mathcal{F} = \{f_\xi : \sum_{j=2}^n E|\xi_j(X_j)|^{k/(k-1)} = 1\}$, we have

$$(V_n^k)^{1/k} = \sup_{f\in\mathcal{F}}\left|\sum_{i=1}^{n-1} f_\xi(\mathbf{X}_i)\right|,$$

and moreover, by separability of the L^p spaces of finite measures, $\mathcal{F}$ can be replaced by a countable subset $\mathcal{F}_0$. Therefore, we can apply Talagrand's inequality for non-i.i.d. random variables (Theorem 3.3.16) to estimate the size of V_n^k.

The bound on the tail probabilities of U_n will be obtained by bounding the variables V_n^k on sets of large probability using Talagrand's inequality and then using Lemma 3.4.6 on these sets by means of optional stopping. In the case of a single f (i.e., $f^i = f$ for all $f = (f^1,\dots,f^n) \in \mathcal{F}$ in Theorem 3.3.16), and with the same transformations as in the first part of the proof of Theorem 3.4.3, the exponential inequality (3.128) becomes, for X_i independent (not necessarily identically distributed), $\mathcal{F}$ a countable collection of measurable functions such that all $f \in \mathcal{F}$ are centred ($Ef(X_i) = 0$ for all $1 \le i \le n$) and $\|f\|_\infty \le U$,

$$\Pr\left\{\left\|\sum_{k=1}^n f(X_k)\right\|_{\mathcal{F}} \ge (1+\varepsilon)E\left\|\sum_{k=1}^n f(X_k)\right\|_{\mathcal{F}} + \sqrt{2\mathcal{V}_n x} + \frac{2+3\varepsilon}{2\varepsilon}Ux\right\} \le e^{-x},$$

for all $x \ge 0$ and $\varepsilon > 0$, where $\mathcal{V}_n = \sup_f \sum_{k=1}^n Ef^2(X_k)$. Thus we obtain

$$\Pr\left\{(V_n^k)^{1/k} \ge (1+\varepsilon)E(V_n^k)^{1/k} + \sqrt{2\mathcal{V}_k x} + \kappa(\varepsilon)b_k x\right\} \le e^{-x} \tag{3.150}$$

for

$$\mathcal{V}_k = \sup_{\sum_{j=2}^n E|\xi_j(X_j)|^{k/(k-1)}=1} \sum_{i=1}^{n-1} E\left[\sum_{j=i+1}^{n} E_j\left(h_{ij}(X_i,X_j)\xi_j(X_j)\right)\right]^2 \tag{3.151}$$

and

$$b_k = \sup_{\sum_{j=2}^n E|\xi_j(X_j)|^{k/(k-1)}=1} \max_i \sup_x \left|\sum_{j=i+1}^{n} E_j\left(h_{ij}(x,X_j)\xi_j(X_j)\right)\right|, \tag{3.152}$$

where the suprema are extended over all $\xi_j \in L^{k/(k-1)}(P_j)$, $j = 2,\dots,n$, satisfying the stated condition.

This gives the following lemma:

Lemma 3.4.7 *For every $u \geq 0$, with $\mathcal{V}_k$ and b_k defined by (3.151) and (3.152), respectively, we have*

$$\Pr\bigcup_{k=2}^{\infty}\Big\{(V_n^k)^{1/k} \geq (1+\varepsilon)(EV_n^k)^{1/k} + \sqrt{2\mathcal{V}_k ku} + \kappa(\varepsilon)b_k ku\Big\} \leq \frac{1+\sqrt{5}}{2}e^{-u} \leq 1.62e^{-u}. \quad (3.153)$$

Proof With the change of variables $x = ku$ in the preceding exponential inequality for $(V_n^k)^{1/k}$, we obtain that the probability on the left side of inequality (3.153) is dominated by

$$1 \wedge \sum_{k=2}^{\infty} e^{-ku} \leq 1 \wedge \frac{1}{e^u(e^u-1)} = \left(e^u \wedge \frac{1}{e^u-1}\right)e^{-u} \leq \frac{1+\sqrt{5}}{2}e^{-u}. \quad \blacksquare$$

Interchanging the first supremum in the definition of b_k with the sum and using Hölder, we obtain

$$b_k \leq \max_i \sup_x \left[\sum_{j=i+1}^{n} E_j\left|h_{ij}(x,X_j)\xi_j(X_j)\right|^k\right]^{1/k} \leq (B^2A^{k-2})^{1/k}, \quad (3.154)$$

where A and B are as defined in (3.143) and (3.144). For $\mathcal{V}_k$, we have, again using duality,

$$\begin{aligned}
\mathcal{V}_k^{1/2} &= \sup_{\substack{\sum_{j=2}^n E|\xi_j(X_j)|^{k/(k-1)}=1 \\ \sum_{i=1}^{n-1} E\zeta_i^2(X_i)=1}} \sum_{i=1}^{n-1} E_i\left[\sum_{j=i+1}^{n} E_j(h_{ij}(X_i,X_j)\xi_j(X_j)\zeta_i(X_i))\right] \\
&= \sup_{\substack{\sum_{j=2}^n E|\xi_j(X_j)|^{k/(k-1)}=1 \\ \sum_{i=1}^{n-1} E\zeta_i^2(X_i)=1}} \sum_{j=2}^{n} E_j\left[\sum_{i=1}^{j-1} E_i(h_{ij}(X_i,X_j)\xi_j(X_j)\zeta_i(X_i))\right] \\
&= \sup_{\sum_{i=1}^{n-1} E\zeta_i^2(X_i)=1} \left[\sum_{j=2}^{n} E_j\left|\sum_{i=1}^{j-1} E_i(h_{ij}(X_i,X_j)\zeta_i(X_i))\right|^k\right]^{1/k} \\
&\leq (B^{k-2})^{1/k} \sup_{\sum_{i=1}^{n-1} E\zeta_i^2(X_i)=1} \left[\sum_{j=2}^{n} E_j\left|\sum_{i=1}^{j-1} E_i(h_{ij}(X_i,X_j)\zeta_i(X_i))\right|^2\right]^{1/k} \\
&= (B^{k-2}D^2)^{1/k},
\end{aligned} \quad (3.155)$$

where D is defined in (3.145).

Now notice that for all $\theta_1, \theta_2 \geq 0$ and $0 < \varepsilon \leq 1$, by convexity,

$$\left(\frac{\theta_1+\theta_2}{1+\varepsilon}\right)^k \leq \left(\frac{\theta_1}{1+\varepsilon} + \frac{\varepsilon\theta_2}{1+\varepsilon}\right)^k \leq \frac{1}{1+\varepsilon}\theta_1^k + \frac{\varepsilon}{1+\varepsilon}\theta_2^k,$$

so

$$(\theta_1+\theta_2)^k \leq (1+\varepsilon)^{k-1}\theta_1^k + \varepsilon(1+\varepsilon)^{k-1}\theta_2^k \leq (1+\varepsilon)^{k-1}\theta_1^k + (1+\varepsilon^{-1})\theta_2^k.$$

By symmetry, this inequality holds for all $\varepsilon \geq 0$; that is, for all $\theta_1, \theta_2, \varepsilon \geq 0$,

$$(\theta_1 + \theta_2)^k \leq \varepsilon)^{k-1}\theta_1^k + (1+\varepsilon^{-1})\theta_2^k.$$

Using this inequality twice and the bounds (3.155) and (3.154), we have, for $u > 0$,

$$\begin{aligned}
&\left[(1+\varepsilon)(EV_n^k)^{1/k} + \sqrt{2V_k ku} + \kappa(\varepsilon)b_k ku\right]^k \\
&\quad \leq \left[1+\varepsilon)(EV_n^k)^{1/k} + (B^{k-2}D^2)^{1/k}\sqrt{2ku} + (B^2A^{k-2})^{1/k}\kappa(\varepsilon)ku\right]^k \\
&\quad \leq (1+\varepsilon)^{2k-1}EV_n^k + (1+\varepsilon^{-1})^{k-1}\left[(B^{k-2}D^2)^{1/k}\sqrt{2ku} + (B^2A^{k-2})^{1/k}\kappa(\varepsilon)ku\right]^k \\
&\quad \leq (1+\varepsilon)^{2k-1}EV_n^k + (1+\varepsilon^{-1})^{k-1}(1+\varepsilon)^{k-1}B^{k-2}D^2(2ku)^{k/2} \\
&\qquad + (1+\varepsilon^{-1})^{2k-2}B^2A^{k-2}\kappa(\varepsilon)^k(ku)^k.
\end{aligned}$$

Thus, setting

$$\begin{aligned}
w_n^k &:= (1+\varepsilon)^{2k-1}EV_n^k + (2+\varepsilon+\varepsilon^{-1})B^{k-2}D^2(2ku)^{k/2} \\
&\quad + (1+\varepsilon^{-1})^{2k-2}B^2A^{k-2}(ku)^k\kappa(\varepsilon)^k(ku)^k, \qquad (3.156)
\end{aligned}$$

we have, by Lemma 3.4.7,

$$\Pr\left\{V_n^k \leq w_n^k \text{ for all } k \geq 2\right\} \geq 1 - 1.62e^{-u}, \qquad (3.157)$$

where we leave implicit the dependence of w_n^k on $u > 0$.

Inequalities (3.148), (3.149) and (3.156) will combine to produce the following theorem, which is the analogue of Bernstein's inequality for canonical U-statistics of order 2:

Theorem 3.4.8 *Let U_n be a U-statistic as defined by (3.141), and assume that the functions h_{ij} are uniformly bounded and canonical for $X_1, \ldots, X_n$, that is, that they satisfy equations (3.142). Let A, B, C, D be as defined by (3.143), (3.144) and (3.145). For $\varepsilon > 0$, define*

$$\begin{aligned}
&\kappa(\varepsilon) = 3/2 + 1/\varepsilon, \quad \eta(\varepsilon) = \sqrt{2}(2+\varepsilon+\varepsilon^{-1}), \\
&\beta(\varepsilon) = e(1+\varepsilon^{-1})^2\kappa(\varepsilon) + [\sqrt{2}(2+\varepsilon+\varepsilon^{-1}) \vee (1+\varepsilon)^2/\sqrt{2}], \\
&\gamma(\varepsilon) = [e(1+\varepsilon^{-1})^2\kappa(\varepsilon)] \vee (1+\varepsilon)^2/3.
\end{aligned}$$

Then, for all $\varepsilon, u > 0$,

$$\Pr\left\{U_n \geq 2(1+\varepsilon)^{3/2}C\sqrt{u} + \eta(\varepsilon)Du + \beta(\varepsilon)Bu^{3/2} + \gamma(\varepsilon)Au^2\right\} \leq e^{1-u}. \qquad (3.158)$$

For example, with $\varepsilon = 1/2$, inequality (3.4.8) becomes

$$\Pr\left\{U_n \geq \frac{3\sqrt{3}}{\sqrt{2}}C\sqrt{u} + \frac{9\sqrt{2}}{2}Du + \frac{63e + 9\sqrt{2}}{2}Bu^{3/2} + \frac{63e}{2}Au^2\right\} \leq e^{1-u}. \qquad (3.159)$$

Proof Let

$$T + 1 := \inf\left\{\ell \in \mathbb{N} : V_\ell^k \geq w_n^k \text{ for some } k \geq 2\right\}.$$

Then the event $\{T \leq \ell\}$ depends only on $X_1, \ldots, X_\ell$ for all $\ell \geq 1$, so T is a stopping time for the filtration $\mathcal{G}_\ell$, and therefore, $U_\ell^T = U_{\ell \wedge T}$, $\ell = 0, 1, \ldots, n$, is a martingale with respect to

$\{\mathcal{G}_\ell\}$ with $U_0^T = U_0 = 0$ and $U_1^T = U_1 = 0$ ($V_1^k = 0$, whereas $w_n^k > 0$; hence, $T \geq 1$ a.s.). Note that $U_j^T - U_{j-1}^T = U_j - U_{j-1}$ if $T \geq j$ and is zero otherwise and that $\{T \geq j\}$ is $\mathcal{G}_{j-1}$ measurable. Then the angle brackets of this martingale admit the following bound:

$$\begin{aligned}
A_n^k(U^T) &= \sum_{j=2}^{n} E(U_j^T - U_{j-1}^T)^k | \mathcal{G}_{j-1}) \\
&\leq V_n^k(U^T) := \sum_{j=2}^{n} E(|U_j^T - U_{j-1}^T|^k | \mathcal{G}_{j-1}) \\
&= \sum_{j=2}^{n} E(|U_j - U_{j-1}|^k | \mathcal{G}_{j-1}) I_{T \geq j} \\
&= \sum_{j=2}^{n} E_j \left| \sum_{i=1}^{j-1} h_{ij}(X_i, X_j) \right|^k I_{T \geq j} \\
&= \sum_{j=2}^{n-1} V_j^k I_{T=j} + V_n^k I_{T \geq n} \\
&\leq w_n^k \left(\sum_{j=2}^{n} I_{T=j} + I_{T \geq n} \right) \leq w_n^k,
\end{aligned}$$

since, by definition of T, $V_j^k \leq w_n^k$ for all k on $\{T \geq j\}$. Hence, Lemma 3.4.6 applied to the martingale U_n^T implies that

$$Ee^{\lambda U_n^T} \leq \exp\left(\sum_{k \geq 2} \frac{\lambda^k}{k} w_n^k \right).$$

Also, since V_n^k is nondecreasing in n for each k, inequality (3.157) implies that

$$\Pr\{T < n\} = \Pr\{V_n^k \geq w_n^k \text{ for some } k \geq 2\} \leq 1.62e^{-u}.$$

We thus have, for all $s \geq 0$,

$$\Pr\{U_n \geq s\} \leq \Pr\left\{U_n^T \geq s, T \geq n\right\} + \Pr\{T < n\} \leq e^{-\lambda s} \exp\left(\sum_{k \geq 2} \frac{\lambda^k}{k} w_n^k \right) + 1.62e^{-u}. \quad (3.160)$$

Finally, we will simplify the right-hand side of this inequality. Plugging in the definition of w_n^k into (3.160), we need to estimate

$$\begin{aligned}
\sum_{k \geq 2} \frac{\lambda^k}{k} w_n^k &= \sum_{k \geq 2} \frac{\lambda^k}{k} (1+\varepsilon)^{2k-1} EV_n^k + \sum_{k \geq 2} \frac{\lambda^k}{k} (2+\varepsilon+\varepsilon^{-1})^{k-1} B^{k-2} D^2 (2ku)^{k/2} \\
&\quad + \sum_{k \geq 2} \frac{\lambda^k}{k} (1+\varepsilon^{-1})^{2k-2} A^{k-2} B^2 \kappa(\varepsilon)^k (ku)^k \\
&:= \alpha + \beta + \gamma.
\end{aligned}$$

To simplify the third term γ, we use the elementary inequality $k \geq (k/e)^k$. To see it, just note that

$$\log k = \sum_{\ell=1}^{k} \log \ell \geq \int_1^k \log x\, dx = k \log k - k + 1 \geq \log (k/e)^k.$$

Replacing k by $(k/e)^k$, the series-defining γ becomes geometric, and its sum gives, with the notation $\delta(\varepsilon) := e(1+\varepsilon^{-1})^2\kappa(\varepsilon)$,

$$\gamma \leq \frac{(\delta(\varepsilon)Bu)^2\lambda^2}{1 - A\delta(\varepsilon)u\lambda}, \tag{3.161}$$

for $\lambda < (A\delta(\varepsilon)u)^{-1}$. To simplify β, we use the inequality $k \geq k^{k/2}$. Since $(k/e)^k > k^{k/2}$ for $k \geq e^2$, the argument immediately preceding (3.161) gives the inequality for $k > 7$; for $k \leq 7$, the inequality follows by direct verification. Then, setting $\eta(\varepsilon) = \sqrt{2}(2+\varepsilon+\varepsilon^{-1})$ and using $2+\varepsilon+\varepsilon^{-1} \geq 4$, we have, again by summing a geometric series,

$$\beta \leq \frac{\lambda^2 D^2 \eta^2(\varepsilon)u/4}{1 - B\eta(\varepsilon)\sqrt{u}\lambda}, \tag{3.162}$$

for $\lambda < (B\eta(\varepsilon)\sqrt{u})^{-1}$.

Next, we consider the term α. Recall that

$$EV_n^k = E\sum_{j=2}^{n} E_j \left|\sum_{i=1}^{j-1} h_{ij}(X_i,X_j)\right|^k = \sum_{j=2}^{n} E_j \left[E\left(\left|\sum_{i=1}^{j-1} h_{ij}(X_i,X_j)\right|^k \middle| X_j \right)\right].$$

Thus, setting $C_j := \sum_{i=1}^{j-1} h_{ij}(X_i,X_j)$, we have

$$\alpha = \frac{1}{1+\varepsilon}\sum_{j=2}^{n} E_j \left[E\left(e^{\lambda(1+\varepsilon)^2|C_i|} \middle| X_j \right) - \lambda(1+\varepsilon)^2 E(|C_i||X_j) - 1 \right].$$

Now we symmetrise: since $e^x - x - 1 \geq 0$ for all x, and since $e^{a|x|} + e^{-a|x|} = e^{ax} + e^{-ax}$, adding $E\left(e^{-\lambda(1+\varepsilon)^2|C_i|}\middle|X_j\right) + \lambda(1+\varepsilon)^2 E(|C_i||X_j) - 1$ to α, we obtain

$$\alpha \leq \frac{1}{1+\varepsilon}\sum_{j=2}^{n} E_j \left[E\left(e^{\lambda(1+\varepsilon)^2 C_i} \middle| X_j \right) - 1 + E\left(e^{-\lambda(1+\varepsilon)^2 C_i} \middle| X_j \right) - 1 \right].$$

Conditionally on X_j, C_i is a sum of $j-1$ independent centred random variables bounded in absolute value by A and with sum of variances

$$v_j(X_j) = \sum_{i=1}^{j-1} E_i h_{ij}^2(X_i,X_j) \leq B^2, \quad \sum_{j=2}^{n} E_j v_j^k(X_j) \leq C^2 B^{2(k-1)},$$

where A, B, C are the parameters defined by (3.143) and (3.144), respectively. Then, using Bernstein's inequality (Theorem 3.1.7), we obtain, for $\lambda < [(1+\varepsilon)^2(A/3+B/\sqrt{2})]^{-1}$,

$$\alpha \leq \frac{2}{1+\varepsilon}\sum_{j=2}^{n} E_j \left(\exp\left(\frac{\lambda^2(1+\varepsilon)^4 v_j(X_j)}{2 - 2A\lambda(1+\varepsilon)^2/3} \right) - 1 \right)$$

$$\leq \frac{2}{1+\varepsilon}\sum_{k=1}^{\infty} \frac{\lambda^{2k}(1+\varepsilon)^{4k} C^2 B^{2(k-1)}}{(2 - 2A\lambda(1+\varepsilon)^2/3)^k}$$

$$
\begin{aligned}
&= \frac{(1+\varepsilon)^3C^2\lambda^2}{1-\lambda A(1+\varepsilon)^3/3-\lambda^2(1+\varepsilon)^4B^2/2} \\
&\le \frac{(1+\varepsilon)^3C^2\lambda^2}{1-(1+\varepsilon)^2\lambda(A/3+B/\sqrt{2})}. \qquad (3.163)
\end{aligned}
$$

Putting together (3.163), (3.162) and (3.161), we obtain

$$
\sum_{k\ge 2}\frac{\lambda^k}{k}w_n^k \le \exp\left(\frac{\lambda^2W^2}{1-\lambda c}\right),
$$

for

$$
W = (1+\varepsilon)^{3/2}C + \eta(\varepsilon)D\sqrt{u}/2 + \delta(\varepsilon)Bu
$$

and

$$
c = \max\left[(1+\varepsilon)^2(A/3+B/\sqrt{2}), \eta(\varepsilon)B\sqrt{u}, \delta(\varepsilon)Au\right].
$$

Plugging this estimate in (3.160) and taking $s = 2W\sqrt{u}+cu$ and $\lambda = \sqrt{u}/(W+c\sqrt{u})$ in this inequality yield

$$
\Pr\left\{U_n \ge 2\sqrt{u}+cu\right\} \le 2.62e^{-u}.
$$

For $u \ge 1$, $cu \le \left((1+\varepsilon)^2/3 \vee \delta(\varepsilon)\right)Au^2 + \left((1+\varepsilon)^2/\sqrt{2} \vee \eta(\varepsilon)\right)Bu^{3/2}$, and this last inequality gives the theorem in this case. For $u < 1$, the inequality trivially holds if we replace the coefficient 2.62 by e. ■

The estimation of the quantity α in the preceding proof used symmetrisation, and this is the reason we have a spurious factor of 2 in front of $C\sqrt{u}$ in inequality (3.158). Bernstein's inequality usually gives best results when it is used in the Gaussian range. This is also true for this inequality: it produces best results when the dominating term among the four summands of the tail range, $2(1+\varepsilon)^{3/2}C\sqrt{u}$, is largest, in which case inequality (3.158) prescribes a Gaussian tail probability for U_n. Note also that this inequality is not useful for $u \le 1$.

Exercises

3.4.1 Let X_i be independent and with respective probability laws P_i, let $k > 1$ and consider the space $\mathbb{H} = \{(f^1(X_1),\dots,f^n(X_n)) : f^i \in L^k(P_i)\}$. Show that the duality of L^p spaces and the independence of the variables X_i imply that the pseudo-norm $\left(\sum_{i=1}^n E|f^i(X_i)|^k\right)^{1/k}$ satisfies

$$
\left(\sum_{i=1}^n E|f^i(X_i)|^k\right)^{1/k} = \sup_{\sum_{i=1}^N E|\xi_i(X_i)|^{k/(k-1)}=1} \sum_{i=1}^m E(f^i(X_i)\xi_i(X_i)),
$$

where the sup runs over $\xi_i \in L^{k/(k-1)}(P_i)$. Note that if $F(i,\omega) = f^i(X_i(\omega))$, this pseudo-norm is just the $L^k(\mu\times\Pr)$ norm of F, where μ is counting measure on $\{1,\dots,m\}$.

3.4.2 Prove versions of Theorems 3.4.3 and 3.4.5 for non-i.i.d. random variables.

3.4.3 Show that $\sum_{j=1}^{\infty} e^{-sq^{\alpha j}} \le \frac{1}{q^{\alpha}-1}\frac{1}{s}e^{-s}$. *Hint*: The left-hand side can be written as

$$\frac{q^{\alpha}}{q^{\alpha}-1}\sum_{j=1}^{\infty} q^{-\alpha j}e^{-sq^{\alpha j}}(q^{\alpha j}-q^{\alpha(j-1)}),$$

and $\sum_{j=1}^{\infty} e^{-sq^{\alpha j}}(q^{\alpha j}-q^{\alpha(j-1)})$ is a Riemann sum.

3.4.4 Sometimes it is handier to have the exponential inequality in Theorem 3.4.8 inverted, that is, for $\Pr\{U_n \ge t\}$. Show that inequality (3.158) implies that, for all $t \ge 0$,

$$\Pr\{U_n \ge t\}$$
$$\le \exp\left[1-\left(\left(\frac{t}{8(1+\varepsilon)^{3/2}C}\right)^2 \wedge \frac{t}{18\sqrt{2}D} \wedge \left(\frac{t}{(126e+18\sqrt{2})B}\right)^{2/3} \wedge \left(\frac{t}{126eA}\right)^{1/2}\right)\right].$$

3.5 Metric Entropy Bounds for Suprema of Empirical Processes

Clearly, to make effective use of the exponential inequalities in Section 3.3, we should have available good estimates for the mean of the supremum of an empirical process $E\|P_n - P\|_{\mathcal{F}}$. This section and the next are devoted to this important subject.

3.5.1 Random Entropy Bounds via Randomisation

By Theorem 3.1.21, we can randomise the empirical process by Rademacher multipliers. The resulting process is sub-Gaussian conditionally on the data X_i, and therefore, the metric entropy bounds in Section 2.3, in particular, Theorem 2.3.7, apply to it. This simple procedure produces a bound that will turn out to be very useful because there are many important classes of functions that have very good $L^2(Q)$ metric entropy bounds, actually, *uniform in* Q, as we will see in the next section. Here we just record the result, and we will wait until the next section for its application to meaningful examples.

For any $n \in \mathbb{N}$, let P_n denote the empirical measure corresponding to n i.i.d. S-valued random variables X_i of law P. Then, for any measurable real functions f, g on S, we let $e_{n,2}(f,g)$ denote their $L^2(P_n)$ (pseudo)distance, that is,

$$e_{n,2}^2(f,g) = \frac{1}{n}\sum_{i=1}^{n}(f-g)^2(X_i).$$

Note that this is a random (pseudo)distance. These random distances give rise to *random or empirical metric entropies*: given a class of measurable functions $\mathcal{F}$ on S, the empirical metric entropies of $\mathcal{F}$ are defined as $\log N(\mathcal{F}, e_{n,2}, \tau)$ for any $\tau > 0$ (recall the definition of the covering numbers $N(T,d,\tau)$ from Section 2.3). Often we will write $N(\mathcal{F}, L^2(P_n), \tau)$ for $N(\mathcal{F}, e_{n,2}, \tau)$. Recall also the packing numbers $D(T,d,\tau)$ and their relationship with covering numbers: for all $\tau > 0$

$$N(T,d,\tau) \le D(T,d,\tau) \le N(T,d,\tau/2). \tag{3.164}$$

There is a formal advantage to using the packing numbers $D(\mathcal{F}, L^2(P_n), \tau)$ instead of the covering numbers $N(\mathcal{F}, L^2(P_n), \tau)$: by definition, for all m,

$$D(\mathcal{F}, L^2(P_n), \tau) \geq m \Leftrightarrow \sup_{f_1,\ldots,f_m \in \mathcal{F}} \min_{1 \leq i \neq j \leq m} P_n^2(f_i, f_j)^2 > \varepsilon^2;$$

hence, it follows that if $\mathcal{F}$ is countable, then $D(\mathcal{F}, L^2(P_n), \tau)$ is a random variable.

Theorem 3.5.1 *In the preceding notation, assuming $\mathcal{F}$ countable and $0 \in \mathcal{F}$,*

$$E\left[\sqrt{n}\|P_n - P\|_{\mathcal{F}}\right] \leq 8\sqrt{2}E\left[\int_0^{\sqrt{\|P_n f^2\|_{\mathcal{F}}}} \sqrt{\log 2D(\mathcal{F}, L^2(P_n), \tau)}\, d\tau\right] \tag{3.165}$$

and, for all $\delta > 0$,

$$E\left[\sqrt{n} \sup_{f,g \in \mathcal{F}: P_n|f-g|^2 \leq \delta^2} |(P_n - P)(f-g)|\right]$$
$$\leq 2(16\sqrt{2}+2)E\left[\int_0^{\delta} \sqrt{\log 2D(\mathcal{F}, L^2(P_n), \tau)}\, d\tau\right]. \tag{3.166}$$

Proof The integrals in the preceding inequalities are Riemann integrals because D is monotone. Hence, since $D(\mathcal{F}, L^2(P_n), \tau)$ is a random variable for each τ, these integrals are also measurable. By Theorem 3.1.21 and Fubini, dropping as usual the subindex $\mathcal{F}$ from the supremum norms, we have

$$E\sqrt{n}\|P_n - P\| \leq 2E\left\|\frac{1}{\sqrt{n}}\sum_{i=1}^n \varepsilon_i f(X_i)\right\| = 2E_X E_\varepsilon \left\|\frac{1}{\sqrt{n}}\sum_{i=1}^n \varepsilon_i f(X_i)\right\|.$$

Since the process $\sum_{i=1}^n a_i \varepsilon_i$, $(a_1, \ldots, a_n) \in \mathbb{R}^n$, is separable for the Euclidean distance (see Definition 2.1.2) and is sub-Gaussian for this distance (Definition 2.3.5 and the paragraph following it), Theorem 2.3.7 applies to the process $(1/\sqrt{n})\sum_{i=1}^n \varepsilon_i f(X_i)$ conditionally on the variables X_i. Thus, noting that

$$\frac{1}{n}E_\varepsilon\left[\sum_{i=1}^n \varepsilon_i(f(X_i) - g(X_i))\right]^2 = \frac{1}{n}\sum_{i=1}^n (f-g)^2(X_i) = \|f-g\|_{L^2(P_n)}^2$$

and recalling (3.164), the entropy bound (2.41) gives

$$E_\varepsilon\left\|\frac{1}{\sqrt{n}}\sum_{i=1}^n \varepsilon_i f(X_i)\right\| \leq 4\sqrt{2}\int_0^{\sqrt{\|P_n f^2\|}} \sqrt{\log 2D(\mathcal{F}, L^2(P_n), \tau)}\, d\tau,$$

which, combined with the preceding randomisation inequality implies the first bound in the theorem. The second bound follows in the same way using (2.42). ∎

Remark 3.5.2 Note that except for measurability, the random packing numbers in the bounds (3.165) and (3.166) can be replaced by the random covering numbers $N(\mathcal{F}, L^2(P_n), \tau)$. In fact, if $N^*(\mathcal{F}, L^2(P_n), \tau) \geq N(\mathcal{F}, L^2(P_n), \tau)$, is a random variable for each $\tau > 0$ and is nondecreasing in τ, then the bound (3.165) can be replaced by

$$E\left[\sqrt{n}\|P_n - P\|_{\mathcal{F}}\right] \leq 8\sqrt{2}E\left[\int_0^{\sqrt{\|P_n f^2\|_{\mathcal{F}}}} \sqrt{\log 2N^*(\mathcal{F}, L^2(P_n), \tau)}\, d\tau\right], \tag{3.167}$$

and likewise for (3.166).

The bound in Theorem 3.5.1 is mostly useful when the random entropies $\log N(\mathcal{F}, L^2(P_n), \tau)$ admit good bounds that are uniform in P_n and satisfy some regularity such as, for example, being regularly varying at zero. As we see in the next section, there are many classes of functions $\mathcal{F}$, denoted by *Vapnik-Červonenkis classes of functions*, whose covering numbers admit the bound

$$N(\mathcal{F}, L^2(Q), \tau\|F\|_{L^2(Q)}) \leq \left(\frac{A}{\tau}\right)^v, \quad 0 < \tau \leq 1, \tag{3.168}$$

for some A, v positive and finite and for all probability measures Q on $(S, \mathcal{S})$, where F is a measurable function such that $|f| \leq F$, for all $f \in \mathcal{F}$. If a measurable function F satisfies this property, we say that F is a *measurable envelope* (or just *envelope*) of the class of functions $\mathcal{F}$. The next theorem will cover in particular the Vapnik-Červonenkis case.

For ease of notation, we set, for all $0 < \delta < \infty$,

$$J(\mathcal{F}, F, \delta) := \int_0^\delta \sup_Q \sqrt{\log 2N(\mathcal{F}, L_2(Q), \tau\|F\|_{L_2(Q)})}\, d\tau, \tag{3.169}$$

where the supremum is taken over all discrete probabilities with a finite number of atoms and rational weights (in particular, over all possible empirical measures), and we assume that our class of functions $\mathcal{F}$ satisfies $J(\mathcal{F}, F, \delta) < \infty$ for some $\delta > 0$ (hence for all). The integrand of J is denoted as the *Koltchinskii-Pollard entropy of* $\mathcal{F}$. Before establishing a bound for $E\|P_n - P\|_{\mathcal{F}}$, it is convenient to single out several concavity properties of the function J.

Lemma 3.5.3 *Let $G(x) = \int_0^x g(t)dt$, $0 < x < \infty$, where $g : (0,\infty) \mapsto [0,\infty)$ is locally integrable, nonnegative and nonincreasing. Then*

(a) *G is concave, nondecreasing, and $G(cx) \leq cG(x)$, for all $c \geq 1$ and all $x > 0$,*
(b) *The function $x \mapsto xG(1/x)$ is nondecreasing and*
(c) *The function of two variables $(x,t) \mapsto \sqrt{t}G(\sqrt{x/t})$, $(x,t) \in (0,\infty) \times (0,\infty)$, is concave.*

Proof Part (a) is obvious because G' is nonincreasing and nonnegative. For part (b), note that $G(y)/y = \frac{1}{y}\int_0^y g(t)dt$ is the average over $(0,y)$ of a nonincreasing function, so it is nonincreasing in y and hence nondecreasing in $x = 1/y$. Part (c) is better proved in two parts. First, we claim that the function of two variables $H(x,t) = tG(x/t)$ is concave on $(0,\infty) \times (0,\infty)$: if $0 < \lambda < 1$ and $0 < x_i < \infty$, $0 < t_i < \infty$, $i = 1,2$,

$$\begin{aligned} H(\lambda(x_1,t_1) + (1-\lambda)(x_2,t_2)) &= (\lambda t_1 + (1-\lambda)t_2) G\left(\frac{\lambda x_1 + (1-\lambda)x_2}{\lambda t_1 + (1-\lambda)t_2}\right) \\ &= (\lambda t_1 + (1-\lambda)t_2) G\left(\frac{\lambda t_1}{\lambda t_1 + (1-\lambda)t_2}\frac{x_1}{t_1} + \frac{(1-\lambda)t_2}{\lambda t_1 + (1-\lambda)t_2}\frac{x_2}{t_2}\right) \\ &\geq \lambda t_1 G(x_1/t_1) + (1-\lambda)t_2 G(x_2/t_2). \end{aligned}$$

Thus, $H(x,t)$ is concave as a function of two variables and is also nondecreasing in each coordinate separately (by (a) and (b)). Using these two properties, one sees that $H(\sqrt{x}, \sqrt{t})$ is also concave: by the monotonicity in each coordinate and the concavity of $(\cdot)^{1/2}$, we have

$$H(\sqrt{\lambda x_1 + (1-\lambda)t_1}, \sqrt{\lambda x_2 + (1-\lambda)t_2}) \geq H(\lambda\sqrt{x_1} + (1-\lambda)\sqrt{t_1}, \lambda\sqrt{x_2} + (1-\lambda)\sqrt{t_2}),$$

and now part (c) follows form the concavity of H. ■

This lemma obviously applies to the function J, and we are now ready to obtain a bound on the expected value of the empirical process in terms of J.

Theorem 3.5.4 *Let $\mathcal{F}$ be a countable class of measurable functions with $0 \in \mathcal{F}$, and let F be a strictly positive envelope for $\mathcal{F}$. Assume that*

$$J(\mathcal{F}, F, \delta) < \infty, \quad \text{for some (for all) } \delta > 0, \tag{3.170}$$

where J is defined in (3.169). Given $X_1, \ldots, X_n$ independent identically distributed S-valued random variables with common law P such that $PF^2 < \infty$, let P_n be the corresponding empirical measure and $\nu_n(f) = \sqrt{n}(P_n - P)(f)$, $f \in \mathcal{F}$, the normalised empirical process indexed by $\mathcal{F}$. Set $U = \max_{1 \le i \le n} F(X_i)$, $\sigma^2 = \sup_{f \in \mathcal{F}} Pf^2$ and $\delta = \sigma / \|F\|_{L^2(P)}$. Then, for all $n \in \mathbb{N}$,

$$E\|\nu_n\|_{\mathcal{F}} \le \max\left[A_1 \|F\|_{L^2(P)} J(\mathcal{F}, F, \delta), \frac{A_2 \|U\|_{L^2(P)} J^2(\mathcal{F}, F, \delta)}{\sqrt{n}\delta^2}\right], \tag{3.171}$$

where we can take

$$A_1 = 8\sqrt{6} \quad \text{and} \quad A_2 = 2^{15} 3^{5/2}. \tag{3.172}$$

Proof Let us write $J(t)$ for $J(\mathcal{F}, F, t)$. Set $\sigma_n^2 = \|P_n f^2\|_{\mathcal{F}}$, and note that the diameter of $\mathcal{F}$ for the $L^2(P_n)$ random pseudo-norm is dominated by $2\sigma_n$. We randomise by Rademacher variables and recall that, as in Theorem 3.5.1, by the metric entropy bound for sub-Gaussian processes (2.41) in Theorem 2.3.7, we have

$$\begin{aligned}
E_\varepsilon \|\nu_{n,\mathrm{rad}}\|_{\mathcal{F}} := E_\varepsilon \left\| \frac{1}{\sqrt{n}} \sum_{i=1}^n \varepsilon_i f(X_i) \right\|_{\mathcal{F}} &\le 4\sqrt{2} \int_0^{\sigma_n} \sqrt{\log 2N(\mathcal{F}, e_{n,2}, \tau)}\, d\tau \\
&= 4\sqrt{2} \|F\|_{L^2(P_n)} \int_0^{\sigma_n / \|F\|_{L^2(P_n)}} \sqrt{\log 2N(\mathcal{F}, e_{n,2}, \tau \|F\|_{L^2(P_n)})}\, d\tau \\
&\le 4\sqrt{2} \|F\|_{L^2(P_n)} J(\sigma_n / \|F\|_{L^2(P_n)}).
\end{aligned} \tag{3.173}$$

Then, by Fubini's theorem and the concavity of $\sqrt{t}J(\sqrt{x/t})$ (Lemma 3.5.3, part (c)), we have

$$E\|\nu_{n,\mathrm{rad}}\|_{\mathcal{F}} \le 4\sqrt{2} \|F\|_{L^2(P)} J(\|\sigma_n\|_{L^2(P)} / \|F\|_{L^2(P)}). \tag{3.174}$$

Now we estimate $\|\sigma_n\|_{L^2(P)}$ by means of Corollary 3.2.2 (a consequence of the comparison theorem for Rademacher processes), followed by the Cauchy-Schwarz inequality and by Hoffmann-Jørgensen's inequality (3.42) on comparison between the first and second moments of a sum of independent random vectors, to obtain

$$\begin{aligned}
n\|\sigma_n\|_{L^2(P)}^2 &= E\left\| \sum_{i=1}^n f^2(X_i) \right\|_{\mathcal{F}} \\
&\le n\sigma^2 + 8E\left[U \left\| \sum_{i=1}^n \varepsilon_i f(X_i) \right\|_{\mathcal{F}} \right]
\end{aligned}$$

$$\leq n\sigma^2 + 8\|U\|_{L^2(P)} \left\| \left\| \sum_{i=1}^n \varepsilon_i f(X_i) \right\|_{\mathcal{F}} \right\|_{L^2(P)}$$

$$\leq n\sigma^2 + 2^5 \cdot 3^{3/2} \|U\|_{L^2(P)} \left(2^4 E \left\| \sum_{i=1}^n \varepsilon_i f(X_i) \right\|_{\mathcal{F}} + \|U\|_{L^2(P)} \right). \quad (3.175)$$

Setting $Z = E\|\nu_{n,\mathrm{rad}}\|_{\mathcal{F}}$, we may then write

$$\|\sigma_n\|^2_{L^2(P)} \leq \max\left[3\sigma^2, 2^9 \cdot 3^{5/2} n^{-1/2} \|U\|_{L^2(P)} Z, 2^5 \cdot 3^{5/2} n^{-1} \|U\|^2_{L^2(P)}\right]. \quad (3.176)$$

If the largest of the three terms on the right-hand side of (3.176) is the first, then plugging this estimate into (3.174) and using Lemma 3.5.3(a), we obtain

$$Z \leq 4\sqrt{2}\|F\|_{L^2(P)} J(\sqrt{3}\delta) \leq 4\sqrt{6}\|F\|_{L^2(P)} J(\delta).$$

If the largest term on the right-hand side of (3.176) is the second, then we have in particular that $\sqrt{3}\delta \leq (2^9 \cdot 3^{5/2} n^{-1/2} \|U\|_{L^2(P)} Z)^{1/2} / \|F\|_{L^2(P)}$, and denoting this last quantity by L, the inequalities (3.176) and (3.174) give, using Lemma 3.5.3(a) and (b), that

$$Z \leq 4\sqrt{2}\|F\|_{L^2(P)} L J(L)/L \leq 4\sqrt{2}\|F\|_{L^2(P)} L J(\delta)/\delta = 2^7 \cdot 3^{5/4} n^{-1/4} \|U\|^{1/2}_{L^2(P)} \sqrt{Z} J(\delta)/\delta;$$

that is,

$$Z \leq 2^{14} 3^{5/2} n^{-1/2} \|U\|_{L^2(P)} J^2(\delta)/\delta^2.$$

Finally, if it is the third term that dominates the right-hand side of (3.176), then, since in particular $\sqrt{3}\delta = \sqrt{3}\sigma/\|F\|_{L^2(P)} \leq 2^{5/2} \cdot 3^{5/4} n^{-1/2} \|U\|_{L^2(P)} / \|F\|_{L^2(P)}$, if we denote this last quantity by $3^{1/2} M$, and since $J(\delta)/\delta \geq \sqrt{\log 2}$,

$$Z \leq 4\sqrt{2}\sqrt{3} M J(M)/M \leq 4\sqrt{2}\sqrt{3} M J(\delta)/\delta$$
$$\leq 2^5 \cdot 3^{5/4} (\log 2)^{-1/2} n^{-1/2} (\|U\|_{L^2(P)} / \|F\|_{L^2(P)}) J^2(\delta)/\delta^2.$$

Now the theorem follows by taking the maximum of these three estimates of Z, given that by the basic Rademacher randomisation inequality $E\|\nu_n\|_{\mathcal{F}} \leq 2E\|\nu_{n,\mathrm{rad}}\|_{\mathcal{F}} = 2Z$. ■

Remark 3.5.5 It is convenient to single out the following simple consequence of the metric entropy bound (3.173): since by Hölder's inequality $E\|F\|_{L^2(P_n)} \leq \|F\|_{L^2(P)}$, and since J is nondecreasing, it follows from (3.173) and the randomisation inequality that under the assumptions of the preceding theorem,

$$E\|\nu_n\|_{\mathcal{F}} \leq 8\sqrt{2} J(1) \|F\|_{L^2(P)}. \quad (3.177)$$

This bound is only interesting when $\|F\|_{L^2(P)}$ is similar in magnitude to σ.

When the Koltchinskii-Pollard entropy admits as upper bound a regularly varying function, then the integral over $(0,\delta]$ defining J is dominated by a constant times the value of this function at δ (just as with the integral of a power of x). Since in this case the resulting bound for the expected value of the empirical process becomes particularly simple and applies in many situations including the Vapnik-Červonenkis case – see (3.168) – we make it explicit in the next theorem.

Theorem 3.5.6 *Let $\mathcal{F}$ be a countable class of functions with $0 \in \mathcal{F}$, let F be an envelope for $\mathcal{F}$ and let $H : [0,\infty) \mapsto [0,\infty)$ be a function equal to $\log 2$ for $0 < x \le 1$ and such that*

(a) *$H(x)$ is nondecreasing for $x > 0$, and so is $xH^{1/2}(1/x)$ for $0 < x \le 1$, and*
(b) *there exists C_H finite such that $\int_0^c \sqrt{H(1/x)}\,dx \le C_H c H^{1/2}(1/c)$ for all $0 < c \le 1$.*

Assume that

$$\sup_Q \log[2N(\mathcal{F}, L^2(Q), \tau\|F\|_{L^2(Q)})] \le H\left(\frac{1}{\tau}\right), \quad \text{for all } \tau > 0, \tag{3.178}$$

where the supremum is taken over all discrete probability measures Q with a finite number of atoms and with rational weights. Then

$$E\|\nu_n\|_{\mathcal{F}} \le \max\left[A_1 C_H \sigma\sqrt{H(\|F\|_{L^2(P)}/\sigma)}, A_2 C_H^2 \|U\|_{L^2(P)} H(\|F\|_{L^2(P)}/\sigma)/\sqrt{n}\right], \tag{3.179}$$

where A_1 and A_2 are the constants in (3.172).

Proof By definition and by property (b) of H,

$$J(\mathcal{F}, F, \delta) \le \int_0^\delta \sqrt{H(1/\tau)}d\tau \le C_H \delta \sqrt{H(1/\delta)} = C_H \frac{\sigma}{\|F\|_{L^2(P)}} \sqrt{H(\|F\|_{L^2(P)}/\sigma)},$$

and the theorem follows by applying this bound for $J(\delta)$ in inequality (3.171). ■

Similarly, if we set $D_H = \int_0^1 \sqrt{H(1/\tau)}d\tau$, inequality (3.177) becomes

$$E\|\nu_n\|_{\mathcal{F}} \le 8\sqrt{2} D_H \|F\|_{L^2(P)}. \tag{3.180}$$

The uniformly bounded case in the preceding two theorems admits a more elementary proof based only on randomisation and the entropy bound (i.e., neither Hoffmann-Jørgensen's inequality nor comparison of Rademacher processes is required) which yields better constants. This is illustrated here for the second theorem.

Corollary 3.5.7 *Assume that the hypotheses of Theorem 3.5.6 are satisfied and that, moreover, the functions in $\mathcal{F}$ are bounded in absolute value by a constant u. Then*

$$E\|\nu_n\|_{\mathcal{F}} \le 8\sqrt{2} C_H \sigma \sqrt{H(\|F\|_{L^2(P)}/\sigma)} + 2^7 C_H^2 u H(\|F\|_{L^2(P)}/\sigma)/\sqrt{n}. \tag{3.181}$$

Proof It suffices to prove the bound for $u = 1/2$. Using $J(c) \le C_H c H^{1/2}(1/c)$, that H is monotone nondecreasing, and that

$$\sigma^2 = \|Pf^2\| = \|PP_n f^2\| \le P\|P_n f^2\| = \|\sigma_n\|_{L^2(P)}^2,$$

inequality (3.174) gives

$$\begin{aligned} Z = E\|\nu_{n,\mathrm{rad}}\|_{\mathcal{F}} &\le 4\sqrt{2} C_H \|\sigma_n\|_{L^2(P)} H^{1/2}(\|F\|_{L^2(P)}/\|\sigma_n\|_{L^2(P)}) \\ &\le 4\sqrt{2} C_H \|\sigma_n\|_{L^2(P)} H^{1/2}(\|F\|_{L^2(P)}/\sigma) =: B. \end{aligned} \tag{3.182}$$

Now we will obtain a bound on B by estimating $\|\sigma_n\|_{L^2(P)}$ in terms of B and solving the resulting inequation. First, we observe that, by Rademacher randomisation,

$$\|\sigma_n\|^2_{L^2(P)} = E\|P_n f^2\|_{\mathcal{F}} \leq \|Pf^2\|_{\mathcal{F}} + E\|(P_n - P)f^2\|_{\mathcal{F}}$$

$$\leq \sigma^2 + 2E\left\|\frac{1}{n}\sum_{i=1}^n \varepsilon_i f^2(X_i)\right\|_{\mathcal{F}}, \tag{3.183}$$

where the second term is the expected value of an empirical process to which we will apply the very same inequality (3.182). Since $|f| \leq 1/2$ for all $f \in \mathcal{F}$, it follows that for $f, g \in \mathcal{F}$, $P_n(f^2 - g^2)^2 \leq P_n(f - g)^2 = e^2_{n,2}(f,g)$, and therefore, if we set $\mathcal{F}^2 = \{f^2 : f \in \mathcal{F}\}$, we have that $N(\mathcal{F}^2, e_{n,2}, \tau) \leq N(\mathcal{F}, e_{n,2}, \tau)$. Then, proceeding as in the derivation of inequality (3.174) followed by (3.182), we obtain

$$E\left\|\frac{1}{\sqrt{n}}\sum_{i=1}^n \varepsilon_i f^2(X_i)\right\|_{\mathcal{F}} \leq 2\sqrt{2}C_H\|\sigma_n\|_{L^2(P)}H^{1/2}(\|F\|_{L^2(P)}/\sigma) \leq B/2.$$

Combined with (3.183), we have

$$\|\sigma_n\|^2_{L^2(P)} \leq \sigma^2 + B/\sqrt{n},$$

which, by the definition of B in (3.182), gives the following inequation for B:

$$B^2 \leq 2^5 C_H^2 H(\|F\|_{L^2(P)}/\sigma)\left(\sigma^2 + B/\sqrt{n}\right).$$

Hence, B is bounded by the largest solution of the corresponding second-degree equation, that is,

$$B \leq 2^5 C_H^2 H(\|F\|_{L^2(P)}/\sigma)/\sqrt{n} + 2^{5/2} C_H \sigma H^{1/2}(\|F\|_{L^2(P)}/\sigma).$$

Inequality (3.181) now follows by the basic randomisation and (3.182), which together give $E\|\nu_n\|_{\mathcal{F}} \leq 2Z \leq 2B$, and by the bound on B. ■

Corollary 3.5.8 *Suppose that* $\sup_Q N(\mathcal{F}, L^2(Q), \varepsilon\|F\|_{L^2(Q)}) \leq (A/\varepsilon)^v$, *for* $0 < \varepsilon < A$, *for some* $v \geq 1$ *and* $A \geq 2$, *the supremum extending over all Borel probability measures* Q, *and let* $u = \|F\|_\infty$. *Then*

$$E\|\nu_n\|_{\mathcal{F}} \leq 8\sqrt{2}C_A\sigma\sqrt{2v\log\frac{A\|F\|_{L^2(P)}}{\sigma}} + 2^8 C_A \frac{1}{\sqrt{n}} uv\log\frac{A\|F\|_{L^2(P)}}{\sigma}, \tag{3.184}$$

where $C_A = 2\log A/(2\log A - 1)$.

Proof The proof follows from the preceding corollary, taking $H(x) = 2v\log(Ax)$ for $x \geq 1$ (and $H(x) = \log 2$ for $0 < x < 1$). To compute C_H, note that, by differentiation, $\int_0^c (\log(A/x)))^{1/2}(1 - 2^{-1}(\log(A/x))^{-1} = c(\log(A/c))^{1/2}$, from which it follows that, for all $0 < c \leq 1$,

$$\int_0^c (\log(A/x))^{1/2} dx \leq \frac{2\log A}{2\log A - 1} c(\log(A/c))^{1/2},$$

so $C_H = C_A = (2\log A)/(2\log A - 1)$. ■

Perhaps the main observation regarding Theorem 3.5.6 is that if

$$n\sigma^2/\|U\|_{L^2(P)}^2 \gtrsim H(\|F\|_{L^2(P)}/\sigma),$$

then the bound (3.179) becomes, disregarding constants,

$$E\left\|\sum_{i=1}^n (f(X_i) - Pf)\right\|_{\mathcal{F}} \lesssim \sqrt{n\sigma^2 H\left(\frac{2\|F\|_{L^2(P)}}{\sigma}\right)},$$

which means that if $n\sigma^2$ is not too small, then the 'price' one pays for considering the expectation of the supremum of infinitely many sums instead of just one is the factor $(H(\|F\|_{L^2(P)}/\sigma))^{1/2}$. Since, as we see next, this bound is best possible, we single out this observation in the following corollary:

Corollary 3.5.9 *Under the hypotheses of Theorem 3.5.6 and with the same notation, if, moreover, for some $\lambda \geq 1$,*

$$\frac{n\sigma^2}{\|U\|_{L^2(P)}^2} \geq \left(\frac{A_2 C_H}{\lambda A_1}\right)^2 H(\|F\|_{L^2(P)}/\sigma), \tag{3.185}$$

then

$$E\left\|\sum_{i=1}^n (f(X_i) - Pf)\right\|_{\mathcal{F}} \leq \lambda A_1 C_H \sqrt{n\sigma^2 H\left(\frac{2\|F\|_2}{\sigma}\right)} \leq \frac{\lambda^2 A_1^2}{A_2}\frac{n\sigma^2}{\|U\|_{L^2(P)}}, \tag{3.186}$$

where A_1 and A_2 are defined in (3.172). In the uniformly bounded case, if

$$\frac{n\sigma^2}{u} \geq \frac{2^7 C_H^2}{(\lambda-1)^2} H(\|F\|_{L^2(P)}/\sigma),$$

then

$$E\left\|\sum_{i=1}^n (f(X_i) - Pf)\right\|_{\mathcal{F}} \leq 8\sqrt{2}\lambda C_H \sqrt{n\sigma^2 H\left(\frac{2\|F\|_2}{\sigma}\right)} \leq \lambda^2 \frac{n\sigma^2}{u}.$$

Note that the bound in this corollary resembles the bound (2.61) for Gaussian processes when the metric entropy is tightly majorised by a function of regular variation. We show now that as in the case of Gaussian processes, when condition (3.185) is satisfied, the expectation bound (3.186) is two sided. The proof is based on the Sudakov-type bound for Rademacher processes given in Theorem 3.2.9. We need a definition just to describe how the function H must also be, up to constants, a lower bound for the metric entropy of $\mathcal{F}$.

Definition 3.5.10 A class of functions $\mathcal{F}$ that satisfies the hypotheses of Theorem 3.5.6 and such that $|f| \leq 1$ for all $f \in \mathcal{F}$ is full for H and P if, moreover, there exists $c > 0$ such that

$$\log N(\mathcal{F}, L^2(P), \sigma/2) \geq cH\left(\frac{\|F\|_{L^2(P)}}{\sigma}\right), \tag{3.187}$$

for a measurable envelope F of $\mathcal{F}$.

Usually the function F in this definition will be the 'smallest possible' measurable envelope of F, that is, one that satisfies that if $\overline{F}$ is another measurable cover, then $F \le \overline{F}$ P a.s. Such an envelope will be called the *P-measurable cover of* $\mathcal{F}$. The P-measurable cover is unique P a.s., and it exists as soon as $\sup_{f\in\mathcal{F}}|f| < \infty$ P a.s. (see Section 3.7.1)

Theorem 3.5.11 *Let $\mathcal{F}$, H and F be as in Theorem 3.5.6 but further assume that the functions in $\mathcal{F}$ take values in $[-1,1]$, let P_n, $n \in \mathbb{N}$, be the empirical measure corresponding to samples from a probability measure P on $(S,\mathcal{S})$ and suppose as well that $Pf = 0$ for all $f \in \mathcal{F}$. Assume that*

$$n\sigma^2 \ge (2^{15} \vee (2^{22}K^2C_H^2))H(6\|F\|_2/\sigma) \quad \text{and} \quad n^2\sigma^2 \ge 32\sqrt{2}D_H/(3e^{1/2}), \tag{3.188}$$

where $K \ge 1$ is as in Theorem 3.2.9. Then

$$E\left\|\sum_{i=1}^n f(X_i)\right\|_{\mathcal{F}} \ge \frac{\sqrt{n}\,\sigma}{32K}\sqrt{\log N(\mathcal{F}, L^2(P), \sigma/2)}. \tag{3.189}$$

If, moreover, the class $\mathcal{F}$ is full for H, P and F with constant c, then

$$\frac{c}{32K}\sqrt{n\sigma^2 H\left(\frac{\|F\|_{L^2(P)}}{\sigma}\right)} \le E\left\|\sum_{i=1}^n f(X_i)\right\|_{\mathcal{F}} \le 8\sqrt{22}\sqrt{n\sigma^2 H\left(\frac{2\|F\|_{L^2(P)}}{\sigma}\right)} \tag{3.190}$$

(fullness is only required for the left-hand side inequality).

Proof Application of Theorem 3.2.9 for $T = \{(f(X_1(\omega)), \dots, f(X_n(\omega))) : f \in \mathcal{F}\}$ (keeping with regular usage, we will not show the variable ω) gives that for a universal constant $K \ge 1$, if

$$E_\varepsilon\left\|\frac{1}{\sqrt{n}}\sum_{i=1}^n \varepsilon_i f(X_i)\right\|_{\mathcal{F}} \le \frac{\sqrt{n}\sigma^2}{64K}, \tag{3.191}$$

then

$$E_\varepsilon\left\|\frac{1}{\sqrt{n}}\sum_{i=1}^n \varepsilon_i f(X_i)\right\|_{\mathcal{F}} \ge \frac{\sigma}{8K}\sqrt{\log N(\mathcal{F}, L^2(P_n), \sigma/8)}. \tag{3.192}$$

The proof of the theorem will consists in finding an upper bound for the left-hand side of (3.191) and a lower bound for the right-hand side of (3.192), both holding with large probability. We start with the latter. Let $D := D(\mathcal{F}, L^2(P), \sigma/2)$, and let $f_1, \dots, f_D$ be $\sigma/2$-separated in $L^2(P)$. By the law of large numbers, we have that almost surely

$$P_n(f_i - f_j)^2 \to P(f_i - f_j)^2, \quad 1 \le i,j \le D, \ \|F\|_{2,n} \to \|F\|_2;$$

hence, given $\varepsilon > 0$, there exist n and ω such that $(1-\varepsilon)\|f_i - f_j\|_{L^2(P_n(\omega))} \ge \|f_i - f_j\|_2$, for $i,j \le D$ and $\|F\|_{L^2(P_n(\omega))} \le (1+\varepsilon)\|F\|_2$. Thus, using (3.164),

$$D(\mathcal{F}, L^2(P), \sigma/2) \le N(\mathcal{F}, L^2(P_n(\omega)), (1-\varepsilon)\sigma/4),$$

and therefore, taking $\varepsilon = 1/5$, we obtain, by the hypothesis (3.178) on the random entropies,

$$D(\mathcal{F}, L^2(P), \sigma/2) \le e^{H(6\|F\|_2/\sigma)}. \tag{3.193}$$

Now, since $P(f_i - f_j)^4 \le 4P(f_i - f_j)^2 \le 16 \max P f_i^2 \le 16\sigma^2$, Bernstein's inequality in the form given by (3.24) gives

$$\Pr\left\{\max_{1\le i\ne j\le D}\left[n(P(f_i - f_j)^2 - \sum_{k=1}^{n}(f_i - f_j)^2(X_k)\right] \ge \frac{x}{3} + \sqrt{32n\sigma^2 x}\right\} \le D^2 e^{-x}.$$

Hence, taking $x = \delta n\sigma^2$ (for some $\delta > 0$), since $P(f_i - f_j)^2 \ge \sigma^2/4$, we have by (3.193) that

$$\Pr\left\{\frac{\sigma^2}{4} - \min_{1\le i\ne j\le D} P_n(f_i - f_j)^2 \ge \frac{\delta\sigma^2}{3} + \sqrt{32\delta\sigma^4}\right\} \le e^{-\delta n\sigma^2 + 2H(6\|F\|_2/\sigma)}$$

or, taking $\delta = 1/(32\cdot 4^4)$,

$$\Pr\left\{\min_{1\le i\ne j\le D} P_n(f_i - f_j)^2 \le \frac{\sigma^2}{16}\right\} \le e^{-n\sigma^2/(32\cdot 4^4) + 2H(6\|F\|_2/\sigma)}.$$

We have thus proved that the event A_1 defined by

$$\begin{aligned} N(\mathcal{F}, L^2(P_n), \sigma/8) &\ge D(\mathcal{F}, L^2(P_n), \sigma/4) \\ &\ge D = D(\mathcal{F}, L^2(P), \sigma/2) \\ &\ge N(\mathcal{F}, L^2(P), \sigma/2), \end{aligned} \tag{3.194}$$

has probability

$$\Pr(A_1) \ge 1 - e^{-n\sigma^2/(32\cdot 4^4) + 2H(6\|F\|_2/\sigma)}. \tag{3.195}$$

We now turn to estimation of the left-hand side of (3.191). The starting point is the metric entropy bound for sub-Gaussian processes (Theorem 2.3.7) applied to the Rademacher empirical process, namely (using (3.178)),

$$E_\varepsilon \left\|\frac{1}{\sqrt{n}}\sum_{i=1}^{n}\varepsilon_i f(X_i)\right\|_{\mathcal{F}} \le 4\sqrt{2}\int_0^{\sigma_n}\sqrt{H(\|F\|_{2,n}/\tau)}\,d\tau, \tag{3.196}$$

and we must show that with large probability σ_n can be replaced by σ and $\|F\|_{2,n}$ by $\|F\|_2$ up to multiplicative constants. If

$$A_2 = \{\|F\|_{2,n} \le 2\|F\|_2\},$$

then Bernstein's inequality (3.23) gives

$$\Pr(A_2^c) = \Pr\left\{\sum_{i=1}^{n}(F^2(X_i) - PF^2) \ge 3nPF^2\right\} \le e^{-9n\|F\|_2^2/4}. \tag{3.197}$$

Now we will apply Talagrand's inequality to

$$|\sigma_n^2 - \sigma^2| \le \frac{1}{n}\left\|\sum_{i=1}^{n}(f^2(X_i) - Pf^2)\right\|_{\mathcal{F}},$$

and this requires some preparation. A key observation is that since for any probability measure Q,

$$Q([f^2 - g^2]^2) = Q\left[(f-g)^2(f+g)^2\right] \le 4Q(f-g)^2,$$

it follows that, for $Q = P_n$ and setting $\mathcal{F}^2 = \{f^2 : f \in \mathcal{F}\}$,

$$\log N(\mathcal{F}^2, e_{n,2}, \varepsilon) \le \log N(\mathcal{F}, e_{n,2}, \varepsilon/2) \le H(2\|F\|_{2,n}/\varepsilon). \tag{3.198}$$

Note also that $\sup_{f\in\mathcal{F}} Pf^4 \le \sigma^2$. Hence, we can apply Corollary 3.5.9 to $\mathcal{F}^2$ with envelope $2F$ instead of F to obtain

$$E\left\|\sum_{i=1}^n (f^2(X_i) - Pf^2)\right\|_{\mathcal{F}} \le 2n\sigma^2. \tag{3.199}$$

Therefore, Bousquet's version of Talagrand's inequality, concretely inequality (3.101), gives

$$\Pr\left\{\left\|\sum_{i=1}^n (f^2(X_i) - Pf^2)\right\|_{\mathcal{F}} \ge 2n\sigma^2 + \sqrt{6n\sigma^2 t} + t/3\right\} \le e^{-t},$$

which, with $t = 6n\sigma^2$, becomes

$$\Pr\left\{\left\|\sum_{i=1}^n (f^2(X_i) - Pf^2)\right\|_{\mathcal{F}} \ge 10n\sigma^2\right\} \le e^{-6n\sigma^2}.$$

Thus, the event

$$A_3 = \left\{\left\|\sum_{i=1}^n f^2(X_i)\right\|_{\mathcal{F}} < 11n\sigma^2\right\}$$

has probability

$$\Pr(A_3) \ge 1 - e^{-6n\sigma^2}. \tag{3.200}$$

Then, combining the bounds (3.197) and (3.200) with inequality (3.196) and using the properties of H, we obtain that, on the event $A_2 \cap A_3$,

$$E_\varepsilon \left\|\frac{1}{\sqrt{n}}\sum_{i=1}^n \varepsilon_i f(X_i)\right\|_{\mathcal{F}} \le 4\sqrt{2}\int_0^{\sqrt{11}\sigma} \sqrt{H(4\|F\|_2/\tau)}\,d\tau$$

$$\le 4\sqrt{22}C_H\sigma H^{1/2}(4\|F\|_2/\sqrt{11}\sigma) < \sqrt{n}\sigma^2/64K. \tag{3.201}$$

Hence, by Theorem 3.2.9, inequality (3.192) holds on the event $A_2 \cap A_3$. Thus, on the intersection of this event with A_1, we can replace the random entropy in this inequality by the $L^2(P)$ entropy. Integrating with respect to the X variables, we then obtain

$$E\left\|\sum_{i=1}^n \varepsilon_i f(X_i)\right\|_{\mathcal{F}} \ge \frac{\sqrt{n}\sigma}{8K}\sqrt{\log N(\mathcal{F}, L^2(P), \sigma/2)}\Pr(A_1 \cap A_2 \cap A_3).$$

By (3.195), (3.197) and (3.200),

$$\Pr(A_1 \cap A_2 \cap A_3) \ge 1 - e^{-n\sigma^2/(32\cdot 4^4) + 2H(6\|F\|_2/\sigma)} - e^{-9n\|F\|_2^2/4} - e^{-6n\sigma^2} \ge 1/2,$$

as is easily seen using (3.188) and that $H(u) \ge \log 2$ for $u \ge 1$. Inequality (3.189) follows now by desymmetrisation (Theorem 3.1.21).

The right-hand-side inequality in (3.190) follows by integrating in (3.201), and it does not require $\mathcal{F}$ to be full. The left-side inequality in (3.190) is a consequence of (3.189) and the fullness of $\mathcal{F}$. ■

For simple examples of computation of random entropies, see the exercises at the end of this section, and more importantly, see Sections 3.6 and 3.7.

3.5.2 Bracketing I: An Expectation Bound

When a class of functions admits regularly varying tight metric entropy bounds that are uniform in P, the preceding subsection provides expectation bounds for the empirical process that are good up to constants, and the next section will display a wealth of such classes. In this subsection we consider classes that are not necessarily small in $L^2(P)$ uniformly in P but that may be small only for one probability measure P, albeit in a stronger sense than metric entropy.

For any $\varepsilon > 0$, the $L^p(P)$-bracketing number $N_{[]}(\mathcal{F}, L^p(P), \varepsilon)$ of a class of functions $\mathcal{F} \subset L^p(P)$ is defined, if it exists, as the smallest cardinality of any partition $B_1, \ldots, B_N$ of $\mathcal{F}$ such that, for each $i = 1, \ldots, N$,

$$P\Delta_i^p := P\left[\left(\sup_{f,g\in B_i} |f-g|\right)^*\right]^p \leq \varepsilon^p.$$

Here, for a nonnegative, not necessarily measurable function g, g^* denotes its *measurable cover*, that is, a measurable function g^* such that a) $g^* \geq g$ P a.s. and b) $g^* \leq h$ P a.s for any measurable function h such that $h \geq g$ P a.s. See Proposition 3.7.1 for the existence of g^*. If $\mathcal{F}$ is countable or separable, then $\sup_{f,g\in B_i} |f-g|$ is measurable, and the asterisk is not required in the definition of Δ_i.

An alternate definition of the bracketing covering numbers is as follows: $N = N_{[]}(\mathcal{F}, L^p(P), \varepsilon)$ if there exist N pairs of functions $\underline{f}_i \leq \overline{f}_i$ such that (a) $P(\overline{f}_i - \underline{f}_i)^p \leq \varepsilon^p$, (b) for every $f \in \mathcal{F}$ there is $i \leq N$ such that $\underline{f}_i \leq f \leq \overline{f}_i$ and (c) N is the smallest number of pairs of functions satisfying properties (a) and (b). The sets $[\underline{f}_i, \overline{f}_i] = \{f : \underline{f}_i \leq f \leq \overline{f}_i\}$ are called *$L^p(P)$ brackets of size ε*. See Exercise 3.5.4 for the equivalence (but not identity) of the two definitions. Unless otherwise stated, we will use the first definition.

The main result in this subsection is a bound on the expected value of the supremum of the empirical process based on an independent sample of probability law P and indexed by $\mathcal{F}$, given in terms of the *$L^2(P)$-bracketing integral*

$$\int_0^1 \sqrt{\log N_{[]}(\mathcal{F}, L^2(P), \varepsilon)}\, d\varepsilon.$$

We will estimate in subsequent sections the bracketing numbers of classes of functions defined by their smoothness properties and of classes of sets with smooth boundaries. As a first example as usual, consider $\mathcal{F} = \{I_{(-\infty,t]} : t \in \mathbb{R}\}$: it is easy to see that $N_{[]}(\mathcal{F}, L^p(\mathbb{R}), \varepsilon) \leq 2/\varepsilon^p$. We now proceed to prove the bracketing bounds for empirical processes and will conclude the section by estimating the bracketing numbers for the class of all monotone nondecreasing functions on $\mathbb{R}$ with uniformly bounded absolute values.

The proof of the bracketing theorem is based on chaining using the interplay between an exponential inequality and the entropy numbers, as in other chaining arguments. However, randomisation does not seem to offer any advantages in this setting, which means that the Gaussian exponential inequality is not available, and we must use Bernstein's instead. Hence, we must truncate at each step of the chain, somewhat complicating the chaining argument. We start by stating the maximal inequality associated to Bernstein's inequality, a combination of Theorem 3.1.10(b) and Theorem 3.1.5.

Lemma 3.5.12 *Let X, X_i, $i = 1,\dots,n$, be independent S-valued random variables with common probability law P, and let $f_1,\dots,f_N$ be measurable real functions on S such that $\max_r \|f_r - Pf_r\|_\infty \le c < \infty$ and $\sigma^2 = \max_r \mathrm{Var}(f_r(X))$. Then*

$$E\max_{r\le N}\left|\sum_{i=1}^n (f_r(X_i) - Pf_r)\right| \le \sqrt{2n\sigma^2\log(2N)} + \frac{c}{3}\log(2N). \tag{3.202}$$

Proof For $g = f_r - Pf_r$ or $g = -f_r + Pf_r$, we have, by Theorem 3.1.5,

$$Ee^{\lambda\sum_{i=1}^n g(X_i)/c} \le \exp\left(\frac{n\sigma^2}{c^2}(e^\lambda - 1 - \lambda)\right).$$

Then Theorem 3.1.10(b) applied to the $2N$ functions $f_r - Pf_r, -f_r + Pf_r, r = 1,\dots,N$, gives

$$E\max_{r\le N}\frac{1}{c}\left|\sum_{i=1}^n (f_r(X_i) - Pf_r)\right| \le \sqrt{\frac{2n\sigma^2}{c^2}\log(2N)} + \frac{1}{3}\log(2N),$$

which is (3.202). ■

In the proof of the bracketing theorem, the truncation levels at each step of the chaining will be precisely those that balance the two summands at the right of inequality (3.202), that is, the largest for which the 'Gaussian' part of the bound (which is the first summand) dominates.

Theorem 3.5.13 *Let P be a probability measure on $(S,\mathcal{S})$ and for any $n\in\mathbb{N}$, and let $X_1,\dots,X_n$ be an independent sample of size n from P. Let $\mathcal{F}$ be a class of measurable functions on S that admits a P-square integrable envelope F and satisfies the $L^2(P)$-bracketing condition*

$$\int_0^2 \sqrt{\log(N_{[]}(\mathcal{F}, L^2(P), \|F\|_2\tau))}\,d\tau < \infty,$$

where we write $\|F\|_2$ for $\|F\|_{L^2(P)}$. Set $\sigma^2 := \sup_{f\in\mathcal{F}} Pf^2$ and

$$a(\delta) = \frac{\delta}{\sqrt{32\log(2N_{[]}(\mathcal{F}, L^2(P), \delta/2))}}.$$

Then, for any $\delta > 0$,

$$E\left\|\sum_{i=1}^n (f(X_i) - Pf)\right\|_{\mathcal{F}}^* \le 56\sqrt{n}\int_0^{2\delta}\sqrt{\log(2N_{[]}(\mathcal{F}, L^2(P), \tau)}\,d\tau \tag{3.203}$$

$$+ 4nP[FI(F > \sqrt{n}a(\delta))] + \sqrt{n\sigma^2\log(2N_{[]}(\mathcal{F}, L^2(P), \delta))}.$$

Proof Assume that $\mathcal{F}$ satisfies the bracketing integral condition, and fix $n, j \in \mathbb{N}$. For $k \ge j$, set $N_k := N_{[]}(\mathcal{F}, L^2(P), 2^{-k})$, let $\{T_{k,i}\}_{i=1}^{N_k}$ be a partition of $\mathcal{F}$ such that

$$E\left(\left(\sup_{f,g\in T_{k,i}} |f - g|\right)^*\right)^2 \le 2^{-2k}, \quad \text{for all } 1 \le i \le N_k,\ k \in \mathbb{N},$$

and note that the bracketing condition implies that

$$\sum_{k=j}^{\infty} 2^{-k}\sqrt{\log N_k} \le 2\int_0^{2^{-j}} \sqrt{\log N_{[]}(\mathcal{F}, L^2(P), \tau)}\, d\tau < \infty.$$

We would like the partitions to be nested and still satisfy these two conditions. This can be achieved as follows: for each $s = (s_j, \ldots, s_k)$, $s_\ell \in \{1, \ldots, N_\ell\}$, $\ell = j, \ldots, k$, take $A_{k,s} = \cap_{\ell=j}^{k} T_{\ell, s_\ell}$. Then, for each k, the collection of sets $\{A_{k,s} : s = (s_j, \ldots, s_k), 1 \le s_\ell \le N_\ell, \ell = j, \ldots, k\}$ is obviously a partition (which may contain some empty sets). For each k, the number of $A_{k,s}$ is dominated by the product $N_j \cdots N_k$, and

$$\begin{aligned}\sum_{k=j}^{\infty} 2^{-k}\sqrt{\log(N_j \cdots N_k)} &\le \sum_{k=j}^{\infty} 2^{-k} \sum_{i=j}^{k} (\log N_i)^{1/2} = \sum_{i=j}^{\infty} (\log N_i)^{1/2} \sum_{k \ge i} 2^{-k} \\ &= 2\sum_{i=j}^{\infty} 2^{-i} (\log N_i)^{1/2} \le 4\int_0^{2^{-j}} \sqrt{\log N_{[]}(\mathcal{F}, L^2(P), \tau)}\, d\tau < \infty.\end{aligned}$$

To ease notation, let us enumerate the indices $= (s_j, \ldots, s_k)$ corresponding to nonempty sets $A_{k,s}$ and denote these sets by $A_{k,i}$, $1 \le i \le \tilde{N}_k$, where $\tilde{N}_k \le N_j \cdots N_k$. Thus, we have a collection of partitions of $\mathcal{F}$, $\{A_{k,i} : 1 \le i \le \tilde{N}_k\}$, $k \ge j$, such that

(a) $\sum_{k=j}^{\infty} 2^{-k}\sqrt{\log \tilde{N}_k} \le 4\int_0^{2^{-j}} \sqrt{\log N_{[]}(\mathcal{F}, L^2(P), \tau)}\, d\tau < \infty$,
(b) $E\left(\left(\sup_{f,g \in A_{k,i}} |f - g|\right)^*\right)^2 \le 2^{-2k}$, for all $1 \le i \le \tilde{N}_k$, $k \ge j$, and
(c) the partitions $\{A_{k,i} : 1 \le i \le \tilde{N}_k\}$ are nested; that is, if $j \le \ell < k$, for each $A_{k,i}$, there is a unique r such that $A_{k,i} \subseteq A_{\ell,r}$.

For each $f \in \mathcal{F}$, let $i_k(f)$ be the index i such that $f \in A_{k,i}$. Then we have, by nestedness,

$$A_{k, i_k(f)} \subseteq A_{k-1, i_{k-1}(f)} \subseteq \cdots \subseteq A_{j, i_j(f)}, \tag{3.204}$$

and in particular, $A_{k, i_k(f)}$ determines $A_{\ell, i_\ell(f)}$, $\ell \le k$; hence, the number of chains (3.204) is just $\tilde{N}_k$.

Pick up $f_{k,i} \in A_{k,i}$, and set, for $i = 1, \ldots, \tilde{N}_k$, $k \ge j$,

$$\pi_k f = f_{k,i} \quad \text{and} \quad \Delta_k(f) = \left(\sup_{h,g \in A_{k,i}} |g - h|\right)^*, \quad \text{if } f \in A_{k,i}.$$

The varying truncation levels will be

$$\alpha_{n,k} := \sqrt{n}\, a_k := \frac{\sqrt{n}}{2^{k+1}\sqrt{\log(2\tilde{N}_{k+1})}}. \tag{3.205}$$

Define

$$\tau f := \tau_{j,n}(f, x) = \min\{k \ge j : \Delta_k f(x) > \alpha_{n,k}\}$$

with the convention $\min \emptyset = \infty$, and notice that

$$\{\tau f = j\} = \{\Delta_j f > \alpha_{n,j}\}$$

and that, for $k > j$,

$$\{\tau f = k\} = \{\Delta_j f \le \alpha_{n,j}, \ldots, \Delta_{k-1} f \le \alpha_{n,k-1}, \Delta_k f > \alpha_{n,k}\} \subset \{\Delta_k f > \alpha_{n,k}, \Delta_{k-1} f \le \alpha_{n,k-1}\},$$
$$\times \{\tau f \ge k\} = \{\Delta_j f \le \alpha_{n,j}, \ldots, \Delta_{k-1} f \le \alpha_{n,k-1}\} \subseteq \{\Delta_k f \le \Delta_{k-1} f \le \alpha_{n,k-1}\}.$$

Note that from the point of view of controlling $\Delta_k f$, the sets $\{\tau f = k\}$ and $\{\tau f \ge k\}$ are 'good', whereas the sets $\{\tau f < k\}$ are 'bad'. Next, we will see how, starting in a natural way with the chain decomposition

$$f - \pi_j f = f - \pi_r f + \sum_{k=j+1}^{r} (\pi_k f - \pi_{k-1} f)$$

and then decomposing the kth link according to $\tau f < k$ or $\tau f \ge k$, we arrive at a decomposition of $f - \pi_j f$ that contains no bad sets. We have

$$f - \pi_j f = f - \pi_r f + \sum_{k=j+1}^{r} (\pi_k f - \pi_{k-1} f) I_{\tau f < k} + \sum_{k=j+1}^{r} (\pi_k f - \pi_{k-1} f) I_{\tau f \ge k}.$$

Now the 'bad' sets telescope, that is,

$$\begin{aligned}
\sum_{k=j+1}^{r} (\pi_k f - \pi_{k-1} f) I_{\tau f < k} &= \sum_{k=j+1}^{r} \pi_k f I_{\tau f < k} - \sum_{k=j+1}^{r} \pi_{k-1} f I_{\tau f = k-1} - \sum_{k=j+1}^{r} \pi_{k-1} f I_{\tau f < k-1} \\
&= \pi_r f I_{\tau f < r} - \pi_j f I_{\tau f < j} - \sum_{k=j+1}^{r} \pi_{k-1} f I_{\tau f = k-1} \\
&= \pi_r f I_{\tau f < r} - \sum_{k=j+1}^{r} \pi_{k-1} f I_{\tau f = k-1}.
\end{aligned}$$

We can further use $-\pi_r f + \pi_r f I_{\tau f < r} = -\pi_r f I_{\tau f \ge r}$ to finally obtain

$$\begin{aligned}
f - \pi_j f &= f - \pi_r f I_{\tau f \ge r} - \sum_{k=j+1}^{r} \pi_{k-1} f I_{\tau f = k-1} + \sum_{k=j+1}^{r} (\pi_k f - \pi_{k-1} f) I_{\tau f \ge k} \\
&= (f - \pi_j f) I_{\tau f = j} + (f - \pi_r f) I_{\tau f \ge r} + \sum_{k=j+1}^{r-1} (f - \pi_k f) I_{\tau f = k} \\
&\quad + \sum_{k=j+1}^{r} (\pi_k f - \pi_{k-1} f) I_{\tau f \ge k}.
\end{aligned} \tag{3.206}$$

We now proceed to estimate the expected value of the empirical process over each of the four terms in the decomposition (3.206). The simple observation that

$$|f| \le g \quad \text{implies} \quad |(P_n - P) f| \le P_n g + P g \le (P_n - P) g + 2Pg$$

will be used repeatedly and without further mention.

First term: We have

$$|f - \pi_j f| I_{\tau f = j} = |f - \pi_j f| I(\Delta_j f > \sqrt{n} a_j) \le 2F I(2F > \sqrt{n} a_j),$$

so

$$E\|\sqrt{n}(P_n - P)((f - \pi_j f)I_{\tau f=j})\|^*_{\mathcal{F}} \le 4\sqrt{n}P\big[FI(2F > \sqrt{n}a_j)\big]. \tag{3.207}$$

Second term: We show that

$$\lim_{r\to\infty} E\|\sqrt{n}(P_n - P)((f - \pi_r f)I_{\tau f\ge r})\|_{\mathcal{F}} = 0, \tag{3.208}$$

so we will be able to ignore the second term as long as we let the sums in the third and fourth terms run up to infinity (instead of only up to r). By definition of τf,

$$|f - \pi_r f|I_{\tau f\ge r} \le (\Delta_r f)I(\Delta_r f < \alpha_{n,r-1}),$$

and therefore,

$$\begin{aligned}\|\sqrt{n}(P_n - P)((f - \pi_r f)I_{\tau f\ge r})\|_{\mathcal{F}} &\le \|\sqrt{n}(P_n - P)((\Delta_r f)I(\Delta_r f < \alpha_{n,r-1}))\|_{\mathcal{F}} \\ &\quad + 2\sqrt{n}\|P((\Delta_r f)I(\Delta_r f < \alpha_{n,r-1}))\|_{\mathcal{F}} \\ &:= (I) + (II).\end{aligned}$$

We use inequality (3.202) on the first term: in this term the empirical process is applied to $\tilde{N}_r$ functions whose variances are dominated by 2^{-2r} (by b)) and whose sup norms are dominated by $\alpha_{n,r-1} = \sqrt{n}/\left(2^r\sqrt{\log\tilde{N}_r}\right)$, so

$$E(I)^* \le 2^{-r}\sqrt{\log(2\tilde{N}_r)} + \frac{\sqrt{n}}{3\cdot 2^r\sqrt{\log(2\tilde{N}_r)}}\frac{1}{\sqrt{n}}\log(2\tilde{N}_r) = \frac{4}{3}2^{-r}\sqrt{\log(2\tilde{N}_r)} \to 0,$$

if $r \to \infty$, by (a). The second term is obviously bounded by

$$(II) \le 2\sqrt{n}\|P(\Delta_r f)^2\|^{1/2}_{\mathcal{F}} \le 2\sqrt{n}2^{-r} \to 0 \text{ as } r \to \infty,$$

so (3.208) follows.

Third term: Since $|f - \pi_k f|I_{\tau f=k} \le (\Delta_k f)I_{\tau f=k}$, we have

$$\begin{aligned}E\left\|\sum_{k=j+1}^{\infty}\sqrt{n}(P_n - P)(f - \pi_k f)I_{\tau f=k}\right\|^*_{\mathcal{F}} &\le \sum_{k=j+1}^{\infty}\Big(E\left\|\sqrt{n}(P_n - P)\left((\Delta_k f)I_{\tau f=k}\right)\right\|_{\mathcal{F}} \\ &\quad + 2\sqrt{n}\left\|P\left((\Delta_k f)I_{\tau f=k}\right)\right\|_{\mathcal{F}}\Big).\end{aligned} \tag{3.209}$$

In order to estimate this expectation using the maximal inequality (3.202), we note that (1) there are $\tilde{N}_k$ different functions $(\Delta_k f)I_{\tau f=k}$, $f \in \mathcal{F}$ (the number of chains (3.204) is $\tilde{N}_k$), (2) on $\{\tau f = k\}$ we have $\Delta_k f \le \Delta_{k-1} f \le \alpha_{n,k-1} = \sqrt{n}/\left(2^k\sqrt{\log(2\tilde{N}_k)}\right)$, (3)

$$\begin{aligned}\mathrm{Var}(\Delta_k f I_{\tau f=k}) &\le P(\Delta_k f)^2 I(\Delta_k f \le \alpha_{n,k-1}, \Delta_k f > \alpha_{n,k}) \\ &\le \alpha_{n,k-1}P(\Delta_k f)I(\Delta_k f > \alpha_{n,k}) \le \frac{\alpha_{n,k-1}}{\alpha_{n,k}}E(\Delta_k f)^2 \\ &\le \frac{2\sqrt{\log(2\tilde{N}_{k+1})}}{\sqrt{\log(2\tilde{N}_k)}}2^{-2k} \le \frac{2\log(2\tilde{N}_{k+1})}{\log(2\tilde{N}_k)}2^{-2k},\end{aligned}$$

since $\tilde{N}_{k+1} \geq \tilde{N}_k$, and (4)

$$P\big((\Delta_k f)I_{\tau f=k}\big) \leq P(\Delta_k f)I(\Delta_k f > \alpha_{n,k}) \leq \frac{E(\Delta_k f)^2}{\alpha_{n,k}} \leq \frac{2^{-k+1}\sqrt{\log(2\tilde{N}_{k+1})}}{\sqrt{n}}.$$

Using (1)–(4) and (3.202) in (3.209), we obtain

$$E\left\|\sum_{j=1}^{\infty}\sqrt{n}(P_n-P)(f-\pi_k f)I_{\tau f=k}\right\|_{\mathcal{F}}^{*}$$
$$\leq \sum_{k=j+1}^{\infty}\left[\sqrt{\frac{2^{-2k+1}\log(2\tilde{N}_{k+1})}{\log(2\tilde{N}_k)}}\sqrt{\log(2\tilde{N}_k)}+\frac{\sqrt{n}}{3\cdot 2^k\sqrt{\log(2\tilde{N}_k)}}\frac{1}{\sqrt{n}}\log(2\tilde{N}_k)\right]$$
$$+\sum_{k=j+1}^{\infty}2^{-k+2}\sqrt{\log(2\tilde{N}_{k+1})}.$$

Bounding the sums by integrals as earlier in the proof, we obtain

$$E\left\|\sum_{j=1}^{\infty}\sqrt{n}(P_n-P)(f-\pi_k f)I_{\tau f=k}\right\|_{\mathcal{F}}^{*}$$
$$\leq 4(2\sqrt{2}+3^{-1}+8)\int_0^{2^{-j}}\sqrt{\log(2N_{[]}(\mathcal{F},L^2(P),\tau)}\,d\tau. \tag{3.210}$$

Fourth term: Again, the number of functions $(\pi_k f-\pi_{k-1}f)I_{\tau f\geq k}$, $f\in\mathcal{F}$, is just $\tilde{N}_k$ (all the functions f in $A_{k,i}$ have the same $\pi_k f$, $\pi_{k-1}f$, $\Delta_j f,\dots,\Delta_k f$). The variance of $(\pi_k f-\pi_{k-1}f)I_{\tau f\geq k}$ is dominated by

$$P|\pi_k f-\pi_{k-1}f|^2 \leq P(\Delta_{k-1}f)^2 \leq 2^{-2(k-1)}$$

and its sup norm by

$$|(\pi_k f-\pi_{k-1}f)I_{\tau f\geq k}| \leq 2(\Delta_{k-1}f)I(\Delta_{k-1}\leq\alpha_{n,k-1}) \leq \frac{2\sqrt{n}}{2^k\sqrt{\log(2\tilde{N}_k)}}.$$

Hence, applying inequality (3.202) as before, we obtain

$$E\left\|\sum_{k=j+1}^{\infty}\sqrt{n}(P_n-P)((\pi_k f-\pi_{k-1}f)I_{\tau f\geq k})\right\|_{\mathcal{F}}^{*}$$
$$\leq \sum_{k=j+1}^{\infty}\left(2^{-(k-1)}\sqrt{\log(2\tilde{N}_k)}+\frac{2\sqrt{n}}{3\cdot 2^k\sqrt{\log(2\tilde{N}_k)}}\frac{1}{\sqrt{n}}\log(2\tilde{N}_k)\right)$$
$$\leq 4(2+2/3)\int_0^{2^{-j-1}}\sqrt{\log(2N_{[]}(\mathcal{F},L^2(P),\tau)}\,d\tau. \tag{3.211}$$

Combining the bounds (3.207), (3.208), (3.210) and (3.211) with the decomposition (3.206) of $f - \pi_j f$, we obtain

$$E\left\|\sqrt{n}(P_n - P)(f - \pi_j f)\right\|_{\mathcal{F}}^* \leq 55.314 \int_0^{2^{-j}} \sqrt{\log(2N_{[]}(\mathcal{F}, L^2(P), \tau)}\, d\tau + 4\sqrt{n}P[FI(2F > \sqrt{n}a_j)]. \quad (3.212)$$

This inequality will turn out to be useful to prove a central limit theorem in a subsequent section, but now, to obtain a bound for $E\|\sqrt{n}(P_n - P)\|_{\mathcal{F}}^*$, we need to combine this bound with a bound for $E\|\sqrt{n}(P_n - P)(\pi_j f)\|_{\mathcal{F}}^*$. This is the expected value of the maximum of $2N_j$ random variables. To apply Lemma 3.5.12, we must truncate, that is,

$$E\|\sqrt{n}(P_n - P)(\pi_j f)\|_{\mathcal{F}} \leq E\|\sqrt{n}(P_n - P)(\pi_j f I(F \leq \sqrt{n}a_j))\|_{\mathcal{F}} + 2\sqrt{n}P(FI(F > \sqrt{n}a_j).$$

Then, by (3.202), the preceding expectation admits the bound

$$\sigma\sqrt{2\log(2N_j)} + \frac{a_j}{3}\log(2N_j) \leq \sigma\sqrt{2\log(2N_j)} + \frac{2}{3}\int_0^{2^{-j-1}} \sqrt{\log(2N_{[]}(\mathcal{F}, L^2(P), \tau)}\, d\tau,$$

where $\sigma^2 = \sup_{f \in \mathcal{F}} Pf^2$. Hence,

$$E\left\|\sum_{i=1}^n (f(X_i) - Pf)\right\|_{\mathcal{F}}^* \leq 56\sqrt{n}\int_0^{2^{-j}} \sqrt{\log(2N_{[]}(\mathcal{F}, L^2(P), \tau)}\, d\tau + 4nP[FI(2F > \sqrt{n}a_j)] + \sqrt{2n\sigma^2\log(2N_j)}.$$

Now the theorem follows for any fixed $\delta > 0$ by taking $j = j(\delta)$ such that $2^{-j-1} \leq \delta\|F\|_2 \leq 2^{-j}$ and using that $N_{[]}$ is nonincreasing. ■

Remark 3.5.14 If in addition to the hypotheses in Theorem 3.5.13 we also have $\sigma \leq \delta$, then the third summand in the bound (3.203) can be assimilated into the first to obtain

$$E\left\|\sum_{i=1}^n (f(X_i) - Pf)\right\|_{\mathcal{F}}^* \leq 58\sqrt{n}\int_0^{2\delta} \sqrt{\log(2N_{[]}(\mathcal{F}, L^2(P), \varepsilon)}\, d\varepsilon + 4nP[FI(F > \sqrt{n}a(\delta))]. \quad (3.213)$$

If we take $\delta = 4\|F\|_2$ (we always have $\sigma \leq \|F\|_2$), we have $a(4\|F\|_2) = \|F\|_2/\sqrt{2\log 2}$, and the first summand dominates the second because

$$\begin{aligned} 4nP[FI(F \geq \sqrt{n}a(4\|F\|_2))] &\leq 4nP[F^2 I(F \geq \sqrt{n}a(4\|F\|_2))]/(\sqrt{n}a(4\|F\|_2)) \\ &\leq 4\sqrt{2\log 2}\sqrt{n}\|F\|_2. \end{aligned}$$

This shows that for any class of functions satisfying the $L^2(P)$-bracketing integral condition, we have

$$E\left\|\sum_{i=1}^n (f(X_i) - Pf)\right\|_{\mathcal{F}}^* \leq 59\sqrt{n}\int_0^{8\|F\|_2} \sqrt{\log(2N_{[]}(\mathcal{F}, L^2(P), \varepsilon)}\, d\varepsilon. \quad (3.214)$$

If the class $\mathcal{F}$ is *uniformly bounded*, then the bound (3.213) can be improved as follows:

Proposition 3.5.15 *Assume that* $\|F\|_\infty < \infty$ *and* $Pf^2 \le \delta^2$ *for all* $f \in \mathcal{F}$ *for some* $\delta > 0$. *Then*

$$E\left\|\sum_{i=1}^n (f(X_i) - Pf)\right\|_{\mathcal{F}}^* \le \left(58\sqrt{n} + \frac{\|F\|_\infty}{3\delta^2}\int_0^{2\delta}\sqrt{\log(2N_{[]}(\mathcal{F}, L^2(P), \varepsilon)}\,d\varepsilon\right)$$
$$\times \int_0^{2\delta}\sqrt{\log(2N_{[]}(\mathcal{F}, L^2(P), \varepsilon)}\,d\varepsilon. \qquad (3.215)$$

Proof Assume that $\delta = 2^{-j}$ for some $j \in \mathbb{N}$ (otherwise take j as indicated at the end of the proof of Theorem 3.5.13). Inspection of the proof of this theorem shows that the term we wish to cancel, $\sqrt{n}P(FI(F > \sqrt{n}a(\delta))$, comes only from the evaluation of the *first term* in the decomposition (3.206) and of $E\|\sqrt{n}(P_n - P)(\pi_j f)\|_{\mathcal{F}}$. Each of these two terms can be estimated in a more precise way than in the preceding proof when the class of functions is uniformly bounded by invoking Lemma 3.5.12, Bernstein's maximal inequality. For the last term, this inequality gives

$$E\|\sqrt{n}(P_n - P)(\pi_j f)\|_{\mathcal{F}} \le \sqrt{2\delta^2\log(2N)} + \frac{2\|F\|_\infty}{3\sqrt{n}}\log(2N),$$

where $N := N_{[]}(\mathcal{F}, L^2(P), \delta)$. As for the first term, again by the same maximal inequality, we have

$$\begin{aligned} E\|\sqrt{n}(P_n - P)(f - \pi_j f)I(\Delta_j f > \sqrt{n}a_j)\|_{\mathcal{F}} &\le E\|\sqrt{n}(P_n - P)((f - \pi_j f)(\Delta_j f)\|_{\mathcal{F}} \\ &\quad + \sqrt{n}\|P(\Delta_j f)I(\Delta_j f > \sqrt{n}a_j)\|_{\mathcal{F}} \\ &\le \sqrt{2\delta^2\log(2N)} + \frac{2\|F\|_\infty}{3\sqrt{n}}\log(2N) + \frac{P(\Delta_j f)^2}{a_j}, \end{aligned}$$

where, by definition of a_j and monotonicity of $N(\varepsilon)$,

$$P(\Delta_j f)^2/a_j \le 2\delta\sqrt{\log(2N)} \le \int_0^{2\delta}\sqrt{\log(2N_{[]}(\mathcal{F}, L^2(P), \varepsilon))}\,d\varepsilon.$$

Likewise, $\log(2N)$ is dominated by the square of the entropy integral divided by $4\delta^2$, and the result follows from this and the preceding bounds together with (3.208), (3.210) and (3.211) in the preceding proof. ■

See Exercise 3.5.5 for a similar bound without assuming that $\mathcal{F}$ is uniformly bounded.

Remark 3.5.16 The expectation bounds in the preceding two subsections can be combined with the moment inequalities in Exercise 3.3.4 or in Theorem 3.4.3, or even with Hoffmann-Jørgensen's inequality, to obtain estimates of higher moments of the empirical process.

We conclude this section with estimation of the bracketing numbers of classes of monotone functions, both as an example and because of its usefulness. See also Section 3.7 for some extensions.

Proposition 3.5.17 *The class $\mathcal{F}$ of monotone functions on $\mathbb{R}$ satisfying $a \le f(x) \le b$, for all $x \in \mathbb{R}$ and some $-\infty < a < b < \infty$, admits the following bound for its $L^2(P)$-bracketing numbers, uniform on all the Borel probability measures P on $\mathbb{R}$:*

$$\log N_{[]}(\mathcal{F}, L^2(P), \varepsilon) \le K/\varepsilon, \quad 0 < \varepsilon \le b - a,$$

where $K < \infty$ is a numerical constant that depends only on a and b.

Proof The proof consists of several steps. We use the second definition of bracket $[\underline{f}, \overline{f}]$ in the proof.

Step 1: Reductions. Since the minimal number of brackets covering the union of two classes of functions is dominated by their sum, and since reflection about $x = 0$ is a one-to-one correspondence between monotone nondecreasing and monotone nonincreasing functions that sends $L^2(P)$ brackets into $L^2(\tilde{P})$ brackets of the same size, where $\tilde{P}$ is the reflection of P, it suffices to consider the class of monotone nondecreasing functions. Also, a and b can be replaced, respectively, by 0 and 1 just by considering the class $\{(f - a)/(b - a) : f \text{ nondecreasing}, a \le f \le b\}$. Thus, the proposition needs only be proved for the class $\mathcal{H}$ of monotone nondecreasing functions taking values in $[0, 1]$. Define now, given a function h, the *left bracket* of $L^2(P)$-size δ as $LB(h, \delta) = \{f \in \mathcal{H} : f \ge h, \|f - h\|_{L^2(P)} \le \delta\}$. Right brackets $RB(h, \delta)$ are defined analogously, with reversal of the first inequality. Then, since if $f \in LB(h_1, \delta) \cup RB(h_2, \delta)$, f belongs to the $L^2(P)$ bracket $[h_1, h_2]$ with size $\|h_2 - h_1\|_{L^2(P)} \le 2\delta$, it suffices to consider one-sided brackets.

Let

$$\mathcal{G} = \{g : \mathbb{R} \mapsto [0, 1], g \text{ monotone nondecresing}\}.$$

We now reduce proving the proposition to showing that for some fixed $K < \infty$,

$$\log N_{LB}(\mathcal{G}, L^2(\mu), \delta) \le K/\delta, \quad 0 < \delta \le 1,$$

where μ is Lebesgue measure, and $N_{LB}(\mathcal{G}, L^2(\mu), \delta)$ is the smallest number of left brackets of $L^2(\mu)$-size δ needed to cover $\mathcal{G}$. To see this, given a probability measure P on $\mathbb{R}$, let F be its cumulative distribution function and $F^{-1}(u) = \inf\{t : F(t) \ge u\}$ its quantile function, which are monotone nondecreasing. The class $\mathcal{H} \circ F^{-1} = \{f \circ F^{-1} : f \in \mathcal{F}\}$ is contained in $\mathcal{G}$, and we recall that $F^{-1} \circ F(t) \le t$ and $u \le F \circ F^{-1}(u)$ for all $u, t \in \mathbb{R}$. Let h define a left bracket for $f \circ F^{-1}$ of $L^2(\mu)$-size δ so that $h \le f \circ F^{-1}$ and $\|f \circ F^1 - h\|_{L^2(\mu)} \le \delta$. Also note that we can assume h to be nondecreasing because it can be replaced by $\inf\{g \circ F^{-1} : \|g \circ F^{-1} - h\|_{L^2(\mu)} \le \delta, g \circ F^{-1} \ge h, g \in \mathcal{H}\}$, which is monotone nondecreasing. Then, since f and h are nondecreasing, we have $h \circ F \le f \circ F^{-1} \circ F \le f$, so also $h \circ F \circ F^{-1} \le f \circ F^{-1}$, as well as $h \circ F \circ F^{-1} \ge h$. Hence, also,

$$\|f - h \circ F\|_{L^2(P)} = \|f \circ f^{-1} - h \circ F \circ F^{-1}\|_{L^2(\mu)} \le \|f \circ F^{-1} - h\|_{L^2(\mu)} \le \delta.$$

That is, $h \circ F$ defines a left bracket of $L^2(P)$-size δ for f. This shows that for all Borel probability measures on $\mathbb{R}$, $N_{LB}(\mathcal{F}, L^2(P), \delta) \le N_{LB}(\mathcal{G}, L^2(\mu), \delta)$.

Step 2: Construction of the brackets. Let $0 < \delta < 1$ and $g \in \mathcal{G}$. We now construct a left bracket for g by means of a step function h, and in subsequent steps we will count the number and estimate the size of the brackets. We start by building the interval partition of

$[0,1]$ to define of h. The partition is reached as the last of a nested sequence of partitions as follows. $\pi_0 = \pi_0(g) = \{[0,1]\}$, and then we construct partitions $\pi_i(g)$ recursively. Given $\pi_i = \{\Delta_j\}_{i=1}^{n_i}$, where

$$\Delta_j = [x_{j-1}, x_j),\ 1 \le j < n_i, \quad \Delta_{n_i} = [x_{n_i-1}, x_{n_i}], \quad x_0 = 0, \quad x_{n_i} = 1,$$

we set

$$|\Delta_j| = x_j - x_{j-1},\ J(\Delta_j) = g(x_i) - g(x_{i-1}),\ \delta_i^2 = \max_{1\le j\le n_i} |\Delta_j| J^2(\Delta_j);$$

then we obtain π_{i+1} by subdividing into two subintervals of equal length those intervals Δ_i such that

$$|\Delta_j| J^2(\Delta_j) \ge \delta_i^2/2 \tag{3.216}$$

and keeping those Δ_j for which (3.216) does not hold. It is convenient to denote by s_i the number of intervals $\Delta_j \in \pi_i$ satisfying (3.216), so $n_{i+1} = n_i + s_i$. Observe also that $s_i \ne 0$ for all i. Next, we define

$$\tilde{\delta}_i^2 = \min_{0\le j\le i} \{2^{-(i-j)} n_j^{-3}\} \tag{3.217}$$

and continue the process of subdivision, $\pi_0 \subset \pi_1 \subset \cdots \subset \pi_k$, up to the smallest positive integer $k = k(g)$ such that $\tilde{\delta}_k^2 \le \delta^3$. (Note that the process terminates because $\tilde{\delta}_i^2 \le \tilde{\delta}_{i-1}^2/2$.) To construct the function defining the left bracket for g, we start with $g_0 = 0$ and define recursively a function g_i constant on the intervals Δ_j of π_i as follows: given g_{i-1}, define g_i on $\Delta_j \in \pi_i$ by

$$g_i(x) = g_{i-1}(x_{j-1}) + \ell_j^i \frac{\tilde{\delta}_i}{|\Delta_j|^{1/2}}, \quad x \in \Delta_j, \tag{3.218}$$

where $\ell_j^i \ge 0$ is the largest integer such that $g_i \le g$, that is, such that $g_i(x_{j-1}) \le g(x_j -)$. Then the left bracket for g is defined by the function $g_{k(g)}$, with size $\|g - g_k\|_{L^2(P)}$, to be estimated next.

Step 3: Bracket size. First, we relate δ_i, $\tilde{\delta}_i$ and n_i. If $\Delta_j \in \pi_i$ satisfies (3.216), for $j = 1, \dots, s_i$, we have, by Cauchy-Schwarz,

$$s_i(\delta_i^2/2)^{1/3} \le \sum_{j=1}^{s_i} |\Delta_j|^{1/3} J^{2/3}(\Delta_j) \le \left(\sum_{j=1}^{s_i} |\Delta_j|\right)^{1/3} \left(\sum_{j=1}^{s_i} J(\Delta_j)\right)^{2/3} \le 1,$$

so $\delta_i^2 \le 2/s_i^3$. Since $\delta_i^2 \le \delta_{i-1}^2/2$, we also have, for $i \ge j$, $\delta_i^2 \le 2^{-(i-j)}\delta_j^2 \le 2^{-(i-j-1)} s_j^{-3}$. Hence,

$$\begin{aligned} n_i = 1 + \sum_{j=0}^{i-1} s_j &\le 1 + \sum_{j=0}^{i-1} 2^{-(i-j-1)/3} \delta_i^{-2/3} \\ &< 1 + 5\delta_i^{-2/3} < 6\delta_i^{-2/3}. \end{aligned}$$

Thus, $n_i^{-3} \ge 6^{-3}\delta_j^2 \ge 6^{-3} 2^{i-j} \delta_i^2$, for $i \ge j$, which implies, by the definition (3.217) of $\tilde{\delta}_i$, that $\delta_i^2 \le 6^3 \tilde{\delta}_i^2$. This and (3.217) give

$$n_i \le \tilde{\delta}_i^{-2/3}, \quad n_i \le 6\delta_i^{-2/3}, \quad \delta_i^2 \le 6^3 \tilde{\delta}_i^2. \tag{3.219}$$

Now, by (3.218) and the definition of ℓ_j^i in (3.218), we have

$$(g(x_j-)-g_i(x_{i-1}))|\Delta_j|^{1/2} \le \tilde{\delta}_i,$$

but, by the definition of δ_i and (3.219),

$$(g(x)-g_i(x_{j-1})|\Delta_j|^{1/2} \le \delta_i + \tilde{\delta}_i \le (6^{3/2}+1)\tilde{\delta}_i, \quad x \in \Delta_j,$$

so, by (3.219),

$$\|g-g_i\|^2_{L^2(\mu)} \le n_1(6^{3/2}+1)^2\tilde{\delta}_i^{\,2} \le 241\tilde{\delta}_i^{\,2-2/3}.$$

Since for $i = k(g)$ we have $\delta^2_{k(g)} \le \delta^3$, it follows that

$$\|g-g_{k(g)}\|^2_{L^2(\mu)} \le 241\delta^2; \tag{3.220}$$

that is, the brackets are of the right size: their size is bounded by a fixed multiple of δ.

Step 4: Number of brackets. Every bracket function is built from a sequence of partitions $\pi_0(g) \subset \cdots \subset \pi_{k(g)}(g)$, which determines the quantities $\tilde{\delta}_i$ and the partition points, and from k vectors of ℓ_j^i, $(\ell_1^1, \ldots, \ell_{n_1}^1), \ldots, (\ell_1^k, \ldots, \ell_{n_k}^k)$ (see (3.218) for $i = k$). By definition, we have $n_k \le 2n_{k-1} \le 2\tilde{\delta}_{k-1}^{-2/3} < 2/\delta$. The number of possible choices for the sequences $\{n_i\}_{i=1}^k$ is just the number of ways we can write $n_k - 1$ into a sum of $k-1$ positive integers s_i, $\binom{n_k-2}{k-2} < 2^{n_k-2}$. Give one such sequence $\{n_i\}$, for each $i \le k$, the number of partitions π_{i+1} we can construct given that a partition π_i is bounded by the number of ways we can choose $s_i = n_{i+1} - n_i$ among n_i intervals, $\binom{n_i}{s_i} < 2^{n_i}$. Hence, the number of partitions corresponding to a sequence $\{n_i\}$ is dominated by $2^{\sum_{i=0}^{k-1} n_i}$. But by (3.217), $n_i \le 2^{-(k-1-i)/3}\tilde{\delta}_{k-1}^{-2/3}$, for $i \le k-1$, and, by definition of $k(g)$, $\delta_{k-1}^{-2/3} \le \delta^{-1}$, which gives

$$\sum_{i=0}^{k-1} n_i \le \frac{1}{\delta}\sum_{i=0}^{\infty} 2^{-1/3} \le \frac{5}{\delta}. \tag{3.221}$$

We conclude that the number of possible partitions N in the definition of the brackets is dominated by

$$N \le 2^{2/\delta-2}2^{5/\delta} = \frac{1}{4}2^{7/\delta}.$$

To determine the number of vectors of numbers ℓ_j^i, we note that, by definition, $g_{i-1}(x_{j-1}) + \ell_j^i\tilde{\delta}_i/|\Delta_j|^{1/2} \le g(x)$, for all $x \in \Delta_j$, hence, $\le g(x_{i-1})$, and recall from step 3 that if $\tilde{\Delta}_j$ is the interval from π_{i-1} that contains Δ_i from π_i, then $g(x_{j-1}) - g_{i-1}(x_{j-1}) \le (6^{3/2}+1)\tilde{\delta}_{i-1}|\tilde{\Delta}_j|^{-1/2}$. Therefore, we must have

$$\ell_j^i \le \frac{(6^{3/2}+1)\tilde{\delta}_{i-1}|\tilde{\Delta}_j|^{-1/2}}{\tilde{\delta}_i|\Delta_j|^{-1/2}} + 1.$$

Now $|\Delta_j| \le |\tilde{\Delta}_j|$ and $\tilde{\delta}_{i-1}/\tilde{\delta}_i \le \sqrt{8}$: the first assertion is obvious, and the second follows because $\tilde{\delta}_i^{\,2} = (\delta^2_{i-1}/2) \wedge n_i^{-3} \ge (\delta^2_{i-1}/2) \wedge (2n_{i-1})^{-3}$ and $\tilde{\delta}_{i-1}^{\,2} \le n_{i-1}^{-3}$. This yields $\ell_j^i \le (6^{3/2}+1)\sqrt{8}+1 < 50$. Hence, for each sequence of partitions, we have at most $50^{\sum_{i=1}^k n_i}$ possible brackets. Thus, by (3.221) and the estimate for N, the number of possible brackets does not exceed

$$N \cdot 50^5/\delta \leq \frac{1}{4} 2^{7/\delta} 50^{5/\delta} < 100^{7/\delta}.$$

Summarising, we obtain

$$\log N_{LB}(\mathcal{G}, L^2(\mu), \quad \sqrt{241\delta}) < \frac{7 \log 100}{\delta},$$

which, by step 1, proves the proposition. ■

See Chapter 4 for estimates on the bracketing numbers (via L^∞-metric entropy bounds) for classes of smooth functions.

3.5.3 Bracketing II: An Exponential Bound for Empirical Processes over Not Necessarily Bounded Classes of Functions

Bernstein's inequality (Proposition 3.1.8) provides exponential bounds for sums of independent random variables with exponential tail probabilities, and a slight modification of the preceding proof should do the same for empirical processes, as we see in this subsection. This is interesting because, although the expectation bounds in the preceding subsection will suffice for the central limit theorem, they only combine with Talagrand's exponential bound in the case of uniformly bounded processes. As it always seems to be the case with chaining, the constants in these bounds are far from optimal, so in this subsection we will not even bother with explicit constants and will be concerned only with the order of the bounds.

We begin by modifying the way the size of a bracket is measured: instead of the $L^2(P)$-norm, we will use a quantity $\rho_k(f)$ (or $\rho_K(g-f)$) that is neither a norm nor a distance but that is adequate for Bernstein's inequality. Given a probability measure P on $(S, \mathcal{S})$ and a positive constant K, we set

$$\rho_K^2(f) = 2K^2 E(e^{|f(X)|/K} - 1 - |f(X)|/K) = 2K^2 \sum_{k=2}^{\infty} \frac{E|f(X)|^k}{K^k k}$$

where X has probability law P, and define the *Bernstein size* of f as the nonnegative square root of $\rho_K^2(f)$. Note that $\rho_K(f) < \infty$ if and only if $Ee^{|f(X)|/K} < \infty$ and that if this holds for some K, then $\lim_{K\to\infty} \rho_K(f) = Ef^2(X)$. Here are some properties of this function that can be easily checked:

Lemma 3.5.18

(a) *$\rho_K(f)$ is nonincreasing in K and $\rho_K(\lambda f) = |\lambda| \rho_{K/|\lambda|}(f)$.*
(b) *$\rho_K^2(f+g) \leq 2\rho_{K/2}^2(f) + 2\rho_{K/2}^2(g)$.*
(c) *If $\rho_K(f) \leq R$, then $E|f(X)|^k \leq kK^{k-2}R^2/2$, for all $k \geq 2$.*
(d) *If $E|f(X)|^k \leq kK^{k-2}R^2/2$, for all $k \geq 2$, then $\rho_{2K}^2(f) \leq 2R^2$.*
(e) *If $\|f\|_\infty \leq K$ and $Ef(X)^2 \leq R^2$, then $\rho_{2K}^2 \leq 2R^2$.*

The version of Bernstein's inequality in Proposition 3.1.8 together with part (c) of the preceding lemma gives the following corollary, which provides the motivation for the definition of ρ_K:

Corollary 3.5.19 *Let X_i be independent identically distributed random variables with law P, and let P_n be the corresponding empirical measure. Assume that $Ef(X) = 0$ and $\rho_K(f) \le R$. Then, given $C > 0$,*

$$\Pr\left\{\sqrt{n}|(P_n - P)(f)| \ge t\right\} \le 2\exp\left(-\frac{t^2}{2(C+1)R^2}\right), \quad \textit{for all } t \le C\sqrt{n}R^2/K. \tag{3.222}$$

This inequality could be stated for the whole range of $t \ge 0$, but it is only stated for t in the 'Gaussian range' because we will only use Bernstein's inequality for this range of t (just as in the proof in the preceding subsection).

Definition 3.5.20 Let $\mathcal{F}$ be a class of measurable functions $f : S \mapsto \mathbb{R}$ such that $Ee^{|f(X)|/K} < \infty$. For each $\varepsilon > 0$, the *B*(K,P)-bracketing number $N_{BK}(\mathcal{F},P,\varepsilon)$ is defined as the smallest N for which there exists a partition of the class $\mathcal{F}$ into N subsets $B_1,\ldots,B_N$ such that, letting $\Delta_i := \left(\sup_{f,g\in B_i}|f-g|\right)^*$, we have $\rho_K(\Delta_i) \le \varepsilon$, for $1 \le i \le N$. For each $\varepsilon > 0$, the *B*(K,P)-bracketing entropy of $\mathcal{F}$ is defined as $H_{BK}(\mathcal{F},P,\varepsilon) = \log N_{BK}(\mathcal{F},P,\varepsilon)$.

Here is the main result of this subsection. Its proof has the same structure as the proof of Theorem 3.5.13 based on chaining combined with a different truncation at each step of the chain (see the decomposition (3.206), which will also be used in the next proof).

Theorem 3.5.21 *Let P be a probability measure on $(S,\mathcal{S})$, and for each n, let P_n be the empirical measure corresponding to n independent identically distributed random variables with law P. Let $\mathcal{F}$ be a class of measurable functions such that $\rho_K(f) \le R$, for all $f \in \mathcal{F}$. Given $C_1 < \infty$, for all C sufficiently large and C_0 satisfying*

$$C_0^2 \ge C^2(C_1+1), \tag{3.223}$$

and for $n \in \mathbb{N}$ and $t > 0$ satisfying

$$C_0\left(R \vee \int_{t/(2^6\sqrt{n})}^{R} \sqrt{H_{B,K}(\mathcal{F},P,\varepsilon)}d\varepsilon\right) \le t \le \sqrt{n}((8R)\wedge(C_1R^2/K)), \tag{3.224}$$

we have

$$\Pr\left\{\sqrt{n}\|P_n - P\|_{\mathcal{F}} \ge t\right\} \le C\exp\left(-\frac{t^2}{C^2(C_1+1)R^2}\right). \tag{3.225}$$

Proof In the proof of Theorem 3.5.13 redefine N_k as $N_k = N_{BK}(\mathcal{F},P,2^{-k}R))$, for $k = 0,1,\ldots$, and $\{T_{k,i}\}_{i=1}^{N_k}$ by a partition of $\mathcal{F}$ such that

$$\rho_K\left(\left(\sup_{f,g\in T_{k,i}}|f-g|\right)^*\right) \le 2^{-k}, \quad \text{for } 1 \le i \le N_k,\ 0 \le k < \infty.$$

Given t satisfying condition (3.224), define

$$L = \min\left\{k \ge 0 : 2^{-k} \le \frac{t}{2^4\sqrt{n}R}\right\}.$$

Then $t \le 8\sqrt{n}R$ implies $L \ge 1$, and we also have, by the definition of L, $R > 2^{-(L+1)}R > t/(2^6\sqrt{n})$, so

$$\frac{1}{2}\sum_{k=0}^{L} 2^{-k}RH_k^{1/2} = \sum_{k=1}^{L+1} 2^{-k}RH_{k-1}^{1/2} \le \sum_{k=1}^{L+1}\int_{2^{-k}R}^{2^{-k+1}R} H_{BK}^{1/2}(u)du$$
$$= \int_{2^{-(L+1)}R}^{R} H_{BK}^{1/2}(u)du \le \int_{t/2^6\sqrt{n}}^{R} H_{BK}^{1/2}(u)du \le t/C_0.$$

Also, as in the preceding proof,

$$\sum_{k=0}^{L} 2^{-k}(\log N_0\cdots N_k)^{1/2} \le 2\sum_{i=0}^{L} 2^{-i}(\log N_i)^{1/2}.$$

Thus, we can assume that we have a collection of partitions $\{A_{k,i} : 1 \le i \le \tilde{N}_i\}$, $\tilde{N}_k = N_0\cdots N_k$, $k \ge 0$, such that

(a) $\sum_{k=0}^{L} 2^{-k}R\sqrt{\log\tilde{N}_k} \le 4\int_{t/2^6\sqrt{n}}^{R}\sqrt{H_{B,K}(\mathcal{F},P,\varepsilon)}d\varepsilon \le 4t/C_0$

(b) $\rho_K\left(\left(\sup_{f,g\in A_{k,i}}|f-g|\right)^*\right) \le 2^{-k}R, \quad \text{for } 1 \le i \le N_k,\ 0 \le k \le L.$

(c) The partitions $\{A_{k,i} : 1 \le i \le \tilde{N}_k\}$ are nested; that is, if $j \le \ell < k$, for each $A_{k,i}$ there is one and only one r such that $A_{k,i} \subseteq A_{\ell,r}$.

As in the preceding bracketing proof, for each f and $k = 0,\ldots,L$, we have a nested sequence of partition sets $A_{\ell,i_\ell(f)}$, $\ell = k,\ldots,L$, each containing f, and the number of such chains is just $\tilde{N}_k$.

For each pair k,i such that $0 \le k \le L$ and $i = 1,\ldots,\tilde{N}_k$, we pick up a function $f_{k,i} \in A_{k,i}$ and set

$$\pi_k f = f_{k,i} \quad \text{and} \quad \Delta_k(f) = \left(\sup_{h,g\in A_{k,i}}|g-h|\right)^*, \quad \text{if } f \in A_{k,i}.$$

The varying truncation levels are also similar to those in the preceding proof, but with different constants. Setting $\log\tilde{N}_k = \tilde{H}_k$, define

$$\eta_k = \max\left(2^{-(k+3)}\tilde{H}_k^{1/2}C_0R/t, 2^{-(k+3)}\sqrt{k}\right)$$

so that $\sum_{k=0}^{L}\eta_k \le 1$,

$$\alpha_k = \sqrt{n}a_k = \frac{2^4\sqrt{n}R^2}{2^{2k}\eta_{k+1}t}, \qquad k = 0,\ldots,L-1,$$

and $\tau f := \tau(f,x) = \min\{0 \le k \le L-1 : \Delta_k f(x) > \alpha_k\}$ if this set is not empty, and $\tau f = L$ otherwise. In analogy with the preceding proof, we also have the following properties of τf:

$$\{\tau f = k\} \subset \{\Delta_k f > \alpha_k\}, \quad \text{for } 0 \le k \le L-1, \quad \{\tau f = k\} \subset \{\Delta_{k-1}f \le \alpha_{k-1}\}, \quad \text{for } 1 \le k \le L,$$
$$\times\{\tau f \ge k\} \subseteq \{\Delta_k f \le \Delta_{k-1}f \le \alpha_{k-1}\}, \quad 1 \le k \le L.$$

Finally, since $\{\tau f = L\} = \{\tau f \geq L\}$, the decomposition (3.206) becomes for all $f \in \mathcal{F}$

$$f = \pi_0 f + \sum_{k=0}^{L} (f - \pi_k f) I_{\tau f = k} + \sum_{k=1}^{L} (\pi_k f - \pi_{k-1} f) I_{\tau f \geq k}. \tag{3.226}$$

Now we can proceed with application of Bernstein's inequality combined with the entropy bound to obtain upper bounds for the tail probabilities of the empirical process applied to each of these three summands.

For the first term, we note that $\rho_K(\pi_0 f) \leq R$ because $\pi_0 f \in \mathcal{F}$ and that by (3.224)

$$C_1 \sqrt{n} R^2 / K \geq t \geq C_0 H_0^{1/2} (R - t/2^6 \sqrt{n}) \geq C_0 H_0^{1/2} R/2,$$

so Bernstein's inequality in the form of Corollary 3.5.19 does apply and gives

$$\begin{aligned} \Pr\left\{ \sup_{f \in \mathcal{F}} |\sqrt{n}(P_n - P)(\pi_0 f)| \geq \frac{t}{4} \right\} &\leq 2e^{H_0} \exp\left(- \frac{t^2/2^4}{2(C_1/4 + 1)R^2} \right) \\ &\leq 2 \exp\left(\frac{4t^2}{C_0^2 R^2} - \frac{t^2}{2^5 (C_1 + 1) R^2} \right) \\ &\leq 2 \exp\left(- \frac{t^2}{2^6 (C_1 + 1) R^2} \right), \end{aligned} \tag{3.227}$$

if we take $C_0^2 \geq 2^8 (C_1 + 1)$.

For the third term of (3.226), which corresponds to the fourth term in (3.206), we first note that the number of functions $(\pi_k f - \pi_{k-1} f) I_{\tau f \geq k}$, $f \in \mathcal{F}$, is just $\tilde{N}_k$. Next, we note that the variance under P of each $(\pi_k f - \pi_{k-1} f) I_{\tau f \geq k}$ is dominated by the second moment of $\Delta_{k-1}(f)(X_i)$, which, by Lemma 3.5.18(c) is, in turn, dominated by $\rho_K^2(\Delta_{k-1}(f)) \leq 2^{-2(k-1)} R^2$, and that its supremum norm is dominated by α_{k-1} (by the properties of τf). Thus, we may apply to these variables either Bernstein's inequality for bounded variables in Theorem 3.1.7 or Corollary 3.5.19 combined with Lemma 3.5.18(e). We apply the former, which for i.i.d. variables bounded by c in absolute value and variance σ^2 is given by $\Pr\{|S_n|/\sqrt{n} \geq t\} \leq 2 \exp(-t^2/(2(1+\lambda)\sigma^2))$, if $t \leq 3\lambda \sqrt{n} \sigma^2 / c$. Then, since by the definitions of α_k and η_k we have

$$\frac{\eta_k t}{4} = \frac{4}{3} \frac{3\sqrt{n} 2^{-2(k-1)} R^2}{\alpha_{k-1}}$$

and

$$\tilde{H}_k = \sum_{k=0}^{L} H_k \leq \eta_k^2 2^{2(k+3)} t^2 / C_0^2 R^2,$$

we conclude, assuming $C_0^2 \geq 3 \cdot 2^{12}$, that

$$\Pr\left\{\sup_{f\in\mathcal{F}}\left|\sqrt{n}(P_n-P)\left(\sum_{k=1}^{L}(\pi_k f-\pi_{k-1}f)I_{\tau f\geq k}\right)\right|\geq\frac{t}{4}\right\}$$

$$\leq\sum_{k=1}^{L}\Pr\left\{\sup_{f\in\mathcal{F}}\left|\sqrt{n}(P_n-P)\left((\pi_k f-\pi_{k-1}f)I_{\tau f\geq k}\right)\right|\geq\frac{\eta_k t}{4}\right\}$$

$$\leq 2\sum_{k=1}^{L}\exp\left(\sum_{k=1}^{L}H_k-\frac{t^2\eta_k^2/2^4}{6\cdot 2^{-2(k-1)}R^2}\right)\leq 2\sum_{k=1}^{L}e^{-2^{2k}\eta_k^2 t^2/2^9R^2}.$$

Now, since $\eta_k \geq 2^{-k-3}\sqrt{k}$ and $t \geq C_0 R$, we have, for $C_0^2 \geq 2^{16}$,

$$\sum_{k=1}^{L}e^{-2^{2k}\eta_k^2 t^2/2^9R^2}\leq\sum_{k=1}^{L}e^{-t^2k/2^{15}R^2}<\frac{1}{1-e^{-C_0^2/2^{15}}}e^{-t^2/2^{15}R^2}\leq 2e^{-t^2/2^{15}R^2};$$

hence,

$$\Pr\left\{\sup_{f\in\mathcal{F}}\left|\sqrt{n}(P_n-P)\left(\sum_{k=1}^{L}(\pi_k f-\pi_{k-1}f)I_{\tau f\geq k}\right)\right|\geq\frac{t}{4}\right\}\leq 4e^{-t^2/2^{15}R^2}. \tag{3.228}$$

Finally, we consider the middle term in the decomposition (3.226), similar to the third term in decomposition (3.206). First, we note that given that $|(f-\pi_k f)I_{\tau f=k}| \leq \Delta_k(f)I_{\tau f=k}$, we have

$$\left|(P_n-P)((f-\pi_k f)I_{\tau f=k})\right|\leq(P_n-P)(\Delta_k(f)I_{\tau f=k})+2P(\Delta_k(f)I_{\tau f=k}).$$

Now, since $\Delta_k(f) > \alpha_k$ if $\tau f = k$ for $k \leq L-1$, and since $\rho_K(\Delta_k(f)) \leq 2^{-k}R$, we have, for $k \leq L-1$,

$$P(\Delta_k(f)I_{\tau f=k})\leq\frac{P(\Delta_k(f))^2}{\alpha_k}\leq\frac{\rho_K^2(\Delta_k(f))}{\alpha_k}\leq\frac{2^{-2k}R^2}{\alpha_k}\leq\frac{\eta_{k+1}t}{2^4\sqrt{n}}$$

and, by the definition of L,

$$P(\Delta_L(f)I_{\tau f=L})\leq(P(\Delta_L^2(f)))^{1/2}\leq 2^{-L}R\leq\frac{t}{2^4\sqrt{n}}.$$

Then, collecting the last three inequalities (recall that $\sum\eta_k \leq 1$), we get

$$\left|(P_n-P)\left(\sum_{k=0}^{L}(f-\pi_k f)I_{\tau f=k}\right)\right|\leq(P_n-P)\left(\sum_{k=0}^{L}\Delta_k(f)I_{\tau f=k}\right)+\frac{t}{4\sqrt{n}}.$$

Hence,

$$\Pr\left\{\left|\sqrt{n}(P_n-P)\left(\sum_{k=0}^{L}(f-\pi_k f)I_{\tau f=k}\right)\right|\geq\frac{t}{2}\right\}\leq\Pr\left\{\sqrt{n}(P_n-P)\left(\sum_{k=0}^{L}\Delta_k(f)I_{\tau f=k}\right)>\frac{t}{4}\right\}$$

$$\leq\Pr\left\{\sqrt{n}(P_n-P)\left(\sum_{k=1}^{L}\Delta_k(f)I_{\tau f=k}\right)>\frac{t}{8}\right\}$$

$$+\Pr\left\{\sqrt{n}(P_n-P)\left(\Delta_0(f)I_{\tau f=0}\right)>\frac{t}{8}\right\}.$$

For the last summand, we can apply Corollary 3.5.19 as we do in (3.227) with the preceding first term, which, since $\rho_K(\Delta_0) \le R$, $t \le C_1\sqrt{n}R^2/K$ and $H_0 \le 16t^2/C_0^2R^2$, gives

$$\Pr\left\{\sqrt{n}(P_n - P)\left(\Delta_0(f)I_{\tau f=0}\right) > \frac{t}{8}\right\} \le 2\exp\left(-\frac{t^2}{2^8(C_1+1)R^2}\right)$$

if we take $C_0^2 \ge 2^{12}(C_1+1)$. The preceding summand can be bounded using Bernstein's inequality for bounded variables just as is done in (3.228) with the second term of the decomposition (3.226) because

$$E\Delta_k^2(f) \le 2^{-2k}R^2 \quad \text{and} \quad \Delta_k(f)I_{\tau f=k} \le \Delta_k(f)I_{\tau f=k} \le \alpha_{k-1}.$$

Thus, we obtain

$$\Pr\left\{\sqrt{n}(P_n - P)\left(\sum_{k=1}^{L}\Delta_k(f)I_{\tau f=k}\right) > \frac{t}{8}\right\} \le 4e^{-t^2/2^{17}R^2},$$

assuming $C_0^2 \ge 2^{18}$. Now the theorem follows by collecting the last two estimates, (3.227) and (3.228) ∎

Exercises

3.5.1 Let $\mathcal{F} = \{I_{(-\infty,t]} : t \in \mathbb{R}\}$. Show that for every Borel probability measure P on $\mathbb{R}$, $N(\mathcal{F}, L^2(P), \tau) \le \tau^{-2} + 1 \le 2\tau^{-2}$, for $0 < \tau < 1$. Hence, since also the measurable cover of $\mathcal{F}$ is $F = 1$, we can take H in Theorem 3.5.6 to be $H(x) = \log(4x^2)$ for $x \ge 1$. Then, by differentiating $xH^{1/2}(1/x)$ and by noting that the derivative of this function, $H^{1/2}(1/x)(1 - 1/H(1/x))$, is larger than or equal to $H^{1/2}(1/x)(1 - 1/\log 4)$ for $x < 1$, it follows that H satisfies condition (ii) in the theorem for $C_H = \log 4/(\log 4 - 1)$. Deduce that for all Borel probability measures P on $\mathbb{R}$, if F and F_n are, respectively, the cumulative distribution function (cdf) of P and the empirical cdf corresponding to an independent sample from P, then $E\|F_n - F\|_\infty \le C/\sqrt{n}$. This bound is of the right order but for the constant much larger than 4. Recall from Exercise 3.1.7 that $E\|F_n - F\|_\infty \le 4/\sqrt{n}$.

3.5.2 (a) Repeat Exercise 3.5.1 for $\mathcal{F} = \{f_t : f_s \le f_t \text{ for } s \le t \in \mathbb{R},\ f_{-\infty+} \ge 0, f_{+\infty-} \le 1\}$. In particular, this applies to the classes of sets in $\mathbb{R}^d$, $\mathcal{F}_i = \{I_{x_i \le t} : t \in \mathbb{R}\}$, $1 \le i \le d$. (b) Observe that if $|f_i|, |g_i| \le 1$, then for any probability measure P, $P(\prod_{i=1}^d f_i - \prod_{i=1}^d g_i)^2 \le d\sum_{i=1}^d P(f_i - g_i)^2$, and use this and part (a) to give an estimate for the $L^2(P)$-covering numbers of $\mathcal{G} = \{I_{x_1 \le t_1, \dots, x_d \le t_d} : (t_1, \dots, t_d) \in \mathbb{R}^d\}$. (c) Conclude that for every $d \in \mathbb{N}$ there exists $C_d < \infty$ such that if for any Borel probability measure P on $\mathbb{R}^d$, F is its cdf and F_n the empirical cdf corresponding to an independent sample from P, then we have $\|F_n - F\|_\infty \le C_d/\sqrt{n}$.

3.5.3 The classes of functions considered in the preceding two exercises are linearly ordered by the usual relation of order between functions (or of inclusion between sets). Show that, as a consequence, the estimates of the covering numbers of the classes of indicators in these exercises also hold true for their bracketing covering numbers (any P) and hence that in these cases the bracketing expectation bounds produce the same results (up to constants, none optimal).

3.5.4 Let $N^1_{[]}(\mathcal{F}, L^p(P), \varepsilon)$ and $N^2_{[]}(\mathcal{F}, L^p(P), \varepsilon)$ denote the bracketing numbers of $\mathcal{F}$ according to the first and second definitions of brackets given at the beginning of Section 3.5.2. Show that

$$N^2_{[]}(\mathcal{F}, L^p(P), 2\varepsilon) \le N^1_{[]}(\mathcal{F}, L^p(P), \varepsilon) \le N^2_{[]}(\mathcal{F}, L^p(P), \varepsilon).$$

Hint: If, for $B \subset \mathcal{F}$, $\Delta_B = (\sup_{f,g \in B}|f - g|)^*$ and $f \in B$, then B is contained in the bracket $[f - \Delta, f + \Delta]$ of L^p size $2\|\Delta_B\|_{L^p}$. In the other direction, if $B = [\underline{f}, \overline{f}]$, then $\Delta_B = \overline{f} - \underline{f}$.

3.5.5 Prove that there exists a constant $K < \infty$ such that if $\mathcal{F}$ is such that $\rho_1(f) \le \delta$, for all $f \in \mathcal{F}$, where ρ_1 is the Bernstein size, then

$$E\left\|\sum_{i=1}^{n}(f(X_i)-Pf)\right\|_{\mathcal{F}}^{*} \le K\left(1+\frac{1}{\delta^2}\int_0^\delta \sqrt{H_{B,1}(\mathcal{F},P,\varepsilon)}\,d\varepsilon\right)\int_0^\delta \sqrt{H_{B,1}(\mathcal{F},P,\varepsilon)}\,d\varepsilon.$$

Hint: Improve on some of the estimates in the proof of Theorem 3.5.21 along the lines of the proof of Proposition 3.5.15.

3.5.6 Modify the proof of Proposition 3.5.17 to show that if $\mathcal{F}$ is a collection of uniformly bounded monotone functions on $\mathbb{R}$, then uniformly in P, for all $p \ge 1$, $\log N_{[]}(\mathcal{F}, L^p(P), \delta) \le K/\delta$.

3.6 Vapnik-Červonenkis Classes of Sets and Functions

Many classes of sets and functions used in statistical applications of empirical processes have the remarkable property of admitting bounds for their $L^2(P)$ covering numbers $N(\mathcal{F}, L^2(P), \varepsilon)$ of the order of ε^{-s}, for some $s < \infty$, *uniformly* in $P \in \mathcal{P}(S)$, where $\mathcal{P}(S)$ is the set of all probability measures on $(S, \mathcal{S})$. Consider as a first example linearly ordered uniformly bounded classes of functions, as in the exercises from the preceding section. Any empirical process indexed by these classes admits excellent bounds for the expectation of its supremum, and these, combined with Talagrand's inequality, produce exponential bounds that are of Gaussian type in large portions of their range, that is, best possible at least up to constants. It will be seen in later sections that the usual limit theorems do hold uniformly in P for these classes. These classes of sets were discovered by Vapnik and Červonenkis and their entropy properties by Dudley who called them *VC classes*. The entropy properties of *VC* classes of sets are inherited by related classes of functions that go by the name of *VC subgraph classes* of functions. Other related classes of functions such as *VC*-major and *VC*-hull classes admit larger but still manageable uniform bounds. In this section, *VC* classes of sets and the related class functions are defined and their $L^2(P)$ metric entropy properties established.

3.6.1 Vapnik-Červonenkis Classes of Sets

Let $\mathcal{C}$ be a class of subsets of a set S. Let $A \subseteq S$ be a finite set. The *trace* of $\mathcal{C}$ on A is the collection of all the subsets of A obtained by intersection of A with sets C from $\mathcal{C}$. Denote by $\Delta^{\mathcal{C}}(A)$ the cardinality of the trace of the class $\mathcal{C}$ on A. Then $\Delta^{\mathcal{C}}(A) \le 2^{\mathrm{Card}(A)}$, and we say that $\mathcal{C}$ *shatters* A if $\Delta^{\mathcal{C}}(A) = 2^{\mathrm{Card}(A)}$, that is, if every subset of A is the intersection of A with some set $C \in \mathcal{C}$. Let

$$m^{\mathcal{C}}(k) = \sup_{\substack{A \subseteq S \\ \mathrm{Card}(A)=k}} \Delta^{\mathcal{C}}(A).$$

Definition 3.6.1 A collection of sets $\mathcal{C}$ is a Vapnik-Červonenkis class ($\mathcal{C}$ is VC) if the quantity

$$v(\mathcal{C}) := \begin{cases} \min\{k : m^{\mathcal{C}}(k) < 2^k\} & \text{if } m^{\mathcal{C}}(k) < 2^k \text{ for some } k < \infty \\ \infty & \text{otherwise} \end{cases}$$

is finite, that is, if there exists $k < \infty$ such that $\mathcal{C}$ does not shatter any subsets of S of cardinality k. The VC index of the class $\mathcal{C}$ is defined as $v(\mathcal{C})$.

For instance, the collection of half-lines $\mathcal{C} = \{(-\infty, t] \ : \ t \in \mathbb{R}\}$ is VC and $v(\mathcal{C}) = 2$; if $x_1 < x_2$, then $\Delta^{\mathcal{C}}(x_1, x_2) = 3 < 2^2$ because $\mathcal{C}$ cannot pick out $\{x_2\}$.

It is easy to see that $v(\mathcal{C}) = 0$ if and only if $\mathcal{C}$ is empty and $v(\mathcal{C}) = 1$ if and only if $\mathcal{C}$ consists of only one set (Exercise 3.6.1). In the first case, $m^{\mathcal{C}}(n) = 0$ for all n, and in the second, $m^{\mathcal{C}}(n) = 1$ for all $n \geq 1$.

The following remarkable combinatorial theorem asserts that *either $m^{\mathcal{C}}(k) = 2^k$ for all k or $m^{\mathcal{C}}(k)$ grows only polynomially in k*. The main results on VC classes of sets and functions are based on this theorem.

Theorem 3.6.2 *Let $\mathcal{C}$ be a non-empty VC class, and let $v = v(\mathcal{C}\Rightarrow$. Then, for any finite set $A \subseteq S$,*

$$\Delta^{\mathcal{C}}(A) \leq \mathrm{Card}\{B \subseteq A : \ \mathrm{Card}(B) < v\} = \sum_{j=0}^{v-1} \binom{\mathrm{Card}(A)}{j},$$

and therefore,

$$m^{\mathcal{C}}(n) \leq \sum_{j=0}^{v-1} \binom{n}{j},$$

a polynomial in n of degree $v-1$. In particular, there is a constant $B(v) < \infty$ depending only on v such that $m^{\mathcal{C}}(n) \leq B(v)n^{v-1}$, for all $n \geq v-1$.

The result follows from the following proposition by letting $\mathcal{U} = \{C \cap A : \ C \in \mathcal{C}\}$:

Proposition 3.6.3 *Let A be a finite set, and let $\mathcal{U}$ be a class of subsets of A. Then*

$$\mathrm{Card}(\mathcal{U}) \leq \mathrm{Card}\{B \subseteq A : \ B \text{ is shattered by } \mathcal{U}\}.$$

Proof Say that a class of subsets of A, $\mathcal{U}'$, is hereditary if when a set B is in $\mathcal{U}'$ then all the subsets of B are also in $\mathcal{U}'$. By definition, if $\mathcal{U}'$ is hereditary, then $\mathcal{U}'$ shatters all the sets it contains, and therefore,

$$\mathrm{Card}(\mathcal{U}') \leq \mathrm{Card}\{B \subseteq A : \ B \text{ is shattered by } \mathcal{U}'\}.$$

Hence, the proposition will be proved if we construct a collection $\mathcal{U}'$ from $\mathcal{U}$ such that

(i) $\mathcal{U}'$ is hereditary,
(ii) $\mathrm{Card}(\mathcal{U}) = \mathrm{Card}(\mathcal{U}')$, and
(iii) $\mathcal{U}$ shatters at least as many sets as $\mathcal{U}'$ shatters.

We will obtain $\mathcal{U}'$ by repeated application of a simple transformation T_x that with each application renders the class of sets a step closer to being hereditary. Given $x \in A$ and a collection $\mathcal{V}$ of subsets of A, let $T_x^{\mathcal{V}}(\mathcal{V}) := T_x(\mathcal{V}) := \{T_x(V) : \ V \in \mathcal{V}\}$, where, for $V \in \mathcal{V}$,

$$T_x(V) = \begin{cases} V \setminus \{x\} & \text{if } x \in V \text{ and } V \setminus \{x\} \notin \mathcal{V} \\ V & \text{otherwise.} \end{cases}$$

Thus all the sets $V \in T_x(\mathcal{V})$ satisfy that if $x \in V$ then $V \setminus \{x\} \in T_x(\mathcal{V})$, which is clearly a step towards transforming $\mathcal{V}$ into a hereditary set, as mentioned earlier. Another property of $T_x(\mathcal{V})$ to be used later that follows directly from the definition is that if $V \in T_x(\mathcal{V})$ and $x \in V$, then $V \in \mathcal{V}$, $V \setminus \{x\} \in \mathcal{V}$ and $T_x(V) = V$.

Define, on classes $\mathcal{V}$ of subsets of A, the functional $w(\mathcal{V}) := \sum_{V\in\mathcal{V}} \mathrm{Card}(V)$. Then, since $\mathrm{Card}(T_x^{\mathcal{V}}(V))$ is either $\mathrm{Card}(V)$ or $\mathrm{Card}(V) - 1$, it follows that $w \circ T_x \leq w$. This implies that the minimum of w among all the collections $\mathcal{U}'$ obtained from $\mathcal{U}$ by finitely many repeated applications of the maps T_x, $x \in A$, is attained. Let $\mathcal{U}'$ be one of the possibly more than one such collections on which w is minimal. Then $w(\mathcal{U}') = w(T_x^{\mathcal{U}'}(\mathcal{U}'))$, for all $x \in A$. But this implies that $\mathcal{U}'$ is hereditary if $\mathcal{U}'$ were not hereditary, then there would exist a set $B \in \mathcal{U}'$ and a point $x \in A$ such that $B \setminus \{x\} \notin \mathcal{U}'$, implying that $\mathrm{Card}(T_x(B)) < \mathrm{Card}(B)$ and hence also that $w(\mathcal{U}') < w(T_x(\mathcal{U}')$, a contradiction. (Repeated applications of the maps T_x means, for example, that $T_x^{T_y\mathcal{U}}(T_y^{\mathcal{U}}(B))$.)

Since $\mathcal{U}'$ is obtained from $\mathcal{U}$ by a finite number of operations T_x, to prove that $\mathcal{U}'$ satisfies preceding properties (ii) and (iii), it suffices to see that they are satisfied by $T_x(\mathcal{U})$. Thus, we must show that $\mathrm{Card}(T_x(\mathcal{U})) = \mathrm{Card}(\mathcal{U})$ and that $\mathcal{U}$ shatters at least as many sets as $T_x(\mathcal{U})$ does. To prove the first property, we observe that T_x is one to one: the operation T_x keeps unchanged those sets $U \in \mathcal{U}$ which contain x and are such that $U \setminus \{x\}$ is in $\mathcal{U}$ as well as those sets $U \in \mathcal{U}$ which do not contain x, and T_x replaces U by $U \setminus \{x\}$ if U contains x but $U \setminus \{x\}$ is not in $\mathcal{U}$. This is obviously a one-to-one replacement. To see the second property, we must show that if $T_x(\mathcal{U})$ shatters B, then $\mathcal{U}$ also shatters B. Suppose that $T_x(\mathcal{U})$ shatters B. If $x \notin B$, then $\mathcal{U}$ shatters B because $U \in \mathcal{U}$ differs from $T_x(U) \in T_x(\mathcal{U})$ at most by x. If $x \in B$, then for every $B' \subseteq B\setminus\{x\}$ there exists $V \in T_x(\mathcal{U})$ such that $V \cap B = B' \cup \{x\}$ because $T_x(\mathcal{U})$ shatters B. This implies that $x \in V$ and therefore that V is also in $\mathcal{U}$, thus showing that $U \cap B = B' \cup \{x\}$ for some $U \in \mathcal{U}$; moreover, $V \setminus \{x\} \in \mathcal{U}$ and $(V \setminus \{x\}) \cap B = B'$. Hence, $\mathcal{U}$ shatters B. ■

The following proposition will simplify the bound for $m^{\mathcal{C}}(n)$ in the preceding theorem.

Proposition 3.6.4 *Let k and n be nonnegative integers such that $n \geq k + 2$. Then*

$$\sum_{j=0}^{k} \binom{n}{j} \leq \frac{1.5n^k}{k}. \tag{3.229}$$

Proof We recall the nonasymptotic Stirling's formula (see Feller, vol. I, 1968, p. 52)

$$(n/e)^n \sqrt{2\pi n} \leq n \leq e^{1/12n}(n/e)^n \sqrt{2\pi n}, \quad n \in \mathbb{N}, \tag{3.230}$$

to be used at a crucial step in this proof. We also recall the 'Pascal triangle' property of the quantities $C_{n,\leq k} := \sum_{j=0}^{k} \binom{n}{j}$, simply inherited from the same property of the combinatorial numbers $\binom{n}{k}$

$$C_{n,\leq k} = C_{n-1,\leq k} + C_{n-1,\leq k-1}, \quad k, n \in \mathbb{N},\ k < n. \tag{3.231}$$

Another inequality we need is the following simple one: for $n, k \geq 1$,

$$\frac{n^k}{k} \geq \frac{(n-1)^k}{k} + \frac{(n-1)^{k-1}}{(k-1)}, \tag{3.232}$$

which follows because, by the binomial theorem, $n^k \geq (n-1)^k + k(n-1)^{k-1}$. Suppose that the proposition is true for $C_{n,\leq n-2}$ for all n and that, given K, the proposition is also true for $C_{n,\leq k}$, for all (n,k) with $0 \leq k \leq K - 1$ and $n \geq k + 2$. Then, using (3.231) and (3.232),

$$C_{K+3,\leq K} = C_{K+2,\leq K} + C_{K+2,\leq K-1} \leq \frac{1.5(K+2)^K}{K} + \frac{1.5(K+2)^{K-1}}{(K-1)} \leq \frac{1.5(K+3)^K}{K}.$$

Consequently,

$$C_{K+4,\leq K} = C_{K+3,\leq K} + C_{K+3,\leq K-1} \leq \frac{1.5(K+4)^K}{K}$$

and, in general, $C_{n,\leq K} \leq 1.5n^K/K$, for all $n \geq K+2$. Hence, the proposition will follow by induction on k if we show that it holds for the pairs (n,k) such that $k=0$ and $n \geq 2$, $k=1$ and $n \geq 3$ and $(n, n-2)$ for all $n \geq 3$. For $k=0$, it follows because $C_{n,\leq 0} = 1$, and for $k=1$ and $n \geq 3$, it follows because $1+n \leq 1.5n$. For $k = n-2$, we must show that

$$2^n - 1 - n \leq \frac{1.5n^{n-2}}{(n-2)} = \frac{1.5(n-1)n^{n-1}}{n}.$$

This can be checked directly for $n = 3, \ldots, 6$. For $n \geq 7$, we will prove the slightly stronger inequality $2^n n \leq 1.5(n-1)n^{n-1}$. By Stirling's formula, it suffices to show that

$$2^n (n/e)^n (2\pi n)^{1/2} e^{\frac{1}{12n}} \leq 1.5(n-1)n^{n-1}.$$

Taking logarithms and then derivatives, it is easy to see that $(e/2)^n \geq 2n^{1/2}$ for all $n \geq 7$ (in fact, $n \geq 5$). But then the preceding inequality follows from

$$\sqrt{2\pi}\, e^{\frac{1}{12n}} \leq 3(1 - 1/n),$$

which does hold for all $n \geq 7$ ($\sqrt{2\pi} \leq 2.51$, $18/7 \geq 2.57$ and the exponential is very close to 1). ■

The preceding proposition and theorem give the following:

Corollary 3.6.5 *If $\mathcal{C}$ is a non-empty VC class of sets and $v = v(\mathcal{C})$ is its VC index, then*

$$m^{\mathcal{C}}(n) \leq \frac{1.5n^{v-1}}{(v-1)}, \quad \textit{for } n \geq v+1.$$

For $n = v$, $m^{\mathcal{C}}(n) \leq 2^v - 1$, and for $n < v$, $m^{\mathcal{C}}(n) \leq 2^n < 2^v - 1$. In particular,

$$m^{\mathcal{C}}(n) \leq 2n^{v-1}, \quad \textit{for all } n \geq 1.$$

Whereas the inequalities in this corollary are not sharp, the inequality in Theorem 3.6.2 is (see Exercise 3.6.2). The preceding inequality follows easily for $v(\mathcal{C}) \geq 2$, and it follows from Exercise 3.6.1 for $v=0$ and $v=1$ (in these cases, $m^{\mathcal{C}}(n)$ is, respectively, 0 and 1).

It is not always easy to prove that a class of sets is *VC*. The following two propositions can be very helpful:

Proposition 3.6.6 *If $\mathcal{G}$ is a finite-dimensional vector space of real functions on S, then the class of sets $\mathcal{C} := [\{g \geq 0\} : g \in \mathcal{G}]$ is VC with $v(\mathcal{C}) = \dim \mathcal{G} + 1$. The same is true for $[\{g > 0\} : g \in \mathcal{G}]$.*

Proof Let $v - 1 = \dim \mathcal{G}$, and let $\{s_1, \ldots, s_v\}$ be v distinct points of S. (The result clearly holds if S contains fewer than v points.) Let $L : \mathcal{G} \mapsto \mathbb{R}^v$ be given by $L(g) = (g(s_1), \ldots, g(s_v))$. Then $\dim L(\mathcal{G}) \leq v-1$. Let $w = (w_1, \ldots, w_v)$ be a nonzero vector orthogonal to $L(\mathcal{G})$. Then

$$\sum w_i I(w_i \geq 0) g(s_i) = -\sum w_i I(w_i < 0) g(s_i),$$

and we can assume that the set of i with $w_i < 0$ is not empty (otherwise, we can replace w by $-w$). If there existed g for which $\{g \geq 0\} \cap \{s_1,\dots,s_v\} = \{s_i : w_i \geq 0\}$, then the left side of the preceding equation for this g would be larger than or equal to zero, whereas the right side would be negative, which is impossible. Thus, there is a subset of $\{s_1,\dots,s_v\}$ that is not the intersection of $\{s_1,\dots,s_v\}$ with any set $\{g \geq 0\}$. Hence, the class $\mathcal{C}$ is VC, and $v(\mathcal{C}) \leq \dim \mathcal{G} + 1$.

Now let $m = v - 1 = \dim(G)$. Consider the vector space $G'_S = \{\sum a_i \delta_{x_i} : a_i \in \mathbb{R},\ x_i \in S\}$, where δ_x is unit mass at x, and note that the natural inclusion $G \subset (G'_S)'$ is injective (if two functions are different, they are also different as linear functionals on G'_S). Hence, the dimension of G'_S, which equals that of its dual, is at least m. Since $\{\delta_x : x \in S\}$ generates G'_S, it follows that there exist m points $x_i \in S$ such that $\delta_{x_1},\dots,\delta_{x_m}$ are linearly independent as elements of G'_S. Therefore, $\{(g(x_1),\dots,g(x_m)) : g \in G\} = \mathbb{R}^m$, which implies that every subset of the set $\{x_1,\dots,x_m\}$ is the intersection of this set with $\{g \geq 0\}$ for some $g \in G$. Thus, $v(\mathcal{C}) \geq m + 1 = v$. The second assertion follows in a similar way. ■

For example, this proposition immediately gives that the class of all closed half-spaces of $\mathbb{R}^d$ is VC and so is the class of all the open half-spaces. The same is true for the class of all the closed balls (or all open balls) of $\mathbb{R}^d$.

Exercise 3.6.4 and the following proposition are examples of permanence properties of VC classes.

Proposition 3.6.7

(i) *If $\mathcal{C}$ is VC, then $\mathcal{C}^c := \{C^c : C \in \mathcal{C}\}$ is VC.*

(ii) *If $\mathcal{C}$ and $\mathcal{D}$ are VC, then $\mathcal{C} \cup \mathcal{D}$ and $\mathcal{C} \cap \mathcal{D}$ are VC. Here $\mathcal{C} \cup \mathcal{D} = \{C \cup D : C \in \mathcal{C}, D \in \mathcal{D}\}$ and $\mathcal{C} \cap \mathcal{D} = \{C \cap D : C \in \mathcal{C}, D \in \mathcal{D}\}$.*

(iii) *If $\mathcal{C}$ is a collection of subsets of S and $\mathcal{D}$ is a collections of subsets of T and both are VC, then $\mathcal{C} \times \mathcal{D}$ is also VC, where $\mathcal{C} \times \mathcal{D} = \{C \times D : C \in \mathcal{C}, D \in \mathcal{D}\}$.*

(iv) *If $\mathcal{C} \subset \mathcal{D}$ and $\mathcal{D}$ are VC, then $\mathcal{C}$ is VC.*

Proof Obviously, $\mathcal{C}$ shatters A if and only if $\mathcal{C}^c$ does, and we also have $v(\mathcal{C}) = v(\mathcal{C}^c)$. The trace of $\mathcal{C}$ on a set of n points consists of a number of sets not exceeding $1.5n^{v(\mathcal{C})-1}$, and the trace of $\mathcal{D}$ on each of these subsets does not exceed $1.5n^{v(\mathcal{D})-1}$; hence, the trace of $\mathcal{C} \cap \mathcal{D}$ does not exceed $2.25n^{v(\mathcal{C})+v(\mathcal{D})-2}$, smaller than 2^n for n large enough. This also proves $\mathcal{C} \cup \mathcal{D}$ is VC by taking complements and applying the first property. If $A \subset S \times T$ has n points, then clearly $\Delta^{\mathcal{C} \times T}(A) = \Delta^{\mathcal{C}}(\pi_1(A))$, where $\pi_1(s,t) = s$, and likewise for $S \times \mathcal{D}$, showing that these two classes are VC; hence, so is their intersection $\mathcal{C} \times \mathcal{D}$. ■

The preceding two propositions show, for example, that the class of all polygons of $\mathbb{R}^2$ of less than k sides, any fixed k, is VC, and likewise in $\mathbb{R}^d$ for piecewise polynomial regions of a fixed finite number of pieces and degrees not exceeding a fixed number k.

We conclude this subsection by showing in the most naive way possible how Theorem 3.6.2 can be used to evaluate empirical processes indexed by VC classes of sets. The reader will find it meaningful because we already have a relatively large collection of examples of VC classes. Let $\mathcal{C}$ be a countable, not empty VC class of subsets of S of index v, and consider the empirical processes indexed by $\mathcal{C}$ and based on an i.i.d. sample $\{X_i\}$ from

P. Let

$$\tilde{P}_n(C) = \frac{1}{n}\sum_{i=1}^{n} \varepsilon_i \delta_{X_i}(C) = \frac{1}{n}\sum_{i=1}^{n} \varepsilon_i I_{X_i \in C}, \quad C \in \mathcal{C},$$

be the empirical process randomised by an independent Rademacher sequence $\{\varepsilon_i\}$ independent of the variables X_j. Observe that if $\{X_1,\dots,X_n\} \cap C = \{X_1,\dots,X_n\} \cap D$, then $\tilde{P}_n(C) = \tilde{P}_n(D)$. This implies by Corollary 3.6.5 that $\sup_{C\in\mathcal{C}} |\tilde{P}_n(C)|$ is in fact the maximum over at most $m^{\mathcal{C}}(n) \le 1.5n^{v-1}/(v-1)$ random variables of the form $n^{-1}\sum_{i=1}^{n} \varepsilon_i a_i$, with $a_i = 0$ or $a_i = 1$. But these random variables are sub-Gaussian with $\sigma^2 \le 1/n$. Hence, the simple maximal inequality for finite collections of sub-Gaussian variables in Lemma 2.3.4 together with randomisation yields, for $n \ge 2$,

$$E\|P_n - P\|_{\mathcal{C}} \le 2E\|\tilde{P}_n\|_{\mathcal{C}} = 2E_X E_\varepsilon \|\tilde{P}_n\|_{\mathcal{C}} \le 2\sqrt{2\log\frac{1.5n^{v-1}}{(v-1)}}/\sqrt{n} \to 0,$$

a uniform Glivenko-Cantelli theorem for these classes. The first identity here requires measurability, and this is why we are assuming $\mathcal{C}$ to be countable, an assumption that can be relaxed but not altogether ignored. This is the celebrated Vapnik-Červonenkis law of large numbers, which does go far beyond the multivariate Glivenko-Cantelli theorem. Strictly speaking, the Glivenko-Cantelli theorem states a.s. convergence, but here, once convergence to zero in probability is established, one simply uses the reverse submartingale convergence theorem to show convergence a.s. (Exercise 3.6.8).

For more sophisticated applications, a bound on the $L^2(P)$-covering numbers of VC classes of sets is needed. This is given in the next subsection, directly for VC subgraph classes of functions, which include in particular VC classes of sets.

3.6.2 *VC* Subgraph Classes of Functions

In this subsection we consider classes of functions on S that are related in different ways to VC classes of sets. The most interesting are probably the VC subgraph classes. The main result is Theorem 3.6.9, which shows that the $L^p(P)$-covering numbers of these classes of functions admit small bounds, of the order of $\varepsilon^{-(v-1)p}$, uniformly in P.

Definition 3.6.8 The subgraph of a real function f on S is the set

$$G_f = \{(s,t) : \ s \in S, \ t \in \mathbb{R}, \ t \le f(s)\}.$$

A class of functions $\mathcal{F}$ is *VC subgraph* if the class of sets $\mathcal{C} = \{G_f : \ f \in \mathcal{F}\}$ is VC.

The family of indicator functions of the sets in a VC class is a VC subgraph class of functions. More generally, if f is a function on S and $\mathcal{C}$ is a VC class of sets, then the class of functions $\mathcal{F} = \{fI_C : C \in \mathcal{C}\}$ is VC subgraph. To see this, we just note that for a subset $\{(s_1,t_1),\dots,(s_n,t_n)\}$ of $S \times \mathbb{R}$ to be shattered by the subgraphs of functions in $\mathcal{F}$, it is necessary that the S coordinates $s_1,\dots,s_n$ be all different and that the set $\{s_1,\dots,s_n\}$ be shattered by $\mathcal{C}$.

Hence, the following key proposition for VC subgraph classes of functions applies to VC classes of sets and beyond. First, here is some notation. Let $\mathcal{F}$ be a class of functions in

$L^p(S,\mathcal{S},P)$, $0<p<\infty$. Then, letting $e_{p,P}=e_p$ denote the $L^p(P)$ pseudo-distance $e_p(f,g)=\|f-g\|_p=\int|f-g|^p dP$ for $0<p<1$, $e_p(f,g)=\|f-g\|_p=\left[\int|f-g|^p dP\right]^{1/p}$ for $p\geq 1$, we write, as usual,

$$D(\mathcal{F},L^p(P),\varepsilon):=D(\mathcal{F},e_p,\varepsilon)$$

and likewise for N_p. We recall that given a class of functions $\mathcal{F}$, a *measurable envelope* or an *envelope* F of $\mathcal{F}$ is any *everywhere finite, measurable* function F such that

$$\sup_{f\in\mathcal{F}}|f(s)|\leq F(s), \quad s\in S.$$

Of course, a class $\mathcal{F}$ may not admit any measurable envelopes. Finally, we should emphasise that although $\mathcal{F}$ is always a subset of measurable functions on a probability space, we do not identify functions which are a.s. equal.

Theorem 3.6.9 (Dudley-Pollard) *Let $\mathcal{F}$ be a non-empty VC subgraph class of functions admitting an envelope $F\in L^p(S,\mathcal{S},P)$ for some $0<p<\infty$ which is assumed, without loss of generality, bounded away from zero. Suppose that the class $\mathcal{C}$ of subgraphs of the functions in $\mathcal{F}$ has index v. Set $m_{v,w}=\max\{m\in\mathbb{N}:\log m\geq m^{1/(v-1)-1/w}\}$ for $w>v-1$. Then we have*

$$D(\mathcal{F},L^p(P),\varepsilon\|F\|_p)\leq m_{v,w}\vee\left[2^{w/(v-1)}\left(\frac{2^{p+1}}{\varepsilon^p}\right)^w\right], \quad \text{for all } w>v-1 \text{ if } p\geq 1, \tag{3.233}$$

and

$$D(\mathcal{F},L^p(P),\varepsilon\|F\|_p)\leq m_{v,w}\vee\left[2^{w/(v-1)}\left(\frac{2}{\varepsilon^{1/p}}\right)^w\right], \quad \text{for all } w>v-1 \text{ if } p<1. \tag{3.234}$$

Hence, the same bounds apply to $N(\mathcal{F},L^p(P),\varepsilon\|F\|_p)$.

Proof The theorem is proved if $D(\mathcal{F},L^p(P),\varepsilon\|F\|_p)\leq 1$, so we may assume the packing numbers of the class $\mathcal{C}$ of subgraphs to be larger than 1 and, in particular, $v\geq 2$. First, we consider the case $p\geq 1$. Let $f_1,\dots,f_m$ be a maximal collection of functions in $\mathcal{F}$ for which

$$P|f_i-f_j|^p>\varepsilon^p PF^p, \quad i\neq j,$$

so that $m=D(\mathcal{F},L^p(P),\varepsilon\|F\|_p)$. For k, to be specified later, let (s_i,t_i), $1\leq i\leq k$, be independent identically distributed $(S\times\mathbb{R})$ random vectors with law

$$\Pr\{(s,t)\in A\times[a,b]\}=\frac{\int_A\lambda[(-F(s))\vee a,F(s)\wedge b]F^{p-1}(s)dP(s)}{2PF^p},$$

for $A\in\mathcal{S}$ and real numbers $a<b$, where λ is Lebesgue measure. (That is, independently for each i, s_i is chosen according to the law $P_F(A)=P(I_AF^p)/PF^p$, $A\in\mathcal{S}$, and given s_i, t_i is chosen uniformly on $[-F(s_i),F(s_i)]$.) Let C_i denote the subgraph of f_i. The probability that

at least two graphs have the same intersection with the sample $\{(s_i,t_i),\ i \le k\}$ is at most

$$
\begin{aligned}
&\binom{m}{2} \max_{i\neq j} \Pr\{C_i \text{ and } C_j \text{ have the same intersection with the sample}\}\\
&= \binom{m}{2} \max_{i\neq j} \prod_{r=1}^{k} \Pr\{(s_r,t_r) \notin C_i \Delta C_j\}\\
&= \binom{m}{2} \max_{i\neq j} \prod_{r=1}^{k} \Big[1 - \Pr\{(s_r,t_r) \in C_i \Delta C_j\}\Big]\\
&= \binom{m}{2} \max_{i\neq j} \prod_{r=1}^{k} \Big[1 - \Pr\big\{(s_r,t_r) : t_r \text{ is between } f_i(s_r), f_j(s_r)\big\}\Big]\\
&= \binom{m}{2} \max_{i\neq j} \left[1 - \frac{1}{\|F\|_p^p} \int \frac{|f_i - f_j|}{2F} F^p \, dP\right]^k\\
&\le \binom{m}{2} \max_{i\neq j} \left[1 - \frac{1}{\|F\|_p^p} \int \frac{|f_i - f_j|^p}{2^p} \, dP\right]^k\\
&\le \binom{m}{2} \left[1 - \frac{\varepsilon^p}{2^p}\right]^k\\
&\le \binom{m}{2} \exp\Big(-\frac{\varepsilon^p k}{2^p}\Big).
\end{aligned}
$$

Let k be such that this probability is strictly less than 1. Then there exists a set of k elements of S such that graphs $C_i \in \mathcal{C}$, $1 \le i \le m$, intersect different subsets of this set, which implies that $m^{\mathcal{C}}(k) \ge m$. Since $2^p(\log 2)/\varepsilon^p > 1$ for $p \ge 1$ and $0 < \varepsilon \le 1$, the smallest integer k such that $\binom{m}{2} \exp\Big(-\frac{\varepsilon^p k}{2^p}\Big) < 1$ satisfies $1 \le k \le (2^{p+1}/\varepsilon^p)\log m$. Therefore, by Corollary 3.6.5, we have

$$
m \le m^{\mathcal{G}}(k) \le 2k^{v-1} \le 2\left(\frac{2^{p+1}}{\varepsilon^p} \log m\right)^{v-1}.
$$

Then, given $w > v-1$ and setting $\tau = 1/(v-1) - 1/w$, either $m \le m_{v,w}$ or $m \le 2\left((2^p/\varepsilon^p)m^\tau\right)^{v-1}$, proving (3.233).

The proof for $p < 1$ is similar, with the following changes: the functions f_i satisfy $P|f_i - f_j|^p \ge \varepsilon P F^p$, and one uses the following estimate of $\Pr\{(s_r,t_r) \in C_i \Delta C_j\}$ in the preceding string of inequalities:

$$
\begin{aligned}
\Pr\{(s_r,t_r) \in C_i \Delta C_j\} &= P_F\Big(\frac{|f_i(s_r) - f_j(s_r)|}{2F(s_r)}\Big)\\
&\ge \left[P_F\Big(\frac{|f_i(s_r) - f_j(s_r)|}{2F(s_r)}\Big)^p\right]^{1/p}\\
&= \Big(\frac{P|f_i - f_j|^p}{2^p P F^p}\Big)^{1/p} \ge \frac{\varepsilon^{1/p}}{2}. \quad \blacksquare
\end{aligned}
$$

It follows from the preceding theorem that if $\mathcal{F}$ is *VC* subgraph, then, for any probability measure P on $(S,\mathcal{S})$ and $\tau \le \|F\|_{L^2(P)}$,

$$\log 2N(\mathcal{F},L^2(P),\tau) \le \log(2m_{v,w}) \vee \log\left(\frac{2^{3+1/(v-1)+1/w}\|F\|^2_{L^2(P)}}{\tau^2}\right)^w \le 2w\log\left(\frac{A\|F\|_{L^2(P)}}{\tau}\right), \tag{3.235}$$

where $w > v$ and $A = A_{v,w} = \max\left(2^{(3+1/(v-1)+1/w)/2}, (2m_{v,w})^{1/2w}\right)$.

As this observation shows, Theorem 3.6.9 is precisely what makes *VC* subgraph classes of functions useful in empirical process theory and practice. One can combine *VC* subgraph classes of functions by, for example, addition to obtain classes that are no longer *VC* subgraph but whose covering numbers still enjoy uniform bounds of the same type, and they are equally useful. For example, if $\mathcal{G}$ and $\mathcal{G}$ are *VC* subgraph, then so are $\mathcal{F}+\mathcal{G} = \{f+g : f \in \mathcal{F}, g \in \mathcal{G}\}$ and $\mathcal{F}-\mathcal{G}$ (if f_i and g_j are the centres of balls of radius ε covering, respectively, $\mathcal{F}$ and $\mathcal{G}$, then the balls with centres $f_i + g_j$ and radius 2ε cover $\mathcal{F}+\mathcal{G}$). This justifies the following definition:

Definition 3.6.10 A class of measurable functions is of *VC* type with respect to a measurable envelope F of $\mathcal{F}$ if there exist finite constants A, v such that for all probability measures Q on $(S,\mathcal{S})$

$$N(\mathcal{F},L^2(Q),\varepsilon\|F\|_{L^2(Q)}) \le (A/\varepsilon)^v.$$

Next, we consider an example of *VC* subgraph classes of functions that is particularly relevant in density estimation. To motivate it, let us anticipate that a class of functions that naturally arises in the analysis of density estimators based on convolution kernels is

$$\mathcal{K} = \{K((t-\cdot)/h) : t \in \mathbb{R}, h > 0\},$$

where K is a function of bounded variation, and that, in the case of wavelet density estimators, the corresponding class of functions (projection kernels) is

$$\mathcal{F}_\phi = \left\{\sum_{k\in\mathbb{Z}} \phi(2^j y - k)\phi(2^j(\cdot) - k) : y \in \mathbb{R}, j \in \mathbb{N}\cup\{0\}\right\},$$

where ϕ is an α-Hölder continuous function with bounded support for some $\alpha \in (0,1]$ (ϕ is the scaling function of a Daubechies wavelet). Many properties of convolution kernel or of wavelet projection density estimators require good estimates on the size of the empirical process indexed by each of these classes, and these estimates follow as a direct consequence of the fact that these classes of functions are of *VC* type. We now prove that this is indeed the case. It is convenient to recall the following classical definition: given $1 \le p < \infty$, a function $f : \mathbb{R} \mapsto \mathbb{R}$ is of *bounded p-variation* if the quantity

$$v_p(f) := \sup\left\{\sum_{i=1}^n |f(x_i) - f(x_{i-1})|^p : -\infty < x_0 < \cdots < x_n < \infty, n \in \mathbb{N}\right\}$$

is finite. In this case, the total p-variation of f is defined as $v_p(f)$, and the p-variation function of f is defined as $v_{p,f}(x) = v_p(fI_{(-\infty,x]})$, $x \in \mathbb{R}$. The functions of bounded 1-variation are precisely the functions of bounded variation. Note also that if f is α-Hölder continuous with compact support, then f is of bounded $1/\alpha$-variation. These functions are relevant in density estimation because most convolution kernels are of bounded variation, and the scaling functions of Daubechies wavelets, being α-Hölder for some $\alpha \in (0,1]$, are functions of bounded $(1/\alpha)$-variation (see Chapter 4).

Functions of bounded p-variation admit the following decomposition:

Lemma 3.6.11 *Let f be a function of bounded p-variation. Then $f = g \circ h$, where h is nondecreasing and $0 \le h(x) \le v_p(f)$, and g is $1/p$-Hölder continuous on the interval $[0, v_p(f)]$ and $\|g\|_\infty = \|f\|_\infty$.*

Proof Set $h = v_{p,f}$, and let $R_h \subseteq [0, v_p(f)]$ denote its range. By definition, for any $x < y$, $|f(y) - f(x)|^p \le h(y) - h(x)$, showing that f is constant on the level sets $h^{-1}\{u\}$ of h for any $u \in R_h$. For $u \in R_h$, define $g(u)$ as the value of f on any of the points of $h^{-1}\{u\}$. Then, for $u, v \in R_h$ and x, y such that $h(x) = u$ and $h(y) = v$, $|g(u) - g(v)| = |f(x) - f(y)| \le |h(x) - h(y)|^{1/p} = |u - v|^{1/p}$. Thus, g is $1/p$-Hölder continuous on $R_h \subseteq [0, v_p(f)]$, and $\|g\|_\infty = \|f\|_\infty$. Then, by the Kirszbraun-McShane extension theorem (see Exercise 3.6.13), g admits an extension to $[0, v_p(f)]$ with the same modulus of continuity and uniform bound. By construction, $f = g \circ h$. ■

It is easy to see that the set of dilations and translations of a nondecreasing function is a *VC*-type class of functions, and as a consequence of the preceding decomposition, this property is also shared by functions of bounded p-variation. This is the content of the next proposition.

Proposition 3.6.12 *Let f be a function of bounded p-variation, $p \ge 1$. Then the collection $\mathcal{F}$ of translations and dilations of f*

$$\mathcal{F} = \{x \mapsto f(tx - s) : t > 0, s \in \mathbb{R}\}$$

is of VC type; concretely, for all $0 < \varepsilon \le v_p^{1/p}(f)$ and $w > 6$, there exists $A_{w,p} < \infty$ such that

$$N\left(\mathcal{F}, L^2(Q), \varepsilon v_p^{1/p}(f)\right) \le \left(\frac{A_{w,p}}{\varepsilon}\right)^{(p\vee 2)w}, \quad 0 < \varepsilon \le 1.$$

If, moreover, f is right (or left) continuous, then $\mathcal{F}$ is of VC type, and the $L^2(Q)$ ε-covering numbers of $\mathcal{F}$ admit the uniform bound $(A_{w,p}/\varepsilon)^{(p\vee 2)w}$ for any $w > 3$.

Proof Assume that f is right continuous, and set $M^p = v_p(f)$. By Lemma 3.6.11, $f = g \circ h$, where g is $1/p$-Hölder continuous on $[0, M^p]$ and h is nondecreasing, right continuous (see Exercise 3.6.16) and $0 \le h(x) \le M^p$ for all $-\infty < x < \infty$. Then, by Exercise 3.6.15, letting h^{-1-} denote the left-continuous generalised inverse of h, we have that for every $s \in \mathbb{R}$ and $t > 0$,

$$\begin{aligned}\{(x,u) \in \mathbb{R} \times [0, M^p] : u \le h(tx - s)\} &= \{(x,u) \in \mathbb{R} \times [0, M^p] : h^{-1-}(u) \le tx - s\} \\ &= \{(x,u) \in \mathbb{R} \times [0, M^p] : h^{-1-}(u) - tx + s \le 0\} \subseteq \mathcal{G},\end{aligned}$$

where $\mathcal{G}$ is the *negativity set* of the vector space of real functions on $\mathbb{R} \times [0, M^p]$ spanned by the three functions $(x,u) \mapsto h^{-1-}(u)$, $(x,u) \mapsto x$ and 1. In particular, $\mathcal{G}$ is a *VC* class of index 4 by Proposition 3.6.6. Then the class $\mathcal{C}_0 = \{\{(x,u) : 0 \le u \le h(tx - s)\} : t > 0, s \in \mathbb{R}\}$ is also

VC of index at most 4, and consequently, the same is true, as is easy to check, of the class $\mathcal{C}_0 = \{\{(x,u) : -\infty < u \le h(tx-s)\} : t > 0, s \in \mathbb{R}\}$. Therefore, the class of functions

$$\mathcal{M} := \{x \mapsto h(tx-s) : t > 0, s \in \mathbb{R}\}$$

is VC subgraph of index at most 4.

Let $1 \le p \le 2$. Then $2/p \ge 1$, and the first inequality in Theorem 3.6.9 gives that there exists $A_{w,p}$ such that for every Borel probability measure Q on $\mathbb{R}$,

$$N(\mathcal{M}, L^{2/p}(Q), \varepsilon M^p) \le \left(\frac{A_{w,p}}{\varepsilon^{2/p}}\right)^w, \quad 0 < \varepsilon \le 1,$$

for all $w > 3$ (in fact, for $w = 3$, see the notes at the end of the chapter). Now, since g is $1/p$-Hölder continuous, we have that if $m_1, m_2 \in \mathcal{M}$ and $\|m_1 - m_2\|_{L^{2/p}(Q)} \le \tau$, then

$$\left[\int (g(m_1) - g(m_2))^2 dQ\right]^{1/2} \le \left[\int |m_1 - m_2|^{2/p} dQ\right]^{1/2} \le \tau^{1/p},$$

showing that any τ-covering of $\mathcal{M}$ for the $L^{2/p}(Q)$ distance induces a $\tau^{1/p}$ covering of $\mathcal{F}$ in $L^2(Q)$. Combining this observation with the preceding estimate on the covering numbers of $\mathcal{M}$, we obtain

$$N\left(\mathcal{F}, L^2(Q), \varepsilon M\right) \le \left(\frac{A_{w,p}}{\varepsilon^2}\right)^w.$$

The case $p \ge 2$ follows in an analogous way, the only difference being that now one uses the second inequality in Theorem 3.6.9, valid for $L^{2/p}(Q)$-covering numbers with $2/p < 1$. This proves the proposition for f right continuous and, by analogy, for f left continuous (just use right-continuous inverses).

Without continuity assumptions, using part (b) of Exercise 3.6.15, one obtains

$$\begin{aligned}\{(x,u) \in \mathbb{R} \times [0, M^p] : u \le h(tx-s)\} &= \{(x,u) \in \mathbb{R} \times [0, M^p] : h^{-1^-}(u) \le tx - s, u \in R_1\} \\ &\times \cup \{(x,u) \in \mathbb{R} \times [0, M^p] : h^{-1^-}(u) < tx - s, u \in R_2\}\end{aligned}$$

for a convenient partition $\{R_1, R_2\}$ of $[0, M^p]$, and the arguments in the first part of the proof then show that $\mathcal{M}$ is VC subgraph, but now the collection of its subgraphs is the union of two VC classes each of index bounded by 4, that is, by the proof of Proposition 3.6.7, for a VC class $\mathcal{V}$ such that $m^{\mathcal{V}}(n) \le cn^6$ for some $c < \infty$ and all $n \in \mathbb{N}$. Then Theorem 3.6.9 gives, for example, in the case $2/p \ge 1$, $N(\mathcal{M}, L^{2/p}(Q), \varepsilon M^p) \le A_{w,p}/\varepsilon^{2w}$, for any $w > 6$. ■

3.6.3 VC Hull and VC Major Classes of Functions

Other types of classes of functions related to the VC property, but with sensibly larger yet still manageable uniform bounds for their $L^2(Q)$-metric entropies, are the VC hull and the VC major classes. The result developed in this section shows that if a class of functions admits a bound on its $L^2(Q)$ covering numbers of the order ε^w uniformly in Q, then the covering numbers of its convex hull admit a uniform bound of the order $e^{\varepsilon^{-2w/(2+w)}}$.

Definition 3.6.13 Given a class of functions $\mathcal{F}$, $\mathrm{co}(\mathcal{F})$ is defined as the convex hull of $\mathcal{F}$, that is,

$$\mathrm{co}(\mathcal{F}) = \left\{ \sum_{f\in\mathcal{F}} \lambda_f f : f \in \mathcal{F}, \sum_f \lambda_f = 1, \lambda_f \geq 0, \lambda_f \neq 0 \text{ only for finitely many } f \right\},$$

and $\overline{\mathrm{co}}(\mathcal{F})$ is defined as the pointwise sequential closure of $\mathrm{co}(\mathcal{F})$, that is, $f \in \overline{\mathrm{co}}(\mathcal{F})$ if there exist $f_n \in \mathrm{co}(\mathcal{F})$ such that $f_n(x) \to f(x)$ for all $x \in S$ as $n \to \infty$. If the class $\mathcal{F}$ is *VC* subgraph, then we say that $\overline{\mathrm{co}}(\mathcal{F})$ is a *VC* hull class of functions.

Example 3.6.14 Let $\mathcal{F}$ be the class of monotone nondecreasing functions $f : \mathbb{R} \mapsto [0,1]$. Then $\mathcal{F} \subseteq \overline{\mathrm{co}}(\mathcal{G})$, where $\mathcal{G} = \{I_{(x,\infty)}, I_{[x,\infty)} : x \in \mathbb{R}\}$. To see this, just note that if

$$f_n = \frac{1}{n}\sum_{i=1}^{n-1} I_{\{f>i/n\}} = \sum_{j=0}^{n-1} \frac{j}{n} I_{\{j/n<f\leq(j+1)/n\}},$$

then $\sup_{x\in\mathbb{R}} |f_n(x) - f(x)| \leq 1/n$ and that the sets $\{f > i/n\}$ are half-lines, so $I_{\{f>i/n\}} \in \mathcal{G}$.

The *VC* major classes constitute a generalisation of this example: $\mathcal{F}$ is a *VC major class* if the collection of sets $\{f \geq t\}$ for $t \in \mathbb{R}$ and $f \in \mathcal{F}$ is a *VC* class of sets. Proceeding as in the preceding example, we see that if $\mathcal{F}$ is *VC* major and the functions in $\mathcal{F}$ take values in $[0,1]$, then $\mathcal{F}$ is *VC* hull. If $\mathcal{F}$ is just uniformly bounded, then it is a multiple of a *VC* hull class. We will not consider these classes any further.

Now we prove the main result on *VC* hull classes. We begin with a fundamental lemma that estimates the covering numbers of convex hulls of finite classes in terms of their cardinality and their diameter.

Lemma 3.6.15 *Let Q be a probability measure on $(S,\mathcal{S})$, and let $\mathcal{F} = \{f_1,\dots,f_n\}$ be a collection of n functions in $\mathcal{L}^2(Q)$. Then, for all $\varepsilon > 0$,*

$$N(\mathrm{co}(\mathcal{F}), L^2(Q), \varepsilon(\mathrm{diam}\ \mathcal{F})) \leq (e + en\varepsilon^2)^{2/\varepsilon^2}.$$

Proof Given $f \in \mathrm{co}(\mathcal{F})$, let $\lambda_1,\dots,\lambda_n$ be nonnegative numbers adding up to one such that $f = \sum_{j=1}^n \lambda_j f_j$, and let λ be the discrete probability measure on $\mathcal{F}$ that assigns mass λ_j to $f_j, j = 1,\dots,n$. Let $Y_1,\dots,Y_k$ be independent $\mathcal{F}$-valued random variables with common law λ, that is, such that $\Pr\{Y_i = f_j\} = \lambda_j$, $j = 1,\dots,n$. Then, letting E denote expectation with respect to Pr, we have $f = EY_1$, and moreover, if we set $\bar{Y} = (1/k)\sum_{i=1}^k Y_i$, we obtain, using Fubini's theorem and independence,

$$\begin{aligned} E\|\bar{Y} - f\|_{L^2(Q)}^2 &= \int E\left(\frac{1}{k}\sum_{i=1}^k (Y_i - EY_i)\right)^2 dQ = \frac{1}{k^2}\int \sum_{i=1}^k E(Y_i - EY_i)^2 dQ \\ &= \frac{1}{k} E\|Y_1 - EY_1\|_{L^2(Q)}^2 \leq \frac{1}{k}(\mathrm{diam}(\mathcal{F}))^2. \end{aligned}$$

Hence, at least one realization of $\bar{Y}$ must be at $L^2(Q)$-distance not exceeding $(\mathrm{diam}(\mathcal{F}))/\sqrt{k}$ from f. Now, independently of λ and hence of $f \in \mathrm{co}(\mathcal{F})$, every such realisation has the form $\sum_{i=1}^k g_i/k$, $g_i \in \mathcal{F}$. These sums only depend on the number x_j of $g_i = f_j$, for $j = 1,\dots,n$;

that is, the number of different such averages does not exceed the number of nonnegative integer solutions of the equation $x_1 + \cdots + x_n = k$, namely, $\binom{k+n-1}{k}$. Therefore,

$$N\left(\mathrm{co}(\mathcal{F}), L^2(Q), \mathrm{diam}(\mathcal{F})/\sqrt{k}\right) \le \binom{k+n-1}{k}.$$

Estimating $\binom{n+k}{k}$ by means of Stirling's formula (3.230) and using that $(1+k/n)^n < e^k$ for all $n \in \mathbb{N}$, it follows that $\binom{k+n-1}{k} \le e^k(1+n/k)^k$. This proves the lemma for $\varepsilon \le 1$ just by taking k to be the smallest integer such that $k^{-1} \le \varepsilon^2$ in the last inequality. For $\varepsilon > 1$, the covering number is 1, and the lemma holds as well. ■

The next lemma gives the first step of an induction procedure that will yield the result. The proof of this lemma is itself by induction, and Lemma 3.6.15 plays a basic role in it.

Lemma 3.6.16 *Let Q be a probability measure on $(S, \mathcal{S})$, and let $\mathcal{F}$ be a collection of measurable functions with envelope $F \in L^2(Q)$ such that*

$$N(\mathcal{F}, L^2(Q), \varepsilon \|F\|_{L^2(Q)}) \le C\varepsilon^{-w}, \quad 0 < \varepsilon \le 1.$$

Set $u = 1/2 + 1/w$ and $L = C^{1/w}\|F\|_{L^2(Q)}$. For each $n \in \mathbb{N}$, let $\mathcal{F}_n$ be a maximal $Ln^{-1/w}$-separated subset of $\mathcal{F}$ for the $L^2(Q)$-norm. Then there exists $C_1 < \infty$ depending only on C and w such that

$$\log N(\mathrm{co}(\mathcal{F}_n), L^2(Q), C_1 L n^{-u}) \le n, \quad n \in \mathbb{N}. \tag{3.236}$$

Proof The proof is by induction on n. Given n_0 fixed, the entropy in the statement is zero for all $n \le n_0$ as soon as C_1 satisfies $C_1 C^{1/w} n_0^{-u} \ge 2$. n_0 and C_1 satisfying this condition will be specified later. Let now, with these choices, $n > n_0$ and $m = n/d$, for $d > 1$ large enough, also to be conveniently chosen later (d will slightly depend on n, just enough to ensure that m is an integer). For each $f \in \mathcal{F}_n$ we choose one and only one function $\pi_m f \in \mathcal{F}_m$ at $L^2(Q)$-distance at most $Lm^{-1/w}$ from f: $\pi_m f$ exists by the definition of $\mathcal{F}_m$. Then the decomposition $f = \pi_m f + (f - \pi_m f)$ induces a decomposition of any $g = \sum_{f \in \mathcal{F}_n} \lambda_f f \in \mathrm{co}(\mathcal{F}_n)$, where $\sum \lambda_f = 1$ and $\lambda_f > 0$, as

$$g = \sum_{f \in \mathcal{F}_m} \mu_f f + \sum_{f \in \mathcal{F}_n} \lambda_f (f - \pi_m f), \tag{3.237}$$

where $\mu_f \ge 0$ and $\sum \mu_f = 1$.

By definition of L and the hypothesis, the cardinality of $\mathcal{F}_n$ does not exceed n. Hence, the set of functions $\mathcal{G}_n = \{f - \pi_m f : f \in \mathcal{F}_n\}$ has cardinality at most n. Moreover, since $\|f - \pi_m f\|_{L^2(Q)} \le Lm^{-1/w}$, the diameter of $\mathcal{G}_n$ is dominated by $2Lm^{-1/w}$. Then, if we apply Lemma 3.6.15 to $\mathcal{G}_n$ with ε such that $2m^{-1/w}\varepsilon = (1/2)C_1 n^{-u}$, it follows that we can cover $\mathrm{co}(\mathcal{G}_n)$ by a collection of balls of radius at most $2^{-1}C_1 L n^{-u}$ whose cardinality does not exceed $\left(e + eC_1^2/16d^{2/w}\right)^{32d^{2/w}C_1^{-2}n}$. Let $\mathcal{K}_1$ be the collection of these centres.

The induction hypothesis on $\mathcal{F}_m$ implies that there exists a covering of $\mathrm{co}(\mathcal{F}_m)$ consisting of (at most) e^m balls of radius at most $C_1 L m^{-u}$. Since $\mathcal{F}_m$ has m elements, its linear span is a subspace H_m of $L^2(Q)$ of dimension at most m; hence, each of the e^m balls of the covering of $\mathrm{co}(\mathcal{F}_m)$ is in fact a ball of radius at most $C_1 L m^{-u}$ in an m-dimensional Hilbert space. By, for example, Exercise 3.6.5, each such ball admits a covering by balls of radius $C_1 L n^{-u}/2$ of cardinality at most $((3C_1Lm^{-u})/(C_1Ln^{-u}/2))^m = (6d^u)^{n/d}$. Let $\mathcal{K}_2$ be the union of the centres of the balls of these 2^m coverings.

Then, by (3.237), the collection of balls with centres at the functions in the set $\mathcal{K} := \{f+g : f \in \mathcal{K}_1, g \in \mathcal{K}_2\}$ and radius C_1Ln^{-u} cover $\mathrm{co}(\mathcal{F}_n)$, and by the considerations in the preceding two paragraphs, the cardinality of this cover is at most

$$e^{n/d}(6d^u)^{n/d}\left(e+\frac{eC_1^2}{16d^{2/w}}\right)^{32d^{2/w}C_1^{-2}n} = \exp\left[n\left(\frac{1+u\log(6d)}{d}+\frac{32d^{2/w}}{C_1^2}\left(1+\log\frac{C_1^2}{16d^{2/w}}\right)\right)\right].$$

Now we take d_0 large enough so that $(1+u\log(12d_0))/d_0 \le 1/2$, $n_0 \ge 2d_0$, and C_1 large enough so that both $C_1C^{1/w}n_0^{-u} \ge 2$ and $(64d_0^{2/w}/C_1^2)\left(1+\log(C_1^2/16d_0^{2/w})\right) \le 1/2$ (for this last choice, note that $x^{-1}\log(1+x) \to 0$ as $x \to \infty$, decreasing for $x \ge 2$). With these choices, for each $n \ge n_0$, we take $d \in [d_0, 2d_0]$ such that $n/d \in \mathbb{N}$ (which is possible because $n/d_0 - n/2d_0 \ge n_0/d_0 - n_0/2d_0 \ge 1$). For these $d = d_n$, C_1 and $n \ge n_0$, the cardinality of the cover of $\mathrm{co}(\mathcal{F}_n)$ just constructed is at most e^n, which completes the induction argument (note that the choice of C_1 and n_0 ensure the validity of (3.236) for $n \le n_0$). ■

We are now ready to prove the following.

Theorem 3.6.17 *Let Q be a probability measure on $(S,\mathcal{S})$, and let $\mathcal{F}$ be a collection of measurable functions with envelope $F \in L^2(Q)$ such that*

$$N(\mathcal{F}, L^2(Q), \varepsilon\|F\|_{L^2(Q)}) \le C\varepsilon^{-w}, \quad 0 < \varepsilon \le 1. \tag{3.238}$$

Then there exists a constant K depending only on C and w such that

$$\log N(\overline{\mathrm{co}}(\mathcal{F}), L^2(Q), \varepsilon\|F\|_{L^2(Q)}) \le K\varepsilon^{-2w/(w+2)}, \quad 0 < \varepsilon \le 1. \tag{3.239}$$

Proof Suppose that the theorem holds for finite collections of functions satisfying (3.238), and let $\mathcal{F}$ satisfy (3.238). Then, given $0 < \varepsilon < 1$, there exists a $\varepsilon\|F\|_{L^2(Q)}$-dense subset $\mathcal{G}$ of $\mathcal{F}$ that is finite. $\mathcal{G}$ obviously satisfies (3.238) and therefore also (3.239). But by convexity of the $L^2(Q)$-norm, any covering of $\mathrm{co}(\mathcal{G})$ by balls of radius $\varepsilon\|F\|_{L^2(Q)}$ (or less) induces a covering of $\mathrm{co}(\mathcal{F})$ by balls of radius $2\varepsilon\|F\|_{L^2(Q)}$ (and the same centres) so that, for this ε,

$$\log N(\overline{\mathrm{co}}(\mathcal{F}), L^2(Q), \varepsilon\|F\|_{L^2(Q)}) \le N(\mathrm{co}(\mathcal{G}), L^2(Q), \varepsilon\|F\|_{L^2(Q)}) \le K2^{w/(w+2)}\varepsilon^{-2w/(w+2)}.$$

Thus, we may assume that $\mathcal{F}$ is finite.

Set $u = (w+2)/2w = 1/2 + 1/w$ and $L = C^{1/w}\|F\|_{L^2(Q)}$. By assumption for $(C/n)^{1/w} < 1$ and trivially for $(C/n)^{1/w} \ge 1$, $\mathcal{F}$ can be covered by n or fewer balls of radius at most $Ln^{-1/w}$, and we let $\mathcal{F}_n$ denote the collection of the centres of such a covering, $n \in \mathbb{N}$. In particular, for each n, $\mathcal{F}_n$ consists of at most n functions. The theorem will be proved if we show that there exist constants C_k, D_k such that $\sup_{k\in\mathbb{N}} \max(C_k, D_k) < \infty$, and $q > 1$, satisfying

$$\log N(\mathrm{co}(\mathcal{F}_{nk^q}), L^2(Q), C_kLn^{-u}) \le D_kn, \quad n,k \ge 1. \tag{3.240}$$

(Note that given n, there exists $k < \infty$ such that $\mathcal{F}_{nk^q} = \mathcal{F}$.)

Lemma 3.6.16 proves (3.240) for $k=1$ and all n with $C_1 < \infty$ and $D_1 = 1$. To proceed by induction, we assume that (3.240) holds for $k-1$ and all n and for $q \ge 3w$. Proceeding as in the proof of Lemma 3.6.16, we have

$$\mathrm{co}(\mathcal{F}_{nk^q}) \subseteq \mathrm{co}(\mathcal{F}_{n(k-1)^q}) + \mathrm{co}(\mathcal{G}_{n,k}),$$

where $\mathcal{G}_{n,k}$ is a collection of at most nk^q functions of $L^2(Q)$-norm at most $L(n(k-1)^q)^{-1/w}$. Thus, Lemma 3.6.15 applied to $\mathcal{G}_{n,k}$ for $\varepsilon = Lk^{-2}n^{-u}/(2L(n(k-1)^q)^{-1/w})$ shows the existence a covering of $\mathcal{G}_{n,k}$ of cardinality at most $(e + ek^{2q/w+q-4}/4)^{2^{3+2q/w}k^{4-2q/w}n}$ by balls of radius not larger than $Lk^{-2}n^{-u}$. The induction hypothesis applied to $\mathcal{F}_{n(k-1)^q}$ yields the existence of a covering of $\mathrm{co}(\mathcal{F}_{n(k-1)^q})$ of cardinality at most $C_{k-1}Ln^{-u}$ by balls of radius no larger than $e^{D_{k-1}n}$. Then the sums $f+g$ of centres f from the balls covering $\mathcal{G}_{n,k}$ and centres g from the balls covering $\mathrm{co}(\mathcal{F}_{n(k-1)^q})$ are the centres of at most

$$\exp\left[n\left(D_{k-1} + \frac{2^{3+2q/w}\log(1+k^{2q/w+q-4}/4}{k^{2q/w-4}}\right)\right]$$

balls of radius at most $C_{k-1}Ln^{-u} + Lk^{-2}n^{-u}$ covering $\mathrm{co}\mathcal{F}_{nk^q}$. Thus, inequality (3.240) is proved for k and for all n with C_k and D_k given by

$$C_k = C_{k-1} + \frac{1}{k^2} \quad \text{and} \quad D_k = D_{k-1} + \frac{2^{3+2q/w}\log(1+k^{2q/w+q-4}/4}{k^{2q/w-4}},$$

which satisfy $\sup_{k\in\mathbb{N}} \max(C_k, D_k) < \infty$ given that $q \geq 3w$. ■

Exercises

3.6.1 Prove that: $v(\mathcal{C}) = 0$ if and only if $\mathcal{C} = \emptyset$. $v(\mathcal{C}) = 1$ if and only if $\mathcal{C}$ consists of only one set. *Hint*: $\mathcal{C}$ shatters the empty set if and only if $\mathcal{C}$ contains at least one set. If $\mathcal{C}$ contains two different sets A and B, then it shatters any set $\{x\}$ for $x \in A\Delta B \neq \emptyset$.

3.6.2 Prove that: If S is an infinite set and $\mathcal{C}$ is the collection of all subsets of S of cardinality k, then all sets of cardinality not exceeding k are shattered by $\mathcal{C}$, but no set of cardinality $k+1$ is shattered. This implies that the bound for $m^{\mathcal{C}}(n)$ in Theorem 3.6.2 is attained.

3.6.3 Prove that: If $\mathcal{C}$ is ordered by inclusion, then $v(\mathcal{C}) = 2$: $\mathcal{C}$ cannot shatter any set consisting of two distinct points. The same is true if $\mathcal{C}$ consists of disjoint sets.

3.6.4 Let S and T be sets, and let $F : S \mapsto T$ be a function. Let $\mathcal{C}$ be a collection of subsets of T, and let $F^{-1}(\mathcal{C}) = \{F^{-1}(C) : C \in \mathcal{C}\}$. Prove that $v(F^{-1}(\mathcal{C})) \leq v(\mathcal{C})$.

3.6.5 Let $B(a,r) = \{x \in \mathbb{R}^d : |x-a| \leq r\}$ be a ball of radius R and centre a in $\mathbb{R}^d$. Prove the following bound for the packing number of $B(a,r)$ with respect to Euclidean distance $d(x,y) = |x-y|$, that is,

$$D(B(a,r), d, \varepsilon) \leq \left(\frac{3r}{\varepsilon}\right)^d, \quad 0 < \varepsilon \leq r.$$

Hint: See Proposition 4.3.34.

3.6.6 Let $\mathcal{F}$ be a class of (measurable) functions ordered by the relation $f \leq g$ iff $f(x) \leq g(x)$, for all $x \in S$, and assume that $0 \leq f \leq 1$ for all $f \in \mathcal{F}$. Show that $N(\mathcal{F}, L^2(Q), \varepsilon) \leq 2\varepsilon^{-2}$ for all $0 < \varepsilon \leq 1$ and probability measures Q on $(S, \mathcal{S})$. *Hint*: Each set in the partition of $\mathcal{F}$ $A_k = \{(k-1)\varepsilon^2 \leq f < k\varepsilon^2 : f \in \mathcal{F}\}$, $1 \leq k \leq \varepsilon^{-2}+1$, is contained in a $L^2(Q)$-ball of radius at most ε (for $g \leq f$ both in A_k, $Q(f-g)^2 \leq Q(f-g) \leq \varepsilon^2$). *Note*: Theorem 3.6.9 falls short of implying this simple result (it gives a bound for the covering numbers of the order of ε^{-w} for any $w > 2$; see the notes at the end of the chapter).

3.6.7 Use Exercise 3.6.6, Example 3.6.14 and Theorem 3.6.17 to show that if $\mathcal{F}$ is the class of monotonically nondecreasing functions $f : \mathbb{R} \mapsto [0,1]$, then there exist $K < \infty$ such that

$$\log N(\mathcal{F}, L^2(Q), \varepsilon) \leq K/\varepsilon, \quad 0 < \varepsilon < 1.$$

Show that the same is true for the class of monotonically nondecreasing functions and for the class of functions of bounded variation on $\mathbb{R}$ taking values on taking values on $[0,1]$ and of variation bounded by one. [See also Corollary 3.7.50]. *Hint*: Recall that functions of bounded variation are differences of monotone functions.

3.6.8 If $\mathcal{F}$ is countable and $PF < \infty$, where $F = \sup_{f\in\mathcal{F}}|f|$, then $\|P_n - P\|_{\mathcal{F}}$ converges almost surely and in L^1. *Hint*: Assume that the variables X_i that make the empirical process are the coordinate functions on an infinite product probability space $(S^{\mathbb{N}}, \mathcal{S}^{\mathbb{N}}, P^{\mathbb{N}})$. Let $\mathcal{S}_n$ be the smallest σ-algebra that contains the sets of $P^{\mathbb{N}}$-measure zero and the sets in

$$\{A \in \mathcal{S}^{\mathbb{N}} : I_A(x) = I_A(\sigma_n x) \text{ for any permutation } \sigma_n \text{ of the first } n \text{ coordinates}\}.$$

Then, if $A \in \mathcal{S}_n$ and $i,j \le n$, $\int_A f(X_i)d\Pr = \int_A f(X_j)d\Pr = \int_A P_n(f)d\Pr$, and therefore, since, moreover, $\|\cdot\|_{\mathcal{F}}$ is convex,

$$E(\|P_{n-1} - P\|_{\mathcal{F}}|\mathcal{S}_n) \ge \|E((P_{n-1} - P)(f)|\mathcal{S}_n)\|_{\mathcal{F}} = \|P_n - P\|_{\mathcal{F}}.$$

This fact requires no measurability other than that the X_i being the coordinates of a product probability space (see Proposition 3.7.8). Thus, $\|P_n - P\|_{\mathcal{F}}$ is a reverse submartingale if $EF < \infty$. Now apply reverse submartingale convergence.

3.6.9 (Vapnik-Červonenkis Glivenko-Cantelli theorem.) Use the preceding exercise, the bound for the packing numbers of *VC* subgraph classes in this section and Theorem 3.5.6 to show that if $\mathcal{F}$ is a countable *VC* subgraph class of functions such that if $PF < \infty$, then $\|P_n - P\|_{\mathcal{F}} \to 0$ a.s. Deduce that the same is true if $\mathcal{F}$ is *VC* hull or *VC* major (assuming that $PF < \infty$).

3.6.10 Produce versions of Theorem 3.6.17 in the following cases: a) $\overline{\text{co}}\mathcal{F}$, where $\mathcal{F} = \{\sum_{i=1}^k g_i : g_i \in \mathcal{F}_i\}$, where $\mathcal{F}_i$ are *VC* subgraph classes and $k < \infty$; (b) $\mathcal{F}$ just as in (a), but the convex hull of $\mathcal{F}$ is replaced by its *symmetric convex hull* $\overline{\text{sco}}\mathcal{F}$ – same definition except that $\sum_f |\lambda_f| \le 1$ – and (c) for $M\overline{\text{co}}\mathcal{F}$ or $M\overline{\text{sco}}\mathcal{F}$ for any M finite.

3.6.11 Prove that: Any finite dimensional space of functions is *VC* subgraph.

3.6.12 Show that if $\mathcal{F}$ is *VC* subgraph, then so are $\mathcal{F}+g$, $\mathcal{F}\cdot g$, $\mathcal{F}\circ g$, for any function g, and $g\circ\mathcal{F}$ if g is monotone.

3.6.13 Let $K : \mathbb{R}^d \mapsto \mathbb{R}$ be a finite linear combination of measurable functions k whose subgraphs $\{(s,u) : k(s) \ge u\}$ can be represented as a finite number of Boolean operations of sets of the form $\{(s,u) : p(s,u) \ge \phi(u)\}$, where p is a polynomial and ϕ is an arbitrary measurable function. Prove that the collection of functions

$$\left\{K\left(\frac{t-\cdot}{h}\right) : t \in \mathbb{R}^d, h > 0\right\}$$

is *VC* type. Examples: $K(x) = L(\|x\|)$, where L is of bounded variation, and $K = I_{[-1,1]^d}$.

3.6.14 (Kirszbraun-McShane extension theorem.) Let (T,d) be a metric space and $S \subset T$. Let $f : S \mapsto \mathbb{R}$ be bounded by M in absolute value and admit a modulus of continuity φ, that is, $|f(s)| \le M$ for all $s \in S$ and $|f(t) - f(s)| \le \varphi(d(s,t))$ for all $s,t \in S$, where $\varphi : [0,\infty) \mapsto (0,\infty)$ satisfies $\varphi(0) = 0$ and $0 \le \varphi(x) \le \varphi(x+y) \le \varphi(x) + \varphi(y)$ for $x,y \ge 0$. Prove that there exists an extension g of f defined on all of T that admits on T the bound M (for its absolute value) and the modulus of continuity φ. *Hint*: Prove that if $h(t) := \inf_{s\in S}[f(s) + \varphi(d(s,t))]$, $t \in T$, then the function $g(t) = \max[\min(h(t),M), -M]$ satisfies the prescribed properties. (See, e.g., Dudley (2002).)

3.6.15 (Quantile functions.) Let f be a monotone nondecreasing function on $\mathbb{R}$, with $-\infty < a = f(-\infty+) < f(+\infty-) = b$, and let f^{-1-} denote its left-continuous generalised inverse $f^{-1-}(t) = \inf\{x : f(x) \ge t\}$. (a) Prove that if f is right continuous, then, for $x \in \mathbb{R}$ and $t \in (a,b)$, $f(x) \ge t$ if and only if $x \ge f^{-1-}(t)$. (b) Prove that (a) does not hold without the continuity assumption. (c) Show that for any given $t \in (a,b)$, either $\{x : f(x) \ge t\} = \{x : x \ge f^{-1-}(t)\}$ or

$\{x : f(x) \geq t\} = \{x : x > f^{-1-}(t)\}$. (When f is a cumulative distribution function, the function f^{-1-} is the quantile function.)

3.6.16 If f of bounded p-variation is right continuous, then so is its p-variation function $v_{f,p}$. *Hint*: Check, if needed, lemma 3.28 in Folland (1999, p. 104).

3.6.17 Let $BL_1([a,b])$ be the collection of functions $f : [a,b] \mapsto [-1,1]$ with Lipschitz constant 1, that is, such that $|f(x) - f(y)| \leq |x - y|$, $x, y \in [a,b]$. Show that there exists $c < \infty$ such that $N(BL_1([a,b]), \|\cdot\|_\infty, \varepsilon) \leq e^{c|b-a|/\varepsilon}$, $0 < \varepsilon < 1$. *Hint*: This is done in Corollary 4.3.38, but here is a hint for a different proof. Assume for simplicity that $\varepsilon = 1/k$ and $b = a + \ell$ for some $k, \ell \in \mathbb{N}$, and make a grid on the rectangle $[a,b] \times [-1,1]$ with the lines $x = a + i\varepsilon$, $i = 1, \ldots, \ell/\varepsilon$, and $y = j\varepsilon$, $-k/\varepsilon \leq j \leq k/\varepsilon$. Show that the collection of continuous functions starting at $(a, j\varepsilon)$ with constant slope 1 or -1 for $a + i\varepsilon < x < a + (i+1)\varepsilon$, $i = 0, \ldots, \ell - 1$, is ε-dense in supremum norm on $BL_1([a,b])$.

3.7 Limit Theorems for Empirical Processes

Whereas the first sections of this chapter dealt with finite sample inequalities, here we consider the asymptotic properties of empirical processes, precisely the law of large numbers and the central limit theorem. These two subjects (as well as the law of the iterated logarithm) have a long history: let us just mention the Glivenko-Cantelli theorem regarding the law of large numbers (Glivenko (1933), Cantelli (1933)) and the Kolmogorov (1933a), Doob (1949), Donsker (1952) and Dudley (1966) theorems on the central limit theorem, both for the empirical distribution function. These theorems, respectively, state that if F is the cumulative distribution function of a probability measure P on the line and F_n is the cumulative empirical distribution function corresponding to an independent sample from P, then $\|F_n - F\|_\infty \to 0$ a.s., and the processes $\sqrt{n}(F_n(t) - F(t))$, $t \in \mathbb{R}$, converge *in law in* $\ell_\infty(\mathbb{R})$ to a centred Gaussian process G_P with the same covariance. The notion of convergence in law took some time to reach its final form (see later), and this convergence implies, then, by the continuous mapping theorem, that the sequence of random variables $\sqrt{n}\|F_n - F\|_\infty$ converges in distribution to $\|G_P\|_\infty$. The same is true for any other continuous functional on $\ell_\infty(\mathbb{R})$, and this makes this notion of convergence in law very powerful. Here, letting X_i to be independent identically distributed S-valued random variables with law P, $F_n(t) = P_n(-\infty, t] = \sum_{i=1}^n I_{X_i \leq t}/n$, $t \in \mathbb{R}$, is replaced by $P_n(f) = \sum_{i=1}^n f(X_i)/n$, $f \in \mathcal{F}$, where $\mathcal{F}$ is an infinite collections of measurable functions on $(S, \mathcal{S})$, a general measurable space. These analogues of the Givenko-Cantelli law of large numbers and the Kolmogorov-Doob-Donsker-Dudley central limit theorem were the first results obtained within the modern general framework of empirical processes indexed by general classes of functions, and they constitute an invaluable tool in asymptotic statistics.

The first section deals with some unavoidable measurability questions, and we have tried hard to be brief on this subject. We continue with a section devoted to the law of large numbers. Then we set up the framework for the central limit theorem (CLT) for the empirical process indexed by a class $\mathcal{F}$ of functions by carefully defining convergence in law of processes with bounded paths, that is, random elements defined on the space $\ell_\infty(\mathcal{F})$ of all bounded functions $H : \mathcal{F} \mapsto \mathbb{R}$, equipped with the supremum norm, measurable only with respect to the σ-algebra generated by the cylinder sets. $\ell_\infty(\mathcal{F})$ is a nonseparable metric space (unless $\mathcal{F}$ is finite), and in order to recover the usual and crucial uniform tightness property associated to convergence in law, the definition asks for the limiting process to have a tight Borel probability law in this space. Besides uniform tightness (asymptotic equicontinuity), a generalisation to this framework of the Skorokhod representation is also discussed. The next three subsections deal with the central limit theorem for empirical processes: permanence properties and extension by convexity; the two main general criteria, namely, *VC* type classes and random entropies, and bracketing; classes of functions that satisfy the central limit theorem uniformly in the law P of the data; and an introduction to a general approach obtained by relating the CLT property of the empirical process indexed by $\mathcal{F}$ to the existence of the limiting Gaussian process G_P.

3.7.1 Some Measurability

We have been able to avoid measurability considerations in preceding sections by restricting attention to countable classes of sets and functions, although some results, for example, from Section 3.5, do extend to uncountable classes. It turns out, however, that the definition of convergence in law in the nonseparable space $\ell_\infty(\mathcal{F})$, needed for the central limit theorem uniform in $f \in \mathcal{F}$, does require the notion of outer expectation as soon as $\mathcal{F}$ is infinite, whether countable or not. In this subsection we collect the few facts about the calculus of nonmeasurable functions that are needed in the rest of this section and in some subsequent ones.

Let (Ω, Σ, P) be a probability space, and let $A \subset \Omega$ be a not necessarily measurable set. The *outer probability* $P^*(A)$ of $A \subseteq \Omega$ is defined as

$$P^*(A) = \inf\{P(C) : A \subseteq C, C \in \Sigma\}, \tag{3.241}$$

which coincides with $P(A)$ if A is measurable. Likewise, with the notation $Eg := \int g dP$ for g measurable, if $f : \Omega \mapsto [-\infty, \infty]$ is not measurable, we may also define its *outer expectation* or integral as

$$\int^* f dP = E^* f = \inf\{Eg : g \geq f,\ g \text{ measurable, } [-\infty, \infty]\text{–valued}\}, \tag{3.242}$$

except that $E^* f$ is undefined if there exist a measurable function $g \geq f$ such that $Eg^+ = Eg^- = \infty$ and no measurable function $g \geq f$ such that $Eg = -\infty$. Here we say that Eg exists if at most one of Eg^+ and Eg^- is infinite, and then Eg is defined as their difference, that is, $Eg = Eg^+ - Eg^-$ (recall that $g+ = \max(g, 0)$ and $g^- = (-g)^+$). Set

$$\mathcal{C}_A = \{C : A \subseteq C, C \in \Sigma\},\ \mathcal{G}_f = \{g \geq f :\ g \text{ measurable and } [-\infty, \infty]\text{–valued}\},$$

and note that $\Omega \in \mathcal{C}_A$ and $\infty \in \mathcal{G}_f$, so outer probabilities always exist and outer expectations exist or are undefined. The following proposition shows that the infimum in (3.241) and (3.242) are, respectively, attained at a P a.s. unique set in $\mathcal{C}_A$ and a P a.s. unique function in $\mathcal{G}_f$:

Proposition 3.7.1 *(a) For every set $A \subset \Omega$, the infimum in the definition (3.241) of $P^*(A)$ is attained at a measurable set $A^* \in \mathcal{C}_A$ which is P a.s. uniquely determined. In particular, $P^*(A) = P(A^*)$. (b) For every function $f : \Omega \mapsto \overline{R}$, there exists a P a.s. unique function $f^* \in \mathcal{G}_f$ such that $f^* \leq g$ P a.s. for every $g \in \mathcal{G}_f$. Then, if either of $E^* f$ or $E f^*$ is defined,* *both are equal, as is the case, for example, if f is bounded above or below. (c) For any set $A \subset \Omega$, $(I_A)^* = I_{A^*}$ a.s. and hence $P^*(A) = E^*(I_A)$.*

Proof (a) Note that $\mathcal{C}_A$ is closed by intersections. Hence, there exists a decreasing sequence C_n of sets in $\mathcal{C}_A$ such that $P(C_n) \le P^*(A) + 1/n$. Thus, setting $A^* = \cap_n C_n \in \mathcal{C}_A$, we have $P(A^*) = P^*(A)$. If this infimum is attained at another set $C \in \mathcal{C}_A$, then $A^* \cap C \in \mathcal{C}_A$ and $P(A^* \cap C) = P(A^*) = P(C)$, which implies that $P(A^* \Delta C) = 0$.

(b) The collection of functions $\mathcal{G}_f$ is obviously closed by pointwise minima; that is, $g, h \in \mathcal{G}_f$ implies that $g \wedge h \in \mathcal{G}_f$, where $(g \wedge h)(x) := \min(g(x), h(x))$ for all $x \in \Omega$. Then $-\pi/2 \le \alpha := \inf\{E(\tan^{-1} g) : g \in \mathcal{G}_f\}$ is attained: if $h_n \downarrow$, $h_n \in \mathcal{G}_f$ and $E(\tan^{-1} h_n) \le \alpha + 1/n$, it follows that $f^* := \lim h_n \in \mathcal{G}_f$ and, by dominated convergence, $E(\tan^{-1} f^*) = \alpha$. Now, if $g \in \mathcal{G}_f$, since $g \wedge f^* \in \mathcal{G}_f$, we have $\alpha \le E(\tan^{-1}(g \wedge f^*)) \le E(\tan^{-1} f^*) = \alpha$, which implies that $g \wedge f^* = f^*$ a.s., and therefore, $f^* \ge g$ a.s. for all $g \in \mathcal{G}_f$. If either $E(f^*)$ or $E^* f$ exists, then the definition and the fact that $f^* \ge g$ a.s. for all $g \in \mathcal{G}_f$ imply that $E^* f = E f^*$.

(c) If we identify $\mathcal{C}_A$ with the collection of its indicator functions, then $\mathcal{C}_A \subset \mathcal{G}_{I_A}$, so $P^* A \ge E^*(I_A)$. However, if $g \in \mathcal{G}_{I_A}$, then $I_{g \ge 1} \in \mathcal{G}_{I_A}$, $Eg \ge EI_{g \ge 1}$ and $\{g \ge 1\} \in \mathcal{C}_A$, which readily implies that $P^* A = E^*(I_A)$. ■

The set A^* and the function f^* are called the *P-measurable covers*, respectively, of the set A and function f. It will also be convenient to call a function F a *P-measurable envelope of f* if $F \ge f^*$ P a.s. and likewise for sets. Note that if P and Q are mutually absolutely continuous, the P- and Q-measurable covers of f coincide and likewise for measurable envelopes. Here are a few simple but useful facts on measurable covers:

Proposition 3.7.2 *(a) For any two functions $f, g : \Omega \mapsto (-\infty, \infty]$, we have*

$$(f+g)^* \le f^* + g^* \text{ a.s.} \quad \text{and} \quad (f-g)^* \ge f^* - g^*,$$

where the second inequality requires that both sides be defined. (b) For $f : \Omega \mapsto \mathbb{R}$, $t \in \mathbb{R}$ and $\varepsilon > 0$,

$$P^*\{f > t\} = P\{f^* > t\} \quad \text{and} \quad P^*\{f \ge t\} \le P\{f^* \ge t\} \le P^*\{f \ge t - \varepsilon\}.$$

(c) If B is a vector space with a pseudo-norm $\|\cdot\|$, then for any functions $f, g : \Omega \mapsto B$,

$$\|f+g\|^* \le \|f\|^* + \|g\|^* \text{ a.s.} \quad \text{and} \quad \|cf\|^* = |c| \|f\|^* \text{ a.s.}$$

Proof The first inequality in (a) is obvious because $f^* + g^*$, which is measurable, dominates $f + g$. The second inequality in (a) obviously holds at all points x where $f^*(x) - g^*(x) = -\infty$; for other x, $g^*(x)$ is finite, and so is $|g(x)|$, and we can write $f(x) = (f(x) - g(x)) + g(x)$ and hence $f^* I_{g^* < \infty} \le (f-g)^* I_{g^* < \infty} + g^* I_{g^* < \infty}$ a.s. (see Exercise 6.4.1(c)).

For any t, $\{f > t\} \subseteq \{f^* > t\}$ and $\{f \ge t\} \subseteq \{f^* \ge t\}$, so $P\{f > t\}^* \le P\{f^* > t\}$, and the same is true replacing $>$ by $\ge$. If $\{f > t\} \subset C \in \Sigma$ (e.g., $C = \{f > t\}^*$), then $f \le t$ on C^c and hence also $f^* \le t$ a.s. on C^c (otherwise, we could replace f^* by $f^* I_C + (f^* \wedge t) I_{C^c}$ and contradict the definition of f^*); hence, $P\{f^* > t\} \le PC$, implying that $P\{f^* > t\} \le P^*\{f > t\}$, and the first part of (b) and the first inequality in the second part of (b) follow. For the remaining inequality in (b), note that by the first inequality in (b), if $0 \le \tau < \varepsilon$, then

$$\Pr\{f^* > t - \tau\} = P^*\{f > t - \tau\} \le P^*\{f > t - \varepsilon\},$$

which completes the proof of (b) by taking limits as $\tau \to 0$.

The first inequality in (c) follows by the triangle inequality and (a) and the second from $\|cf\| = |c|\|f\|$ and $\mathcal{G}_{|c|\|f\|} = |c|\mathcal{G}_{\|f\|}$. ■

The following one-sided Fubini-Tonelli theorem is an important tool in the calculus of nonmeasurable functions and it will be used often:

Proposition 3.7.3 *Let $(X \times Y, \mathcal{A} \otimes \mathcal{B}, P \times Q)$ be a product probability space. Let $f : X \times Y \mapsto [0,\infty)$, and let f^* be its measurable cover with respect to $P \times Q$. Let E_P^* and E_Q^* denote, respectively, the outer expectations with respect to P and Q. Then*

$$E_P^* E_Q^* f \le E(f^*), \quad E_Q^* E_P^* f \le E(f^*).$$

If, moreover, Q is discrete and $\mathcal{B}$ is the collection of all the subsets of Y, then

$$E_P^* E_Q f \le E(f^*) = E_Q E_P^* f.$$

Proof We may apply the usual Fubini-Tonelli theorem to f^* to the effect that $Ef^* = E_P E_Q f^*$ and that $E_Q f^*(x,\cdot)$ is $\mathcal{A}$-measurable for each x. To estimate this last integral, we just observe that if f_x^* is the measurable cover of $f(x,\cdot)$ with respect to Q for each fixed $x \in X$, then $f_x^*(y) \le f^*(x,y)$ a.s. because this last function is $\mathcal{B}$-measurable for each x. The first inequality, in this proposition will follow from this observation and the fact that, by Proposition 3.7.1, $E_Q^* f_x = E_Q f_x^*$: these two inequalities give that for each x, $E_Q f^*(x,\cdot) \ge E_Q f_x^* = E_Q^* f_x$. For the second inequality, interchange P and Q in the first.

Next, if Q is discrete, then $E_Q^* = E_Q$ because all the functions are Q-measurable, so $E_P^* E_Q f \le E(f^*)$ follows from the first part of the proof. The equality follows from the Fubini-Tonelli theorem because $f_y^*(x) = f^*(x,y)$ a.s., where f_y^* is the measurable cover with respect to P of the function $f(\cdot,y)$. To prove this assertion, just note that $f^*(x,y) = \sum_i f^*(x,y_i) I_{y=y_i}$ a.s. if $Q = \sum_{i \in I} \delta_{y_i}$, where $I \subseteq \mathbb{N}$, and that, as seen earlier, $f_y^*(x) \le f^*(x,y)$ P a.s. for each y and hence $f^*(x,y) \ge \sum_i f_y^*(x) I_{y=y_i} = \sum_i f_{y_i}^*(x) I_{y=y_i} \ge f(x,y)$ a.s., but the middle term is a $\mathcal{A} \otimes \mathcal{B}$ measurable function and equals $f_y^*(x)$; hence, $f_y^*(x) = f^*(x,y)$ a.s. It then follows that $E_P(f^*(\cdot,y)) = E_P(f_y^*) = E_P^* f$ Q a.s. and $Ef^* = E_Q E_P f^* = E_Q E_P^* f$. ■

Example 3.7.4 A first application of the calculus for nonmeasurable functions consists in extending the Lévy and Hofmann-Jørgensen's inequalities in Section 3.1.3 and the symmetrisation and randomisation inequalities in Section 3.1.4, which are proved for sample bounded processes with *countable* index set, to general, not necessarily countable index set at the expense of replacing expectations and probabilities by outer expectations and probabilities. In the case of these two subsections, the proofs of the extended results follow from the proofs that assume measurability with not much more work than just adding stars to E and P (note that, by Lemma 3.7.2, the functional $\|\cdot\|^*$ is convex), being careful to use only the valid directions of the Fubini-Tonelli theorem. To illustrate this point, here is a proof of the randomisation inequalities in Theorem 3.1.21, namely, that if Y_i are independent centred sample bounded processes indexed by a not necessarily countable set T, and if $\{\varepsilon_i\}$ is a Rademacher sequence independent of $\{Y_i\}$ in the strong sense that $\{Y_1,\ldots,Y_{2n},\varepsilon_1,\ldots,\varepsilon_n\}$

are the coordinate functions of a product probability space, then

$$2^{-p}E^*\left\|\sum_{i=1}^n \varepsilon_i Y_i\right\|^p \le E^*\left\|\sum_{i=1}^n Y_i\right\|^p \le 2^p E^*\left\|\sum_{i=1}^n \varepsilon_i(Y_i - c_i)\right\|^p$$

for any functions $c_i = c_i(t)$, where we recall the notation $\|z\| := \sup_{t\in T}|z(t)|$ for any function $z(t)$.

By Theorem 3.7.1, convexity of $(\|\cdot\|^p)^*$ (Lemma 3.7.2 plus the easy-to-prove fact that $(\|\cdot\|^*)^p = (\|\cdot\|^p)^*$) and Proposition 3.7.3, if A and B are disjoint sets of indices, then

$$E^*\left\|\sum_{i\in A} Y_i\right\|^p = E_A^*\left\|\sum_{i\in A} Y_i + E\sum_{i\in B} Y_i\right\|^p \le E_A^* E_B^*\left\|\sum_{i\in A\cup B} Y_i\right\|^p \le E^*\left\|\sum_{i\in A\cup B} Y_i\right\|^p,$$

where E_A denotes integration with respect to Y_i, $i \in A$, and likewise for E_B. Hence,

$$E^*\left\|\sum \varepsilon_i Y_i\right\|^p = E_\varepsilon E_Y^*\left\|\sum_{i:\varepsilon_i=1} Y_i - \sum_{i:\varepsilon_i=-1} Y_i\right\|^p \le 2^p E_\varepsilon E_Y^*\left\|\sum Y_i\right\|^p = 2^p E^*\left\|\sum Y_i\right\|^p,$$

where E_ε and E_X denote integration with respect to the Rademachaer and the Y variables, respectively. In the other direction,

$$\begin{aligned} E^*\left\|\sum_{i=1}^n Y_i\right\|^p &= E^*\left\|\sum_{i=1}^n (Y_i - EY_{n+i})\right\|^p \le E^*\left\|\sum_{i=1}^n (Y_i + c_i) - \sum_{i=1}^n (Y_{n+i} + c_i)\right\|^p \\ &= E_\varepsilon E_Y^*\left\|\sum_{i=1}^n \varepsilon_i(Y_i + c_i - Y_{n+i} - c_i)\right\|^p \le 2^p E^*\left\|\sum_{i=1}^n \varepsilon_i(Y_i + c_i)\right\|^p, \end{aligned}$$

where the last equality follows because $P_1 \times \cdots \times P_n \times P_1 \times \cdots \times P_n$ is invariant by permutations of the coordinates i and $i+n$ for each $i \le n$, and the remaining inequalities follow by previous arguments.

Finally, we introduce a concept that will be useful when extending to the nonmeasurable setting Skorokhod's theorem about almost sure convergent representations of sequences of random variables that converge in distribution. Let $\phi : (\tilde{X}, \tilde{\mathcal{A}}) \mapsto (X, \mathcal{A})$ be measurable, let $\tilde{P}$ be a probability measure on $\tilde{\mathcal{A}}$ and let $\tilde{P}\circ\phi^{-1}$ be the probability law of ϕ. Then, if $f : X \mapsto \mathbb{R}$ is arbitrary, we have $f^*\circ\phi \ge f\circ\phi$, where f^* is the $\tilde{P}\circ\phi^{-1}$-measurable cover of f and hence $f^*\circ\phi$ is $\tilde{P}$-measurable and therefore $f^*\circ\phi \ge (f\circ\phi)^*$ $\tilde{P}$ a.s.

Definition 3.7.5 A measurable map $\phi : \tilde{X} \mapsto X$ is $\tilde{P}$-perfect if $f^*\circ\phi = (f\circ\phi)^*$ $\tilde{P}$ a.s. for every bounded function $f : X \mapsto \mathbb{R}$, where $(f\circ\phi)^*$ is the $\tilde{P}$-measurable cover of $f\circ\phi$ and f^* is the $P\circ\phi^{-1}$-measurable cover of f.

Then, if ϕ is perfect and f is bounded,

$$\begin{aligned} E_{\tilde{P}}^*(f\circ\phi) &= \int (f\circ\phi)^* d\tilde{P} = \int f^*\circ\phi\, d\tilde{P} = \int f^* d(\tilde{P}\circ\phi^{-1}) \\ &= \int^* f\, d(\tilde{P}\circ\phi^{-1}) = E_{\tilde{P}\circ\phi^{-1}}^* f, \end{aligned} \tag{3.243}$$

or, for indicators, $\tilde{P}^*\{\phi\in A\} = (\tilde{P}\circ\phi^{-1})^*(A)$ for any $A \subset X$. It is this property that will make perfectness useful.

Example 3.7.6 Coordinate projections in product probability spaces are perfect. Let $\pi_1 : (X \times Y, \mathcal{A} \otimes \mathcal{B}, P \times Q) \mapsto (X, \mathcal{A}, P)$ be the projection onto X, $\pi_1(x,y) = x$, and let $f : X \mapsto \mathbb{R}$ be a bounded function. It suffices to prove that $(f \circ \pi_1)^* \geq f^* \circ \pi_1$. Let $h : X \times Y \mapsto \mathbb{R}$ be a measurable function such that $h(x,y) \geq f(x)$ $P \times Q$ a.s. Then, by Fubini's theorem, Q a.s. we have $h(x,y) \geq f(x)$ P a.s. But then, since the sections $h(\cdot,y)$ are P-measurable, we also have that Q a.s., $h(x,y) \geq f^*(x)$ P a.s., and applying Fubini's theorem once more, we obtain that $h(x,y) \geq f^*(x)$ $P \times Q$ a.s.

Example 3.7.7 Here is a related example that will be useful later. Let $(\Omega, \mathcal{A}, Q)$ be a measurable space, let $\Omega_k \in \mathcal{A}$ with $Q(\Omega_k) > 0$, $\mathcal{A}_k = \mathcal{A} \cap \Omega_k$ and $Q_k(\cdot) = Q(\cdot \,|\Omega_k)$, $k = 1, \ldots, r$, and let $(\Omega_0, \mathcal{A}_0, Q_0)$ be another probability space. Consider the product probability space $(\Omega_0 \times \cdots \times \Omega_r, \mathcal{A}_0 \otimes \cdots \otimes \mathcal{A}_r, Q_0 \times \cdots \times Q_r)$ and on it the function $\phi = \sum_{i=1}^k (I_{A_{0,i}} \circ \pi_0)\pi_i$, where π_i are the coordinate projections and $A_{0,i} \in \mathcal{A}_0$ are disjoint sets. Then ϕ is perfect. To see this, note that for $f : \Omega \mapsto \mathbb{R}$, $f \circ \phi(\omega_0, \ldots, \omega_r) = \sum_{i=1}^r I_{A_{0,i}}(\omega_0) f(\omega_i)$. Then, since for each fixed $\omega_0, \omega_2, \ldots, \omega_k$, $(f \circ \phi)^*$ is measurable in ω_1, we have $(f \circ \phi)^*(\omega_0, \ldots, \omega_r) \geq I_{A_{0,1}}(\omega_0)(f_{|\Omega_1} \circ \pi_1)^*(\omega_1) + \sum_{i=2}^r I_{A_{0,i}}(\omega_0) f(\omega_i)$, and recursively and by perfectness of projections, with $\omega = (\omega_0, \ldots, \omega_r)$,

$$
\begin{aligned}
(f \circ \phi)^*(\omega) &\geq \sum_{i=1}^r I_{A_{0,i}}(\omega_0)(f_{|\Omega_i} \circ \pi_i)^*(\omega) \\
&\geq \sum_{i=1}^n I_{A_{0,i}}(\omega_0)(f_{|\Omega_i})^* \circ \pi_i(\omega) \\
&= \sum_{i=1}^n I_{A_{0,i}}(\omega_0) f^* \circ \pi_i(\omega) \\
&= f^* \circ \phi(\omega),
\end{aligned}
$$

where the first identity follows from the fact that $Q_{|\mathcal{A}_i}$ and $Q(\cdot \,|\Omega_i)$ are mutually absolutely continuous. The reversed inequality, $(f \circ \phi)^*(\omega) \leq \sum_{i=1}^r I_{A_{0,i}}(\omega_0)(f_{|\Omega_i} \circ \pi_i)^*(\omega)$, is obvious because the second of these two functions is measurable in the product space.

3.7.2 Uniform Laws of Large Numbers (Glivenko-Cantelli Theorems)

Given as usual the coordinates X_i, $i \in \mathbb{N}$, on $(\Omega, \Sigma, \Pr) := (S, \mathcal{S}, P)^{\mathbb{N}}$, the product of countably many copies of $(S, \mathcal{S}, P)$ and a collection of real-valued measurable functions $\mathcal{F}$ on S, we are now interested in obtaining conditions on $\mathcal{F}$ and P ensuring that the law of large numbers holds uniformly in $f \in \mathcal{F}$, that is, so that

$$\lim_{n\to\infty} \|P_n - P\|_{\mathcal{F}}^* = 0 \text{ a.s.},$$

where, as usual, $P_n = \sum_{i=1}^n \delta_{X_i}/n$ is the empirical measure based on X_i, $1 \leq i \leq n$, $n \in \mathbb{N}$. Let F be the P-measurable cover of the function $x \mapsto \sup_{f\in\mathcal{F}} |f(x)|$. With some abuse of notation, we call this function the *measurable cover of* $\mathcal{F}$. Here is a first useful observation:

Proposition 3.7.8 *If $PF < \infty$, then the sequence $\{\|P_n - P\|_{\mathcal{F}}^*\}_{n=1}^\infty$ converges a.s. and in L^1 to a finite limit.*

Proof We just verify that the usual reversed submartingale proof of the law of large numbers extends to this setting. Define for each n the σ-algebra Σ_n as the smallest σ-algebra that contains the sets of $P^{\mathbb{N}}$ measure zero in Σ as well as

$$\{A \in \Sigma : I_A(x) = I_A(\sigma_n x), \text{ for any permutations } \sigma_n \text{ of the first } n \text{ coordinates}\}.$$

We show that the collection $\{(\|P_n - P\|^*, \Sigma_n) : n \in \mathbb{N}\}$ is a reversed submartingale. First, $\|P_n - P\|^* \le P_n F + PF$, so $E\|P_n - P\|^* \le 2PF < \infty$, for all n. Also, since neither $P_n - P$ nor $P^{\mathbb{N}}$ is changed by permutations σ_n of the first n coordinates, it follows that for each n, $\|P_n - P\|^*$ is Σ_n-measurable. Set $P_{n,i} = \sum_{j \ne i, j \le n+1} \delta_{X_j}/n$. Now $P_{n,i} - P$ becomes $P_n - P$ by a permutation of the first $n+1$ coordinates, and any such permutation transforms the infinite product space into itself. Then, if $\|P_{n,n+1} - P\|^* = \|P_n - P\|^* = H(X_1, \ldots, X_n)$ for a measurable function H of n variables (see Exercise 3.7.2), we have that $\|P_{n,i} - P\|^* = H(X_1, \ldots, X_{i-1}, X_{i+1}, \ldots, X_{n+1})$, and the invariance of the σ-algebra Σ_{n+1} with respect to permutations of the first $n+1$ coordinates then gives that the conditional expectations $E(\|P_{n,i} - P\|^* | \Sigma_{n+1})$, $i = 1, \ldots, n+1$, are all a.s. equal and hence equal to $E(\|P_n - P\|^* | \Sigma_{n+1})$. Therefore, since

$$\|P_{n+1} - P\|^* = \frac{1}{n+1} \left\| \sum_{i=1}^{n+1} (P_{n,i} - P) \right\|^* \le \frac{1}{n+1} \sum_{i=1}^{n+1} \|P_{n,i} - P\|^* \text{ a.s.}$$

and $\|P_{n+1} - P\|^*$ is Σ_{n+1}-measurable, it follows that

$$\|P_{n+1} - P\|^* = E(\|P_{n+1} - P\|^* | \Sigma_{n+1}) \le E(\|P_n - P\|^* | \Sigma_{n+1}) \text{ a.s.},$$

proving that $\{\|P_n - P\|^*\}$ is a reversed submaringale with respect to the σ-algebras Σ_n. Now the lemma follows by the convergence theorem for reversed submartingales. ■

The limit in the preceding proposition may not be zero: if, for example, P gives mass zero to all finite sets of $\mathbb{R}$ and $\mathcal{F}$ is the collection of indicators of all finite sets in $\mathbb{R}$, then $\|P_n - P\|_{\mathcal{F}} = \left\|(1/n)\sum_{l=1}^n \delta_{X_i}(\{X_1, \ldots, X_n\})\right\|_{\mathcal{F}} = 1$. However, it has the following useful corollary:

Corollary 3.7.9 *If $PF < \infty$ and $\{\|P_n - P\|_{\mathcal{F}}^*\}$ converges in probability to zero, then it converges a.s. to zero.*

In other words, under integrability of the measurable cover of the class $\mathcal{F}$, the weak law of large numbers uniform in $f \in \mathcal{F}$ implies the uniform strong law. The following definition is given by analogy with the classical Glivenko-Cantelli theorem for the empirical distribution function in $\mathbb{R}$, which is just the law of large numbers for the empirical process over $\mathcal{F} = \{I_{(-\infty,x]} : x \in \mathbb{R}\}$.

Definition 3.7.10 A class of functions $\mathcal{F}$ is a P-Glivenko-Cantelli class if $\|P_n - P\|_{\mathcal{F}}^* \to 0$ a.s., where P_n is the empirical process based on the coordinate projections X_i, $i = 1, \ldots, n$, $n \in \mathbb{N}$, of the product probability space $(S, \mathcal{S}, P)^{\mathbb{N}}$.

The theorem we are about to prove requires that the empirical process indexed by the class $\mathcal{F}$ satisfy a measurability condition. The problem is that although without measurability assumptions we may still compare $\|P_n - P\|_{\mathcal{F}}$ with $\left\|n^{-1}\sum_{i=1}^n \varepsilon_i f(X_i)\right\|_{\mathcal{F}}$ as shown in Example 3.7.4, without measurability of these suprema, Fubini's theorem only works in one direction (see Proposition 3.7.3), and we cannot take full advantage of the fact that, conditionally on the variables X_i, the randomised sum is sub-Gaussian. We state and give a name to the property we need.

Definition 3.7.11 A class of functions $\mathcal{F}$ is P-measurable, or P-empirically measurable, or just measurable, if for each $\{a_i, b\} \subset \mathbb{R}$ and $n \in \mathbb{N}$, the quantity $\left\|\sum_{i=1}^n a_i f(X_i) + bPf\right\|_{\mathcal{F}}$ is measurable for the completion of P^n.

For example, if $\mathcal{F}$ is countable, then it is P-measurable for every P. If $\mathcal{F}_0 \subset \mathcal{F}$ is P-measurable and for each $\{a_i, b\} \subset \mathbb{R}$ and $n \in \mathbb{N}$,

$$\Pr^*\left\{\left\|\sum_{i=1}^n a_i f(X_i) + bPf\right\|_{\mathcal{F}} \neq \left\|\sum_{i=1}^n a_i f(X_i) + bPf\right\|_{\mathcal{F}_0}\right\} = 0, \tag{3.244}$$

then $\mathcal{F}$ is P-measurable; for instance, if for every $f \in \mathcal{F}$ there exist $f_n \in \mathcal{F}_0$ such that $f_n(x) \to f(x)$ for all $x \in S$ and $Pf_n \to Pf$, then $\mathcal{F}$ is P-measurable. If the processes $f \mapsto \sum_{i=1}^n a_i f(X_i) + bPf$, $f \in \mathcal{F}$, are separable for the supremum norm (Definition 2.1.2), then $\mathcal{F}$ is P-measurable.

Example 3.7.12 Examples of measurable classes are $\mathcal{F} = \{I_{u_i \le x_i, 1 \le i \le d}(u) : x \in \mathbb{R}^d\}$ and $\mathcal{F} = \{K((x - \cdot)/h) : x \in \mathbb{R}, h > 0\}$ if K is right (or left) continuous. For the first, $\mathcal{F}_0$ consists of the functions in the class corresponding to $x \in \mathbb{Q}^d$, and for the second, $\mathcal{F}_0$ is the subclass corresponding to x and h rational.

Example 3.7.13 A more complicated example of P-measurable class for every P is the set $BV_{p,M_1,M_2}(\mathbb{R})$ of the bounded functions of bounded p-variation on $\mathbb{R}$ with $p \ge 1$ with supremum norm bounded by M_1 and total p-variation norm bounded by M_2, for some $p \ge 1$ (see immediately before Lemma 3.6.11 for definitions). We may assume without loss of generality that $M_1 = M_2 = 1$. By Lemma 3.6.11, if $f \in BV_{p,1,1}(\mathbb{R})$, then $f = g \circ h$, where h, the p-variation function of f, is nondecreasing and takes values on $[0,1]$, and g is $1/p$-Hölder on $[0, v_p(f)]$ with supremum norm and Hölder constants both bounded by 1. We may extend g to $[0,1]$ by making $g(x) = g(v_p(f))$ on $(v_p(f), 1]$. By the Arzelà-Ascoli theorem, this set of Hölder functions is compact for the supremum norm, and in particular, it has a countable dense subset, say, D_p. Also, as seen in Example 3.6.14, h is uniformly approximated by $h_n(x) = n^{-1}\sum_{i=1}^{n-1} I_{h > i/n}(x)$, where $\{h > i/n\}$ is an open or closed half-line (x, ∞) of $[x, \infty)$. Hence, by right or left continuity of these indicators, the functions h_n are limits of finite linear combinations of indicators of half-lines with rational points, concretely, of functions in the countable set

$$\mathcal{H} = \left\{n^{-1}\sum_{i=1}^{n-1}(I_{(r_i,\infty)} + \tau_i I_{\{r_i\}}) : n \in \mathbb{N}, r_i \in \mathbb{Q}, \tau_i \in \{0,1\}, 1 \le i < n\right\}.$$

Then any function in $BV_{p,1,1}(\mathbb{R})$ is the pointwise limit of a sequence of functions $g_n \circ h_n$, with $g_n \in D_p$ and $h_n \in \mathcal{H}$ as $|g \circ h(x) - g_n \circ h_n(x)| \le \|g - g_n\|_\infty + |h(x) - h_n(x)|^{1/p}$. Thus, in this case, $\mathcal{F}_0 = \{g \circ h : g \in D_p, h \in \mathcal{H}\}$.

In the measurable case, the Glivenko-Cantelli property for $\mathcal{F}$ can be characterised by a condition on the metric entropies of $\mathcal{F}$ with respect to the $L^p(P_n)$ pseudo-metrics, for any $0 < p \le \infty$. These metric entropies are random, so the result does not constitute a complete solution to the problem, but it does simplify it, as we will see in a couple of corollaries later.

Recall from Section 3.5 the definition of the empirical L^p pseudo-distances $e_{n,p}(f,g) = \|f-g\|_{L^p(P_n)}$, that is, in the case $p=\infty$, $e_{n,\infty}(f,g) = \max_{1\le i\le n}|f(X_i)-g(X_i)|$, and for $0<p<\infty$, $e_{n,p}(f,g) = \left[\sum_{i=1}^n |f(X_i)-g(X_i)|^p\right]^{1/(p\vee 1)}$. Recall the notation $N(T,d,\varepsilon)$ and $D(T,d,\varepsilon)$ for the covering numbers and packing numbers of (T,d). The following notation is also convenient: given a class of functions $\mathcal{F}$ and a positive number M, we set

$$\mathcal{F}_M = \{fI_{F\le M} : f\in\mathcal{F}\},$$

where F is the P-measurable cover of $\mathcal{F}$ (determined only P a.s.).

Theorem 3.7.14 *Let $\mathcal{F}$ be class of functions with an everywhere finite measurable cover F and such that $\mathcal{F}_M$ is P-measurable for all $M\le\infty$. Assume also that $\mathcal{F}$ is $L^1(P)$-bounded, that is, $\sup_{f\in\mathcal{F}} P|f|<\infty$. Then the following are equivalent:*

(a) $\mathcal{F}$ is a P-Glivenko-Cantelli class of functions.
(b) $PF<\infty$ and $\|P_n-P\|_{\mathcal{F}}\to 0$ in probability.
(c) $PF<\infty$, and for all $M<\infty$, $\varepsilon>0$ and $p\in(0,\infty]$, $(\log N^*(\mathcal{F}_M,e_{n,p},\varepsilon))/n\to 0$ in probability (in L^r for any $0<r<\infty$).
(d) $PF<\infty$, and for all $M<\infty$ and $\varepsilon>0$ and for some $p\in(0,\infty]$, $(\log N^*(\mathcal{F}_M,e_{n,p},\varepsilon))/n\to 0$ in probability (in L^r for any $0<r<\infty$).
(e) $PF<\infty$, and for all $M<\infty$ and $\varepsilon>0$,

$$E\left(1\wedge(1/\sqrt{n})\int_0^{2M}\sqrt{\log N^*(\mathcal{F}_M,e_{n,2},\tau)}d\tau\right)\to 0.$$

Proof (b) implies (a) by Corollary 3.7.9. (d) for any $p>0$ implies (d) for $p=1$ because $e_{n,p}\ge e_{n,1}$ for all $p\in[1,\infty]$ and because for any $0<p<1$ and $f,g\in\mathcal{F}_M$, $e_{n,1}(f,g)\le (2M)^{1-p}e_{n,p}(f,g)$. Thus, to prove that (d) implies (b), it suffices to prove that (d) for $p=1$ and with convergence in probability implies (b). First, we see that since $EF<\infty$, if ε_i are i.i.d. Rademacher variables independent of the variables X_i (we take all these variables as coordinates in an infinite product probability space), we have

$$E\left\|\frac{1}{n}\sum_{i=1}^n\varepsilon_i f(X_i)\right\|_{\mathcal{F}} \le E\left\|\frac{1}{n}\sum_{i=1}^n\varepsilon_i f(X_i)I(F(X_i)\le M)\right\|_{\mathcal{F}} + E\left\|\frac{1}{n}\sum_{i=1}^n\varepsilon_i f(X_i)I(F(X_i)>M)\right\|_{\mathcal{F}}.$$

The last summand is dominated by

$$\frac{1}{n}E\left(\sum_{i=1}^n F(X_i)I(F(X_i)>M)\right) = E(FI(F>M))\to 0, \quad \text{as } M\to\infty.$$

Hence, this and the Rademacher randomisation Lemma 3.1.21 imply that the statement in (b) will be proved if we show that

$$E\left\|\frac{1}{n}\sum_{i=1}^n\varepsilon_i f(X_i)\right\|_{\mathcal{F}_M}\to 0,$$

for all $M<\infty$. To prove that this last statement follows from the metric entropy condition, we will use the fact that conditionally on the variables X_i, these Rademacher averages are sub-Gaussian variables for each f. Fix the variables X_i. Given that $\varepsilon>0$, let $f_1,\dots,f_N$ be the centres of $N = N(\mathcal{F}_M, e_{n,1}, \varepsilon)$ $e_{n,1}$-balls of radius less than or equal to ε covering $\mathcal{F}_M$. Then, for each f in $\mathcal{F}_M$, there is $k(f) \le N$ such that $e_{n,1}(f_{k(f)}, f) \le \varepsilon$, and we have, letting E_ε denote expectation with respect to the Rademacher variables only and using the maximal inequality (2.35) for sub-Gaussian variables,

$$\begin{aligned} E_\varepsilon \left\| \frac{1}{n}\sum_{i=1}^n \varepsilon_i f(X_i) \right\|_{\mathcal{F}_M} &\le E_\varepsilon \left\| \frac{1}{n}\sum_{i=1}^n \varepsilon_i (f - f_{k(f)})(X_i) \right\|_{\mathcal{F}_M} + E_\varepsilon \max_{1\le k\le N} \left| \frac{1}{n}\sum_{i=1}^n \varepsilon_i f_k(X_i) \right| \\ &\le \sup_{f\in\mathcal{F}_M} e_{n,1}(f, f_{k(f)}) + \sqrt{2\log 2N} \times \max_{1\le k\le N} \left(\frac{\sum_{i=1}^n f_k^2(X_i)}{n^2} \right)^{1/2} \\ &\le \varepsilon + M\sqrt{\frac{2\log 2 + 2\log N}{n}}. \end{aligned}$$

Hence, by Fubini,

$$E\left\| \frac{1}{n}\sum_{i=1}^n \varepsilon_i f(X_i) \right\|_{\mathcal{F}_M} \le \varepsilon + M\sqrt{\frac{2\log 2}{n}} + \sqrt{2} M E^* \sqrt{\frac{\log N(\mathcal{F}_M, e_{n,1}, \varepsilon)}{n}}. \tag{3.245}$$

(Note that in the absence of measurability, this would not follow from the nonmeasurable form of the Fubini-Tonelli theorem, Proposition 3.7.3.) Now, since for each $f \in \mathcal{F}_M$, $(f(X_1), \dots, f(X_n)) \in [-M, M]^n$ and $[-M, M]^n$ can be covered by fewer than $(1 + M/\varepsilon)^n$ hypercubes of the form $\{x : \max_{i\le n} |x_i - x_i^0| \le \varepsilon\}$, $x^0 \in [-M, M]^n$, it follows that for $\varepsilon \le M$,

$$N(\mathcal{F}_M, e_{n,1}, \varepsilon) \le N(\mathcal{F}_M, e_{n,\infty}, \varepsilon) \le \left(\frac{2M}{\varepsilon} \right)^n.$$

Hence, $n^{-1}\log N(\mathcal{F}_M, e_{n,1}, \varepsilon) \le \log(2M/\varepsilon)$, so if $n^{-1}\log N^*(\mathcal{F}_M, e_{n,1}, \varepsilon) \to 0$ in probability, then, by bounded convergence,

$$E\left[n^{-1}\log N^*(\mathcal{F}_M, e_{n,1}, \varepsilon)\right]^r \to 0,$$

for all $r > 0$. Therefore, condition (d) in probability and for $p = 1$ gives, by (3.245), that

$$\limsup_n E\left\| \frac{1}{n}\sum_{i=1}^n \varepsilon_i f(X_i) \right\|_{\mathcal{F}_M} \le \varepsilon, \quad \text{for all } \varepsilon > 0,$$

so this limit is zero, proving condition (b). The preceding argument also shows that for all $p \le \infty$, convergence in probability to zero of $n^{-1}\log N^*(\mathcal{F}_M, e_{n,p}, \varepsilon)$ is equivalent to its convergence to zero in L^r for all $r < \infty$. So far we have proved that (d) in probability for $p = 1$ implies (b), which implies (a), and it also implies (c) for any $0 < p < \infty$ in probability and in L^r for any $r < \infty$. Next we prove that condition (a) implies condition (d) in L^r (for any $r < \infty$) for $p < \infty$ as well as condition (d).

Supose that (a) holds. Then

$$\begin{aligned} \frac{1}{n}\|f(X_n) - Pf\|_{\mathcal{F}} &= \left\| \frac{1}{n}\sum_{i=1}^n (f(X_i) - Pf) - \frac{1}{n}\sum_{i=1}^{n-1} (f(X_i) - Pf) \right\| \\ &\le \frac{n-1}{n}\|P_{n-1} - P\|_{\mathcal{F}} + \|P_n - P\|_{\mathcal{F}} \to 0 \text{ a.s.} \end{aligned}$$

Hence, by the Borel-Cantelli lemma, $\sum_{n=1}^{\infty}\Pr\{\|f(X_n)-Pf\|_{\mathcal{F}}>n\}<\infty$, and therefore, $E\|f(X_1)-Pf\|_{\mathcal{F}}<\infty$, which by the $L^1(P)$ boundedness of $\mathcal{F}$ implies that $PF=E\|f(X_1)\|_{\mathcal{F}}<\infty$, proving that the first part of condition (d) holds. The main step in proving the rest of (c) and (d) consists in showing that the metric entropy condition holds for $p=2$, and this will be achieved by randomising the empirical process with standard normal multipliers and then applying Sudakov's minorisation conditionally on the variables and X_i. Let g_i be i.i.d. standard normal variables and ε_i i.i.d. Rademacher variables, all independent from the sequence $\{X_i\}$. The randomisation inequality in Proposition 3.1.26 gives that, for $0<n_0<n$,

$$E\left\|\frac{1}{n}\sum_{i=1}^{n}g_if(X_i)\right\|_{\mathcal{F}}\leq n_0\frac{PF}{n}E\max_{1\leq i\leq n}|g_i|+\frac{1}{\sqrt{n}}\Lambda_{2,1}(g_1)\max_{n_0<k\leq n}\times E\left\|\frac{1}{\sqrt{k}}\sum_{i=n_0+1}^{k}\varepsilon_if(X_i)\right\|_{\mathcal{F}}. \tag{3.246}$$

Now (a) $E\max_{i\leq n}|g_i|$ is dominated by a constant times $\sqrt{\log n}$ (see (2.35)), (b) $\Lambda_{2,1}(g_1)=\int_0^\infty\sqrt{\Pr\{|g|>t\}}dt<3$ and (c) since $PF<\infty$ and (a) holds, it follows from Corollary 3.7.9 that $E\|P_n-P\|_{\mathcal{F}}\to 0$ and, by the Rademacher randomisation inequality (3.46), that

$$\lim_{n\to\infty}E\left\|\frac{1}{n}\sum_{i=1}^{n}\varepsilon_if(X_i)\right\|_{\mathcal{F}}=0$$

(note that $E\left\|\sum_{i=1}^{n}\varepsilon_iPf/n\right\|_{\mathcal{F}}=\|Pf\|_{\mathcal{F}}E\left|\sum_{i=1}^{n}\varepsilon_i\right|/n=\|Pf\|_{\mathcal{F}}/\sqrt{n}\to 0$.). Therefore, (3.246) implies that

$$\lim_{n\to\infty}E\left\|\frac{1}{n}\sum_{i=1}^{n}g_if(X_i)\right\|_{\mathcal{F}}=0.$$

This, in turn, implies by Sudakov's inequality (Theorem 2.4.12) that

$$\lim_{n\to\infty}n^{-1/2}E^*\left[\sup_{\varepsilon>0}\varepsilon\sqrt{\log N(\mathcal{F},e_{n,2},\varepsilon)}\right]=0. \tag{3.247}$$

Since, for all $M<\infty$, $N(\mathcal{F}_M,e_{n,2},\varepsilon)\leq N(\mathcal{F},e_{n,2},\varepsilon)$ and the first of these two covering numbers is bounded by $(2M/\varepsilon)^n$, it then follows from (3.247) that for $r\geq 1/2$,

$$E^*\left[\frac{1}{n}\log N(\mathcal{F}_M,e_{n,2},\varepsilon)\right]^r\leq(\log(2M/\varepsilon))^{r-1/2}E^*\left[\frac{1}{n}\log N(\mathcal{F}_M,e_{n,2},\varepsilon)\right]^{1/2}\to 0,$$

proving condition (d) in L^r, $r<\infty$ for $p=2$, and hence also for $0<p<\infty$ (and all r). For condition (e), note that, as observed earlier in this proof ($e_{n,2}\leq e_{n,\infty}$), $E\sqrt{n^{-1}\log N^*(\mathcal{F}_M,e_{n,2},\varepsilon)}\leq\sqrt{\log(2M/\varepsilon)}<\infty$ and that $E\sqrt{n^{-1}\log N^*(\mathcal{F}_M,e_{n,2},\varepsilon)}\to 0$ by (3.247) both for all $\varepsilon>0$, and hence condition (e) follows by the dominated convergence theorem.

Now we complete the proof of statement (d) by proving the case $p=\infty$. Let us assume that the variables X_i are fixed. Fix $0<\varepsilon<M$, and let $0<\alpha<\varepsilon$. Let $\pi:\mathcal{F}_M\mapsto\mathcal{F}_M$ satisfy (a) $e_{n,1}(f,\pi f)\leq\alpha\varepsilon/2$ and (b) $\mathrm{Card}\{\pi f: f\in\mathcal{F}_M\}=N(\mathcal{F}_M,e_{n,1},\alpha\varepsilon/2)$. Such a function π exists:

just disjointify an optimal covering of $\mathcal{F}_M$ by $e_{n,1}$-balls of radius $\alpha\varepsilon/2$ to obtain a partition of $\mathcal{F}_M$ into $N(\mathcal{F}_M, e_{n,1}, \alpha\varepsilon/2)$ sets $\mathcal{Q}_i$ each contained in an $e_{n,1}$-ball of radius $\alpha\varepsilon/2$ and centre f_i, and set $\pi f = f_i$ for $f \in \mathcal{Q}_i$. Set $\mathcal{G} = \{f - \pi f : f \in \mathcal{F}_M\}$. Since for $f, g \in \mathcal{Q}_i$, $e_{n,\infty}(f,g) = e_{n,\infty}(f - \pi f, g - \pi g)$, it follow that for each ε and $\mathcal{Q}_i$, $N(\mathcal{Q}_i, e_{n,\infty}, \varepsilon) \le N(\mathcal{G}, e_{n,\infty}, \varepsilon)$, and we have

$$N(\mathcal{F}_M, e_{n,\infty}, \varepsilon) \le N(\mathcal{F}_M, e_{n,1}, \alpha\varepsilon/2) N(\mathcal{G}, e_{n,\infty}, \varepsilon). \tag{3.248}$$

We need to estimate the last covering number. By definition of πf, if $g \in \mathcal{G}$, then $\sum_{j=1}^n |g(X_j)| \le \alpha n\varepsilon/2$, and this implies that there are at most $[\alpha n]$ subindices j such that $|g(X_j)| > \varepsilon/2$. Now let $\mathcal{H}$ be the family of functions $f : \{X_1, \ldots, X_n\} \mapsto \mathbb{R}$ such that $f(X_i) = 0$ for $n - [\alpha n]$ subindices and takes values $k\varepsilon/2$, $k \in \mathbb{Z}$, $|k| \le 4M/\varepsilon$, for the remaining $[\alpha n]$. Then every function in $\mathcal{G}$ is at most at $e_{n,\infty}$-distance $\varepsilon/2$ from a function in $\mathcal{H}$, and therefore,

$$N(\mathcal{G}, e_{n,\infty}, \varepsilon) \le \mathrm{Card}(\mathcal{H}) \le \binom{n}{[\alpha n]} (1 + 8M/\varepsilon)^{[\alpha n]}.$$

For each $\alpha \in (0,1)$, by Stirling's formula, the logarithm of this bound divided by n is asymptotically as $n \to \infty$ of the order of

$$\begin{aligned} \frac{1}{n} \log \mathrm{Card}(\mathcal{H}) &\le \frac{1}{n} \log\left(\binom{n}{[\alpha n]} (1 + 8M/\varepsilon)^{[\alpha n]} \right) := h(n, \alpha) \\ &\asymp (1-\alpha)|\log(1-\alpha)| + \alpha|\log\alpha| + \alpha\log(1 + 8M/\varepsilon) - \frac{\log(2\pi n\alpha(1-\alpha))}{2n}, \end{aligned}$$

so $\lim_{\alpha\to 0}\lim_{n\to\infty} h(n,\alpha) = 0$. This and (3.248) imply that if $n^{-1}\log N^*(\mathcal{F}_M, e_{n,1}, \alpha\varepsilon/2) \to 0$ in probability or in L^r for any $r < \infty$, then we also have $n^{-1}\log N^*(\mathcal{F}_M, e_{n,\infty}, \varepsilon) \to 0$, proving that statement (d) for $p = \infty$ follows from statement (d) for $p = 1$ (which holds if statement (a) does by the preceding paragraph).

Finally, to prove that condition (e) implies condition (b), we just apply Rademacher randomisation in probability (Corollary 3.1.25 with $Y_i - a_i = \{f(X_i)/n : f \in \mathcal{F}_M\}$ and $\sigma^2 \le M^2/n^2$) and the metric entropy bound for sub-Gaussian processes (Theorem 2.3.7) to obtain that for all $\varepsilon > 0$ and $n \ge 2M^2/\varepsilon^2$,

$$\begin{aligned} \Pr\left\{\|P_n - P\|_{\mathcal{F}_M} > 4\varepsilon\right\} &\le 4\Pr\left\{ \frac{1}{n}\left\| \sum_{i=1}^n \varepsilon_i f(X_i) \right\|_{\mathcal{F}_M} > \varepsilon \right\} \\ &\le 4E_X\left(1 \wedge \frac{1}{\varepsilon} E_\varepsilon \left\| \frac{1}{n}\sum_{i=1}^n \varepsilon_i f(X_i) \right\|_{\mathcal{F}_M} \right) \\ &\le \frac{4M}{\sqrt{n}\varepsilon} + 4E\left(1 \wedge \frac{4\sqrt{2}}{\sqrt{n}\varepsilon} \int_0^M \sqrt{\log N^*(\mathcal{F}_M, e_{n,2}, \tau)}\, d\tau \right). \quad \blacksquare \end{aligned}$$

The following is a corollary to the preceding proof:

Corollary 3.7.15 *Let $\mathcal{F}$ be an $L^1(P)$-bounded, P-measurable class of functions, and let F be its P-measurable cover. Then $\mathcal{F}$ is P-Glivenko-Cantelli if and only if*

(a) *$PF < \infty$, and*
(b) *$(1/n)\log N^*(\mathcal{F}, e_{n,2}, \varepsilon) \to 0$ in probability (or in $L^{1/2}$).*

For classes of sets $\mathcal{C}$, recall the definition of $\Delta^{\mathcal{C}}(A)$ for finite sets A in Section 3.6.1, $\Delta^{\mathcal{C}}(A) = \text{Card}\{A \cap C : C \in \mathcal{C}\}$, and note that for $A(\omega) = \{X_1(\omega), \dots, X_n(\omega)\}$, if $C \cap \{X_1, \dots, X_n\} = D \cap \{X_1, \dots, X_n\}$, then $e_{n,p}(C,D) = 0$ for all $0 < p \leq \infty$ and that $e_{n,p}(C,D) \geq n^{-1/(p\vee 1)}$ otherwise. Hence, $N(\mathcal{C}, e_{n,p}, \varepsilon) \leq \Delta^{\mathcal{C}}(X_1, \dots, X_n)$ for all $\varepsilon > 0$, with equality for $0 < \varepsilon \leq n^{-1/(p\vee 1)}$. This observation and Theorem 3.7.14 for $p = \infty$ give the following result for classes of sets:

Corollary 3.7.16 *Let $\mathcal{C}$ be a P-measurable class of sets. Then $\|P_n - P\|_{\mathcal{C}}^* \to 0$ a.s. if and only if*

$$\lim_{n\to\infty} \frac{1}{n} \log (\Delta^{\mathcal{C}}(X_1, \dots, X_n))^* = 0 \textit{ in probability}$$

(or in L^r for any $r < \infty$).

Next, combining Corollary 3.7.15 with Theorem 3.6.9 about the empirical metric entropy properties of *VC* type classes of functions, we obtain the following uniform law of large numbers (see also exercise 3.6.9):

Corollary 3.7.17 *Let P be any probability measure on $(S, \mathcal{S})$, and let $\mathcal{F}$ be a P-measurable class of functions whose measurable cover F is P-integrable. Assume that*

(a) *$\mathcal{F}$ is VC subgraph or, more generally, of VC type, or*
(b) *$\mathcal{F}$ is VC hull.*

Then $\mathcal{F}$ is P-Glivenko-Cantelli.

Proof The result for *VC* type classes of functions follows directly from Corollary 3.7.15 and the definition of *VC* type (Definition 3.6.10). That *VC* subgraph classes of functions are *VC* type follows from Theorem 3.6.9, and that the uniform law of large numbers also holds for *VC* hull classes follows immediately from the fact that $\|P_n - P\|_{\mathcal{F}} = \|P_n - P\|_{\overline{\text{co}}\mathcal{F}}$, which is obvious (so the entropy estimate for *VC* hull classes, Theorem 3.6.17, is not needed here, although, of course, it also gives the result). ■

Remark 3.7.18 Since, as mentioned in the preceding proof, $\|P_n - P\|_{\mathcal{F}} = \|P_n - P\|_{\overline{\text{co}}\mathcal{F}}$, it follows that the Glivenko-Cantelli property is preserved by taking pointwise closures of convex hulls; that is, $\mathcal{F}$ is *P*-Glivenko-Cantelli if and only if $\overline{\text{co}}\mathcal{F}$ is.

Example 3.7.19 The preceding corollary includes the Glivenko-Cantelli theorem for distribution functions on $\mathbb{R}^d$, $\|F_n - F\|_\infty \to 0$ a.s. To see this, just note that $F_n(x) - F(x) = (P_n - P)(-\infty, x]$, $x \in \mathbb{R}^d$, where $(-\infty, x] := \{y \in \mathbb{R}^d : y_i \leq x_i : i = 1, \dots d\}$, and $x = (x_1, \dots, x_d)$ and likewise for y. We show that $\mathcal{C} = \{(-\infty, x] : x \in \mathbb{R}^d\}$ is a *VC* class of sets as follows: for each $i = 1, \dots, d$, the class of half-spaces $\mathcal{C}_i = \{\{y \in \mathbb{R}^d : y_i \leq a\} : a \in \mathbb{R}\}$ is *VC* because it is ordered by inclusion (see Exercise 3.6.3), but $\mathcal{C} \subset \{C_1 \cap \cdots \cap C_d : C_i \in \mathcal{C}_i\}$, and Proposition 3.6.7 shows that $\mathcal{C}$ is a *VC* class. Also, $\|F_n - F\|_\infty = \sup_{x \in \mathbb{Q}^d} |(P_n - P)(-\infty, x]|$, so the class $\mathcal{C}$ is *P*-measurable for any Borel probability measure P on $\mathbb{R}^d$. Hence, by Corollary 3.7.17, we have

$$\|F_n - F\|_\infty \to 0 \text{ a.s.}$$

See also Exercise 3.6.9.

If a class of functions is uniformly bounded and is of VC type, then it is P-Glivenko-Cantelli for all probablity measures P on $(S, \mathcal{S})$. For larger classes, one may use the random entropies in Corollary 3.7.15 and Theorem 3.7.14; however, the following criterion for the Glivenko-Cantelli property based on $L^1(P)$ bracketing is more user friendly when it applies:

Theorem 3.7.20 *If* $\mathcal{F} \subset L^1(S, \mathcal{S}, P)$ *and* $N_{[]}(\mathcal{F}, L^1(P), \varepsilon) < \infty$ *for all* $\varepsilon > 0$, *then*

$$\|P_n - P\|_{\mathcal{F}}^* \to 0 \text{ a.s.}$$

Proof Let, for given ε, $[\underline{f}_i, \overline{f}_i]$ be $N = N_{[]}(\mathcal{F}, L^1(P), \varepsilon)$ $L^1(P)$-brackets of size ε (or less) covering $\mathcal{F}$. Recall that $[\underline{f}_i, \overline{f}_i] = \{h \in L^1(P) : \underline{f}_i \leq h \leq \overline{f}_i\}$ and that $P(\overline{f}_i - \underline{f}_i) \leq \varepsilon$. We have, for $f \in [\underline{f}_i, \overline{f}_i]$,

$$|(P_n - P)(f)| = |(P_n - P)\underline{f}_i + (P_n - P)(f - \underline{f}_i)| \leq |(P_n - P)\underline{f}_i| + P_n(\overline{f}_i - \underline{f}_i) + P(\overline{f}_i - \underline{f}_i).$$

Hence,

$$\|P_n - P\|_{\mathcal{F}}^* \leq \max_{1 \leq i \leq N} |(P_n - P)\underline{f}_i| + \max_{1 \leq i \leq N} P_n(\overline{f}_i - \underline{f}_i) + \max_{1 \leq i \leq N} P(\overline{f}_i - \underline{f}_i).$$

By definition of the brackets, the last summand is dominated by ε, and by the law of large numbers in $\mathbb{R}$, both the first summand tends to zero a.s. and the limsup of the second is dominated by ε a.s. Take $\varepsilon = 1/m$, and let $m \to \infty$ to immediately obtain the result. ■

It is easy to see that this theorem implies the classical Glivenko-Cantelli theorem. It also implies the law of large numbers in separable Banach spaces. For a random variable X in a Banach space B, the expectation EX is defined in the Bochner sense (see before Lemma 2.6.3).

Corollary 3.7.21 (Mourier law of large numbers) *Let B be a separable Banach space, and let X, X_i be i.i.d. B-valued random vectors such that $E\|X\| < \infty$. Then*

$$\frac{1}{n}\sum_{i=1}^{n} X_i \to EX \text{ a.s.}$$

Proof Let B^* denote the topological dual of B. It suffices to show that $\mathcal{F} := \{f \in B^* : \|f\| \leq 1\}$ is a P-Glivenko-Cantelli class of functions over B, where $P = \mathcal{L}(X)$, because

$$\left\|\frac{1}{n}\sum_{i=1}^{n} X_i - EX\right\| = \sup_{f \in \mathcal{F}} \left|\frac{1}{n}\sum_{i=1}^{n} (f(X_i) - Ef(X))\right|.$$

Since $\int \|x\| dP = E\|X\| < \infty$, it follows that the set function $A \mapsto \int_A \|x\| dP(x)$, defined on the Borel sets of B, is a finite Borel measure. By tightness of finite Borel measures on complete separable metric spaces, given $\varepsilon > 0$, there exists a compact subset K of B such that $\int_{B \setminus K} \|x\| dP(x) < \varepsilon/4$. Also, K is bounded, say, $K \subset \{x : \|x\| \leq C\}$. Now, if $f \in \mathcal{F}$, then $|f(x)| \leq \|f\| \|x\| \leq C$ for all $x \in K$, and moreover, $|f(x) - f(y)| \leq \|x - y\|$. Hence, $\mathcal{F}$ is a uniformly bounded and equi-continuous set of functions on $C(K)$, the Banach space of continuous functions on the metric space $K \subset B$, and hence precompact by the

Arzelà-Ascoli theorem and thus hence totally bounded. Thus, there exist $m < \infty$ functions from $\mathcal{F}$, $f_1, \ldots, f_m$, such that for all $f \in \mathcal{F}$, $\sup_{x \in K} |f(x) - f_i(x)| < \varepsilon/4$ for some $i \le m$. Define, for $i = 1, \ldots, m$,

$$\underline{f}_i(x) = \begin{cases} f_i(x) - \varepsilon/4 & \text{for } x \in K \\ -\|x\| & \text{for } x \notin K, \end{cases} \quad \overline{f}_i(x) = \begin{cases} f_i(x) + \varepsilon/4 & \text{for } x \in K \\ \|x\| & \text{for } x \notin K. \end{cases}$$

Then

$$0 \le P(\overline{f}_i - \underline{f}_i) = (\varepsilon/2)P(K) + 2\int_{B \setminus K} \|x\| dP(x) < \varepsilon,$$

that is, $N_{[]}(\mathcal{F}, L^1(P), \varepsilon) \le m < \infty$, and the result follows from the preceding theorem. ■

The Mourier law of large numbers also follows from Theorem 3.7.14 (see Exercise 3.7.6).

3.7.3 Convergence in Law of Bounded Processes

If $f \mapsto f(x) - Pf$ is a bounded functional on $\mathcal{F}$, for example, if $\sup_{f \in \mathcal{F}} |f(x)| < \infty$ for all x and $\sup_{f \in \mathcal{F}} |Pf| < \infty$, the empirical process $(P_n - P)(f)$, $f \in \mathcal{F}$, is a process with bounded sample paths, that is, a random element taking values in the space $\ell_\infty(\mathcal{F})$, the space of bounded real functions on the set $\mathcal{F}$. Since we are interested in particular in limit theorems for $\|P_n - P\|_{\mathcal{F}}$, we need to consider the supremum norm $\|H\|_{\mathcal{F}} = \sup_{f \in \mathcal{F}} |H(f)|$ in $\ell^\infty(\mathcal{F})$. Unless $\mathcal{F}$ is finite, this Banach space is not separable, and the law of the empirical process $f \mapsto (P_n - P)(f)$, $f \in \mathcal{F}$, which is a probability measure on the cylindrical σ-algebra of $\ell_\infty(\mathcal{F})$, does not extend to a tight Borel probability measure (see Exercise 3.7.7). Thus, the classical theory of convergence in law on complete separable metric spaces needs to be extended to include empirical processes. It turns out that this theory extends nicely if the limit laws are assumed to be tight Borel probability measures on $\ell_\infty(\mathcal{F})$. If this is the case, then, as shown later, (a) convergence in law of a sequence of sample bounded processes is equivalent to weak convergence of the finite-dimensional probability laws together with asymptotic equi-continuity, a condition that is expressed in terms of probability inequalities, and (b) the Skorokhod almost-sure representation of sequences that converge in law extends very nicely in this context. As a consequence of the latter, one can show that if the empirical process indexed by $\mathcal{F}$ satisfies the central limit theorem, then so does the empirical process indexed by the class of convex combinations of functions in $\mathcal{F}$. This section is devoted to these two basic results. A sort of metrisability of convergence in law is also briefly considered.

The extension of convergence in law just mentioned may be better described in the general context of bounded processes or processes with bounded sample paths. A bounded process (or a process with bounded sample paths) X of index set T defined on a measure space $(\Omega, \mathcal{A}, \Pr)$ is a measurable map $(\Omega, \mathcal{A}) \mapsto (\ell_\infty(T), \Sigma)$, where Σ is the cylindrical σ-algebra of $\ell_\infty(T)$. Hence, in general, even if $H : \ell_\infty(T) \mapsto \mathbb{R}$ is continuous, the random element $H(X)$ needs not be measurable. However, if the finite-dimensional probability distributions of a sample bounded process X are those of a tight Borel probability measure on $\ell_\infty(T)$, then there is a version of X that defines a Borel measurable map $\Omega \mapsto \ell_\infty(T)$ with σ-compact range (hence separable for the supremum norm), and in particular, $H(X)$ is measurable for any continuous function H.

Definition 3.7.22 Let $X(t)$, $t \in T$, be a bounded process whose finite-dimensional laws correspond to the finite-dimensional projections of a tight Borel probability measure on $\ell_\infty(T)$, and denote by $\tilde{X}$ a measurable version of X with separable range. Let $X_n(t)$, $t \in T$, be bounded processes. Then we say that X_n converge in law to X in $\ell_\infty(T)$, or uniformly in $t \in T$, or that

$$X_n \to_{\mathcal{L}} X, \quad in\ \ell_\infty(T)$$

if

$$E^* H(X_n) \to EH(\tilde{X}),$$

for all functions $H : \ell_\infty(T) \mapsto \mathbb{R}$ bounded and continuous, where E^* denotes outer expectation.

In general, we will still denote by X (and not by $\tilde{X}$) its measurable version with separable range unless confusion may arise. As with regular convergence in law, if H is a continuous function on $\ell_\infty(T)$ with values in another metric space, and if $H(X_n)$ is measurable, then convergence in law of X_n to X implies that $H(X_n) \to_{\mathcal{L}} H(X)$ in the usual way, and this makes the concept of convergence in law in $\ell_\infty(T)$ quite useful.

Recall from Proposition 2.1.7 that if the probability law of X is a tight Borel measure on $\ell_\infty(T)$, then X is sample continuous with respect to a metric d on T that makes T totally bounded. The main result in this subsection is the following theorem. It reduces convergence in law in $\ell_\infty(T)$ to maximal inequalities, which are tractable with the techniques presented in the preceding sections. This theorem will be referred to as the *asymptotic equi-continuity criterion* for convergence in law in $\ell_\infty(T)$.

Theorem 3.7.23 *Let $X_n(t)$, $t \in T$, $n \in \mathbb{N}$, be a sequence of bounded processes. Then the following statements are equivalent:*

(a) The finite-dimensional distributions of the processes X_n converge in law, and there exists a pseudo-metric d on T such that (T,d) is totally bounded, and

$$\lim_{\delta\to 0} \limsup_{n\to\infty} \Pr{}^*\Big\{ \sup_{d(s,t)\le\delta} |X_n(t) - X_n(s)| > \varepsilon \Big\} = 0, \tag{3.249}$$

for all $\varepsilon > 0$.

(b) There exists a process X whose law is a tight Borel probability measure on $\ell_\infty(T)$ and such that

$$X_n \to_{\mathcal{L}} X, \quad \text{in } \ell_\infty(T).$$

Moreover, if (a) holds, then the process X in (b) has a version with bounded uniformly continuous paths for d, and if (b) holds and X has a version with almost all of its trajectories in $C_u(T,\rho)$ for a pseudo-distance ρ such that (T,ρ) is totally bounded, then the distance d in (a) can be taken to be ρ.

Proof Let us assume that (a) holds. Clearly, the limit laws of the finite-dimensional distributions of the processes X_n are consistent and thus define a stochastic process X on T. Let T_0 be a countable d-dense subset of (T,d), and let T_k, $k \in \mathbb{N}$, be finite sets increasing to T_0. Then, by one of the conditions equivalent to convergence in law in $\mathbb{R}^d$ (or, in general, in complete, separable metric spaces, Portmanteau theorem), for all $k \in \mathbb{N}$ and $\delta > 0$,

$$\Pr\left\{\max_{d(s,t)\le\delta,\ s,t\in T_k} |X(t)-X(s)| > \varepsilon\right\} \le \liminf_{n\to\infty}\Pr\left\{\max_{d(s,t)\le\delta,\ s,t\in T_k} |X_n(t)-X_n(s)| > \varepsilon\right\}$$

$$\le \liminf_{n\to\infty}\Pr\left\{\max_{d(s,t)\le\delta,\ s,t\in T_0} |X_n(t)-X_n(s)| > \varepsilon\right\}.$$

Hence, taking limits as $k \to \infty$ and using condition (3.249), it follows that there exists a sequence $\delta_r \searrow 0$, $\delta_r > 0$, such that

$$\Pr\left\{\sup_{d(s,t)\le\delta_r,\ s,t\in T_0} |X(t)-X(s)| > 2^{-r}\right\} \le 2^{-r},$$

and by the Borel–Cantelli lemma, there exists $r(\omega) < \infty$ a.s. such that

$$\sup_{d(s,t)\le\delta_r,\ s,t\in T_0} |X(t,\omega)-X(s,\omega)| \le 2^{-r},$$

for all $r > r(\omega)$. Hence, $X(t,\omega)$ is a d-uniformly continuous function of t for almost every ω. Also, since T is totally bounded, $X(t,\omega)$ is also bounded for those ω for which it is d-uniformly continuous. The extension to T by uniform continuity of the restriction of $X(\omega)$ to T_0 for all these ω produces a version of X with all its sample paths in $C_u(T,d)$, and this shows, by Proposition 2.1.7, that the law of X admits a tight extension to the Borel σ–algebra of $\ell_\infty(T)$.

Fix $\tau > 0$. Let $t_1,\dots,t_{N(\tau)}$, $N(\tau) < \infty$ be a τ-dense subset of (T,d) (such a set exists for each τ because (T,d) is totally bounded), and let $\pi_\tau : T \mapsto \{t_1,\dots,t_{N(\tau)}\}$ be a mapping satisfying that $d(\pi_\tau(t),t) < \tau$. We then define processes $X_{n,\tau}$, $n \in \mathbb{N}$, and X_τ as

$$X_{n,\tau}(t) = X_n(\pi_\tau(t)) \quad \text{and} \quad X_\tau(t) = X(\pi_\tau(t)), \quad t \in T.$$

For each τ, these approximations of X_n and X are in fact $\mathbb{R}^{N(\tau)}$-valued random variables, and convergence of the finite-dimensional distributions of X_n to those of X implies their convergence in law in finite dimensional space and hence also that

$$X_{n,\tau} \to_{\mathcal{L}} X_\tau, \quad \text{in } \ell_\infty(T), \tag{3.250}$$

as can be seen from Definition 3.7.22 (formally, if $H : \ell_\infty(T) \mapsto \mathbb{R}$ is bounded and continuous, then so is $H \circ I : \mathbb{R}^{N(\tau)} \mapsto \mathbb{R}$, where I is the isometric imbeding $\mathbb{R}^{N(\tau)} \mapsto \ell_\infty(T)$ that assigns to each point $(a_1,\dots,a_{N(\tau)})$ the function $t \mapsto \sum_{i=1}^{N(\tau)} a_i I_{\pi_\tau(t)=t_i}$, and $E(H \circ I(X_n(t_1),\dots,X_n(t_{N(\tau)}))) = EH(X_{n,\tau})$ and likewise for X_τ, so convergence in law of the vectors $(X_n(t_1),\dots,X_n(t_{N(\tau)}))$ implies convergence in law of the processes $X_{n,\tau}$). Moreover, by uniform continuity of the sample paths of X,

$$\lim_{\tau\to 0} \|X - X_\tau\|_T = 0 \text{ a.s.}, \tag{3.251}$$

where we uses the notation $\|x\|_T = \sup_{t\in T}|x(t)|$ for $x \in \ell_\infty(T)$. Now let $H : \ell_\infty(T) \mapsto \mathbb{R}$ be a bounded continuous function. We may write

$$\begin{aligned} |E^*H(X_n) - EH(X)| &\le \left|E^*H(X_n) - EH(X_{n,\tau})\right| \\ &\quad + \left|EH(X_{n,\tau}) - EH(X_\tau)\right| + |EH(X_\tau) - EH(X)| \qquad (3.252) \\ &:= I_{n,\tau} + II_{n,\tau} + III_\tau. \end{aligned}$$

By the definition of convergence in law, it follows from (3.250) that

$$\lim_{\tau\to 0}\limsup_{n\to\infty} II_{n,\tau} = 0,$$

and by sample continuity of X, precisely (3.251), and the dominated convergence theorem, we also have

$$\lim_{\tau\to 0} III_\tau = 0.$$

Hence, proving that the double limit of $I_{n,\tau}$ is zero will complete the proof of (b). Given $\varepsilon > 0$, let $K \subset \ell_\infty(T)$ be a compact set such that $\Pr\{X \in K^c\} < \varepsilon/(12\|H\|_\infty)$. Given such a set K, let $\delta > 0$ be such that

$$\|u - v\|_T < \delta, \quad u \in K, \quad v \in \ell_\infty(T) \implies |H(u) - H(v)| < \varepsilon/6,$$

which exists by Exercise 3.7.8. Given such a $\delta > 0$, let $\tau_1 > 0$ be such that $\Pr\{\|X_\tau - X\|_T \ge \delta/2\} < \varepsilon/(12\|H\|_\infty)$ for all $\tau < \tau_1$, which exists by (3.251). Let $K_{\delta/2} = \{v \in \ell_\infty(T) : \inf_{u\in K}\|v - u\|_T < \delta/2\}$ denote the open neighbourhood of the set K of 'radius' $\delta/2$ for the sup norm, and note that these choices of K, δ and τ imply (a) if $X_{n,\tau} \in K_{\delta/2}$ and $\|X_n - X_{n,\tau}\|_T < \delta/2$, then there exists $u \in K$ such that $\|u - X_{n,\tau}\|_T < \delta/2$ and hence $\|u - X_n\|_T < \delta$, and (b) if $u \in K$ and $\|u - v\|_T < \delta$, then $|H(u) - H(v)| < \varepsilon/6$. We thus have

$$\begin{aligned}
\left|E^*H(X_n) - EH(X_{n,\tau})\right| &\le 2\|H\|_\infty\left[\Pr{}^*\left\{\|X_n - X_{n,\tau}\|_T \ge \frac{\delta}{2}\right\} + \Pr\left\{X_{n,\tau} \in \left(K_{\delta/2}\right)^c\right\}\right]\\
&\quad + 2\sup\{|H(u) - H(v)| : u \in K, \|u - v\|_T < \delta\} \qquad (3.253)\\
&\le 2\|H\|_\infty\left[\Pr{}^*\left\{\|X_n - X_{n,\tau}\|_T \ge \frac{\delta}{2}\right\} + \Pr\left\{X_{n,\tau} \in \left(K_{\delta/2}\right)^c\right\}\right] + \frac{2\varepsilon}{6}.
\end{aligned}$$

Now, by (3.250), we have for $\tau < \tau_1$ that

$$\begin{aligned}
\limsup_{n\to\infty}\Pr\left\{X_{n,\tau} \in \left(K_{\delta/2}\right)^c\right\} &\le \Pr\left\{X_\tau \in \left(K_{\delta/2}\right)^c\right\} \le \Pr\{X \in K^c\} + \Pr\{\|X_\tau - X\|_T \ge \delta/2\}\\
&\le \frac{\varepsilon}{6\|H\|_\infty},
\end{aligned}$$

and by the asymptotic equi-continuity hypothesis (3.249), there exists $\tau_2 > 0$ such that

$$\limsup_{n\to\infty}\Pr{}^*\left\{\|X_{n,\tau} - X_n\|_T \ge \frac{\delta}{2}\right\} < \frac{\varepsilon}{6\|H\|_\infty},$$

for all $\tau < \tau_2$. Combining these two inequalities with (3.253) gives that for all $\tau < \tau_1 \wedge \tau_2$,

$$\limsup_{n\to\infty}\left|E^*H(X_n) - EH(X_{n,\tau})\right| < \varepsilon.$$

This proves that $\lim_{\tau\to 0}\limsup_{n\to\infty} I_{n,\tau} = 0$, thus completing the proof that (a) implies (b).

Suppose now that (b) holds. Then, by Proposition 2.1.7, there exists a pseudo-distance d on T for which (T,d) is totally bounded and such that X has a version (that we still denote by X) with all its sample paths in $C_u(T,d)$. Set, for $\varepsilon, \delta > 0$, $F_{\delta,\varepsilon} = \left\{u \in \ell_\infty(T) : \sup_{d(s,t)\le\delta}|u(s) - u(t)| \ge \varepsilon\right\}$, which is a closed set in $\ell_\infty(T)$. Then convergence in law of X_n to X implies, by

Exercise 3.7.7, that $\limsup_{n\to\infty} \Pr^*\{X_n \in F_{\delta,\varepsilon}\} \le \Pr\{X \in F_{\delta,\varepsilon}\}$ for all $\varepsilon, \delta > 0$, and this and the fact that $X \in C_u(T,d)$ yield

$$\lim_{\delta\to 0}\limsup_{n\to\infty}\Pr^*\left\{\sup_{d(s,t)\le\delta}\left|X_n(t)-X_n(s)\right|\ge\varepsilon\right\}\le\lim_{\delta\to 0}\Pr\left\{\sup_{d(s,t)\le\delta}\left|X(t)-X(s)\right|\ge\varepsilon\right\}=0,$$

for all $\varepsilon > 0$, proving (a) for d (convergence of the finite-dimensional distributions of X_n to those of X follows from the definition of convergence in law in $\ell_\infty(T)$). ■

The useful fact that weak convergence of Borel probability measures in complete separable metric spaces (B,d) is metrisable by a norm extends to convergence in law of bounded processes. For instance, the (dual) bounded-Lipschitz distance $\beta_{B,d}(\mu,\nu)$ between Borel probability measures μ, ν on B, defined as the supremum of $\left|\int f d(\mu-\nu)\right|$ over all the functions $f : B \mapsto \mathbb{R}$ such that $\|f\|_\infty \le 1$ and $\sup_{x\ne y}|f(x)-f(y)|/d(x,y) \le 1$ metrises weak convergence of Borel probability measures, already encountered in Chapter 1. We now develop an extension of this fact in the nonseparable context.

Let $BL_1\left(\ell_\infty(T)\right)$ denote the set of real functionals H on $\ell_\infty(T)$ such that

$$\sup_{x\in\ell_\infty(T)}|H(x)|+\sup_{x\ne y,x,y\in\ell_\infty(T)}|H(y)-H(x)|/\|y-x\|_T\le 1.$$

(BL_1 or $BL(1)$ stands for the unit ball of the space of bounded Lipschitz functions as in Section 1.1.) If Y is a process on T with almost all its trajectories bounded, X is a process whose law is a tight Borel measure on $\ell_\infty(T)$, and we also denote by X one of its versions almost all of whose sample paths are in $C_u(T,d)$ for some distance d for which (T,d) is separable, we define

$$d_{BL}(Y,X) := d_{BL(T)}(Y,X) := \sup\{|E^*H(Y)-EH(X)| : H \in BL_1(\ell_\infty(T))\}, \tag{3.254}$$

the *(dual) bounded Lipschitz distance* between X and Y. (We write d_{BL} instead of $d_{BL(T)}$ if no confusion may arise.)

Proposition 3.7.24 *If X is a $C_u(T,d)$-valued random variable, where (T,d) is a separable metric or pseudo-metric space, and if $X_n(t)$, $t\in T$, are processes with bounded sample paths, then $X_n \to_{\mathcal{L}} X$ in $\ell_\infty(T)$ if and only if $d_{BL(T)}(X_n,X)\to 0$.*

Proof We keep the notation from the proof of Theorem 3.7.23. Let us assume convergence in law of X_n to X, and let us consider the decomposition from the preceding proof, that is,

$$|E^*H(X_n)-Ef(X)|\le I_{n,\tau}(H)+II_{n,\tau}(H)+III_\tau(H),$$

for H bounded Lipschitz, specifically, $H \in BL_1(\ell_\infty(T))$. (In particular, $|H(x)-H(y)| \le \min(\|x-y\|_T, 2)$.) Since, for τ fixed, $X_{n,\tau} \to_{\mathcal{L}} X_\tau$ as random vectors in $\mathbb{R}^{N(\tau)}$, and since d_{BL} metrises this convergence, it follows that

$$\lim_{n\to\infty}\sup_{H\in BL_1(\ell_\infty(T))}|II_{n,\tau}(H)|=0,$$

for all $\tau > 0$ (since if H is Lipschitz on $\ell_\infty(T)$ and I is as after (3.250), then $H\circ I$ is Lipschitz on $\mathbb{R}^{N(\tau)}$). Since $\|X_\tau - X\|_T \to 0$ a.s.,

$$\lim_{\tau\to 0}\sup_{H\in BL_1(\ell_\infty(T))}|III_\tau(H)|\le\lim_{\tau\to 0}E(\|X_\tau-X\|_T\wedge 2)=0$$

by the dominated convergence theorem. Finally,

$$\lim_{\tau\to 0}\limsup_{n\to\infty}\sup_{H\in BL_1(\ell_\infty(T))}|I_{n,\tau}(H)| \le \lim_{\tau\to 0}\limsup_{n\to\infty}E^*\left[\sup_{d(s,t)\le\tau}(|X_n(t)-X_n(s)|\wedge 2)\right]$$
$$\le 2\lim_{\tau\to 0}\limsup_{n\to\infty}\Pr{}^*\left\{\sup_{d(s,t)\le\tau}|X_n(t)-X_n(s)|>\varepsilon\right\}+2\varepsilon$$
$$=2\varepsilon,$$

for all $\varepsilon>0$, by Theorem 3.7.23 (3.249). Thus, $d_{BL}(X_n,X)\to 0$.

For the converse, assume that $d_{BL}(X_n,X)\to 0$, and set, for $\delta>0$ fixed and all $\varepsilon>0$,

$$A_\varepsilon(\delta)=A_\varepsilon=\left\{x\in\ell_\infty(T):\sup_{d(s,t)\le\delta}|x(t)-x(s)|\ge\varepsilon\right\}.$$

Then, if $x\in A_\varepsilon$ and $y\in A^c_{\varepsilon/2}$, we have $\|x-y\|_T\ge\varepsilon/5$: there exist s and t with $d(s,t)\le\delta$ such that $|x(t)-x(s)|>9\varepsilon/10$; hence, if, for example, $|x(t)-y(t)|<\varepsilon/5$, then

$$9\varepsilon/10<|x(t)-x(s)|\le|x(t)-y(t)|+|y(t)-y(s)|$$
$$+|y(s)-x(s)|<\varepsilon/5+\varepsilon/2+|y(s)-x(s)|,$$

that is, $|x(s)-y(s)|>\varepsilon/5$. Therefore, the restriction to the set $A_\varepsilon\cap A^c_{\varepsilon/2}$ of the indicator function I_{A_ε} is Lipschitz with constant bounded by $5/\varepsilon$. Hence, by the Kirzbraun–McShane extension theorem (Exercise 3.6.14), there exists a bounded Lipschitz function H defined on all of $\ell_\infty(T)$ which is 0 on $A^c_{\varepsilon/2}$ and 1 on A_ε, nonnegative and bounded by 1 and whose Lipschitz constant is bounded by $5/\varepsilon$. For such a function H, we have $I_{A_\varepsilon}\le H\le I_{A_{\varepsilon/2}}$ and $|E^*H(X_n)-EH(X)|\le(5/\varepsilon)d_{BL}(X_n,X)$. Hence, the hypothesis implies that

$$\limsup_{n\to\infty}\Pr{}^*\{X_n\in A_\varepsilon\}\le\limsup_{n\to\infty}E^*H(X_n)=EH(X)$$
$$\le\Pr\{X\in A_{\varepsilon/2}\}=\Pr\left\{\sup_{d(s,t)\le\delta}|X(t)-X(s)|\ge\frac{\varepsilon}{2}\right\}.$$

But since $X\in C_u(T,d)$ a.s., we have

$$\lim_{\delta\to 0}\Pr\left\{\sup_{d(s,t)\le\delta}|X(t)-X(s)|\ge\frac{\varepsilon}{2}\right\}=0,$$

which, combined with the preceding inequality, yields

$$\lim_{\delta\to 0}\limsup_{n\to\infty}\Pr{}^*\{X_n\in A_\varepsilon(\delta)\}=0,$$

that is, the asymptotic equicontinuity condition (3.249). Then, by Theorem 3.7.23, $X_n\to_{\mathcal{L}} X$ in $\ell_\infty(T)$. ■

Another quite useful general property of convergence in law in the present general setting is the analogue of Skorokhod's representation theorem:

Theorem 3.7.25 *Let $(\Omega_n, \mathcal{A}_n, Q_n) \mapsto \ell_\infty(T)$, $n \in \mathbb{N} \cup \{\infty\}$, be probability spaces, and let $X_n : \Omega_n \mapsto \ell_\infty(T)$ be bounded processes such that X_∞ is Borel measurable and has separable range in $\ell_\infty(T)$. Then $X_n \to_{\mathcal{L}} X_\infty$ in $\ell_\infty(T)$ if and only if there exists a probability space $(\tilde{\Omega}, \tilde{\mathcal{A}}, \tilde{Q})$ and perfect maps $\phi_n : \tilde{\Omega} \mapsto \Omega_n$ such that $Q_n = \tilde{Q} \circ \phi_n^{-1}$, $n \leq \infty$, and $\lim_{n\to\infty} \|\tilde{X}_n - \tilde{X}_\infty\|_T^* = 0$ $\tilde{Q}$ a.s. as $n \to \infty$, where $\tilde{X}_n := X_n \circ \phi_n$.*

Proof For necessity of the condition note that by Exercise 3.7.9, if $\|X_n \circ \phi_n - X_\infty \circ \phi_\infty\|_T^* \to 0$ a.s., then $X_n \circ \phi_n \to_{\mathcal{L}} X_\infty \circ \phi_\infty$ in $\ell_\infty(T)$ and that by perfectness of ϕ_n, $E_{\tilde{Q}}^* H(\tilde{X}_n) = E_{Q_n}^* H(X_n)$, $n \leq \infty$, for every $H : \ell_\infty(T) \mapsto \mathbb{R}$ bounded and continuous (see the end of Section 3.7.1) and thus that $X_n \to_{\mathcal{L}} X_\infty$ in $\ell_\infty(T)$.

The proof of sufficiency is much more elaborate. Let $C \subset \ell_\infty(T)$ be the range of X_∞, and let $\{x_i\}_{i=1}^\infty$ be a dense subset of C. Since, for each $x \in \ell_\infty(T)$, all but at most a countable number of open balls $B(x,r)$ of centre x and radius r (for the supremum norm over T) are continuity sets for the law of X_∞ (meaning that $Q_\infty\{X_\infty \in \partial B(x,r)\} = 0$), for every $\varepsilon > 0$, there exists a collection $B(x_i, r_i) \subset \ell_\infty(T)$ of open balls with radii $\varepsilon/3 < r_i < \varepsilon/2$, $i \in \mathbb{N}$, that are continuity sets for the law of X_∞. These balls cover C and, by subtracting from each such ball the union of the previous ones (set $B_1^{(\varepsilon)} = B(x_1, r_1)$, $B_2^{(\varepsilon)} = B(x_1, r_1) \setminus B(x_2, r_2)$, and so on), we obtain a countable collection $\left\{B_i^{(\varepsilon)}\right\}_{i=1}^\infty$ of disjoint continuity sets for X_∞ such that $Q_\infty\left\{X_\infty \in \cup_{i=1}^\infty B_i^{(\varepsilon)}\right\} = 1$ (note that the boundary of a finite union of sets is contained in the union of their boundaries). Let $k_\varepsilon < \infty$ be such that $\sum_{i>k_\varepsilon} Q_\infty\left\{X_\infty \in B_i^{(\varepsilon)}\right\} < \varepsilon$, and set $B_0^{(\varepsilon)} = \ell_\infty(T) \setminus \cup_{i=1}^{k_\varepsilon} B_i^{(\varepsilon)}$. We have thus constructed a partition of $\ell_\infty(T)$ into a finite number of X_∞-continuity sets $\{B_i^{(\varepsilon)}\}_{i=0}^{k_\varepsilon}$ such that

$$Q_\infty\left\{X_\infty \in B_0^{(\varepsilon)}\right\} < \varepsilon, \quad \operatorname{diam}(B_i^{(\varepsilon)}) < \varepsilon, \quad \text{for } i = 1, \dots, k_\varepsilon.$$

We may also assume that $Q_\infty\left\{X_\infty \in B_i^{(\varepsilon)}\right\} > 0$ for all $i = 1, \dots, k_\varepsilon$ by discarding sets $B_i^{(\varepsilon)}$, $i \geq 1$, incorporating them into $B_0^{(\varepsilon)}$, if necessary, and renumbering the rest.

Since $X_n \to_{\mathcal{L}} X_\infty$ in $\ell_\infty(T)$, by the Portmanteau theorem (Exercise 3.7.7), for each $\varepsilon > 0$ and $i = 0, 1, \dots, k_\varepsilon$, we have that $\lim_{n\to\infty} (Q_n)_*\left\{X_n \in B_i^{(\varepsilon)}\right\} = Q_\infty\left\{X_\infty \in B_i^{(\varepsilon)}\right\}$. Hence, given $\varepsilon_m = 1/(m+1)^2$, there exist $n_m \nearrow \infty$ such that

$$(Q_n)_*\left\{X_n \in B_i^{(\varepsilon_m)}\right\} \geq (1 - \varepsilon_m) Q_\infty\left\{X_\infty \in B_i^{(\varepsilon_m)}\right\}, \quad i = 0, 1, \dots, k_{\varepsilon_m}, \quad \text{and} \quad n \geq n_m.$$

Take $\eta_n = \varepsilon_m$ for $n_m \leq n < n_{m+1}$, $m \in \mathbb{N}$, and by discarding the first $n_1 - 1$, X_n and renumbering, set $n_1 = 1$. Note that $\eta_n \searrow 0$ and that the range of the sequence $\{\eta_n\}$ is contained in $\{1/(m+1)^2 : m \in \mathbb{N}\}$. The preceding inequality then becomes

$$(Q_n)_*\left\{X_n \in B_i^{(\eta_n)}\right\} \geq (1 - \eta_n) Q_\infty\left\{X_\infty \in B_i^{(\eta_n)}\right\}, \quad i = 0, 1, \dots, k_{\eta_n}, \quad \text{and} \quad n \in \mathbb{N}. \quad (3.255)$$

Let now $A_i^n \in \mathcal{A}_n$ be such that $A_i^n \subseteq \{X_n \in B_i^{(\eta_n)}\}$ and $Q_n(A_i^n) = (Q_n)_*\{X_n \in B_i^{(\eta_n}\}$, $i = 1, \dots, k_{\eta_n}$, which exist by the definition of inner probability (see Exercise 3.7.4 and Proposition 3.7.1), and set $A_0^n = \Omega_n \setminus \cup_{i=1}^{k_{\eta_n}} A_i^n$. Let $N_1 = \{n \in \mathbb{N} : Q_\infty\{X_\infty \in B_0^{(\eta_n)}\} = 0\}$, and let

$N_2^n = \mathbb{N} \setminus N_1$. Define

$$\tilde{\Omega} = \Omega_\infty \times \prod_{n \in N_1} \left[\Omega_n \times \prod_{i=1}^{k_{\eta n}} A_i^n \right] \times \prod_{n \in N_2} \left[\Omega_n \times \prod_{i=0}^{k_{\eta n}} A_i^n \right] \times [0,1],$$

$$\tilde{\mathcal{A}} = \mathcal{A}_\infty \otimes \prod_{n \in N_1} \left[\mathcal{A}_n \otimes \prod_{i=1}^{k_{\eta n}} (\mathcal{A}_n \cap A_i^n) \right] \otimes \prod_{n \in N_2} \left[\mathcal{A}_n \otimes \prod_{i=0}^{k_{\eta n}} (\mathcal{A}_n \cap A_i^n) \right] \otimes \mathcal{B},$$

$$\tilde{Q} = Q_\infty \times \prod_{n \in N_1} \left[\mu_n \times \prod_{i=1}^{k_{\eta n}} Q_n(\cdot\,|A_i^n) \right] \times \prod_{n \in N_2} \left[\mu_n \times \prod_{i=0}^{k_{\eta n}} Q_n(\cdot\,|A_i^n) \right] \times \lambda,$$

where $Q_n(\cdot\,|A_i^n)$ denotes conditional Q_n probability given A_i^n, λ is Lebesgue measure, $\mathcal{B}$ is the Borel σ-algebra of $[0,1]$ and μ_n is the probability measure on $\mathcal{A}_n$ given by

$$\mu_n(A) = \frac{1}{\eta_n} Q_n(A \cap A_0^n) + \frac{1}{\eta_n} \sum_{i=1}^{k_{\eta n}} Q_n(A|A_i^n) \Big[Q_n(A_i^n) - (1-\eta_n) Q_\infty\{X_\infty \in B_i^{(\eta_n)}\} \Big],$$

for $n \in N_1$, and

$$\mu_n(A) = \frac{1}{\eta_n} \sum_{i=0}^{k_{\eta n}} Q_n(A|A_i^n) \Big[Q_n(A_i^n) - (1-\eta_n) Q_\infty\{X_\infty \in B_i^{(\eta_n)}\} \Big],$$

for $n \in N_2$. Note that μ_n is a probability measure for each n because of the inequalities (3.255). With the notation

$$\tilde{\omega} = (\omega_\infty, \dots, \omega_n, \omega_{n,1}, \dots, \omega_{n,k_{\eta n}}, \dots, \omega_{n'}, \omega_{n',0}, \dots, \omega_{n',k_{\eta_{n'}}}, \dots, \xi)$$

for $n \in N_1$ and $n' \in N_2$, we define ϕ_n, $1 \le n \le \infty$, by

$$\phi_\infty(\tilde{\omega}) = \omega_\infty$$

$$\phi_n(\tilde{\omega}) = \begin{cases} \tilde{\omega}_n & \text{if } \xi > 1 - \eta_n \\ \tilde{\omega}_{n,i} & \text{if } \xi \le 1 - \eta_n \text{ and } X_\infty(\omega_\infty) \in B_i^{(\eta_n)}, \end{cases}$$

where i starts at 1 if $n \in N_1$ and otherwise at 0. Finally, define $\tilde{X}_n = X_n \circ \phi_n$, $1 \le n \le \infty$.

We now prove that this construction gives the sufficiency part of the theorem. Suppose that $\tilde{X}_\infty \notin B_0^{(\eta_n)}$ and $\xi \le 1 - \eta_n$. Then $\tilde{X}_\infty$ is in some $B_i^{(\eta_n)}$, $1 \le i \le k_{\eta_n}$, and the definitions of ϕ_n and A_i^n then give that $\tilde{X}_n$ is in the same $B_i^{(\eta_n)}$ and therefore that $\|\tilde{X}_n - \tilde{X}_\infty\|_T \le \eta_n$. We thus have, for any $m \in \mathbb{N}$,

$$\begin{aligned} \tilde{Q}^* \left\{ \sup_{n \ge n_m} \|\tilde{X}_n - \tilde{X}_\infty\|_T > \eta_n \right\} &\le \tilde{Q}\Big(\cup_{n: \eta_n \le \varepsilon_m} [\{\tilde{X}_\infty \in B_0^{(\eta_n)}\} \cup \{\xi > 1 - \eta_n\}] \Big) \\ &= \tilde{Q}\Big(\cup_{\ell \ge m} \{\tilde{X}_\infty \in B_0^{(\varepsilon_\ell)}\} \cup \{\xi > 1 - \varepsilon_m\} \Big) \\ &\le \sum_{\ell \ge m} \frac{1}{(\ell+1)^2} + \frac{1}{(m+1)^2} \to 0 \text{ as } m \to \infty. \end{aligned}$$

Since $\eta_n \to 0$, we have proved that $\|\tilde{X}_n - \tilde{X}_\infty\|_T^* \to 0$ a.s. (note Exercise 3.7.1 and Proposition 3.7.2). Next, we see that $Q_n = \tilde{Q} \circ \phi_n^{-1}$, $1 \le n \le \infty$. This is obvious for $n = \infty$. For $n < \infty$ and $A \in \mathcal{A}_n$, we have

$$\tilde{Q}_n\{\phi_n \in A\} = (1-\eta_n)\sum_{i=0}^{k_{\eta n}} \tilde{Q}\left\{\omega_{n,i} \in A, X_\infty(\omega_\infty) \in B_i^{(\eta_n)}\right\} + \eta_n\mu_n(A),$$

and note that, for any $n < \infty$,

$$\begin{aligned}(1-\eta_n)\sum_{i=0}^{k_{\eta n}} \tilde{Q}\left\{\omega_{n,i} \in A, X_\infty(\omega_\infty) \in B_i^{(\eta_n)}\right\} &= (1-\eta_n)\sum_{i=0}^{k_{\eta n}} Q_n(A|A_i^n)Q_\infty(X_\infty(\omega_\infty) \in B_i^{(\eta_n)}\} \\ &= -\eta_n\mu_n(A) + Q_n(A),\end{aligned}$$

proving that $Q_n = \tilde{Q} \circ \phi_n^{-1}$. Finally, we show that the random variables ϕ_n are perfect. $\phi_\infty = \pi_\infty$ is a projection in a product probability space; hence it is perfect by Example 3.7.6. Here we denote π_ξ, π_n and $\pi_{n,i}$, respectively, the projections of $\tilde{\Omega}_n$ onto $[0,1]$, Ω_n and A_i^n. With this notation, we have

$$\phi_n = I_{\pi_\xi \le 1-\eta_n} \sum_{i=0}^{k_{\eta n}} I_{X_\infty(\omega_\infty) \in B_i^{(\eta_n)}} \pi_{n,i} + I_{\pi_\xi > 1-\eta_n}\pi_n,$$

for $n \in N_2$, and the same expression with the sum starting at $i = 1$ for $n \in N_1$, and the perfectness of ϕ_n follows from Example 3.7.7. ■

3.7.4 Central Limit Theorems for Empirical Processes I: Definition and Some Properties of Donsker Classes of Functions

As usual, we let $X_i : S^{\mathbb{N}} \mapsto S$, $i \in \mathbb{N}$, be the coordinate functions on the infinite product probability space $(\Omega, \Sigma, \Pr) := (S^{\mathbb{N}}, \mathcal{S}^{\mathbb{N}}, P^{\mathbb{N}})$ (in particular, then, the functions X_i are independent identically distributed S-valued random variables with probability law P) and $\mathcal{F}$ a class of measurable functions $f : S \mapsto \mathbb{R}$. We also denote as X an S-valued random variable with law P. In this section we assume that $\mathcal{F}$ consists of P-square integrable functions and that

$$\sup_{f\in\mathcal{F}} |f(x) - Pf| < \infty, \quad \forall x \in S. \tag{3.256}$$

With this condition, the centred empirical process based on $\{X_i\}$ and indexed by $\mathcal{F}$,

$$f \mapsto (P_n(\omega) - P)(f) = \frac{1}{n}\sum_{i=1}^{n}(f(X_i(\omega)) - Pf),$$

is a bounded map $\mathcal{F} \mapsto \mathbb{R}$; that is, the centred empirical process $P_n - P$ has all its sample paths bounded, and the results from the preceding section apply to it. We may impose, instead of condition (3.256), the more restrictive conditions

$$\sup_{f\in\mathcal{F}} |f(x)| < \infty, \quad \forall x \in S, \quad \text{and} \quad \sup_{f\in\mathcal{F}} |Pf| < \infty \tag{3.257}$$

so that the uncentred empirical process $\{P_n(f), f \in \mathcal{F}\}$ is also bounded. In this subsection we introduce the central limit theorem for empirical processes. The framework and the results constitute a far-reaching generalisation of the classical work of Kolmogorov, Doob, Donsker and others on the central limit theorem for the empirical distribution function (the invariance principle), of great value in asymptotic statistics.

If $\mathcal{F}$ is in $L^2(P)$, by the central limit theorem in finite dimensions, the finite-dimensional distributions of the process

$$\nu_n(f) := \sqrt{n}(P_n - P)(f), \quad f \in \mathcal{F},$$

converge in law to the corresponding finite-dimensional distributions $(G_P(f_1), \ldots, G_P(f_k))$ of a centred Gaussian process $\{G_P(f) : f \in \mathcal{F}\}$ with covariance that of $f(X) - Pf$, $f \in \mathcal{F}$; that is,

$$\mathcal{L}\left(\frac{1}{\sqrt{n}} \sum_{l=1}^{n} (f_1(X_i) - Pf_1, \ldots, f_k(X_i) - Pf_k)\right) \to_w \mathcal{L}(G_P(f_1), \ldots, G_P(f_k)), \quad f_i \in \mathcal{F},\ k \in \mathbb{N},$$

where $G_P(f)$, $f \in \mathcal{F}$, is a centred Gaussian process with the same covariance as the process $\{f(X) : f \in \mathcal{F}\}$,

$$E(G_P(f)G_P(g)) = E[(f(X) - Pf)(g(X) - Pg)] = P[(f - Pf)(g - Pg)], \tag{3.258}$$

and where $\to_w$ denotes weak convergence of probability measures (in this case in $\mathbb{R}^n$). When P is Lebesgue measure on $[0,1]$ and $\mathcal{F} = \{I_{[0,x]} : x \in [0,1]\}$, then

$$E(G(x)G(y)) := E(G(I_{[0,x]})G(I_{[0,y]})) = x(1-y),$$

for $0 \le x \le y \le 1$; that is, G is a Brownian bridge. We may refer to G_P as the *P-bridge process indexed by* $\mathcal{F}$.

Definition 3.7.26 We say that the class of functions $\mathcal{F}$ is P-pre-Gaussian if the P-bridge process $G_P(f)$, $f \in \mathcal{F}$, admits a version whose sample paths are all bounded and uniformly continuous for its intrinsic L^2-distance $d_P^2(f,g) = P(f-g)^2 - (P(f-g))^2$, $f,g \in \mathcal{F}$.

Remark 3.7.27 By Proposition 2.1.5 and Sudakov's theorem (Corollary 2.4.13), if $\mathcal{F}$ is P-pre-Gaussian, then the pseudo-metric space $(\mathcal{F}, d_P)$ is totally bounded, and the law of G_P is a tight Borel probability measure on the Banach space $C_u(\mathcal{F}, d_P)$ (and, in particular, on $\ell_\infty(\mathcal{F})$). Thus, we will be able to apply Definition 3.7.22 to the convergence in law in $\ell_\infty(\mathcal{F} \Rightarrow$ of the empirical process $\nu_n = \sqrt{n}(P_n - P)$ to G_P.

Before considering the central limit theorem for the empirical process indexed by the class $\mathcal{F}$, it is convenient to examine the linearity of G_P. Since its covariance structure is that of $f(X)$, G_P inherits some of the linearity of the map $f \mapsto f(X)$. By computing covariances, it is clear that $\sum_{i=1}^r \lambda_i G_P(f_i) = 0$ a.s. whenever $r < \infty$, $\lambda_i \in \mathbb{R}$ and $f_i \in \mathcal{F}$ are such that $\sum_{i=1}^r \lambda_i f_i = 0$ pointwise, but in principle, the set of probability 1 where this happens may depend on the functions f_i and the constants λ_i. We may thus ask whether G_P has a version whose sample paths satisfy this linearity property besides being bounded and uniformly continuous.

Formally, if $\mathcal{F}$ is a subset of a vector space and $g : \mathcal{F} \mapsto \mathbb{R}$ satisfies $\sum_{i=1}^r \lambda_i g(f_i) = 0$ whenever $\sum_{i=1}^r \lambda_i f_i = 0$ pointwise, for $r < \infty$, $\lambda_i \in \mathbb{R}$ and $f_i \in \mathcal{F}$, we say that g is *prelinear* (on $\mathcal{F}$). If g is prelinear, then it extends uniquely to a linear function on the linear span of $\mathcal{F}$, and such an extension exists only if g is prelinear (Exercise 3.7.10). The answer to the question in the preceding paragraph is given by the following theorem:

Theorem 3.7.28 *Let $\mathcal{F}$ be a P-pre-Gaussian class of functions, and let G_P be a version of the P-bridge that is also a Borel measurable $C_u(\mathcal{F},d_P)$-valued random variable. Then, for almost all ω, the function $f \mapsto G_P(\omega)(f)$, $f \in \mathcal{F}$, is prelinear and therefore extends uniquely to a linear functional on the linear span of $\mathcal{F}$ and is bounded and uniformly d_P-continuous on the symmetric convex hull* $\mathrm{sco}(\mathcal{F})$ *of $\mathcal{F}$. Hence, it extends as a prelinear bounded uniformly continuous function on the d_P-closure $\overline{\mathrm{sco}}(\mathcal{F})$ of* $\mathrm{sco}(\mathcal{F})$.

Proof From Remark 3.7.27, the metric space $(\mathcal{F},d_X)$ is totally bounded, and therefore, $C_u(\mathcal{F},d_X)$, which is isomorphic to $C(\overline{\mathcal{F}},d_X)$, where $\overline{\mathcal{F}}$ is the completion for d_X of $\mathcal{F}$, which is compact, is a separable Banach space (see the discussion before Proposition 2.1.5). Thus, the Karhunen-Loève expansion, Theorem 2.6.10, applies to give a sequence of independent standard normal random variables k_i, with k_i and $G_P(f)$, $j \in \mathbb{N}$, $f \in \mathcal{F}$, jointly normal, such that

$$G_P = \sum_{i=1}^{\infty} [E(k_i G_P)] k_i \text{ a.s.,}$$

where the series converges in the norm of $C_u(\mathcal{F},d_X)$. That is, $G_P(f) = \sum_{i=1}^{\infty} [E(k_i G_P(f))] k_i$ uniformly in f almost surely. If, for $\lambda_j \in \mathbb{R}$ and $f_j \in \mathcal{F}$, $1 \le j \le r < \infty$, $\sum \lambda_j f_j = 0$, then $E\left(\sum \lambda_j G_P(f_j)\right)^2 = P\left(\sum \lambda_j f_j\right)^2 - \left(P\left(\sum \lambda_j f_j\right)\right)^2 = 0$, and therefore, $\sum_j \lambda_j E(k_i G(f_j)) = 0$ for each i; that is, for almost all ω, the function $\mathcal{F} \mapsto \mathbb{R}$, $f \mapsto G_P(f)(\omega)$ is prelinear, and hence it has a unique linear extension to the linear span of $\mathcal{F}$ given by $G_P\left(\sum \lambda_i f_i\right)(\omega) := \sum \lambda_i G(f_i)(\omega)$.

Let us continue denoting as G_P the just-constructed linear extension of the original process to the span of $\mathcal{F}$. Consider now the symmetric convex hull of $\mathcal{F}$,

$$\mathrm{sco}(\mathcal{F}) := \left\{ \sum_{j=1}^{r} \lambda_j f_j : f_j \in \mathcal{F}, \sum_{j=1}^{r} |\lambda_j| \le 1, r \in \mathbb{N} \right\},$$

and note that if $f,g \in \mathrm{sco}(\mathcal{F})$, we still have $E(G_P(f) - G_P(g))^2 = P(f-g)^2 - (P(f-g))^2 = d_P^2(f,g)$ so that for each k_i, $|E(k_i G_P(f)) k_i(\omega) - E(k_i G_P(g)) k_i(\omega)| \le |k_i(\omega)| d_P(f,g)$ by Hölder's inequality. Hence, the terms of the series $\sum_{i=1}^{\infty} [E(k_i G_P)] k_i$ are a.s. in $C_u(\mathrm{sco}(\mathcal{F}), d_P)$. Moreover,

$$\sup_{f \in \mathrm{sco}(\mathcal{F})} \left| \sum_{i=n}^{m} E(k_i G_P(f)) k_i(\omega) \right| = \sup_{f \in \mathcal{F}} \left| \sum_{i=n}^{m} E(k_i G_P(f)) k_i(\omega) \right|.$$

We thus conclude that

$$G_P(f) = \sum_{i=1}^{\infty} [E(k_i G_P(f))] k_i \in C_u(\mathrm{sco}(\mathcal{F}), d_P) \text{ a.s.}$$

Finally, $G_P(f)$ extends by uniform continuity to a uniformly continuous function on the closure $\overline{\mathrm{sco}}(\mathcal{F})$ of $\mathrm{sco}(\mathcal{F})$ with linearity preserved. ∎

If $\mathcal{F}$ is P-pre-Gaussian, any versions of the P-bridge whose sample paths are bounded, uniformly d_P-continuous and prelinear will be called *suitable*.

We now come to the definition of the central limit theorem uniform in $\mathcal{F}$. The finite-dimensional distributions of the (centred normalised) empirical process $\nu_n = \sqrt{n}(P_n - P)$ converge in law to those of the P-Brownian bridge G_P by the central limit theorem in $\mathbb{R}^n$ as long as $\mathcal{F} \subset L^2(S, \mathcal{S}, P)$. Moreover, if $\mathcal{F}$ is P-pre-Gaussian, then the law of G_P is a tight Borel measure in $\ell_\infty(\mathcal{F})$ (and in $C_u(\mathcal{F}, d_P)$), and hence, it is a possible limit in the definition of convergence in law for bounded processes, Definition 3.7.22. Hence, the following definition is natural:

Definition 3.7.29 We say that the class of functions $\mathcal{F} \subset L^2(S, \mathcal{S}, P)$ satisfying the boundedness condition (3.256) is a P-Donsker class or that $\mathcal{F}$ satisfies the central limit theorem for P, $\mathcal{F} \in CLT(P)$ for short, if $\mathcal{F}$ is P-pre-Gaussian and the P-empirical processes indexed by $\mathcal{F}$, $\nu_n(f) = \sqrt{n}(P_n - P)(f)$, $f \in \mathcal{F}$, converge in law in $\ell_\infty(\mathcal{F})$ to the Gaussian process G_P as $n \to \infty$.

Note that the envelope condition (3.256) is natural because ν_n is not stochastically bounded in $\ell_\infty(\mathcal{F})$ if $\sup_{f\in\mathcal{F}} |f(X) - Pf| = \infty$ with positive P-probability (see Exercise 3.7.23).

Remark 3.7.30 (Central limit theorem for Banach space–valued random variables) Let B be a separable Banach space, let P be a weakly centred Borel probability measure on B, meaning that $\int f(x)dP(x) = 0$ for all $f \in B^*$, the topological dual of B, and let X_i be i.i.d.(P) B-valued random variables. If

$$\mathcal{L}\left(\frac{1}{\sqrt{n}}\sum_{i=1}^{n} X_i\right) \to_w \mathcal{L}(Z), \tag{3.259}$$

where Z is a centred Gaussian B-valued random variable, we say that the central limit theorem (CLT) holds for P on B. If the CLT holds for P on B, then the class of functions $\mathcal{F} = B_1^*$, the unit ball of B^*, is a P-Donsker class, and conversely, if B_1^* is P-Donsker for a Borel probability measure P on B, then the CLT for P holds on B. To prove this, first observe that the map $i : B \mapsto \ell_\infty(B_1^*)$ sending $x \in B$ to evaluation at x of $f \in B_1^*$ is a linear isometric imbedding, $i(B)$ is closed in $\ell_\infty(B_1^*)$ and $i\left(\sum_{i=1}^n X_i/\sqrt{n}\right) = \sqrt{n}P_n$. Recall also that the dual bounded Lipschitz norm metrises weak convergence of probability measures on complete separable metric spaces. If $H : \ell_\infty(B_1^*) \mapsto \mathbb{R}$ is bounded Lipschitz, then so is $H \circ i : B \mapsto \mathbb{R}$ and with the same supremum and Lipschitz norms. If the CLT holds for P on B, this observation immediately yields $d_{BL}(\sqrt{n}P_n, i(Z)) \le \beta_{(B,\|\cdot\|)}\left(\mathcal{L}\left(\sum_{i=1}^n X_i/\sqrt{n}\right), \mathcal{L}(Z)\right) \to 0$, and therefore, B_1^* is P-Donsker by Proposition 3.7.24. For the converse, we need two additional observations: the first is that if $H : B \mapsto \mathbb{R}$ is bounded Lipschitz, then $i^{-1}H$ is bounded Lipschitz (with the same norms) on $i(B)$, and it extends by the Kirszbraun-McShane extension theorem (Exercise 3.6.13) to a bounded Lipschitz function $\tilde{H}$ with the same supremum and Lipschitz norms over all of $\ell_\infty(B_1^*)$, and the second is that if B_1^* is P-Donsker, then since $\sqrt{n}P_n \in i(B)$ and $i(B)$ is closed, $G_P \in i(B)$ by the Portmanteau theorem (Exercise 3.7.7); that is, $G_P = i(Z)$ for some B-valued centred Gaussian random variable Z. These two observations imply that if B_1^* is P-Donsker, then $\beta_{(B,\|\cdot\|)}\left(\mathcal{L}\left(\sum_{i=1}^n X_i/\sqrt{n}\right), \mathcal{L}(Z)\right) = d_{BL}(\sqrt{n}P_n, G_P) \to 0$ by Proposition 3.7.24, and therefore, the CLT holds for P on B.

Theorem 3.7.23 together with the central limit theorem in finite dimensions immediately yields the following *asymptotic equi-continuity criterion* for $\mathcal{F}$ to be a P-Donsker class.

Recall that by Remark 3.7.27, if $\mathcal{F}$ is P-Donsker, then the pseudo-metric space $(\mathcal{F}, d_P)$ is totally bounded.

Theorem 3.7.31 *Assume that $\mathcal{F} \subset L^2(S, \mathcal{S}, P)$ and satisfies condition (3.256). Then the following conditions are equivalent:*

(a) $\mathcal{F}$ is a P-Donsker class.
(b) The pseudo-metric space $(\mathcal{F}, d_P)$ is totally bounded, and

$$\lim_{\delta\to 0}\limsup_{n\to\infty}\Pr^*\left\{\sup_{\substack{f,g\in\mathcal{F}\\ d_P(f,g)\le\delta}}|\sqrt{n}(P_n-P)(f-g)|>\varepsilon\right\}=0, \tag{3.260}$$

for all $\varepsilon > 0$.
(c) There exists a pseudo-distance e on $\mathcal{F}$ such that $(\mathcal{F}, e)$ is totally bounded, and

$$\lim_{\delta\to 0}\limsup_{n\to\infty}\Pr^*\left\{\sup_{\substack{f,g\in\mathcal{F}\\ e(f,g)\le\delta}}|\sqrt{n}(P_n-P)(f-g)|>\varepsilon\right\}=0, \tag{3.261}$$

for all $\varepsilon > 0$.

A typical distance e in condition (c) is $e(f,g) = e_P(f,g) = \|f-g\|_{L^2(P)}$, but it is not the only one that we will use.

With this theorem, proving the central limit theorem for the empirical process essentially reduces to proving a maximal inequality, which is the subject of several of the preceding sections. Here is an application of Theorem 3.7.31 to an important necessary integrability condition for $\mathcal{F}$ to be P-Donsker: if $\mathcal{F}$ is P-Donsker, the measurable cover of F is in weak-$L^2(P)$, in particular, $PF^\alpha < \infty$, for all $0 < \alpha < 2$.

Proposition 3.7.32 *Let $\mathcal{F} \subset L^2(P)$ be a P-Donsker class satisfying the conditions (3.257), and let F be its measurable cover. Then*

$$\lim_{t\to\infty} t^2 \Pr\{F > t\} = 0.$$

If the P-Donsker class $\mathcal{F}$ only satisfies condition (3.256), and $\overline{F}$ is the measurable cover of the centred class $\{f - Pf : f \in \mathcal{F}\}$, then

$$\lim_{t\to\infty} t^2 \Pr\left\{\overline{F} > t\right\} = 0.$$

Proof Since in the first case $\sup_{f\in\mathcal{F}}|f(x) - Pf| + \sup_{f\in\mathcal{F}}|Pf| \ge F(x)$ and the second summand is a finite number, it suffices to prove the second part of the proposition. Given $\tau > 0$, let $f_1, \ldots, f_{N(\tau)}$ be a subset of $\mathcal{F}$ τ-dense for the pseudo-metric d_P, which exists by Theorem 3.7.31, and define the functionals $Y_i(f) = f(X_i) - Pf$, $Y_{i,\tau}(f) = Y_i(\pi_\tau f))$, where $\pi_\tau : \mathcal{F} \mapsto \{f_1, \ldots, f_{N(\tau)}\}$ is a mapping satisfying $d_P(f, \pi_\tau(f)) \le \tau$. Then

$$n\Pr^*\left\{\overline{F}(X_1) > 2n^{1/2}\right\} \le n\Pr^*\left\{\|Y_1 - Y_{1,\tau}\|_{\mathcal{F}} > n^{1/2}\right\} + n\Pr\left\{\|Y_{1,\tau}\|_{\mathcal{F}} > n^{1/2}\right\}.$$

Since $E\|Y_{1,\tau}\|_{\mathcal{F}}^2 = E\max_{i\le N(\tau)}|f_i(X_1) - Pf_i|^2 \le N(\tau)\max_{i\le N(\tau)} Pf_i^2 < \infty$, it follows that

$$\lim_{n\to\infty} n\Pr\left\{\|Y_{1,\tau}\|_{\mathcal{F}} > n^{1/2}\right\} = 0, \quad \text{for all } \tau > 0,$$

and we only need to prove that the first summand tends to zero as we let first $n \to \infty$ and then τ to zero. To this end, note that for $x \geq 0$, $x/(1+x) = 1 - 1/(1+x) \leq 1 - e^{-x}$. Then, using Lévy's inequality (3.34) (together with the calculus of nonmeasurable functions, concretely Exercises 3.7.1 and 3.7.3 and Proposition 3.7.2), we obtain

$$\frac{n\Pr^*\left\{\|Y_1 - Y_{1,\tau}\|_{\mathcal{F}} > n^{1/2}\right\}}{1 + n\Pr^*\left\{\|Y_1 - Y_{1,\tau}\|_{\mathcal{F}} > n^{1/2}\right\}} \leq 1 - \exp\left(-n\Pr\left\{\|Y_1 - Y_{1,\tau}\|^*_{\mathcal{F}} > n^{1/2}\right\}\right)$$

$$\leq 1 - \prod_{i=1}^{n}\left[1 - \Pr\left\{\|Y_i - Y_{i,\tau}\|^*_{\mathcal{F}} > n^{1/2}\right\}\right]$$

$$\leq \Pr^*\left\{\max_{i\leq n}\|Y_i - Y_{i,\tau}\|_{\mathcal{F}} > n^{1/2}\right\}$$

$$\leq 2\Pr^*\left\{\left\|\sum_{i=1}^{n}\varepsilon_i(Y_i - Y_{i,\tau})\right\|_{\mathcal{F}} > n^{1/2}\right\}$$

$$\leq 2\Pr^*\left\{\sup_{\substack{f,g\in\mathcal{F}\\ d_P(f,g)\leq\tau}}\left|\sum_{i=1}^{n}\varepsilon_i((f-g)(X_i) - P(f-g))\right| > n^{1/2}\right\} := (I_n), \tag{3.262}$$

where ε_i and X_j are all independent, in fact, coordinates in a product probability space, and $\Pr\{\varepsilon_i = 1\} = \Pr\{\varepsilon_i = -1\} = 1/2$ (Rademacher variables). Now, by Proposition 3.1.23 (and Exercise 3.7.3),

$$(I_n) \leq 6\max_{1\leq j\leq n}\Pr^*\left\{\sup_{\substack{f,g\in\mathcal{F}\\ d_P(f,g)\leq\tau}}|\nu_j(f-g)| > \frac{n^{1/2}}{9j^{1/2}}\right\}.$$

It is easy to see that in Theorem 3.7.31, condition (b), limsup over n can in fact be replaced by supremum over n; hence, Theorem 3.7.31 shows that $\lim_{\tau\to 0}\sup_n (I_n) = 0$. Since also $x/(1+x) < \eta$ iff $x < \eta/(1-\eta)$ for $x, \eta > 0$, this last limit and inequality (3.262) yield

$$\lim_{\tau\to 0}\limsup_{n} n\Pr^*\left\{\|Y_1 - Y_{1,\tau}\|_{\mathcal{F}} > n^{1/2}\right\} = 0. \quad \blacksquare$$

Next, we consider two of the most important permanence properties of Donsker classes.

Proposition 3.7.33 *If $\mathcal{F}_1$ and $\mathcal{F}_2$ are P-Donsker classes of functions, then so is $\mathcal{F}_1 \cup \mathcal{F}_2$.*

Proof Recall the notation $\nu_n = \sqrt{n}(P_n - P)$. Since

$$\sup_{\substack{f,g\in\mathcal{F}_1\cup\mathcal{F}_2\\ d_P(f,g)\leq\delta}}|\nu_n(f-g)| \leq \sum_{i=1}^{2}\sup_{\substack{f,g\in\mathcal{F}_i\\ d_P(f,g)\leq\delta}}|\nu_n(f-g)| + \sup_{\substack{f\in\mathcal{F}_1,g\in\mathcal{F}_2\\ d_P(f,g)\leq\delta}}|\nu_n(f-g)|,$$

by Theorem 3.7.31 applied to $\mathcal{F}_1$ and $\mathcal{F}_2$ and to their union, it suffices to prove that

$$\lim_{\delta\to 0}\limsup_{n\to\infty}\Pr^*\left\{\sup_{\substack{f\in\mathcal{F}_1,g\in\mathcal{F}_2\\ d_P(f,g)\leq\delta}}|\nu_n(f-g)| > \varepsilon\right\} = 0, \tag{3.263}$$

for all $\varepsilon > 0$. Note also that since the classes $\mathcal{F}_i$ are P-pre-Gaussian, Sudakov's minorisation for sample continuous Gaussian processes (Corollary 2.4.14) implies that

$$\lim_{\delta\to 0}\delta\sqrt{\log N(\mathcal{F}_i, d_P, \delta)} = 0, \quad i = 1,2. \tag{3.264}$$

Fix $\varepsilon > 0$. Then, given $\tau > 0$, there exists $\delta_0 > 0$ such that both

$$\limsup_{n\to\infty} \Pr{}^*\left\{ \sup_{\substack{f,g\in\mathcal{F}_i\\ d_P(f,g)\le\delta_0}} |\nu_n(f-g)| > \varepsilon/3 \right\} < \tau/3, \quad i = 1,2, \tag{3.265}$$

and

$$\frac{9}{\varepsilon}\delta_0\sqrt{2\log[2(N(\mathcal{F}_1,d_P,\delta_0)+N(\mathcal{F}_2,d_P,\delta_0))^2]} < \tau/3. \tag{3.266}$$

Let $\mathcal{G}_i \subset \mathcal{F}_i$ be δ_0-dense in $\mathcal{F}_i$ and with cardinality $N(\mathcal{F}_i, d_P, \delta_0)$, $i = 1,2$. For each $f \in \mathcal{F}_i$, let $\pi_i f \in \mathcal{G}_i$ be such that $d_P(f, \pi_i f) \le \delta_0$. Let $\mathcal{G} = \mathcal{G}_1 \cup \mathcal{G}_2$. Then, for $f \in \mathcal{F}_1$ and $g \in \mathcal{F}_2$, the decomposition

$$f - g = (f - \pi_1 f) + (\pi_1 f - \pi_2 g) + (\pi_2 g - g)$$

together with the inequalities (3.265) give, for $0 < \delta \le \delta_0$,

$$\begin{aligned}\limsup_{n\to\infty} \Pr{}^*\left\{ \sup_{\substack{f\in\mathcal{F}_1, g\in\mathcal{F}_2\\ d_P(f,g)\le\delta}} |\nu_n(f-g)| > \varepsilon \right\} &< 2\tau/3\\ &+ \limsup_{n\to\infty} \Pr{}^*\left\{ \sup_{\substack{f,g\in\mathcal{G}\\ d_P(f,g)\le 3\delta_0}} |\nu_n(f-g)| > \varepsilon/3 \right\}.\end{aligned} \tag{3.267}$$

Let $d = \mathrm{Card}(\mathcal{G}) \le N(\mathcal{F}_1, d_P, \delta_0) + N(\mathcal{F}_2, d_P, \delta_0)$, and denote by $x = (x_f : f \in \mathcal{G})$ the points in $\mathbb{R}^d$. By the central limit theorem in $\mathbb{R}^d$, we have that the $\mathbb{R}^d$-valued random vectors $\nu_n|_{\mathcal{G}} = (\nu_n(f) : f \in \mathcal{G})$ converge in law to the normal variable $G_P|_{\mathcal{G}} = (G_P(f) : f \in \mathcal{G})$. In particular, for any closed set F of $\mathbb{R}^d$, we have by the Portmanteau theorem in finite dimensions (see also Exercise 3.7.7) that $\limsup_n \Pr\{\nu_n|_{\mathcal{G}} \in F\} \le \Pr\{G_P|_{\mathcal{G}} \in F\}$. Let $F_{\delta_0} = \{x : \max_{d_P(f,g)\le 3\delta_0} |x_f - x_g| \ge \varepsilon/3\} \subset \mathbb{R}^d$, and note that the set F_{δ_0} is closed and that, by, for example, Lemma 2.3.4,

$$\Pr\{G_P|_{\mathcal{G}} \in F_{\delta_0}\} \le \frac{9\delta_0}{\varepsilon}\sqrt{2\log(2d^2)},$$

which is smaller than $\tau/3$ by inequality (3.266) and the definition of d. Therefore,

$$\limsup_{n\to\infty} \Pr{}^*\left\{ \sup_{\substack{f,g\in\mathcal{G}\\ d_P(f,g)\le 3\delta_0}} |\nu_n(f-g)| \ge \varepsilon/3 \right\} \le \Pr\left\{ \sup_{\substack{f,g\in\mathcal{G}\\ d_P(f,g)\le 3\delta_0}} |G_P(f) - G_P(g)| \ge \varepsilon/3 \right\} < \tau/3.$$

The limit (3.263) now follows from this inequality and (3.267). ■

For the next property, we need a definition. Recall that the *symmetric convex hull of* $\mathcal{F}$, $\mathrm{sco}(\mathcal{F})$, is defined as the set of functions g of the form $g = \sum_{i=1}^k \lambda_i f_i$, where $k \in \mathbb{N}$, $f_i \in \mathcal{F}$

and $\lambda_i \in \mathbb{R}$ with $\sum_{k=1}^{n} |\lambda_i| \le 1$. We denote as $H(\mathcal{F},P)$ the sequential closure, both pointwise and in $L^2(P)$, of the symmetric convex hull of $\mathcal{F}$, that is,

$$H(\mathcal{F},P) = \left\{ g : S \mapsto \mathbb{R} : g(x) = \lim_n g_n(x)\ \forall x \in S \text{ and } \lim_n P(g_n - g)^2 = 0,\ g_n \in \mathrm{sco}(\mathcal{F}) \right\}.$$

The result holds as well with the term *sequential* removed (see Exercise 3.7.14).

Proposition 3.7.34 *If $\mathcal{F}$ is P-Donsker, so is $H(\mathcal{F},P)$.*

Proof To prove (a), we apply Theorem 3.7.25 on almost-sure representations for convergence in law for $(\Omega_n, \mathcal{A}_n, \mathcal{Q}_n) = (S^n, \mathcal{S}^n, P^n)$, $n < \infty$, and for $(\Omega_\infty, \mathcal{A}_\infty, \mathcal{Q}_\infty)$ a probability space where a suitable version of G_P is defined over $\overline{\mathrm{sco}}(\mathcal{F})$, for $X_n = \nu_n$, $n < \infty$, and $X_\infty = G_P$ (a suitable version), and for $T = \mathcal{F}$ and, in a second instance, $T = H(\mathcal{F},P)$. Note that the range of G_P, $C_u(\mathrm{sco}\mathcal{F})$, is separable. By Theorem 3.7.25, there exist a probability space $(\tilde{\Omega}, \tilde{\mathcal{A}}, \tilde{P})$ and perfect maps $\phi_n : \tilde{\Omega}_n \mapsto \Omega_n$, $n \le \infty$, such that $P^n = \tilde{\mathcal{Q}} \circ \phi_n^{-1}$ for $n < \infty$, $\mathcal{Q}_\infty = \tilde{\mathcal{Q}} \circ \phi_\infty^{-1}$ and the processes

$$\tilde{\nu}_n(f,\omega) := \nu_n \circ \phi_n(f,\omega) = \frac{1}{n} \sum_{i=1}^{n} (f(\phi_n(\omega)_i) - Pf),\ \tilde{G}_P(f,\omega) := G_P(f, \phi_\infty(\omega))$$

satisfy $\sup_{f \in \mathcal{F}} |\tilde{\nu}_n(f) - \tilde{G}_P(f)|^* \to 0$ a.s., where $\phi_n(\omega)_i$ is the ith coordinate of $\phi_n(\omega) \in S^n$. Note that $\tilde{\nu}_n$ and $\tilde{G}_P$ are versions of ν_n and G_P. But, by linearity and continuity,

$$\sup_{f \in \mathcal{F}} |\tilde{\nu}_n(f) - \tilde{G}_P(f)| = \sup_{f \in H(\mathcal{F},P)} |\tilde{\nu}_n(f) - \tilde{G}_P(f)|,$$

so $\sup_{f \in H(\mathcal{F},P)} |\tilde{\nu}_n(f) - \tilde{G}_P(f)|^* \to 0$ a.s. Hence, another application of Theorem 3.7.25 then gives $\nu_n \to_w G_P$ in $\ell_\infty(H(\mathcal{F},P))$. ■

It is clear that if $\mathcal{F}$ is P-Donsker, so is $\lambda\mathcal{F} = \{\lambda f : f \in \mathcal{F}\}$, for all $\lambda \in \mathbb{R}$. Thus, as a corollary to the preceding two permanence properties of the Donsker property, we have the following one:

Corollary 3.7.35 *If $\mathcal{F}_i$, $1 \le k < \infty$, are P-Donsker, so is $\mathcal{F} = \left\{ \sum_{i=1}^{k} f_i : f_i \in \mathcal{F}_i \right\}$.*

3.7.5 Central Limit Theorems for Empirical Processes II: Metric and Bracketing Entropy Sufficient Conditions for the Donsker Property

The expectation bounds based on random entropies (hence, in particular, for VC type classes) or on bracketing entropy given in Section 3.5 immediately provide what are probably the most useful central limit theorems for empirical processes. In the case of random entropies, we need some measurability, so we will assume our classes of functions to be P-measurable (see Definition 3.7.11). Recall the notation $e_{n,p}^p(f,g) = P_n|f-g|^p$ for $p \ge 1$ and $e(f,g) = e_P(f,g) = \|f-g\|_{L^2(P)}$.

Theorem 3.7.36 *Let $\mathcal{F}$ be a class of measurable functions satisfying condition (3.257) and with a measurable envelope F in $L^2(P)$. Assume that the classes of functions*

$\mathcal{F}$, $\mathcal{G} := \{(f-g)^2 : f,g \in \mathcal{F}\}$ and $\mathcal{F}'_\delta := \{f-g : f,g \in \mathcal{F}, \|f-g\|_{L^2(P)} \le \delta\}$ for all $\delta > 0$ are all P-measurable. Then, if

$$\lim_{\delta\to 0}\limsup_{n\to\infty} E\left[1 \wedge \int_0^\delta \sqrt{\log N^*(\mathcal{F}, e_{n,2}, \varepsilon)}\, d\varepsilon\right] = 0, \tag{3.268}$$

the class $\mathcal{F}$ is P-Donsker.

Note that we take the finiteness of the double limit (3.268) to mean in particular that the expected values involved are finite for all $n \in \mathbb{N}$ and $\delta > 0$.

Proof We will apply the random entropy bound (3.166) for the expected value of the empirical process (with the packing number D replaced by the measurable cover N^* of N) in combination with the asymptotic equi-continuity criterion, Theorem 3.7.23. We begin by showing that $\mathcal{G}$ is P-Glivenko-Cantelli. First, $G = \sup_{g\in\mathcal{G}}|g| \le 4F^2 \in L^1(P)$. Next, note that for each $M < \infty$,

$$\begin{aligned}|(f_1-g_1)^2-(f_2-g_2)^2|I_{G\le M} &= |f_1-g_1+f_2-g_2||f_1-g_1-f_2+g_2|I_{G\le M}\\ &\le 2M^{1/2}(|f_1-f_2|+|g_2-g_1|),\end{aligned}$$

which implies that

$$N(\mathcal{G}_M, e_{n,2}, \varepsilon) \le N^2(\mathcal{F}, e_{n,2}, \varepsilon/4M^{1/2}),$$

where $\mathcal{G}_M := \{gI_{G\le M} : g \in \mathcal{G}\}$. Then

$$\begin{aligned}n^{-1/2}\int_\delta^M \sqrt{\log N^*(\mathcal{G}_M, e_{n,2}, \varepsilon)}\, d\varepsilon &\le Mn^{-1/2}\sqrt{2\log N^*(\mathcal{F}, e_{n,2}, \delta/4M^{1/2})}\\ &\le \frac{4M^{3/2}}{n^{1/2}\delta}\int_0^{\delta/4M^{1/2}} \sqrt{2\log N^*(\mathcal{F}, e_{n,2}, \tau)}\, d\tau,\end{aligned}$$

which is dominated, for all $n \ge 16M^3/\delta^2$, by just the last integral. Hence, for these values of n and δ, we have

$$n^{-1/2}\int_0^M \sqrt{\log N^*(\mathcal{G}_M, e_{n,2}, \varepsilon)}\, d\varepsilon \le 2\int_0^{\delta/4M^{1/2}} \sqrt{2\log N^*(\mathcal{F}, e_{n,2}, \tau)}\, d\tau.$$

Now, condition (e) in Theorem 3.7.14 for $\mathcal{G}$ follows from condition (3.268) for $\mathcal{F}$, proving that $\mathcal{G}$ is a Glivenko-Cantelli class. Next, we show that $(\mathcal{F}, e_P)$ is totally bounded, where $e_P(f,g) = \|f-g\|_{L^2(P)}$. On the one hand, the hypothesis of the theorem implies that for each n and ε there is a set of Pr measure 1 where $N(\mathcal{F}, e_{n,2}, \varepsilon^{1/2}) < \infty$. On the other hand, since $\mathcal{G}$ is P-GC, there is n such that the event $\{\sup_{g\in\mathcal{G}}|P_n(g) - P(g)| \le \varepsilon\}$ has positive probability. Thus, for ω in the intersection of the two events, if $f_i \in \mathcal{F}$ are the centres of a covering of $\mathcal{F}$ by $N(\mathcal{F}, e_{n,2}(\omega), \varepsilon^{1/2})$ balls for the $e_{n,2}(\omega)$ pseudo-metric, they are also the centres of a covering of $\mathcal{F}$ by balls of radius $(2\varepsilon)^{1/2}$ in the $L^2(P)$ pseudo-metric: $P_n(\omega)(f-f_i)^2 \le \varepsilon$ implies, for these ω, that $P(f-f_i)^2 \le \varepsilon + P_n(\omega)(f-f_i)^2 \le 2\varepsilon$.

Finally, we prove (3.261) for the metric $\tau = e_P$, which will conclude the proof of the theorem by the asymptotic equi-continuity criterion for Donsker classes. We have

$$\Pr\left\{\|\nu_n(f-g)\|_{\mathcal{F}'_\delta}| > \varepsilon\right\} \le \Pr\left\{\sup_{\substack{f,g\in\mathcal{F}\\ e_{P,n}(f,g)\le\sqrt{2}\delta}} |\nu_n(f-g)| > \varepsilon\right\}$$

$$+\Pr\left\{\sup_{f,g\in\mathcal{F}} |e^2_{P,n}(f,g) - e^2_P(f,g)| \ge \delta^2\right\}$$

$$\le E_X\left[1 \wedge \frac{2}{\varepsilon}E_\varepsilon\left(\sup_{\substack{f,g\in\mathcal{F}\\ e_{P,n}(f,g)\le\sqrt{2}\delta}} \left|\frac{1}{\sqrt{n}}\sum_{i=1}^n \varepsilon_i(f-g)(X_i)\right|\right)\right]$$

$$+\Pr\left\{\sup_{g\in\mathcal{G}} |P_n(g) - P(g)| > \delta^2\right\}.$$

The second probability tends to zero as $n \to \infty$ for all $\delta > 0$ because $\mathcal{G}$ is P-Glivenko-Cantelli, whereas the first is dominated, by Chebyshev's inequality and (2.42) Theorem 2.3.7, by

$$\frac{2(16\sqrt{2}+2)}{\varepsilon}E\left[1 \wedge \int_0^{\sqrt{2}\delta} \sqrt{\log N^*(\mathcal{F}, e_{n,2}, \tau)}\, d\tau\right],$$

which, by (3.268), tends to zero when we take first $\limsup_{n\to\infty}$ and then $\lim_{\delta\to 0}$. This proves the asymptotic equi-continuity of the empirical process indexed by $\mathcal{F}$ for the metric e_P, for which $\mathcal{F}$ is totally bounded. Hence, $\mathcal{F}$ is P-Donsker by Theorem 3.7.23. ■

Here is the main application of this theorem (see Exercise 3.7.13 for another application):

Theorem 3.7.37 *Let $\mathcal{F}$ satisfy the P measurability conditions in Theorem 3.7.36, and assume that (a) the P-measurable cover F of $\mathcal{F}$ is in $L^2(P)$ and (b) for some $a > 0$ there exists a function $\lambda : [0,a) \mapsto \mathbb{R}$ integrable on $[0,a)$ for Lebesgue measure such that*

$$\sup_Q \sqrt{\log N(\mathcal{F}, L^2(Q), \varepsilon\|F\|_{L^2(Q)})} \le \lambda(\varepsilon), \quad 0 \le \varepsilon \le a, \tag{3.269}$$

where the supremum is over all discrete probability measures Q on S with a finite number of atoms and rational weights on them. Then $\mathcal{F}$ is P-Donsker. In particular, if $\mathcal{F}$ is VC subgraph, VC type, VC hull, or a finite union or sum (in the sense of Corollary 3.7.35) of such classes, and if $F \in L^2(P)$, then $\mathcal{F}$ is P-Donsker.

Proof Assume that $\mathcal{F}$ satisfies condition (3.269) and, without loss of generality, that the measurable cover $F \in L^2(P)$ satisfies $F(s) \ge 1$ for all $s \in S$. Since $\sqrt{\log N^*(\mathcal{F}, e_{n,2}(\omega), \varepsilon\|F\|_{L^2(P_n)})} \le \lambda(\varepsilon)$ for all n and ω and $0 < \varepsilon \le a$, we have, for $\delta \le a$,

$$\int_0^\delta \sqrt{\log N^*(\mathcal{F}, e_{n,2}, \varepsilon)}\, d\varepsilon \le (P_nF^2)^{1/2}\int_0^\delta \lambda(\tau)d\tau = (P_nF^2)^{1/2}O(\delta). \tag{3.270}$$

Then the law of large numbers for F proves condition (3.268), and $\mathcal{F}$ is P-Donsker by Theorem 3.7.36. The consequence for VC type classes follows from the uniform metric entropy bounds they satisfy (Theorems 3.6.9 and 3.6.17 and Definition 3.6.10) and Proposition 3.7.33 and Corollary 3.7.35. ∎

Next, we present the *bracketing CLT*, based on (the proof of) Theorem 3.5.13.

Theorem 3.7.38 *Let $\mathcal{F}$ be a class of measurable functions on S with measurable cover F in $L^2(P)$ and satisfying the $L^2(P)$-bracketing condition*

$$\int_0^{2\|F\|_2} \sqrt{\log(N_{[]}(\mathcal{F},L^2(P),\tau)}\, d\tau < \infty,$$

where we write $\|F\|_2$ for $\|F\|_{L^2(P)}$. Then $\mathcal{F}$ is P-Donsker.

Proof In the context of the proof of Theorem 3.5.13, consider the collection of nested partitions $\{A_{k,i} : 1 \le i \le \tilde{N}_k\}$, $k \ge 1$, where $\tilde{N}_k = N_1,\dots,N_k$, $N_k = N(\mathcal{F},L^2(P),2^{-k})$. Define a pseudo-distance on $\mathcal{F}$ by $\rho(f,g) = 1/2^k$ if k is the largest integer such that both f and g are in the same set $A_{k,i}$ from the kth partition, and $\rho(f,g) = 0$ if no such k exists (in which case $f = g$ a.s.). Then $(\mathcal{F},\rho)$ is totally bounded: just note that each partition set $A_{k,i}$ is contained in a closed ρ-ball of radius $1/2^k$ and that there are just $\tilde{N}_k$ of them, a finite number by the bracketing condition. Recall the projections $\pi_k f$ from the same proof ($\pi_k f$ is a predetermined function $f_{k,i} \in A_{k,i}$ if $f \in A_{k,i}$). Then

$$\sup_{\substack{\rho(f,g)\le 2^{-j}\\ f,g\in\mathcal{F}}} |\nu_n(f-g)| \le 2\|\nu_n(f-\pi_j f)\|_{\mathcal{F}}$$

because, if $\rho(f,g) \le 2^{-j}$, then $\pi_j f = \pi_j g$ and $|\nu_n(f-g) \le |\nu_n(f-\pi_j f)| + |\nu_n(g-\pi_j g)|$. We then obtain, by inequality (3.212),

$$E\sup_{\substack{\rho(f,g)\le 2^{-j}\\ f,g\in\mathcal{F}}} |\nu_n(f-g)| \le 111\int_0^{2^{-j}} \sqrt{\log(N_{[]}(\mathcal{F},L^2(P),\tau)}\, d\tau + 8\sqrt{n}P(FI_{2F>\sqrt{n}a_j}),$$

where $a_j^{-1} = 2^{j+1}\sqrt{\log \tilde{N}_{j+1}}$. The first summand does not depend on n and tends to zero as $j \to \infty$ by the bracketing entropy hypothesis, whereas the second is dominated, for each j fixed, by $16a_j^{-1}P[F^2 I_{2F>\sqrt{n}a_j}] \to 0$ as $n \to \infty$ because $PF^2 < \infty$. In conclusion, the pseudo-metric space $(\mathcal{F},\rho)$ is totally bounded, and

$$\lim_{j\to\infty}\limsup_n E\sup_{\substack{\rho(f,g)\le 2^{-j}\\ f,g\in\mathcal{F}}} |\nu_n(f-g)| = 0,$$

so $\mathcal{F} \in CLT(P)$ by the asymptotic equi-continuity criterion, Theorem 3.7.31 ($\mathcal{F}$ satisfies (3.7.31) for ρ). ∎

Either of the preceding theorems implies the classical Donsker-Kolmogorov-Dudley theorem for the empirical distribution function of an i.i.d. sample in $\mathbb{R}^d$. For instance, using Example 3.7.19, we obtain the following:

Corollary 3.7.39 *Let $X_1,\dots,X_n$ be i.i.d. $\mathbb{R}^d$-valued random variables from a law P with distribution function $F(t) = P((-\infty,t])$, and let $F_n = P_n((-\infty,t])$. Then, as $n \to \infty$,*

$$\sqrt{n}(F_n - F) \to_w G_P \text{ in } \ell_\infty(\mathbb{R}^d). \tag{3.271}$$

In fact, the convergence in this corollary is uniform in P. We investigate such uniformity results in the next section.

3.7.6 Central Limit Theorems for Empirical Processes III: Limit Theorems Uniform in P and Limit Theorems for P-Pre-Gaussian Classes

We consider now two additional questions on the CLT for empirical processes, related only in that both require deep use of Gaussian process theory. The first is, on what classes of functions $\mathcal{F}$ does the empirical process hold uniformly in P? The answer to this question is relevant in statistics because in inference problems one does not usually know the distribution of the data. We could ask the same question for the law of large numbers, but this subject will not be treated here (see the notes at the end of this chapter). The second question is more theoretical: if $\mathcal{F}$ is P-Donsker, then it is P-pre-Gaussian (since a necessary condition for the limit theorem to hold is that the limit exists), and determining whether a class of functions is P pre-Gaussian is simpler than showing that it is P-Donsker. This then begs the following question: what additional conditions (if any) should a P-pre-Gaussian class of functions satisfy in order for it to be P-Donsker?

Randomisation by Rademacher or by normal multipliers is very convenient when dealing with these (and other) questions, so we examine randomisation briefly before considering the two main subjects of this section. To efficiently apply randomisation, we must impose two mild conditions on the class $\mathcal{F}$, namely, that it be *measurable* and that it be $L^1(P)$-*bounded*. Thus, we assume the boundedness conditions (3.257) instead of the slightly weaker (3.256). Consider the Rademacher randomisation of the empirical process,

$$\nu_{n,\mathrm{rad}}(f) = \frac{1}{\sqrt{n}}\sum_{i=1}^{n} \varepsilon_i f(X_i), \quad f \in \mathcal{F}, \tag{3.272}$$

where the random variables ε_i, X_j are all coordinates in a infinite product probability space, and $\Pr(\varepsilon_i = 1\} = \Pr\{\varepsilon_i = -1\} = 1/2$, and the Gaussian randomisation

$$\nu_{n,g}(f) = \frac{1}{\sqrt{n}}\sum_{i=1}^{n} g_i f(X_i), \quad f \in \mathcal{F}, \tag{3.273}$$

where the random variables g_i, X_j are also all coordinates in a infinite product probability space, but the variables g_i are standard normal. Both processes have covariance $E(f(X)g(X))$ or, what is the same,

$$E(\nu_{n,\mathrm{rad}}(f) - \nu_{n,\mathrm{rad}}(h))^2 = P(f-h)^2 = e_P(f,h),$$

and the same is true for $\nu_{n,g}$. We will denote by $Z_P(f)$, $f \in \mathcal{F}$, the centred Gaussian process with this covariance, $E(Z_P(f)Z_P(h)) = P(fh)$, and hence with intrinsic metric $e_P^2(f,h) = E(Z_P(f) - Z_P(h))^2 = P(f-h)^2$. It is related to G_P, the P-bridge, by the fact that the process $G_P(f) + gP(f)$, where g is standard normal independent of G_P, is a version of Z_P, as can be seen by computing covariances, so we will call it the *P-Brownian motion* or just *P-motion*. We express this relationship by the equation

$$Z_P(f) = G_P(f) + gP(f), \quad f \in \mathcal{F}, \tag{3.274}$$

which, in fact, makes sense for all $f \in L^2(P)$. Let us also recall the notation $\mathcal{F}'_\delta = \{f - g : f, g \in \mathcal{F}, e_P(f,g) \le \delta\}$.

Theorem 3.7.40 *Assume that $\mathcal{F}$ and $\mathcal{F}'_\delta$ are measurable classes of functions for every $\delta > 0$ and that $\mathcal{F}$ satisfies the boundedness conditions (3.257). Then the following conditions are equivalent:*

(a) $\mathcal{F}$ is P-Donsker.
(b) The process $Z_P(f)$, $f \in \mathcal{F}$, admits a version with bounded and e_P-equi-continuous sample paths, and $\nu_{n,\mathrm{rad}} \to_{\mathcal{L}} Z_P$ in $\ell_\infty(\mathcal{F})$.

If (a) or (b) hold, Z_P admits a suitable version.

Proof If $(\mathcal{F}, e_P)$ is totally bounded, then so is $(\mathcal{F}, d_P)$ because $d_P \le e_P$, and if $\mathcal{F}$ is L^1-bounded and $(\mathcal{F}, d_P)$ is totally bounded, then so is $(\mathcal{F}, e_P)$. To see the last assertion, if the $L^1(P)$ bound of $\mathcal{F}$ is M, we divide $\mathcal{F}$ into $[4M/\varepsilon] + 1$ subsets $\mathcal{H}_i$ such that if f and g belong to one of these subsets; then $|Pf - Pg| \le \varepsilon/2$. Next, for each $i \le [4M/\varepsilon] + 1$, we find a $\varepsilon/2$-dense subset of $\mathcal{H}_i$ in the sense of d_P: the union of these sets, which is finite, is ε-dense in $\mathcal{F}$ for e_P because $e(f,g) \le d(f,g) + |Pf - Pg|$.

Suppose that $\mathcal{F}$ is P-Donsker. Then, by Theorem 3.7.31, $(\mathcal{F}, d_P)$ is totally bounded, and the asymptotic equi-continuity condition (3.260) with respect to d_P holds for ν_n. Also, replacing d_P by e_P in this condition results in a weaker statement, so we have

$$\lim_{\delta \to 0} \sup_{n \ge 1} \Pr{}^* \left\{ \|\nu_n\|_{\mathcal{F}'_\delta} > \varepsilon \right\} = 0,$$

for all $\varepsilon > 0$ (the limsup over n in (b) and (c), Theorem 3.7.31, can be replaced by supremum over n). Now, by Proposition 3.1.23 and Exercise 3.7.3,

$$\Pr{}^* \left\{ \|\nu_{n,\mathrm{rad}}\|_{\mathcal{F}'_\delta} > \varepsilon \right\} \le 3 \max_{j \le n} \Pr \left\{ \|\nu_j\|_{\mathcal{F}'_\delta} > \frac{\sqrt{n}\varepsilon}{9\sqrt{j}} \right\},$$

for all n and $\varepsilon > 0$. Hence,

$$\lim_{\delta \to 0} \sup_{n \ge 1} \Pr{}^* \left\{ \|\nu_{n,\mathrm{rad}}\|_{\mathcal{F}'_\delta} > \varepsilon \right\} = 0. \tag{3.275}$$

This, together with the fact that $(\mathcal{F}, e_P)$ is totally bounded by the observation at the beginning of the proof, implies by Theorem 3.7.31 that Z_P admits versions with bounded e_P-uniformly continuous sample paths and that $\nu_{n,\mathrm{rad}} \to_{\mathcal{L}} Z_P$ in $\ell^\infty(\mathcal{F})$.

Assume now that (b) holds. Then $(\mathcal{F}, e_P)$ is totally bounded by Sudakov's lower bound (Corollary 2.4.13), and hence, Theorem 3.7.31 implies the asymptotic equi-continuity (3.275). But then, by Corollary 3.1.25,

$$\Pr \left\{ \|\nu_n\|_{\mathcal{F}'_\delta} > \varepsilon \right\} \le 4 \Pr{}^* \left\{ \|\nu_{n,\mathrm{rad}}\|_{\mathcal{F}'_\delta} > (\varepsilon - \sqrt{2\delta})/2 \right\},$$

which gives the asymptotic equi-continuity of ν_n with respect to e_P. Theorem 3.7.31 now implies that $\mathcal{F}$ is P-Donsker.

If (a) or (b) hold, then, $\mathcal{F}$ being P-Donsker, G_P admits a suitable version, call it G_P, but continuity with respect to d_P implies continuity with respect to e_P, and $f \mapsto gPf$ is also linear, bounded on $\mathcal{F}$, and e_P uniformly continuous; therefore, the process $Z_P(f) = G_P(f) + gPf$, $f \in \mathcal{F}$, has prelinear, bounded and e_P-uniformly continuous sample paths. ∎

See Exercises 3.7.15 and 3.7.16 for additional necessary and sufficient conditions for $\mathcal{F}$ to be a P-Donsker class also based on randomisation.

Uniform Donsker and Uniformly Pre-Gaussian Classes of Functions

We begin with the definitions of uniformly pre-Gaussian and uniform Donsker classes of functions, with some discussion. Let $\mathcal{P}(S)$ be the set of all probability measures on $(S,\mathcal{S}\Rightarrow$, and let $\mathcal{P}_f(S)$ be the set of probability measures on S that are discrete and have a finite number of atoms. For $P=\sum_{i=1}^m \alpha_i\delta_{x_i}\in\mathcal{P}_f(S)$, the P-bridge G_P and the P-motion Z_P admit the explicit versions

$$G_P=\sum_{i=1}^m \alpha_i^{1/2} g_i(\delta_{x_i}-P) \quad\text{and}\quad Z_P=\sum_{i=1}^m \alpha_i^{1/2} g_i\delta_{x_i}, \tag{3.276}$$

where g_i are independent standard normal random variables, and we continue denoting by G_P and Z_P. If $\mathcal{F}$ is uniformly bounded, then both $(\mathcal{F},d_P)$ and $(\mathcal{F},e_P)$ are totally bounded, and these versions of G_P and Z_P have all their sample paths bounded and, respectively, d_P- and e_P-uniformly continuous (e.g., for Z_P: the set $\{(f(x_1),\dots,f(x_m)) : f\in\mathcal{F}\}$ is a bounded set of $\mathbb{R}^m$ and $e_P^2(f,h)=\sum_{i=1}^m \alpha_i(f(x_i)-h(x_i))^2\le \max_{1\le i\le m}|f(x_i)-h(x_i)|^2$, so $(\mathcal{F},e_P)$ is totally bounded, and by Cauchy-Schwarz's inequality, $|Z_P(f)-Z_P(h)|\le\left(\sum_{i=1}^m g_i^2\right)^{1/2} e_P(f,h)$). Z_P is somewhat simpler than G_P, and if $\mathcal{F}$ is L^1-bounded, then the sizes of $\mathcal{F}$ for the Z_P and G_P pseudo-distances are comparable (see the proof of Theorem 3.7.40). Thus, we use Z_P in the following definition instead of the perhaps more natural G_P. In this section, when $P\in\mathcal{P}_f(S)$, we take G_P and Z_P to mean precisely their versions in (3.276).

Definition 3.7.41 A class $\mathcal{F}$ of measurable functions on $(S,\mathcal{S}\Rightarrow$ is finitely uniformly pre-Gaussian, $\mathcal{F}\in UPG_f$ for short, if both

$$\sup_{P\in\mathcal{P}_f(S)} E\|Z_P\|_{\mathcal{F}}<\infty \quad\text{and}\quad \lim_{\delta\to 0}\sup_{P\in\mathcal{P}_f(S)} E\|Z_P\|_{\mathcal{F}'_{\delta,P}}=0, \tag{3.277}$$

where $\mathcal{F}'_{\delta,P}=\{f-g : f,g\in\mathcal{F}, e_P(f,g)\le\delta\}$.We say that $\mathcal{F}$ is uniformly pre-Gaussian, $\mathcal{F}\in UPG$, if the probability law of Z_P is a tight Borel measure on $\ell_\infty(\mathcal{F}\Rightarrow$ for all $P\in\mathcal{P}_f(S)$ and $\mathcal{F}$ satisfies the conditions (3.277) uniformly in $\mathcal{P}(S)$, that is, with $\mathcal{P}_f(S)$ replaced by $\mathcal{P}(S)$.

Note that if $\mathcal{F}\in UPG_f$, then $\mathcal{F}$ is a uniformly bounded class: for $P_x=\delta_x$, $x\in S$, $Z_{P_x}(f)=f(x)g$, g standard normal, so $E\|Z_{P_x}\|_\infty=\sqrt{2/\pi}\,|f(x)|$, and the supremum of these expectations is bounded by definition of the UPG_f property.

Example 3.7.42 If $\mathcal{F}$ is a uniformly bounded *VC* subgraph, *VC* type or *VC* hull class, then the uniform bounds on the metric entropy of these classes (Theorems 3.6.9 and 3.6.17) together with the metric entropy bound for Gaussian processes in Theorem 2.3.7 imply that $\mathcal{F}$ is *UPG*, so, in particular, UPG_f. In more generality and for the same reasons, if $\mathcal{F}$ is uniformly bounded and satisfies the entropy condition of Theorem 3.7.37 in terms of the Koltchinski-Pollard entropy

$$\int_0^\infty \sup_Q\sqrt{\log N(\mathcal{F},e_Q,\varepsilon)}d\varepsilon<\infty \tag{3.278}$$

with the supremum extended over all the discrete probability measures Q with a finite number of atoms and rational weights on them, then $\mathcal{F}$ is UPG_f. Thus, many classes of functions are UPG_f (see Proposition 3.7.49 for a concrete important example).

Example 3.7.43 Let $\mathcal{F} = \{f_k\}_{k=2}^{\infty}$ with $\|f_k\|_\infty = o(1/(\log k)^{1/2})$. We show that $\mathcal{F} \in UPG_f$. We have $\beta_k := \sqrt{\log k}\|f_k\|_\infty \to 0$ and $\overline{\beta}_N := \sup_{k\geq N}\beta_k \to 0$. If $\overline{g}_k$ are $N(0,1)$ random variables, not necessarily independent, then Exercise 2.3.8 shows that

$$E\sup_{k\geq N}|\beta_k\overline{g}_k|/\sqrt{\log k} \leq c\overline{\beta}_N.$$

Then, if $P = \sum_{i=1}^m \alpha_i\delta_{x_i}$ we have

$$Z_P(f_k) = \sum_{i=1}^m \alpha_i^{1/2} f_k(x_i)g_i = \left(\sum_{i=1}^m \alpha_i f_k^2(x_i)\right)^{1/2} \overline{g}_k^P = (Pf_k^2)^{1/2}\overline{g}_k^P,$$

where g_k^P are $N(0,1)$ random variables. Hence,

$$E\|Z_P\|_{\mathcal{F}} = E\sup_k (Pf_k^2)^{1/2}|\overline{g}_k^P| \leq E\sup_k \beta_k|\overline{g}_k^P|/\sqrt{\log k} \leq c\overline{\beta}_2$$

independently of P. Given $\delta > 0$ and $2 \leq N < \infty$, if for $k \leq N$ the set $A_{k,N,\delta} = \{\ell > N : e_P(f_k, f_\ell) \leq \delta\}$ is not empty, choose $\ell_k \in A_{k,N,\delta}$, and observe that $\sup_{\ell\in A_{k,N,\delta}}|Z_P(f_k - f_\ell)| \leq |Z_P(f_k - f_{\ell_k})| + \sup_{\ell,r>N}|Z_P(f_\ell - f_r)|$. Hence,

$$\|Z_P\|_{\mathcal{F}'_{\delta,P}} = \max\left[\max_{k,\ell\leq N, e_P(f_k,f_\ell)\leq\delta}|Z_P(f_k - f_\ell)|,\ \max_{\substack{k\leq N\\ A_{k,N,\delta}\neq\emptyset}}\left(|Z_P(f_k - f_{\ell_k})| + \sup_{\ell,r>N}|Z_P(f_\ell - f_r)|\right),\right.$$
$$\left.\sup_{\ell,r>N}|Z_P(f_\ell - f_r)|\right]$$
$$\leq \max_{k,\ell\leq N, e_P(f_k,f_\ell)\leq\delta}|Z_P(f_k - f_\ell)| + \max_{\substack{k\leq N\\ A_{k,N,\delta}\neq\emptyset}}|Z_P(f_k - f_{\ell_k})| + \sup_{\ell,r>N}|Z_P(f_\ell - f_r)|.$$

Therefore,

$$E\|Z_P\|_{\mathcal{F}'_{\delta,P}} \leq \delta N^2 + \delta N + 2c\overline{\beta}_N$$

independently of P. Hence, $\limsup_{\delta\to 0}\sup_{P\in\mathcal{P}_f(S)} E\|Z_P\|_{\mathcal{F}'_{\delta,P}} \leq 2c\overline{\beta}_N$ for all N, and letting $N \to \infty$, $\lim_{\delta\to 0}\sup_{P\in\mathcal{P}_f(S)} E\|Z_P\|_{\mathcal{F}'_{\delta,P}} = 0$.

The two main reasons behind Definition 3.7.41 are that (1) as we will immediately see, empirical processes indexed by UPG_f classes satisfy very strong uniformity in P-limiting properties, and (2) Gaussian processes are sufficiently well understood so as to make it feasible, in general, to decide whether a given class satisfies the UPG_f property, and in fact, as the preceding examples show, there are many classes that satisfy this property.

Recall that X_i are the coordinate functions $S^{\mathbb{N}} \mapsto S$ and that $\nu_n = \sqrt{n}(P_n - P)$, where $P_n = n^{-1}\sum_{i=1}^n \delta_{X_i}$. Since we simultaneously consider all the Borel probability measures P, it is convenient to write ν_n^P for ν_n, and we will do so in this subsection: for any probability Q on S, $\nu_n^Q = \sum_{i=1}^n(\delta_{X_i} - Q)/\sqrt{n}$ is defined on $(S^{\mathbb{N}}, \mathcal{S}^{\mathbb{N}}, Q^{\mathbb{N}})$: the same product space for all Q, but the probability measure on it depends on Q so as to make the coordinates X_i of

$S^{\mathbb{N}}$ independent and with law Q. Let us also recall the definition of the bounded Lipschitz distance, which depends on P:

$$d_{BL}(\nu_n^P, G_P) = d_{BL(\mathcal{F})}(\nu_n^P, G_P)$$
$$= \sup\left\{\left|\int^* H(\nu_n^P)dP^{\mathbb{N}} - \int H(G_P)d\Pr\right| : H : \ell_\infty(\mathcal{F}) \mapsto \mathbb{R} \text{ with } \|H\|_\infty \le 1, \|H\|_{\text{Lip}} \le 1\right\}.$$

Note that the superindex P on ν_n^P determines the probability measure in the first integral. Recall that d_{BL} metrises convergence in law in $\ell_\infty(\mathcal{F})$ (see Proposition 3.7.24).

Definition 3.7.44 A class of functions $\mathcal{F}$ is uniform Donsker if $\mathcal{F}$ is uniformly pre-Gaussian and $\lim_{n\to\infty} \sup_{P\in\mathcal{P}(S)} d_{BL(\mathcal{F})}(\nu_n^P, G_P) = 0$.

In particular, this definition of uniform Donsker class implies uniform boundedness of the class by the comment following the definition of the UPG_f property. Before stating and proving the main theorem, we look first at the uniform CLT in finite dimensions:

Lemma 3.7.45 *Let $\mathcal{P}_M^d$ be the collection of Borel probability measures on $\mathbb{R}^d$ with support in the unit ball of radius $M < \infty$. For $P \in \mathcal{P}_M^d$, let ξ_i^P be i.i.d.(P), and let $\Phi_P = \mathrm{Cov}(P)$. Then*

$$\lim_{n\to\infty} \sup_{P\in\mathcal{P}_M^d} d_{BL}\left[\mathcal{L}\left(\sum_{i=1}^n (\xi_i^P - E\xi_i^P)/\sqrt{n}\right), N(0,\Phi_P)\right] = 0,$$

where $N(0,\Phi_P)$ denotes the centred normal law of $\mathbb{R}^d$ with covariance Φ_P.

Proof This follows from standard results on speed of convergence in the multidimensional CLT, but an elementary proof obtains along the following lines: for notational convenience, we give the proof only in dimension 1 (as the proof in higher dimensions is only formally different). Let $\xi_i = \xi_i^P - E\xi_i^P$, and let ζ_i be standard normal variables with variance $\sigma_2 = E\xi_i^2$, all independent. Let f be a bounded function with the first three derivatives bounded, with $\|f^{(3)}\|_\infty \le m_3$. Now, following the classical Lindeberg proof of the CLT, we estimate $\left|Ef\left(\sum_{i=1}^n \xi_i/\sqrt{n}\right) - Ef\left(\sum_{i=1}^n \zeta_i/\sqrt{n}\right)\right|$ by subtracting and adding $Ef(\sum_{i=1}^{n-1} \xi_i/\sqrt{n} + \zeta_n/\sqrt{n}),\ldots,$ $Ef\left(\xi_1/\sqrt{n} + \sum_{i=2}^n \zeta_i/\sqrt{n}\right)$ (each term in this sequence is obtained from the preceding one by replacing an ξ_i variable by a ζ_i variable, one at a time). Then it follows by the triangle inequality that the preceding difference is bounded by the sum of n terms of the form $\left|Ef(U_i + \xi_i/\sqrt{n}) - Ef(U_i + \zeta_i/\sqrt{n})\right|$, where U_i, ξ_i and ζ_i are independent. Deleting the subindex i, using independence and that $E\xi = E\zeta = 0$, $E\xi^2 = E\zeta^2 = \sigma^2$, $E|\xi|^3 \le M^3$ and $E|\zeta|^3 = \sqrt{8/\pi}\sigma^3 \le \sqrt{8/\pi}M^3$, a limited Taylor development about U gives

$$\left|Ef(U + \xi/\sqrt{n}) - Ef(U + \zeta/\sqrt{n})\right| \le \frac{1}{3}\left(1 + \sqrt{8/\pi}\right) m_3 M^3/n^{3/2}.$$

That is,

$$\left|Ef\left(\sum_{i=1}^n \xi_i/\sqrt{n}\right) - Ef\left(\sum_{i=1}^n \zeta_i/\sqrt{n}\right)\right| \le \frac{1}{3}\left(1 + \sqrt{8/\pi}\right) m_3 M^3/n^{1/2} \to 0$$

as $n \to \infty$ uniformly in f bounded and with bounded derivatives such that $\|f^{(3)}\|_\infty \le m_3$.

For f such that $\|f\|_\infty \le 1$ and $\|f\|_{\rm Lip} \le 1$, and for $\varepsilon > 0$, define

$$f_\varepsilon(x) = \frac{1}{(2\pi)^{1/2}} \int f(x - \varepsilon y) e^{-y^2/2} dy = \frac{1}{(2\pi\varepsilon^2)^{1/2}} \int f(v) e^{-(v-x)^2/2\varepsilon^2} dv,$$

and note that

$$\|f - f_\varepsilon\|_\infty \le \frac{1}{(2\pi)^{1/2}} \int (2 \wedge \varepsilon|y|) e^{-y^2/2} dy \le c\varepsilon,$$

where c_1 is an absolute constant, whereas $\|f_\varepsilon^{(k)}\|_\infty \le \varepsilon^{-k} \int |\varphi^{(k)}(x)| dx =: c_k/\varepsilon^k$, where φ is the density of the standard normal law. Then, if ζ is $N(0, \sigma^2)$,

$$\left| Ef\left(\sum_{i=1}^n \xi_i/\sqrt{n}\right) - Ef(\zeta)\right| \le 2c\varepsilon + \left| Ef_\varepsilon\left(\sum_{i=1}^n \xi_i/\sqrt{n}\right) - Ef_\varepsilon(\zeta)\right| \le 2c_1\varepsilon + \frac{(1+\sqrt{8/\pi})c_3}{3\varepsilon^3 n^{1/2}}.$$

Choosing $\varepsilon = n^{-1/8}$ proves the lemma. ■

The same idea can be used to prove the following:

Lemma 3.7.46 *Let Φ and Ψ be two covariance operators on $\mathbb{R}^d \times \mathbb{R}^d$, and let $N(0, \Phi)$ and $N(0, \Psi)$ denote the centred normal laws with these covariances. Set $\|\Phi - \Psi\|_\infty := \max_{1 \le i,j \le d} |\Phi(e_i, e_j) - \Psi(e_i, e_j)|$, where e_i is the canonical basis of $\mathbb{R}^d$. Then*

$$d_{BL}[N(0, \Phi), N(0, \Psi)] \le c(d)\|\Phi - \Psi\|_\infty^{1/3},$$

where $c(d)$ depends only on d.

Proof Let X_i, Y_i, $i = 1, \ldots, n$, be independent normal variables with laws $N(0, \Phi)$ for the X_i and $N(0, \Psi)$ for the Y_i respectively. Then the variables $X^{(n)} = \sum_{i=1}^n X_i/\sqrt{n}$, $Y^{(n)} = \sum_{i=1}^n Y_i/\sqrt{n}$ have, respectively, laws $N(0, \Phi)$ and $N(0, \Psi)$ for all n. Let $f : \mathbb{R}^d \mapsto \mathbb{R}$ be a uniformly bounded function with uniformly bounded partial derivatives of order at least 3. Let M_i, $i = 0, \ldots, 3$, be uniform bounds for f (M_0) and for all the derivatives of order i, $i \le 3$. We use Lindeberg's procedure of estimating $|Ef(X^{(n)}) - Ef(Y^{(n)})|$ step by step, replacing X_i by Y_i one at a time and doing a Taylor expansion up to the third term just as in the preceding proof. The linear term at each of the n steps is zero, the quadratic term at each step is dominated by $d^2 M_2 \|\Phi - \Psi\|_\infty/(2n)$ and the third term is dominated by a constant times $d^3 M_3/n^{3/2}$. Multiplying these bounds by n and taking limits as $n \to \infty$, noting that $|Ef(X^{(n)}) - Ef(Y^{(n)})|$ does not depend on n, we obtain $|Ef(X^{(n)}) - Ef(Y^{(n)})| \le d^2 M_2 \|\Phi - \Psi\|_\infty/2$, for all n. If H is bounded Lipschitz with BL norm equal to 1, then we may convolve with a Gaussian approximate identity (again, just as in the preceding proof) and obtain the result: as earlier, the convolution H_ε of H with $N(0, \varepsilon^2 I)$ satisfies $\|H - H_\varepsilon\|_\infty \le c\varepsilon$, and H_ε has uniformly bounded partial derivatives of all orders, and in particular, $\|(H_\varepsilon)''_{i,j}\|_\infty \le c_2/\varepsilon^2$ for a universal constant $c_3 < \infty$. This, together with the bound for differentiable functions, gives

$$|EH(X^{(n)}) - EH(Y^{(n)})| \le c\varepsilon + d^2 c_2 \|\|\Phi - \Psi\|_\infty/(2\varepsilon^2),$$

and the result follows by taking ε to be a constant times $\|\Phi - \Psi\|_\infty^{1/3}$. ■

No claim for optimality is made on the exponent of $\|\Phi - \Psi\|_\infty$ in the preceding lemma (any power would do in the proof of the next theorem). The theorem that follows requires measurability of the uniformly bounded class $\mathcal{F}$ for every $P \in \mathcal{P}(S)$. A general condition is that there exists a countable class $\mathcal{F}_0 \subset \mathcal{F}$ such that every f in $\mathcal{F}$ is a pointwise limit of functions in $\mathcal{F}_0$. If $\mathcal{F}$ satisfies this property, we say that $\mathcal{F}$ satisfies the *pointwise countable approximation property*. Note that if $\mathcal{F}$ is uniformly bounded and satisfies the pointwise countable approximation property, then, in particular, it is measurable in the sense of Definition 3.7.11. Another useful general condition is that $\mathcal{F}$ be *image admissible Suslin* (see Dudley (2014)). We will assume the first condition, although the second would work as well. Some notation: $\mathcal{F}\mathcal{F} = \{f^2, f : f \in \mathcal{F}\}$, and $\mathcal{R}(\ell_\infty(\mathcal{F}))$ denotes the set of tight Borel probability measures on $\ell_\infty(\mathcal{F})$.

Theorem 3.7.47 *Let $\mathcal{F}$ be a uniformly bounded class of measurable functions on S satisfying the pointwise countable approximation property. Then the following conditions are equivalent:*

(a) $\mathcal{F} \in UPG_f$.

(b) $(\mathcal{F}, e_P)$ is uniformly totally bounded, and $\lim_{\delta\to 0} \limsup_n \sup_{P\in\mathcal{P}(S)} P^{\mathbb{N}}\{\|\nu_n^P\|_{\mathcal{F}'_{\delta,P}} > \varepsilon\} = 0$, *for all $\varepsilon > 0$.*

(c) $\mathcal{F} \in UPG$, and the same uniformity extends to G_P; that is, for each P, G_P admits a suitable version, and for these versions, $\sup_{P\in\mathcal{P}(S)} E\|G_P\|_{\mathcal{F}} < \infty$ *and* $\lim_{\delta\to 0} \sup_{P\in\mathcal{P}(S)} E\|G_P\|_{\mathcal{F}'_{\delta,P}} = 0$.

(d) $\mathcal{F}$ is uniform Donsker.

Moreover, if either of these conditions holds, then the map $G : (\mathcal{P}(S), \|\cdot\|_{\mathcal{F}\mathcal{F}}) \mapsto (\mathcal{R}(\ell_\infty(\mathcal{F})), d_{BL})$ given by $G(P) = \mathcal{L}(G_P)$ is uniformly continuous.

Proof We can assume that $\mathcal{F}$ is uniformly bounded by 1.

Claim 1. Set $\mathcal{G} = \{f, f^2, f - g, (f - g)^2 : f, g \in \mathcal{F}\}$. Assume that the class $\mathcal{F}$ is UPG_f and satisfies the required measurability. Let ε_i be independent Rademacher variables and g_i independent $N(0,1)$ variables such that ε_i, g_j, X_k are all coordinates in a product probability space and hence independent. Then

$$\sup_{P\in\mathcal{P}(S)} E_P\|P_n - P\|_{\mathcal{G}} = O(n^{-1/2}), \quad \sup_{P\in\mathcal{P}(S)} E_{P,\varepsilon}\left\|\frac{1}{n}\sum_{i=1}^n \varepsilon_i h(X_i)\right\|_{\mathcal{G}} = O(n^{-1/2}),$$

$$\sup_{P\in\mathcal{P}(S)} E_{P,g}\left\|\frac{1}{n}\sum_{i=1}^n g_i h(X_i)\right\|_{\mathcal{G}} = O(n^{-1/2}), \tag{3.279}$$

where $E_{P,\varepsilon}$ and $E_{P,g}$ indicate, respectively, expected value with respect to $P^{\mathbb{N}} \times ((\delta_1 + \delta_{-1})/2)^{\mathbb{N}}$ and $P^{\mathbb{N}} \times N(0,1)^{\mathbb{N}}$

Proof. It suffices to prove the claim for $\mathcal{H} = \{(f - g)^2 : f, g \in \mathcal{F}\}$ because the proof for $\mathcal{G}$ is essentially a subset of the proof for $\mathcal{H}$, and we will avoid repetition. The usual randomisation for expectations (Theorem 3.1.21 and the easy part of Proposition 3.1.26) gives

$$E_P\|P_n - P\|_{\mathcal{H}} \le 2E_{P,\varepsilon}\left\|\frac{1}{n}\sum_{i=1}^n \varepsilon_i h(X_i)\right\|_{\mathcal{H}} \le \sqrt{\frac{2\pi}{n}}E_{P,g}\left\|\frac{1}{\sqrt{n}}\sum_{i=1}^n g_i h(X_i)\right\|_{\mathcal{H}}.$$

Thus, it suffices to prove the claim for the Gaussian randomisation of the empirical process. Next, we see that we can dominate the preceding expectation by the same one with $\mathcal{H}$ replaced by $\mathcal{F}' = \{f - g : f, g \in \mathcal{F}\}$ using the Slepian-Fernique comparison theorem for Gaussian processes, Corollary 2.4.10. For X_i fixed, consider the two Gaussian processes on $\mathcal{F}'$ given by

$$Z_1(f) = \frac{1}{\sqrt{n}}\sum_{i=1}^n g_i f^2(X_i), \quad Z_2(f) = \frac{1}{\sqrt{n}}\sum_{i=1}^n g_i f(X_i), \quad f \in \mathcal{F}'.$$

Then (i) $E_g(Z_1(f_1) - Z_1(f_2))^2 = P_n(f_1^2 - f_2^2)^2 \le 4P_n(f_1 - f_2)^2 = 4E_g((Z_2(f_1) - Z_2(f_2))^2$. (ii) Both processes have bounded and uniformly continuous sample paths with respect to their corresponding L^2-distances. (iii) Both attain the value zero at some $f \in \mathcal{F}'$ (trivially because $0 \in \mathcal{F}'$). Therefore, we can apply Corollary 2.4.10 to both $\{Z_1, Z_2\}$ and $\{-Z_1, -Z_2\}$, use that by (iii) $E_g\|Z_1\|_{\mathcal{F}'} \le E_g \sup_{f\in\mathcal{F}'} Z_1(f) + E_g \sup_{f\in\mathcal{F}'}(-Z_1(f))$ and likewise for Z_2 and obtain

$$E_g\|Z_1\|_{\mathcal{F}'} \le 16E_g\|Z_2\|_{\mathcal{F}'}.$$

Integrating with respect to the variables X_i we further obtain

$$\begin{aligned}
\sqrt{\frac{1}{n}}E_{P,g}\left\|\frac{1}{\sqrt{n}}\sum_{i=1}^n g_i h(X_i)\right\|_{\mathcal{H}} &\le 16\sqrt{\frac{1}{n}}E_{P,g}\left\|\frac{1}{\sqrt{n}}\sum_{i=1}^n g_i h(X_i)\right\|_{\mathcal{F}'} \\
&\le 32\sqrt{\frac{1}{n}}E_{P,g}\left\|\frac{1}{\sqrt{n}}\sum_{i=1}^n g_i h(X_i)\right\|_{\mathcal{F}} \\
&= 32\sqrt{\frac{1}{n}}E_P E_g\|Z_{P_n}\|_{\mathcal{F}} \le 32\sqrt{\frac{1}{n}} \sup_{Q\in\mathcal{P}_f(S)} E\|Z_Q\|_{\mathcal{F}} < \infty.
\end{aligned}$$

This proves claim (a). Note how Gaussian randomisation reduces properties of the empirical process to properties of the simple Gaussian processes Z_Q: this is the idea of proof of the whole theorem.

Claim 2. (a) implies (b).

Proof Since $\mathcal{F}$ is UPG_f, we have in particular

$$\sup\left\{E\|Z_{P_n(\omega)}\|_{\mathcal{F}} : P \in \mathcal{P}(S),\ \omega \in S^{\mathbb{N}},\ n \in \mathbb{N}\right\} < \infty,$$

and therefore, Sudakov's lower bound (Theorem 2.4.12) gives that there is $c < \infty$ such that for all $\varepsilon > 0$,

$$\sup_{P,n,\omega} \log N(\mathcal{F}, e_{P_n(\omega)}, \varepsilon) < \frac{c}{\varepsilon^2}.$$

Now, by claim 1,

$$\sup_{f,g\in\mathcal{F}} |e^2_{P_n(\omega)}(f,g) - e^2_P(f,g)| \to 0$$

in $P^{\mathbb{N}}$ probability. Then we have convergence along a subsequence for at least one ω for each fixed P, which together with the Sudakov estimate of the $e_{P_n(\omega)}$ covering numbers implies that

$$\sup_{P\in\mathcal{P}(S)} \log N(\mathcal{F}, e_P, \varepsilon) < \frac{2c}{\varepsilon^2}, \quad \text{for all } \varepsilon > 0, \tag{3.280}$$

proving that the pseudo-metric spaces $(\mathcal{F}, e_P)$ are totally bounded uniformly in $P \in \mathcal{P}(S)$. To prove the uniform asymptotic equi-continuity, we first note that the symmetrisation inequality in Theorem 3.1.25 implies, for $\delta \le 2\tau$, that

$$\begin{aligned} P^{\mathbb{N}}\left\{\|v_n^P\|_{\mathcal{F}'_{\delta,P}} > 4\tau\right\} &\le 4\mathrm{Pr}_{P,\varepsilon}\left\{\|v^P_{n,\mathrm{rad}}\|_{\mathcal{F}'_{\delta,P}} > \tau\right\} \\ &\le 4\mathrm{Pr}_{P,\varepsilon}\left\{\sup_{\substack{f,g\in\mathcal{F} \\ e_{P_n}(f,g)\le 2^{1/2}\delta}} |v^P_{n,\mathrm{rad}}| > \tau\right\} \\ &\quad + 4P^{\mathbb{N}}\left\{\sup_{f,g\in\mathcal{F}} |e^2_{P_n(\omega)}(f,g) - e^2_P(f,g)| > \delta^2\right\} \\ &:= I_{P,n} + II_{P,n}, \end{aligned}$$

where $\mathrm{Pr}_{P,\varepsilon}$ is $P^{\mathbb{N}} \times ((\delta_1 + \delta_{-1})/2)^{\mathbb{N}}$. Now claim 1 directly gives

$$\lim_{n\to\infty} \sup_{P\in\mathcal{P}(S)} II_{P,n} = 0.$$

Set, for each ω fixed, that we leave implicit $\mathcal{H}_{\delta,n} = \{h \in \mathcal{H} : P_n h \le \delta^2\}$, where we recall that $\mathcal{H} = \{(f-h)^2 : f, h \in \mathcal{F}\}$, and note that by comparison of Gaussian processes as in the proof of claim 1 and with the same notation, $E_g\|Z_1\|_{\mathcal{F}'_{\delta,P_n}} \le 16 E_g\|Z_2\|_{\mathcal{F}'_{\delta,P_n}}$. This gives

$$\begin{aligned} E_g\left\|\frac{1}{\sqrt{n}}\sum_{i=1}^n g_i h(X_i)\right\|_{\mathcal{H}_{\delta,n}} &\le 16 E_g\left\|\frac{1}{\sqrt{n}}\sum_{i=1}^n g_i h(X_i)\right\|_{\mathcal{F}'_{\delta,P_n}} \\ &= 16 E_g\|Z_{P_n}\|_{\mathcal{F}'_{\delta,P_n}} \le 16 \sup_{Q\in\mathcal{P}(S)} E\|Z_Q\|_{\mathcal{F}'_{\delta,Q}}, \end{aligned}$$

which tends to zero as $\delta \to 0$ by definition of the UPG_f property. Taking expectation with respect to the X_i variables, we thus obtain

$$\lim_{\delta\to 0} \limsup_n I_{P,n} = 0$$

because, as observed in the proof of claim 1, $E_\varepsilon\|v_{n,\mathrm{rad}}\|_{\mathcal{H}_{\delta,n}} \le \sqrt{\pi/2} E_g\|v_{n,g}\|_{\mathcal{H}_{\delta,n}}$. Claim 2 is proved.

Claim 3. (b) implies (c) and hence (a).

Proof Since the process $G_P + gP$, where g is $N(0,1)$ independent of G_P, is a version of Z_P for all P, and $\|P\|_{\mathcal{F}} \le 1$ and $\|P\|_{\mathcal{F}'_{\delta,P}} \le \delta$, it suffices to prove the claim for G_P. By claim 2, $\mathcal{F}$ is P-Donsker for all P. Hence, the Portmanteau theorem (Exercise 3.7.7) and claim 2 give

$$\limsup_{\delta\to 0} \sup_{P\in\mathcal{P}(S)} \Pr\left\{\|G_P\|_{\mathcal{F}'_{\delta,P}} > \varepsilon\right\} \le \limsup_{\delta\to 0} \sup_{P\in\mathcal{P}(S)} \liminf_{n\to\infty} P^{\mathbb{N}}\left\{\|\nu_n^P\|_{\mathcal{F}'_{\delta,P}} > \varepsilon\right\} = 0,$$

for all $\varepsilon > 0$. The integrability of Gaussian processes and the Paley-Zygmund argument (Exercises 2.1.2 and 2.1.4) give that there is a universal constant K such that for all $0 < \lambda < 1$ and any measurable pseudo-norm $\|\cdot\|$ satisfying $\Pr\{\|G_P\| < \infty\} > 0$, we have $\Pr\{\|G_P\| > \lambda E\|G_P\|\} \ge K(1-\lambda)^{1/2}$. Hence, by the preceding limits, for every $\varepsilon > 0$, there is $\tau > 0$ such that for all $0 < \delta < \tau$ and $P \in \mathcal{P}(S)$, $E\|G_P\|_{\mathcal{F}'_{\delta,P}}/2 < \varepsilon$; that is,

$$\lim_{\delta\to 0} \sup_{P\in\mathcal{P}(S)} E\|G_P\|_{\mathcal{F}'_{\delta,P}} = 0.$$

Now

$$\sup_{P\in\mathcal{P}(S)} E\|G_P\|_{\mathcal{F}} < \infty$$

follows from the preceding limit and the fact that $(\mathcal{F}, e_P)$ is totally bounded uniformly in $P \in \mathcal{P}(S)$ (Claim 2). The claim is proved.

Claim 4. (b) implies (d) (which implies a)).

Proof By claim 3, $\mathcal{F}$ is *UPG*, so it suffices to prove that

$$\lim_{n\to\infty} \sup_{P\in\mathcal{P}} d_{BL(\mathcal{F})}\left(\nu_n^P, G_P\right) = 0.$$

In the decomposition of $|E_P^* H(\nu_n^P) - EH(G_P)|$ into $I_{n,\tau} + II_{n,\tau} + III_\tau$ from the proof of Proposition 3.7.24 for H Lipschitz in $\ell^\infty(\mathcal{F})$, the three terms can be estimated uniformly in P, the first, by condition (b); the second, by Lemma 3.7.45, and the third by condition (c) (which holds by claim 3). We omit the details to avoid repetition.

Claims 2 to 4 yield the equivalence of the four conditions (a)–(d).

Claim 5. If the equivalent conditions (a)–(d) hold, then the map G defined in the statement of the theorem is uniformly continuous.

Proof Let $P, Q \in \mathcal{P}(S)$ and $\tau > 0$. The uniform entropy bound (3.280) shows that there is a universal constant $a < \infty$ and $N(\tau, P, Q) \le e^{a/\tau^2}$ disjoint subsets A_i of $\mathcal{F}$ whose union covers $\mathcal{F}$ and that for each of them there is $f_i \in \mathcal{F}$ such that $A_i \subseteq \{f : e_P(f, f_i) \vee e_Q(f, f_i) \le \tau\}$. Let $H : \ell_\infty(\mathcal{F}) \to \mathbb{R}$ be bounded Lipschitz with supremum and Lipschitz norms bounded by 1. As in several other proofs, define the Gaussian processes $Z_{P,\tau}(f) = Z_P(f_i)$ if $f \in A_i$, and likewise, define $Z_{Q,\tau}$. These processes are in fact centred normal random vectors in $\mathbb{R}^{[\exp(a/\tau^2)]+1}$. If $\Phi_{P,\tau}$ and $\Phi_{Q,\tau}$ are their covariances, we have

$$\begin{aligned} E\left|H(Z_P) - H(Z_Q)\right| &\le E\left|H(Z_P) - H(Z_{P,\tau})\right| + E\left|H(Z_Q) - H(Z_{Q,\tau})\right| + E\left|H(Z_{P,\tau}) - H(Z_{Q,\tau})\right| \\ &\le E\|Z_P\|_{\mathcal{F}'_{\tau,P}} + E\|Z_P\|_{\mathcal{F}'_{\tau,P}} + c\left(e^{a/\tau^2}\right)\left\|\Phi_{P,\tau} - \Phi_{Q,\tau}\right\|_\infty^{1/3}, \end{aligned}$$

where the last inequality follows from Lemma 3.7.46. Since

$$\left\|\Phi_{P,\tau} - \Phi_{Q,\tau}\right\|_\infty = \max_{i,j\le N(\tau,P,Q)} \left|\Phi_{P,\tau}(f_i, f_j) - \Phi_{Q,\tau}(f_i, f_j)\right| \le \|P - Q\|_{\mathcal{F}\mathcal{F}}$$

independently of τ, uniform continuity of the map G follows from the fact that $\mathcal{F}$ is *UPG*. Claim 5 is proved, and so is the theorem. ■

The uniform continuity of the map G in the preceding theorem is particularly appropriate for the bootstrap: it shows that the central limit theorem for empirical processes over UPG_f classes can be bootstrapped in many different ways. Although the bootstrap is not studied in this book, it is nevertheless worthwhile to point out how to use this property.

Corollary 3.7.48 *Let $\mathcal{F}$ be a UPG_f class of functions satisfying the pointwise countable approximation property, and let Q_n be random probability measures on $(S,\mathcal{S})$ such that, for some $P \in \mathcal{P}(S)$,*

$$\lim_{n\to\infty} \|Q_n - P\|_{\mathcal{F}\mathcal{F}} = 0 \text{ a.s. (in pr.)},$$

where we assume $\|Q_n - P\|_{\mathcal{F}\mathcal{F}}$ to be measurable. Then

$$\lim_{n\to\infty} d_{BL(\mathcal{F})}\left(\nu_n^{Q_n}, G_P\right)^* = 0 \text{ a.s. (in outer pr.)}.$$

In resampling (as in different kinds of bootstraps), Q_n depends on the observations; that is, Q_n is defined on the probability space $(\Omega_1, \Sigma_1, \Pr_1) = (S^{\mathbb{N}}, \mathcal{S}^{\mathbb{N}}, P^{\mathbb{N}})$ as a $(\mathcal{P}(S), \mathcal{F}\mathcal{F})$-valued random variable and depends only on the first n coordinates. Then, for each $\omega_1 \in \Omega_1$, $\nu_n^{Q_n(\omega_1)}(f) = \sum_{i=1}^n (f(X_i^*) - Q_n(\omega_1)(f))/\sqrt{n}$, that is, the variables X_i^* are sampled according to the law $Q_n(\omega_1)$, or conditionally on ω_1, the X_i^* are i.i.d. with law $Q_n(\omega_1)$: they constitute the *bootstrap sample*. This corollary asserts that if Q_n tends to P uniformly over the class $\mathcal{F}\mathcal{F}$ a.s. or in pr., then the empirical process based on Q_n, conditionally on the 'original sample $\omega_1 = (X_1,\dots,X_n)$', has almost surely (or in probability) the same limit law as P_n, G_P.

Typically, $Q_n = P_n$ (*Efron's bootstrap*), or $Q_n = P_n * \lambda_n$, where λ_n is an approximate identity (*smoothed bootstrap*), or $Q_n = P_{\theta_n}$, where $\theta_n = \theta_n(X_1,\dots,X_n)$ is an estimator of a parameter θ, and $(P_\theta : \theta \in \Theta)$ is a parametric model (*parametric bootstrap*) for the data, etc.

Proof The proof is basically a triangle inequality. We prove only the a.s. version because the version for convergence in probability follows from it by taking subsequences. Assume that the first limit holds. Just note that

$$d_{BL}\left(\nu_n^{Q_n}, G_P\right)^* \le d_{BL}\left(\nu_n^{Q_n}, G_{Q_n}\right)^* + d_{BL}\left(G_{Q_n}, G_P\right)^*.$$

Since $\mathcal{F}$ is uniform Donsker (conclusion (d)) in Theorem 3.7.47), there exist $c_n \to 0$ such that $d_{BL}\left(\nu_n^{Q_n(\omega_1)}, G_{Q_n(\omega_1)}\right)^* \le c_n$, independently of ω_1. By the continuity of the map G in Theorem 3.7.47, given $\varepsilon > 0$, there is $\delta > 0$ such that, for all n,

$$d_{BL}\left(G_{Q_n}, G_P\right) \le \varepsilon I_{\|Q_n - P\|_{\mathcal{F}\mathcal{F}} < \delta} + 2 I_{\|Q_n - P\|_{\mathcal{F}\mathcal{F}} \ge \delta}$$

(recall that d_{BL} is bounded by 2). For each n, the right side is a measurable random variable, and the limsup of these random variables as $n \to \infty$ is dominated by ε by the hypothesis of the corollary. Hence, $d_{BL}\left(\nu_n^{Q_n}, G_P\right)^* \to 0$ a.s. ■

We end this subsection with an important example.

Corollary 3.7.49 *For $1 \le p < 2$, the collection of functions of bounded p-variation on $\mathbb{R}$ with supremum norm and total p-variation norm bounded by $M < \infty$, $BV_{p,M}(\mathbb{R})$, is uniform Donsker.*

Proof We may assume that $M = 1$. $BV_{p,1}(\mathbb{R})$ satisfies the measurability condition in Theorem 3.7.47, as shown in Example 3.7.13. We will show that $BV_{p,1}(\mathbb{R})$ also satisfies the uniform metric entropy bound (3.278), so the result will then follow by Theorems 2.3.7 and 3.7.47. By Lemma 3.6.11, if $f \in BV_{p,1}(\mathbb{R})$, then $f = g \circ h$, where $h \in \mathcal{H}_1$, the collection of nondecreasing functions taking values in $[0,1]$, and $g \in \mathcal{G}_1$, the collection of p^{-1}-Holder continuous functions on $[0,1]$ with supremum norm and Hölder constant both bounded by 1 (the Holder constant of g is $\sup_{x\neq y} |g(x)-g(y)|^{1/p}/|x-y|$). Now, by Corollary 4.3.38, there exists $C_1 < \infty$ such that for all $0 < \varepsilon < 1$, $N(\mathcal{G}_1, \|\cdot\|_\infty, \varepsilon) \le e^{C_1/\varepsilon^p}$, and by Exercise 3.6.7, there exists C_2 such that for all $P \in \mathcal{P}(\mathbb{R})$, $N(\mathcal{H}_1, e_P, \varepsilon) \le e^{C_2/\varepsilon}$, for all $0 < \varepsilon < 1$. Let G_ε be ε-dense in $\mathcal{G}_1$ for the supremum norm and of cardinality $N(\mathcal{G}_1, \|\cdot\|_\infty, \varepsilon)$, and let H_ε be ε^p-dense in $\mathcal{H}_1$ for the $L_2(P)$ pseudo-distance and of cardinality $N(\mathcal{H}_1, e_P, \varepsilon^p)$. Then the set $\mathcal{V}_\varepsilon = \{g\circ h : g \in G_\varepsilon, h \in H_\varepsilon\}$ has cardinality bounded by $e^{(C_1+C_2)/\varepsilon^p}$ and is 2ε-dense in $BV_{p,1}(\mathbb{R})$ in e_P. The first assertion is obvious, and to see the second, note that if $\|g-\bar{g}\|_\infty \le \varepsilon$ (with $g, \bar{g} \in \mathcal{G}_1$) and $\|h-\bar{h}\|_{L^2(P)} \le \varepsilon^p$, then

$$
\begin{aligned}
\|g\circ h - \bar{g}\circ\bar{h}\|_{L^2(P)} &= \|g\circ h - \bar{g}\circ h + \bar{g}\circ h - \bar{g}\circ\bar{h}\|_{L^2(P)} \\
&\le \|g-\bar{g}\|_\infty + \||h-\bar{h}|^{1/p}\|_{L^2(P)} \le \varepsilon + \left(\int |h-\bar{h}|^{2/p} dP\right)^{1/2} \\
&\le \varepsilon + \left(\int |h-\bar{h}|^2 dP\right)^{1/2p} \le 2\varepsilon.
\end{aligned}
$$

Hence, $\int_0^\infty \sup_{P\in\mathcal{P}_f(\mathbb{R})} \sqrt{\log N(BV_{p,1}(\mathbb{R}), e_P, \varepsilon)}\,d\varepsilon \le 2\sqrt{C_1+C_2}\int_0^\infty d\varepsilon/\varepsilon^{p/2} < \infty$ for $p < 2$; that is, (3.278) holds for $BV_{p,1}(\mathbb{R})$. ■

We record for further use the entropy bound obtained in the preceding proof.

Corollary 3.7.50 *For $p \ge 1$, there exists $C_p < \infty$ such that for all Borel probability measures P on $\mathbb{R}$ and $0 < \varepsilon \le 1$,*

$$N(BV_{p,1}(\mathbb{R}), L^2(P), \varepsilon) \le e^{C_p/\varepsilon^p}.$$

Actually, Proposition 3.5.17 on the $L^2(P)$-bracketing numbers for the class of monotone functions provides, by the same argument as in the proof of the preceding two corollaries, a bound for the $L^2(P)$-bracketing numbers of $BV_{p,M}(\mathbb{R})$, which in fact contains Corollary 3.7.50.

Corollary 3.7.51 *For $p \ge 1$, there exists $C_p < \infty$ such that for all Borel probability measures P on $\mathbb{R}$ and $0 < \varepsilon \le 1$,*

$$N_{[]}(BV_{p,1}(\mathbb{R}), L^2(P), \varepsilon) \le e^{C_p/\varepsilon^p}.$$

Proof Just note that balls for the supremum norm are in fact brackets for any $L^p(P)$-norms, $p \le \infty$, and that if $G_{\varepsilon,\infty}$ is a supremum norm bracket of $\mathcal{G}_1$ of size ε and $H_{\varepsilon,P}$ is an $L^2(P)$-bracket of $\mathcal{H}_1$ of size ε^p (where $\mathcal{G}_1$ and $\mathcal{H}_1$ are as in the proof of Corollary 3.7.49), then

$$\left\| \sup_{\substack{g,\bar{g}\in G_{\varepsilon,\infty} \\ h,\bar{h}\in H_{\varepsilon,P}}} |g\circ h - \bar{g}\circ\bar{h}| \right\|_{L^2(P)} \le \left\| \sup_{g,\bar{g}\in G_{\varepsilon,\infty}} \|g-\bar{g}\|_\infty + \sup_{h,\bar{h}\in H_{\varepsilon,P}} |h-\bar{h}|^{1/p} \right\|_{L^2(P)} \le 2\varepsilon.$$

Now the result follows from Corollary 4.3.38 and Proposition 3.5.17. ∎

Central Limit Theorem for Pre-Gaussian Classes of Functions

If a uniformly bounded class of functions $\mathcal{F}$ is known to be P-pre-Gaussian, then checking that it is P-Donsker is somewhat easier than otherwise; concretely, if $\mathcal{F}$ is P-pre-Gaussian, then automatically the oscillations of the randomised process $\sum_{i=1}^n \varepsilon_i f(X_i)$ in the range $n^{-1/4} \le e_P(f,g) \le \delta$ are already small. We prove this fact and a consequence for the central limit theorem for classes of sets.

Some notation: given a class of functions $\mathcal{F}$, define

$$\mathcal{F}'_{\varepsilon,n} = \mathcal{F}'_{\varepsilon^{1/2}n^{-1/4}} = \left\{f - g : f,g \in \mathcal{F} : P(f-g)^2 \le \varepsilon n^{-1/2}\right\}.$$

Theorem 3.7.52 *Let P be a probability measure on $(S,\mathcal{S})$, and let $\mathcal{F}$ be a uniformly bounded class of measurable functions on S satisfying the countable pointwise approximation property. Then the following conditions are equivalent:*

(a) $\mathcal{F}$ is P-Donsker.

(b) $\mathcal{F}$ is P-pre-Gaussian, and

$$\lim_{\varepsilon\to 0}\limsup_n \Pr\left\{\left\|\sum_{i=1}^n \varepsilon_i f(X_i)/n^{1/2}\right\|_{\mathcal{F}'_{\varepsilon,n}} \ge \gamma\right\} = 0, \tag{3.281}$$

for all $\gamma > 0$.

Proof We may assume that $\|f\|_\infty \le 1$ for all $f \in \mathcal{F}$. Condition (a) implies condition (b) by the definition of P-Donsker class, and Theorem 3.7.31: By part c) of that theorem the class $\tilde{\mathcal{F}} = (rf(x) : f \in \mathcal{F})$ defined on the sample space $\{-1,1\} \times S$ is Q-Donsker for the law $Q = R \times P$, where R is the Rademacher law. Then $\tilde{\mathcal{F}}$ is Q-centred, and so the expression in (3.281) converges to zero by Theorem 3.7.31b). For the reverse implication, we first note that, as in the proof of claim 2 in Theorem 3.7.47, by Sudakov's inequality and boundedness of $\mathcal{F}$, the fact that $\mathcal{F}$ is P-pre-Gaussian implies that $(\mathcal{F}, e_P)$ is totally bounded. Hence, by Theorem 3.7.40 and the asymptotic equi-continuity condition for processes that converge in law, Theorem 3.7.23, it suffices to prove that

$$\lim_{\delta\to 0}\limsup_n \Pr\left\{\|\nu_{n,rad}\|_{\mathcal{F}'_\delta} \ge \gamma\right\} = 0, \tag{3.282}$$

for all $\gamma > 0$ (recall the definition (3.272) of $\nu_{n,\text{rad}}$). Let $\mathcal{H} = \mathcal{H}(\varepsilon,n)$ be a maximal collection of function $h_1,\dots,h_m \in \mathcal{F}$ such that $e_P^2(h_i,h_j) = P(h_i - h_j)^2 > \varepsilon/n^{1/2}$ for all $i \ne j$, and notice that then

$$\sup_{f\in\mathcal{F}}\min_{h_i\in\mathcal{H}} e_P^2(f,h_i) \le \varepsilon/n^{1/2}.$$

If $e_P(f,g) \le \delta$, $e_P^2(f,h_i) \le \varepsilon/n^{1/2}$ and $e_P^2(g,h_j) \le \varepsilon/n^{1/2}$, then $e_P(h_i,h_j) \le 2\delta$, for all $n \ge 2^4\varepsilon^2/\delta^4$. Hence, for n sufficiently large,

$$\Pr\left\{\|\nu_{n,\text{rad}}\|_{\mathcal{F}'_\delta} \ge 3\gamma\right\} \le 2\Pr\left\{\|\nu_{n,\text{rad}}\|_{\mathcal{F}'_{\varepsilon,n}} \ge \gamma\right\} + \Pr\left\{\|\nu_{n,\text{rad}}\|_{\mathcal{H}'_{2\delta}} \ge \gamma\right\}.$$

Thus, by the limit in condition (b), it suffices to show that for all $\gamma > 0$,

$$\lim_{\delta\to 0}\limsup_n \Pr\left\{\|\nu_{n,\text{rad}}\|_{\mathcal{H}'_{2\delta}} \ge \gamma\right\} = 0. \tag{3.283}$$

Define

$$\mathcal{K}(\varepsilon,n)=\{h_i-h_j: h_i,h_j\in\mathcal{H}(\varepsilon,n), i\neq j\},\ A(\varepsilon,n)=\left\{\sup_{f\in\mathcal{H}(\varepsilon,n)}\frac{\sum_{i=1}^n f^2(X_i)}{nPf^2}\leq 4\right\},$$

and decompose the preceding probability as follows:

$$\Pr\left\{\left\|\nu_{n,\mathrm{rad}}\right\|_{\mathcal{H}'_{2\delta}}\geq\gamma\right\}\leq \Pr(A^c)+\gamma^{-1}E_XE_\varepsilon\left(\left\|\nu_{n,\mathrm{rad}}\right\|_{\mathcal{H}'_{2\delta}}I_A\right)$$
$$:=(I)+(II). \tag{3.284}$$

Since $\mathcal{F}$ is P-pre-Gaussian, uniformly bounded and satisfies enough measurability, it follows by Theorem 3.7.40 that the P-motion Z_P has a suitable version (i.e., with prelinear, bounded and uniformly e_P-continuous sample paths). Hence, the refinement of Sudakov's bound for sample continuous processes, Corollary 2.4.14, gives

$$\lim_{\lambda\to 0}\lambda^2\log N(\mathcal{F},e_P,\lambda)=0. \tag{3.285}$$

Since by construction $\mathcal{H}(\varepsilon,n)$ has cardinality $D(\mathcal{F},e_P,\varepsilon^{1/2}/n^{1/4})\leq N(\mathcal{F},e_P,\varepsilon^{1/2}/2n^{1/4})$, it follows that the cardinality of $\mathcal{H}(\varepsilon,n)$ is dominated by $\exp(c_nn^{1/2}/\varepsilon)$, where $c_n=c_n(\varepsilon)\to 0$. Combined with Bernstein's inequality (Theorem 3.1.7), this gives

$$\begin{aligned}\limsup_n \Pr(A(\varepsilon,n)^c)&\leq\limsup_n(\mathrm{Card}(\mathcal{H}))^2\sup_{f\in\mathcal{K}(\varepsilon,n)}\Pr\left\{\sum_{i=1}^n f^2(X_i)\geq 4nPf^2\right\}\\&\leq\limsup_n\exp(2c_nn^{1/2}/\varepsilon)\sup_{f\in\mathcal{K}(\varepsilon,n)}\exp\left(-\frac{9n^2(Pf^2)^2}{2nPf^4+2nPf^2}\right)\\&\leq\limsup_n\exp\left(2c_nn^{1/2}/\varepsilon-9n^{1/2}\varepsilon/4\right)=0,\end{aligned}$$

where in the last inequality we use that $P(h_i-h_j)^2>\varepsilon/n^{1/2}$. Thus, we have $\limsup_n(I)=0$, and we now consider (II) in (3.284). By the first inequality in (3.50), we can replace the Rademacher variables ε_i by standard normal g_i in (II). Then the process in the resulting expression is Gaussian when conditioned on the variables X_i. Thus, for each $n\in\mathbb{N}$ and $\omega\in A(\varepsilon,n)$, we consider the Gaussian process

$$Z_{\omega,n}(h)=\sum_{i=1}^n g_ih(X_i(\omega))/n^{1/2},\quad h\in\mathcal{H}(\varepsilon,n),$$

and note that since $\omega\in A(\varepsilon,n)$,

$$E_g(Z_{\omega,n}(h)-Z_{\omega,n}(h'))^2=P_n(\omega)(h-h')^2\leq 4P(h-h')^2=E(2Z_P(h)-2Z_P(h'))^2.$$

Then, by the result on comparison of moduli of continuity in Exercise 2.4.12, there exists $K<\infty$ such that for all $\omega\in A(\varepsilon,n)$, $\varepsilon>0$ and $n\in\mathbb{N}$,

$$E_g\left\|\nu_{n,\mathrm{rad}}\right\|_{\mathcal{H}'_{2\delta}}\leq K\left[\sup_{f\in\mathcal{F}}E\sup_{h\in\mathcal{F}:e_P(f,h)\leq 2\delta}|Z_P(f)-Z_P(h)|+\delta\sqrt{\log N(\mathcal{F},e_P,\delta)}\right].$$

Since $\|Z_P\|_{\mathcal{F}'_\delta}\to 0$ a.s. by uniform continuity with respect to e_P, and since $E\|Z_P\|_\infty<\infty$ (by, e.g., Theorem 2.1.20), the first term at the right-hand side of this inequality tends to zero as

$\delta \to 0$. The second term also tends to zero as $\delta \to 0$ by (3.285) (Sudakov's inequality). These two terms are independent of $\omega \in A(\varepsilon, n)$ and $n \in \mathbb{N}$, thus showing that $\lim_{\delta\to 0} \limsup_n (II) = 0$. Plugging the limits of (I) and (II) into (3.284) proves the asymptotic equi-continuity condition (3.283) and hence the theorem. ■

Note that in this proof Bernstein's inequality is used at the limit of its Gaussian range and that the cutoff $\varepsilon/n^{1/2}$ for $P(f-g)^2$ in the asymptotic equi-continuity condition is obtained by balancing this limit with the size of $\mathcal{H}$ as estimated by Sudakov's bound.

Next, we apply this theorem to obtain conditions for classes of sets to be P-Donsker not covered by results presented earlier in this chapter. The proof requires a second ingredient, namely, a probability estimate of the supremum norm of the empirical process over a class of bounded *positive* functions that combines an exponential inequality with the random entropy of the process over the 'square root' class by means of an elaborate use of randomisation. Note that for a single function f, setting $r = \|f\|_\infty$ and $\sigma^2 = Pf^2$, since $Pf^4 \le r^2\sigma^2$, Bernstein's inequality gives, for all $u \ge 2n\sigma^2$,

$$\Pr\left\{\sum_{i=1}^n f^2(X_i) \ge u\right\} \le 2\exp\left(-\frac{u^2/4}{2nr^2\sigma^2 + r^2u/3}\right) \le 2\exp\left(-\frac{3u}{16r^2}\right).$$

Here is what can be obtained for the supremum of these sums over a class of functions. For the next lemma, recall the notation $e_{n,2}^2(f,g) = P_n(f-g)^2$.

Lemma 3.7.53 (Square root trick) *Let $\mathcal{F}$ be a class of functions satisfying the pointwise countable approximation hypothesis. Let $\sigma^2 = \sup_{f\in\mathcal{F}} Pf^2$ and $r = \sup_{f\in\mathcal{F}} \|f\|_\infty$. For $u > 4n\sigma^2$ and $0 < 2^{5/2}\rho < u^{1/2} - 2n^{1/2}\sigma$, set*

$$\lambda = \frac{1}{2}(u^{1/2} - 2n^{1/2}\sigma - 2^{5/2}\rho)^2.$$

Then, for all $n, m \in \mathbb{N}$ and $u > 4n\sigma^2$,

$$\Pr\left\{\left\|\sum_{i=1}^n f^2(X_i)\right\|_{\mathcal{F}} > u\right\} \le 4\Pr^*\left\{N(\mathcal{F}, e_{n,2}, \rho/n^{1/2}) > m\right\} + 16me^{-\lambda/2r^2}. \tag{3.286}$$

Proof We begin with some notation. Given a Rademacher sequence ε_i, $i \le n$, independent from the sequence $\{X_i\}$ (as usual, all the variables defined as coordinates in a product probability space), set, for $f \in \mathcal{F}$,

$$N_+(f) = \sum_{i\le n:\varepsilon_i=1} f^2(X_i), \; N_-(f) = \sum_{i\le n:\varepsilon_i=-1} f^2(X_i).$$

Then N_+ and N_- are equi-distributed, they are conditionally independent given $\{\varepsilon_i\}$ and

$$N_+(f) - N_-(f) = \sum_{i=1}^n \varepsilon_i f^2(X_i), \; N_+(f) + N_-(f) = \sum_{i=1}^n f^2(X_i)$$

and $E(N_-^{1/2}(f))^2 \le nPf^2 \le n\sigma^2$. We can use these properties together with the symmetrisation inequality in Proposition 3.1.24 part (b), applied conditionally on the Rademacher variables to $Y = N_+^{1/2}$ and $Y' = N_-^{1/2}$, and obtain, using Fubini's theorem,

$$\begin{aligned}
\Pr\left\{\left\|\sum_{i=1}^n f^2(X_i)\right\|_{\mathcal{F}} > u\right\} &\le 2\Pr\left\|N_+^{1/2}\|_{\mathcal{F}} \ge 2^{-1/2}u^{1/2}\right\} \\
&\le 4E_\varepsilon \Pr_X\left\{\left\|N_+^{1/2} - N_-^{1/2}\right\|_{\mathcal{F}} > 2^{-1/2}u^{1/2} - (2n\sigma^2)^{1/2}\right\} \\
&\le 4\Pr^*\left\{N(\mathcal{F}, e_{n,2}, \rho/n^{1/2}) > m\right\} \\
&\quad + 4E_X \Pr_\varepsilon\left\{\left\|N_+^{1/2} - N_-^{1/2}\right\|_{\mathcal{F}} > 2^{-1/2}u^{1/2} - (2n\sigma^2)^{1/2}, N(\mathcal{F}, e_{n,2}, \rho/n^{1/2}) \le m\right\}.
\end{aligned}$$

Now we notice that by the triangle inequality for the Euclidean norm, $|N_+^{1/2}(f) - N_+^{1/2}(g)| \le \left[\sum_{i:\varepsilon_i=1}(f(X_i) - g(X_i))^2\right]^{1/2}$, and likewise for N_-, so

$$|(N_+^{1/2}(f) - N_-^{1/2}(f)) - (N_+^{1/2}(g) - N_-^{1/2}(g))| \le 2\left[\sum_{i=1}^n (f(X_i) - g(X_i))^2\right]^{1/2} = 2n^{1/2}e_{n,2}(f, g).$$

Thus, if for $X_1, \ldots, X_n$ fixed $\mathcal{D}$ is a $\rho/n^{1/2}$-dense subset of $\mathcal{F}$ for the $e_{n,2}$ pseudo-distance of minimal cardinality, we have

$$\begin{aligned}
&\Pr_\varepsilon\left\{\left\|N_+^{1/2} - N_-^{1/2}\right\|_{\mathcal{F}} > 2^{-1/2}u^{1/2} - (2n\sigma^2)^{1/2}, N(\mathcal{F}, e_{n,2}, \rho/n^{1/2}) \le m\right\} \\
&\le m \max_{f \in \mathcal{D}} \Pr_\varepsilon\left\{|N_+^{1/2}(f) - N_-^{1/2}(f)| > 2^{-1/2}u^{1/2} - (2n\sigma^2)^{1/2} - 2\rho\right\}.
\end{aligned}$$

If $0 \in \mathcal{D}$, we replace the function 0 by a function in $\mathcal{D} \cap \{f \in \mathcal{F} : P_n f^2 \le \rho/n^{1/2}\}$, and the resulting set, which we still call $\mathcal{D}$, is at least $2\rho/n^{1/2}$-dense in $\mathcal{F}$, so if we use it in the preceding probability, we must subtract 4ρ instead of 2ρ to obtain instead the bound

$$m \max_{f \in \mathcal{D}} \Pr_\varepsilon\left\{|N_+^{1/2}(f) - N_-^{1/2}(f)| > \lambda^{1/2}\right\}.$$

Next, we note that

$$\begin{aligned}
|N_+(f) - N_-(f)| &= |N_+^{1/2}(f) - N_-^{1/2}(f)|(N_+^{1/2}(f) + N_-^{1/2}(f)) \\
&\ge |N_+^{1/2}(f) - N_-^{1/2}(f)|\left(\sum_{i=1}^n f^2(X_i)\right)^{1/2} \\
&\ge |N_+^{1/2}(f) - N_-^{1/2}(f)|\left(\sum_{i=1}^n f^4(X_i)\right)^{1/2} / r.
\end{aligned}$$

Hence, Hoeffding's inequality ((3.9) in Theorem 3.1.2) gives

$$\max_{f \in \mathcal{D}} \Pr_\varepsilon\left\{|N_+^{1/2}(f) - N_-^{1/2}(f)| > \lambda^{1/2}\right\} \le \Pr\left\{\frac{\left|\sum_{i=1}^n \varepsilon_i f^2(X_i)\right|}{\left(\sum_{i=1}^n f^4(X_i)\right)^{1/2}} > \frac{\lambda^{1/2}}{r}\right\} \le 2e^{-\lambda/2r^2}.$$

The lemma follows by collecting the preceding probability bounds. ■

Here is how the preceding lemma can be used to estimate the small oscillations of the empirical process in Theorem 3.7.52 by means of random entropies. The resulting condition is *necessary* for classes of sets $\mathcal{F}$ to be P-Donsker

Theorem 3.7.54 *Let $\mathcal{F}$ be a uniformly bounded class of functions satisfying the pointwise countable approximation hypothesis. If $\mathcal{F}$ is P-pre-Gaussian and, for some $c > 0$ and all $\tau > 0$,*

$$\lim_{\varepsilon\to 0}\limsup_{n} \Pr^* \left\{ \frac{\log N(\mathcal{F}'_{\varepsilon,n}, L^1(P^n), \tau/n^{1/2})}{n^{1/2}} > c\tau \right\} = 0, \tag{3.287}$$

then $\mathcal{F}$ is P-Donsker. Conversely, if $\mathcal{F}$ is a collection of indicator functions and is P-Donsker, then $\mathcal{F}$ is P-pre-Gaussian and satisfies condition (3.287).

Proof We prove the direct part first. We can assume without loss of generality that $\|f\|_\infty \le 1/2$, for all $f \in \mathcal{F}'_{\varepsilon,n}$. By Theorem 3.7.52, we only need to prove that the limit (3.281) holds for all $\gamma > 0$. Somewhat as earlier, we write, for $\alpha, \beta > 0$ to be chosen later,

$$\begin{aligned} \Pr\left\{ \|\nu_{n,\mathrm{rad}}\|_{\mathcal{F}'_{\varepsilon,n}} > \gamma \right\} &\le \Pr^*\left\{ N(\mathcal{F}'_{\varepsilon,n}, e_{n,1}, \beta/n^{1/2}) > m \right\} + \Pr\left\{ \left\| \sum_{i=1}^n f^2(X_i) \right\|_{\mathcal{F}'_{\varepsilon,n}} > \alpha n^{1/2} \right\} \\ &+ E_X^* \Pr_\varepsilon \left\{ \|\nu_{n,\mathrm{rad}}\|_{\mathcal{F}'_{\varepsilon,n}} > \gamma, N(\mathcal{F}_{\varepsilon,n}, e_{n,1}, \beta/n^{1/2}) \le m, \left\| \sum_{i=1}^n f^2(X_i) \right\|_{\mathcal{F}'_{\varepsilon,n}} \le \alpha n^{1/2} \right\}. \end{aligned} \tag{3.288}$$

For $X_1, \ldots, X_n$ fixed, let $\mathcal{D}$ be a $\beta/n^{1/2}$-dense subset of $\mathcal{F}'_{\varepsilon,n}$ for the $e_{n,1}$ distance, and note that $\left|\sum_{i=1}^n \varepsilon_i (f(X_i) - g(X_i)\right| \le e_{n,1}(f,g)$. Thus, if we take

$$\alpha = (\gamma - \beta)/2 > 0$$

(hence $\beta < \gamma$), the $\Pr_\varepsilon$ probability in the third summand is dominated by

$$m \sup_{f\in\mathcal{D}} \Pr_\varepsilon \left\{ |\nu_{n,\mathrm{rad}}| > (\gamma - \beta), \sum_{i=1}^n f^2(X_i) \le (\gamma - \beta)n^{1/2}/2 \right\} \le 2me^{-(\gamma-\beta)n^{1/2}}$$

by Hoeffding's inequality (Theorem 3.1.2).

Next, we apply Lemma 3.7.53 to the second summand in the decomposition (3.288). We have $\sigma^2 = \sup Pf^2 \le \varepsilon/n^{1/2}$, $r = 1$, $u = \alpha n^{1/2} = (\gamma - \beta)n^{1/2}/2$, and the lemma requires $(\gamma - \beta)/2 > 4\varepsilon$, and then ρ must satisfy $0 < \rho < 2^{-5/2}\left[2^{-1/2}(\gamma - \beta)^{1/2} - 2\varepsilon^{1/2}\right] n^{1/4}$. Also observe that since the functions in $\mathcal{F}'_{\varepsilon,n}$ are bounded by 1, we have on this class that $e_{n,2} \le 2^{1/2} e_{n,1}^{1/2}$, so $N(\mathcal{F}'_{\varepsilon,n}, e_{n,2}, \rho/n^{1/2}) \le N(\mathcal{F}'_{\varepsilon,n}, e_{n,1}, \rho^2/2n)$. Thus, assuming that the conditions imposed on $\gamma, \beta, \varepsilon, \rho$ are met, Lemma 3.7.53 implies that

$$\Pr\left\{ \left\| \sum_{i=1}^n f^2(X_i) \right\|_{\mathcal{F}'_{\varepsilon,n}} > \alpha n^{1/2} \right\} \le 4\Pr^*\left\{ N(\mathcal{F}'_{\varepsilon,n}, e_{n,1}, \rho^2/2n) > m \right\} + 16m \exp(-\lambda/2),$$

where λ is defined as in the lemma, $\lambda^{1/2} = 2^{-1/2}\left(\left[2^{-1/2}(\gamma - \beta)^{1/2} - 2\varepsilon^{1/2}\right] n^{1/4} - 2^{5/2}\rho\right)$ (in fact, the inequality holds for any λ not exceeding this value).

Now consider only $\varepsilon < \gamma/2^6$, and take $\beta = \gamma/2$, $\alpha = \gamma/4$, $\rho = \gamma^{1/2}n^{1/4}/2^{11/2}$ and $\lambda = \gamma n^{1/2}/2^7$ to obtain from the preceding estimates that

$$\Pr\left\{\left\|\nu_{n,\mathrm{rad}}\right\|_{\mathcal{F}'_{\varepsilon,n}} > \gamma\right\} \le 5\Pr^*\left\{N(\mathcal{F}'_{\varepsilon,n}, e_{n,1}, \gamma/(2^{13}n^{1/2})) > m\right\} + 18me^{-\gamma n^{1/2}/2^8}. \quad (3.289)$$

Thus, taking $m = e^{\gamma n^{1/2}/2^{13}}$ in this bound, we see that the limit as n tends to infinity is zero for all $\varepsilon < \gamma/2^6$, proving (3.287) for $c = 1$ and all $\gamma > 0$ (a little more has been proved, namely, that the $\limsup_n$ of these probabilities is zero for all $0 < \varepsilon \le \nu_0(\gamma)$).

For the converse for classes of sets, we assume that $\mathcal{F}$ is P-Donsker and must prove that (3.287) holds. If $t_{n,\varepsilon} = \inf\left[t : \Pr\left\{\|\nu_{n,\mathrm{rad}}\|_{\mathcal{F}'_{\varepsilon,n}} > 1/8\right\}\right]$, Hoffmann-Jørgensen's inequality (3.39) shows that

$$E\|\nu_{n,\mathrm{rad}}\|_{\mathcal{F}'_{\varepsilon,n}} \le C(n^{-1/2} + t_{n,\varepsilon}).$$

But the fact that $\mathcal{F}$ is P-Donsker implies that $\lim_{\varepsilon\to 0}\limsup_n t_{n,\varepsilon} = 0$ by Theorem 3.7.52 and hence that

$$\lim_{\varepsilon\to 0}\limsup_n E\|\nu_{n,\mathrm{rad}}\|_{\mathcal{F}'_{\varepsilon,n}} = 0.$$

Now, since by Proposition 3.1.26 (and convexity) and Lemma 2.3.4, for $0 \le n_0 < n$ and all n,

$$E\|\nu_{n,g}\|_{\mathcal{F}'_{\varepsilon,n}} \le C\left(\frac{n_0(\log n)^{1/2}}{n^{1/2}} + \max_{n_0<k\le n} E\|\nu_{k,\mathrm{rad}}\|_{\mathcal{F}'_{\varepsilon,n}}\right),$$

it follows that

$$\lim_{\varepsilon\to 0}\limsup_n E\|\nu_{n,g}\|_{\mathcal{F}'_{\varepsilon,n}} = 0$$

by taking, for example, $n_0 = \sqrt{n}/\log n$. (This also follows from Exercises 3.7.15 and 3.7.16, and the preceding arguments are only given for completeness.) Then Sudakov's inequality (Theorem 2.4.12) gives

$$\lim_{\varepsilon\to 0}\limsup_n E^*\left[\sup_{\lambda>0}\lambda\left(\log N(\mathcal{F}'_{\varepsilon,n}, e_{n,2}, \lambda)\right)^{1/2}\right] = 0.$$

But $\mathcal{F}$ being a class of indicators, any $f \in \mathcal{F}'_{\varepsilon,n}$ takes on only the values 1, 0 and -1, and therefore, $e_{n,1}(f,g) \le e^2_{n,2}(f,g)$, for all $f, g \in \mathcal{F}'_{\varepsilon,n}$, so $N(\mathcal{F}'_{\varepsilon,n}, e_{n,1}, \lambda^{1/2}) \le N(\mathcal{F}'_{\varepsilon,n}, e_{n,2}, \lambda)$ and hence,

$$\lim_{\varepsilon\to 0}\limsup_n E^*\left[\sup_{\lambda>0}\left(\lambda \log N(\mathcal{F}'_{\varepsilon,n}, e_{n,2}, \lambda)\right)^{1/2}\right] = 0.$$

This implies (3.287). ■

Now, since if $\{f_i : 1 \le i \le N\}$ is $\varepsilon/2$-dense in $\mathcal{F}$ in $L^1(P_n)$, then $\{f_i - f_j : 1 \le i,j \le N\}$ is ε-dense in $\mathcal{F}'_{\varepsilon,n}$, and we can find a subset of $\mathcal{F}'_{\varepsilon,n}$ of cardinality at most N^2 that is ε-dense in $\mathcal{F}'_{\varepsilon,n}$. It follows that $N(\mathcal{F}'_{\varepsilon,n}, e_{n,1}, \varepsilon) \le (N(\mathcal{F}, e_{n,1}, \varepsilon/2))^2$. In the case of indicators of sets, it is convenient to note that all the possible values of $P_n(|I_A - I_B|)$ are $\{k/n : 0 \le k \le n\}$ and that $P_n(\omega)(|I_A - I_B|) = 0$ if and only if $A \cap \{X_1(\omega), \ldots, X_n(\omega)\} = B \cap \{X_1(\omega), \ldots, X_n(\omega)\}$. Thus, if $\mathcal{F} = \{I_C : C \in \mathcal{C}\}$, then $N(\mathcal{F}, e_{n,1}, \varepsilon) \le \Delta^{\mathcal{C}}(X_1, \ldots, X_n)$, with equality for $0 \le \varepsilon \le 1/n$. (See Section 3.6.1 for the definition of $\Delta^{\mathcal{C}}$.) This and the preceding theorem yield the following:

Theorem 3.7.55 (P-Donsker class of sets: necessary and sufficient conditions) *Let $\mathcal{C}$ be a class of sets satisfying the pointwise countable approximation property. If $\mathcal{C}$ is P-pre-Gaussian and*

$$\frac{\log \Delta^{\mathcal{C}}(X_1,\ldots,X_n)}{n^{1/2}} \to 0 \quad \textit{in outer probability}, \tag{3.290}$$

then $\mathcal{C}$ is a P-Donsker class.

The last two theorems may be thought of as the analogues for the central limit theorem of Theorem 3.7.14 for the law of large numbers for classes of sets. In fact, the converse of Theorem 3.7.55 does hold, and for general classes of functions $\mathcal{F}$, there are necessary and sufficient conditions for $\mathcal{F}$ to be a P-Donsker class that combines pre-Gaussianness and $e_{n,1}$ conditions. This subject will not be pursued. See the notes at the end of this chapter.

Here are two interesting examples. Parts of the proofs will be relegated to the exercises.

Example 3.7.56 We will show that *the collection $\mathcal{C}$ of all the subsets of $\mathbb{N}$ is P-Donsker for $P = \sum_{k=1}^{\infty} p_k \delta_k$ if and only if $\sum_{k=1}^{\infty} p_k^{1/2} < \infty$, in particular*, by Exercise 3.7.21, *if and only if $\mathcal{C}$ is P-pre-Gaussian.* First, $\mathcal{C}$ satisfies the pointwise countable approximation property because indicators of countable sets can be pointwise approximated by indicators of finite sets. As seen in Exercise 3.7.21, the condition is necessary and sufficient for $\mathcal{C}$ to be P-pre-Gaussian; hence, it is necessary for $\mathcal{C}$ to be P-Donsker, and in order to prove that it is sufficient, it suffices to show that if it holds, then condition (3.290) holds. Assume that $\{p_k\}$ is nonincreasing so that, in particular, $p_k < 1/k^2$ for all k large enough and $\sum_{k=r}^{\infty} p_k = p_r^{1/2} \sum_{k=r}^{\infty} p_k^{1/2} = o(1/r)$ as $r \to \infty$. Then

$$\begin{aligned}
\Pr^*\left\{\frac{\log \Delta^{\mathcal{C}}(X_1,\ldots,X_n)}{n^{1/2}} > \varepsilon \log 2\right\} &= \Pr^*\left\{\Delta^{\mathcal{C}}(X_1,\ldots,X_n) > 2^{\varepsilon n^{1/2}}\right\} \\
&\le \Pr\left\{\text{the number of distint } X_1,\ldots,X_n \text{ exceeds } \varepsilon n^{1/2}\right\} \\
&\le \Pr\left\{\text{at least } [\varepsilon n^{1/2}/2]\, X_i' \text{ out of } X_1,\ldots,X_n \text{ exceed } \varepsilon n^{1/2}/2\right\}.
\end{aligned}$$

By Exercise 3.1.8, this probability is bounded by

$$\left(\frac{en \Pr\{X > \varepsilon n^{1/2}/2\}}{[\varepsilon n^{1/2}/2]}\right)^{[\varepsilon n^{1/2}/2]} = \left(\frac{o(en/[\varepsilon n^{1/2}/2])}{[\varepsilon n^{1/2}/2]}\right)^{[\varepsilon n^{1/2}/2]} \to 0$$

as $n \to \infty$ for all $\varepsilon > 0$, and the result follows from Theorem 3.7.55.

Example 3.7.57 We will show that $BL_1(\mathbb{R})$, *the collection of bounded Lipschitz functions on $\mathbb{R}$ with supremum norm and Lipschitz constants not exceeding 1 is P-Donsker if and only if*

$$\sum_{j=1}^{\infty} (P\{j-1 < |x| \le j\})^{1/2} < \infty, \tag{3.291}$$

which happens if and only if $BL_1(\mathbb{R})$ is P-pre-Gaussian. To prove that condition (3.291) is necessary for $BL_1(\mathbb{R})$ to be P-pre-Gaussian, consider the functions $f_{3j}(x) = 1$, for $x \in [3j, 3j+1]$, and $f_{3j}(x) = 0$, for $x \in [3j-1, 3j+2]^c$, and linear in between. Then for the subclass of bounded Lipschitz functions $\{\sum_{j=-\infty}^{\infty} \tau_j f_{3j} : \tau_j = \pm 1\}$ to be P-pre-Gaussian, it is necessary that the series $\sum_{j=-\infty}^{\infty} |g_{3j}|$ converge a.s., where g_{3j} are independent centred normal variables with variance equal to the second moment of $f_{3j}(X)$, where X is a random variable with law P (to see this, just argue as in Exercise 3.7.21). But $Pf_{3j}^2(X) \geq P(3j,3j+1]$, and the series $\sum_{j=-\infty}^{\infty}|g_{3j}|$ converges if and only if $\sum_{j=-\infty}^{\infty}(P(3j,3j+1])^{1/2} < \infty$. Convergence of the other two-thirds of the series $\sum_{j=1}^{\infty}(P(j-1,j])^{1/2}$ follows by applying the same reasoning to the translation by one unit to the right and one to the left of the same set of functions. To prove that condition (3.291) is sufficient for $BL_1(\mathbb{R})$ to be P-Donsker, let us first note that $BL_1(\mathbb{R})$ is separable for the usual metric of uniform convergence on compact sets (e.g., because $BL([a,b])$ is compact in $C([a,b])$ $-\infty < a < b < \infty$) and that this implies that $BL_1(\mathbb{R})$ satisfies the pointwise countable approximation property. Next, we show that $BL_1(\mathbb{R})$ is P-pre-Gaussian if P satisfies condition (3.291). For $j \in \mathbb{Z}$, let $\tilde{Z}_j(f)$, $f \in BL_1(\mathbb{R})$, be a centred Gaussian process with covariance $E(\tilde{Z}_j(f)\tilde{Z}_j(g)) = E(f(X)g(X)I(j-1 < X \leq j))$. Then

$$e_j^2(f,g) := E(\tilde{Z}_j(f) - \tilde{Z}_j(g))^2 \leq \|f-g\|_\infty^2 P(j-1,j],$$

and it follows from the metric entropy estimate in Exercise 3.6.17 that

$$N(BL_1(\mathbb{R})), e_j, \varepsilon) \leq N(BL_1([j-1,j]), \|\cdot\|_\infty, \varepsilon/(P(j-1,j])^{1/2}) \leq e^{c(P(j-1,j])^{1/2})/\varepsilon},$$

for all $0 < \varepsilon < (P(j-1,j]))^{1/2}$. Then, by Theorem 2.3.7, $\tilde{Z}_j$ admits a version Z_j with bounded and uniformly continuous sample paths in $(BL_1(\mathbb{R}), L^2(P))$ such that

$$E\|Z_j(f)\|_{BL_1(\mathbb{R})} \leq 4\sqrt{2}(P(j-1,j])^{1/2}\int_0^1 \sqrt{1+c/\varepsilon}\,d\varepsilon =: K(P(j-1,j])^{1/2},$$

for $K < \infty$ independent of j. Take the processes Z_j , $j \in \mathbb{Z}$, to be independent. Then, by Lévy's inequality,

$$\begin{aligned}\lim_{n\to\infty}\Pr\left\{\sup_{k\geq n}\left\|\sum_{n\leq|j|\leq k} Z_j\right\|_{BL_1(\mathbb{R})} > \varepsilon\right\} &= \lim_{n\to\infty}\lim_{m\to\infty}\Pr\left\{\sup_{n\leq k\leq m}\left\|\sum_{n\leq|j|\leq k} Z_j\right\|_{BL_1(\mathbb{R})} > \varepsilon\right\}\\ &\leq 2\lim_{n\to\infty}\lim_{m\to\infty}\Pr\left\{\left\|\sum_{n\leq|j|\leq m} Z_j\right\|_{BL_1(\mathbb{R})} > \varepsilon\right\}\\ &\leq 2K\varepsilon^{-1}\lim_n\sum_{j\geq n}(P(j-1,j])^{1/2} = 0;\end{aligned}$$

that is, the series $Z(f) = \sum_{j=-\infty}^{\infty} Z_j(f)$ converges uniformly a.s. in $BL_1(\mathbb{R})$, and therefore, Z has bounded and uniformly continuous paths in $(BL_1(\mathbb{R}), L^2(P))$. Since $E(Z(f)Z(g)) = E(f(X)g(X))$, we have proved that $BL_1(\mathbb{R})$ is P-pre-Gaussian. By Theorem 3.7.52, to prove that $BL_1(\mathbb{R})$ is P-Donsker, it suffices to show that condition (3.281) holds for this class. Since by Exercise 3.6.17, as mentioned earlier, for all $r < \infty$ and for any probability measure Q,

$$N(BL_1([-r,r]), L^2(Q), \varepsilon) \leq N(BL_1([-r,r]), Q^{1/2}[-r,r]\|\cdot\|_\infty, \varepsilon) \leq \exp\left(2crQ^{1/2}[-r,r]/\varepsilon\right),$$

it follows from Theorem 3.7.37 that $BL_1([-r,r])$ is Q-Donsker for every probability measure Q (in fact, uniform Donsker). Hence, by Theorem 3.7.52, we have that for all P, $r > 0$ and $\gamma > 0$,

$$\lim_{\varepsilon\to 0}\limsup_n \Pr\left\{\sup_{\substack{f,g\in BL_1(\mathbb{R})\\ P(f-g)^2\le\varepsilon/n^{1/2}}} |\nu_{n,\mathrm{rad}}((f-g)I_{[-r,r]})|\ge\gamma\right\}=0. \tag{3.292}$$

By the proof of Theorem 3.7.54, in fact, inequality (3.289), for all $\gamma>0$ and $\varepsilon<\gamma/2^6$, gives

$$\Pr\left\{\sup_{\substack{f,g\in BL_1(\mathbb{R})\\ P(f-g)^2\le\varepsilon/n^{1/2}}} |\nu_{n,\mathrm{rad}}((f-g)I_{[-r,r]^c})|\ge\gamma\right\}$$
$$\le 5\Pr^*\left\{\log N(BL_1([-r,r]^c),L^1(P_n),\gamma/(\tau n^{1/2}))\ge\gamma n^{1/2}/\tau\right\}+o(1), \tag{3.293}$$

where $\tau=2^{14}$ and $o(1)\to 0$ as $n\to\infty$ and depends only on γ. To compute this random entropy, set $C_{r,n}=\sum_{j=r+1}^{\infty}P_n^{1/2}(j-1,j]$ and $I_j=(j-1,j]$. Then, for $f,g\in BL_1([-r,r]^c)$,

$$\begin{aligned}P_n|f-g| &= \sum_{j=r+1}^{\infty}P_n(|f-g|I_{I_j})\le\sum_{j=r+1}^{\infty}\|(f-g)I_{I_j}\|_\infty P_n(I_j)\\ &=\sum_{j=r+1}^{\infty}\frac{P_n^{1/2}(I_j)}{C_{r,n}}(\|(f-g)I_{I_j}\|_\infty P_n^{1/2}(I_j)C_{r,n}),\end{aligned}$$

so if $\|(f-g)I_{I_j}\|_\infty P_n^{1/2}(I_j)C_{r,n}\le\eta$ for each $j>r$, then $P_n|f-g|\le\eta$. Therefore, for all $\eta>0$,

$$\begin{aligned}N(BL_1([-r,r]^c,L^1(P_n),\eta) &\le\prod_{j=r+1}^{\infty}N(BL(I_j),\|\cdot\|_\infty P_n^{1/2}(I_j)C_{r,n},\tau)\\ &\le\exp\left(c\eta^{-1}\left(\sum_{j=r+1}^{\infty}P_n^{1/2}(I_j)\right)^2\right).\end{aligned}$$

This gives

$$\begin{aligned}\Pr^*\left\{\log N(BL_1([-r,r]^c,L^1(P_n),\gamma/(\tau n^{1/2}))>\gamma n^{1/2}/\tau\right\} &\le\Pr\left\{\sum_{j=r+1}^{\infty}P_n^{1/2}(I_j)\ge\frac{\gamma}{\tau c^{1/2}}\right\}\\ &\le\frac{\tau c^{1/2}}{\gamma}\sum_{j=r+1}^{\infty}E(P_n^{1/2}(I_j))\\ &\le\frac{\tau c^{1/2}}{\gamma}\sum_{j=r+1}^{\infty}P^{1/2}(j-1,j],\end{aligned}$$

which tends to zero as $r\to\infty$ independently of n. Plugging this into (3.293), we see that $\lim_r\limsup_n(3.293)=0$, which together with (3.292) shows that condition (3.281) is satisfied and therefore that $BL_1(\mathbb{R})$ is P-Donsker by Theorem 3.7.52.

Exercises

3.7.1 Prove that: (a) $(\max(f,g))^* = \max(f^*,g^*)$ a.s.; in fact, the same is true for a countable number of functions. (b) For $f \geq 0$ and $p > 0$, $(f^p)^* = (f^*)^p$ a.s. (c) If $g \geq 0$ is a measurable function, then $(fg)^* = f^*g$ a.s.

3.7.2 Let $(X \times Y, \mathcal{A} \otimes \mathcal{B}, P \times Q)$ be a product probability space, and let $g(x,y) = f(x)$ be a function of only the first coordinate. Prove that $g^*_{P\times Q} = f^*_P$ $P \times Q$-a.s. *Hint*: $g(x_1,x_2) \leq f^*_P(x_1)$ and, however, $f(x_1) \leq g^*_{P\times Q}(x,y)$, and this last function is P-measurable for every y.

3.7.3 State and prove versions of Lévy's, Ottaviani's and Hoffmann-Jørgensen's inequalities (Theorems 3.1.11, 3.1.15 and 3.1.16 and Proposition 3.1.12) in terms of outer probabilities and expectations for sample bounded processes indexed by not necessarily countable sets T. Do the same for the randomisation and symmetrisation inequalities in probability from Propositions 3.1.23, 3.1.24 and 3.1.26 and Corollary 3.1.25.

3.7.4 Define $E_*f = \sup\{Eg : g \text{ measurable}, g \leq f\}$, the inner integral of f. As for outer integrals, show that if f_* is the essential supremum of the set of measurable $g \leq f$, then $E_*f = Ef_*$ whenever either is defined. Show also that $f_* = -((-f)^*)$ and $E_*f = -E^*(-f)$.

3.7.5 Prove that the Glivenko-Cantelli theorem in $\mathbb{R}^d$ follows from Theorem 3.7.20.

3.7.6 Let B be a separable Banach space, let $\mathcal{F} := \{f \in B^* : \|f\| \leq 1\}$, where B^* is the dual of B, and let X, X_i, be i.i.d. B-valued random variables with $E\|X\| < \infty$. Prove that there exists $L < \infty$ such that $\lim_{n\to\infty} \Pr^*\{N(\mathcal{F}, e_{n,1}, \varepsilon) > L\} = 0$ for all $\varepsilon > 0$. *Hint*: Proceed in a way similar to the proof of Corollary 3.7.21. This shows that the Mourier law of large numbers is also a consequence of Theorem 3.7.14.

3.7.7 (Portmanteau theorem.) Prove that for bounded processes X_n and a bounded process X with tight Borel probability law, convergence in law in $\ell_\infty(T)$ of X_n to X is equivalent to each of the following three conditions: (a) for every closed set $F \subset \ell_\infty(T)$, $\limsup_{n\to\infty} \Pr^*\{X_n \in F\} \leq \Pr\{X \in F\}$; (b) for every open set $G \subset \ell_\infty(T)$, $\liminf_{n\to\infty} \Pr_*\{X_n \in G\} \geq \Pr\{X \in G\}$; and (c) if $A \subset \ell_\infty(T)$ is a continuity set for the law of X (meaning that the probability that X is in the boundary of A is zero), then $\liminf_{n\to\infty} \Pr_*\{X_n \in A\} = \limsup_{n\to\infty} \Pr^*\{X_n \in A\} = \Pr\{X \in A\}$. *Hint for (a)*: Apply the definition to bounded continuous functions H approximating I_F ($0 \leq H(x) \leq 1$, $g(x) = 1$, for $x \in F$, and $g(x) = 0$, for x outside of a neighborhood of F). *Hint for (c) to imply convergence in law*: For any bounded continuous function H, all but a countable number of sets of the form $\{H < t\}$ are continuity sets for X, so H can be uniformly approximated with any accuracy by simple functions based on continuity sets for X, to which (c) applies.

3.7.8 (A simple extension of uniform continuity on compact sets, used in the proof of Theorem 3.7.23.) Let (S,d) be a metric space, let $K \subset S$ be a compact set and let $f : S \mapsto \mathbb{R}$ be a continuous bounded function. Then, for every $\varepsilon > 0$, there exists $\delta > 0$ such that

$$d(u,v) < \delta, \quad u \in K, \quad v \in S \implies |f(u) - f(v)| < \varepsilon.$$

3.7.9 Let f_n and f_∞ be (sample) bounded processes indexed by a set T, and assume the probability law of f_∞ to be a tight Borel measure. Prove that $\|f_n - f_\infty\|^*_T \to 0$ a.s. $\implies$ $\|f_n - f_\infty\|^*_T \to 0$ in probability $\implies$ $f_n \to_{\mathcal{L}} f_\infty$ in $\ell_\infty(T)$. *Hint*: The first assertion is clear. For the second, let d be a distance on T such that (T,d) is totally bounded and $f_\infty \in C_u(T,d)$ (Proposition 2.1.7), and observe that

$$\left(\sup_{d(s,t)\leq\delta} |f_n(t) - f_n(s)|\right)^* \leq 2(\|f_n - f_\infty\|_T)^* + \sup_{d(s,t)\leq\delta} |f_\infty(t) - f_\infty(s)|.$$

Now the asymptotic equi-continuity condition (3.249) for $\{f_n\}$ follows from convergence in outer probability of f_n to f_∞ and uniform continuity of f_∞.

3.7.10 Let $C \subset V$, where V is a vector space. Let $G : C \mapsto \mathbb{R}$ be a prelinear map; that is, whenever $\lambda_i \in \mathbb{R}$ and $c_i \in C$ satisfy $\sum \lambda_i c_i = 0$, one has $\sum \lambda_i G(c_i) = 0$ (finite sums). Prove that G extends as a linear map on the linear span of C by the formula $g\left(\sum \lambda_i c_i\right) = \sum \lambda_i g(c_i)$. Conversely, if $G : C \mapsto \mathbb{R}$ admits a linear extension to the linear span of C, then G is prelinear.

3.7.11 In the proof of Theorem 3.7.28 we use the Karhunen-Loève expansion for the reason that it was proved in Chapter 2; however, any expansion with respect to a complete ortho-normal system of the range of G_P will do. Concretely, if $\mathcal{F}$ is P-pre-Gaussian and G_P is a $C_u(\mathcal{F}, d_X)$-valued version of the P-bridge, then $G = \{G_P(f) : f \in \mathcal{F}\}$ is a precompact subset of $\mathcal{L}^2$ (because $(\mathcal{F}, d_P)$ is totally bounded), so we can take a countable dense subset of G and ortho-normalize it by Gram-Schmidt to obtain an i.i.d. sequence g_i, $i \in \mathbb{N}$, of standard normal random variables which constitute an ortho-normal basis of the linear span of G. Prove that a.s. $G_P = \sum_{i=1}^{\infty} [E(g_i G_P)] g_i$ uniformly in $f \in \mathcal{F}$. *Hint*: For all $f \in G$, $G_P(f) = \sum [E(g_i G_P(f))] g_i$ with convergence in L^2 and hence, by Lévy's theorem on equivalence of forms of convergence for series of independent random variables, also with a.s. convergence. Because of this, because $S_n = \sum_{i=1}^{n} E(g_i G_P) g_i \in C_u(\mathcal{F}, d_x)$ a.s. and because the law of G_P is determined by $G_P(f)$, $f \in \mathcal{F}$, to prove the claim, it suffices to prove that a.s., the sequence of partial sums $S_n(\omega)$ forms a relatively compact subset of $C_u(\mathcal{F}, d_X)$, just as in the proof of Theorem 2.6.10. Now proceed as in that proof, where one can take as the elements v_i of B' simply a countable dense subset of $\mathcal{F}$ for d_P. (As mentioned in the proof of Theorem 2.6.10, a slicker proof using Ito-Nisio's theorem is possible, but this proof is more elementary.)

3.7.12 Let $\mathcal{F}_1$ and $\mathcal{F}_2$ be P-pre-Gaussian classes of functions, and let G_P be a version of the P-bridge which is in $C_u(\mathcal{F}_i, d_X)$ for $i = 1, 2$. Show that there exists an i.i.d. sequence of standard normal random variables which is dense in $(\mathcal{F}_1 \cup \mathcal{F}_2, d_X)$ and that $G_P = \sum_{i=1}^{\infty} [E(g_i G_P)] g_i$ a.s. both as a $C_u(\mathcal{F}_1, d_X)$ and as a $C_u(\mathcal{F}_2, d_X)$ random variable, with uniform convergence in the sup norm for each and hence uniform in $f \in \mathcal{F}_1 \cup \mathcal{F}_2$. Show that, moreover, the functional $f \mapsto E(g_i G_P(f)) g_i$ is in $C_u(\mathcal{F}_1 \cup \mathcal{F}_2)$ and that therefore the process $f \mapsto G_P(f)$, $f \in \mathcal{F}_1 \cup \mathcal{F}_2$, is bounded and uniformly d_P-continuous. Conclude that $\mathcal{F}_1 \cup \mathcal{F}_2$ is P-pre-Gaussian. *Hint*: Proof similar to that of Exercise 3.7.11.

3.7.13 (The Jain-Marcus CLT.) Let (T, ρ) be a compact metric space, and let $X(t)$, $t \in T$, be a stochastic process satisfying $EX(t) = 0$, $EX^2(t) < \infty$, for all $t \in T$, and $|X(t, \omega) - X(s, \omega)| \leq M(\omega)\rho(s, t)$, for all $s, t \in T$, $\omega \in \Omega$, where $EM^2 < \infty$ and $\int_0^{\infty} \sqrt{\log N(T, \rho, \varepsilon)}\, d\varepsilon < \infty$. Let X_i be i.i.d. with the same law as X (the coordinates on $C(T, d)^{\mathbb{N}}$). Prove that $n^{-1/2} \sum_{i=1}^{n} X_i \to_{\mathcal{L}} G$ as random variables taking values in $C(T, d)$, where G is Gaussian. *Hint*: Set $S = C(T, d)$, define $f_t(X) = X(t)$ and apply Theorem 3.7.36 to $\mathcal{F} = \{f_t : t \in T\}$ and $P = \mathcal{L}(X)$. Note that $e_{n,2}(s, t) \leq \left(n^{-1} \sum_{i=1}^{n} M_i^2\right)^{1/2} \rho(s, t)$, and use the law of large numbers for M_i^2.

3.7.14 Consider in $L^2(P)$ the topology τ_1 of pointwise convergence (with neighbourhood base $N(f, x_1, \ldots, x_r, \varepsilon) = \{g \in L^2 : |f(x_i) - g(x_i)| \leq \varepsilon, 1 \leq i \leq r\}$, $f \in L^2, x_i \in S, r \in \mathbb{N}$), and denote by τ_2 the topology given by the L^2-pseudo-norm. Let $\tau = \tau_1 \vee \tau_2$ be the coarsest topology finer than τ_1 and τ_2 (τ is the collection of arbitrary unions of $A_1 \cap A_2$, where $A_i \in \tau_i$, $i = 1, 2$, and recall that a map is continuous in the τ-topology if it is continuous in either one of the topologies τ_1 or τ_2). Prove that if $\mathcal{F}$ is P-Donsker, then so is $\tilde{H}(\mathcal{F}, P)$, the closure of $\mathcal{F}$ in $L^2(P)$ for the τ-topology. (Thus, we are replacing 'sequential closure' by 'closure' in Proposition 3.7.34.) *Hint*: Set $\mathcal{G} = \mathrm{sco}(\mathcal{F})$ and $\mathcal{H} = \tilde{\mathcal{H}} \Leftarrow \mathcal{F}, \mathcal{P} \Rightarrow$. Use almost-sure representations as in the proof of Proposition 3.7.34 together with the fact that if L_i are linear functionals defined on the linear span of $\mathcal{F}$ that are continuous in either of the two topologies (just as $\delta_{X_i(\omega)}$, P and a suitable version of G_P are), then $\left\|\sum_{i=1}^{n} L_i\right\|_{\mathcal{G}} = \left\|\sum_{i=1}^{n} L_i\right\|_{\mathcal{H}}$.

3.7.15 Prove that in Theorem 3.7.40 the processes $\nu_{n,\mathrm{rad}}$ in condition (b) can be replaced by the processes $\nu_{n,g}$; that is, it is possible to randomise in the central limit theorem for empirical processes not only by Rademacher but also by standard normal random variables. In fact, it is possible to randomise by i.i.d. symmetric variables ξ_i, independent of the X_j, such that $\Lambda_{2,1}(\xi_1) < \infty$ (see (3.49)). *Hint*: Use the randomisation inequality for general multipliers, Proposition 3.1.26.

3.7.16 Prove that in Theorem 3.7.40 the equivalence of (a) and (b) extends to the following four conditions which are therefore equivalent to $\mathcal{F}$ being P-Donsker:

(c) $(\mathcal{F}, e_P)$ is totally bounded, and $\lim_{\delta\to 0}\limsup_n \Pr\left\{\|\nu_{n,\mathrm{rad}}\|_{\mathcal{F}'_\delta} > \varepsilon\right\} = 0$, for all $\varepsilon > 0$.

(d) Same as (c) with $\nu_{n,g}$ replacing $\nu_{n,\mathrm{rad}}$.

(e) $(\mathcal{F}, e_P)$ is totally bounded, and $\lim_{\delta\to 0}\limsup_n E\|\nu_{n,rad}\|_{\mathcal{F}'_\delta} = 0$.

(f) Same as (e) with $\nu_{n,g}$ replacing $\nu_{n,\mathrm{rad}}$.

Hint: Use Hoffmann-Jørgensen' s inequality.

3.7.17 Show that if $\mathcal{F}$ is P-Donsker for every $P \in \mathcal{P}_f(S)$, that is, if $\mathcal{F}$ is *universal Donsker*, then $\sup_{f\in\mathcal{F}}(\sup_{x\in S} f(x) - \inf_{x\in S} f(x)) < \infty$, so the class $\{f - \inf f : f \in \mathcal{F}\}$ is uniformly bounded. *Hint*: If not, there exist $x_n, y_n \in S$ and $f_n \in \mathcal{F}$ such that $f_n(x_n) - f_n(y_n) > 2^n$ for all n. Show that if $P_n = \sum_{n=1}^{\infty}(\delta_{x_n} + \delta_{y_n})/2^{n+1}$, then $E_{P_n}(f_n - Ef_n)^2 \to \infty$, which implies that $\sup_{f\in\mathcal{F}} EG_P^2(f) = \infty$ and hence that $\mathcal{F}$ is not P-pre-Gaussian.

3.7.18 Use the asymptotic equi-continuity criterion to prove the central limit theorem in Hilbert space: Let H be a separable Hilbert space, and let X, X_i, $i \in \mathbb{N}$, be independent identically distributed H-valued random variables such that $EX = 0$ and $E\|X\|^2 < \infty$; then the sequence $\left\{T_n := n^{-1/2}\sum_{i=1}^n X_i\right\}_{n=1}^{\infty}$ converges in law on H. *Hint*: As shown in Remark 3.7.30, this statement is equivalent to the class of functions $H_1 = \{\langle h, \cdot\rangle : h \in H, \|h\| \le 1\}$ being P-Donsker, where $P = \mathcal{L}(X)$. Choose an ortho-normal basis $\{e_i\}$ of H such that the coordinates of X in this basis are uncorrelated, and assume without loss of generality that $X^i = \langle X, e_i\rangle$ is not P a.s. zero. Given $N \in \mathbb{N}$, consider the complementary orthogonal projections $P_N x = \sum_{i=1}^N \langle x, e_i\rangle e_i = \sum_{i=1}^N x^i e_i$ and $I - P_N$. Since $E\|X\|^2 = \sum_{i=1}^{\infty} E(X^i)^2 < \infty$, we have $E\|(I - P_N)(X)\|^2 \to 0$ as $N \to \infty$. Use this fact to show that (H_1, e_P) is totally bounded by reducing the problem to an easy finite-dimensional one. For the asymptotic equi-continuity notice that if, given $\varepsilon > 0$, N_ε is such that $E\|(I - P_{N_\varepsilon})(X)\|^2 \le \varepsilon^2$, then

$$
\begin{aligned}
E \sup_{\substack{\|h\|\le 2\\ E\langle h,X\rangle^2\le\delta}} \langle h, T_n\rangle^2 &\le 2E \sup_{\substack{\|h\|\le 2\\ E\langle P_{N_\varepsilon} h,X\rangle^2\le\delta}} \langle P_{N_\varepsilon} h, T_n\rangle^2 + 2E\left\|(I - P_{N_\varepsilon})(T_n)\right\|^2 \\
&\le 2\sum_{j=1}^{N_\varepsilon}\left(1 \wedge \left(\delta / \sum_{j=1}^{N_\varepsilon} P(X^j)^2\right)\right) P(X^j)^2 + 2E\left\|(I - P_{N_\varepsilon})(X)\right\|^2 \\
&\le 2N_\varepsilon(\delta \wedge E\|X\|^2) + 2\varepsilon^2.
\end{aligned}
$$

3.7.19 Let H be a separable Hilbert space, let $S = \{x_k : k \in \mathbb{N}\} \subset H$ with $\|x_k\| \to 0$, where $\|\cdot\|$ denotes Hilbert space norm, and let $\mathcal{F} = H_1$ be the unit ball centred at 0 of H, with $z \in \mathcal{F}$ acting on S by inner product, as in the preceding exercise. Prove that: $\mathcal{F} \in UPG$. *Hint*: Let $P = \sum_{k=1}^{\infty}\alpha_k\delta_{x_k}$ with $\sum \alpha_k = 1, \alpha_k \ge 0$. Then, with g_k i.i.d. $N(0,1)$, we have

$$
E\|Z_P\|_{\mathcal{F}} = E\left\|\sum \alpha_k^{1/2} g_k x_k\right\| \le \left(E\left\|\sum \alpha_k^{1/2} x_k g_k\right\|^2\right)^{1/2} = \left(\sum \alpha_k \|x_k\|^2\right)^{1/2} \le \sup_k \|x_k\|,
$$

independently of P. Hence, $\sup_P E\|Z_P\|_{\mathcal{F}} < \infty$. Also, since $e_P(z,z')^2 = \sum \alpha_k \langle x_k, z - z'\rangle^2$,

$$
\begin{aligned}
E\|Z_P\|_{\mathcal{F}'_{\delta,P}} &= E \sup_{\sum \alpha_k \langle x_k, z\rangle^2 \le \delta^2, \|z\| \le 2} \left|\sum \alpha_k^{1/2} \langle x_k, z\rangle g_k\right| \\
&\le E \sup_{\sum \alpha_k \langle x_k, z\rangle^2 \le \delta^2} \left|\sum_{k=1}^n \alpha_k^{1/2} \langle x_k, z\rangle g_k\right| + 2E \sup_{\|z\| \le 1} \left|\sum_{k=n+1}^\infty \alpha_k^{1/2} \langle x_k, z\rangle g_k\right| \\
&\le \delta E \left(\sum_{k=1}^n g_k^2\right)^{1/2} + 2E \left\|\sum_{k=n+1}^\infty \alpha_k^{1/2} g_k x_k\right\| \\
&\le \delta n^{1/2} + 2\left(\sum_{k=1}^\infty \alpha_k \|x_k\|^2\right)^{1/2} \le \delta n^{1/2} + 2 \sup_{k>n} \|x\|_k,
\end{aligned}
$$

independently of P. Thus, $\mathcal{F}$ is *UPG*. (Note that it can be shown that there are sequences $\{x_k\}$ for which this example satisfies $\sup_P \int_0^\infty \sqrt{\log N(\mathcal{F}, L^2(P), \varepsilon)}d\varepsilon = \infty$ (see Giné and Zinn (1991)), so this example does not follow from uniform metric entropy bounds as Example 3.7.42.)

3.7.20 If in Exercise 3.7.19 x_k does not tend to zero, then $\mathcal{F}$ may not be UPG_f: let $x_k = e_k$, $k \in \mathbb{N}$, be an ortho-normal basis of H, and take $Q_N = \sum_{i=1}^N \delta_{e_i}/N$; using a reverse Hölder inequality (Exercise 2.1.2), show that $E\|Z_{P_N}\|_{\mathcal{F}'_{\delta,P_N}} \ge c(2 \wedge \sqrt{\delta N})$. Use Exercise 3.7.18 to show that $\mathcal{F}$ is universal Donsker (here $S = \{e_i\}_{i=1}^n$, so the probability measures on S integrate $\|x\|^2$). That is, there are universal Donsker classes that are not uniform Donsker.

3.7.21 Let $\mathcal{F} = \{I_A : A \subset \mathbb{N}\}$ be the collection of (the indicator functions of) all the subsets of $\mathbb{N}$, and let $P = \sum_{i=1}^n p_k \delta_k$ be a probability measure on $\mathbb{N}$. Prove that $\mathcal{F}$ is P-pre-Gaussian if and only if $\sum_{k=1}^\infty p_k^{1/2} < \infty$. *Hint*: Since $\mathcal{F}$ is uniformly bounded, $\mathcal{F}$ is P-pre-Gaussian iff $Z_P(I_A) = \sum_{k \in A} p_k^{1/2} g_k$, where g_k are i.i.d. and $N(0,1)$ is a bounded e_P-uniformly continuous process. Suppose that it is. Then so is the linear extension of Z_P to the symmetric convex hull of $\{I_A\}$, and hence, so is the process $(A,B) \mapsto Z_P(I_A) - Z_P(I_B)$. Therefore,

$$
\infty > E \sup_{A,B \subset \mathbb{N}} |Z_P(I_A) - Z_P(I_B)| = E \sup_{\tau_k = \pm 1} \left|\sum \tau_k p_k^{1/2} g_k\right| = E \sum p_k^{1/2} |g_k|.
$$

Conversely, suppose that $\sum p_k^{1/2} < \infty$, and assume without loss of generality that the sequence $\{p_k\}$ is nonincreasing. Set $k_\delta = \sup\{k \in \mathbb{N} : p_k^{1/2} > \delta^2\}$, and note that $e_P(I_A, I_B) \le \delta$ implies $A \Delta B \subset (k_\delta, \infty)$. Thus, $E \sup_{e_P(I_A, I_B) \le \delta} |Z_P(A) - Z_P(B)| \le E \sum_{k \ge k_\delta} p_k^{1/2} |g_k| \to 0$.

3.7.22 Let $\mathcal{F}$ be a P-pre-Gaussian bounded class satisfying the pointwise countable approximation property. Prove that if, for all $\varepsilon > 0$,

$$
\lim_n E^* \left[1 \wedge \int_0^{n^{1/4}} \sqrt{\log N(\mathcal{F}'_{\varepsilon,n}, L^2(P), \lambda)}\, d\lambda\right] = 0,
$$

then $\mathcal{F}$ is P-Donsker. *Hint*: Split the probability $\Pr\left\{\|\nu_{n,\mathrm{rad}}\|_{\mathcal{F}'_{\varepsilon,n}} > \tau\varepsilon\right\}$ according to whether $\left\|\sum_{i=1}^n f^2(X_i)\right\|_{\mathcal{F}'_{\varepsilon,n}} > n^{1/2}/4$ or $\le n^{1/2}/4$, and apply the square root trick bound on the first part and the usual metric entropy integral bound, conditionally on the X_i on the second; then invoke Theorem 3.7.54.

3.7.23 Prove that: If the sequence $\sqrt{n}\|P_n - P\|_{\mathcal{F}}^*$ is stochastically bounded, then $\|f(X_n) - Pf\|_{\mathcal{F}}^*/\sqrt{n}$ is also stochastically bounded. Hence, necessarily, $\sup_{f \in \mathcal{F}} |f(X) - Pf| < \infty$ P-almost surely. *Hint*: Note that $\|f(X_n) - Pf\|_{\mathcal{F}}^*/\sqrt{n} \le \sqrt{n}\|\nu_n\|_{\mathcal{F}}^* + \sqrt{(n-1)/n}\|\nu_{n-1}\|_{\mathcal{F}}^*$.

3.7.24 Let H be a separable Hilbert space, let (S,d) be a metric space and let $T : H \mapsto C_b(S)$ be a continuous linear map. Prove that the set of functions $T(\mathcal{U})$, where $\mathcal{U}$ is the unit ball of H, is P-Donsker for all Borel probability measures P on $(S,\mathcal{S})$. *Hint*: Let $\tilde{e}_P(Tf,Tg) = \|f-g\|$, for $f,g \in H$. Show that the metric space $(T(\mathcal{U}),\tilde{e})$ is separable. Hence, it suffices to prove the asymptotic equi-continuity condition (3.261) with $e = \tilde{e}_P$. Using Chebyshev's inequality and linearity, that condition holds if $\limsup_n E\sup_{f\in\mathcal{U}}((P_n-P)(Tf))^2 < \infty$, where P_n is the empirical measure corresponding to n i.i.d. random variables with law P. To prove that this limit is finite, observe first that if $\{e_k\}_{k=1}^{\infty}$ is an ortho-normal basis of H, then $\left\|\sum_k a_k(Te_k)(x)\right\|_\infty \le \|T\|\left(\sum a_k^2\right)^{1/2}$ for all $x \in S$, which implies that $\sum_k (Te_k)(x))^2 \le \|T\|^2$. Thus, $\sum_k P(Te_k)^2 \le \|T\|^2 < \infty$, and therefore, for every $i < \infty$, $\sup_n \sum_{k=1}^{i} E(\sqrt{n}(P_n - P)(Te_k))^2 \le \|T\|^2 < \infty$. Finally, argue that

$$\sup_{\sum a_k^2\le 1} E\left[\sqrt{n}(P_n-P)\left(\sum_{k=i+1}^{\infty} a_k(Te_k)\right)\right]^2 \le \sup_n \sum_{k=i+1}^{\infty} E\left(\sqrt{n}(P_n-P)(Te_k)\right)^2$$

$$\le \sum_{k=i+1}^{\infty} P(Te_k)^2 \to 0 \text{ as } i \to \infty.$$

3.7.25 Let (T,d) be a separable metric or pseudo-metric space, and let X be a $C_u(T,d)$-valued random variable. Let X_n and Y_n, $n \in \mathbb{N}$, be sample bounded processes on T such that $X_n \to_{\mathcal{L}} X$ in $\ell_\infty(T)$ and $X_n - Y_n \to 0$ in outer probability. Then $Y_n \to_{\mathcal{L}} X$ in $\ell_\infty(T)$. *Hint*: The second part of the proof of Theorem 3.7.24 shows that $Z_n \to_{\mathcal{L}} X$ in $\ell_\infty(T)$ if and only if $E^*H(Z_n) \to EH(X)$ for all $H : \ell_\infty(T) \mapsto \mathbb{R}$ bounded Lipschitz. Without loss of generality, assume that $\|H\|_{BL} \le 1$. Then $(Y_n - X_n)^* \to 0$ in probability implies that

$$|E(H(Y_n))^* - E(H(X_n))^*| \le E(|H(Y_n) - H(X_n)|^*) \le E(2 \wedge \|Y_n - X_n\|_T^*) \to 0,$$

so $E^*H(Y_n) \to EH(X)$.

3.8 Notes

Section 3.1 The modern yet classical references to the inequalities of Hoeffding, Bennett, Prokhorov and Bernstein are Bennett (1962) and Hoeffding (1963). The present exposition of these inequalities borrows also from McDiarmid (1989), Rio (2009) and Boucheron, Lugosi and Massart (2013), particularly the proof of Lemma 3.1.1, which improves on Hoeffding's original proof, follows the last reference. The convexity argument for the maximal inequalities in (3.30) is due to Pisier (1983), but the use of $\Phi(x) = e^{\lambda x}$ (instead of, e.g., $\Phi(x) = e^{ux^2}$ in the sub-Gaussian case), which gives very good constants, comes from Boucheron, Lugosi and Massart (2013), and so does the statement and the proof of the exponential inequality for Gaussian quadratic forms, which, without specific constants, is due to Hanson and Wright (1971). For more on Gaussian quadratic and multilinear forms, including two-sided tail estimates, see Latała (2006).

The present proof of Lévy's classical inequalities in infinite dimensions is basically taken from Kahane (1968). Montgomery-Smith (1994) obtained a similar reflection principle for independent, identically distributed processes. The Lévy-Ottaviani inequality dates back to Ottaviani (1939), and its consequence, inequality (3.36), was observed by Kwapień and Woyczynski (1992). Hoffmann-Jørgensen's inequality, with larger constants than those given in the text, is due to Hoffmann-Jørgensen (1974) and, with the constants and proof given earlier, to Kwapień and Woyczynski (1992). Theorem 3.1.16 with constants of the order $p/\log p$, which are of the best type for any given q, was discovered by Johnson, Schechtman and Zinn (1985) in $\mathbb{R}$ and by Talagrand (1989) in Banach spaces, with much less elementary proofs than the one given here. Another important complement to Hoffmann-Jørgensen's inequality is that the pth norm in the inequality for $q = 1$ can be replaced by the Orlicz exponential norm ψ_α of order α for $0 < \alpha \leq 1$; see Ledoux and Talagrand (1991).

The contraction principle for Rademacher series was discovered by Kahane (1968) but in the present form by Hoffmann-Jørgensen (1974), and this important article contains versions of Theorems 3.1.17 and 3.1.21. The proof of Theorem 3.1.17 borrows from van der Vaart and Wellner (1996). The symmetrisation inequalities for tail probabilities are classical; see, for example, Alexander (1984) for a version of Proposition 3.1.24 with fewer measurability requirements and Giné and Zinn (1984) for Exercise 3.1.5. Inequality (3.50) is due to G. Pisier or X. Fernique (private communication to one of the authors in 1977, but none of the persons involved remembers exactly), published, with a different proof, in Giné and Zinn (1984) (see also Giné (1996)). Ledoux and Talagrand (1986) proved that the integrability condition on ξ cannot in general be relaxed.

Exercise 3.1.1 comes from Boucheron, Lugosi and Massart (2013), Exercise 3.1.4 from Dudley (1978), Exercise 3.1.9 from Giné and Zinn (1983) and Exercise 3.1.10 from Einmahl and Mason (1996).

Section 3.2 The comparison principle for Rademacher processes comes from Ledoux and Talagrand (1989), and the proof given here follows that in Ledoux and Talagrand (1991). In their book, they attribute the result to Talagrand. Talagrand's inequality for the convex distance comes from Talagrand (1995), with the more specialised concentration inequality in the cube given in Exercise 3.2.2 in Talagrand (1988a). See also Ledoux (2001), McDiarmid (1998) and Ledoux and Talagrand (1991). The Khinchin-Kahane inequalities for a single vector t are due to Khinchin (1923) (see also Littlewood (1930) and Paley and Zygmund (1930)) and were extended to Banach space–valued coefficients by Kahane (1964). With the constants $((p-1)/(q-1))^{1/2}$, they are due to Bonami (1970), Gross (1975), Beckner (1975) and Borell (1979) for Banach-valued coefficients. The best constant in the comparison between the first and second was obtained by Szarek (1976) in the real case and Latała and Oleskiewicz (1994) in the Banach case, and Haagerup (1982) obtained the best constants for comparison of any moments with the second. See also de la Peña and Giné (1999) for an exposition about the Khinchin-Kahane inequalities and their extensions.

The Sudakov-type metric entropy lower bound for Rademacher processes is due to Talagrand; see Ledoux and Talagrand (1991). The proof presented here comes from this reference. For further work on this subject, see this reference, Talagrand (1994a) and, more recently, Bednorz and Latala (2014).

Section 3.3 Talagrand's inequality, with unspecified constants, was proved in Talagrand (1996) and a weaker one-sided version of it in Talagrand (1994). Ledoux (1997) gave a simpler proof of the upper tail in Talagrand's inequality using the entropy method, and later Samson (2000) showed that the lower-tail inequality is also accessible by the same method. Ledoux's method uses tensorisation of entropy for functions of several independent random variables, somehow allowing effective use at the exponential level of the cancellation due to independence in order to deduce log-Sobolev-type integrable differential inequalities for the Laplace transform of the supremum of the empirical process. Massart (2000), Rio (2001, 2002, 2012), Bousquet (2003), Klein (2002) and Klein and Rio (2005) have the best results about constants; in particular, Rio (2012) contains a version of the upper tail of Talagrand's inequality where the term $2ES_n$ in the definition of v_n is replaced by a smaller function of ES_n. The exposition here follows Bousquet (2003), Klein (2002) and Rio (2009), complemented at some points by Boucheron, Lugosi and Massart (2013).

Hoeffding's inequality dates back to Hoeffding (1963), and its 'extension' for random variables with bounded differences is due to McDiarmid (1989). This inequality has a simple martingale proof, but it seemed fit, to unify the exposition, to transcribe here a proof based on the entropy method that belongs to A. Maurer (unpublished) and that we have learned from Boucheron, Lugosi and Massart (2013); see also Maurer (2012). The concentration inequality for self-bounding random variables was obtained by Boucheron, Lugosi and Massart (2000). The examples on the application of the bounded differences inequality are taken form Devroye (1991). The sub-Gaussianness of self-normalised sums was observed in Giné and Mason (1998).

We have considered only bounded empirical processes. Regarding unbounded empirical processes, we refer to Section 3.5.3, and further results can be found in Adamczak (2008). For a more exhaustive account of the concentration-of-measure phenomenon, including the earlier history of the subject, see Ledoux (2001), Massart's (2007) lecture notes and the above-mentioned monograph by Boucheron, Lugosi and Massart (2013).

Section 3.4 The moment inequality at the beginning of Section 3.4.1 is due to Pinelis (1994), and the constants come from Latała (1997) and Giné, Latała and Zinn (2000). The moment inequality for empirical processes (3.137) in the first subsection was proved by Giné, Latała and Zinn (2000) up to an undetermined factor A^p, and Boucheron et al. (2005) obtained it with reasonable constants but by a somewhat more specialised proof that does not depend on Talagrand's inequality.

The idea of using Rademacher complexities to derive data-driven inequalities belongs to Koltchinskii (2001, 2006) and to Bartlett, Boucheron and Lugosi (2002), and Theorem 3.4.5 comes from Koltchinskii (2006). Giné and Nickl (2010a) introduced the weak variance σ^2 in this type of inequality, and the result presented here, Theorem 3.4.3, was obtained by Lounici and Nickl (2011).

The Bernstein-type inequality for canonical U-processes of order 2 was proved in Giné, Latała and Zinn (2000) up to an unspecified multiplicative constant using Talagrand's inequality by way of inequality (3.137) and 'decoupling'. The inequality with specific constants as well as its proof come from Houdré and Reynaud-Bouret (2003), and here we have a slight improvement resulting from the use of a tighter empirical process bound. The analogue of this inequality for canonical U-statistics of order larger than 2 was obtained by Adamczak (2006).

Section 3.5 Randomisation by Rademacher variables is a technique used in probability in Banach spaces at least since Kahane (1968) and, in particular, to prove central limit theorems since Jain and Marcus (1975). It was introduced in empirical process theory by Pollard (1981), and it was put to intensive use, together with Gaussian and even Poisson randomisation, by Giné and Zinn (1984). The expectation bound in Theorem 3.5.6 was proved for Vapnik-Červonenkis (VC) classes of sets by Talagrand (1994) and for bounded VC classes of functions by Einmahl and Mason (2000), Giné and Guillou (2001) and Giné and Koltchinskii (2006), who prove a version of Corollary 3.5.7; see also Giné and Mason (2007) for a proof similar to the one given here and for an extension to U-processes. The final extension to classes with square integrable envelope, Theorem 3.5.4, belongs to Chernozhukov, Chetverikov and Kato (2014) with elements of proof from van der Vaart and Wellner (2011). The partial converse inequality, Theorem 3.5.11, is due to Giné and Koltchinskii (2006).

Bracketing was introduced as a condition to control empirical processes first by Blum (1955) and DeHardt (1971) for the uniform law of large numbers and by Dudley (1978, 1984) for the central limit theorem for classes of sets and uniformly bounded classes of functions. Bass (1984) introduced the basic idea of truncation at each step of the chain to prove the law of the iterated logarithm under a bracketing condition for (unbounded) partial-sum processes. Ossiander (1987) used this idea to prove the central limit theorem for (unbounded) classes of functions, a work sharpened in some respects by Andersen et al. (1988). The proof of the expectation inequality (Theorem 3.5.13) given here originated in an adaptation by Arcones and Giné (1983) of these authors' proof and in van der Vaart and Wellner's (1996) replacement of probability estimates with expectation estimates. The precise statement of

Theorem 3.5.13 is due to van der Vaart and Wellner (1996) and the bracketing exponential bound in Theorem 3.5.21 to Birgé and Massart (1993) and to van de Geer (1995, 2000). Proposition 3.5.15 and Exercise 3.5.5 come from van der Vaart and Wellner (1996); see also van de Geer (1995). The bound on the bracketing numbers for classes of uniformly bounded monotone functions on $\mathbb{R}$ is due to van de Geer (1991), following work by Birman and Solomjak (1967); the present exposition follows van de Geer's proof but has also benefitted from an adaptation in van der Vaart and Wellner (1996) that includes L^p-bracketing numbers.

Section 3.6 Vapnik-Červonenkis classes of sets were introduced in the seminal paper of Vapnik and Červonenkis (1971) in connection with their version of the Glivenko-Cantelli theorem, where the class of half-lines is replaced by a *VC* class of sets. They also proved a slightly weaker version of the basic combinatorial lemma, Theorem 3.6.2, which is due in the present form to Sauer (1972) and Shelah (1972). The present proof using Proposition 3.6.3 follows this last author. Proposition 3.6.6 comes from the Vapnik and Červonenkis (1974) book (in Russian). The *VC* property for positivity (negativity) sets of finite-dimensional spaces of functions was observed by Dudley (1978). Dudley (1978) also obtained the basic relation between the *VC* property and metric entropy, which was extended to *VC* subgraph classes of functions by Pollard (1982), Theorem 3.6.9. With considerably more work, it was proved by Haussler (1995) that the exponent of ε in Theorem 3.6.9 can be improved to $-vp$ (instead of $-wp$ for any $w > v$). We have chosen to present the simpler yet somewhat weaker original version of Dudley and Pollard in part because it is close enough to the best for most purposes and in part because of the beauty and transparency of the Dudley-Pollard proof. Lemma 3.6.11, the decomposition of functions of bounded p-variation, is classical (Love and Young (1937) via Dudley (1992)), and the *VC* property of translations and dilations of these functions comes from Giné and Nickl (2009) for $p \neq 1$ and Nolan and Pollard (1987) for $p = 1$.

The metric entropy bound for convex hulls of classes of functions, Theorem 3.6.17, was obtained, in a slightly weaker form, by Dudley (1987) and then sharpened by Ball and Pajor (1990) under additional assumptions on $\mathcal{F}$ and, in the present definitive version, by van der Vaart and Wellner (1996) and Carl (1997). We follow van der Vaart and Wellner in our presentation. The result was shown by Dudley (1987) to be best possible (up to multiplication constants). The key Lemma 3.6.16 appears in Pisier (1981), where it is attributed to B. Maurey. Exercise 3.6.17 comes from Kolmogorov and Tikhomirov (1961).

Section 3.7 The calculus of nonmeasurable random elements, including measurable covers and envelopes and perfect maps, was developed, in the context of empirical processes, by Dudley (1966, 1967a, 1985), Dudley and Philipp (1983), Hoffmann-Jørgensen (1984, 1985), and Andersen (1985). The present definition of convergence in law for sample bounded processes is due to Hoffmann-Jørgensen (1984), with previous contributions by Dudley (1966, 1967). The asymptotic equi-continuity condition, which adapts Prohorov's criterion for uniform tightness in $C(S)$, is due to Dudley (1978) and, with different pseudo-metrics, to Andersen and Dobrić (1987), and its present statement and proof (Theorem 3.7.23) come from Giné and Zinn (1986). Giné and Zinn (1990) observed and used the bounded Lipschitz distance characterisation of convergence in law (Proposition 3.7.24) in connection with the bootstrap, and so did Dudley (1990). Theorem 3.7.25 extending Skorokhod's theorem on almost-sure convergent versions of sequences converging in law is due to Dudley (1985), and the proof here follows the exposition in van der Vaart and Wellner (1996).

The law of large numbers for the empirical process indexed by classes of sets and classes of bounded functions was obtained by Vapnik and Čěrvonenkis in their seminal 1971 paper and in its sequel in 1981. The present version for unbounded classes was obtained by Giné and Zinn (1984), who introduced Gaussian randomisation and the use of Sudakov's bound, whereas the submartingale proof for the equivalence of a.s. and in probability convergence comes from Nolan and Pollard (1987) (see also Strobl (1995)). The law of large numbers under bracketing conditions is due to Blum (1955), deHardt (1971) and Dudley (1984) at increasing degrees of generality, whereas Mourier (1953) proved the law of large numbers in Banach spaces.

The existence of suitable versions for P-bridges and the fact that the convex hull of a P-Donsker class is P-Donsker are due to Dudley (1985) (the extension in Exercise 3.7.14 coming from Giné and Nickl (2008)), that the union of P-Donsker classes is P-Donsker is due to Alexander (1987) and the present proof of this last result is ascribed by Dudley (2014) to Arcones. The necessary integrability condition for $\mathcal{F}$ to be P-Donsker comes from Giné and Zinn (1986).

The random entropy integral sufficient condition for $\mathcal{F}$ to be P-Donsker comes from Giné and Zinn (1984), whereas Condition (3.278) is due to Pollard (1982), see Koltchinskii (1981) for a precedent. The bracketing entropy central limit theorem is due to Ossiander (1987); see Dudley (1978, 1984) for earlier versions and Andersen, Giné, Ossiander and Zinn (1988) for a strictly stronger version. The proof presented here comes from this last article, as adapted to the present situation by Arcones and Giné (1993) and by van der Vaart and Wellner (1996). Van der Vaart and Wellner replaced probability inequalities by expectation inequalities, which makes for a more streamlined presentation.

The uniform in P central limit theorem comes from Giné and Zinn (1991); Sheehy and Wellner (1992) consider uniformity on classes of probability measures P other than all of them. The same question for the law of large numbers was answered by Dudley, Giné and Zinn (1991), but this is not considered here. The limit theorems for pregaussian classes of functions in the second part of Subsection 3.7.6, Theorems 3.7.52, 3.7.54 (which contains a necessary and sufficient condition for a class of sets $\mathcal{C}$ to be P-Donsker) and 3.7.55, were obtained by Giné and Zinn (1984). The 'square root trick' inequality, also from the same article, is based on a technique of Le Cam (1986) (see the proof of his Lemma 6 on page 546). Such results have been useful in the theory of smoothed empirical processes discussed in Section 5.2, see Giné and Nickl (2008)) – Nickl and Reiß (2012) and Nickl et al. (2015) use it to prove Donsker type theorems for Lévy measures. It is also worth mentioning that the approach to limit theorems using pregaussianness was further developed by Talagrand, who obtained necessary and sufficient conditions in Gaussian and $L^1(P_n)$ terms for general classes of functions $\mathcal{F}$ to be P-Donsker in Talagrand (1987). See also Ledoux and Talagrand (1991) and, for a different exposition of his result, Giné and Zinn (1986). It is worth noting as well that the $\Delta^{\mathcal{C}}$ condition in Theorem 3.7.55 is also necessary for a class of sets $\mathcal{C}$ to be P-Donsker (Talagrand (1988)).

The uniform Donsker property for the class of uniformly bounded functions with uniformly bounded p-variation was proved (differently) by Dudley (1992), the characterization of the probability measures on $\mathbb{R}$ for which $2^{\mathbb{N}}$ is P-Donsker was obtained by Borisov (1981) and Durst and Dudley (1981), and the present proof comes from Giné and Zinn (1984), whereas central limit theorem for the empirical process over the bounded Lipschitz functions comes from Giné and Zinn (1986a); see this last article for alternative ways to prove the last two mentioned results. For more applications to limit theorems of the results in this section, see Giné and Zinn (1984) and Rhee (1986).

4

Function Spaces and Approximation Theory

This chapter presents core materials from the theory of function spaces that will serve as building blocks for the statistical models considered subsequently in this book. Some classical materials from functional analysis, such as Sobolev spaces and approximate identities based on convolution kernels or ortho-normal decompositions of L^2, are reviewed. A particular emphasis is placed on wavelet theory, including the complete construction of ortho-normal wavelet bases with compact support in time or frequency domain. The main aspects of the theory of Besov spaces as a unifying scale of function spaces are developed. As a consequence, sharp results on approximation of smooth functions from finite-dimensional function spaces and related metric entropy results are obtained. The main presentation is for functions of one variable, defined on $\mathbb{R}$ or on a subinterval of it, but it is shown how the techniques generalise without difficulty to various multivariate settings. The proofs rely only on basic real and Fourier analysis techniques, the most important of which are briefly reviewed at the beginning of the chapter.

4.1 Definitions and Basic Approximation Theory

4.1.1 Notation and Preliminaries

We recall that for $1 \leq p < \infty$ and A a Borel-measurable subset of $\mathbb{R}$, the space

$$L^p(A) = \left\{ f : A \to \mathbb{R} : \int_A |f(x)|^p dx < \infty \right\}$$

consists of p-fold Lebesgue-integrable functions and is normed by

$$\| f \|_p := \| f \|_{L^p(A)} = \left(\int_A |f(x)|^p dx \right)^{1/p}.$$

We write L^p only when no confusion about A may arise, and ℓ_p for the usual sequence spaces when $A \subset \mathbb{Z}$ is equipped with counting measure dx. The Hilbert space $L^2(A)$ carries the natural inner product

$$\langle f, g \rangle = \int_A f(x)\overline{g(x)}dx.$$

We shall say that a function $f : \mathbb{R} \mapsto \mathbb{R}$ is *locally integrable* if it satisfies $\int_B |f(x)|dx < \infty$ for every bounded (Borel) subset $B \subset \mathbb{R}$. The symbol $L^\infty(A)$, $L^\infty \equiv L^\infty(\mathbb{R})$, denotes the space of bounded measurable functions on A, normed by the usual supremum norm $\| \cdot \|_\infty$, $C(A)$ denotes the subspace of continuous functions and $C_u(A)$ the subspace of uniformly continuous functions. When A is a half-open bounded interval with endpoints a,b, then $C_{\mathrm{per}}(A)$ denotes the space of continuous periodic functions on A, $f(a)=f(b)$. Throughout δ_x denotes the Dirac-δ point probability measure at x, whereas δ_{kl} denotes the Kronecker-δ (i.e., $\delta_{kl}=1$ when $k=l$ and 0 otherwise).

Regularity properties of a function f may be measured through the size in L^p of the derivatives of f. In the L^p-setting, it is natural to consider weakly differentiable functions. For $A\subseteq\mathbb{R}$ an interval, a function $f\in L^p(A)$ is said to be *weakly differentiable* if there exists a locally integrable function Df – the weak derivative of f – such that

$$\int_A f(u)\phi'(u)du=-\int_A Df(u)\phi(u)du, \tag{4.1}$$

for every infinitely differentiable function ϕ of compact support in the interior of A. It follows from integration by parts that any classically differentiable function f on A – with derivatives understood to be one sided if A includes its endpoints – has a weak derivative that coincides almost everywhere with the classical derivative. To distinguish the two concepts when necessary, we shall write D^α for the weak differential operator of order α and $f^{(\alpha)}$ for the classical derivative of f of order α. If the weak derivative Df is almost everywhere equal to a continuous function $f^{(1)}$, then we always identify Df with $f^{(1)}$, and f is then classically differentiable by the fundamental theorem of calculus. One thus defines the *Sobolev spaces* of weakly differentiable functions on A: for $1\le p<\infty$, the L^p-Sobolev space of order $m\in\mathbb{N}$ is defined as

$$H^m_p(A)=\left\{f\in L^p: D^jf\in L^p(A)\ \forall j=1,\dots,m: \|f\|_{H^m_p(A)}\equiv\|f\|_p+\|D^mf\|_p<\infty\right\},$$

and we shall suppress A when no confusion may arise in the notation. One further defines, for $m\in\mathbb{N}$,

$$C^m(A)=\left\{f\in C_u(A): f^{(j)}\in C_u(A)\ \forall j=1,\dots,m: \|f\|_{C^m(A)}\equiv\|f\|_\infty+\|f^{(m)}\|_\infty<\infty\right\},$$

where again $C^m\equiv C^m(\mathbb{R})$. We further define $C^\infty(A)$ to be the space of infinitely differentiable functions defined on A, with derivatives defined to be one sided at the endpoints if A is not open. The subspace $C^\infty_0(A)$ consists of all $\phi\in C^\infty(A)$ that have compact support in the interior of A. The Schwartz space $\mathcal{S}(\mathbb{R})$ consists of all functions $f\in C^\infty(\mathbb{R})$ such that all derivatives $f^{(\alpha)},\alpha\ge 0$, exist and decay at $\pm\infty$ faster than any inverse polynomial.

Much of the theory in this chapter will be based on exploiting the symmetry of the group action of translation on $\mathbb{R}$ via the Fourier transform. To prepare this, we review here some standard facts on convolutions and Fourier transforms that can be found in any real analysis book (see the notes at the end of this chapter). All the preceding spaces make sense for complex-valued functions, too, and unless necessary, we shall not distinguish in the notation whether the functions involved are real or complex valued.

For two measurable functions f,g defined on $\mathbb{R}$, their *convolution* is

$$f*g(x)\equiv\int_{\mathbb{R}}f(x-y)g(y)dy,\quad x\in\mathbb{R},$$

whenever the integral exists. If $f\in L^p(\mathbb{R}),g\in L^q(\mathbb{R})$ with $1\le p,q\le\infty$ such that $1/p+1/q=1$, then $f*g$ defines an element of $C(\mathbb{R})$, and Hölder's inequality implies that

$$\|f*g\|_\infty\le\|f\|_p\|g\|_q. \tag{4.2}$$

Furthermore, if $f \in L^p(\mathbb{R}), g \in L^1(\mathbb{R}), 1 \le p \le \infty$, then the function $f * g$ is well defined a.e., and

$$\|f * g\|_p \le \|f\|_p \, \|g\|_1, \tag{4.3}$$

in view of Minkowski's inequality for integrals. In the case where $f \in C(\mathbb{R})$, $f * g$ is in fact defined everywhere and itself contained in $C(\mathbb{R})$. We can also define

$$f * \mu(x) = \int_{\mathbb{R}} f(x-y) d\mu(y), \quad x \in \mathbb{R}, \tag{4.4}$$

for $\mu \in M(\mathbb{R})$, where $M(\mathbb{R})$ denotes the spaces of finite signed measures on $\mathbb{R}$, and one has likewise $\|f * \mu\|_p \le \|f\|_p |\mu|(\mathbb{R})$, where $|\mu|$ is the total variation measure of μ.

For a function $f \in L^1(\mathbb{R})$, we define the Fourier transform

$$\mathcal{F}[f](u) \equiv \hat{f}(u) = \int_{\mathbb{R}} f(x) e^{-iux} dx, \quad u \in \mathbb{R}.$$

If $f \in L^1(\mathbb{R})$ is such that $\hat{f} \in L^1(\mathbb{R})$, then the Fourier inversion theorem states that

$$\mathcal{F}^{-1}(\hat{f}) \equiv \frac{1}{2\pi} \int_{\mathbb{R}} e^{iu\cdot} \hat{f}(u) du = f \ \ a.e., \tag{4.5}$$

and f can be modified on a set of Lebesgue measure zero to equal a continuous function for which the inversion formula holds everywhere. One immediately has

$$\|\hat{f}\|_\infty \le \|f\|_1, \quad \|f\|_\infty \le \frac{1}{2\pi} \|\hat{f}\|_1,$$

and the (inverse) Fourier transform is injective from L^1 to L^∞. Moreover, if $f \in L^1 \cap L^2$, Plancherel's theorem states that

$$\|f\|_2 = \frac{1}{\sqrt{2\pi}} \|\hat{f}\|_2, \quad \langle f, g \rangle = \frac{1}{2\pi} \langle \hat{f}, \hat{g} \rangle,$$

and $\sqrt{2\pi}\mathcal{F}$ extends continuously to an isometry from L^2 to L^2. Some further basic properties of the Fourier transform are the following:

$$\mathcal{F}[f(\cdot - k)](u) = e^{-iku} \hat{f}(u), \tag{4.6}$$

$$\mathcal{F}[f(a \cdot)](u) = a^{-1} \hat{f}(u/a), \quad a > 0, \tag{4.7}$$

$$\mathcal{F}[f * g](u) = \hat{f}(u)\hat{g}(u), \quad \mathcal{F}[f * \overline{f(-\cdot)}](u) = |\hat{f}(u)|^2, \tag{4.8}$$

and finally, for every $N \in \mathbb{N}$,

$$\frac{d^N}{(du)^N} \hat{f}(u) = \int_{\mathbb{R}} f(x)(-ix)^N e^{-ixu} dx, \ (iu)^N \hat{f}(u) = \mathcal{F}[D^N f](u) \tag{4.9}$$

whenever $|f|$ and $|\hat{f}|$ integrate $|\cdot|^N$, respectively. Conclude in particular that the Fourier transform $\mathcal{F}$ maps the Schwartz space $\mathcal{S}(\mathbb{R})$ into itself.

If instead of $\mathbb{R}$ the group is $(0, 2\pi]$ with addition modulo 2π, we have similar results. In particular, any 2π-periodic $f \in L^2((0, 2\pi])$ decomposes into its Fourier series

$$f = \sum_{k \in \mathbb{Z}} c_k e^{ik\cdot} \ \text{ in } L^2((0, 2\pi]), \quad c_k \equiv c_k(f) = \frac{1}{2\pi} \int_0^{2\pi} f(x) e^{-ikx} dx, \quad \{c_k\} \in \ell_2; \tag{4.10}$$

in fact, $f \mapsto \{c_l\}$ gives a Hilbert space isometry between $L^2((0,2\pi])$ and ℓ_2. If, further, $\{c_k\} \in \ell_1$, then the Fourier series of f converges a.e. on $(0,2\pi]$ (pointwise if f is continuous).

Fourier inversion and Fourier series can be linked to each other by the *Poisson summation formula*: if $f \in L^1(\mathbb{R})$, then the periodised sum

$$S(x) = \sum_{l\in\mathbb{Z}} f(x+2\pi l), \quad x \in (0,2\pi],$$

converges a.e., belongs to $L^1((0,2\pi])$ and the Fourier coefficients of S are given by

$$c_k(S) = \frac{1}{2\pi}\hat{f}(k) = \mathcal{F}^{-1}[f](-k). \tag{4.11}$$

More generally, for $a > 0$, the series $\sum_l f(\cdot + la)$ converges in $L^1((0,a])$ and has Fourier coefficients

$$a^{-1}\int_0^a \sum_l f(x+la)e^{-ixk2\pi/a}dx = a^{-1}\hat{f}(2\pi k/a). \tag{4.12}$$

We finally introduce generalised functions. Recall the space $\mathcal{S} = \mathcal{S}(\mathbb{R})$ of infinitely differentiable functions on $\mathbb{R}$ such that all derivatives $f^{(\alpha)}, \alpha \geq 0$, exist and decay at $\pm\infty$ faster than any inverse polynomial. We define a countable family of seminorms on $\mathcal{S}(\mathbb{R})$ by

$$\|f\|_{m,r} = \max_{\alpha\leq r} \|(1+|\cdot|^2)^m f^{(\alpha)}\|_\infty, \quad m,r \in \mathbb{N}\cup\{0\};$$

these seminorms provide a metrisable locally convex topology on $\mathcal{S}(\mathbb{R})$; in fact, $\mathcal{S}(\mathbb{R})$ is complete, and the set $C_0^\infty(\mathbb{R})$ is dense in $\mathcal{S}(\mathbb{R})$ for this topology. Henceforth when we speak of $\mathcal{S}(\mathbb{R})$, we will always endow it with this topology.

We define $\mathcal{S}^* \equiv \mathcal{S}(\mathbb{R})^*$ to be the topological dual space of $\mathcal{S}(\mathbb{R})$, that is, all continuous linear forms on $\mathcal{S}(\mathbb{R})$. The space $\mathcal{S}(\mathbb{R})^*$ is known as the space of *tempered distributions*, or *Schwartz distributions*, equipped with the weak topology: $T_n \to T$ in $\mathcal{S}^*$ if $T_n(\phi) \to T(\phi)$ for every $\phi \in \mathcal{S}$. Weak differentiation defined through (4.1) is then a continuous operation from $\mathcal{S}^*$ to $\mathcal{S}^*$. If μ is a finite measure, then the action of μ on $\mathcal{S}(\mathbb{R})$ via integration gives rise to an element of $\mathcal{S}(\mathbb{R})^*$ because

$$\left|\int_{\mathbb{R}} f d\mu\right| \leq |\mu|(\mathbb{R})\|f\|_{0,0}.$$

Moreover, any signed measure of at most polynomial growth at $\pm\infty$, that is, such that $|\mu|(x : |x| < R) \lesssim (1+|R|^2)^l$ for all $R > 0$, some $l \in \mathbb{N}$, defines an element of $\mathcal{S}(\mathbb{R})^*$. Likewise, one shows that any $f \in L^p$ acting on $\mathcal{S}(\mathbb{R})$ by integration $\phi \mapsto \int f\phi$ defines an element of $\mathcal{S}^*(\mathbb{R})$.

As remarked earlier, the Fourier transform maps $\mathcal{S}(\mathbb{R})$ onto $\mathcal{S}(\mathbb{R})$. We thus can define the Fourier transform of $T \in \mathcal{S}^*$ as the element $\mathcal{F}T$ of $\mathcal{S}^*$ whose action on $\mathcal{S}$ is given by

$$\phi \mapsto \mathcal{F}T(\phi) = T(\hat{\phi}), \tag{4.13}$$

which, for $T = f \in L^1$, returns the usual definition since, by Fubini's theorem,

$$\int_{\mathbb{R}} \hat{f}(u)\phi(u)dx = \int_{\mathbb{R}}\int_{\mathbb{R}} e^{-iux}f(x)\phi(u)dudx = \int_{\mathbb{R}} f(u)\hat{\phi}(u)du.$$

In particular, the Fourier transform maps $\mathcal{S}^*$ continuously onto itself, and $\mathcal{F}^{-1}[\mathcal{F}T] = T$ in $\mathcal{S}^*$.

We can by the same principles define periodic Schwartz distributions. For A any interval $(0,a]$, we let $C^\infty_{\text{per}}(A)$ denote the space of infinitely differentiable periodic real-valued functions on A and let $\mathcal{D}^* = \mathcal{D}^*(A)$ denote its topological dual space. The preceding duality arguments can be carried through in the same way; in particular, the Fourier coefficients $\langle f, e_k\rangle$ of any $f \in \mathcal{D}^*$ with respect to suitably scaled trigonometric polynomials e_k are well defined, and the Fourier series $\sum_k \langle f, e_k\rangle e_k$ converges in $\mathcal{D}^*$.

4.1.2 Approximate Identities

Convolution with Kernels

For $f : \mathbb{R} \to \mathbb{R}$, we can define the convolution

$$K_h * f(x) = \int_{\mathbb{R}} K_h(x-y) f(y) dy = \int_{\mathbb{R}} f(x-y) K_h(y) dy = f * K_h(x) \tag{4.14}$$

of f with a suitably 'localised' kernel function

$$K_h(x) = \frac{1}{h} K\left(\frac{x}{h}\right), \quad h > 0, x \in \mathbb{R},$$

where K is typically chosen to be bounded and integrable and in particular satisfies $\int_{\mathbb{R}} K(x)dx = 1$. The parameter h governs the degree of 'localisation': if we decrease h, then K_h is more concentrated near zero. For instance, if $K = 1_{[-1/2,1/2]}$, then $K_{0.5} = 2 \cdot 1_{[-1/4,1/4]}$ and $K_{0.1} = 10 \cdot 1_{[-1/20,1/20]}$, so as $h \to 0$ the function K_h looks more and more like a point mass δ_0 at 0. Intuitively, then, convolution with K_h should approximately behave as

$$f * K_h \sim f * \delta_0 = \int_{\mathbb{R}} f(x-y) d\delta_0(y) = f(x),$$

which is why K_h can be called an *approximate identity* (convolution with δ_0 being an *identity operator*). A simple and basic result that formalises these ideas is as follows.

Proposition 4.1.1 *Let $f : \mathbb{R} \to \mathbb{R}$ be a measurable function, and let $K \in L^1$ satisfy $\int_{\mathbb{R}} K(x)dx = 1$.*

(i) If f is bounded on $\mathbb{R}$ and continuous at $x \in \mathbb{R}$, then $K_h * f(x)$ converges to $f(x)$ as $h \to 0$.
(ii) If f is bounded and uniformly continuous on $\mathbb{R}$, then $\|K_h * f - f\|_\infty \to 0$ as $h \to 0$.
(iii) If $f \in L^p$ for some $1 \le p < \infty$, then $\|K_h * f - f\|_p \to 0$ as $h \to 0$.

Proof By (4.3), the integral $K_h * f$ defines an element of L^p if $f \in L^p$. We have from the substitution $(x-y)/h \mapsto u$ with derivative $|du/dy| = 1/h$, and since K integrates to 1,

$$\begin{aligned} K_h * f(x) - f(x) &= \int_{\mathbb{R}} \frac{1}{h} K\left(\frac{x-y}{h}\right) f(y) dy - f(x) \\ &= \int_{\mathbb{R}} K(u) f(x-uh) du - f(x) \\ &= \int_{\mathbb{R}} K(u)(f(x-uh) - f(x)) du. \end{aligned}$$

For (i), by continuity, $f(x-uh)-f(x)\to 0$ as $h\to 0$ for every u, and since f is bounded, the last quantity in the last display converges to zero by the dominated convergence theorem. For (ii), let $\varepsilon>0$ be given. By integrability of K, outside a fixed interval $[-A,A]$ the quantity in the last display is bounded by $2\|f\|_\infty \int_{[-A,A]^c}|K(u)|du<\varepsilon/2$, and for $|u|\le A$ and h small enough, we can use uniform continuity of f to obtain the same bound $\varepsilon/2$, which implies the result by adding both bounds. For (iii), by Minkowski's inequality for integrals,

$$\left\|\int_{\mathbb{R}} K(u)(f(\cdot-uh)-f(\cdot))du\right\|_p \le \int_{\mathbb{R}} |K(u)|\|f(\cdot-uh)-f\|_p du.$$

By continuity of translation in L^p, we see that $\|f(\cdot-uh)-f\|_p\to 0$ as $h\to 0$ for every u, and since $\|f(\cdot-uh)-f\|_p\le 2\|f\|_p$ by invariance of translation of Lebesgue measure, the quantity in the last display converges to zero again by the dominated convergence theorem. ■

We see that $K_h*(\cdot)$ acts asymptotically like δ_0 on any L^p-space with $1\le p<\infty$ and on the space of bounded uniformly continuous functions equipped with the $\|\cdot\|_\infty$-norm. It also reconstructs a function locally at any continuity point. Note that the condition that f be bounded in Proposition 4.1.1 can be relaxed if the integral defining $K_h*f(x)$ converges locally near x.

Orthogonal Series and L^2-Projection Kernels

Let $\mathcal{L}\subset\mathbb{Z}$ be an index set. A family of functions $\{e_l: l\in\mathcal{L}\}\subset L^2(A)$ is called an *ortho-normal system* if $\langle e_k,e_l\rangle=0$ whenever $k\neq l$ and $\langle e_l,e_l\rangle=\|e_l\|_2^2=1$ otherwise. The family $\{e_l: l\in\mathcal{L}\}$ is further called *complete* if the linear span

$$\left\{\sum_{l\in\mathcal{L}} c_l e_l : c_l\in\mathbb{R}\right\}$$

is norm-dense in $L^2(A)$. Any complete ortho-normal system is called an *ortho-normal basis of* $L^2(A)$. By completeness of the system, we conclude from basic Hilbert space theory that an arbitrary $f\in L^2(A)$ can be decomposed into the series

$$f=\sum_{l\in\mathcal{L}}\langle f,e_l\rangle e_l,$$

with convergence in $L^2(A)$ and where

$$\|f\|_2^2=\sum_{l\in\mathcal{L}}|\langle f,e_l\rangle|^2,\quad \langle f,g\rangle=\sum_k\langle f,e_k\rangle\langle g,e_k\rangle, \tag{4.15}$$

known as *Parseval's identity*.

We thus can reconstruct an arbitrary function $f\in L^2(A)$ in terms of the fixed family of basis functions e_l and the coefficients $\langle f,e_l\rangle$. This constitutes an alternative to approximating functions f by convolutions. If we denote by V the closed subspace of $L^2(A)$ generated by the linear span of $\{e_l: l\in\mathcal{L}'\}$ for some subset $\mathcal{L}'\subset\mathcal{L}$, then the ortho-normal projection of any $f\in L^2(A)$ onto V is given by

$$\pi_V(f)=\sum_{l\in\mathcal{L}'}\langle f,e_l\rangle e_l,$$

with convergence in $L^2(A)$. Standard Hilbert space theory implies that $\pi_V(f)$ is the best L^2-approximation of f from the subspace V. Moreover, by definition of the $L^2(A)$ inner product

$$\pi_V(f)(x) = \sum_{l\in\mathcal{L}'}\langle f, e_l\rangle e_l(x) = \sum_{l\in\mathcal{L}'}\int_A f(y)e_l(y)e_l(x)dy = \int_A \sum_{l\in\mathcal{L}'} e_l(x)e_l(y)f(y)dy$$

under suitable conditions that allow summation and integration to be interchanged. If we define the *projection kernel*

$$K_V(x,y) = \sum_{l\in\mathcal{L}'} e_l(x)e_l(y), \tag{4.16}$$

then

$$\pi_V(f)(x) = \int_A K_V(x,y)f(y)dy, \tag{4.17}$$

a representation that resembles (4.14) with the convolution kernel $K_h(x,y) = K_h(x-y)$ replaced by the general kernel function $K_V(x,y)$, not necessarily of translation type.

We shall now discuss some classical examples of ortho-normal bases of L^2, including the trigonometric system and some basic historical examples of *wavelet bases*, which will be introduced in full generality later.

The Trigonometric Basis

If $A = (0,1]$, then the trigonometric basis of $L^2((0,1])$ consists of the complex trigonometric polynomials

$$\{e_l = e^{2\pi il\cdot} = \cos(2\pi l\cdot) + i\sin(2\pi l\cdot) : l \in \mathbb{Z}\}.$$

For intervals $A = (0,a]$, we only modify the phase and take $e_l = a^{-1/2}e^{(2\pi/a)il}$; for $a = 2\pi$ we obtain the basis used in (4.10); and for intervals (a,b), we suitably translate the e_l from $(0, b-a)$ to (a,b). These functions form an ortho-normal system in $L^2(A)$ because

$$\langle e_l, e_k\rangle = \int_0^1 e_l(x)\overline{e_k(x)}dx = \int_0^1 e^{2\pi ilx}e^{-2\pi ikx} = \int_0^1 e^{2\pi i(l-k)x}dx = \delta_{lk}$$

equals 1 for $l = k$ and 0 otherwise. The trigonometric polynomials are further dense in $C_{\mathrm{per}}(A)$ for $\|\cdot\|_\infty$ by the Stone-Weierstrass theorem, and $C_{\mathrm{per}}(A)$ is dense in $L^2(A)$ by standard approximation arguments, so the $\{e_l\}_{l\in\mathbb{Z}}$ indeed form an ortho-normal basis of $L^2(A)$. Thus, any $f \in L^2(A)$ can be decomposed into its Fourier series

$$f = \sum_{l\in\mathbb{Z}}\langle f, e_l\rangle e_l,$$

with convergence in $L^2(A)$. The partial sums can be represented as

$$S_N(f)(x) = \sum_{|l|\le N}\langle f, e_l\rangle e_l(x) = \int_0^1 D_N(x-y)f(y)dy = D_N * f(x),$$

where

$$D_N(x) = \sum_{|l|\le N} e^{2\pi ilx} = \frac{\sin((2N+1)\pi x)}{\sin(\pi x)} \tag{4.18}$$

is the *Dirichlet* kernel. We see that in this particular case the L^2-projection kernel is in fact of convolution type. However, the sequence D_N is, in contrast to $K_h = h^{-1}K(\cdot/h)$, with $K \in L^1$, not bounded uniformly in $L^1(A)$; some simple calculus shows that $\|D_N\|_1$ is of greater order of magnitude than $\log N$ as $N \to \infty$. Thus, a proof along the lines of the one of Proposition 4.1.1 cannot be used to study $S_N(f) - f$. Indeed, convergence of $S_N(f) \to f$ in $L^p(A), p \neq 2$, or in $C_u(A)$ does not hold in general (see Exercise 4.1.2 for some facts). One way around this problem is based on the Fejér kernel

$$F_m = \frac{1}{m+1}\sum_{k=0}^{m} D_k, \tag{4.19}$$

with corresponding Fejér (or Cesàro) sums

$$f * F_m = \frac{1}{m+1}\sum_{k=0}^{m} S_k(f),$$

which can be shown to converge uniformly to $f \in C_{\text{per}}((0,1))$ (see Exercise 4.1.3). This averaging over partial Fourier sums, however, has no simple interpretation in terms of ortho-normal bases.

The Haar Basis

The drawbacks of the Fourier basis in $L^p, p \neq 2$, motivated the first construction of an ortho-normal basis of L^2 for which an analogue of Proposition 4.1.1 can be proved. We partition $\mathbb{R}$ into dyadic intervals $(k/2^j, (k+1)/2^j]$, where $j \in \mathbb{N} \cup \{0\}$, $k \in \mathbb{Z}$. For $\phi = 1_{(0,1]}$, the normalised indicator functions

$$\{\phi_{jk} \equiv 2^{j/2}\phi(2^j(\cdot) - k), k \in \mathbb{Z}\}, \quad j \in \mathbb{N} \cup \{0\},$$

form an ortho-normal system in L^2 simply because they all have disjoint support. The L^2 projection kernel equals

$$K_j(x,y) = 2^j K(2^j x, 2^j y) = \sum_{k \in \mathbb{Z}} 2^j \phi(2^j x - k)\phi(2^j y - k) = \sum_{k \in \mathbb{Z}} \phi_{jk}(x)\phi_{jk}(y), \tag{4.20}$$

and the best L^2-approximation of f in L^2 by a function piecewise constant on the dyadic intervals $(k/2^j, (k+1)/2^j]$ is given by $K_j(f) = \int_{\mathbb{R}} K_j(x,y) f(y)dy$, which is of a similar form to the convolution approximation from (4.14) if we convert 2^{-j} to h, but with a kernel

$$K(x,y) = \sum_{k \in \mathbb{Z}} \phi(x-k)\phi(y-k),$$

which is now *not* of convolution type. It still has some comparable approximation properties, however.

Proposition 4.1.2 *Let $f : \mathbb{R} \to \mathbb{R}$ be a measurable function, and let K be the Haar projection kernel from (4.20).*

(i) If f is bounded on $\mathbb{R}$ and continuous at $x \in \mathbb{R}$, then $K_j(f)(x)$ converges to $f(x)$ as $j \to \infty$.

(ii) *If f is bounded and uniformly continuous on $\mathbb{R}$, then $\|K_j(f)-f\|_\infty \to 0$ as $j\to\infty$.*
(iii) *If $f\in L^p$ for some $1\le p<\infty$, then $\|K_j(f)-f\|_p\to 0$ as $j\to\infty$.*

Proof From the partition properties, we clearly have

$$\int_{\mathbb{R}} K(x,y)dy = \sum_{k\in\mathbb{Z}} 1_{(0,1]}(x-k) = 1 \quad \forall x\in\mathbb{R}.$$

Moreover,

$$\sup_{x\in\mathbb{R}} |K(x,x-u)| \le 1_{[-1,1]}(u)$$

because ϕ is supported in $[0,1]$ and bounded by 1. We substitute $2^j y \mapsto 2^j x - u$ with derivative $|du/dy| = 2^j$ to obtain

$$\begin{aligned}
|K_j(f)(x)-f(x)| &= \left|\int_{\mathbb{R}} 2^j K(2^jx,2^jy)f(y)dy - f(x)\right| \\
&= \left|\int_{\mathbb{R}} K(2^jx,2^jx-u)f(x-u2^{-j})du - f(x)\right| \\
&= \left|\int_{\mathbb{R}} K(2^jx,2^jx-u)(f(x-u2^{-j})-f(x))du\right| \\
&\le \int_{-1}^{1} |f(x-u2^{-j})-f(x)|du. \qquad (4.21)
\end{aligned}$$

The rest of the proof is as in the proof of Proposition 4.1.1, replacing $h\to 0$ by $2^{-j}\to 0$ for $j\to\infty$ and $|K(u)|$ by $1_{[-1,1]}(u)$ everywhere. ∎

Having established the preceding, we can telescope the projections

$$K_j(f) = K_0(f) + \sum_{l=0}^{j-1} (K_{l+1}(f) - K_l(f)), \qquad (4.22)$$

and an elementary computation shows that

$$K_{l+1}(f) - K_l(f) = \sum_{k\in\mathbb{Z}} \langle \psi_{lk}, f\rangle \psi_{lk},$$

where $\psi = 1_{[0,1/2]} - 1_{(1/2,1]}$, $\psi_{lk}(x) = 2^{l/2}\psi(2^l x - k)$. We see that the ψ_{lk} and $\phi(\cdot - k)$ form an ortho-normal system, and since $K_j(f)\to f$ in L^p for any $f\in L^p, 1\le p<\infty$, by Proposition 4.1.2, we can take L^p-limits in (4.22) to conclude, writing $\phi_k = \phi(\cdot - k)$, that

$$f = \sum_{k\in\mathbb{Z}} \langle \phi_k, f\rangle \phi_k + \sum_{l=0}^{\infty}\sum_{k\in\mathbb{Z}} \langle \psi_{lk}, f\rangle \psi_{lk},$$

with convergence in $L^p, 1\le p<\infty$, and with uniform convergence if $f\in C_u(\mathbb{R})$. In particular, the family

$$\{\phi_k, \psi_{lk} : k\in\mathbb{Z}, l\in\mathbb{N}\cup\{0\}\}$$

forms an ortho-normal basis of L^2 known as the *Haar basis*.

The Shannon Basis

Consider a function $f \in \mathcal{V}_\pi$, where $\mathcal{V}_\pi$ is the space of continuous functions $f \in L^2$, which have (distributional) Fourier transform $\hat{f}$ supported in $[-\pi, \pi]$. One also says that f is a *band-limited* function (with band limit π). As discussed earlier, the functions $\{(2\pi)^{-1/2} e^{ik\cdot}\}_{k \in \mathbb{Z}}$ form an ortho-normal system in $L^2([-\pi,\pi])$, and we thus can represent $\hat{f}$ by its Fourier series

$$\hat{f} = \sum_{k \in \mathbb{Z}} c_k e^{ik(\cdot)}, \quad \text{in } L^2([-\pi,\pi]),$$

with Fourier coefficients given by

$$c_k = c_k(\hat{f}) = \frac{1}{2\pi} \int_{-\pi}^{\pi} e^{-iku} \hat{f}(u) du = f(-k),$$

the last identity following from (4.5) if $f \in L^1$. In this case, by Fourier inversion, for every $x \in \mathbb{R} \setminus \mathbb{Z}$, and if $\{c_k : k \in \mathbb{Z}\} \in \ell_1$,

$$\begin{aligned} f(x) &= \frac{1}{2\pi} \int_{-\infty}^{\infty} e^{iux} \hat{f}(u) du \\ &= \frac{1}{2\pi} \int_{-\pi}^{\pi} e^{iux} \sum_{k \in \mathbb{Z}} c_k e^{iku} du \\ &= \frac{1}{2\pi} \sum_{k \in \mathbb{Z}} c_k \int_{-\pi}^{\pi} e^{iu(k+x)} du \\ &= \sum_{k \in \mathbb{Z}} f(k) \frac{\sin \pi(x-k)}{\pi(x-k)}, \end{aligned}$$

and this argument extends to all $f \in \mathcal{V}_\pi$ by standard approximation arguments (using that $\sum_l |c_l|^2 = \|f\|_2^2 < \infty$). Conclude that such f can be *exactly* reconstructed by its 'sampled' values $f(k), k \in \mathbb{Z}$. This result is known as the *Shannon sampling theorem*. Setting $\phi(x) = \sin(\pi x)/(\pi x)$, we have $\hat{\phi} = 1_{[-\pi,\pi]}, \phi \in \mathcal{V}_\pi$, so, in particular, by Plancherel's theorem and (4.6), the integer translates of the function ϕ are ortho-normal in L^2, and the family $\{\phi_k = \phi(\cdot - k) : k \in \mathbb{Z}\}$ is an ortho-normal basis for the space $\mathcal{V}_\pi$. Since

$$\widehat{\phi(2^j \cdot)} = 2^{-j} 1_{[-2^j\pi, 2^j\pi]},$$

we see, moreover, that the functions $\{\phi_{jk} = 2^{j/2}\phi(2^j(\cdot) - k) : k \in \mathbb{Z}\}$ span $\mathcal{V}_{2^j\pi}$. The projection of $f \in L^2(\mathbb{R})$ onto $\mathcal{V}_{2^j\pi}$ is thus

$$\Pi_{\mathcal{V}_{2^j\pi}}(f) = \sum_k \langle \phi_{jk}, f \rangle \phi_{jk}.$$

As $j \to \infty$, the projections $\Pi_{\mathcal{V}_{2^j\pi}}(f)$ converge to f in L^2 because they are the best approximations in L^2 from the spaces $(\mathcal{V}_{2^j\pi})_{j \geq 0}$ which are dense in $L^2(\mathbb{R})$ (noting that $\|\hat{f} - \hat{f} 1_{[-2^j\pi, 2^j\pi]}\|_2 \to 0$ as $j \to \infty$, so $\mathcal{F}^{-1}(\hat{f} 1_{[-2^j\pi, 2^j\pi]})$ converges to f by Plancherel's theorem). As for the Haar basis, we can telescope these projections: set

$$\psi = \mathcal{F}^{-1}[1_{[-2\pi,-\pi]} + 1_{[\pi,2\pi]}];$$

then the functions $\{\psi_{lk} = 2^{l/2}\psi(2^l \cdot -k) : k \in \mathbb{Z}\}$ form an ortho-normal basis for $W_l = V_{2^l\pi} \ominus V_{2^{l-1}\pi}$. This is proved by a similar sampling formula as earlier (with $[-\pi, \pi]$ enlarged to $[-2\pi, 2\pi]$, noting that $\psi = 2\phi(2\cdot) - \phi$ and that ϕ vanishes at the integers, so one can ignore half-integers in the sampling formula). We thus can expand every $f \in L^2(\mathbb{R})$ into the series

$$f = \sum_k \langle \phi_k, f\rangle \phi_k + \sum_{l=0}^{\infty}\sum_k \langle \psi_{lk}, f\rangle \psi_{lk},$$

and the ortho-normal 'Shannon' basis

$$\{\phi_k, \psi_{lk} : k \in \mathbb{Z}, l \in \mathbb{N} \cup \{0\}\}$$

of L^2 can be regarded as the frequency domain analogue of the Haar basis.

4.1.3 Approximation in Sobolev Spaces by General Integral Operators

We now consider the general framework of integral operators

$$f \mapsto K_h(f) = \int_{\mathbb{R}} K_h(\cdot, y) f(y)dy = \frac{1}{h}\int_{\mathbb{R}} K\left(\frac{\cdot}{h}, \frac{y}{h}\right) f(y)dy, \quad h > 0, \tag{4.23}$$

under general conditions on the kernel $K : \mathbb{R} \times \mathbb{R} \to \mathbb{R}$. Such operators are sometimes called *Calderon-Zygmund operators* in the theory of singular integrals. They accommodate both convolution and projection kernels, with the obvious notational conversion $h = 2^{-j}$.

The following proposition is proved exactly in the same way as Proposition 4.1.2:

Proposition 4.1.3 *Let $f : \mathbb{R} \to \mathbb{R}$ be a measurable function, let K_h be as in (4.23) and suppose that $\int_{\mathbb{R}} \sup_{v\in\mathbb{R}} |K(v, v-u)|du < \infty$, $\int_{\mathbb{R}} K(x,y)dy = 1$ for every $x \in \mathbb{R}$. Then we have*

(i) If f is bounded on $\mathbb{R}$ and continuous at $x \in \mathbb{R}$, then $K_h(f)(x)$ converges to $f(x)$ as $h \to 0$.
(ii) If f is bounded and uniformly continuous on $\mathbb{R}$, then $\|K_h(f) - f\|_\infty \to 0$ as $h \to 0$.
(iii) If $f \in L^p$ for some $1 \le p < \infty$, then $\|K_h(f) - f\|_p \to 0$ as $h \to 0$.

To investigate further approximation properties, we shall impose the following condition.

Condition 4.1.4 *Let K be a measurable function $K(x,y) : \mathbb{R} \times \mathbb{R} \to \mathbb{R}$. For $N \in \mathbb{N}$, assume that*

***(M)**: $c_N(K) \equiv \int_{\mathbb{R}} \sup_{v\in\mathbb{R}} |K(v, v-u)||u|^N du < \infty$.*
***(P)**: For every $v \in \mathbb{R}$ and $k = 1, \ldots, N-1$,*

$$\int_{\mathbb{R}} K(v, v+u)du = 1 \quad \text{and} \quad \int_{\mathbb{R}} K(v, v+u)u^k du = 0.$$

We wish to study quantitative approximation properties of $K_h(f)$ for functions $f \in L^p$. For spaces of L^p-differentiable functions, we have the following basic approximation result.

Proposition 4.1.5 *Let K be a kernel that satisfies Condition 4.1.4 for some $N \in \mathbb{N}$, let $K_h(f)$ be as in (4.23) and let*

$$c(m, K) = c_m(K)\int_0^1 \frac{(1-t)^{m-1}}{(m-1)}dt,$$

for any integer $m \le N$.

(i) *If* $f \in H_p^m(\mathbb{R}), 1 \le p < \infty$, *then*

$$\|K_h(f) - f\|_p \le c(m,K)\|D^m f\|_p h^m,$$

(ii) *If* $f \in C^m(\mathbb{R})$, *then*

$$\|K_h(f) - f\|_\infty \le c(m,K)\|f^{(m)}\|_\infty h^m.$$

Proof By Taylor's theorem (for part (i), cf. Exercise 4.1.1), for any z,x,

$$f(z) - f(x) = \sum_{k=1}^{m-1} \frac{D^k f(x)}{k!}(z-x)^k + r_m f(z,x),$$

with the convention $r_1 f(z,x) = f(z) - f(x)$ and otherwise with remainder

$$r_m f(z,x) = \int_0^1 (z-x)^m \frac{(1-t)^{m-1}}{(m-1)!} D^m f(x + t(z-x))dt.$$

Using the substitution $y/h \mapsto (x/h) - u$ with derivative $|dy/du| = h$, the Taylor expansion with $z = x - uh$ and Condition 4.1.4, we thus may write

$$\begin{aligned}
K_h(f)(x) - f(x) &= \int \frac{1}{h} K\left(\frac{x}{h}, \frac{y}{h}\right) f(y)dy - f(x) \\
&= \int K\left(\frac{x}{h}, \frac{x}{h} - u\right) f(x - uh)du - f(x) \\
&= \int K\left(\frac{x}{h}, \frac{x}{h} - u\right)(f(x - uh) - f(x))du \\
&= \int K\left(\frac{x}{h}, \frac{x}{h} - u\right) \sum_{k=1}^{m-1} \frac{D^k f(x)}{k!}(-uh)^k du + \int K\left(\frac{x}{h}, \frac{x}{h} - u\right) r_m(x - uh, x)du \\
&= \int \int_0^1 K\left(\frac{x}{h}, \frac{x}{h} - u\right)(-u)^m h^m \frac{(1-t)^{m-1}}{(m-1)!} D^m f(x + tuh)dtdu.
\end{aligned}$$

Now, for $1 \le p < \infty$, by Minkowski's inequality for integrals, and directly for $p = \infty$, we have

$$\|K_h(f) - f\|_p \le h^m \int \sup_v |K(v, v-u)||u|^m du \int_0^1 \frac{(1-t)^{m-1}}{(m-1)!} dt \, \|D^m f\|_p,$$

which completes the proof. ■

The basic example for an operator satisfying Condition 4.1.4 is to take a translation kernel $K(x,y) = K(x-y)$, where $K \in L^1(\mathbb{R})$ integrates to 1. If K is also symmetric, then $\int_{\mathbb{R}} K(u)udu = 0$, that is, Condition 4.1.4 **(P)** with $N = 2$, and Condition 4.1.4 **(M)** reduces to a standard moment condition on K satisfied, for instance, for any compactly supported function K. Higher-order kernels of compact support exist for any $N \in \mathbb{N}$, as the following result shows.

Proposition 4.1.6 *For every $N \in \mathbb{N}$ there exists a bounded measurable function $K^{(N)} : \mathbb{R} \to \mathbb{R}$ supported in $[-1,1]$ such that $\int_{\mathbb{R}} K^{(N)}(u)du = 1$ and $\int_{\mathbb{R}} K^{(N)}(u)u^k du = 0$, for every $k = 1,\ldots,N-1$.*

Proof Let $\{\phi_m\}_{m\in\mathbb{N}}$ be the ortho-normal system in $L^2([-1,1])$ of Legendre polynomials defined by

$$\phi_0(x) := 2^{-1/2}, \quad \phi_m(x) = \sqrt{\frac{2m+1}{2}}\frac{1}{2^m m!}\frac{d^m}{dx^m}[(x^2-1)^m],$$

for $x \in [-1,1]$ and $m \in \mathbb{N}$. Define

$$K^{(N)}(u) = \sum_{m=0}^{N-1} \phi_m(0)\phi_m(u)1\{|u| \le 1\},$$

which is bounded and supported in $[-1,1]$. The $\{\phi_q\}_{q\le k}$ form a basis for the space of polynomials of degree at most k, so we can find coefficients b_{qk}, $q = 1,\ldots,k$, such that $u^k = \sum_{q=0}^k b_{qk}\phi_q(u) =: L(u)$. Therefore, for $k \le N-1$, we conclude, using ortho-normality of the ϕ_m in $L^2([-1,1])$, that

$$\begin{aligned}\int_{\mathbb{R}} K^{(N)}(u)u^k du &= \int_{-1}^1 \sum_{q=0}^k b_{qk}\phi_q(u)\sum_{m=0}^{N-1}\phi_m(0)\phi_m(u)du \\ &= \sum_{q=0}^k\sum_{m=0}^{N-1} b_{qk}\phi_m(0)\int_{-1}^1 \phi_q(u)\phi_m(u)du \\ &= \sum_{q=0}^k b_{qk}\phi_q(0) = L(0),\end{aligned}$$

which equals 1 if $k = 0$ and 0 otherwise and completes the proof. ■

For projection-type kernels, such results are not necessarily as simple. The proof of Proposition 4.1.2 implies the following:

Proposition 4.1.7 *The Haar kernel $K(x,y)$ from (4.20) satisfies Condition 4.1.4 with $N = 0$.*

For $N \ge 1$, the Haar kernel does not satisfy Condition 4.1.4: one sees immediately that

$$\int_{\mathbb{R}} K(1, 1+u)udu \ne 0.$$

However, if ϕ is the Shannon-basis function so that $\mathcal{F}[\phi] = 1_{[-\pi,\pi]}$, then from the point of view of Schwartz distributions, we can formally write

$$\int_{\mathbb{R}} x^k\phi(x)dx = (-i)^{-k}D^k\mathcal{F}[\phi](0) = 0 \quad \forall k \in \mathbb{N},$$

but this does not make sense in the L^p setting because $\phi \notin L^1$. We would like to construct ortho-normal bases of L^2 that are in a sense 'interpolating' between the Haar and Shannon bases, and this is what leads to wavelet theory, as we shall see later.

4.1.4 Littlewood-Paley Decomposition

The main idea behind the Haar and Shannon bases of L^2 was a partition of unity either in the time or the frequency domain. The functions used in the partition are, however, indicators of intervals and thus not smooth, which poses a difficulty either in approximating very regular functions well or in approximating functions in norms other than L^2. The main idea behind Littlewood-Paley theory is to construct partitions of unity that consist of smooth functions, relaxing the requirement of orthogonality of the functions involved.

Take $\phi \in \mathcal{S}(\mathbb{R})$ to be any symmetric function such that

$$\hat{\phi} \in C_0^\infty(\mathbb{R}), \quad supp(\hat{\phi}) \in [-1,1], \quad \hat{\phi} = 1 \text{ on } \left[-\frac{3}{4}, \frac{3}{4}\right]. \tag{4.24}$$

Define, moreover,

$$\hat{\psi} = \hat{\phi}\left(\frac{\cdot}{2}\right) - \hat{\phi} \quad \text{equivalent to} \quad \psi = 2\phi(2\,\cdot) - \phi$$

so that $\hat{\psi}$ is supported in $\{2^{-1} \leq |u| \leq 2\}$. If we set $\psi_{2^{-j}} = 2^j\psi(2^j\,\cdot)$, then $\widehat{\psi_{2^{-j}}} = \hat{\psi}(\,\cdot\,/2^j)$, and by a telescoping sum, for every $u \in \mathbb{R}$,

$$\hat{\phi}(u) + \sum_{j=0}^{\infty} \hat{\psi}(u/2^j) = \lim_{J\to\infty}\left(\hat{\phi}(u) + \sum_{j=0}^{J-1} \hat{\psi}(u/2^j)\right) = \lim_J \hat{\phi}(u/2^J) = 1. \tag{4.25}$$

Note that for fixed u, the preceding sums are all finite. Thus, for f with Fourier transform $\hat{f}(u)$ and every $u \in \mathbb{R}$,

$$\hat{f}(u) = \hat{f}(u)\hat{\phi}(u) + \sum_{j=0}^{\infty} \hat{\psi}(u/2^j)\hat{f}(u), \tag{4.26}$$

which by (4.8) is formally the same as

$$f = f * \phi + \sum_{j=0}^{\infty} f * \psi_{2^{-j}} = \lim_{J\to\infty} f * \phi_{2^{-J}}, \tag{4.27}$$

where $\phi_{2^{-J}} = 2^J\phi(2^J\,\cdot)$. Since $\hat{\phi}(0) = 1$, we see that $\int \phi = 1$, and since $\phi \in \mathcal{S}(\mathbb{R})$, we conclude from Proposition 4.1.1 that the last limit holds in L^p whenever $f \in L^p$ ($f \in C_u(\mathbb{R})$ when $p = \infty$). Moreover, $\int_{\mathbb{R}} x^k\phi(x)dx$ equals zero for every $k \in \mathbb{N}$ because $D^k\mathcal{F}[\phi](0)$ does, so Proposition 4.1.5 applies for every N with $h = 2^{-j}$. We have thus succeeded in decomposing f into a sum of infinitely many band-limited functions, known as the *Littlewood-Paley decomposition of* f. It shares in some sense the good properties of both the Haar and Shannon bases but itself does not constitute an ortho-normal basis of L^2.

Exercises

4.1.1 Let f be a function that has N weak derivatives $D^j f, j = 1, \dots, N$, on an interval $I \subset \mathbb{R}$ containing x. Prove that f can be redefined on a set of measure zero such that the following properties hold: if $N = 1$ and Df is integrable on I, then, for $y \geq x, y \in I$,

$$f(y) - f(x) = \int_x^y Df(t)dt;$$

in particular, f is continuous on I. Moreover, for $N \geq 1$, we have Taylor's formula

$$f(y) = f(x) + \sum_{k=1}^{N-1} \frac{D^k f(x)}{k!}(y-x)^k + \int_0^1 (y-x)^N \frac{(1-t)^{N-1}}{(N-1)!} D^N f(x+t(y-x))dt.$$

Hint: If f has a weak derivative Df, then use

$$f(y) = f(x) + \int_0^1 Df(x+t(y-x))(y-x)dt$$

recursively.

4.1.2 For D_m the Dirichlet kernel, let

$$S_m(f)(x) = f * D_m(x)$$

be the mth Fourier partial sum of $f \in C_{\text{per}}((0,1])$, and let $x \in (0,1]$ be arbitrary. Prove that there exists $f \in C_{\text{per}}((0,1])$ such that $S_m(f)(x)$ does not converge. Prove that in fact the set of $f \in C_{\text{per}}((0,1])$ such that $S_m(f)(x)$ converges is *nowhere dense* for the uniform-norm topology of $C_{\text{per}}((0,1])$. *Hint*: Prove first that $\|D_m\|_1 \to \infty$ as $m \to \infty$. Then use the fact that $f \mapsto S_m(f)(x)$ for x fixed is a continuous linear form on $C_{\text{per}}((0,1])$ of operator (dual) norm $\|D_m\|_1$, and invoke the uniform boundedness principle (Banach-Steinhaus theorem) from functional analysis to deduce a contradiction.

4.1.3 Let

$$F_m(x) = \frac{1}{m+1} \sum_{k=0}^{m} D_k(x)$$

be the Fejér kernel. Show that for every $f \in C_{\text{per}}((0,1])$, $\|f * F_m - f\|_\infty \to 0$ as $m \to \infty$. *Hint*: Prove first that F_m is a probability density function on $(0,1]$, and then apply a variant of Proposition 4.1.1 to the periodic extension of f.

4.2 Orthonormal Wavelet Bases

In this section we present some of the basic ingredients of wavelet theory, the main goal being the construction of smooth wavelet bases that have compact support in time or frequency domain and which outperform the classical Haar and Shannon bases in many respects, particularly from an approximation-theoretic perspective. We focus on the one-dimensional theory – multivariate wavelets can be constructed from univariate ones by a simple tensor-product method that will be introduced and used in Section 4.3.6.

4.2.1 Multiresolution Analysis of L^2

The abstract framework in which wavelets naturally arise and which unifies some of the ideas of the preceding section is the one of a *multiresolution analysis* of L^2.

Definition 4.2.1 We say that $\phi \in L^2(\mathbb{R})$ is the scaling function of a multiresolution analysis (MRA) of $L^2(\mathbb{R})$ if it satisfies the following conditions:

(a) The family

$$\{\phi(\cdot - k) : k \in \mathbb{Z}\}$$

is an ortho-normal system in $L^2(\mathbb{R})$; that is, $\langle\phi(\cdot-k),\phi(\cdot-l)\rangle$ equals 1 when $k=l$ and 0 otherwise.

(b) The linear spaces

$$V_0=\left\{f=\sum_{k\in\mathbb{Z}}c_k\phi(\cdot-k),\{c_k\}_{k\in\mathbb{Z}}:\sum_{k\in\mathbb{Z}}c_k^2<\infty\right\},\ldots,$$

$$V_j=\{h=f(2^j(\cdot)):f\in V_0\},\ldots,$$

are nested; that is, $V_{j-1}\subset V_j$ for every $j\in\mathbb{N}$.

(c) The union $\cup_{j\geq 0}V_j$ is dense in L^2.

We note that under (a) and (b), it is immediate that the functions

$$\phi_{jk}=2^{j/2}\phi(2^j(\cdot)-k),\quad k\in\mathbb{Z},$$

form an ortho-normal basis of the space $V_j, j\in\mathbb{N}$.

Examples for ϕ generating a multiresolution analysis exist, for example, the Haar function $\phi=1_{(0,1]}$ from the preceding section, where V_j equals the space of functions that are piecewise constant on the dyadic intervals $(k/2^j,(k+1)/2^j]$. Another example from the preceding section is the function $\phi(x)=\sin(\pi x)/(\pi x)$ from the Shannon basis, where $V_j\equiv\mathcal{V}_{2^j\pi}$. These examples are in some sense opposite extremes because the function ϕ generating the Haar basis is localised in time but not in frequency, and the function ϕ generating the Shannon basis is localised in frequency but not in time. A question that has been essential to the development of wavelet theory and, more generally, to time-frequency analysis was, among other things, whether good localisation properties of ϕ could be achieved in time and frequency simultaneously, in the flavour of a Littlewood-Paley decomposition, but without loosing the ortho-normal basis property.

Before we answer this question in the positive, let us first spell out some simple properties of a multiresolution analysis of L^2 that follow immediately from its definition. Since the spaces V_j are nested, there are nontrivial subspaces of L^2 obtained from taking the orthogonal complements $W_l:=V_{l+1}\ominus V_l$ in Hilbert space: we can 'telescope' these orthogonal complements to see that the space V_j can be written as

$$V_j=V_0\oplus\left(\bigoplus_{l=0}^{j-1}W_l\right).\tag{4.28}$$

If we want to find the orthogonal L^2-projection of $f\in L^2$ onto V_j, then the preceding decomposition of V_j tells us that we can describe this projection as the projection of f onto V_0 plus the sum of the projections of f onto W_l from $l=0$ to $j-1$. The projection of f onto V_0 is

$$K_0(f)(x)=\sum_{k\in\mathbb{Z}}\langle\phi_k,f\rangle\phi_k(x),$$

where we write $\phi_k=\phi(\cdot-k)$. To describe the projections onto W_l, we would like to find basis functions that span the spaces W_l, and this is where the *wavelet* function ψ enters the stage: assume that there exists a fixed $\psi\in L^2(\mathbb{R})$ such that, for every $l\in\mathbb{N}\cup\{0\}$,

$$\{\psi_{lk}:=2^{l/2}\psi(2^l(\cdot)-k):k\in\mathbb{Z}\}$$

is an ortho-normal set of functions that spans W_l. Again, such ψ exists: if $\phi = 1_{(0,1]}$, the corresponding *Haar wavelet* is $\psi = 1_{[0,1/2]} - 1_{(1/2,1]}$; for the Shannon basis, we have $\psi = \mathcal{F}^{-1}[1_{[-2\pi,-\pi]} + 1_{[\pi,2\pi]}]$. The projection of f onto W_l is

$$\sum_k \langle \psi_{lk}, f \rangle \psi_{lk},$$

and adding things up, we see that the projection $K_j(f)$ of f onto V_j has the two identical presentations

$$K_j(f)(x) = \sum_{k\in\mathbb{Z}} \langle \phi_{jk}, f \rangle \phi_{jk}(x) \tag{4.29}$$

$$= \sum_{k\in\mathbb{Z}} \langle \phi_k, f \rangle \phi_k(x) + \sum_{l=0}^{j-1} \sum_{k\in\mathbb{Z}} \langle \psi_{lk}, f \rangle \psi_{lk}(x). \tag{4.30}$$

This projection is the partial sum of what is called the *wavelet series* of a function $f \in L^2$: to understand this, note that, if $\cup_{j\geq 0} V_j$ is dense in L^2, then (4.28) implies that the space L^2 can be decomposed into the direct sum

$$L^2 = V_0 \oplus \left(\bigoplus_{l=0}^{\infty} W_l \right),$$

so the set of functions

$$\{\phi(\cdot - k), 2^{l/2}\psi(2^l(\cdot) - k) : k \in \mathbb{Z}, l \in \mathbb{N} \cup \{0\}\} \tag{4.31}$$

is an ortho-normal *wavelet* basis of the Hilbert space L^2. *It will often be convenient in later chapters to denote the scaling functions ϕ_k as the 'first' wavelets' ψ_{-1k} – this way we can abbreviate the wavelet basis as consisting of functions $\{\psi_{lk}\}$, which expedites notation.* As a consequence, every $f \in L^2$ has the *wavelet series* expansion

$$f = \sum_{k\in\mathbb{Z}} \langle \phi_k, f \rangle \phi_k + \sum_{l=0}^{\infty} \sum_{k\in\mathbb{Z}} \langle \psi_{lk}, f \rangle \psi_{lk} \tag{4.32}$$

$$= \sum_{k\in\mathbb{Z}} \langle \phi_{jk}, f \rangle \phi_{jk}(x) + \sum_{l=j}^{\infty} \sum_{k\in\mathbb{Z}} \langle \psi_{lk}, f \rangle \psi_{lk}$$

$$= \sum_{l\geq -1} \sum_{k\in\mathbb{Z}} \psi_{lk} \langle \psi_{lk}, f \rangle,$$

where convergence is guaranteed at least in the space L^2.

Now the question arises as to whether functions ϕ and ψ other than those given by the Haar and Shannon bases exist such that the class of functions (4.31) forms a multiresolution analysis of L^2. The following theorem gives Fourier-analytical conditions for the first two relevant properties in Definition 4.2.1 and also gives a generic construction of the wavelet function ψ.

Theorem 4.2.2 *Let $\phi \in L^2(\mathbb{R}), \phi \neq 0$.*

(a) The family of functions $\{\phi(\cdot - k) : k \in \mathbb{Z}\}$ forms an ortho-normal system in L^2 if and only if

$$\sum_{k \in \mathbb{Z}} |\hat{\phi}(u + 2\pi k)|^2 = 1 \quad a.e. \tag{4.33}$$

(b) Suppose that the $\{\phi(\cdot - k) : k \in \mathbb{Z}\}$ form an ortho-normal system in L^2. Then the corresponding spaces $(V_j)_{j \in \mathbb{Z}}$ are nested if and only if there exists a 2π-periodic function $m_0 \in L^2((0, 2\pi])$ such that

$$\hat{\phi}(u) = m_0\left(\frac{u}{2}\right)\hat{\phi}\left(\frac{u}{2}\right) \quad a.e. \tag{4.34}$$

(c) Let ϕ be a scaling function satisfying properties (a) and (b) of Definition 4.2.1, and let m_0 satisfy (4.34). If $\psi \in L^2$ satisfies

$$\hat{\psi}(u) = m_1\left(\frac{u}{2}\right)\hat{\phi}\left(\frac{u}{2}\right) \quad a.e., \tag{4.35}$$

where $m_1(u) = \overline{m_0(u + \pi)}e^{-iu}$, then ψ is a wavelet function; that is, $\{\psi(\cdot - k) : k \in \mathbb{Z}\}$ forms an ortho-normal basis of $W_0 = V_1 \ominus V_0$, and any $f \in V_1$ can be uniquely decomposed as $\sum_k c_k \phi(\cdot - k) + \sum_k c'_k \psi(\cdot - k)$ for sequences $\{c_k\}, \{c'_k\} \in \ell_2$.

Proof

(a) If we let $\tilde{\phi} = \overline{\phi(-\cdot)}, q = \phi * \tilde{\phi}$, then

$$S(u) \equiv \sum_k |\hat{\phi}(u + 2\pi k)|^2 = \sum_k \hat{q}(u + 2\pi k).$$

Since $\phi \in L^2$, we have $\hat{q} \in L^1$ by Plancherel's theorem and the Cauchy-Schwarz inequality, so the Poisson summation formula (before (4.11)) implies that the series defining S converges a.e. on $(0, 2\pi]$ and has Fourier coefficients $c_k = \mathcal{F}^{-1}[\hat{q}](-k) = q(-k)$. Now, if δ_{kl} is the Kronecker-δ, we need to show that, for every $k, l \in \mathbb{Z}$,

$$\int_{\mathbb{R}} \phi(x - k)\overline{\phi(x - l)}dx = \delta_{kl}$$

or, equivalently, $\forall k \in \mathbb{Z}$,

$$\int_{\mathbb{R}} \phi(x)\tilde{\phi}(k - x)dx = q(k) = \delta_{0k}.$$

The Poisson summation formula yields

$$\sum_k \hat{q}(u + 2\pi k) = \sum_k c_k e^{iku} = \sum_k q(k)e^{-iku}, \tag{4.36}$$

where we note that $\{c_k\} \in \ell_1$ because either $S = 1$ or $q(k) = \delta_{0k}$. In fact, if $q(k) = \delta_{0k}$, then the preceding sum equals 1, and if the preceding sum equals 1, then $q(k) = \delta_{0k}$ by uniqueness of Fourier series.

(b) By definition of the spaces V_j, it suffices to prove that $V_0 \subseteq V_1$. We first assume that $V_0 \subseteq V_1$, so $\phi \in V_1$, and hence for coefficients $h_k = \sqrt{2}\int \phi(x)\overline{\phi(2x-k)}dx$, $\{h_k\} \in \ell_2$; we can represent ϕ in the basis of V_1, that is,

$$\phi(x) = \sqrt{2}\sum_k h_k \phi(2x-k).$$

Taking Fourier transforms on both sides, we see that

$$\hat{\phi}(u) = 2^{-1/2}\sum_k h_k e^{-iuk/2}\hat{\phi}(u/2) \equiv m_0(u/2)\hat{\phi}(u/2)$$

almost everywhere.

For the converse, we first need an auxiliary result that will be useful repeatedly in what follows.

Lemma 4.2.3 *Let $\{\phi(\cdot - k) : k \in \mathbb{Z}\}$ be an ortho-normal system in $L^2(\mathbb{R})$. Every 2π-periodic function $m_0 \in L^2((0,2\pi])$ satisfying (4.34) also satisfies*

$$|m_0(u)|^2 + |m_0(u+\pi)|^2 = 1 \quad a.e.,$$

in particular, any such m_0 is bounded on $(0,2\pi]$.

Proof From (4.34), we have

$$|\hat{\phi}(2u+2\pi k)|^2 = |m_0(u+\pi k)|^2 |\hat{\phi}(u+\pi k)|^2.$$

Summing this identity, using part (a), splitting the summation indices into odd and even integers (possible by absolute convergence) and using periodicity of m_0 and part (a) again, we have

$$\begin{aligned}
1 &= \sum_k |m_0(u+\pi k)|^2 |\hat{\phi}(u+\pi k)|^2 \\
&= \sum_l |m_0(u+2\pi l)|^2 |\hat{\phi}(u+2\pi l)|^2 + \sum_l |m_0(u+2\pi l+\pi)|^2 |\hat{\phi}(u+2\pi l+\pi)|^2 \\
&= \sum_l |m_0(u)|^2 |\hat{\phi}(u+2\pi l)|^2 + \sum_l |m_0(u+\pi)|^2 |\hat{\phi}(u+2\pi l+\pi)|^2 \\
&= |m_0(u)|^2 + |m_0(u+\pi)|^2,
\end{aligned}$$

completing the proof. ■

To proceed, any $f \in V_0$ has Fourier transform

$$\hat{f}(u) = \hat{\phi}(u)\sum_k c_k e^{-iuk} \equiv \hat{\phi}(u)m(u), \quad \{c_k\} \in \ell_2,$$

where m is a 2π-periodic square integrable function on $(0,2\pi]$, and any $m \in L^2((0,2\pi])$ gives rise to such f by the Fourier isometry of $L^2((0,2\pi])$ with ℓ_2. Conclude that the images $\hat{V}_0, \hat{V}_1$ of V_0, V_1, respectively, under the Fourier transform are

$$\hat{V}_0 = \{m(u)\hat{\phi}(u) : m \text{ is } 2\pi-\text{periodic}, m \in L^2((0,2\pi])\}$$

and

$$\hat{V}_1 = \{m(u/2)\hat{\phi}(u/2) : m \text{ is } 2\pi - \text{periodic}\ , m \in L^2((0,2\pi])\}.$$

The identity (4.34) then implies that any element of $\hat{V}_0$ can be written as $m(u)m_0(u/2)\hat{\phi}(u/2)$, which belongs to $\hat{V}_1$ as the boundedness and 2π-periodicity of m imply that $m(2\cdot)m_0$ is also 2π-periodic and in L^2. By injectivity of the Fourier transform, we conclude that $V_0 \subseteq V_1$.
(c) We finally turn to construction of the wavelet function, and we proceed in three steps.

(i) To show that the translates of ψ form an ortho-normal system in L^2, we verify (4.33) with ψ in place of ϕ. Indeed, using the hypothesis on $\hat{\psi}$, 2π-periodicity and boundedness of m_0, (4.33) for ϕ and Lemma 4.2.3, we see that

$$\begin{aligned}
\sum_k |\hat{\psi}(u+2\pi k)|^2 &= \sum_k \left|m_1\left(\frac{u}{2}+\pi k\right)\right|^2 \left|\hat{\phi}\left(\frac{u}{2}+\pi k\right)\right|^2 \\
&= \sum_k \left|m_0\left(\frac{u}{2}+\pi+\pi k\right)\right|^2 \left|\hat{\phi}\left(\frac{u}{2}+\pi k\right)\right|^2 \\
&= \sum_l \left|m_0\left(\frac{u}{2}+\pi+2\pi l+\pi\right)\right|^2 \left|\hat{\phi}\left(\frac{u}{2}+2\pi l+\pi\right)\right|^2 \\
&\quad + \sum_l \left|m_0\left(\frac{u}{2}+\pi+2\pi l\right)\right|^2 \left|\hat{\phi}\left(\frac{u}{2}+2\pi l\right)\right|^2 \\
&= \left|m_0\left(\frac{u}{2}\right)\right|^2 + \left|m_0\left(\frac{u}{2}+\pi\right)\right|^2 = 1 \ a.e.
\end{aligned}$$

(ii) We further need to show that the translates of ψ are orthogonal to ϕ, that is, $\langle\phi(\cdot - k), \psi(\cdot - l)\rangle = 0\ \forall k,l$, or equivalently, recalling the notation $\tilde{\psi} = \overline{\psi(-\cdot)}$,

$$g(k) \equiv \langle\phi, \psi(\cdot - k)\rangle = \phi * \tilde{\psi}(k) = 0, \quad k \in \mathbb{Z}.$$

We have $\hat{g} = \hat{\phi}\hat{\tilde{\psi}} = \hat{\phi}\overline{\hat{\psi}} \in L^1$ by the Cauchy-Schwarz inequality, and the Poisson summation formula (4.11) applied to the 2π-periodic function $S = \sum_k \hat{g}(\cdot + 2\pi k)$ gives Fourier coefficients $g(-k)$ of S. Therefore, $g(k) = 0\ \forall k$ is equivalent to $S(u) = 0\ a.e.$ on $(0, 2\pi]$, or reinserting the definition of $\hat{g}$,

$$\sum_k \hat{\phi}(u+2\pi k)\overline{\hat{\psi}(u+2\pi k)} = 0 \ a.e., \tag{4.37}$$

which it remains to verify. Using the definition of $\hat{\psi}$, (4.34), part (a), periodicity and boundedness of m_0, m_1 (Lemma 4.2.3), the left-hand side in the preceding display equals

$$\begin{aligned}
&\sum_k \hat{\phi}\left(\frac{u}{2}+\pi k\right) m_0\left(\frac{u}{2}+\pi k\right) \overline{\hat{\phi}\left(\frac{u}{2}+\pi k\right) m_1\left(\frac{u}{2}+\pi k\right)} \\
&= \sum_k \left|\hat{\phi}\left(\frac{u}{2}+\pi k\right)\right|^2 m_0\left(\frac{u}{2}+\pi k\right) \overline{m_1\left(\frac{u}{2}+\pi k\right)}
\end{aligned}$$

$$= \sum_l \left|\hat{\phi}\left(\frac{u}{2}+2\pi l\right)\right|^2 m_0\left(\frac{u}{2}+2\pi l\right)\overline{m_1\left(\frac{u}{2}+2\pi l\right)}$$

$$+ \sum_l \left|\hat{\phi}\left(\frac{u}{2}+2\pi l+\pi\right)\right|^2 m_0\left(\frac{u}{2}+2\pi l+\pi\right)\overline{m_1\left(\frac{u}{2}+2\pi l+\pi\right)}$$

$$= m_0\left(\frac{u}{2}\right)\overline{m_1\left(\frac{u}{2}\right)} + m_0\left(\frac{u}{2}+\pi\right)\overline{m_1\left(\frac{u}{2}+\pi\right)}$$

It thus suffices to prove that $m_0\overline{m_1}+m_0(\cdot+\pi)\overline{m_1(\cdot+\pi)}$ equals zero almost everywhere, which follows because, by definition of m_1, this quantity equals

$$m_0(u)e^{iu}m_0(u+\pi)+m_0(u+\pi)m_0(u)e^{iu+i\pi} = m_0(u)m_0(u+\pi)e^{iu}(1+e^{i\pi}) = 0.$$

(iii) It remains to establish the unique decomposition of $f \in V_1$ as desired. Expand f into a series with respect to the basis $\{\sqrt{2}\phi(2(\cdot)-k) : k \in \mathbb{Z}\}$ of V_1, which, in the Fourier domain, and proceeding as in the proof of part (a), reads as

$$\hat{f}(u) = q\left(\frac{u}{2}\right)\hat{\phi}\left(\frac{u}{2}\right) \quad \text{with} \quad q(u) = \frac{1}{\sqrt{2}}\sum_k q(k)e^{-iuk}. \tag{4.38}$$

Now, using the formulas for $\hat{\phi}, \hat{\psi}$, we see that

$$\overline{m_0\left(\frac{u}{2}\right)}\hat{\phi}(u) = \left|m_0\left(\frac{u}{2}\right)\right|^2\hat{\phi}\left(\frac{u}{2}\right), \quad \overline{m_1\left(\frac{u}{2}\right)}\hat{\psi}(u) = \left|m_1\left(\frac{u}{2}\right)\right|^2\hat{\phi}\left(\frac{u}{2}\right),$$

and summing these identities,

$$\hat{\phi}\left(\frac{u}{2}\right)\left[\left|m_0\left(\frac{u}{2}\right)\right|^2+\left|m_1\left(\frac{u}{2}\right)\right|^2\right] = \overline{m_0\left(\frac{u}{2}\right)}\hat{\phi}(u)+\overline{m_1\left(\frac{u}{2}\right)}\hat{\psi}(u) \ \ a.e.$$

By definition of m_1, we see $|m_1(u/2)|^2 = |m_0(\pi+u/2)|^2$, so, by Lemma 4.2.3, $\hat{\phi}(u/2) = \overline{m_0(u/2)}\hat{\phi}(u)+\overline{m_1(u/2)}\hat{\psi}(u)$, which when substituted into (4.38) gives the decomposition

$$\hat{f}(u) = q\left(\frac{u}{2}\right)\overline{m_0\left(\frac{u}{2}\right)}\hat{\phi}(u)+q\left(\frac{u}{2}\right)\overline{m_1\left(\frac{u}{2}\right)}\hat{\psi}(u).$$

Taking inverse Fourier transforms gives the result. ∎

Theorem 4.2.2 gives sufficient conditions to construct ϕ, ψ in the Fourier domain. In the time domain, these properties imply the following representations of ϕ, ψ:

Corollary 4.2.4 *For ϕ a scaling function generating a multiresolution analysis of L^2, we have*

$$\phi(x) = \sqrt{2}\sum_k h_k\phi(2x-k) \ a.e., \quad h_k = \sqrt{2}\int_{\mathbb{R}}\phi(x)\overline{\phi(2x-k)}dx \tag{4.39}$$

and

$$\psi(x) = \sqrt{2}\sum_k \lambda_k\phi(2x-k) \ a.e., \quad \lambda_k = (-1)^{k+1}\bar{h}_{1-k}. \tag{4.40}$$

Moreover, if $\int_{\mathbb{R}}\phi(x)dx = 1$, then

$$\sum_k \bar{h}_k h_{k+2l} = \delta_{0l}, \quad \frac{1}{\sqrt{2}}\sum_k h_k = 1. \tag{4.41}$$

Proof The proof of the direct part of Theorem 4.2.2, part (b), already established (4.39), as well as that $m_0(u) = 2^{-1/2}\sum_k h_k e^{-iku}$. If $\hat{\phi}(0) = \int \phi(x)dx = 1$, we have necessarily $m_0(0) = 1$ from (4.34), which already implies the second claim in (4.41). Using this representation of m_0, we see that

$$\overline{m_0\left(\frac{u}{2}+\pi\right)} = \overline{\frac{1}{\sqrt{2}}\sum_k h_k e^{-ik((u/2)+\pi)}} = \frac{1}{\sqrt{2}}\sum_k \bar{h}_k(-1)^k e^{ik(u/2)},$$

and from Theorem 4.2.2, part (c) we further have

$$\begin{aligned}\hat{\psi}(u) &= \frac{1}{\sqrt{2}}\sum_k \bar{h}_k(-1)^k e^{i(k-1)(u/2)}\hat{\phi}\left(\frac{u}{2}\right) = \sqrt{2}\sum_{k'} \bar{h}_{k'}(-1)^{k'+1}e^{-ik'(u/2)}\frac{1}{2}\hat{\phi}\left(\frac{u}{2}\right)\\ &= \sqrt{2}\sum_{k'}\lambda_k e^{-ik'(u/2)}\frac{1}{2}\hat{\phi}\left(\frac{u}{2}\right),\end{aligned}$$

so (4.40) follows from taking inverse Fourier transforms. Finally, the conclusion of Lemma 4.2.3 gives

$$m_0(u)\overline{m_0(u)} + m_0(u+\pi)\overline{m_0(u+\pi)} = 1;$$

thus,

$$\begin{aligned}1 &= \frac{1}{2}\sum_{k,k'}\bar{h}_k h_{k'} e^{-iu(k'-k)} + \frac{1}{2}\sum_{k,k'}\bar{h}_k h'_k e^{-iu(k'-k)-i(k'-k)\pi}\\ &= \frac{1}{2}\sum_{k,k'} e^{-iu(k-k')}[1+e^{-i(k'-k)\pi}] = \sum_l\sum_k \bar{h}_k h_{k+2l}e^{-2iul},\end{aligned}$$

implying the first property of (4.41). ■

4.2.2 Approximation with Periodic Kernels

The preceding results give conditions to verify (a) and (b) from Definition 4.2.1 in the Fourier domain. They do not, however, address the question of whether the $\{V_j\}_{j\geq 0}$ are dense in L^2. This can be established by showing that the projection kernel $K(x,y) = \sum_k \phi(x-k)\phi(y-k)$ satisfies the conditions of Proposition 4.1.3, or in fact the stronger Condition 4.1.4, so that then $K_{2^{-j}}(f) \to f$ in L^2 as $j \to \infty$. Convergence then will in fact hold in any $L^p, 1 \leq p < \infty$, and in $C_u(\mathbb{R})$ whenever f is in any of these spaces, with quantitative bounds on the approximation errors depending on regularity properties of f.

We consider general projection kernels of the form

$$K(x,y) \equiv K_\phi(x,y) = \sum_{k\in\mathbb{Z}}\phi(x-k)\phi(y-k), \tag{4.42}$$

where $\phi \in L^2$ is a fixed real-valued function such that the preceding sum converges pointwise. Note that in this subsection we do *not* require that ϕ generate a multiresolution analysis, although the results are tailor-made for such situations. If ϕ is a bounded function of compact support in the interval $[-a,a]$ say, then

$$\sup_x \sum_{k\in\mathbb{Z}} |\phi(x-k)| \le \|\phi\|_\infty \sum_k 1_{\{k:-2a\le k\le 2a\}}(x) \le (4a+1))\|\phi\|_\infty, \tag{4.43}$$

$$|K(v,v-u)| \le C(\|\phi\|_\infty, a)1_{[-2a,2a]}(u) \quad \forall u\in\mathbb{R}, \tag{4.44}$$

and this generalises to noncompactly supported ϕ in the sense that good localisation properties of ϕ imply good localisation properties of the kernel K. Note that the conditions of the following proposition exclude too spread-out functions ϕ such as the function ϕ generating the Shannon basis (where $\phi \notin L^1$):

Proposition 4.2.5 *Assume that for some nonincreasing function* $\Phi \in L^\infty([0,\infty)) \cap L^1([0,\infty))$ *we have* $|\phi(u)| \le \Phi(|u|)\ \forall u\in\mathbb{R}$. *Then*

(a) $\sum_{k\in\mathbb{Z}} |\phi(\cdot - k)| \in L^\infty(\mathbb{R})$ *and*
(b) $\sup_{v\in\mathbb{R}} |K(v,v-u)| \le c_1\Phi(c_2|u|)$ *for some* $0<c_1,c_2<\infty$ *and every* $u\in\mathbb{R}$.

Proof The function $\sum_{k\in\mathbb{Z}}|\phi(x-k)|$ is one periodic, so we can restrict to bounding it on $[0,1]$. For $x\in[0,1], |k|\ge 2$, we have $|x-k|\ge |k|/2$, so by monotonicity, $\Phi(|x-k|)\le \Phi(|k|/2)$ for those x,k. Thus, again using monotonicity of Φ,

$$\sum_{k\in\mathbb{Z}}|\phi(x-k)| \le \sum_{k\in\mathbb{Z}}\Phi(|x-k|) \le \sum_{|k|\le 1}\Phi(|x-k|) + \sum_{|k|\ge 2}\Phi(|k|/2)$$
$$\le 4\Phi(0) + \int_{\mathbb{R}} \Phi(|u|/2)du < \infty.$$

As mentioned earlier, part (b) is immediate for compactly supported ϕ (see (4.44)), and the general case is proved using similar arguments as in part (a), (see Exercise 4.2.1). ∎

Verifying Condition 4.1.4 **(P)** for K is more delicate; finer properties of the function ϕ are required. The following theorem allows us to derive wavelet projection kernel analogues of Proposition 4.1.6 from properties of $\hat{\phi}$ only.

Proposition 4.2.6 *Assume the conditions of Proposition 4.2.5 with* Φ *such that* $\int_{\mathbb{R}}\Phi(|u|)|u|^N du<\infty$ *for some* $N\in\mathbb{N}\cup\{0\}$. *Assume, moreover, that, as* $u\to 0$,

$$|\hat{\phi}(u)|^2 = 1 + o(|u|^N),\ \hat{\phi}(u+2\pi k) = o(|u|^N) \quad \forall k\neq 0. \tag{4.45}$$

Then $\left|\int_{\mathbb{R}}\phi(x)dx\right| = 1$, *and for every* $l=1,\dots,N$ *and almost every* $x\in\mathbb{R}$, *we have*

$$\int_{\mathbb{R}} K(x,x+u)du = 1 \quad \text{and} \quad \int_{\mathbb{R}} K(x,x+u)u^l = 0. \tag{4.46}$$

Proof The assumption $|\hat{\phi}(0)|^2=1$ implies that $\left|\int_{\mathbb{R}}\phi(x)dx\right|=1$. To prove (4.46), note that by a change of variables it suffices to prove that

$$\int_{\mathbb{R}} K(x,y)(y-x)^l dy = \delta_{0l},$$

where δ_{0l} is the Kronecker delta. The integral in question equals, by the binomial theorem and a change of variables,

$$\sum_m \int_{\mathbb{R}} \phi(x-m)\phi(y-m)(y-m+m-x)^l dy$$

$$= \sum_{j=0}^{l} (-1)^{l-j} C_j^l \int_{\mathbb{R}} y^j \phi(y) dy \sum_m \phi(x-m)(x-m)^{l-j}, \tag{4.47}$$

where the interchange of integration and summation is permitted in view of Fubini's theorem and Proposition 4.2.5. Note next that

$$(D^j\hat{\phi})(2\pi k) = \int_{\mathbb{R}} \phi(y)(-iy)^j e^{-i2\pi ky} dy, \quad k \in \mathbb{Z}, \tag{4.48}$$

in view of (4.9). The Poisson summation formula (4.12) with $a = 1$ implies that

$$(D^j\hat{\phi})(2\pi k) = \int_0^1 \sum_{m\in\mathbb{Z}} \phi(x-m)(-i(x-m))^j e^{-i2\pi kx} dx \equiv \langle D_j, e_k \rangle.$$

Since $\hat{\phi}(u+2\pi k) = o(|u|^N)\ \forall k \neq 0$, we see $(D^j\hat{\phi})(2\pi k) = 0\ \forall k \neq 0$ for $j \leq N$, so the Fourier coefficients $(k \neq 0)$ of $D_j(x) = \sum_{m\in\mathbb{Z}} \phi(x-m)(-i(x-m))^j$ all vanish, and $D_j(x) = D_j = (D^j\hat{\phi})(0)$ is thus constant. However, (4.48) at $k = 0$ gives $(D^j\hat{\phi})(0) = \int_{\mathbb{R}} \phi(y)(-iy)^j dy$, so the quantity in (4.47) equals, again by the binomial theorem,

$$\sum_{j=0}^{l} (-1)^{l-j} C_j^l \int_{\mathbb{R}} y^j \phi(y) dy \int_{\mathbb{R}} z^{l-j}\phi(z) dz = \int_{\mathbb{R}}\int_{\mathbb{R}} \sum_{j=0}^{l} C_j^l y^j (-z)^{l-j} \phi(y)\phi(z) dy dz$$

$$= \int_{\mathbb{R}}\int_{\mathbb{R}} (y-z)^l \phi(y)\phi(z) dy dz$$

$$= \int_{\mathbb{R}} (-t)^l (\phi * \tilde{\phi})(t) dt,$$

where we recall the notation $\tilde{\phi} = \phi(-\cdot)$. This last expression equals $i^{-l}(D^l|\hat{\phi}|^2)(0)$, which equals 1 for $l = 0$ and 0 otherwise in view of the assumption $|\hat{\phi}(u)|^2 = 1 + o(|u|^N)$, completing the proof. ■

It is easily shown that the conclusions of the preceding proposition hold for *every* $x \in \mathbb{R}$ if the scaling function ϕ involved is continuous. If the conditions of the preceding proposition are satisfied for some $N \geq 0$, we conclude, combining Propositions 4.1.3, 4.2.5 and 4.2.6, that $\|K_{2^{-j}}(f) - f\|_p \to 0$ whenever $f \in L^p, 1 \leq p < \infty$. In particular, if $K_{2^{-j}}$ is the projector onto V_j, then the $\{V_j\}_{j\geq 0}$ are dense in L^p. The same remarks apply to $p = \infty$ if L^p is replaced by C_u and if ϕ is continuous.

If the function ϕ in the preceding proposition comes from a multiresolution analysis, we further have that the wavelet function ψ corresponding to ϕ is automatically 'orthogonal to polynomials' up to degree N, proved as follows.

Proposition 4.2.7 *Let ϕ, ψ be as in Theorem 4.2.2, part (c), and suppose that ϕ satisfies the conditions of Proposition 4.2.6 for some N. Then*

$$\int_{\mathbb{R}} \psi(x)x^l dx = 0 \quad \forall l = 0, \dots, N.$$

Proof The integral in question is proportional to $(D^l \hat{\psi})(0)$ – it thus suffices to show that

$$\hat{\psi}(u) = m_1\left(\frac{u}{2}\right)\hat{\phi}\left(\frac{u}{2}\right) = o(|u|^N),$$

as $u \to 0$. By definition of m_1 (see Theorem 4.2.2, part (c)) and boundedness of $\hat{\phi}$ (noting $|\phi| \le \Phi \in L^1$), it thus suffices to show that $m_0(u+\pi) = o(|u|^N)$ as $u \to 0$. By (4.33), the preceding proposition and Exercise 4.2.2 we necessarily have $\hat{\phi}(\pi + 2\pi k_0) \neq 0$ for some k_0, and letting $k' = 2k_0 + 1$, we see, using the periodicity of m_0, that

$$\hat{\phi}(u + 2\pi k') = m_0\left(\frac{u}{2} + \pi k'\right)\hat{\phi}\left(\frac{u}{2} + \pi k'\right) = m_0\left(\frac{u}{2} + \pi\right)\hat{\phi}\left(\frac{u}{2} + \pi + 2\pi k_0\right).$$

Since $\hat{\phi}(u + 2\pi k') = o(|u|^N)$, by hypothesis, we deduce, as $u \to 0$, that

$$m_0\left(\frac{u}{2} + \pi\right) = o(|u|^N),$$

concluding the proof. ■

We conclude this subsection with the following basic but important result on the relation between the L^p-norms of translation averages $\sum_{k\in\mathbb{Z}} c_k 2^{l/2}\phi(2^l x - k)$ for functions ϕ satisfying $\sum_k |\phi(\cdot - k)| \in L^\infty$ and the ℓ_p-norms of $\{c_k : k \in \mathbb{Z}\}$. For scaling functions ϕ, the result holds with equality for $p = 2$ by Parseval's identity; for $p \neq 2$, the result holds as a two-sided inequality with universal scaling constants.

Proposition 4.2.8 *Suppose that $\phi \in L^1(\mathbb{R})$ is such that $\sup_{x\in\mathbb{R}} \sum_{k\in\mathbb{Z}} |\phi(x-k)| \equiv \kappa < \infty$. Let $c \equiv \{c_k : k \in \mathbb{Z}\} \in \ell_p, 1 \le p \le \infty$. Then, for every $l \ge 0$ and some constant $K = K(\kappa, \|\phi\|_1, p)$, we have*

$$\left\| \sum_{k\in\mathbb{Z}} c_k 2^{l/2}\phi(2^l \cdot - k) \right\|_p \le K\|c\|_p 2^{l(1/2-1/p)}.$$

If, moreover, the set $\{\phi(\cdot - k) : k \in \mathbb{Z}\}$ is ortho-normal in L^2, then, for some constant $K' = K'(\kappa, \|\phi\|_1, p)$,

$$\left\| \sum_{k\in\mathbb{Z}} c_k 2^{l/2}\phi(2^l \cdot - k) \right\|_p \ge K'\|c\|_p 2^{l(1/2-1/p)}.$$

Proof For $c \in \ell_p \subset \ell_\infty$, the series $\sum_k c_k \phi(\cdot - k)$ converges absolutely uniformly; in particular, the case $p = \infty$ is immediate in the first inequality. For $p < \infty$, we consider first $l = 0$ and have, for q such that $1 = 1/p + 1/q$,

$$\left| \sum_k c_k \phi(x-k) \right| \le \sum_k |c_k| |\phi(x-k)|^{1/p} |\phi(x-k)|^{1/q},$$

so, by Hölder's inequality,

$$\left\|\sum_k c_k \phi(\cdot - k)\right\|_p^p \le \int_{\mathbb{R}} \sum_k |c_k|^p |\phi(x-k)| \left(\sum_k |\phi(x-k)|\right)^{p/q} dx$$
$$\le \kappa^{p/q} \|c\|_p^p \|\phi\|_1,$$

which gives the desired result for $l = 0$. The case $l > 0$ follows directly from the last estimate and $\|\phi(2^l \cdot)\|_1 = 2^{-l}\|\phi\|_1$.

For the converse, we again restrict to $l = 0$ as the general case is the same up to scaling. Note that if we define $h(x) = \sum_k c_k \phi(x-k)$, then the c_k necessarily equal the inner products $\langle h, \phi(\cdot - k)\rangle$. Using Hölder's inequality and writing $\phi_k = \phi(\cdot - k)$ as usual, the estimate

$$\|\langle \phi_\cdot, h\rangle\|_p^p \le \sum_k \left(\int |h(x)||\phi_k(x)|^{1/p}|\phi_k(x)|^{1/q} dx\right)^p$$
$$\le \sum_k \int |h(x)|^p |\phi_k(x)| dx \left(\int |\phi_k(x)| dx\right)^{p/q}$$
$$\le \|\phi\|_1^{p/q} \|h\|_p^p \kappa \tag{4.49}$$

gives

$$\|c\|_p = \|\langle \phi_\cdot, h\rangle\|_p \le \|\phi\|_1^{1/q} \kappa^{1/p} \left\|\sum_k c_k \phi(\cdot - k)\right\|_p,$$

the desired result for $p < \infty$. The case $p = \infty$ is again obvious. ∎

4.2.3 Construction of Scaling Functions

We now want to construct concrete scaling functions ϕ that generate a multiresolution analysis of $L^2(\mathbb{R})$ as in Definition 4.2.1 and whose projection kernels have good approximation properties in the sense that they satisfy Condition 4.1.4. We will focus here on the construction of two main examples of scaling functions and wavelets – one where ϕ, ψ have compact support and one where $\hat{\phi}, \hat{\psi}$ have compact support – and in both cases such that the relevant functions are well localised in both the frequency and the time domains, avoiding the shortcomings of Haar and Shannon wavelets. Note that it is impossible to have both ϕ and $\hat{\phi}$ of compact support (loosely speaking, this is Heisenberg's uncertainty principle). As the construction of band-limited wavelets (i.e., with compactly supported Fourier transform) is significantly easier, we start with this case.

Band-Limited Wavelets

One of the first examples of a wavelet basis was the *Meyer* scaling function ϕ defined by

$$\hat{\phi}_\nu(u) = \begin{cases} 1 & |u| \le \dfrac{2\pi}{3} \\ \cos\left[\dfrac{\pi}{2}\nu\left(\dfrac{3}{2\pi}|u| - 1\right)\right] & \dfrac{2\pi}{3} \le |u| \le \dfrac{4\pi}{3} \\ 0 & \text{otherwise,} \end{cases} \tag{4.50}$$

where $\nu : [0,\infty) \to \mathbb{R}$ is any infinitely differentiable function that satisfies $\nu(0) = 0, \nu(x) = 1\ \forall x \geq 1$ and $\nu(x) + \nu(1-x) = 1$. Using the results from the preceding subsection, one can check directly that ϕ generates a multiresolution analysis. More generally, we construct band-limited wavelets as follows: take μ any probability measure supported in a closed subinterval of $[-\pi/3, \pi/3]$, and define ϕ by

$$\hat{\phi}(u) = \sqrt{\int_{u-\pi}^{u+\pi} d\mu}, \tag{4.51}$$

which is supported in $[-4\pi/3, 4\pi/3]$ and equals 1 on $(-2\pi/3, 2\pi/3)$; in particular, $\int \phi = \hat{\phi}(0) = 1$. Suitable choices of μ give the Meyer function and other examples, including the Shannon scaling function ($\mu = \delta_0$). We have

$$\sum_{k\in\mathbb{Z}} |\hat{\phi}(u+2\pi k)|^2 = \sum_{k\in\mathbb{Z}} \int_{u+(2k-1)\pi}^{u+(2k+1)\pi} d\mu = \int_{\mathbb{R}} d\mu = 1,$$

which checks the first condition of Definition 4.2.1 in view of Theorem 4.2.2, part (a). Next, we set, for $u \in [-2\pi, 2\pi]$,

$$m_0(u/2) = \begin{cases} \hat{\phi}(u) & |u| \leq 4\pi/3 \\ 0 & 4\pi/3 < |u| \leq 2\pi \end{cases} \tag{4.52}$$

extended periodically to the real line so that the second condition of Definition 4.2.1 follows in view of Theorem 4.2.2, part (b). Since $\hat{\phi}$ is identically 1 near the origin, we trivially have $|\hat{\phi}(u)|^2 = 1 + o(|u|^N)$ for every N, and since $\hat{\phi}$ is supported in $[-4\pi/3, 4\pi/3]$, we also have, for $|u|$ small enough and every N, $\hat{\phi}(u + 2\pi k) = 0 = o(|u|^N)$ whenever $k \neq 0$, so Proposition 4.2.6 can be used. By Theorem 4.2.2, part (c), the wavelet ψ is seen to equal

$$\hat{\psi}(u) = e^{-iu/2} \sqrt{\int_{|u/2|-\pi}^{|u|-\pi} d\mu}, \tag{4.53}$$

which is supported in $\{u : |u| \in [2\pi/3, 8\pi/3]\}$; in particular, $\int \psi = \hat{\psi}(0) = 0$. Finally, by taking μ to have a suitably regular density, we can take $\hat{\phi}, \hat{\psi} \in C^\infty$ with compact support, so ϕ and ψ are contained in the Schwartz space $\mathcal{S}$ and have dominating functions Φ, Ψ with moments of arbitrary order; thus, Proposition 4.2.5 applies. Summarising, we have proved the following result, which joins the forces of the Littlewood-Paley decomposition and the Shannon basis.

Theorem 4.2.9 *There exists a band-limited ortho-normal multiresolution wavelet basis*

$$\{\phi_k = \phi(\cdot - k), \psi_{lk} = 2^{l/2}\psi(2^l(\cdot) - k) : k \in \mathbb{Z}, l \in \mathbb{N} \cup \{0\}\}$$

of $L^2(\mathbb{R})$ with scaling function $\phi \in \mathcal{S}(\mathbb{R})$, $\int \phi = 1$, wavelet $\psi \in \mathcal{S}(\mathbb{R})$, $\int \psi = 0$ and projection kernel $K(x,y) = \sum_{k\in\mathbb{Z}} \phi(x-k)\phi(y-k)$ such that

(a) $supp(\hat{\phi}) \subset \{u : |u| \leq 4\pi/3\}$, $\quad supp(\hat{\psi}) \subset \{u : |u| \in [2\pi/3, 8\pi/3]\}$,
(b) $\int_{\mathbb{R}} \psi(u)u^l du = 0\ \forall l \in \mathbb{N} \cup \{0\}$ *and, for all* $v \in \mathbb{R}$, $l \in \mathbb{N}$,

$$\int_{\mathbb{R}} K(v, v+u)du = 1, \qquad \int_{\mathbb{R}} K(v, v+u)u^l du = 0,$$

(c) $\sum_{k\in\mathbb{Z}}|\phi(\cdot-k)|\in L^\infty(\mathbb{R})$, $\sum_{k\in\mathbb{Z}}|\psi(\cdot-k)|\in L^\infty(\mathbb{R})$, *and*
(d) *For* $\kappa(x,y)$ *equal to either* $K(x,y)$ *or* $\sum_k \psi(x-k)\psi(y-k)$,

$$\sup_{v\in\mathbb{R}}|\kappa(v,v-u)|\le c_1\Phi(c_2|u|) \quad \text{for some } 0<c_1,c_2<\infty \text{ and every } u\in\mathbb{R},$$

for some bounded function $\Phi:[0,\infty)\to\mathbb{R}$ *that decays faster than any inverse polynomial at* $+\infty$.

Daubechies Wavelets

We now construct a scaling function ϕ and wavelet ψ that have compact support and, unlike the Haar wavelet, possess a prescribed number of derivatives. These wavelets are called *Daubechies wavelets* because I. Daubechies gave the first construction of such a remarkable basis.

Theorem 4.2.10 *For every* $N\in\mathbb{N}$ *there exists an ortho-normal multiresolution wavelet basis*

$$\{\phi_k=\phi(\cdot-k),\ \psi_{lk}=2^{l/2}\psi(2^l(\cdot)-k): k\in\mathbb{Z}, l\in\mathbb{N}\cup\{0\}\}$$

of $L^2(\mathbb{R})$ *with scaling function* $\phi\equiv\phi^{(N)}$, $\int\phi=1$, *wavelet* $\psi\equiv\psi^{(N)}$, $\int\psi=0$ *and projection kernel* $K(x,y)=\sum_{k\in\mathbb{Z}}\phi(x-k)\phi(y-k)$ *such that*

(a) $supp(\phi)\subset\{x:0\le x\le 2N-1\}$, $supp(\psi)\subset\{x:-N+1\le x\le N\}$,
(b) $\int_{\mathbb{R}}\psi(u)u^l du=0$ $\forall l=0,1,\dots,N-1$, *and, for all* $v\in\mathbb{R}$,

$$\int_{\mathbb{R}}K(v,v+u)du=1, \qquad \int_{\mathbb{R}}K(v,v+u)u^l du=0, \quad \forall l=1,\dots,N-1,$$

(c) $\sum_{k\in\mathbb{Z}}|\phi(\cdot-k)|\in L^\infty(\mathbb{R})$, $\sum_{k\in\mathbb{Z}}|\psi(\cdot-k)|\in L^\infty(\mathbb{R})$,
(d) *For* $\kappa(x,y)$ *equal to either* $K(x,y)$ *or* $\sum_k\psi(x-k)\psi(y-k)$,

$$\sup_{v\in\mathbb{R}}|\kappa(v,v-u)|\le c_1\Phi(c_2|u|)$$

for some $0<c_1,c_2<\infty$ *and every* $u\in\mathbb{R}$, *for some bounded and compactly supported function* $\Phi:[0,\infty)\to\mathbb{R}$, *and*
(e) *For* $N\ge 2$, *both functions* ϕ,ψ *are elements of* $C^{[\lambda(N-1)]}(\mathbb{R})$ *for some* $\lambda\ge 0.18$.

Remark 4.2.11 We recall for part (e) that the spaces C^m for $m\in\mathbb{N}$ have been defined at the beginning of this chapter; when $[\lambda N]=0$, we understand C^0 as the space of bounded uniformly continuous functions. In fact, the conclusion in (e) holds as well with λN replacing its integer part $[\lambda N]$ if we use the more general definition of C^s, real $s>0$, from (4.111). Also, the smoothness estimate can be slightly improved to $\lambda\ge 0.193$ (but not much beyond that).

Proof For $N=1$, the theorem follows from taking $\phi^{(1)}$ equal to the Haar scaling function (Section 4.1.2). We thus consider only $N\ge 2$. The rough idea is as follows: since, trigonometric polynomials have inverse Fourier transforms of compact support, one starts by constructing a trigonometric polynomial m_0 from scratch that satisfies the conclusion of Lemma 4.2.3. The recursion (4.34) then motivates us to define

$$\hat\phi=\Pi_{j=1}^\infty m_0(2^{-j}\cdot), \quad \phi=\mathcal{F}^{-1}\hat\phi.$$

Since trigonometric polynomials are Fourier transforms of Dirac measures, the function ϕ can be understood as an infinite convolution product of certain discrete signed measures. Remarkably, such ϕ can be shown to be in L^2, to have compact support and in fact to generate a multiresolution analysis. The proof of this fact, however, is neither short nor simple. The numerical computation of ϕ as an inverse Fourier transform of $\hat{\phi}$ is possible by efficient algorithms – we discuss this in the notes at the end of this chapter.

Step I: Construction of m_0 We wish to construct a trigonometric polynomial m_0 on $(0, 2\pi]$ (periodically extended to $\mathbb{R}$) such that

$$|m_0(u)|^2 + |m_0(u+\pi)|^2 = 1 \quad \forall u. \tag{4.54}$$

For reasons that will become apparent later, we also want m_0 to factorise, for the given N, as

$$m_0(u) = \left(\frac{1+e^{-iu}}{2}\right)^N \mathcal{L}(u), \tag{4.55}$$

where $\mathcal{L}$ is the trigonometric polynomial to be found. If we write $M_0(u) = |m_0(u)|^2, L(u) = |\mathcal{L}(u)|^2$, this is the same as requiring, for all u,

$$M_0(u) + M_0(u+\pi) = 1 \quad \text{and} \quad M_0(u) = \left(\cos^2\frac{u}{2}\right)^N L(u).$$

Note that both M and L are now polynomials in $\cos u$, and if we rewrite L as a polynomial P in $\sin^2(u/2) = (1-\cos u)/2$, then

$$M_0(u) = \left(\cos^2\frac{u}{2}\right)^N P\left(\sin^2\frac{u}{2}\right),$$

so the desired equation becomes

$$(1-y)^N P(y) + y^N P(1-y) = 1 \ \forall y \in [0,1]. \tag{4.56}$$

The proof of the following lemma is not difficult using Bezout's theorem; see Exercise 4.2.3.

Lemma 4.2.12 *For any $N \in \mathbb{N}$, the polynomial*

$$P_N(y) = \sum_{k=0}^{N-1} \binom{N-1+k}{k} y^k$$

is a solution of equation (4.56).

Note that there are other solutions than P_N which could be used as well, but this shall not concern us here. The solution P_N uniquely defines $|m_0(u)|^2$. Moreover, any positive polynomial in $\cos(u)$ has a square root that is itself a trigonometric polynomial of the same degree, with real coefficients (Exercise 4.2.4). Applying this to $L(u) = P_N(\sin^2(u/2))$, we conclude that there exists, for every $N \in \mathbb{N}$, a trigonometric polynomial

$$m_0 = m_0^{(N)} = \frac{1}{\sqrt{2}} \sum_{k=0}^{2N-1} h_k e^{-ik\cdot}, \quad h_k = h_k^{(N)}, \tag{4.57}$$

satisfying the desired properties (4.54) and (4.55). (The scaling by $1/\sqrt{2}$ will be convenient later.) We finally notice that $L(0)=1$ implies that $|m_0(0)|^2=1$, and since $m_0(0)$ is real valued, we can always take the positive square root in the preceding argument so that

$$m_0(0)=1, \quad \frac{1}{\sqrt{2}}\sum_k h_k = 1. \tag{4.58}$$

Step II: Definition of $\phi^{(N)}, \psi^{(N)}$ and their support properties Given the integer $N \geq 2$, let $m_0^{(N)}$ be the function from (4.57), and define

$$\hat{\phi}(u) = \hat{\phi}^{(N)}(u) = \prod_{j=1}^{\infty} m_0(2^{-j}u), \quad u \in \mathbb{R}. \tag{4.59}$$

The infinite product is well defined: since $m_0(0)=1$ and the polynomial m_0 is Lipschitz continuous, we have $|m_0(2^{-j}u)| \leq 1 + |m_0(2^{-j}u) - m_0(0)| \leq 1 + K2^{-j}u$ for some constant K, hence,

$$\prod_{j=1}^{\infty} |m_0(2^{-j}u)| \leq \exp\left\{K\sum_{j=1}^{\infty} 2^{-j}|u|\right\} \leq e^{K|u|} < \infty \tag{4.60}$$

and thus the infinite product converges uniformly on compact subsets of $\mathbb{R}$. Moreover, the functions

$$f_k(u) \equiv 1_{[-\pi,\pi]}(2^{-k}u)\prod_{j=1}^{k} m_0(2^{-j}u) \tag{4.61}$$

converge to $\hat{\phi}(u)$ pointwise as $k \to \infty$. The following observation on the integrals

$$\int_{\mathbb{R}} |f_k(u)|^2 e^{inu}\,du = \int_{-2^k\pi}^{2^k\pi} |f_k(u)|^2 e^{inu}\,du, \quad n \in \mathbb{Z},$$

will be useful: using 2π-periodicity of m_0 and (4.54), the first 'half' of this integral equals, per substitution $v = u + 2^k\pi$,

$$\begin{aligned}
\int_{-2^k\pi}^{0} \prod_{j=1}^{k} |m_0(2^{-j}u)|^2 e^{inu}\,du &= \int_0^{2^k\pi} \prod_{j=1}^{k} |m_0(2^{-j}v - 2^{k-j}\pi)|^2 e^{in(v-2^k\pi)}\,dv \\
&= \int_0^{2^k\pi} \prod_{j=1}^{k-1} |m_0(2^{-j}v)|^2 |m_0(2^{-j}v+\pi)|^2 e^{inv}\,dv \\
&= \int_0^{2^k\pi} \prod_{j=1}^{k-1} |m_0(2^{-j}v)|^2 e^{inv}\,dv - \int_0^{2^k\pi} \prod_{j=1}^{k} |m_0(2^{-j}v)|^2 e^{inv}\,dv.
\end{aligned}$$

Hence, using again the periodicity of m_0,

$$\int |f_k(u)|^2 e^{inu}\,du = \int |f_{k-1}(u)|^2 e^{inu}\,du = 2\pi\,\delta_{0n} \quad \forall n \in \mathbb{Z}, k \geq 2, \tag{4.62}$$

with the last identity following from

$$\int |f_1(u)|^2 e^{inu} du = \int_{-2\pi}^{0} |m_0(u/2)|^2 e^{inu} du + \int_{0}^{2\pi} |m_0(u/2)|^2 e^{inu} du$$
$$= 2\int_0^{\pi} \left(|m_0(v)|^2 + |m_0(v+\pi)|^2\right) e^{i2nv} dv$$
$$= \int_0^{2\pi} e^{inv} dv = 2\pi\,\delta_{0n}.$$

We conclude, in particular, for $n=0$ and from Fatou's lemma combined with $|f_k|^2 \to |\hat{\phi}|^2$ pointwise, that $\hat{\phi} \in L^2$; in fact,

$$\|\hat{\phi}\|_2^2 \le \limsup_k \|f_k\|_2^2 \le 2\pi.$$

We can therefore define, by Plancherel's theorem,

$$\phi = \phi^{(N)} = \mathcal{F}^{-1}(\hat{\phi}^{(N)}) \tag{4.63}$$

as an element of L^2 satisfying $\|\phi\|_2 \le 1$. We immediately have from $m_0(0)=1$ that $\hat{\phi}(0) = \int \phi = 1$.

Having defined ϕ, we next show that it has compact support contained in $[0, 2N-1]$, where the isomorphism of $\mathcal{S}^*(\mathbb{R})$ under the distributional Fourier transform is helpful: from (4.57) and (4.58) we know that m_0 is the Fourier transform of the discrete probability measure

$$\frac{1}{\sqrt{2}}\sum_{k=0}^{2N-1} h_k \delta_k,$$

and thus the (distributional) inverse Fourier transform $\mathcal{F}^{-1}(m_0(\cdot/2^j))$ has support contained in $[0, 2^{-j}(2N-1)]$ for $j \ge 0$. Moreover, we established earlier that, as $J \to \infty$,

$$\prod_{j=1}^{J} m_0\left(\frac{\cdot}{2^j}\right) \to \prod_{j=1}^{\infty} m_0\left(\frac{\cdot}{2^j}\right) = \hat{\phi}$$

uniformly on compact subsets of $\mathbb{R}$, which implies, in particular, convergence in $\mathcal{S}^*$. Since $\mathcal{F}^{-1}$ is a continuous isomorphism of $\mathcal{S}^*$, we conclude, as $J \to \infty$, that

$$\mathcal{F}^{-1}\left[\prod_{j=1}^{J} m_0\left(\frac{\cdot}{2^j}\right)\right] = \mathcal{F}^{-1}\left[m_0\left(\frac{\cdot}{2}\right)\right] * \cdots * \mathcal{F}^{-1}\left[m_0\left(\frac{\cdot}{2^J}\right)\right] \to \phi$$

in $\mathcal{S}^*$. The support of the convolution products $\mathcal{F}^{-1}[m_0(\cdot/2)] * \cdots * \mathcal{F}^{-1}m_0(\cdot/2^J)$ is contained in $[0, (2N-1)\sum_{j=1}^{J} 2^{-j}] \subset [0, (2N-1)]$ uniformly in J, so $supp(\phi) \subset [0, 2N]$ follows from taking limits (and noting that convergence in $\mathcal{S}^*$ implies convergence of the support sets by checking all integrals against suitably supported test functions).

Trusting that ϕ will give rise to a scaling function, it is now natural to define, in light of (4.35),

$$\hat{\psi}(u) = \hat{\psi}^{(N)} = m_1\left(\frac{u}{2}\right)\hat{\phi}\left(\frac{u}{2}\right), \tag{4.64}$$

where $m_1(u) = \overline{m_0(u+\pi)e^{-iu}}$. As m_1 is bounded by 1 (by (4.54)), and since $\hat{\phi} \in L^2$, we immediately conclude that $\hat{\psi} \in L^2$, and we can define the Daubechies wavelet function by

$$\psi = \psi^{(N)} = \mathcal{F}^{-1}(\hat{\psi}^{(N)}), \tag{4.65}$$

which is also in L^2 by Plancherel's theorem. From (4.54) and $m_0(0) = 1$, we obtain $\hat{\psi}(0) = \int \psi = 0$. Moreover, as in the proof of Corollary 4.2.4, one shows that

$$\psi(x) = \sqrt{2} \sum_k \lambda_k \phi(2x-k),$$

with λ_k, h_k as in that Corollary for $\phi = \phi^{(N)}$. By the just established support property of ϕ, we necessarily have that the preceding sum extends over finitely many k, namely, $0 \le 1-k \le 2N-1$ (as only these $\lambda_k \neq 0$) and $0 \le 2x-k \le 2N-1$, which combined gives that x is in the support of ψ whenever

$$1 - 2N + 1 \le 2x \le 1 + 2N - 1 \iff -N+1 \le x \le N.$$

Step III: Verifying ortho-normality Although we have constructed ϕ, we do not yet know if it is the scaling function of a multiresolution analysis. It remains to establish the ortho-normality of the translates of ϕ; the nestedness of the V_j spaces and the ortho-normality properties of ψ then follow immediately from Theorem 4.2.2, noting that the property (4.34) is immediate from the definition of $\hat{\phi}$. Let us thus prove that for $\phi = \phi^{(N)}$; we have, for all $n \in \mathbb{Z}$,

$$\int \phi(x)\overline{\phi(x-n)}dx = \int |\hat{\phi}(u)|^2 e^{imu} du = 2\pi\,\delta_{0n}, \tag{4.66}$$

using Plancherel's theorem in the first identity. We recall the functions f_k from (4.61), for which we have already shown $|f_k|^2 \to |\hat{\phi}|^2$ pointwise. Moreover, in (4.62) we proved that

$$\int |f_k|^2 e^{inu} = 2\pi\,\delta_{0n}, \quad \forall n \in \mathbb{Z}, k \ge 1. \tag{4.67}$$

We want to integrate the limit $|f_k|^2 \to |\hat{\phi}|^2$ to deduce (4.66) from (4.67), and to achieve this, we will construct a dominating function for $\{|f_k|^2 : k \ge 1\}$ that is integrable: from the explicit representation of m_0 from step 1, we have $\inf_{|u|\le\pi/2} |m_0(u)|^2 \ge \cos^{2N}(\pi/4) > 0$ (using $L(u) \ge 1$), and thus, for some $C > 0$, every $j \ge 1$,

$$\inf_{|u|\le\pi} |m_0(2^{-j}u)| \ge C.$$

Moreover, again by $m_0(0) = 1$ and by Lipschitz continuity of m_0, for some C',

$$|m_0(u)| \ge 1 - C'|u|.$$

We next choose j_0 large enough that $2^{-j_0}C'|u| < 1/2$ for $|u| \leq \pi$. Then, using $1-x \geq e^{-2x}$ for $0 \leq x \leq 1/2$, we obtain, for all $|u| \leq \pi$,

$$
\begin{aligned}
|\hat{\phi}(u)| &= \prod_{j=1}^{j_0} |m_0(2^{-j}u)| \prod_{j=j_0+1}^{\infty} |m_0(2^{-j}u)| \\
&\geq C^{j_0} \prod_{j=j_0+1}^{\infty} e^{-2C'2^{-j}|u|} \\
&\geq C^{j_0} e^{-C'2^{-j_0+1}\pi} \equiv C'' > 0
\end{aligned}
$$

or, equivalently, $1_{[-\pi,\pi]}(u) \leq |\hat{\phi}(u)|^2 (C'')^{-2}$ for all u. Thus, by (4.59),

$$
\begin{aligned}
|f_k(u)|^2 &= \prod_{j=1}^{k} |m_0(2^{-j}u)|^2 1_{[-\pi,\pi]}(2^{-k}u) \\
&\leq (C'')^{-2} \prod_{j=1}^{k} |m_0(2^{-j}u)|^2 |\hat{\phi}(u)(2^{-k}u)|^2 \\
&= (C'')^{-2} |\hat{\phi}(u)|^2.
\end{aligned}
$$

Since we already know $|\hat{\phi}|^2 \in L^1$ from the preceding step, we deduce from $|f_k(u)|^2 \to |\hat{\phi}(u)|^2$ pointwise, (4.67) and the dominated convergence theorem that (4.66) indeed holds true.

Step IV: Regularity of ϕ, ψ We next establish property (e) – once this is achieved, the claims (c) and (d) of the theorem follow from Proposition 4.2.5 because ϕ, ψ are then bounded and continuous (being C^γ for some $\gamma > 0$) and compactly supported, so a bounded dominating function Φ of compact support exists.

To establish $\phi \in C^\gamma$, it suffices to show, by Fourier inversion and (4.9), that $\int |\hat{\phi}(u)|(1+|u|)^\gamma < \infty$ and therefore to establish the estimate

$$
|\hat{\phi}(u)| \leq C(1+|u|)^{-\gamma-1-\delta}, \tag{4.68}
$$

for some $C, \delta > 0$ and all u. From (4.64) and since m_1 is bounded, we see that any such proof will also establish $\psi \in C^\gamma$; we thus restrict to ϕ. We will use the factorisation $m_0(u) = [(1+e^{-iu})/2)]^N \mathcal{L}(u)$ from step I, which implies that

$$
|\hat{\phi}(u)|^2 = \prod_{j=1}^{\infty} \left|\cos\frac{u}{2^{j+1}}\right|^{2N} \prod_{j=1}^{\infty} \left|\mathcal{L}\left(\frac{u}{2^j}\right)\right|^2 = \frac{|\sin(u/2)|^{2N}}{|u/2|^{2N}} \prod_{j=1}^{\infty} \left|\mathcal{L}\left(\frac{u}{2^j}\right)\right|^2, \tag{4.69}
$$

where we have used the identity

$$
\frac{\sin x}{x} = \prod_{j=1}^{\infty} \cos\frac{x}{2^j},
$$

which follows easily from $\sin 2x = 2\cos x \sin x$. We can thus reduce the problem to estimating the uniform norms of the polynomial $\mathcal{L}(u)$. To obtain useful results, some care is necessary.

Lemma 4.2.13 *Let*

$$q_l = \sup_u \left| \prod_{k=0}^{l-1} \mathcal{L}\left(\frac{u}{2^k}\right) \right|, \quad \kappa_l = \frac{\log q_l}{l \log 2},$$

and suppose that $\kappa_\ell < N - 1 - \gamma$ for some $\ell \in \mathbb{N}$. Then $\phi \in C^\gamma$.

Proof Setting $\mathcal{L}_\ell(u) = \prod_{k=0}^{\ell-1} \mathcal{L}(2^{-k}u)$, we can write

$$\prod_{j=1}^{\infty} \mathcal{L}\left(\frac{u}{2^j}\right) = \prod_{j=0}^{\infty} \prod_{k=0}^{\ell-1} \mathcal{L}(2^{-k-\ell j-1}u) = \prod_{j=0}^{\infty} \mathcal{L}_\ell(2^{-\ell j-1}u).$$

We know that $\mathcal{L}(0) = \mathcal{L}_\ell(0) = 1$, and since $\mathcal{L}_\ell$ is a polynomial, it is locally Lipschitz, so $|\mathcal{L}_\ell(u)| \leq 1 + C|u|$, for all $|u| \leq 1$ and some fixed $0 < C < \infty$. Thus, as in (4.60), we see, for all $|u| \leq 1$, that

$$\prod_{j=0}^{\infty} |\mathcal{L}_\ell(2^{-\ell j-1}u)| \leq e^C.$$

For $|u| \geq 1$, we choose J large enough that $2^{J-1} \leq |u| \leq 2^J$ and observe that

$$\begin{aligned}
\prod_{j=0}^{\infty} |\mathcal{L}_\ell(2^{-\ell j-1}u)| &= \prod_{j=0}^{J} |\mathcal{L}_\ell(2^{-\ell j-1}u)| \prod_{j=0}^{\infty} |\mathcal{L}_\ell(2^{-J}2^{-\ell j-1}u)| \\
&\leq q_\ell^J e^C = 2^{J\left(\frac{\log q_\ell}{\ell \log 2}\right)} e^C \\
&= 2^{J\kappa_\ell} e^C \leq 2^{J(N-1-\gamma-\delta)} e^C = O(|u|^{N-1-\gamma-\delta}),
\end{aligned}$$

for some $\delta > 0$, which combined with (4.68) and (4.69) gives the result. ∎

Now estimates on κ_ℓ for $\ell \geq 1$ lead to smoothness estimates for ϕ. Considering $\ell = 2$ and making the dependence of $\mathcal{L} = \mathcal{L}_N$ on N explicit in the notation for the moment, we need to bound

$$q_2^2 = \sup_u |\mathcal{L}_N(u)\mathcal{L}_N(2u)|^2 = \sup_{y \in [0,1]} |P_N(y)P_N(4y(1-y))|,$$

recalling that $|\mathcal{L}_N(u)|^2 = P_N(u)$ with $P = P_N$ from Lemma 4.2.12 and using the trigonometric identity $\sin^2 u = 4\sin^2(u/2)(1 - \sin^2(u/2))$. Simple computations, using that P_N solves (4.56), establish that

$$P_N(y) \leq 2^{N-1} \max(1, 2y)^{N-1}, \quad y \in [0,1],$$

so for such y further satisfying either $y \leq (1/2) - \sqrt{2}/4$ or $y \geq (1/2) + \sqrt{2}/4$ (so that $4y(1-y) \leq 1/2$), we have

$$\sqrt{|P_N(y)P_N(4y(1-y))|} \leq 2^{3(N-1)/2}.$$

For the remaining $y \in [(1/2) - (\sqrt{2}/4), (1/2) + (\sqrt{2}/4)]$, we have likewise, since $y^2(1-y) \leq 4/27$ for $y \in [0,1]$, that

$$\sqrt{|P_N(y)P_N(4y(1-y))|} \leq 2^{N-1}[16y^2(1-y)]^{(N-1)/2} \leq 2^{3(N-1)}(4/27)^{(N-1)/2},$$

a bound which strictly dominates the preceding one and which results in

$$q_2 \le 2^{4(N-1)} 3^{-3(N-1)/2}, \quad \kappa_2 \le (N-1)\left[2 - \frac{3}{4}\frac{\log 3}{\log 2}\right].$$

Hence, to establish that $\phi \in C^{(N-1)\lambda}$, we need to check $[2-(3/4)(\log 3/\log 2)] < 1-\lambda$, true, for instance, for $\lambda = 0.18$. (Better values can be obtained from estimating $\ell > 2$.)

Step V: Cancellation properties We finally establish (b) by verifying the hypotheses of Proposition 4.2.6 so that this proposition and Proposition 4.2.7 imply the desired result. A dominating function Φ with arbitrary moments exists by the preceding step, so it remains to verify (4.45) with N replaced by $N-1$. We note from step I and since P is bounded on any compact set,

$$\begin{aligned} |m_0(u+\pi)|^2 &= \left|\cos\frac{u+\pi}{2}\right|^{2N} P\left(\sin^2\frac{u+\pi}{2}\right) \\ &= O(|u|^{2N}) = o(|u|^{2(N-1)}), \quad \text{as } |u| \to 0. \end{aligned} \tag{4.70}$$

We thus have, for $k \in \mathbb{Z}, k \neq 0, q \geq 0$, such that $k = 2^q k'$, k' odd, that, as $u \to 0$,

$$\begin{aligned} \hat{\phi}(u+2k\pi) &= \hat{\phi}\left(\frac{u}{2}+k\pi\right) m_0\left(\frac{u}{2}+k\pi\right) \\ &= \hat{\phi}\left(\frac{u}{2^{q+1}}+k'\pi\right) m_0\left(\frac{u}{2^{q+1}}+k'\pi\right) \cdots m_0\left(\frac{u}{2}+k\pi\right) = o(|u|^{N-1}) \end{aligned}$$

because $\hat{\phi}, m_0$ are bounded and because, by the periodicity of m_0,

$$m_0\left(\frac{u}{2^{q+1}}+k'\pi\right) = m_0\left(\frac{u}{2^{q+1}}+\pi\right) = o(|u|^{N-1})$$

in view of (4.70). Next, (4.54) and (4.70) imply that $|m_0(u)|^2 = 1 + o(|u|^{2(N-1)})$, and so, as $u \to 0$,

$$|\hat{\phi}(u)|^2 = \left|\hat{\phi}\left(\frac{u}{2}\right)\right|^2 \left|m_0\left(\frac{u}{2}\right)\right|^2 = \left|\hat{\phi}\left(\frac{u}{2}\right)\right|^2 (1 + o(|u|^{2(N-1)})).$$

Moreover, as $\hat{\phi}$ is the Fourier transform of a compactly supported continuous function, it possesses derivatives of all orders, and so then does $|\hat{\phi}|^2$. Since $|\hat{\phi}(0)|^2 = 1$ has already been established, we know that $|\hat{\phi}|^2$ has a Taylor expansion

$$|\hat{\phi}(u)|^2 = 1 + \sum_{k=1}^{N-1} b_k u^k + o(|u|^{N-1}), \quad b_k \in \mathbb{R},$$

near the origin. Summarising, for all u near 0,

$$1 + \sum_{k=1}^{N-1} b_k u^k + o(|u|^{N-1}) = (1 + o(|u|^{2(N-1)}))\left(1 + \sum_{k=1}^{N-1} b_k \left(\frac{u}{2}\right)^k + o(|u|^{N-1})\right),$$

so all the $b_k, 0 \le k \le N-1$, necessarily must equal zero, implying that

$$|\hat{\phi}(u)|^2 = 1 + o(|u|^{N-1}),$$

verifying the second hypothesis in Proposition 4.2.6 and completing step V and thereby also the proof of the theorem. ■

S-Regular Wavelet Bases

For several results we shall not be needing a particular wavelet basis, but any that satisfies the following key properties. Meyer and Daubechies wavelets provide examples, but other wavelet bases (not treated here in detail) satisfy the conditions as well.

Definition 4.2.14 A multiresolution wavelet basis

$$\{\phi_k = \phi(\cdot - k), \psi_{lk} = 2^{l/2}\psi(2^l(\cdot) - k) : k \in \mathbb{Z}, l \in \mathbb{N} \cup \{0\}\}$$

of $L^2(\mathbb{R})$ with projection kernel $K(x,y) = \sum_k \phi(x-k)\phi(y-k)$ is said to be S-regular for some $S \in \mathbb{N}$ if the following conditions are satisfied:

(a) $\int_{\mathbb{R}} \psi(u)u^l du = 0 \quad \forall l = 0, 1, \ldots, S-1, \quad \int_{\mathbb{R}} \phi(x)dx = 1$, and, for all $v \in \mathbb{R}$,

$$\int_{\mathbb{R}} K(v, v+u)du = 1, \qquad \int_{\mathbb{R}} K(v, v+u)u^l du = 0 \quad \forall l = 1, \ldots, S-1,$$

(b) $\sum_{k\in\mathbb{Z}} |\phi(\cdot - k)| \in L^\infty(\mathbb{R}), \quad \sum_{k\in\mathbb{Z}} |\psi(\cdot - k)| \in L^\infty(\mathbb{R})$, and
(c) For $\kappa(x,y)$ equal to either $K(x,y)$ or $\sum_k \psi(x-k)\psi(y-k)$,

$$\sup_{v\in\mathbb{R}} |\kappa(v, v-u)| \le c_1 \Phi(c_2|u|), \quad \text{for some } 0 < c_1, c_2 < \infty \text{ and every } u \in \mathbb{R},$$

for some bounded integrable function $\Phi : [0,\infty) \to \mathbb{R}$ such that $\int_{\mathbb{R}} |u|^S \Phi(|u|)du < \infty$.

When using such an S-regular wavelet basis, we see from Proposition 4.1.3 and the discussion after Definition 4.2.1 that the associated wavelet series

$$f = \sum_{k\in\mathbb{Z}} \langle \phi_k, f \rangle \phi_k + \sum_{l=0}^{\infty} \sum_{k\in\mathbb{Z}} \langle \psi_{lk}, f \rangle \psi_{lk} \tag{4.71}$$

converges not only in L^2 but also in fact in L^p or $C_u(\mathbb{R})$ whenever $f \in L^p(\mathbb{R}), 1 \le p < \infty$, or $f \in C_u(\mathbb{R})$, respectively.

Exercises

4.2.1 Prove Proposition 4.2.5(b). *Hint*: Show that for any u, v, we have

$$\Phi(|v|)\Phi(|v-u|) \le \Phi(|u|/2)\max(\Phi(|v|), \Phi(|v-u|))$$

because either $|v| \ge |u|/2$ or $|v-u| \ge |u|/2$; hence

$$K(v, v-u)| \le 2 \left\| \sum_k \Phi(\cdot - k) \right\|_\infty \Phi(|u|/2).$$

4.2.2 Suppose that $\sum_k |\hat{\phi}(\cdot + 2\pi k)|^2 = 1$ almost everywhere and that $\sum_k |\phi(\cdot - k)| \in L^\infty$. Prove that

$$\sum_k |\hat{\phi}(u + 2\pi k)|^2 = 1 \quad \forall u \in \mathbb{R}.$$

Hint: Define

$$g_u(x) = \sum_n \phi(x+n)e^{-iu(x+n)},$$

and use the Poisson summation formula as well as Parseval's identity to show that $\sum_k |\hat{\phi}(u + 2k\pi)|^2 = \int_0^1 |g_u(x)|^2 dx$ for every $u \in \mathbb{R}$.

4.2.3 Prove that if p_1, p_2 are two polynomials of degree n_1, n_2, respectively, with no common zeros, then there exist unique polynomials q_1, q_2 of degree $n_2 - 1, n_1 - 1$, respectively, such that $p_1q_1 + p_2q_2 = 1$ (this is known as *Bezout's theorem*). Deduce that there exist unique polynomials q_1, q_2 of degree less than or equal to $N - 1$ such that

$$(1-y)^N q_1(y) + y^N q_2(y) = 1,$$

and use this to prove Lemma 4.2.12.

4.2.4 (Riesz' lemma.) Let

$$A(u) = \sum_{m=0}^{M} a_m \cos(mu), \quad a_m \in \mathbb{R},$$

be a positive trigonometric polynomial. Prove that there exists a trigonometric polynomial

$$B(u) = \sum_{m=0}^{M} b_m e^{imu}, \quad b_m \in \mathbb{R}, \quad s.t. \ |B(u)|^2 = A(u).$$

4.2.5 If a multiresolution basis is S-regular for some $S > 0$, prove that, for all $1 \le p < \infty$, there exist $c_p > 0$, $d_p < \infty$ such that, for all x,

$$c_p \le \int |K(x,y)|^p dx \le d_p.$$

Hint: For the upper bound, use the domination of K by a majorising kernel Φ. The lower bound is clear for $p = 1$ because $\int K(v, v+u)du = 1$.

4.2.6 Let V_j be a multiresolution analysis of $L^2(\mathbb{R})$ based on $\phi, \psi \in C^S(\mathbb{R})$. Prove that, for every $\alpha < S, 1 \le p \le \infty$ and $f \in V_j$, we have the following *Bernstein-type inequality*:

$$\|D^\alpha f\|_p \le C 2^{|\alpha| j} \|f\|_p,$$

for some finite constant $C > 0$. *Hint*: Reduce to $j = 0$, and argue as in the proof of Proposition 4.2.8.

4.3 Besov Spaces

There are several ways to measure regularity properties of functions. A common and meaningful one is in terms of quantitative bounds on the L^p-norms of the derivatives of a function or, more generally, on the L^p-size of its local oscillations. This paradigm encompasses classical Hölder, Lipschitz and Sobolev smoothness conditions but also relates well to the notion of p-variation of a function. The fact that derivatives and moduli of continuity are defined in terms of the translation operator gives rise to powerful 'harmonic analysis' characterisations of the function spaces defined through such regularity properties. In this section we construct a unifying scale of function spaces – the *Besov spaces* – that allows us to measure the regularity of functions in a general and flexible way. The classical smoothness function spaces will be seen to be contained in these spaces as special cases. We shall use Besov spaces for the construction of high- and infinite-dimensional statistical models in subsequent chapters.

4.3.1 Definitions and Characterisations

Besov spaces can be defined in various ways, and we give several of these definitions and then establish their equivalence. We first develop the theory for Besov spaces defined on the real line and turn to Besov spaces over different domains of definition in Sections 4.3.4 to 4.3.6.

When studying embeddings between function spaces, the following remark will be used repeatedly:

Remark 4.3.1 An imbedding $X \subseteq Y$ of a normed space into another is continuous if the identity map $id : X \to Y$ is continuous, in this case, by linearity, $\|x\|_Y \le C\|x\|_X$ for some universal constant $0 < C < \infty$. Thus, if a normed space X is continuously imbedded into another normed space Y, and vice versa, then $X = Y$, and the norms of X, Y are equivalent. If X, Y is any pair of function spaces in which norm convergence implies convergence almost everywhere along a subsequence, and if $X \subseteq Y$, then the identity map $id : X \to Y$ has a closed graph and is thus automatically continuous by the closed-graph theorem from functional analysis. In particular, establishing the set inclusions $X \subset Y, Y \subset X$ and thus $X = Y$ for such spaces automatically implies equivalence of the norms $\|\cdot\|_X, \|\cdot\|_Y$.

Definition by Moduli of Smoothness

For a function f defined on a subinterval A of $\mathbb{R}$, possibly $A = \mathbb{R}$, we let $\tau_h, h \in \mathbb{R}$, denote the translation operator; that is, $\tau_h(f)(x) = f(x+h)$ whenever $x+h$ is in the domain of f. The difference operator then equals $\Delta_h = \tau_h - id$, so $\Delta_h(f) = f(\cdot + h) - f$. Inductively, we define, for $r \in \mathbb{N}$, $\Delta_h^r \equiv \Delta_h[\Delta_h^{r-1}] = (\tau_h - id)^r$, which by the binomial theorem equals

$$\Delta_h^r(f)(x) = \sum_{k=0}^{r} \binom{r}{k} (-1)^{r-k} f(x+kh). \tag{4.72}$$

If $A = \mathbb{R}$, then this operator is defined everywhere, and if $A = [a,b], h > 0$, its domain of definition is $A_{rh} = [a, b - rh]$, in which case we set, by convention, $\Delta_h^r(f)(x) \equiv 0$, for $x \in A \setminus A_{rh}$.

For $f \in L^p(A), 1 \le p \le \infty$, we define the rth modulus of smoothness

$$\omega_r(f,t) \equiv \omega_r(f,t,p) = \sup_{0<h\le t} \|\Delta_h^r(f)\|_p, \quad t > 0, \tag{4.73}$$

occasionally not reflecting the dependence on p when no confusion may arise.

If the weak derivative Df exists and is locally integrable, then (cf. Exercise 4.1.1)

$$\Delta_h^1(f)(x) = \int_x^{x+h} Df(u)du = \int_{\mathbb{R}} Df(u) 1_{[0,1]}\left(\frac{u-x}{h}\right) du,$$

and by a simple induction argument on r we show more generally that

$$\Delta_h^r(f)(x) = h^r \int_{\mathbb{R}} D^r f(u) N_{r,h}(u-x)du, \quad x \in A_{rh}, \tag{4.74}$$

whenever $D^r f$ exists and is locally integrable, where $N_{r,h}(x) = h^{-1}N_r(\cdot/h)$, and N_r is the $r-1$-fold convolution of $1_{[0,1]} = N_1$ with itself. Note that as $h \to 0$, we have approximately

$$h^{-r}\Delta_h^r(f)(x) \sim D^r f(x)$$

since $\{N_{r,h}\}_{h \searrow 0}$ is an approximate identity (cf. Proposition 4.1.1) – so higher differences $\Delta_h^r(f)$ encode precise quantitive information about higher derivatives of a function f. Moreover, we see, by Minkowski's inequality for integrals and since $\|N_{r,h}\|_1 = 1$, that

$$\omega_r(f,t,p) \le t^r \|D^r f\|_p \tag{4.75}$$

and, similarly,

$$\omega_{r+r'}(f,t,p) \le t^{r}\omega_{r'}(D^{r}f,t,p). \tag{4.76}$$

Let now $s>0$ be given, and let $r>s$ be an integer. For $1\le q\le\infty, 1\le p\le\infty$, we define the Besov space

$$B^s_{pq} \equiv B^s_{pq}(A) = \begin{cases} \{f\in L^p(A): \|f\|_{B^s_{pq}} \equiv \|f\|_p + |f|_{B^s_{pq}} < \infty\}, & 1\le p<\infty, \\ \{f\in C_u(A): \|f\|_{B^s_{pq}} \equiv \|f\|_p + |f|_{B^s_{pq}} < \infty\}, & p=\infty, \end{cases} \tag{4.77}$$

where

$$|f|_{B^s_{pq}} \equiv |f|_{B^s_{pq}(A)} = \begin{cases} \left(\int_0^\infty \left[\frac{\omega_r(f,t,p)}{t^s}\right]^q \frac{dt}{t}\right)^{1/q}, & 1\le q<\infty, \\ \sup_{t>0}\frac{\omega_r(f,t,p)}{t^s}, & q=\infty, \end{cases} \tag{4.78}$$

is the Besov seminorm. One shows (Exercise 4.3.2) that an equivalent norm on B^s_{pq} is obtained if the integral/supremum over t in the preceding display is restricted to $(0,1)$ and also if ω_r is replaced by ω_k for any $k>r$. In fact, Theorem 4.3.2 will imply that this definition is independent of the choice of r as long as $r>s$.

The integral in the preceding norm can be 'discretised' in a natural way. Since $\omega_r(f,t,p)$ is monotone increasing in t, we show, using also the hint for Exercise 4.3.2, that

$$2^{-r}\frac{\omega_r(f,2^{-k})}{2^{-ks}} \le \frac{\omega_r(f,t)}{t^s} \le 2^s\frac{\omega_r(f,2^{-k})}{2^{-ks}}, \quad \forall t\in[2^{-k-1},2^{-k}],$$

so, up to constants independent of f,k,

$$\left(\int_{2^{-k-1}}^{2^{-k}} \left[\frac{\omega_r(f,t)}{t^s}\right]^q \frac{dt}{t}\right)^{1/q} \sim \frac{\omega_r(f,2^{-k})}{2^{-sk}}, \quad k\in\mathbb{Z}.$$

We thus obtain the equivalent norm

$$\|f\|_{B^s_{pq}} \sim \begin{cases} \|f\|_p + \left(\sum_{k=0}^\infty [2^{ks}\omega_r(f,2^{-k},p)]^q\right)^{1/q}, & 1\le q<\infty, \\ \|f\|_\infty + \sup_{k\ge 0}(2^{ks}\omega_r(f,2^{-k},p)), & q=\infty, \end{cases} \tag{4.79}$$

on B^s_{pq}. An immediate conclusion is that $B^s_{pq}\subset B^s_{pq'}$ whenever $0<q\le q'\le\infty$.

Other equivalent norms exist, in a similar flavour of measuring the L^p-size of certain moduli of smoothness. In particular, we can restrict to second differences only if we consider the last existing derivative of f (we refer to the notes at the end of this chapter for some references).

Definition by Low-Frequency Approximations

In this section we give an alternative definition of Besov spaces on the real line that will be seen to be equivalent to the one given in the preceding subsection and which realises Besov spaces as consisting of functions that have a prescribed rate of approximation by band-limited functions (the continuous analogue of Fourier series). For $t>0, 1\le p\le\infty$, we introduce the spaces

$$\mathcal{V}^p_t = \{f\in L^p(\mathbb{R}), supp(\hat f)\in\{u:|u|\le t\}\}$$

of functions $f \in L^p$ whose Fourier transform $\hat{f}$ (in the distributional sense of (4.13) in case $p > 2$) is band limited by t. We may approximate arbitrary $f \in L^p$ by elements of $\mathcal{V}_t^p$ with approximation errors

$$\sigma_p(t, f) = \inf_{g \in \mathcal{V}_t^p} \|f - g\|_p, \tag{4.80}$$

and define Besov spaces, for $1 \le p \le \infty, 1 \le q \le \infty, s > 0$, as

$$B_{pq}^{s,V} \equiv \begin{cases} \{f \in L^p(\mathbb{R}) : \|f\|_{B_{pq}^{s,V}} \equiv \|f\|_p + |f|_{B_{pq}^{s,V}} < \infty\}, & 1 \le p < \infty, \\ \{f \in C_u(\mathbb{R}) : \|f\|_{B_{pq}^{s,V}} \equiv \|f\|_p + |f|_{B_{pq}^{s,V}} < \infty\}, & p = \infty, \end{cases} \tag{4.81}$$

where

$$|f|_{B_{pq}^{s,V}} \equiv \begin{cases} \left(\int_1^\infty \left[t^s \sigma_p(t, f)\right]^q \frac{dt}{t}\right)^{1/q}, & 1 \le q < \infty, \\ \sup_{t \ge 1} t^s \sigma_p(t, f), & q = \infty, \end{cases} \tag{4.82}$$

is the relevant Besov seminorm. Extending the integral/supremum over all $t > 0$ also gives an equivalent norm (since $f \in L^p$). Also, as in the preceding section, the integral can be discretised by estimating $\sigma_p(t, f)$ from above and below on intervals $[2^j, 2^{j+1}]$ to give an equivalent seminorm

$$|f|_{B_{pq}^{s,V}} \sim \begin{cases} \left(\sum_{j=1}^\infty [2^{js}\sigma_p(2^j, f)]^q\right)^{1/q}, & 1 \le q < \infty, \\ \sup_{j \ge 1} 2^{js}\sigma_p(2^j, f), & q = \infty. \end{cases} \tag{4.83}$$

Definition by Littlewood-Paley Theory

The ideas from Section 4.1.4 can be used to give another definition of Besov spaces which foreshadows the wavelet definition from the next subsection and which will be independently useful in what follows. We have seen in (4.27) that any $f \in L^p(\mathbb{R})$ for $p < \infty$ and every $f \in C_u(\mathbb{R})$ for $p = \infty$ can be written as

$$f = f * \phi + \sum_{j=0}^{\infty} f * \psi_{2^{-j}}, \quad \text{in } L^p, \quad 1 \le p \le \infty, \tag{4.84}$$

where $\phi, \psi \in L^1$ are smooth functions of compactly supported Fourier transform, $\psi_{2^{-j}} = 2^j \psi(2^j \cdot)$. By (4.3), we have $\|f * \phi\|_p \le \|\phi\|_1 \|f\|_p < \infty$, and Besov spaces can be defined by requiring sufficient geometric decay of the ℓ^q-sequence space norms of $\{\|f * \psi_{2^{-j}}\|_p\}_{j \ge 0}$. Formally for $1 \le p \le \infty, 1 \le q \le \infty, s > 0$, we set

$$B_{pq}^{s,LP} \equiv \begin{cases} \{f \in L^p(\mathbb{R}) : \|f\|_{B_{pq}^{s,LP}} < \infty\}, & 1 \le p < \infty, \\ \{f \in C_u(\mathbb{R}) : \|f\|_{B_{pq}^{s,LP}} < \infty\}, & p = \infty, \end{cases} \tag{4.85}$$

where the Littlewood-Paley norm is given, for $s \in \mathbb{R}$, by

$$\|f\|_{B_{pq}^{s,LP}} \equiv \begin{cases} \|f * \phi\|_p + \left(\sum_{j=0}^\infty 2^{jsq} \|f * \psi_{2^{-j}}\|_p^q\right)^{1/q}, & 1 \le q < \infty, \\ \|f * \phi\|_p + \sup_{j \ge 0} 2^{js} \|f * \psi_{2^{-j}}\|_p, & q = \infty. \end{cases} \tag{4.86}$$

Theorem 4.3.2 will imply that this definition is independent of the choice of ϕ, ψ as long as ϕ, ψ satisfy the requirements in and after (4.24).

Definition by Wavelet Coefficients

We will finally give the for our purposes perhaps most useful definition of Besov spaces in terms of wavelet expansions. We shall use a wavelet basis of regularity $S > s, S \in \mathbb{N}$, in the sense of Definition 4.2.14, satisfying in addition that $\phi, \psi \in C^S(\mathbb{R})$ with $D^S\phi$, $D^S\psi$ dominated by some integrable function. For instance, we can take band-limited or sufficiently regular Daubechies wavelets. Starting from the wavelet series

$$f = \sum_{k\in\mathbb{Z}} \langle \phi_k, f\rangle \phi_k + \sum_{l=0}^{\infty}\sum_{k\in\mathbb{Z}} \langle \psi_{lk}, f\rangle \psi_{lk}, \quad \text{in } L^p, \quad 1 \le p \le \infty, \tag{4.87}$$

of $f \in L^p(\mathbb{R})$ $(p < \infty)$ and of $f \in C_u(\mathbb{R})$ $(p = \infty)$, the idea is to use the decay, as $l \to \infty$, of the L^p norms

$$\left\| \sum_k \langle f, \psi_{lk}\rangle \psi_{lk} \right\|_p \simeq 2^{l(\frac{1}{2}-\frac{1}{p})} \|\langle f, \psi_{l\cdot}\rangle\|_p$$

to describe the regularity of a function f (recalling Proposition 4.2.8 for the preceding display). Formally for $1 \le p \le \infty, 1 \le q \le \infty, 0 < s < S$, we set

$$B^{s,W}_{pq} \equiv \begin{cases} \{f \in L^p(\mathbb{R}) : \|f\|_{B^{s,W}_{pq}} < \infty\}, & 1 \le p < \infty \\ \{f \in C_u(\mathbb{R}) : \|f\|_{B^{s,W}_{pq}} < \infty\}, & p = \infty \end{cases} \tag{4.88}$$

with wavelet-sequence norm given, for $s \in \mathbb{R}$, by

$$\|f\|_{B^{s,W}_{pq}} \equiv \begin{cases} \|\langle f, \phi_{\cdot}\rangle\|_p + \left(\sum_{l=0}^{\infty} 2^{ql(s+\frac{1}{2}-\frac{1}{p})} \|\langle f, \psi_{l\cdot}\rangle\|_p^q\right)^{1/q}, & 1 \le q < \infty, \\ \|\langle f, \phi_{\cdot}\rangle\|_p + \sup_{l\ge 0} 2^{l(s+\frac{1}{2}-\frac{1}{p})} \|\langle f, \psi_{l\cdot}\rangle\|_p, & q = \infty. \end{cases} \tag{4.89}$$

Theorem 4.3.2 implies that the definition is independent of the wavelets $\phi, \psi \in C^S$ used as long as they are S-regular.

Equivalence of All Definitions

We now prove that, on the real line, all the preceding definitions of Besov spaces coincide and that the respective norms are equivalent. Since $B^{s,V}_{pq}$ does not depend on any free parameters, the result implies, in particular, that the preceding definitions are independent of the particular choice of ϕ, ψ, r. Equivalence results for domains different from A (including in particular $A = [0,1]$) will be discussed in Sections 4.3.4 and 4.3.5.

Theorem 4.3.2 *Let $1 \le p \le \infty, 1 \le q \le \infty, s > 0$. Then we have*

$$B^s_{pq}(\mathbb{R}) = B^{s,V}_{pq} = B^{s,LP}_{pq} = B^{s,W}_{pq}. \tag{4.90}$$

Moreover, the norms $\|\cdot\|_{B^s_{pq}}, \|\cdot\|_{B^{s,V}_{pq}}, \|\cdot\|_{B^{s,LP}_{pq}}, \|\cdot\|_{B^{s,W}_{pq}}$ are all pairwise equivalent.

Proof The spaces $B^s_{pq}(\mathbb{R}), B^{s,V}_{pq}, B^{s,LP}_{pq}, B^{s,W}_{pq}$ are all Banach spaces of real-valued functions defined on $\mathbb{R}$ in which norm convergence implies convergence almost everywhere along a subsequence. This is obvious for the first two norms as these imply L^p-convergence, which, in turn, implies convergence almost everywhere along a subsequence. It follows likewise for the third and fourth norms, noting that the respective $\|\cdot\|_{s,p,q}$-norms, $s > 0$, dominate the

$\|\cdot\|_p$-norm, using the L^p-identities (4.84), (4.87) (and also Proposition 4.2.8 in the wavelet case). By Remark 4.3.1, it thus remains to prove the set-theoretic identities (4.90) to prove the theorem. This will be organised into two separate steps.

We start with the following simple result relating low frequency approximations to Littlewood-Paley decompositions. Recall the functions ϕ, ψ from Section 4.1.4, where in the following proposition we write $\phi_{2^{-j}} = 2^j\phi(2^j\cdot)$ in slight abuse of (wavelet) notation.

Proposition 4.3.3 *For $f \in L^p$ and s,p,q as in Theorem 4.3.2, the following are equivalent:*

(a) $f \in B^{s,V}_{pq}$,
(b) $\|\phi_{2^{-j}} * f - f\|_p = c_j 2^{-js}, \ \forall j \geq 0$, *for some nonnegative sequence* $\{c_j\} \in \ell_q$, *and*
(c) $\|\psi_{2^{-j}} * f\|_p = c_j 2^{-js}, \ \forall j \geq 0$, *for some nonnegative sequence* $\{c_j\} \in \ell_q$.

Proof (a) $\iff$ (b): By definition, $\phi_{2^{-j}}$ has a Fourier transform supported in $[-2^j, 2^j]$, and thus, $\phi_{2^{-j}} * f \in \mathcal{V}^p_{2^j}$ by (4.8), (4.3), so

$$\sigma_p(2^j, f) \leq \|\phi_{2^{-j}} * f - f\|_p,$$

and (4.83) gives one direction. Conversely, for $h \in \mathcal{V}^p_{2^j}$ we have $\phi_{2^{-j+1}} * h = h$ since $\mathcal{F}[\phi_{2^{-j+1}}] = 1$ on the support $[-2^j, 2^j]$ of $\hat{h}$. We can then write $\phi_{2^{-j+1}} * f - f = \phi_{2^{-j+1}} * (f-h) + h - f$, so

$$\|\phi_{2^{-j+1}} * f - f\|_p \leq \|\phi_{2^{-j+1}} * (f-h)\|_p + \|f-h\|_p \leq (1 + \|\phi\|_1)\|f-h\|_p$$

by (4.3). Since h was arbitrary, we can take the infimum over $\mathcal{V}^p_{2^j}$ to see that, for every j,

$$\frac{1}{1+\|\phi\|_1}\|\phi_{2^{-j+1}} * f - f\|_p \leq \sigma_p(2^j, f) \equiv 2^{-js} c_j,$$

where $\{c_j\} \in \ell_q$ in view of (4.83).
(b) $\iff$ (c): We have from the definitions that

$$\|\psi_{2^{-j}} * f\|_p = \|(\phi_{2^{-j+1}} - \phi_{2^{-j}}) * f\|_p \leq \|\phi_{2^{-j}} * f - f\|_p + \|\phi_{2^{-j+1}} * f - f\|_p,$$

and, conversely,

$$\|\phi_{2^{-j}} * f - f\|_p \leq \sum_{l \geq j} \|\psi_{2^{-l}} * f\|_p \leq \sum_{l \geq j} c_l 2^{-ls} \leq 2^{-js} \sum_{l \geq j} c_l 2^{-|l-j|s} = c'_j 2^{-js}$$

using Exercise 4.3.1. ∎

To complete the proof of Theorem 4.3.2, we first need the following lemma on band-limited functions $f \in \mathcal{V}^p_t$.

Lemma 4.3.4 (Bernstein) *Let $f \in \mathcal{V}^p_t$. Then, for some fixed constant $0 < C < \infty$ that depends only on n,*

(a) $D^n f \in \mathcal{V}^p_t$, *for every* $n \in \mathbb{N}$,
(b) $\|D^n f\|_p \leq C t^n \|f\|_p$, *and*
(c) $\|D^n f\|_p \leq C t^{n-k} \|D^k f\|_p$.

Proof By (4.9),

$$\mathcal{F}[D^n f](u) = \hat{f}(u)(iu)^n,$$

which implies the first claim (a) modulo showing $D^n f \in L^p$ (part (b)). Take $\Phi \in \mathcal{S}(\mathbb{R})$ such that $0 \le \hat{\Phi} \le 1$, $\hat{\Phi}(u) = 1$, for $|u| \le 1$, supported in $\{u : |u| \le 2\}$ and such that $\|D^n \Phi\|_1 < \infty$. We then have $\hat{f}(u) = \hat{f}(u)\hat{\Phi}(u/t)$ or, in other words, $f = f * \Phi_{1/t}$, where we recall the notation $\Phi_{1/t} = t\Phi(t\,\cdot\,)$. Now $D^n(\Phi_{1/t}) = t^n(D^n\Phi)_{1/t}$ and, interchanging differentiation and integration,

$$D^n f = t^n f * ((D^n \Phi)_{1/t}),$$

so (b) follows with $C = \|D^n\Phi\|_1$ from (4.3). Part (c) follows likewise, writing $D^n f = D^k f * D^{n-k}(\Phi_{1/t})$ and proceeding as earlier. ■

The key result is now the following, in which the Littlewood-Paley approach to the definition of Besov spaces turns out to be very helpful:

Proposition 4.3.5 *For $f \in L^p$ and s, p, q as in Theorem 4.3.2, the following are equivalent:*

(a) $\|\psi_{2^{-j}} * f\|_p = c_j 2^{-js}$, *for some nonnegative sequence* $\{c_j\} \in \ell_q$,
(b) There exists $N \in \mathbb{N}, N > s$, *functions* $\{u_l\} \in L^p$ *and a nonnegative sequence* $\{c_l\} \in \ell_q$ *such that*

$$f = \sum_{l=0}^{\infty} u_l \quad \text{in } L^p, \quad \|u_l\|_p \le c_l 2^{-ls}, \quad \|D^N u_l\|_p \le c_l 2^{l(N-s)},$$

(c) $f \in B^s_{pq}(\mathbb{R})$, *and*
(d) $f = \sum_{k \in \mathbb{Z}} \langle \phi_k, f \rangle \phi_k + \sum_{l=0}^{\infty} \sum_{k \in \mathbb{Z}} \langle \psi_{lk}, f \rangle \psi_{lk}$ *in* L^p, *and for some nonnegative* $\{c_l\} \in \ell_q$,

$$\|\langle \phi_\cdot, f \rangle\|_p < \infty, \|\langle \psi_{l\cdot}, f \rangle\|_p \le c_l 2^{-l(s+1/2-1/p)}.$$

Proof (a) ⇒ (b): letting $u_0 = \phi * f, u_l = \psi_{2^{-l}} * f$, we see that $f = \sum_l u_l$ from (4.27). Clearly,

$$\|u_0\|_p \le \|\phi\|_1 \|f\|_p, \quad \|u_l\|_p = \|\psi_{2^{-l}} * f\|_p = c_l 2^{-ls}.$$

Moreover, $u_l \in \mathcal{V}^p_{2^{l+1}}$ by definition of ϕ, ψ, so Lemma 4.3.4 gives, for any $N \in \mathbb{N}$,

$$\|D^N u_l\|_p \le \|\phi\|_1 2^{(l+1)N} \|u_l\|_p,$$

which, combined with the preceding display, gives the last bound in (b).

(b) $\Rightarrow$ (c): Using (4.3) and (4.74) and setting $r = N$, we can write, for any j,

$$\|\Delta_h^r(f)\|_p = \left\|\sum_{l=0}^{\infty}\Delta_h^r(u_l)\right\|_p \leq \sum_{l=0}^{\infty}\|\Delta_h^r(u_l)\|_p \leq \sum_{l=0}^{\infty}\min(2^r\|u_l\|_p, |h|^r\|D^r u_l\|_p)$$

$$\leq \sum_{l=0}^{\infty}\min(2^r c_l 2^{-ls}, |h|^r c_l 2^{l(r-s)}) \leq |h|^r \sum_{l=0}^{j} c_l 2^{l(r-s)} + 2^r \sum_{l>j} c_l 2^{-ls}$$

$$= |2^j h|^r 2^{-js}\sum_{l=0}^{j} c_l 2^{-|j-l|(r-s)} + 2^r 2^{-js}\sum_{l>j} c_l 2^{-|j-l|s},$$

so we can conclude from Exercise 4.3.1 that

$$\omega_r(f, 2^{-j}, p) \leq c_j' 2^{-js},$$

for some $\{c_j'\} \in \ell_q$. The result now follows from the equivalence (4.79).
(c) $\Rightarrow$ (a): Since the support of $\hat{\psi}$ is contained in $\{1/2 \leq |u| \leq 2\}$, we see that

$$\widehat{H}_r(u) \equiv \frac{\hat{\psi}(u)}{(e^{iu} - 1)^r}$$

is contained in $\mathcal{S}(\mathbb{R})$. Note, moreover, that $\widehat{\Delta_h^r(f)} = (e^{ih\cdot} - 1)^r \hat{f}$, so

$$\mathcal{F}[\psi_{2^{-j}} * f](u) = \hat{\psi}(u/2^j)\hat{f}(u) = \frac{\hat{\psi}(u/2^j)}{(e^{iu/2^j} - 1)^r}(e^{iu/2^j} - 1)^r \hat{f}(u) = \mathcal{F}\left[(H_r)_{2^{-j}} * \Delta_{2^{-j}}^r(f)\right](u).$$

Thus, we see, using (4.3), that

$$\|\psi_{2^{-j}} * f\|_p \leq \|H_r\|_1 \omega_r(f, 2^{-j}, p),$$

which completes the proof using the equivalence (4.79).
(d) $\Rightarrow$ (b): Setting

$$u_0 = \sum_{k\in\mathbb{Z}}\langle\phi_k, f\rangle\phi_k, \quad u_l = \sum_{k\in\mathbb{Z}}\langle\psi_{lk}, f\rangle\psi_{lk},$$

the identity $f = \sum_l u_l$ in L^p follows directly from (4.87). Moreover, for $l > 0$, by Proposition 4.2.8,

$$\|u_l\|_p \leq 2^{-ls} 2^{l(s+\frac{1}{2}-\frac{1}{p})}\|\langle\psi_{l\cdot}, f\rangle\|_p \equiv 2^{-ls} c_l,$$

where $\{c_l\} \in \ell_q$ in view of (4.89), and a similar estimate holds for u_0. Setting $N = S$, we know by hypothesis that $D^N\psi$ is bounded, continuous and dominated by some integrable function. Thus, interchanging differentiation and summation, we have, from the chain rule,

$$D^N u_l = \sum_{k\in\mathbb{Z}}\langle\psi_{lk}, f\rangle D^N(\psi_{lk}) = 2^{lN}\sum_{k\in\mathbb{Z}}\langle\psi_{lk}, f\rangle 2^{l/2}(D^N\psi)(2^l(\cdot) - k),$$

so Proposition 4.2.8 applied with $\phi = D^N\psi$ (possible in view of Proposition 4.2.5, using the dominating function for $D^N\psi$) allows us to proceed as in the preceding estimate to establish $\|D^N u_l\|_p \leq c_l 2^{l(N-s)}$.
(b) $\Rightarrow$ (d): By what has already been proved, we can use without loss of generality the Littlewood-Paley decomposition $f = \sum_l u_l$, where $u_0 = \phi * f, u_l = \psi_{2^{-l}} * f$ (with slight

abuse of notation for ϕ), and we let $K_j(f)$ be the wavelet projection of f onto V_j at resolution level j. Then, using the majorising kernel Φ for K_j (Definition 4.2.14), we see that

$$K_j(f) - f = \sum_l (K_j(u_l) - u_l)$$

and

$$\|K_j(f) - f\|_p \leq \sum_l \|K_j(u_l) - u_l\|_p \leq \sum_{l \leq j} \|K_j(u_l) - u_l\|_p + \sum_{l > j} (\|\Phi\|_1 + 1)\|u_l\|_p.$$

We know that $\|u_l\|_p \leq 2^{-ls} c_l$ for some $\{c_l\} \in \ell_q$, so the tail of the second term in the preceding expression is of order $2^{-js} c_j$ by Exercise 4.3.1. Moreover, the support of $\hat{u}_l$ is in $[-2^{l+1}, 2^{l+1}]$, so by Lemma 4.3.4, we see, for $N = S$, that

$$\|D^N u_l\|_p \leq C 2^{Nl} \|u_l\|_p \leq c_l' 2^{(N-s)l}, \quad \{c_l'\} \in \ell_q.$$

We thus have, by Proposition 4.1.5 with $h = 2^{-j}$, that

$$\|K_j(u_l) - u_l\|_p \leq c 2^{-jN} \|D^N u_l\|_p \leq c_l'' 2^{(N-s)l} 2^{-jN}, \quad \{c_l''\} \in \ell_q,$$

and the sum over $l \leq j$ of these terms is bounded by $c_j''' 2^{-js}$ for some $\{c_j'''\} \in \ell_q$ using Exercise 4.3.1. Summarising, for every $j \geq 0$,

$$\|K_j(f) - f\|_p \leq d_j 2^{-js}, \quad \{d_j\} \in \ell_q.$$

This implies, for every $j \geq 0$, that

$$\left\| \sum_k \langle \psi_{jk}, f \rangle \psi_{jk} \right\|_p = \|K_{j+1}(f) - K_j(f)\|_p \leq \|K_{j+1}(f) - f\|_p + \|K_j(f) - f\|_p \leq 2 d_j 2^{-js}.$$

To translate this into an estimate on the wavelet coefficients themselves, we use Proposition 4.2.8 to see that

$$\|\langle \psi_{j\cdot}, f \rangle\|_p \leq C(\psi) 2^{j(1/p - 1/2)} \left\| \sum_k \langle \psi_{jk}, f \rangle \psi_{jk} \right\|_p \leq d_j' 2^{j(1/p - 1/2 - s)}, \quad \{d_j'\} \in \ell_q,$$

which proves the implication (b) $\Rightarrow$ (d) (noting that the estimate for $\|\langle \phi_k, f \rangle\|_p$ is the same). The proof of Proposition 4.3.5 is complete. ■

The two preceding propositions complete the proof of Theorem 4.3.2. ■

First Basic Properties of Besov Spaces

Proposition 4.3.6 *Let $s, s' > 0, p, p', q, q' \in [1, \infty]$. Then the following continuous imbeddings hold:*

(i) $B^s_{pq} \subset B^s_{pq'}$ *whenever $q \leq q'$,*
(ii) $B^s_{pq} \subset B^{s'}_{pq'}$ *whenever $s > s'$, and*
(iii) $B^s_{pq} \subset B^{s'}_{p'q}$ *whenever $p \leq p'$, $s' - (1/p') = s - (1/p)$.*

Proof These assertions are immediate for the wavelet definition of Besov spaces using the imbedding $\ell_r \subset \ell_{r'}$ for $r \leq r'$, so the result follows from Theorem 4.3.2. ■

Approximation Operators and Besov Spaces

The essence of the proof of Theorem 4.3.2 can be distilled into a general theorem that is quite independent of Littlewood-Paley or wavelet theory and that applies to general approximation operators satisfying certain conditions. By the preceding proofs, the Littlewood-Paley convolution operator $P_j(f) = \phi_{2^{-j}} * f$ or the wavelet projection operator $P_j(f) = K_j(f)$ is admissible in the following theorem, but other approximation schemes could be used.

Theorem 4.3.7 *Let* $1 \le p \le \infty$, *and let*

$$\{P_j : L^p \to L^p\}_{j=0}^{\infty}$$

be a family of operators on L^p *such that, for some fixed constant* $0 < C < \infty$, *some* $N \in \mathbb{N}$,

(i) $\sup_j \sup_{f:\|f\|_p \le 1} \|P_j(f)\|_p \le C$,
(ii) $\|P_j(f) - f\|_p \le C2^{-jN}\|D^N f\|_p \quad \forall j \ge 0, \quad f \in H_p^N$, *and*
(iii) *For* $Q_j = P_{j+1} - P_j$ *and every* $f \in L^p, j \ge 0$, *we have*

$$\|D^N(Q_j(f))\|_p \le C2^{jN}\|Q_j(f)\|_p, \quad \|D^N(P_0(f))\|_p \le C\|f\|_p.$$

Then, for $f \in L^p$ *and* $0 < s < N, 1 \le q \le \infty$, *we have*

$$f \in B^s_{pq}(\mathbb{R}) \iff \|P_j(f) - f\|_p \le c'_j 2^{-js}; \quad \textit{for some sequence } \{c'_j\} \in \ell_q.$$

Proof We prove the result by showing the equivalence of the hypotheses on $P_j(f)$ with part (b) of Proposition 4.3.5. First, if $u_j = Q_j(f)$ for $j > 0$, $u_0 = P_0(f)$; from a telescoping series we have, as usual,

$$f = P_0(f) + \sum_{j=0}^{\infty} Q_j(f)$$

in L^p, under the hypothesis $\|P_j(f) - f\|_p \le c'_j 2^{-js} \to 0$ as $j \to \infty$. Moreover,

$$\|u_j\|_p = \|Q_j(f)\|_p \le \|P_j(f) - f\|_p + \|P_{j+1}(f) - f\|_p = c_j 2^{-js}, \quad \{c_j\} \in \ell_q,$$

and similarly, $\|u_0\|_p \le c_0$. Moreover, by assumption (iii),

$$\|D^N(Q_j(f))\|_p \le C2^{jN}\|Q_j(f)\|_p \le Cc_j 2^{j(N-s)},$$

with $\{c_j\} \in \ell_q$, so $f \in B^s_{pq}$ by Proposition 4.3.5. Conversely, if $f \in B^s_{pq}$, we can decompose it in L^p as $f = \sum_{k=0}^{\infty} u_k$ from Proposition 4.3.5, part (b). Then $P_j(f) - f = \sum_j (P_j u_k - u_k)$, so, using assumption (ii) and Exercise 4.3.1

$$
\begin{aligned}
\|P_j(f)-f\|_p &\le \sum_{k=0}^{\infty}\|P_j u_k - u_k\|_p \le \sum_{k=0}^{\infty}\min(2C\|u_k\|_p, C2^{-jN}\|D^N u_k\|_p)\\
&\le C\sum_{k=0}^{\infty}\min(2c_k 2^{-ks}, 2^{-jN}c_k 2^{k(N-s)})\\
&\le C\sum_{k=0}^{j}2^{-jN}c_k 2^{k(N-s)} + 2C\sum_{k=j+1}^{\infty}c_k 2^{-ks} \qquad (4.91)\\
&= C2^{-js}\sum_{k=0}^{j}2^{-|j-k|(N-s)}c_k + 2C2^{-js}\sum_{k=j+1}^{\infty}c_k 2^{-|k-j|s}\\
&= c_j' 2^{-js},
\end{aligned}
$$

completing the proof. ■

We can now refine Proposition 4.1.5, which was restricted to the spaces H_p^m and C^m, to cover approximation in the full scale of Besov spaces. We recall the integral operators

$$
f \mapsto K_h(f) = \int_{\mathbb{R}} K_h(\cdot,y)f(y)dy = \frac{1}{h}\int_{\mathbb{R}} K\left(\frac{\cdot}{h},\frac{y}{h}\right) f(y)dy, \quad h>0,
$$

which cover approximate identities arising from either convolution or projection kernels (using the conversion $h = 2^{-j}$ in the latter case).

The following result is almost trivial for wavelet projection kernels $K_h = K_{2^{-j}}$ in view of the wavelet definition of the Besov norm when the basis is N-regular *and* $\phi, \psi \in C^N$. While sufficient smoothness of ϕ, ψ is necessary to *characterise* Besov spaces, for approximation-theoretic results, it is not because N-regular wavelet bases may satisfy condition (B) in the following proposition without requiring $\phi \in C^N$ (cf. Definition (4.2.14)).

Proposition 4.3.8 *Let $N \in \mathbb{N}$, and let $K(\cdot,\cdot)$ be a kernel such that either*

(Ai) $\sup_{h>0}\sup_{f:\|f\|_p\le 1}\|K_h(f)\|_p \le C$ *and*
(Aii) $\|K_h(f)-f\|_p \le Ch^N\|D^N f\|_p \quad \forall h>0, f\in H_p^N$ *or*
(B) Condition 4.1.4 is satisfied for this N.

If $f \in B^s_{pq}$ for some $0<s<N, 1\le p,q\le\infty$, then, for some constant $0<C(K)<\infty$,

$$
\|K_h(f)-f\|_p \le C(K)\|f\|_{B^s_{pq}}h^s.
$$

Proof First, by Proposition 4.1.5, condition (B) implies condition (A), so it suffices to prove the result for the latter assumption. Moreover, since $B^s_{pq} \subset B^s_{p\infty}$, it suffices to prove $q=\infty$. By Proposition 4.3.5, we can decompose $f = \sum_k u_k$ with $2^{ks}\|u_k\|_p = 2^{ks}\|f * \psi_{2^{-k}}\|_p$ bounded by $c\|f\|_{B^s_{p\infty}}$ for some $c>0$ and every k. The result now follows from the estimate (4.91) with $c_k = c\|f\|_{B^s_{p\infty}}$. ■

4.3.2 Basic Theory of the Spaces B^s_{pq}

Besov Spaces, L^p-Spaces and Sobolev Imbeddings

Fix a wavelet basis of regularity $S > 0$. If $f \in L^p, 1 \le p < \infty$, then, by (4.87), the partial sums

$$\sum_{k\in\mathbb{Z}} \langle f, \phi_k \rangle \phi_k + \sum_{l=0}^{J-1} \sum_{k\in\mathbb{Z}} \langle f, \psi_{lk} \rangle \psi_{lk} \tag{4.92}$$

converge to f in L^p as $J \to \infty$, and by completeness of L^p, any function f that is the L^p-limit of such partial sums belongs to L^p. There exists a proper subspace of L^p of those $f \in L^p$ for which the partial sums in l converge absolutely, that is, for which

$$\sum_{l=0}^{\infty} \left\| \sum_k \langle f, \psi_{lk} \rangle \psi_{lk} \right\|_p < \infty.$$

Recalling Proposition 4.2.8, these functions are precisely those for which

$$\sum_{l=0}^{\infty} 2^{l(\frac{1}{2}-\frac{1}{p})} \| \langle f, \psi_{l\cdot} \rangle \|_p < \infty.$$

Formally, we define this set of functions, for $1 \le p < \infty$, to be

$$B^0_{p1} = \left\{ f \in L^p : \| \langle f, \phi_\cdot \rangle \|_p + \sum_{l=0}^{\infty} 2^{(1/2-1/p)} \| \langle f, \psi_{l\cdot} \rangle \|_p < \infty \right\}.$$

Denoting this space by B^0_{p1} suggests itself from definition of the norm (4.89) for general values of s. Likewise, for $p = \infty$, we define

$$B^0_{\infty 1} = \left\{ f \in C_u(\mathbb{R}) : \| \langle f, \phi_\cdot \rangle \|_\infty + \sum_{l=0}^{\infty} 2^{l/2} \| \langle f, \psi_{l\cdot} \rangle \|_\infty < \infty \right\}.$$

The continuous imbeddings

$$B^0_{p1}(\mathbb{R}) \subset L^p(\mathbb{R}), \quad 1 \le p < \infty, \quad B^0_{\infty 1}(\mathbb{R}) \subset C_u(\mathbb{R}) \tag{4.93}$$

now follow directly from

$$\|f\|_p = \left\| \sum_{k\in\mathbb{Z}} \langle f, \phi_k \rangle \phi_k + \sum_{l=0}^{\infty} \sum_k \langle \psi_{lk}, f \rangle \psi_{lk} \right\|_p \lesssim \|f\|_{B^0_{p1}}.$$

Moreover, when $p = q = 2$, then Parseval's identity (4.15) implies directly that

$$B^0_{22}(\mathbb{R}) = L^2(\mathbb{R}).$$

Combined with Proposition 4.3.6, we obtain the following 'Sobolev' imbedding proposition for Besov spaces:

Proposition 4.3.9 *Let $s \ge 0, p, p', q, \in [1, \infty]$. Then the following continuous imbeddings hold:*

(i) $B^s_{pq}(\mathbb{R}) \subset C_u(\mathbb{R})$ whenever $s > 1/p$ or $s = 1/p, q = 1$,

(ii) $B^s_{pq}(\mathbb{R}) \subset L^{p'}(\mathbb{R})$ *whenever* $p \le p'$ *and* $s > 1/p - 1/p'$ *or* $s = 1/p - 1/p'$, $q = 1$,

(iii) $B^s_{pq}(\mathbb{R}) \subset L^2(\mathbb{R})$ *whenever* $p \le 2$ *and* $s > 1/p - 1/2$ *or* $s = 1/p - 1/2$, $q \le 2$.

This result can be used to show that pointwise products of $f, g \in B^s_{pq}$ also belong to B^s_{pq} whenever the Sobolev imbedding into L^∞ holds (see Exercise 4.3.4).

The preceding motivates that, for $q > 1$, the spaces B^0_{pq} could be defined likewise, simply by requiring finiteness of the $\|\cdot\|_{B^0_{pq}}$-norm. For $p \neq 2$ or $q > 2$, however, a problem arises, as it is not a fortiori clear that convergence of the partial sums (4.92) for the norm $\|\cdot\|_{B^0_{pq}}$ implies that the full wavelet series defines an element of L^p. More precisely, for $q > 1$, the Besov spaces need to contain elements which are not functions in order to be Banach spaces. Indeed, considering the example of Dirac measure δ_0 acting on the wavelet coefficients by integration (i.e., evaluation at zero), we see that

$$\|\delta_0\|_{B^0_{1\infty}} = \sum_k |\phi(k)| + \sup_{l \ge 0} 2^{-l/2} 2^{l/2} \sum_k |\psi(k)| < \infty$$

by Definition 4.2.14, so any reasonable definition of $B^0_{1\infty}$ will be such that $\delta_0 \in B^0_{1\infty}$. In fact, by the same argument, if we let $M(\mathbb{R})$ be the space of finite signed measures, then

$$\|\mu\|_{B^0_{1\infty}} \le C|\mu|(\mathbb{R}) \quad \forall \mu \in M(\mathbb{R}), \quad M(\mathbb{R}) \subset B^0_{1\infty}(\mathbb{R}), \tag{4.94}$$

where $|\mu|$ is the total variation measure of μ, and $0 < C < \infty$ is some universal constant. Thus, for general $q > 1, s \le 0, 1 \le p \le \infty$, Besov spaces need to be interpreted as spaces of 'generalised functions', that is, as elements of the space $\mathcal{S}^*$ of *tempered distributions*.

Besov Spaces of Tempered Distributions

We shall now define Besov spaces for general $s \in \mathbb{R}$ as spaces of tempered distributions. For $1 \le p \le \infty, 1 \le q \le \infty, s \in \mathbb{R}$, we set

$$B^s_{pq} \equiv \{f \in \mathcal{S}^* : \|f\|_{B^s_{pq}} < \infty\}, \tag{4.95}$$

where the Besov norms

$$\|\cdot\|_{B^s_{pq}} \equiv \|\cdot\|_{B^{s,W}_{pq}}$$

are given as in (4.89) using the band-limited wavelet basis from Theorem 4.2.9. The inner products $\langle f, \phi_k \rangle, \langle f, \psi_{lk} \rangle$ are now interpreted as the action of f on $\phi_k, \psi_{lk} \in \mathcal{S}$ in the sense of tempered distributions. Alternatively, we require the Littlewood-Paley norms $\|\cdot\|_{B^{s,LP}_{pq}}, s \in \mathbb{R}$, from (4.86) to be finite, where the convolutions $f * (\cdot)(x)$ are defined as the action of f on $\phi(\cdot - x)$ and on $\psi_{2^{-l}}(\cdot - x)$. Since at each level l the wavelet and Littlewood-Paley norms only involve L^p-functions, we show as in the proof of Theorem 4.3.2 that the Littlewood-Paley and wavelet definitions coincide, with equivalent norms. Moreover, this definition reproduces the definition of the space $B^s_{pq}, s > 0$, and of B^0_{p1}, by identifying $f \in L^p$ with the tempered distribution $\phi \mapsto \int_{\mathbb{R}} \phi f$. To see this, note that finiteness of $\|f\|_{B^s_{pq}}$ implies that $f \in \mathcal{S}^*$ is the L^p/uniform limit of L^p/uniformly continuous functions and thus must itself lie in L^p/C_u. Moreover, we shall see in the next subsection that B^{-s}_{pq} for $s > 0$ can be interpreted as a dual space of a Besov space of index s. As the definition of the latter space is independent of the wavelet basis used, so is the definition of B^{-s}_{pq}.

It is now immediate that Proposition 4.3.6 holds without the restriction $s,s' > 0$.

Proposition 4.3.10 *Let $s,s' \in \mathbb{R}, p,p',q,q' \in [1,\infty]$. Then the following continuous imbeddings hold:*

(i) $B^s_{pq} \subset B^s_{pq'}$ whenever $q \leq q'$,
(ii) $B^s_{pq} \subset B^{s'}_{pq'}$ whenever $s > s'$, and
(iii) $B^s_{pq} \subset B^{s'}_{p'q}$ whenever $p \leq p'$, $s' - 1/p' = s - 1/p$.

Moreover, the spaces $B^0_{p\infty}$ always contain the L^p-spaces.

Proposition 4.3.11 *Let $1 \leq p \leq \infty$. We have the continuous imbedding $L^p \subset B^0_{p\infty}$.*

Proof The result follows immediately from the definition (4.89) of the Besov norm and, for $p < \infty$, from the second part of Proposition 4.2.8 combined with the estimate

$$\left\| \sum_k \langle f, \psi_{lk} \rangle \psi_{lk} \right\|_p \leq 2 \sup_l \|K_l(f)\| \leq C\|f\|_p. \quad \blacksquare$$

We note again that for $p = 2$, we have $L^2 = B^0_{22}$.

Duality of Besov Spaces

Besov spaces for $s < 0$, alternatively, can be defined as the dual spaces of the Besov spaces with $s > 0$. Let us first consider the easiest case, $p = q = 2$, so that B^s_{22} is a Hilbert space. Since $\mathcal{S} \subset B^s_{22}$, we can naturally view the topological dual space $(B^s_{22})^*$ as a subset of $\mathcal{S}^*$, and viewed in such a way, this alternative definition coincides with the one from (4.95).

Proposition 4.3.12 *Let $s > 0$. We have*

$$(B^s_{22})^* = B^{-s}_{22},$$

and the norms are equivalent:

$$\|f\|_{(B^s_{22})^*} \equiv \sup_{g:\|g\|_{B^s_{22}} \leq 1} \left| \int_{\mathbb{R}} g(x) f(x) \right| \simeq \|f\|_{B^{-s}_{22}}.$$

Proof We write throughout $\phi_k = \psi_{-1,k}$ to shorten notation. If $f \in L^2$ and ϕ is any element of $B^s_{22} \subset L^2$, then, by Parseval's identity and the Cauchy-Schwarz inequality,

$$\int_{\mathbb{R}} f\phi = \sum_{l,k} \langle f, \psi_{lk} \rangle \langle \phi, \psi_{lk} \rangle = \sum_l \sum_k 2^{-ls} \langle f, \psi_{lk} \rangle 2^{ls} \langle \phi, \psi_{lk} \rangle$$
$$\leq \|f\|_{B^{-s}_{22}} \|\phi\|_{B^s_{22}} = C\|\phi\|_{B^s_{22}},$$

for

$$C = \|f\|_{B^{-s}_{22}} \leq \|f\|_2.$$

Thus, any $f \in L^2$, acting by integration on B^s_{22}, belongs to $(B^s_{22})^*$. By standard approximation arguments (using that $\mathcal{S}$ is dense in B^s_{22} and that L^2 is dense in B^{-s}_{22}), we see that any tempered distribution $f \in B^{-s}_{22}$ defines an element of $(B^s_{22})^*$. Conversely, let $L \in (B^s_{22})^*$. In view of the wavelet definition of Besov spaces, we see that B^s_{22} is isometric to the Hilbert space $\ell_2(\mu_s)$ of sequences $\{c_{lk}\}_{k\in\mathbb{Z},l\geq -1}$ that are square summable with regard to the weighted counting measure $d\mu_s(k,l) = 2^{2ls}dldk$. Thus, L defines a continuous linear form on $\ell_2(\mu_s)$, and by the Riesz-representation theorem for continuous linear functionals on $\ell_2(\mu_s)$, there exists $g_L \in \ell_2(\mu_s)$ such that

$$L(\phi) = \sum_{k,l} 2^{2ls}\langle g_L, \psi_{lk}\rangle\langle\phi, \psi_{lk}\rangle \quad \forall\phi \in B^s_{22}, \quad \|g_L\|_{\ell_2(\mu_s)} = \|L\|_{(B^s_{22})^*}.$$

Now, since $\mathcal{S} \subset B^s_{22}$, we have $L \in \mathcal{S}^*$ and, using the last representation for $L(\psi_{lk})$ and ortho-normality of the $\psi_{lk} \in \mathcal{S}$,

$$\|L\|_{B^{-s}_{22}} = \sum_{l,k} 2^{-2ls}|L(\psi_{lk})|^2 = \sum_{l,k} 2^{2ls}|\langle g_L, \psi_{lk}\rangle|^2 = \|L\|_{(B^s_{22})^*} < \infty,$$

so $L \in B^{-s}_{22}$, completing the proof. ∎

The same result holds for $s \leq 0$ as the spaces involved are Hilbert spaces and hence reflexive. Moreover, these arguments extend without conceptual difficulty to $1 < p,q < \infty$, and one shows that

$$(B^s_{pq})^* = B^{-s}_{p'q'}, \quad 1/p + 1/p' = 1, \quad 1/q + 1/q' = 1, \tag{4.96}$$

in this situation. The proofs, which follow from the same arguments as in the proof of the preceding proposition combined with standard duality theory of ℓ_p-spaces, are left to the reader.

Since B^s_{pq} can be defined using a wavelet basis of regularity $r > s$ with equivalent norms, we see that we can define $B^{-s}_{p'q'}$ for these values of p,q also using such a wavelet basis, and there is no need to restrict ourselves to $\phi, \psi \in \mathcal{S}$ as long as the wavelets are S-regular, $s < S$.

The limiting cases where p or q take values in $\{1,\infty\}$ deserve separate attention. In essence, we shall be interested in the following estimate:

Proposition 4.3.13 *Let $s > 0$. Then*

(a) Every $f \in L^\infty$ defines a tempered distribution in $B^{-s}_{\infty 1}$, and

$$\sup_{g:\|g\|_{B^s_{1\infty}}\leq 1}\left|\int_{\mathbb{R}} f(x)g(x)dx\right| \leq \|f\|_{B^{-s}_{\infty 1}}.$$

(b) Every finite signed measure μ defines a tempered distribution in $B^{-s}_{1\infty}$, and

$$\sup_{g:\|g\|_{B^s_{\infty 1}}\leq 1}\left|\int_{\mathbb{R}} g(x)d\mu(x)\right| \leq \|\mu\|_{B^{-s}_{1\infty}}.$$

Proof (a) We have

$$f \in L^\infty \subset B^0_{\infty\infty} \subset B^{-s}_{\infty 1}, \quad g \in B^s_{1\infty} \subset B^0_{11} \subset L^1$$

by Propositions 4.3.9, 4.3.10 and 4.3.11. The wavelet series of g converges in L^1, and thus, by dominated convergence,

$$\int_{\mathbb{R}} fg = \sum_{l,k}\int f(x)\langle g, \psi_{lk}\rangle\psi_{lk}(x)dx = \sum_{l,k}\langle g, \psi_{lk}\rangle\langle f, \psi_{lk}\rangle,$$

so

$$\left|\int_{\mathbb{R}} fg\right| \leq \sup_l 2^{l(s-1/2)} \|\langle g, \psi_{l\cdot}\rangle\|_1 \cdot \sum_l 2^{l(-s+1/2)} \|\langle f, \psi_{l\cdot}\rangle\|_\infty = \|g\|_{B^s_{1\infty}} \|f\|_{B^{-s}_{\infty 1}},$$

which gives part (a). Part (b) is proved in exactly the same way, noting that the wavelet series $g \in B^s_{\infty 1} \subset C_u$ converges uniformly and replacing $f(x)dx$ by $d\mu(x)$ everywhere. ■

Unlike in the cases $1 < p,q < \infty$, one cannot conclude, however, that $(B^s_{1\infty})^* = B^{-s}_{\infty 1}$ and $(B^s_{\infty 1})^* = B^{-s}_{1\infty}$ because $\mathcal{S}$ is not dense in B^s_{pq} for $\max(p,q) = \infty$ (so the approximation arguments from before cannot be used). Rather, the spaces $B^s_{1\infty}, B^s_{\infty 1}$ have to replaced by the completion of $\mathcal{S}$ for the corresponding norms. We do not pursue this further as it will not be relevant in what follows.

Approximation of Functions in Weak Norms and for Integral Functionals

In Propositions 4.1.5 and 4.3.8 we studied properties of kernel-type approximation schemes in L^p-type distance functions. These bounds clearly imply the same approximation rates in weaker distance functions than L^p-loss, such as B^{-r}_{pq}-loss for $r > 0$, simply by the continuous imbeddings of $L^p \subset B^{-r}_{pq}$. However, the bounds can be quantitatively improved in such situations, and this can be most easily understood for wavelet approximation schemes.

Proposition 4.3.14 *Let $K_j = 2^j K(2^j \cdot, 2^j \cdot)$ be the projection kernel of an S-regular wavelet basis of $L^2(\mathbb{R})$, and let* $\max(s,r) < S$, $1 \leq p,q \leq \infty$. *Suppose that $f \in B^s_{pq}(\mathbb{R})$. Then, for every $s \geq 0, r \geq 0, j \geq 0$, and some constant C that depends only on the wavelet basis,*

$$\|K_j(f) - f\|_{B^{-r}_{pq}} \leq C\|f\|_{B^s_{pq}} 2^{-j(s+r)}.$$

Proof For notational simplicity, we prove only $q = \infty$; the general case is the same. Using the wavelet characterisation of the Besov norm, we have

$$\begin{aligned}\|K_j(f) - f\|_{B^{-r}_{p\infty}} &= \sup_{l \geq j} 2^{l(-r+1/2-1/p)} \|\langle f, \psi_{l\cdot}\rangle\|_p \\ &\leq 2^{-j(s+r)} \sup_{l \geq j} 2^{l(s+1/2-1/p)} \|\langle f, \psi_{l\cdot}\rangle\|_p \\ &\leq 2^{-j(s+r)} \|f\|_{B^{s,W}_{p\infty}}\end{aligned}$$

so the result follows from equivalence of the different Besov norms. ■

This result, in view of the duality theory of the preceding section, can be used to bound the approximation errors of integrals

$$\left|\int_{\mathbb{R}} g(x)(K_j(f) - f)(x)dx\right| \lesssim \|K_j(f) - f\|_{B^{-s}_{pq}} \|g\|_{B^r_{p'q'}} \leq C\|f\|_{B^s_{pq}} \|g\|_{B^r_{p'q'}} 2^{-j(s+r)}, \quad (4.97)$$

for $f \in L^p, g \in L^q$, $1/p + 1/p' = 1, 1/q + 1/q' = 1$ and $0 < C < \infty$ some universal constant, showing that for smooth integral functionals the precision of approximation of f by $K_j(f)$ can be quantitatively better than in L^p. In particular, these arguments can be applied to g contained in the spaces H^m_p, C^m by using their imbeddings into suitable B^r_{pq} spaces, as discussed in the next section.

For some functions, direct estimates are preferable over duality arguments. We illustrate this for indicator functions $1_{(-\infty,t]}$ which are bounded but smooth in L^1 only, so the duality theory via Besov spaces is not efficient. The following proposition generalises immediately to classes $\mathcal{G}$ of functions of bounded variation (defined in the next section) and then also to the space H_1^1 and, more generally, to the spaces H_1^m (see Exercise 4.3.5).

Proposition 4.3.15 *Let $K_j = 2^j K(2^j\cdot, 2^j\cdot)$ be the projection kernel of an S-regular wavelet basis of $L^2(\mathbb{R})$. Assume that $f \in B^s_{\infty\infty} \cap L^1$ for some $s < S$. Let $\mathcal{G} = \{1_{(-\infty,t]} : t \in \mathbb{R}\}$. Then, for some constant C depending only on the wavelet basis and every $l \geq 0$,*

$$\sup_{g\in\mathcal{G}} \|\langle g, \psi_{l\cdot}\rangle\|_1 \leq C2^{-l/2} \tag{4.98}$$

and

$$\sup_{g\in\mathcal{G}} \left|\int_{\mathbb{R}} (K_j(f) - f)g\right| \leq C\|f\|_{B^{s,W}_{\infty\infty}} 2^{-j(s+1)}. \tag{4.99}$$

Proof Since the wavelet series of f converges in L^1, we have

$$K_j(f) - f = -\sum_{l=j}^{\infty}\sum_k \langle f, \psi_{lk}\rangle\psi_{lk} \text{ in } L^1,$$

and since $g = 1_{(-\infty,s]} \in L^\infty$, we can interchange integration and summation to see that

$$\int (K_j(f) - f)g = \sum_{l=j}^{\infty}\sum_k \langle f, \psi_{lk}\rangle\langle g, \psi_{lk}\rangle.$$

We now have

$$\begin{aligned}
\|\langle g, \psi_{l\cdot}\rangle\|_1 &\leq \int \sum_k |(K_{l+1} - K_l)(g)(x)||\psi_{lk}(x)|dx \\
&\leq 2^{l/2}\left\|\sum_k |\psi(2^l(\cdot) - k)|\right\|_\infty \|K_{l+1}(g) - K_l(g)\|_1 \qquad (4.100)\\
&\leq c2^{l/2}(\|K_{l+1}(g) - g\|_1 + \|K_l(g) - g\|_1).
\end{aligned}$$

To bound the right-hand side, we have by Fubini's theorem, using a majorising kernel Φ for $K(x,y)$,

$$\begin{aligned}
\int\left|\int 2^l K(2^l y, 2^l x)g(x)dx - g(y)\right|dy &= \int\left|\int 2^l K(2^l y, 2^l u + 2^l y)(g(u+y) - g(y))du\right|dy \\
&\leq \int\int 2^l\Phi(2^l u)|g(u+y) - g(y)|dudy
\end{aligned}$$

$$= \int \Phi(u) \int |g(2^{-l}u+y) - g(y)| dy du$$
$$= \int \Phi(u) \left| \int_{s-2^{-l}u}^{s} dy \right| du$$
$$\leq 2^{-l} \int \Phi(u)|u| du$$

to conclude that (4.98) holds true. For the wavelet coefficients of f, we have by the wavelet definition of the Besov norm that $\|\langle f, \psi_{l\cdot}\rangle\|_\infty \leq C2^{-l(s+1/2)}$. Combining these two bounds gives (4.99). ■

A similar phenomenon occurs for convolution kernel approximations

$$\int_{\mathbb{R}} g(x)(K_h * f - f)(x), \quad K_h = h^{-1}K(\cdot/h),$$

where we cannot use the wavelet characterisation of Besov spaces directly, however. Instead, we notice the following:

Lemma 4.3.16 *Let $K \in L^1(\mathbb{R}), \int K = 1,\ f \in L^p(\mathbb{R}), g \in L^q(\mathbb{R})$, with $1/p + 1/q = 1$. Then, writing $\bar{f} = f(-\cdot)$, we have $\bar{f} * g \in C(\mathbb{R})$, and*

$$\int_{\mathbb{R}} g(x)(K_h * f - f)(x)\, dx = \int_{\mathbb{R}} K(t)[\bar{f} * g(ht) - \bar{f} * g(0)]\, dt. \tag{4.101}$$

Proof By (4.2) and (4.3), the properties of $\bar{f} * g$ follow and also that $(f * K_h)g, fg \in L^1$. By substitution and Fubini's theorem,

$$\frac{1}{h} \int_{\mathbb{R}} \int_{\mathbb{R}} (f(x-y) - f(x))K\left(\frac{y}{h}\right) dy g(x)\, dx$$
$$= \int_{\mathbb{R}} \int_{\mathbb{R}} (f(x-th) - f(x))K(t) dt g(x)\, dx$$
$$= \int_{\mathbb{R}} K(t) \left[\int_{\mathbb{R}} f(x-th)g(x)dx - \int_{\mathbb{R}} f(x)g(x)dx \right] dt$$
$$= \int_{\mathbb{R}} K(t) \left[\int_{\mathbb{R}} \bar{f}(th-x)g(x)dx - \int_{\mathbb{R}} \bar{f}(0-x)g(x)dx \right] dt$$
$$= \int_{\mathbb{R}} K(t)[\bar{f} * g(ht) - \bar{f} * g(0)] dt. \quad ■$$

Using Taylor expansion arguments as in the case $p = \infty$ in Proposition 4.1.5, we now see that the regularity of the convolution $\bar{f} * g$ near zero governs the rate of convergence to zero of the quantity in the preceding lemma. More precisely, if we can bound the $C^m(\mathbb{R})$-norm of $\bar{f} * g$, then we will obtain a bound of the order of h^m for $\left|\int_{\mathbb{R}} (K_h * f - f)g\right|$. To obtain sharp results, one notices that, intuitively speaking, the smoothness of $\bar{f} * g$ will be the sum of the smoothness degrees of the individual functions f, g, paralleling the role of $s + r$ in Proposition 4.3.14.

To investigate this further, we start with the following simple lemma, for which we note that a locally integrable function f may have a weak derivative $Df = \nu_f$ equal to a signed measure by interpreting $Df(u)du$ in (4.1) as $d\nu_f(u)$.

Lemma 4.3.17 *(a) Let $f \in C(\mathbb{R})$ be such that $Df \in L^\infty(\mathbb{R})$, and let $\nu \in M(\mathbb{R})$ be a finite signed measure. Then, for every $x \in \mathbb{R}$, $D(f * \nu)(x)$ exists, and*

$$D(f * \nu)(x) = (Df * \nu)(x).$$

(b) Let $g \in C(\mathbb{R})$, let $f \in L^\infty(\mathbb{R})$ be such that $Df \in M(\mathbb{R})$ and suppose that $g * f(x)$ is defined for every $x \in \mathbb{R}$. Then, for every $x \in \mathbb{R}$, $D(g * f)(x)$ exists, and

$$D(g * f)(x) = (g * \nu_f)(x),$$

where ν_f is the finite signed measure defined by $\nu_f((a,b]) = \tilde{f}(b) - \tilde{f}(a), \tilde{f}(x) \equiv Df((-\infty,x])$.

Proof For part (a), note that by the mean value theorem and boundedness of Df, $h^{-1}[f(x-y+h) - f(x-y)]$ is uniformly bounded; hence, by the dominated convergence theorem, we have

$$\begin{aligned} D(f * \nu)(x) &= \lim_{h\to 0} h^{-1} \int_{\mathbb{R}} (f(x-y+h) - f(x-y))\, d\nu(y)dy \\ &= \int_{\mathbb{R}} \lim_{h\to 0} h^{-1}[f(x-y+h) - f(x-y)] d\nu(y)dy \\ &= \int_{\mathbb{R}} Df(x-y) d\nu(y)dy = (Df * \nu)(x), \end{aligned}$$

the last integral being convergent for every $x \in \mathbb{R}$ because Df is bounded.

For part (b), we have $f = \tilde{f}$ almost everywhere (Exercise 4.3.6), and thus, $(g * f)(x) = (g * \tilde{f})(x)$ holds for every $x \in \mathbb{R}$, so we have

$$\begin{aligned} D(g * f)(x) &= \lim_{h\to 0} h^{-1} \int_{\mathbb{R}} \left(\tilde{f}(x-y+h) - \tilde{f}(x-y)\right) g(y)dy \\ &= \lim_{h\to 0} h^{-1} \int_{\mathbb{R}} \int_{x-y}^{x-y+h} d\nu_f(t) g(y)dy \\ &= \lim_{h\to 0} h^{-1} \int_{\mathbb{R}} \int_{x-t}^{x-t+h} g(y)dy d\nu_f(t) \\ &= \int_{\mathbb{R}} \lim_{h\to 0} h^{-1} \int_{x-t}^{x-t+h} g(y)dy d\nu_f(t) \\ &= \int_{\mathbb{R}} g(x-t) d\nu_f(t) = g * \nu_f(x), \end{aligned}$$

for every x. The first two equalities are obvious. The third is Fubini's theorem, and the fourth equality follows from $g \in C(\mathbb{R})$ and the dominated convergence theorem. The fifth equality follows from the fundamental theorem of calculus. The integral in the last line converges for every $x \in \mathbb{R}$ by the boundedness of g. ■

In the $p = 2$ setting we have the following:

Lemma 4.3.18 *Assume that* $f \in H_2^s(\mathbb{R}), g \in H_2^r(\mathbb{R})$ *for nonnegative integers* $s, r \geq 0$ *(where* $H_2^0 = L^2$*). Then* $f * g \in C^{s+r}(\mathbb{R})$*, and*

$$\|f * g\|_{C^{s+r}(\mathbb{R})} \leq 2\pi \|f\|_{H_2^s(\mathbb{R})} \|g\|_{H_2^r(\mathbb{R})},$$

for some universal constant C.

Proof We have $\|f * g\|_\infty \leq \|f\|_2 \|g\|_2$, and using the Fourier-analytical tools reviewed in Section 4.1, as well as the Cauchy-Schwarz inequality, we get

$$\begin{aligned}
\|D^{s+r}(f * g)\|_\infty &\leq \int_{\mathbb{R}} |u|^{s+r} |\hat{f}(u)||\hat{g}(u)| du \\
&= \int_{\mathbb{R}} |\mathcal{F}[D^s f](u)||\mathcal{F}[D^r g](u)| du \\
&\leq \|\mathcal{F}[D^s f]\|_2 \|\mathcal{F}[D^r g]\|_2 \leq 2\pi \|f\|_{H_2^s} \|g\|_{H_2^r},
\end{aligned}$$

giving the result. ■

We can conclude that if $f \in C^s(\mathbb{R}), g \in H_1^r(\mathbb{R})$ or $f \in H_2^s(\mathbb{R}), g \in H_2^r(\mathbb{R})$, then

$$D^{s+r}(f * g) = D^s f * D^r g \Rightarrow f * g \in C^{s+r}(\mathbb{R}) \tag{4.102}$$

from the preceding lemmas (applied iteratively in the first case). In particular, if K is a kernel that satisfies Condition 4.1.3 for $N = s + r$, then

$$\left| \int_{\mathbb{R}} g(x)(K_h * f - f)(x) \right| \leq C h^{s+r}, \tag{4.103}$$

where C equals a universal constant times $\|f\|_{C^s} \|g\|_{H_1^r}$ or $\|f\|_{H_2^s} \|g\|_{H_2^r}$, respectively. To obtain similar results for $p \notin \{1, 2, \infty\}$ and $s, r \notin \mathbb{N}$ for convolution kernel approximations is possible but requires more sophisticated arguments; we give some references in the notes at the end of this chapter. Recall, however, that for wavelet approximations one can immediately use Proposition 4.3.14 without difficulty for such values of s, r, p.

The Differential Operator on Besov Spaces

Since Besov spaces model regularity properties, it is natural to expect that the differentiation operator $f \mapsto D^n f$ acts on the scale B^s_{pq} by decreasing the s index by n. This holds true in general ($s \in \mathbb{R}$), where derivatives are understood in the distributional sense of the space $\mathcal{S}^*$ if necessary.

Proposition 4.3.19 *Let* $s \in \mathbb{R}, p, q \in [1, \infty]$*. For any* $n \in \mathbb{N}$*, the mapping* $f \mapsto D^n f$ *is linear and continuous from* $B^s_{pq}(\mathbb{R})$ *to* $B^{s-n}_{pq}(\mathbb{R})$*.*

Proof We estimate the Littlewood-Paley norm

$$\|D^n f\|_{B^{s-n}_{pq}} = \|(D^n f) * \phi\|_p + \left(\sum_{j=0}^{\infty} 2^{j(s-n)q} \|(D^n f) * \psi_{2^{-j}}\|_p^q \right)^{1/q}$$

with obvious notational modifications if $q = \infty$. We use Lemma 4.3.4 and the fact that $f * \phi$ has compactly supported Fourier transform to see that

$$\|(D^n f) * \phi\|_p = \|D^n(f * \phi)\|_p \leq C\|f * \phi\|_p.$$

Moreover, again by Lemma 4.3.4 and since the support of the Fourier transform of $f * \psi_{2^{-j}}$ is contained in an interval of width of order 2^j, we have, from the same argument,

$$\sum_{j=0}^{\infty} 2^{j(s-n)q} \|(D^n f) * \psi_{2^{-j}}\|_p^q \leq \sum_{j=0}^{\infty} 2^{jsq} \|f * \psi_{2^{-j}}\|_p^q,$$

which implies that

$$\|D^n f\|_{B^{s-n}_{pq}} \leq \|f\|_{B^s_{pq}},$$

completing the proof. ∎

4.3.3 Relationships to Classical Function Spaces

We have seen that Besov spaces are the maximal spaces for which the accuracy of approximation from a variety of common approximation schemes has a prescribed degree of precision. We show in this section that many of the more familiar classical spaces of smooth functions (i) either coincide with or (ii) are contained in a suitable Besov space. In the former cases, this implies, by virtue of Theorem 4.3.2, that the classical spaces, which themselves are usually defined by more intuitive regularity conditions, have powerful characterisations by their wavelet coefficients. In the latter case, it allows us to use approximation-theoretic results such as Proposition 4.3.8 for these Besov subspaces as well. The results in this subsection particularly establish that for the purposes of constructing statistical models for functions, Besov spaces are in a sense the right general framework.

Spaces of Differentiable Functions

We recall, for $m \in \mathbb{N}$, $1 \leq p < \infty$, that

$$H^m_p \equiv H^m_p(\mathbb{R}) = \left\{ f \in L^p : D^j f \in L^p \; \forall j = 1, \dots, m : \|f\|_{H^m_p} := \|f\|_p + \|D^m f\|_p < \infty \right\},$$

and for $p = \infty$,

$$C^m \equiv C^m(\mathbb{R}) = \{ f \in C_u(\mathbb{R}) : f^{(j)} \in C_u(\mathbb{R}) \; \forall j = 1, \dots, m : \|f\|_{C^m} := \|f\|_\infty + \|f^{(m)}\|_\infty < \infty \}.$$

These spaces can be related to Besov spaces $B^s_{pq}(\mathbb{R})$ in a natural way.

Proposition 4.3.20 *The following continuous imbeddings hold: for every $1 \leq p < \infty, m \in \mathbb{N}$,*

$$B^m_{p1}(\mathbb{R}) \subset H^m_p(\mathbb{R}) \subset B^m_{p\infty}(\mathbb{R}), \quad \text{if } 1 \leq p < \infty, \quad B^m_{\infty 1}(\mathbb{R}) \subset C^m(\mathbb{R}) \subset B^m_{\infty\infty}(\mathbb{R}), \tag{4.104}$$

as well as

$$B^m_{22}(\mathbb{R}) = H^m_2(\mathbb{R}), \tag{4.105}$$

with equivalent norms.

Proof In view of (4.75) and the modulus of continuity definition of B^s_{pq}, we immediately deduce $H^m_p(\mathbb{R}) \subseteq B^m_{p\infty}(\mathbb{R}), C^m(\mathbb{R}) \subset B^m_{\infty\infty}(\mathbb{R})$. Moreover, using Proposition 4.3.5, part (b), we see that for $f \in B^m_{p1}(\mathbb{R})$ we have $f \in L^p$ and, interchanging differentiation and summation,

$$\|D^m f\|_p = \left\| D^m \sum_l u_l \right\|_p \leq \sum_l \|D^m u_l\|_p \leq \sum_l c_l < \infty,$$

so (4.104) follows.

To establish (4.105), assume first that $f, D^m f \in L^2$, and let us show that the Littlewood-Paley Besov norm (4.86) is finite. By Plancherel's theorem,

$$2^{2jm}\|\psi_{2^{-j}} * f\|_2^2 = \frac{2^{2jm}}{2\pi}\int_{2^{-j-1}}^{2^{-j+1}} |\hat{\psi}(2^{-j}u)|^2 |\hat{f}(u)|^2 |u|^{2m}|u|^{-2m} du$$
$$\leq c \int |\hat{\psi}(2^{-j}u)|^2 |\mathcal{F}[D^m f](u)|^2 du,$$

and this bound is summable because

$$\sum_j |\hat{\psi}(2^{-j}u)|^2 \leq 2\|\hat{\psi}\|_\infty^2, \qquad \int |\mathcal{F}[D^m f](u)|^2 du \leq \|D^m f\|_2^2,$$

using that $\hat{\psi}(2^{-j}u)$ and $\hat{\psi}(2^{-j'}u)$ have disjoint support as soon as $|j - j'| \geq 2$. Thus, $H^m_2 \subset B^m_{22}$ follows.

Conversely, assume that $f \in B^m_{22}$. Hence, $f \in L^2$, and it suffices to show that $\|D^m f\|_2 < \infty$. It is easy to see that the function ψ in the Littlewood-Paley decomposition can be chosen such that

$$\inf_{u \in \mathbb{R}} \sum_j |\hat{\psi}(2^{-j}u)|^2 \geq c > 0,$$

for some universal constant c so that, by Plancherel's theorem,

$$\sum_j 2^{2jm}\|\psi_{2^{-j}} * f\|_2^2 = \frac{1}{2\pi}\sum_j 2^{2jm}\int_{2^{-j-1}}^{2^{-j+1}} |\hat{\psi}(2^{-j}u)|^2 |\hat{f}(u)|^2 |u|^{2m}|u|^{-2m} du$$
$$\geq c' \int \sum_j |\hat{\psi}(2^{-j}u)|^2 |\mathcal{F}[D^m f](u)|^2 du$$
$$\geq c'' \|D^m f\|_2^2,$$

so $D^m f \in L^2$, and the result follows. ■

Bounded Variation Spaces

The space H^1_1 is the space of functions $f \in L^1$ with weak derivative Df in L^1. Since Df is understood in the sense of (4.1), the requirement $Df \in L^1$ can be weakened further to require only that $Df \in M(\mathbb{R})$, that is, that Df is a finite signed measure on $\mathbb{R}$; in this case, we speak of a function of bounded variation

$$BV \equiv BV(\mathbb{R}) = \{f \in L^1(\mathbb{R}) : Df \in M(\mathbb{R})\} \tag{4.106}$$

equipped with the norm

$$\|f\|_{BV} = \|f\|_1 + |Df|(\mathbb{R}),$$

where $|Df|$ is the total variation of Df.

Proposition 4.3.21 *We have the continuous imbeddings*

$$B^1_{11}(\mathbb{R}) \subset H^1_1(\mathbb{R}) \subset BV(\mathbb{R}) \subset B^1_{1\infty}(\mathbb{R}).$$

Proof Clearly, $H^1_1 \subseteq BV$ as any $Df \in L^1$ is the density of a finite signed measure, so in particular, $B^1_{11} \subset BV$ by (4.104). It remains to prove the third imbedding. We can write

$$f(x) = f(-\infty) + \int_{\mathbb{R}} 1_{(-\infty,x]}(u)dDf(u),$$

and thus from Fubini's theorem and since $\int \psi_{lk} = 0$, we have

$$\langle f, \psi_{lk}\rangle = \int_{\mathbb{R}} \langle 1_{[u,\infty)}, \psi_{lk}\rangle dDf(u).$$

The estimate (4.98) now implies that

$$\sup_{l\geq 0} 2^{l/2} \sum_k |\langle f, \psi_{lk}\rangle| \lesssim |Df|(\mathbb{R}),$$

and since $\|\langle f, \phi_\cdot\rangle\|_1 \leq C\|f\|_1$, we conclude that $\|f\|_{B^1_{1\infty}} < \infty$, implying the result. ■

In a similar vein, if instead of requiring $f \in L^1$ in the definition of $BV(\mathbb{R})$ we require $f \in L^\infty$, then the norm $\|f\|_\infty + |Df|(\mathbb{R})$ of this space can be estimated by $\|f\|_\infty + \|Df\|_{0,1,\infty}$. We recall that so-defined bounded variation spaces are related to the classical notion of BV spaces as follows: if $\mathcal{P}$ denotes the set of all finite dissections $\{x_i\}_{i=1}^n$ of $\mathbb{R}$, then one shows (Exercise 4.3.6) that any function $f \in BV(\mathbb{R})$ is a.e. equal to a function f which satisfies

$$\sup_{\{x_i\}\in\mathcal{P}} \sum_i |f(x_{i+1}) - f(x_i)| < \infty, \tag{4.107}$$

and conversely, any f for which the last supremum is finite defines an element of $BV(\mathbb{R})$. A similar remark holds for the relationship between H^1_1 and the classical space of absolutely continuous functions.

Finally, if in the preceding definition pth powers are used, we obtain the spaces of functions of bounded p-variation, that is,

$$BV_p(\mathbb{R}) = \left\{ f : \mathbb{R} \to \mathbb{R} : v_p(f) \equiv \sup_{\{x_i\}\in\mathcal{P}} \sum_i |f(x_{i+1}) - f(x_i)|^p < \infty \right\}, \quad 1 \leq p < \infty.$$

An important case is $p = 2$, the space of functions of finite *quadratic variation*.

Proposition 4.3.22 *Let* $1 \leq p < \infty$. *Then we have*

$$BV_p(\mathbb{R}) \cap L^p(\mathbb{R}) \subset B^{1/p}_{p\infty}(\mathbb{R}).$$

Proof The case $p = 1$ is already established, and we only prove $p = 2$; the general case is the same in view of Lemma 3.6.11. Since $f \in L^2$, it suffices to show that the squared Besov norm

$$\sup_{t>0} \frac{\omega_1^2(f,t,2)}{t} = \sup_{t>0} \frac{\int (f(x+t) - f(x))^2 dx}{t}$$

from (4.78) with $r = 1 > 1/2$ is finite; in fact, since $f \in L^2$, we can restrict the supremum and the integral to a bounded set K including the origin. By Lemma 3.6.11, we have $f(x) = g(m(x))$, where $m(x)$ is a nondecreasing function with range contained in $[0, v_2(f)]$ and g is $1/2$-Hölder continuous on $[0, v_2(f)]$. In particular, $m \in BV$ has a weak derivative of variation at most $v_2(f)$, and thus, the quantity in the preceding display can be bounded, using Fubini's theorem and Exercise 4.3.6, by

$$\sup_{t\in K} \frac{\int_K |m(x+t) - m(x)|dx}{t} = \sup_{t\in K} \frac{1}{t} \int_K \int_x^{x+t} |dDm(u)|dx \leq C(K)v_2(f),$$

completing the proof. ∎

Using more sophisticated techniques from interpolation theory, one can also prove a 'converse' of the preceding proposition in the sense that

$$B^{1/p}_{p1}(\mathbb{R}) \subset BV_p(\mathbb{R}) \tag{4.108}$$

(see the notes at the end of this chapter for references). The easier case $p = 2$ is hinted at in Exercise 4.3.10.

Spaces Defined by Hölder-Type Conditions

For A an arbitrary measurable subset of $\mathbb{R}$, the classical Lipschitz space is defined as

$$BL(A) = \left\{ f \in C_u(A) : \|f\|_\infty + \sup_{x\neq y, x,y\in A} \frac{|f(x) - f(y)|}{|x-y|} < \infty \right\}. \tag{4.109}$$

For A an interval possibly equal to $\mathbb{R}$, the space $BL(A)$ contains, by the mean value theorem, the space $C^1(A)$, and this containment is strict (since $|\cdot| \in BL(\mathbb{R}) \setminus C^1(\mathbb{R})$). Lipschitz spaces have an obvious generalisation to noninteger $s \in (0,1)$ as

$$C^s(A) = \left\{ f \in C_u(A) : \|f\|_\infty + \sup_{x\neq y, x,y\in A} \frac{|f(x) - f(y)|}{|x-y|^s} < \infty \right\}, \tag{4.110}$$

and then, for $s > 0$, any noninteger real number with integer part $[s]$,

$$C^s(A) = \left\{ f \in C_u(A) : \|f\|_{C^s(A)} \equiv \|f\|_{C^{[s]}(A)} + \sup_{x\neq y, x,y\in A} \frac{|D^{[s]}f(x) - D^{[s]}f(y)|}{|x-y|^{s-[s]}} \right\}. \tag{4.111}$$

For noninteger s, the spaces $C^s(A)$ are known as *Hölder spaces*. It is immediate from (4.78) that, for A an interval,

$$C^s(A) = B^s_{\infty\infty}(A), \quad 0 < s < 1,$$

and (4.76) gives $C^s(A) \subset B^s_{\infty\infty}(A)$ for any $s \notin \mathbb{N}$. Finally, Exercise 4.3.3 shows that also $B^s_{\infty\infty}(A) \subset C^s(A)$. Summarising:

Proposition 4.3.23 *We have*

$$C^s(A) = B^s_{\infty\infty}(A), \quad 0 < s < \infty, s \notin \mathbb{N},$$

with equivalent norms.

Remarkably, the norm of the nonseparable (see Exercise 4.3.7) space $C^s(A), s \notin \mathbb{N}$, has several useful equivalent characterisations by wavelet or Littlewood-Paley bases (cf. Theorem 4.3.2). Combined with the imbedding $C^m \subset B^m_{\infty\infty}, m \in \mathbb{N}$, we see that, for any $s > 0$,

$$f \in C^s(\mathbb{R}) \Rightarrow \sup_{k\in\mathbb{Z}} |\langle f, \psi_{lk}\rangle| \leq K\|f\|_{C^s(\mathbb{R})} 2^{-l(s+1/2)}, \tag{4.112}$$

for some constant K that depends only on the (S-regular, $S > s$) wavelet basis. Thus, smoothness of f translates into faster decay of the wavelet coefficients, and for $s \notin \mathbb{N}$, this implication is in fact an equivalence.

The preceding definitions of $C^s(A)$ and $BL(A)$ leads to potential ambiguities when s is an integer. For convenience of the reader, we discuss here what is known without going into detailed proofs, and we refer you to the notes at the end of this chapter for references. The classical definition of C^m, and the one we use, is in terms of derivatives, given at the beginning of this chapter. Alternatively, one can generalise the Lipschitz spaces and define, for $s \in \mathbb{N}$,

$$BL(s,A) = \left\{ f \in C_u(A) : \|f\|_{C^s(A)} \equiv \|f\|_\infty + \sup_{x\neq y, x,y\in A} \frac{|D^{(s-1)}f(x) - D^{(s-1)}f(y)|}{|x-y|} \right\}. \tag{4.113}$$

We have already noted that $C^m \subset B^m_{\infty\infty}$, and one may show that this imbedding is strict, $C^m \neq B^m_{\infty\infty}$ for $m \in \mathbb{N}$ as the latter space contains nondifferentiable functions (e.g., $|\cdot|$, as is easily seen). One may next ask whether $BL(m,A) = B^m_{\infty\infty}(A)$ holds true. The answer is also negative; the Besov norm in terms of higher differences gives rise to a space that is still larger than an s-Lipschitz space – in some sense, the $B^m_{\infty\infty}$ are the largest spaces that still have L^∞-regularity m. The spaces $B^m_{\infty\infty}$ are sometimes studied separately under the name of *Hölder-Zygmund* or just *Zygmund spaces*, and are then denoted by $\mathcal{C}^m$. For these spaces, (4.112) is always an equivalence, including the case $s \in \mathbb{N}$. Summarising, we note that

$$C^m(\mathbb{R}) \subsetneq BL(m,\mathbb{R}) \subsetneq B^m_{\infty\infty}(\mathbb{R}) \equiv \mathcal{C}^m(\mathbb{R}), \quad m \in \mathbb{N}.$$

Similar remarks apply to the $p < \infty$ case when considering the integrated Hölder-type conditions ($s \notin \mathbb{N}$) that define B^s_{pq}. For noninteger s, another approach to define the spaces H^s_p exists. Note first that in the definition of $H^m_2, m \in \mathbb{N}$, we require

$$f + D^m f \in L^2 \iff \hat{f} + (iu)^m \hat{f} = (1 + (iu)^m)\hat{f} \in L^2 \tag{4.114}$$

by Plancherel's theorem and (4.9). One thus can construct an equivalent norm on H^m_2 by $\|f\| = \|\langle u\rangle^m \hat{f}\|_2$, where $\langle u\rangle \equiv (1+|u|^2)^{m/2}$. This motivates the definition, for general $s \geq 0, 1 \leq p < \infty$,

$$\hat{H}^s_p = \{f \in L^p : \|\mathcal{F}^{-1}\langle u\rangle^s \hat{f}\|_p < \infty\},$$

with Fourier transform understood in the sense of (4.13) if $p > 2$. The preceding arguments and Plancherel immediately give $\hat{H}^m_2 = H^m_2 = B^m_{22}$ for $m \in \mathbb{N}$, and the proof of

$$\hat{H}^s_2 = B^s_{22} \quad \forall s > 0, \tag{4.115}$$

with equivalent norms, is only slightly more involved, using the spectral synthesis of the translation operator (4.6). A deeper fact is

$$B^s_{pp} = \hat{H}^s_p, \quad 1 < p < \infty, \tag{4.116}$$

which can be proved using Fourier multiplier arguments that we do not develop here.

Local Wavelet Reconstruction of Hölderian Functions

If the wavelet basis is sufficiently localised, then the estimate (4.112) is in fact a local phenomenon. For instance, take an S-regular Daubechies wavelet ψ, and recall that the support of ψ_{lk} is $[(-N+1+k)/2^l, (N+k)/2^l]$. Fix a point $x_0 \in \mathbb{R}$, let $\delta > 0$, let $A_{x_0,\delta}$ be the δ-neighborhood of x_0 and assume that $f \in C^s(A_{x_0,\delta})$ for some $s > 0$. If k_0 is any integer such that the support set

$$[(-N+1+k_0)/2^l, (N+k_0)/2^l]$$

of ψ_{lk_0} is contained in $A_{x_0,\delta}$, then

$$\int_{A_{x_0,\delta}} \psi_{lk_0}(x) x^\ell dx = \int_{\mathbb{R}} \psi_{lk_0} x^\ell dx = 0,$$

for every $\ell = 0, \dots, S-1$, and

$$\langle f, \psi_{lk_0} \rangle = \int_{\mathbb{R}} f(x)\psi_{lk_0}(x)dx = 2^{l/2} \int_{A(x_0,\delta)} (f(x) - f(x_0))\psi(2^l x - k_0)dx$$

so that, after the usual Taylor expansion arguments (as in Proposition 4.1.5),

$$\left| \langle f, \psi_{lk_0} \rangle \right| \le C 2^{-l(s+1/2)}, \tag{4.117}$$

where C depends only on S and the local Hölder constant of f. This result is in fact uniform in all k_0 for which ψ_{lk_0} is supported in $A_{x,\delta}$.

4.3.4 Periodic Besov Spaces on $[0,1]$

The function spaces considered so far consisted of functions defined on all of $\mathbb{R}$, the only exception being the definition of Besov and Hölder spaces on subintervals A of $\mathbb{R}$ by (integrated) moduli of smoothness (e.g., (4.77)). The restriction to $\mathbb{R}$ was not arbitrary, as the group structure of the real line was exploited heavily in the proofs via the Fourier transform. In this section we show how a similar theory can be developed on a fixed interval $(a,b]$ when one restricts to periodic functions so that the group operation of translation modulo 1 can be used. In the case where periodicity is inappropriate, one needs to introduce a boundary correction that we discuss in the next section.

We restrict to the case where $(a,b] = (0,1]$ as the general case consists only in more cumbersome notation. Perodic Besov spaces on $(0,1]$ can be defined via integrated moduli of continuity as in (4.77) with the choice of $A = [0,1]$, and if the translation $\cdot + h$ is understood 'modulo 1' in the definition of the translation operator, $\Delta_h(f) = f(\cdot + h) - f$ (whose domain of definition is then all of $(0,1]$ instead of the restricted set A_{rh} from after (4.72)). Note that continuous functions are then necessarily periodic, $f(0) = f(1)$. *Throughout this section,*

$B^s_{pq}((0,1])$ *will thus stand for the Besov space defined as in (4.77) with the translation operator adapted to the periodic setting.*

Alternatively, we can define Besov spaces through approximation properties of their elements from 'band-limited functions', as in (4.80). Since the group characters of $(0,1]$ with addition modulo 1 are given by the Fourier basis

$$\{e_k = e^{2\pi i(\cdot)k} : k \in \mathbb{Z}\}$$

of $L^2 = L^2((0,1])$, these spaces consist of all trigonometric polynomials of degree less than $t \in \mathbb{N}$; more precisely,

$$V_t = \left\{ f = \sum_{k:|k|\le t} e_k c_k : \; c_k \in \mathbb{R} \right\}. \tag{4.118}$$

Now if

$$\sigma_p(t,f) = \inf_{g \in V_t} \|f - g\|_p \tag{4.119}$$

is the best approximation error of $f \in L^p$ from V_t, then, for $1 \le p \le \infty, 1 \le q \le \infty, s > 0$, we define

$$B^{s,V}_{pq}((0,1]) \equiv \begin{cases} \{f \in L^p((0,1]) : \|f\|_{B^{s,V}_{pq}} \equiv \|f\|_p + |f|_{B^{s,V}_{pq}} < \infty\}, & 1 \le p < \infty \\ \{f \in C_{\mathrm{per}}((0,1]) : \|f\|_{B^{s,V}_{pq}} \equiv \|f\|_p + |f|_{B^{s,V}_{pq}} < \infty\}, & p = \infty, \end{cases} \tag{4.120}$$

where

$$|f|_{B^{s,V}_{pq}} \equiv \begin{cases} \left(\sum_{j=1}^{\infty} [2^{js}\sigma_p(2^j,f)]^q]\right)^{1/q}, & 1 \le q < \infty \\ \sup_{j \ge 1} 2^{js}\sigma_p(2^j,f) & q = \infty \end{cases} \tag{4.121}$$

is the relevant Besov seminorm.

The main purpose of this section is to show that

$$B^{s,V}_{pq}((0,1]) = B^s_{pq}((0,1])$$

and to give a wavelet characterisation of the periodic Besov space. To achieve this, we next introduce periodic wavelets.

Periodised Wavelets on the Unit Circle

One can periodise a wavelet basis of $L^2(\mathbb{R})$ to construct a basis on $L^2((0,1])$. This allows us to characterise spaces of smooth *periodic* functions on $(0,1]$ by their wavelet expansions. We note that in all that follows one may replace $(0,1]$ by the unit circle $\mathbb{T}$, isomorphic to $(0,1]$ when addition is modulo 1, or by $(a,b]$ when considering functions that are $b-a$ periodic.

For ϕ, ψ an S-regular wavelet basis of $L^2(\mathbb{R})$ in the sense of Definition 4.2.14, let V_j be the multiresolution ladder from Definition 4.2.1 associated to ϕ. Denote by V_j^∞ the completion of V_j for the weak topology on L^∞ generated by all integrals against L^1-functions. Then one shows by standard arguments that $f \in V_0^\infty \iff \sum_k c_k \phi(\cdot - k)$ for some $\{c_k\} \in \ell_\infty$, and one still has

$$f \in V_0^\infty \iff f(2^j \cdot) \in V_j^\infty.$$

Let P_j be the subspace of V_j^{∞} consisting of 1-periodic functions. Then P_0 consists only of the constant functions. Indeed, any $f \in P_0$ is of the form

$$f = \sum_{k \in \mathbb{Z}} c_k \phi(\cdot - k)$$

where, for any $k \in \mathbb{Z}$,

$$c_k = \int_{\mathbb{R}} f(x)\phi(x-k)dx = \int_{\mathbb{R}} f(y)\phi(y)dy = const,$$

and $\sum_k \phi(x-k) = 1$ (recall that $\int \phi = 1, \int K(x,y)dy = 1$). If $j > 0$, then, again by periodicity, for any $f \in P_j$,

$$f = \sum_{k \in \mathbb{Z}} c_k \phi(2^j x - k), \quad c_k = 2^j \int_{\mathbb{R}} f(x)\phi(2^j x - k)dx = c_{k+2^j},$$

so f is determined by 2^j-many coefficients c_k. Conclude that P_j has dimension 2^j. Moreover $\cup_j P_j$ is dense in the space $C_{\text{per}}((0,1])$ for the uniform norm: for $f \in C_{\text{per}}((0,1])$, by the preceding arguments, the projection $K_j(f) = \sum_k \langle f, \phi_{jk} \rangle \phi_{jk}$ is seen to belong to P_j, and it converges to f uniformly on $(0,1]$, proved just as Proposition 4.1.3 with supremum restricted to $x \in (0,1]$. Since $C_{\text{per}}(0,1])$ is dense in $L^2((0,1])$, we deduce that the nested sequence $(P_j : j = 0, 1, \dots)$ forms a multiresolution analysis of $L^2((0,1])$ that is comparable to those of $L^2(\mathbb{R})$ encountered so far.

To construct a wavelet basis for $\cup_{j \geq 0} P_j$, note that the periodisations

$$\phi_j^{(\text{per})} \equiv \sum_{k \in \mathbb{Z}} 2^{j/2} \phi(2^j(\cdot - k)) \tag{4.122}$$

are contained in P_j. Clearly,

$$\phi^{(\text{per})} \equiv \phi_0^{(\text{per})} = \sum_{k \in \mathbb{Z}} \phi(\cdot - k) = 1$$

is a basis of P_0. For $j > 0$, likewise,

$$\left\{ \phi_{jm}^{(\text{per})} = \phi_j^{(\text{per})}(\cdot - 2^{-j} m) : 0 \leq m < 2^j, m \in \mathbb{Z} \right\}$$

forms an ortho-normal basis of P_j for the $L^2((0,1])$ inner product. Indeed, since these functions are all in P_j and $2^j = dim(P_j)$-many, it suffices to prove that the $\phi_{jm}^{(\text{per})}$ are ortho-normal in $L^2((0,1])$: changing variables from (k,ℓ) to (k,r) via $r = \ell - k$ and then $t = 2^j(x-k)$ for each k, we have

$$\begin{aligned}
\langle \phi_{jm}^{(\text{per})}, \phi_{jm'}^{(\text{per})} \rangle &= 2^j \sum_{k,\ell} \int_0^1 \phi(2^j x - 2^j k - m)\phi(2^j x - 2^j \ell - m')dx \\
&= 2^j \sum_r \sum_k \int_0^1 \phi(2^j(x-k) - m)\phi(2^j(x-k) - 2^j r - m')dx \\
&= \sum_r \int_{\mathbb{R}} \phi(t-m)\phi(t - 2^j r - m')dt.
\end{aligned}$$

By ortho-normality of the unperiodised wavelets, these integrals are all zero except for $m = m'$ and $r = 0$ (recall that $0 \leq m, m' < 2^j$), in which case they equal $\|\phi_{jm}^{(\mathrm{per})}\|_2^2 = 1$. The change in order of summation is justified by the finiteness of $\|\sum|\phi(\cdot - k)|\|_\infty$.

Now, if $Q_j = P_{j+1} \ominus P_j$ and

$$\psi_j^{(\mathrm{per})} \equiv \sum_{k\in\mathbb{Z}} 2^{j/2}\psi(2^j(\cdot - k)), \tag{4.123}$$

then one proceeds as earlier to show that the family

$$\left\{\psi_{jm}^{(\mathrm{per})} = \psi_j^{(\mathrm{per})}(\cdot - 2^{-j}m) : 0 \leq m < 2^j, m \in \mathbb{Z}\right\}$$

forms an ortho-normal basis of Q_j for the $L^2((0,1])$ inner product. Using the same arguments as after Definition 4.2.1, we can then decompose

$$L^2((0,1]) = P_0 \oplus \left(\bigoplus_{l=0}^{\infty} Q_l\right) = P_j \oplus \left(\bigoplus_{l=j}^{\infty} Q_l\right)$$

so that

$$\left\{1, \psi_{lm}^{(\mathrm{per})} = \psi_l^{(\mathrm{per})}(\cdot - 2^{-l}m) : 0 \leq m < 2^l, m \in \mathbb{Z}, l \in \mathbb{N} \cup \{0\}\right\} \tag{4.124}$$

forms an ortho-normal basis of $L^2((0,1])$. We can thus expand any $f \in L^2((0,1])$ into its orthogonal wavelet series

$$\begin{aligned} f &= \langle f, 1\rangle + \sum_{l=0}^{\infty}\sum_{m=0}^{2^l-1} \langle f, \psi_{lm}^{(\mathrm{per})}\rangle\, \psi_{lm}^{(\mathrm{per})} \qquad (4.125)\\ &= \sum_{m=0}^{2^j-1} \langle f, \phi_{jm}^{(\mathrm{per})}\rangle \phi_{jm}^{(\mathrm{per})} + \sum_{l=j}^{\infty}\sum_{m=0}^{2^l-1} \langle f, \psi_{lm}^{(\mathrm{per})}\rangle\, \psi_{lm}^{(\mathrm{per})} \end{aligned}$$

with convergence holding at least in L^2. We now investigate the general convergence properties of the partial sums

$$K_{j,\mathrm{per}}(f)(x) \equiv \sum_{m=0}^{2^j-1} \langle f, \phi_{jm}^{(\mathrm{per})}\rangle \phi_{jm}^{(\mathrm{per})}(x), \quad x \in (0,1], \tag{4.126}$$

of this wavelet series. We recall the spaces $C^m((0,1])$ and Sobolev spaces H_p^k from the beginning of this chapter.

Proposition 4.3.24 *Let ϕ be the scaling function of an S-regular wavelet basis of $L^2(\mathbb{R})$, and let $\phi_{jm}^{(\mathrm{per})}, j \geq 0, m = 0, \ldots, 2^j - 1$, be the associated periodised basis functions in $L^2((0,1])$. For $f \in L^1((0,1])$, let $K_{j,\mathrm{per}}(f)$ be the projection (4.126).*

(a) If f is in $C_{\mathrm{per}}((0,1])$, then, as $j \to \infty$,

$$K_{j,\mathrm{per}}(f) \to f \quad \textit{uniformly on (0,1].}$$

If, moreover, for some $m \in \mathbb{N}$, $m < S$, $f \in C^m((0,1])$ with all derivatives $D^\alpha f, 0 < \alpha \leq m$ periodic, then for every $j \geq 0$ and some finite constant $C = C(\phi)$,

$$\|K_{j,per}(f) - f\|_{L^\infty((0,1])} \leq C\|f\|_{C^m((0,1])}2^{-jm}.$$

(b) If $f \in L^p((0,1])$, $1 \le p < \infty$, then, as $j \to \infty$,

$$K_{j,\mathrm{per}}(f) \to f \quad \text{in } L^p((0,1]).$$

If, moreover, $f \in H_p^k$ with $D^\alpha f$, $0 \le \alpha < k$ periodic, for some $k \in \mathbb{N}$, $k < S$, then

$$\|K_{j,\mathrm{per}}(f) - f\|_{L^p((0,1])} \le C\|f\|_{H_p^k((0,1])} 2^{-jk}.$$

Proof Identifying f with its 1-periodic extension to $\mathbb{R}$ and denoting by K_j the standard projection operator onto $V_j \subset L^2(\mathbb{R})$ spanned by the (nonperiodised) ϕ_{jk}, the key observation is that

$$K_{j,\mathrm{per}}(f)(x) = K_j(f)(x), \quad x \in (0,1], \tag{4.127}$$

in view of the identities

$$\begin{aligned}
K_j(f)(x) &= 2^j \sum_{k\in\mathbb{Z}} \phi(2^j x - k) \int_{\mathbb{R}} f(t)\phi(2^j t - k)dt \\
&= 2^j \sum_{m=0}^{2^j-1} \sum_{\ell\in\mathbb{Z}} \phi(2^j x - 2^j l - m) \int_{\mathbb{R}} f(t-\ell)\phi(2^j t - 2^j \ell - m)dt \\
&= 2^j \sum_{m=0}^{2^j-1} \int_{\mathbb{R}} f(v)\phi(2^j v - m)dv \cdot \sum_{\ell\in\mathbb{Z}} \phi(2^j(x-l) - m) \\
&= 2^{j/2} \sum_{m=0}^{2^j-1} \sum_{l\in\mathbb{Z}} \int_{-l}^{-l+1} f(v+l)\phi(2^j v - m)dv \cdot \phi_{jm}^{\mathrm{per}}(x) \\
&= 2^{j/2} \sum_{m=0}^{2^j-1} \sum_{l\in\mathbb{Z}} \int_0^1 f(z)\phi(2^j z - 2^j l - m)dz \cdot \phi_{jm}^{\mathrm{per}}(x) \\
&= \sum_{m=0}^{2^j-1} \langle \phi_{jm}^{\mathrm{per}}, f\rangle \phi_{jm}^{\mathrm{per}}(x) = K_{j,\mathrm{per}}(f)(x),
\end{aligned}$$

where we have used 1-periodicity of f, the substitutions $t - \ell = v, v + l = z$, and $\phi \in \cap_q L^q(\mathbb{R}), \sum_k |\phi(\cdot - k)| \in L^\infty(\mathbb{R})$. The first claims in (a) and (b) now follow from the same proof as that of Proposition 4.1.3 (in fact, Proposition 4.1.1), with the L^p-norms, $1 \le p \le \infty$, in the proof restricted to $(0,1]$ (and noting that the translation uh can be restricted to u in a compact set by the moment condition on K). Moreover, since any f from part a), when periodically extended to $\mathbb{R}$, belongs to $C^m(\mathbb{R}) \subset B^m_{\infty\infty}(\mathbb{R})$ with the same norm, we have, from Proposition 4.3.8,

$$\|K_{j,\mathrm{per}}(f) - f\|_{L^\infty((0,1])} \le \|K_j(f) - f\|_\infty \le C'\|f\|_{B^m_{\infty\infty}(\mathbb{R})} 2^{-jm} \le C\|f\|_{C^m((0,1])} 2^{-jm}.$$

Part (b) follows likewise, as in the proof of Proposition 4.1.5, using that the L^p-norms considered at the end of the proof can be restricted to $(0,1]$ (and noting again that the translation tuh in $D^m f(x + tuh)$ can be restricted to u in a compact set by the moment condition on the projection kernel K). ■

The preceding proposition implies decay estimates for wavelet coefficients of functions f satisfying the preceding approximation bounds. To see this, we note that one can establish an analogue of Proposition 4.2.8 in the periodic setting. Since

$$\|\psi_{lm}^{\mathrm{per}}\|_1 \le 2^{l/2}\int_0^1 \sum_k |\psi(2^l(x-k)-m)|dx = 2^{l/2}\sum_k \int_{-k}^{-k+1} |\psi(2^l u - m)|du = 2^{-l/2}\|\psi\|_1,$$

one proves as in Proposition 4.2.8 that, for every $l \ge 0$,

$$\begin{aligned}\|\langle f, \psi_{l\cdot}^{\mathrm{per}}\rangle\|_p &\simeq 2^{l(1/p-1/2)}\left\|\sum_m \langle f, \psi_{lm}^{\mathrm{per}}\rangle \psi_{lm}^{\mathrm{per}}\right\|_p \\ &\lesssim 2^{l(1/p-1/2)}(\|K_{l+1,\mathrm{per}}(f)-f\|_p + \|K_{l,\mathrm{per}}(f)-f\|_p).\end{aligned} \tag{4.128}$$

Remark 4.3.25 (Comparison to classical Fourier series) If we use band-limited wavelets ϕ, ψ from Theorem 4.2.9 in the preceding construction, we see from the Poisson summation formula (4.12) that the Fourier coefficients of $\psi_{lm}^{(\mathrm{per})}$ vanish for all k large enough; more precisely,

$$\langle \psi_{lm}^{(\mathrm{per})}, e_k\rangle = 0 \quad \text{whenever} \quad |k| \notin [2^l/3, 2^l(4/3)], \tag{4.129}$$

so the ψ_{lm} are again band limited in the (discrete) Fourier domain. In other words, such periodised wavelets consist of *finitely many* (but growing in l) linear combinations of elements of the standard Fourier basis. Thus, remarkably, having replaced the standard trigonometric polynomials at frequencies $l > 1$ by suitable finite linear combinations of them has made the series in (4.125) converge uniformly for any $f \in C_{\mathrm{per}}((0,1])$, whereas Fourier series fail for 'almost all' elements of $C_{\mathrm{per}}((0,1])$ (see Exercise 4.1.2). The approximation properties of Fourier series are investigated further in Proposition 4.3.29.

Wavelet Characterisation of Periodic Besov Spaces

If $\phi, \psi \in C^S(\mathbb{R})$ generate an S-regular wavelet basis of L^2, and if $\psi_{lm}^{(\mathrm{per})}$ are the associated periodised wavelets constructed earlier, then for $1 \le p \le \infty, 1 \le q \le \infty, 0 < s < S$ (with possibly $S = \infty$ if one uses Meyer wavelets), we define periodic Besov spaces

$$B_{pq}^{s,\mathrm{per}}((0,1]) \equiv \begin{cases}\{f \in L^p((0,1]) : \|f\|_{B_{pq}^{s,\mathrm{per}}} < \infty\}, & 1 \le p < \infty \\ \{f \in C_{\mathrm{per}}((0,1]) : \|f\|_{B_{pq}^{s,\mathrm{per}}} < \infty\}, & p = \infty\end{cases} \tag{4.130}$$

with wavelet-sequence norm given by

$$\begin{aligned}&\|f\|_{B_{pq}^{s,\mathrm{per}}((0,1])} \\ &\equiv \begin{cases}|\langle f,1\rangle| + \left(\sum_{l=0}^{\infty} 2^{ql(s+(1/2)-(1/p))}\left(\sum_{m=0}^{2^l-1}|\langle f, \psi_{lm}^{(\mathrm{per})}\rangle|^p\right)^{q/p}\right)^{1/q}, & 0 < q < \infty \\ |\langle f,1\rangle| + \sup_{l\ge 0} 2^{l(s+(1/2)-(1/p))}\left(\sum_{m=0}^{2^l-1}|\langle f, \psi_{lm}^{(\mathrm{per})}\rangle|^p\right)^{1/p} & q = \infty.\end{cases}\end{aligned} \tag{4.131}$$

When $p = \infty$, the ℓ_p-sequence norms in this display have to be replaced by the supremum norm of ℓ_∞.

In the following theorem we show that the preceding wavelet definition coincides with the definition of $B_{pq}^s((0,1])$ from (4.77) adapted to the periodic situation (discussed at the beginning of this section) and with the approximation-theoretic definition of $B^{s,V}_{pq}((0,1])$ from (4.120).

Theorem 4.3.26 *Let $s > 0, p, q \in [1,\infty]$. Then the spaces*

$$B^s_{pq}((0,1]),\ B^{s,V}_{pq}((0,1]) \quad and \quad B^{s,\mathrm{per}}_{pq}((0,1])$$

coincide, and their norms are all pairwise equivalent.

Proof The proof is similar to Theorem 4.3.2 but needs some modifications because the dilation $x \mapsto 2^l x, l \geq 0$, has no convenient representation modulo 1. The following 'Fourier multiplier' lemma will be useful to deal with this. It implies in particular the classical Bernstein inequality ((4.133); cf. also Lemma 4.3.4) on the circle. All L^p-norms are over $(0,1]$ unless explicitly indicated otherwise.

Lemma 4.3.27 *Let $f \in V_t$ be a trigonometric polynomial of degree at most $t \in \mathbb{N}$, let $m : \mathbb{R} \to \mathbb{C}$ be infinitely differentiable and let $\Phi \in \mathcal{S}(\mathbb{R})$ such that $\Phi = 1$ on $[-1,1]$ and zero outside of $[-2,2]$. We then have, for every $1 \leq p \leq \infty$,*

$$\left\| \sum_{k\in\mathbb{Z}:|k|\leq t} \langle f, e_k\rangle m(k) e_k \right\|_p \leq c \left\| \mathcal{F}^{-1}[m] * \mathcal{F}^{-1}[\Phi(\cdot/t)] \right\|_{L^1(\mathbb{R})} \|f\|_p, \tag{4.132}$$

where $c > 0$ is a universal constant, and $\mathcal{F}^{-1}$ denotes the usual inverse Fourier transform on $\mathbb{R}$. In particular, for any $n \in \mathbb{N}$, D^n the differential operator and every $f \in V_t$, we have

$$\|D^n f\|_p \leq C t^n \|f\|_p, \tag{4.133}$$

where $C = C(n) > 0$ is a fixed constant.

Remark 4.3.28 The proof in fact only requires that m is regular enough so that its Fourier inverse and its convolution with $\mathcal{F}^{-1}[\Phi]$ are defined and regular enough that the operations in the following proof are justified.

Proof The function $\Phi(\cdot/t)$ is supported in $[-2t,2t]$ and identically 1 on $[-t,t]$. Define the function

$$M(u) = m(u)\Phi(u/t), \quad u \in \mathbb{R},$$

which is in $C^\infty(\mathbb{R})$ with support in $[-2t,2t]$ and coincides with $m(u)$ on $[-t,t]$. Then $\mathcal{F}^{-1}M$ defines a continuous function on $\mathbb{R}$ for which Fourier inversion $\mathcal{F}(\mathcal{F}^{-1}M)(k) = M(k)$ holds pointwise. For $c_k = \langle f, e_k\rangle$, we need to estimate the $L^p((0,1])$-norm of

$$\begin{aligned}
\sum_{|k|\leq t} c_k m(k) e_k &= \sum_{|k|\leq t} c_k M(k) e_k = \sum_{|k|\leq t} c_k \mathcal{F}(\mathcal{F}^{-1}M)(k) e_k \\
&= \sum_{|k|\leq t} \int_{\mathbb{R}} (\mathcal{F}^{-1}M)(x) e^{-ikx} c_k e_k dx \\
&= 2\pi \sum_{|k|\leq t} \int_{\mathbb{R}} (\mathcal{F}^{-1}M)(2\pi y) c_k e^{2\pi i(\cdot - y)k} dy \\
&= 2\pi \int_{\mathbb{R}} (\mathcal{F}^{-1}M)(2\pi y) f(\cdot - y) dy,
\end{aligned}$$

where we have used Fubini's theorem twice. Now, applying the $L^p((0,1])$-norm to this identity, using Minkowski's inequality for integrals and the Cauchy-Schwarz inequality, we obtain, for positive constants C, C',

$$\left\| \sum_{|k|\le t} c_k m(k) e_k \right\|_p \le C\|f\|_p \int_{\mathbb{R}} |(\mathcal{F}^{-1}M)(y)|dy.$$

The first inequality establishes the first claim of the lemma because $\mathcal{F}^{-1}M = \mathcal{F}^{-1}m * F^{-1}[\Phi(\cdot/t)]$. The second inequality also follows because we have for $m(u) = (2\pi iu)^n$ that

$$F^{-1}m * F^{-1}[\Phi(\cdot/t)] = t^n D^n[(\mathcal{F}^{-1}\Phi)(t\cdot)]$$

and because $\mathcal{F}^{-1}\Phi \in L^1$. ■

Now, to prove the theorem, we take functions $\hat{\phi}, \hat{\psi}$ generating a Littlewood-Paley decomposition as in (4.24) and define

$$u_0(x) = \sum_{k\in\mathbb{Z}} \langle f, e_k\rangle \hat{\phi}(k) e_k(x), \; u_l(x) = \sum_{k\in\mathbb{Z}} \langle f, e_k\rangle \hat{\psi}(k/2^l) e_k(x), \quad x \in (0,1],$$

so that

$$f = \lim_{j\to\infty} \sum_{k\in\mathbb{Z}} \langle f, e_k\rangle \hat{\phi}(k/2^j) e_k(x) = \lim_{j\to\infty} \sum_{l\le j} u_l.$$

Since $\phi \in \mathcal{S}, \phi(0) = 1$ and $\sup_j \|2^j\varphi(2^j\cdot)\|_{L^1(\mathbb{R})} < \infty$ for any $\varphi \in L^1(\mathbb{R})$, we deduce from Lemma 4.3.27 and the dominated convergence theorem that the last series converges in L^p whenever $f \in L^p$. The proof of the implication (c) $\Rightarrow$ (a) in Proposition 4.3.5 combined with first part of Lemma 4.3.27 applied to

$$m = \frac{\hat{\psi}(\cdot/2^l)}{(e^{i\cdot/2^l} - 1)^r}, \quad \mathcal{F}^{-1}m = 2^l \tilde{m}_r(2^l\cdot), \quad \tilde{m}_r \in \mathcal{S},$$

gives

$$f \in B^s_{pq}((0,1]) \Rightarrow \|u_l\|_p \le c_l 2^{-ls}, \quad \{c_l\} \in \ell_q.$$

Since $u_l \in V_{c2^l}$ for some c, we deduce further from the second claim of Lemma 4.3.27 that

$$\|D^N u_l\|_p \le C' 2^{lN} \|u_l\|_p \le c'_l 2^{l(N-s)}, \quad \{c'_l\} \in \ell_q.$$

We have thus proved an $(0,1]$-analogue of the key decomposition in Proposition 4.3.5, part (b), which characterises $B^s_{pq}((0,1])$ by the decay of $\|u_l\|_p$ as $l \to \infty$, noting that the converse implication (b) $\Rightarrow$ (c) from that proposition follows just as well given Lemma 4.3.27. The rest of the proof of this theorem is now the same as for Theorem 4.3.2, using Proposition 4.3.24 and also (4.128) to compare wavelet sequence norms with approximation errors. We leave the details as Exercise 4.3.8. ■

Relationship to Classical Periodic Function Spaces

Using the preceding theorem and (4.128), we see that for these spaces one has the same imbedding relationships as in Section 4.3 if all spaces involved are replaced by their periodic counterparts. In particular,

$$B^{s,\mathrm{per}}_{pq}((0,1]) \subset C_{\mathrm{per}}((0,1]) \quad \text{whenever} \quad s > 1/p \quad \text{or} \quad s = 1/p, q = 1, \tag{4.134}$$

arguing as before Proposition 4.3.9. We also note that since the support of the functions involved is now bounded,

$$B^{s,per}_{pq}((0,1]) \subseteq B^{s,per}_{p'q}((0,1]),$$

for any $p \geq p', q \in [1,\infty]$. Also from Proposition 4.3.24, we deduce that

$$H^{m,per}_p((0,1]) \equiv H^m_p((0,1]) \cap \{D^\alpha f \text{ periodic } \forall 0 \leq \alpha < m\} \subset B^{m,per}_{p\infty}((0,1]), \quad m \in \mathbb{N}, \tag{4.135}$$

as well as

$$C^{s,per}((0,1]) \equiv C^s((0,1])) \cap \{D^\alpha f \text{ periodic } \forall 0 \leq \alpha \leq [s]\} \subseteq B^{s,per}_{\infty\infty}((0,1]), \quad s > 0. \tag{4.136}$$

One shows further, as in Section 4.3.3, that the last set inclusion can be replaced by an equality if $s \notin \mathbb{N}$. It also follows directly from the definition of $B^{s,V}_{pq}((0,1])$ that the classical periodic Sobolev spaces

$$H^s \equiv \left\{ f \in L^2((0,1]) : \sum_{l \in \mathbb{Z}} (1+|l|)^{2s} |\langle f, e_l \rangle|^2 < \infty \right\}, \quad s > 0,$$

for the trigonometric basis $\{e_l\}$ are equal to $B^s_{22}((0,1])$. We finally remark that the duality theory for Besov spaces with $s \leq 0$ from Section 4.3.2 can be developed for the periodic spaces in exactly the same way, replacing tempered distributions by the space $\mathcal{D}^*$ of periodic Schwartz distributions with only notational changes. In particular, we can define general Besov spaces

$$B^{s,\mathrm{per}}_{pq}((0,1]) \equiv \{f \in \mathcal{D}^* : \|f\|_{B^{s,\mathrm{per}}_{pq}} < \infty\}, \quad s \in \mathbb{R}, p, q \in [1,\infty], \tag{4.137}$$

with norms as in (4.131) with the duality $\langle f, \cdot \rangle$ replaced by the action $T_f(\,\cdot\,)$ of Schwartz distributions.

Approximation Properties of Classical Fourier Series in Besov Spaces

The preceding wavelet techniques give a powerful tool to approximate $f \in L^p$ by its periodised wavelet series. As indicated in Remark 4.3.25, the Meyer wavelet partial sums are trigonometric polynomials in $V_{c2^l}, c > 0$, which uniformly approximate any continuous function $f : (0,1] \to \mathbb{R}$, thus outperforming the standard L^2-projection onto the trigonometric basis of V_{c2^l} which need not converge uniformly for continuous f (see Exercise 4.1.2). A deeper reason behind this fact is that the L^1-norm of the Dirichlet projection kernel D_n from (4.18) diverges as $n \to \infty$; in fact, one can show that

$$\|D_n\|_1 \simeq \log n \tag{4.138}$$

as $n \to \infty$ (see Exercise 4.3.9). From this we can deduce that the approximation properties of Fourier partial sums in periodic B^s_{pq} spaces are off the optimal rate by at most a logarithmic term.

Proposition 4.3.29 *Let $s > 0, p \in [1,\infty]$ and $f \in B^s_{p\infty}((0,1])$. For $e_k = e^{2\pi ik(\cdot)}$, define*

$$S_n(f) = D_n * f = \sum_{|k|\le n} \langle e_k, f\rangle e_k$$

to be the nth Fourier partial sum of f. Then, for every $n \in \mathbb{N}$, there exists a constant c independent of n such that

$$\|S_n(f) - f\|_p \le cn^{-s}\log n.$$

Proof From the definition of $B^{s,V}_{p\infty} = B^s_{p\infty}$ there exists $f_n \in V_n$ such that

$$\|f_n - f\|_p \le c'n^{-s},$$

for some $c' > 0$. Then $S_n(f_n) = f_n$, and so, since $\|S_n(h)\|_p \le \|D_n\|_1\|h\|_p$ for any $h \in L^p$, we obtain

$$\|S_n(f) - f\|_p \le \|S_n(f - f_n)\|_p + \|f_n - f\|_p \lesssim \log n \|f_n - f\|_p,$$

implying the result. ■

The classical way to deal with the suboptimal approximation properties of Fourier series is to consider Fejér sums $F_n * f$ (recall Exercise 4.1.3), for which the Fourier series converges uniformly and in any L^p. Note, however, that Fejér sums have other approximation-theoretic shortcomings when compared to periodised wavelets: they do not in general optimally approximate functions of higher smoothness than 1, they do not give rise to an ortho-normal basis, and they do not characterise general Besov spaces by the decay of their approximation errors. In contrast, if one replaces $S_n(f)$ by the wavelet projection $K_j(f)$ onto the Meyer basis, then (for $p \neq 2$) we have a uniform improvement on $S_n(f)$ simply by removing the $\log n$ term.

4.3.5 Boundary-Corrected Wavelet Bases*

The main problem in defining Besov spaces on a subset of the real line is how to measure regularity of functions at the boundary points. Clearly, any function of compact support in $\mathbb{R}$ can be viewed as a function defined on the whole space, but this gives possibly excessive attention to irregularities of the function at the boundaries of the support set. For periodic functions, one can proceed as in the preceding section. However, it is natural to say that the function $x1_{[0,1]}$ is very regular on $[0,1]$, whereas an expansion in periodic wavelets (or in a basis for $L^2(\mathbb{R})$) will give coefficients pertaining to the right edge point that reflect a step discontinuity. In this subsection we describe a construction of a wavelet basis of $L^2([0,1])$ that allows us to measure the regularity at the boundary points in a correct way even for nonperiodic functions. We rigorously present the main ideas of the construction but refer to the notes at the end of this chapter for references where one can find complete numerical details.

For $N = 1$, we can use the Haar basis directly to approximate functions up to regularity 1 on $[0,1]$. For smoother basis functions, we need a separate construction. Let $N \ge 2$, and let ϕ, ψ be the Nth Daubechies scaling function and wavelet, respectively, from Theorem 4.2.10. Note that $\int_{\mathbb{R}} K(v, v+u)u^l du = \delta_{0l}$, for $l = 0, \dots, N-1, v \in \mathbb{R}$, implies that

$$\int_{\mathbb{R}} K(x,y)y^l dy = \sum_k \langle \phi_k, (\cdot)^l \rangle \phi_k(x) = x^l \quad \forall x.$$

In other words, all polynomials of degree less than or equal to $N-1$ are generated by the ϕ_k in the sense that any such polynomial can be exactly reconstructed by linear combinations of the ϕ_k. The main idea now is to first retain all the scaling functions supported in the interior of $[0,1]$ and then to add ad hoc edge functions so that the resulting basis generates polynomials of degree $\leq N-1$ on $[0,1]$. One then constructs wavelets ψ that are orthogonal in $L^2([0,1])$ to the basis functions and thus also to all polynomials of degree $\leq N-1$, which can be used to describe regularity properties of functions by the decay of the wavelet coefficients.

It is convenient to first construct such a basis of $L^2([0,\infty))$; taking 'mirror images', this then also solves the problem on $[0,1]$. The Daubechies wavelet $\psi = \psi^{(N)}$ is supported in $[-N+1, N]$, and we translate the scaling function $\phi = \phi^{(N)}$ by $-N+1$ to be supported in this interval as well. For notational convenience, we still denote, in this subsection, the translated scaling function by ϕ (it generates the same multiresolution analysis). As ϕ, ψ are continuous, we necessarily have $\phi(-N+1) = \phi(N) = \psi(-N+1) = \psi(N) = 0$, and the translates $\phi_k = \phi(\cdot - k)$, for $k \geq N-1$, are compactly supported in $[0,\infty)$. Considering an example to start with, to reproduce constants on $[0,\infty)$, one can define

$$\phi^0(x) = 1 - \sum_{k=N-1}^{\infty} \phi(x-k).$$

Since $\sum_{k\in\mathbb{Z}} \phi(x-k) = 1$ (recall that $\int \phi = \int K(x,y)dy = 1$), we see that, for $0 \leq x < \infty$,

$$\phi^0(x) = \sum_{k=-\infty}^{N-2} \phi(x-k) = \sum_{k=-N+1}^{N-2} \phi(x-k), \tag{4.139}$$

so ϕ^0 itself has compact support and, as a finite sum of ϕ_k with $k < N-1$, is orthogonal to all $\phi_k, k \geq N-1$. By construction, the functions $\phi^0, \phi_k, k \geq N-1$ generate the constants on $[0,\infty)$.

To construct projection kernels that reproduce polynomials up to a given degree without leaving the multiresolution framework requires a little more care.

Proposition 4.3.30 *For $N \geq 2$, let $\phi = \phi^{(N)}$ be the Nth Daubechies scaling functions translated such that its support is in $[-N+1, N]$. For $k = 0, 1, \dots, N-1$, define*

$$\tilde{\phi}^{(k)}(x) = \sum_{n=k}^{2N-2} \binom{n}{k} \phi(x+n-N+1), \quad x \geq 0.$$

Then the $\tilde{\phi}^{(k)}$ are linearly independent, supported in $[0, 2N-1-k]$ and orthogonal in $L^2([0,\infty))$ to the $\phi_m, m \geq N$. The family

$$\left\{ \tilde{\phi}^{(k)}, \phi_m : k = 0, \dots, N-1, m \geq N \right\}$$

generates all polynomials on $[0,1]$ up to degree $N-1$.

Proof The support property of $\tilde{\phi}^{(k)}$ is immediate from the definitions, and since the support sets are strictly nested, the linear independence also follows. Since $\tilde{\phi}^{(k)}$ consists of a finite sum of ϕ_m, $m < N$, orthogonality to any $\phi_m, m \geq N$ follows as well (noting that $\int_0^\infty \tilde{\phi}^{(k)} \phi_m = \int_{\mathbb{R}} \tilde{\phi}^{(k)} \phi_m$ for those m).

We next show that this basis reproduces polynomials up to order $N-1$. Note that $\binom{n}{k}$ is, as a function of n, a polynomial of degree k, and as is easy to see, these polynomials $(k=0,1,\ldots,N-1)$ can be triangularly transformed into the basic polynomials $1, n, \ldots, n^{N-1}$. Therefore, the linear span of the functions

$$\{\tilde{\phi}^{(k)}, \phi_m : k=0,\ldots,N-1, m \geq N\}$$

contains all polynomials of degree $\leq N-1$ if

$$\{\bar{\phi}^{(k)}, \phi_m : k=0,\ldots,N-1, m \geq N\}$$

does, where

$$\bar{\phi}^{(k)} = \sum_{k=0}^{2N-2} n^k \phi(x+n-N+1), \quad k=0,\ldots,N-1. \tag{4.140}$$

As in the proof of Proposition 4.2.6, using the Poisson summation formula and that $\hat{\phi}(2\pi k) = \delta_{0k}$ (cf. the last step of the proof of Theorem 4.2.10), we see that, for any $k=0,\ldots,N-1$,

$$\sum_{n\in\mathbb{Z}} (x-n)^k \phi(x-n) = \int_{\mathbb{R}} x^k \phi(x) \equiv C_k, \tag{4.141}$$

where we note that $C_0 = 1$. Combined with (4.140) and the binomial theorem, this gives

$$\begin{aligned} p_k(x) \equiv \sum_{n\in\mathbb{Z}} n^k \phi(x-n-N+1) &= \sum_{m\in\mathbb{Z}} [x-N+1-(x-m)]^k \phi(x-m) \\ &= \sum_{l=0}^{k} \binom{k}{l} (-1)^l (x-N+1)^{k-l} C_l. \end{aligned}$$

Thus, p_k is a degree k polynomial with leading term $C_0 x^k = x^k$ – conclude that the $\{p_k : k=0,\ldots,N-1\}$ generate all polynomials of degree less than or equal to $N-1$. Each p_k is represented in the basis $\{\bar{\phi}^{(k)}, \phi_m : k=0,\ldots,N-1, m \geq N\}$ as

$$p_k(x) = (-1)^k \bar{\phi}^{(k)}(x) + \sum_{n=1}^{\infty} n^k \phi(x-n-N+1),$$

the last sum being a finite linear combination of ϕ_m due to the compact support of ϕ. Conclude that this basis indeed generates all polynomials up to degree $N-1$. ■

We note that the $\{\tilde{\phi}^{(k)} : k=0,\ldots,N-1\}$ constructed in the preceding proposition, despite being orthogonal to the ϕ_m, $m \geq N$, are not orthonormal among themselves. Since they are linearly independent, we can apply a Gram-Schmidt procedure to ortho-normalise them. If we start the process at $\tilde{\phi}^{(N-1)}$ and proceed downwards to $k=0$, we can obtain ortho-normal functions $\phi_k^{\text{left}}, k=0,\ldots,N-1$, with support contained in $[0, N+k]$. Summarising, the family

$$\{\phi_k^{\text{left}}, \phi_m : k=0,\ldots,N-1, m \geq N\}$$

is an ortho-normal system in $L^2([0,\infty))$ and generates all polynomials of degree less than or equal to $N-1$.

If we are interested in a similar system in $L^2([0,1])$, we first use the preceding construction with ϕ replaced by $2^{J/2}\phi(2^J(\cdot))$ everywhere, where $J = J(N)$ is such that $2^J \geq 2N$. This ensures that the boundaries do not interact in the sense that the ϕ_{Jk}^{left} are then supported away from 1 (more precisely, in $[0,(2N-1)/2^J]$). We then repeat the preceding procedure on $(-\infty,1]$ (or on $(-\infty,0]$ and then shift everything by 1) to construct ortho-normal edge basis functions $\{\phi_{Jk}^{\text{right}}, k=-N,\dots,-1\}$ near the endpoint 1, supported in $[1-(2N-1)/2^J,1]$. This gives in total 2^J-2N standard Daubechies wavelets ϕ_{Jm} supported in the interior of $[0,1]$ and $2N$ edge basis functions $\phi_{Jk}^{\text{left}}, \phi_{Jk}^{\text{right}}$ which together reproduce polynomials up to degree $N-1$ on $[0,1]$. In particular, the family

$$\left\{\phi_{Jk}^{\text{left}}, \phi_{Jk'}^{\text{right}}, \phi_{Jm} : k=0,\dots N-1, k'=-1,\cdots -N, m=N,\dots,2^J-N-1\right\}, \tag{4.142}$$

which we denote henceforth by

$$\{\phi_{Jk}^{bc} : k=0,\dots,2^J-1\},$$

forms an ortho-normal system in $L^2([0,1])$ whose linear span contains all polynomials on $[0,1]$ up to degree $N-1$.

We now turn to the construction of corresponding wavelet functions, restricting attention first to $[0,\infty)$. Define

$$\tilde{\psi}^{(k)} = \sqrt{2}\phi_k^{\text{left}}(2\cdot) - \sum_{m=0}^{N-1}\langle\sqrt{2}\phi_k^{\text{left}}(2\cdot), \phi_{0,m}^{\text{left}}\rangle\phi_{0,m}^{\text{left}}, \quad k=0,\dots,N-1, \tag{4.143}$$

which are orthogonal to all the $\psi_m = \psi(\cdot - m), m \geq N$, and which are supported in $[0, N+k]$. These can be transformed, after some simple but technical computations, including a Gram-Schmidt ortho-normalisation step, into the ortho-normal system

$$\{\psi_k^{\text{left}}, \psi_m : k=0,\dots,N-1, m \geq N\}$$

of $L^2([0,\infty))$, where the ψ_k^{left} are the ortho-normal boundary-correction wavelets. Repeating this process symmetrically, starting with dilated wavelets $\psi_{Jk} = 2^{J/2}\psi(2^J\cdot)$, $2^J \geq N$, we obtain an ortho-normal system

$$\left\{\psi_{Jk}^{\text{left}}, \psi_{Jk'}^{\text{right}}, \psi_{Jm} : k=0,\dots N-1, k'=-1,\cdots -N, m=N,\dots,2^J-N-1\right\} \tag{4.144}$$

in $L^2([0,1])$. For $j \geq J$, we now define

$$\psi_{jk}^{\text{left}} = 2^{(j-J)/2}\psi_{Jk}^{\text{left}}(2^{j-J}\cdot), \psi_{jk}^{\text{right}} = 2^{(j-J)/2}\psi_{Jk}^{\text{right}}(2^{j-J}\cdot).$$

One then shows, again by simple but technical manipulations, that the family

$$\left\{\phi_{Jk}^{\text{left}}, \phi_{Jk'}^{\text{right}}, \phi_{Jm} : k=0,\dots N-1, k'=-N,\cdots -1, m=N,\dots,2^J-N-1\right\}$$
$$\bigcup\left\{\psi_{lk}^{\text{left}}, \psi_{lk'}^{\text{right}}, \psi_{lm} : k=0,\dots N-1, k'=-N,\cdots -1, m=N,\dots,2^l-N-1, l=J,J+1,\dots\right\}$$
$$\equiv \{\phi_{Jk}^{bc}, \psi_{lm}^{bc} : k=0,\dots,2^J-1, m=0,\dots,2^l-1, l=J,J+1,\dots\} \tag{4.145}$$

forms an ortho-normal system in $L^2([0,1])$. This system is immediately seen to form an actual basis of $L^2([0,1])$ because already the interior standard wavelets $\{\phi_{Jm}, \psi_{lk}\}$ are dense

$L^2([0,1])$ (noting that the support of the ψ_{lm} closest to the boundaries equals $[2^{-l}, 2^{-l}N]$ or $[1-2^{-l}N, 1-2^{-l}]$ and hence approaches the boundary points as $l \to \infty$). Therefore, any $f \in L^2$ can be expanded into its wavelet series

$$f = \sum_{k=0}^{2^J-1} \langle \phi_{Jk}^{bc}, f \rangle \phi_{Jk}^{bc} + \sum_{l=J}^{\infty} \sum_{k=0}^{2^l-1} \langle \psi_{lk}^{bc}, f \rangle \psi_{lk}^{bc}, \tag{4.146}$$

with convergence holding at least in L^2. In fact, as we show now, the series converges uniformly on $[0,1]$ for $f \in C([0,1])$, and more regular $f \in C^s([0,1])$ will give rise to faster decay of the $|\langle f, \psi_{lk}^{bc} \rangle|$.

To see this, note that since the ϕ_{Jk}^{bc} reproduce polynomials of degree up to $N-1$, we necessarily must have

$$\int_0^1 x^\ell \psi_{lk}^{bc}(x) dx = 0 \quad \forall \ell = 0, \dots, N-1, \quad \forall l, k, \tag{4.147}$$

as the ψ_{lk}^{bc} are orthogonal in $L^2([0,1])$ to all the ϕ_{JK}^{bc} (which, in turn, generate the x^ℓ). One thus shows for $f \in C^s([0,1]), 0 < s \leq N$, that

$$\max_k |\langle f, \psi_{lk}^{bc} \rangle| \leq c \|f\|_{C^s([0,1])} 2^{-l(s+1/2)}. \tag{4.148}$$

The last estimate is immediate for interior wavelets, arguing as before (4.117), and follows for the boundary wavelets as well: for instance, at the 0-boundary and assuming $s = N$ for notational simplicity, using the support and dilation properties of ψ_{lk}^{left}, (4.147) and a Taylor expansion, we have, for some $0 < \zeta < 1$,

$$\begin{aligned}
\left| \int_0^1 \psi_{lk}^{\text{left}}(x) f(x) dx \right| &= \left| \int_0^1 \psi_{lk}^{\text{left}}(x)(f(x) - f(0)) dx \right| \\
&= \left| \int_0^1 \psi_{lk}^{\text{left}}(x) \frac{1}{s} D^s f(\zeta x) x^s \right| \\
&\leq C \|f\|_{C^s([0,1])} 2^{-l(s+1/2)} \int_0^{2^{-J}(N+k)} |\psi_{Jk}^{\text{left}}(u)| |u|^s du \\
&\leq C(s, J, N) \|f\|_{C^s([0,1])} 2^{-l(s+1/2)}.
\end{aligned}$$

For $f \in C([0,1])$, these arguments imply that

$$\left| \int \psi_{lk}^{\text{left}}(x) f(x) dx \right| \leq C(J, N) 2^{-l/2} \sup_{|x| \leq 2^{-l}} |f(x) - f(0)| \leq \varepsilon 2^{-l/2}$$

for l large enough. Combined with a similar estimate for the coefficients involving ψ_{lk}^{right}, using uniform convergence of the standard wavelet series in the interior of $[0,1]$ and noting that the number of boundary-correction wavelets stays the same at all levels l, we conclude that the partial sums of the wavelet series in (4.146) converge uniformly whenever $f \in C([0,1])$. Similar results can be proved in L^p when considering Sobolev-type smoothness conditions $f \in H_p^m([0,1])$.

Motivated by these observations, we can define general Besov spaces on $[0,1]$ that will reflect Sobolev-Hölder-type smoothness conditions on $[0,1]$ through the decay of their wavelet coefficients. For $1 \le p \le \infty, 1 \le q \le \infty, 0 < s < S$, take a boundary-corrected Daubechies wavelet basis of regularity S and such that $\phi, \psi \in C^S([0,1])$, and define

$$B^{s,W}_{pq}([0,1]) \equiv \begin{cases} \{f \in L^p([0,1]) : \|f\|_{B^{s,W}_{pq}} < \infty\}, & 1 \le p < \infty \\ \{f \in C([0,1]) : \|f\|_{B^{s,W}_{pq}} < \infty\}, & p = \infty, \end{cases} \tag{4.149}$$

with wavelet-sequence norm, for $|s| < S$, given by

$$\|f\|_{B^{s,W}_{pq}([0,1])}$$

$$\equiv \begin{cases} \left(\sum_{k=0}^{2^J-1} |\langle f, \phi^{bc}_{Jk}\rangle|^p\right)^{1/p} + \left(\sum_{l=J}^{\infty} 2^{ql(s+1/2-1/p)} \left(\sum_{m=0}^{2^l-1} |\langle f, \psi^{bc}_{lm}\rangle|^p\right)^{q/p}\right)^{1/q}, & q < \infty \\ \left(\sum_{k=0}^{2^J-1} |\langle f, \phi^{bc}_{Jk}\rangle|^p\right)^{1/p} + \sup_{l \ge J} 2^{l(s+1/2-1/p)} \left(\sum_{m=0}^{2^l-1} |\langle f, \psi^{bc}_{lm}\rangle|^p\right)^{1/p} & q = \infty, \end{cases}$$

where in case $p = \infty$ the ℓ_p-sequence norms have to be replaced by the maximum norms $\|\cdot\|_\infty$. One then shows, as in previous sections, that the Besov spaces $B^s_{pq}([0,1])$ defined in terms of moduli of continuity in (4.77) coincide with the space $B^{s,W}_{pq}([0,1])$ defined here, with equivalent norms. In particular, using (4.147) and Proposition 4.2.8 (in fact, a simple modification thereof for the boundary-corrected wavelet basis), one shows, by estimating directly the size of $\|\langle f, \psi^{bc}_{l\cdot}\rangle\|_p$ as earlier, that

$$B^{s,W}_{pq}([0,1]) \subset C([0,1]), \quad \text{for } s > 1/p \text{ or } s = 1/p, q = 1, \tag{4.150}$$

that

$$H^m_p([0,1]) \subset B^{m,W}_{p\infty}([0,1]) \quad \forall m \in \mathbb{N}, \tag{4.151}$$

and that

$$C^s([0,1]) \subseteq B^{s,W}_{\infty\infty}([0,1]) \quad \forall s > 0. \tag{4.152}$$

Again, in the last display we have in fact equality when $s \notin \mathbb{N}$. As in the periodic case, the duality theory for Besov spaces with $s \le 0$ from Section 4.3.2 can be developed for the spaces on $[0,1]$ as well, with only formal changes. We also define general-order Besov spaces

$$B^s_{pq}([0,1]) \equiv \left\{ f = \sum_{k=0}^{2J-1} a_k \phi^{bc}_{Jk} + \sum_{l=J}^{\infty} \sum_{k=0}^{2^l-1} c_{lk} \psi^{bc}_{lk} : \|f\|_{B^{s,W}_{pq}([0,1])} < \infty \right\},$$

$$s \in \mathbb{R}, p, q \in [1, \infty], \tag{4.153}$$

consisting of all wavelet series from (4.146) whose coefficients give rise to finite Besov norms $\|\cdot\|_{B^{s,W}_{pq}([0,1])}$. The elements of $B^s_{pq}([0,1])$ for $s < 0$ may be interpreted as Schwartz distributions $f \in \mathcal{S}^*$ whose support is contained in $[0,1]$ (i.e., $f(\varphi) = 0$ for all compactly supported $\varphi \in C^\infty([0,1]^c)$) *plus* a fixed finite linear combination of edge functions.

4.3.6 Besov Spaces on Subsets of $\mathbb{R}^d$

We consider next the situation where the functions f whose regularity one wishes to measure are defined on a general Euclidean space $\mathbb{R}^d$ or a subset thereof. We denote the Euclidean norm of an element x of $\mathbb{R}^d$ by $|x|$. The standard definitions from the beginning of this chapter, such as those of L^p-spaces, obviously generalise to the multivariate case with obvious notation. If $\alpha = (\alpha_1, \dots, \alpha_d), \alpha_i \in \mathbb{N} \cup \{0\}$, is a multi-index of length $|\alpha| = \sum_{j=1}^d \alpha_j$ (noting the slight abuse of the $|\cdot|$ notation), then $x^\alpha = x_1^{\alpha_1} \dots x_d^{\alpha_d}$ whenever $x \in \mathbb{R}^d$, and

$$D^\alpha = \frac{\partial^{|\alpha|}}{\partial x_1^{\alpha_1} \dots \partial x_d^{\alpha_d}}$$

is the mixed partial weak differential operator of order $|\alpha|$. If $f : \mathbb{R}^d \to \mathbb{R}$ is classically differentiable at x, then we set $D^\alpha f = f^{(\alpha)}$ equal to the classical mixed partial derivative. We define

$$H_p^m(\mathbb{R}^d) = \left\{ f \in L^p(\mathbb{R}^d) : D^\alpha f \in L^p(\mathbb{R}^d)\ \forall |\alpha| \le m : \|f\|_{H_p^m(\mathbb{R}^d)} \equiv \|f\|_p + \sum_{|\alpha|=m} \|D^\alpha f\|_p < \infty \right\},$$

$$C^m(\mathbb{R}^d) = \left\{ f \in C_u(\mathbb{R}^d) : f^{(\alpha)} \in C_u(\mathbb{R}^d)\ \forall |\alpha| \le m : \|f\|_{C^m(\mathbb{R}^d)} \equiv \|f\|_\infty + \sum_{|\alpha|=m} \|f^{(\alpha)}\|_\infty < \infty \right\}.$$

As in (4.111), the latter spaces can be generalised directly to $s \notin \mathbb{N}$ with integer part $[s]$

$$C^s(\mathbb{R}^d) = \left\{ f \in C_u(\mathbb{R}^d) : \|f\|_{C^s(\mathbb{R}^d)} \equiv \|f\|_{C^{[s]}(\mathbb{R}^d)} + \sum_{|\alpha|=[s]} \sup_{x \neq y, x,y \in \mathbb{R}^d} \frac{|D^\alpha f(x) - D^\alpha f(y)|}{|x-y|^{s-[s]}} \right\}.$$

Again, we shall occasionally write $H_p^m(\mathbb{R}^d) = H_p^m, C^s(\mathbb{R}^d) = C^s$.

These spaces measure the regularity of the functions f in an *isotropic* way, since the same regularity degree m is assumed to hold for all coordinate directions in $\mathbb{R}^d$. We shall restrict ourselves to the isotropic case in what follows, as the theory then only requires mostly straightforward adaptation from the univariate case. The main ideas for the anisotropic case are, in principle, also similar but require approximation schemes with coordinate-dependent bandwidths/resolution levels, which result in a somewhat cumbersome notation.

Approximation by Multivariate Kernel-Type Operators

We first extend the basic kernel-based approximation schemes from Section 4.1 to the multivariate case. Consider

$$f \mapsto K_h(f) = \int_{\mathbb{R}^d} K_h(\cdot, y) f(y) dy = \frac{1}{h^d} \int_{\mathbb{R}^d} K\left(\frac{\cdot}{h}, \frac{y}{h}\right) f(y) dy, \quad h > 0, \tag{4.154}$$

where $K : \mathbb{R}^d \times \mathbb{R}^d \to \mathbb{R}$ and where $x/h = (x_1/h, \dots, x_d/h)$ for any scalar $h \neq 0$. These correspond to multivariate approximate identities.

Proposition 4.3.31 *Let $f : \mathbb{R}^d \to \mathbb{R}$ be a measurable function, and let K_h as in (4.154) satisfy $\int_{\mathbb{R}^d} \sup_{v \in \mathbb{R}^d} |K(v, v-u)| du < \infty, \int_{\mathbb{R}^d} K(x,y) dy = 1\ \forall x \in \mathbb{R}^d$.*

(i) If f is bounded on $\mathbb{R}^d$ and continuous at $x \in \mathbb{R}^d$, then $K_h(f)(x) \to f(x)$ as $h \to 0$.

(ii) If f is bounded and uniformly continuous on $\mathbb{R}^d$, then $\|K_h(f)-f\|_\infty \to 0$ as $h \to 0$.
(iii) If $f \in L^p(\mathbb{R}^d)$ for some $1 \le p < \infty$, then $\|K_h(f)-f\|_p \to 0$ as $h \to 0$.

The proof is the same as that for Proposition 4.1.3 with elementary modifications pertaining to the multivariate case.

Condition 4.3.32 *Let K be a measurable function $K(x,y):\mathbb{R}^d \times \mathbb{R}^d \to \mathbb{R}$. For $N \in \mathbb{N}$, assume that*

***(M)**: $c_N(K) \equiv \int_{\mathbb{R}^d} \sup_{v\in\mathbb{R}^d} |K(v,v-u)||u|^N du < \infty$, and*
***(P)**: For every $v \in \mathbb{R}^d$ and multiindex α such that $|\alpha| = 1,\dots,N-1$,*

$$\int_{\mathbb{R}^d} K(v,v+u)du = 1 \quad \text{and} \quad \int_{\mathbb{R}^d} K(v,v+u)u^\alpha = 0.$$

The proof of the following proposition is then analogous to that for Proposition 4.1.5 using Taylor's theorem for functions of several variables.

Proposition 4.3.33 *Let K be a kernel that satisfies Condition 4.3.32 for some $N \in \mathbb{N}\cup\{0\}$, and let $K_h(f)$ be as in (4.154). We then have, for c a constant depending only on $m \le N, K$,*

(i) $f \in H_p^m(\mathbb{R}^d), 1 \le p < \infty \Rightarrow \|K_h(f)-f\|_p \le c\|f\|_{H_p^m(\mathbb{R}^d)} p h^m$, and
(ii) $f \in C^m(\mathbb{R}^d) \Rightarrow \|K_h(f)-f\|_\infty \le c\|f\|_{C^m(\mathbb{R}^d)} h^m$.

This result immediately applies, for instance, to product kernels $K(x) = \prod_{i=1}^d k(x_i)$, where k is a kernel satisfying Condition 4.1.4.

Multivariate Tensor Product Wavelet Bases of $L^2(\mathbb{R}^d)$

With the case $d=1$ already established, one can easily construct tensor product wavelet bases of $L^2(\mathbb{R}^d)$ that generate a multiresolution analysis of that space. If ϕ is a scaling function of an S-regular wavelet basis of $L^2(\mathbb{R})$, then the function

$$\Phi(x) = \phi(x_1)\dots\phi(x_d), \quad x = (x_1,\dots,x_d),$$

obviously satisfies

$$\int_{\mathbb{R}^d} \Phi(x)dx = \prod_{i=1}^d \int_{\mathbb{R}} \phi(x)dx = 1, \quad \sum_{k\in\mathbb{Z}^d} \Phi(x-k) = \prod_{i=1}^d \sum_{k_i\in\mathbb{Z}} \phi(x-k_i) = 1. \tag{4.155}$$

The translates $\Phi_k, k \in \mathbb{Z}^d$, are ortho-normal, and if we set

$$K(x,y) = \sum_{k\in\mathbb{Z}^d} \Phi(x-k)\Phi(y-k),$$

then, for every $v \in \mathbb{R}^d$ and multi-index α such that $|\alpha| = 1,\dots,S-1$,

$$\int_{\mathbb{R}^d} K(x,y)dy = 1 \quad \text{and} \quad \int_{\mathbb{R}^d} K(v,v+u)u^\alpha = 0. \tag{4.156}$$

Conclude that the family $\{\Phi_k \equiv \Phi(\cdot - k) : k \in \mathbb{Z}^d\}$ constitutes an ortho-normal system in $L^2(\mathbb{R}^d)$ that reproduces polynomials up to degree $S-1$. The corresponding wavelets are defined as follows: for $\mathcal{I}$ equal to the set of $2^d - 1$ sequences $\iota = (\iota_1, \ldots, \iota_d)$ of zeros and ones, excluding $\iota = 0$, we define

$$\Psi^\iota(x) = \psi^{\iota_1}(x_1)\ldots\psi^{\iota_d}(x_d), \quad \Psi^\iota_{lk} = 2^{ld/2}\Psi^\iota(2^l x - k), \quad l \in \mathbb{N}\cup\{0\}, \quad k \in \mathbb{Z}^d, \tag{4.157}$$

where $\psi^0 = \phi, \psi^1 = \psi$. If we define by $\mathbf{V}_j$ the linear span of the translates

$$\Phi_{jk} = 2^{jd/2}\Phi(2^j x - k), k \in \mathbb{Z}^d,$$

and by

$$\mathbf{W}_j = \mathbf{V}_j \ominus \mathbf{V}_{j-1},$$

then one sees easily that the

$$\left\{\Psi^\iota_{lk} : \iota \in \mathcal{I}, k \in \mathbb{Z}^d, l \in \mathbb{N}\cup\{0\}\right\} \tag{4.158}$$

form an ortho-normal basis of $\mathbf{W}_j$. For example, when $d = 2$, we can write

$$\begin{aligned}\mathbf{V}_j &= V_j \otimes V_j = (V_{j-1}\oplus W_{j-1})\otimes(V_{j-1}\oplus W_{j-1})\\ &= \mathbf{V}_{j-1}\oplus\big((W_{j-1}\otimes V_{j-1})\oplus(V_{j-1}\otimes W_{j-1})\oplus(W_{j-1}\otimes W_{j-1})\big)\\ &\equiv \mathbf{V}_{j-1}\oplus\mathbf{W}_{j-1},\end{aligned}$$

and the wavelets spanning $\mathbf{W}_0$ are the translates of the three functions

$$\psi(x_1)\phi(x_2), \quad \phi(x_1)\psi(x_2), \quad \psi(x_1)\psi(x_2).$$

Moreover, the $\mathbf{V}_j$ are nested, and they are also dense in $L^2(\mathbb{R}^d)$ because the L^2-projections

$$K_j(f) \equiv \int_{\mathbb{R}^d}\sum_{k\in\mathbb{Z}^d}\Phi_{jk}(y)f(y)dy\ \Phi_{jk} \tag{4.159}$$

of any $f \in L^2$ onto $\mathbf{V}_j$ converge to f in L^2 in view of Proposition 4.3.31 with $h = 2^{-j}$ and (4.156). We thus can decompose

$$L^2(\mathbb{R}^d) = \mathbf{V}_0 \oplus \left(\bigoplus_{l=0}^{\infty}\mathbf{W}_l\right) = \mathbf{V}_j \oplus \left(\bigoplus_{l=j}^{\infty}\mathbf{W}_l\right), \tag{4.160}$$

and any $f \in L^2(\mathbb{R}^d)$ can be decomposed into its wavelet series

$$f = \sum_{k\in\mathbb{Z}^d}\langle f, \Phi_k\rangle\Phi_k + \sum_{l=0}^{\infty}\sum_{k\in\mathbb{Z}^d,\iota\in\mathcal{I}}\langle\Psi^\iota_{lk}, f\rangle\Psi^\iota_{lk} \tag{4.161}$$

$$= \sum_{k\in\mathbb{Z}^d}\langle f, \Phi_{jk}\rangle\Phi_{jk} + \sum_{l=j}^{\infty}\sum_{k\in\mathbb{Z}^d,\iota\in\mathcal{I}}\langle\Psi^\iota_{lk}, f\rangle\Psi^\iota_{lk}. \tag{4.162}$$

The series converges in fact uniformly on $\mathbb{R}^d$ whenever $f \in C_u(\mathbb{R}^d)$, using again Proposition 4.3.31 and (4.156). But more is true: by (4.156), the wavelet projection kernel $K(x,y)$ of this tensor product wavelet basis satisfies the conditions of Proposition 4.3.33 with $h = 2^{-j}$, and we can therefore approximate arbitrary $f \in C^m(\mathbb{R}^d)$ or $f \in H^m_p(\mathbb{R}^d)$ by $K_j(f)$ at precision 2^{-jm} whenever $m \le S$. As with univariate wavelets, the decay of the wavelet coefficients for a suitably regular wavelet basis in fact characterises containment of a function in a Besov space on $\mathbb{R}^d$. For $1 \le p \le \infty, 1 \le q \le \infty, 0 < s < S$ and a tensor product wavelet basis based on S-regular $\phi, \psi \in C^S([0,1])$, we define

$$B^{s,W}_{pq}(\mathbb{R}^d) \equiv \begin{cases} \{f \in L^p(\mathbb{R}^d) : \|f\|_{B^{s,W}_{pq}} < \infty\}, & 1 \le p < \infty \\ \{f \in C_u(\mathbb{R}^d) : \|f\|_{B^{s,W}_{pq}} < \infty\}, & p = \infty, \end{cases} \tag{4.163}$$

with wavelet-sequence norm given, for $s \in \mathbb{R}$, by

$$\|f\|_{B^{s,W}_{pq}(\mathbb{R}^d)} \equiv \begin{cases} \left(\sum_{k\in\mathbb{Z}^d} \langle f, \Phi_k\rangle|^p\right)^{1/p} + \left(\sum_{l=0}^{\infty} 2^{ql(s+d/2-d/p)} \left(\sum_{k\in\mathbb{Z}^d, \iota\in\mathcal{I}} |\langle f, \Psi^\iota_{lk}\rangle|^p\right)^{q/p}\right)^{1/q}, & q < \infty \\ \left(\sum_{k\in\mathbb{Z}^d} |\langle f, \Phi_k\rangle|^p\right)^{1/p} + \sup_{l\ge 0} 2^{l(s+d/2-d/p)} \left(\sum_{k\in\mathbb{Z}^d, \iota\in\mathcal{I}} |\langle f, \Psi^\iota_{lk}\rangle|^p\right)^{1/p} & q = \infty, \end{cases}$$

where in case $p = \infty$ the ℓ_p-sequence norms have to be replaced by the supremum norms $\|\cdot\|_\infty$. One can then show, just as in the univariate case, that a Besov space $B^s_{pq}(\mathbb{R}^d)$ defined in terms of multivariate moduli of continuity (similar to (4.77)) coincides with $B^{s,W}_{pq}(\mathbb{R}^d)$ defined here with equivalent norms. We also have the Sobolev imbeddings

$$B^{s,W}_{pq}(\mathbb{R}^d) \subset C_u(\mathbb{R}^d), \quad \text{if } s > d/p \text{ or } s = d/p, q = 1, \tag{4.164}$$

and

$$H^m_p(\mathbb{R}^d) \subset B^{m,W}_{p\infty}(\mathbb{R}^d), \quad C^m(\mathbb{R}^d) \subset B^{m,W}_{\infty\infty}(\mathbb{R}^d) \quad \forall m \in \mathbb{N}, \tag{4.165}$$

and the multidimensional identities

$$C^s(\mathbb{R}^d) = B^{s,W}_{\infty\infty}(\mathbb{R}^d), \quad s \notin \mathbb{N}, \quad H^m_2(\mathbb{R}^d) = B^{m,W}_{22}(\mathbb{R}^d), \quad m \in \mathbb{N}.$$

Since we only consider isotropic function spaces, the proofs of these claims need only formal modifications compared to the univariate case, as we can treat each axis of $\mathbb{R}^d$ individually in the tensor product basis. We thus leave the details to the reader.

Besov Spaces on $[0,1]^d$

Just as in the case of $\mathbb{R}^d$, if we start with a wavelet basis $\{\phi_k, \psi_{lk}\}$ of $L^2([0,1])$, such as the periodic ($\phi_k = 1, \psi_{lk} = \psi^{\mathrm{per}}_{lk}$) or boundary-corrected ($\phi_k = \phi^{bc}_{Jk}, \psi_{lk} = \psi^{bc}_{lk}$) ones from the preceding section, we obtain a tensor product wavelet basis $\{\Phi_k, \Psi^\iota_{lk}\}$ in $L^2([0,1]^d)$, where $\iota \in \mathcal{I}, |\mathcal{I}| = 2^d - 1$, and at the lth level, there are now $O(2^{ld})$ wavelets ψ^ι_{lk} indexed by $k \in \mathcal{K}(l)$. Any such basis is complete in $L^2([0,1]^d)$ because the functions $\{f_1(x_1)\cdots f_d(x_d) : f_i \in L^2([0,1])\}$ are dense in $L^2([0,1]^d)$.

We then can define Besov spaces via wavelet norms as usual:

$$B^{s,W}_{pq}([0,1]^d) \equiv \begin{cases} \{f \in L^p([0,1]^d) : \|f\|_{B^{s,W}_{pq}([0,1]^d)} < \infty\}, & 1 \le p < \infty \\ \{f \in C_u([0,1]^d) : \|f\|_{B^{s,W}_{pq}([0,1]^d)} < \infty\}, & p = \infty, \end{cases} \tag{4.166}$$

with wavelet-sequence norm given, for $s \in \mathbb{R}$, by

$$\|f\|_{B^{s,W}_{pq}([0,1]^d)} \equiv \begin{cases} \left(\sum_k \langle f, \Phi_k\rangle|^p\right)^{1/p} + \left(\sum_l 2^{ql(s+d/2-d/p)} \left(\sum_{k\in\mathcal{K}(l), \iota\in\mathcal{I}} |\langle f, \Psi^\iota_{lk}\rangle|^p\right)^{q/p}\right)^{1/q}, & q < \infty \\ \left(\sum_k |\langle f, \Phi_k\rangle|^p\right)^{1/p} + \sup_l 2^{l(s+d/2-d/p)} \left(\sum_{k\in\mathcal{K}(l), \iota\in\mathcal{I}} |\langle f, \Psi^\iota_{lk}\rangle|^p\right)^{1/p}, & q = \infty, \end{cases}$$

where in case $p = \infty$ the ℓ_p-sequence norms have to be replaced by the maximum norms $\|\cdot\|_\infty$. If the wavelets used in the definition are the periodic ones from (4.130), we shall indicate this by writing $(0,1]^d$ instead of $[0,1]^d$ in the preceding definition. In this case, the periodic n-torus analogues $C^{s,\mathrm{per}}((0,1]^d), H_p^{m,\mathrm{per}}((0,1]^d)$ of the isotropic spaces $C^s(\mathbb{R}^d), H_p^m(\mathbb{R}^d)$ from earlier satisfy

$$C^s((0,1]^d) \subset B^{s,W}_{\infty\infty}((0,1]^d), \quad H_p^m((0,1]^d) \subset B^{m,W}_{p\infty}((0,1]^d), \tag{4.167}$$

with identity when $s \notin \mathbb{N}$. Likewise, $H_2^m((0,1]^d) = B^{m,W}_{22}((0,1]^d)$, and

$$B^{s,W}_{pq}((0,1]^d) \subset C_{\mathrm{per}}((0,1]^d) \quad \text{if } s > d/p \text{ or } s = d/p, q = 1. \tag{4.168}$$

The proofs of these facts proceed as in preceding subsections.

We note that the spaces $C^{s,\mathrm{per}}((0,1]^d), H_p^{m,\mathrm{per}}((0,1]^d)$ of differentiable functions can be naturally defined on the n-torus due to periodicity of the functions involved. When considering nonperiodic boundary-corrected wavelets, one can define $B^{s,W}_{pq}$ just as earlier, but the spaces $C^s([0,1]^d)$ for $s \geq 1$ and $d > 1$ need some interpretation as it is not clear how to define the derivative of a function f on $[0,1]^d$ at the boundary of the unit cube. One may assume Df to exist in the interior of $[0,1]^d$, and the interior tensor product Daubechies wavelets will then have the desired decay behaviour for such functions, arguing as in the preceding subsection. Another possibility is to avoid derivatives altogether, and define Besov spaces only by higher multivariate moduli of smoothness, as in (4.77), which can be shown to be equivalent to the Besov space generated by boundary-corrected tensor product wavelets.

Besov Spaces on General Domains $\Omega \subset \mathbb{R}^d$

We finally consider the situation where we wish to measure the regularity of functions defined on general domains $\Omega \subset \mathbb{R}^d$. Several approaches can be taken here.

The definition of a Besov space $B^s_{pq}(\Omega)$ by integrated moduli of continuity (cf. (4.77), generalised to the multivariate situation in a natural way) is perhaps the most intuitive one. To relate this definition to spaces of differentiable functions, one needs to establish some geometric properties of Ω to define what differentiation means. As soon as some geometric properties of Ω are available, it may seem, in principle, preferable to construct directly basis functions for $L^2(\Omega)$ to approximate functions on Ω. For instance, if Ω is a compact Riemannian manifold or a manifold with some boundary conditions, it is natural to take the eigenfunctions of a Laplace-Beltrami operator as a starting point of general wavelet based on Ω replacing the tools from commutative Fourier analysis from the preceding section by appropriate (usually noncommutative) tools from geometric analysis. This in itself constitutes a separate field of mathematics that we do not introduce here, and we refer to the notes at the end of this chapter for some references.

If no geometric structure of Ω is available, or if one wishes a practical definition that avoids geometric considerations, one may define a Besov space $B^s_{pq}(\Omega)$ simply as the restrictions $g|\Omega$ of all elements $g \in B^s_{pq}(\mathbb{R}^d)$ to the domain Ω. The space

$$B^s_{pq}(\Omega) = \{f : \Omega \to \mathbb{R}, f = g|\Omega \text{ on } \Omega,\ g \in B^s_{pq}(\mathbb{R}^d)\} \tag{4.169}$$

with natural quotient norm

$$\inf\{\|g\|_{B^s_{pq}(\mathbb{R}^d)} : g \in B^s_{pq}(\mathbb{R}^d), g|\Omega = f\} \tag{4.170}$$

is a natural way to measure smoothness on a domain of Ω: it amounts to saying that the functions $f \in B^s_{pq}(\Omega)$ can be extended to functions on $\mathbb{R}^d$ of the same regularity degree. If Ω is bounded, we can take g to be of bounded support in the preceding definition so that the Besov space consists of restrictions to Ω of functions supported in a fixed compact subset of $\mathbb{R}^d$. Moreover, as soon as Ω has sufficiently smooth boundaries, this definition can be shown to coincide with intrinsic definitions that arise from considering integrated moduli of continuity. We do not pursue this further here but discuss references in the notes at the end of this chapter.

4.3.7 Metric Entropy Estimates

Recall that for K a (relatively) compact subset of a metric space $X = (X,d)$, its metric entropy is

$$H(K,d,\varepsilon) = \log N(K,d,\varepsilon),$$

where $N(K,d,\varepsilon)$ denotes the minimal number of closed d-balls of radius $\leq \varepsilon$, and with centres in K, required to cover K. We note the simple properties that if (X,d) is a normed linear space, then, for any scalar $\lambda \neq 0$,

$$H(\lambda K,d,\varepsilon) = H(K,d,\varepsilon/|\lambda|) \tag{4.171}$$

and, for

$$T : (X, \|\cdot\|_X) \to (Y, \|\cdot\|_Y)$$

a continuous linear map between two normed spaces X, Y of operator norm $\|T\|$,

$$H(T(K), \|\cdot\|_Y, \varepsilon) \leq H(K, \|\cdot\|_X, \varepsilon/\|T\|,), \tag{4.172}$$

properties we shall use repeatedly later.

As such, entropy is a quantitative way of measuring the degree of compactness (on the log scale) of a subset of a metric space. In this section we establish bounds for the metric entropy of balls in Besov spaces defined over $[0,1]$ when viewed as subsets of $X = L^q([0,1])$. By the relationships of Besov spaces to classical function spaces, this gives entropy bounds for these spaces as well. Generalisations to Besov spaces defined on subsets of $\mathbb{R}$ or $\mathbb{R}^d$ will be discussed too.

Our proofs proceed by using wavelet theory to reduce the problem to a finite-dimensional one. We therefore start with some entropy estimates for subsets of finite-dimensional spaces, which are of independent interest.

Entropy of ℓ_p-Balls in ℓ_q-Norms

We start with the following standard estimate of the entropy of a ball in the finite-dimensional space:

$$\ell^n_p = (\mathbb{R}^n, \|\cdot\|_{\ell^n_p}), \quad \|x\|^p_{\ell^n_p} = \sum_{i=1}^n |x_i|^p, \quad 1 \leq p \leq \infty.$$

We denote the ball of radius r in ℓ^n_p by $b^n_p(r)$, and write $b^n_p \equiv b^n_p(1)$ for the unit ball.

Proposition 4.3.34 *Let $1 \le p \le \infty$. Then*

$$n\log\left(\frac{r}{2\varepsilon}\right) \le H(b_p^n(r), \|\cdot\|_{\ell_p^n}, \varepsilon) \le n\log\left(\frac{3r}{\varepsilon}\right) \quad \forall\, 0 < \varepsilon < r. \tag{4.173}$$

Proof We only have to prove $r = 1$ in view of (4.171). If vol denotes the Euclidean volume of any measurable subset K of $\ell_p^n = \mathbb{R}^n$, then $\mathrm{vol}(x_0 + \lambda K) = |\lambda|^n \mathrm{vol}(K)$. Given ε, let $y_1, \dots, y_{M(\varepsilon)}$ be a maximal set of points in b_p^n such that $\min_{i \ne j} \|y_i - y_j\|_{\ell_p^n} > \varepsilon$. By maximality, the balls of radius $\le \varepsilon$ centred at the y_i cover b_p^n. Comparing their Euclidean volumes, we see that

$$\mathrm{vol}(b_p^n) \le \mathrm{vol}(b_p^n)\varepsilon^n M(\varepsilon).$$

Also, balls with the same centres y_i and radius $\varepsilon/2$ are disjoint and contained in $b_p^n(3/2)$, so

$$\mathrm{vol}(b_p^n)(\varepsilon/2)^n M(\varepsilon) \le \mathrm{vol}(b_p^n)(3/2)^n,$$

which implies that $(1/\varepsilon)^n \le M(\varepsilon) \le 3^n(1/\varepsilon)^n$. Now the result follows from $M(2\varepsilon) \le N(b_p^n, \|\cdot\|_{\ell_p^n}, \varepsilon) \le M(\varepsilon)$, after taking logarithms. ∎

This result in fact does not depend on p at all; as the proof shows, it just estimates the covering number of the unit ball of an arbitrary finite-dimensional normed space. A more delicate situation arises when the ℓ_q-norm on $\mathbb{R}^n$ is different from the ℓ_p-norm describing the unit ball. For this situation, the following result will be useful:

Proposition 4.3.35 *Let $1 \le p < q \le \infty$, $1/t = 1/p - 1/q$. Then, for some constant $C = C(p,q)$ and every $0 < \varepsilon < 1$,*

$$H(b_p^n, \|\cdot\|_{\ell_q^n}, \varepsilon) \le \begin{cases} C\varepsilon^{-t}\log(2n\varepsilon^t), & \text{if } \varepsilon \ge n^{-1/t} \\ Cn\log(2/(n\varepsilon^t)), & \text{if } \varepsilon \le n^{-1/t}. \end{cases} \tag{4.174}$$

Proof Consider first $q = \infty$, so $t = p$. For the given ε, let $k \in \mathbb{N}$ be such that $(k+1)^{-1} < \varepsilon \le k^{-1}$. Consider all the vectors $z = (z_1, \dots, z_n) \in b_p^n$ whose entries are of the form $z_i = \nu/k$, for some $\nu = 0, \pm 1, \dots, \pm k$. The ℓ_∞-balls of radius ε centred at these vectors cover b_p^n, and the number $N(\varepsilon, p, n)$ of such vectors that lie in b_p^n does not exceed the number of nonnegative integer solutions $(\nu_1, \dots, \nu_n)$ of the inequality $\sum_{i=1}^n |\nu_i|^p \le k^p$. Basic combinatorial arguments, including the standard inequality $m > (m/e)^m$, yield that

$$N(\varepsilon, p, n) \le \left(\frac{2e(n + k^p)}{k^p}\right)^{k^p}.$$

Now, if $\varepsilon \ge n^{-1/t} = n^{-1/p}$, we have $(2\varepsilon)^{-p} \le k^p \le \varepsilon^{-p} \le n$, so

$$H(b_p^n, \|\cdot\|_{\ell_\infty^n}, \varepsilon) \le c\log N(\varepsilon, p, n) \le C\varepsilon^{-p}\log(2n\varepsilon^p). \tag{4.175}$$

For $\varepsilon = n^{-1/p}$, this gives $H(b_p^n, \|\cdot\|_{\ell_\infty^n}, n^{-1/p}) \le C'n$. Now any set $A \subset \ell_\infty^n$ satisfies, for $1 \le \lambda \le [\lambda] + 1 = \bar{\lambda}$, any $\varepsilon' > 0$,

$$\begin{aligned} H(A, \|\cdot\|_{\ell_\infty^n}, \varepsilon'/\lambda) &\le H(A, \|\cdot\|_{\ell_\infty^n}, \varepsilon'/\bar{\lambda}) \le \log(\bar{\lambda}^n) + H(A, \|\cdot\|_{\ell_\infty^n}, \varepsilon') \\ &\le n\log(2\lambda) + H(A, \|\cdot\|_{\ell_\infty^n}, \varepsilon'); \end{aligned}$$

hence, when $\varepsilon \le n^{-1/p} = \varepsilon', \lambda = \varepsilon^{-1} n^{-1/p}$, we deduce

$$H(b_p^n, \|\cdot\|_{\ell_\infty^n}, \varepsilon) \le n \log(2/(\varepsilon n^{1/p})) + C'n \le C''n(\log(2/(\varepsilon n^{1/p})), \tag{4.176}$$

which, combined with (4.175), proves the proposition for $q = \infty$.

When $p < q$, consider any η-covering $\{x_i\}$ of b_p^n by N balls in the ℓ_∞^n-metric with centres in b_n^p. For $x \in b_p^n$ and x_i such that $\|x - x_i\|_\infty \le \eta$, we have

$$\|x - x_i\|_q \le \|x - x_i\|_\infty^{1-p/q} \|x - x_i\|_p^{p/q} \le \eta^{1-p/q} 2^{p/q} \le 2\eta^{1-p/q}.$$

Setting $\eta = \varepsilon^{q/(p-q)}$, we see that

$$H(b_p^n, \|\cdot\|_{\ell_q^n}, 2\varepsilon) \le H(b_p^n, \|\cdot\|_{\ell_\infty^n}, \varepsilon^{q/(q-p)}) = H(b_p^n, \|\cdot\|_{\ell_\infty^n}, \varepsilon^{t/p}),$$

completing the proof by using the result for $q = \infty$ just established. ■

A General Entropy Bound for Classes of Smooth Functions

The main result of this subsection is the following bound on the entropy of Besov balls. It implies the same bound for many of the common classical function spaces, as we discuss later. We can restrict the second Besov index to ∞ in view of $B^s_{pr} \subset B^s_{p\infty}$ for every r.

Theorem 4.3.36 *Let $1 \le p,q \le \infty$, and assume that $s > \max(1/p - 1/q, 0)$. Let $\mathcal{B} \equiv \mathcal{B}(s,p,M)$ be the norm-ball of radius M in either $B^{s,W}_{p\infty}([0,1])$ or $B^{s,\mathrm{per}}_{p\infty}((0,1])$. Then $\mathcal{B}$ is relatively compact in $L^q([0,1])$ and satisfies*

$$H(\mathcal{B}, L^q([0,1]), \varepsilon) \le K \left(\frac{M}{\varepsilon}\right)^{1/s} \quad \forall \varepsilon > 0, \tag{4.177}$$

where the constant K depends only on s,p,q.

Remark 4.3.37 One may show that this bound cannot be improved as a function of ε. The result implies the remarkable fact that the degree of compactness of a ball in $B^s_{pq}([0,1])$ in L^q is the same for all permitted values of p,q. For example (see the following corollary), an s-Sobolev ball with $s > 1/2$ has, up to constants, the same entropy in L^∞ as an s-Hölder ball.

Proof In this proof, all spaces are defined over $[0,1]$, and we omit to mention it in the notation. We set $M = 1$ in view of (4.171), and since $B^s_{p\infty} \subset B^s_{q\infty}$ whenever $p \ge q$, we can restrict to $p \le q$. Under the maintained hypotheses on s,p,q, we have $B^s_{pq} \subset L^q$ (as in Proposition 4.3.9), so $\mathcal{B}$ is a bounded subset of L^q. It hence suffices to prove (4.177) for ε small enough. Also, we may replace in the following arguments ε by $c\varepsilon$ for any fixed constant $c > 0$ at the expense of increasing the constant K in (4.177).

The first step is the reduction to sequence space. Taking ϕ, ψ generating an S-regular, $S > s$, periodic or boundary-corrected wavelet basis of L^2 from Section 4.3.4 or 4.3.5, we can expand any $f \in \mathcal{B}$ as

$$f = \sum_{l \ge J-1} \sum_{k=0}^{2^l - 1} \langle f, \psi_{lk} \rangle \psi_{lk},$$

with the convention that $\psi_{J-1,k} = \phi_k$ – here $J = 0$ in the periodic case and $J \ge 2N$ in the boundary-corrected case. We note that Proposition 4.2.8 holds for the wavelet coefficient sequence norms in $\ell_p^{2^l}$ as well – for the periodised wavelets, we recall (4.128) – and follows likewise for the boundary-corrected wavelets because the number of boundary-correction wavelets is constant in l. Moreover, we notice that $f \in \mathcal{B}$ implies, by definition of the wavelet norm on the Besov space,

$$2^{l(s+1/2-1/p)} \|\langle f, \psi_{l\cdot}\rangle\|_p \le 1 \iff 2^{l(s+1/2-1/p)} \{\langle f, \psi_{l\cdot}\rangle\} \in b_p^{2^l}, \tag{4.178}$$

where we recall that b_p^n is the unit ball of ℓ_p^n. Combining these observations, we thus may estimate the L^q-distance of any two $f, g \in \mathcal{B}$ by

$$\begin{aligned} \|f-g\|_q &\le \sum_l \left\| \sum_{k=0}^{2^l-1} \langle f-g, \psi_{lk}\rangle \psi_{lk} \right\|_q \\ &\lesssim \sum_l 2^{l(1/2-1/q)} \|\langle f-g, \psi_{l\cdot}\rangle\|_q \\ &= \sum_l 2^{l(-s+1/p-1/q)} 2^{l(s+1/2-1/p)} \|\langle f-g, \psi_{l\cdot}\rangle\|_q. \end{aligned} \tag{4.179}$$

If we set

$$\varepsilon_l' = \varepsilon_l 2^{l(s-1/p+1/q)},$$

this means that any sequence of ε_l'-coverings of $b_p^{2^l}$ in ℓ_q-norm will induce an $\varepsilon = \sum_l \varepsilon_l$-covering of $\mathcal{B}$ in L^q norm or, in compact notation,

$$H(\mathcal{B}, L^q, \varepsilon) \le \sum_l H(b_p^{2^l}, \ell_q^{2^l}, \varepsilon_l'). \tag{4.180}$$

Let now $\varepsilon > 0$ be given, and define

$$l_0 = \left[\frac{1}{s}\log_2 \frac{1}{\varepsilon}\right] \tag{4.181}$$

and

$$\varepsilon_l = 2^{-sl_0 - A|l-l_0|}, \tag{4.182}$$

for A a positive constant to be chosen. Since

$$\sum_l \varepsilon_l \le c 2^{-sl_0} \le c'\varepsilon,$$

any such sequence of ε_l'-coverings will give rise to a radius ε covering of $\mathcal{B}$ in L^q-distance. It thus remains to estimate the right-hand side of (4.180) for this choice of ε_l.

Case $p = q$. Choose $A = 1$. We see from (4.178) that the tail of the series (4.179) is of order

$$\sum_{l \ge l_0} 2^{-ls} \le c(s) 2^{-l_0 s} \le c(s)' \varepsilon.$$

This implies that in (4.180) we can restrict the sum to $l \le l_0$, since any ε-covering of the partial sums also covers the full series (with ε increased by a constant factor). Now, to sum the terms $l \le l_0$, we use Proposition 4.3.34 to the effect that $H(b_p^n, \|\cdot\|_{\ell_p^n}, \delta) \lesssim n\log(1/\delta)$; hence,

$$\begin{aligned}
\sum_{l\le l_0} H(b_p^{2^l}, \|\cdot\|_{\ell_p^{2^l}}, \varepsilon_l') &\lesssim \sum_{l\le l_0} 2^l \log\frac{1}{\varepsilon_l 2^{ls}} \\
&\lesssim (s+1)\sum_{l\le l_0} 2^l(l_0-l) \\
&\lesssim (s+1)2^{l_0}\sum_{l\le l_0} 2^{l-l_0}(l_0-l) \\
&\lesssim 2^{l_0} \lesssim (1/\varepsilon)^{1/s},
\end{aligned}$$

completing the proof of this case.

Case $p<q$. Let $1/t \equiv 1/p - 1/q$, and choose $0 < A < s - t^{-1}$, possible by our assumptions on s, p, q. Note that $\varepsilon_l' = \varepsilon_l 2^{l(s-t^{-1})}$, so

$$\log_2\left(2^l(\varepsilon_l')^t\right) = t(s(l-l_0) - A|l-l_0|). \tag{4.183}$$

For $l \le l_0$, this equals $t(s+A)(l-l_0) \le 0$; in particular, the quantity inside the logarithm must be less than or equal to 1, so the second estimate from (4.174) applies. Thus, as in the preceding step,

$$\begin{aligned}
\sum_{l\le l_0} H(b_p^{2^l}, \|\cdot\|_{\ell_q^{2^l}}, \varepsilon_l') &\lesssim \sum_{l\le l_0} 2^l \log\frac{1}{2^l(\varepsilon_l')^t} \\
&\lesssim t(s+A)\sum_{l\le l_0} 2^l(l_0-l) \\
&\le c2^{l_0} \lesssim (1/\varepsilon)^{1/s}.
\end{aligned}$$

In the range $l > l_0$, the first bound from (4.174) applies. Since the right-hand side in (4.183) becomes

$$t(s(l-l_0) - A|l-l_0|) = t(s-A)(l-l_0),$$

for $l > l_0$, we obtain

$$\begin{aligned}
\sum_{l> l_0} H(b_p^{2^l}, \|\cdot\|_{\ell_q^{2^l}}, \varepsilon_l') &\lesssim \sum_{l> l_0} (\varepsilon_l')^{-t}(l-l_0) \\
&\lesssim \sum_{l> l_0} 2^l 2^{(At-st)(l-l_0)}(l-l_0) \\
&\lesssim 2^{l_0}\sum_{j=0}^{\infty} 2^j 2^{(At-st)j} j \\
&\lesssim 2^{l_0} \lesssim (1/\varepsilon)^{1/s},
\end{aligned}$$

since $1 - st + At < 0$ by our choice of A. This completes the proof. ■

This proof generalises without difficulty to the multivariate setting if the Besov space $B_{pq}^s([0,1]^d)$ is defined as in (4.166). Noting that the dimension of the spaces spanned by the ψ_{lk}^{ι} is then of order 2^{ld} instead of 2^l, and from formal adaptations of the proof of Theorem 4.3.36, we obtain that the norm ball $\mathcal{B} \equiv \mathcal{B}(s,p,d,M)$ of $B^s_{p\infty}([0,1]^d)$ of radius M satisfies, for $s > \max(d/p - d/q, 0)$,

$$H(\mathcal{B}, L^q([0,1]^d), \varepsilon) \le K\left(\frac{M}{\varepsilon}\right)^{d/s} \quad \forall \varepsilon > 0, \tag{4.184}$$

where the constant K depends only on s,p,q,d.

The generalisation of Theorem 4.3.36 to Besov spaces defined on unbounded subsets of $\mathbb{R}^d$, such as $\mathbb{R}$ or $\mathbb{R}^d$ itself, is more difficult. The reason is that compactness in $L^q(\mathbb{R})$ is not driven only by smoothness of the functions in a given class but also by their decay at infinity. For instance, the unit ball of $B^s_{\infty\infty}(\mathbb{R})$ is not compact in L^∞ for any s. This means that L^q-entropy estimates for Besov-type function classes will need some additional uniform decay conditions at $\pm\infty$ on the functions considered, or alternatively, one needs to weaken the L^q-norms by re-weighting them by a fixed function that vanishes at $\pm\infty$. We do not pursue this further but refer to the notes at the end of this chapter for relevant references.

The preceding estimate for Besov classes of functions implies immediately estimates for balls in Sobolev-Hölder-type spaces on $[0,1]$, retrieving, in particular, Exercise 3.6.17. Note that when $q = \infty$, this gives rise to similar L^2-bracketing metric entropy bounds from Chapter 3 because supremum-norm coverings automatically induce bracketing coverings of $L^2(P)$-size no larger than in sup-norm.

Corollary 4.3.38 *Let $q \in [1,\infty]$, and let $\mathcal{B}_s$ be the unit ball of either (a) $H_2^{s,\mathrm{per}}((0,1]), H_2^s([0,1])$, for $s \in \mathbb{N}, s > \max(0, 1/2 - 1/q)$, or of (b) $C^{s,\mathrm{per}}((0,1]), C^s([0,1])$ for $s > 0$. Then*

$$H(\mathcal{B}_s, L^q([0,1]), \varepsilon) \le K\left(\frac{1}{\varepsilon}\right)^{1/s} \quad \forall \varepsilon > 0, \tag{4.185}$$

where the constant K depends only on s,q.

A similar result holds for periodic Sobolev-Hölder spaces on the d-dimensional unit cube using (4.184). Again, extensions to function spaces defined on unbounded sets need qualitatively stronger assumptions (see the notes to at the end of this chapter).

Exercises

4.3.1 (Discrete convolution inequality.) Let $c = \{c_l\} \in \ell_q$, and let $c' = \{c'_l\} \in \ell_1$. Define

$$(c * c')_k = \sum_l c_l c_{k-l}.$$

Prove that $c * c' \in \ell_q$ and

$$\|c * c'\|_q \le \|c'\|_1 \|c\|_q.$$

In particular, $\{c'_l\} = \{2^{-ls}\} \in \ell_1$ for any $s > 0$, so $\|c * c'\|_q \le C\|c\|_q$ for some constant C that depends only on s.

4.3.2 Show that an equivalent norm on $B^s_{pq}(\mathbb{R})$ is obtained if the integral/supremum over t in (4.78) is restricted to $(0,1)$ and also if ω_r is replaced by ω_k for any $k > r$. *Hint*: Prove first the auxiliary inequalities

$$\omega_r(f,t,p) \le 2^{r-k}\omega_k(f,t),\ \omega_r(f,\lambda t,p) \le (\lambda+1)^r \omega_r(f,t,p),$$

for $\lambda > 0$, and use $\omega_r(f,t,p) \le 2^r \|f\|_p$.

4.3.3 Show that an equivalent norm on $B^s_{pq}(\mathbb{R})$ is obtained by replacing $\omega_r(f,t,p)$ by $\omega_1(D^{[s]}f,t,p)$ in (4.78). *Hint*: Use (4.74).

4.3.4 (Multiplication algebras.) Let $f,g \in B^s_{pq}(A)$, where $A \subset \mathbb{R}^d$, and assume that s,p,q are such that $B^s_{pq} \subset L^\infty$ (e.g., for $A = \mathbb{R}$ whenever $s > 1/p$ or $s = 1/p, q = 1$ in view of Proposition 4.3.9 and for other A by the corresponding Sobolev imbeddings obtained in this section). Show that the pointwise product

$$h(x) = f(x)g(x), \quad x \in A,$$

satisfies $h \in B^s_{pq}(A)$. When A is a bounded interval, show that $f \in B^s_{pq}(A)$ and $\inf_{x\in A}|f(x)| \geq \zeta > 0$ imply that $1/f \in B^s_{pq}(A)$. *Hint*: Use the modulus of continuity definition of $B^s_{pq}(A)$ and that all functions involved are bounded.

4.3.5 Show that Proposition 4.3.15 still holds true when $\mathcal{G}$ is replaced by a ball in the space $BV(\mathbb{R})$ or by

$$\mathcal{G}(M) = \left\{ g \text{ right-continuous} : \lim_{x\to-\infty} g(x) = 0, \|g\|_\infty + |Dg|(\mathbb{R}) \leq M \right\}, \quad M > 0.$$

Hint: Write $g(x) = \int_{\mathbb{R}} 1_{(-\infty,x]}(t)dDg(t)$. Show next that any ball in $H^1_1(\mathbb{R})$ is contained in $\mathcal{G}(M)$ for some M. Next, generalise Proposition 4.3.15 to hold for $\mathcal{G}$ equal to a ball in $H^m_1, m \in \mathbb{N}$, with approximation error $2^{-j(m+s)}$.

4.3.6 Let $\tilde{f} \in BV(\mathbb{R})$. Show that $\tilde{f}$ is almost everywhere equal to a function f for which the quantity in (4.107) is finite and that, for almost every x,y,

$$f(x) - f(y) = \int_y^x dDf(u).$$

4.3.7 (Separability.) Recall that a Banach space is called *separable* if it contains a countable dense subset. Let A equal either $[0,1]$ or $\mathbb{R}$. Show that the spaces $B^s_{pq}(A)$ are separable for their norm whenever $\max(p,q) < \infty$. Show, moreover, that $B^s_{\infty q}([0,1])$ is separable for any $1 \leq q < \infty$ but that $B^s_{\infty\infty}([0,1])$ is not separable. Finally, show that $B^s_{\infty q}(\mathbb{R})$ is not separable for any q. *Hint*: Use the wavelet characterisations of these spaces to reduce the problem to sequence space.

4.3.8 Complete the proof of Theorem 4.3.26, proceeding as in Theorem 4.3.2. *Hint*: Establish analogues of the conditions of Proposition 4.3.5 for the periodised Meyer wavelet series.

4.3.9 Let D_n be the Dirichlet kernel on $(0,1]$. Prove the inequalities (4.138); that is, find constants $0 < c < C < \infty$ such that

$$c \log n \leq \int_0^1 |D_n(x)| dx \leq C \log n$$

holds for every $n \in \mathbb{N}$. *Hint*: For the lower bound, use the representation (4.18). For the upper bound, show first that $|D_n(x)| \leq C \min(n, |x|^{-1})$ for some $C > 0$, and split the integral into $[0,1/n], [1/n,1]$.

4.3.10 Prove (4.108) when $p = 2$. *Hint*: Since $B^{1/2}_{21}(\mathbb{R}) \subset C_u(\mathbb{R}) \cap L^2$, it suffices, by approximation, to prove the imbedding for compactly supported continuous functions $f \in B^{1/2}_{21}$. Then write

$$\sum_i (f(x_{i+1}) - f(x_i))^2 = \sum_i \frac{(f(x_{i+1}) - f(x_i))^2}{x_{i+1} - x_i}(x_{i+1} - x_i),$$

and use Riemann integrability of f as well as the modulus of smoothness definition of the Besov norm.

4.4 Gaussian and Empirical Processes in Besov Spaces

In this section we use the techniques from this chapter to study the connection between certain Gaussian and empirical processes and the Besov spaces $B^s_{pq}([0,1]), s \in \mathbb{R}$. Throughout this section we let $\{\psi_{lk}\}$ be an S-regular, S sufficiently large, wavelet basis of $L^2([0,1])$, for instance, the periodised basis or the boundary-corrected basis from Sections 4.3.4 and 4.3.5, so that it generates all the Besov spaces $B^s_{pq}([0,1]), |s| < S$, considered. We adopt again the convention that the scaling functions ϕ_k equal the 'first' wavelets $\psi_{J-1,k}$, where $J = 0$ in the periodic case and $J \in \mathbb{N}$ large enough in the boundary-corrected case, and we recall that there are 2^l wavelets ψ_{lk} at level $l \geq 0$.

4.4.1 Random Gaussian Wavelet Series in Besov Spaces

Consider the Gaussian white noise or isonormal Gaussian process on $L^2([0,1])$, that is, the mean zero Gaussian process given by

$$\mathbb{W}(g) \sim N(0, \|g\|_2^2), \quad E\mathbb{W}(g)\mathbb{W}(g') = \langle g, g'\rangle, \quad g, g' \in L^2([0,1]). \tag{4.186}$$

This process generates an infinite sequence of standard Gaussian random variables $g_{lk} = \mathbb{W}(\psi_{lk}) \sim N(0,1)$, where $\{\psi_{lk}\}$ can be any ortho-normal basis of L^2. The process $\mathbb{W}$ can be viewed as a generalised function (or element of $\mathcal{S}^*$) simply by considering the action of the random wavelet series

$$\sum_{l \geq J-1} \sum_k g_{lk} \psi_{lk}$$

on test functions. We can then ask whether $\mathbb{W}$ defines a random variable in some B^s_{pq}, a question that can be reduced to checking convergence of the Besov sequence norms of (g_{lk}). A similar question can be asked for the Brownian bridge process

$$\mathbb{G}(g) \sim N\left(0, \left\|g - \int_0^1 g\right\|_2^2\right), \quad E\mathbb{G}(g)\mathbb{G}(g') = \langle g, g'\rangle - \int_0^1 g \int_0^1 g',$$
$$g, g' \in L^2([0,1]), \tag{4.187}$$

which satisfies $G(c) = 0$ for any constant c but otherwise has the same distribution as $\mathbb{W}$ because all the ψ_{lk} have integral $\int_0^1 \psi_{lk} = 0, l \geq J$. The proofs for both processes thus are effectively the same.

We first consider the simplest case, where $p = q < \infty$ in the Besov indices. The following result is then immediate:

Proposition 4.4.1 *The white noise process $\mathbb{W}$ and the Brownian bridge process $\mathbb{G}$ define tight Gaussian Borel random variables in $B^{-s}_{pp}([0,1])$ for any $s > 1/2$ and $1 \leq p < \infty$.*

Proof For $e_p = E|g_{11}|^p$, we have, from Fubini's theorem, that

$$E\|\mathbb{W}\|^p_{B^{-s}_{pp}} = \sum_l 2^{pl(-s+1/2-1/p)} \sum_k E|g_{lk}|^p = e_p \sum_l 2^{pl(1/2-s)} < \infty,$$

so $\mathbb{W} \in B^{-s}_{pp}$ almost surely, measurable for the cylindrical σ-algebra. Since B^{-s}_{pp} is separable (Exercise 4.3.7) and complete, $\mathbb{W}$ is Borel measurable and the result follows from the Oxtoby-Ulam theorem (Proposition 2.1.4). The Brownian bridge case is the same. ∎

This proof leaves some space for improvement if one considers *logarithmic* Besov spaces, defined as

$$B^{s,\delta}_{pp} \equiv \left\{ f : \|f\|^p_{B^{s,\delta}_{pp}} \equiv \sum_l 2^{pl(s+1/2-1/p)} \max(l,1)^{p\delta} \sum_k |\langle \psi_{lk}, f \rangle|^p < \infty \right\}, \quad \delta, s \in \mathbb{R}. \quad (4.188)$$

Note that $B^{s,0}_{pp} = B^s_{pp}$, but otherwise we can decrease or increase the regularity of the functional space on the logarithmic scale. By the same proof as in the preceding proposition, we obtain

Proposition 4.4.2 *The white noise process* $\mathbb{W}$ *and the Brownian bridge process* $\mathbb{G}$ *define tight Gaussian Borel random variables in* $B^{-1/2,-\delta}_{pp}([0,1])$ *for any* $1 \le p < \infty, \delta > 1/p$.

A way to say that $\mathbb{W}$ has exact smoothness $-1/2$ can be obtained from setting the q-index equal to ∞.

Theorem 4.4.3 *For any* $1 \le p < \infty$, *the random variables* $\|\mathbb{W}\|_{B^{-1/2}_{p\infty}([0,1])}$ *and* $\|\mathbb{G}\|_{B^{-1/2}_{p\infty}([0,1])}$ *are finite almost surely.*

Proof For every M large enough and $e_p = E|g_{11}|^p$, from a union bound and Chebyshev's inequality,

$$\begin{aligned}
\Pr\left(\|\mathbb{W}\|_{B^{-1/2}_{p\infty}} > M \right) &= \Pr\left(\sup_l 2^{-l} \sum_k |g_{lk}|^p > M^p \right) \\
&\le \sum_l \Pr\left(2^{-l} \sum_k (|g_{lk}|^p - e_p) > M^p - e_p \right) \\
&\le \frac{1}{(M^p - e_p)^2} \sum_l 2^{-l} e_{2p},
\end{aligned}$$

so for M large enough, we deduce

$$\Pr(\|\mathbb{W}\|_{B^{-1/2}_{p\infty}} < \infty) > 0.$$

The result now follows from the 0-1 law for Gaussian measures (Corollary 2.1.14) and because the Besov norm as a countable supremum of finite-dimensional ℓ_p-norms is measurable for the cylindrical σ-algebra $\mathcal{C}$. The Brownian bridge case is again the same. ■

Note that by the Borell-Sudakov-Tsirelson inequality (see Exercise 2.1.1) the random variables

$$\|\mathbb{W}\|_{B^{-1/2}_{p\infty}([0,1])}, \quad \|\mathbb{G}\|_{B^{-1/2}_{p\infty}([0,1])}$$

are actually sub-Gaussian. However, unlike in the case $\max(p,q) < \infty$ in Theorem 4.4.3, we cannot infer that $\mathbb{W}, \mathbb{G}$ are tight random variables in $B^{-1/2}_{p\infty}$ because these spaces are nonseparable (Exercise 4.3.7). In fact, $\mathbb{W}, \mathbb{G}$ are not tight in these spaces (see Exercise 4.4.3).

In the case $p = \infty$, we consider certain weighting schemes to obtain sharp results. Define, for $w = (w_l), w_l \ge 1$, a weighting sequence

$$B^{s,w}_{\infty\infty}([0,1]) \equiv \left\{ f : \|f\|_{B^{s,w}_{\infty\infty}} \equiv \sup_l 2^{l(s+1/2)} w_l^{-1} \max_k |\langle \psi_{lk}, f \rangle| < \infty \right\}, \quad s \in \mathbb{R}. \quad (4.189)$$

The preceding spaces can be defined for $q \in [1,\infty)$ too, and a version of the following theorem holds for such q and suitable w as well (with obvious modifications of the proof).

Theorem 4.4.4 *(a) For $\omega = (\omega_l) = (\sqrt{l})$, we have*

$$\Pr\left(\|\mathbb{W}\|_{B^{-1/2,\omega}_{\infty\infty}([0,1])} < \infty\right) = 1.$$

(b) For any w such that $(w_l/\sqrt{l}) \uparrow \infty$ as $l \to \infty$, the white noise process $\mathbb{W}$ defines a tight Gaussian Borel random variable in the closed subspace $B^{-1/2,w}_{\infty\infty,0}$ of $B^{-1/2,w}_{\infty\infty}$ consisting of coefficient sequences satisfying

$$\lim_{l\to\infty} w_l^{-1} \max_k |\langle f, \psi_{lk}\rangle| = 0.$$

(c) The preceding statements remain true if $\mathbb{W}$ is replaced by $\mathbb{G}$.

Proof Since there are 2^l standard Gaussians g_{lk} at the lth level, we have from Lemma 2.3.4

$$E\max_k |g_{lk}| \le C\sqrt{l},$$

for some universal constant C. To prove (a), we have from a union bound and the Borell-Sudakov-Tsirelson inequality (Theorem 2.5.8), for M large enough and some universal constant $c > 0$,

$$\begin{aligned}\Pr\left(\|\mathbb{W}\|_{B^{-1/2,\omega}_{\infty\infty}} > M\right) &= \Pr\left(\sup_l l^{-1/2} \max_k |g_{lk}| > M\right)\\ &\le \sum_l \Pr\left(\max_k |g_{lk}| - E\max_k |g_{lk}| > \sqrt{l}M - E\max_k |g_{lk}|\right)\\ &\le 2\sum_l \exp\{-c(M-C)^2 l\},\end{aligned}$$

so (a) follows from the 0-1 law for Gaussian measures as in the proof of the preceding theorem. The claim (b) follows from the preceding estimate, which implies the desired decay estimate for $|\langle \mathbb{W}, \psi_{lk}\rangle|$, and the fact that $B^{-1/2,-1/2}_{\infty\infty,0}(w)$ is a separable space (isomorphic to c_0). The proof of (c) is identical to the preceding one. ■

Just as before, the random variable $\|\mathbb{W}\|_{B^{-1/2,\omega}_{\infty\infty}}$ is actually sub-Gaussian in view of the Borell-Sudakov-Tsirelson inequality, but $\mathbb{W}$ in (a) does *not* define a tight Gaussian measure on $B^{-1/2,\omega}_{\infty\infty}$ (see Exercise 4.4.3 and the notes at the end of this chapter).

4.4.2 Donsker Properties of Balls in Besov Spaces

For P a probability measure on A and $\mathcal{B}$ a subset of a Besov space $B^s_{pq}(A)$, one may ask whether $\mathcal{B}$ is P-pre-Gaussian or even P-Donsker in the sense of Definitions 3.7.26 and 3.7.29, similar to results in Corollary 3.7.49 and Example 3.7.57 for classes of p-variation and bounded Lipschitz functions. We investigate this question in this subsection for the case $A = [0,1]$, and the case $A = \mathbb{R}$ is discussed in the notes at the end of this chapter. Certain Besov balls will be shown to be P-pre-Gaussian but *not* P-Donsker, displaying a gap between the uniform central limit theorem $\sqrt{n}(P_n - P) \to^d G_P$ and the existence of the limit 'experiment' G_P.

Besov Balls with $s > 1/2$

The metric entropy bound Theorem 4.3.36 and the uniform entropy central limit theorem (CLT) from Theorem 3.7.37 imply the following result:

Theorem 4.4.5 *Let $1 \le p, q \le \infty$, and assume that $s > \max(1/p, 1/2)$. Then any bounded subset $\mathcal{B}$ of $B^s_{pq}([0,1])$ is a uniform Donsker class. In particular, bounded subsets of Sobolev spaces $H^s([0,1])$ and Hölder spaces $C^s([0,1])$ are P-Donsker for $s > 1/2$ and any P.*

Proof For any $\varepsilon > 0$, Theorem 4.3.36 gives an ε-covering of $\mathcal{B}$ in the uniform norm. Since

$$\|f - g\|_{L^2(Q)} \le \|f - g\|_\infty,$$

this induces an ε-covering of $\mathcal{B}$ of the same cardinality in the $L^2(Q)$-norm for any probability measure Q on $[0,1]$. Since

$$\int_0^a (1/\varepsilon)^{1/2s} d\varepsilon < \infty,$$

for any a whenever $s > 1/2$, the result follows from Theorem 3.7.37. ■

To be precise, we note that since we require $s > 1/p$, we can and do view $\mathcal{B}$ as a family of continuous functions in the preceding theorem (possible by the Sobolev imbedding).

This result implies in particular that $\mathcal{B}$ is P-pre-Gaussian for any P. If it is known that P has a bounded density on $[0,1]$, then we can strengthen the preceding result in terms of the P-pre-Gaussian property when $p \in [1,2)$.

Proposition 4.4.6 *If P has a bounded Lebesgue density on $[0,1]$, then any bounded subset $\mathcal{B}$ of $B^s_{pq}([0,1])$ for $1 \le p, q \le \infty$ and $s > 1/2$ is P-pre-Gaussian.*

Proof This follows from Theorem 2.3.7 combined with Theorem 4.3.36, which gives an L^2 covering and then for P with bounded density also an $L^2(P)$ covering of log-cardinality $(1/\varepsilon)^{1/s}$ for the class $\mathcal{B}$. The $L^2(P)$-entropy is thus integrable at zero whenever $s > 1/2$. ■

In the preceding theorem, $\mathcal{B}$ need not in general consist of continuous functions, and to be precise, we note that subsets of B^s_{pq} are then only equivalence classes $[f]$ of functions f. The result holds either for $\mathcal{B}$ consisting of arbitrary selections from each equivalence class or for the Gaussian process $[f] \mapsto G_P(f)$, which is a version of the same process because the covariance function of G_P is constant on each equivalence class for P possessing a Lebesgue density.

An interesting gap between Theorem 4.4.5 and Proposition 4.4.6 arises when $1 \le p < 2$ and P indeed has a bounded density, as Theorem 4.4.5 then requires $s > 1/p$, whereas Proposition 4.4.6 only needs $s > 1/2$. This gap is *real* and provides examples for P-pre-Gaussian classes of functions that are not P-Donsker, as we show in the following proposition for the representative case $p = 1, q = \infty$ (the cases $1 \le p < 2, 1/2 < s \le 1/p, q > 1$ are proved in a similar way using Exercise 4.4.2).

Proposition 4.4.7 *Suppose that P has a bounded Lebesgue density on $[0,1]$, and let $1/2 < s \le 1$. The unit ball $\mathcal{B}$ of $B^s_{1\infty}([0,1])$ is P-pre-Gaussian but not P-Donsker.*

Remark 4.4.8 Again, $B^s_{1\infty}$ does, for $s \le 1$, consist of not necessarily continuous functions and hence has to be viewed as a space Lebesgue-equivalence class of functions. Empirical processes are not defined on equivalence classes of functions but on functions. The set of all a.e. modifications of a fixed (e.g., the constant) function can easily be shown not to be P-Donsker, so to avoid triviality, the preceding statement should be understood as holding for $\mathcal{B}$ equal to *any* class of functions constructed from selecting one element f from each equivalence class $[f]$ in the unit ball of $B^s_{1\infty}([0,1])$, which is what we now prove.

Proof We already know that $\mathcal{B}$ is P-pre-Gaussian; hence, it remains to prove that it is not P-Donsker. The idea is to show that $\mathcal{B}$ contains a fixed unbounded function and then also arbitrary rational translates of it, leading to an almost everywhere unbounded envelope which contradicts the P-Donsker property for absolutely continuous P.

We give the construction of an unbounded function explicitly here for $s < 1$ (the limiting case $s = 1$ is left as Exercise 4.4.2). Take J large enough and a 1-regular Daubechies wavelet ψ translated by a suitable integer such that for every $l \ge J$ the dilated function $\psi(2^l\cdot)$ is supported in $[1/4, 3/4]$, which is possible by Theorem 4.2.10. Since the support is interior to $[0,1]$, the Besov norms of $\psi(2^l\cdot)$ on $A = \mathbb{R}$ coincide with those on $A = [0,1]$. For $\bar{J} \in \mathbb{N}$ and $\epsilon = 1 - s$, define

$$\Psi_{\bar{J}} = \sum_{J \le l \le \bar{J}} 2^{l\epsilon}\psi(2^l\cdot),$$

which is a bounded and continuous function for every fixed $\bar{J}$. Since $\int_0^1 |\psi(2^l\cdot)|dx = 2^{-l}\|\psi\|_1 \le 2^{-l}$, we have $\|\Psi_{\bar{J}}\|_1 \le C$ for some fixed constant C. Moreover, by the wavelet definition of the Besov norm, we have $\|\Psi_{\bar{J}}\|_{B^s_{1\infty}} = 1$, for all $\bar{J} \in \mathbb{N}$. Since $L^\infty \subset B^0_{\infty\infty}$, we have, for some $c > 0$,

$$\|\Psi_{\bar{J}}\|_\infty \ge c\|\Psi_{\bar{J}}\|_{B^0_{\infty\infty}} = \sup_{l \in [J, \bar{J}]} 2^{l\epsilon}.$$

Conclude that by continuity for any given U, there exists $\bar{J}$ large enough and $x_0 \in [1/4, 3/4]$ such that $|\Psi_{\bar{J}}(x)| > U$ for all x in a neighborhood of x_0. For $z \in (0, 1/4)$, any translate $\Psi_{\bar{J}}(\cdot - z)$ is supported in $[0,1]$ and contained in the unit ball of $B^s_{1\infty}$: indeed, the $B^{s,V}_{1\infty}$-norm is unchanged since $\mathcal{F}[f(\cdot - z)] = e^{-izu}\hat{f}(u)$, and by Theorem 4.3.2, the wavelet norm thus can increase by at most a universal constant (by which, if it exceeds 1, we can renormalise). Now let $\{z_i : i = 1, 2, \dots\}$ be an enumeration of the rationals in $(0, 1/4)$, and set

$$\Psi_{i,\bar{J}} = \Psi_{\bar{J}}(\cdot - z_i),$$

which are all contained in the unit ball of $B^s_{1\infty}([0,1])$ and unbounded near $x_0 + z_i \in (0,1)$. We can modify each of these function on a set N_i of Lebesgue measure zero to accommodate the selection from each equivalence class (cf. Remark 4.4.8), and $N = \cup_i N_i$ is still a set of zero Lebesgue measure. For $x \in (x_0, x_0 + 1/4) \setminus N$ arbitrary, we can find, by density of the rationals, z_i such that $(x - z_i)$ is near x_0. Thus,

$$\sup_{f \in \mathcal{B}} |f(x)| \ge \sup_i |\Psi_{i,\bar{J}}(x)| = \sup_i |\Psi_{\bar{J}}(x - z_i)| > U.$$

Since U was arbitrary, we conclude that necessarily $\sup_{f\in\mathcal{B}}|f(x)|=\infty$ for every $x\in(x_0,x_0+1/4)\setminus N$. Since $\sup_{f\in\mathcal{B}}\left|\int f dP\right|\le C\|dP\|_\infty$, we have

$$\sup_{f\in\mathcal{B}}|f(x)-Pf|=\infty$$

for Lebesgue and then also P almost every $x\in[x_0,x_0+1/4]$, so $\mathcal{B}$ cannot be P-Donsker in view of Exercise 3.7.23. ■

Donsker Properties for Critical Values of s

We start with the following result:

Proposition 4.4.9 *Bounded subsets of* $B^{1/p}_{p1}(A)$, $1\le p<2$, *are uniform Donsker classes for* A *any interval in (possibly equal to)* $\mathbb{R}$.

Proof The result follows from the imbedding (4.108) and since p-variation balls are uniform Donsker for $p<2$ (see Corollary 3.7.49). ■

As soon as q is increased above 1, balls in $B^{1/p}_{pq}, p<\infty$, contain unbounded functions (Exercise 4.4.2) and thus by the same proof as in Proposition 4.4.7 can be shown not to form P-Donsker classes. We note that for $s<1/2$, none of the Besov classes are even P-pre-Gaussian for, for example, P equal to the uniform distribution, since $\|G_P\|_{\mathcal{B}}=\infty$ almost surely, arguing as in Section 4.4.1 and using duality arguments.

The critical boundary $s=1/p$ can be analysed more closely. Let us investigate this in the Hilbert space case $p=q=2$: balls in $B^{1/2,\delta}_{22}([0,1])$ defined in (4.188) can be shown to be P-Donsker if (and only if) $\delta>1/2$. Proofs of this fact for arbitrary P can be given using the explicit Hilbert space structure of these spaces (see Exercise 3.7.24 and also the notes at the end of this chapter). When P has a bounded density, a simple direct proof of this fact can be given using the wavelet techniques from this chapter.

Theorem 4.4.10 *For* $\delta>1/2$, *any bounded subset* $\mathcal{B}$ *of* $B^{1/2,\delta}_{22}([0,1])$ *consists of uniformly bounded continuous functions and is P-Donsker for any P with bounded Lebesgue density on* $[0,1]$.

Proof Without loss of generality, we can set $\mathcal{B}$ equal to the unit ball of $B^{1/2,\delta}_{22}$. The uniform boundedness and continuity of elements of $\mathcal{B}$ follow from

$$\|f\|_\infty\lesssim\|f\|_{B^0_{\infty 1}}=\sum_l 2^{l/2}\max_k|\langle f,\psi_{lk}\rangle|$$

$$\le\sqrt{\sum_l l^{-2\delta}}\left(\sum_l 2^l l^{2\delta}\sum_k|\langle f,\psi_{lk}\rangle|^2\right)^{1/2}=C\|f\|_{B^{1/2,\delta}_{22}}$$

in view of the Cauchy-Schwarz inequality and the fact that uniform limits of continuous functions are continuous.

By the same arguments as in Proposition 4.3.12, we have the Hilbert space duality

$$\left(B^{1/2,\delta}_{22}\right)^*=B^{-1/2,-\delta}_{22}.$$

Moreover, one shows easily from the definitions that $(B^{s,\delta}_{22})^*$ is isometric to the subspace of $\ell^\infty(\mathcal{B})$ consisting of (pre-)linear mappings from $\mathcal{B}$ to $\mathbb{R}$, equipped with the supremum norm. The proof of Proposition 4.4.2 and the fact that P has a bounded density imply that G_P exists as a tight random variable in $B^{-1/2,-\delta}_{22}$ for every $\delta > 1/2$ and that, in turn, $\mathcal{B}$ is P-pre-Gaussian. Moreover, since $\mathcal{B}$ is uniformly bounded,

$$\sup_{f\in\mathcal{B}}\left|\int f(dP_n - dP)\right| < \infty,$$

so $\sqrt{n}(P_n - P)$ also defines a random variable in $B^{-1/2,-\delta}_{22}$. To prove that $\mathcal{B}$ is P-Donsker, it thus suffices to prove that

$$\nu_n \equiv \sqrt{n}(P_n - P) \to^d G_P \quad \text{in } B^{-1/2,-\delta}_{22}.$$

For J to be chosen later, let V_J be the space spanned by wavelets up to resolution level J, and let π_{V_J} denote the projection operator onto V_J. Further, let β be the bounded Lipschitz metric for weak convergence in $B^{-1/2,-\delta}_{22}$ (cf. Proposition 3.7.24). Writing $\mathcal{L}(X)$ for the law of a random variable X, it suffices to show that $\beta(\mathcal{L}(\nu_n), \mathcal{L}(G_P)) \to 0$ as $n \to \infty$ to prove the theorem. We have

$$\begin{aligned}\beta(\mathcal{L}(\nu_n), \mathcal{L}(G_P)) \le \beta(\mathcal{L}(\nu_n), \mathcal{L}(\nu_n)\circ\pi^{-1}_{V_J}) + \beta(\mathcal{L}(\nu_n)\circ\pi^{-1}_{V_J},\\ \mathcal{L}(G_P)\circ\pi^{-1}_{V_J}) + \beta(\mathcal{L}(G_P), \mathcal{L}(G_P)\circ\pi^{-1}_{V_J}).\end{aligned}$$

Let $\varepsilon > 0$ be given. The second term is less than $\varepsilon/3$ for every J fixed and n large enough because

$$\pi_{V_J}(\sqrt{n}(P_n - P)) \to^d \pi_{V_J}(G_P) \quad \text{in } V_J$$

by the multivariate CLT. For the first term, note that by Fubini, independence, ortho-normality, definition of the bounded Lipschitz metric and the variance bound

$$nE\langle P_n - P, \psi_{lk}\rangle^2 = Var_P(\psi_{lk}) \le \|p\|_\infty,$$

we can find, for every $\varepsilon > 0$, some J large enough such that

$$\begin{aligned}\beta(\mathcal{L}(\nu_n), \mathcal{L}(\nu_n)\circ\pi^{-1}_{V_J}) &\le E\|\sqrt{n}(id - \pi_{V_J})(P_n - P)\|^2_{B^{-1/2,-\delta}_{22}}\\ &= \sum_{l>J} 2^{-l} l^{-2\delta} \sum_{k=0}^{2^l-1} nE\langle P_n - P, \psi_{lk}\rangle^2\\ &\le \|p\|_\infty \sum_{l>J} l^{-2\delta} < \varepsilon/3.\end{aligned}$$

For the third Gaussian term, the same bound holds true, replacing the variance bound before the preceding display by $EG_P(\psi_{lk})^2 \le \|p\|_\infty$, completing the proof. ■

The duality argument in this proof derives convergence in $\ell^\infty(\mathcal{G})$ for $\mathcal{G}$ a unit ball of a Besov space B^s_{pq} from convergence in $B^{-s}_{p'q'}$ and also can be used for other Besov spaces as long as the duality relationship (4.96) is satisfied.

Exercises

4.4.1 Show that the identity mapping $B^{s,\delta}_{\infty\infty} \to B^{s,\delta'}_{\infty\infty}$ for any $\delta < \delta'$ used in Theorem 4.4.4 is a compact operator. *Hint*: Show that any bounded sequence in the domain space is a Cauchy sequence in the image space.

4.4.2 Show that for any x the space $B^{1/p}_{pq}, q > 1, p < \infty$, contains a function unbounded at x.

4.4.3 Show that the white noise process $\mathbb{W}$ does not define a tight Gaussian random variable in the spaces $B^{-1/2}_{2\infty}, B^{-1/2,\omega}_{\infty\infty}, \omega = (\sqrt{l})$. *Hint*: Show first that if $\mathbb{W}$ were tight, it would concentrate on the completion of its RKHS ℓ_2, which consists of all sequences

$$\left\{(x_{lk}) : \lim_l 2^{-l} \sum_k |x_{lk}|^2 = 0\right\}, \quad \left\{(x_{lk}) : \lim_l l^{-1/2} \max_k |x_{lk}| = 0\right\},$$

respectively, and then use standard properties of i.i.d. $g_{lk} \sim N(0,1)$ sequences to show that $\mathbb{W}$ does not concentrate on these spaces.

4.5 Notes

Section 4.1 All topics in this section consist of classical material. Good references for the background in real analysis are, for example, Folland (1999) and Dudley (2002). Chapters 8 and 9 in Folland (1999) can be particularly recommended for the basic Fourier analysis reviewed here, as well as for historical references on that subject. There are manifold other reference works that can be consulted as well.

Trigonometric series have been studied extensively in the last two centuries, a classical reference is the reprint Zygmund (2002) of the first two editions (1935, 1968). The Haar basis was introduced in Haar (1910), and the Shannon sampling theorem is due to Shannon (1949). The main ideas of a Littlewood-Paley decomposition first occurred in Littlewood and Paley (1931, 1936), who used it to characterise the spaces $L^p((0,1]), 1 < p < \infty$, by different means than Fourier series (which fail when $p \neq 2$).

Section 4.2 Wavelet theory is by now a vast independent subject at the intersection of applied mathematics and harmonic analysis, and there is no space to review its history here. In particular, it should be emphasised that this chapter only attempts to develop some of the by now classical ideas from first principles, with a focus on what is statistically relevant, and does by no means reflect the wide range of activities until today in the field of wavelet theory.

Key contributions in the early days of wavelet theory are the construction of compactly supported wavelets by Daubechies (1988), the rigorous treatment by Mallat (1989) of the multiresolution analysis approach first formulated by Y. Meyer in the mid-1980s, and the landmark monographs by Meyer (1992) and Daubechies (1992) on the subject. The current exposition draws heavily from these references, as well as from the presentation in Härdle et al. (1998). Wavelets usually cannot be expressed in closed analytical form because they are defined as Fourier transforms of possibly quite intricate functions, but efficient algorithms for their computation exist; we refer to Daubechies (1992) for a treatment of these topics.

Section 4.3 The systematic study of function spaces is a subject on its own, and a comprehensive monograph on many aspects of the theory is Triebel (1983). The field developed progressively in the early twentieth century, with a focus on the Lebesgue L^p-spaces, the spaces C^m of continuous and differentiable functions and spaces of functions of bounded variation. The study of Sobolev spaces H^m_2 and of their fruitful interactions with functional analysis was initiated in the 1930s by Sobolev

(1935, 1938). Zygmund's (1945) seminal study of function spaces, where derivatives are replaced by higher-order differences, led to the study of spaces of generalised L^p-moduli of smoothness by Besov (1959, 1961), who introduced the classical definition of Besov spaces. Both Sobolev and Besov spaces were initially (and still are) of interest in the study of regularity questions concerning solutions of elliptic partial differential equations. Several important techniques originated in the work of Hardy and Littlewood (see, e.g., Hardy, Littlewood and Polya (1967)).

The unifying Littlewood-Paley theory approach to Besov spaces is due to Petree (1976) and Triebel (1983), see also Bergh and Löfström (1976). We refer to these references for more on Fourier multiplier theory and for a treatment of interpolation theory that is omitted here. The wavelet approach to Besov spaces is inspired by the main ideas of Littlewood-Paley theory and is rigorously developed in Meyer (1992) and Frazier, Jawerth and Weiss (1991).

The connection between Besov spaces and approximation theory is intimate; we refer to the extensive monographs by DeVore and Lorentz (1993), Lorentz, Golitschek and Makovoz (1996) and also Härdle et al. (1998), which inspired our treatment of the wavelet aspect. Approximation theory often uses piecewise polynomials to approximate functions – for dyadic piecewise polynomials, the usual B-spline bases are 'almost' wavelet bases (up to a Gram-Schmidt ortho-normalisation step). Piecewise polynomials are particularly useful to deal with functions defined on closed intervals. The boundary-corrected wavelets introduced in Cohen, Daubechies and Vial (1993) provide a multiresolution alternative to spline methods on intervals. For approximation in weak metrics, wavelet approximation schemes are most powerful through the duality theory for B^s_{pq}. For convolution kernels, analogues of Proposition 4.3.14 for $s, r \notin \mathbb{N}$ and general p, q also can be proved by using Fourier multipliers techniques; see Giné and Nickl (2008, lemma 12).

The approximation-theoretic perspective is particularly interesting for metric entropy arguments and for the related problem of studying the spectral distributions of eigenvalues of compact operators: the classical papers by Kolmogorov and Tihomirov (1959) and Birman and Solomjak (1967) contain Corollary 4.3.38, part (b), and Theorem 4.3.36, respectively. For general results in $B^s_{pq}(\mathbb{R}^d)$, we refer to the monograph by Edmunds and Triebel (1996) – many authors often study entropy numbers rather than metric entropy, these are related by an inverse relationship to each other. See chapter 15 in Lorentz, Golitschek and Makovoz (1996) for references and results with a focus on metric entropy bounds rather than entropy numbers; see also Nickl and Pötscher (2007) for the multivariate case.

Exhaustive references for further relationships between classical function spaces and Besov spaces, including the issue of the definition of C^s for integer s as well as Fourier analytical characterisations of Sobolev spaces for $p \neq 2$, are Triebel (1983) and DeVore and Lorentz (1993). The relationships between the V_p spaces of functions of bounded p-variation and Besov spaces are further investigated in Bourdaud, de Cristoforis and Sickel (2006), who use interpolation-theoretic ideas from Peetre (1976). A different, more classical approach to p-variation spaces is Dudley and Norvaisa (2011). Recent references for Besov spaces that are defined on general geometric objects (such as manifolds or Dirichlet spaces) are Geller and Pesenson (2011) and Coulhon, Kerkyacharian and Petrushev (2012). Results for when Besov spaces on domains that are defined by restriction coincide with intrinsic definitions can be found in Triebel (1983).

Section 4.4 The results on regularity of white noise processes in negative-order Besov spaces are related to results about the regularity properties of trajectories of Brownian motion, which are in some sense 'integrated versions' of the results here. See Ciesielski, Kerkyacharian and Roynette (1992) as well as Meyer, Sellan and Taqqu (1999). The process $\mathbb{W}$ in $B^{-1/2,\omega}_{\infty\infty}$ for $\omega = (\sqrt{l})$ is an example of a Gaussian random variable in a Banach space that is not tight (Exercise 4.4.3). The cylindrically defined law of $\mathbb{W}$ is in fact a 'degenerate' Gaussian measure that does (assuming the continuum hypothesis) not admit an extension to a Borel measure on $B^{-1/2,\omega}_{\infty\infty}$ (see definition 3.6.2 and proposition 3.11.5 in Bogachev (1998)) and has further unusual properties: $\mathbb{W}$ has a 'hole', that is, $\|\mathbb{W}\|_{\mathcal{M}(\omega)} \in [c, \infty)$ almost surely (see Ciesielski, Kerkyacharian and Roynette (1992)), and depending on finer properties of the sequence ω, the distribution of $\|\mathbb{W}\|_{\mathcal{M}}$ may not be absolutely continuous, and its absolutely continuous part may have infinitely many modes; see Hoffmann-Jorgensen, Shepp and Dudley (1979).

Balls in Hölder and Sobolev spaces were among the first examples of classes of functions for which the P-Donsker property was established; see Strassen and Dudley (1969), Giné (1975) and Marcus (1985). The results for general Besov spaces in this section are based on Nickl (2006) and Nickl and Pötscher (2007), who also deal with the more general case of Besov spaces on $\mathbb{R}$ or even $\mathbb{R}^d$, where moment conditions on P need to be added whenever the Besov index p exceeds 2. The observation that certain balls in Besov spaces are P-pre-Gaussian but not P-Donsker (Proposition 4.4.7) is due to Nickl (2006). Marcus (1985) showed that balls in Sobolev spaces $B^s_{22}(\mathbb{R}^d)$ are P-Donsker for every P on $\mathbb{R}^d$ if and only if $s > d/2$, by using the theory of Hilbert-Schmidt imbeddings (see also Exercise 3.7.24). This also provides another proof of Theorem 4.4.10 (not treated in Marcus (1985)) but does not generalise to the non-Hilbertian setting of Besov spaces B^s_{pq} where either p or q is different from 2.

5

Linear Nonparametric Estimators

In this chapter we consider the classical linear estimators of densities and functions observed under white noise: they are some of the simplest nonparametric estimators and constitute the building blocks for more complex estimators and other statistical procedures such as tests and confidence intervals. For densities, these are the convolution kernel density estimators and the projection-based density estimators, particularly those based on projections of the empirical measure over the subspaces of wavelet multiresolution analyses of $L^2(\mathbb{R})$ or $L^2([0,1])$. For functions observed under white noise, we look at projections of the observation Y_f over the subspaces of multiresolution analyses of $L^2([0,1])$.

In the first section we derive upper bounds for the moments and tail probabilities of $\|f_n - f\|_p$, $1 \le p \le \infty$, where f_n is a linear estimator of f, and apply them to obtain asymptotic convergence rates almost surely and in probability. The rates are optimal in the minimax sense (see Chapter 6) but slower than $1/\sqrt{n}$ and depend on the smoothness of f through the bias. In the second section we look at the approximation of f by f_n in norms weaker than the supremum norm or the L^p-norms, such as multiscale sequence spaces (or negative-order Besov spaces) and norms defined by the supremum of $\int g(f_n - f)$ taken over g in special subsets of the unit balls $\{\|g\|_p \le M\}$, such as the indicators of half-lines (the distribution function), Sobolev balls and bounded variation balls: in these cases, slightly increasing the order of the kernel or the wavelet basis produces estimators of f that attain the optimal rates both in L^p-norms and in these weaker norms. Finally, in the third section we consider the additional topics of contaminated observations, concretely deconvolution, and estimation of nonlinear functionals of a density such as $\int f^2$ and, more generally, integrals $\int \phi(f(x),x)f(x)dx$ for smooth functions ϕ.

Linear estimators have been widely studied, maybe as some of the simplest examples of nonparametric estimation, and there is a wealth of important results about them, impossible to summarise in a book chapter. Far from being exhaustive, the choice of topics here is in part dictated by the statistical developments in subsequent chapters.

5.1 Kernel and Projection-Type Estimators

Given i.i.d. random variables X, X_i, $1 \le i \le n$, $n \in \mathbb{N}$, with law of density f on $\mathbb{R}$, and given a convolution kernel K that we take to be a bounded integrable function which integrates to 1 and a bandwidth or resolution parameter $h = h_n$, the corresponding estimator of f is

$$f_n(x) = P_n * K_h(x) = \frac{1}{n}\sum_{i=1}^{n} K_h(x - X_i) = \frac{1}{nh}\sum_{i=1}^{n} K\left(\frac{x - X_i}{h}\right), \tag{5.1}$$

and clearly, $Ef_n(x) = EK_h(x - X) = \int K_h(x-y)f(y)dy = K_h * f(x)$, which, under quite general conditions (depending on properties of K and the smoothness of f), approximates f pointwise and in different L^p-norms as $h_n \to 0$ (see Section 4.1.2, the definition in (4.14) and Proposition 4.1.1). We call these *convolution kernel density estimators*.

Another kind of estimator based on sampling from f is the projection-type estimators. If $K(x,y)$ is the kernel of an orthogonal projection onto a subspace V of $L^2(\mathbb{R})$ or, more generally, $L^2(A)$, $A \subseteq \mathbb{R}$, if f is a density on $\mathbb{R}$ (or on A), and if X_i are i.i.d. with law of density f as earlier, the corresponding projection estimator is defined as

$$f_n(x) = \int K(x,y)dP_n(y) = \frac{1}{n}\sum_{i=1}^{n} K(x,X_i).$$

Note that $Ef_n(x) = \int K(x,y)f(y)dy = \pi_V(f)$. See (4.16) and (4.17). We are mostly interested, for $A = \mathbb{R}$, in the special case when $V = V_j$ is one of the nested subspaces in a multiresolution analysis of $\mathbb{R}$ (Definition 4.2.1) admitting an S-regular wavelet basis $\{\phi_k, \psi_{\ell k} : k \in \mathbb{Z}, \ell \geq 0\}$, where ϕ is a scaling function and ψ the corresponding wavelet function (Definition 4.2.14). We take the resolution $j = j_n \to \infty$, so $K_{j_n}(f) \to f$ (as in Proposition 4.1.3). Then the kernel of the projection onto V_j is defined by any of the two expressions in (4.29)

$$K_j(x,y) = 2^j K(2^j x, 2^j y), \quad or \quad K_j(x,y) = K(x,y) + \sum_{\ell=0}^{j-1}\sum_{k\in\mathbb{Z}} \psi_{\ell k}(x)\psi_{\ell k}(y), \quad j = 0,1,\ldots, \tag{5.2}$$

where

$$K(x,y) = K_0(x,y) = \sum_{k\in\mathbb{Z}} \phi(x-k)\phi(y-k).$$

(We may equally take $K_j(x,y) = K_J(x,y) + \sum_{\ell=J}^{j-1}\sum_{k\in\mathbb{Z}} \psi_{\ell k}(x)\psi_{\ell k}(y)$, for $J < j$.) The two expressions of K_j in (5.2) are obviously a.s. equal, and they are pointwise equal, for instance, if ϕ and ψ define the Haar basis ($S = 1$) or if ϕ and ψ are uniformly continuous. Under these assumptions and with this notation, the wavelet density estimators f_n of f are defined as

$$f_n(x) = \int K_{j_n}(x,y)dP_n(y) = \frac{1}{n}\sum_{i=1}^{n} K_{j_n}(x,X_i). \tag{5.3}$$

It is convenient to note that $f_n \in V_{j_n}$. This follows because (i) the partial sums of the series defining $K_j(x,x_0)$ for every fixed x_0 are in V_j and (ii) this series is Cauchy in L^2 by part (b) of Definition 4.2.14.

If f is defined on $[0,1]$ or $(0,1]$, then we may take instead of K_{j_n} in (5.3) the projection kernels corresponding to boundary-corrected Nth Daubechies wavelets in Section 4.3.5 or to periodized wavelets (see (4.126)) with respective projection kernels

$$K_{j,bc}(x,y) = \sum_{k=0}^{2^j-1} \phi_{jk}^{bc}(x)\phi_{jk}^{bc}(y), \quad K_{j,\mathrm{per}}(x,y) = \sum_{k=0}^{2^j-1} \phi_{jk}^{\mathrm{per}}(x)\phi_{jk}^{\mathrm{per}}(y), \tag{5.4}$$

and f_n is then defined as in (5.3) but using these kernels instead.

Given $f \in L^2([0,1])$, suppose that f is observed under white noise; that is, we observe

$$dY_f \equiv dY_f^{(n)}(t) = f(t)dt + \frac{\sigma}{\sqrt{n}}dW(t), \quad t \in [0,1], \tag{5.5}$$

totally or partially, as in (1.5) or equivalently (1.8). Then the estimator f_n of f is defined as the orthogonal projection of Y_f onto V_{j_n} for some resolution j_n depending on n; that is,

$$f_n(x) = \int_0^1 K_{j_n,bc}(x,y)dY_f^{(n)}(y) = \int_0^1 K_{j_n,bc}(x,y)f(y)dy + \frac{\sigma}{\sqrt{n}}\int_0^1 K_{j_n,bc}(x,y)dW(y). \tag{5.6}$$

Writing $g_{jk} = \int \phi_{jk}^{bc}(y)dW(y)$ but just g_k if no confusion may arise, this estimator becomes

$$f_n(x) = K_{j_n,bc}(f) + \frac{\sigma}{\sqrt{n}}\sum_{k=0}^{2^{j_n}-1} g_k\phi_{j_nk}^{bc}(x), \quad g_k \text{ i.i.d. } N(0,1), \tag{5.7}$$

or, equivalently, writing $\tilde{g}_{jk} = \int \psi_{jk}^{bc}(y)dW(y)$, which are also i.i.d. $N(0,1)$,

$$f_n(x) = K_{j_n,bc}(f) + \frac{\sigma}{\sqrt{n}}\left[\sum_{k=0}^{2^J-1} g_{Jk}\phi_{Jk}^{bc}(x) + \sum_{l=J}^{j_n-1}\sum_{k=0}^{2^l-1}\tilde{g}_{lk}\psi_{lk}^{bc}(x)\right]. \tag{5.8}$$

If $A = (0,1]$ or $f(0) = f(1)$, $K_{j_n,bc}$ can be replaced by $K_{j_n,\text{per}}$ and ϕ_{jk}^{bc}, ψ_{jk}^{bc} by ϕ_{jk}^{per}, ψ_{jk}^{per}.

This section first develops moment bounds and exponential bounds (concentration and deviation inequalities) for the deviations of these estimators from their means and from the true f, measured in the L^p-norms, $1 \le p \le \infty$, with application to the almost-sure asymptotics of these deviations. We also present the asymptotic distribution of the uniform deviations $\|f_n - Ef_n\|_\infty$ (suitably centred and normalised).

5.1.1 Moment Bounds

The convolution and projection kernels to be used throughout this section will satisfy the following conditions:

For conciseness, in what follows, under Condition 5.1.1(a), we take $h_n = 2^{-j_n}$ and write $K_j = K_{j_n}$ for K_{h_n}.

Condition 5.1.1 *Let $A = \mathbb{R}, A = [0,1]$ or $A = (0,1]$. The sequence of operators*

$$K_j(x,y) = 2^jK(2^jx, 2^jy), \quad x,y \in A, j \ge 0,$$

is called an admissible approximating sequence if it satisfies one of the following conditions:

(a) (Convolution kernel case, $A = \mathbb{R}$): $K(x,y) = K(x-y)$, where $K \in L^1(A) \cap L^\infty(A)$ integrates to 1 and is of bounded p-variation for some finite $p \ge 1$ and right (or left) continuous.

(b) (Multiresolution projection case, $A = \mathbb{R}$):

$$K(x,y) = \sum_{k\in\mathbb{Z}}\phi(x-k)\phi(y-k),$$

with K_j as earlier or $K_j(x,y) = K(x,y) + \sum_{\ell=0}^{j-1}\sum_k \psi_{\ell k}(x)\psi_{\ell k}(y)$, where $\phi, \psi \in L^1(\mathbb{R}) \cap L^\infty(\mathbb{R})$ define an S-regular wavelet basis, have bounded p variation for some $p \geq 1$ and are uniformly continuous, or define the Haar basis.

(c) *(Multiresolution case, $A = [0,1]$): $K_{j,bc}(x,y)$ is the projection kernel at resolution j of a boundary corrected wavelet basis, defined in Chapter 4.*

(d) *(Multiresolution case, $A = (0,1]$): $K_{j,\mathrm{per}}(x,y)$ is the projection kernel at resolution j of the periodisation of a scaling function satisfying condition (b); see (4.126) and (4.127).*

See Lemma 3.6.11 and the text immediately preceding it and Proposition 3.6.12 for the definition and metric entropy properties of functions of bounded p-variation and Proposition 4.3.21 and thereafter for their relationship to Besov spaces. It is also convenient to recall two very useful properties of S-regular wavelet bases, namely, that both functions $\sum_k |\phi(\cdot - k)|$ and $\sum_k |\psi(\cdot - k)|$ are bounded and that there exists a nonnegative measurable function $\Phi \in L^1(\mathbb{R}) \cap L^\infty(\mathbb{R})$ such that $|K(x,y)| \leq \Phi(|x-y|)$, for all $x, y \in \mathbb{R}$; that is, K is dominated by a bounded and integrable convolution kernel Φ.

Some of the properties of the scaling function and the projection kernels from Condition 5.1.1(b) translate into similar properties under Conditions 5.1.1(c) and (d), and we will point them out as needed. For instance, for periodised wavelets as in Conditions 5.1.1(d), recall that $\phi_{jm}^{\mathrm{per}}(x) = 2^{j/2}\sum_{k\in\mathbb{Z}}\phi(2^j x - 2^j k - m)$ (see just after (4.122)), and set $C_{jm}(x) := \sum_{k\in\mathbb{Z}} |\phi(2^j x - 2^j k - m)|$, $m = 0, \ldots, 2^j - 1$. Then, since $\left\|\sum_{k\in\mathbb{Z}} |\phi(x-k)|\right\|_\infty =: \kappa < \infty$ for the scaling functions satisfying Condition 5.1.1(b), we have

$$2^{-j/2}\left\|\sum_{m=0}^{2^j-1} |\phi_{jm}^{\mathrm{per}}(x)|\right\|_\infty \leq \left\|\sum_{m=0}^{2^j-1} C_{jm}\right\|_\infty = \kappa < \infty, \tag{5.9}$$

and for each m, with the changes of variables $y = x - k$, $k \in \mathbb{Z}$, and then $z = 2^j y - m$,

$$\begin{aligned}\int_0^1 C_{jm}(x)dx &= \int_0^1 \sum_{k\in\mathbb{Z}} |\phi(2^j(x-k) - m)|dx \\ &= \sum_{k\in\mathbb{Z}} \int_k^{k+1} |\phi(2^j y - m)|dy \\ &= \int_{\mathbb{R}} |\phi(2^j y - m)|dy = \|\phi\|_1 2^{-j}.\end{aligned} \tag{5.10}$$

Under Condition 5.1.1(c), that is, for boundary-corrected wavelets based on the Nth Daubechies scaling and wavelet functions, the level j multiresolution subspace V_j also has dimension 2^j, but it consists of three orthogonal components, the linear spans of the N left-edge ortho-normal functions $\phi_{jk}^{bc}(x) = \phi_{jk}^{\mathrm{left}}(x) = 2^{j/2}\phi_k^{\mathrm{left}}(2^j x)$, $k = 0, \ldots, N-1$, and the right-edge ortho-normal functions $\phi_{jk}^{bc}(x) = 2^{j/2}\phi_k^{\mathrm{right}}(2^j x)$, $k = 2^j - N, \ldots, 2^j - 1$, where ϕ_k^{left} and ϕ_k^{right} are as smooth as ϕ, bounded and of bounded support, and the linear span of the $2^j - 2N$ interior functions $\phi_{jk}^{bc} = \phi_{j,k}(x) = 2^{j/2}\phi(2^j x - k)$, $k = N, \ldots, 2^j - N - 1$ ((4.145)). To keep a similar notation, set $D_{jm}(x) = 2^{-j/2}|\phi_{jm}(x)|$, $m = 0, \ldots, 2^j - 1$. Then, since N is fixed and these functions have compact support and are bounded, we have, just as earlier,

$$\left\| \sum_{m=0}^{2^j-1} D_{jm} \right\|_\infty =: \kappa' < \infty \quad \text{and} \quad \int_0^1 D_{jm}(x)dx = c2^{-j}, \tag{5.11}$$

where c and κ' are constants that depend only on ϕ.

We consider first-moment bounds for the estimator (5.6) of a function observed in the presence of additive white noise: as usual, the Gaussian case provides a benchmark for the types of results to be obtained in the sampling case and is easier to handle.

Proposition 5.1.2 *Assume Condition 5.1.1(c) or (d) and that $f \in L^2([0,1])$ (with $f(0) = f(1)$ under Condition 5.1.1(d)), and let Y_f be as in (5.5). Let K_j denote either $K_{j,bc}$ and then $2^j \geq 2N$ or $K_{j,\text{per}}$. Let $f_n(x) = \int_0^1 K_j(x,y)dY_f(y)$. Then there exists $C(\phi,p) < \infty$ such that*

$$E\|f_n - Ef_n\|_p = E\left\| \frac{\sigma}{\sqrt{n}} \int K_j(\cdot,t)dW(t) \right\|_p \leq C(\phi,p)\sigma\sqrt{2^j/n}, \tag{5.12}$$

for $1 \leq p < \infty$, *and*

$$E\|f_n - Ef_n\|_\infty \leq C(\phi,\infty)\sigma\sqrt{2^j(j+1)/n}, \tag{5.13}$$

for all n and j ($2^j > 2N$ in the case of boundary-corrected wavelets).

Proof For concreteness we take the kernel based on periodised wavelets, but the proof for boundary corrected Daubechies scaling functions is just the same. The case $1 \leq p \leq 2$ reduces to $p = 2$ for which we have, using (5.7), (5.9) and (5.10),

$$\begin{aligned}
(n/\sigma^2)E\|f_n - Ef_n\|_2^2 &= \int_0^1 E\left(\sum_{m=0}^{2^j-1} g_m\phi_{jm}^{per}(t)\right)^2 dt \\
&\leq 2^j \int_0^1 \sum_{m=0}^{2^j-1} C_{jm}^2(t)dt \\
&\leq 2^j\kappa \sum_{m=0}^{2^j-1} \int_0^1 C_{jm}(t)dt \leq 2^j\kappa\|\phi\|_1.
\end{aligned}$$

For $2 < p < \infty$ we also use Hoffmann-Jørgensen's inequality (Theorem 3.1.22 with T a singleton set and $q = 2$) to obtain

$$\begin{aligned}
(\sqrt{n}/\sigma)E\|f_n - Ef_n\|_p &\leq \left(\int_0^1 E\left|\sum_{m=0}^{2^j-1} g_m\phi_{jm}^{per}(t)\right|^p dt\right)^{1/p} \\
&\lesssim \left(\int_0^1 \left(\sum_{m=0}^{2^j-1} 2^jC_{jm}^2(t)\right)^{p/2} dt\right)^{1/p} + \left(\int_0^1 E\max_m (|g_m|2^{j/2}C_{jm}(t))^p dt\right)^{1/p} \\
&\lesssim 2^{j/2}\int_0^1 \sum_{m=0}^{2^j-1} C_{jm}(t)dt + 2^{j/2}\left(\sum_{m=0}^{2^j-1}\int_0^1 E|g_m|^pC_{jm}^p(t)dt\right)^{1/p} \lesssim 2^{j/2},
\end{aligned}$$

where the implicit multiplicative constants depend only on p and on ϕ through κ and $\|\phi\|_1$. For (5.13), just note that, by (2.35),

$$E\left\|\sum_{m=0}^{2^j-1} g_m \phi_{jm}^{\text{per}}\right\|_\infty \leq 2^{j/2}\left\|\sum_{m=0}^{2^j-1} C_{jm}\right\|_\infty E\left(\max_{0\leq m\leq 2^j-1}|g_m|\right) \leq \kappa 2^{j/2}\sqrt{2\log 2^{j+1}}. \quad \blacksquare$$

This proof gives in fact a bound for the pth moment of $\|f_n - Ef_n\|_p$. We may ask whether these bounds are best possible up to constants (see Exercises 5.1.5 and 5.1.6).

With some more work, we will now obtain the same estimates, up to constants, for density estimators based on sampling. The following lemma will help for $1 \leq p < 2$ as a replacement to Young's inequalities. Given a measure ν on $\mathbb{R}$, we use the notations $L^1(\nu) = L^1(\mathbb{R}, \mathcal{B}, \nu)$ and $\|f\|_{L^1(\nu)} = \int_{\mathbb{R}} |f| d\nu$.

Lemma 5.1.3 *Let* $0 < r < 1$, $t \geq 0$. *If* f *and* k *are two nonnegative functions on* $\mathbb{R}$ *such that* $f, k \in L^1(\mu_{s+t/r})$ *for some* $s > (1-r)/r$, *where* $d\mu_{s+t/r}(x) = (1+|x|)^{s+t/r}dx$, *then, for any* $0 < b < \infty$,

$$\sup_{h\in(0,b]}\int (k_h * f)^r(y)(1+|y|)^t dy \leq C\left(\|f\|_{L^1(\mu_{s+t/r})}\|k\|_{L^1(\mu_{s+t/r})}\right)^r,$$

where $C = (2(1-r)/(sr-(1-r)))^{1-r}(1\vee b)^{sr+t}$.

Proof Let $u = sr/(1-r)$, and note that $v(y) := (1+|y|)^{-u}$ is integrable. Then, by Jensen's inequality with respect to the probability measure $v(y)dy/\|v\|_1$, we have

$$\int (k_h * f)^r(y)(1+|y|)^t dy = \int v(y)^{-1}(k_h * f)^r(y)(1+|y|)^t v(y)dy$$
$$\leq \|v\|_1^{1-r}\left(\int (1+|y|)^{s+t/r}(k_h * f)(y)dy\right)^r.$$

Since $(1+|u+v|) \leq (1+|u|)(1+|v|)$ and $1+|hu| \leq (1\vee h)(1+|u|)$, for $h > 0$, we also have

$$\int (1+|y|)^{s+t/r}(k_h * f)(y)dy \leq \int\int (1+|y-x|)^{s+t/r}(1+|x|)^{s+t/r}k_h(y-x)f(x)dxdy$$
$$= \|f\|_{L^1(\mu_{s+t/r})}\int (1+|v|)^{s+t/r}h^{-1}k(v/h)dv$$
$$\leq (1\vee h^{s+t/r})\|f\|_{L_1(\mu_{s+t/r})}\|k\|_{L_1(\mu_{s+t/r})}.$$

The Lemma follows from these inequalities. ■

The case $p = \infty$ also may benefit from a few remarks before the proof of the main theorem.

Remark 5.1.4 For $p = \infty$, the variable of interest is $\|(P_n - P)(K_j(x,\cdot))\|_\infty$. Under Condition 5.1.1(a), this variable takes the form

$$h\|(P_n - P)(K_h(x,\cdot))\|_\infty = \sup_{x\in\mathbb{R}}\left|(P_n - P)\left(K\left(\frac{x-\cdot}{h}\right)\right)\right|.$$

If K is a bounded function of bounded p-variation for some $p \geq 1$, then, by Proposition 3.6.12, the collection of functions $\mathcal{K} = \{K((x-\cdot)/h) : x \in \mathbb{R}, h > 0\}$ is of *VC* type; in fact, for all $0 < \varepsilon \leq v_p^{1/p}(K)$ and $w > 6$ ($w > 3$ if K is right or left continuous), there exists $A_{w,p} < \infty$ such that

$$N(\mathcal{K}, L^2(Q), \varepsilon v_p^{1/p}(K)) \leq (A_{w,p}/\varepsilon)^{(p\vee 2)w},$$

for all Borel probability measures Q on $\mathbb{R}$. Then we can apply the moment bounds, for example, from Corollary 3.5.8. Now let us consider the same variables under Condition 5.1.1(b). Since for S-regular wavelets $\kappa := \left\|\sum_k |\phi(x-k)|\right\|_\infty < \infty$, we have

$$\begin{aligned}\|2^{-j}(P_n - P)(K_j(x,\cdot))\|_\infty &= \left\|\sum_k \phi(2^j x - k)(P_n - P)(\phi(2^j \cdot -k))\right\|_\infty \\ &\leq \kappa \sup_k |(P_n - P)(\phi(2^j \cdot -k))|, \qquad (5.14)\end{aligned}$$

and the same comment as for convolution kernels applies if the scaling function ϕ is of finite p-variation, $p \geq 1$; see Proposition 3.6.12. This holds as well under Condition 5.1.1(c), although, strictly speaking, under Conditions 5.1.1(c) and (d), the variables $\|2^{-j}(P_n - P)(K_j(x,\cdot))\|_\infty$ can be estimated without resorting to empirical process theory, just by means of Bernstein's inequality.

Theorem 5.1.5 *Let f be a density in $L^p(A)$, assume any of Conditions 5.1.1(a), (b), (c) or (d) for K and let $f_n(x) = \int K_j(x,y)dP_n(y)$, as defined in (5.3). For $1 \leq p < 2$ in the cases Conditions 5.1.1(a) and (b), assume further that f, K^2 and Φ^2 integrate $1 + |t|^s$ for some $s > (2-p)/p$. Then, for $1 \leq p < \infty$, there exist constants L_p, depending on p, K or Φ and f, such that, for all $j \geq 0$ if $p \leq 2$ and for all j such that $2^j < n$ if $p > 2$,*

$$E\|f_n - Ef_n\|_p \leq L_p\sqrt{2^j/n}. \qquad (5.15)$$

If $p = \infty$ and f is bounded, there exists a constant L_∞ depending on K and f in the case of Condition 5.1.1(a) and on ϕ, Φ and f in the other cases such that for all j satisfying $2^j j < n$, we have

$$E\|f_n - Ef_n\|_\infty \leq L_\infty\sqrt{2^j(j+1)/n}. \qquad (5.16)$$

Remark 5.1.6 The proof of Theorem 5.1.5 will produce the following upper bounds for the constants L_p in (5.15) and (5.16): under Conditions 5.1.1(a) and (b), setting $\Phi = K$ in case (a),

$$L_p \leq \left(\frac{2(2-p)}{sp-(2-p)}\right)^{1/p-1/2} \|\Phi^2\|_{L^1(\mu_s)}^{1/2}\|f\|_{L^1(\mu_s)}^{1/2} \quad \text{for } 1 \leq p < 2,$$

$$L_p \leq (p-1)^{1+1/p}2^{4+3/p}\|\Phi\|_2\|f\|_{p/2}^{1/2} + (p-1)^{1+1/p}2^{1-2/p}\|\Phi\|_p \quad \text{for } 2 < p < \infty,$$

$L_p \leq \|\Phi\|_2$ for $p = 2$ and $L_\infty \leq C(1 \vee \|f\|_\infty)^{1/2}$ if $p = \infty$, where C is a numerical constant depending only on K in case (a) and on ϕ and Φ in case (b). Under Conditions 5.1.1(c) and (d), these bounds are $C(\phi)(1 \vee \|f\|_{p/2})^{1/2}$ for $p < \infty$ and $C(1 \vee \|f\|_\infty)^{1/2}$ for $p = \infty$, where the constants $C(\phi)$ depend only on ϕ.

Proof Let $\Phi = K$ in the convolution case and Φ equal to the majorising kernel for the wavelet projection kernel K in the multiresolution case.

Case 1: $1 \le p < 2$ under Conditions 5.1.1(a) and (b). Cauchy-Schwarz's inequality gives, for $j \ge 0$,

$$E\left\|\sum_{i=1}^{n}(K_j(\cdot,X_i) - EK_j(\cdot,X))\right\|_p \le \left(\int E\left|\sum_{i=1}^{n}(K_j(t,X_i) - EK_j(t,X))\right|^p dt\right)^{1/p}$$

$$\le \left(\int \left(E\left(\sum_{i=1}^{n}(K_j(t,X_i) - EK_j(t,X))\right)^2\right)^{p/2} dt\right)^{1/p}$$

$$\le \left\|n^{1/2}(EK_j^2(\cdot,X))^{1/2}\right\|_p$$

$$\le \left\|n^{1/2}(E\Phi_j^2(\cdot - X))^{1/2}\right\|_p$$

$$= (n2^j)^{1/2}\left(\int ((\Phi^2)_j * f)^{p/2}(t)dt\right)^{1/p}.$$

Since $p/2 \le 1$, we cannot use Young's inequality, but since both f and Φ^2 belong to $L^1(\mu_s)$ for some $s > (2-p)/p$, it follows from Lemma 5.1.3 that the $(1/p)$th power of this integral is dominated by $C^{1/p}(\|f\|_{L_1(\mu_s)}\|\Phi^2\|_{L_1(\mu_s)})^{1/2}$, where C is as in the lemma. This proves the proposition for $p < 2$ under Conditions 5.1.1(a) and (b).

Case 2: $p > 2$ under Conditions 5.1.1(a) and (b). Hoffmann-Jørgensen's inequality (Theorem 3.1.22) allows us to write

$$E\left\|\sum_{i=1}^{n}(K_j(\cdot,X_i) - EK_j(\cdot,X))\right\|_p \le \left(\int E\left|\sum_{i=1}^{n}(K_j(t,X_i) - EK_j(t,X))\right|^p dt\right)^{1/p}$$

$$\le (p-1)^{1+1/p}2^{1+2/p}\left(\int \left(2^{3+1/p}\left(nEK_j^2(t,X)\right)^{1/2}\right.\right.$$

$$\left.\left. + \left(E\max_{i\le n}|K_j(t,X_i)|^p\right)^{1/p}\right)^p dt\right)^{1/p}$$

$$\le (p-1)^{1+1/p}2^{1+2/p}\left(2^{3+1/p}\left(\int \left(nEK_j^2(t,X)\right)^{p/2} dt\right)^{1/p}\right.$$

$$\left. + \left(\int E\max_{i\le n}|K_j(t,X_i)|^p dt\right)^{1/p}\right)$$

$$\le (p-1)^{1+1/p}2^{4+3/p}\left(\int (n2^j)^{p/2}((\Phi^2)_j * f)^{p/2}(t)dt\right)^{1/p}$$

$$+ (p-1)^{1+1/p}2^{1+2/p}\left(\int n2^{j(p-1)}((\Phi^p)_j * f)(t)\right) dt.$$

We have, by Young's inequality, that

$$\left[\int ((\Phi^2)_j * f)^{p/2}(t)dt\right]^{2/p} \le \|f\|_{p/2}\|(\Phi^2)_j\|_1 = \|f\|_{p/2}\|\Phi\|_2^2 < \infty,$$

since $f \in L^p$ and $\Phi \in L^2$. Thus, the first summand in the preceding integral is of the right order. Since $\int ((\Phi^p)_j * f)(t)dt = \int \Phi^p(u)du$, the second term in the preceding integral raised to the power $1/p$ is of the order $n^{1/p}2^{j(p-1)/p}$, which, for $2^j < n$, is dominated by $(n2^j)^{1/2}$, again the right bound. Case 2 under Conditions 5.1.1(a) and (b) is thus proved.

Case 3: $p = 2$, Conditions 5.1.1(a) and (b). Same as case 2, with simplifications because Hoffmann-Jørgensen's inequality is not needed.

Case 4: $p = \infty$, Conditions 5.1.1(a) and (b). By Remark 5.1.4 under Condition 5.1.1(a), the class of functions $\{2^{-j}K_j(x,\cdot) : j \in \mathbb{N},\ x \in A\}$ is of *VC* type; that is, its L^2-covering numbers satisfy the uniform in Q bounds in that remark, and we can apply Corollary 3.5.8 since, by continuity, $\|f_n - Ef_n\|_\infty$ is in fact a countable supremum in all cases (Exercise 5.1.3); we take $F = u = v_p^{1/p}(K) \vee \|K\|_\infty$ and $EK^2((x-X)/h) = h\int K^2(-z)f(x+hz)dz \le \|f\|_\infty \|K\|_2^2 h$, and the bound given by this corollary is

$$E\|f_n - Ef_n\|_\infty \le \overline{C}(\sqrt{2^j(j+1)/n} + 2^j(j+1)/n) \le C\sqrt{2^j(j+1)/n},$$

since $n > 2^j j$, where $C = C(K)\|f \vee 1\|_\infty^{1/2}$. Under Condition 5.1.1(b), if ϕ is of bounded p-variation, we use (5.14) and Corollary 3.5.8 on the class of translations and dilations of ϕ with $F = u = v_p^{1/p}(\phi) \vee \phi_\infty$ and $E\phi^2(2^jX - k) \le 2^{-j}\|f\|_\infty$ to obtain a similar bound.

Case 5: $1 \le p < \infty$, Conditions 5.1.1(c) and (d). By the computations in cases 1 and 2, we only need to show that $\|n^{1/2}(EK^2_{j,\mathrm{per}}(\cdot, X))^{1/2}\|_p$ is of the order of $(n2^j)^{1/2}$ and that, moreover, for $p > 2$, the quantity

$$\left(\int_0^1 E\max_{1\le i\le n} |K_{j,\mathrm{per}}(t, X_i)|^p d(t)\right)^{1/p}$$

is of the same order and the same for $K_{j,bc}$. Using (5.9) and (5.10) and the ortho-normality of the functions ϕ_{jm}^{per}, $m = 0\ldots, 2^{j-1}$, w.r.t. Lebesgue measure on $[0,1]$, and letting q be the conjugate of p, we have, for $1 \le p \le 2$,

$$\begin{aligned}
\|n^{1/2}(EK^2_{j,\mathrm{per}}(\cdot, X))^{1/2}\|_p^p &\le n^{p/2}\left(\int_0^1 E\left(\sum_{m=0}^{2^j-1}\phi_{jm}^{\mathrm{per}}(X)\phi_{j,m}^{per}(x)\right)^2 dx\right)^{p/2} \\
&= n^{p/2}\left(\int_0^1 \sum_{m=0}^{2^j-1} E(\phi_{jm}^{\mathrm{per}}(X))^2 dx\right)^{p/2} \\
&\le n^{p/2}2^{jp/2}\left(\int_0^1 \sum_{m=0}^{2^j-1} C_{jm}^2(x)f(x)dx\right)^{p/2} \\
&\le n^{p/2}2^{jp/2}\kappa^p,
\end{aligned}$$

which is what we wanted for this quantity for $1 < p \le 2$. For $p > 2$, recall that, as a typical application of the Fubini-Tonelli theorem and Hölder's inequality, if $Tf(x) = \int_0^1 Q(x,y)f(y)dy$ and $f \in L^{p/2}([0,1])$, then $\|Tf\|_{p/2} \le C\|f\|_{p/2}$, where $C = \left\|\int_0^1 |Q(x,y)|dy\right\|_\infty \vee \left\|\int_0^1 |Q(x,y)|dx\right\|_\infty$. Applying this observation with $Q = K^2_{j,\mathrm{per}}$, we have, by ortho-normality in $L^2([0,1])$ of the functions ϕ_{jm}^{per},

$$\left\|n^{1/2}(EK_{j,\mathrm{per}}^2(\cdot,X))^{1/2}\right\|_p^2 = n\left[\int_0^1\left(\int_0^1 K_{j,\mathrm{per}}^2(x,y)f(y)dy\right)^{p/2}dx\right]^{2/p}$$

$$\leq n\|f\|_{p/2}\sup_{x\in(0,1]}\sum_{m=0}^{2^j-1}(\phi_{jm}^{\mathrm{per}}(x))^2$$

$$\leq \kappa^2\|f\|_{p/2}n2^j$$

as desired. Still, for $p>2$, respectively, by hypothesis and a simple computation, we have

$$n\leq n^{p/2}2^{j(1-p/2)}\quad\text{and}\quad\int_0^1 EK_{j,\mathrm{per}}^2(y,X)dy=\int_0^1\sum_{m=0}^{2^j-1}(\phi_{jm}^{\mathrm{per}}(y))^2f(y)dy\leq 2^j\kappa^2$$

and, by (5.9),

$$\|K_{j,\mathrm{per}}\|_\infty\leq 2^j\left\|\sum_{m=0}^{2^j-1}C_{jm}\right\|_\infty\max_m\|C_{jm}\|_\infty\leq\kappa^2 2^j.$$

Hence, we have

$$\int_0^1 E\max_{1\leq i\leq n}|K_{j,\mathrm{per}}(t,X_i)|^p dt\leq n\|K_{j,\mathrm{per}}\|_\infty^{p-2}\int_0^1 EK_{j,\mathrm{per}}^2(t,X)dt$$

$$\leq n^{p/2}2^{j(1-p/2)}(\kappa^2 2^j)^{p-2}2^j\kappa^2$$

$$=\kappa^{2(p-1)}n^{p/2}2^{jp/2},$$

which is also of the desired order.

The proof for boundary-corrected wavelets is not different, just using (5.11) instead of (5.9) and (5.10).

Case 6: $p=\infty$, Conditions 5.1.1(c) and (d). As in the preceding case, it suffices to consider the theorem for periodised wavelets. By (5.9),

$$E\|f_n-Ef_n\|_\infty=E\left\|\sum_{m=0}^{2^j-1}\phi_{jm}^{\mathrm{per}}(\cdot)(P_n-P)(\phi_{jm}^{\mathrm{per}})\right\|_\infty$$

$$\leq\left\|\sum_{m=0}^{2^j-1}\phi_{jm}^{\mathrm{per}}\right\|_\infty E\max_{0\leq m<2^j}|(P_n-P)(\phi_{jm}^{\mathrm{per}})|$$

$$\leq\kappa 2^j E\max_{0\leq m<2^j}|(P_n-P)(C_{jm})|.$$

We apply Bernstein's inequality in expectation form, (3.202) in Lemma 3.5.12, which, since by (5.9) and (5.10)

$$\|C_{jm}\|_\infty\leq\kappa\quad\text{and}\quad EC_{jm}^2(X)\leq\kappa EC_{jm}(X)\leq\kappa\|f\|_\infty\|\phi\|_1 2^{-j},$$

gives

$$nE \max_{0\le m<2^j} |(P_n - P)(C_{jm})| \le \sqrt{2n\kappa \|f\|_\infty \|\phi\|_1 2^{-j}\log(2^{j+1})} + \frac{\kappa}{3}\log(2^{j+1}).$$

Collecting terms and using that $2^j j < n$, we thus obtain

$$E\|f_n - Ef_n\|_\infty \le \left[2\kappa^{3/2}\|\phi\|_1^{1/2}\|f\|_\infty^{1/2} + \frac{\kappa^2}{4}\right]\sqrt{\frac{2^j j}{n}}. \quad ■$$

Note that the proof of this theorem provides concrete reasonable expressions for the numerical factors C depending on K, ϕ or Φ in the definition of L_p in Remark 5.1.6. This result can be combined with approximation bounds for $K_j(f) - f$. If we use, for instance, an S-regular wavelet basis (as in Definition 4.2.14, or in (4.124) or (4.145)) with $S > s$, then, for $f \in B^s_{p\infty}(A)$, we have the bound (cf. Proposition 4.3.8)

$$\|K_j(f) - f\|_p \le c\|f\|_{B^s_{p\infty}(A)} 2^{-js}, \quad 1 \le p \le \infty.$$

For $f_n(j,x) = P_n(K_j(x,\cdot))$ with K_j as in Condition 5.1.1, and under the conditions of Theorem 5.1.5, this leads to the bound

$$E\|f_n - f\|_p \le L_p\sqrt{\frac{2^j}{n}} + c\|f\|_{B^s_{p\infty}(A)} 2^{-js}, \quad 1 \le p < \infty. \tag{5.17}$$

This holds as well for $f_n - f$ in the Gaussian noise model case with the numerical constants from Proposition 5.1.2. If B is a bound for $\|f\|_{B^s_{p\infty}(A)}$, we can balance the antagonistic terms in (5.17) by choosing

$$2^{j_n} = (cB/L_p)^{2/(2s+1)} n^{1/(2s+1)}, \tag{5.18}$$

giving a rate of convergence for f_n to f in L^p-risk. For $p = \infty$, similar remarks apply, replacing n by $n/\log n$ in the definition of 2^{j_n}. Moreover, this remark applies as well to convolution kernel estimators, with the order of the kernel playing the role of the regularity of the basis. We say that a convolution kernel K is of order S if it satisfies Condition 4.1.4; that is, if

$$\int_{\mathbb{R}} |K(u)||u|^S du < \infty \quad \text{and} \quad \int_{\mathbb{R}} K(u)u^k du = \delta_{0k}, \quad k = 0,\ldots,S-1.$$

Proposition 4.3.8 clearly applies to these kernels, and the bound (5.17) follows as well for convolution kernel estimators based on these kernels, with $A = \mathbb{R}$. Thus, for both the white noise and the sampling case, we obtain the following rates:

Proposition 5.1.7 *Let f_n be a wavelet projection estimator based on an S-regular wavelet basis, $S > s$, or a convolution kernel density estimator based on a kernel of order S, and let j_n be such that $2^{j_n} \simeq (cB/L_p)^{2/(2s+1)} n^{1/(2s+1)}$. Then, for $1 \le p < \infty$, if $\|f\|_{B^s_{p\infty}(A)} \le B$,*

$$E\|f_n(j_n) - f\|_p \le 2L_p^{2s/(2s+1)}(cB)^{1/(2s+1)} n^{-s/(2s+1)}, \quad 1 \le p < \infty, \tag{5.19}$$

and if the convolution kernel in one case or the scaling function in the other is, of bounded r-variation for some $r \ge 1$, if $\|f\|_{B^s_{\infty\infty}(A)} \le B$ and if $2^{j_n} \simeq (cB/L_p)^{2/(2s+1)}(n/\log n)^{1/(2s+1)}$, then

$$E\|f_n(j_n) - f\|_\infty \le 2L_\infty^{2s/(2s+1)}(cB)^{1/(2s+1)}(n/\log n)^{-s/(2s+1)}, \tag{5.20}$$

where L_p is as in Theorem 5.1.5. Bounds of the same order hold as well for function estimation under additive white noise, when the periodised or boundary-corrected wavelets are S-regular (and, for $p = \infty$, of bounded r-variation).

The constants L_p, c are often known or can be estimated (cf. Remark 5.1.6). The constants s, B are usually unknown and cannot be estimated, leading to an adaptation problem that will be the subject of Chapter 8. For the moment, we note that one can simply replace B by a fixed constant, say, 1, in the definition of 2^{j_n}, producing the same rate of convergence $n^{-s/(2s+1)}$ for the estimators if s is known. We shall show in Chapter 6 that this rate is minimax optimal for any given s.

Empirical Wavelet Coefficients

Theorem 5.1.5 also applies to moment bounds for the deviation from their expectations of the empirical wavelet coefficients, that is, the wavelet coefficients of the wavelet projection estimator f_n of a density f. We will illustrate this under the assumption that the scaling function ϕ and wavelet function ψ satisfy Condition 5.1.1(b). Recall that the projection $K_j(f)$ has two expressions (see (4.29)):

$$K_j(f)(x) = \sum_{k\in\mathbb{Z}} \langle \phi_{jk}, f\rangle \phi_{jk}(x) = \sum_{k\in\mathbb{Z}} \langle \phi_k, f\rangle \phi_k(x) + \sum_{\ell=0}^{j-1}\sum_{k\in\mathbb{Z}} \langle \psi_{\ell k}, f\rangle \psi_{\ell k}(x),$$

where $\phi_k = \phi_{0k}$, and that, by (5.3), if $j = j_n$, then

$$f_n(x) = P_n(K_j(x,\cdot)) = \sum_{k\in\mathbb{Z}} P_n(\phi_{jk})\phi_{jk}(x) = \sum_{k\in\mathbb{Z}} P_n(\phi_k)\phi_k(x) + \sum_{\ell=0}^{j-1}\sum_{k\in\mathbb{Z}} P_n(\psi_{\ell k})\psi_{\ell k}(x).$$

To ease notation, define

$$\alpha_{jk}(f) = \langle \phi_{jk}, f\rangle = E\phi_{jk}(X), \quad \beta_{jk}(f) = \langle \psi_{jk}, f\rangle = E\psi_{jk}(X), \quad j \geq 0,\ k \in \mathbb{Z}, \tag{5.21}$$

where the distribution of X has density f, and for the same values of k and j,

$$\hat{\alpha}_{jk} = P_n(\phi_{jk}), \quad \hat{\beta}_{jk} = P_n(\psi_{jk}). \tag{5.22}$$

If no confusion may arise, we write α_{jk}, β_{jk} for $\alpha_{jk}(f)$ and $\beta_{jk}(f)$. Also, we write α_k and $\hat{\alpha}_k$ for α_{0k} and $\hat{\alpha}_{0k}$.

Since by the comments following (5.2), $K_{j+1}(x,y) - K_j(x,y) = \sum_{r\in\mathbb{Z}} \psi_{jr}(x)\psi_{jr}(y)$, for all x and y, if we set $h(x) := \sum_r (\hat{\beta}_{jr} - \beta_{jr})\psi_{jr}(x)$, we have, on the one hand,

$$h(x) = (P_n - P)(K_{j+1}(x,\cdot) - K_j(x,\cdot))$$

and, on the other, by ortho-normality of the ψ_{jk} functions

$$\hat{\beta}_{jk} - \beta_{jk} = \int \sum_r (\hat{\beta}_{jr} - \beta_{jr})\psi_{jr}(x)\psi_{jk}(x)dx = \int h(x)\psi_{jk}(x)dx.$$

To justify these identities, note that $\{\beta_{jr} : r \in \mathbb{Z}\} \subset \ell_1(\mathbb{Z})$ and $\{\psi_{jr}(y) : r \in \mathbb{Z}\} \subset \ell_1(\mathbb{Z})$, for all y, since $\sum_r |\psi(x-r)| \in L^1(\mathbb{R}) \cap L^\infty(\mathbb{R})$. Therefore,

$$\sup_k |\hat{\beta}_{jk} - \beta_{jk}| \leq \|h\|_\infty \sup_k \|\psi_{jk}\|_1 \leq 2^{-j/2}\|\psi\|_1 \left\|(P_n - P)(K_{j+1} - K_j)\right\|_\infty. \tag{5.23}$$

Likewise,

$$\sup_k |\hat{\alpha}_{jk} - \alpha_{jk}| \le 2^{-j/2} \|\phi\|_1 \left\|(P_n - P)(K_j)\right\|_\infty. \tag{5.24}$$

These observations apply as well to the empirical wavelet coefficients under Conditions 5.1.1(c) and (d). Then a direct application of Theorem 5.1.5 gives the following:

Proposition 5.1.8 *With the notation and assumptions of Theorem 5.1.5, assuming further that $\|f\|_\infty < \infty$ and $2^j j < n$, the empirical wavelet coefficients $\hat{\alpha}_{jk}$, $\hat{\beta}_{jk}$ defined by (5.22), under Conditions 5.1.1(b), (c) or (d), satisfy*

$$E \sup_k |\hat{\alpha}_{jk} - \alpha_{jk}| \le \|\phi\|_1 L_\infty \sqrt{j/n}, \quad E \sup_k |\hat{\beta}_{jk} - \beta_{jk}| \le 2\|\phi\|_1 L_\infty \sqrt{j/n}. \tag{5.25}$$

Derivatives of Densities

We briefly consider here linear estimators of the derivatives of a density, which are interesting because they appear as part of different statistics such as, for example, the Fisher information of a location parameter. The methods to obtain inequalities for their estimators do not differ from the ones just developed for density estimators; hence, to avoid repetition, we will only indicate how to obtain analogues of Theorem 5.1.5 and Proposition 5.1.7, leaving to the reader the development of exponential and higher-moment inequalities.

Let K and f be, respectively, a convolution kernel and a density on $\mathbb{R}$, both in $C^m(\mathbb{R})$. Then, by integration by parts, we have

$$K_h * f^{(m)}(x) = \frac{1}{h^{m+1}} \int_{\mathbb{R}} K^{(m)}\left(\frac{x-y}{h}\right) f(y)dy = \frac{1}{h^m}(K^{(m)})_h * f(x). \tag{5.26}$$

Thus, since $K_h * f^{(m)}$ is a good approximation of $f^{(m)}$, it makes sense to estimate $f^{(m)}$ by the plug-in estimator of the expression on the right-hand side of this identity, namely,

$$f_n^{(m)}(x) = \frac{1}{h^{m+1}} \sum_{i=1}^n K^{(m)}\left(\frac{x - X_i}{h}\right), \tag{5.27}$$

where X_i are i.i.d. samples from f, and $h = h_n \to 0$ as $n \to \infty$. The notation $f_n^{(m)}$ is adequate because clearly the mth derivative of the kernel density estimator $f_n = P_n * K_h$ coincides with the statistic defined by (5.27). Then the proofs of cases 1, 2 and 4 in Theorem 5.1.5 give the following:

Proposition 5.1.9 *Let K and f be, respectively, a convolution kernel and a density on $\mathbb{R}$, both in $C^m(\mathbb{R})$, and assume that $K^{(m)}$ and f integrate $(1+|t|)^s$ for some $s > (2-p)/p$ for $1 \le p < 2$, that $K^{(m)}$ is of finite r-variation for some $r \ge 1$ for $p = \infty$, that $h > 1/n$ if $p > 2$ and that $h/\log h^{-1} > 1/n$ for $p = \infty$. Then*

$$E\|f_n^{(m)} - Ef_n^{(m)}\|_p \le L_p/\sqrt{nh^{2m+1}}, \quad \text{for } 1 \le p < \infty,$$

and

$$E\|f_n^{(m)} - Ef_n^{(m)}\|_\infty \le L_\infty \sqrt{\log(h^{-1})/(nh^{2m+1})}, \quad \text{for } p = \infty,$$

where L_p is as in Theorem 5.1.5 with Φ replaced by $|K^{(m)}|$.

Now, by (5.26), $Ef_n^{(m)} = K_h * f^{(m)}$, and by Proposition 4.3.19, $\|f^{(m)}\|_{B^s_{pq}} \le \|f\|_{B^{s+m}_{pq}}$, so, by Proposition 4.3.8,

$$\|Ef_n^{(m)} - f^{(m)}\|_p \le c\|f^{(m)}\|_{B^s_{pq}} h^s \le c\|f\|_{B^{s+m}_{pq}} h^s.$$

Hence, we obtain the following analogue of Proposition 5.1.7:

Proposition 5.1.10 *Under the same hypotheses as in Corollary 5.1.9, if K is a kernel of order S, $s+m<S$ and $h^{-1} \simeq (cB/L_p)^{2/(2(s+m)+1)} n^{1/(2(s+m)+1)}$, then for $1 \le p < \infty$,*

$$\sup_{f:\|f\|_{B^{s+m}_{p\infty}(\mathbb{R})} \le B} E\|f_n^{(m)} - f^{(m)}\|_p \le 2L_p^{2s/(2(s+m)+1)} (cB)^{(2m+1)/(2(s+m)+1)} n^{-s/(2(s+m)+1)},$$

and if $h^{-1} \simeq (cB/L_p)^{2/(2(s+m)+1)} (n/(\log n)^{1/(2(s+m)+1)}$ and K has bounded r-variation for some $r \ge 1$, then

$$\sup_{f:\|f\|_{B^{s+m}_{\infty\infty}(\mathbb{R})} \le B} E\|f_n^{(m)} - f^{(m)}\|_p \le 2L_\infty^{2s/(2(s+m)+1)} (cB)^{(2m+1)/(2(s+m)+1)} (n/\log n)^{-s/(2(s+m)+1)}.$$

Likewise, one can obtain exponential bounds and higher-moment bounds for $\|f_n^{(m)} - Ef_n^{(m)}\|_p$ in complete analogy with the results given in Theorems 5.1.13 and 5.1.15 for f_n. This is left to the reader.

Next, we consider wavelet projection estimation of derivatives of densities. Let $K(x,y) = \sum_k \phi(x-k)\phi(y-k)$ be the orthogonal projection onto V_0 associated to a scaling function $\phi \in C^m(\mathbb{R})$ such that ϕ and its first m derivatives $\phi^{(r)}$, $r \le m$, are rapidly decaying at $\pm\infty$ (meaning that $|x|^k|\phi^{(r)}(x)| \to 0$ as $|x| \to \infty$ for $0 \le r \le m$ and all $k \in \mathbb{N}$). Then not only the series $\sum_k \phi(x-k)\phi(y-k)$ converges uniformly in x and y, but this series can be differentiated term by term m times with respect to x (or y), again by uniform convergence. Let $f \in C^m(\mathbb{R})$ be a probability density, and observe that, by integration by parts,

$$K_j(f^{(m)}) = \sum_k \langle \phi_{jk}, f^{(m)} \rangle \phi_{jk} = (-1)^m \sum_k \langle \phi_{jk}^{(m)}, f \rangle \phi_{jk}.$$

Then, as in the convolution case, it makes sense to estimate $f^{(m)}$ by the plug-in estimator

$$f_{n,m}(y) = (-1)^m \sum_k (P_n(\phi_{jk}^{(m)}))\phi_{jk}(y), \quad y \in \mathbb{R}, \tag{5.28}$$

where P_n is as usual the empirical measure based on n i.i.d. samples X_i from f, and where $j = j_n \to \infty$. Note that, by the preceding observation,

$$Ef_{n,m} = K_j(f^{(m)})$$

(since $\sum_k |\phi_{jk}^{(m)}(x)|$ is integrable and f is bounded, so we can integrate term by term in (5.28)). Consider the asymmetric kernel

$$\bar{K}(x,y) = (-1)^m \sum_k \phi^{(m)}(x-k)\phi(x-k).$$

For every $R > 0$ there exists $\tilde{C}_R < \infty$ such that $|\phi^{(k)}(x)| \le \tilde{C}_R/(1+|x|^R)$, $k = 0,1,\dots,m$. Then we may proceed as in the proof of Lemma 4.2.5 and Exercise 4.2.1 to conclude that for each $R > 0$ there is $C_R > 0$ such that $\bar{K}(x,y) \leq C_R/(1+|x-y|^R)$, $x,y \in \mathbb{R}$. Thus, we have that

$$\bar{K}(x,y) \leq \Phi(x-y)$$

for a convolution kernel Φ which is even, nonincreasing on $[0,\infty)$, bounded, integrable and such that $\int \Phi(x)(1+|x|^t)dx < \infty$, for all $t \leq R-2$, where $R > 2$ can be chosen at will. Finally, note that

$$\sum_{i=1}^{n} |\bar{K}_j(X_i,y)| = 2^{j(m+1)} \sum_{i=1}^{n} |K(2^j X_i, 2^j y)| \leq 2^{j(m+1)} \sum_{i=1}^{n} \Phi(2^j(X_i - y)).$$

At this point it is clear that we can proceed as in the proof of Theorem 5.1.5 if we choose R large enough in the definition of Φ so that the integrals in the proofs of cases 1, 2 and 4 for this Φ are finite and conclude the following:

Proposition 5.1.11 *Let $\phi \in C^m(\mathbb{R})$ be an MRA scaling function such that ϕ and its first m derivatives $\phi^{(r)}$, $r \leq m$, are rapidly decaying at $\pm\infty$. Let $f \in C^{(m)}(\mathbb{R})$ be a probability density, and let $f_{n,m}$ be the estimator of $f^{(m)}$ defined by (5.28) and based on ϕ. Then, assuming $2^j < n$ if $p > 2$, we have*

$$E\|f_{n,m} - Ef_{n,m}\|_p \leq L_p\sqrt{2^{(2m+1)j}/n}, \quad 1 \leq p < \infty,$$

and if, moreover, $2^j j < n$ and $\phi^{(m)}$ is of finite p-variation for some $p \geq 1$, then also

$$E\|f_{n,m} - Ef_{n,m}\|_\infty \leq L_\infty\sqrt{2^{(2m+1)j}j/n},$$

where the constants L_p are as in Theorem 5.1.5, cases 1, 2 and 4 (with a different majorising convolution kernel Φ).

Now assume in addition that ϕ determines an S-regular wavelet basis, that $f \in B^{s+m}_{p\infty}(\mathbb{R})$ and that $s+m < S$. Then, since $f^{(m)} \in B^s_{p\infty}(\mathbb{R})$ by Proposition 4.3.19, Proposition 4.3.8 gives

$$\|Ef_{n,m} - f^{(m)}\|_p = \|K_j(f^{(m)}) - f^{(m)}\|_p \leq C\|f^{(m)}\|_{B^s_{p\infty}} 2^{-js} \leq \bar{C}\|f\|_{B^{s+m}_{p\infty}} 2^{-js}.$$

Combined with the preceding proposition, this bound yields a result for $E\|f_{n,m} - f^{(m)}\|_p$ completely analogous to the one in Proposition 5.1.10 for kernel estimators, in fact, the very same bounds.

Several Dimensions

Density estimation in several dimensions under smoothness restrictions that are homogeneous across different coordinates is only formally different from dimension 1. Basically, all the preceding results on the L^p-norms of $f_n - Ef_n$ and $f_n - f$ work with a slight change: 2^j changes to 2^{dj} in the bounds for $E\|f_n - Ef_n\|_p$ because $K_j(x,y) = 2^{dj}K(x/d,y/d)$, $x,y \in \mathbb{R}^d$, both in the convolution case $K(x,y) = K(x-y)$ and in the wavelet projection case $K(x,y) = \sum_{k\in\mathbb{Z}^d} \Phi(x-k)\Phi(y-k)$, $\Phi(x_1,\ldots,x_d) = \phi(x_1)\cdots\phi(x_d)$ (see Section 4.3.6). The proofs for $p = \infty$ require metric entropy computations, but these are also analogous to the case $d = 1$, particularly if we also take, in the convolution case, $K(x) = \Phi(|x|)$, Φ

a real function of bounded p-variation, or $K(x) = K(x_1)\cdots K(x_d)$, $K : \mathbb{R} \mapsto \mathbb{R}$ of bounded p-variation. For instance, in the product case, for example, for $d = 2$, the inequality

$$\left\|\phi(\lambda_1\cdot -k_1)\phi(\lambda_2\cdot -k_2) - \phi(\lambda_1'\cdot -k_1')\phi(\lambda_2'\cdot -k_2')\right\|_{L^2(Q)}$$
$$\leq \|\phi\|_\infty \sum_{i=1}^{2} \left\|\phi(\lambda_i\cdot -k_i) - \phi(\lambda_i'\cdot -k_i')\right\|_{L^2(Q)}$$

allows estimation of the $L^2(Q)$ covering numbers for the collection of translations and dilations of $\Phi(x_1, x_2)$ in terms of the known covering numbers of translations and dilations of functions of bounded p-variation in dimension 1, and the same is true for any finite dmension d. In the case $K(x) = \Phi(|x|)$, we see that the subgraphs of the functions $x \mapsto \Phi(|x-y|/h)$, $y \in \mathbb{R}^d$ and $h > 0$, for Φ strictly increasing, are the positivity sets of the finite-dimensional space of functions of x and t generated by x_i^2, x_i, 1 and $(\Phi^{-1}(t))^2$, hence with good uniform entropy bounds (see Propositions 3.6.12 and 3.6.6).

However, in several dimensions we still have $\|K_j(f) - f\|_p \leq c\|f\|_{B^s_{p\infty}(\mathbb{R}^d)} 2^{-js}$ if K is of order S or if ϕ is S-regular for $S > s$ by the analogue in several dimensions of Proposition 4.3.8. Hence, the analogues in several dimensions of the bounds in Proposition 5.1.7 are, for $2^{j_n} \simeq (cB/L_p)^{2/(2s+d)} n^{1/(2s+d)}$,

$$\sup_{f:\|f\|_{B^s_{p\infty}(\mathbb{R}^d)} \leq B} E\|f_n(j_n) - f\|_p \leq 2L_p^{2s/(2s+d)}(cB)^{d/(2s+d)} n^{-s/(2s+d)}, \quad 1 \leq p < \infty,$$

and, for j_n such that $2^{j_n} = (cB/L_p)^{2/(2s+1)}(n/\log n)^{1/(2s+1)}$,

$$\sup_{f:\|f\|_{B^s_{\infty\infty}(\mathbb{R}^d)} \leq B} E\|f_n(j_n) - f\|_\infty \leq 2L_\infty^{2s/(2s+d)}(cB)^{d/(2s+1)}(n/\log n)^{-s/(2s+d)}.$$

The same comment applies to multivariate function estimation under Gaussian noise. In this case, the process W is the iso-normal Gaussian process on $L^2((0,1]^d)$, defined as in (4.186) with $[0,1]$ and Lebesgue measure replaced by $(0,1]^d$ (or $[0,1]^d$) and multivariate Lebesgue measure, so, in particular, $\int_{(0,1]^d} g(y)dW(y) := W(g)$ is $N(0, \|g\|_2^2)$, and the statistical model is

$$dY_f \equiv dY_f^{(n)}(y) = f(y)dy + \sigma dW(y)/\sqrt{n}, \quad y \in [0,1]^d,$$

(see (1.10)). With the notation $k = (k_1, \ldots, k_d) \in \mathbb{Z}^d$, $x = (x_1, \ldots, x_d) \in (0,1]^d$, we take $\Phi_{j,k}(x) = \prod_{i=1}^d \phi_{jk_i}(x_i)$ for $\phi_{jk_i} = \phi_{jk_i}^{\text{per}}$ or $\phi_{jk_i} = \phi_{jk_i}^{bc}$, as in Section 4.3.6, and $j = j_n \to \infty$ and define

$$K_j(x,y) = \sum_{k \in \mathbb{Z}^d : 0 \leq k_i \leq 2^j - 1} \Phi_{j,k}(x)\Phi_{j,k}(y).$$

Then f is estimated by

$$f_n(x) = \int_{(0,1]^d} K_j(x,y)dY_f^{(n)}(y) = K_j(f) + \frac{\sigma}{\sqrt{n}} W(K_j(x,\cdot)).$$

With these definitions, the comments in the preceding two paragraphs for multidimensional density estimators apply verbatim to this estimator of f.

5.1.2 Exponential Inequalities, Higher Moments and Almost-Sure Limit Theorems

The results in the preceding subsection are useful, in particular, to help obtain exponential inequalities for $\|f_n - Ef_n\|_p$, $1 \le p \le \infty$. These, in turn, produce bounds for higher moments. In the white noise case, with, for example, boundary-corrected wavelets, the centred estimator of f,

$$f_n - Ef_n = \frac{1}{\sqrt{n}}\int_0^1 K_j(\cdot,t)\sigma\, dW(t) = \frac{\sigma}{\sqrt{n}}\left[\sum_{k=0}^{2^J-1} g_{Jk}\phi_{Jk}^{bc} + \sum_{l=J}^{j-1}\sum_{k=0}^{2^l-1} \tilde{g}_{lk}\psi_{lk}^{bc}\right],$$

is an L^p-valued centred Gaussian random variable, $1 \le p \le \infty$, where $j = j_n$ depends on n. We first consider $1 \le p \le 2$, and in this case we apply Gaussian concentration for Lipschitz functions. To this end, observe that if $h_1,\dots,h_m$ are ortho-normal functions in $L^2([0,1],\lambda)$, λ being Lebesgue measure, then the function $F(x_1,\dots,x_m) = \left\|\sum_{i=1}^m x_i h_i\right\|_p$ is Lipschitz with constant 1, that is,

$$|F(x) - F(y)| \le \left\|\sum_{i=1}^m (x_i - y_i)h_i\right\|_p \le \left\|\sum_{i=1}^m (x_i - y_i)h_i\right\|_2 \le |x - y|,$$

where $|\cdot|$ denotes Euclidean norm. This applies to $m = 2\cdot 2^J + 2^{J+1} + 2^{J+2} + \cdots + 2^{j-1} = 2^j$ and to the ortho-normal system $\{\phi_{Jk}^{bc},\dots,\psi_{j-1k}^{bc}\}$ in the expression for $f_n - Ef_n$. Then the concentration inequality for the standard Gaussian variable and Lipschitz functions in Euclidean space, (2.69) in Theorem 2.5.7, shows that, for $1 \le p \le 2$,

$$\Pr\left\{\|f_n - Ef_n\|_p \ge E\|f_n - Ef_n\|_p + \sigma t/\sqrt{n}\right\} \le e^{-t^2/2}. \tag{5.29}$$

Case $2 < p \le \infty$ is less direct. For $p < \infty$, if B_0^q is a countable subset of the unit ball of $L^q([0,1])$, q conjugate of p, such that $\|u\|_p = \sup_{v\in B_0^q}\int_0^1 u(x)v(x)dx$, then

$$\|f_n - Ef_n\|_p = \frac{\sigma}{\sqrt{n}}\sup_{v\in B_0^q}\left[\sum_{k=0}^{2^J-1} g_{Jk}\int_0^1 \phi_{Jk}^{bc}(x)v(x)dx + \sum_{l=J}^{j-1}\sum_{k=0}^{2^l-1}\tilde{g}_{lk}\int_0^1 \psi_{lk}^{bc}(x)v(x)dx\right].$$

(Recall that such a set B_0^q exists by separability of L^p and the Hahn-Banach theorem.) Hence, in the language of processes, we may consider $f_n - Ef_n$ as the Gaussian process indexed by B_0^q, $h \mapsto \langle f_n - Ef_n, h\rangle$ for $p < \infty$. For $p = \infty$, we think of $f_n - Ef_n$ as the Gaussian process indexed by $[0,1]$, $t \mapsto f_n(t) - Ef_n(t)$, which is separable by continuity of the functions ϕ_{jk}, ψ_{lk}. In either case, we apply the Borell-Sudakov-Tsirelson inequality (2.70). To this end, we need to estimate the parameter σ in that inequality, which here we denote as τ to avoid confusion. For $p < \infty$, we have

$$\tau^2 := \frac{\sigma^2}{n}\sup_{v\in B_0^q}\left[\sum_{k=0}^{2^J-1}\langle\phi_{Jk}^{bc}, v\rangle^2 + \sum_{l=J}^{j-1}\sum_{k=0}^{2^l-1}\langle\psi_{lk}^{bc}, v\rangle^2\right].$$

Since L^q embeds continuously into the Besov space $B^0_{q\infty}$ by the analogue for $[0,1]$ of Proposition 4.3.11, it follows from the boundary-corrected characterisation of Besov spaces on $[0,1]$ (Section 4.3.5) that $\sum_{k=0}^{2^J-1}|\langle\phi_{Jk}^{bc}, v\rangle|^q \le D_p^q$ and $\sum_{k=0}^{2^l-1}|\langle\psi_{lk}^{bc}, v\rangle|^q \le D_p^q 2^{-lq(1/2-1/q)}$, where D_p denotes the norm of the embedding of L^q into $B^0_{q\infty}$. Then, using that $\left(\sum |a_i|^2\right)^{1/2} \le \left(\sum |a_i|^q\right)^{1/q}$ for $1 \le q \le 2$, we obtain

$$\tau^2 \le \frac{C_p^2 \sigma^2 2^{j\left(1-\frac{2}{p}\right)}}{n}, \quad 2 < p < \infty,$$

where C_p^2 is easy to compute and contains as factors D_p^q and the quantity $2^{2/q-1}/(2^{1-q/2}-1)^{2/q}$ (which tends to infinity as $q \to 2$). For $p = \infty$, using the expression (5.8) for f_n, the bound (5.11) gives

$$\tau^2 := \frac{\sigma^2}{n}\left\| \sum_{k=0}^{2^j-1} (\phi_{jk}^{bc})^2 \right\|_\infty = \kappa'^2 \frac{\sigma^2 2^j}{n}.$$

Similar estimates hold for function estimators based on periodised wavelets (see (5.10) and note that Besov spaces on $[0,1]$ admit characterisations in terms of both boundary-corrected and periodised wavelets). Then the abovementioned Borell-Sudakov-Tsirelson inequality (2.70) yields the exponential inequality for $\|f_n - Ef_n\|_p$, $p > 2$, in the following proposition, whereas inequality (5.29) yields the part of the inequality corresponding to $1 \le p \le 2$.

In this proposition, we let C_p denote the quantity just described for $p > 2$, but $C_p = 1$ for $1 \le p \le 2$ and $C_\infty = \kappa'$ in the boundary-corrected case and similarly in the periodic case.

Proposition 5.1.12 *Assume Condition 5.1.1(c) or (d) and that $f \in L^p([0,1])$, $1 \le p \le \infty$ (with $f(0) = f(1)$ under Condition 5.1.1(d)). Let K_j denote either $K_{j,bc}$ and then $2^j \ge 2N$ or $K_{j,\mathrm{per}}$ as in Proposition 5.1.2. Let $f_n = \int K_j(\cdot,t)dY_f^{(n)}(t)$. Then, for all $x > 0$ and $1 \le p \le \infty$,*

$$\Pr\left\{ \|f_n - Ef_n\|_p \ge E\|f_n - Ef_n\|_p + \sqrt{2C_p^2\sigma^2 2^{j(1-2/(p\vee 2))}x/n} \right\} \le e^{-x}. \tag{5.30}$$

In particular, by Proposition 5.1.2,

$$\Pr\left\{ \|f_n - Ef_n\|_p \ge C(\phi,p)\sigma\sqrt{2^j/n} + \sqrt{2C_p^2\sigma^2 2^{j(1-2/(p\vee 2))}x/n} \right\} \le e^{-x}, \tag{5.31}$$

for $p < \infty$ and with $E\|f_n - Ef_n\|_\infty$ replaced by $C(\phi,\infty)\sigma\sqrt{2^j(j+1)/n}$ for $p = \infty$.

In general, we will only be interested in the upper tail of the concentration of $\|f_n - f\|_p$ about its expectation: for the lower-tail estimate to be practical, we would need lower bounds for $E\|f_n - f\|_p$, which we are not considering.

As in the preceding subsection, the Gaussian noise case provides a model for the sampling case, which is again more complicated. In this case, we apply Talagrand's inequality, Bousquet version, Theorem 3.3.9, instead of the Borel-Sudakov-Tsirelson Gaussian concentration inequality. For $p = \infty$, we will apply this theorem to the class of functions $\mathcal{K} = \mathcal{K}_j = \{K_j(x,\cdot) - K_j(f)(x) : x \in A\}$, whereas for $1 \le p < \infty$, the relevant class of functions is, with B_0^q as earlier

$$\mathcal{K} = \mathcal{K}_j = \left\{ x \mapsto \int_A g(t)K_j(t,x)dt - \int_A g(t)K_j(f)(t)dt : g \in B_0^q \right\}.$$

Then $n\|f_n - Ef_n\|_p = \sup_{H\in\mathcal{K}} \left|\sum_{i=1}^n H(X_i)\right| =: S_n$. Besides ES_n, which is upper bounded in Theorem 5.1.5, the parameters to be estimated (bounded from above) for Talagrand's inequality are $\sigma^2 = \sup_{H\in\mathcal{K}} EH^2(X)$ and $U = \sup_{H\in\mathcal{K}} \|H\|_\infty$. Let us consider first density estimation under Conditions 5.1.1(a) and (b), and as earlier, let us set $\Phi = K$ under Condition 5.1.1(a), and let Φ be the dominating convolution kernel under Condition 5.1.1(b). Then, for $1 \le p < \infty$, given $g \in B_0^q$,

$$\|K_j(g)\|_\infty \le \|\Phi_j * |g|\|_\infty \le \|\Phi_j\|_p \|g\|_q = \|\Phi\|_p 2^{j(1-1/p)},$$

whereas for $p = \infty$, $2^j\Phi(2^j(x-y)) \le \|\Phi\|_\infty 2^j$, so we can take

$$U = 2\|\Phi\|_p 2^{j(1-1/p)} \tag{5.32}$$

in Theorem 3.3.9. Regarding σ^2, we have, for $1 \le p < \infty$,

$$E(\Phi_j * g)^2(X) \le \|f\|_p \|\Phi_j * g\|_{2q}^2 \le \|f\|_p \|g\|_q^2 \|\Phi_j\|_{2p/(1+p)}^2 \le \|f\|_p \|\Phi\|_{2p/(1+p)}^2 2^{j(1-1/p)},$$

where we use first Hölder's and then Young's inequalities (Young's inequality: $\|u * v\|_t \le \|u\|_p \|v\|_q$, for $1 + 1/t = 1/p + 1/q$, $0 \le t,p,q \le \infty$). And for $p = \infty$, $E\Phi_j(x,X)^2 \le \|f\|_\infty \|\Phi\|_2^2 2^j$. Thus, we may take, for $1 \le p \le \infty$,

$$\sigma^2 = \|f\|_p \|\Phi\|_{2p/(1+p)}^2 2^{j(1-1/p)}. \tag{5.33}$$

This estimate can be refined if f is bounded: for $1 \le p < 2$, since $f \in L^\infty \cap L^1 \subset L^{q/(q-2)}$ for q conjugate of p, by Hölder's inequality,

$$E(\Phi_j * g)^2(X) \le \|\Phi_j * g\|_q^2 \|f\|_{q/(q-2)} \le \|\Phi\|_1^2 \|f\|_{q/(q-2)},$$

whereas for $2 \le p < \infty$, again by Hölder's and Young's inequalities,

$$E(\Phi_j * g)^2(X) \le \|f\|_\infty \|\Phi_j * g\|_2^2 \le \|f\|_\infty \|\Phi\|_{2p/(2+p)}^2 \|g\|_q 2^{j(1-2/p)}.$$

Thus, if $\|f\|_\infty < \infty$, we can take

$$\sigma^2 = \|f\|_\infty \|\Phi\|_{2p/(2+p)}^2 2^{j(1-2/(p\vee 2))}. \tag{5.34}$$

See Exercises 5.1.1 and 5.1.2 for similar estimates of U and σ^2 in the cases of periodic and boundary-corrected wavelets on the unit interval.

Inequality (3.101) in Theorem 3.3.9, with $v = n\sigma^2 + 2UnE\|f_n - Ef_n\|_p$, $1 \le p \le \infty$, states that

$$\Pr\left\{n\|f_n - Ef_n\|_p \ge nE\|f_n - Ef_n\|_p + \sqrt{2vx} + Ux/3\right\} \le e^{-x}, \tag{5.35}$$

and we can simplify it somewhat using the standard inequalities $\sqrt{a+b} \le \sqrt{a} + \sqrt{b}$, $\sqrt{ab} \le (a+b)/2$, $a,b \ge 0$. Thus,

$$\Pr\left\{n\|f_n - Ef_n\|_p \ge \frac{3}{2}nE\|f_n - Ef_n\|_p + \sqrt{2n\sigma^2 x} + 7Ux/3\right\} \le e^{-x}. \tag{5.36}$$

Plugging the estimates (5.32), (5.33) and (5.34) for U and σ into this inequality, we obtain the following result:

Theorem 5.1.13 *Under the hypotheses of Theorem 5.1.5, we have that, for all $t > 0$ and $1 \le p \le \infty$,*

$$\Pr\left\{n\|f_n - Ef_n\|_p \ge \frac{3}{2}nE\|f_n - Ef_n\|_p + \sqrt{C_1 n\|f\|_p 2^{j(1-1/p)} x} + C_2 2^{j(1-1/p)} x\right\} \le e^{-x}. \tag{5.37}$$

and if, moreover, the density f is uniformly bounded,

$$\Pr\left\{n\|f_n - Ef_n\|_p \geq \frac{3}{2}nE\|f_n - Ef_n\|_p + \sqrt{C_1 n\|f\|_\infty 2^{j(1-2/(p\vee 2))}x} + C_2 2^{j(1-1/p)}x\right\} \leq e^{-x}, \tag{5.38}$$

where, by Theorem 5.1.5, $nE\|f_n - Ef\|_p \leq L_p\sqrt{n2^j}$ if $1 \leq p < \infty$ and $nE\|f_n - Ef\|_\infty \leq L_\infty\sqrt{n2^j j}$, and where the constants C_1 and C_2 depend only on K (Condition 5.1.1(a)) or Φ (Condition 5.1.1(b)) or κ' (Condition 5.1.1(c)) or κ (Condition 5.1.1(d)).

The constants L_p are precisely those in Theorem 5.1.5 (Remark 5.1.6). The constants C_1 and C_2 are, respectively, $C_1 = 2\|\Phi\|^2_{2p/(1+p)}$ and $C_2 = 14\|\Phi\|_p/3$ under Condition 5.1.1(a) or Condition 5.1.1(b), and similar expressions can be obtained under Conditions 5.1.1(c) and (d) (see Exercises 5.1.1 and 5.1.2).

Remark 5.1.14 Besides Talagrand's inequality, the concentration inequality for bounded differences, Theorem 3.3.14, also applies to density estimators (as indicated in an example in Section 3.3.4; see (3.121)), and it actually obtains results that are somewhat better than those in Theorem 5.1.13 for $p = 1$. Moreover, the bounded differences theorem allows us to effortlessly consider not only $\|f_n - Ef_n\|_p$ but also $\|f_n - f\|_p$. Let

$$g(x_1,\dots,x_n) = \left\|\frac{1}{n}\sum_{i=1}^n K_j(\cdot,x_i) - f(\cdot)\right\|_p .$$

Then

$$\begin{aligned}|g(x_1,\dots,x_n) - g(x_1,\dots,x_{i-1},x_i',x_{i+1},\dots,x_n)| &\leq \frac{1}{n}\|K_j(\cdot,x_i) - K_j(\cdot,x_i')\|_p \\ &\leq \frac{2}{n}\|\Phi_j\|_p = \frac{2\cdot 2^{j(1-1/p)}\|\Phi\|_p}{n},\end{aligned}$$

for all x_i,x_j', $1 \leq i,j \leq n$, for $1 \leq p \leq \infty$. Hence, g has bounded differences with $c_i = 2\cdot 2^{j(1-1/p)}\|\Phi\|_p/n$, and Theorem 3.3.14 gives, after a change of variables,

$$\Pr\left\{\left|\|f_n - f\|_p - E\|f_n - f\|_p\right| \geq \sqrt{2\cdot 2^{2j(1-1/p)}\|\Phi\|_p^2 x/n}\right\} \leq 2e^{-x}, \tag{5.39}$$

for all $p \in [1,\infty]$, as well as the same bound with f replaced by Ef_n. This bound outperforms the bounds (5.37) and (5.38) for $p = 1$, but its dependence on j_n is worse for other values of p.

Now we apply the preceding exponential inequalities to obtain bounds for moments higher than 1 of $\|f_n - Ef_n\|_p$ (and hence also of $\|f_n - f\|_p$). The result is stated only for density estimators, and it is left as an exercise for function estimation under additive Gaussian noise – recall from, for example, Exercise 2.1.2 that all the L^p-norms of suprema of Gaussian processes are equivalent. Combining Exercise 3.3.4 with the bounds (5.32) for U and (5.33) for σ ((5.34), if $\|f\|_\infty < \infty$), we obtain the following result:

Theorem 5.1.15 *Under the hypotheses of Theorem 5.1.5, we have that for all* $1 \le p \le \infty$ *and all* $1 < r < \infty$*, assuming Condition 5.1.1(a) or Condition 5.1.1(b),*

$$\left(E\|f_n - Ef_n\|_p^r\right)^{1/r} \le 2E\|f_n - Ef_n\|_p + 1.24 \cdot 3^{1/r} r^{1/2} \|f\|_p^{1/2} \|\Phi\|_{2p/(1+p)} \sqrt{2^{j(1-1/p)}/n}$$
$$+ 4.26 \cdot 9^{1/r} r \|\Phi\|_p 2^{j(1-1/p)}/n, \tag{5.40}$$

where we may replace $\|f\|_p^{1/2} \|\Phi\|_{2p/(1+p)} \sqrt{2^{j(1-1/p)}/n}$ *by* $\|f\|_\infty^{1/2} \|\Phi\|_{2p/(2+p)} \sqrt{2^{j(1-2/(p\vee 2))}/n}$ *in the second summand if* $\|f\|_\infty < \infty$*. Bounds of the same order but different constants hold under Conditions 5.1.1(c) and (d).*

We could have used Hoffmann-Jørgensen's inequality (Theorem 3.1.22) directly on Theorem 5.1.5, but Exercise 3.3.4 produces better constants. Note that as a consequence of this theorem, assuming $2^j < n$ if $p > 2$ ($2^j j < n$ for $p = \infty$), the bound for $E\|f_n - Ef_n\|_p$ prescribed by Theorem 5.1.5 does hold as well for the rth moments perhaps with different numerical factors: the last two terms on the right-hand side of inequality (5.40) are of smaller order than $\sqrt{2^j/n}$ under this assumption.

Remark 5.1.16 We should also point out that the results in this subsection apply to produce higher-moment bounds and exponential inequalities for empirical wavelet coefficients just by combining Theorems 5.1.13 and 5.1.15 with inequalities (5.24) and (5.23) in complete analogy with Proposition 5.1.8.

A second, very important application of the moment and exponential bounds in this section obtains upper bounds on the asymptotic order of $\|f_n - Ef_n\|_p$, $1 \le p \le \infty$, both a.s. and in probability. Consider sample-based density estimation under the hypotheses of Theorem 5.1.13. Regarding the L_1-norm, here is what Remark 5.1.14 (inequality (5.39)) gives for $\|f_n - f\|_1$ (as well as for $\|f_n - Ef_n\|_1$, which we omit in this proposition):

Proposition 5.1.17 *For any density* f*,*

(a) $\|f_n - f\|_1 - E\|f_n - f\|_1 = O_{pr}(1/\sqrt{n})$*; in particular, if* $\sqrt{n}E\|f_n - f\|_1 \to \infty$*, then* $\|f_n - f\|_1 / E\|f_n - f\|_1 \to 1$ *in probability, and*
(b) $\limsup_n \sqrt{n/\log n}\,|\|f_n - f\|_1 - E\|f_n - f\|_1| \le \sqrt{2}\|\Phi\|_1$*; in particular,* $\lim_{n\to\infty} |\|f_n - f\|_1 / E\|f_n - f\|_1 - 1| = 0$ *a.s. if* $\sqrt{n/\log n}E\|f_n - f\|_1 \to \infty$*.*

Here is another simple consequence of Theorem 5.1.13, in fact, of inequality (5.35) and Theorem 5.1.5, for the a.s. rate of convergence of $\|f_n - Ef_n\|_p$, $1 \le p < \infty$, whose proof is also immediate:

Proposition 5.1.18 *If under the hypotheses of Theorem 5.1.5, with* f *not necessarily bounded, we have* $n2^{-j_n} \to \infty$ *and* $j_n/\log\log n \to \infty$*, then there exists* $C_p \le L_p$ *such that*

$$\limsup_{n\to\infty} \sqrt{\frac{n}{2^{j_n}}} \|f_n - Ef_n\|_p = C_p \text{ a.s.}, \quad 1 \le p < \infty.$$

Regarding the proof of this proposition (that of the preceding proposition is similar but easier), note that the best upper bounds we can come up with for the orders of the three summands on the right-hand side of the probability expression in inequality (5.37) are, respectively, after dividing by n and disregarding constants, as follows: $(2^{j_n}/n)^{1/2}$ for the first, $[(1 + \sqrt{2^{j_n}/n})2^{j_n(1-1/p)}x/n]^{1/2}$ for the second and $2^{j_n(1-1/p)}x/n$ for the third. Thus, if we take $x_n = 2^{j_n/(2p)}$, we have both the first term dominating and the series $\sum e^{-x_n}$ converging. This allows application of the Borel-Cantelli lemma and the 0-1 law to reach the conclusion. Also, note that automatically the rate obtained in this proposition is as good as the bound for the expected value of $\|f_n - Ef_n\|_p$ and hence not improvable in general.

The case $p = \infty$ is slightly more complicated: inequality (5.37) offers too narrow a choice for x if we still want the last two summands on the right-hand side of the probability expression in this inequality of smaller order than the first summand. The problem is solved by 'blocking' as in the typical proofs of the law of the iterated logarithm.

Let $j_n \nearrow \infty$ be a sequence of nonnegative integers satisfying the following conditions:

$$\frac{n}{j_n 2^{j_n}} \to \infty, \qquad \frac{j_n}{\log\log n} \to \infty, \qquad \sup_{n \ge n_0} (j_{2n} - j_n) \le \tau, \tag{5.41}$$

for some $\tau \ge 1$ and some $n_0 < \infty$.

Proposition 5.1.19 *Assume that either of Conditions 5.1.1(a), (b), (c) or (d) holds, let $\{j_n\}$ be a sequence of integers satisfying (5.41), let f be a bounded density on $\mathbb{R}$ and let f_n be the corresponding density estimator (defined by (5.1) or (5.3) with projection kernels as in (5.2) or (5.4)). Then we have*

$$\limsup_{n\to\infty} \sqrt{\frac{n}{j_n 2^{j_n}}} \|f_n - Ef_n\|_\infty = C \ \ a.s., \tag{5.42}$$

where $C^2 \le M^2 2^\tau \|f\|_\infty$ for a constant M that depends only on K or Φ.

Proof Assume Condition 5.1.1(a) or Condition 5.1.1(b). Let $n_k = 2^k$, and to unify notation, denote $K(x,y) = K(x-y)$ under Condition 5.1.1(a). We have, for any $s > 0$,

$$\Pr\left\{\max_{n_{k-1}<n\le n_k} \sup_{y\in\mathbb{R}} \sqrt{\frac{1}{n2^{-j_n}j_n}} \left|\sum_{i=1}^n \left(K(2^{j_n}y, 2^{j_n}X_i) - EK(2^{j_n}y, 2^{j_n}X)\right)\right| > s\right\}$$

$$\le \Pr\left\{\max_{n_{k-1}<n\le n_k} \sup_{\substack{y\in\mathbb{R}\\ j_{n_{k-1}}<j\le j_{n_k}}} \left|\sum_{i=1}^n \left(K(2^j y, 2^j X_i) - EK(2^j y, 2^j X)\right)\right| > s\sqrt{\frac{n_{k-1} j_{n_k}}{2^{j_{n_k}}}}\right\}, \tag{5.43}$$

where $j \in \mathbb{N}$. To estimate this probability, we apply Talagrand's inequality (3.101) for maxima of suprema of partial sums to the classes of functions

$$\mathcal{F}_k = \{K(2^j y, 2^j(\cdot)) - P(K(2^j y, 2^j(\cdot))) : y \in \mathbb{R},\ j_{n_{k-1}} < j \le j_{n_k}\},$$

which, by (5.32) and (5.33), have constant envelope and weak variance, respectively, bounded by $U = 2\|\Phi\|_\infty$ and $\sigma^2 = \|f\|_\infty \|\Phi\|_2^2 2^{-j}$ and satisfy

$$E \sup_{\substack{y\in\mathbb{R}\\ j_{n_{k-1}}<j\le j_{n_k}}} \left|\sum_{i=1}^{n_k} \left(K(2^j y, 2^j X_i) - EK(2^j y, 2^j X)\right)\right| \le L(\Phi)\|f\|_\infty^{1/2} \sqrt{n_k 2^{-j_{n_{k-1}}} j_{n_{k-1}}}$$

for n_k large enough by Remark 5.1.4 and Corollary 3.5.8 as in Theorem 5.1.5. Hence, Talagrand's inequality, simplified as in (5.36), gives

$$\Pr\left\{\max_{n_{k-1}<n\le n_k}\ \sup_{\substack{y\in\mathbb{R}\\ j_{n_{k-1}}<j\le j_{n_k}}}\left|\sum_{i=1}^{n}\left(K(2^jy,2^jX_i)-EK(2^jy,2^jX)\right)\right|\right.$$
$$\left.\times > \frac{3}{2}L(\phi)\|f\|_\infty\sqrt{n_k2^{-j_{n_{k-1}}}j_{n_{k-1}}}+\sqrt{2n_k\|\Phi\|_2^2 2^{-j_{n_{k-1}}}x_k}+14\|\Phi\|_\infty 2^{-j_{n_{k-1}}}x_k/3\right\}\le e^{-x_k}.$$

Setting $x_k=j_{n_{k-1}}$, which satisfies $\sum e^{-x_k}<\infty$, and comparing with (5.43), we see that for $s=M2^{\tau/2}\|f\|_\infty^{1/2}$ for some M depending only on Φ, the Borel-Cantelli lemma gives

$$\Pr\left\{\max_{n_{k-1}<n\le n_k}\sup_{y\in\mathbb{R}}\sqrt{\frac{1}{n2^{-j_n}j_n}}\left|\sum_{i=1}^{n}\left(K(2^{j_n}y,2^{j_n}X_i)-EK(2^{j_n}y,2^{j_n}X)\right)\right|>s\ i.o.\right\}=0.$$

Then the 0-1 law gives the proposition under Conditions 5.1.1(a) and (b). The proof under Conditions 5.1.1(c) and (d) is similar and is omitted. ■

Propositions 5.1.18 and 5.1.19 also hold in dimension d, with the only difference that the factor 2^{j_n} in the norming is replaced by 2^{j_nd}. These propositions also admit analogues for the estimators (5.7) of functions under additive white noise (see Exercise 5.1.7).

Proposition 5.1.19 can be made much more precise, both in one and in several dimensions, at the price of considerable work. See the notes and complements at the end of this chapter.

Combining inequality (5.37) with the lower tail in Talagrand's inequality, Theorem 3.3.10, we also may obtain an upper bound on the rate of a.s. and in probability convergence to zero of $\|f_n-Ef_n\|_p-E\|f_n-Ef_n\|_p$ (see Exercise 5.1.9).

*5.1.3 A Distributional Limit Theorem for Uniform Deviations**

Uniform deviations of f_n from f are easy to visualize and lead to 'confidence bands' for a density f, as in Proposition 6.4.3. In this respect, the result in this section, showing that the limiting distribution of $\|f_n-f\|_\infty$ suitably centred and normalised is the double exponential extreme-value distribution, is theoretically quite interesting (however, its usefulness in practise is hampered by slow speed of convergence). The proof requires use of the famous Komlós-Major-Tusnadý (KMT) approximation of the empirical process by Brownian bridges, a subject that is not developed in this book, as well as limit theorems for the distributions of maxima of stationary and not necessarily stationary Gaussian processes (see Section 2.7). Here is the statement of the KMT theorem.

Theorem 5.1.20 (Komlós-Major-Tusnadý approximation theorem) *There exists a probability space with a sequence $\{\xi_i\}$ of i.i.d. uniform on $[0,1]$ random variables and a sequence of Brownian motions W_n defined on it such that setting*

$$\alpha_n(t)=\frac{1}{\sqrt{n}}\sum_{i=1}^{n}(\delta_{\xi_i}[0,t]-t)$$

and $W_n^\circ(t) = W_n(t) - tW_n(1)$, *then*

$$\Pr\left\{\|\alpha_n - W_n^\circ\|_{[0,1]} > \frac{x + C\log n}{\sqrt{n}}\right\} \le \Lambda e^{-\theta x}, \quad 0 \le x < \infty,\ n \in \mathbb{N}, \tag{5.44}$$

for some universal finite, positive constants C, Λ, θ.

This theorem, combined with the limit theorems for suprema of Gaussian processes in Section 2.7, produces distributional limit theorems for $\|f_n - Ef_n\|_\infty$ in some cases. A large part of this section consists of showing how to use Theorem 5.1.20 to reduce weak convergence of the laws of $\|f_n - Ef_n\|_\infty$ (properly centred and normalised) to weak convergence of certain Gaussian processes. First, we state some conditions.

Our densities will belong to the class of functions

$$\begin{aligned}\mathcal{D} &= \mathcal{D}(\alpha, D, \delta, A, L)\\ &= \left\{f : \mathbb{R} \to \mathbb{R}, \int_{\mathbb{R}} f = 1, f \ge 0 \ on\ \mathbb{R}, f \ge \delta \ on\ A, \|f\|_\infty \le L, \|f\|_{C^\alpha(A)} \le D\right\},\end{aligned} \tag{5.45}$$

where $A = [F_1, F_2]$, $F_1 < 0 < 1 < F_2$ and α, δ are nonnegative. We assume without loss of generality that $[-\delta, 1+\delta] \subseteq A$ by decreasing δ if necessary. Recall the definition of $C^\alpha(A)$ from Section 4.3.3. To avoid triviality, we shall only consider combinations of α, D, δ, A, L such that $\mathcal{D}$ is nonempty.

We let $K : \mathbb{R}^2 \mapsto \mathbb{R}$ be a measurable function satisfying the following properties, of which the first three have appeared earlier in the text and whose verification will be discussed in some concrete examples later.

(K1) K is symmetric in its arguments and bounded, and for all $s \in \mathbb{R}$, $K(s,t)$ is right or left continuous in t for every $s \in \mathbb{R}$,
(K2) $\sup_t \|K(t,\cdot)\|_v := \|K\|_V < \infty$, where $\|\cdot\|_v$ denotes the total variation norm on $\mathbb{R}$, $K(t, -\infty) = 0$ for all t,
(K3) There is a bounded, nonincreasing, exponentially decaying function $\Phi : \mathbb{R}^+ \cup \{0\} \mapsto \mathbb{R}^+ \cup \{0\}$ such that

$$|K(x,y)| \le \Phi(|x-y|),$$

(K4) For all $\lambda \ge 1$, the covering numbers $N(\lambda[F_1,F_2], d, \varepsilon)$ of the intervals $[\lambda F_1, \lambda F_2]$ for the pseudo-distance $d(s,t) = \left(\int_{\mathbb{R}} (K(t,u) - K(s,u))^2 du\right)^{1/2}$ admit the bounds

$$N(\lambda[F_1,F_2], d, \varepsilon) \le \frac{A'\lambda^{v_2}}{\varepsilon^{v_1}},$$

for some $A', v_i < \infty$ independent of ε, λ, and these bounds are valid for all positive ε not exceeding the d diameter of $[\lambda F_1, \lambda F_2]$, and
(K5) There exist $\bar{A}$, $\bar{v}$ finite such that if $\mathcal{K} = \left\{K(2^j t, 2^j(\cdot)) : t \in \mathbb{R}, j \in \mathbb{N} \cup \{0\}\right\}$, and if $\mathcal{Q}$ is the set of Borel probability measures on $\mathbb{R}$, then

$$\sup_{Q \in \mathcal{Q}} N(\mathcal{K}, L^2(Q), \varepsilon) \le \left(\frac{\bar{A}}{\varepsilon}\right)^{\bar{v}}, \tag{5.46}$$

for $0 < \varepsilon \le \|K\|_\infty$.

Let $I = [0,1]$ (we could as well consider $[a,b]$, with $-\infty < a < b < \infty$, with only formal changes). Given a real sequence $j_n \to \infty$, define on I the Gaussian processes

$$Y_n(t) = 2^{j_n/2} \int_{\mathbb{R}} K(2^{j_n}t, 2^{j_n}s)dW(s) = \int_{\mathbb{R}} K(2^{j_n}t, u)dW(u), \tag{5.47}$$

where dW is a standard white noise. It will often be convenient to rewrite $Y_n(t)$ as $Y_n(t) = Y(2^{j_n}t)$, where

$$Y(t) = \int_{-\infty}^{\infty} K(t,s)dW(s). \tag{5.48}$$

Note also that condition *(K4)* ensures that the processes Y_n are sample continuous: for $u, v \in I$,

$$d_n^2(u,v) := E(Y_n(u) - Y_n(v))^2 = \int_{\mathbb{R}} (K(2^{j_n}u,s) - K(2^{j_n}v,s))^2 ds \le d^2(2^{j_n}u, 2^{j_n}v), \tag{5.49}$$

so $N(I, d_n, \varepsilon) \le N(2^{j_n}I, d, \varepsilon)$, and it follows from condition $(K4)$ that the square root of the metric entropy of I with respect to the distance d_n is integrable at zero; hence the claim is an immediate consequence of Dudley's theorem, Theorem 2.3.7. In particular, if we still denote a sample continuous version of Y_n by Y_n, the norms $\|Y_n\|_I = \sup_{t \in I} |Y_n(t)|$ are proper random variables.

Now let

$$\mathcal{F}_n = \cup_{f \in \mathcal{D}} \mathcal{F}_n^f, \quad \mathcal{F}_n^f = \{K(2^{j_n}t, 2^{j_n}\cdot)/\sqrt{f(t)} : t \in I\}. \tag{5.50}$$

Given $f \in \mathcal{D}$, let X_i be i.i.d. with law $dP_f(t) := f(t)dt$, and let, as usual,

$$\nu_n^f = \frac{1}{\sqrt{n}} \sum_{i=1}^{n} (\delta_{X_i} - P_f)$$

be the empirical process based on the sequence X_i. Note that by the properties of K and f, the supremum in $\|\nu_n^f\|_{\mathcal{F}_n^f}$ is countable and hence measurable.

Our first goal is to prove the following proposition, which reduces our problem on empirical processes to a problem about Gaussian processes. For the remainder of this section, $\Pr_f$ will denote the product probability $P_f^{\mathbb{N}}$, but the symbol Pr will denote the probability measure determining the laws of all relevant other random variables (such as Y_n and random variables constructed in the Gaussian coupling in the proof of Proposition 5.1.21).

Proposition 5.1.21 *Let $I = [0,1]$, let K be a function satisfying conditions (K1)–(K5) and let $j_n \to \infty$ as $n \to \infty$. Let $\{A_n\}$ and $\{B_n\}$ be numerical sequences such that $A_n \to \infty$ and*

$$A_n = o\left(\frac{\sqrt{n}}{2^{j_n/2}\log n} \wedge 2^{j_n/2} \wedge \frac{2^{\alpha j_n}}{\sqrt{j_n}}\right), \tag{5.51}$$

for some $0 < \alpha < 1$. Assume that there exists a random variable Z with continuous distribution such that

$$\lim_{n\to\infty} \Pr\{A_n(\|Y_n\|_I - B_n) \le x\} = \Pr\{Z \le x\}, \quad x \in \mathbb{R}, \tag{5.52}$$

where the processes Y_n are defined by (5.47). Let $\mathcal{D}(\alpha, D, \delta, A, L)$ be as in (5.45) for the given α, $A = [F_1, F_2] \supset I$, and δ, D and L, such that $\mathcal{D}$ is not empty. Define, for each $f \in \mathcal{D}$, $\mathcal{F}_n^f$

as in (5.50), and further let ν_n^f, $n \in \mathbb{N}$, be the empirical processes based on the variables X_i. Then, for all $x \in \mathbb{R}$,

$$\lim_{n\to\infty} \sup_{f\in\mathcal{D}} \left| \Pr_f \left\{ A_n(2^{j_n/2}\|\nu_n^f\|_{\mathcal{F}_n^f} - B_n) \le x \right\} - \Pr\{Z \le x\} \right| = 0. \tag{5.53}$$

Proof *Step 1*: Define new random variables $\tilde{X}_i = F_f^{-1}(\xi_i)$, where F_f^{-1} is the left continuous generalised inverse of the distribution function F_f of f, right continuous at zero. For every $f \in \mathcal{D}$, the variables $\tilde{X}_i$ are i.i.d. with law P_f, and we denote by $\tilde{\nu}_n^f$ the associated empirical process. By $(K2)$ and $f \ge \delta$ on I, the functions in $\mathcal{F}_n$ have total variation norm not exceeding $\|K\|_V/\sqrt{\delta}$, and since F_f^{-1} is monotone, it follows that the same bound on the total variation norm (for functions on $[0,1]$) holds for all the functions in the classes

$$\tilde{\mathcal{F}}_n^f = \{h \circ F_f^{-1} : h \in \mathcal{F}_n^f\}, \quad f \in \mathcal{D}, \ n \in \mathbb{N}.$$

Moreover, if g is nonincreasing on $[0,1]$ with $g(0) = 1$ and $g(1) = 0$, then g is the pointwise nondecreasing limit – and, by dominated convergence, also the limit in $L_2([0,1])$ – of convex combinations of indicators $I_{[0,t]}$, $0 \le t \le 1$. Thus, by $(K2)$, both α_n and W_n extend from sets $I_{[0,t]}$ to functions in $\tilde{\mathcal{F}}_n^f$ by linearity and continuity (see Theorem 3.7.28), and so does W_n°. We conclude that, for all $f \in \mathcal{D}$,

$$\|\alpha_n - W_n^\circ\|_{\tilde{\mathcal{F}}_n^f} \le \|K\|_V \delta^{-1/2} \|\alpha_n - W_n^\circ\|_{[0,1]},$$

and, writing $G_{n,f}^\circ(g) = W_n^\circ(g \circ F_f^{-1})$ for $g \in \mathcal{F}_n^f$, that $E(G_{n,f}^\circ(g)G_{n,f}^\circ(\bar{g})) = P_f(g\bar{g}) - (P_f g)(P_f \bar{g})$; that is, $G_{n,f}^\circ$ is a (sample continuous) version of the P_f-Brownian bridge. Since, furthermore,

$$\alpha_n(g \circ F_f^{-1}) = \tilde{\nu}_n^f(g)$$

by construction, (5.44) gives

$$\begin{aligned}
\sup_{f\in\mathcal{D}} \Pr&\left\{ \|\tilde{\nu}_n^f - G_{n,f}^\circ\|_{\mathcal{F}_n^f} > \frac{\|K\|_V \delta^{-1/2}(x + C\log n)}{\sqrt{n}} \right\} \\
&= \sup_{f\in\mathcal{D}} \Pr\left\{ \|\alpha_n - W_n^\circ\|_{\tilde{\mathcal{F}}_n^f} > \frac{\|K\|_V \delta^{-1/2}(x + C\log n)}{\sqrt{n}} \right\} \\
&\le \Pr\left\{ \|\alpha_n - W_n^\circ\|_{[0,1]} > \frac{x + C\log n}{\sqrt{n}} \right\} \le \Lambda e^{-\theta x},
\end{aligned}$$

for all $x \ge 0$ and $n \in \mathbb{N}$. Taking $x = (C' - C)\log n$ for some $C' > C$ in this inequality, we have

$$\sup_{f\in\mathcal{D}} \Pr\left\{ A_n 2^{j_n/2} \|\tilde{\nu}_n^f - G_{n,f}^\circ\|_{\mathcal{F}_n^f} > \frac{\|K\|_V \delta^{-1/2} C' A_n \log n}{\sqrt{n2^{-j_n}}} \right\} \le \frac{\Lambda}{n^{(C'-C)\theta}}. \tag{5.54}$$

In particular, if

$$A_n = o\left(\frac{\sqrt{n}}{2^{j_n/2}\log n} \right), \tag{5.55}$$

then (5.54) implies that there exists a sequence $\varepsilon_n'' \to 0$ such that

$$\lim_n \sup_{f\in\mathcal{D}} \Pr\left\{ A_n 2^{j_n/2} \|\tilde{\nu}_n^f - G_{n,f}^\circ\|_{\mathcal{F}_n^f} \ge \varepsilon_n'' \right\} = 0. \tag{5.56}$$

Consider next the processes $G_{n,f}(g) = W_n(g \circ F_f^{-1})$, $g \in \mathcal{F}_n^f$, which are sample continuous versions of the P_f-Brownian motion W_{P_f} because

$$E(W_n(g \circ F_f^{-1})W_n(\bar{g} \circ F_f^{-1})) = \int_0^1 g \circ F_f^{-1}(x)\bar{g} \circ F_f^{-1}(x)dx = \int_{\mathbb{R}} g(y)\bar{g}(y)f(y)dy.$$

Since $W_n^\circ(g \circ F^{-1}) = W_n(g \circ F^{-1}) - \left(\int_0^1 g \circ F^{-1}(t)dt\right) W_n(1)$ and since, by (K3),

$$\sup_{g \in \mathcal{F}_n^f} \left| \int_0^1 g \circ F^{-1}(t)dt \right| = \sup_{g \in \mathcal{F}_n^f} |P_f g| \leq \delta^{-1/2} \|\Phi\|_1 2^{-j_n}, \tag{5.57}$$

it follows that if

$$A_n = o(2^{j_n/2}), \tag{5.58}$$

then we can replace G_n° by G_n in (5.56); that is, there exists $\varepsilon_n' \to 0$ such that

$$\lim_n \sup_{f \in \mathcal{D}} \Pr\left\{A_n 2^{j_n/2} \|\tilde{\nu}_n^f - G_{n,f}\|_{\mathcal{F}_n^f} \geq \varepsilon_n'\right\} = 0. \tag{5.59}$$

(Note that by Theorem 3.7.28, for all n and f, the process $W_{P_f}(g)$, $g \in \mathcal{F}_n^f$, is sample continuous (hence sample bounded).)

Step 2: To compare $G_{n,f}$ on $\mathcal{F}_n^f$ with Y_n, we must couple in the right way sample continuous versions of both processes. Since the functions in $\mathcal{F}_n^f$ are parametrised by $t \in I$, we will write (in slight abuse of notation) $G_{n,f}(t)$, $t \in I$, for $G_n(g_t)$, $g_t(\cdot) = K(2^{j_n}t, 2^{j_n}\cdot)/\sqrt{f(t)} \in \mathcal{F}_n^f$. First, we observe that the process

$$W\left(K(2^{j_n}t, 2^{j_n}\cdot)\sqrt{f(\cdot)/f(t)}\right), \quad t \in I,$$

where W is Brownian motion acting on functions as described in step 1, is a version of $G_{n,f}$ (both processes have the same covariance). Next, we observe that for λ Lebesgue measure, the isonormal process of $L^2(\mathbb{R}, \lambda)$, $g \mapsto \int_{\mathbb{R}} g\, dW = W(g)$, restricted to the set $\mathcal{G}_n$ defined by

$$\mathcal{G}_n = \left\{2^{j_n/2}K(2^{j_n}t, 2^{j_n}\cdot), K(2^{j_n}t, 2^{j_n}\cdot)\sqrt{f(\cdot)/f(t)} : t \in I\right\},$$

admits a version with bounded uniformly continuous sample paths (for the $L^2(\mathbb{R}, \lambda)$ distance): this follows from the entropy bounds ($K4$) and ($K5$) and the metric entropy theorem for Gaussian processes, Theorem 2.3.7. We call $\tilde{G}_{n,f}(t)$ and $\tilde{Y}_n(t)$ the restrictions of this process to the sets $\{K(2^{j_n}t, 2^{j_n}\cdot)\sqrt{f(\cdot)/f(t)} : t \in I\}$ and $\{2^{j_n/2}K(2^{j_n}t, 2^{j_n}\cdot) : t \in I\}$, respectively. They are versions of $G_{n,f}$ and Y_n, respectively, and, as we see next, we can control the supremum norm of their difference. Set

$$Z_{n,f}(t) = 2^{j_n/2}\tilde{G}_n(t) - \tilde{Y}_n(t) = 2^{j_n/2}\int_{\mathbb{R}} K(2^{j_n}t, 2^{j_n}s)\left(\sqrt{\frac{f(s)}{f(t)}} - 1\right) dW(s), \quad t \in I.$$

We have, for $u, v \in I$,

$$\begin{aligned} d_{Z_{n,f}}(u,v) &:= \left(E(Z_{n,f}(u) - Z_{n,f}(v))^2\right)^{1/2} \\ &\leq \delta^{-1/2}\|K(2^{j_n}u,\cdot)) - K(2^{j_n}v,\cdot))\|_{L_2(P_f)} + d_n(u,v), \end{aligned}$$

where $d_n(u,v) = d(2^{j_n}u, 2^{j_n}v)$ (cf. (5.49)), and using $(K4)$ and $(K5)$, the covering numbers of I for this distance are bounded by

$$N(I, d_{Z_{n,f}}, \varepsilon) \leq N(\mathcal{K}, L^2(P_f), \delta^{1/2}\varepsilon/2) N(2^{j_n}I, d, \varepsilon/2) \leq B2^{v_3 j_n}/\varepsilon^{v_4}, \tag{5.60}$$

for every (small) $\varepsilon > 0$ and constants B, v_3, v_4 independent of n and f. Since if $f \in \mathcal{D}$ then $\|f\|_\infty \leq L$ and its α-Hölder constant on $[-\delta, 1+\delta]$ is at most D, we have, for $t \in I$,

$$\left(\sqrt{f(t-2^{-j_n}u)} - \sqrt{f(t)}\right)^2 \leq LI(|u| > \delta 2^{j_n}) + 4^{-1}\delta^{-1}D^2 2^{-2\alpha j_n} u^{2\alpha} I(|u| \leq \delta 2^{j_n}),$$

and we obtain, for all $t \in I$,

$$\begin{aligned}
E(Z_{n,f}(t))^2 &= 2^{j_n} \int_{\mathbb{R}} K^2(2^{j_n}t, 2^{j_n}s) \left(\sqrt{\frac{f(s)}{f(t)}} - 1\right)^2 ds \\
&\leq 2^{j_n}\delta^{-1} \int_{\mathbb{R}} K^2(2^{j_n}t, 2^{j_n}s) \left(\sqrt{f(s)} - \sqrt{f(t)}\right)^2 ds \\
&\leq \delta^{-1} \int_{\mathbb{R}} \Phi^2(u) \left(\sqrt{f(t-2^{-j_n}u)} - \sqrt{f(t)}\right)^2 du \\
&\leq \delta^{-1} L \|\Phi\|_1 \Phi(\delta 2^{j_n}) + 4^{-1}\delta^{-2} D^2 2^{-2\alpha j_n} \int_{\mathbb{R}} \Phi^2(u) u^{2\alpha} du \leq D_1^2 2^{-2\alpha j_n},
\end{aligned} \tag{5.61}$$

where D_1 is a constant that does not depend on n or f. That is, the diameter of I for the L_2-distance induced by the process $Z_{n,f}$ is at most $2D_1 2^{-\alpha j_n}$. Hence, the metric entropy bound in Theorem 2.3.7, (5.60) and (5.61) give

$$\begin{aligned}
E \sup_{t \in I} \left| 2^{j_n/2} \tilde{G}_{n,f}(t) - \tilde{Y}_n(t) \right| &\lesssim D_1 2^{-\alpha j_n} + \int_0^{D_1 2^{-\alpha j_n}} \sqrt{\log \frac{B2^{v_3 j_n}}{\varepsilon^{v_4}}} d\varepsilon \\
&\lesssim \sqrt{j_n} 2^{-\alpha j_n},
\end{aligned}$$

with unspecified multiplicative constants independent of $f \in \mathcal{D}$ and n. Thus, if, besides (5.55) and (5.58), the sequence $\{A_n\}$ satisfies

$$A_n = o\left(2^{\alpha j_n}/\sqrt{j_n}\right)$$

(hence, if $\{A_n\}$ satisfies (5.51)), then there exists $\varepsilon_n \to 0$ such that

$$\lim_{n \to \infty} \sup_{f \in \mathcal{D}} \Pr\left\{A_n \|2^{j_n/2}\tilde{G}_{n,f} - \tilde{Y}_n\|_I \geq \varepsilon_n\right\} = 0. \tag{5.62}$$

Step 3: We finally combine the bounds obtained. Clearly, $\|\tilde{G}_{n,f}\|_I$ has the same probability law as $\|G_{n,f}\|_{\mathcal{F}_n^f}$, and likewise, $\|\tilde{Y}_n\|_I$ has the same law as $\|Y_n\|_I$. Therefore, under the hypotheses of the proposition, we have, for all $f \in \mathcal{D}$ and $x_n \to x$, $x \in \mathbb{R}$,

$$\begin{aligned}
&[\Pr\{A_n(\|Y_n\|_I - B_n) \leq x_n - \varepsilon_n\} - \Pr\{Z \leq x\}] - \sup_{f \in \mathcal{D}} \Pr\left\{A_n \left\|2^{j_n/2}\tilde{G}_{n,f} - \tilde{Y}_n\right\|_I > \varepsilon_n\right\} \\
&\leq \Pr\left\{A_n\left(2^{j_n/2}\|G_{n,f}\|_{\mathcal{F}_n^f} - B_n\right) \leq x_n\right\} - \Pr\{Z \leq x\} \\
&\leq [\Pr\{A_n(\|Y_n\|_I - B_n) \leq x_n + \varepsilon_n\} - \Pr\{Z \leq x\}] + \sup_{f \in \mathcal{D}} \Pr\left\{A_n \left\|2^{j_n/2}\tilde{G}_{n,f} - \tilde{Y}_n\right\|_I > \varepsilon_n\right\}.
\end{aligned}$$

The left- and rightmost sides of this inequality do not depend on $f \in \mathcal{D}$ and tend to zero by (5.52), the continuity of the probability law of Z and (5.62). Thus, we have

$$\lim_{n\to\infty} \sup_{f\in\mathcal{F}_n^f} \left| \Pr\left\{A_n\left(2^{j_n/2}\|G_{n,f}\|_{\mathcal{F}_n^f} - B_n\right) \le x_n\right\} - \Pr\{Z \le x\}\right| = 0, \tag{5.63}$$

for any sequence $x_n \to x$, any $x \in \mathbb{R}$. Similarly, since the random variables $\|\tilde{\nu}_n^f\|_{\mathcal{F}_n^f}$ and $\|\nu_n^f\|_{\mathcal{F}_n^f}$ have the same law, we have, for any $x \in \mathbb{R}$,

$$\left[\Pr\{A_n(2^{j_n/2}\|G_{n,f}\|_{\mathcal{F}_n^f} - B_n) \le x - \varepsilon_n'\} - \Pr\{Z \le x\}\right] - \sup_{f\in\mathcal{D}} \Pr\left\{A_n 2^{j_n/2}\|\tilde{\nu}_n^f - \tilde{G}_{n,f}\|_{\mathcal{F}_n^f} > \varepsilon_n'\right\}$$

$$\le \Pr_f\left\{A_n\left(2^{j_n/2}\|\nu_n^f\|_{\mathcal{F}_n^f} - B_n\right) \le x\right\} - \Pr\{Z \le x\}$$

$$\le \left[\Pr\{A_n(2^{j_n/2}\|G_{n,f}\|_{\mathcal{F}_n^f} - B_n) \le x + \varepsilon_n'\} - \Pr\{Z \le x\}\right]$$

$$+ \sup_{f\in\mathcal{D}} \Pr\left\{A_n 2^{j_n/2}\|\tilde{\nu}_n^f - \tilde{G}_{n,f}\|_{\mathcal{F}_n^f} > \varepsilon_n'\right\},$$

which, by (5.59) and (5.63) with $x_n = x \pm \varepsilon_n'$, gives (5.53). ∎

Condition $(K3)$ is used only in the preceding equation (5.57) and in (5.61); therefore, it can be relaxed to there is Φ measurable, bounded and satisfying that, for some y_0 and $\eta > 0$ and all $y > y_0$, $\sup_{x\ge y}\Phi(x) \le y^{-1-\eta}$ such that $|K(x,y)|$ is dominated by $\Phi(|x-y|)$.

Here is the first example of application of Proposition 5.1.21. The projection kernel corresponding to the Haar wavelet is

$$K(x,y) = \sum_{k\in\mathbb{Z}} I_{[0,1)}(x-k)I_{[0,1)}(y-k) = I([x] = [y]). \tag{5.64}$$

It obviously satisfies conditions $(K1)$–$(K3)$ ($\|K(t,\cdot)\|_v = 2$, $\Phi(|u|) = I(|u| \le 1)$). Moreover, $d^2(x,y) = \int_{\mathbb{R}}(K(x,u) - K(y,u))^2 du = 0$ if $[x] = [y]$ and 2 otherwise, so

$$N(\lambda[F_1,F_2],d,\varepsilon) \le N(\lambda[F_1,F_2],d,0) \le \lambda(F_2 - F_1) + 2 \le \frac{2\lambda(F_2 - F_1 + 2)}{\varepsilon},$$

for $0 < \varepsilon < 2$ (note that 2 is an upper bound for the d-diameter of any set of real numbers), so $(K4)$ holds. $(K5)$ follows because $\mathcal{K} = \{I_{[k/2^j,(k=1)/2^j)} : k \in \mathbb{Z}, j \ge 0\}$ consists of indicators of a VC class of sets (see Theorem 3.6.9).

Thus, Proposition 5.1.21 applies, and we are led to consider the process (see (5.48))

$$Y(t) = \sum_{k\in\mathbb{Z}} I(t \in [k,k+1)) \int_k^{k+1} dW(s) = \sum_{k\in\mathbb{Z}} I(t \in [k,k+1))g_k,$$

where g_k are i.i.d. $N(0,1)$, and therefore, taking $I = [0,1]$,

$$\sup_{0\le t\le 1} |Y_n(t)| = \sup_{0\le u\le 2^{j_n}} |Y(u)| = \max_{0\le k\le 2^{j_n}} |g_k|.$$

Now Theorem 2.7.1 gives

$$\Pr\left\{A_n\left(\sup_{0\le t\le 1}|Y_n(t)| - B_n\right) \le x\right\} \to e^{-e^{-x}}, \quad \text{for all } x \in \mathbb{R},$$

where $A_n = A(j_n)$, $B_n = B(j_n)$, and

$$A(l) = [2(\log 2)l]^{1/2}, \quad B(l) = A(l) - \frac{\log l + \log(\pi \log 2)}{2A(l)}. \tag{5.65}$$

Combining this with Proposition 5.1.21, we have, recalling the set $\mathcal{D}$ from (5.45), the following:

Proposition 5.1.22 *Let $\mathcal{D} = \mathcal{D}(\alpha, D, \delta, A, L)$, for some $0 < \alpha \leq 1$, $0 < D < \infty$ and where δ, A are admissible. If $j_n \to \infty$ as $n \to \infty$ satisfies $j_n 2^{j_n} = o(n/(\log n)^2)$, and if $f_n := f_n(\cdot, j_n)$ is the Haar wavelet estimator from (5.3) with $\phi = 1_{[0,1)}$, then*

$$\sup_{f\in\mathcal{D}} \left| \Pr_f \left\{ A_n \left(\sqrt{n2^{-j_n}} \left\| \frac{f_n - Ef_n}{\sqrt{f}} \right\|_{[0,1]} - B_n \right) \leq x \right\} - e^{-e^{-x}} \right| \to 0, \quad \text{for all } x \in \mathbb{R},$$

as $n \to \infty$, where A_n and B_n are as before (5.65).

As a second example, we consider (convolution) kernel density estimators. If K is a real-valued function with bounded support, symmetric and Lipschitz continuous, then the kernel $K(x,y) := K(x-y)$ satisfies conditions $(K1)$–$(K4)$ with $\Phi = K$ and $d(s,t)$ proportional to $|s-t|$, and note that $(K5)$ holds as well by Remark 5.1.4. (These are not the only convolution kernels satisfying $(K1)$–$(K5)$; for instance, the Gaussian kernel also satisfies theses conditions.)

Assume now that K is bounded, symmetric, supported by $[-1,1]$ and twice continuously differentiable on $\mathbb{R}$. Writing $Y_n(t) = Y(2^{j_n}t)$ with Y as in (5.48), we have

$$\sup_{t\in[0,1]} |Y_n(t)| = \sup_{0\leq t\leq 2^{j_n}} |Y(t)|.$$

In this case, $Y(t) = \int_{\mathbb{R}} K(t-s)dW(s)$ is a stationary Gaussian process with covariance

$$r(t) := E(Y(t)Y(0)) = \int_{\mathbb{R}} K(t+u)K(u)du = \|K\|_2^2 - Ct^2 + o(t^2),$$

where $C = -2^{-1}\int_{\mathbb{R}} K(u)K''(u)du > 0$ (by integration by parts), and $r(t) = 0$ for $|t| > 2$. Set $\tilde{Y} = Y/\|K\|_2$ and $\tilde{C} = C/\|K\|_2^2$. We apply Theorem 2.7.9: with

$$B(l) = \sqrt{2(\log 2)l} + \frac{\log\sqrt{2\tilde{C}} - \log\pi}{\sqrt{2(\log 2)l}}, \tag{5.66}$$

we have

$$\lim_{n\to\infty} \Pr\left\{ \sqrt{2(\log 2)j_n} \left(\sup_{0\leq t\leq 2^{j_n}} |Y(t)|/\|K\|_2 - B(j_n) \right) \leq x \right\} \to e^{-e^{-x}}, \quad x \in \mathbb{R},$$

which, combined with Proposition 5.1.21, yields the following proposition:

Proposition 5.1.23 *If $K : \mathbb{R} \mapsto \mathbb{R}$ is bounded, symmetric, supported by $[-1,1]$ and twice continuously differentiable, $\mathcal{D}$ and j_n are as in Proposition 5.1.22, $B(l)$ is as in (5.66) and if $f_n := f_n(y, j_n)$ is the kernel estimator from (5.3), then, as $n \to \infty$, for all $x \in \mathbb{R}$*

$$\sup_{f\in\mathcal{D}} \left| \Pr_f \left\{ \sqrt{2(\log 2)j_n} \left(\sqrt{n2^{-j_n}} \left\| \frac{f_n - Ef_n}{\|K\|_2\sqrt{f}} \right\|_{[0,1]} - B(j_n) \right) \leq x \right\} - e^{-e^{-x}} \right| \to 0.$$

Wavelet projection density estimators based on Daubechies, band-limited or spline wavelets satisfy limit theorems similar to these, but the proofs require extreme-value theory for nonstationary processes, such as in Theorem 2.8.3, and will not be treated here. See the notes at the end of this chapter for references.

Exercises

5.1.1 Let $K_{j,\mathrm{per}}$ be as defined in (5.4) for a scaling function ϕ satisfying Condition 5.1.1(b). Prove that for q conjugate to p,

(a) For all $x \in [0,1]$, $\int_0^1 |K_{j,\mathrm{per}}(x,y)|dy \le \kappa \|\phi\|_1 \le \kappa^2$ where $\kappa := \|\sum_k |\phi(\cdot - k)|\|_\infty$;

(b) For all $x \in [0,1], p \ge 1$, $\left(\int_0^1 |K_{j,\mathrm{per}}(x,y)|^p dy\right)^{1/p} \le \kappa^2 2^{j(1-1/p)}$;

(c) For $1 \le p < \infty$ and $\|g\|_q \le 1$, $\|K_{j,\mathrm{per}}(g)\|_\infty \le \sup_x \|K_{j,\mathrm{per}}(x,\cdot)\|_p \le \kappa^2 2^{j(1-1/p)}$;

(d) $\|K_{j,\mathrm{per}}\|_\infty \le \kappa_2 2^j$ and $EK^2_{j,\mathrm{per}}(x,X) \le \|f\|_\infty \kappa^2 2^j$;

(e) For $\|g\|_q \le 1$, $E(K_{j,\mathrm{per}}(g)(X))^2 \le \|f\|_p \|K_{j,\mathrm{per}}(g)\|^2_{2q} \le C_p \|f\|_p \sup_x \|K_{j,per}(x,\cdot)\|^2_{2p/(1+p)}$ $\le C_p \|f\|_p \kappa^4 2^{j(1-1/p)}$, where C_p depends only on p;

(f) For $\|g\|_q \le 1$ and $p > 2$, $E(K_{j,\mathrm{per}}(g)(X))^2 \le \|f\|_\infty \|K_{j,\mathrm{per}}(g)\|_2^2 \le \bar{C}_p \|f\|_\infty \|K\|^2_{2p/(2+p)} \le$ $\bar{C}_p \|f\|_\infty \kappa^4 2^{j(1-2/p)}$, where $\bar{C}_p$ depends only on p;

(g) For $\|g\|_q \le 1$ and $p \le 2$,

$$E(K_{j,\mathrm{per}}(g)(X))^2 \le \|f\|_{p/(2-p)} \|K_j(g)\|_q^2 \le \|f\|_{p/(2-p)} \|K_{j,\mathrm{per}}(x,\cdot)\|_1^2 \le \|f\|_{p/(2-p)} \kappa^2 \|\phi\|_1^2.$$

Hints: For (a), note that

$$\int_0^1 |K_{j,\mathrm{per}}(x,y)|dy \le 2^j \int_{-\infty}^{\infty} \sum_{m=0}^{2^j-1} |\phi(2^j y - m)| \sum_\ell |\phi(2^j(x-\ell) - m)| dy \le \|\phi\|_1 \kappa.$$

For (b), note that, by (a),

$$\left(\int_0^1 |K_{j,\mathrm{per}}(x,y)|^p dy\right)^{1/p} \le \|K_{j,\mathrm{per}}\|_\infty^{1-1/p} (\|\phi\|_1 \kappa)^{1/p} \le 2^{j(1-1/p)} \kappa^{2-1/p}.$$

For most of the rest, what is needed is a generalisation of Young's inequality for operators $Tg = \int_0^1 K(x,y)g(y)dy$: if $\int |K(x,y)|dy \le C$ and $\int |K(x,y)|dx \le C$ a.e., then $\|Tg\|_p \le C\|g\|_p$, for $1 \le p \le \infty$; also, for $1 < p < r < \infty$ and $1 + 1/r = 1/p + 1/q$, if $\|K(x,\cdot)\|_q \le C$ and $\|K(\cdot,y)\|_q \le C$ a.e., then $\|Tg\|_r \le B_p C \|g\|_p$, where B_p is an absolute constant depending only on p. See Folland (1999), Theorems 6.18, 6.36.

5.1.2 Prove an analogue of Exercise 5.1.1 for $K_{j,bc}$ assuming that ϕ is a Daubechies scaling function. *Hint*: one may deal separately with $K^{left}_{j,bc}$, $K^{right}_{j,bc}$ and $\tilde{K}_{j,bc}$, exploiting the facts that $\tilde{K}_{j,bc}$ is dominated by a convolution kernel and that the other two kernels involve only finitely many sums (N if ϕ is the Daubechies Nth scaling function).

5.1.3 Under Condition 5.1.1, prove that $\|f_n - Ef_n\|_\infty = \max_{x \in L} |f_n(x) - Ef_n(x)|$, where L is a countable subset of $\mathbb{R}$. *Hints*: If K is a right-continuous kernel, then the function $f_n(x) - Ef_n(x)$ is also right continuous (note that K is bounded), and we can take $L = \mathbb{Q}$. If ϕ is as in Condition 5.1.1(b), since ϕ has finite p-variation, ϕ is continuous except perhaps on a countable set D of A (as it is the composition of a Hölder continuous with a nondecreasing function, see Lemma 3.6.11); use this and the fact that ϕ has bounded support to argue that one can take L to be the union of $\{2^j y - k : y \in D, k \in \mathbb{Z}\}$ and $\mathbb{Q}$. The remaining two cases are similar.

5.1.4 Use Remark 5.1.14 to show that $|\|f_n - f\|_1 - E\|f_n - f\|_1| = O_{pr}(1/\sqrt{n})$ and that, for all densities, the estimators f_n considered in Theorem 5.1.13 satisfy

$$\limsup_{n\to\infty} \sqrt{\frac{n}{2\log n}}\, |\|f_n - f\|_1 - E\|f_n - f\|_1| \le \|\Phi\|_1.$$

5.1.5 The expectation bounds in Theorem 5.1.5 cannot in general be improved except perhaps for the multiplicative constant L_p because, as mentioned earlier, they lead to unimprovable minimax rates. In some instances, one can directly obtain lower bounds on the expected value of $\|f_n - Ef_n\|_p$ of the same order as the upper bounds. One such instance occurs with the convolution estimator based on $K = I_{[-1/2,1/2]}$ when the density $f(x)$ is bounded from below on an interval. (a) Prove this for $p = \infty$ by using the metric entropy lower bound in Theorem 3.5.11 after randomising (i.e., replacing $nh_n(f_n - Ef_n)$ by $\sum_{i=1}^n \varepsilon_i K((x - X_i)/h_n)$). (b) To obtain the lower bound for $1 \le p < \infty$, use the Minkowski inequality for integrals to bound $E\|f_n - Ef_n\|_p$ from below by

$$(nh_n)^{-1}\left[\int_{f(x)>\delta}\left(E\left|\sum_{i=1}^n (I_{[x-h_n/2,x+h_n/2]}(X_i) - \Pr(|x - X_1| \le h_n/2)\}\right|\right)^p dx\right]^{1/p}.$$

Then argue that by normal approximation and uniform integrability, $\sup_{np_n(1-p_n)>M} |E|Z_n|/\sqrt{np_n(1-p_n)} - E|g||\to 0$ as $M\to\infty$, where g is $N(0,1)$ and Z_n is $Bin(n,p_n)$ This shows there exist constants $C > 0$, $M < \infty$, $\delta > 0$ such that if $nh_n(1 - h_n) > M$ and $h_n < \delta$, then $E\|f_n - Ef_n\|_p \ge C(nh_n)^{-1/2}$.

5.1.6 The expectation bound in Proposition 5.1.2 for $p = \infty$ cannot in general be improved. Show that for the Haar basis of $[0,1]$,

$$E\left\|\sum_{k=0}^{2^j-1} g_k 2^{j/2} I_{[k/2^j,(k+1)/2^j]}(x)\right\|_\infty \ge c\sqrt{2^j j}.$$

5.1.7 Prove analogues of Propositions 5.1.18 and 5.1.19 for the estimators of a function observed under additive white noise. When considering the supremum norm, it is convenient to take the alternate definition of the projection kernel; for example, for the boundary corrected wavelets,

$$K_{j,bc}(x,y) = \sum_{k=0}^{2^J-1} \phi_{Jk}(x)\phi_{Jk}(y) + \sum_{\ell=J}^{j-1}\sum_{k=0}^{2^\ell-1} \psi_{\ell k}^{bc}(x)\psi_{\ell k}^{bc}(y),$$

since then, for j_n increasing, the processes

$$\frac{\sigma}{\sqrt{n}}(f_n - Ef_n) = \sum_{k=0}^{2^J-1} g_{Jk}\phi_{Jk} + \sum_{\ell=J}^{j-1}\sum_{k=0}^{2^\ell-1} \tilde{g}_{\ell k}\psi_{\ell k}^{bc},$$

where the g variables are all independent standard normal and are, in fact, partial sums of independent random processes to which P. Lévy's inequality applies. This allows us to essentially proceed as in the proof of Proposition 5.1.19.

5.1.8 Again, on Proposition 5.1.2, use the fact that the L^p-norms of Banach-valued Gaussian random variables are equivalent and that there exist c_2, d_2 finite and positive such that $c_1 2^{-j} \le \sum_{k=0}^{2^j-1} (\phi_{jk}^{bc}(x))^2 \le c_2 2^{-j}$, for all $x \in [0,1]$, to show that for all $1 \le p < \infty$ there exists $D_p > 0$ such that $E\left\|\sum_{k=0}^{2^j-1} g_k \phi_{jk}^{bc}(\cdot)\right\|_p \ge D_p 2^{j/2}$.

5.1.9 Combining inequality (5.35) with the lower tail in Talagrand's inequality, Theorem 3.3.10, we obtain that for some $0 < C < \infty$,

$$\Pr\left\{\left|\|f_n - Ef_n\|_p - E\|f_n - Ef_n\|_p\right| > C\left(\sqrt{vx/n^2} + Ux/n\right)\right\} \le 2e^{-x}.$$

Use this inequality to show that assuming that $2^{j_n} < n$ and $\|f\|_\infty < \infty$ if $p > 1$, then $\|f_n - Ef_n\|_p - E\|f_n - Ef_n\|_p = O_{pr}(1/\sqrt{n})$ for $1 \le p \le 2$, and $\|f_n - Ef_n\|_p - E\|f_n - Ef_n\|_p = O_{pr}(\sqrt{2^{j(1-2/p)}}/n)$ for $p > 2$. Deduce analogous results by assuming that $\|f\|_p < \infty$ instead of f bounded. Deduce also upper bounds for almost-sure rates. Show also that the same rates hold in the white noise case.

5.2 Weak and Multiscale Metrics

Whereas a probability density can only be estimated at a rate slower than $n^{-1/2}$ (depending on its smoothness), the distribution function can be estimated by the empirical distribution function at the rate $n^{-1/2}$. If $\{f_n\}$ is a sequence of convolution kernel or wavelet linear estimators that approximate f at the best rate in L^p for some $1 \le p \le \infty$, it is shown in this section that the empirical distribution function corresponding to f_n also approaches the true F at the optimal rate $n^{-1/2}$, provided that we take the kernel K or the wavelet basis just of a slightly higher order than needed for only the L^p approximation. This is a consequence of an exponential inequality measuring the closeness of the empirical distribution function and the distribution function of f_n and constitutes an example of a 'plug in' property: f_n approximates f at the best rates simultaneously in the L^p metric and in a weaker metric. The weaker metrics considered in this section go far beyond the metric defining the supremum of a difference between distribution functions of a density, $\|F_{f_n} - F_f\|_\infty = \sup_x \left|\int I_{(-\infty,x]}(f_n - f)\right|$, to include the supremum of $\int g(f_n - f)$ with g running over bounded subsets of several important Banach spaces of functions such as Sobolev spaces, bounded variation spaces and, more generally, some Besov spaces.

5.2.1 Smoothed Empirical Processes

Both in the density estimation case and in the case of function estimation under additive white noise, the rate of approximation of f by f_n given by Proposition 5.1.7, assuming that f has degree of smoothness s and K is the kernel projection of an S-regular wavelet basis with $s < S$, is $n^{-s/(2s+1)}$ in the L^p norms, $p < \infty$, and is $(n/\log n)^{-s/(2s+1)}$ in the supremum norm and, as we will see in Chapter 6, these rates cannot be improved (in the sense that they are minimax). Although f_n does not approximate f at the usual finite-dimensional rate $n^{-1/2}$ in L^2 loss, the question arises as to whether the approximation rate is $n^{-1/2}$ in weaker norms of interest. Typical weaker norms of interest are the supremum over bounded subsets of $L^2([0,1])$ or $L^2(\mathbb{R})$ much smaller than the unit ball, usually consisting of smooth functions or of functions of bounded variation (the latter including $I_{(-\infty,u]}$ for all u, thus being distribution functions based on f_n).

When an estimator f_n of f is minimax in quadratic or supremum norm loss and is also optimal in its action on a class of functions $\mathcal{H} \subset L^2$, then we say that it has *the plug-in property with respect to the class of functions* $\mathcal{H}$. We show in this subsection that convolution kernel and wavelet projection density estimators satisfy the plug-in property not only for $\mathcal{H} = \{I_{(-\infty,t]} : t \in \mathbb{R}\}$ but also for many other classes of functions. Although this property holds even for certain classes of functions that are not P-Donsker (see the notes at the end of this chapter), we will only consider Donsker classes. For a P-Donsker class $\mathcal{H}$, since by the empirical central limit theorem $\sqrt{n}(P_n - P) \to_{\mathcal{L}} G_P$ in $\ell_\infty(\mathcal{H})$, it will suffice to show

$\sup_{g\in\mathcal{F}} \left|\int g f_n - P_n(g)\right| = o_P(n^{1/2})$ to obtain rates for these weak metrics of the order $n^{-1/2}$. In addition, and related to this problem, we will also obtain an exponential bound for $\|F_{f_n} - F_n\|_\infty$, where F_{f_n} is the distribution function corresponding to the measure of density f_n, whereas F_n is the empirical distribution function, that is, the distribution function of the empirical measure P_n.

The Donsker classes to be considered are bounded subsets of the space $BV(A)$ of bounded variation functions and bounded subsets of smooth functions, concretely, of the Sobolev spaces $H_2^s(A) = B_{22}^s(A)$, for $s > 1/2$. We will use the notation

$$H_{2,M}^s(A) = \{f \in C_u(A) : \|f\|_{H_2^s(A)} \leq M\},$$

where we recall the Sobolev imbedding, Proposition 4.3.9, and where A is either of $\mathbb{R}$, $[0,1]$ or $(0,1]$. In the case of functions of bounded variation, for $A = \mathbb{R}$, recall that each function f is almost surely $f(x) = f(-\infty+) + \nu(-\infty,x]$, where $\nu = Df$ is a finite signed measure on $\mathbb{R}$ (see (4.106) and Exercise 4.3.6), and in the case of $A = [0,1]$, $f(x) = \nu[0,x]$, where ν is a finite signed measure on $[0,1]$ (with $f(0) = f(1)$ in the periodic case). Then we set

$$BV_M(\mathbb{R}) := \{f(x) = f(-\infty+) + \nu(-\infty,x], x \in \mathbb{R} : |\nu|(\mathbb{R}) \leq M\}$$

and similarly define $BV_M([0,1])$ and $BV_M((0,1])$.

Here is what we obtain for the estimation of functions under Gaussian white noise for these two classes of sets:

Proposition 5.2.1 *Let K_{j_n} be projection kernels associated to boundary-corrected or periodised wavelet bases of regularity $S > s$. Let $f \in B_{\infty\infty}^s$ for some $s \geq 0$. Let $j = j_n$ be such that $2^{-j_n(s+1)}n^{1/2} \to 0$. Let $Y_f(x)$, $x \in [0,1]$, be as in (5.5), and let $f_n(y) = \int_0^1 K_{j_n}(x,y)dY_f(x)$ be the estimator of f at resolution j_n. Set also $dQ_n(x) = f_n(x)dx$, $dQ(x) = f(x)dx$ and $W(g) = \int_0^1 g(x)dW(x)$ for g square integrable. Then $W|_{BV_M}$ admits a sample continuous version, and*

$$\sqrt{n}(Q_n - Q) \to_{\mathcal{L}} \sigma W, \quad \text{in } \ell_\infty(BV_M),$$

for any $M > 0$, where $BV_M = BV_M([0,1])$ (with the obvious modification in the periodic case). Also, $W|_{H_{2,M}^s}$ for $s > 1/2$ admits a sample continuous version, and

$$\sqrt{n}(Q_n - Q) \to_{\mathcal{L}} \sigma W, \quad \text{in } \ell_\infty(H_{2,M}^s),$$

for any $M > 0$.

Proof We may assume that $\sigma = 1$. To prove this proposition, we will look at the two terms in the decomposition

$$\sqrt{n}\int_0^1 gd(Q_n - Q) = \sqrt{n}\int_0^1 (K_j(f) - f)g + W(K_j(g)),$$

where we write j for j_n. Consider first the space BV. By Example 3.6.14, monotone nondecreasing functions on $[0,1]$ taking values on $[0,1]$ are in the pointwise closure of the convex hull of the set of functions $I_{(x,1]}$ and $I_{[x,1]}$, $x \in [0,1]$; hence, the collection of functions $BV_M([0,1])$ is in M times the pointwise closure of the symmetric convex hull of these indicators. Therefore, a version of Proposition 4.3.15 for boundary-corrected or periodised wavelets applies to the effect that

$$\sup_{g\in BV_M}\left|\int_0^1 (K_j(f)-f)g\right| \le CM\|f\|_{B^s_{\infty\infty}} 2^{-j(s+1)} = o(n^{-1/2}). \tag{5.67}$$

Regarding the second term, since $E(W(K_j(g_1)) - W(K_j(g_2)))^2 = \int_0^1 (K_j(g_1 - g_2))^2 \le \|g_1 - g_2\|_2^2$, Dudley's metric entropy bound (Theorem 2.3.7) and the entropy estimate in Corollary 3.7.50 show that

$$\sup_j E \sup_{g_1,g_2\in BV_M:\|g_1-g_2\|_2\le\delta} |W(K_j(g_1)) - W(K_j(g_2))| \to 0$$

as $\delta \to 0$. Also, the finite-dimensional distributions of the processes $W(K_j(g))$ converge to those of $W(g) = \int_0^1 g dW$ because the projections π_j onto V_j converge to the identity in L^2. We then conclude by Theorem 3.7.23 that

$$\sqrt{n}W(K_j(\cdot)) \to W(\cdot), \quad \text{in } \ell_\infty(BV_M). \tag{5.68}$$

The result for BV_M follows from (5.67) and (5.68). The proof for Sobolev spaces is exactly the same, except that we now invoke the bound in Proposition 4.3.14 (concretely, the remark after it) and the metric entropy estimates in Corollary 4.3.38. ■

Smoothing the Empirical Measure by Convolution Kernels

Density estimators may be considered as smoothed versions of the empirical measure, that is, random probability measures with densities that approximate the true probability P. As in previous sections, we consider both convolution kernel density estimators and wavelet projection estimators. The classes of functions $\mathcal{F}$ for which we will prove the plug-in property will be Sobolev classes of functions and classes of functions of bounded variation as described earlier, and we will use a different method on each.

Let P_n be the empirical measure corresponding to i.i.d. observations from a probability measure P on $\mathbb{R}$ (as usual, coordinate functions). Convolution kernel density estimators have the form $P_n * \mu_n$, where μ_n is the signed measure with density $K_{h_n}(y) = h_n^{-1}K(y/h_n)$, $h_n \to 0$ (and $nh_n \to \infty$). Note that if $d\mu_n(y) = K_{h_n}(y)dy$, then

(a) $\mu_n(\mathbb{R}) = \int K(y)dy = 1$,
(b) For all f bounded and continuous, $\int f d\mu_n \to f(0)$ as $n \to \infty$ and
(c) For all $a > 0$, $\mu_n([-a,a]^c) \to 0$ as $n \to \infty$.

We call a sequence of signed measures $\{\mu_n\}$ satisfying conditions (a) and (b) an *approximate identity* for convolution, and we say that an approximate identity is *proper* if it also satisfies condition (c).

Let $\mathcal{F}$ be a P-Donsker class of measurable functions. Since for any finite signed measure Q (random or not)

$$(Q * \mu_n)(f) = \int\int f(x+y)dQ(x)d\mu_n(y) = \int (f * \bar{\mu}_n)(x)dQ(x),$$

where $\bar{\mu}_n(A) := \mu_n(-A)$ for all Borel sets A, and where it is assumed that $f(x+\cdot) \in L^1(|\mu_n|)$ (here and elsewhere $|\mu_n|$ denotes the total variation measure of μ_n) and that $|f| * |\bar{\mu}_n|$ is

$|Q|$-integrable, a first natural step to establish that $(P_n * \mu_n)(f)$ approximates $P(f)$ uniformly in $\mathcal{F}$ at the rate $1/\sqrt{n}$ will be to prove the following proposition. We say that a class $\mathcal{F}$ of functions is *translation invariant* if $f(\cdot - y) \in \mathcal{F}$, for all $f \in \mathcal{F}$ and $y \in \mathbb{R}$.

Proposition 5.2.2 *Let $\mathcal{F}$ be a translation-invariant P-Donsker class of functions on $\mathbb{R}$, and let $\mathcal{M}$ be a collection of signed Borel measures of finite variation such that $\sup_{\mu \in \mathcal{M}} |\mu|(\mathbb{R}) < \infty$. Assume that for all $f \in \mathcal{F}$ and $\mu \in \mathcal{M}$, the functions*

$$y \mapsto f(x+y), \quad \textit{for all } x \in \mathbb{R}, \quad \textit{and} \quad y \mapsto \|f(\cdot + y)\|_{L^2(P)}$$

are in $L^1(|\mu|)$. Then the class of functions

$$\tilde{\mathcal{F}} := \{f * \mu : f \in \mathcal{F}, \mu \in \mathcal{M}\}$$

is P-Donsker.

Proof Proposition 3.7.34 as extended by Exercise 3.7.14 shows that if a class of functions $\mathcal{F}$ is P-Donsker, then so is the closure in $L^2(P)$ of its convex hull for the topology $\tau = \tau_1 \vee \tau_2$, where τ_1 is the topology of pointwise convergence and τ_2 is the topology of $L^2(P)$. Hence, it suffices to show that for each $f \in \mathcal{F}$ and $\mu \in \mathcal{M}$, every neighbourhood of $f * \mu$ for the τ-topology has a nonvoid intersection with $|\mu|(\mathbb{R})$ times the symmetric convex hull of $\mathcal{F}_f = \{f(\cdot - y) : y \in \mathbb{R}\}$. By definition of the neighbourhood base for τ, it suffices to prove this only for any set of the form

$$A_{x_1,\dots,x_r,\varepsilon} = \left\{g \in L^2(P) : \|f * \mu - g\|_{L^2(P)} < \varepsilon, |f * \mu(x_i) - g(x_i)| < \varepsilon, 1 \le i \le r\right\},$$

where $r < \infty$, $x_i \in \mathbb{R}$ and $\varepsilon > 0$. Define

$$Q = P + \delta_{x_1} + \cdots + \delta_{x_r}$$

and note that the hypotheses of Exercise 5.2.1 are satisfied by Q, $\mu \in \mathcal{M}$ and $f \in \mathcal{F}$. The conclusion of that exercise is that $|\mu|(\mathbb{R})$ times the symmetric convex hull of $\mathcal{F}_f$ intersects any neighbourhood B_ε of $f * \mu$ for the $L^2(Q)$-(pseudo)norm

$$B_\varepsilon = \{g \in L^2(P) : \|f * \mu - g\|_{L^2(Q)} < \varepsilon\}, \quad 0 < \varepsilon < \infty.$$

But, obviously, $B_\varepsilon \subseteq A_{x_1,\dots,x_r,\varepsilon}$, which proves the proposition. ■

Now, if we wish to take advantage of the fact that $\mathcal{F}$ is P-Donsker, we should compare our process of interest $\sqrt{n}(P_n * \mu_n - P)$ with $\sqrt{n}(P_n - P)$, and for this, it is natural to consider the following decomposition:

$$P_n * \mu_n - P_n = (P_n - P) * \mu_n - (P_n - P) + (P * \mu_n - P). \tag{5.69}$$

This decomposition and the preceding lemma will reduce proving convergence in law in $\ell_\infty(\mathcal{F})$ of the uncentred smoothed empirical process $\sqrt{n}(P_n * \mu_n - P)$ to the P-bridge G_P to the verification of two manageable limits as follows:

Proposition 5.2.3 *Let $\mathcal{F}$ be a translation-invariant $\mathbb{P}$-Donsker class of real-valued functions on $\mathbb{R}$, and let $\{\mu_n\}_{n=1}^\infty$ be an approximate convolution identity such that $\mu_n(\mathbb{R}) = 1$ for every* *n. Further, assume that for every n,* $\mathcal{F} \subseteq L^1(|\mu_n|)$ *and* $\int_{\mathbb{R}} \|f(\cdot - y)\|_{L^2(P)} d|\mu_n|(y) < \infty$, *for all* $f \in \mathcal{F}$. *Then the conditions*

$$\sup_{f \in \mathcal{F}} E\left(\int_{\mathbb{R}} (f(X+y) - f(X)) d\mu_n(y)\right)^2 \to_{n\to\infty} 0 \tag{5.70}$$

and

$$\sup_{f \in \mathcal{F}} \sqrt{n}\left|E\int_{\mathbb{R}} (f(X+y) - f(X)) d\mu_n(y)\right| \to_{n\to\infty} 0 \tag{5.71}$$

imply that $\sqrt{n}\|P_n * \mu_n - P_n\|_{\mathcal{F}}$ *converges to zero in outer probability and that*

$$\sqrt{n}(P_n * \mu_n - P) \to_{\mathcal{L}} G_P, \quad \text{in } \ell_\infty(\mathcal{F}). \tag{5.72}$$

Proof Since $(P * \mu_n - P)(f) = E\int (f(X+y) - f(X))\, d\mu_n(y)$, condition (5.71) implies that

$$\|P * \mu_n - P\|_{\mathcal{F}} = o(1/\sqrt{n}),$$

which takes care of the last summand in the decomposition (5.69). For the remaining part of that decomposition, note that

$$((P_n - P) * \mu_n - (P_n - P))(f) = (P_n - P)(\bar{\mu}_n * f - f),$$

where $\bar{\mu}_n(A) = \mu_n(-A)$, A a Borel set. Now $\mathcal{F}$ is P-Donsker by hypothesis, and $\{\bar{\mu}_n * f : f \in \mathcal{F}, n \in \mathbb{N}\}$ is also P-Donsker by Proposition 5.2.2; therefore, the class of functions $\cup_n\{\bar{\mu}_n * f - f : f \in \mathcal{F}\}$ is P-Donsker as well by Corollary 3.7.35 (this is a subclass of $\{f * \bar{\mu}_n + g : f, g \in \mathcal{F}, n \in \mathbb{N}\}$). Thus, since by condition (5.70), $\sup_{f\in\mathcal{F}} P(\bar{\mu}_n * f - f)^2 \to 0$, it follows that

$$\sup_{f\in\mathcal{F}} |(P_n - P)(\bar{\mu}_n * f - f)| = o_P(1/\sqrt{n})$$

by the asymptotic equi-continuity condition (see Theorem 3.7.23), proving the first conclusion of the proposition. Combining the last two estimates with (5.69) gives that $\|\sqrt{n}(P_n * \mu_n - P) - \sqrt{n}(P_n - P)\|_{\mathcal{F}} \to 0$ in outer probability. Now the result follows because $\mathcal{F}$ is a P-Donsker class (use Exercise 3.7.25). ■

Note that the limit in (5.71) ensures that the 'bias' part of the discrepancy between the smoothed empirical measure $P_n * \mu_n$ and the empirical measure proper, P_n, tends to zero, whereas the limit in (5.70) does the same for the 'variance' part of this discrepancy.

As will become apparent later in Remark 5.2.6, it is convenient to extend the definition of kernel of order t to noninteger values. We restrict to *symmetric* kernels.

Definition 5.2.4 A kernel $K : \mathbb{R} \to \mathbb{R}$ of order $r > 0$ is a Lebesgue integrable function, symmetric around the origin, such that

$$\int_{\mathbb{R}} K(y)dy = 1, \quad \int_{\mathbb{R}} y^j K(y)dy = 0, \quad \text{for } j = 1, \ldots, \{r\}, \quad \text{and} \quad \int_{\mathbb{R}} |y|^r |K(y)| dy < \infty,$$

where $\{r\}$ is the largest integer strictly smaller than r.

Then, given such a kernel and a sequence $h_n \to 0$, $h_n > 0$, we take $d\mu_n(x) = K_{h_n}(x)dx$ as our approximate identity.

By Exercise 3.7.24, any bounded subset of a Sobolev space $H_2^s(\mathbb{R})$ with $s > 1/2$ consisting of continuous functions is universal Donsker because $H_2^s(\mathbb{R})$ is a Hilbert space and its imbedding into $C(\mathbb{R})$ is continuous (by Proposition 4.3.9 as $H_2^s(\mathbb{R}) = B_{22}^s(\mathbb{R})$). In particular, the sets $H_{2,M}^s(\mathbb{R})$ defined in the introduction to this section are P-Donsker for all $M > 0$, all $s > 1/2$ and all Borel probability measures P on $\mathbb{R}$. The following theorem shows that under quite general conditions the central limit theorem uniform over $H_{2,M}^s(\mathbb{R})$ holds for $\sqrt{n}(P_n^K - P)$, where

$$dP_n^K(x) = d(P_n * K_{h_n})(x), \quad x \in \mathbb{R}.$$

Recall Proposition 4.1.6 for the existence of such kernels.

Theorem 5.2.5 *Let f_0 be a probability density in $H_2^t(\mathbb{R})$ for some $t \geq 0$, let P_n be the empirical measure corresponding to i.i.d. samples from P, let K be a symmetric kernel of order $r = t + s - k$ for some $s > 1/2$ and $0 \leq k < t + s$ and let $f_n = P_n * K_{h_n}$ be the corresponding kernel density estimators of f, where the window widths $h_n > 0$ are such that $h_n^{t+s-k} n^{1/2} \to 0$. Then the P-bridge G_P indexed by $H_{2,M}^s(\mathbb{R})$ admits a sample continuous version, and*

$$\sqrt{n}(P_n^K - P) \to_{\mathcal{L}} G_P, \quad \text{in } \ell_\infty(H_{2,M}^s(\mathbb{R})),$$

for all $M < \infty$.

Proof We apply Proposition 5.2.3. The class $\mathcal{F} := H_{2,M}^s(\mathbb{R})$ is clearly translation invariant and is P-Donsker, as indicated in the preceding paragraph. Also, by continuity of the Sobolev imbedding, $\mathcal{F}$ is uniformly bounded. Set $d\mu_n(x) = K_{h_n}(x)dx$. Since K is integrable and $\mathcal{F}$ is bounded, we have both $\mathcal{F} \subset L^1(|\mu_n|)$ and $\int \|f(\cdot - y)\|_{L^2(P)} d|\mu_n|(y) < \infty$. To prove (5.70), we first note that $H_2^s \subset C^\alpha(\mathbb{R})$ for some $\alpha > 0$ (actually, for all $\alpha < s - 1/2$): If $\alpha < (s - 1/2) \wedge 1$, we have, for $f \in \mathcal{F}$ with the Sobolev norm $\|\langle u\rangle^s \hat{f}\|_2$ (see before (4.115), and recall the notation $\langle t\rangle^s = (1 + |t|^2)^{s/2}$),

$$\begin{aligned}
\frac{1}{|h|^\alpha}|f(x+h) - f(x)| &= \frac{1}{\sqrt{2\pi}}\left|\int |ht|^{-\alpha}(e^{-iht} - 1)|t|^\alpha e^{-ixt}\hat{f}(t)dt\right| \\
&\leq \frac{2}{\sqrt{2\pi}}\left|\int |t|^\alpha \hat{f}(t)dt\right| = \frac{2}{\sqrt{2\pi}}\left|\int |t|^\alpha (\langle t\rangle^s)^{-1}\langle t\rangle^s \hat{f}(t)dt\right| \\
&\leq \frac{2}{\sqrt{2\pi}}\left(\int \frac{|t|^\alpha}{(1+|t|^2)^s}dt\right)^{1/2} \|f\|_{H_2^s}.
\end{aligned}$$

Now, by Minkowski's inequality for integrals, we have

$$\begin{aligned}
&\sup_{f\in\mathcal{F}}\left(E\left(\int_{\mathbb{R}}(f(X+y) - f(X))d\mu_n(y)\right)^2\right)^{1/2} \\
&\leq \sup_{f\in\mathcal{F}}\int_{|y|\leq\delta}\left(E(f(X+y) - f(X))^2\right)^{1/2} d|\mu_n|(y) \qquad (5.73) \\
&\quad + \sup_{f\in\mathcal{F}}\int_{|y|>\delta}\left(E(f(X+y) - f(X))^2\right)^{1/2} d|\mu_n|(y) = (I)_{n,\delta} + (II)_{n,\delta}.
\end{aligned}$$

By the preceding observation, there is $c < \infty$ such that, for all $f \in \mathcal{F}$,

$$(I)_{n,\delta} \le c \int_{|y|\le\delta} |y|^\alpha d|\mu_n|(y) \le c\delta^\alpha \sup_n \|\mu_n\|_v \to 0, \quad \text{as } \delta \to 0,$$

uniformly in n. However, since the imbedding of H_2^s into L^2 is also continuous and f_0 is uniformly bounded, $\sup_y \sup_{f\in\mathcal{F}} P(f(\cdot+y))^2 \le c\|f_0\|_\infty = D < \infty$, and $\int_{|y|>\delta} |K_{h_n}(y)|dy = \int_{-\delta/h_n}^{\delta/h_n} |K(y)|dy \to 0$ as $n\to\infty$, for all $\delta > 0$, we have

$$\lim_n (II)_{n,\delta} \le \lim_n 2D|\mu_n|\{|y| > \delta\} = 0,$$

for all $\delta > 0$. The variance condition (5.70) in Proposition 5.2.3 is thus proved. Next, we prove the bias condition (5.71). Recall that by Lemma 4.3.16 for $p = q = 2$,

$$E\int (f(X+y) - f(X))d\mu_n(y) = \int (K_{h_n} * f - f)f_0 = \int K(t)(\bar{f} * f_0(h_n t) - \bar{f} * f_0(0)))dt, \tag{5.74}$$

where $\bar{f}(x) = f(-x)$. If s and t are integers, Lemma 4.3.18 gives that for all $f \in \mathcal{F}$, $f * f_0 \in C^{s+t}(\mathbb{R})$, and

$$\sup_{f\in\mathcal{F}} \|f * f_0\|_{C^{s+t}(\mathbb{R})} \le 2\pi \|f\|_{H_2^s(\mathbb{R})} \|f_0\|_{H_2^t(\mathbb{R})},$$

and this is also true for $s,t > 0$ not necessarily integers (proved, e.g., as in the proof of Theorem 5.3.2, where $\alpha = s = t$). Thus, we can apply Taylor development in (5.74) and use the fact that the kernel K is of order $r = s+t-k$. Then, if $t+s-k \notin \mathbb{N}$, we obtain

$$\left|E\int (f(X+y) - f(X))d\mu_n(y)\right|$$
$$= \frac{h_n^{[t+s-k]}}{[t+s-k]} \left|\int K(u)u^{[t+s-k]}[D^{[t+s-k]}(\bar{f} * f_0)(\zeta h_n u) - D^{[t+s-k]}(\bar{f} * f_0)(0)]du\right|,$$

for some $0 \le \zeta \le 1$, and since $\bar{f} * f_0 \in C^{s+t} \subset C^{s+t-k}$, we have that $D^{[t+s-k]}(\bar{f} * f_0)]$ is Hölder continuous of order at least $\alpha = t+s-k-[t+s-k]$, with uniformly bounded C^α norm. We conclude that $\sup_{f\in\mathcal{F}} |E\int (f(X+y) - f(X))d\mu_n(y)| \le Ch_n^{t+s-k}$. Likewise, if $t+s-k$ is an integer, then

$$\left|E\int (f(X+y) - f(X))d\mu_n(y)\right| \le \frac{1}{(t+s-k)} h_n^{t+s-k} \|D^{t+s-k}(\bar{f} * f_0)\|_\infty \int |K(u)|u|^{t+s-k} du.$$

In either case, the assumption on h_n yields $\sqrt{n} \sup_{f\in\mathcal{F}} |E\int (f(X+y) - f(X))d\mu_n(y)| \to 0$. Now the theorem follows from Proposition 5.2.3. ■

Remark 5.2.6 For $f_0 \in B_{2\infty}^t$, $t > 0$, and K of order $S > t$, taking $h_n \simeq n^{-1/(2t+1)}$ gives the optimal rate $n^{-t/(2t+1)}$ for the estimation of f by f_n in L^2-norm (Proposition 5.1.7). Theorem 5.2.5 gives in this case that for $\int ff_n$ to also estimate $\int ff_0$ uniformly in f in the unit ball of $H_2^s(\mathbb{R})$ at the optimal rate of $n^{-1/2}$, we just need to take K of order $S > t+1/2$ (choose $k < s-1/2$). That is, the increase of 1/2 in the order of K yields the plug-in property for f_n with respect to Sobolev balls.

Proposition 5.2.3 also applies to bounded subsets of $BV(\mathbb{R})$; however, we will deduce the central limit theorem for $\sqrt{n}(f_n - f)(x)dx$ over such classes from an *exponential inequality* for the discrepancy between the distribution function of $P_n * \mu_n$ and P_n, because of its independent interest.

Let $F_n(x) = \int_{-\infty}^x dP_n(u),\ x \in \mathbb{R}$, be the empirical distribution function, and let

$$F_n^K(h)(x) = \int_{-\infty}^x P_n * K_h(u)du, \quad x \in \mathbb{R},$$

be the distribution function of the kernel density estimator $P_n * K_h$ with $h = h_n \to 0$. We have the following theorem, where $BV_M(\mathbb{R})$ is defined in the introduction to this section:

Theorem 5.2.7 *Suppose that P has a density f_0 with respect to Lebesgue measure. Assume that f_0 is bounded, in which case we set $t = 0$ in what follows, or that $f_0 \in C^t(\mathbb{R})$ for some $t > 0$. Let $e^{-1} \geq h := h_n \to 0$ as $n \to \infty$ satisfy $h > \log n/n$, and let K be a kernel of order $r = t+1-k$ for some $0 \leq k < t+1$. Then there exist finite positive constants $L := L(\|f_0\|_\infty, K)$ and $\Lambda_0 := \Lambda_0(\|f_0\|_{C^t}, K) \geq 1$ such that, for all $\lambda \geq \Lambda_0 \max(\sqrt{h\log(1/h)}, \sqrt{n}h^{t+1-k})$ and $n > 1$,*

$$\Pr\left(\sqrt{n}\|F_n^K(h) - F_n\|_\infty > \lambda\right) \leq 2\exp\left\{-L\min\left(h^{-1}\lambda^2, \sqrt{n}\lambda\right)\right\}. \tag{5.75}$$

As a consequence, for all $n > e$,

$$\Pr\left(\sqrt{n}\|P_n^K - P_n\|_{BV_M(\mathbb{R})} > 2M\lambda\right) \leq 2\exp\left\{-L\min\left(h^{-1}\lambda^2, \sqrt{n}\lambda\right)\right\}.$$

Proof Note first that $F_n^K(h)(x) - F_n(x)$ is a random variable for each $x \in \mathbb{R}$, and hence, so is $\|F_n^K(h) - F_n\|_\infty$, since, by right continuity, this is in fact a supremum over a countable set. We will also use this observation when we apply Talagrand's inequality later.

We set $\mathcal{F} = \{1_{(-\infty,x]} : x \in \mathbb{R}\}$ throughout the proof and note that $P_n 1_{(-\infty,x]} = F_n(x)$ as well as $(P_n * K_h)1_{(-\infty,x]} = F_n^K(x)$. We will still use the decomposition (5.69), now with $d\mu_n(x) = K_h(x)dx$. For the deterministic bias $P * K_h - P$, we have by Lemma 4.3.16 that, for any given $f \in \mathcal{F}$ with $\bar{f}(x) = f(-x)$,

$$(P * K_h - P)f = \int_{\mathbb{R}} K(u)[f_0 * \bar{f}(hu) - f_0 * \bar{f}(0)]du.$$

First, if $t = 0$, we have for every $x \in \mathbb{R}$ and $u \geq 0$ that, for $f = I_{(-\infty,x]}$,

$$|f_0 * \bar{f}(hu) - f_0 * \bar{f}(0)| = \left|\int_{\mathbb{R}} 1_{(x,x+hu]}(y) f_0(y)dy\right| \leq (hu)^r \frac{hu\|f_0\|_\infty \wedge 1}{(hu)^r}.$$

Hence, since for $a > 0$ and $r = 1 - k \leq 1$ $(a \wedge 1)/a^r \leq 1$, it follows that

$$|(P * K_h - P)f| \leq h^{1-k}\|f_0\|_\infty^r \int |K(u)||u|^{1-k}du,$$

and likewise for $u < 0$. More generally, if $t > 0$, the distribution function F of f_0 is contained in $C^{t+1}(\mathbb{R})$, so by a standard Taylor expansion as in the preceding proof and since the kernel is of order $r = t + 1 - k$, it follows that

$$\sup_{x\in\mathbb{R}} |(P * K_h - P)1_{(-\infty,x]}| = \left|\int_{\mathbb{R}} K(u)[F(x - uh) - F(x)]du\right| \leq dh^{t+1-k}, \tag{5.76}$$

for some constant d depending only on $\|f_0\|_{C^t}$ and $\int_{\mathbb{R}} |K(u)||u|^{t+1-k}du$.

For the remaining part of the decomposition (5.69), observe that, using the symmetry of the kernel,

$$(P_n * K_h - P * K_h - P_n + P)f = (P_n - P)(K_h * f - f),$$

for $f \in \mathcal{F}$. Consequently,

$$\Pr\left(\sqrt{n}\|F_n^K(h) - F_n\|_\infty > \lambda\right) \le \Pr\left(\sqrt{n}\sup_{f\in\mathcal{F}}|(P_n - P)(K_h * f - f)| > \lambda - d\sqrt{n}h^{t+1-k}\right)$$
$$\le \Pr\left(n\sup_{f\in\mathcal{F}}|(P_n - P)(K_h * f - f)| > \frac{\sqrt{n}\lambda}{2}\right), \tag{5.77}$$

by assumption on λ, and we will apply Talagrand's inequality (3.101) in Theorem 3.3.9 to the class

$$\tilde{\mathcal{F}} = \{K_h * f - f - P(K_h * f - f) : f \in \mathcal{F}\}$$

to bound the last probability. For this, we need the following facts:

(a) First, we note that the class of functions $\{K_h * f - f : f \in \mathcal{F}\}$ is uniformly bounded by $2\|K\|_1$, and hence, $\tilde{\mathcal{F}}$ has envelope $U = 4\|K\|_1$.

(b) Also,

$$\sup_{f\in\mathcal{F}}\|K_h * f - f\|_{L^2(P)} \le Ch^{1/2} =: \sigma, \tag{5.78}$$

for $C = \|p_0\|_\infty^{1/2}\int_{\mathbb{R}}|u|^{1/2}|K(u)|$, since

$$E(f(X+y) - f(X))^2 = E|f(X+y) - f(X)| = \int_{-\infty}^{\infty} 1_{[x-y,x)}(u)f_0(u)dx \le \|f_0\|_\infty y$$

if $y > 0$ and similarly if $y < 0$, and therefore, using Minkowski's inequality for integrals,

$$\left(E\left(\int_{\mathbb{R}}(f(X+y) - f(X))K_h(y)dy\right)^2\right)^{1/2} \le \int_{\mathbb{R}}\left(E(f(X+y) - f(X))^2\right)^{1/2}|K_h(y)|dy$$
$$\le \|f_0\|_\infty^{1/2}\int_{\mathbb{R}}|y|^{1/2}|K_h(y)|dy$$
$$= \|f_0\|_\infty^{1/2}h^{1/2}\int_{\mathbb{R}}|u|^{1/2}|K(u)|. \tag{5.79}$$

(c) Moreover, we will need the expectation bound

$$nE\sup_{f\in\mathcal{F}}|(P_n - P)(K_h * f - f)| \le d'\sqrt{nh\log(1/h)}, \tag{5.80}$$

for some constant $0 < d' < \infty$ depending only on $\|f_0\|_\infty$ and K, which is proved as follows. For each $h > 0$, the class $\{K_h * 1_{(-\infty,x]} : x \in \mathbb{R}\}$ is just $\{F^K((x - \cdot)/h) : x \in \mathbb{R}\}$, where $F^K(t) = \int_{-\infty}^t K(s)ds$, since

$$K_h * 1_{(-\infty,x]}(u) = h^{-1}\int_{-\infty}^x K\left(\frac{y-u}{h}\right)dy = \int_{-\infty}^{(x-u)/h} K(t)dt = F^K\left(\frac{x-u}{h}\right),$$

and F^K is of bounded variation since it is the distribution function of a finite signed measure. Similarly, $\{1_{(-\infty,x]}(t): x\in\mathbb{R}\} = \{1_{(-\infty,0]}(t-x): x\in\mathbb{R}\}$, so $\{K_h * 1_{(-\infty,x]} - 1_{(-\infty,x]}: x\in\mathbb{R}\}$ is contained in the set of all translates of the function $F^K(\cdot/h) - 1_{(-\infty,0]}(\cdot)$, which is of bounded variation, and Proposition 3.6.12 hence gives an $L^2(Q)$ metric entropy bound for the class $\{K_h * f - f : f\in\mathcal{F}\}$ independent of h and Q and of the order of a constant times $\log\varepsilon^{-1}$. This entropy bound and the bounds from (a) and (b) now allow us to apply the expectation bound (3.184) in Corollary 3.5.8, yielding (5.80), since $nh \geq \log n > \log h^{-1}$.

Finally, we apply Talagrand's inequality (see (3.101)), with

$$x = L\min\left(\sqrt{n}\lambda, h^{-1}\lambda^2\right)$$

and σ, U as in (a) and (b) to expression (5.77). In the notation of (3.101), we have

(i) $ES_n = nE\sup_{f\in\mathcal{F}}|(P_n-P)(K_h * f - f)| \leq d'\sqrt{nh\log(1/h)} \leq \sqrt{n}\lambda/6$ by (5.80) and the assumption on λ;

(ii) $v_n = 2UES_n + n\sigma^2 \leq C^2nh + 8\|K\|_1 d'\sqrt{nh\log(1/h)} \leq C'nh$ for some constant C' since $h\geq(\log n/n)$, and hence,

$$\sqrt{2v_nx} \leq \sqrt{2C'nhL\min\left(\sqrt{n}\lambda, h^{-1}\lambda^2\right)} \leq \sqrt{2C'L}\sqrt{n}\lambda \leq \frac{\sqrt{n}\lambda}{6},$$

for L small enough.

(iii) Furthermore,

$$Ux/3 \leq (4/3)\|K\|_1 L\min\left(\sqrt{n}\lambda, h^{-1}\lambda^2\right) \leq \frac{\sqrt{n}\lambda}{6}.$$

Summarising, the sum of the terms in (i)–(iii) is smaller than $(\sqrt{n}\lambda/2)$ if L is chosen suitably small, and we obtain from (3.101) for the given choice of x that

$$\Pr\left(n\sup_{f\in\mathcal{F}}|(P_n-P)(K_h * f - f)| > \frac{\sqrt{n}\lambda}{2}\right) \leq 2\exp\{-x\},$$

which implies the theorem for distribution functions.

Let us now consider $BV_M(\mathbb{R})$ with $M=1$ without loss of generality. We recall that by Jordan's decomposition, every function f of variation bounded by 1 that is 0 at $-\infty$ is the difference of two nondecreasing functions whose sum does not exceed 1 and that are 0 at $-\infty$ (the positive and negative variations of f) and that nondecreasing functions on $\mathbb{R}$ taking values on $[0,1]$ are in the pointwise closure of the convex hull of the class $\mathcal{G} = \{I_{(x,\infty)}, I_{[x,\infty)}: x\in\mathbb{R}\}$ (see Example 3.6.14). Hence, $BV_1 \subset \overline{\text{co}}(\mathcal{G})$, and therefore, $\|P_n^K - P_n\|_{BV_1} = \|P_n^K - P_n\|_{\mathcal{G}}$. Since with probability 1, P_n has only atoms of size $1/n$, we also have, almost surely, that

$$\sqrt{n}\|P_n^K - P_n\|_{\mathcal{G}} \leq \sqrt{n}\|F_n^K - F_n\|_\infty + 1/\sqrt{n} \leq \sqrt{n}\|F_n^K - F_n\|_\infty + \lambda.$$

The exponential inequality for BV_M now follows from (5.75) together with these observations. ■

The preceding inequalities may be used to transfer several properties about the empirical process $\sqrt{n}(P_n - P)$ to the smoothed empirical process $\sqrt{n}(P_n * K_{h_n} - P)$, such as a

Dvoretzky-Kiefer-type inequality, or the central limit theorem, or the law of the iterated logarithm, or the Komlós-Major-Tusńady almost-sure approximation by Gaussian processes. We only record the central limit theorem.

Theorem 5.2.8 *Let f_0 and K be as in the preceding theorem, and let h_n be such that $h_n^{t+1-k}n^{1/2} \to 0$. Then the restriction of the P-bridge G_P to BV_M admits a sample continuous version, and*

$$\sqrt{n}(P_n * K_{h_n} - P) \to_{\mathcal{L}} G_P, \quad \textit{in } \ell_\infty(BV_M),$$

for all $0 < M < \infty$; in particular,

$$\sqrt{n}(F_n^K(h_n) - F_P) \to_{\mathcal{L}} G_P, \quad \textit{in } \ell_\infty(\mathbb{R}).$$

Proof Since $BV_1 \subset \overline{\text{co}}\mathcal{G}$ and $\mathcal{G}$ is P-Donsker, so is BV_1 (and BV_M for all M) by Proposition 3.7.34 (or rather Exercise 3.7.14). Now the result follows immediately from the convergence in law in $\ell_\infty(BV_M)$ of the empirical process $\sqrt{n}(P_n - P)$ together with Theorem 5.2.7 (see, e.g., Exercise 3.7.25). ■

As a consequence of this theorem and in analogy with Remark 5.2.6, when $f_0 \in C^t(\mathbb{R})$ and $h_n \simeq n^{-1/(2t+1)}$, if we take the kernel of any order $r > t + 1/2$, then the kernel estimator achieves optimal rates of convergence simultaneously in squared error loss (and in any L^p loss, $1 \le p \le \infty$) and in its action over bounded variation classes.

Smoothing the Empirical Measure by Wavelet Projections

Next, we consider the plug-in property for wavelet projection density estimators. P, f_0, P_n and F_n are as in the preceding section, and we set $f_n(x) = P_n(K_j(\cdot,x))$, $x \in \mathbb{R}$, to be the wavelet projection estimator of f_0, where, for each j, K_j is the projection kernel onto the space V_j corresponding to a wavelet basis $\{\phi_k, \psi_{j,k}\}$ of regularity index $S > 0$. We also define

$$dP_n^W(x) = f_n(x)dx = P_n(K_j(\cdot,x))dx \quad \text{and} \quad F_n^W(x) := \int_{-\infty}^{x} P_n(K_j(\cdot,y))dy, \quad x \in \mathbb{R},$$

where $j = j_n \to \infty$; that is, P_n^W is the probability measure of density f_n, and F_n^W is its cumulative distribution function. We prove the following theorem only for Daubechies wavelets, the reason being that in this case several classes of functions related to the projection kernel are of VC type (see Exercise 5.2.3):

Theorem 5.2.9 *With the notation from the preceding paragraph, assume that the density f_0 is a bounded function – in which case we set $t = 0$ – or that $f_0 \in C^t(\mathbb{R})$ for some t, $0 < t < S+1$. Let j satisfy $2^{-j} > (\log n)/n$. Assume also that the wavelet density projection estimators f_n of f_0, and hence P_n^W and F_n^W, are defined via Daubechies wavelets of regularity $S > s$. Then there exist finite positive constants $L := L(\|f_0\|_\infty, K)$ and $\Lambda_0 := \Lambda_0(\|f_0\|_{C^t}, K) \ge 1$ such that, for all $n > 1$ and $\lambda \ge \Lambda_0 \max(\sqrt{j2^{-j}}, \sqrt{n}2^{-j(t+1)})$, we have*

$$\Pr\left(\sqrt{n}\|F_n^W - F_n\|_\infty > \lambda\right) \le 2\exp\left\{-L\min(2^j\lambda^2, \sqrt{n}\lambda)\right\}$$

and therefore also

$$\Pr\left(\sqrt{n}\|P_n^W - P_n\|_{BV_M(\mathbb{R})} > 2M\lambda\right) \le 2\exp\left\{-L\min(2^j\lambda^2, \sqrt{n}\lambda)\right\},$$

for all $n > 1$ and $M > 0$.

Proof Set $\mathcal{F} = \{1_{(-\infty,s]} : s \in \mathbb{R}\}$, and note the following analogue of the decomposition (5.69):

$$(P_n^W - P_n)(f) = (P_n - P)(K_j(f) - f) + (P(K_j) - P)(f), \tag{5.81}$$

where we use the symbol $P(K_j)$ both for the function and for the finite signed measure that has it as density. Since $C^t(\mathbb{R}) \subset B^t_{\infty\infty}(\mathbb{R})$, Proposition 4.3.14 and the remark after it gives $\sup_{f\in\mathcal{F}} |(P(K_j) - P)(f)| \leq C\|f_0\|_{B^t_{\infty\infty}} 2^{-j(t+1)}$ for $t > 0$, and the proof of this proposition gives the bound $C\|f\|_\infty 2^{-j}$ for $t = 0$ because $\|\langle f_0, \psi_{\ell\cdot}\rangle\|_\infty \leq \|f_0\|_\infty \|\psi\|_1 2^{-\ell/2}$. Hence,

$$\sup_{f\in\mathcal{F}} |(P(K_j) - P)(f)| \leq C\|f_0\|_{C^t} 2^{-j(t+1)}, \quad t \geq 0.$$

The decomposition (5.81) then yields

$$\begin{aligned}\Pr\left(\sqrt{n}\|F_n^W - F_n\|_\infty > \lambda\right) &\leq \Pr\left(\sqrt{n}\sup_{f\in\mathcal{F}} |(P_n - P)(K_j(f) - f)| > \lambda - C\sqrt{n}2^{-j(t+1)}\right)\\ &\leq \Pr\left(n\sup_{f\in\mathcal{F}} |(P_n - P)(K_j(f) - f)| > \frac{\sqrt{n}\lambda}{2}\right)\end{aligned}$$

by assumption on λ (if we take $\Lambda_0 \geq 2C$). As in the preceding proof, we apply Talagrand's inequality (3.101) to the class

$$\tilde{\mathcal{F}} = \{K_j(f) - f - P(K_j(f) - f) : f \in \mathcal{F}\},$$

and for this we need to compute U, σ and ES_n. (Notice that the supremum over $f \in \mathcal{F}$ is in fact over a countable set by right continuity of the functions f, so Talagrand's inequality can be applied.)

(a) First, we note that the class of functions $\{K_j(f) - f : f \in \mathcal{F}\}$ is uniformly bounded by $\|\sum_k |\phi(\cdot - k)|\|_\infty \|\phi\|_1 + 1 = \kappa\|\phi\|_\infty + 1 < \infty$, and hence, $\tilde{\mathcal{F}}$ has constant envelope $U = c\|\phi\|_1$.

(b) Second,

$$\sup_{f\in\mathcal{F}} \|K_j(f) - f\|_{L^2(P)} \leq c' 2^{-j/2} =: \sigma,$$

for some finite positive c', since

$$E(f(X+u) - f(X))^2 \leq E|f(X+u) - f(X)| = \int 1_{[s-u,s)}(x) f_0(x)dx \leq \|f_0\|_\infty u,$$

if $u > 0$ and, similarly, if $u < 0$, and hence, using Minkowski's inequality for integrals and the majorisation of K by Φ, we obtain

$$\begin{aligned}\|K_j(f) - f\|_{L^2(P)} &= \left(E\left(\int (f(X+u) - f(X))K_j(X, u+X)du\right)^2\right)^{1/2}\\ &\leq \int \left(E(f(X+u) - f(X))^2\right)^{1/2} 2^j\Phi(2^j u)du\\ &\leq \|f_0\|_\infty^{1/2} \int |u|^{1/2} 2^j\Phi(2^j u)du\\ &= 2^{-j/2}\|f_0\|_\infty^{1/2} \int |v|^{1/2}\Phi(v)dv.\end{aligned}$$

(c) Finally, consider $S_n = \|n(P_n - P)(K_j f - f)\|_{\mathcal{F}}$. By Exercise 5.2.3, the classes of functions $\{K_j(f) : f \in \mathcal{F}\}$ are of VC type with constants A and v in the metric entropy bound independent of j, hence, since the class $\mathcal{F}$ is also VC, the same is true for the classes of functions $\{K_j(f) - f : f \in \mathcal{F}\}$. Then, by (a) and (b), the expectation bound in Corollary 3.5.8 gives $ES_n \leq \tilde{C}[\sqrt{n2^{-j}j} + j] \leq C\sqrt{n2^{-j}j}$ for a constant C that depends only on ϕ and Φ and where the last inequality follows from the hypothesis on j_n.

The bounds in (a), (b) and (c) are just as in Theorem 5.2.7 with h replaced by 2^{-j}; hence, an application of Talagrand's inequality as in the preceding theorem concludes the proof of the exponential inequality for the cdf. The inequality for BV_M follows exactly as in the proof of the same theorem. ■

As a consequence, and in complete analogy with the convolution kernel case, we thus have the following plug-in property for the wavelet projection estimator.

Theorem 5.2.10 *Let ϕ, ψ and f_0 satisfy the conditions of Theorem 5.2.9 for some $t \geq 0$, and let j_n satisfy $2^{-j_n} \geq (\log n)/n$ for all n and $\sqrt{n}2^{-j_n(t+1)} \to 0$ as $n \to \infty$. If F is the distribution function of P, then*

$$\sqrt{n}(F_n^W - F) \to_{\mathcal{L}} G_P, \quad \text{in } \ell_\infty(\mathbb{R}),$$

where $G_P(x) = G_P(I_{(-\infty,x]})$ is a sample continuous version of the P-bridge. Moreover, $G_P|_{BV_M}$ also admits a version with continuous sample paths, and

$$\sqrt{n}(P_n^W - P) \to_{\mathcal{L}} G_P, \quad \text{in } \ell_\infty(BV_M),$$

for all $M < \infty$.

Next, we prove an analogue of Theorem 5.2.5 for wavelet density estimators. The following proof does not require the scaling or the wavelet functions to have bounded support. More general plug-in results are available; see the notes at the end of this chapter and the remark following this theorem.

Theorem 5.2.11 *Let $s > 1/2$, and let $\mathcal{F} = H^s_{2,M}(\mathbb{R})$ be the set of continuous functions whose Sobolev norm is bounded by $M < \infty$. Let f_0 be a probability density in $H^t_2(\mathbb{R})$ for some $t \geq 0$, let P_n be the empirical measure corresponding to i.i.d. samples from $dP(x) = f_0(x)dx$, let K_j be the projection kernel onto V_j for an S-regular wavelet basis $\{\phi_k, \psi_{\ell k}\}$, $S > t + s$ and let $f_n(x) = P_n(K_{h_n}(\cdot, x))$ be the corresponding kernel density estimators of f and $dP_n^W(x) = f_n(x)dx$, $n \in \mathbb{N}$. Assume that the resolutions $j = j_n \to \infty$ satisfy $2^{-j_n(t+s)}n^{1/2} \to 0$. Then, $G_P|_{\mathcal{F}}$ is sample continuous, and*

$$\sqrt{n}(P_n^W - P) \to_{\mathcal{L}} G_P, \quad \text{in } \ell_\infty(\mathcal{F}).$$

Proof Assume without loss of generality that $M = 1$. We start with the decomposition (5.81). For the bias part, we have, by Proposition 4.3.14 and the inequality following its proof, that

$$\|P(K_{j_n}) - P)\|_{\mathcal{F}} = \sup_{f \in \mathcal{F}} \left| \int (K_{j_n}(f_0) - f_0)(f) \right| \leq C \|f_0\|_{H_2^s} 2^{-j_n(s+t)} = o(1/\sqrt{n})$$

by the assumption on j_n.

For the random term, just note that, by definition, K_j is a contraction in the $B_{22}^{s,W}$-norm of $B_{22}^s = H_2^s$; hence, $\cup_j\{K_j(f) - f : f \in \mathcal{F}\}$ is a bounded subset of H_2^s, and therefore, since $s > 1/2$, it is a P-Donsker class for all P (as observed just before Theorem 5.2.5). Hence, in order to prove that

$$\|P_n - P\|_{\tilde{\mathcal{F}}_{jn}} = o_P(1/\sqrt{n}),$$

where $\tilde{\mathcal{F}}_{j_n} = \{K_{j_n}(f) - f : f \in \mathcal{F}\}$, it suffices to show that $\sup_{f\in\mathcal{F}_{jn}} Pf^2 \to 0$. But this follows because, using that bounded subsets of H_2^s are uniformly bounded by the Sobolev imbedding and that $K_{j_n}(f) \to f$ in L^2, we have, for some $c < \infty$,

$$\sup_{f\in\mathcal{F}} E(K_{j_n}(f) - f)^2 \leq c \int |K_{j_n}(f) - f| f_0 \leq c\|K_{j_n}(f) - f\|_2 \|f_0\|_2 \to 0.$$

Hence, combining these estimates with the decomposition (5.81), we obtain that $\sqrt{n}\|P_n^W - P_n\|_{\mathcal{F}} \to 0$ in probability. Since $\mathcal{F}$ is P-Donsker, the theorem follows from Exercise 3.7.25. ■

Remark 5.2.12 The results in Section 4.4 about the Donsker property for Besov balls on $[0,1]$ (particularly Theorem 4.4.5) can be combined with the bias estimates provided by Proposition 4.3.14 (in particular, by the display following the proof of this proposition) exactly as in the preceding proof to give if $f_0 \in B_{p,q}^t([0,1]) \cup L^p([0,1])$ for some $t > 0$ and $\mathcal{F} = \{f \in C([0,1]) : \|f\|_{B^s_{p'q'}([0,1])} \leq M\}$, where $1 < p, p', q, q' < \infty$, $1/p + 1/q = 1$, $1/p' + 1/q' = 1$ and $s > \max(1/p', 1/2)$, then $G_P|_{\mathcal{F}}$ is sample continuous, and moreover, if the resolutions $j_n \to \infty$ satisfy $2^{-j_n(t+s)} n^{1/2} \to 0$, then also

$$\sqrt{n}(P_n^W - P) \to_{\mathcal{L}} G_P, \quad in\ \ell_\infty(\mathcal{F}),$$

for any Borel probabilty measure P on $[0,1]$. Here P_n^W is the measure with density equal to the wavelet projection estimator of f_0 when boundary-corrected wavelets or periodised wavelets satisfying Condition 5.1.1(c) or Condition 5.1.1(d) are used.

Finally, we should remark that all the results in this subsection admit extensions to dimension d with only formal changes.

5.2.2 Multiscale Spaces

We have seen in Section 4.4 how the white noise $\mathbb{W}$ and empirical process $\sqrt{n}(P_n - P)$ induce (asymptotically) tight random variables in certain negative-order Besov spaces. Using the wavelet characterisation of these spaces, these results are in fact equivalent to tightness results in certain multiscale spaces, where $1/\sqrt{n}$-consistent estimation of statistical parameters is possible. This gives an alternative approach to the analysis of nonparametric procedures where the bias-variance decomposition is done in the frequency domain and will be particularly interesting in the construction of confidence sets in Chapter 6.

For monotone increasing weighting sequences $w = (w_l)$, $w_l \geq 1, l \in \mathbb{N}$, we define multiscale sequence spaces of the following type:

$$\mathcal{M} \equiv \mathcal{M}(w) \equiv \left\{ x = \{x_{lk}\} : \|x\|_{\mathcal{M}(w)} \equiv \sup_l \frac{\max_k |x_{lk}|}{w_l} < \infty \right\}. \tag{5.82}$$

The space $\mathcal{M}(w)$ is a nonseparable Banach space (it is isomorphic to ℓ_∞). A separable closed subspace (isomorphic to c_0) is obtained by defining

$$\mathcal{M}_0 = \mathcal{M}_0(w) = \left\{ x \in \mathcal{M}(w) : \lim_{l\to\infty} \max_k \frac{|x_{lk}|}{w_l} = 0 \right\}, \tag{5.83}$$

consisting of those (weighted) sequences in $\mathcal{M}(w)$ whose l-limit vanishes at infinity.

We notice that $w_l \geq 1$ implies $\|x\|_{\mathcal{M}} \leq \|x\|_{\ell_2}$, so $\mathcal{M}$ always contains ℓ_2. For suitable divergent weighting sequences (w_l), these spaces actually contain objects that are much less regular than ℓ_2-sequences.

Gaussian White Noise in Multiscale Spaces

The following results parallel those from Theorem 4.4.4:

Definition 5.2.13 We call a sequence (w_l) admissible if $w_l/\sqrt{l} \uparrow \infty$ as $l \to \infty$.

Proposition 5.2.14 *For $\{\psi_{lk}\}$ a periodic or boundary-corrected wavelet basis of $L^2([0,1])$ from Sections 4.3.4 and 4.3.5, let $\mathbb{W} = (\int \psi_{lk} dW : l,k) = (g_{lk}), g_{lk} \sim N(0,1)$ be a Gaussian white noise. For $\omega = (\omega_l) = \sqrt{l}$, we have*

$$E\|\mathbb{W}\|_{\mathcal{M}(\omega)} < \infty. \tag{5.84}$$

If $w = (w_l)$ is admissible, then $\mathbb{W}$ defines a tight Gaussian Borel probability measure in the space $\mathcal{M}_0(w)$.

Proof The proof is the same as that for Theorem 4.4.4, using also Theorem 2.1.20 a). ■

We note that an assumption such as admissibility of w is necessary if we want to show that $\mathbb{W}$ is tight in $\mathcal{M}(w)$ (see Exercise 4.4.3 in and the notes at the end of Chapter 4). In particular, it is impossible to converge weakly towards $\mathbb{W}$ in $\mathcal{M}(\omega)$ because weak convergence of probability measures on a complete metric space implies tightness of the limit distribution.

For $f \in L^2$, a trajectory in the white noise model gives rise to a tight Gaussian shift experiment

$$\mathbb{Y}^{(n)} = f + \frac{1}{\sqrt{n}}\mathbb{W} \tag{5.85}$$

in $\mathcal{M}_0(w)$, for any admissible (w_l), in view of Proposition 5.2.14. Then

$$\sqrt{n}(\mathbb{Y}^{(n)} - f) = \mathbb{W}, \quad \text{in } \mathcal{M}_0, \tag{5.86}$$

so $\mathbb{Y}^{(n)}$ is a $1/\sqrt{n}$-consistent estimator of f in $\mathcal{M}_0$.

Smoothed Empirical Processes in Multiscale Spaces

Consider next the situation where we observe $X_1, \dots, X_n$ i.i.d. from P with density f on $[0,1]$. For $\{\psi_{lk}\}$ a periodic or boundary corrected wavelet basis of $L^2([0,1])$ from Chapter 4, a natural estimate of $\langle f, \psi_{lk}\rangle$ is given by

$$P_n\psi_{lk} := \int_0^1 \psi_{lk} dP_n = \frac{1}{n}\sum_{i=1}^n \psi_{lk}(X_i).$$

From the results in Chapter 3, we expect $\sqrt{n}(P_n - P)(\psi_{lk})$ to be approximately equal to a P-Brownian bridge G_P, which defines an element of a multiscale space by the action

$$G_P(\psi_{lk}) \sim N(0, \|\psi_{lk} - E_P\psi_{lk}(X)\|_{2,P}^2), \quad k, l, \tag{5.87}$$

and we have the following result:

Proposition 5.2.15 *Proposition 5.2.14 holds true for the P-Brownian bridge $\mathbb{G}_P$ replacing $\mathbb{W}$ whenever P has a bounded density on* $[0,1]$.

Proof The proof is exactly the same, using the standard bounds $Var(\mathbb{G}_P(\psi_{lk})) \leq \|dP\|_\infty$ and $E\max_k |G_P(\psi_{lk})| \leq C\sqrt{l}\|dP\|_\infty$, where dP denotes the density of P, and where C is a universal constant. ■

We note again that admissibility of the weighting sequence w is necessary to obtain tightness and hence also for any sequence of random variables to converge weakly towards G_P.

Any P with bounded density has coefficients $\langle f, \psi_{lk}\rangle \in \ell_2 \subset \mathcal{M}_0(w)$. We would like to formulate a statement such as $\sqrt{n}(P_n - P) \to^{\mathcal{L}} G_P$ in $\mathcal{M}_0$, paralleling (5.86) in the Gaussian white noise setting. The fluctuations of $\sqrt{n}(P_n - P)(\psi_{lk})$ are indeed at most of order $\sqrt{l}$ for l such that $2^l \leq n$ but for high frequencies can lie in the Poissonian regime of order $2^{l/2}l/\sqrt{n}$. Thus, P_n will not define an element of $\mathcal{M}_0$ for every admissible w. In our setting, we can consider only frequencies up to levels $2^l \leq n$, so let us introduce an appropriate 'projection' of the empirical measure P_n onto V_j associated to the sample

$$P_n(j) = \begin{cases} P_n\psi_{lk}, & \text{if } l \leq j \\ 0, & \text{if } l > j, \end{cases}$$

which obviously defines a tight random element in $\mathcal{M}_0$, and the following theorem shows that the smoothed empirical measure $P_n(j)$ estimates P efficiently in $\mathcal{M}_0$ if $j = j_n$ is chosen appropriately:

Theorem 5.2.16 *Suppose that P has density f in $C^\gamma([0,1])$ for some $\gamma \geq 0$. Let j_n be such that*

$$2^{-j_n(\gamma+1/2)} w_{j_n}^{-1} = o(1/\sqrt{n}), \qquad \frac{2^{j_n} j_n}{n} = O(1).$$

Then we have

$$\sqrt{n}(P_n(j_n) - P) \to^{\mathcal{L}} G_P, \qquad \text{in } \mathcal{M}_0(w),$$

as $n \to \infty$ for any w that is admissible.

Proof For J to be chosen later, let V_J be the subspace of $\mathcal{M}(w)$ consisting of the scales $l \leq J$, and let $\pi_{V_J}(f) = P(J)$ be the projection of P onto V_J. We have, by definition of the Hölder space,

$$\|P - P(j_n)\|_{\mathcal{M}_0} = \sup_{l > j_n} w_l^{-1} \max_k |\langle f, \psi_{lk}\rangle| \lesssim w_{j_n}^{-1} 2^{-j_n(\gamma+1/2)} = o(1/\sqrt{n}), \tag{5.88}$$

so this term is negligible in the limit distribution. Writing

$$\sqrt{n}(P_n(j_n) - P(j_n)) = \nu_n$$

and $\mathcal{L}(X)$ for the law of a random variable X, it suffices to show that $\beta(\mathcal{L}(\nu_n),\mathcal{L}(G_P)) \to 0$, where we denote by β the bounded Lipschitz distance d_{BL} between probability measures on the complete separable metric space $\mathcal{M}_0$, which, we recall, metrises weak convergence (see (3.254) and the paragraph preceding this definition). We have

$$\beta(\mathcal{L}(\nu_n),\mathcal{L}(G_P)) \le \beta(\mathcal{L}(\nu_n),\mathcal{L}(\nu_n)\circ\pi_{V_J}^{-1}) + \beta(\mathcal{L}(\nu_n)\circ\pi_{V_J}^{-1},\mathcal{L}(G_P)\circ\pi_{V_J}^{-1})$$
$$+\beta(\mathcal{L}(G_P),\mathcal{L}(G_P)\circ\pi_{V_J}^{-1}). \tag{5.89}$$

Let $\varepsilon > 0$ be given. The second term is less than $\varepsilon/3$ for every J fixed and n large enough by the multivariate central limit theorem applied to

$$\frac{1}{\sqrt{n}}\sum_{i=1}^{n}(\psi_{lk}(X_i) - E\psi_{lk}(X)), \quad k,l \le J,$$

noting that eventually $j_n > J$. For the first term, we have

$$\beta(\mathcal{L}(\nu_n),\mathcal{L}(\nu_n)\circ\pi_{V_J}^{-1}) \le E\|\sqrt{n}(\pi_{V_j} - \pi_{V_J})(P_n - P)\|_{\mathcal{M}(w)}$$
$$\le \left[\max_{J<l\le j}\frac{\sqrt{l}}{w_l}\right] E\max_{J<l\le j} l^{-1/2}\max_k |\langle\sqrt{n}(P_n - P),\psi_{lk}\rangle|.$$

Thus, for J large enough, this term can be made smaller than $\varepsilon/3$ if we can show that the expectation is bounded by a fixed constant. For M a large enough constant, this expectation is bounded above by M plus

$$\int_M^\infty P\left(\max_{J\le l\le j} l^{-1/2}\max_k |\langle\sqrt{n}(P_n - P),\psi_{lk}\rangle| > u\right)du$$
$$\le \sum_{J\le l\le j,k}\int_M^\infty P\left(|\langle\sqrt{n}(P_n - P),\psi_{lk}\rangle| > \sqrt{l}u\right)du$$
$$\le \sum_{J\le l\le j} 2^l\int_M^\infty e^{-Clu}du \lesssim e^{-C'JM},$$

where the third inequality follows from an application of Bernstein's inequality (see Chapter 3) together with the bounds $P\psi_{lk}^2 \le \|f\|_\infty$ and $\sqrt{l}\|\psi_{lk}\|_\infty \le \sqrt{l}2^{l/2} = O(\sqrt{n})$, for $l \le j_n$, using the assumption on j_n. For the third Gaussian term, we argue similarly using the arguments from the proof of Theorem 4.4.4. ∎

Exercises

5.2.1 Let Q be a finite positive Borel measure on $\mathbb{R}$, let μ be a Borel signed measure of finite variation on $\mathbb{R}$ and let $f:\mathbb{R}\mapsto\mathbb{R}$ be a Borel measurable function. Assume that (a) $f(\cdot+y)\in L^2(Q)$ for all $y\in\mathbb{R}$, (b) $f(x+\cdot)\in L^1(|\mu|)$ for all $x\in\mathbb{R}$ and (c) the function $y\mapsto\|f(\cdot+y)\|_{L^2(Q)}$ is in $L^1(|\mu|)$. Prove that the function

$$h(x) := \int f(x+y)d\mu(y)$$

is in the $L^2(Q)$-closure of $|\mu|(\mathbb{R})$ times the symmetric convex hull of $\mathcal{F}_f := \{f(\cdot+y) : y \in \mathbb{R}\}$. *Hint*: By separability there exists a countable set $\{y_k\} \in \mathbb{R}$ such that the set of functions

$\{f(\cdot+y_k)\}$ is dense in $\mathcal{F}_f$ for the $\mathcal{L}^2$-norm. Given $\varepsilon>0$, let $\varepsilon'=\varepsilon/2\|\mu\|$ define the measurable partition $\{A_k\}_{k=1}^{\infty}$ of $\mathbb{R}$: $A_1=\{y\in\mathbb{R}:\|f(\cdot+y)-f(\cdot+y_1)\|_{L^2(Q)}<\varepsilon'\}$ and, recursively, for all $k\in\mathbb{N}$, $A_k=(\cup_{j=1}^{k-1}A_j)^c\cap\{y\in\mathbb{R}:\|f(\cdot+y)-f(\cdot+y_k)\|_{L^2(Q)}<\varepsilon'\}$. By Minkowski for integrals,

$$\left\|\int f(\cdot+y)d\mu(y)-\sum_{k=1}^{r}f(\cdot+y_k)\mu(A_k)\right\|_{L^2(Q)}$$

$$\leq\sum_{k=1}^{r}\int\|f(x+y)-f(x+y_k)\|_{L^2(Q)}I_{A_k}(y)d|\mu|(y)+\int_{\cup_{k=r+1}^{\infty}A_k}\|f(x+y)\|_{L^2(Q)}d|\mu|(y)$$

$$\leq\varepsilon'\|\mu\|+\int_{\cup_{k=r+1}^{\infty}A_k}\|f(x+y)\|_{L^2(Q)}d|\mu|(y),$$

which proves the statement by letting $r\to\infty$.

5.2.2 Theorem 5.2.7 does not hold for discrete probability measures P. Concretely, let P be a probability measure such that $P\{x_0\}=a>0$ for some $x_0\in\mathbb{R}$. If $K\in L^1(\mathbb{R})$ and $h_n\to 0$ as $n\to\infty$. Prove that $\lim_n\Pr\{\|F_n^K(h)-F_n\|_\infty>a/3\}=1$; in particular, the sequence $\sqrt{n}\|F_n^K(h)-F_n\|_\infty$ is not stochastically bounded. *Hint*: Observe that the continuity of $P*K_h$ implies that $\|P*K_h-P\|_{\mathcal{F}}\geq a/2$, where $\mathcal{F}=\{I_{(-\infty,t]}:t\in\mathbb{R}\}$. Then proceed by analogy to parts of the proof of Theorem 5.2.7 to show that $\sqrt{n}E\|(P_n-P)(K_h*f-f)\|_{\mathcal{F}}=O(1)$, and conclude from decomposition (5.69) that $\sqrt{n}\|F_n^K(h)-F_n\|_\infty\geq\sqrt{n}a/2-O_P(1)$.

5.2.3 Let $\phi:\mathbb{R}\to\mathbb{R}$ be of bounded p-variation for some $1\leq p<\infty$ and with support contained in $(B_1,B_2]$ for some $-\infty<B_1<B_2<\infty$, and let

$$\mathcal{F}_\phi=\left\{\sum_{k\in\mathbb{Z}}\phi(2^jy-k)dy\phi(2^j(\cdot)-k):y\in\mathbb{R},\,j\in\mathbb{N}\right\}$$

and

$$\mathcal{D}_{\phi,j}=\left\{\sum_{k\in\mathbb{Z}}2^j\int_{-\infty}^{t}\phi(2^jy-k)\phi(2^j(\cdot)-k):\ t\in\mathbb{R}\right\},\quad j\in\mathbb{N}\cup\{0\}.$$

Prove that these classes of functions are of *VC* type (uniformly in j for $\mathcal{D}_{\phi,j}$); concretely, if $\mathcal{G}$ is either $\mathcal{F}_\phi$ or $\mathcal{D}_{\phi,j}$, for any j, then for all Borel probability measures Q on $\mathbb{R}$,

$$\sup_Q N(\mathcal{G},L^2(Q),\varepsilon)\leq\left(\frac{A}{\varepsilon}\right)^v,\quad 0<\varepsilon<A,$$

for A,v positive and finite constants depending only on ϕ. In particular, by Theorem 4.2.10, part (e), this applies to Daubechies scaling and wavelet functions. *Hint*: Proposition 3.6.12 applies to the class $\mathcal{M}$ of dilations and translations of ϕ, giving for its covering numbers a uniform in Q bound as earlier. Now, for y,j fixed, the sum $\sum_{k\in\mathbb{Z}}\phi(2^jy-k)\phi(2^j(\cdot)-k)$ consists of at most $[B_2-B_1]+1$ summands, each of which has the form

$$\phi(2^jy-k)\phi(2^j(\cdot)-k)=c_{j,y,k}\phi(2^j(\cdot)-k),$$

where k is a fixed integer satisfying $2^jy-B_2\leq k<2^jy-B_1$, and where $|c_{j,y,k}|\leq\|\phi\|_\infty$. A simple computation on covering numbers then shows that if a class $\mathcal{M}$ has a bound on $L^2(Q)$-covering numbers of the form $(A/\varepsilon)^v$, then so does (for possibly different A, v) any class $\mathcal{F}_\phi$ consisting of linear combinations of a fixed number of elements of $\mathcal{M}$ with coefficients bounded in absolute value by a fixed constant, proving the statement for $\mathcal{F}_\phi$. For $\mathcal{D}_{\phi,j}$, by the support assumption on

ϕ, we have, for every fixed t,

$$\sum_{k\in\mathbb{Z}} 2^j \int_{-\infty}^{t} \phi(2^j y - k)dy\, \phi(2^j(\cdot) - k) = c \sum_{k\le 2^j t - B_2} \phi(2^j(\cdot) - k)$$
$$+ \sum_{2^j t - B_2 < k \le 2^j t - B_1} c_{j,t,k}\phi(2^j(\cdot) - k),$$

where $c = \int_{-\infty}^{\infty} \phi(y)dy$ and $|c_{j,t,k}| \le \|\phi\|_1$. The class of functions

$$\left\{ \sum_{2^j t - B_2 < k < 2^j t - B_1} c_{j,t,k}\phi(2^j(\cdot) - k) : t \in \mathbb{R} \right\}$$

satisfies the bound on covering numbers with A, v independent of j by the argument in the first part of the hint. Each function in the class

$$\left\{ c \sum_{k \le 2^j t - B_2} \phi(2^j(\cdot) - k) : t \in \mathbb{R} \right\}$$

is the difference of two functions, one in each of the classes

$$\left\{ c \sum_{k \le 2^j t - B_2} \phi_+(2^j(\cdot) - k) : t \in \mathbb{R} \right\} \quad \text{and} \quad \left\{ c \sum_{k \le 2^j t - B_2} \phi_-(2^j(\cdot) - k) : t \in \mathbb{R} \right\},$$

where $\phi = \phi_+ - \phi_-$ and $\phi_+, \phi_- \ge 0$. But these classes are linearly ordered, so their subgraphs are ordered by inclusion and therefore are VC subgraphs of index 1. Now a simple computation on covering numbers gives the bound for $\mathcal{D}_{\phi,j}$.

5.2.4 Let f_0 satisfy the hypotheses of Theorem 5.2.7 for some $t \ge 0$, and suppose that its support is contained in $(-1, 1)$. Let K be a symmetric kernel of order $r = t + 1 - k$ for some $0 < k < 1/2$, and let h_n be such that $h_n^{t+1-k} n^{-1/2} \to 0$, e.g., $h_n = n^{-1/(2t+1)}$. Show that if $\mathcal{F} = \{\alpha^{-1}|x|^{\alpha} I_{[-1,1]} : 0 < \alpha < \infty\}$, then $\sqrt{n}(P_n^k - P) \to_{\mathcal{L}} G_P$ in $\ell_\infty(\mathcal{F})$. That is, f_n is minimax in L^p loss, and the empirical moments and absolute moments based on f_n converge simultaneously to those of $f_0(x)dx$ at the rate $1/\sqrt{n}$. Show also that $\|P_n^K - P\|_{BL} = O_P(1/\sqrt{n})$.

5.3 Some Further Topics

The minimax paradigm extends far beyond the basic function estimation problems introduced earlier. In this section we discuss two further topics that arise naturally in the field. In the first subsection we look at functionals of a density consisting of integrals of smooth functions, starting with the integral of the square of a density. In the second subsection we consider the very natural problem of estimating a density or a signal from contaminated observations, concretely, from observations $X_i + \varepsilon_i$ that contain additive i.i.d. errors ε_i. Just as in previous sections in this chapter, we obtain upper bounds and/or limit theorems. As in previous sections, the important subject of optimality can be treated with the tools of Chapter 6 (see the notes at the end of the chapter).

5.3.1 Estimation of Functionals

Even when a statistical parameter space $\mathcal{F}$ is infinite dimensional often one is interested only in a one-dimensional aspect of it. Such situations can be conveniently modeled as functionals

$$\Phi : \mathcal{F} \to X,$$

where X is the range space of the functional: typically, $X = \mathbb{R}$ or $\mathbb{R}^n$ or some other function space. The most natural approach to estimating $\Phi(f)$ is perhaps based on $\Phi(\hat{f}_n)$, where $\hat{f}_n$ is a nonparametric estimator for f, but this need not always be optimal. This is illustrated here with some nonlinear integral functionals such as $\int f^2$ or $\int \phi(f(x),x)f^2(x)dx$, where ϕ enjoys some smoothness.

Estimation of $\|f\|_2^2$ under Gaussian White Noise

Consider an observation in the white noise model (5.5), and suppose that the goal is to estimate $\|f\|_2^2$. If

$$\begin{aligned} f_n(j)(x) &= \sum_{k=0}^{2^J-1}\left(\int_0^1 \phi_{Jk}dY\right)\phi_{Jk} + \sum_{l=J}^{j-1}\sum_{k=0}^{2^l-1}\left(\int_0^1 \psi_{lk}dY\right)\psi_{lk}(x) \\ &= \sum_{J-1\le l<j,k}\left(\int_0^1 \psi_{lk}dY\right)\psi_{lk}(x), \quad x\in[0,1], \end{aligned}$$

denotes the projection of the observation dY onto a wavelet subspace V_j and where, for ease of notation, we recall the notational convention of replacing ϕ_{Jk} by ψ_{J-1k} introduced in (4.32), we can use as our estimate for $\|f\|_2^2$ the statistic

$$T_n = \|f_n(j)\|_2^2 - \frac{\sigma^2 2^j}{n} = \sum_{l\le j,k}\left(\int_0^1 \psi_{lk}(t)dY(t)\right)^2 - \frac{\sigma^2 2^j}{n}. \tag{5.90}$$

(We omit the lower summation index for l in (5.90) because it depends on the starting V_J which needs to be some $J > 0$ for boundary-corrected wavelets but is -1 for periodised wavelets if we start at $l = 0$.) We see that the estimation error is

$$\|f_n(j)\|_2^2 - \frac{\sigma^2 2^j}{n} - \|f\|_2^2 = \frac{2\sigma}{\sqrt{n}}\sum_{l\le j,k} g_{lk}\langle f, \psi_{lk}\rangle + \frac{\sigma^2}{n}\sum_{l\le j,k}(g_{lk}^2 - 1) + \|K_j(f) - f\|_2^2,$$

which implies, for

$$Z = \frac{\sigma}{\sqrt{n}}\sum_{l,k} g_{lk}\langle f, \psi_{lk}\rangle \sim N(0, (\sigma^2/n)\|f\|_2^2),$$

that

$$E|T_n - \|f\|_2^2 - 2Z| \lesssim \frac{\sigma^2 2^{j/2}}{n} + \|K_j(f) - f\|_2^2 + \frac{2\sigma}{\sqrt{n}}\|K_j(f) - f\|_2. \tag{5.91}$$

From this we obtain the following result:

Proposition 5.3.1 *Suppose that $f \in B^r_{2\infty}([0,1])$ for some $r > 1/4$, and let $2^{j_n} \sim n^{1/(2r+1/2)}$. Then*

$$\sqrt{n}(T_n - \|f\|_2^2) \to^d N(0, 4\sigma^2\|f\|_2^2),$$

as $n \to \infty$.

Proof For such $f \in B^r_{2\infty}$, we have $\|K_j(f) - f\|_2^2 = O(2^{-2jr})$, which is balanced with $2^{j/2}/n$ when $2^{j_n} \sim n^{1/(2r+1/2)}$. For $r > 1/4$, we then have $2^{-2j_n r} = o(n^{-1/2})$, and as a consequence,

$$E|T_n - \|f\|_2^2 - 2Z| = o(1/\sqrt{n}) \Rightarrow \sqrt{n}(T_n - \|f\|_2^2) \to^d 2\sqrt{n}Z,$$

completing the proof. ■

For $r \leq 1/4$, the preceding estimates give a rate of convergence

$$E|T_n - \|f\|_2^2| = O(n^{-2r/(2r+1/2)}), \tag{5.92}$$

which can be shown to be optimal in a minimax sense using methods from Chapter 6.

Estimation of $\|f\|_2^2$ Based on U-Statistics

We now consider the sampling analogue to the problem from the preceding subsection. Suppose that one wants to estimate $\int_{\mathbb{R}} f^2(x)dx$ based on a sample $X_1,\dots,X_n$ from f. A natural estimator is

$$T_n(h_n) := \frac{2}{n(n-1)h_n} \sum_{1\leq i<j\leq n} K\left(\frac{X_i - X_j}{h_n}\right), \tag{5.93}$$

where we take K to be a bounded symmetric kernel such that $\int K(u)du = 1$ as well as $\int |K(u)||u|du < \infty$ and $0 < h_n \to 0$. (Note that $T_n(h)$ equals $\int (K_h * P_n)dP_n$ with the diagonal terms removed and with a slight renormalisation: roughly, one replaces f by its kernel density estimator and integration against $f(x)dx$ by integration with respect to the empirical measure.)

It is convenient to recall U-statistics notation. For a symmetric function of two variables $R(x,y)$, we write

$$U_n^{(2)}(R) = \frac{2}{n(n-1)} \sum_{1\leq i<j\leq n} R(X_i,X_j).$$

The two Hoeffding projections of R are

$$\pi_1 R(x) = ER(x,X_1) - ER(X_1,X_2),$$
$$\pi_2(R)(x,y) = R(x,y) - ER(x,X_1) - ER(y,X_1) + ER(X_1,X_2),$$

which induce the Hoeffding decomposition

$$U_n^{(2)}(R) - ER(X_1,X_2) = 2U_n^{(1)}(\pi_1 R) + U_n^{(2)}(\pi_2 R), \tag{5.94}$$

where $U_n^{(1)}(\pi_1 R) = n^{-1}\sum_{i=1}^n (\pi_1 R)(X_i)$. Note that, by orthogonality,

$$E\left(U_n^{(1)}(\pi_1 R)\right)^2 = n^{-1}E((\pi_1 R)(X_1))^2 \leq n^{-1}E(R(X_1,X_2) - ER(X_1,X_2))^2,$$
$$E\left(U_n^{(2)}(\pi_2 R)\right)^2 = \frac{2}{n(n-1)}E((\pi_2 R)(X_1,X_2))^2 \leq \frac{2}{n(n-1)}E(R(X_1,X_2) - ER(X_1,X_2))^2.$$

(See the beginning of Section 3.4.3.)

We will also use the notations $H^\alpha_{2,M}$ and L^∞_L, respectively, for the ball of radius M of $H^\alpha_2 \equiv B^\alpha_{22}(\mathbb{R})$ and for the ball of radius L of $L^\infty(\mathbb{R})$.

Theorem 5.3.2 *Let $f \in H_2^\alpha(\mathbb{R}) \cap L^\infty(\mathbb{R})$ for some $0 < \alpha \le 1/2$.*

(i) We have

$$\sup_{f \in H_{2,M}^\alpha} \left| ET_n(h_n) - \int_{\mathbb{R}} f^2(x)dx \right| \le B(h_n) := c_1(M) h_n^{2\alpha} \tag{5.95}$$

and

$$\sup_{f \in H_{2,M}^\alpha \cap L_L^\infty} E\left(T_n(h_n) - ET_n(h_n) - \frac{1}{n}\sum_{i=1}^n Y_i \right)^2 \le c_2^2(M)\sigma^2(h_n, n) \tag{5.96}$$

$$:= c_2^2(M)\left(\frac{1}{n^2 h_n} \vee \frac{L h_n^{2\alpha}}{n} \right),$$

where $Y_i = 2\left(f(X_i) - \int_{\mathbb{R}} f^2\right)$, and where $c_1(M)$ and $c_2(M)$ are numerical constants depending only on M and the function K.

(ii) As a consequence, taking h_n so that $h_n \approx n^{-2/(4\alpha+1)}$, we have the following:

(a) If $0 < \alpha \le 1/4$, then

$$T_n(h_n) - \int_{\mathbb{R}} f(x)^2 dx = O_P(n^{-4\alpha/(4\alpha+1)}),$$

(b) If $\alpha > 1/4$ and if $\tau^2 = \left[\int_{\mathbb{R}} f^3 - \left(\int_{\mathbb{R}} f^2\right)^2\right]$, then

$$\sqrt{n}\left(T_n(h_n) - \int_{\mathbb{R}} f(x)^2 dx \right) \to_d Z \sim N(0, 4\tau^2).$$

Proof The bias equals

$$\begin{aligned}
ET_n(h_n) - \int f^2 &= \int_{\mathbb{R}}\int_{\mathbb{R}} K_{h_n}(x-y) f(y)dy f(x)dx - \int_{\mathbb{R}} f(x)f(x)dx \\
&= \int_{\mathbb{R}}\int_{\mathbb{R}} K_{h_n}(x-y)[f(y) - f(x)] f(x)dydx \\
&= \int_{\mathbb{R}}\int_{\mathbb{R}} K(u)[f(x-uh_n) - f(x)] f(x)dudx \\
&= \int_{\mathbb{R}} K(u)\left[\int_{\mathbb{R}} \bar{f}(uh_n - x) f(x)dx - \int_{\mathbb{R}} \bar{f}(0-x) f(x)dx \right] du \\
&= \int_{\mathbb{R}} K(u)\left[(\bar{f} * f)(uh_n) - (\bar{f} * f)(0)\right] du,
\end{aligned} \tag{5.97}$$

where $\bar{f}(x) = f(-x)$, and where $*$ denotes convolution. As in Lemma 4.3.18 we have for $f, g \in H_2^\alpha \cap L^1, 0 < \alpha \le 1/2$, for any $x \in \mathbb{R}, t \ne 0$, using Fourier inversion and (4.115),

$$\begin{aligned}\frac{|(f*g)(x+t)-(f*g)(x)|}{|t|^{2\alpha}} &\le |t|^{-2\alpha}\|\mathcal{F}^{-1}\mathcal{F}[(f*g)(\cdot+t)-(f*g)(\cdot)]\|_\infty\\ &\le (2\pi)^{-1}|t|^{-2\alpha}\|F[(f*g)(\cdot+t)-(f*g)(\cdot)]\|_1\\ &= (2\pi)^{-1}|t|^{-2\alpha}\int_{\mathbb{R}}\left|\mathcal{F}(f*g)(u)\left[e^{-iut}-1\right]\right|du\\ &= (2\pi)^{-1}\int_{\mathbb{R}}|\mathcal{F}f||u|^\alpha|\mathcal{F}g||u|^\alpha\frac{|e^{-iut}-e^{-i0}|}{|u|^{2\alpha}|t|^{2\alpha}}du\\ &\le C\|f\|_{H_2^\alpha}\|g\|_{H_2^\alpha},\end{aligned}$$

since $e^{-i(\cdot)}$ is bounded Lipschitz.

To proceed, identity (5.97) now gives, by the conditions on the kernel, that

$$\left|ET_n(h_n)-\int f^2\right|\le c_1 h_n^{2\alpha},$$

where $c_1 = C\|f\|_{H_2^\alpha}\int|K(u)||u|^{2\alpha}du \le CM^2\int|K(u)||u|^{2\alpha}du$, that is, (5.95).

Next, we show (5.97). Setting

$$R(u,v) := K_{h_n}(u-v),$$

we can write, in U-statistic notation, $T_n(h_n) = U_n^{(2)}(R)$ or, if $\tilde{R}(u,v) = R(u,v) - ER(X_1,X_2)$,

$$T_n(h_n) - ET_n(h_n) = U_n^{(2)}(\tilde{R}).$$

Thus, by Hoeffding's decomposition, it remains to estimate the following statistics (note that $\pi_i R = \pi_i\tilde{R}$, $i = 1,2$):

$$U_n^{(2)}(\tilde{R}) - \frac{1}{n}\sum_{i=1}^n Y_i = \left(2U_n^{(1)}(\pi_1 R) - \frac{1}{n}\sum_{i=1}^n Y_i\right) + U_n^{(2)}(\pi_2 R) =: S_1 + S_2.$$

First, we have, by Proposition 4.3.8 (note that K is a kernel of order $1 > \alpha$),

$$\begin{aligned}nES_1^2 &\le E\left[\int 2K_{h_n}(X_1-y)f(y)dy - 2f(X_1)\right]^2\\ &\le 4\|f\|_\infty\|K_{h_n}*f-f\|_2^2\\ &\le C\|f\|_\infty\|f\|_{H_2^\alpha}h_n^{2\alpha}.\end{aligned}\qquad(5.98)$$

Next, since π_2 is a projection of $L^2(f(x)dx)$, it follows from Young's inequalities that

$$\begin{aligned}ES_2^2 &\le \frac{2}{n(n-1)}ER^2 = \frac{2}{n(n-1)}E\left[K_{h_n}(X_1-X_2)\right]^2\\ &= \frac{2}{n(n-1)}\int (K_{h_n}^2 * f)(y)\, f(y)\, dy\\ &\le \frac{2\|f\|_\infty\|K\|_2^2}{n(n-1)h_n}.\end{aligned}\qquad(5.99)$$

Now (5.98) and (5.99) complete the proof of (5.97). The remaining claims in part (ii) follow by the choice of the bandwidth and in case (a) (and hence $\alpha \le 1/4$) noting that we have $n^{-1}\sum_{i=1}^n Y_i = O_P(n^{-1/2}) = O_P(n^{-4\alpha/(4\alpha+1)})$ and in case (b) from the central limit theorem (CLT) for the random variables Y_i. ∎

With a view towards estimating $\int \varphi(f(x),x)dx$ for differentiable functions φ by means of a limited Taylor development, it is convenient to estimate $\int \psi(x) f_0^2(x)dx$ for ψ bounded. The estimator

$$T_n^{\psi,1}(h) := \frac{1}{n(n-1)h} \sum_{1\le i\neq j\le n} K\left(\frac{X_i - X_j}{h}\right)\psi(X_i)$$
$$= \frac{1}{n(n-1)h} \sum_{i<j\le n} K\left(\frac{X_i - X_j}{h}\right)(\psi(X_i) + \psi(X_j)),$$

where the identity follows from the symmetry of K, is the analogue of T_n earlier: both are obtained by deleting the diagonal and re-norming in the expression $\int \psi(x) f_n(x)dP_n(x)$, where f_n is the kernel density estimator and P_n is the empirical measure (T_n corresponds to $\psi \equiv 1$). Application of the bias calculations earlier (see (5.97) and the text after it) to this case only obtains the crucial bound $h_n^{2\alpha}$ if $\|\psi f\|_{H_2^\alpha} < \infty$, but this condition essentially requires ψ to be in a Sobolev space. There is a bias-reduction device that allows us to obtains this bound under only the boundedness assumption on ψ; it consists in subtracting of the original estimator a different one with an overlapping bias, thus cancelling some of it. Another natural estimator of $\int \psi f^2$ is $\int \psi f_n^2$, where f_n is the kernel density estimator of f, again with the diagonal deleted and a slight re-norming (the diagonal terms have different expected values than the other terms in the resulting double sum, and their deletion reduces bias and helps in computations), explicitly,

$$T_n^{\psi,2}(h) = \frac{2}{n(n-1)h} \sum_{i<j\le n} \int K\left(\frac{x - X_i}{h}\right) K\left(\frac{x - X_j}{h}\right)\psi(x)\,dx.$$

Then we observe that with the usual notation $K_h(\cdot) = K(\cdot/h)/h$,

$$ET_n^{\psi,1}(h) = \int (K_h * f)\psi f \quad \text{and} \quad ET_n^{\psi,2}(h) = \int (K_h * f)^2 \psi.$$

Therefore,

$$E\left(2T_n^{\psi,1}(h) - T_n^{\psi,2}(h) - \int \psi f^2\right) = 2\int (K_h * f)\psi f - \int (K_h * f)^2\psi - \int \psi f^2$$
$$= -\int (K_h * f - f)^2 \psi, \tag{5.100}$$

so the absolute value of this expression admits the bound $\|\psi\|_\infty \|K_h * f - f\|_2^2$, of order at most $h^{2\alpha}$, if f is in $B_{2\infty}^\alpha$, just as in Theorem 5.3.2. Thus, we define

$$T_n^{\psi}(h) = 2T_n^{\psi,1}(h) - T_n^{\psi,2}(h) \tag{5.101}$$

and note that, by the preceding argument and Proposition 4.3.8 (recall that K is a kernel of order 1),

$$\left| ET_n^{\psi}(h) - \int \psi f^2 \right| \le \|\psi\|_\infty \|K_h * f - f\|_2^2 \le C\|\psi\|_\infty \|f\|_{B_{2\infty}^\alpha}^2 h^{2\alpha},$$

where C depends only on K and the Besov norm used. Note that the bias bound for this new estimator does work for $0 < \alpha < 1$ (as opposed to $0 < \alpha < 1/2$ in the preceding theorem).

And, although there is not much difference, it seems more natural in this new context to measure smoothness of f in terms of the Besov scale rather than the Sobolev scale.

We now look at the variance part of this estimator, that is, at $T_n^\psi(h) - ET_n^\psi(h)$. We consider the two terms separately. For $T_n^{\psi,1}$, letting

$$R_1(u,v) = \frac{1}{2} K_h(u-v)(\psi(u) + \psi(v)),$$

we have, $T_n^{\psi,1}(h) = 2/(n(n-1)) \sum_{1 \le i < j \le n} R_1(X_i, X_j) =: U_n^{(2)}(R_1)$, so, with analogous notation as in the proof of the preceding theorem and with $Y_i = 2\left(\psi(X_i) f(X_i) - \int \psi f^2\right)$,

$$T_n^{\psi,1}(h) - ET_n^{\psi,1}(h) - \frac{1}{n}\sum_{i=1}^n Y_i = \left(2U_n^{(1)}(\pi_1 R_1) - \frac{1}{n}\sum_{i=1}^n Y_i\right) + U_n^{(2)}(\pi_2 R_1) =: S_1 + S_2.$$

Then, mimicking the preceding proof,

$$\begin{aligned} nES_1^2 &\le E\left[\int K_h(X_1 - y)(\psi(y) + \psi(X_1)) f(y)dy - 2\psi(X_1) f(X_1)\right]^2 \\ &\le 2\|\psi\|_\infty^2 \|f\|_\infty \|K_h * f - f\|_2^2 + 2\|f\|_\infty \|K_h * (\psi f) - \psi f\|_2^2. \end{aligned}$$

This bound is not as small as (5.98), but note that it still converges to zero when h tends to zero: if K is a kernel integrating to 1, and if g is in L^2, Minkowski's inequality for integrals gives

$$\begin{aligned} \left(\int (K_h * g - g)^2(y)dy\right)^{1/2} &= \left(\int \left(\int K(u)(g(y-uh) - g(y))du\right)^2 dy\right)^{1/2} \\ &\le \int |K(u)| \left(\int (g(y-uh) - g(y))^2 dy\right)^{1/2} du, \end{aligned}$$

and this tends to zero as $h \to 0$ by Lebesgue's dominated convergence and by continuity of translations in L^2. We also have

$$nES_2^2 \le \frac{1}{n-1} E[K_h(X_1 - X_2)(\psi(X_1) + \psi(X_2))]^2 \le \frac{2\|\psi\|_\infty^2 \|f\|_\infty \|K\|_2^2}{(n-1)h}$$

by boundedness of ψ and Young's inequality, just as in (5.99). For the variance part of $T_n^{\psi,2}$, we use the U-statistic kernel

$$R_2(u,v) = \int K_h(x-u) K_h(x-v) \psi(x) dx,$$

with Hoeffding projections

$$\pi_2(R_2) = \int (K_h(x-u) - (K_h * f)(x))(K_h(x-v) - (K_h * f)(x))\psi(x)dx,$$

$$\pi_1(R_2) = \int (K_h * f)(x) K_h(x-u)\psi(x)dx - ER_2(X_1, X_2).$$

If we still denote $S_1 = 2U_n^{(1)}(\pi_1 R_2) - (1/n)\sum_{i=1}^n Y_i$ and $S_2 = U_n^{(2)}(\pi_2 R_2)$, we obtain in this case the following two bounds. The first bound makes repeated use of Young's inequality:

$$\frac{n(n-1)}{2} ES_2^2 \le E\left[\int K_h(x-X_1)K_h(x-X_2)\psi(x)dx\right]^2$$
$$\le \|\psi\|_\infty^2 E\int\int |K_h(x-X_1)||K_h(x-X_2)||K_h(y-X_1)||K_h(y-X_2)|dxdy$$
$$= \|\psi\|_\infty^2 \int\int\int\int |K_h(x-u)||K_h(x-v)||K_h(y-u)|$$
$$\times |K_h(y-v)| f(u)f(v)dudvdxdy$$
$$= \|\psi\|_\infty^2 \int\int (K_h * K_h)^2(v-u)f(u)f(v)dudv$$
$$= \|\psi\|_\infty^2 \int\int ((K_h * K_h)^2 * f)(v)f(v)dudv$$
$$\le \|\psi\|_\infty^2 \|f\|_\infty \|K_h * K_h\|_2^2$$
$$\le \|\psi\|_\infty^2 \|f\|_\infty \|K_h\|_1^2 \|K_h\|_2^2 = \|\psi\|_\infty^2 \|f\|_\infty \|K\|_1^2 \|K\|_2^2/h.$$

For the second bound, we start with the decomposition

$$nES_1^2 \le E\left[2\int (K_h * f)(x)K_h(x-X)\psi(x)dx - 2\psi(X)f(X)\right]^2$$
$$\le 8E\left[\int ((K_h * f)(x)K_h(x-X)\psi(x)dx - (K_h * (\psi f))(X)\right]^2$$
$$+8E[(K_h * (\psi f))(X) - \psi(X)f(X)]^2$$

The second summand is just

$$\int (K_h * (\psi f) - \psi f)^2 f \le \|f\|_\infty \|K_h * (\psi f) - \psi f\|_2^2,$$

which tends to zero as $h \to 0$, as observed earlier. As for the first summand, we have, by Young's inequality,

$$E\left[\int (K_h * f)(x)K_h(x-X)\psi(x)dx - (K_h * (\psi f))(X)\right]^2$$
$$= \int (K_h * ((K_h * f)\psi) - K_h * (\psi f))^2 f$$
$$\le \|f\|_\infty \|K_h\|_1^2 \|(K_h * f)\psi - f\psi\|_2^2 \le \|f\|_\infty \|K\|_1 \|\psi\|_\infty^2 \|K_h * f - f\|_2^2.$$

Collecting these bounds, noting that $\|K_h * f - f\|_2$ is bounded by a constant times $\|f\|_{H_2^\alpha} h^\alpha$ for $0 < \alpha < 1$ by Proposition 4.3.8, that $\|K_h * (\psi f) - \psi f\|_2 \to 0$ as $h \to 0$ for ψf in L^2 and that the random variables $\{Y_i\}$ satisfy the CTL, we obtain the following extension of Theorem 5.3.2:

Theorem 5.3.3 *Let f be a bounded density in $B_{2\infty}^\alpha(\mathbb{R})$ for some $0 < \alpha < 1$, let K be a symmetric kernel in $L^1(\mathbb{R}) \cap L^2(\mathbb{R})$ integrating to 1 and such that $\int |K(u)||u|du < \infty$ and let*

ψ be a bounded measurable function. Let X_i, $i \in \mathbb{N}$, be independent, identically distributed with law of density f, and let $Y_i = 2\left(\psi(X_i)f(X_i) - \int \psi f^2\right)$. Let $T_n^\psi(h)$ be the estimator of $\int_{\mathbb{R}} \psi(x) f_0^2(x)dx$ defined by equation (5.101). Then there exists a constant C that depends only on K (and the Besov norm used) such that, for $h > 0$ and $n \geq 2$,

$$E\left(T_n^\psi(h) - \int_{\mathbb{R}} \psi(x) f^2(x)dx - \frac{1}{n}\sum_{i=1}^{n} Y_i\right)^2$$

$$\leq C(K)\left[\|\psi\|_\infty \|f\|_\infty \frac{1}{n^2 h} + \|\psi\|_\infty \|f\|_\infty \frac{\|K_h * f - f\|_2^2}{n}\right.$$

$$\left. + \|f\|_\infty \frac{\|K_h * (\psi f) - \psi f\|_2^2}{n} + \|\psi\|_\infty^2 \|K_h * f - f\|_2^4\right]. \tag{5.102}$$

Hence, taking $h_n = n^{-2/(4\alpha+1)}$, we have

(a) For $0 < \alpha \leq 1/4$,

$$T_n^\psi(h_n) - \int \psi f^2 = O_P(n^{-4\alpha/(4\alpha+1)}),$$

(b) For $\alpha > 1/4$ and $\tau^2 = Var(Y_1)$,

$$\sqrt{n}\left(T_n^\psi(h_n) - \int \psi f^2\right) \to_d Z \sim N(0, \tau^2).$$

The preceding two theorems admit versions for wavelet estimators. With the obvious changes (including significant simplifications due to ortho-normality), the preceding proof can be easily modified to show that for $f \in B_{2\infty}^\alpha$, $0 < \alpha < 1$, and for projection kernels K_j onto the nested spaces V_j from a multiresolution analysis with a wavelet basis $\{\phi_k, \psi_{j,k}\}$ of regularity index $S > 0$, the estimators

$$T_n^W(j_n) = \frac{1}{n(n-1)} \sum_{i \neq j \leq n} K_{j_n}(X_i, X_j)$$

satisfy the conclusions of Theorem 5.3.2, and the estimators

$$T_n^{W,\psi}(j_n) = \frac{2}{n(n-1)} \sum_{k \in \mathbb{Z}, \ell < j_n} \sum_{i \neq j \leq n} \psi_{\ell,k}(X_i)\psi_{\ell,k}(X_j)\psi(X_j) \tag{5.103}$$

$$- \frac{1}{n(n-1)} \sum_{k \in \mathbb{Z}, \ell < j_n} \sum_{k' \in \mathbb{Z}, \ell' \leq j_n} \sum_{i < j \leq n} \psi_{\ell,k}(X_i)\psi_{\ell',k'}(X_j) \int \psi_{\ell,k}\psi_{\ell',k'}\psi$$

satisfy the conclusions of Theorem 5.3.3. Likewise, the densities may be defined on an interval $A \subset \mathbb{R}$ instead of $\mathbb{R}$, and the wavelet basis then may be replaced by a boundary-corrected basis on $L^2(A)$ or by a periodised wavelet basis of $f(0) = f(1)$.

Estimation of General Integral Functionals

Let $\phi(u,x)$ be a differentiable function of two variables defined on $I \times \mathbb{R}$, where I is an open interval containing the range $[a,b]$ of a bounded density f. The object in this subsection is to estimate $T(f) := \int_{\mathbb{R}} \phi(f(x),x)f(x)dx$. An important example of a functional of this type is the Shannon entropy functional, which is considered in Exercise 5.3.4.

To this end, we begin by taking a preliminary estimator $\hat{f}$ of f using only $n_1 \simeq n/\log n$ of the n data and do a Taylor expansion of ϕ about the point $(\hat{f}(x), x)$ for each x. Then, denoting by $\phi_1'(u,v)$, $\phi_1''(u,v)$, $\phi_1^{(j)}(u,v)$ the partial derivatives of ϕ with respect to the first variable and

$$\|\phi_1^{(j)}\|_\infty = \sup_{u\in I, x\in\mathbb{R}} \left|\frac{\partial^j \phi}{\partial u^j}(u,v)\right|$$

and assuming that $\|\phi_1^{(3)}\|_\infty < \infty$, we have

$$\begin{aligned} T(f) = &\int \phi(\hat{f}(x),x)dx + \int \phi_1'(\hat{f}(x),x)(f-\hat{f})(x)dx \\ &+ \frac{1}{2}\int \phi_1''(\hat{f}(x),x)(f-\hat{f})^2(x)dx + \Gamma_n, \end{aligned}$$

where the remainder Γ_n satisfies $|\Gamma_n| \le 1/6\|\phi_1^{(3)}\|_\infty \int |f-\hat{f}|^3$. Collecting the terms that are linear in f, those that are quadratic in f and the remainder, this becomes

$$T(f) = \int G(x)dx + \int H(x)f(x)dx + \int J(x)f^2(x)dx + \Gamma_n,$$

where, writing $\hat{f}$ for $\hat{f}(\cdot)$ to ease notation,

$$G(\cdot) = \phi(\hat{f},\cdot) - \phi_1'(\hat{f},\cdot)\hat{f} + \frac{1}{2}\phi_1''(\hat{f},\cdot)\hat{f}^2, \tag{5.104}$$

$$H(\cdot) = \phi_1'(\hat{f},\cdot) - \hat{f}\phi_1''(\hat{f},\cdot), \tag{5.105}$$

$$J(\cdot) = \frac{1}{2}\phi_1''(\hat{f},\cdot). \tag{5.106}$$

Now we use the remaining $n_2 = n - n_1$ data to estimate $\int H(x)f(x)dx$ and $\int J(x)f(x)dx$, the first by $P_{n_2}H$ and the second by the estimator constructed in the preceding subsection taking $\psi(x) = J(x)$. We assume that $\hat{f}$ depends on the first n_1 data, $X_1, \dots, X_{n_1}$, and that the estimators of H and J depend on $\hat{f}$ and the last n_2 data. Thus, we define

$$\begin{aligned} T_n := &\int G(x)dx + \frac{1}{n_2}\sum_{i=n_1+1}^{n} H(X_i) + \frac{2}{n_2(n_2-1)}\sum_{n_1<i<j\le n} K_{h_n}(X_i - X_j)(J(X_i) + J(X_j)) \\ &- \frac{2}{n_2(n_2-1)h}\sum_{n_1<i<j\le n}\int K_{h_n}(x-X_i)K_{h_n}(x-X_j)J(x)dx, \end{aligned} \tag{5.107}$$

where $h_n \simeq n_2^{-2/(4\alpha+1)}$, and we assume that $f \in B^\alpha_{2\infty}(\mathbb{R})$ and that K is a kernel with the same properties as in Theorem 5.3.3. We could as well use wavelets rather than kernels to define the second and third parts of the estimator (as at the end of the preceding subsection) and replace $\mathbb{R}$ by an interval, for example, $[0,1]$.

We will only sketch the proof of the following theorem, mainly to illustrate the use of Theorem 5.3.3 and give a sense of the computations involved. To make for a cleaner statement, we first list the hypotheses:

$H1$: f is a bounded density, $0 \le a \le f(x) \le b < \infty$ for all $x \in supp(f)$, and $f \in B^{\alpha}_{2\infty}$ for some $1/4 < \alpha < 1$.

$H2$: The preliminary estimator $\hat{f} = \hat{f}_n$ of f is constructed from $X_1, \dots, X_{n_1}$, where $n_1 \simeq n/\log n$, such that (a) for any fixed choice of ε, for all $x \in supp(f), a - \epsilon \le \hat{f}(x) \le b + \varepsilon$ with probability 1 for all n_1 large enough depending on ε; (b) $E\|f - \hat{f}\|_q \to 0$ for $q = 1,2$; hence, $E\|f - \hat{f}\|_q^r \to 0$ for any $r > 0$; and (c) $E\|f - \hat{f}\|_3^6 \le Cn_1^{-\beta}$ for some $\beta > 1$ and $C < \infty$ independent of n (examples: see Proposition 5.1.7 and Theorem 5.1.15).

$H3$: The kernel K is as in Theorem 5.3.2.

$H4$: The function $\phi(u,x)$ is a bounded function of two variables defined on $I \times \mathbb{R}$, where I is an open interval containing the interval $[a - \varepsilon, b + \varepsilon]$ (hence the range of the density f and that of the preliminary estimators $\hat{f}$ for n_1 large enough), and it is three times differentiable with respect to the first variable, with the first three partial derivatives uniformly bounded on $I \times \mathbb{R}$.

Theorem 5.3.4 *Assume $H1$–$H4$ for a given $\alpha \in (1/4,1)$, let $T(f) := \int_{\mathbb{R}} \phi(f(x),x)f(x)dx$ and let T_n be the estimator of $T(f)$ defined by (5.107) for $h_n \simeq n_2^{2/(4\alpha+1)}$ for n_1 and $\hat{f}$ as specified in $H2$ and $n_2 = n - n_1$. Then we have*

$$\sqrt{n}(T_n - T(f)) \to_d Z \sim N(0, C(f,\phi)), \tag{5.108}$$

with convergence of second moments, where

$$C(f,\phi) = \int (\phi_1'(f(x),x)^2 f(x)dx - \left(\int \phi_1'(f(x),x)f(x)dx\right)^2,$$

with $\phi_1'(u,v) = \partial\phi(u,v)/\partial u$.

Proof (Sketch.) By the central limit theorem for $\phi_1'(f(X),X)$, it suffices to prove that

$$\sqrt{n}\big[T_n - T(f) - (P_{n_2} - P)(\phi_1'(f(\cdot),\cdot)\big] \to 0$$

in probability, where P_{n_2} is the empirical measure based on $X_{n_1+1}, \dots, X_n$. Since, as observed earlier, $|\Gamma_n| \le 1/6\|\phi_1^{(3)}\|_\infty \int |f - \hat{f}|^3$, it follows from $H2$ that $E\Gamma_n^2 = O(n_1^{-\beta}) = o(n^{-1})$. Also, denoting by P_{n_2} the empirical measure based on the second portion of the sample, $X_{n_1+1}, \dots, X_n$, and integrating first conditionally on the first n_1 variables,

$$\begin{aligned} nE\Big[(P_{n_2} - P)(\phi_1'(\hat{f},\cdot) - \phi_1'(f,\cdot))\Big]^2 &\le \frac{n}{n_2} E\left[\int (\phi_1'(\hat{f}(x),x) - \phi_1'(f(x),x))^2 f(x)dx\right] \\ &\lesssim 2\|\phi_1'\|_\infty \|\phi_1''\|_\infty \|f\|_\infty E\|\hat{f} - f\|_2^2 \to 0 \end{aligned}$$

by $H2$ and $H4$. These three observations show that it suffices to prove that

$$nE\Big[T_n - T(f) - (P_{n_2} - P)(\phi_1'(\hat{f}(\cdot),\cdot))\Big]^2 \to 0. \tag{5.109}$$

The proof of this limit requires some care: the expression inside the brackets consists of the sum of

$$(P_{n_2} - P)\Big(-\hat{f}\phi_1''(\hat{f},\cdot)\Big) =: \hat{L} - L$$

(where, as earlier, we write $\hat{f}$ for $\hat{f}(\cdot)$) and

$$T_{n_2}^{\psi}(f) - \int \psi f^2 =: \hat{Q} - Q,$$

where T_{n_2} is based on the n_2 variables $X_{n_1+1},\dots,X_n$, and $\psi(x) = (1/2)\phi_1''(\hat{f}(x),x)$, and it turns out that neither of the limits of its three components

$$\lim_{n\to\infty} nE(\hat{L}-L)^2, \quad \lim_{n\to\infty} nE(\hat{Q}-Q)^2 \quad \text{and} \quad \lim_{n\to\infty} nE((\hat{L}-L)(\hat{Q}-Q))$$

is zero, but they cancel each other. We develop some of the computations but not all.

Let E_2 denote conditional expectation given $X_1,\dots,X_{n_1}$. Then

$$\begin{aligned} nE(\hat{L}-L)^2 &= nEE_2(\hat{L}-L)^2 \\ &= \frac{n}{n_2} E\left[\int (\hat{f}\phi_1''(\hat{f},\cdot))^2 f - \left(\int \hat{f}\phi_1''(\hat{f},\cdot)f\right)^2\right]. \end{aligned}$$

Now, $n/n_2 \to 1$, the first summand should clearly converge to $\int f^3(x)\phi_1''(f(x),x))^2 dx$ and the second to $\left(\int f^2(x)\phi_1''(f(x),x)dx\right)^2$, so we should have

$$\lim_{n\to\infty} nE(\hat{L}-L)^2 = \int f^3(x)\phi_1''(f(x),x))^2 dx - \left(\int f^2(x)\phi_1''(f(x),x)dx\right)^2. \tag{5.110}$$

We check convergence of the first summand and that of the second follows in a similar way. Using the hypotheses, we have

$$\begin{aligned} &\left|E\left[\int (\hat{f}\phi_1''(\hat{f},\cdot))^2 f - \int f^3(\phi_1''(f,\cdot))^2\right]\right| \\ &\le \|f\|_\infty E\left[\int |\hat{f}\phi_1''(\hat{f},\cdot) + f\phi_1''(f,\cdot)||\hat{f}\phi_1''(\hat{f},\cdot) - f\phi_1''(f,\cdot)|\right] \\ &\le 2\|f\|_\infty \max(|a-\varepsilon|,|b+\varepsilon|)\|\phi_1''\|_\infty E|\hat{f}\phi_1''(\hat{f},\cdot) - f\phi_1''(f,\cdot)| \\ &\le 2\|f\|_\infty \max(|a-\varepsilon|,|b+\varepsilon|)\|\|\phi_1''\|_\infty(\|f\|_\infty\|\phi_1^{(3)}\|_\infty + \|\phi_1''\|_\infty)E\|\hat{f} - f\|_1 \to 0, \end{aligned}$$

as $n\to\infty$.

To prove that $nE(\hat{Q}-Q)^2$ has the same limit as $nE(\hat{L}-L)^2$, we will use inequality (5.102) in Theorem 5.3.3 with $\psi(x) = (1/2)\phi_1''(\hat{f}(x),x)$ and hence with $Y_i = \phi_1''(\hat{f}(X_i),X_i)f(X_i) - \int \phi_1''(\hat{f},\cdot)f^2$. Since $\|\psi\|_\infty = \|\phi_1''(\hat{f},\cdot)\|_\infty/2 \le \|\phi_1''\|_\infty/2 < \infty$ and $\|f\|_\infty < \infty$, this inequality implies that there exists a finite constant C independent of n such that

$$\begin{aligned} &n_2E_2\left[(\hat{Q}-Q) - \frac{1}{\sqrt{n_2}}\sum_{i=n_1+1}^{n} Y_i\right]^2 \\ &\le C\left[\frac{1}{n_2h} + \|K_h*f-f\|_2^2 + n_2\|K_h*f-f\|_2^4 + \|K_h*(f\phi_1''(\hat{f},\cdot)) - f\phi_1''(\hat{f},\cdot)\|_2^2\right], \end{aligned}$$

where $h = h_n$. As in Theorem 5.3.3, the four terms in this bound tend to zero as $n\to\infty$, but since the fourth term is random, we still need to show that its expected value also tends to zero. By Young's inequalities,

$$\begin{aligned}E\|K_h*(f\phi_1''(\hat f,\cdot))-f\phi_1''(\hat f,\cdot)\|_2^2 &\le E\|K_h*(f\phi_1''(\hat f,\cdot))-K_h*(f\phi_1''(f,\cdot))\|_2^2\\ &+\|K_h*(f\phi_1''(f,\cdot))-f\phi_1''(f,\cdot)\|_2^2+E\|f\phi_1''(f,\cdot)-f\phi_1''(\hat f,\cdot)\|_2^2\\ &\le(1+\|K\|_1)E\|f\phi_1''(f,\cdot)-f\phi_1''(\hat f,\cdot)\|_2^2+\|K_h*(f\phi_1''(f,\cdot))-f\phi_1''(f,\cdot)\|_2^2.\end{aligned}$$

The first summand in the last line tends to zero because $\|f\|_\infty<\infty$, $\|\phi_1^{(3)}\|\le\infty$ and $E\|f-\hat f\|_2^2\to 0$ and the second because $f\phi_1''(f,\cdot))\in L^2$. We thus have proved (recall $n_2/n\to 1$) that

$$nE\left[(\hat Q-Q)-\frac{1}{\sqrt{n_2}}\sum_{i=n_1+1}^{n}Y_i\right]^2\to 0.$$

This and the fact that

$$\begin{aligned}\lim_{n\to\infty}\frac{1}{n_2}E\left[\sum_{i=n_1+1}^{n}Y_i\right]^2 &=\lim_{n\to\infty}E\left[\int(\phi_1''(\hat f,\cdot))^2f^3-\left(\int\phi_1''(\hat f,\cdot))f^2\right)^2\right]\\ &=\int(\phi_1''(f,\cdot))^2f^3-\left(\int\phi_1''(f,\cdot))f^2\right)^2,\end{aligned}$$

which follows by the mean value theorem, boundedness of ϕ_1''' and f and because $E\|\hat f-f\|_2^2\to 0$, imply that

$$\lim_{n\to\infty}nE(\hat Q-Q)^2=\int f^3(x)(\phi_1''(f(x),x))^2dx-\left(\int f^2(x)\phi_1''(f(x),x)dx\right)^2. \qquad (5.111)$$

Computations based on Young's inequality and conditions $H1$–$H4$ similar to those for the limits (5.110) and (5.111) also yield

$$\lim_{n\to\infty}2E(\hat L-L)(\hat Q-Q)=-2\left[\int f^3(x)(\phi_1''(f(x),x))^2dx-\left(\int f^2(x)\phi_1''(f(x),x)dx\right)^2\right],$$

which, together with (5.110) and (5.111) yields the limit (5.109) and concludes the proof of the theorem. ■

Theorem 5.3.4 also holds if the estimator (5.107) is replaced by its wavelet analogue (obtained from $T_{n_2}^{W,\psi}$ (5.103) in the same way as (5.107) is obtained from $T_{n_2}^{\psi}$). See Exercise 5.3.4 for its application to the entropy functional.

5.3.2 Deconvolution

Suppose that we would like to estimate the density of a random variable that cannot be directly observed. A particularly important model for 'indirect observations' is the deconvolution model, which we consider next. In this model, i.i.d. data X_i are contaminated by i.i.d. errors ε_i independent of the variables X_i, and the object is to recover the common density of the variables X_i from these noisy observations, namely, from

$$Y_i=X_i+\varepsilon_i,\quad 1\le i\le n.$$

We denote by f the density of the variables X_i, by φ the probability law of the i.i.d. variables ε_i and by g the density of the variables Y_i,

$$g = f * \varphi.$$

Also, we recall from Chapter 4 the notation $\mathcal{F}[h]$ for the Fourier transform of a function h. Then we have $\mathcal{F}[g] = (\mathcal{F}[f])(\mathcal{F}[\varphi])$, and if $\mathcal{F}[\varphi]$ does not vanish on the support of $\mathcal{F}[g]$, we can solve for $\mathcal{F}[f]$. We could try to estimate $\mathcal{F}[g]$ by dividing the empirical characteristic function based on $\{Y_i\}$ by $\mathcal{F}[\varphi]$ and then apply the inverse Fourier transform; however, for this to work, we would need the quotient to be in $L^1(\mathbb{R})$ or in $L^2(\mathbb{R})$, which is a very vexing hypothesis (we would need to impose not only $|\mathcal{F}[\varphi]|(x) \neq 0$ on all of $\mathbb{R}$ but also $1/\mathcal{F}[\varphi] \in L^1$). We should instead try to apply this strategy not directly to g but to regularisations of g whose Fourier transforms have bounded support. We will do this for kernel and wavelet projection regularisations as usual. Both the kernels K and the scaling and wavelet functions ϕ, ψ will have to be band limited. Meyer-type or band-limited wavelets are discussed in (4.50) and Theorem 4.2.9, and we refer to Exercise 5.3.5 for the existence of band-limited convolution kernels.

Let K be a symmetric function in $L^1(\mathbb{R}) \cap L^2(\mathbb{R})$ integrating to 1 (a kernel) and whose Fourier transform has support contained in $[-a,a]$, let $h > 0$ and assume that $\mathcal{F}[\varphi](x) \neq 0$, for all $x \in \mathbb{R}$. Assume that the density f is bounded and that $\mathcal{F}[f] \in L^1(\mathbb{R})$. Then, by Plancherel's theorem, we have, writing $\tau_x F(\cdot) = F(\cdot - x)$,

$$\begin{aligned}
K_h(f)(x) = K_h * f(x) &= \int \tau_x(K_h)(y) f(y)dy = \frac{1}{2\pi}\int \mathcal{F}[\tau_x(K_h)](t)\overline{\mathcal{F}[f](t)}dt \\
&= \frac{1}{2\pi}\int e^{-ixt}\mathcal{F}[K_h](t)\overline{\mathcal{F}[g](t)/\mathcal{F}[\varphi](t)}dt \\
&= \int g(y)\mathcal{F}^{-1}\left(e^{-ix\cdot}\mathcal{F}[K_h]I_{[-a/h,a/h]}/\overline{\mathcal{F}[\varphi]}\right)(y)dy \qquad (5.112) \\
&= \int g(y)\left[(\tau_x(K_h)) * \mathcal{F}^{-1}(I_{[-a/h,a/h]}/\overline{\mathcal{F}[\varphi]})\right](y)dy \\
&= (g * K_h^*)(x),
\end{aligned}$$

where

$$K_h^* = \left[K_h * \mathcal{F}^{-1}\left[\frac{I_{[-a/h,a/h]}}{\mathcal{F}[\varphi]}\right]\right]^{\sim},$$

with the notation $w^{\sim}(x) = w(-x)$. See Exercise 5.3.6 for a detailed justification of the steps in this computation and to see that $\|K_h^*\|_\infty < \infty$. If Y_i are the i.i.d. noisy observations, with law of density g, and $Q_n = n^{-1}\sum_{i=1}^n \delta_{Y_i}$ are the corresponding empirical measures, it is then natural, in view of (5.112), to estimate the density f by

$$f_n(x) = f_n(x,h) = K_h^* * Q_n(x) = \frac{1}{n}\sum_{i=1}^n K_h^*(x - Y_i), \quad x \in \mathbb{R},\ h > 0, \qquad (5.113)$$

where $h = h_n \to 0$ as $n \to \infty$. Undoing some of the steps in (5.112), we see that

$$\begin{aligned} K_h^*(x - Y_i) &= K_h * \mathcal{F}^{-1}\left[I_{[-a/h,a/h]}/\overline{\mathcal{F}[\varphi]} \right](Y_i - x) \\ &= \mathcal{F}^{-1}\left[\mathcal{F}[K_h] I_{[-a/h,a/h]}/\overline{\mathcal{F}[\varphi]} \right](Y_i - x) = \mathcal{F}^{-1}\left[\mathcal{F}[K_h]/\overline{\mathcal{F}[\varphi]} \right](Y_i - x) \\ &= \frac{1}{2\pi} \int e^{i(Y_i - x)u} \frac{\mathcal{F}[K](hu)}{\overline{\mathcal{F}[\varphi](u)}} du. \end{aligned}$$

Thus, the estimator (5.113) has the alternative expression

$$f_n(x) = f_n(x,h) = \frac{1}{2\pi} \int_{-\infty}^{\infty} e^{-iux} \frac{\mathcal{F}[K](hu)}{\mathcal{F}[\varphi](u)} \frac{1}{n} \sum_{i=1}^{n} e^{iuY_i} du, \quad x \in \mathbb{R},\ h > 0. \tag{5.114}$$

This is the *standard deconvolution kernel density estimator*.

Next, we define the wavelet deconvolution density estimator. We assume that ϕ and ψ are, respectively, the scaling function and the wavelet function of a band-limited ortho-normal multiresolution wavelet basis satisfying the properties specified in Theorem 4.2.9, in particular, such that

$$c_\phi := \left\| \sum_{k \in \mathbb{Z}} |\phi(\cdot - k)| \right\|_\infty < \infty \quad \text{and} \quad c_\psi := \left\| \sum_{k \in \mathbb{Z}} |\psi(\cdot - k)| \right\|_\infty < \infty. \tag{5.115}$$

Let $K_j(x,y) = \sum_k \phi_{jk}(x)\phi_{jk}(y)$, $j = 0, 1, \ldots$, be the wavelet projection kernels for the MRA associated to ϕ and ψ, where $\phi_{0k} = \phi_k$. Then we have, essentially as in (5.112),

$$\begin{aligned} K_j(f)(x) &= 2^j \sum_k \phi(2^j x - k) \int \phi(2^j y - k) f(y) dy \\ &= \sum_k \phi(2^j x - k) \frac{1}{2\pi} \int \overline{\mathcal{F}[\phi_k(2^{-j} t)} \mathcal{F}[f](t) dt \\ &= \sum_k \phi(2^j x - k) \frac{1}{2\pi} \int \overline{\mathcal{F}[\phi_k(2^{-j} t)} \mathcal{F}[g](t) (\mathcal{F}[\varphi](t))^{-1} dt \\ &= 2^j \sum_k \phi(2^j x - k) \int \tilde{\phi}_{jk}(y) g(y) dy = \int K_j^*(x,y) g(y) dy, \end{aligned} \tag{5.116}$$

where the nonsymmetric kernel K_j^* is defined as

$$K_j^*(x,y) = 2^j \sum_k \phi(2^j x - k) \tilde{\phi}_{jk}(y),$$

with

$$\tilde{\phi}_{jk}(x) = \mathcal{F}^{-1}\left[\frac{\mathcal{F}[\phi_k](2^{-j} \cdot)}{2^j \overline{\mathcal{F}[\varphi]}} \right](x) = \phi_{0k}(2^j \cdot) * \mathcal{F}^{-1}\left[\frac{I_{[-a/h,a/h]}}{\overline{\mathcal{F}[\varphi]}} \right](x). \tag{5.117}$$

As earlier, by Young's inequality, $\|\tilde{\phi}_{jk}\|_\infty < \infty$ and, consequently, also $\|K_j^*\|_\infty < \infty$. Then the *wavelet deconvolution density estimator* is defined as

$$f_n(x) = f_n(x,j) = \frac{1}{n} \sum_{i=1}^{n} K_j^*(x, Y_i), \quad x \in \mathbb{R},\ j \geq 0. \tag{5.118}$$

For the next result, which is basic for the application of empirical processes to determine the quality of approximation of f_n to f in the supremum norm, we set

$$\delta_j = \inf_{t\in[-2^j a, 2^j a]} |\mathcal{F}[\varphi](t)|,$$

where $[-a,a]$ is the support of $\mathcal{F}[\phi]$ in the wavelet projection case and that of $\mathcal{F}[K]$ in the convolution kernel case. In this last case, we take, as in other instances, $h = 2^{-j}$ and $K_j := K_{2^{-j}}$. We also set

$$\mathcal{H}_j = \{\delta_j \tilde{\phi}_{jk} : k \in \mathbb{Z}\}, \quad j \in \mathbb{N},$$

in the wavelet projection case and

$$\mathcal{H}_j = \{\delta_j 2^{-j} K_j^*(x - \cdot) : x \in \mathbb{R}\}, \quad j \in \mathbb{N},$$

in the convolution kernel case. Note that since $|\mathcal{F}[\varphi]| \geq 0$ is uniformly continuous, δ_j is strictly positive.

Lemma 5.3.5 *Let φ be probability measure on $\mathbb{R}$ such that $\mathcal{F}[\varphi](x) \neq 0$ for all $x \in \mathbb{R}$. Let K be an even function in $L^1(\mathbb{R}) \cap L^2(\mathbb{R})$ integrating to 1 (a kernel) and whose Fourier transform has support contained in $[-a,a]$ for some $0 < a < \infty$. Let ϕ, ψ be, respectively, the scaling function and the wavelet function of a band-limited ortho-normal multiresolution wavelet basis satisfying the properties specified in Theorem 4.2.9. Let $\mathcal{H}_j$, $j \in \mathbb{N}$, be the collections of functions defined immediately before the lemma via either K or ϕ. Then these classes of functions are uniformly bounded, uniformly in j, and their $L^2(Q)$ covering numbers $N(\mathcal{H}_j, L^2(Q), \varepsilon)$ satisfy*

$$\sup_j \sup_Q N\left(\mathcal{H}_j, L^2(Q), \varepsilon\right) \leq \left(\frac{A}{\varepsilon}\right)^v, \quad 0 < \varepsilon \leq A,$$

for finite positive constants A, v depending only on K in one case and on ϕ and ψ in the other, where the first supremum extends over $j \in \mathbb{N}$ and the second extends over all the Borel probability measures Q on $\mathbb{R}$.

Proof Set

$$\eta_j(x) = \mathcal{F}^{-1}\left[I_{[-2^j a, 2^j a]}/\overline{\mathcal{F}[\varphi]}\right](x),$$

clearly a bounded continuous function. In the convolution kernel case, the functions in $\mathcal{H}_j$ are just translates of the function $2^{-j}\delta_j(K_j * \eta_j)^{\sim}$. This is a bounded function because, by Young's inequality and Plancherel's theorem,

$$2^{-j}\delta_j \|K_j * \eta_j\|_\infty \leq 2^{-j}\delta_j \|K_j\|_2 \|\eta_j\|_2 \leq \sqrt{a/\pi}\,\|K\|_2 < \infty.$$

We will show that this function has finite quadratic variation, and the result will follow from Proposition 3.6.12. Similarly, in the wavelet projection case,

$$\begin{aligned}\tilde{\phi}_{jk}(x) &= (\phi_{0k}(2^j \cdot) * \eta_j)(x) = \int \phi(2^j x - 2^j y - k)\eta_j(y)dy \\ &= \int 2^{-j/2}\phi_{j0}(x - y - 2^{-j}k)\eta_j(y)dy = 2^{-j/2}(\phi_{j0} * \eta_j)(x - 2^{-j}k),\end{aligned}$$

so the functions in $\mathcal{H}_j$ are translates of the function $2^{-j/2}\phi_{j0} * \eta_j$, a function bounded by

$$2^{-j/2}\delta_j\|\phi_{j0} * \eta_j\|_\infty \le \sqrt{a/\pi}.$$

The facts that $2^{-j/2}\phi_{j0}(x) = \phi(2^j x)$ and $2^{-j}K_j(x) = K(2^j x)$ and that both functions ϕ and K have Fourier transforms with support on $[-a,a]$ allow for a unified treatment of the remainder of the proof (assume that $a \ge 4\pi/3$ in the wavelet case). We denote by L either of ϕ or K, that is, a function in $L^1 \cap L^2$ with Fourier transform supported by $[-a,a]$, and consider the function $H_j = \delta_j L(2^j\cdot) * \eta_j$. To prove that this is a function of finite quadratic variation, we use that the Besov space $B^{1/2}_{21}(\mathbb{R})$ is continuously imbedded into the space $V_2(\mathbb{R})$ of functions of bounded 2-variation on $\mathbb{R}$ (see (4.108)). The Besov norm admits the following Littlewood-Paley characterisation (see (4.86), with the functions $\phi, \psi_{2^{-j}}$ there re-labelled here to θ, γ_l, respectively, to avoid confusion):

$$\|h\|_{B^{1/2}_{21}} = \|h * \theta\|_2 + \sum_{\ell \ge 0} 2^{\ell/2}\|\mathcal{F}^{-1}[\gamma_\ell \mathcal{F}[h]]\|_2.$$

Thus, it suffices to show that $\|H_j\|_{B^{1/2}_{21}}$ is bounded uniformly in $j \in \mathbb{N}$. First, we have, using Plancherel's theorem,

$$\|H_j * \theta\|_2^2 \le \|H_j\|_2^2 = \delta_j^2 \int_{-2^j a}^{2^j a} \left|\mathcal{F}[L(2^j\cdot)/\overline{\mathcal{F}[\varphi]}\right|^2 \lesssim \|L(2^j\cdot)\|_2^2 \lesssim 2^{-j/2} \lesssim 1,$$

with uniform constants. For the second part of the norm, by Plancherel's theorem and with the notation $\langle u\rangle = (1+u^2)^{1/2}$, we have

$$\begin{aligned}
&\delta_j \sum_\ell 2^{\ell/2}\left\|\mathcal{F}^{-1}\left[\gamma_\ell \mathcal{F}[L(2^j\cdot) * \eta_j]\right]\right\|_2 \\
&= \frac{1}{\sqrt{2\pi}} 2^{-j}\delta_j \sum_\ell 2^{\ell/2}\left\|\gamma_\ell \mathcal{F}[L](2^{-j}\cdot)I_{[-2^j a, 2^j a]}(\overline{\mathcal{F}[\varphi]})^{-1}\langle u\rangle^{1/2}\langle u\rangle^{-1/2}\right\|_2 \\
&\le C 2^{-j}\delta_j \sum_\ell \left\|\gamma_\ell \mathcal{F}[L](2^{-j}\cdot)I_{[-2^j a, 2^j a]}(\overline{\mathcal{F}[\varphi]})^{-1}\langle u\rangle^{1/2}\right\|_2 \\
&\le C(a)\sum_\ell \|\mathcal{F}^{-1}[\gamma_\ell 2^{-j/2}\mathcal{F}[L](2^{-j}\cdot)]\|_2 \le c(a)\|2^{j/2}L(2^j\cdot)\|_{B^{1/2}_{21}}.
\end{aligned}$$

Note that since the supports of the Meyer wavelet ψ and L are, respectively, contained in $[2\pi/3, 4\pi/3]$ and $[-a,a]$, by Plancherel,

$$\langle \psi_{\ell k}, 2^{j/2}L(2^j\cdot)\rangle = \frac{2^{(\ell+j)/2}}{2\pi}\langle \hat\psi(2^{-\ell}\cdot), \hat L(2^{-j}\cdot)\rangle = 0,$$

for $\ell > j + c$ with $c = \log_2 a - \log_2(4\pi/3)$. Hence, by the wavelet definition (4.89) of the Besov norm (with Meyer wavelets), we can use Theorem 4.3.2 to bound $\|2^{j/2}L(2^j\cdot)\|_{B^{1/2}_{21}}$ by a constant multiple of

$$\|2^{j/2}L(2^j\cdot)\|_{B^{1/2,W}_{21}} = \sqrt{\sum_k |\langle \phi_{0k}, 2^{j/2}L(2^j\cdot)\rangle^2} + \sum_{\ell=0}^{j+c-1}\sqrt{\sum_k |\langle \psi_{\ell k}, 2^{j/2}L(2^j\cdot)\rangle^2}.$$

The first term is the norm of the orthogonal projection of the function $2^{j/2}L(2^j\cdot)$ onto $V_0 \subset L^2$; hence, it is dominated by $\|2^{j/2}L(2^j\cdot)\|_2 = \|L\|_2 < \infty$. For the second term, we note first that the change of variables $2^jx = u$ and the boundedness condition (5.115) give

$$\begin{aligned}\sum_k \langle \psi_{\ell k}, 2^{j/2}L(2^j\cdot)\rangle^2 &= \sum_k \left(2^{\ell/2}2^{j/2}\int \psi(2^\ell x - k)L(2^jx)dx\right)^2 \\ &= \sum_k \left(2^{\ell/2}2^{-j/2}\int \psi(2^{\ell-j}u - k)L(u)du\right)^2 \\ &\le 2^{\ell-j}c_\psi \|L\|_1 \sup_k \left|\int \psi(2^{\ell-j}u-k)L(u)du\right| \\ &\le \tilde{C}(\psi,L)2^{\ell-j},\end{aligned}$$

for some $\tilde{C}(\psi,L)$ depending only on these two functions, where in the last inequality we use that ψ is bounded and L is integrable. Hence,

$$\sum_{\ell=0}^{j+c-1}\sqrt{\sum_k \langle \psi_{\ell k}, 2^{j/2}L(2^j\cdot)\rangle^2} \le \tilde{C}(\psi,L)2^{-j/2}\sum_{\ell=0}^{j+c-1}2^{\ell/2} \le C(\psi,L),$$

for some constant C depending only on ψ and L. This shows that the functions in both definitions of $\mathcal{H}_j$ are uniformly bounded and have uniformly bounded quadratic variation. Then Proposition 3.6.12 proves the lemma. ■

This lemma together with the moment bound for empirical processes in Corollary 3.5.8 or Talagrand's exponential inequality in Theorem 3.3.9 immediately give moment and exponential bounds for the uniform deviation of f_n from its mean, in complete analogy with the bounds in Theorems 5.1.5 and 5.1.15 and Proposition 5.1.13 for the linear kernel and wavelet density estimators.

Theorem 5.3.6 *Under the same assumptions as in Lemma 5.3.5 and with the same notation, if f is a bounded density on $\mathbb{R}$ and $f_n(x,j)$ are the deconvolution density estimators defined by (5.113) with $h = 2^{-j}$ or by (5.118) for $j \ge 0$, and if $2^jj \le Cn$ for some $C < \infty$, then there exist constants L', L'' depending only on K, $p \in [1,\infty)$ and C in the convolution kernel case or on ϕ, ψ, p and C in the wavelet projection case such that*

$$\left[E\|f_n(\cdot,j) - Ef_n(\cdot,j)\|_\infty^p\right]^{1/p} \le \frac{L'}{\delta_j}(\|g\|_\infty^{1/2} \vee 1)\sqrt{2^j(j\vee 1)/n} \tag{5.119}$$

and

$$\Pr\left\{\|f_n(\cdot,j) - Ef_n(\cdot,j)\|_\infty \ge \frac{L''}{\delta_j}\left((\|g\|_\infty^{1/2}\vee 1)\sqrt{(1+u)2^j(j\vee 1)/n} + (1+u)2^j(j\vee 1)/n\right)\right\}$$
$$\le e^{-(1+u)(j\vee 1)}, \tag{5.120}$$

for all $u > 0$.

Proof In the convolution case, the supremum in

$$\|f_n(\cdot,j)-Ef_n(\cdot,j)\|_\infty = \frac{1}{n}\sup_x\left|\sum_{i=1}^n (K_j^*(Y_i-x)-EK_j^*(Y_1-x))\right|$$

is countable because $K_j^*(x)$ is continuous (as it is the value at $-x$ of the convolution of the integrable function K_j with the continuous function η_j). The same is true in the wavelet case because, since $\tilde{\phi}_{jk}$ is bounded and since $\phi \in \mathcal{S}(\mathbb{R})$, the series of continuous functions defining K_j^* converges uniformly. This observation is needed to apply Talagrand's inequality, and moreover, it renders unnecessary the use of outer expectations and probabilities.

In the convolution case, by Lemma 5.3.5, we can apply the moment inequality (3.184) to the empirical processes indexed by $\mathcal{H}_j$ and based, for example, on $\{Y_i\}$. This requires evaluation of the two parameters U and σ. In the convolution case, from the preceding proof, we can take $U=\sqrt{a/\pi}\|K\|_2$, for all j (since $\sup_{h\in\mathcal{H}_j}\|h\|_\infty \le \sqrt{a/\pi}\|K\|_2$), whereas for $\sigma^2 \ge \sup_{h\in\mathcal{H}_j} Eh^2(Y)$, where the distribution of Y has density g, the estimate

$$\begin{aligned}
Eh^2(Y) &= \delta_j^2 2^{-2j}\int (K_j^*(x-y))^2 g(y)dy \le \delta_j^2 2^{-2j}\|g\|_\infty \|K_j^*\|_2^2\\
&= \frac{1}{2\pi}\delta_j^2 2^{-2j}\|g\|_\infty \int_{-2^j a}^{2^j a} |\mathcal{F}[K_j]|^2 |\mathcal{F}[\varphi]|^{-2} du\\
&\le \frac{1}{2\pi}2^{-2j}\|g\|_\infty \int K^2(u/2^j)du = \frac{1}{2\pi}2^{-j}\|g\|_\infty\|K\|_2^2,
\end{aligned} \tag{5.121}$$

valid for all $h\in\mathcal{H}_j$, $j\in\mathbb{N}$, implies that we can take $\sigma^2 = 2^{-j}\|g\|_\infty\|K\|_2^2/(2\pi)$. Now the bound (5.119) for $p=1$ follows from Corollary 3.5.8 because U is constant, σ is of the order $\|g\|_\infty^{1/2}2^{-j/2}$ and $2^j j \le n$. The bound for $p>1$ follows from the one for $p=1$ and Theorem 3.4.1 and the exponential inequality (5.120) from (3.101) in Theorem 3.3.9 (with $x=(1+u)(j\vee 1)$).

In the wavelet case, we note that, by the definitions of f_n, $\tilde{\phi}_{j,k}$ and c_ϕ,

$$\|f_n(\cdot,j)-Ef_n(\cdot,j)\|_\infty \le c_\phi 2^j \sup_{k\in\mathbb{Z}}\left|\frac{1}{n}\sum_{i=1}^n\left(\tilde{\phi}_{jk}(Y_i)-E\tilde{\phi}_{jk}(Y)\right)\right|.$$

Now the theorem in this case follows by applying the moment and exponential inequalities for empirical processes to the classes of functions $\mathcal{H}_j=\{\delta_j\phi_{jk}: k\in\mathbb{Z}\}$, which is possible by Lemma 5.3.5 as well. The estimation of σ^2 works in complete analogy with the convolution case and yields the same value for σ up to a constant. Also, in this case, as shown in the proof of the preceding lemma, U is again a constant independent of j. Hence, the same moment and exponential bounds hold in the wavelet projection case. ■

To obtain a rate of approximation of f by f_n, it remains to consider the bias. Since by the computations in (5.112) and (5.116), both in the convolution kernel and in the wavelet projection cases, we have

$$f(x)-Ef_n(x,j) = f(x)-K_j(f)(x), \quad x\in\mathbb{R}, \tag{5.122}$$

Proposition 4.3.8 applies (just as for linear wavelet estimators). Using this proposition, in the next theorem we obtain rates of convergence of f_n to f in the supremum norm for Hölder continuous densities f under typical decay conditions on $\mathcal{F}[\varphi]$. Two cases are considered: the 'severely ill-posed' case, corresponding to exponential rates of decay of $\mathcal{F}[\varphi]$, and the 'moderately ill-posed' case, where $\mathcal{F}[\varphi]$ decays polynomially. Recall the definition of the order of a convolution kernel, just before Proposition 5.1.7. In the following theorem, f_n refers to the deconvolution kernel and wavelet projection estimators defined, respectively, by (5.113) (or equivalently by (5.114)) and (5.118).

Theorem 5.3.7 *Assume that the band-limited kernel $K \in L^1 \cap L^2$ is of order $S \geq 1$. In the wavelet case, assume that ϕ, ψ are the scaling and wavelet functions defining a band-limited ortho-normal multiresolution wavelet basis satisfying the properties in Theorem 4.2.9. Let $B(s,L)$ denote the ball of radius L about the origin in the Besov space $B^s_{\infty\infty}$. Then*

(a) If $|\mathcal{F}[\varphi](t)| \geq Ce^{-c_0|t|^\alpha}$ for all $t \in \mathbb{R}$ and some C, c_0, α positive, and if $j_n = \alpha^{-1}\log_2(\nu \log n)$ for some ν such that $c_0 a^\alpha \nu < 1/2$, (where a is such that $[-a,a]$ contains supp($\hat{K}$) or supp($\hat{\phi}$)), then there exists a constant L' depending only on s, L, c_0, C, α, ν and K or ϕ and ψ such that, for all $n \geq 2$ and all $s > 0$ in the wavelet case and $s < S$ in the convolution kernel case,

$$\sup_{f \in B(s,L)} E\|f_n(\cdot,j_n) - f(\cdot)\|_\infty \leq L'\left(\frac{1}{\log n}\right)^{s/\alpha}.$$

(b) If $|\mathcal{F}[\varphi](t)| \geq C(1+t^2)^{-w/2}$ for all $t \in \mathbb{R}$ and some C, w nonnegative, and if j_n is such that $2^{j_n} \simeq (n/\log n)^{1/(2s+2w+1)}$, then there exists a constant L'' depending only on s, L, C, w and K or ϕ and ψ such that, for all $n \geq 2$ and all $s > 0$ in the wavelet case and $0 < s < S$ in the convolution kernel case,

$$\sup_{f \in B(s,L)} E\|f_n(\cdot,j_n) - f(\cdot)\|_\infty \leq L'\left(\frac{\log n}{n}\right)^{s/(2s+2w+1)}.$$

Proof Part (a): The convolution kernel K satisfies Condition 4.1.4 for $N = S$, and therefore, in view of (5.122), Proposition 4.3.8 implies that for $0 < s < S$, under the assumptions in (a),

$$\|f(\cdot) - Ef_n(\cdot,j_n)\|_\infty \leq C(K)\|f\|_{B^s_{\infty\infty}} 2^{-j_n s} = C(K)\|f\|_{B^s_{\infty\infty}}\left(\frac{1}{\nu \log n}\right)^{s/\alpha}.$$

If K is the wavelet projection kernel associated to ϕ, then, by Theorem 4.2.9, K satisfies Condition 4.1.4 for any $N \in \mathbb{N}$, which implies, also by Proposition 4.3.8, that this bias bound holds for every $s > 0$ when f_n is the deconvolution wavelet density estimator of f. Moreover, Theorem 5.3.6 gives

$$\begin{aligned} E\|f_n(\cdot,j_n) - Ef_n(\cdot,j_n)\|_\infty &\leq \frac{L'}{C} e^{c_0 a^\alpha 2^{\alpha j_n}} (\|g\|_\infty \vee 1)\sqrt{2^{j_n}(j_n \vee 1)/n} \\ &\leq \frac{L'}{C} n^{c_0 a^\alpha \nu - 1/2}\sqrt{(\nu \log n)^{1/\alpha}(1 \vee \alpha^{-1}\log_2(\nu \log n)} = o(n^{-\delta}), \end{aligned}$$

for any $0 < \delta < c_0 a^\alpha \nu - 1/2$. Then the conclusion in part (a) of the theorem follows by these two bounds and the triangle inequality.

Part (b): The proof of part (b) is very similar to that of part (a) and is omitted. ■

In the severely ill posed problem and for the supremum norm, this theorem obtains only logarithmic rates of recovery of f, which may be impractical. This is unfortunate because the severely ill posed case includes some very common candidates for φ, such as the normal or the stable distributions. It is interesting to observe that if the density f is assumed to be 'supersmooth', then these rates become faster than any power of the logarithm, in fact, even polynomial for choices of ϕ related to the smoothness of f. We illustrate this phenomenon in the next proposition (see also Exercise 5.3.9).

For α, s, L positive, the class $\mathcal{A}_{\alpha,s}(L)$ of supersmooth densities is defined as

$$\mathcal{A}_{\alpha,s}(L) = \left\{ f : \mathbb{R} \mapsto [0,\infty), \int f = 1, \int |\mathcal{F}[f](t)|^2 e^{2\alpha|t|^s} \leq 2\pi L \right\}.$$

Recall that if ϕ, ψ are as in Theorem 4.2.9, there exist a, a' positive such that the support of $\mathcal{F}[\phi]$ is contained in $[-a,a]$ and the support of $\mathcal{F}[\psi]$ has null intersection with $[-a',a']$.

Proposition 5.3.8 *For $s > 0$, $\alpha > 0$, $\beta \geq 0$, $w \geq 0$, $L > 0$ and $b > 0$, let $f \in \mathcal{A}_{\alpha,s}(L)$, and assume that $\mathcal{F}[\varphi]$ satisfies*

$$\mathcal{F}[\varphi](t) \geq b(1+t^2)^{-w/2} e^{-\beta|t|^r}, \quad t \in \mathbb{R},$$

for some $r > 0$. Assume that $\beta = 0$ or $r = s$. Let ϕ and ψ be as in Theorem 4.2.9, and let $f_n(x,j)$ be the corresponding deconvolution wavelet density estimator of f. Then there exist $c', c > 0$ depending, respectively, on ϕ, ψ, α, s and on ϕ and ψ such that, for all $n, j > 0$ satisfing $2^j j < n$,

$$E\|f_n(\cdot,j) - f\|_\infty \leq c'\sqrt{L} e^{-\alpha(a')^s 2^{js}} 2^{j(1-s)/2} + \frac{c(\|g\|_\infty \vee 1)}{\delta_j}\sqrt{\frac{2^j j}{n}}.$$

Consequently, taking $2^{j_n} = [1/2(\alpha(a')^s + \beta a^s)\log n]^{1/s}$, we have, for $n \geq e^e$,

$$E\|f_n(\cdot,j_n) - f\|_\infty \leq C \frac{(\log n)^{(w+1/2)/s}(\log\log n)^{1/2}}{n^{\alpha(a')^s/(2\alpha(a')^s + 2\beta a^s)}},$$

where C depends on ϕ, ψ, α, β, s, w, b and $\|g\|_\infty$.

Proof The second part follows from the first by just replacing j_n by its value. The bound in the first part consists of a bias bound that we obtain later and of the 'variance' bound (5.119).

The bias bound is as follows: by (5.122), Plancherel and the fact that f is supersmooth, we have

$$\begin{aligned}\|Ef_n(\cdot,j) - f(\cdot)\|_\infty &= \|K_j(f) - f\|_\infty \leq c_\psi \sum_{\ell \geq j} 2^{\ell/2} \sup_k |\beta_{\ell k}(f)| \\ &\leq c' \sum_{\ell \geq j} 2^{\ell/2} \sup_k \int |\mathcal{F}[\psi_{\ell k}](u)||\mathcal{F}[f](u)| du\end{aligned}$$

$$
\begin{aligned}
&= c' \sum_{\ell \ge j} \int |\mathcal{F}[\psi](2^{-\ell}u)||\mathcal{F}[f](u)|du \\
&\le c' \|\psi\|_1 \sum_{\ell \ge j} \int_{[-2^\ell a', 2^\ell a']^c} |\mathcal{F}[f](u)| e^{\alpha |u|^s} e^{-\alpha |u|^s} du \\
&\le c'' \|\psi\|_1 \sqrt{L} \sum_{\ell > j} \left(\int_{2^\ell a'}^{\infty} e^{-2\alpha |u|^s} du \right)^{1/2} \\
&\le c''' \sqrt{L} \sum_{\ell \ge j} (2^\ell a')^{(1-s)/2} e^{-2\alpha (2^\ell a')^s / 2} \le c^{iv} \sqrt{L} (2^j a')^{(1-s)/2} e^{-2\alpha (2^j a')^s / 2},
\end{aligned}
$$

where the next to last inequality follows, for example, by L'Hôpital's rule. ■

Note that the supremum norm recovery rate of a supersmooth density is, up to logarithmic terms, of the order $n^{-1/2}$ in the moderately ill-posed case and slower but still of the order n to a negative power in the severely ill-posed case.

Next, we look at the size of the MISE of f_n. We could as well consider moments other than 2, that is, $\|f_n - f\|_p$ for $1 \le p < \infty$, but given the developments in this subsection and those in Section 5.1.1, this would be repetitive. And for $p = 2$, again to avoid repetition, the exponential bound is left as an exercise, and only the moment bound is developed here. Whereas for supremum norm bounds it is natural to assume the density in $B^s_{\infty\infty}$ (or in the Hölder spaces $C^s \subseteq B^s_{\infty\infty}$), it is more natural for L^2 bounds to assume f in $B^s_{2\infty}$ (recall that $B^s_{2\infty}$ contains the Sobolev-Hilbert space H^s_2).

Theorem 5.3.9 *Under the same notation for f, φ, g, K, ϕ, ψ and $f_n(x,j)$ and the assumptions as in Lemma 5.3.5, both for deconvolution kernel and wavelet density estimators, we have*

$$E\|f_n - Ef_n\|_2^2 \le C \frac{2^j}{\delta_j^2 n},$$

where $C = 1/2\pi$ in the wavelet case and $C = \|K\|_2^2/2\pi$ in the convolution case. Now let $B(s,L)$ denote the ball of radius L about the origin in the Besov space $B^s_{2\infty}$ ($s > 0$). Then the following hold:

(a) If $|\mathcal{F}[\varphi](t)| \ge Ce^{-c_0|t|^\alpha}$ for all $t \in \mathbb{R}$ and some C, c_0, α positive, and if $j_n = \alpha^{-1}\log_2(v \log n)$ for some v such that $c_0 a^\alpha v < 1/2$, then there exists a constant L' depending only on s, L, c_0, C, α, v and K or ϕ and ψ such that, for all $n \ge 2$ and all $s > 0$ in the wavelet case and $s < S$ in the convolution kernel case,

$$\sup_{f \in B(s,L)} E\|f_n(\cdot, j_n) - f(\cdot)\|_2^2 \le L' \left(\frac{1}{\log n} \right)^{2s/\alpha},$$

and

(b) If $|\mathcal{F}[\varphi](t)| \ge C(1+t^2)^{-w/2}$ for all $t \in \mathbb{R}$ and some C, w nonnegative, and if j_n is such that $2^{j_n} \simeq (n/\log n)^{1/(2s+2w+1)}$, then there exists a constant L'' depending only on s, L, C, w and K or ϕ and ψ such that, for all $n \ge 1$ and all $s > 0$ in the wavelet case and

$0 < s < S$ in the convolution kernel case,

$$\sup_{f\in B(s,L)} E\|f_n(\cdot,j_n)-f(\cdot)\|_2^2 \le L'\left(\frac{1}{n}\right)^{2s/(2s+2w+1)}.$$

Proof Using (5.121) in the convolution kernel case, we obtain, for the 'variance part' of the mean integrated squared error,

$$\begin{aligned}E\|f_n(\cdot,j)-Ef_n(\cdot,j)\|_2^2 &= \frac{1}{n}\int E(K_j^*(x-Y)-EK_j^*(x-Y))^2dx \le \frac{1}{n}\int E(K_j^*(x-Y))^2dx\\ &= \frac{1}{n}\int\int (K_j^*(x-y))^2 g(y)dydx = \frac{1}{n}\int (K_j^*(u))^2du\\ &\le \frac{\|K\|_2^2}{2\pi}\frac{2^j}{\delta_j^2 n},\end{aligned}$$

whereas for the bias part, using the observation (5.122) and the bound in Proposition 4.3.8, we have, for $s < S$,

$$\|f - Ef_n\|_2 \le C(K)\|f\|_{B^s_{2\infty}} 2^{-j_n s}.$$

Combining these two estimates, we obtain the results for deconvolution kernel estimators.

In the wavelet case, the bias inequality holds for all $s > 0$, again by Proposition 4.3.8 and because the projection kernel in this case is of order S for all S (Theorem 4.2.9). The variance inequality is similar to the preceding case, and left to the reader. ■

Note the improvement of the L^2 rate over the L^∞ rate by a logarithmic factor when the deconvolution problem is only moderately ill posed, just as for the linear density estimator in the absence of contamination. The rate of decay of $E\|f_n - f\|_2$ for supersmooth densities, as in the case of the supremum norm, also can be made to decay at polynomial rates, even in the severely ill-posed case (see Exercise 5.3.10).

Exercises

5.3.1 Complete the proof of Theorem 5.3.4. *Hint*: Decompose Q as $2Q_1 - Q_2$ and likewise for $\hat{Q}$, based on the definition of T_n^ψ.

5.3.2 Prove versions of Theorems 5.3.2 and 5.3.3 for the wavelet estimators T_n^W and $T_n^{W,\psi}$ defined by (5.103), including the case of boundary-corrected wavelets on $[0,1]$.

5.3.3 (Continued from Exercise 5.3.2.) Use the estimators from Exercise 5.3.2 to define an estimator of $\int_0^1 \phi(f(x),x)f(x)dx$ analogue to T_n in (5.107) satisfying the central limit theorem in Theorem 5.3.4.

5.3.4 Use the preceding exercise to construct an estimator of the Shannon entropy functional $\int_0^1 f\log f$ for a density f on $[0,1]$ bounded from below by some positive constant and such that $f \in B^\alpha_{2,\infty}([0,1]) \cap B^r_{\infty,\infty}([0,1])$ for some $\alpha \in (1/4,1)$ and some $r > 0$. *Hint*: Start with the Taylor expansion $\int f\log f = -\frac{1}{2}\int\hat{f} + \int\log(\hat{f})f + \frac{1}{2}\int\frac{f^2}{\hat{f}} + \Gamma_n$.

5.3.5 Let $\ell = I_{[-1/2,1/2]}$. Then $\ell*\ell$ is the tent function on $[-1,1]$, prove that $\mathcal{F}^{-1}[\ell](x) = (\sin x)/(\pi x)$, $\mathcal{F}^{-1}[\ell*\ell](x) = (\sin x)^2/(\pi x)^2$ and that therefore dividing the function $(\sin x)^2/(\pi x)^2$ by its L^1-norm, we obtain a positive function $K(x)$ that integrates to 1 and whose Fourier transform has compact support.

5.3.6 Justify (5.112), and show that the kernel K_h^* is bounded. *Hints*: (a) The second identity in the first line follows by Plancherel's identity; (b) $\mathcal{F}[g] \in L^1 \cap L^2$ (by Young's inequality and since $\|\mathcal{F}[g]\|_1 \le \|\mathcal{F}[f]\|_1$ because $\|\mathcal{F}[\varphi]\|_\infty \le 1$), $\mathcal{F}[K_h]$ is bounded and has compact support, and $\mathcal{F}[\varphi]$ being continuous (φ is probability density) and nonzero on it is bounded away form zero on any compact set; hence, both $\mathcal{F}[g]$ and $e^{-ix\cdot}\mathcal{F}[K_h]I_{[-a/h,a/h]}/\mathcal{F}[\varphi]$ are in $L^1 \cap L^2$, which justifies the identity between the second and third lines; (c) K_h^* is bound by Young's inequality.

5.3.7 (Uniform fluctuations of the empirical wavelet coefficients in the deconvolution wavelet density estimator.) Show that the entropy bounds obtained for the classes $\mathcal{H}_j$ in Lemma 5.3.5 hold as well for $\mathcal{H}_j(\psi) = \{\delta_j \tilde{\psi}_{jk} : k \in \mathbb{Z}\}$, where $\tilde{\psi}_{\ell k}$ is defined as $\tilde{\phi}_{\ell k}$ in (5.117) with ϕ replaced by ψ. Use this and methods from this section to show that if $\beta_{\ell k} = \beta_{\ell k}(f)$ and $\hat{\beta}_{\ell k} = 2^{\ell/2}\sum_{i=1}^n \tilde{\psi}_{\ell k}(Y_i)/n$, then, under the assumptions of Theorem 5.3.6, we have both

$$\left(E\sup_k |\tilde{\beta}_{\ell k} - \beta_{\ell k}|^p\right)^{1/p} \le L\frac{1}{\delta_\ell}\left((\|g\|_\infty \vee 1)\sqrt{(\ell \vee 1)/n} + 2^{\ell/2}(\ell \vee 1)/n\right),$$

for all $1 \le p < \infty$, and

$$\Pr\left\{\sup_k |\tilde{\beta}_{\ell k} - \beta_{\ell k}| \ge \frac{D}{\delta_\ell}\left((\|g\|_\infty \vee 1)\sqrt{((1+u)\ell \vee 1)/n} + (1+u)2^{\ell/2}(\ell \vee 1)/n\right)\right\}$$
$$\le e^{-(1+u)(\ell \vee 1)},$$

for all $u > 0$ and for some constants $L = L(p, \phi, \psi)$ and $D = D(\phi, \psi)$. (One cannot proceed to prove these inequalities as in Proposition 5.1.8 due to the lack of symmetry of the convolution 'projection kernel' in the deconvolution case.)

5.3.8 Complement Theorem 5.3.9 with the corresponding exponential bounds for $\|f_n - Ef_n\|_2$ for deconvolution density estimators.

5.3.9 If $r \neq s$ in Proposition 5.3.8, one still obtains fast rates: if $s < r$ (and $\beta \neq 0$), take $2^{j_n} \simeq ((3\beta a^r)^{-1}\log n)^{1/r}$ to obtain a rate faster than any negative power of the logarithm for the bias term and a faster rate for the 'variance' term; if $s > r$, take $2^{j_n} \simeq ((2\alpha(a')^s)^{-1}\log n)^{1/s}$ to obtain a rate faster than $n^{-\tau}$ for any $0 < \tau < 1/2$ for the variance and a faster rate for the bias.

5.3.10 Define K_j^* by its Fourier transform, $\mathcal{F}[K_j^*] = I_{[-2^j,2^j]}/\mathcal{F}[\varphi]$, and let $f_n(x,j) = \sum_{i=1}^n K_j^*(x - Y_i)/n$ as in (5.113) so that $Ef_n(x,j) = K_j^* * g(x)$ (notation from the section on deconvolution). (a) Show that if the density f is in $A_{\alpha,r}(L)$, then $\|Ef_n(\cdot,j) - f(\cdot)\|_2^2 = (1/2\pi)\int_{-2^j}^{2^j}|\mathcal{F}[f]|^2(t)dt \le Le^{-2\alpha 2^{rj_n}}$. (b) If $|\mathcal{F}[\varphi](u) \ge b(1 + |u|^2)^{-w/2}e^{-\beta|u|^s}$, $u \in \mathbb{R}$, then $E\|f_n(\cdot,j) - Ef_n(\cdot,j)\|_2^2 \le (\|g\|_\infty^2/n)\int_{-2^j}^{2^j} du/|\mathcal{F}[\varphi])|^2(u) \le C(\|g\|_\infty^2/n)2^{2w+s-1}e^{2\beta 2^{sj}}$ for some $C < \infty$ depending only on b, s, w. (c) Use (a) and (b) to obtain, for suitably chosen j_n, rates of decay for $E\|f_n(\cdot,j) - Ef_n(\cdot,j)\|_2$ slightly better than for $E\|f_n(\cdot,j) - Ef_n(\cdot,j)\|_\infty$ in Proposition 5.3.8 and in Exercise 5.3.11.

5.3.11 Prove the analogue of Theorem 5.3.9 for the pth moment of the deconvolution wavelet density estimator. (Use methods from Sections 5.1.1 and 5.3.2.)

5.4 Notes

Section 5.1 Convolution kernel density estimators were introduced by Akaike (1954), Rosenblatt (1956) and Parzen (1962) and orthogonal projection estimators by Čencov (1962). The moment and exponential inequalities for L_p-norms of density estimators in Theorems 5.1.5 and 5.1.13 are mostly taken from Giné and Nickl (2011) and its supplementary material (doi:10.1214/11-AOS924SUPP), with exceptions as follows: the moment inequality in the case $1 \le p < 2$ for convolutions comes from Giné and Mason (2007), and the moment and exponential inequalities in the case $p = \infty$ for convolutions come from Giné and Guillou (2002) and from Giné and Nickl (2009) for wavelet projection estimators. Lemma 5.1.3 comes from Giné and Mason (2007), and it is based on work in Devroye (1987, 1992). Devroye (1991) showed that the exponential inequality for bounded differences produces best possible results for the L^1 norm of $f_n - f$, and we base Remark 5.1.14 on his observation. Kerkyacharian and Picard (1992) showed that Besov spaces provide a natural framework for function estimation in L^p norms ($p < \infty$); see also Chapter 6.

The stability result for the L^1-norm of $f_n - f$ was proved by Devroye (1991) and Devroye and Lugosi (2001). Stute (1982, 1984) proved exact a.s. limit theorems for the supremum norm over an interval $\|f_n - Ef_n\|_{[a,b]^d}$ in one and several dimensions and for convolution kernel estimators: see Einmahl and Mason (2000) for a different approach using modern empirical processes. These were extended to limits for suprema over the whole of $\mathbb{R}^d$ by Giné and Guillou (2002) and by Deheuvels (2000) for $d = 1$. Giné and Nickl (2009) extended these results to wavelet density estimators. Here is the exact a.s. limit theorem, in one dimension, both for convolution kernel estimators and for wavelet projection estimators: write $K(x,y)$ for $K(x-y)$ in the convolution kernel case.

Theorem 5.4.1 *Under the assumptions in Proposition 5.1.19 (hence for both wavelet and convolution estimators) and with the same notation,*

$$\lim_{n\to\infty} \sqrt{\frac{n}{2(\log 2) j_n 2^{j_n}}} \sup_{y\in\mathbb{R}} \left| \frac{f_n(y) - Ef_n(y)}{\left(\int_{\mathbb{R}} K^2(2^{j_n}y,x)dx\right)^{1/2}} \right| = \|f\|_\infty^{1/2} \ a.s.$$

Note that the denominator equals $\|K\|_2$ in the convolution kernel case and $\sqrt{\sum_k \phi^2(2^{j_n}y-k)}$ in the wavelet cases. When the supremum is taken over an interval where f is bounded from below above zero, then we may replace $\|f\|_\infty$ by 1 and have $\sqrt{f(y)}$ multiplying in the denominator within the absolute value. For other weights besides $\sqrt{f(y)}$, valid over the whole line, see Giné, Koltchinskii and Zinn (2004). There are versions of this theorem in the convolution kernel case with a limsup bound holding uniformly over bandwidths $(\log n)/n \le h \le 1$; see Einmahl and Mason (2005).

Bickel and Rosenblatt (1973) proved a distributional limit theorem for the supremum norm of $f_n - f$ over an interval similar to Proposition 5.1.23 (under stronger assumptions) – this work was preceded by Smirnov (1950), who obtained a similar result for a histogram density estimator. Giné and Nickl (2010) show how the KMT representation together with Theorem 2.8.3 for cyclo-stationary Gaussian processes implies an analogue for wavelet density estimators, and this approach combined with technical results in Giné, Günturk and Madych (2011) and Bull (2013) obtains the limit theorem for wavelet density estimators based on the Battle-Lemarié spline wavelets and on Daubechies wavelets, respectively. Proposition 5.1.21 is from Giné, Koltchinskii and Sakhanenko (2004) for convolution kernels and is adapted to wavelet projection kernels in Giné and Nickl (2010). Rio (1994) extended to several dimensions the Bickel-Rosenblatt result as a consequence of a KMT-type Gaussian approximation for empirical processes over *VC* classes. See also the more recent work of Chernozhukov, Chetverikov and Kato (2014), where approximation of the empirical process by a Gaussian process is replaced by approximation of the supnorm of one process by the supnorm of the other, with better rates of approximation. A full proof of the KMT theorem can be found in the second edition of Dudley (1999) (Dudley (2014)).

For limit theorems for the L^p norm of density estimators with $1 \le p < \infty$ – a topic not treated here – see Bickel and Rosenblatt (1973), Csörgö and Hórvath (1988), Beirlant and Mason (1995), Giné, Mason and Zaitsev (2003) and Giné and Madych (2014).

Section 5.2 The plug-in property was put forward by Bickel and Ritov (2003), and they also made the observation that the plug-in property for the distribution function requires a kernel of larger order than for the minimax rate in L^2 loss. The setup of Section 5.2.1 with the distinction between weak and strong metrics originates in Nickl (2007). Both these papers contain applications to interesting statistical problems. See Chapter 7 for derivations of such results for maximum likelihood estimators. Van der Vaart (1994) proved Lemma 5.2.2 and Proposition 5.2.3, whereas the rest of the results given in this subsection for kernel density estimators come from Giné and Nickl (2008, 2009a). In particular, the first of these articles shows that any increase of more than 1/2 in the order of the kernel suffices for the plug-in property with respect to both Sobolev balls for $s > 1/2$ and bounded variation balls (thus including distribution functions). The second article contains the exponential inequality for the discrepancy between P_n^K and P_n, which is shown to be sharp. The same inequality for wavelet density estimators is given in Giné and Nickl (2009), and in fact, the plug-in property for wavelet density estimators over most Besov balls that are P-Donsker is also proved in this article (Section 5.2.1 contains only part of this result). Giné and Nickl (2008) show as well that under certain conditions on f_0 and h_n, $\sqrt{n}(P_n^K - P)$ converges in law uniformly over the unit ball of B_{11}^s, $1/2 < s < 1$, which is P-pre-Gaussian but not P-Donsker (see Section 4.4). The proof of this last result uses a variation of Theorem 3.7.52 in combination with sharp Fourier analytical computations. Radulović and Wegkamp (2009) also consider convergence in law of smoothed empirical processes over P-pre-Gaussian not necessarily P-Donsker classes. See Yukich (1992) for earlier results on smoothed empirical processes. The results on smoothed empirical processes on multiscale spaces come from Castillo and Nickl (2014).

Section 5.3 The problem of estimating $\int f^2$ (and even $\int (f^{(k)})^2$) has been considered by several authors, motivated by the fact that these functionals appear in the asymptotic variance of important statistics and also because they constitute relatively simple examples of nonlinear functionals. Bickel and Ritov (1988) obtained optimal rates for this problem by means of (a version of) the debiasing device described above for $\int \psi f^2$. Giné and Nickl (2008a) proved Theorem 5.3.2 showing in particular that in the case $\psi = 1$ there is no need for debiasing: carefully using the smoothness of $f * f$ instead of only that of f, they prove that the bias of the simple estimator (5.93) is already small enough. Estimation of integrals of smooth functions has been considered by Ibragimov and Khasminskii (1978), Levit (1978), Ibragimov, Nemirovski and Khasminskii (1987), Donoho and Nussbaum (1990), Nemirovski (1990, 2000), Birgé and Massart (1995), Laurent (1996), Cai and Low (2005), Klemela (2006), among others. The exposition above adapts Laurent's – who works with orthogonal projection estimators – to kernel estimators.

Deconvolution problems have been extensively studied at least since the 1980's. See Meister's (2009) monograph and Cavalier's (2008) survey paper on inverse problems. The deconvolution kernel density estimator studied here comes from Stefanski and Carroll (1990), where it is proven to be pointwise consistent. The recovery rates in L^2 norm for this estimator were obtained by Fan (1991) and shown to be minimax in Fan (1993), when the density is assumed to belong to a Sobolev space. The observation that for supersmooth densities the rates are better than logarithmic even in the extremely ill posed case, that includes the important case when the errors are normal, was made by Pensky and Vidakovic (1999), by means of deconvolution wavelet density estimators base on Meyer wavelets, and Butucea and Tsybakov (2008a) constructed kernel estimators also achieving these rates (exercise 6 is based on their article), which they prove in (2008b) to be optimal even with respect to constants. The sup norm rates and the key Lemma 5.3.5 were obtained in Lounici and Nickl (2011) with (band limited) wavelet estimators, and it is shown here how their result extends as well to kernel estimators (see Ray (2010) for another approach to deconvolution with kernels). The exposition above is close to the Lounici and Nickl article for the supremum norm and to the Pensky and Vidakovic article for the MISE.

One shows, just as in Chapter 6 for the regular function estimation problems (density estimation and white noise), that the convergence rates for functional estimation and deconvolution problems obtained in Propositions/Theorems 5.3.2, 5.3.3, 5.3.7, 5.3.8, 5.3.9 are optimal from a minimax point of view. The ideas are similar to those in Chapter 6 and full proofs can be found in Birge and Massart (1995), Laurent (1996), Butucea and Tsybakov (2008ab), Lounici and Nickl (2011). Likewise, using now classical ideas in semi-parametric statistics (e.g., Chapter 25 in van der Vaart (1998)), one can show that the limiting covariances obtained in Propositions/Theorems 5.3.1, 5.3.2, 5.3.3 , 5.3.4 are optimal in the sense that they attain the Cramer-Rao lower bounds for these estimation problems. See Ibragimov, Nemirovski and Khasminskii (1987), Nemirovski (1990), and also the appendix in Laurent (1996) for some specific calculations.

6

The Minimax Paradigm

In this chapter we consider a statistical experiment $\mathcal{E}$ whose distribution P_f is indexed by an infinite-dimensional parameter $f \in \mathcal{F}$. We concentrate on the cases of i.i.d. observations $X_1,\ldots,X_n$, where P_f is equal to the infinite product probability measure $P_f^{\mathbb{N}}$ on some product sample space, or, alternatively, on an observation dY in Gaussian white noise of law P_f^Y. In both situations, the goal is to make statistical inference on the unknown function f generating the observations. We shall analyse this problem under the assumption that f satisfies a natural nonparametric regularity condition such that it is contained in a Hölder or Sobolev space. The minimax paradigm provides a coherent setting for the statistical theory of such nonparametric models – it searches for best possible procedures that have guaranteed performance for all elements simultaneously in the given parameter space. In other words, the minimax paradigm takes into account that results about the performance of a statistical procedure should *not* depend on the particular true function f but should hold *no matter what* f is, as long as f belongs to the model used. Given this (maximal) constraint, one searches for the statistical procedure with minimal risk – hence the name *minimax*. In some sense, the minimax paradigm goes hand in hand with the general nonparametric 'philosophy' that statistical procedures should work for 'many' f instead of just for a few specific ones. Following this paradigm, we develop in this chapter a basic theory of nonparametric inference, naturally dividing our treatment into separate but closely related sections on nonparametric testing problems, estimation problems and the construction of nonparametric confidence sets. The proofs rely on the techniques developed in previous chapters, particularly wavelet theory and concentration inequalities.

6.1 Likelihoods and Information

The complexity of a statistical problem and the information it contains can be measured by the difficulty to test hypotheses that arise in it. Let us start by illustrating this by a simple example. For f_0, f_1 two elements in a parameter space $\mathcal{F}$ indexing the laws $(P_f : f \in \mathcal{F})$ of a statistical experiment $\mathcal{E}$, consider testing the hypothesis

$$H_0 : f = f_0 \quad \text{against} \quad H_1 : f = f_1,$$

based on observations Y from the experiment. Let $\Psi(Y)$ be any such test, that is, a measurable function of Y that takes values 0 or 1. We adopt the notational convention that $\Psi(Y) = 0$ means to accept H_0 and $\Psi(Y) = 1$ means to reject H_0. The sum of type 1 and type 2 errors of such a test is then

$$E_{f_0}\Psi(Y)+E_{f_1}(1-\Psi(Y)) = P_{f_0}(\Psi=1)+P_{f_1}(\Psi=0).$$

A trivial upper bound for this sum is 1, corresponding to the case where one takes a test that always accepts one of the hypotheses or where one flips a coin to accept H_0 or not. A good test should allow us to control the preceding sum at any given level $0<\alpha<1$. Whether or not this is possible depends strongly on the nature of the hypotheses. Assume that P_{f_1} is absolutely continuous with respect to P_{f_0}, or otherwise replace E_{f_1} later by the expectation $E^a_{f_1}$ with respect to the absolutely continuous part $P^a_{f_1}$ of P_{f_1}, and use $E_{f_1}(1-\Psi)\geq E^a_{f_1}(1-\Psi)$. We can lower bound the sum of type 1 and type 2 errors, for any $\eta>0$, by

$$\begin{aligned}E_{f_0}\Psi(Y)+E_{f_1}(1-\Psi(Y)) &= E_{f_0}\left[\Psi(Y)+(1-\Psi(Y))\frac{dP_{f_1}}{dP_{f_0}}(Y)\right]\\ &\geq (1-\eta)P_{f_0}\left(\frac{dP_{f_1}}{dP_{f_0}}(Y)\geq 1-\eta\right),\end{aligned}\tag{6.1}$$

where dP_{f_1}/dP_{f_0} is the likelihood ratio (Radon-Nikodym derivative) of the probability measure P_{f_1} with respect to P_{f_0}. Intuitively, a likelihood ratio close to 1 indicates that observations drawn from P_{f_1} look very much like those drawn from P_{f_0} and that hence it is difficult to distinguish f_0 and f_1 based on observations drawn this way. Formally, in situations where the preceding probability can be made close to 1 for small enough $\eta>0$ we see that *no* test can ever distinguish between H_0 and H_1 except at a trivial level $\alpha=1$.

Lower bounds for the performance of statistical tests, and more generally for statistical procedures that imply the existence of certain tests, can be obtained through an application of the preceding method (and refinements thereof). This motivates a thorough study of likelihood ratios and their properties.

Consider first the case where i.i.d. random variables $X=(X_1,\dots,X_n)$ are drawn either from P_{f_1} or P_{f_0} on a measurable space $(\mathcal{X},\mathcal{A})$, with $P^n_f=\otimes_{i\leq n}P_f$ denoting the corresponding product measures on $\mathcal{X}^n$. Suppose further that μ is a common-dominating measure on $\mathcal{X}$, and let $\mu^n=\otimes_{i\leq n}\mu$. By the Radon-Nikodym theorem, the likelihood ratio equals the ratio of the product densities: for $x=(x_1,\dots,x_n)\in\mathcal{X}^n$, we have, μ^n-almost surely,

$$\frac{dP^n_{f_1}/d\mu^n}{dP^n_{f_0}/d\mu^n}(x) = \frac{\prod_{i=1}^n \frac{dP_{f_1}}{d\mu}(x_i)}{\prod_{i=1}^n \frac{dP_{f_0}}{d\mu}(x_i)}.\tag{6.2}$$

For Gaussian nonparametric models, the interpretation of such ratios needs some more care, and we clarify this in the next subsection.

6.1.1 *Infinite-Dimensional Gaussian Likelihoods*

In this section we develop tools to compute likelihood ratios in the infinite-dimensional Gaussian models introduced in Chapter 1. This includes, in particular, the construction of tractable sample spaces on which these models can be realised.

The Gaussian White Noise Model

Let, as usual, $L^2 = L^2([0,1])$ denote the Hilbert space of square-integrable functions on $[0,1]$, with inner product $\langle \cdot, \cdot \rangle$. We consider the functional Gaussian white noise model

$$dY(t) = dY_f^{(n)}(t) = f(t)dt + \frac{\sigma}{\sqrt{n}} dW(t), \quad t \in [0,1], \ \sigma > 0, \ n \in \mathbb{N}, \tag{6.3}$$

for different choices of $f \in L^2$. This model was already introduced informally in (1.5). Observing a trajectory dY means that we can observe all integrals $\int g(t)dY(t)$ against L^2-functions g or, what is the same, that we observe a realisation of the Gaussian process $\mathbb{Y}$ given by the shift experiment

$$\mathbb{Y}(g) = \mathbb{Y}_f^{(n)}(g) = \langle f, g \rangle + \frac{\sigma}{\sqrt{n}} \mathbb{W}(g), \quad g \in L^2([0,1]), \tag{6.4}$$

where $\mathbb{W}$ is the standard white noise or the isonormal process on $L^2([0,1])$ (see Example 2.1.11),

$$\mathbb{W}(g) \sim N(0, \|g\|_2^2), \quad E\mathbb{W}(g)\mathbb{W}(g') = \langle g, g' \rangle, \quad g, g' \in L^2([0,1]). \tag{6.5}$$

In the terminology of Section 2.1, both processes $\mathbb{Y}$ and $\mathbb{W}$ define Gaussian random variables in the 'path' space $\mathbb{R}^{L^2([0,1])}$, measurable for the cylindrical σ-algebra. For likelihood ratio computations, this nonseparable space is not convenient. To circumvent this problem, we now show that we can take the separable Banach space of continuous functions $C([0,1])$ as the underlying sample space for the model (6.3) and obtain a formula for the likelihood ratio on $C([0,1])$. Another approach, which is based on the sequence space isometry of L^2, is discussed at the end of the next subsection.

Let P^W be a Wiener measure, that is, the Gaussian Borel probability measure on $C([0,1])$ corresponding to the law of a standard Brownian motion $(W(x) : x \in [0,1])$ (see Exercise 2.3.2 and Example 2.6.7). The covariance metric of W is

$$\rho_W(x,y) = \sqrt{E|W(x) - W(y)|^2} = \sqrt{|x-y|} = \|1_{(x,y]}\|_2.$$

Hence, if for $0 \le x \le y \le 1$ we define the Gaussian variables $\tilde{\mathbb{W}}(1_{(x,y]}) = W(y) - W(x)$, it follows that, by independence of the increments of Brownian motion, for any finite number of disjoint intervals I_i and coefficients a_i, we have

$$E\left(\sum_i a_i \tilde{\mathbb{W}}(I_i)\right)^2 = \sum_i a_i^2 |I_i| = \int \left(\sum_i a_i I_i(x)\right)^2 dx.$$

Any $g \in L^2([0,1])$ can be approximated in L^2, as $N \to \infty$, by linear combinations $g_N = \sum_{i \le N} a_{i,N} 1_{I_{i,N}}$ of indicator functions of disjoint intervals $I_{i,N}$, and since these g_N form a Cauchy sequence in $L^2([0,1])$, the preceding identity implies that the variables

$$\tilde{\mathbb{W}}(g_N) := \sum_{i \le N} a_{i,N} \tilde{\mathbb{W}}(1_{I_{i,N}})$$

form a Cauchy sequence in $L^2(P^W)$. Thus, by completeness of $L^2(P^W)$, we can define $\tilde{\mathbb{W}}(g)$ to equal the $L^2(P^W)$-limit of the Gaussian random variables $\tilde{\mathbb{W}}(g_N)$ (which are independent of the approximating sequence g_N used). Then, by construction, the Gaussian process

$g \mapsto \tilde{\mathbb{W}}(g)$ defines an isometry between $L^2([0,1])$ and the closed linear span of $(\tilde{\mathbb{W}}(g) : g \in L^2)$ in $L^2(P^W)$; in other words, just like $\mathbb{W}$, $\tilde{\mathbb{W}}$ is a version of the isonormal process of $L^2([0,1])$. We thus may take $C([0,1])$ as the sample space underlying (6.5), with P^W determining the law of $\mathbb{W}$. Note that the preceding construction resembles a 'stochastic integral', obtained from taking limits of random integrals of simple functions g_N, explaining the notation $\mathbb{W}(g) = \int g dW$.

The arguments from the preceding paragraph work as well for the scaled white noise process $(\sigma/\sqrt{n})\mathbb{W}$, whose law is determined by the tight Gaussian Borel probability measure $P^W_{n,\sigma}$ on $C([0,1])$ arising from a scaled Brownian motion $(\sigma/\sqrt{n})W$. The process $\mathbb{Y}$ from (6.4), likewise, can be realised in this way as the law of the random variable $Y \in C([0,1])$, that is,

$$Y(x) = \mathbb{Y}(1_{[0,x]}) = F(x) + \frac{\sigma}{\sqrt{n}} W(x), \quad x \in [0,1],$$

with shift equal to the absolutely continuous function

$$F(x) = \int_0^x f(t)dt, \quad x \in [0,1].$$

We denote by $P^Y_{f,n,\sigma}$ the corresponding Gaussian Borel probability measure on $C([0,1])$. Note that $P^Y_{0,n,\sigma} = P^W_{n,\sigma}$. The likelihood ratio between the measures $P^Y_{f,n,\sigma}$ and $P^Y_{0,n,\sigma}$ on the separable Banach space $C([0,1])$ can be obtained from the Cameron-Martin Theorem 2.6.13 as follows.

Proposition 6.1.1 *Let $f \in L^2$. Then $P^Y_{f,n,\sigma}$ is absolutely continuous with respect to $P^Y_{0,n,\sigma}$ on $C([0,1])$, and the likelihood ratio, for $Y \sim P^Y_{0,n,\sigma}$, is given by*

$$\frac{dP^Y_{f,n,\sigma}}{dP^Y_{0,n,\sigma}}(Y) = \exp\left\{\frac{n}{\sigma^2}\mathbb{Y}_0^{(n)}(f) - \frac{n}{2\sigma^2}\|f\|_2^2\right\}. \tag{6.6}$$

Remark 6.1.2 From the preceding arguments, a draw $Y \sim P^Y_{0,n,\sigma}$ generates a draw from the process

$$\mathbb{Y}_0^{(n)}(g) = \frac{\sigma}{\sqrt{n}}\mathbb{W}(g), \quad g \in L^2([0,1]).$$

Scaling this equation, we see that the distribution of $(n/\sigma^2)\mathbb{Y}_0^{(n)}(f)$ coincides, in the integral notation dW from the standard white noise model (6.3), with the one of $(\sqrt{n}/\sigma)\int_0^1 f(t)dW(t)$. Therefore, the preceding likelihood ratio is often rewritten as

$$\exp\left\{\frac{\sqrt{n}}{\sigma}\int_0^1 f(t)dW(t) - \frac{n}{2\sigma^2}\|f\|_2^2\right\},$$

which is the density of $P^Y_{f,n,\sigma}$ with respect to Wiener measure P^W.

Remark 6.1.3 As in Remark 2.6.14, whenever $f, g \in L^2$, we can obtain a formula for the likelihood ratios by writing

$$\frac{dP^Y_{f,n,\sigma}}{dP^Y_{g,n,\sigma}}(Y) = \frac{dP^Y_{f,n,\sigma}}{dP^Y_{0,n,\sigma}}(Y)\frac{dP^Y_{0,n,\sigma}}{dP^Y_{g,n,\sigma}}(Y)$$

and using (6.6) for each factor.

Proof Note that by definition F is absolutely continuous with $F' = f \in L^2$ and satisfies $F(0) = 0$; hence, it is contained in the reproducing kernel Hilbert space of Brownian motion (see Example 2.6.7). Moreover, as in that example, one shows that the reproducing kernel Hilbert space of a Brownian motion scaled by $(\sigma/\sqrt{n})$ is the same as the unscaled one, with norm $\|\cdot\|_H$ of F equal to

$$\|F\|_H^2 = \frac{n}{\sigma^2}\int_0^1 (F'(x))^2 dx = \frac{n}{\sigma^2}\|f\|_2.$$

The result thus follows from the Cameron-Martin Theorem 2.6.13, with Radon-Nikodym derivative of $P^Y_{f,n,\sigma}$ with respect to $P^Y_{0,n,\sigma}$ given by

$$\frac{dP^Y_{f,n,\sigma}}{dP^Y_{0,n,\sigma}}(y) = \exp\left\{\phi^{-1}(F)(y) - \frac{\|F\|_H^2}{2}\right\} = \exp\left\{\phi^{-1}(F)(y) - \frac{n}{2\sigma^2}\|f\|_2^2\right\}.$$

As at the beginning of the proof of Theorem 2.6.13, we see that $\phi^{-1}(F)(Y)$ is a centred normal random variable with variance equal to the reproducing kernel Hilbert space norm $\|F\|_H^2 = (n/\sigma^2)\|f\|_2^2$ and hence is equal in distribution, under $Y \sim P^Y_{0,n,\sigma}$, to

$$\frac{n}{\sigma^2}\mathbb{Y}_0^{(n)}(f) \sim N\left(0, \frac{n}{\sigma^2}\|f\|_2^2\right),$$

so the result follows. ■

The Gaussian Sequence Space Model

For $f \in L^2([0,1])$ and $\{e_k : k \in \mathbb{Z}\}$ an ortho-normal basis of $L^2([0,1])$, consider an observation in the Gaussian sequence space model

$$Y_k = \langle f, e_k\rangle + \frac{\sigma}{\sqrt{n}} g_k,\ g_k \sim^{i.i.d.} N(0,1), \quad k \in \mathbb{Z}. \tag{6.7}$$

To study likelihood ratios in this model, we note that by Parseval's equality

$$(f_k : k \in \mathbb{Z}) \equiv \langle f, e_\cdot\rangle \in \ell_2,$$

but the white noise part $(g_k : k \in \mathbb{Z})$ of (6.7) does not define an element of ℓ_2 because $\|g_\cdot\|_{\ell_2} = \infty$ almost surely. Rather, $(g_k : k \in \mathbb{Z})$ defines a Gaussian product probability measure $\prod_k N(0,1)$ on $\mathbb{R}^{\mathbb{Z}}$, and so does the random vector $(Y_k : k \in \mathbb{Z})$. Kakutani's theorem for Gaussian product measures (Proposition 2.6.16) immediately implies the following result:

Proposition 6.1.4 *Denote by $P^Y_{f,n,\sigma}$ the product law of the Gaussian vector $(Y_k : k \in \mathbb{Z})$ on the cylindrical σ-algebra $\mathcal{C}$ of $\mathbb{R}^{\mathbb{Z}}$. If $(f_k : k \in \mathbb{Z}) \in \ell_2$, then $P^Y_{f,n,\sigma}$ is absolutely continuous with respect to $P^Y_{0,n,\sigma}$, and the likelihood ratio, for $Y \sim P^Y_{0,n,\sigma}$, is given by*

$$\frac{dP^Y_{f,n,\sigma}}{dP^Y_{0,n,\sigma}}(Y) = \exp\left\{\frac{n}{\sigma^2}\sum_{k\in\mathbb{Z}} f_k Y_k - \frac{n}{2\sigma^2}\|f_\cdot\|_{\ell_2}^2\right\}. \tag{6.8}$$

Similar to the remark after Proposition 6.1.1, we note that the preceding likelihood ratio can be written, for $(g_k : k \in \mathbb{Z})$ a sequence of i.i.d. standard normal random variables, as

$$\exp\left\{\frac{\sqrt{n}}{\sigma}\sum_{k\in\mathbb{Z}} f_k g_k - \frac{n}{2\sigma^2}\|f_\cdot\|_{\ell_2}^2\right\}.$$

Note in particular that $E[\sum_k f_k g_k]^2 = \sum_k f_k^2 < \infty$ for any $(f_k : k \in \mathbb{Z}) \in \ell_2$, and hence, the random Gaussian series $\sum_{k\in\mathbb{Z}} f_k Y_k$ converges almost surely. Moreover, a chain rule formula as in (6.1.3) holds true as well.

Whereas for many purposes the preceding formula for the likelihood ratios is sufficient, sometimes it is important to interpret this result in a separable Banach space in which $(Y_k : k \in \mathbb{Z})$ can be realised. This can be done using a Hilbert-Schmidt embedding of ℓ_2 into a larger sequence space: for some monotone decreasing strictly positive weighting sequence $\{w_k : k \in \mathbb{Z}\} \in \ell_2$, define

$$\ell_2(w) = \left\{ x : \|x\|_{\ell_2(w)}^2 \equiv \sum_k x_k^2 w_k^2 < \infty \right\},$$

which is a separable Hilbert space. By Fubini's theorem,

$$E\|g_\cdot\|_{\ell_2(w)}^2 = \sum_k E g_k^2 w_k^2 = \sum_k w_k^2 < \infty,$$

so $(g_k : k \in \mathbb{Z})$ is almost surely in $\ell_2(w)$ and hence defines a tight Gaussian Borel probability measure in this space (using that in separable spaces the cylindrical and Borel σ-algebra coincide and that Borel probability measures are always tight in such spaces; see Chapter 2). Note that the covariance operator of $(g_k : k \in \mathbb{Z})$ defines a Hilbert-Schmidt operator for the inner product of $\ell_2(w)$ (see Exercise 2.6.14). The Cameron-Martin Theorem 2.6.13 now gives the following, in the same way as in Proposition 6.1.1 and arguing as in Example 2.6.2:

Proposition 6.1.5 *Let $(f_k : k \in \mathbb{Z}) \in \ell_2$. Let $P^Y_{f,n,\sigma}$ be the Gaussian Borel probability measure on $\ell_2(w), w \in \ell_2$, representing the law of*

$$Y_k = f_k + \frac{\sigma}{\sqrt{n}} g_k, \quad g_k \sim^{i.i.d.} N(0,1), \quad k \in \mathbb{Z}.$$

Then $P^Y_{f,n,\sigma}$ is absolutely continuous with respect to $P^Y_{0,n,\sigma}$, and the likelihood ratio, for $(Y_k : k \in \mathbb{Z}) \sim P^Y_{0,n,\sigma}$, is given by

$$\frac{dP^Y_{f,n,\sigma}}{dP^Y_{0,n,\sigma}}(Y) = \exp\left\{ \frac{n}{\sigma^2} \sum_{k\in\mathbb{Z}} f_k Y_k - \frac{n}{2\sigma^2} \|f\|_{\ell_2}^2 \right\}. \tag{6.9}$$

The preceding proof would, unlike that of Proposition 6.1.4, work for any Gaussian measure and does not use the fact that $P^Y_{0,n,\sigma}$ is a product probability measure.

The preceding two propositions also give another proof for the likelihood ratio formula in Proposition 6.1.1. Given observations $(Y_k : k \in \mathbb{Z})$ in (6.7), we can define, for $g \in L^2([0,1])$, the Gaussian random variable $\tilde{\mathbb{Y}}(g)$ as the mean square limit of the Gaussian random variables

$$\tilde{\mathbb{Y}}(g_N) = \sum_{k:|k|\le N} Y_k \langle g, e_k \rangle,$$

so, using Parseval's identity,

$$E\tilde{\mathbb{Y}}(g) = \lim_{N\to\infty} \sum_k \langle f, e_k \rangle \langle g_N, e_k \rangle = \langle f, g \rangle, \quad \operatorname{Var} \tilde{\mathbb{Y}}(g) = \frac{\sigma^2}{n} \lim_{N\to\infty} \|g_N\|_2^2 = \frac{\sigma^2}{n} \|g\|_2^2;$$

hence, $(\tilde{\mathbb{Y}}(g) : g \in L^2)$ defines a version of $(\mathbb{Y}(g) : g \in L^2)$. We thus can compute likelihood ratios in the model (6.7) and transfer the result to the model (6.3) using the sequence space isometry of L^2 with ℓ_2. This exactly returns the formula (6.6).

Sample Splitting in the Gaussian White Noise Model

The preceding observations can be particularly useful if one wishes to create two independent copies of an observation in a white noise model, which can be of interest to mimick 'sample splitting' procedures from the i.i.d. observation model. Given an observation dY in the model (6.3), take an ortho-normal basis $\{e_k : k \in \mathbb{Z}\}$ of L^2 and the associated random coefficients

$$\left(Y_k = \int_0^1 e_k dY : k \in \mathbb{Z}\right).$$

We can generate i.i.d. Gaussians $(\tilde{g}_k : k \in \mathbb{Z})$ independent of all the g_k in

$$Y_k = \langle f, e_k \rangle + \frac{\sigma}{\sqrt{n}} g_k,$$

and set

$$Y_k' = Y_k + \frac{\sigma}{\sqrt{n}} \tilde{g}_k, \quad Y_k'' = Y_k - \frac{\sigma}{\sqrt{n}} \tilde{g}_k. \tag{6.10}$$

Then each Y_k', Y_k'' is Gaussian with mean $EY_k' = EY_k'' = \langle f, e_k \rangle$ and variance $2\sigma^2/n$. Moreover,

$$\mathrm{Cov}(Y_k', Y_k'') = \frac{\sigma^2}{n} E(g_k + \tilde{g}_k)(g_k - \tilde{g}_k) = 0,$$

so the Y_k, Y_k' are independent. We can now take dY', dY'' equal to the laws of

$$\sum_{k \in \mathbb{Z}} e_k Y_k', \quad \sum_{k \in \mathbb{Z}} e_k Y_k'',$$

respectively, defining independent white noise experiments of mean $f(t)dt$ and with σ increased by a factor of 2.

6.1.2 Basic Information Theory

In lower bounding the likelihood ratio $Z \equiv dP_{f_1}/dP_{f_0}$ in (6.1), we may use Markov's inequality to see that

$$P_{f_0}\left(\frac{dP_{f_1}}{dP_{f_0}}(Y) \geq 1 - \eta\right) \geq 1 - \frac{E_{f_0}|Z-1|}{\eta} \geq 1 - \frac{\sqrt{E_{f_0}(Z-1)^2}}{\eta}.$$

If μ is a common dominating measure, the quantities

$$E_{f_0}|Z-1| = \int \left| \frac{dP_{f_1}}{d\mu} - \frac{dP_{f_0}}{d\mu} \right| d\mu \equiv \|dP_{f_1} - dP_{f_0}\|_{1,\mu}$$

and

$$E_{f_0}(Z-1)^2 \equiv \chi^2(P_{f_1}, P_{f_0})$$

are called the $L^1(\mu)$- and χ^2-distance between P_{f_0} and P_{f_1}, respectively. For many lower bounds, such as those in Section 6.2, the χ^2-distance will be seen to be convenient to work with. In some other situations, more refined techniques are needed, based on the idea to estimate (6.1), for P_{f_0} absolutely continuous with respect to P_{f_1}, by

$$\begin{aligned} P_{f_0}\left(\frac{dP_{f_1}}{dP_{f_0}}(Y) \geq 1-\eta\right) &= P_{f_0}\left(\frac{dP_{f_0}}{dP_{f_1}}(Y) \leq \frac{1}{1-\eta}\right) \\ &= 1 - P_{f_0}\left(\log\frac{dP_{f_0}}{dP_{f_1}}(Y) \geq \log\frac{1}{1-\eta}\right) \\ &\geq 1 - \frac{1}{\log(1/(1-\eta))}E_{f_0}\left|\log\frac{dP_{f_0}}{dP_{f_1}}(Y)\right| \end{aligned} \tag{6.11}$$

using Markov's inequality. The last term is closely related to the information-theoretic concept of the Kullback-Leibler 'distance' between two probability measures, which we introduce next.

Total Variation and Kullback-Leibler Distance

Let P,Q be two probability measures on a measurable space $(\mathcal{X},\mathcal{A})$, and let μ be a common dominating measure, that is, a measure on $(\mathcal{X},\mathcal{A})$ such that both P and Q are absolutely continuous with respect to μ, with densities $dP/d\mu = p, dQ/d\mu = q$. For instance, we can take $\mu = P+Q$. Also, let us write $P \ll Q$ if P is absolutely continuous with respect to Q.

The *total variation distance* between P and Q equals

$$\|P-Q\|_{TV} \equiv \sup_{A\in\mathcal{A}} |P(A)-Q(A)|. \tag{6.12}$$

It is easy (see Exercise 6.1.1) to check that

$$\|P-Q\|_{TV} = \frac{1}{2}\int_{\mathcal{X}} |p(x)-q(x)|d\mu(x) = \frac{1}{2}\|p-q\|_{1,\mu}. \tag{6.13}$$

An alternative notion of 'distance' between probability measures is the following:

Definition 6.1.6 The *Kullback-Leibler* distance between P and Q is defined as

$$K(P,Q) = \begin{cases} \int_{\mathcal{X}} \log\left(\frac{dP}{dQ}(x)\right) dP(x), & \text{if } P \ll Q \\ +\infty, & \text{elsewhere.} \end{cases}$$

Note that this definition coincides with the one of entropy in Definifion 2.5.1 with $f = dP/d\mu$ because then $\log \int (dP/d\mu)d\mu = 0$. The following inequalities are sometimes known as *Pinsker's* or *Kullback-Czisar inequalities*.

Proposition 6.1.7 *For any probability measures P,Q on a measurable space $(\mathcal{X},\mathcal{A})$ and μ a common-dominating measure, we have*

(a) $\|P-Q\|_{TV} \leq \sqrt{K(P,Q)/2}$, *and*
(b) If $P \ll Q$, then

$$\int_{pq>0}\left|\log\frac{p}{q}\right| p d\mu \leq K(P,Q) + \sqrt{2K(P,Q)}.$$

Proof (a) Assume that $P \ll Q$; otherwise, the result is trivial. If we set $h(x) = x\log x - x + 1$ with $h(0) = 1$, then simple calculus shows that

$$\left(\frac{4}{3} + \frac{2}{3}x\right)h(x) \geq (x-1)^2, \quad x \geq 0,$$

and so, using (6.13) and the Cauchy-Schwarz inequality,

$$\begin{aligned}
\|P - Q\|_{TV} &= \frac{1}{2}\int_{q>0} |(p/q) - 1| q \leq \frac{1}{2}\int_{q>0} q\sqrt{\left(\frac{4}{3} + \frac{2p}{3q}\right)h(p/q)} \\
&\leq \frac{1}{2}\sqrt{\int\left(\frac{4q}{3} + \frac{2p}{3}\right)}\sqrt{\int_{q>0} qh(p/q)} \\
&= \sqrt{\frac{1}{2}\int_{pq>0} p\log\frac{p}{q}} = \sqrt{K(P,Q)/2}.
\end{aligned}$$

(b) By Exercise 6.1.2 and part (a), we have

$$\begin{aligned}
\int_{pq>0} p\left|\log\frac{p}{q}\right| &= \int_{pq>0} p(\log(p/q))_+ + \int_{pq>0} p(\log(p/q))_- \\
&= K(P,Q) + 2\int_{pq>0} p(\log(p/q))_- \leq K(P,Q) + 2\|P - Q\|_{TV} \\
&\leq K(P,Q) + \sqrt{2K(P,Q)},
\end{aligned}$$

completing the proof. ∎

Application to Nonparametric Likelihoods

We now show how the preceding concepts apply to the likelihoods relevant in this book. If we draw i.i.d. samples $X_1, \dots, X_n$ from law P_f as before (6.2), then the Kullback-Leibler distance between the product measures P^f_n, P^n_g tensorises; that is,

$$K(P^n_f, P^n_g) = nK(P_f, P_g). \tag{6.14}$$

In view of Proposition 6.1.7, we deduce that the total variation distance of the product measures associated to samples of size n from P_f, P_g can be controlled by $\sqrt{n}$ times the square root of the Kullback-Leibler distance

$$2\|P^n_f - P^n_g\|_{TV} \leq \sqrt{2nK(P_f, P_g)}. \tag{6.15}$$

The $L^1(\mu)$ distance of P^n_f, P^n_g can be controlled likewise using (6.13). Intuitively speaking, samples arising from laws P_f, P_g with small $K(P_f, P_g)$ will contain similar statistical information. We shall develop tools to control $K(P_f, P_g)$ in various settings later.

In the Gaussian white noise model, the Kullback-Leibler distance has a simple interpretation: for $f, g \in L^2$, let $P^Y_f \equiv P^Y_{f,n,\sigma}$ be the probability measures on $C([0,1])$ from Proposition 6.1.1, and denote by E^Y_f expectation under P^Y_f. Using Proposition 6.1.1 and

Remark 2.6.14, we obtain

$$
\begin{aligned}
K(P_f^Y, P_g^Y) &= \int \log \frac{dP_f^Y}{dP_g^Y}(y) dP_f^Y(y) \\
&= \frac{n}{\sigma^2} E_f^Y \mathbb{Y}(f) - \frac{n}{2\sigma^2}\|f\|_2^2 - \frac{n}{\sigma^2} E_f^Y \mathbb{Y}(g) + \frac{n}{2\sigma^2}\|g\|_2^2 \\
&= \frac{n}{2\sigma^2}\left(\|f\|_2^2 - 2\langle f, g\rangle + \|g\|_2^2\right) \\
&= \frac{n}{2\sigma^2}\|f-g\|_2^2.
\end{aligned} \tag{6.16}
$$

The total variation distance can be controlled likewise, using the first Pinsker inequality. We see that the information-theoretic distance of two white noise experiments equals a constant multiple of $n\sigma^2$ times the squared L^2-distance between f and g. Likewise, for $P_f^Y = P_{f,n,\sigma}^Y$ the law of the Gaussian vector $(Y_k : k \in \mathbb{Z})$, we have

$$
K(P_f^Y, P_g^Y) = \frac{n}{2\sigma^2} \sum_k \langle f - g, e_k\rangle^2 = \frac{n}{2\sigma^2}\|f-g\|_2^2 \tag{6.17}
$$

from Proposition 6.1.4 and the arguments leading to (6.16).

Exercises

6.1.1 Prove (6.13) and, in fact, that $\|P-Q\|_{TV} = 1 - \int \min(p,q) d\mu$. *Hint*: For $A = \{x \in \mathcal{X} : q(x) \geq p(x)\}$, write

$$
\int |p-q| d\mu = 2\int_A (q-p) d\mu.
$$

6.1.2 Let $a_- = \max(0, -a), a_+ = \max(0, a)$. If $P \ll Q$, prove that

$$
\int \log(dP/dQ)_- dP \leq \|P-Q\|_{TV}
$$

and hence

$$
K(P,Q) = \int_{pq>0} p[\log(p/q)]_+ d\mu - \int_{pq>0} p[\log(p/q)]_- d\mu.
$$

6.2 Testing Nonparametric Hypotheses

We start with some definitions and terminology. Consider statistical experiments $\mathcal{E}_n$ giving rise to observations $Y = Y^{(n)}$ on the measurable space $(\mathcal{Y}_n, \mathcal{A}_n), n \in \mathbb{N}$. The model for the distribution of Y consists of probability laws P_f indexed by $f \in \mathcal{F}$ (these laws also may, in principle, depend on n, but we suppress it in the notation unless necessary). A *statistical hypothesis* H_0 is a subset of the parameter space $\mathcal{F}$. A *statistical test* for H_0 is a function

$$
\Psi_n \equiv \Psi(Y^{(n)}) : \mathcal{Y}_n \to \{0, 1\}
$$

of the observations that takes value $\Psi_n = 0$ to accept H_0 and $\Psi_n = 1$ to reject it. Mathematically speaking, Ψ_n is the indicator function of a measurable subset of $\mathcal{Y}_n$. If we do not specify an alternative hypothesis H_1 explicitly, we can always take it to be the complement of H_0 in the whole parameter space $\mathcal{F}$. The performance of a test is naturally measured by the sum of its type 1 and type 2 errors, that is, by the probability of rejecting H_0 when it was true plus the probability of accepting H_0 when it was false. If d is a metric on $\mathcal{F}$, we can ask how well H_1 needs to be separated from H_0 for good tests to exist.

Definition 6.2.1 Let r_n, ρ_n be sequences of nonnegative real numbers, and let $H_0 \subset \mathcal{F}$ be a statistical hypothesis. Let d be a metric on $\mathcal{F}$. The sequence $(\rho_n : n \in \mathbb{N})$ is called the *minimax d-separation rate* for testing the hypotheses

$$f \in H_0 \quad \text{vs.} \quad f \in H_1 = H_1(d, r_n) = \left\{ f \in \mathcal{F}, \inf_{g \in H_0} d(f,g) \geq r_n \right\} \tag{6.18}$$

if the following two requirements are met:

(i) For every $\alpha > 0$ there exists a test Ψ_n such that for every $n \in \mathbb{N}$ large enough,

$$\sup_{f \in H_0} E_f \Psi_n + \sup_{f \in H_1(d,\rho_n)} E_f(1 - \Psi_n) \leq \alpha. \tag{6.19}$$

(ii) For any sequence $r_n = o(\rho_n)$, we have

$$\liminf_n \inf_{\Psi_n} \left[\sup_{f \in H_0} E_f \Psi_n + \sup_{f \in H_1(d,r_n)} E_f(1 - \Psi_n) \right] > 0, \tag{6.20}$$

where the infimum extends over all measurable functions $\Psi_n : \mathcal{Y}_n \to \{0,1\}$.

A few remarks on this definition are in order: it is typically satisfactory to control the type 1 and type 2 errors of a test at any given level $\alpha > 0$, with the test depending on α. In some situations, $\alpha = \alpha_n$ can be taken to converge to 0; that is, the limit as $n \to \infty$ of the left-hand side of (6.19) equals 0 – in this case we say that the test Ψ_n is *consistent*. Moreover, we say that the hypotheses H_0, H_1 from (6.18) are *asymptotically indistinguishable* if the limit inferior in (6.20) equals 1. We note that all the following results on the d-separation rate are in fact nonasymptotic in nature in the sense that the inequalities in (6.19) and (6.20) hold for every $n \in \mathbb{N}$. However, due to possibly large constants involved in the statements, the separation rate usually has a meaningful interpretation only for sufficiently large n.

Following the approach laid out in Definition 6.2.1, nonparametric testing theory will be seen to depend on which minimal assumptions on the parameter space $\mathcal{F}$ one is wishing to make, as well as on the choice of the metric d. For a fully nonparametric signal, such as a completely unknown probability distribution, one can only test for fairly weak aspects of f, such as the cumulative distribution function or other integral functionals, corresponding to 'weak' separation metrics d. If one is willing to assume further regularity properties of the signal at hand, such as that it has some derivatives (or, more generally, lies in a Besov ball), then much finer features of f can be tested, corresponding in a sense to stronger separation metrics d.

The following basic multiple-testing lower bound will be useful in determining the separation rates of testing problems as earlier. It generalises the even simpler bound from (6.1) to multiple alternatives and shows how the complexity of the alternative hypothesis can be encoded in certain averages of likelihood ratios arising from the statistical model. Consider a singleton $H_0 \colon f = f_0$ and a finite family $\mathcal{M} \subset \mathcal{F}$ of cardinality $|\mathcal{M}| = M$ describing the alternative $H_1 \colon f \in \mathcal{M}$. Assume that the P_{f_m} are all absolutely continuous with respect to P_{f_0}, or otherwise replace E_{f_m} later by the expectation $E^a_{f_m}$ with respect to the absolutely continuous part $P^a_{f_m}$ of P_{f_m}, and use $E_{f_m}(1-\Psi) \geq E^a_{f_m}(1-\Psi)$. Then we have, for every $\eta > 0$,

$$E_{f_0}\Psi + \sup_{f\in\mathcal{M}} E_f(1-\Psi) \geq E_{f_0}(1\{\Psi=1\}) + \frac{1}{M}\sum_{m=1}^{M} E_{f_m}(1-\Psi)$$
$$\geq E_{f_0}(1\{\Psi=1\}+1\{\Psi=0\}Z) \geq (1-\eta)\Pr_{f_0}(Z \geq 1-\eta), \quad (6.21)$$

where

$$Z = \frac{1}{M}\sum_{m=1}^{M} \frac{dP_{f_m}}{dP_{f_0}}$$

is an average of likelihood ratios. By Markov's inequality,

$$\Pr_{f_0}(Z \geq 1-\eta) \geq 1 - \frac{E_{f_0}|Z-1|}{\eta} \geq 1 - \frac{\sqrt{E_{f_0}(Z-1)^2}}{\eta}. \quad (6.22)$$

This lower bound is independent of Ψ, and hence,

$$\inf_{\Psi}\left[E_{f_0}\Psi + \sup_{f\in\mathcal{M}} E_f(1-\Psi)\right] \geq (1-\eta)\left(1 - \frac{\sqrt{E_{f_0}(Z-1)^2}}{\eta}\right). \quad (6.23)$$

Before we apply these lower bound techniques, let us introduce some concrete test procedures.

6.2.1 Construction of Tests for Simple Hypotheses

Simple hypotheses are those where H_0 consists of a singleton $\{f_0\}$ subset of $\mathcal{F}$. In this case, it is typically easy to find a test Ψ_n that performs well under H_0, and the challenge is to find tests for which the type 2 errors can be controlled for large sets of alternatives. We discuss several approaches to this problem in this subsection that are based on a combination of tools from Chapters 3 and 4, and that will lead to minimax optimal procedures. A completely different approach to nonparametric testing, based on likelihood methods and the Hellinger distance, will be discussed in Chapter 7.

Kolmogorov-Smirnov Tests

Suppose that we observe i.i.d. random variables $X_1,\dots,X_n$ from an unknown distribution function F on $\mathbb{R}^d$ and wish to test whether F equals a particular distribution function F_0. This corresponds to the null hypothesis $H_0 : F = F_0$, with default alternative hypothesis $H_1 : \mathcal{F}\setminus\{F_0\}$, where $\mathcal{F}$ is a subset of the set of all probability distribution functions. If $F_n(t) = (1\backslash n)\sum_{i=1}^n 1_{(-\infty,t]}(X_i)$ is the empirical distribution function of the sample, then a natural test statistic is

$$T_n = \sup_{t\in\mathbb{R}^d} |F_n(t) - F_0(t)|, \quad (6.24)$$

known as the *Kolmogorov-Smirnov statistic for testing*

$$H_0 : F = F_0 \quad vs. \quad H_1 : F \neq F_0.$$

A natural test is based on rejecting H_0 whenever

$$T_n \geq z_\alpha/\sqrt{n},$$

where z_α is a suitable constant depending on the desired level α of the test. For instance, we can take z_α to be the α-quantile constant of the distribution of $\sup_{t\in\mathbb{R}^d}|G_F(t)|$, where G_F is the F-Brownian bridge (see Section 3.7). In case $d=1$, the distribution of $\|G_F\|_\infty$ equals, for every continuous F, the one of $\max_{t\in[0,1]}|G(t)|$, where G is a standard Brownian bridge, so z_α is a distribution-free quantile constant (see Exercise 6.2.1). The type 1 errors of the test then satisfy, by Donsker's theorem, the continuous mapping theorem, and since the random variable $\|G\|_\infty$ is absolutely continuous (Exercise 2.4.4),

$$E_{F_0}\Psi_n = P_{F_0}^{\mathbb{N}}(\sqrt{n}\|F_n - F_0\|_\infty > z_\alpha) \to \Pr(\|G\|_\infty \geq z_\alpha) = \alpha,$$

as $n\to\infty$, so this test has exact asymptotic level α under H_0. Alternatively, by choosing z_α such that the tail of the Dvoretzky-Kiefer-Wolfowitz inequality (Exercise 3.3.3) equals α, one constructs a test that can be seen to have this level for every $n\in\mathbb{N}$ and uniformly in all distribution functions F_0 (see Exercise 6.2.1).

How does this test perform under the alternative H_1? For continuous distribution functions F for which

$$\sup_{t\in\mathbb{R}}|F(t) - F_0(t)| \geq C/\sqrt{n},$$

for some constant C, we can bound the type 2 errors by

$$E_F(1-\Psi_n) \leq P_F^{\mathbb{N}}(\|F_n - F\|_\infty \geq (C - z_\alpha)/\sqrt{n}) \to \Pr(\|G\|_\infty \geq C - z_\alpha) \leq \beta,$$

whenever $C - z_\alpha \geq z_\beta$. Again, the Dvoretzky-Kiefer-Wolfowitz inequality allows for a nonasymptotic and uniform in F bound on the type 2 errors. In other words, for alternatives that are sufficiently separated from F_0 at the $1/\sqrt{n}$ scale, this test will we able to distinguish between H_0 and H_1. Considering the generic case of F_0 equal to the uniform distribution, we will show later that in this generality this simple test procedure cannot be improved in terms of the separation rate $C/\sqrt{n}$. We will, however, also show that if it is reasonable to assume more structure on F, then the preceding procedure can be substantially improved.

Plug-in Tests Based on Estimators

The preceding Kolmogorov-Smirnov procedures are simply based on the idea that if we can estimate an arbitrary F at rate $1/\sqrt{n}$ in supnorm loss by F_n, then we can use this estimator to construct tests for simple hypotheses $H_0 : F = F_0$ that are accurate for alternatives separated away from F_0 by at least the rate $1/\sqrt{n}$ of estimation. If the statistical model satisfies some nonparametric regularity constraints, such as F_0 possessing a density function f_0 that is contained in a Sobolev, Hölder or Besov ball, then we can use linear nonparametric estimators from Chapter 5 in precisely the same way. More generally, suppose that for some metric space (S,d) that contains the statistical model $\mathcal{F}$ we can construct an estimator $\hat{f}_n$ such that

$$\sup_{f\in\mathcal{F}} E_f d(\hat{f}_n, f) \leq r_n, \tag{6.25}$$

where $(r_n : n\in\mathbb{N})$ is a sequence of nonnegative real numbers (the estimation rate). To test $H_0 : f = f_0$, we define the test statistic

$$T_n := d(\hat{f}_n, f_0)$$

and, for $0 < \alpha < 1$, the test

$$\Psi_n = 1\{T_n > 2r_n/\alpha\}. \tag{6.26}$$

Proposition 6.2.2 *Let $f_0 \in \mathcal{F}$ and $\alpha > 0$ be given, and consider testing hypotheses*

$$H_0 : f = f_0 \quad vs. \quad H_1 \subset \mathcal{F} \cap \{f : d(f, f_0) \geq cr_n\}, \quad c \geq 4/\alpha.$$

Then, for Ψ_n given in (6.26), we have, for every $n \in \mathbb{N}$,

$$E_{f_0}\Psi_n + \sup_{f\in H_1} E_f(1-\Psi_n) \leq \alpha.$$

Proof Under the null hypothesis, we have, using Markov's inequality and (6.25),

$$\begin{aligned} E_{f_0}\Psi_n &= P_{f_0}^{\mathbb{N}}\left(d(\hat{f}_n, f_0) > 2r_n/\alpha\right) \\ &\leq \frac{\alpha}{2r_n} Ed(\hat{f}_n, f_0) \leq \alpha/2. \end{aligned}$$

Under the alternatives, we have, by the triangle inequality and the same arguments as earlier,

$$\begin{aligned} E_f(1-\Psi_n) &= P_f^{\mathbb{N}}\left(d(\hat{f}_n, f_0) \leq 2r_n/\alpha\right) \\ &\leq P_f^{\mathbb{N}}\left(d(f, f_0) - d(\hat{f}_n, f) \leq 2r_n/\alpha\right) \\ &\leq P_f^{\mathbb{N}}\left(d(\hat{f}_n, f) \geq (c-2/\alpha)r_n\right) \leq \alpha/2, \end{aligned}$$

for $c - 2/\alpha \geq 2/\alpha$, completing the proof. ■

For example, if we consider the white noise or i.i.d. sampling model and take $\mathcal{F}$ equal to a ball in the Besov space $B^r_{\infty\infty}(A)$, $A = [0, 1]$ or $A = \mathbb{R}$, equipped with the supremum norm metric d, then the separation rate in Proposition 6.2.2 equals

$$r_n \simeq \left(\frac{\log n}{n}\right)^{r/(2r+1)}, \tag{6.27}$$

in view of Proposition 5.1.7. Similarly, if $\mathcal{F}$ consists of a ball in the Besov space $B^r_{p\infty}, p < \infty$, with $d(f,g) = \|f-g\|_p$, we have the separation rate $r_n \simeq n^{-r/(2r+1)}$ in L^p-norm.

The preceding test and separation constant $c \geq 4/\alpha$ can be improved by studying more exact distributional properties of the estimator $\hat{f}_n$ – earlier we only relied on (6.25) and Markov's inequality. One can obtain concentration bounds on the type 1 and type 2 errors simply by using concentration inequalities for L^p-norms of centred linear estimators as in Chapter 5. In fact, ideally, we would base the preceding test on a tight *confidence set* centred at $\hat{f}_n$, and the theory of nonparametric confidence sets provides another approach to nonparametric testing problems that we shall investigate systematically in Section 6.4.

If we leave the question of tight constants aside, then Proposition 6.2.2 essentially gives what is possible for plug-in tests in terms of order of magnitude of r_n. For several situations, the so-obtained separation boundaries will be seen to be minimax optimal in the sense of Definition 6.2.1 (see the next section). But, interestingly, for several other relevant situations, plug-in tests are not best possible, and more refined techniques are necessary.

Tests Based on χ^2-Statistics

If the testing problem involved has some geometric properties expressed, for instance, in the Hilbert space L^2, then one can often take advantage of this. To illustrate this, consider first an observation in the white noise model (6.3), and suppose that we want to test whether $f = f_0$ for some fixed function $f_0 \in L^2$. We recall that we assume that σ is known – if this is not the case, it may be replaced by a suitable estimate.

Consider a preliminary wavelet estimate of f_0 at resolution level $j \in \mathbb{N}$, that is,

$$\hat{f}_n(x) = f_n(j)(x) = \sum_{l \le j-1} \sum_k \psi_{lk}(x) \int_0^1 \psi_{lk}(t) dY(t), \tag{6.28}$$

where $\{\psi_{lk}\}$ is a wavelet basis of $L^2([0,1])$ from Section 4.3.4 or section 4.3.5 (and where as in (4.32) the scaling functions ϕ_k equal, by notational convention, the ψ_{lk} at the first resolution level $J_0 = J - 1$ or $J_0 = -1$ in the periodic case). Then, by Parseval's identity,

$$\|\hat{f}_n - f_0\|_2^2 = \sum_{l \le j-1} \sum_k \left(\int_0^1 \psi_{lk}(t) dY(t) - \langle f_0, \psi_{lk} \rangle \right)^2 + \sum_{l \ge j} \sum_k \langle f_0, \psi_{lk} \rangle^2. \tag{6.29}$$

Under the null hypothesis $H_0 : f = f_0$, the first summand equals, for $g_{lk} \sim^{i.i.d.} N(0,1)$,

$$\frac{\sigma^2}{n} \sum_{l \le j-1} \sum_k g_{lk}^2, \quad \text{which has expectation} \quad \frac{\sigma^2}{n} \sum_{l \le j-1} \sum_k 1 \equiv \frac{\sigma^2 2^j}{n}.$$

If we can control the second term in (6.29) – equal to the approximation error $\|K_j(f_0) - f_0\|_2^2$ of f_0 by its wavelet projection $K_j(f_0) = \sum_{l \le j,k} \langle f_0, \psi_{lk} \rangle \psi_{lk}$ – this motivates the test statistic

$$T_n(f_0) = \|f_n(j) - K_j(f_0)\|_2^2 - \frac{\sigma^2 2^j}{n} = \sum_{l \le j-1} \sum_k \left(\int_0^1 \psi_{lk}(t) dY(t) - \langle f_0, \psi_{lk} \rangle \right)^2 - \frac{\sigma^2 2^j}{n} \tag{6.30}$$

to test

$$H_0 : f = f_0 \quad vs. \quad H_1 : f \neq f_0$$

by

$$\Psi_n = 1\{|T_n(f_0)| \ge \tau_n\}, \quad \tau_n = \sigma^2 L \frac{2^{j/2}}{n}, \tag{6.31}$$

for a positive constant L to be chosen. The performance of this test is the subject of the following result.

Proposition 6.2.3 *Let $\alpha > 0$, and consider testing hypotheses*

$$H_0 : f = f_0 \quad vs. \quad H_1 \subset \{\|f - f_0\|_2 \ge \rho_n\}, \quad \rho_n \ge 0,$$

based on observations $dY \sim P_f^Y$ in the white noise model (6.3). Let $B(j), j \in \mathbb{N}$, be a nonincreasing sequence such that

$$\sup_{f \in H_1} \|K_j(f - f_0) - (f - f_0)\|_2^2 \le B(j), \tag{6.32}$$

and for $j = j_n \in \mathbb{N}$, let

$$\rho_n^2 = c^2 \max\left(\sigma^2 L \frac{2^{j_n/2}}{n}, B(j_n) \right), \quad c > 0. \tag{6.33}$$

Then, for Ψ_n from (6.31), there exist $L = L(\alpha)$ and $c = c(\alpha, \sigma)$ large enough such that, for all $n \in \mathbb{N}$,

$$E_{f_0}\Psi_n + \sup_{f \in H_1} E_f(1 - \Psi_n) \leq \alpha.$$

Proof We set $\sigma = 1$ for notational simplicity. Under the null hypothesis, we have, by Chebyshev's inequality and independence,

$$E_{f_0}\Psi_n = \Pr\left(\left|\frac{1}{n}\sum_{l \leq j_n - 1}\sum_k (g_{lk}^2 - 1)\right| \geq L\frac{2^{j_n/2}}{n}\right) \leq \frac{Eg_{11}^4}{L^2} \leq \alpha/2,$$

for L large enough. For the alternatives, we notice, from Parseval's identity and the definition of $B(j)$, ρ_n, that for c large enough,

$$\left|\sum_{l \leq j_n - 1, k} (f_{lk} - f_{0,lk})^2\right| \geq \|f - f_0\|_2^2 - B(j_n) \geq \|f - f_0\|_2^2/2. \tag{6.34}$$

Writing $f_{lk} = \langle f, \psi_{lk}\rangle$, we can hence bound the type 2 errors $E_f(1 - \Psi_n), f \in H_1$, by definition of ρ_n and for c large enough by

$$\begin{aligned}
&P_f^Y\left(\left|\sum_{l \leq j_n - 1}\sum_k (f_{lk} - f_{0,lk} + \frac{1}{\sqrt{n}}g_{lk})^2 - \frac{1}{n})\right| < L\frac{2^{j_n/2}}{n}\right) \\
&\leq \Pr\left(\left|\sum_{l \leq j_n - 1, k} (f_{lk} - f_{0,lk})^2\right| - L\frac{2^{j_n/2}}{n} < \left|\frac{1}{n}\sum_{l \leq j_n, k} (g_{lk}^2 - 1)\right| + \left|\frac{2}{\sqrt{n}}\sum_{l \leq j_n - 1, k} g_{lk}(f_{lk} - f_{0,lk})\right|\right) \\
&\leq \Pr\left(\left|\sum_{l \leq j_n - 1, k} (g_{lk}^2 - 1)\right| > L2^{j_n/2}\right) \\
&\quad + \Pr\left(\left|\frac{2}{\sqrt{n}}\sum_{l \leq j_n - 1, k} g_{lk}(f_{lk} - f_{0,lk})\right| > \frac{1}{4}\sum_{l \leq j_n - 1, k} (f_{lk} - f_{0,lk})^2\right).
\end{aligned}$$

The first term in the last bound is less than $\alpha/4$, arguing as for H_0 (c large enough). Next, by independence and again Parseval's identity,

$$\mathrm{Var}\left|\frac{2}{\sqrt{n}}\sum_{l \leq j_n - 1, k} g_{lk}(f_{lk} - f_{0,lk})\right| \leq \frac{4\|K_{j_n}(f - f_0)\|_2^2}{n},$$

so the last probability is bounded, using Chebyshev's inequality, (6.34) and the definition of ρ_n, by

$$\frac{1}{n}\frac{c''}{\|K_j(f - f_0)\|_2^2} \leq \frac{1}{c^2} \times O(1) \leq \alpha/4,$$

again for c large enough, completing the proof. ■

A tight choice of the constant L can be obtained, for instance, from the χ^2-concentration inequality in Theorem 3.1.9 applied to the probability constituting the type 1 error. Using this inequality, one can establish nonasymptotic concentration bounds for the type 1 and type 2 errors in Proposition 6.2.3. We refer to the general Theorem 6.2.14 (with $H_0 = \{f_0\}$) for such results.

The typical bounds for the approximation errors $B(j)$ over Sobolev or Besov balls of functions imply geometric decay in j. For instance, for $f \in B^r_{2\infty}([0,1])$ we know from Chapter 4 that

$$\|K_j(f) - f\|_2^2 \leq C\|f\|^2_{B^r_{2\infty}} 2^{-2jr} \tag{6.35}$$

using boundary-corrected or periodic r-regular wavelet bases of $L^2([0,1])$ (see Sections 4.3.4 and 4.3.5). If B is a bound for $\{\|f\|_{B^r_{2\infty}([0,1])} : f \in H_0 \cup H_1\}$, then we can take

$$B(j) = 4CB^2 2^{-2jr}, \quad 2^{j_n} \simeq B^{2/(2r+1/2)} n^{1/(2r+1/2)}$$

in the preceding proposition to obtain the separation rate

$$\rho_n = c' \max(1,B)^{1/(4r+1)} n^{-r/(2r+1/2)}, \tag{6.36}$$

for some constant $c' = c'(\alpha,\sigma)$. We shall see in Theorem 6.2.11 for the specific choice $f_0 = 0$ that this separation rate is optimal for alternatives consisting of a ball in $B^r_{2\infty}([0,1])$. Note that we need to know r, B in this case to implement the test – adaptation to unknown r, B is treated in Chapter 8.

Remark 6.2.4 For computational purposes, the test statistic in (6.30) is often based simply on Haar wavelets $\{\psi_{lk}\}$. At first sight, this seems to induce the restriction $r \leq 1$ on the alternatives (to guarantee (6.35)), but a lower bound of the form $\|K_j(f)\|_2 \geq C\rho_n$ whenever $\|f\|_2 \geq \rho_n$ also can be shown to hold for general $r > 1$ (see Exercise 6.2.6).

Tests Based on U-Statistics

We now show how an analogue of Proposition 6.2.3 can be proved in the i.i.d. sampling model. In the following, the observations $X_1, \ldots, X_n$ are assumed to take values in $[0,1]$. Generalisations to random samples on general Euclidean spaces $\mathbb{R}^d$ are possible (see Exercise 6.2.3). The analogue of $\hat{f}_n$ from (6.28) is

$$\tilde{f}_n(x) = f_n(j)(x) = \sum_{l \leq j-1} \sum_k \psi_{lk}(x) \int_0^1 \psi_{lk}(t) dP_n(t) = \frac{1}{n} \sum_{i=1}^n \sum_{l \leq j-1} \sum_k \psi_{lk}(X_i)\psi_{lk}(x),$$

where $P_n = (1/n)\sum_{i=1}^n \delta_{X_i}$ is the empirical measure associated to the sample and where the ψ_{lk} form a wavelet basis of $L^2([0,1])$ just as the one used in (6.28). The squared L^2-norm $\|f_n(j) - K_j(f_0)\|_2^2$ can be estimated unbiasedly by the U-statistic

$$U_n(f_0) = \frac{2}{n(n-1)} \sum_{i<i' \leq n} \sum_{l \leq j-1} \sum_k (\psi_{lk}(X_i) - \langle f_0, \psi_{lk} \rangle)(\psi_{lk}(X_{i'}) - \langle f_0, \psi_{lk} \rangle).$$

See Section 3.4.3 and Chapter 5 for some basic properties of U-statistics – removing the diagonal ($i = i'$) terms from the average corresponds to the subtraction of $\sigma^2 2^{j/2}/n$ in (6.30), resulting in a variance reduction. If the densities f involved in the testing problem are all uniformly bounded, we have a complete sampling analogue of the preceding proposition for the test

$$\Psi_n = 1\{|U_n(f_0)| \geq \tau_n\}, \quad \tau_n = L\frac{2^{j/2}}{n}. \tag{6.37}$$

Proposition 6.2.5 *Let $\alpha > 0$, and for $f_0 : [0,1] \to [0,\infty)$ a bounded probability density consider testing hypotheses*

$$H_0 : f = f_0 \quad vs. \quad H_1 \subset \{\|f - f_0\|_2 \geq \rho_n, \|f\|_\infty \leq U\}, \quad U > 0,\ \rho_n \geq 0,$$

based on observations $X_1, \dots, X_n \sim^{i.i.d.} f$. Let $B(j), j \in \mathbb{N}$, be a nonincreasing sequence such that

$$\sup_{f \in H_1} \|K_j(f - f_0) - (f - f_0)\|_2^2 \leq B(j), \tag{6.38}$$

and for $j = j_n \in \mathbb{N}$ let

$$\rho_n^2 = c^2 \max\left(L\frac{2^{j_n/2}}{n}, B(j_n)\right), \quad c > 0. \tag{6.39}$$

Then, for Ψ_n from (6.37), there exist $L = L(\alpha, U)$ and $c = c(U, \alpha)$ large enough that, for all $n \in \mathbb{N}$,

$$E_{f_0}\Psi_n + \sup_{f \in H_1} E_f(1 - \Psi_n) \leq \alpha.$$

Proof We assume that w.l.o.g. $\|f_0\|_\infty \leq U$. Under the null hypothesis, the factors $\psi_{lk}(X_i) - \langle f_0, \psi_{lk}\rangle$ of the summands in the U-statistic $U_n(f_0)$ are all centred, and hence, using independence and orthonormality of the wavelet basis,

$$\begin{aligned}\mathrm{Var}_{f_0}(U_n(f_0)) &\leq \frac{2}{n(n-1)} \int \left(\sum_{l \leq j_n - 1} \sum_k \psi_{lk}(x)\psi_{lk}(y)\right)^2 f_0(x) f_0(y)\, dx\, dy \\ &\leq \frac{2\|f_0\|_\infty^2}{n(n-1)} \sum_{l \leq j_n - 1, k} 1 \leq \frac{2^{j_n+1}\|f_0\|_\infty^2}{n(n-1)},\end{aligned}$$

so, by Chebyshev's inequality, for L large enough

$$E_{f_0}\Psi_n = \Pr(|U_n(f_0)| \geq \tau_n) \leq \frac{\mathrm{Var}_{f_0}(U_n(f_0))}{L^2 2^{j_n}/n^2} \leq \alpha/2.$$

For the alternatives $f \in H_1$, we have

$$\begin{aligned}(\psi_{lk}(X_i) - \langle\psi_{lk}, f_0\rangle)(\psi_{lk}(X_j) - \langle\psi_{lk}, f_0\rangle) &= (\psi_{lk}(X_i) - \langle\psi_{lk}, f\rangle)(\psi_{lk}(X_j) - \langle\psi_{lk}, f\rangle) \\ &\quad + (\psi_{lk}(X_i) - \langle\psi_{lk}, f\rangle)\langle\psi_{lk}, f - f_0\rangle + (\psi_{lk}(X_j) - \langle\psi_{lk}, f\rangle)\langle\psi_{lk}, f - f_0\rangle + \langle\psi_{lk}, f - f_0\rangle^2,\end{aligned}$$

so, by the triangle inequality, writing

$$L_n(f) = \frac{2}{n}\sum_{i=1}^n \sum_{l \leq j_n - 1} \sum_k (\psi_{lk}(X_i) - \langle\psi_{lk}, f\rangle)\langle\psi_{lk}, f - f_0\rangle, \tag{6.40}$$

we conclude that

$$\begin{aligned}|U_n(f_0)| &\geq \sum_{l \leq j_n - 1} \sum_k \langle\psi_{lk}, f - f_0\rangle^2 - |U_n(f)| - |L_n(f)| \\ &= \|K_{j_n}(f - f_0)\|_2^2 - |U_n(f)| - |L_n(f)|, \end{aligned} \tag{6.41}$$

for every $f \in H_1$. For c large enough, we have

$$\left| \sum_{l \le j_n - 1, k} (f_{lk} - f_{0,lk})^2 \right| \ge \|f - f_0\|_2^2 - B(j_n) \ge \|f - f_0\|_2^2/2. \tag{6.42}$$

We thus can bound the type 2 errors $E_f(1 - \Psi_n)$, for c large enough and some $c'' > 0$, by

$$\begin{aligned} P_f^{\mathbb{N}}(|U_n(f_0)| < \tau_n) &\le P_f^{\mathbb{N}}(|U_n(f)| + |L_n(f)| > \|K_j(f - f_0)\|_2^2 - \tau_n) \\ &\le P_f^{\mathbb{N}}(|U_n(f)| > \sqrt{2} L 2^{j_n/2}/n) + P_f^{\mathbb{N}}(L_n(f) > \|K_{j_n}(f - f_0)\|_2^2/4). \end{aligned}$$

The first term is less than $\alpha/4$ by the arguments used for H_0. By independence, orthogonality of the ψ_{lk} and Parseval's identity, the linear term has variance

$$\begin{aligned} \mathrm{Var}_f(|L_n(f)|) &\le \frac{4}{n} \int \left(\sum_{l \le j_n - 1, k} \psi_{lk}(x) \langle \psi_{lk}, f - f_0 \rangle \right)^2 f(x)\,dx \\ &\le \frac{4\|f\|_\infty \|K_{j_n}(f - f_0)\|_2^2}{n}, \end{aligned} \tag{6.43}$$

so the last probability can be bounded, using Chebyshev's inequality and (6.42), by

$$\frac{c''}{n} \frac{B}{\|K_{j_n}(f - f_0)\|_2^2} \le \frac{1}{c^2} \times O(1) \le \alpha/4,$$

for $c > 0$ large enough, completing the proof. ■

Similar remarks as after Proposition 6.2.3 apply – in particular, we obtain the separation rate $\rho_n \simeq n^{-r/(2r+1/2)}$ for alternatives contained in balls of $B^r_{2\infty}([0,1])$ if we choose $2^{j_n} \simeq n^{1/(2r+1/2)}$. This rate will be seen to be optimal in Theorem 6.2.9. To choose L in sharp dependence of α, we can use the exponential inequality for U-statistics in Theorem 3.4.8 – this then also yields nonasymptotic concentration bounds on the type 1 and type 2 errors of Ψ_n (see Theorem 6.2.17 (with $H_0 = \{f_0\}$)).

If no upper bound U for $\|f\|_\infty$ is known, we can replace it by a preliminary estimate such as $\|f_n(\bar{j}_n)\|_\infty$ where $2^{\bar{j}_n} \sim (\log n)^2/n$ and use that this estimate is consistent for $\|f\|_\infty$ (see Proposition 5.1.7).

Finally, we may use convolution kernels instead of wavelet projections in the preceding test statistics, but discretisation of the L^2-norm on the basis is obviously very convenient, both for proofs and computationally.

6.2.2 Minimax Testing of Uniformity on $[0,1]$

We now ask: in which sense are the separation rates obtained for the tests suggested in the preceding subsection optimal? We investigate this first in the perhaps most basic nonparametric testing problem, where the null hypothesis is that the unknown density function is uniform on $[0,1]$ and where the alternatives either consist of all distribution functions or are constrained to lie in a Hölder or Sobolev ball. Note that any simple testing problem $H_0 = \{F_0\}$ can be reduced to testing for uniformity by transforming the observations via the quantile transform F_0^{-1}.

Suppose that we are given a sample of n i.i.d. random variables $X_1,\dots,X_n$ on $[0,1]$ with common distribution function F or probability density function f. All statistical hypotheses in this subsection are automatically intersected with the set of probability distribution/density functions. The uniform density is identically 1 on $[0,1]$, with distribution function $F(t)=t$ on $[0,1]$. Suppose that we wish to test whether the probability distribution/density that generated the observations indeed is uniform or not; that is, we want to test

$$H_0: f=1 \quad vs. \quad H_1: f\neq 1. \tag{6.44}$$

Optimality of the Kolmogorov-Smirnov Test for General Alternatives

If $F_n(t) = (1/n)\sum_{i=1}^n 1_{[0,t]}(X_i)$ is the empirical distribution function, then the Kolmogorov-Smirnov test statistic is

$$T_n = \sup_{t\in[0,1]} |F_n(t)-t|. \tag{6.45}$$

As discussed after (6.24), the test $\Psi_n = 1\{T_n > z_\alpha/\sqrt{n}\}$ has level α under the null hypothesis and has power against the distribution functions F that are separated in supremum norm from the null hypothesis by at least a constant multiple of $1/\sqrt{n}$. The following proposition shows that for general alternatives of distribution functions, the separation rate $C/\sqrt{n}$ cannot be improved:

Proposition 6.2.6 *The minimax separation rate for testing the hypotheses*

$$H_0 \colon F(t) = t\,\forall t\in[0,1] \quad vs. \quad H_1: F\in\left\{\sup_{t\in[0,1]}|F(t)-t|\geq r_n\right\}$$

in the sense of Definition 6.2.1 based on observations $X_1,\dots,X_n \sim^{i.i.d.} F$ *is equal to*

$$(1/n)^{r/(2r+1/2)},$$

where C *is a constant that depends only on the desired level* α *of the test.*

Proof The upper bound follows from the reasoning after (6.24) (see also Exercise 6.2.1). For the lower bound, let $f_0=1$ be the uniform density, and let ψ be any bounded function supported in $[0,1]$ such that $\|\psi\|_1\leq\|\psi\|_2=1$ and $\int_0^1\psi=0$ – for example, Daubechies wavelets from Chapter 4, but much simpler examples are constructed easily. For $r_n=o(n^{-1/2})$, let $\psi_n=r_n\psi$; then $f_1=f_0+\psi_n$ is a positive probability density for n large enough (depending only on $\|\psi\|_\infty$), and its distribution function F_1 satisfies

$$\|F_0-F_1\|_\infty = r_n\sup_t\left|\int_0^t\psi(x)dx\right| = Cr_n$$

and hence is contained in H_1 for all n large enough. We use the reduction (6.23) with $M=1$ and need to control the χ^2-distance between P_{f_1} and P_{f_0}. For $Z=dP_{f_1}/dP_{f_0}$, using the properties of ψ and $1+x \le e^x$, we have, by independence,

$$\begin{aligned}
E_{f_0}(Z-1)^2 &= \int_{[0,1]^n} \left((\Pi_{i=1}^n f_1(x_i) - 1)\right)^2 dx \\
&= \int_{[0,1]^n} \Pi_{i=1}^n (1 + r_n \psi(x_i))^2 dx - 1 \\
&= \left(\int_{[0,1]} (1 + r_n \psi(x))^2 dx\right)^n - 1 \\
&= (1 + r_n^2)^n - 1 \le e^{r_n^2 n} - 1 \to 0,
\end{aligned}$$

so the result follows from (6.23). ■

Beyond Kolmogorov-Smirnov: Spiky Lipschitz Alternatives

Proposition 6.2.6 shows that testing problems for distribution functions have a natural separation boundary of $1/\sqrt{n}$ and that the Kolmogorov-Smirnov test is optimal in this sense. This is not unexpected – basic information theory shows that perturbations of the uniform distribution of constant size $1/\sqrt{n}$ cannot be detected by any test, paralleling the separation boundary in standard finite-dimensional models.

It pays to take a second look at the problem. Separating alternatives on the level of distribution functions is in some sense restrictive: small departures from $f_0 = 1$ on small intervals look even smaller after computing a cumulative integral. More abstractly speaking, the topology, or metric, in which we separate should play a role in our infinite-dimensional situation. Consider an example: let ψ be the zigzag function such that $\psi(0) = \psi(1/2) = \psi(1) = 0, \psi(1/4) = 1, \psi(3/4) = -1$, that is, 0 outside of $[0,1]$ and linear between $0, 1/4, 1/2, 3/4$ and 1. Define $\psi_n = \epsilon n^{-1/4} \psi(n^{1/4} \cdot)$ for some $\epsilon > 0$. We can think of ψ_n as of two tiny spikes of size $\pm \epsilon n^{-1/4}$ supported in an interval of length $n^{-1/4}$. Note that ψ_n integrates to 0; thus, $f_{\psi_n} = 1 + \psi_n$ is a probability density with distribution function F_{ψ_n} satisfying, for any $t \in [0,1]$,

$$\sup_t |F_{\psi_n}(t) - t| = \epsilon n^{-1/4} \sup_t \left| \int_0^t \psi(n^{1/4} x) dx \right| = \epsilon n^{-1/2} \sup_t \left| \int_0^{n^{1/4} t} \psi(v) dv \right| \le \epsilon n^{-1/2}.$$

Thus, for ϵ small, F_{ψ_n} is not contained in the alternatives H_1 covered by Proposition 6.2.6.

However, f_{ψ_n} is a Lipschitz function of Lipschitz constant equal to ϵ for all n, and its distance to f_0 in supremum- and L^2-norm is $\|f_{\psi_n} - 1\|_\infty = \|\psi_n\|_\infty = \epsilon n^{-1/4} \|\psi\|_\infty$ and $\|f_{\psi_n} - 1\|_2 = \epsilon n^{-3/8} \|\psi\|_2$, respectively, both of greater magnitude than $1/\sqrt{n}$. In particular, the test procedures from Propositions 6.2.2 and 6.2.5 can detect such departures from uniformity (recalling the remarks after these propositions and noting that Lipschitz functions are contained in $B^1_{\infty\infty} \cap B^1_{2\infty}$, see Section 4.3.3). The example of ψ_n illustrates that how large a spike has to be for it to be detectable depends, in a subtle way, on the separation metric: for distribution functions, the distance $d(f_0, f_{\psi_n})$ is of order $1/\sqrt{n}$; in L^∞, it is $n^{-1/4}$; and in L^2, it is $n^{-3/8}$. We investigate these questions in a general minimax setting in the following subsections.

A Minimax Test for Smooth L^∞-Separated Alternatives

Motivated by the preceding subsection, we now consider the testing problem (6.44) with the parameter space constrained to lie in a fixed ball of smooth functions in the Besov space $B^r_{\infty\infty}([0,1]), r > 0$. Recall from the discussion in Section 4.3 that these spaces model r-Hölderian functions and include Lipschitz balls for $r = 1$. We want to find the minimax separation rate in the sense of Definition 6.2.1 for the supremum-norm metric.

Theorem 6.2.7 *For any $r > 0, B > 1$, the minimax separation rate for testing the hypotheses*

$$H_0 \colon f = 1 \quad vs. \quad H_1(r_n) \colon \ f \in \left\{ \|f\|_{B^r_{\infty\infty}([0,1])} \leq B, \|f-1\|_\infty \geq r_n \right\}$$

in the sense of Definition 6.2.1 based on observations $X_1, \dots, X_n \sim^{i.i.d.} f$ equals

$$\rho_n = C(1/n)^{r/(2r+1/2)},$$

where C is a constant that depends only on r, B and on the desired level α of the test.

Remark 6.2.8 The proof implies that the hypotheses H_0 and H_1 are in fact asymptotically indistinguishable for any $r_n = o(\rho_n)$ (cf. after Definition 6.2.1).

Proof The upper bound follows from Proposition 6.2.2 and (6.27). To prove the lower bound, let $f_0 = 1$ on $[0,1]$, and let ψ be a Daubechies wavelet from Theorem 4.2.10 translated by N so that its support is $[1, 2N]$. We assume that the wavelet basis associated to ψ is sufficiently regular so that it generates the Besov space $B^r_{\infty\infty}(\mathbb{R})$ from Chapter 4. For $m \in \mathbb{Z}, j \in \mathbb{N}$, we write, as usual,

$$\psi_{jm} = 2^{j/2}\psi(2^j(\cdot) - m).$$

We can choose $j \in \mathbb{N}$ large enough that ψ_{jm} is supported in the interior of $[0,1]$ for every $m = 1, \dots, M, c_0 2^j \leq M < 2^j$ and some $c_0 > 0$ depending only on the regularity of the wavelet basis. Define, for $\epsilon > 0$, the functions

$$f_m := f_0 + \epsilon 2^{-j(r+1/2)}\psi_{jm}, \quad m = 1, \dots, M; \quad \mathcal{M} = \{f_m : m = 1, \dots, M\}. \tag{6.46}$$

Since $\int_0^1 \psi = 0$, we have $\int_0^1 f_m = 1$ for every m and also $f_m > 0 \ \forall m$ if $\epsilon > 0$ is chosen small enough depending only on $\|\psi\|_\infty$. We have $\|f_0\|_{B^r_{\infty\infty}([0,1])} = \|f_0\|_\infty = 1$ for every $r > 0$, and the Besov norm of the perturbation is, in view of the interior support of the ψ_{jm},

$$\|\epsilon 2^{-j(r+1/2)}\psi_{jm}\|_{B^r_{\infty\infty}([0,1])} = \|\epsilon 2^{-j(r+1/2)}\psi_{jm}\|_{B^r_{\infty\infty}(\mathbb{R})} \leq c_1\epsilon, \quad m = 1, \dots, M, \tag{6.47}$$

for some $c_1 > 0$, by equivalence of the wavelet and the Besov norms (Theorem 4.3.2). For small enough ϵ, we thus have, by the triangle inequality, $\|f_m\|_{B^r_{\infty\infty}([0,1])} \leq B$ for every m. Moreover,

$$\|f_m - 1\|_\infty = \epsilon\|\psi\|_\infty 2^{-jr}.$$

Summarising, we see that

$$\mathcal{M} \subset H_1\left(\epsilon\|\psi\|_\infty 2^{-jr}\right),$$

for every j large enough. Now, for $r_n = o(\rho_n)$, if j_n^* is defined such that $\rho_n \simeq 2^{-j_n^* r}$, we can find $j = j_n > j_n^*$ such that

$$r_n \leq \epsilon\|\psi\|_\infty 2^{-j_n r} = o(\rho_n), \tag{6.48}$$

and the so-chosen f_m are all contained in $H_1(r_n)$. Hence, for any test Ψ,

$$E_{f_0}\Psi + \sup_{f\in H_1} E_f(1-\Psi) \geq E_{f_0}\Psi + \sup_{f\in\mathcal{M}} E_f(1-\Psi),$$

and in view of Definition 6.2.1, the proof thus will be complete if we show that

$$\liminf_n \inf_\Psi \left(E_{f_0}\Psi + \sup_{f\in\mathcal{M}} E_f(1-\Psi)\right) \geq 1. \tag{6.49}$$

Using (6.23), it suffices to bound $E_{f_0}(Z-1)^2$ where

$$Z = \frac{1}{M}\sum_{m=1}^{M} \frac{dP_m^n}{dP_0^n},$$

with P_m^n the product probability measures induced by a sample of size n from density f_m. Writing

$$\gamma_n = \epsilon 2^{-j_n(r+1/2)},$$

using independence, ortho-normality of the ψ_{jm} and $\int \psi_{jm} = 0$ repeatedly, as well as $(1+x) \leq e^x$, we see that

$$\begin{aligned}
E_{f_0}(Z-1)^2 &= \frac{1}{M^2}\int_{[0,1]^n}\left(\sum_{m=1}^{M}\left(\prod_{i=1}^{n} f_m(x_i) - 1\right)\right)^2 dx \\
&= \frac{1}{M^2}\sum_{m=1}^{M}\int_{[0,1]^n}\left(\prod_{i=1}^{n}(f_m(x_i)) - 1\right)^2 dx \\
&= \frac{1}{M^2}\sum_{m=1}^{M}\left(\int_{[0,1]^n}\prod_{i=1}^{n} f_m^2(x_i)dx - 1\right) \\
&= \frac{1}{M^2}\sum_{m=1}^{M}\left(\left(\int_{[0,1]}(1+\gamma_n\psi_{j_n m}(x))^2 dx\right)^n - 1\right) \\
&= \frac{1}{M}((1+\gamma_n^2)^n - 1) \leq \frac{e^{n\gamma_n^2}-1}{M}.
\end{aligned}$$

Now, using (6.48) and the definition of ρ_n, we see $n\gamma_{j_n}^2 = o(\log n)$, so $e^{n\gamma_j^2} = o(n^\kappa)$, for every $\kappa > 0$, whereas $M \simeq 2^{j_n} \geq 2^{j_n^*} \simeq \rho_n^{1/r}$ still diverges at a fixed polynomial rate in n. Conclude that the preceding quantity converges to 0, which proves (6.49) because η was arbitrary. ■

L^2-Separation Conditions

We now investigate what happens if the L^∞-norm is replaced by the L^2-norm as separation metric. It is then natural to work with a more general Besov-Sobolev B_{2q}^r constraint on H_1 consisting of bounded densities.

Theorem 6.2.9 *For any $r>0, B>1, 1\leq q\leq\infty$, the minimax separation rate for testing the hypotheses*

$$H_0\colon f=1 \quad vs. \quad H_1(r_n)\colon f\in\left\{\max(\|f\|_\infty, \|f\|_{B_{2q}^r([0,1])}) \leq B, \|f-1\|_2 \geq r_n\right\}$$

in the sense of Definition 6.2.1 based on observations $X_1, \dots, X_n \sim^{i.i.d.} f$ *is equal to*

$$\rho_n = C\,(1/n)^{r/(2r+1/2)},$$

where C *is a constant that depends only on* r, B *and on the desired level* α *of the test.*

Remark 6.2.10 The proof implies that the hypotheses H_0 and H_1 are in fact asymptotically indistinguishable for any $r_n = o(\rho_n)$ (cf. after Definition 6.2.1).

Proof The upper bound follows from Proposition 6.2.5 using the bias bound (6.35) combined with $B^r_{2q} \subset B^r_{2\infty}$, $\|1\|_{B^r_{2q}} = 1$ and the assumptions on H_1 to obtain the desired separation rate by choosing j_n as before (6.36). It remains to prove the lower bound. Let $f_0 = 1$. As in the proof of Theorem 6.2.7, we take Daubechies wavelets ψ_{jk}, j large enough that are supported in the interior $[0,1]$. Recall that at the jth level there are $c_0 2^j$, of these interior wavelets, where $c_0 < 1$ is a fixed positive constant. We denote by $\mathcal{Z}_j, |\mathcal{Z}_j| = c_0 2^j$ the index set of those k at level j. For $\beta_m = (\beta_{mk} : k \in \mathcal{Z}_j)$, any point in the discrete hypercube $\{-1,1\}^{|\mathcal{Z}_j|}$ and $\epsilon > 0$ a small constant, define functions

$$f_m(x) = f_0 + \epsilon 2^{-j(r+1/2)} \sum_{k \in \mathcal{Z}_j} \beta_{mk} \psi_{jk}(x), \quad m = 1, \dots, 2^{|\mathcal{Z}_j|} \equiv M.$$

All the ψ_{jk} integrate to 0 on $[0,1]$, and moreover,

$$\|f_m - f_0\|_\infty \le \epsilon 2^{-jr} \sup_{x \in \mathbb{R}} \sum_{k \in \mathbb{Z}} |\psi(2^j x - k)| = c(\epsilon, \psi) 2^{-jr},$$

so, for ϵ small enough, the f_m are all uniformly bounded positive probability densities. The Besov norm of f_0 equals $\|f_0\|_2 = 1$, and for the f_m we have from the wavelet characterisation of Besov spaces (Section 4.3) and since all the ψ_{jk} are supported in the interior of $[0,1]$ that

$$\|f_m - f_0\|^2_{B^r_{2q}([0,1])} = \|f_m - f_0\|^2_{B^r_{2q}(\mathbb{R})} = \epsilon^2 2^{2jr} 2^{-2j(r+1/2)} \sum_{k \in \mathcal{Z}_j} 1 = c_0 \epsilon^2;$$

hence, by the triangle inequality and for ϵ small enough, the f_m are all contained in a B-ball of $B^r_{2q}([0,1])$. Finally,

$$\|f_m - f_0\|_2^2 = \sum_{l,k} \langle f_m, \psi_{lk} \rangle^2 \le \epsilon^2 2^{-2jr},$$

so if $r_n = o(\rho_n)$, then if j_n^* is defined such that $\rho_n \simeq 2^{-j_n^* r}$, we can find $j_n > j_n^*$ such that

$$r_n \le \epsilon 2^{-j_n r} = o(\rho_n),$$

and the so-chosen f_m are all contained in $H_1(r_n)$. Now to prove the lower bound from Definition 6.2.1, we use (6.23) and have to bound $E_{f_0}(Z-1)^2$ where

$$Z = \frac{1}{M} \sum_{m=1}^{M} \frac{dP^n_m}{dP^n_0} = \frac{1}{M} \sum_{m=1}^{M} \prod_{i=1}^{n} f_m(X_i) \equiv \frac{1}{M} \sum_{m=1}^{M} Z_m$$

is the corresponding likelihood ratio. Now, using $\int f_m - f_0 = 0$, for all m, independence and ortho-normality of the ψ_{jk}, we see that

$$\begin{aligned}
E_{f_0}[Z^2] &= \frac{1}{M^2}\sum_{m,m'} E_{f_0}\prod_{i=1}^{n}[f_m f_{m'}(X_i)] \\
&= \frac{1}{M^2}\sum_{m,m'}\left(\int_0^1 f_m(x)f_{m'}(x)dx\right)^n \\
&= \frac{1}{M^2}\sum_{m,m'}\left(1+\epsilon^2 2^{-j_n(2r+1)}\sum_k \beta_{mk}\beta_{m'k}\right)^n \qquad (6.50) \\
&= E[(1+\epsilon^2 2^{-j_n(2r+1)}Y_{j_n})^n],
\end{aligned}$$

where $Y_{j_n} \equiv \sum_{k=1}^{|\mathcal{Z}_{j_n}|} R_k$, for i.i.d. Rademacher ± 1 random variables R_k. Set

$$\gamma_n = \epsilon^2 n 2^{-j_n(2r+1/2)},$$

which, as $n \to \infty$, is $o(1)$ by the assumption on j_n. Using $1+x \le e^x$ and the expansion

$$\cosh(z) = \cos(iz) = 1 + \frac{z^2}{2} + o(z^2), \quad \text{as } |z| \to 0,$$

the quantity in (6.50) can further be bounded by

$$\begin{aligned}
E[\exp(\epsilon^2 n 2^{-j_n(2r+1)}Y_{j_n})] &= E[\exp(\gamma_n 2^{-j_n/2}Y_{j_n})] = E\exp\left(\gamma_n 2^{-j_n/2}\sum_{k=1}^{|\mathcal{Z}_{j_n}|}R_k\right) \\
&= \left(\frac{e^{\gamma_n 2^{-j_n/2}} + e^{-\gamma_n 2^{-j_n/2}}}{2}\right)^{|\mathcal{Z}_{j_n}|} \\
&= \cosh\left(\gamma_n 2^{-j_n/2}\right)^{|\mathcal{Z}_{j_n}|} \\
&= \left(1+\gamma_n^2 2^{-j_n-1}(1+o(1))\right)^{c_0 2^{j_n}} \\
&\le \exp\left(c_0\gamma_n^2(1+o(1))/2\right) \\
&\le 1+\delta^2,
\end{aligned}$$

for any $\delta > 0$. Conclude that, for any $\delta > 0$,

$$E_{f_0}(Z-1)^2 = E_{f_0}[Z^2] - 1 \le \delta^2$$

whenever n is large enough, completing the proof. ■

We remark that in this proof we could have taken $j_n = j_n^*$, and the lower bound on the testing errors still would be true on choosing ϵ small enough in the last step.

The theorem shows that the L^2-separation rates are of a smaller order of magnitude as those in L^∞. The case where one separates in L^p for general $1 \le p \le \infty$ is discussed in Exercise 6.2.5 and in the notes at the end of this chapter. In particular, decreasing p below 2 does not improve the separation rate further.

6.2.3 Minimax Signal-Detection Problems in Gaussian White Noise

Consider observing a function $f \in L^2([0,1])$ in white noise, that is,

$$dY(t) = dY_f^{(n)}(t) = f(t)dt + \frac{\sigma}{\sqrt{n}}dW(t), \quad t \in [0,1],\ \sigma > 0,\ n \in \mathbb{N},$$

as in (6.3). A natural problem is to test whether there has been a signal f at all or whether the observations are just pure white noise. That is, we want to test

$$H_0 : f = 0 \quad vs. \quad H_1 : f \neq 0,$$

which can be considered the white noise model analogue of testing for uniformity of a sampling density. The situation here is quite similar to, in fact slightly simpler than, the one in the preceding section. A Kolmogorov-Smirnov type of test is based on

$$\frac{\sqrt{n}}{\sigma} \sup_{t\in[0,1]} \left| \int_0^1 dY(t) \right| = \frac{\sqrt{n}}{\sigma} \sup_{t\in[0,1]} |\mathbb{Y}(1_{[0,t]})|.$$

Under H_0, this statistic exactly equals the maximum of a standard Brownian motion. Plug-in tests based on nonparametric estimators in the white noise model can be used likewise, and for L^2-separation results, we replace the U-statistic arguments by χ^2-statistics. We summarise the results in the following theorem, which shows that the signal detection problem in white noise is similar to the one of testing for uniformity of a probability density on $[0,1]$.

Theorem 6.2.11 *Consider the signal detection problem based on observations $dY \sim P_f^Y$ in the white noise model (6.3), and let $B > 0, 1 \leq q \leq \infty$. The minimax rates ρ_n of separation for testing the following hypotheses in the sense of Definition 6.2.1 are given as follows:*

(a) H_0: $f = 0$ *vs.* H_1: $f \in \left\{ \sup_{t\in[0,1]} \left| \int_0^t f(x)dx \right| \geq r_n \right\} \Rightarrow \rho_n = \frac{c}{\sqrt{n}}$,
where c is a constant that depends on σ and the desired level α of the test.

(b) H_0: $f = 0$ *vs.* H_1: $f \in L^2 \cap \left\{ \|f\|_{B^r_{\infty\infty}([0,1])} \leq B, \|f\|_\infty \geq r_n \right\} \Rightarrow \rho_n = c\left(\frac{\log n}{n}\right)^{\frac{r}{2r+1}}$,
where c is a constant that depends on r, B, σ and on the desired level α of the test.

(c) H_0: $f = 0$ *vs.* H_1: $f \in \left\{ \|f\|_{B^r_{2q}([0,1])} \leq B, \|f\|_2 \geq r_n \right\} \Rightarrow \rho_n = c\left(\frac{1}{n}\right)^{\frac{r}{2r+1/2}}$,
where c is a constant that depends on r, B, σ and on the desired level α of the test.

Remark 6.2.12 The proof implies that the hypotheses H_0 and H_1 are in fact asymptotically indistinguishable for any $r_n = o(\rho_n)$ (cf. after Definition 6.2.1).

Proof We start with proofs of the lower bounds, which are similar to those of Proposition 6.2.6 and Theorems 6.2.7 and 6.2.9 up to the likelihood ratio calculations.

(a) Assume the separation rate is $r_n = o(1/\sqrt{n})$. Take $\psi_n = r_n\psi$ as in the proof of Proposition 6.2.6. We can assume $\sigma = 1$ (otherwise renormalise so that $\|\psi\|_2 = \sigma$). Then $\int_0^t \psi_n(x)dx = Cr_n$ hence $\psi_n \in H_1$. By (6.23) with $M = 1$ and Proposition 6.1.1 applied with $f = \psi_n$ so that

$$n\mathbb{Y}_0^{(n)}(\psi_n) = \sqrt{n}r_n \int_0^1 \psi dW, \quad \int_0^1 \psi dW \sim N(0,1),$$

we can lower bound, for every $\eta > 0$, the relevant infimum in Definition 6.2.1 by $(1-1/\eta)$ times the square root of

$$E_0^Y(Z-1)^2 = \int_{\mathbb{R}} \left(e^{\sqrt{n}r_n x - nr_n^2/2} - 1\right)^2 \frac{1}{\sqrt{2\pi}} e^{-x^2/2} dx - 1$$
$$= \frac{1}{\sqrt{2\pi}} e^{-nr_n^2} \int_{\mathbb{R}} e^{2\sqrt{n}r_n x} e^{-x^2/2} dx - 1 = e^{-nr_n^2 + 2nr_n^2} - 1$$
$$= e^{nr_n^2} - 1 \to 0$$

as $n \to \infty$ using standard properties of standard normal random variables, including $Ee^{uX} = e^{u^2/2}$.

(b) Assume that the separation rate is $r_n = o(\rho_n)$. Take functions

$$f_m := \epsilon 2^{-j(r+1/2)} \psi_{jm}, \quad m = 1, \dots, M; \quad \mathcal{M} = \{f_m : m = 1, \dots, M\},$$

as in (6.46) but with $f_0 = 0$, for $\epsilon > 0$ and $j_n \in \mathbb{N}$ chosen as before (6.48), so that $f_m \in H_1$ for every m and such r_n. Again, we can take $\sigma = 1$ without loss of generality (otherwise scale ψ such that $\|\psi\|_2 = \sigma$). By (6.23), to prove the lower bound, we need to bound $E_{P^Y_{0,n,1}}(Z-1)^2$, where

$$Z = \frac{1}{M} \sum_{m=1}^{M} \frac{dP^Y_{f_m,n,1}}{dP^Y_{0,n,1}} = \frac{1}{M} \sum_{m=1}^{M} e^{n\mathbb{Y}_0^{(n)}(f_m) - n\|f_m\|_2^2/2}$$
$$= \frac{1}{M} \sum_{m=1}^{M} e^{\epsilon\sqrt{n}2^{-j_n(r+1/2)} g_m} e^{-\epsilon^2 n 2^{-j_n(2r+1)}/2}.$$

in view of Proposition 6.1.1, with $g_m \sim^{i.i.d.} N(0,1)$. Now, writing $\bar{\gamma}_n = \epsilon\sqrt{n}2^{-j_n(r+1/2)}$ and using independence of the g_m, $Ee^{ug_1} = e^{u^2/2}$, we have

$$E^Y_{0,n,1}(Z-1)^2 = E\left(\frac{1}{M}\sum_{m=1}^{M} e^{\bar{\gamma}_n g_m} e^{-\bar{\gamma}_n^2/2} - 1\right)^2$$
$$= \frac{1}{M} E\left(e^{\bar{\gamma}_n g_1} e^{-\bar{\gamma}_n^2/2} - 1\right)^2$$
$$= \frac{1}{M}\left(Ee^{2\bar{\gamma}_n g_1} e^{-\bar{\gamma}_n^2} - 1\right) = \frac{e^{\bar{\gamma}_n^2} - 1}{M}.$$

The proof is now completed exactly as at the end of the proof of Theorem 6.2.7.

(c) Assume that the separation rate is $r_n = o(\rho_n)$, and for $\beta_m = (\beta_{mk} : k \in \mathcal{Z}_j)$ a point in the discrete hypercube $\{-1,1\}^{|\mathcal{Z}_j|}$, consider functions

$$f_m(x) = \epsilon 2^{-j(r+1/2)} \sum_{k \in \mathcal{Z}_j} \beta_{mk} \psi_{jk}(x), \quad m = 1, \dots, 2^{|\mathcal{Z}_j|} \equiv M,$$

as in the proof of Theorem 6.2.9 (with $f_0 = 0$). We can set $\sigma = 1$ (otherwise renormalise so that $\|\psi\|_2 = \sigma$). We let $j = j_n$ as in the proof of Theorem 6.2.9 so that all these f_m are contained in H_1. By (6.23), to prove the lower bound, we need to bound $E_{P^Y_{0,n,1}}(Z-1)^2$, where, in view of Proposition 6.1.1,

$$
\begin{aligned}
Z &= \frac{1}{M}\sum_{m=1}^{M}\frac{dP^Y_{f_m,n,1}}{dP^Y_{0,n,1}} \\
&= \frac{1}{M}\sum_{m=1}^{M} e^{n\mathbb{Y}_0^{(n)}(f_m)-n\|f_m\|_2^2/2} \\
&= \frac{1}{M}\sum_{m=1}^{M} e^{\epsilon\sqrt{n}2^{-j_n(r+1/2)}\sum_k \beta_{mk}g_k} e^{-(n\epsilon^2 2^{-j_n(2r+1)}\sum_k 1)/2} \\
&= \frac{1}{M}\sum_{m=1}^{M}\prod_k \left[e^{\sqrt{\gamma_n'}\beta_{mk}g_k} e^{-\gamma_n'/2}\right],
\end{aligned}
$$

for $\gamma_n' \equiv \epsilon^2 n 2^{-j_n(2r+1)}$. Now, since the g_k are i.i.d. and from $Ee^{ug_1} = e^{u^2/2}$, we have

$$
\begin{aligned}
E_0 Z^2 &= E\left(\frac{1}{M}\sum_{m=1}^{M}\prod_k \left[e^{\sqrt{\gamma_n'}\beta_{mk}g_k} e^{-\gamma_n'/2}\right]\right)^2 \\
&= \frac{1}{M^2}\sum_{m,m'}\prod_k E e^{\sqrt{\gamma_n'}(\beta_{mk}+\beta_{m'k})g_k-\gamma_n'} \\
&= \frac{1}{M^2}\sum_{m,m'}\prod_k e^{\frac{\gamma_n'}{2}(\beta_{mk}+\beta_{m'k})^2-\gamma_n'} \\
&= \frac{1}{M^2}\sum_{m,m'} e^{\gamma_n'\sum_k \beta_{mk}\beta_{m'k}} \\
&= E[\exp(\gamma_n' Y_{jn})],
\end{aligned}
$$

where $Y_{j_n} = \sum_{k\in\mathcal{Z}_{j_n}} R_k$, with R_k i.i.d. ± 1 Rademachers. The proof from now on is the same as after (6.50), noting that $\gamma_n' = \gamma_n 2^{-j_n/2}$.

We now turn to upper bounds. For (a), we can argue as before Proposition 6.2.6, even without invoking Donsker's theorem, using that

$$
E_0^Y \sup_{t\in[0,1]}\left|\int_0^1 dY(t)\right| = (\sigma/\sqrt{n})E\|W\|_\infty = O(1/\sqrt{n}),
$$

where W is a standard Brownian motion. Part (b) follows from Proposition 6.2.2 and (6.27) after it, and Part (c) follows from Proposition 6.2.3 and the discussion after it (see (6.35) and (6.36)) using also $B^r_{2q} \subset B^r_{2\infty}$. ■

6.2.4 Composite Testing Problems

In this section we turn to general testing problems where the null hypothesis itself is possibly *composite*, that is, consists of more than just one element $\{f_0\}$. A classical application is as a *goodness-of-fit test* designed to test whether a given parametric model $\{P_\theta : \theta \in \Theta\}, \Theta \subset \mathbb{R}^p$, of probability distributions fits the data or not. But such testing problems are important beyond this, particularly for multiple testing situations as well as for the adaptive inference procedures discussed in Chapter 8.

The composite testing theory is more difficult as it depends on the geometry, size and other properties of H_0. One can view such problems as multiple testing problems where we want to control the type 1 errors for many simple hypotheses $\{f_0\}$ simultaneously. A natural test statistic is thus obtained from accepting H_0 as soon as one of the elements of H_0 has been accepted by a test designed for a simple testing problem, such as the χ^2- or U-statistic tests from Section 6.2.1. Mathematically, this amounts to studying the infimum over all elements of H_0 of the individual test statistics.

For null hypotheses that are not too complex measured via an entropy condition, we show that such *infimum tests* give rise to consistent tests. In proving this, we establish nonasymptotic exponential concentration bounds for the type 1 and type 2 errors for the relevant tests, which are of independent interest even for the case of a singleton null hypothesis. We then also show that in specific settings where H_0 is a very large set but with nice geometric properties, tailor-made tests can outperform minimum-H_0 tests.

Plug-in Approach

We start again with the Kolmogorov-Smirnov approach. Suppose that we are given a sample of n i.i.d. random variables $X_1,\dots,X_n$ from law P_F on $\mathbb{R}^d$ with distribution function F. Consider a general, possibly composite null hypothesis H_0 equal to an arbitrary set $\mathcal{F}_0$ of probability distribution functions on $\mathbb{R}^d$. For F_n the empirical distribution function of the observations, define the test statistic

$$T_n = \inf_{G \in H_0} \|F_n - G\|_\infty, \tag{6.51}$$

searching for the minimal uniform distance from the 'observations' F_n to the null hypothesis H_0. A test for the null hypothesis is given by

$$\Psi_n = 1\{T_n \geq z_\alpha/\sqrt{n}\},$$

where z_α are some α-quantile constants. The type 1 errors of this test satisfy, for $F \in H_0$,

$$E_F \Psi_n \leq P_F^{\mathbb{N}}\left(\sqrt{n}\|F_n - F\|_\infty > z_\alpha\right)$$

and hence behave as in the situation of a simple null hypothesis, with corresponding choices for z_α. We can consider separated alternatives

$$H_1 \subset \left\{F : \inf_{G \in H_0} \|F - G\|_\infty \geq C/\sqrt{n}\right\},$$

for which we can, as before Proposition 6.2.6, control the type 2 errors.

This approach works in generality for any model in which a good estimator $\hat{f}_n$ is available, similar to Proposition 6.2.2. Consider a statistical model $\mathcal{F}$ contained in some metric space (S,d) such that we can construct an estimator $\hat{f}_n$ for which

$$\sup_{f \in \mathcal{F}} E_f d(\hat{f}_n, f) \leq r_n, \tag{6.52}$$

where $(r_n : n \in \mathbb{N})$ is a sequence of nonnegative real numbers (the estimation rate). To test $H_0 \subset \mathcal{F}$, we define the test statistic

$$T_n := \inf_{h \in H_0} d(\hat{f}_n, h) = d(\hat{f}_n, H_0)$$

and, for $0 < \alpha < 1$, the test

$$\Psi_n = 1\{T_n > 2r_n/\alpha\}. \tag{6.53}$$

Proposition 6.2.13 *Let $\alpha > 0$ be given, and consider testing hypotheses*

$$H_0 \subset \mathcal{F} \quad vs. \quad H_1 \subset \mathcal{F} \cap \{f : d(f, H_0) \geq c r_n\}, \ c \geq 4/\alpha.$$

Then, for Ψ_n given in (6.53), we have, for every $n \in \mathbb{N}$,

$$\sup_{f \in H_0} E_f \Psi_n + \sup_{f \in H_1} E_f (1 - \Psi_n) \leq \alpha.$$

Proof The proof in exactly the same as Proposition 6.2.2 using the following two observations: for $f \in H_0$, we have $T_n \leq d(\hat{f}_n, f)$, and for the alternatives, we use $\inf_{h \in H_0} d(f, h) \geq c r_n$. ■

As after Proposition 6.2.2, we obtain the nonparametric estimation rates as separation rates. We also note that we can obtain exponential concentration bounds for the type 1 and type 2 errors of the preceding plug-in tests simply by using the concentration inequalities for centred linear estimators from Chapter 5.

Minimum χ^2 Tests in the White Noise Model

Consider observing a function $f \in L^2([0,1])$ in the white noise model (6.3). For an arbitrary $H_0 \subset L^2$, consider testing hypotheses

$$H_0 \quad vs. \quad H_1 \subset \{f : \|f - H_0\|_2 \geq \rho_n\},$$

where $\rho_n \geq 0$ is a separation sequence. For $f_n(j)$ as in (6.30) and

$$K_j(f) = \sum_{l \leq j-1, k} \psi_{lk} \langle f, \psi_{lk} \rangle \tag{6.54}$$

the wavelet approximation of f at level $j \in \mathbb{N}$ associated with a wavelet basis $\{\psi_{lk}\}$ of $L^2([0,1])$, consider the minimum χ^2 test statistic

$$T_n = \inf_{h \in H_0} |T_n(h)| = \inf_{h \in H_0} \left| \|f_n(j_n) - K_{j_n}(h)\|_2^2 - \frac{2^{j_n}\sigma^2}{n} \right| \tag{6.55}$$

and, assuming that the preceding infimum is measurable (see Exercise 6.2.2), the associated test

$$\Psi_n = 1\{T_n \geq \tau_n\}, \tag{6.56}$$

where τ_n are some critical values. This test is based on minimising the simple χ^2 test from Proposition 6.2.3 over all points $h \in H_0$.

Searching over multiple null hypotheses in the test statistic T_n has some cost that we can measure by the increments of the noise process on certain classes of functions associated with the wavelet basis used in the construction of the test. For $m \in \mathbb{Z}, j \in \mathbb{N}, f \in L^2$, define the class of functions

$$\mathcal{G}_{j,m}(H_0, f) = \left\{ \sum_{l \le j-1,k} \psi_{lk}(\cdot)\langle \psi_{lk}, f - h\rangle : h \in H_0,\ \|K_j(f-h)\|_2^2 \le 2^{m+1} \right\}. \tag{6.57}$$

If $\mathbb{W}$ is a standard white noise process acting on L^2, and if g_{lk} are i.i.d. $N(0,1)$, we have

$$\|\mathbb{W}\|_{\mathcal{G}_{j,m}(H_0,f)} = \sup_{h\in H_0, \|K_j(f-h)\|_2^2 \le 2^{m+1}} \left| \sum_{l\le j-1,k} g_{lk}(f_{lk} - h_{lk}) \right|. \tag{6.58}$$

The following theorem gives a general nonasymptotic concentration bound for the infimum χ^2 test Ψ_n on general hypotheses H_0 and H_1. We shall discuss several consequences of it after the proof.

Theorem 6.2.14 *Let H_0 be a bounded subset of L^2, and consider testing the hypotheses*

$$H_0 \quad vs. \quad H_1 \subset \left\{ f \in L^2 : \|f - H_0\|_2 \ge \rho_n \right\}, \qquad \rho_n \ge 0,$$

based on observations $dY \sim P_f^Y$ in the white noise model (6.3). Let $B(j)$, $j \in \mathbb{N}$, be a nonincreasing sequence of positive real numbers such that for K_j from (6.54)

$$\sup_{f_0\in H_0, f \in H_1} \|K_j(f - f_0) - (f - f_0)\|_2^2 \le B(j), \tag{6.59}$$

and for c, L positive constants and $j_n \in \mathbb{N}, d_n \ge 1$ sequences of nonnegative numbers, let

$$\tau_n = L\sigma^2 d_n \frac{2^{j_n/2}}{n}, \qquad \rho_n^2 \ge c^2 \max\left(L\sigma^2 d_n \frac{2^{j_n/2}}{n}, B(j_n) \right).$$

Assume, moreover, that for all $n \in \mathbb{N}$,

$$\sup_{f\in H_1} E\|\mathbb{W}\|_{\mathcal{G}_{j_n,m}(H_0,f)} \le \frac{\sqrt{n}2^m}{16}, \qquad \text{for all } m \in \mathbb{Z} \text{ s.t. } 2^m \ge \frac{\rho_n^2}{2}. \tag{6.60}$$

Then there exist positive constants $L = L(\psi)$ and $c = c(L,\sigma)$ such that the test Ψ_n from (6.56) satisfies, for every $n \in \mathbb{N}$ and constants $c_i, i = 1, \dots, 4$ depending only on c, L, σ,

$$\sup_{f\in H_0} E_f \Psi_n + \sup_{f \in H_1} E_f(1 - \Psi_n) \le c_1 \exp\left\{ -c_2 \frac{d_n^2}{1 + d_n 2^{-j_n/2}} \right\} + c_3 \exp\{-c_4 n \rho_n^2\}. \tag{6.61}$$

Proof We set $\sigma^2 = 1$ for notational simplicity. We first control the type 1 errors. Then $f \in H_0$, and we see that

$$E_f \Psi_n = P_f^Y \left\{ \inf_{h\in H_0} |T_n(h)| > \tau_n \right\} \le P_f^Y \{|T_n(f)| > \tau_n\}. \tag{6.62}$$

The last probability equals, for g_{lk} i.i.d. $N(0,1)$,

$$\Pr\left(\left| \frac{1}{n} \sum_{l \le j_n - 1} \sum_k (g_{lk}^2 - 1) \right| \ge L d_n \frac{2^{j_n/2}}{n} \right) \le C \exp\left\{ -\frac{1}{C} \frac{L^2 d_n^2}{1 + L d_n 2^{-j_n/2}} \right\},$$

using Theorem 3.1.9 and hence can be absorbed into the first bound in (6.61).

For the alternatives, let us write $f_{lk} = \langle f, \psi_{lk} \rangle$ for $f \in L^2$ and $h^* \in H_0$ for a minimiser in h of the test statistic (see Exercise 6.2.2). We have, from the definition of $\rho_n, B(j)$ and for c large enough, uniformly in H_1,

$$\left| \sum_{l \le j_n - 1, k} (f_{lk} - h^*_{lk})^2 \right| \ge \|f - h^*\|_2^2 - 2B(j) \ge \frac{1}{2} \inf_{h \in H_0} \|f - h\|_2^2 \ge \rho_n^2/2 \tag{6.63}$$

and hence can bound $E_f(1 - \Psi_n)$ by

$$\Pr\left(\inf_{h \in H_0} \left| \sum_{l \le j_n - 1} \sum_k (f_{lk} - h_{lk} + \frac{1}{\sqrt{n}} g_{lk})^2 - \frac{1}{n}) \right| \le \tau_n \right)$$

$$\le \Pr\left(\left| \sum_{l \le j_n - 1, k} (f_{lk} - h^*_{lk})^2 \right| - L d_n \frac{2^{j_n/2}}{n} \le \left| \frac{1}{n} \sum_{l \le j_n - 1, k} (g_{lk}^2 - 1) \right| + \left| \frac{2}{\sqrt{n}} \sum_{l \le j_n - 1, k} g_{lk} (f_{lk} - h^*_{lk}) \right| \right)$$

$$\le \Pr\left(\left| \sum_{l \le j_n - 1, k} (g_{lk}^2 - 1) \right| \ge L d_n 2^{j_n/2} \right)$$

$$+ \Pr\left(\left| \frac{2}{\sqrt{n}} \sum_{l \le j_n - 1, k} g_{lk} (f_{lk} - h^*_{lk}) \right| > \frac{1}{4} \left| \sum_{l \le j_n - 1, k} (f_{lk} - h^*_{lk})^2 \right| \right),$$

for $c > 0$ large enough. The first term is bounded as under H_0 and, combined with the H_0 bound, gives rise to the first term in the inequality (6.61). The second term needs a more careful treatment. It can be bounded by

$$\Pr\left(\sup_{h \in H_0} \frac{\left| \sum_{l \le j_n - 1, k} g_{lk} (f_{lk} - h_{lk}) \right|}{\|K_{j_n}(f - h)\|_2^2} > \frac{\sqrt{n}}{8} \right). \tag{6.64}$$

The variances of the numerator in this probability are bounded, using ortho-normality of the ψ_{lk} and independence, by

$$\mathrm{Var} \left| \sum_{l \le j_n - 1, k} g_{lk} (f_{lk} - h_{lk}) \right| \le \|K_{j_n}(f - h)\|_2^2. \tag{6.65}$$

Define, for $h \in H_0$,

$$\sigma^2(h) := \|K_{j_n}(f - h)\|_2^2.$$

By definition of σ^2, boundedness of H_0 in L^2 and $\|K_j(g)\|_2 \le \|g\|_2 \; \forall g \in L^2$, we can realise the supremum in the probability in (6.64) as the maximum over all slices

$$\{h \in H_0 : 2^m \le \sigma^2(h) \le 2^{m+1}\},$$

for $(\rho_n^2/2) \le 2^m \le C$ with fixed constant C; more precisely, (6.64) is less than or equal to

$$\Pr\left\{\max_{m\in\mathbb{Z}:\rho_n^2/2\le 2^m\le C}\ \sup_{h\in H_0:2^m\le\sigma^2(h)\le 2^{m+1}} \frac{\left|\sum_{l\le j_n-1,k} g_{lk}(f_{lk}-h_{lk})\right|}{\sigma^2(h)} > \frac{\sqrt{n}}{8}\right\}$$

$$\le \sum_{m\in\mathbb{Z}:\rho_n^2/2\le 2^m\le C} \Pr\left\{\|\mathbb{W}\|_{\mathcal{G}_{j_n,m}(H_0,f)} > \frac{\sqrt{n}}{8}2^m\right\}$$

$$\le \sum_{m\in\mathbb{Z}:\rho_n^2/2\le 2^m\le C} \Pr\left\{\|\mathbb{W}\|_{\mathcal{G}_{j_n,m}(H_0,f)} - E\|\mathbb{W}\|_{\mathcal{G}_{j_n,m}(H_0,f)} > \sqrt{n}2^{m-3} - E\|\mathbb{W}\|_{\mathcal{G}_{j_n,m}(H_0,f)}\right\},$$

where we use the notation from (6.58). Using (6.60), we bound the last expression by

$$\sum_{m\in\mathbb{Z}:\rho_n^2/2\le 2^m\le C} \Pr\left\{\|\mathbb{W}\|_{\mathcal{G}_{j_n,m}(H_0,f)} - E\|\mathbb{W}\|_{\mathcal{G}_{j_n,m}(H_0,f)} > \sqrt{n}2^{m-4}\right\}.$$

To this expression we can apply Theorem 2.5.8, noting that the supremum over $\mathcal{G}_{j,m}(H_0, f)$ can be realised, by continuity, as one over a countable subset of H_0. By (6.65), the last probability thus is bounded by

$$\sum_{m\in\mathbb{Z}:\rho_n^2/2\le 2^m\le C} c' \exp\left\{-c''\frac{n2^{2m}}{2^m}\right\} \le c_3 e^{-c_4 n\rho_n^2},$$

completing the proof. ■

We now give some corollaries to demonstrate the usefulness of the preceding theorem. The separation rate is driven by similar tradeoffs as in Proposition 6.2.3, with the additional requirement (6.60), which we discuss now.

Since H_0 is bounded in L^2, the class $\mathcal{G}_{j_n,m}(H_0, f)$ from (6.57) varies, for j_n fixed, in a ball of radius $2^{m/2}$ in the finite-dimensional space V_{j_n} spanned by wavelets up to resolution level j_n. These spaces have dimension 2^{j_n}, and their balls of radius $2^{m/2}$ have $L^2([0,1])$-covering numbers of order $(3\cdot 2^{m/2}/\varepsilon)^{2^{j_n}}$ for all $0 < \varepsilon < 2^{m/2}$ (Proposition 4.3.34 combined with the fact that the $L^2([0,1])$ metric coincides with the Euclidean metric on V_{j_n} by Parseval's identity.) Using Dudley's metric entropy bound (Theorem 2.3.7) for the Gaussian process $\mathbb{W}$ indexed by $\mathcal{G}_{j_m,m}(H_0, f)$, we see that, for every $n \in \mathbb{N}$,

$$E\|\mathbb{W}\|_{\mathcal{G}_{j_n,m}(H_0,f)} \lesssim 2^{j_n/2}\int_0^{2^{(m+1)/2}} \sqrt{\log\frac{3\cdot 2^{m/2}}{\varepsilon}}\,d\varepsilon \lesssim 2^{j_n/2}2^{m/2} \tag{6.66}$$

always holds. This bound is somewhat crude but does not require any assumptions whatsoever on H_0 other than that it is bounded in L^2. It verifies (6.60) whenever ρ_n^2 is of larger order than $2^{j_n}/n$ (with c large enough). Taking $d_n = 2^{j_n/2}$ in Theorem 6.2.14 then results in the following

Corollary 6.2.15 *Consider $H_0, H_1, B(j), j_n$ as in Theorem 6.2.14 with separation rate*

$$\rho_n^2 \ge c^2 \max\left(L\sigma^2\frac{2^{j_n}}{n}, B(j_n)\right).$$

Consider the test Ψ_n from (6.56). For every $D > 0$, there exist $L = L(\psi, D)$ and $c = c(L, \sigma, D)$ large enough such that

$$\sup_{f \in H_0} E_f \Psi_n + \sup_{f \in H_1} E_f(1 - \Psi_n) \le D \exp\left\{-D 2^{j_n}\right\}. \tag{6.67}$$

If $H_0 \cup H_1$ is a bounded subset of $B^r_{2\infty}([0,1])$, then ρ_n^2 can be taken of order $c^2 n^{-2r/(2r+1)}$, and then, for j_n such that $2^{j_n} \simeq n^{1/(2r+1)}$,

$$\sup_{f \in H_0} E_f \Psi_n + \sup_{f \in H_1} E_f(1 - \Psi_n) \le D \exp\left\{-D n \rho_n^2\right\}. \tag{6.68}$$

Proof The first part follows from Theorem 6.2.14 with $d_n = 2^{j_n/2}$ and (6.60) verified by (6.66), noting that a constant multiple of $2^{j_n/2} 2^{m/2}$ can be made less than $\sqrt{n} 2^m$ for all $2^m \ge \rho_n^2$ by choosing c large enough. The second result then follows from combining the preceding and the approximation bound

$$\sup_{f_0 \in H_0, f \in H_1} \|K_j(f - f_0) - (f - f_0)\|_2^2 \le C 2^{-2jr} \equiv B(j) \tag{6.69}$$

under the regularity assumptions on $H_0 \cup H_1$ (via the results from Sections 4.3.4 and 4.3.5). ■

Since ρ_n is effectively the L^2-distance between the hypotheses H_0 and H_1, the second assertion of the preceding corollary controls the type 1 and type 2 errors at rate $e^{-Dn\|H_0 - H_1\|_2^2}$ if the distance between H_0 and H_1 is at least of order $n^{-r/(2r+1)}$.

In the bound (6.66), no information on H_0 was used other than that it was bounded in L^2. If further structure is available, the separation rate can be substantially improved, as is already clear from Proposition 6.2.3 for simple H_0 and H_1 bounded in $B^r_{2\infty}([0,1])$, where the rate is

$$n^{-r/(2r+1/2)} = o(n^{-r/(2r+1)}).$$

If we want to retrieve this better rate from Theorem 6.2.14, we need to take $d_n = const$ and $2^{j_n} \sim n^{1/(2r+1/2)}$ in view of (6.69). If we can strengthen the simple entropy bound used in (6.66), for instance, by assuming that

$$\log N(\mathcal{G}_{j,m}(H_0, f), L^2([0,1]), \varepsilon) \le \left(\frac{A}{\varepsilon}\right)^{1/s}, \quad 0 < \varepsilon < A, \tag{6.70}$$

for every $m \in \mathbb{Z}, j \in \mathbb{N}$ and positive constants s, A independent of m, j, then a separation rate improving on $n^{-r/(2r+1)}$ also can be attained in the composite case, at least if H_0 is not 'too large'.

Corollary 6.2.16 *Consider H_0, H_1 as in Theorem 6.2.14 such that $H_0 \cup H_1$ are bounded subsets of $B^r_{2\infty}([0,1])$, for some $r > 0$, such that (6.70) holds for some $s > 1/2$ and with separation rate*

$$\rho_n^2 \sim c^2 \max\left(n^{-2s/(2s+1)}, n^{-r/(2r+1/2)}\right), \quad c > 0. \tag{6.71}$$

Consider the test Ψ_n from (6.56). For every $\alpha > 0$, there exist $L = L(\psi, \alpha)$ and $c = c(L, \sigma, \alpha)$ large enough such that, for all $n \in \mathbb{N}$,

$$\sup_{f \in H_0} E_f \Psi_n + \sup_{f \in H_1} E_f(1 - \Psi_n) \le \alpha. \tag{6.72}$$

Proof For the choices of d_n, j_n indicated earlier, we retrieve the desired separation rate from Theorem 6.2.14 and (6.69) if we can verify (6.60). From Dudley's metric entropy bound (Theorem 2.3.7) and (6.70) and (6.65), we have

$$E\|\mathbb{W}\|_{\mathcal{G}_{j_n,m}(H_0,f)} \leq C\int_0^{2^{(m+1)/2}} (A/\varepsilon)^{1/2s}\,d\varepsilon \leq C'2^{m(1/2-1/4s)}. \tag{6.73}$$

We see that

$$E\|\mathbb{W}\|_{\mathcal{G}_{m,j_n}} \leq \sqrt{n}2^{m-4}$$

is equivalent to $2^m \geq c'''n^{-2s/(2s+1)}$, for some sufficiently large $c''' > 0$, which is satisfied for c large enough because $2^m \geq \rho_n^2/2 \geq (c/2)n^{-2s/(2s+1)}$. ■

Examples for H_0 satisfying condition (6.70) can be found in Chapters 3 and 4. For instance, if H_0 is a bounded subset of $B^s_{p\infty}([0,1])$, then it satisfies (6.70) for any $s > 1/2$ in view of Theorem 4.3.36 and the wavelet characterisation of the norm of $B^s_{2\infty}([0,1])$ (Sections 4.3.4 and 4.3.5). Note that the $\|\cdot\|_{B^s_{p\infty}}$-norm of a function $\sum_{l\leq j_n,k}\langle h,\psi_{lk}\rangle\psi_{lk}$ cannot exceed $\|h\|_{B^s_{p\infty}}$ and that the translation by f extends the entropy for a bounded subset of $B^s_{p\infty}$ to one for the set $\mathcal{G}_{j,m}(H_0,f)$ that is uniform in m,j,f. We discuss some concrete examples after Proposition 6.2.18.

Infimum Tests Based on U-Statistics

We now consider the composite testing problem from the preceding subsection in the sampling setting. Let $X_1,\dots,X_n$ be i.i.d. with common bounded probability density function f on $[0,1]$). For $H_0 \subset L^2([0,1])$, $\{\psi_{lk}\}$ a wavelet basis of $L^2([0,1])$ (with scaling function equal to the wavelets at the initial resolution level) , $j \in \mathbb{N}$ and $h \in H_0$, define the U-statistic

$$U_n(h) = \frac{2}{n(n-1)}\sum_{i<i'}\sum_{l\leq j_n-1,k}(\psi_{lk}(X_i) - \langle\psi_{lk},h\rangle)(\psi_{lk}(X_{i'}) - \langle\psi_{lk},h\rangle).$$

For τ_n some thresholds to be chosen later, and following the ideas from (6.37), define the test

$$\Psi_n = 1\left\{\inf_{h\in H_0}|U_n(h)| > \tau_n\right\}. \tag{6.74}$$

For $f \in L^2, m \in \mathbb{Z}, j \in \mathbb{N}$, we consider again the classes

$$\mathcal{G}_{j,m}(H_0,f) = \left\{\sum_{l\leq j-1,k}\psi_{lk}(\cdot)\langle\psi_{lk},f-h\rangle : h \in H_0,\ \|K_j(f-h)\|_2^2 \leq 2^{m+1}\right\}$$

from the preceding subsection. The role of the white noise process $\mathbb{W}$ is naturally replaced by the empirical process $P_n - P$, and we write

$$\|P_n - P\|_{\mathcal{G}_{j,m}(H_0,f)} = \sup_{h\in H_0,\|K_j(f-h)\|_2^2\leq 2^{m+1}}\left|\frac{1}{n}\sum_{i=1}^n\sum_{l\leq j_n-1,k}(\psi_{lk}(X_i) - \langle f,\psi_{lk}\rangle)(f_{lk} - h_{lk})\right|. \tag{6.75}$$

The proof of the following theorem replaces the Gaussian process tools from Theorem 6.2.14 by appropriate empirical process tools. For the degenerate part of the U-statistic, we use the concentration inequality from Section 3.4.3, which naturally leads to four different concentration regimes. To simplify expressions, we restrict to sequences d_n, j_n which do not grow too quickly (but still cover all the applications we have in mind), leading to a pure Gaussian tail inequality. Other regimes can be obtained likewise from the proof.

Theorem 6.2.17 *Let H_0 be a family of probability densities on $[0,1]$ that are uniformly bounded by U. Consider testing the hypotheses*

$$H_0 \quad vs. \quad H_1 \subset \{f : \|f\|_\infty \le U, \|f - H_0\|_2 \ge \rho_n\}; \quad \rho_n \ge 0,$$

based on i.i.d. observations $X_1, \dots, X_n$ from density f. Let $B(j), j \in \mathbb{N}$, be a nonincreasing sequence of positive real numbers such that

$$\sup_{f_0 \in H_0, f \in H_1} \|K_j(f - f_0) - (f - f_0)\|_2^2 \le B(j), \tag{6.76}$$

and, for c, L positive constants and $j_n \in \mathbb{N}, d_n$ sequences of nonnegative numbers satisfying

$$1 \le d_n \le \min(2^{j_n/2}, n^{-1/4}), \quad 2^{j_n/2} d_n^3 \le n,$$

let

$$\tau_n = L d_n \frac{2^{j_n/2}}{n}, \quad \rho_n^2 \ge c^2 \max\left(L d_n \frac{2^{j_n/2}}{n}, B(j_n)\right).$$

Assume, moreover, that for all $n \in \mathbb{N}$,

$$\sup_{f \in H_1} E_f \|P_n - P\|_{\mathcal{G}_{j_n,m}(H_0,f)} \le \frac{2^m}{16}, \quad \textit{for all } m \in \mathbb{Z} \textit{ s.t. } 2^m \ge \frac{\rho_n^2}{2}. \tag{6.77}$$

Then there exist $L = L(U, \psi)$ and $c = c(L, U)$ large enough such that the test Ψ_n from (6.74) satisfies, for every $n \in \mathbb{N}$ and constants $c_i, i = 1, \dots, 4$, depending only on c, L,

$$\sup_{f \in H_0} E_f \Psi_n + \sup_{f \in H_1} E_f(1 - \Psi_n) \le c_1 \exp\left\{-c_2 d_n^2\right\} + c_3 \exp\{-c_4 n \rho_n^2\}. \tag{6.78}$$

Proof We first control the type 1 errors. For $f \in H_0$, we have

$$E_f \Psi_n = P_f^{\mathbb{N}} \left\{ \inf_{h \in H_0} |U_n(h)| > \tau_n \right\} \le P_f^{\mathbb{N}} \left\{ |U_n(f)| > L d_n \frac{2^{j_n/2}}{n} \right\}. \tag{6.79}$$

Now $U_n(f)$ is a U-statistic with kernel

$$R_f(x,y) = \sum_{l \le j_n - 1, k} (\psi_{lk}(x) - \langle \psi_{lk}, f \rangle)(\psi_{lk}(y) - \langle \psi_{lk}, f \rangle)$$

which satisfies $ER_f(x, X_1) = 0$, for every x, since $E_f(\psi_{lk}(X) - \langle \psi_{lk}, f \rangle) = 0$, for every k, l. Consequently, $T_n(f)$ is a degenerate U-statistic of order 2, and we can apply Theorem 3.4.8 to it, which we shall do with the choice $u = d_n^2$. We thus need to bound the constants A, B, C and D occurring in that theorem (cf. (3.146) and (3.147)) in such a way that

$$\frac{1}{n(n-1)}(C d_n + D d_n^2 + B d_n^3 + A d_n^4) \lesssim d_n \frac{2^{j_n/2}}{n}, \tag{6.80}$$

with constants in that inequality depending only on U, ψ, so the bound

$$c_1 \exp\left\{-c_2 d_n^2\right\} \tag{6.81}$$

for the type 1 errors follows from choosing $L = L(U, \psi)$ large enough. The bounds for A, B, C and D are obtained as follows: first, since R_f is fully centred we can bound $ER_f^2(X_1, X_2)$ by the second moment of the uncentred kernel, and thus, using the ortho-normality of ψ_{lk},

$$ER_f^2(X_1,X_2) \le \int_0^1 \int_0^1 \left(\sum_{l \le j_n - 1, k} \psi_{lk}(x)\psi_{lk}(y) \right)^2 f(x)f(y)dx\,dy$$

$$\le \|f\|_\infty^2 \sum_{l \le j_n - 1, k} \int_0^1 \psi_{lk}^2(x)dx \int_0^1 \psi_{lk}^2(y)\,dy$$

$$\le 2^{j_n} U^2.$$

We obtain $C^2 \le n(n-1)2^{j_n - 1}U^2$, and it follows that

$$\frac{Cd_n}{n(n-1)} \lesssim d_n \frac{2^{j_n/2}U}{n},$$

which precisely matches the right-hand side of (6.80). For the second term, note that, using the Cauchy-Schwarz inequality and that K_j is an L^2-projection operator,

$$\left| \int\int \sum_{l \le j_n - 1, k} \psi_{lk}(x)\psi_{lk}(y)\zeta(x)\xi(y)f(x)f(y)dx\,dy \right| = \left| \int K_{j_n}(\zeta f)(y)\xi(y)f(y)\,dy \right|$$

$$\le \|K_{j_n}(\zeta f)\|_2 \|\xi f\|_2 \le \|f\|_\infty^2 \le U^2$$

and, similarly, that

$$|E[E_{X_1}[K_{j_n}(X_1,X_2)]\zeta(X_1)\xi(X_2)]| \le \|f\|_\infty^2, \quad |EK_{j_n}(X_1,X_2)| \le \|f\|_\infty^2.$$

Thus, $E[R_f(X_1,X_2)\zeta(X_1)\xi(X_2)] \le 4U^2$, and hence,

$$\frac{Dd_n^2}{n(n-1)} \le \frac{2Ud_n^2}{n-1} \lesssim d_n \frac{2^{j_n/2}}{n}$$

is satisfied because $d_n \le 2^{j_n/2}$. For the third term, using the decomposition

$$R_f(x_1,x) = (r(x_1,x) - E_{X_1}r(X,x)) + (E_{X,Y}r(X,Y) - E_Y r(x_1,Y)),$$

for $r(x,y) = \sum_{l \le j_n - 1, k} \psi_{lk}(x)\psi_{lk}(y)$, the inequality $(a+b)^2 \le 2a^2 + 2b^2$ and again ortho-normality, we have that, for every $x \in \mathbb{R}$,

$$|E_{X_1} R_f^2(X_1,x)| \le 2\left[\|f\|_\infty \sum_{l \le j_n - 1, k} \psi_{lk}^2(x) + \|f\|_\infty \|K_{j_n}(f)\|_2^2 \right],$$

so, using $\|\sum_k \psi_{lk}^2\|_\infty \lesssim 2^l$ by regularity of the wavelet basis, we have

$$\frac{Bd_n^3}{n(n-1)} \lesssim \frac{2^{j_n/2}d_n^3}{n} \frac{1}{\sqrt{n}} \lesssim d_n \frac{2^{j_n/2}}{n}$$

because $d_n^2 \le \sqrt{n}$ by assumption. Finally, for the fourth term, we have $A = \|R_f\|_\infty \lesssim 2^{j_n}$ by regularity of the wavelet basis, and hence, by assumption on j_n, d_n,

$$\frac{Ad_n^4}{n(n-1)} \lesssim \frac{2^{j_n}d_n^4}{n^2} \lesssim d_n \frac{2^{j_n/2}}{n},$$

verifying (6.79).

Second, we now turn to the type 2 errors. In this case, for $f \in H_1$,

$$E_f(1-\Psi_n) = P_f^{\mathbb{N}}\left\{\inf_{h \in H_0} |U_n(h)| \le \tau_n\right\}. \tag{6.82}$$

Using (6.41), we have, writing

$$L_n(h) = \frac{2}{n}\sum_{i=1}^{n} \sum_{l \le j_n - 1, k} (\psi_{lk}(X_i) - \langle \psi_{lk}, f\rangle)\langle \psi_{lk}, f-g\rangle \tag{6.83}$$

for every $h \in H_0$,

$$|U_n(h)| \ge \|K_{j_n}(f-h)\|_2^2 - |U_n(f)| - |L_n(g)|. \tag{6.84}$$

We can find random $h_n^* \in H_0$ such that $\inf_{h \in H_0} |T_n(h)| = |T_n(h_n^*)|$ (see Exercise 6.2.2) and hence bound the probability in (6.82), using (6.84), by

$$P_f^{\mathbb{N}}\left\{|L_n(h_n^*)| > \frac{\|K_{j_n}(f-h_n^*)\|_2^2 - \tau_n}{2}\right\} + P_f^{\mathbb{N}}\left\{|U_n(f)| > \frac{\|K_{j_n}(f-h_n^*)\|_2^2 - \tau_n}{2}\right\}.$$

Since $h_n^* \in H_0$, we have uniformly in H_1

$$\left|\sum_{l \le j_n - 1, k} (f_{lk} - h_{lk}^*)^2\right| \ge \inf_{h \in H_0} \|f-h\|_2^2 - 2B(j) \ge \frac{1}{2}\inf_{h \in H_0} \|f-h\|_2^2 \ge \rho_n^2/2 \ge 2\tau_n, \tag{6.85}$$

for c large enough by definition of ρ_n. We thus can bound the sum of the last two probabilities by

$$P_f^{\mathbb{N}}\{|L_n(h_n^*)| > \|K_{J_n}(f-h_n^*)\|_2^2/4\} + P_f^{\mathbb{N}}\{|U_n(f)| > \tau_n\}.$$

For the second degenerate part, the exponential bound from of the first step applies as well (only boundedness of f by U was used there), giving the first bound in (6.78).

The proof concludes by treating the linear part. We have

$$P_f^{\mathbb{N}}\{|L_n(h_n^*)| > \|K_{j_n}(f-h_n^*)\|_2^2/4\} \le P_f^{\mathbb{N}}\left\{\sup_{h \in H_0} \frac{|L_n(h)|}{\|K_{j_n}(f-h)\|_2^2} > \frac{1}{4}\right\}. \tag{6.86}$$

Arguing as in (6.43), the variance of the linear process from (6.83) is bounded, for fixed $h \in H_0$, by

$$\mathrm{Var}_f(|L_n(h)|) \le \frac{4\|f\|_\infty \|K_{j_n}(f-h)\|_2^2}{n}, \tag{6.87}$$

so the supremum in (6.86) is one of a self-normalised ratio-type empirical process. For $h \in H_0$, define

$$\sigma^2(h) := \|K_{j_n}(f-h)\|_2^2 \ge \rho_n^2/2,$$

the inequality holding in view of (6.85). We bound the last probability in (6.86), for a suitable finite constant C (using $H_0 \cup H_1$ is bounded in L^2), by

$$
\begin{aligned}
P_f^{\mathbb{N}} & \left\{ \max_{m\in\mathbb{Z}:\rho_n^2/2\le 2^m\le C} \ \sup_{h\in H_0:2^m\le\sigma^2(h)\le 2^{m+1}} \frac{|L_n(h)|}{\sigma^2(h)} > \frac{1}{4} \right\} \\
& \le \sum_{m\in\mathbb{Z}:\rho_n^2/2\le 2^m\le C} P_f^{\mathbb{N}} \left\{ \sup_{h\in H_0:\sigma^2(h)\le 2^{m+1}} |L_n(h)| > 2^{m-2} \right\} \\
& \le \sum_{m\in\mathbb{Z}:\rho_n^2/2\le 2^m\le C} P_f^{\mathbb{N}} \left\{ \|P_n - P\|_{\mathcal{G}_{jn,m}} - E_f\|P_n - P\|_{\mathcal{G}_{jn,m}} > 2^{m-3} - E_f\|P_n - P\|_{\mathcal{G}_{jn,m}} \right\} \quad (6.88) \\
& \le \sum_{m\in\mathbb{Z}:\rho_n^2/2\le 2^m\le C} P_f^{\mathbb{N}} \left\{ n\|P_n - P\|_{\mathcal{G}_{jn,m}} - nE_f\|P_n - P\|_{\mathcal{G}_{jn,m}} > n2^{m-4} \right\},
\end{aligned}
$$

where we have used (6.77) and written $\mathcal{G}_{jn,m}$ as shorthand for $\mathcal{G}_{jn,m}(H_0, f)$. We note that the classes $\mathcal{G}_{jn,m}(H_0, f)$ are uniformly bounded by a constant multiple of U (using regularity of the wavelet basis and $|\langle\psi_{lk}, f - h\rangle| \le 2^{-l/2}2U$). Hence, we can apply Talagrand's inequality to the preceding expression, noting that the supremum over $\mathcal{G}_{m,j_n}$ can be realised as one over a countable subset of H_0. Renormalising by U and using (3.100) in Theorem 3.3.9 as well as (6.87) and (6.77), we can bound the expression in the preceding display, up to multiplicative constants, by

$$
\begin{aligned}
\sum_{m\in\mathbb{Z}:C'\rho_n^2\le 2^m\le C} \exp\left\{ -c_1 \frac{n^2(2^m)^2}{n2^m + nE_f\|P_n - P\|_{\mathcal{G}_{jn,m}} + n2^m} \right\} & \le \sum_{m\in\mathbb{Z}:C'\rho_n^2\le 2^m\le C} e^{-c_2 n 2^m} \\
& \le c_3 e^{-c_4 n\rho_n^2},
\end{aligned}
$$

which completes the proof. ∎

We obtain separation and concentration rates similar to those obtained in the corollaries after Proposition 6.2.14. To verify (6.77), we may use bounds for the moments of suprema of empirical processes indexed by the classes $\mathcal{G}_{jn,m}(H_0, f)$ (see Section 3.5). The standard entropy condition (6.70) then needs to be strengthened to bracketing or uniform metric entropy bounds. We may use results from Section 3.6 to derive results, paralleling Corollaries 6.2.15 and 6.2.16. For instance, if we assume for any probability measure P that

$$
\log N(\mathcal{G}_{jn,m}(H_0, f), L^2(P), \varepsilon) \le \left(\frac{A}{\varepsilon}\right)^{1/s}, \quad 0 < \varepsilon < A,
$$

for constants A, s independent of m, n, P, and if $H_0 \cup H_1$ lie in a fixed ball of $B^r_{2\infty}([0,1])$, we obtain the separation rate

$$
\rho_n^2 \sim \max\left(n^{-2s/(2s+1)}, n^{-2r/(2r+1)}\right)
$$

for the infimum – U-statistic test (6.74). One example is to take H_0 a ball in $B^s_{2\infty}([0,1])$ for some $s > 1/2$, for which the preceding entropy bound is satisfied in view of Theorem 4.3.36 and because a bound on the Besov norm of elements of H_0 carries over to the same bound on the Besov norm of elements of $\mathcal{G}_{jn,m}(H_0, f)$.

Minimax Theory for Some Composite Testing Problems

For L^∞-separation, the minimax separation rate for composite problems often can be shown to equal the estimation rate (as in Theorem 6.2.7), and results such as Proposition 6.2.13 then can be used directly. We give some references in the notes to this chapter.

Minimax theory for composite testing problems with L^2-separation appears to be more difficult partly because the χ^2- and U-statistic tools from Theorems 6.2.14 and 6.2.17 require bounds on the complexity of H_0 that may harm the separation rates. In particular, we cannot expect the infimum test to be optimal in all situations. But let us start with some conditions for when infimum tests are minimax optimal.

Proposition 6.2.18 *Let $r, B > 0$ and suppose that $H_0 \subset L^2$ satisfies (6.70) for some $s \geq 2r$, contains $f_0 = 0$ and is contained in a ball of $B^t_{2\infty}([0,1])$ for some $t > r$. Then the minimax separation rate for testing the hypotheses*

$$H_0 \quad vs. \quad H_1: \ f \in \left\{\|f\|_{B^r_{2\infty}([0,1])} \leq B, \|f - H_0\|_2 \geq r_n\right\}, \tag{6.89}$$

in the sense of Definition 6.2.1 based on observations $dY \sim P^Y_f$ in the white noise model, equals

$$\rho_n = c\left(\frac{1}{n}\right)^{r/(2r+1/2)},$$

where c is a positive constant.

Proof For the lower bound, we notice that the alternatives from the proof of Theorem 6.2.11 are

$$f_m(x) = \epsilon 2^{-j_n(r+1/2)} \sum_k \beta_{mk}\psi_{j_n k}(x), \quad m = 1, \dots, M,$$

for $\epsilon > 0$ a small constant, $\beta_{ik} = \pm 1$, and with j_n such that $2^{j_n} \simeq n^{1/(2r+1/2)}$. For all $\varepsilon > 0$, some $c > 0$ and n large enough,

$$\inf_{h\in H_0} \|f_m - h\|_2 \geq \sqrt{\sum_{l\geq j_n,k} \langle f_m, \psi_{lk}\rangle^2} - \sup_{h\in H_0}\sqrt{\sum_{l\geq j_n,k}\langle h, \psi_{lk}\rangle^2} \geq c\epsilon n^{-r/(2r+1/2)} \tag{6.90}$$

in view of

$$\sup_{h\in H_0}\sqrt{\sum_{l\geq j_n,k}\langle h, \psi_{lk}\rangle^2} = O(2^{-j_n t}), \quad t > r.$$

Hence, the f_m are contained in H_1, and

$$\sup_{f\in H_0} E_f\Psi + \sup_{f\in H_1} E_f(1-\Psi) \geq E_0\Psi + \sup_{f_m:m=1,\dots,M} E_{f_m}(1-\Psi)$$

implies that the proof of Theorem 6.2.9 applies to yield the lower bound.

For the upper bound, we use Corollary 6.2.15 and note that for $s \geq 2r$, the second term in (6.71) dominates. ■

A similar result can be proved in the sampling setting using Theorem 6.2.17. We see that the minimum χ^2- or U-statistic test is optimal at least in situations where the null hypothesis is not too complex ($s \geq 2r$) and consists of functions that are smoother than the alternative ($t > r$). For classical goodness-of-fit testing problems where H_0 equals a fixed parametric (finite-dimensional) model, we can typically apply the preceding proposition. Instead of going into the details, we consider some more difficult examples where H_0 itself can be infinite dimensional.

Consider, for instance, testing a null hypothesis that lies in a fixed ball of functions of bounded variation on $[0,1]$ (see Section 4.3.3). This includes the important example of testing for monotonicity of a bounded function f, that is,

$$H_0 = \{f : [0,1] \to (-\infty, M],\ f \text{ is nondecreasing}\},$$

but also other examples, such as testing whether f is piecewise constant or not.

Corollary 6.2.19 *Let $0 < r < 1/2$ and $B, M > 0$. Let H_0 be any subset of*

$$\{f : [0,1] \to \mathbb{R}, \|f\|_{BV} \leq M\}$$

that contains $f_0 = 0$. The minimax separation rate for testing the hypotheses

$$H_0 \quad vs. \quad H_1\colon\ f \in \left\{\|f\|_{B^r_{2\infty}([0,1])} \leq B, \|f - H_0\|_2 \geq r_n\right\}, \tag{6.91}$$

in the sense of Definition 6.2.1 based on observations $dY \sim P_f^Y$ in the white noise model, equals

$$\rho_n = c(1/n)^{r/(2r+1/2)},$$

where c is a positive constant.

Proof Any function $f \in H_0$ is contained in a ball of $B^1_{1\infty}$ (see Proposition 4.3.21) and then also in a ball of $B^{1/2}_{2\infty}$ (see Proposition 4.3.6). Theorem 4.3.36 then verifies (6.70), with $s = 1 \geq 2r$, so Proposition 6.2.18 applies with $t = 1/2 > r$. ∎

The infimum χ^2 test from (6.56) is hence minimax in the setting of Corollary 6.2.19. The restriction $r < 1/2$ in this corollary is natural in the sense that then $H_0 \subset B^r_{2\infty}$, so the testing problem is a nested one. In fact, inspection of the proof shows that $r = 1/2$ is also admissible at least if M is fixed and if B is sufficiently large.

Another example to which Proposition 6.2.18 applies would be to take H_0 itself equal to a ball in the space $B^s_{2\infty}([0,1])$, $s > r$, which satisfies (6.70) for the given s, as discussed after Corollary 6.2.16. This amounts to testing whether the signal f observed is of regularity s or r. For $s \geq 2r$, Proposition 6.2.18 implies, via (6.71), that the infimum test is minimax optimal, but for $r < s < 2r$, the interesting question arises whether the first term in this separation rate

$$\max\left(n^{-s/(2s+1)}, n^{-r/(2r+1/2)}\right)$$

should appear or not. On a more conceptual level, this is equivalent to the question of whether the complexity of the null hypothesis should affect the minimax separation rate or not. The following theorem shows that this is not the case and that the rate in the preceding display is suboptimal for $r < s < 2r$. The construction of an optimal test in the proof relies strongly on the geometry of the hypotheses H_0 and H_1 in this particular example.

Theorem 6.2.20 *For any $s > r > 0, B > 0$, the minimax separation rate for testing the hypotheses*

$$H_0: f \in \{\|f\|_{B^s_{2\infty}([0,1])} \le B\} \quad vs. \quad H_1: f \in \left\{\|f\|_{B^r_{2\infty}([0,1])} \le B, \|f - H_0\|_2 \ge r_n\right\}, \quad (6.92)$$

in the sense of Definition 6.2.1 based on observations $dY \sim P^Y_f$ in the white noise model, equals

$$\rho_n = c\left(\frac{1}{n}\right)^{r/(2r+1/2)},$$

where c is a positive constant that depends on r, B, σ and on the desired level α of the test.

Proof Noting that $f_0 = 0 \in H_0$, for f_m as in Proposition 6.2.18, we have

$$\sup_{f\in H_0} E_f\Psi + \sup_{f\in H_1} E_f(1-\Psi) \ge E_0\Psi + \sup_{f_m: m=1,\dots,M} E_{f_m}(1-\Psi)$$

using (6.90) with $t = s$, so the proof of Theorem 6.2.9 applies to yield the lower bound on the separation rate.

For the upper bound, set $\sigma = 1$ without loss of generality. Note from Sections 4.3.4 and 4.3.5 that the wavelet characterisation of the Besov norm of f gives

$$\|f\|^2_{B^s_{2\infty}} = \max_{J_0 \le l < \infty} 2^{2ls} \sum_{k\in\mathcal{Z}_l} \langle f, \psi_{lk}\rangle^2. \quad (6.93)$$

We construct a test that rejects the null hypothesis as soon as any of the scales l indicate a too-large $\|f\|_{B^s_{2\infty}}$ norm. Writing $\hat{f}_{lk} = \int_0^1 \psi_{lk}(t)dY(t) = \langle f, \psi_{lk}\rangle + (1/\sqrt{n})g_{lk}$, for $g_{lk} \sim^{i.i.d.} N(0,1)$, and recalling $|\mathcal{Z}_l| = 2^l$, define

$$T_n(l) = \sum_{k\in\mathcal{Z}_l} \hat{f}^2_{lk} - \frac{2^l}{n},$$

which estimates the contribution of the lth scale to the Besov norm of the signal. Let j_n be such that $2^{j_n} \sim n^{1/(2r+1/2)}$, and define thresholds

$$\tau_n(l) = \frac{B^2}{2^{2ls}} + 2C\frac{B}{2^{ls}}\frac{2^{(l+j_n)/8}}{\sqrt{n}} + C^2\frac{2^{(l+j_n)/4}}{n} = \left(\frac{B}{2^{ls}} + C\frac{2^{(l+j_n)/8}}{\sqrt{n}}\right)^2,$$

for a constant C to be chosen later. Take

$$\Psi_n = 1 - \prod_{J_0 \le l \le j_n - 1} 1\{T_n(l) < \tau_n(l)\}$$

as the test for H_0.

Under the null hypothesis, we know, by definition of the Besov norm, that

$$\sup_{J_0 \le l \le j_n - 1}\left(\sum_{k\in\mathcal{Z}_l} \langle f, \psi_{lk}\rangle^2 - \frac{B^2}{2^{2ls}}\right) \le 0;$$

hence,

$$
\begin{aligned}
E_f\Psi_n &= P_f^Y(T_n(l) > \tau_n(l) \text{ for some } J_0 \le l \le j_n - 1)\\
&\le \sum_{l\le j_n-1} \Pr\left(\sum_{k\in\mathcal{Z}_l} \hat f_{lk}^2 - \frac{2^l}{n} > \tau_n(l)\right)\\
&= \sum_{l\le j_n-1} \Pr\left(\sum_k \langle f,\psi_{lk}\rangle^2 - \frac{B^2}{2^{2ls}} + \frac{2}{\sqrt{n}}\sum_k \langle f,\psi_{lk}\rangle g_{lk} + \frac{1}{n}\sum_k (g_{lk}^2-1) > \tau_n(l) - \frac{B^2}{2^{2ls}}\right)\\
&\le \sum_{l\le j_n-1} \Pr\left(\left|\frac{2}{\sqrt{n}}\sum_k \langle f,\psi_{lk}\rangle g_{lk}\right| > 2C\frac{B}{2^{ls}}\frac{2^{(l+j_n)/8}}{\sqrt{n}}\right)\\
&\quad + \sum_{l\le j_n-1} \Pr\left(\left|\frac{1}{n}\sum_k (g_{lk}^2-1)\right| > C^2\frac{2^{(j+l)/4}}{n}\right)\\
&\le C^{-2}2^{-j_n/4}\sum_{l\le j_n-1} 2^{-l/4} + C^{-4}c'2^{-j/2}\sum_{l\le j_n-1} 2^{l/2} \le c''C^{-2} \le \alpha/2,
\end{aligned}
$$

for C large enough, using Chebyshev's inequality and the variance bounds

$$
\mathrm{Var}\left(\sum_k \langle f,\psi_{lk}\rangle g_{lk}\right) \le \|\langle f,\psi_{l\cdot}\rangle\|_2^2 \le B^2 2^{-2ls} \quad \text{and} \quad \mathrm{Var}\left(\sum_k (g_{lk}^2-1)\right) \le c'2^l.
$$

We now turn to the alternatives and start with the following preliminary observation, where π_W denotes the L^2-projection operator onto the linear spaces $V_j, W_j = V_{j+1}\setminus V_j$ spanned by the wavelet basis. By the triangle inequality, for all j,

$$
\rho_n^2 \le \inf_{h\in H_0} \|f-h\|_2^2 \le \inf_{h\in H_0} \|\pi_{V_j}(f) - h\|_2^2 + B2^{-2jr},
$$

so, for c large enough,

$$
\inf_{h\in H_0} \|\pi_{V_{j_n}}(f) - h\|_2^2 \ge 3\rho_n^2/4. \tag{6.94}
$$

Next,

$$
\begin{aligned}
\inf_{h\in H_0} \|\pi_{V_{j_n}}(f) - h\|_2 &\le \inf_{h\in H_0} \sum_{l\le j_n-1} \|\pi_{W_l}(f) - h\|_2\\
&= \inf_{h_{lk}:2^{ls}\|g_{l\cdot}\|_2\le B} \sum_{l\le j_n-1} \sqrt{\sum_k (f_{lk}-h_{lk})^2}\\
&= \sum_{l\le j_n-1} \inf_{h_{lk}:2^{ls}\|g_{l\cdot}\|_2\le B} \sqrt{\sum_k (f_{lk}-h_{lk})^2}\\
&\le \sum_{l\le j_n-1} \max\left(0, \|f_{l\cdot}\|_2 - \frac{B}{2^{ls}}\right),
\end{aligned}
$$

where we note that the minimisation problems do not interact across the scale indices l. Summarising, for any $L > 1$, our choice of ρ_n with c large enough, and writing

$$
t_n(l) = C^2 2^{(l+j_n)/4}/n,
$$

we have

$$L \sum_{l \le j_n - 1} \sqrt{t_n(l)} \le 3\rho_n/4 \le \sum_{l \le j_n - 1} \left(\|f_{l\cdot}\|_2 - \frac{B}{2^{ls}} \right), \tag{6.95}$$

so at least one summand on the right-hand side, say, the $\bar{l}$th, needs to exceed or equal the corresponding $\sqrt{t_n(\bar{l})}$; hence,

$$\sqrt{Lt_n(\bar{l})} \le \|f_{\bar{l}\cdot}\|_2 - \frac{B}{2^{\bar{l}s}} \Rightarrow \|f_{\bar{l}\cdot}\|_2^2 \ge \left(\sqrt{Lt_n(\bar{l})} + \frac{B}{2^{\bar{l}s}} \right)^2, \tag{6.96}$$

for some $\bar{l} \le j_n - 1$. To bound the type 2 errors, we thus have, for $f \in H_1$ and L large enough

$$\begin{aligned} E_f(1 - \Psi_n) &= P_f^Y\left(T_n(l) \le \tau_n(l) \text{ for all } J_0 \le l \le j_n - 1\right) \le P_f^Y\left(T_n(\bar{l}) \le \tau_n(\bar{l})\right) \\ &= \Pr\left(\sum_k \langle f, \psi_{\bar{l}k} \rangle^2 + \frac{2}{\sqrt{n}} \sum_k \langle f, \psi_{\bar{l}k} \rangle g_{\bar{l}k} + \frac{1}{n} \sum_k (g_{\bar{l}k}^2 - 1) \le \tau_n(\bar{l}) \right) \\ &\le \Pr\Bigg(-\frac{2}{\sqrt{n}} \sum_k \langle f, \psi_{\bar{l}k} \rangle g_{\bar{l}k} - \frac{1}{n} \sum_k (g_{\bar{l}k}^2 - 1) \ge Lt_n(\bar{l}) \\ &\quad + 2\sqrt{Lt_n(\bar{l})} \frac{B}{2^{\bar{l}s}} + \frac{B^2}{2^{2\bar{l}s}} - \tau_n(\bar{l}) \Bigg) \\ &\le \Pr\left(\left| \frac{2}{\sqrt{n}} \sum_k \langle f, \psi_{\bar{l}k} \rangle g_{\bar{l}k} \right| \ge 2(\sqrt{L} - 1) C \frac{B}{2^{ls}} \frac{2^{(l+j_n)/8}}{\sqrt{n}} \right) \\ &\quad + \Pr\left(\left| \frac{1}{n} \sum_k (g_{\bar{l}k}^2 - 1) \right| \ge (L-1) C^2 \frac{2^{(j+l)/4}}{n} \right) \le \alpha/2 \end{aligned}$$

by the same arguments as at the end of the type 1 errors, completing the proof. ∎

It is possible to prove a sampling analogue of the preceding result (see Exercise 6.2.4).

Exercises

6.2.1 (Kolmogorov-Smirnov test.) For F any continuous distribution function on $\mathbb{R}$ and G_F the F-Brownian bridge, show that the distribution of $\sup_{t \in \mathbb{R}} |_F(t)|$ equals the one of the maximum $\sup_{t \in [0,1]} |G(t)|$ of a standard Brownian bridge. *Hint*: Use that the quantile transform F^{-1} maps $[0,1]$ onto $\mathbb{R}$ and that $F^{-1}(U)$ has law F for $U \sim U(0,1)$. Moreover, use the Dvoretzky-Kiefer-Wolfowitz inequality (Exercise 3.3.3) to find a numerical value z_α such that

$$\sup_F P_F^{\mathbb{N}} P(\|F_n - F\|_\infty > z_\alpha) \le \alpha \quad \forall n \in \mathbb{N},$$

where the supremum extends over all distribution functions F.

6.2.2 (Existence of infimum tests.) (a) Let H_0 be a totally bounded subset of L^2. Then its closure $\bar{H}_0$ is compact, and hence, for any continuous mapping defined on L^2,

$$\inf_{h \in H_0} |L_f(h)| = |L_f(h^*)|,$$

for some $h^* \in \bar{H}_0$. *Hint*: Apply standard continuity and weak compactness arguments from real analysis. (b) Use the preceding to establish the existence of the infimum test statistics considered in Theorems 6.2.14 and 6.2.17 for such H_0, including, in particular, balls in Besov spaces $B^r_{2\infty}([0,1])$. (c) Noting that h^* in (b) is random, establish its measurability under the maintained assumptions.

6.2.3 (U-statistic tests on the real line and Euclidean space.) Proposition 6.2.5 and Theorem 6.2.17 generalise to more general sample spaces A than $[0,1]$ simply by replacing the basis functions ψ_{lk} used there by appropriate basis functions of $L^2(A)$. For instance, when $A=\mathbb{R}^d$, take a tensor wavelet basis from Section 4.3.6 based on S-regular wavelets on $\mathbb{R}$, and prove an analogue of Theorem 6.2.17. *Hint*: The proof is very similar – since the sums in k are now not necessarily finite anymore, in the variance estimate for ER^2_f, use that $\sum_{l\le j-1,k}\psi_{lk}(x)\psi_{lk}(y)$ is majorised by a nicely integrable convolution kernel $K(x-y)$ (in view of Definition 4.2.14). The variances are then of order $2^{jd/2}/n$, giving rise to separation rates $\rho_n \sim n^{-r/(2r+d/2)}$.

6.2.4 Prove a sampling analogue of Proposition 6.2.20. *Hint*: Replace $\int \psi_{lk}(t)dY(t)$ in the proof by the empirical wavelet coefficients $\int \psi_{lk}(t)dP_n(t)$.

6.2.5 Show that the alternatives f_m in the proof of Theorem 6.2.9 satisfy

$$\|f_m-1\|_1 \ge cn^{-r/(2r+1/2)}$$

and that thus the minimax separation rate is not changed if the $\|\cdot\|_2$-norm is replaced by $\|\cdot\|_1$ in H_1 in that theorem. *Hint*: Use the results from Chapter 4 to obtain

$$\|f_m-1\|_1 \ge c'\|f_m-1\|_{B^0_{11}} \simeq 2^{-j_n/2}\|\langle f_m,\psi_{j_n\cdot}\rangle\|_1 = 2^{-j_n r}, \quad c'>0,$$

for the wavelet sequence norm of B^0_{11}. For upper bounds, notice that $\rho_n \le \|f-1\|_1 \le \|f-1\|_2$.

6.2.6 Let K_j be the Haar wavelet projection kernel. Show that for $f\in B^r_{2q}, 1\le q\le\infty$, we have

$$\|K_j(f)\|_2 \ge c_1\|f\|_2 - c_2\|f\|_{B^r_{2,q}}2^{-jr}.$$

Deduce that in Proposition 6.2.3, expression (6.34) with the middle inequality ommitted still holds true for f satisfying (6.35) with $r>1$ and that, hence, condition (6.32) can be ommitted in Proposition 6.2.3. This implies that for nonparametric testing problems, we can always use test statistics computed from the Haar basis only. See proposition 2.16 in Ingster and Suslina (2003).

6.3 Nonparametric Estimation

Consider again statistical experiments $\mathcal{E}_n$ giving rise to observations $Y=Y^{(n)}$ on the measurable space $(\mathcal{Y}_n,\mathcal{A}_n), n\in\mathbb{N}$, of law P_f indexed by f varying in a parameter space $\mathcal{F}$ (again, these laws may depend in principle on n). We now consider the problem of estimating the parameter f directly from the observations $Y^{(n)}$. That is, we want to find a (measurable) function $\hat f_n:\mathcal{Y}_n\to\mathcal{F}$ that is close to f in the case where $Y^{(n)}$ is indeed drawn from f. To measure closeness, we will endow $\mathcal{F}$ with a metric d, and consider the *d-risk*

$$E_f d(\hat f_n, f),$$

where E_f is the expectation operator corresponding to P_f. The minimax paradigm requires the risk $E_f d(\hat f_n, f)$ to be controlled independently of which $f\in\mathcal{F}$ has generated $Y^{(n)}$; that is, we are looking for bounds for

$$\sup_{f\in\mathcal{F}} E_f d(\hat f_n, f).$$

We can study the last quantity for particular choices of $\hat{f}_n$ – some results of this kind were given in Chapter 5. A fundamental statistical property of the triple $(\mathcal{E}_n, \mathcal{F}, d)$ is the *minimal uniform*, or *minimax risk*,

$$\inf_{\tilde{f}_n} \sup_{f \in \mathcal{F}} E_f d(\hat{f}_n, f)$$

that can be achieved by the 'best' estimator $\tilde{f}_n$.

Definition 6.3.1 Let $(P_f : f \in \mathcal{F})$ be a statistical model for the law of observations $(Y^{(n)} : n \in \mathbb{N})$ in the measurable space $(\mathcal{Y}_n, \mathcal{A}_n)$. Let $\mathcal{F}$ be a subset of a metric space (S, d) equipped with its Borel σ-field. The sequence $(r_n : n \in \mathbb{N})$ is called the *minimax rate of estimation in d-risk over* $\mathcal{F}$ if the following two requirements are met:

(i) There exists a measurable function $\hat{f}_n : \mathcal{Y}_n \to S$ and a universal constant $\overline{C}$ such that, for every $n \in \mathbb{N}$ large enough,

$$r_n^{-1} \sup_{f \in \mathcal{F}} E_f d(\hat{f}_n, f) \leq \overline{C}. \tag{6.97}$$

(ii) There exists a universal constant $\underline{C}$ such that

$$\liminf_n r_n^{-1} \inf_{\tilde{f}_n} \sup_{f \in \mathcal{F}} E_f d(\hat{f}_n, f) \geq \underline{C}, \tag{6.98}$$

where the infimum extends over all measurable functions $\tilde{f}_n : \mathcal{Y}_n \to S$.

As in the testing case from Definition 6.2.1, we remark that the following bounds we obtain on the minimax risk typically hold for every $n \in \mathbb{N}$ but are informative usually for large n. Moreover, we will show that the lower bounds also hold if expectations in (6.98) are replaced by weaker probability statements (see (6.99)).

Upper bounds for the minimax rate of estimation in function estimation problems were studied in detail in Chapter 5, and in this section we complement these upper bounds by appropriate lower bounds, thus characterising the minimax rate of estimation in a variety of nonparametric statistical models. We have already seen in Propositions 6.2.2 and 6.2.13 that estimators satisfying (6.97) solve certain testing problems, and we thus can attempt to prove lower bounds for estimation by a reduction to testing problems. Some refinements compared to the preceding section will be needed, taking into account that an estimator does not solve only one testing problem H_0 versus H_1 but in fact can be used to solve many such testing problems, depending on the complexity of the parameter space $(\mathcal{F}, d)$. Intuitively, the complexity of the metric space $(\mathcal{F}, d)$ will govern the estimation rate, and for minimax theory on Besov bodies, the wavelet techniques from Chapter 4 will be particularly useful.

6.3.1 Minimax Lower Bounds via Multiple Hypothesis Testing

We demonstrate in this subsection how minimax estimation lower bound results can be reduced to lower bounds of certain testing problems that involve several hypotheses $H_0, H_1, \ldots, H_m$ in the parameter space and how information-theoretic tools from Section 6.1 can be used to obtain quantitative lower bounds for such testing problems.

A General Reduction Principle

Markov's inequality implies that

$$r_n^{-1} \inf_{\tilde{f}_n} \sup_{f \in \mathcal{F}} E_f d(\tilde{f}_n, f) \geq \inf_{\tilde{f}_n} \sup_{f \in \mathcal{F}} P_f(d(\tilde{f}_n, f) > r_n) \geq \inf_{\tilde{f}_n} \max_{m=0,\dots,M} P_{f_m}(d(\tilde{f}_n, f_m) > r_n), \quad (6.99)$$

for any finite set $(f_m : m = 0, \dots, M)$ in $\mathcal{F}$. Suppose that the f_m are $2r_n$-separated from each other; that is,

$$d(f_m, f_{m'}) \geq 2r_n \quad \forall\, m \neq m'. \quad (6.100)$$

Any estimator $\tilde{f}_n$ can be used to test among the $M+1$-many hypotheses f_m simply by choosing the f_m closest to $\tilde{f}_n$: formally, let

$$\Psi_n : \mathcal{Y}^{(n)} \to \{0, \dots, M\}$$

be such that

$$d(\tilde{f}_n, f_{\Psi_n}) = \min_{m=0,\dots,M} d(\tilde{f}_n, f_m),$$

where in case of a tie we may choose any of the minimisers. The errors of this test are bounded by

$$P_{f_m}(\Psi_n \neq m) \leq P_{f_m}(d(\tilde{f}_n, f_{\bar{m}}) \leq d(\tilde{f}_n, f_m) \text{ for some } \bar{m}).$$

On the event in the preceding probability and by the triangle inequality,

$$d(\tilde{f}_n, f_m) \geq d(f_m, f_{\bar{m}}) - d(\tilde{f}_n, f_{\bar{m}}) \geq d(f_m, f_{\bar{m}}) - d(\tilde{f}_n, f_m),$$

so, by the separation hypothesis (6.100),

$$2d(\tilde{f}_n, f_m) \geq d(f_m, f_{\bar{m}}) \ \Rightarrow d(\tilde{f}_n, f_m) \geq r_n,$$

and we conclude that

$$P_{f_m}(\Psi_n \neq m) \leq P_{f_m}(d(\tilde{f}_n, f_m) \geq r_n) \quad \forall m = 0, 1, \dots, M. \quad (6.101)$$

The inequality is preserved by taking maxima over m and infima over $\tilde{f}_n, \Psi_n$, and we thus have from (6.99)

$$r_n^{-1} \inf_{\tilde{f}_n} \sup_{f \in \mathcal{F}} E_f d(\tilde{f}_n, f) \geq \inf_{\Psi_n} \max_{m=0,\dots,M} P_{f_m}(\Psi_n \neq m) \quad (6.102)$$

whenever the hypotheses $(f_m : m = 0, \dots, M) \subset \mathcal{F}$ are $2r_n$-separated as in (6.100).

Information-Theoretic Lower Bounds for Multiple Hypothesis Testing Problems

To bound the right-hand side of (6.102) further, we can use the information-theoretic tools from Section 6.1.2, in particular, the Kullback-Leibler distance $K(P, Q)$ between two probability measures. The following general-purpose result will be used repeatedly:

Theorem 6.3.2 *Suppose that* $\mathcal{F}$ *contains*

$$\{f_m : m = 0, 1, \dots, M\}, \quad M \geq 1,$$

which are $2r_n$ separated as in (6.100) and such that the P_{f_m} are all absolutely continuous with respect to P_{f_0}. Set $\bar{M} = \max(e, M)$, and assume that, for some $\alpha > 0$,

$$\frac{1}{M}\sum_{m=1}^{M} K(P_{f_m}, P_{f_0}) \le \alpha \log \bar{M}. \tag{6.103}$$

Then the minimax risk from Definition 6.3.1 is lower bounded by

$$\inf_{\tilde{f}_n}\sup_{f\in\mathcal{F}} E_f d(\tilde{f}_n, f) \ge r_n \frac{\sqrt{\bar{M}}/3}{1+\sqrt{\bar{M}}}\left(1 - 2\alpha - \sqrt{\frac{8\alpha}{\log \bar{M}}}\right). \tag{6.104}$$

Remark 6.3.3 The constant 8 can be replaced by 2 at the expense of a slightly longer proof. For applications that follow, this improvement will be irrelevant.

Proof We prove the result under the assumption that all the P_{f_m} are mutually absolutely continuous to each other; the general case needs only minor modifications. In view of (6.102), it is sufficient to lower bound

$$\inf_{\Psi_n}\max_{m=0,\dots,M} P_{f_m}(\Psi_n \ne m) \tag{6.105}$$

by the right-hand side of (6.104). For Ψ any measurable function from $\mathcal{Y}^{(n)}$ to $\{0,1,\dots,M\}$, we can write $\{\Psi \ne 0\} = \cup_{1\le m\le M}\{\Psi = m\}$, a union of disjoint events. Define the events

$$A_m = \left\{\frac{dP_{f_0}}{dP_{f_m}} \ge 1 - \eta\right\}.$$

We have, for every $0 < \eta < 1$,

$$\begin{aligned}
P_{f_0}(\Psi \ne 0) &= \sum_{m=1}^{M} P_{f_0}(\Psi = m) \ge \sum_{m=1}^{M} E_{f_0}\left[1\{\Psi = m\}1_{A_m}\right] \\
&= \sum_{m=1}^{M} E_{f_m}\left[1\{\Psi = m\}1_{A_m}\frac{dP_{f_0}}{dP_{f_m}}\right] \\
&\ge (1-\eta)\sum_{m=1}^{M} P_{f_m}(\Psi = m) - (1-\eta)\sum_{m=1}^{M} P_{f_m}(A_m^c).
\end{aligned}$$

Now, writing

$$p_0 = \frac{1}{M}\sum_{m=1}^{M} P_{f_m}(\Psi = m), \qquad L = \frac{1}{M}\sum_{m=1}^{M} P_{f_m}(A_m^c),$$

we have, for every $0 < \eta < 1$,

$$\begin{aligned}
\max_{m=0,\dots,M} P_{f_m}(\Psi_n \ne m) &= \max\left(P_{f_0}(\Psi \ne 0), \max_{1\le m\le M} P_{f_m}(\Psi \ne m)\right) \\
&\ge \max\left((1-\eta)\sum_{m=1}^{M} P_{f_m}(\Psi = m) - (1-\eta)\sum_{m=1}^{M} P_{f_m}(A_m^c), \frac{1}{M}\sum_{m=1}^{M} P_{f_m}(\Psi \ne m)\right)
\end{aligned}$$

$$
\begin{aligned}
&= \max((1-\eta)M(p_0 - L), 1 - p_0) \\
&\geq \inf_{0 \leq p \leq 1} \max((1-\eta)M(p-L), 1-p) \\
&= \frac{(1-\eta)M}{1+(1-\eta)M} \frac{1}{M} \sum_{m=1}^{M} P_{f_m}\left(\frac{dP_{f_0}}{dP_{f_m}} \geq 1-\eta\right)
\end{aligned}
$$

because the infimum over p is attained when the two terms in the maximum are equal. As in (6.11) and by Proposition 6.1.7, each of the summands in the preceding expression can be further bounded below by

$$
1 - \frac{1}{\log(1/(1-\eta))}\left[K(P_{f_m}, P_{f_0}) + \sqrt{2K(P_{f_m}, P_{f_0})}\right],
$$

and, by Jensen's inequality,

$$
\frac{1}{M}\sum_{m=1}^{M}\sqrt{K(P_{f_m}, P_{f_0})} \leq \sqrt{\frac{1}{M}\sum_{m=1}^{M} K(P_{f_m}, P_{f_0})},
$$

so, combined with the preceding bound and using the hypothesis on the Kullback-Leibler distance, we obtain, for every $\eta > 0$,

$$
\max_{m=0,\ldots,M} P_{f_m}(\Psi_n \neq m) \geq \frac{(1-\eta)M}{1+(1-\eta)M}\left(1 - \frac{1}{\log\frac{1}{1-\eta}}\left(\alpha \log \bar{M} + \sqrt{2\alpha \log \bar{M}}\right)\right). \tag{6.106}
$$

Choosing $\eta = 1 - \bar{M}^{-1/2}$ gives the desired result. ■

6.3.2 Function Estimation in L^∞ Loss

We consider in this section the problem of estimating an unknown function f in supnorm loss based on observations in the sampling or Gaussian white noise model. A first basic observation is that when estimating an unknown distribution function F, the rate $1/\sqrt{n}$ obtained from the empirical distribution function F_n cannot be improved on.

Proposition 6.3.4 *Denote by $\mathcal{F}$ the set of all probability distribution functions on* $[0,1]$. *The minimax rate of estimation over $\mathcal{F}$ in $\|\cdot\|_\infty$ risk in the sense of Definition 6.3.1 based on observations $X_1, \ldots X_n \sim^{i.i.d.} F$ equals*

$$
r_n = \frac{1}{\sqrt{n}}.
$$

Proof By Exercise 3.1.7 we have $E\|F_n - F\|_\infty \leq 4/\sqrt{n}$, for some constant C and every $n \in \mathbb{N}$, so the upper bound follows. For the lower bound, suppose that the rate r_n is faster than $1/\sqrt{n}$. Then we can construct an estimator for which Proposition 6.2.2 implies the existence of a test that is consistent against r_n-separated alternatives. This implies a contradiction to Proposition 6.2.6. ■

This result can be easily generalised to the case of probability distribution functions on $\mathbb{R}^d$ (see Exercise 6.3.1). Note further that the proof of Proposition 6.2.6 in fact implies that the preceding result remains true when $\mathcal{F}$ consists of all probability distribution functions that have S derivatives bounded in supremum norm by a fixed constant M, where S, M are arbitrary. Hence, adding regularity constraints on $\mathcal{F}$ does not improve the rate if one is interested in supnorm loss on distribution functions. For estimating densities, or functions in white noise, such additional regularity constraints, however, fundamentally influence the minimax rates. We investigate this in the next subsections.

L^∞-Minimax Rates in Gaussian White Noise

Consider observing dY in (6.3) from some f that belongs to a Besov space $B^r_{\infty\infty}([0,1])$ modelling r-Hölderian functions. In Proposition 5.1.7, we obtained the rate of estimation

$$(n/\log n)^{-r/(2r+1)},$$

and we may wonder whether this is optimal. Similar to the proof of Proposition 6.3.4, we can use Proposition 6.2.2 and the discussion after it combined with the lower bound in Theorem 6.2.11 to show that the rate in the preceding display cannot be improved. Here is another, more direct proof based on Theorem 6.3.2:

Theorem 6.3.5 *Let $B, r > 0$. The minimax rate of estimation over*

$$\mathcal{F} = \{f : \|f\|_{B^r_{\infty\infty}([0,1])} \le B\}$$

in $\|\cdot\|_\infty$ risk in the sense of Definition 6.3.1 based on observations $dY_f \sim P^Y_f$ in the Gaussian white noise model equals

$$r_n = C\left(\frac{\log n}{n}\right)^{r/(2r+1)},$$

where the constant C depends on B, r, σ.

Proof The upper bound follows from Proposition 5.1.7. For the lower bound, take $f_0 = 0$ on $[0,1]$, and for $S > r$, let ψ be an S-regular Daubechies wavelet from Theorem 4.2.10, translated by $N \equiv N(S)$ so that its support is $[1, 2N]$. For $j \in \mathbb{N}$ large enough, we can take $c_0 2^j = M$ wavelets $\psi_{jm} = 2^{j/2}\psi(2^j(\cdot) - m)$ with disjoint support contained in the interior of $[0,1]$, with c_0 a fixed positive constant that depends only on N. Define, for $\epsilon > 0$, the functions

$$f_m := \epsilon 2^{-j(r+1/2)}\psi_{jm}, \quad m = 1, \dots, M. \tag{6.107}$$

We have $\|f_0\|_{B^r_{\infty\infty}([0,1])} = 0$ and, using Theorem 4.3.2,

$$\|f_m\|_{B^r_{\infty\infty}([0,1])} = \epsilon 2^{-j(r+1/2)}\|\psi_{jm}\|_{B^r_{\infty\infty}(\mathbb{R})} \le c_1\epsilon, \tag{6.108}$$

for some constant $c_1 = c_1(r)$, so for every j and $\epsilon \le B/c_1$ we have $f_m \in \mathcal{F}$ for every m. Moreover, in view of the disjoint support of the f_m, we see that

$$\|f_m - f_{m'}\|_\infty = \epsilon 2^{-jr}\|\psi\|_\infty,$$

so if we choose $j = j_n$ such that

$$2^{j_n} \simeq (n/\log n)^{1/(2r+1)},$$

then the f_m are all $2\underline{C}r_n$-separated from each other in supremum norm for $\underline{C}$ small enough depending only on $\epsilon, \|\psi\|_\infty$. The Kullback-Leibler distance between P_{f_m} and P_{f_0} equals, by (6.16), for $m = 1,\dots,M$,

$$K(P_{f_m}, P_{f_0}) = \frac{n}{2\sigma^2}\|f_m\|_2^2 = \frac{n}{2\sigma^2}\epsilon^2 2^{-j_n(2r+1)} \le \epsilon^2 C(\sigma, c_0)\log M,$$

and thus,

$$\frac{1}{M}\sum_{m=1}^{M} K(P_{f_m}, P_{f_0}) \le \epsilon^2 C(\sigma, c_0)\log M,$$

so the result follows from Theorem 6.3.2 after choosing ϵ small enough depending only on c_0, σ, S. ■

Remark 6.3.6 Inspection of this proof, combined with Proposition 5.1.7, shows further that the dependence of the constant C on B is of the form $C = cB^{1/(2r+1)}$, where c does not depend on B.

L^∞-Minimax Rates in Density Estimation

Consider next observing i.i.d. random variables $X_1,\dots,X_n$ from probability density function f, and denote the joint distribution of the observations by P_f^n. The following theorem gives lower bounds for the univariate situation; the multivariate situation is treated in Exercise 6.3.2. The dependence of the constant C on B is as in Remark 6.3.6

Theorem 6.3.7 *Let $B > 1, r > 0$, let A equal either $[0,1]$ or $\mathbb{R}$ and let $\mathcal{F}$ consist of all probability density functions in*

$$\{f : \|f\|_{B^r_{\infty\infty}(A)} \le B\}.$$

The minimax rate of estimation over $\mathcal{F}$ in $\|\cdot\|_\infty$-risk in the sense of Definition 6.3.1 based on observations $X_1,\dots X_n \sim^{i.i.d.} f$ equals

$$r_n = C\left(\frac{\log n}{n}\right)^{r/(2r+1)},$$

where the constant C depends on B, r.

Proof The upper bound follows from Proposition 5.1.7. The proof of the lower bound proceeds differently for $A = [0,1]$ and $A = \mathbb{R}$.

Case $A = [0,1]$: The proof considers similar alternatives as in Theorem 6.3.5. Let $f_0' = 1$ be the uniform density, and let

$$f_m' = 1 + \epsilon 2^{-j_n(r+1/2)}\psi_{jm} = f_0' + f_m, \quad m = 1,\dots,M,$$

where $\psi_{j_n m}, j_n, f_m$ are as in and before (6.107). We have $\|f_0'\|_{B^r_{\infty\infty}} = 1$ for all r, and as in Theorem 6.3.5, we show, using $B > 1$, that for ϵ small enough, the f_m' are all positive and in $\mathcal{F}$ and $2\underline{C}r_n$-separated from each other in supremum norm. By (6.14), the Kullback-Leibler distance between the product measures $P^n_{f'_m}, P^n_{f'_0}$ describing the law of the $X_1,\dots,X_n$ under f'_m, f'_0 is bounded by

$$\begin{aligned}
K(P^n_{f'_m}, P^n_{f'_0}) &= n\int \log(f'_m(x)) f'_m(x)\,dx \\
&= n\int \log(1+\epsilon 2^{-j_n(r+1/2)}\psi_{j_n m}(x)) f'_m(x)\,dx \\
&\le \epsilon n 2^{-j_n(r+1/2)}\int \psi_{j_n m}(x)(1+\epsilon 2^{-j_n(r+1/2)}\psi_{j_n m}(x))\,dx \\
&= \epsilon^2 n 2^{-j_n(2r+1)} \\
&\le \epsilon^2 c\log M,
\end{aligned}$$

where we have used $\log(1+x)\le x$ for $x>-1$ and $\int \psi_{j_n m}=0$ for all m. The result now follows from Theorem 6.3.2 after choosing ϵ small enough.

Case $A=\mathbb{R}$: The proof needs minor modifications since $f_0 = 1$ is not a smooth density on the whole real line. Instead, we take f_0 equal to a normal density with large enough variance such that $\|f_0\|_{B^r_{\infty\infty}} < B$ (possible in view of $\|f_0\|_\infty \le 1$, $B>1$, (4.104) and Proposition 4.3.23), and define

$$f_m = f_0 + \epsilon 2^{-j_n(r+1/2)}\psi_{jm},$$

which, arguing as earlier, are all in $\mathcal{F}$ for ϵ small enough and $2\underline{C}r_n$-separated from each other. The Kullback-Leibler divergence is bounded, using again $\log(1+x)\le x$, for $x>-1$ and $\int\psi_{j_n m}=0$, as follows:

$$\begin{aligned}
K(P^n_{f_m}, P^n_{f_0}) &= n\int \log\left(\frac{f_m(x)}{f_0(x)}\right) f_m(x)\,dx \\
&= n\int \log\left(1+\epsilon 2^{-j_n(r+1/2)}\frac{\psi_{j_n m}(x)}{f_0(x)}\right) f_m(x)\,dx \\
&\le \epsilon n 2^{-j_n(r+1/2)}\int \frac{\psi_{j_n m}(x)}{f_0(x)}(f_0(x)+\epsilon 2^{-j_n(r+1/2)}\psi_{j_n m}(x))\,dx \\
&\le \left(\inf_{x\in[0,1]} f_0(x)\right)^{-1}\epsilon^2 n 2^{-j_n(2r+1)} \\
&\le \epsilon^2 c\log M,
\end{aligned}$$

for a constant c that can be taken to depend on r only. The result now follows again from Theorem 6.3.2 after choosing ϵ small enough. ■

6.3.3 Function Estimation in L^p-Loss

Inspection of the proof of Theorem 6.3.5 reveals that the main idea behind the lower bound was to construct functions that are separated in the L^∞-norm at the estimation rate but whose L^2-distance is of much smaller order $\sqrt{(\log n)/n}$, corresponding to the 'spikes' discussed before Theorem 6.2.7. Since the Kullback-Leibler divergence between two white noise experiments is driven by the L^2-distance of their drift coefficients, this means that functions that are different on a small interval cannot be reliably estimated in uniform loss. One may wonder whether the estimation rate improves if $\|\cdot\|_\infty$-loss is replaced by $\|\cdot\|_2$-loss, since, after all, then the loss function coincides with the information-theoretic distance on the experiment. This would parallel the improvement of the testing rate in Theorem 6.2.11 when L^∞-separation is replaced by L^2-separation. Somewhat surprisingly perhaps, such an improvement does not occur in L^2-loss, except for the removal of the $\log n$ term in Theorem 6.3.5. The proof of this fact, unlike Theorem 6.3.5, cannot be derived directly from testing lower bounds from Section 6.2 but requires the more refined techniques from Theorem 6.3.2.

The proof techniques for L^2-risk in fact imply the same estimation rates for the L^p-risks whenever $1 \le p < \infty$. Remarkably, the minimax estimation rates in L^p over Besov bodies in $B^r_{p\infty}$ do not depend on p. This can change when the p-parameter of the Besov body is not matched with the p-parameter of the L^p-risk (see the notes at the end of this chapter for some discussion).

We finally remark that the natural exhaustive classes for minimax estimation are balls of the spaces $B^r_{p\infty}$. The lower (and trivially also the upper) bounds in this section hold in fact for any value $q \in [1,\infty]$ in B^r_{pq}, showing that the choice of the q-index is not important for estimation problems. Moreover, we could set $p = \infty$, and all the lower bounds would still remain true (see Exercise 6.3.3).

L^p-Minimax Rates in Gaussian White Noise

In Gaussian white noise with L^p-risk, $L^p = L^p([0,1]), 1 \le p < \infty$, and we have the following theorem. Note that Remark 6.3.6 on the dependence of the constant C on B applies here as well.

Theorem 6.3.8 *Let $B, r > 0$ and $1 \le p < \infty, 1 \le q \le \infty$. The minimax rate of estimation over*

$$\mathcal{F} = \{f : \|f\|_{B^r_{pq}([0,1])} \le B\}$$

in $\|\cdot\|_p$-risk in the sense of Definition 6.3.1 based on observations $dY_f \sim P^Y_f$ in the Gaussian white noise model equals

$$r_n = Cn^{-r/(2r+1)},$$

where the constant C depends on B, r, p, σ.

Proof The upper bound follows from Proposition 5.1.7 and $B^r_{pq} \subset B^r_{p\infty}$. For the lower bound, let $S > r$, and take S-regular Daubechies wavelets ψ_{jk} from the proof of Theorem 6.3.5 that are supported in the interior $[0,1]$. Recall that at the jth level there are $c_0 2^j$ of these interior wavelets, where $c_0 \equiv c_0(S) \le 1$ is a fixed positive constant. Set $f_0 = 0$, and for $m \ge 1$, $\beta_m = (\beta_{mk})$ any point in the discrete hypercube $\{-1,1\}^{c_0 2^j}$ and $\epsilon > 0$ a small constant, define functions

$$f_m(x) = \epsilon 2^{-j(r+1/2)} \sum_{k=1}^{c_0 2^j} \beta_{mk} \psi_{jk}(x).$$

By the wavelet characterisation of the Besov norm in Theorem 4.3.2, we have

$$\|f_m\|_{B^r_{pq}} = 2^{j(r+1/2-1/p)}\epsilon 2^{-j(r+1/2)}\left(\sum_k |\beta_{mk}|^p\right)^{1/p} \leq \epsilon,$$

so the f_m are all contained in $\mathcal{F}$. By Parseval's identity, the f_m are L^2-separated by

$$\|f_m - f_{m'}\|_2^2 = \epsilon^2 2^{-2j(r+1/2)} \sum_k (\beta_{mk} - \beta_{m'k})^2 \tag{6.109}$$

and more generally in L^p, from Proposition 4.2.8, by

$$\|f_m - f_{m'}\|_p \geq K' 2^{j(\frac{1}{2}-\frac{1}{p})}\epsilon 2^{-j(r+1/2)}\left(\sum_k |\beta_{mk} - \beta_{m'k}|^p\right)^{1/p}. \tag{6.110}$$

To obtain suitably separated f_m, we need to separate points in the hypercube $\{-1,1\}^{c_0 2^j}$: by Example 3.1.4 (the Varshamov-Gilbert bound) and for j large enough, there exist universal constants $c_1, c_2 > 0$ and a subset $\mathcal{M}$ of $\{-1,1\}^{c_0 2^j}$ of cardinality $M = 2^{c_1 2^j}$ such that

$$\sum_k |\beta_{mk} - \beta_{m'k}|^p \geq c_2 2^p 2^j$$

whenever $m \neq m'$. Hence,

$$\|f_m - f_{m'}\|_p \geq c(\epsilon, c_2) 2^{-jr}, \quad m \neq m', m, m' \in \mathcal{M}, \tag{6.111}$$

and choosing $j = j_n$ such that $2^{j_n} \simeq n^{1/(2r+1)}$, the $\{f_m : m \in \mathcal{M}\}$ are $2Cr_n$-separated and contained in $\mathcal{F}$. By (6.16), the Kullback-Leibler distances $K(P_{f_m}, P_{f_0}), m \geq 1$, are bounded by

$$\frac{n}{2\sigma^2}\|f_m\|_2^2 = \frac{n}{2\sigma^2}\epsilon^2 2^{-2j(r+1/2)} \sum_{k=1}^{c_0 2^j} 1 \leq \epsilon^2 c(c_0, c_1, \sigma^2) \log M, \tag{6.112}$$

so the result follows from Theorem 6.3.2 by choosing ϵ small enough. ■

L^p-Minimax Rates in Density Estimation

The following theorem gives a sampling analogue of Theorem 6.3.8 when $A = [0,1]$. More general sample spaces are discussed later.

Theorem 6.3.9 *Let $B > 1, r > 0, 1 \leq p < \infty, 1 \leq q \leq \infty$, and let $\mathcal{F}$ consist of all probability density functions in*

$$\{f : \|f\|_{B^r_{pq}([0,1])} \leq B\}.$$

The minimax rate of estimation over $\mathcal{F}$ in $\|\cdot\|_p$-risk in the sense of Definition 6.3.1 based on observations $X_1, \dots X_n \sim^{i.i.d.} f$ on $[0,1]$ equals

$$r_n = Cn^{-r/(2r+1)},$$

where the constant C depends on B, r.

Proof The upper bound follows from Proposition 5.1.7 and $B^r_{pq} \subset B^r_{q\infty}$. For the lower bound, we take the uniform density $f'_0 = 1$ and let

$$f'_m = f'_0 + \epsilon 2^{-j(r+1/2)} \sum_k \beta_{mk} \psi_{jk}(x) = f'_0 + f_m,$$

with $\beta_{mk}, \psi_{jk}, f_m$ as in the proof of Theorem 6.3.8. Arguing as in that proof and using

$$\|f'_0\|_{B^r_{pq}} = \|f'_0\|_p = \|f'_0\|_p = 1, \qquad \int \psi_{jk} = 0, \|f'_m - f'_0\|_\infty < 1,$$

for j large enough, we show that the f'_m are positive probability densities contained in $\mathcal{F}$. Moreover, as before (6.111), we can find a subset $\mathcal{M}$ of $\{-1,1\}^{c_0 2^j}$ of cardinality $M = 2^{c_1 2^j}$ such that the $\{f'_m : m \in \mathcal{M}\}$ are $2Cr_n$-separated in L^p-distance. Using $\log(1+x) \le x$ for $x > -1$, the Kullback- Leibler distances $K(P_{f'_m}, P_{f'_0}), m \ge 1$, are bounded by

$$\begin{aligned} K(P^n_{f'_m}, P^n_{f'_0}) &= n \int \log(f'_m(x)) f'_m(x)\,dx \\ &= n \int \log(1+f_m)(1+f_m(x))(x)\,dx \\ &\le n \int f_m + n \int f^2_m(x)\,dx \\ &= n\|f_m\|_2^2, \end{aligned}$$

so the result follows from the estimate (6.112) (with $\sigma = 1$) and Theorem 6.3.2. ∎

We now turn to the case where the observations $X_1, \ldots, X_n$ take values on the real line. The multivariate situation is treated in Exercise 6.3.2.

Theorem 6.3.10 *Let $B > 1, r > 0, 1 \le p < \infty, 1 \le q \le \infty, s > (2-p)/p$, and let $\mathcal{F}$ consist of all probability density functions in*

$$\{f : \|f\|_{B^r_{pq}(\mathbb{R})} \le B\}.$$

If $1 \le p < 2$, intersect, $\mathcal{F}$ further with $\{f : \int_{\mathbb{R}} f(t)(1+|t|^s) \le B\}$. The minimax rate of estimation over $\mathcal{F}$ in $\|\cdot\|_p$-risk in the sense of Definition 6.3.1 based on observations $X_1, \ldots X_n \sim^{i.i.d.} f$ on $\mathbb{R}$ equals

$$r_n = Cn^{-r/(2r+1)},$$

where the constant C depends on B, r.

Proof The upper bound follows from Proposition 5.1.7. The lower bound is proved as in Theorem 6.3.9, but with $f_0 = 1$ there replaced by a normal density with variance large enough that $\|f_0\|_{B^r_{\infty\infty}} < B$ and adapting the proof as in the case $A = \mathbb{R}$ in Theorem 6.3.7 in a straightforward way. The details are left to the reader. ∎

Exercises

6.3.1 Prove an analogue of Proposition 6.3.4 for the set $\mathcal{F}$ of probability distribution functions on $\mathbb{R}^d$.

6.3.2 Let A equal $[0,1]^d$ or $\mathbb{R}^d$. Prove analogues of the theorems in this section for minimax lower bounds when estimating functions in L^p-risk, $1 \le p < \infty$, and L^∞-risk over a Besov ball $B^r_{p\infty}(A)$. *Hint*: The rates are of the order

$$n^{-r/(2r+d)} \quad \text{and} \quad (n/\log n)^{-r/(2r+d)}$$

in L^p and L^∞, respectively. The proofs are the same as earlier using the results from Section 4.3.6 and the alternatives $\epsilon 2^{-j(r+1/2)} \sum_k \beta_{mk} \psi_{jk}(x)$ and $\epsilon 2^{-j(r+1/2)} \psi_{jm}$, with the ψ_{jm} forming a wavelet basis of $L^2(A)$ so that at each resolution level j there are now 2^{jd} wavelet coefficients.

6.3.3 Show that the functions f_m in the lower-bound proofs of Theorems 6.3.8 and 6.3.9 are in fact contained in balls of $B^r_{\infty\infty}([0,1])$ and that hence the minimax rate of convergence over such (Hölder) balls in $L^p([0,1])$ loss, $1 \le p < \infty$, is $n^{-r/(2r+1)}$.

6.4 Nonparametric Confidence Sets

Consider statistical experiments $\mathcal{E}_n$ giving rise to observations $Y = Y^{(n)}$ on the measurable space $(\mathcal{Y}_n, \mathcal{A}_n), n \in \mathbb{N}$. Suppose that the parameter space $\mathcal{F}$ describing the laws $(P_f : f \in \mathcal{F})$ that could have generated Y is a subset of a metric space (S,d) and that the minimax rate of estimation over $\mathcal{F}$ in d-risk in the sense of Definition 6.3.1 is equal to r_n. In this case, we know that an estimator $\tilde{f}_n$ based on $Y^{(n)}$ exists that attains the minimax rate of convergence: for n large enough and some $\overline{C} < \infty$,

$$\sup_{f \in \mathcal{F}} E_f d(\tilde{f}_n, f) \le \overline{C} r_n. \tag{6.113}$$

A key challenge in statistics is to go beyond the mere construction of an optimal decision rule $\tilde{f}_n = \tilde{f}(Y^{(n)})$ by *providing a data-driven quantification of the uncertainty in the estimate of f*. The reason why this is important is that a point estimate $\tilde{f}_n$ alone will be quite useless for the purposes of statistical inference if we cannot guarantee that $\tilde{f}_n$ is within a certain known distance to f with high probability. For instance, to give an extreme example, if we cannot be sure that $\overline{C} r_n$ is of smaller order than the diameter of $\mathcal{F}$, we will not want to trust the estimate $\tilde{f}_n$ in practice. A statistical approach to uncertainty quantification is based on the idea of using the observations $Y^{(n)}$ to construct a random subset C_n of $\mathcal{F}$ which contains f with large probability whenever $Y^{(n)}$ was indeed generated from P_f. Such sets C_n provide a quantification of uncertainty through their size and are called *confidence sets*. We will show that a natural goal is to construct C_n such that its d-diameter $|C_n|_d$ satisfies $|C_n|_d = O_P(r_n)$, reflecting the accuracy of estimation in (6.113).

The theory of confidence sets comprises at least two main challenges: the first is that in nonparametric situations, r_n often depends on several unknown quantities that need to be estimated themselves, and this can pose fundamental difficulties. We will touch on this issue to a certain extent in what follows, and it will occur more prominently in the adaptive setting of Section 8.3. The second challenge arises even in those situations where r_n is known: for example, the basic confidence set $C_n = [\tilde{f}_n - Lr_n, \tilde{f}_n + Lr_n]$ requires the choice of the constant L in dependence of the desired coverage probability $P_f(f \in C_n)$. If we want sharp results, this can be a delicate matter that depends on the precise probabilistic properties of $\tilde{f}_n$ in the loss function in which we want C_n to be small. We present the minimax approach to this problem in this section – a related Bayesian approach is discussed in Section 7.3.4.

6.4.1 Honest Minimax Confidence Sets

To give a general definition, we notice that a confidence set need not necessarily be centred at an estimator $\tilde{f}_n$ but could be any random subset C_n of the parameter space $\mathcal{F}$ that contains f with prescribed P_f probability. Any such C_n then does not only provide an estimate of f (by taking $\tilde{f}_n$ equal to an arbitrary element of C_n) but also an estimate of the uncertainty about f through the diameter of the set C_n.

Paralleling the situation of nonparametric testing and estimation, it is natural to require, from a minimax point of view, that the *coverage probability*

$$P_f(f \in C_n)$$

of a confidence set be controlled at any given level $1-\alpha, 0<\alpha<1$, *uniformly* in all $f \in \mathcal{F}$. The level α has the usual interpretation of being chosen by the statistician; C_n thus is allowed to depend on α. To fix ideas, let us give the following definition:

Definition 6.4.1 Let $(P_f : f \in \mathcal{F})$ be a statistical model for the law of observations $(Y^{(n)} : n \in \mathbb{N})$ in the measurable space $(\mathcal{Y}_n, \mathcal{A}_n)$. Let $\mathcal{F}$ be a subset of a metric space (S,d) equipped with its Borel σ-field $\mathcal{B}_S$. Given $0<\alpha<1$, a honest level $1-\alpha$ confidence set C_n for $\mathcal{F}$ is a random subset $C_n = C(\alpha, Y^{(n)}) : \mathcal{Y}_n \to \mathcal{B}_S$ of (S,d) such that, for some sequence $e_n = o(1)$,

$$\inf_{f \in \mathcal{F}} P_f(f \in C_n) \geq 1-\alpha+e_n. \tag{6.114}$$

The confidence set C_n is said to be of *exact asymptotic level* $1-\alpha$ if

$$\inf_{f \in \mathcal{F}} |P_f(f \in C_n) - (1-\alpha)| = o(1). \tag{6.115}$$

Note that such confidence sets are *honest* in the sense that there exists an index n_0 depending only on the model such that from then onwards, coverage holds for all f in the model. Clearly, for the uncertainty quantification provided by C_n to be as informative as possible, we want C_n to be as small as possible – otherwise we could take $C_n = \mathcal{F}$, which is not an interesting confidence set. Denote the d-diameter of C_n by

$$|C_n|_d = \sup\{d(f,g) : f,g \in C_n\}; \tag{6.116}$$

if d arises from an L^p-metric, we simply write $|C_n|_p$ in slight abuse of notation. We argue now that a natural minimax optimality criterion for confidence sets is to require that $|C_n|_d$ shrinks at the minimax rate of d-estimation over $\mathcal{F}$ in the sense of Definition 6.3.1.

First, we note that on the 'coverage' events $\{f \in C_n\}$ we can find random $\tilde{f}_n \in C_n$ depending only on $Y^{(n)}, \alpha$ such that

$$\{|C_n|_d \leq r_n, f \in C_n\} \subseteq \{d(\tilde{f}_n, f) \leq r_n\}.$$

Negating this inclusion, we have

$$\{|C_n|_d > r_n\} \cup \{f \notin C_n\} \supseteq \{d(\tilde{f}_n, f) > r_n\},$$

so

$$P_f(d(\tilde{f}_n, f) > r_n) \leq P_f(|C_n|_2 > r_n) + P_f(f \notin C_n) \leq P_f(|C_n|_2 > r_n) + \alpha.$$

Hence, if $|C_n|_d = o_P(r_n)$ uniformly in $\mathcal{F}$, then $d(\tilde{f}_n, f)$ would have rate of convergence $o(r_n)$ uniformly in $f \in \mathcal{F}$ on an event of probability if at least $1-\alpha$. A lower bound for the estimation rate from the preceding section thus gives a lower bound for the size $|C_n|_d$ of *any* confidence set. Conversely, if we are given an estimator $\tilde{f}_n$ satisfying (6.113) with risk bound $\overline{C}r_n$, we can consider the confidence set

$$C_n = [\tilde{f}_n - L\overline{C}r_n, \tilde{f}_n + L\overline{C}r_n].$$

Note that this confidence set is statistically feasible only if $\overline{C}r_n$ are known quantities. Assuming that this is the case for the moment, we see that C_n obviously has diameter $|C_n|_d = O_P(r_n)$ and satisfies, for $L = L(\alpha)$ large enough, by Markov's inequality,

$$\inf_{f\in\mathcal{F}} P_f(f \in C_n) \geq 1 - \frac{\sup_{f\in\mathcal{F}} E_f d(\tilde{f}_n, f)}{L\overline{C}r_n} \geq 1 - \alpha, \tag{6.117}$$

so C_n is honest with level $1-\alpha$. The following definition hence is sensible:

Definition 6.4.2 An honest confidence set C_n from Definition 6.4.1 is minimax optimal if its d-diameter $|C_n|_d$ shrinks at the minimax rate of estimation r_n in d-risk over $\mathcal{F}$ from Definition 6.3.1, that is, if for all $\alpha' > 0$ there exists $M = M(\alpha')$ such that, for all $n \in \mathbb{N}$,

$$\sup_{f\in\mathcal{F}} P_f(|C_n|_d > Mr_n) < \alpha'. \tag{6.118}$$

6.4.2 Confidence Sets for Nonparametric Estimators

As just shown, any minimax optimal estimator gives rise to an honest minimax optimal confidence set if the minimax rate of estimation r_n is known (so that r_n can be used in the construction of C_n). However, even in this ideal situation, the bound via Markov's inequality used in (6.117) is very crude, and we describe in this subsection some more exact constructions of minimax and near-minimax confidence sets based on the probabilistic tools from preceding chapters.

From a descriptive point of view, perhaps the most useful confidence sets for nonparametric functions defined on a subset A of $\mathbb{R}$ are *confidence bands*. In dimension one, these are confidence sets of functions that create a band in the Euclidean plane to which the graph of f belongs with probability $1-\alpha$. Formally, they can be described as a family of random intervals $[L(x), U(x)], x \in A$, such that $f(x) \in [L(x), U(x)]$ for all x simultaneously, providing a clear visual description of the confidence set. The simplest example for a confidence band is an L^∞-ball. Other confidence sets with less obvious geometric structure, such as L^2-balls, are also of interest.

Most of what follows will be based on an a priori control of the approximation error of the functions $f \in \mathcal{F}$, in particular, we shall assume that $\mathcal{F}$ is contained in an r-regular Besov class with r known. As mentioned earlier, such knowledge is usually not available – this leads to more fundamental adaptation questions which will be addressed in Section 8.3.

Kolmogorov-Smirnov Confidence Bands

We can construct asymptotic confidence bands for distribution functions by using Donsker's theorem. If $X_1,\ldots,X_n$ are i.i.d. F on the real line, and if

$$F_n(x) = \frac{1}{n}\sum_{i=1}^{n} 1_{(-\infty,x]}(X_i), \quad x \in \mathbb{R},$$

is the empirical distribution function, then by Corollary 3.7.39 and the continuous mapping theorem, for continuous F,

$$\sqrt{n}\|F_n - F\|_\infty \to^d \max_{x\in[0,1]} |G(x)|, \tag{6.119}$$

where G is the standard Brownian bridge on $[0,1]$. If we denote by z_α the α-quantile of that limit distribution, then we can construct the confidence band

$$C_n = \left[F_n(x) - \frac{z_\alpha}{\sqrt{n}}, F_n(x) + \frac{z_\alpha}{\sqrt{n}}\right], \quad x \in \mathbb{R},$$

for which, by (6.119),

$$P_F^{\mathbb{N}}(F \in C_n) = P_F^{\mathbb{N}}\left(\|F_n - F\|_\infty \leq \frac{z_\alpha}{\sqrt{n}}\right) \to 1 - \alpha$$

as $n \to \infty$. This convergence is uniform in F in view of Examples 3.7.19 and 3.7.42, and the L^∞-diameter of C_n shrinks at a rate $z_\alpha/\sqrt{n}$, so this confidence band is minimax optimal in view of Proposition 6.3.4. We can also use the Dvoretzky-Kiefer-Wolfowitz inequality (Exercise 3.3.3) to give a nonasymptotic coverage result for this confidence set with slightly enlarged z_α.

Undersmoothed Confidence Bands via Exact Asymptotics for Linear Estimators

The construction of Kolmogorov-Smirnov confidence sets can be carried over to other nonparametric estimators $\tilde{f}_n$ if we can obtain the exact distribution of the random variable $r_n^{-1} d(\tilde{f}_n, f)$. This is typically untractable for fixed n but may be possible in the large sample limit, perhaps after a suitable centring and scaling.

In Proposition 5.1.7, we constructed linear estimators $\tilde{f}_n$ that attain the minimax rate in supremum-norm loss over a given Besov ball $B^r_{p\infty}$ by using the decomposition

$$\tilde{f}_n - f = \tilde{f}_n - E_f \tilde{f}_n + E_f \tilde{f}_n - f$$

and by balancing the stochastic size of $\tilde{f}_n - E_f \tilde{f}_n$ with the deterministic approximation error $E_f \tilde{f}_n - f$. A classical approach to nonparametric confidence sets consists in simply *undersmoothing* the estimate slightly so that $d(E_f \tilde{f}_n, f)$ is asymptotically negligible compared to the random term whose asymptotic distribution we can obtain. These confidence sets are only 'near minimax' in the sense that an undersmoothing penalty is paid in the rate for $|C_n|_d$.

We illustrate the theory using Gaussian extreme value theory in the white noise model (Theorem 2.7.1) to construct L^∞-type simultaneous confidence bands. Suppose that we observe $dY \sim P_f^Y$ and estimate f by the Haar-projection estimator

$$f_n(j,x) = \sum_{k=0}^{2^j-1} \phi_{jk}(x) \int_0^1 \phi_{jk}(t)dY(t), \quad \phi_{jk}(x) = 2^{j/2} 1_{(k/2^j,(k+1)/2^j]}(x), \quad x \in [0,1]. \tag{6.120}$$

The uniform deviations $\|f_n(j) - E_f f_n(j)\|_\infty$ amount to maxima over an increasing ($j \to \infty$) number of i.i.d. Gaussian random variables. By Theorem 2.7.1, when suitably centred and scaled, such maxima have a Gumbel limiting distribution. The following result then gives 'undersmoothed' confidence bands for f that are near minimax optimal within a $\log n$ factor of the optimal rate over Besov balls $B^r_{\infty\infty}([0,1]), 0 < r \leq 1$ (cf. Theorem 6.3.5).

Proposition 6.4.3 *Let $B > 0, 0 < r \leq 1$, and consider observing $dY \sim P_f^Y$ in the Gaussian white noise model (6.3) with $\sigma = 1$ and where f is contained in*

$$\mathcal{F} \equiv \{f : \|f\|_{B^r_{\infty\infty}([0,1])} \leq B\}.$$

Let $\bar{j}_n$ be such that

$$2^{\bar{j}_n} \simeq (n/\log n)^{1/(2r+1)},$$

and let $j_n = \bar{j}_n + u_n$ for some undersmoothing sequence u_n satisfying $2^{u_n} \sim (\log n)^2$. For this j_n and $f_n(j_n,x)$ as in (6.120), we have, for all $t \in \mathbb{R}$,

$$\limsup_n \sup_{f \in \mathcal{F}} \left| P_f^Y \left(a_n \left(\sqrt{\frac{n}{2^{j_n}}} \|f_n(j_n) - f\|_\infty - b_n \right) \leq t \right) - \exp(-e^{-t}) \right| = 0, \tag{6.121}$$

where

$$a_n = \sqrt{j_n 2 \log 2} \quad \text{and} \quad b_n = a_n - \frac{\log j_n + \log(\pi \log 2)}{2a_n}.$$

Consequently, for z_α such that $1 - \alpha = \exp(-e^{-z_\alpha})$, the family of intervals

$$C_n \equiv C_n(\alpha,x) = \left[f_n(j_n,x) \pm \sqrt{\frac{2^{j_n}}{n}} \left(\frac{z_\alpha}{a_n} + b_n \right) \right], \quad x \in [0,1],$$

defines an honest confidence band C_n satisfying

$$\inf_{f \in \mathcal{F}} |P_f^Y(f \in C_n) - (1-\alpha)| = o(1), \tag{6.122}$$

with L^∞-diameter $|C_n|_\infty$ of the order

$$|C_n|_\infty = O_P\left(\left(\frac{\log n}{n} \right)^{r/(2r+1)} \right) 2^{u_n/2} \tag{6.123}$$

uniformly in $\mathcal{F}$.

Proof Note that

$$\|E_f f_n(j) - f\|_\infty \lesssim B 2^{-jr},$$

for all $j \geq 0$ and $0 < r \leq 1$, by the approximation properties of the Haar wavelet basis (Proposition 4.3.8). Hence,

$$\left| \sqrt{\frac{n}{2^{j_n}}} \|f_n(j_n) - f\|_\infty - \sqrt{\frac{n}{2^{j_n}}} \|f_n(j_n) - E_f f_n(j_n)\|_\infty \right|$$
$$\lesssim B \sqrt{\frac{n}{2^{j_n}}} 2^{-j_n r} = 2^{-u_n(r+1/2)} \sqrt{\log n} = o(a_n^{-1})$$

uniformly in $f \in \mathcal{F}$. Moreover, under P_f^Y, we have, for $g_k \sim^{i.i.d.} N(0,1)$,

$$\sqrt{\frac{n}{2^{j_n}}} \|f_n(j_n) - E_f f_n(j_n)\|_\infty = \sup_{x \in [0,1]} \left| \sum_{k=0}^{2^{j_n}-1} 1_{(k/2^{j_n},(k+1)/2^{j_n}]}(x) g_k \right| = \max_{k=0,\dots,2^{j_n}-1} |g_k|.$$

Combining the last two displays with Theorem 2.7.1 gives (6.121), from which the remaining claims follow immediately, using the definition of j_n, $b_n = O(\sqrt{\log n})$,

$$P_f^Y(f \in C_n) = P_f^Y\left(\|f_n(j_n) - f\|_\infty \le \sqrt{\frac{2^{j_n}}{n}}\left(\frac{z_\alpha}{a_n} + b_n\right)\right)$$

and

$$\sqrt{\frac{2^{j_n}}{n}}\left(\frac{z_\alpha}{a_n} + b_n\right) = O\left(\sqrt{\frac{2^{j_n} j_n}{n}}\right) = O_P\left(\left(\frac{\log n}{n}\right)^{r/(2r+1)}\right) 2^{u_n/2}. \quad \blacksquare$$

Inspection of this proof shows that the result holds in fact for any estimator for which

$$\limsup_n \sup_{f \in \mathcal{F}} \left| P_f^Y\left(a_n\left(\sqrt{\frac{n}{2^{j_n}}}\|f_n(j_n) - E_f f_n(j_n)\|_\infty - b_n\right) \le t\right) - \exp(-e^{-t})\right| = 0$$

can be established for suitable sequences a_n, b_n. This includes more general projection estimators and similar results for the sampling situation. See Section 5.1.3 for such results and the notes at the end of this chapter for some discussion. By using more regular projection kernels, this also generalises to $r > 1$.

The undersmoothing penalty $2^{u_n/2}$ is of order $\log n$ in the preceding result. This is in some sense an artefact of the proof and can be removed at least if we are not interested in exact limit distributions. For instance, we can use a good concentration inequality for $\|f_n(j_n) - E_f f_n(j_n)\|_\infty - b_n$, which is the supremum of a Gaussian process minus its expectation, so Theorem 2.5.8 applies. We can then arrive at undersmoothing penalties of arbitrarily slow divergent order.

Nonasymptotic Multivariate Confidence Tubes via Rademacher Complexities

Let us next discuss a general nonasymptotic approach to nonparametric confidence sets based on ideas from Section 3.4.2 on Rademacher complexities. We investigate both the case of confidence bands for multivariate distribution functions as well as the case of densities.

Consider first observing $X_1, \dots, X_n$ i.i.d. from distribution F on $\mathbb{R}^d$. When $d = 1$, we can use the preceding Kolmogorov-Smirnov-type procedures from (6.119) without difficulty. In the multivariate situation, the limit distribution $\sup_{t \in \mathbb{R}^d} |G_F(t)|$ is much less tractable, and the nonasymptotic approach via the Dvoretzky-Kiefer-Wolfowitz inequality is also restricted to dimension one. An alternative is provided by the Rademacher complexity approach to concentration inequalities. We still centre the confidence band

$$C_n = [F_n(x) \pm R_n], \quad x \in \mathbb{R}^d,$$

at the empirical distribution function

$$F_n(x) = \frac{1}{n}\sum_{i=1}^n 1_{(-\infty,x_1]\times\cdots\times(-\infty,x_d]}(X_i) \equiv \frac{1}{n}\sum_{i=1}^n 1_{(-\infty,x]}(X_i), \quad x = (x_1, \dots, x_d),$$

but now with random width given by the sum of a Rademacher complexity and a Gaussian tail deviation term

$$R_n = \frac{2}{n}\sup_{x \in \mathbb{R}^d}\left|\sum_{i=1}^n \varepsilon_i 1_{(-\infty,x]}(X_i)\right| + 3\sqrt{\frac{2\log(4/\alpha)}{n}}, \quad 0 < \alpha < 1.$$

Alternatively, we can replace R_n by its Rademacher expectation $E_\varepsilon R_n$, a stochastically more stable quantity. This confidence tube has nonasymptotic coverage: for every $n \in \mathbb{N}$,

$$P_F^{\mathbb{N}}(F \in C_n) = 1 - P_F^{\mathbb{N}}(\|F_n - F\|_\infty > R_n) \geq 1 - \alpha \tag{6.124}$$

in view of Theorem 3.4.5 with $\mathcal{F} = \{1_{(-\infty,x]} : x \in \mathbb{R}^d\}$. By desymmetrisation (Theorem 3.1.21 and $E\|P_n - P\|_\mathcal{F} \lesssim 1/\sqrt{n}$ (using (3.177) and Example 3.7.19)), we also have $ER_n \lesssim 1/\sqrt{n}$, so the diameter of this confidence band is minimax optimal.

This idea generalises to density estimation problems, where, however, we need more refined concentration inequalities for Rademacher processes. Consider $X_1, \dots, X_n$ i.i.d. from density f on $\mathbb{R}^d$, and let

$$f_n(h,x) = \frac{1}{nh^d}\sum_{i=1}^n K\left(\frac{x - X_i}{h}\right) \equiv \int_\mathbb{R} K_h(x-y)dP_n(y), \quad P_n = \frac{1}{n}\sum_{i=1}^n \delta_{X_i},$$

be a kernel density estimator based on a kernel $K : \mathbb{R}^d \to \mathbb{R}$ that is of bounded variation, integrates to 1 and is contained in $L^1 \cap L^\infty$. We then know that, for $f \in B^r_{\infty\infty}(\mathbb{R}^d)$,

$$E \sup_{x\in\mathbb{R}^d} |f_n(h,x) - f(x)| \lesssim \|f\|_\infty^{1/2}\sqrt{\frac{\log(1/h)}{nh^d}} + \|f\|_{B^r_{\infty\infty}(\mathbb{R}^d)} 2^{-jr} \tag{6.125}$$

from the results in Chapters 4 and 5, and we wish to construct a corresponding multivariate confidence 'tube' around the estimator $f_n(h)$. All that follows works for general (wavelet-projection) kernels $K(x,y)$, too, if they are suitably regular, with obvious modifications, using the techniques from Chapter 4.

Define a Rademacher process and the associated Rademacher complexity

$$f_n(h,x) = \left\{\frac{1}{n}\sum_{i=1}^n \varepsilon_i K_h(x - X_i)\right\}_{x\in\mathbb{R}^d}, \qquad R_n(h) := E_\varepsilon \sup_{x\in\mathbb{R}^d}\left|\frac{1}{n}\sum_{i=1}^n \varepsilon_i K_h(x - X_i)\right|,$$

with $(\varepsilon_i)_{i=1}^n$ an i.i.d. Rademacher sequence, independent of the X_i. $R_n(h)$ can be easily computed in practice: first, simulate n i.i.d. random signs, apply these signs to the summands $K_h(x - X_i)$ of the kernel density estimator and then maximise the resulting function over $\mathbb{R}^d$. Finally, take the expectation $E_\varepsilon R_n(h)$ of $R_n(h)$ with respect to the Rademacher variables only. This last step could be skipped but gives rise to a stochastically more stable quantity and hence slightly better constants in the following results.

Let us assume that an upper bound U for $\|f\|_\infty$ is known in what follows; otherwise, replace U by the consistent estimate $\|f_n(\bar{h}_n)\|_\infty$, where $\bar{h}_n \sim (\log n)^2/n^{1/d}$. Define the random variable

$$\sigma^R(n,h,z) = 3R_n(h) + 4\sqrt{\frac{2U\|K\|_2^2 z}{nh^d}} + \frac{50\|K\|_\infty z}{3nh^d}. \tag{6.126}$$

We construct a confidence band

$$\bar{C}_n = [f_n(h,\cdot) - \sigma^R(n,h,z), f_n(h,\cdot) + \sigma^R(n,h,z)] \tag{6.127}$$

for the mean Ef_n of f_n.

Proposition 6.4.4 *Let $f_n(x,h)$ be as earlier, and let $X_1,\dots,X_n$ be i.i.d. $f \in L^\infty$. Then we have, for every $n \geq 1$, every $h > 0$ and every $z > 0$, that*

$$P_f^{\mathbb{N}}\left\{\sup_{x\in\mathbb{R}^d} |f_n(h,x) - Ef_n(x,h)| \geq \sigma^R(n,h,z)\right\} \leq 2e^{-z}.$$

Moreover, the expected diameter of $\bar{C}_n$ is bounded by

$$2E_f\sigma^R(n,h,z) \leq C\left(\sqrt{\frac{\log(1/h)}{nh^d}} + \frac{\log(1/h)}{nh^d}\right),$$

for every $z > 0$, every $n \in \mathbb{N}$, every $h > 0$ and some constant C depending only on U,K,z,d.

Proof We use Theorem 3.4.3; in fact, we use the remark after it. We can write in empirical process notation

$$\|f_n(x,h) - E_f f_n(x,h)\|_\infty = \|P_n - P\|_{\mathcal{K}}. \tag{6.128}$$

where $\mathcal{K} = \{K_h(x-\cdot) : x \in \mathbb{R}^d\}$. This class has envelope $h^{-d}\|K\|_\infty$, and hence, the class

$$\mathcal{G} \equiv \{h^d K_h(x-\cdot)/2\|K\|_\infty : x \in \mathbb{R}^d\}$$

is uniformly bounded by $1/2$. For the weak variances, we have the estimate

$$\sup_{g\in\mathcal{G}} Eg^2(X) \leq \sup_{x\in\mathbb{R}^d} \frac{h^{2d}}{4\|K\|_\infty^2} \int_{\mathbb{R}^d} K_h^2(x-y)f(y)dy \leq \frac{U\|K\|_2^2}{4\|K\|_\infty^2} h^d \equiv \sigma^2.$$

Then, by (3.139) with $\tilde{S}_n$ as in the proof of Theorem 3.4.3, we have

$$P_f^{\mathbb{N}}\left(\|f_n(h,\cdot) - Ef_n(h,\cdot)\|_\infty \geq 3R_n(h) + 4\sqrt{\frac{2U\|K\|_2^2 z}{nh^d}} + \frac{50\|K\|_\infty z}{3nh^d}\right)$$

$$= P_f^{\mathbb{N}}\left(\|P_n - P\|_{\mathcal{G}} \geq 3E_\varepsilon \tilde{S}_n + 4\sqrt{\frac{2\sigma^2 z}{n}} + \frac{25z}{3n}\right) \leq 2e^{-z},$$

which proves the first claim. For the second claim of the proposition, we only have to show that $ER_n(h)$ has, up to constants, the required order as a function of h,n. But this follows readily from desymmetrisation (Theorem 3.1.21) and (6.125). ■

The constants in the choice of σ^R are the best we can obtain from the concentration of measure tools developed in Chapter 3. A 'practical' and perhaps optimistic choice for moderate sample sizes may be to replace 3 by 1 in front of R_n and to ignore the third 'Poissonian' term in (6.126) altogether, reflecting a pure Gaussian tail inequality.

Combined with undersmoothing as in Proposition 6.4.3, we can show that $\bar{C}_n$ is indeed a confidence set for the unknown function f which shrinks at the near-minimax rate of estimation (see Exercise 6.3.2).

Corollary 6.4.5 *Consider observing $X_1,\dots,X_n$ i.i.d. f on $\mathbb{R}^d$. Suppose that f is an element of the class $\mathcal{F}$ consisting of all probability densities in*

$$\{f : \|f\|_{B^r_{\infty\infty}(\mathbb{R}^d)} \leq B\}.$$

Let $\bar{h}_n$ be such that $h \simeq (n/\log n)^{-1/(2r+d)}$, and let $h_n = \bar{h}_n u_n$ for some undersmoothing sequence $u_n \to 0$ as $n \to \infty$. Take $\bar{C}_n$ as in (6.127) with $h = h_n$, $z = z_\alpha$ such that $2e^{-z_\alpha} = \alpha$ and based on a kernel of order r. Then

$$\inf_{f \in \mathcal{F}} P_f^Y(f \in C_n) \geq 1 - \alpha + e_n, \tag{6.129}$$

where $e_n = o(1)$, and the L^∞-diameter $|\bar{C}_n|_\infty$ is of the order

$$|C_n|_\infty = O_P\left(\left(\frac{\log n}{n}\right)^{r/(2r+d)}\right) u_n^{-\gamma} \tag{6.130}$$

uniformly in $\mathcal{F}$ for some $\gamma > 0$.

Proof Since

$$\|E_f f_n(h) - f\|_\infty \lesssim Bh^r$$

for all $h > 0$, the result follows from $h_n^r = o(\sigma^R(n, h_n, z))$ and the preceding proposition to control the coverage probabilities. ■

If explicit control of the bias is available, we can give a nonasymptotic version of this corollary as well by incorporating the bias bound into the diameter of the confidence set.

Minimax Confidence Sets via Unbiased Risk Estimation

Another approach to nonparametric confidence sets is based on the idea of estimating the risk $d(\tilde{f}_n, f)$ of a nonparametric estimator $\tilde{f}_n$ directly using sample splitting. This is particularly useful when d is a metric that has some averaging structure, such as L^2-loss.

Let us illustrate these ideas first in the sampling setting, where we have at hand i.i.d. observations X_i from some probability density function f on $[0,1]$. Let us assume that the sample size is $2n, n \in \mathbb{N}$, for notational convenience. Split the sample into two halves, with index sets $\mathcal{S}^1, \mathcal{S}^2$ of equal size n, write $P_f^{(i)}, E_f^{(i)}, i = 1, 2$, for the corresponding probabilities and expectation operators. Let $\tilde{f}_n$ be some preliminary estimator of f based on the sample $\mathcal{S}^1$. For a wavelet basis $\{\psi_{lk}\}$ of $L^2([0,1])$ from Section 4.3.4 or Section 4.3.5 with associated projection operator $K_j, j \in \mathbb{N}$, define the U-statistic

$$U_n(\tilde{f}_n) = \frac{2}{n(n-1)} \sum_{i<i', i,i' \in \mathcal{S}^2} \sum_{l \leq j-1, k} (\psi_{lk}(X_i) - \langle \psi_{lk}, \tilde{f}_n \rangle)(\psi_{lk}(X_{i'}) - \langle \psi_{lk}, \tilde{f}_n \rangle) \tag{6.131}$$

which has expectation

$$E_f^{(2)} U_n(\tilde{f}_n) = \sum_{l \leq j-1, k} \langle \psi_{lk}, f - \tilde{f}_n \rangle^2 = \|K_j(f - \tilde{f}_n)\|_2^2,$$

so $U_n(\tilde{f}_n)$ is an unbiased estimate of the jth wavelet approximation of the squared L^2-estimation error $\|f - \tilde{f}_n\|_2^2$. The idea is to take $U_n(\tilde{f}_n)$ as a proxy for the true risk $\|\tilde{f}_n - f\|_2^2$, suggesting heuristically a confidence set of the form

$$C_n = \{f : \|\tilde{f}_n - f\|_2^2 \leq U_n(\tilde{f}_n) + z_{n,\alpha}\},$$

where $z_{n,\alpha}$ are suitable quantile constants controlling the estimation and approximation error $U_n(\tilde{f}_n) - E_f^{(2)} U_n(\tilde{f}_n)$ and $\|K_j(f - \tilde{f}_n) - (f - \tilde{f}_n)\|_2^2$, respectively.

Theorem 6.4.6 *Consider i.i.d. observations $X_1,\dots,X_{2n}$ on $[0,1]$ from a probability density $f \in L^2([0,1])$. Let $\tilde f_n \in L^2$ be a preliminary estimator of f based on the subsample $\mathcal{S}^1 = \{X_1,\dots,X_n\}$. Let $\mathcal{F}$ be a family of probability densities in $L^\infty([0,1])$ such that*

$$\sup_{f\in\mathcal{F}} \|K_j(f-\tilde f_n)-(f-\tilde f_n)\|_2^2 \le B(j),$$

for some sequence $(B(j): j\in\mathbb{N})$. Let further z_α be a numerical quantile constant, and let

$$\tau_n^2(f)=\frac{2^{j+1}\|f\|_\infty^2}{n(n-1)}+\frac{4\|f\|_\infty}{n}\|K_j(f-\tilde f_n)\|_2^2.$$

Define the confidence set

$$C_n=\left\{f\in\mathcal{F}: \|f-\tilde f_n\|_2 \le \sqrt{z_\alpha \tau_n(f)+U_n(\tilde f_n)+B(j)}\right\}, \tag{6.132}$$

where $U_n(\tilde f_n)$ is as in (6.131) based on the second subsample $\mathcal{S}^2=\{X_{n+1},\dots,X_{2n}\}$. Then, for every $n\in\mathbb{N}$ and z_α large enough,

$$\inf_{f\in\mathcal{F}} P_f^{\mathbb{N}}(f\in C_n)\ge 1-\alpha. \tag{6.133}$$

Proof Writing $P_f^{(2)}$ for the joint law of the X_i in $\mathcal{S}^2$, we have from Chebyshev's inequality

$$\begin{aligned}
&P_f^{(2)}\left\{U_n(\tilde f_n)-\|f-\tilde f_n\|_2^2 \ge -B(j)-z(\alpha)\tau_n(f)\right\}\\
&\ge P_f^{(2)}\left\{U_n(\tilde f_n)-\|K_{j_n}(f-\tilde f_n)\|_2^2 \ge -z(\alpha)\tau_n(f)\right\}\\
&\ge 1-\frac{\mathrm{Var}^{(2)}(U_n(\tilde f_n)-E_f^{(2)}U_n(\tilde f_n))}{(z(\alpha)\tau_n(f))^2},
\end{aligned} \tag{6.134}$$

for all $f\in\mathcal{F}$. The Hoeffding decomposition for the centred U-statistic $U_n(\tilde f_n)-E_f^2U_n(\tilde f_n)$ with kernel

$$R(x,y)=\sum_{l\le j_n-1,k}(\psi_{lk}(x)-\langle\psi_{lk},\tilde f_n\rangle)(\psi_{lk}(y)-\langle\psi_{lk},\tilde f_n\rangle)$$

is (cf. Section 3.4.3 or before (6.40))

$$U_n(\tilde f_n)-E_2U_n(\tilde f_n)=\frac{2}{n}\sum_{i\in\mathcal{S}^2}(\pi_1R)(X_i)+\frac{2}{n(n-1)}\sum_{i<i'}(\pi_2R)(X_i,X_{i'})\equiv L_n+D_n,$$

with linear

$$(\pi_1R)(x)=\sum_{l\le j_n-1,k}(\psi_{lk}(x)-\langle\psi_{lk},f\rangle)\langle\psi_{lk},f-\tilde f_n\rangle$$

and degenerate kernel

$$(\pi_2R)(x,y)=\sum_{l\le j_n-1,k}(\psi_{lk}(x)-\langle\psi_{lk},f\rangle)(\psi_{lk}(y)-\langle\psi_{lk},f\rangle).$$

The variance of $U_n(\tilde{f}_n) - E_f^{(2)} U_n(\tilde{f}_n)$ is the sum of the variances of the two terms in the Hoeffding decomposition. For the linear term we bound the variance $\mathrm{Var}^{(2)}(L_n)$ by the second moment, using ortho-normality of the ψ_{lk}, that is,

$$\frac{4}{n}\int \left(\sum_{l \le j_n - 1, k} \psi_{lk}(x)\langle \psi_{lk}, \tilde{f}_n - f\rangle \right)^2 f(x)dx \le \frac{4\|f\|_\infty}{n} \sum_{l \le j_n - 1, k} \langle \psi_{lk}, \tilde{f}_n - f\rangle^2,$$

which equals the second term in the definition of $\tau_n^2(f)$. For the degenerate term we can bound $Var^{(2)}(D_n)$ analogously by the second moment of the uncentred kernel, i.e., by

$$\frac{2}{n(n-1)}\int \left(\sum_{l \le j_n - 1, k} \psi_{lk}(x)\psi_{lk}(y) \right)^2 f(x)dx f(y)dy \le \frac{2^{j_n+1}\|f\|_\infty^2}{n(n-1)},$$

using ortho-normality of the wavelet basis. For z_α large enough independent of the X_is in $\mathcal{S}^1$, the last term in (6.134) thus is bounded from below by $1-\alpha$, completing the proof by integrating the inequality with respect to $P_f^{(1)}$. ■

Sharp choices for the constant z_α can be obtained from applying the concentration of measure tools for U-statistics from Section 3.4.3 and Bernstein's inequality, to the U-statistic appearing in the preceding proof.

The preceding theorem only proves coverage of the confidence set, and the important question arises whether C_n is optimal in the sense of Definition 6.4.2. We now show that this is the case if $\mathcal{F}$ equals a subset of a ball in $B^r_{2\infty}([0,1])$ with r known and when the preliminary estimator is

$$\tilde{f}_n = f_n(\bar{j}_n, x) = \frac{1}{n} \sum_{l \le \bar{j}_n - 1, k, i \in \mathcal{S}^1} \psi_{lk}(x)\psi_{lk}(X_i), \tag{6.135}$$

constructed on the same wavelet basis, assumed to be S-regular, $S > r$ and for suitable $\bar{j}_n$.

Corollary 6.4.7 *Consider i.i.d. observations $X_1, \dots, X_{2n}$ on $[0,1]$ from probability density f contained in*

$$\mathcal{F} = \{f : \max(\|f\|_\infty, \|f\|_{B^r_{2\infty}([0,1])}) \le B\}.$$

Let $\tilde{f}_n = f_n(\bar{j}_n, x)$ be as in (6.135), where $\bar{j}_n$ is such that $2^{\bar{j}_n} \simeq n^{1/(2r+1)}$. Moreover, let $\tau_n(f), U_n(\tilde{f}_n), C_n$ be as in Theorem 6.4.6, with $j = j_n \ge \bar{j}_n$ such that $2^{j_n} \simeq n^{1/(2r+1/2)}$ and $B(j_n) = d \log n \, 2^{-2j_n r}$ for some $d > 0$. Then, for all $n \ge n_0(B,d)$ large enough,

$$\inf_{f \in \mathcal{F}} P_f^{\mathbb{N}}(f \in C_n) \ge 1 - \alpha, \tag{6.136}$$

and if $|C_n|_2$ is the L^2-diameter of C_n, then for every $\alpha' > 0$ there exists $L > 0$ such that, for all $n \in \mathbb{N}$,

$$\sup_{f \in \mathcal{F}} P_f^{\mathbb{N}}(|C_n|_2 \ge L n^{-r/(2r+1)}) \le \alpha'. \tag{6.137}$$

Proof Using the standard approximation bound

$$\|K_j(f) - f\|_2 \lesssim B 2^{-jr}$$

for $f \in \mathcal{F}$, we deduce the coverage result from Theorem 6.4.6 for n large enough depending on B only, noting also that

$$\|K_{j_n}(\tilde{f}_n) - \tilde{f}_n\|_2 = \|\tilde{f}_n - \tilde{f}_n\|_2 = 0$$

since $\bar{j}_n \le j_n$ and since K_j is a L^2-projector. The diameter of C_n equals

$$\sqrt{z_\alpha \tau_n(f) + U_n(\tilde{f}_n) + B(j_n)}$$

and can be bounded as follows: the nonrandom terms are of order

$$\sqrt{\log n}B2^{-j_n r} + 2^{j_n/4}n^{-1/2} \lesssim \sqrt{\log n}\, n^{-r/(2r+1/2)} = o(n^{-r/(2r+1)}).$$

The random component of $\tau_n(f)$ has expectation

$$\|f\|_\infty^{1/4} n^{-1/4} E^{(1)} \|K_{j_n}(\tilde{f}_n - f)\|_2^{1/2} = o(n^{-r/(2r+1)})$$

since K_{J_n} is a projection operator and since $E\|\tilde{f}_n - f\|_2 = O(n^{-r/(2r+1)})$ by Proposition 5.1.7. Moreover, by that proposition and again the projection properties,

$$EU_n(\tilde{f}_n) = E^{(1)}\|K_{j_n}(\tilde{f}_n - f)\|_2^2 \le E^{(1)}\|\tilde{f}_n - f\|_2^2 \le cn^{-2r/(2r+1)}.$$

The term in this display is the leading term in our bound for the diameter of the confidence set, completing the proof by Markov's inequality. ■

The preceding confidence set is a genuine minimax confidence set in the sense of Definition 6.4.2. Note that we are still undersmoothing (through the choice of $B(j_n)$), but by exploiting the L^2-structure, this bias term is seen to be of smaller order than the estimation rate. In fact, in the preceding proof there is some 'space' left in terms of how fast the bias has to approach zero – this observation will be important in Section 8.3. This phenomenon is tied to the L^2-situation, and the construction of exact minimax confidence sets in L^∞ (without undersmoothing penalty) when the radius B is unknown poses fundamental difficulties. This, again, will be discussed in Section 8.3.

These results have an analogue in the Gaussian white noise model. The proof of Theorem 6.4.6 relies on splitting an i.i.d. sample into two halves. In the Gaussian white noise model, we cannot directly split the sample, but as discussed in the paragraph surrounding (6.10), given an observation dY, we can always create two independent observations dY_1, dY_2 of the same white noise model with variance increased by a factor of 2. We can compute a preliminary estimator $\tilde{f}_n$ from the first observation dY_1 and estimate its L^2-risk based on the second observation dY_2 by

$$T_n(\tilde{f}_n) = \sum_{l \le j-1, k} \left(\int_0^1 \psi_{lk} dY_2 - \langle \psi_{lk}, \tilde{f}_n \rangle \right)^2 - 2\sigma^2 \frac{2^j}{n}, \tag{6.138}$$

where the ψ_{lk} form an S-regular, $S > r$ wavelet basis of $L^2([0,1])$ as in Section 4.3.4 or Section 4.3.5 such that $\sum_{k,l \le j-1} 1 = 2^j$. In this way, we obtain the following white noise analogue of Theorem 6.4.6:

Theorem 6.4.8 *Consider two independent white noise experiments dY_1, dY_2 generated from (6.3) with $f \in \mathcal{F} \subset L^2([0,1])$, and let $\tilde{f}_n$ be a preliminary estimator of f based on the first experiment. Suppose that*

$$\sup_{f\in\mathcal{F}} \|K_j(f-\tilde{f}_n)-(f-\tilde{f}_n)\|_2^2 \le B(j)$$

for some sequence $(B(j): j \in \mathbb{N})$. Let further z_α be a quantile constant, $g \sim N(0,1)$, and let

$$\tau_n^2(f) = \frac{8\sigma^2}{n}\|K_j(f-\tilde{f}_n)\|_2^2 + 4Eg^4\sigma^4\frac{2^j}{n^2}.$$

Define the confidence set

$$C_n = \left\{ f : \|f-\tilde{f}_n\|_2 \le \sqrt{z_\alpha \tau_n(f) + T_n(\tilde{f}_n) + B(j)} \right\}, \tag{6.139}$$

where $T_n(\tilde{f}_n)$ is as in (6.138) based on dY_2. Then, for every $n \in \mathbb{N}$ and z_α large enough,

$$\inf_{f\in\mathcal{F}} P_f^Y(f \in C_n) \ge 1-\alpha. \tag{6.140}$$

Proof We have from Chebyshev's inequality

$$\begin{aligned}
&\inf_{f\in\mathcal{F}} P_f^{Y_2}\left\{T_n(\tilde{f}_n) - \|f-\tilde{f}_n\|_2^2 \ge -B(j) - z(\alpha)\tau_n(f)\right\}\\
&\ge \inf_{f\in\mathcal{F}} P_f^{Y_2}\left\{T_n(\tilde{f}_n) - \|K_{j_n}(f-\hat{f}_n)\|_2^2 \ge -z(\alpha)\tau_n(f)\right\}\\
&\ge 1 - \sup_{f\in\mathcal{F}} \frac{Var_2(T_n(\tilde{f}_n) - E_f^{(2)}T_n(\tilde{f}_n))}{(z(\alpha)\tau_n(f))^2}.
\end{aligned}$$

The result now follows, bounding the variances of the centred χ^2-statistic $T_n(\tilde{f}_n) - E_f^{(2)}T_n(\tilde{f}_n)$. We can decompose, for $g_{lk} \sim^{i.i.d.} N(0,1)$,

$$(T_n(\tilde{f}_n) - E_f^{(2)}T_n(\tilde{f}_n)) = 2\frac{\sqrt{2}\sigma}{\sqrt{n}}\sum_{l\le j-1,k} g_{lk}\langle\psi_{lk}, f-\tilde{f}_n\rangle + \frac{2\sigma^2}{n}\sum_{l\le j-1,k}(g_{lk}^2-1).$$

The variance of the first term is bounded by

$$\frac{8\sigma^2}{n}\|K_j(f-\tilde{f}_n)\|_2^2$$

using ortho-normality of the wavelet basis. The variance of the second term is bounded by $4Eg_{lk}^4\sigma^4 2^j/n^2$, completing the proof. ■

A confidence set as in Corollary 6.4.7 now can be obtained likewise, with the same bounds on $|C|_2$. This is left as Exercise 6.4.1. Sharp choices of z_α can be obtained from replacing Chebyshev's inequality by the Gaussian concentration tools from Chapter 2, in particular, Theorem 3.1.9.

Confidence Sets and Multiscale Inference

Another approach to constructing nonparametric confidence sets is based on the theory for linear estimators in weak metrics developed in Section 5.2. Of particular interest is the approach via multiscale spaces from Section 5.2.2 that we investigate now.

Consider the Gaussian white noise model with f contained in a ball of radius B in the Hölder space $B^r_{\infty\infty}([0,1])$. Recalling (5.86), we have that the estimator $\mathbb{Y}^{(n)}$ satisfies

$$\sqrt{n}(\mathbb{Y}^{(n)} - f) = \mathbb{W} \text{ in } \mathcal{M}_0(w), \tag{6.141}$$

for any admissible sequence w such that $w_l/\sqrt{l} \uparrow \infty$ (recalling Definition 5.2.13), and where white noise $\mathbb{W}$ defines a tight Borel Gaussian random variable on the Banach space $\mathcal{M}_0$.

Then we use this result to construct a confidence set

$$W_n \equiv \left\{f : \|\mathbb{Y}^{(n)} - f\|_{\mathcal{M}_0(w)} \le z_\alpha/\sqrt{n}\right\},$$

where z_α are the α-quantiles of the distribution of the random variable $\|\mathbb{W}\|_{\mathcal{M}_0(w)}$. (A Bayesian approach to 'bootstrap' these quantiles will be presented in Theorem 7.3.23.) The multiscale approach consists in starting with the universal confidence set W_n and further intersecting it with qualitative information about f. In the present case, this information is a bound on the smoothness of f, and we define

$$C_n = W_n \cap B_n, \quad B_n \equiv \{f : \|f\|_{B^r_{\infty\infty}} \le u_n\}, \tag{6.142}$$

where $u_n \to \infty$ as $n \to \infty$ is an undersmoothing sequence accommodating the fact that a bound on the Hölder norm of f is usually not available. More precise (and less ad hoc) choices for u_n are discussed in Exercise 6.4.2.

Theorem 6.4.9 *Let $B, r > 0$, and consider observing $dY \sim P^Y_f$ in the Gaussian white noise model (6.3) with $\sigma = 1$ and where f is contained in*

$$\mathcal{F} \equiv \{f : \|f\|_{B^r_{\infty\infty}([0,1])} \le B\}.$$

Let j_n be such that $2^{j_n} \simeq (n/\log n)^{1/(2r+1)}$, and let C_n be the random subset of $B^r_{\infty\infty}$ given in (6.142) with w chosen such that $w_{j_n}/\sqrt{j_n} = O(u_n)$. Then

$$\sup_{f\in\mathcal{F}} |P^Y_f(f \in C_n) - (1-\alpha)| = o(1), \tag{6.143}$$

with L^∞-diameter $|C_n|_\infty$ of the order

$$|C_n|_\infty = O_P\left(\left(\frac{\log n}{n}\right)^{r/(2r+1)} u_n\right), \tag{6.144}$$

uniformly in $\mathcal{F}$.

Proof For n large enough, $\|f\|_{B^r_{\infty\infty}} \le u_n$, and then $P^Y_f(f \in B_n) = 1$. From this and (6.141) we infer, for n large enough, that

$$P^Y_f(f \in C_n) = P^Y_f(f \in W_n) = 1 - \alpha.$$

To bound the L^∞-diameter of C_n, pick any $f,g \in C_n$, and let $h = f - g$. We have for any S-regular wavelet basis, $S > (1/2, r)$,

$$\|h\|_\infty \lesssim \sum_l 2^{l/2} \max_k |\langle h, \psi_{lk}\rangle|.$$

For the high frequencies $l \geq j_n$, we have from the wavelet characterisation of the Besov norm

$$\begin{aligned}\sum_{l>j_n} 2^{l/2} \max_k |\langle h, \psi_{lk}\rangle| &= \sum_{l>j_n,k} 2^{-lr} 2^{l(r+1/2)} \max_k |\langle h, \psi_{lk}\rangle| \\ &\leq \|h\|_{B^r_{\infty\infty}} 2^{-j_n r} \\ &= O_P\left((n/\log n)^{-r/(2r+1)} u_n\right).\end{aligned}$$

For the low frequencies, we have, using $\|h\|_{\mathcal{M}_0(w)} = O_P(1/\sqrt{n})$,

$$\begin{aligned}\sum_{l\leq j_n} 2^{l/2} \max_k |\langle h, \psi_{lk}\rangle| &= \sum_{l\leq j_n} 2^{l/2} w_l/w_l \max_k |\langle h, \psi_{lk}\rangle| \\ &\lesssim \|h\|_{\mathcal{M}_0(w)} 2^{j_n/2} \sqrt{j_n}(w_{j_n}/\sqrt{j_n}) \\ &= O_P\left((n/\log n)^{-2r/(2r+1)} u_n\right),\end{aligned}$$

completing the proof. ■

The perceding confidence set consists of the simple thresholding rule

$$C_n = \left\{ f : |\langle f, \psi_{lk}\rangle| \leq \min\left(\frac{z_\alpha 2^{l/2} w_l}{\sqrt{n}}, 2^{-l(r+1/2)} u_n \right) \right\}. \tag{6.145}$$

The proof shows that instead of undersmoothing in an additive bias-variance decomposition, we can alternatively 'undersmooth in the frequency domain'. An interesting feature of this multiscale approach is that the confidence set is constructed for the full function f in a weak loss function where parametric rates can be obtained. The intersection with a ball in $B^r_{\infty\infty}$ then yields sufficient regularity of the confidence set that its diameter shrinks at a near-minimax optimal rate.

In the sampling setting, we can use the results from Section 5.2 in a similar fashion, using Theorem 5.2.16 for the projected empirical measure $P_n(j_n)$, and replacing W_n by

$$V_n = \left\{ P : \|P - P_n(j_n)\|_{\mathcal{M}_0(w)} \leq z_\alpha/\sqrt{n} \right\},$$

where z_α are the α-quantiles of the multiscale norm of the P-Brownian bridge. Assuming that P has a bounded density $f \in B^r_{\infty\infty}([0,1])$, we then prove an immediate analogue of Theorem 6.4.9, choosing $2^{j_n} \sim (n/\log n)^{1/(2r+1)}$ and intersecting with a growing ball in $B^r_{\infty\infty}([0,1])$.

Exercises

6.4.1 Prove an analogue of Corollary 6.4.7 in the Gaussian white noise model.

6.4.2 Show that if in (6.142), we replace

$$\{f : \|f\|_{B^r_{\infty\infty}} \leq u_n\}$$

by

$$\{f : \|f\|_{B^{r,\gamma}_{\infty\infty}} \leq \|\hat{f}_n\|_{B^{r,\gamma}_{\infty\infty}} + \delta'\}$$

for $\delta', \gamma > 0$ arbitrary, then Theorem 6.4.9 still remains true. *Hint*: Show that $\|\hat{f}_n\|_{B^{r,\gamma}_{\infty\infty}}$ estimates $\|f\|_{B^{r,\gamma}_{\infty,\infty}}$ consistently for $f \in B^r_{\infty\infty}$, and adapt the interpolation argument in the proof of Theorem 6.4.9 to the logarithmically weakened Besov norm.

6.5 Notes

Section 6.1 Most of the materials in this section are basic. The realisation of the Gaussian white noise model via the path space of Brownian motion in Ibragimov and Khasminski (1981) is based on the basic theory of stochastic integrals $\int f dW$. Our construction of the white noise process $\mathbb{W}$ relies only on basic Gaussian process tools and the Cameron-Martin Theorem 2.6.13. Yet another way to interpret $\int g dW$ is as a P^W-measurable linear functional on $C([0,1])$; see p. 83ff. in Bogachev (1998). The Gaussian sequence space analogues of these results are particularly simple as a basic application of Kakutani's theorem for infinite product measures. The Kullback-Leibler distance was systematically used in information theory, we refer to Kullback (1967), where also a proof of the first Pinsker inequality can be found. Pinsker (1964) obtained a slightly weaker version of it.

Section 6.2 Nonparametric testing theory has been widely used in the theory of goodness-of-fit tests for parametric models, where they serve the purpose of providing a 'sanity check' for a given parametric model in use. Next to the Kolmogorov-Smirnov tests (Smirnov (1939)), a common procedure is the Cramér–von Mises statistic, based on replacing $\|F_n - F\|_\infty$ by $\|F_n - F\|_2$, which is the distribution function analogue of the U-statistic approach considered in this section. Tests based on χ^2-ideas can be traced back to Pearson (1900) and have been in use in particular when the basis functions of L^2 used are those of the Haar basis, where computation of the involved test statistics is simple. Chapter 1 of Ingster and Suslina (2003) contains an extensive review of the classical ideas in the field and further historical remarks.

The nonparametric minimax perspective on testing problems was investigated in the landmark work by Yuri Ingster in the 1980s and 1990s; see, in particular, Ingster (1982, 1986, 1993) and the monograph by Ingster and Suslina (2003). Ingster concentrated mostly on minimax theory for simple hypotheses such as $H_0 = \{0\}$ in white noise and $H_0 = \{1\}$ in the sampling model on $[0,1]$. He noted the fact that Kolmogorov-Smirnov-type procedures are too crude for several important nonparametric problems and that the geometry of the separation metric influences the testing rate. The L^2-separation rate $n^{-r/(2r+1/2)}$ can be shown to be connected to the L^∞-separation rate $n^{-r/(2r+1)}$ (ignoring $\log n$ terms) by the unified rate

$$\rho_n \simeq n^{-r/(2r+1-p^{-1})}, \quad 2 \leq p \leq \infty;$$

see Ingster (1986, 1993). For $1 \leq p < 2$, no further improvement is possible; see Exercise 6.4. Beyond separation rates, we can ask for exact separation constants for tests, which is a more delicate matter. In the Gaussian white noise model, such results are available; for L^2-separation, see Ermakov (1990), and for L^∞-separation, see Lepski and Tsybakov (2000).

Composite nonparametric testing problems have been studied in the context of minimax goodness-of-fit tests, mostly in the setting of $1/\sqrt{n}$ separation rates and where the null hypothesis is a finite-dimensional parametric class see Pouet (2002) and Fromont and Laurent (2006), for instance. The case of general composite hypotheses defined by qualitative constraints on the functions involved, such as monotonicity or convexity, has been considered in Dümbgen and Spokoiny (2001) and Baraud,

Huet and Laurent (2005), among others. For L^∞-separation, the rates found there are typically minimax, but for L^2-separation conditions, the situation is more delicate – Corollary 6.2.19 gives a sharp result of that kind. Theorem 6.2.14 on general infimum χ^2-tests is a Gaussian adaptation of the more difficult Theorem 6.2.17, which in essence is due to Bull and Nickl (2013) (see also Gayraud and Pouet (2005) for some results). Theorem 6.2.17 can be extended to cover situations of hypotheses that are not uniformly bounded if the complexity of the null hypothesis is not too large, as shown in the (related) high-dimensional regression setting in Nickl and van de Geer (2013). Theorem 6.2.20 was proved in Carpentier (2015). Some composite testing theory for L^p-separation, $2 \le p \le \infty$, is implicit in the work Carpentier (2013). The minimax theory of composite nonparametric testing remains a field with several open problems.

Section 6.3 We have presented here only those very basic results on minimax lower bounds for function estimation that are relevant for this book. Using wavelet theory and Kullback-Leibler distances leads to proofs that are particularly short and transparent.

The general reduction principle leading to Theorem 6.3.2 was pioneered by Ibragimov and Khasminskii (1977, 1981) and developed further in Korostelev and Tsybakov (1993), Tsybakov (2009). Our exposition closely follows the one in Tsybakov (2009), to whom Theorem 6.3.2 is due. Other key references include Cencov (1972), Bretagnolle and Huber (1979), Stone (1980, 1982), Ibragimov and Khasminskii (1982), Nemirovski (1985).

Birgé (1983) showed that a general connection exists between the metric entropy of the parameter space $(\mathcal{F}, d)$ and the minimax estimation rate over $\mathcal{F}$ in d-risk. This connection can be successfully exploited by wavelet theory, as shown in this section.

The wavelet approach to minimax lower bounds over Besov spaces was developed by Kerkyacharian and Picard (1992), Donoho, Johnstone, Kerkyacharian and Picard (1996), Donoho and Johnstone (1998). See also Härdle, Kerkyacharian, Picard and Tsybakov (1998) for a general treatment and many further references, as well as a treatment of the theory for L^q-loss over $B^r_{p\infty}$-balls where $p \neq q$, where new 'nonlinear' phenomena can arise. Several other approaches to lower bound proofs exist and Chapter 2 in Tsybakov (2009) gives an excellent account of the general theory and many further references.

Section 6.4 Kolmogorov-Smirnov-type nonparametric confidence sets for a distribution function F can be obtained directly from the classical results of Kolmogorov (1933a) and Smirnov (1939). Nonparametric confidence sets for densities or regression functions require more elaborate constructions. Apparently, Smirnov (1950) was the first to realise the relevance of extreme value theory: he proved a sampling analogue of Proposition 6.4.3 based on histogram estimators. This was generalised to kernel density estimators using Gaussian approximation techniques and extreme value for stationary Gaussian processes by Bickel and Rosenblatt (1973), who were unaware of Smirnov's (1950) work. For regression settings, this approach was developed further by Claeskens and van Keilegom (2003), who also considered related bootstrap methods. The general projection kernel extreme value theory that also allows for wavelet kernels was developed in Giné and Nickl (2010). The difficulty in the general case arises from the lack of stationarity of wavelet-driven white noise integrals $x \mapsto \int K(x,y)dW(y)$. The conditions of the general theorem in Giné and Nickl (2010) for such processes are verified in that paper for spline-based wavelets, and numerical proofs show that the same is true for Daubechies wavelets; see Bull (2013) and the notes for Section 2.7.

The undersmoothing approach to nonparametric confidence sets was used in Smirnov (1950) and Bickel and Rosenblatt (1973) and investigated systematically in Hall (1992), where it is argued that undersmoothing can be more efficient than obtaining an estimate of the bias. The Rademacher symmetrisation approach was introduced in the setting of empirical risk minimisation by Koltchinskii (2006) and in the setting of nonparametric confidence sets by Lounici and Nickl (2011) and

Kerkyacharian, Nickl and Picard (2012). The unbiased risk estimation approach to L^2-confidence sets is adapted from Robins and van der Vaart (2006), and related ideas also can be found in Juditsky and Lambert-Lacroix (2003). The multiscale approach to nonparametric inference was developed in Davies and Kovac (2001), Dümbgen and Spokoiny (2001) and Davies, Kovac and Meise (2009). The functional multiscale space approach to confidence sets presented here is also implicit in corollary 3 in Nickl (2007) and was used in the Bayesian setting (to be discussed in the next chapter) by Castillo and Nickl (2013, 2014).

7

Likelihood-Based Procedures

Consider observations $X = X^{(n)}$ from law P_f indexed by a parameter space $\mathcal{F}$. From a very basic perspective, statistical inference is about finding the value of f that is 'most likely' to have generated the observed values $X = x$. This perspective can be transformed into a rigorous, principled approach to any statistical problem and has resulted in the development of two paradigms of statistical inference that rely on the concept of the *likelihood function*.

The first approach follows the well-known *maximum likelihood principle*, which takes the preceding perspective literally and attempts to maximise a likelihood function which represents the joint distribution of the data as a function of f over the parameter space $\mathcal{F}$. The second approach, to be introduced in more detail later, starts with a probability distribution Π on the parameter space $\mathcal{F}$, often called the *prior distribution*, makes the assumption that $X \sim P_f$ *conditional* on f having been drawn from Π and then computes the conditional *posterior distribution* of f given the observations X, which is a reweighted version of the likelihood function. As the last 'updating' step is often based on an application of Bayes' rule for conditional probabilities, this approach is called the *Bayesian approach* to statistical inference.

In this chapter we develop some basic aspects of the theory of likelihood-based inference for infinite-dimensional models $\mathcal{F}$. A central role will be played by the Hellinger distance – a metric that is naturally compatible with likelihood techniques in the i.i.d. sampling model – and by the corresponding L^2-distance in the Gaussian white noise model. We start with nonparametric testing problems and show that certain likelihood ratio–based procedures allow for general results in the sampling model, replacing the analytic assumptions on the functions employed in Chapter 6 by general Hellinger-distance compactness conditions. We then study the maximum likelihood principle and give a general rate of convergence result using a bracketing version of these Hellinger compactness conditions. We illustrate the theory for two concrete nonparametric maximum likelihood estimators in some detail: the cases where $\mathcal{F}$ equals a ball in a Sobolev space and where $\mathcal{F}$ equals the set of monotone decreasing densities. We will derive convergence rate results in Hellinger and related metrics, and we shall prove an infinite-dimensional version of the classical asymptotic normality result for maximum likelihood estimators. We then lay out the main ideas of the Bayesian approach to nonparametric inference. We shall first prove general contraction results for posterior distributions in the Hellinger and L^2-distance and give applications to Gaussian process priors. In the white noise setting with product priors we conduct a finer asymptotic analysis of posterior distributions. In particular, we prove nonparametric Bernstein–von Mises theorems, which establish asymptotic normality of the posterior distribution in infinite-dimensional settings and which can be used to give a frequentist justification of Bayesian methods to construct nonparametric confidence sets.

7.1 Nonparametric Testing in Hellinger Distance

We now develop the main ingredients of the theory of nonparametric testing using the Hellinger distance on probability densities, which is based on the idea of endowing the space of probability measures with a canonical Hilbert space structure. For any pair of probability measures P,Q on a measurable space $(\mathcal{X},\mathcal{A})$ with densities p,q with respect to a dominating measure μ, the *Hellinger distance* is defined as

$$h^2(p,q) \equiv h^2(P,Q) = \int_{\mathcal{X}} \left(\sqrt{p}-\sqrt{q}\right)^2 d\mu. \tag{7.1}$$

As discussed in Section 6.1.2, such a μ always exists, and one shows further that the Hellinger distance is independent of the choice of μ. We always have $h^2(p,q) \leq 2$. The *Hellinger affinity* is defined as

$$\rho(p,q) \equiv 1 - h^2(p,q)/2 \tag{7.2}$$

and satisfies

$$\rho(p,q) = E_P\sqrt{\frac{q}{p}(X)} = \langle \sqrt{p},\sqrt{q}\rangle_{L^2(\mu)}, \quad \log\rho(p,q) \leq -\frac{1}{2}h^2(p,q), \tag{7.3}$$

where $\langle\cdot,\cdot\rangle_{L^2(\mu)}$ is the usual $L^2(\mu)$ inner product on root densities.

Based on i.i.d. observations $X_1,\dots,X_n$ in $\mathcal{X}$, consider the testing problem of whether the X_i have been generated by the density p or by q. Write $P^{\mathbb{N}},Q^{\mathbb{N}}$ for the infinite product measures arising from the samples from densities p and q, respectively, and E_P,E_Q for the corresponding expectation operators. A natural likelihood ratio test is

$$\Psi_n = 1\left\{\prod_{i=1}^n \frac{q}{p}(X_i) > 1\right\}. \tag{7.4}$$

Proposition 7.1.1 *For two μ-densities p,q on $\mathcal{X}$, consider the test Ψ_n from (7.4) for the problem*

$$H_0 = \{p\} \quad vs. \quad H_1 = \{q\}.$$

Then, for every $n \in \mathbb{N}$,

$$E_P\Psi_n \leq e^{-\frac{1}{2}nh^2(p,q)}, \quad E_Q(1-\Psi_n) \leq e^{-\frac{1}{2}nh^2(p,q)}. \tag{7.5}$$

Proof For the type 1 errors, using Markov's inequality and independence,

$$\begin{aligned}
P^{\mathbb{N}}\left(\prod_{i=1}^n \frac{q}{p}(X_i) > 1\right) &= P^{\mathbb{N}}\left(\prod_{i=1}^n \sqrt{\frac{q}{p}}(X_i) > 1\right)\\
&\leq E_P\left(\sqrt{q/p(X)}\right)^n\\
&= \exp\{n\log\rho(p,q)\}\\
&\leq \exp\left\{-nh^2(p,q)/2\right\}
\end{aligned}$$

and likewise, for the alternatives,

$$Q^{\mathbb{N}}\left(\prod_{i=1}^{n}\frac{q}{p}(X_i)\leq 1\right)=Q^{\mathbb{N}}\left(\prod_{i=1}^{n}\sqrt{\frac{p}{q}}(X_i)\geq 1\right)$$
$$\leq E_Q\left(\sqrt{p/q(X)}\right)^n$$
$$\leq \exp\left\{-nh^2(p,q)/2\right\},$$

completing the proof. ■

This test allows us to distinguish two fixed densities p and q if their Hellinger distance is at least a large constant times $1/\sqrt{n}$.

Let us now generalise this approach to the situation where one wishes to test *composite* hypotheses. The classical likelihood ratio test for composite hypotheses would compare the maxima of both likelihood functions over the corresponding hypotheses, and in the nonparametric situation, this leads to some mathematical difficulties that require control of likelihood ratios uniformly in infinite-dimensional sets. Whereas we shall be able to address some of these difficulties in the next section, for testing problems between two Hellinger balls, there exists an elegant way around this problem using the explicit Hilbert space structure induced by the Hellinger distance.

Theorem 7.1.2 *Let p,q be probability densities on a measurable space $(\mathcal{X},\mathcal{A})$ with respect to a dominating measure μ, $dP=pd\mu, dQ=qd\mu$. Let $h(p,q)=d>0$, and define Hellinger balls of μ-densities*

$$B(p)=\{r:h(r,p)\leq d/4\},\quad B(q)=\{r:h(r,q)\leq d/4\}. \tag{7.6}$$

Then there exists a measurable function $\Psi:\mathcal{X}\to[0,\infty)$ such that, for $dR=rd\mu$,

$$E_R\Psi(X)\leq 1-\frac{h^2(p,q)}{12}\quad \forall r\in B(p);\quad E_R(1/\Psi)(X)\leq 1-\frac{h^2(p,q)}{12}\quad \forall r\in B(q).$$

Proof Define $v_0=\sqrt{p}, v_1=\sqrt{q}$, and denote by V the two-dimensional linear subspace of $L^2(\mu)$ spanned by v_0,v_1. Further, let $V^1=\{v\in V:\|v\|_{L^2(\mu)}=1\}$ denote the unit circle of that space so that $v_i\in V^1, i=0,1$. Let ω be such that

$$d^2=h^2(p,q)=2(1-\cos(\omega)),\quad \omega\in(0,\pi/2],$$

or, what is the same, such that

$$\rho(p,q)=\langle v_0,v_1\rangle_{L^2(\mu)}=\cos(\omega).$$

For $\beta\in[0,2\pi/\omega)$, define v_β to be the rotation of v_0 on V^1 by the angle $\beta\omega$ in V. Then v_β^2 is a probability density too, with

$$\langle v_\alpha,v_\beta\rangle=\cos((\alpha-\beta)\omega)=\rho(v_\alpha^2,v_\beta^2) \tag{7.7}$$

and

$$v_\beta=\frac{\sin(\beta\omega)v_1+\sin((1-\beta)\omega)v_0}{\sin\omega}.$$

We take $\alpha = 2/3, \beta = 1/3$, so that

$$\sqrt{s} \equiv v_{2/3} = \frac{\sin(2\omega/3)v_1 + \sin(\omega/3)v_0}{\sin\omega}, \quad \sqrt{t} \equiv v_{1/3} = \frac{\sin(\omega/3)v_1 + \sin(2\omega/3)v_0}{\sin\omega}$$

and

$$\langle v_{2/3}, v_{1/3}\rangle_{L^2(\mu)} = \cos(\omega/3) = \rho(v_{2/3}^2, v_{1/3}^2) = \rho(s,t). \tag{7.8}$$

Now define

$$\Psi(x) \equiv \frac{v_{2/3}}{v_{1/3}}(x) = \sqrt{\frac{s}{t}(x)}, \tag{7.9}$$

where we use the convention $0/0 = 1$. Since

$$\int r\sqrt{s/t}\,d\mu = \int \sqrt{s/t}(\sqrt{r} - \sqrt{t})^2 d\mu + 2\int \sqrt{sr}\,d\mu - \int \sqrt{st}\,d\mu$$

for any density r and since, using $\sin(2x) = 2\sin x\cos x$,

$$\sqrt{\frac{s}{t}} = \frac{\sin(2\omega/3)v_1 + \sin(\omega/3)v_0}{\sin(\omega/3)v_1 + \sin(2\omega/3)v_0} \le \frac{\sin(2\omega/3)}{\sin(\omega/3)} = 2\cos(\omega/3),$$

we see that

$$E_R\Psi(X) \le 2\cos(\omega/3)h^2(r,t) + 2\rho(r,s) - \rho(s,t). \tag{7.10}$$

We can decompose $\sqrt{r} = u + \theta v_\gamma$ for some $v_\gamma \in V^1$ and $\theta \in [0,1], \gamma \in [0, 2\pi/\omega)$ and u orthogonal in $L^2(\mu)$ to the space V such that $\theta^2 + \|u\|^2_{L^2(\mu)} = 1$. As a consequence,

$$\rho(r,t) = \langle\sqrt{r}, v_{1/3}\rangle_{L^2(\mu)} = \theta\cos((\gamma - 1/3)\omega), \quad \rho(r,s) = \theta\cos((\gamma - 2/3)\omega). \tag{7.11}$$

Feeding this observation and (7.8) into (7.10) gives

$$\begin{aligned} E_R\Psi(X) &\le 4\cos(\omega/3)[1 - \theta\cos((\gamma - 1/3)\omega)] + 2\theta\cos((\gamma - 2/3)\omega) - \cos(\omega/3) \\ &= 2\theta[\cos((\gamma - 2/3)\omega) - 2\cos(\omega/3)\cos((\gamma - 1/3)\omega))] + 4\cos(\omega/3) - \cos(\omega/3) \\ &= 3\cos(\omega/3) - 2\theta\cos(\gamma\omega), \end{aligned}$$

where we have used elementary trigonometric identities (first for $\cos(x+y)$ and then for $\sin x \sin y$ and $\cos x \cos y$) to simplify the term in brackets. Since $\rho(t,p) = \langle v_{1/3}, v_0\rangle_{L^2(\mu)} = \cos(\omega/3)$ and $\rho(r,p) = \theta\cos(\gamma\omega)$, we thus conclude that

$$\begin{aligned} E_R\Psi(X) &\le 3\rho(t,p) - 2\rho(r,p) = 1 - (3/2)h^2(t,p) + h^2(r,p) \\ &\le 1 - (1/6)h^2(p,q) + h^2(r,p), \end{aligned} \tag{7.12}$$

where we have used, in the last inequality,

$$\frac{1}{2}h^2(t,p) = 1 - \cos(\omega/3) = 2\sin^2(\omega/6), \frac{1}{2}h^2(p,q) = 1 - \cos(\omega) = 2\sin^2(\omega/2)$$

and that $(\sin x)/x$ is decreasing on $[0, \pi/2]$ to see that $h^2(t,p) \ge (1/9)h^2(p,q)$. Finally, $h(r,p) \le d/4 = h(p,q)/4$ implies that $h^2(r,p) \le h^2(p,q)/12$, which gives the first inequality of the theorem. The second inequality follows from interchanging the roles of $v_{1/3}, v_{2/3}$, noting that the argument is entirely symmetric. ■

Corollary 7.1.3 *Based on i.i.d. observations $X_1,\dots,X_n$ in $\mathcal{X}$ and for $B(p),B(q)$ as in (7.6), consider the testing problem*

$$H_0\colon r \in B(p) \quad vs. \quad H_1 : r \in B(q), \quad d = h(p,q) > 0. \tag{7.13}$$

For Ψ from the preceding theorem, define the test

$$\Psi_n = 1\left\{\prod_{i=1}^{n} \Psi(X_i) > 1\right\}. \tag{7.14}$$

Then, writing $dR = r d\mu$, we have, for every $n \in \mathbb{N}$,

$$\sup_{r\in H_0} E_R \Psi_n \le e^{-\frac{1}{12}nd^2}, \quad \sup_{r\in H_1} E_R(1-\Psi_n) \le e^{-\frac{1}{12}nd^2}. \tag{7.15}$$

Proof In view of Markov's inequality, the preceding theorem and $\log(1-x) \le -x$, for any $r \in B(p)$,

$$\begin{aligned} R^{\mathbb{N}}\left(\prod_{i=1}^{n} \Psi(X_i) \ge 1\right) &\le (E_R\Psi(X))^n = \exp\{n \log E_R \Psi(X)\} \\ &\le \exp\left\{-n\left(\frac{h^2(p,q)}{12}\right)\right\}, \end{aligned}$$

and likewise, for any $r \in B(q)$, we have

$$R^{\mathbb{N}}\left(\prod_{i=1}^{n} \Psi(X_i) \le 1\right) \le \exp\left\{-n\left(\frac{h^2(p,q)}{12}\right)\right\},$$

completing the proof. ■

This result allows us to construct tests for general problems

$$H_0 : p = p_0 \quad vs. \quad H_1 : p \in \mathcal{P} \cap \{p : h(p,p_0) > \varepsilon\}, \tag{7.16}$$

where p_0 is a fixed density, $dP_0 = p_0 d\mu$, $\varepsilon > 0$, and where $\mathcal{P}$ is a collection of probability densities that is totally bounded for the Hellinger metric. More precisely, we will assume a bound on the ε-covering numbers $N(\mathcal{P},h,\varepsilon)$ required to cover suitable shells of $\mathcal{P}$ by Hellinger balls of radius ε. We can then decompose the problem into testing H_0 against a collection of fixed Hellinger balls that cover H_1 and use the preceding corollary to sum error probabilities.

Theorem 7.1.4 *Let $\mathcal{P}$ be a collection of μ-densities on $(\mathcal{X},\mathcal{A})$, and suppose that for some nonincreasing function $N(\varepsilon)$, some $\varepsilon_0 > 0$ and all $\varepsilon > \varepsilon_0$, we have*

$$N\left(\{p \in \mathcal{P} : \varepsilon < h(p,p_0) \le 2\varepsilon\}, h, \varepsilon/4\right) \le N(\varepsilon).$$

Then, for every $\varepsilon > \varepsilon_0$, there exist tests Ψ_n for the problem (7.16) s.t. for all $n \in \mathbb{N}$ such that $n\varepsilon^2 \ge c_0$ and constant $K = K(c_0)$

$$E_{P_0}(\Psi_n = 1) \le \frac{N(\varepsilon)}{K} e^{-Kn\varepsilon^2}, \qquad \sup_{p\in\mathcal{P}:h(p,p_0)>\varepsilon} E_P(\Psi_n = 0) \le e^{-Kn\varepsilon^2}.$$

Proof Choose a finite set $\mathcal{S}_j'$ of points in each shell

$$\mathcal{S}_j = \{p \in \mathcal{P} : \varepsilon j < h(p,p_0) \leq \varepsilon(1+j)\}, \quad j \in \mathbb{N},$$

such that every $p \in \mathcal{S}_j$ is within distance $j\varepsilon/4$ of at least one of these points. By hypothesis, for j fixed, there are at most $N(j\varepsilon)$ such points $q_{jl} \in \mathcal{S}_j'$, and from the preceding corollary for each of them there exists a test $\Psi_{n,jl}$ such that

$$E_{P_0}\Psi_{n,jl} \leq e^{-Cnj^2\varepsilon^2}, \qquad \sup_{p\in\mathcal{S}_j, h(p_0,q_{jl})>j\varepsilon/4} E_P(1-\Psi_{n,jl}) \leq e^{-Cnj^2\varepsilon^2}$$

for some universal constant $C > 0$. Let Ψ_n be the maximum of all these tests. By a union bound for the maximum, we have

$$P_0^{\mathbb{N}}(\Psi_n = 1) \leq \sum_j \sum_l \exp\{-Cnj^2\varepsilon^2\} \leq \sum_j N(j\varepsilon)\exp\{-Cnj^2\varepsilon^2\} \tag{7.17}$$

and

$$\sup_{p\in\cup_j\mathcal{S}_j} E_P(\Psi_n = 0) \leq \exp\{-Cn\varepsilon^2\}, \tag{7.18}$$

which implies the result (noting also that $N(j\varepsilon) \leq N(\varepsilon)$). ∎

The remarkable feature of this result is that in contrast to results from Chapter 6, the Hellinger approach to nonparametric testing needs no qualitative assumptions (such as smoothness) on the unknown densities but works under a complexity bound on the alternative space H_1 alone. At the same time, without having imposed any uniform boundedness of the densities p involved, we obtain excellent exponential error bounds on the type 1 and type 2 errors.

7.2 Nonparametric Maximum Likelihood Estimators

Consider an i.i.d. sample of n real random variables $X_1,\dots,X_n$ from some unknown law P of density function p with respect to some measure μ on a measurable space $(\mathcal{X},\mathcal{A})$. We denote by $P^{\mathbb{N}}$ the infinite product measure associated with the random experiment $(\mathcal{X}^{\mathbb{N}},\mathcal{A}^{\mathbb{N}})$. The joint probability density of the observations is

$$\prod_{i=1}^n p(x_i), \quad x_i \in \mathcal{X}.$$

If we consider a model $\mathcal{P}$, of probability densities p, we can evaluate this joint density at the observation points X_i and view the resulting statistic as a function of p only – this defines the *likelihood function*

$$L_n(p) \equiv \prod_{i=1}^n p(X_i), \tag{7.19}$$

which is a function of the argument $p \in \mathcal{P}$ that is random through the variables X_i. The maximum likelihood approach suggests that we maximise the function L_n over $\mathcal{P}$. Equivalently, we maximise the (normalised) *log-likelihood function*

$$\ell_n(p) = \frac{1}{n}\sum_{i=1}^n \log p(X_i). \tag{7.20}$$

We set $\log 0 = -\infty$ throughout so that ℓ_n takes values in the extended real line $[-\infty,\infty)$ (endowed with its usual topology). If the X_i have been drawn from a fixed law P_0 with density p_0, and if we assume that $E_{P_0}|\log p(X)| < \infty$, for all $p \in \mathcal{P}$, we can define the limiting log-likelihood function

$$\ell(p) = \int_{\mathcal{X}} \log p(x) dP_0(x) \tag{7.21}$$

which satisfies, by a version of Jensen's inequality, for every probability density p that is absolutely continuous with respect to $p_0(x)dx$,

$$\ell(p) - \ell(p_0) = \int_{\mathcal{X}} \log \frac{p}{p_0} dP_0 \leq \log 1 = 0$$

with equality only if $p = p_0$ μ a.e. and, hence,

$$\ell(p) < \ell(p_0) \quad \forall p \in \mathcal{P}, \quad p \neq p_0. \tag{7.22}$$

This gives an intuitive justification of the maximum likelihood approach as it attempts to maximise the empirical version $\ell_n(p)$ of $\ell(p)$ over $\mathcal{P}$. However, for nonparametric models $\mathcal{P}$ such as the set of all probability density functions on a given interval or even the set of all infinitely differentiable densities, this problem has no solution because we can make $\ell_n(p)$ as large as desired by taking density functions that have very large peaks at the observation points. From an intuitive point of view, the maximiser could be taken to be the empirical measure $P_n = (1/n)\sum_{i=1}^n \delta_{X_i}$, which for problems of nonparametric density estimation is, however, not satisfactory. Indeed, for a nonparametric maximum likelihood estimator to exist, the complexity of the possibly infinite-dimensional set $\mathcal{P}$ has to be restricted. Under suitable constraints, such as a bound on the metric entropy of $\mathcal{P}$, nonparametric maximum likelihood estimators can be shown to exist and even to give minimax optimal estimation procedures.

7.2.1 Rates of Convergence in Hellinger Distance

We consider a measurable space $(\mathcal{X},\mathcal{A})$ and a family $\mathcal{P}$ of probability density functions $p : \mathcal{X} \to [0,\infty)$ with respect to the common σ-finite dominating measure μ on $\mathcal{A}$. Further, let $X_1,\dots,X_n$ be i.i.d. from common density $p_0 \in \mathcal{P}$, $dP_0 = p_0 d\mu$, and let μ_0 equal μ restricted to the support of p_0. Throughout, $\hat{p}_n$ denotes a nonparametric maximum likelihood estimator (NPMLE) assumed to satisfy the relationship

$$\sup_{p\in\mathcal{P}} \ell_n(p) = \ell_n(\hat{p}_n) \tag{7.23}$$

for some model $\mathcal{P}$ of probability densities on $\mathcal{X}$ which contains p_0. Sometimes *sieved* maximum likelihood estimators are of interest, where the maximisation is performed over a sequence of models $\mathcal{P}_n$ that increase with n and whose limit set $\mathcal{P}$ is assumed to contain p_0.

Before we study some concrete examples in more detail, we shall prove a general rate of convergence result for such estimators that is based on a measure of the complexity of the classes $\mathcal{P},\mathcal{P}_n$ only. The result follows from empirical process techniques developed in Section 3.5.3. When applied in the right way, these techniques allow us to control likelihood ratios uniformly in Hellinger balls. The conditions obtained are conceptually related to the Hellinger metric entropy techniques encountered in Theorem 7.1.4 but require additional bracketing conditions to deal with the local behaviour of the empirical processes involved.

We first consider the unsieved case and shall assume for now that $\hat{p}_n$ exists as an element of $\mathcal{P}$. We will measure the distance between two μ-densities p, q in the Hellinger metric which, as we recall from (7.1), equals the L^2-distance on the root densities; more precisely,

$$h^2(p,q) \equiv \int_{\mathcal{X}} (\sqrt{p} - \sqrt{q})^2 d\mu.$$

The Hellinger-distance ε-bracketing entropy of a class $\mathcal{F}$ of probability densities equals the L^2-bracketing metric entropy $\log N_{[]}(\mathcal{F}^{1/2}, L^2(\mu), \varepsilon)$ from Section 3.5.2 of the class of root densities

$$\mathcal{F}^{1/2} = \{\sqrt{f} : f \in \mathcal{F}\}.$$

To accommodate the behaviour of elements in $\mathcal{P}$ near zero, it will be useful to consider classes

$$\bar{\mathcal{P}} = \left\{\bar{p} \equiv \frac{p + p_0}{2} : p \in \mathcal{P}\right\}, \tag{7.24}$$

and we set

$$\bar{\mathcal{P}}^{1/2} \equiv \{\sqrt{\bar{p}} : p \in \mathcal{P}\}, \tag{7.25}$$

for which we will require a bound on the bracketing entropy integral

$$J(\delta) \equiv J(\bar{\mathcal{P}}^{1/2}, \delta) = \int_0^\delta \sqrt{\log N_{[]}(\bar{\mathcal{P}}^{1/2}, L^2(\mu), \varepsilon)}\, d\varepsilon \vee \delta, \quad 0 < \delta \leq 1. \tag{7.26}$$

We can consider slightly smaller quantities than $J(\delta)$ which cover local entropies and divergent entropy integrals – this will not be relevant in the examples we consider later but will be discussed briefly after the proof of the following theorem.

Theorem 7.2.1 *Suppose that $p_0 \in \mathcal{P}$, and let $\hat{p}_n$ solve (7.23). Take $\mathcal{J}(\delta) \geq J(\bar{\mathcal{P}}^{1/2}, \delta)$ such that $\mathcal{J}(\delta)/\delta^2$ is a non increasing function of $\delta \in (0,1]$. Then there exists a fixed number $c > 0$ such that for any δ_n satisfying*

$$\sqrt{n}\delta_n^2 \geq c\mathcal{J}(\delta_n) \quad \forall n \in \mathbb{N} \tag{7.27}$$

we have, for all $\delta \geq \delta_n$,

$$P_0^{\mathbb{N}}(h(\hat{p}_n, p_0) \geq \delta) \leq c \exp\left(-n\delta^2/c^2\right). \tag{7.28}$$

Proof Throughout this proof, the expectation operator, when applied to quantities involving the MLE, has to be understood as $P(g(\hat{p}_n)) = \int g(\hat{p}_n(x))dP(x)$ for measurable functions g (so conditional on the value of $\hat{p}_n$). We start with the following basic inequality, for which we recall the usual notation $P_n = (1/n)\sum_{i=1}^n \delta_{X_i}$ for empirical measures:

Lemma 7.2.2 *For*

$$g_p \equiv \frac{1}{2}\log\frac{\bar{p}}{p_0} = \frac{1}{2}\log\frac{p + p_0}{2p_0}, \tag{7.29}$$

we have

$$h^2((\hat{p}_n + p_0)/2, p_0) \le 2[(P_n - P_0)(g_{\hat{p}_n})].$$

Proof By concavity of log, we see, on the set $A = \{p_0 > 0\}$,

$$\log \frac{\hat{p}_n + p_0}{2p_0} \ge \frac{1}{2} \log \frac{\hat{p}_n}{p_0},$$

so, by definition of $\hat{p}_n$,

$$0 \le \frac{1}{4} \int_A \log \frac{\hat{p}_n}{p_0} dP_n \le \frac{1}{2} \int_A \log \frac{\hat{p}_n + p_0}{2p_0} dP_n$$
$$= (P_n - P_0)(g_{\hat{p}_n}) + \frac{1}{2} \int_A \log \frac{\hat{p}_n + p_0}{2p_0} dP_0. \tag{7.30}$$

Now, since $(1/2)\log x \le \sqrt{x} - 1$, for $x \ge 0$, we have, for any density q, that

$$\frac{1}{2} \log \frac{q}{p_0} \le \sqrt{\frac{q}{p_0}} - 1,$$

and then, using (7.3), also

$$\frac{1}{2} \int_A \log \frac{q}{p_0} dP_0 \le \int_A \sqrt{\frac{q}{p_0}} dP_0 - 1 = -h^2(q, p_0)/2, \tag{7.31}$$

which applied to $q = (\hat{p}_n + p_0)/2$ in (7.30) implies the lemma. ■

We also have for any density p and some universal constant $C > 0$ that

$$h^2(p, p_0) \le C h^2(\bar{p}, p_0); \tag{7.32}$$

see Exercise 7.2.1. Therefore, intersecting the event inside the probability in (7.28) with the inequality of the preceding lemma and (7.32), we can bound the probability in (7.28) by

$$P_0^{\mathbb{N}}\left(\sqrt{n}(P_n - P_0)(g_{\hat{p}_n}) - \sqrt{n} h^2((\hat{p}_n + p_0)/2, p_0) \ge 0, h^2((\hat{p}_n + p_0)/2, p_0) \ge \delta^2/C\right)$$
$$\le P_0^{\mathbb{N}}\left(\sup_{p \in \mathcal{P}, h^2(\bar{p}, p_0) \ge \delta^2/C} \left[\sqrt{n}(P_n - P_0)(g_p) - \sqrt{n} h^2(\bar{p}, p_0)\right] \ge 0\right)$$
$$\le \sum_{s=0}^{S} P_0^{\mathbb{N}}\left(\sup_{p \in \mathcal{P}, h^2(\bar{p}, p_0) \le 2^{s+1}\delta^2/C} \sqrt{n}|(P_n - P_0)(g_p)| \ge \sqrt{n} 2^s \delta^2/C\right),$$

where we have estimated the supremum by the maximum of all slices

$$\left\{2^s \delta^2/C \le h^2(\bar{p}, p_0) \le 2^{s+1}\delta^2/C\right\}, \quad 0 \le s \le S,$$

noting that S is finite because $h^2(p, q) \le 2$ is always true. We apply the exponential inequality Theorem 3.5.21 to sum these probabilities: We can bound the Bernstein norms with $K = 1$; thereby,

$$\rho_1(g_p) \le c' h(\bar{p}, p_0) \le c'' 2^{(s+1)/2} \delta \equiv R$$

since the g_p are bounded from below by $-(\log 2)/2$, and since

$$2(e^{|x|}-1-|x|) \le c(D)(e^x-1)^2, \quad x \ge -D,\ D>0. \tag{7.33}$$

Similarly, the bracketing entropy $H_{B,1}$ of $\bar{\mathcal{P}}$ from Definition 3.5.20 is bounded by the regular bracketing metric entropy $\log N_{[]}(\bar{\mathcal{P}}^{1/2}, L^2(\mu_0), \varepsilon)$; see Exercise 7.2.2. In Theorem 3.5.21 with $K=1$ and R as earlier we choose $t=\sqrt{n}2^s\delta^2/2^4$, which is admissible in view of our choice of $\mathcal{J}$, giving the bound

$$\sum_{s=0}^{S} C' \exp\left\{-\frac{2^{2s}n\delta^2}{C'}\right\} \le c\exp\{-n\delta^2/c\},$$

completing the proof. ■

Inspection of this proof shows that instead of (7.26), we can consider, more generally, bounds for the entropy integral of classes

$$\bar{\mathcal{P}}^{1/2}(\delta) \equiv \left\{\sqrt{\bar{p}} : p \in \mathcal{P}, h(\bar{p}, p_0) \le \delta\right\}$$

bounded away from zero, that is,

$$J'(\bar{\mathcal{P}}^{1/2}, \delta) = \int_{\delta^2/2^{13}}^{\delta} \sqrt{\log N_{[]}(\bar{\mathcal{P}}^{1/2}(\delta), L^2(\mu_0), \varepsilon)}d\varepsilon \vee \delta, \quad 0<\delta\le 1,$$

to obtain the same result.

Let us next consider the case of sieved maximum likelihood estimators, where $\hat{p}_n$ solves

$$\sup_{p\in\mathcal{P}_n} \ell_n(p) = \ell_n(\hat{p}_n) \tag{7.34}$$

for a sequence $\mathcal{P}_n$ of models of probability densities called the *sieve*. Assume that we can construct an approximating sequence $p_n^* \in \mathcal{P}_n$ such that, for some $0<U<\infty$,

$$\frac{p_0}{p_n^*} \le U^2 \quad \forall n \in \mathbb{N}. \tag{7.35}$$

Consider the classes

$$\bar{\mathcal{P}}_{*,n}^{1/2} = \left\{\sqrt{\frac{p+p_n^*}{2}}, p \in \mathcal{P}_n\right\}, \tag{7.36}$$

and require a bound on the bracketing entropy integrals

$$J_*(\delta) \equiv J(\bar{\mathcal{P}}_{*,n}^{1/2}, \delta) = \int_0^{\delta} \sqrt{\log N_{[]}(\bar{\mathcal{P}}_{*,n}^{1/2}, L^2(\mu), \varepsilon)}d\varepsilon \vee \delta, \quad 0<\delta\le 1. \tag{7.37}$$

The following theorem shows that the convergence rate in Hellinger distance is similar to the unsieved case if the approximation errors $h(p_n^*, p_0)$ can be controlled at a suitably fast rate, paralleling the 'bias-variance' tradeoff from Chapter 6:

Theorem 7.2.3 *Let $\hat{p}_n, p_n^*$ be as in (7.34) and (7.35) respectively. Take $\mathcal{J}(\delta) \ge J_*(\bar{\mathcal{P}}_{*,n}^{1/2}, \delta)$ such that $\mathcal{J}(\delta)/\delta^2$ is a nonincreasing function of $\delta \in (0,1]$ for every n large enough. Then there exists a fixed number $c>0$ such that for any δ_n satisfying*

$$\sqrt{n}\delta_n^2 \ge c\mathcal{J}(\delta_n) \quad \forall n \in \mathbb{N} \text{ large enough} \tag{7.38}$$

we have

$$h(\hat{p}_n, p_0) = O_{P_0^{\mathbb{N}}}(\delta_n + h(p_n^*, p_0)). \tag{7.39}$$

Proof As in the proof of Theorem 7.2.1, we have, with all integrals in the following displays over $\{p_n^* > 0\}$,

$$\begin{aligned} 0 &\le \int \log \frac{\hat{p}_n + p_n^*}{2p_n^*} d\mathbb{P}_n \\ &\le \int \log \frac{\hat{p}_n + p_n^*}{2p_n^*} (d\mathbb{P}_n - dP_0) - 2 \int \left(1 - \sqrt{\frac{\hat{p}_n + p_n^*}{2p_n^*}}\right) dP_0, \end{aligned}$$

and the term subtracted can be further bounded using (7.2), (7.35) and the Cauchy-Schwarz inequality, for some constant $C = C(U)$, by

$$\begin{aligned} &\int \left(1 - \sqrt{\frac{\hat{p}_n + p_n^*}{2p_n^*}}\right) p_n^* d\mu + \int \left(1 - \sqrt{\frac{\hat{p}_n + p_n^*}{2p_n^*}}\right) (p_0 - p_n^*) d\mu \\ &= \frac{1}{2} h^2((\hat{p}_n + p_n^*)/2, p_n^*) + \int \left(\sqrt{p_n^*} - \sqrt{\frac{\hat{p}_n + p_n^*}{2}}\right) (\sqrt{p_0} - \sqrt{p_n^*}) \left(1 + \sqrt{\frac{p_0}{p_n^*}}\right) d\mu \\ &\ge \frac{1}{2} h^2((\hat{p}_n + p_n^*)/2, p_n^*) - C h((\hat{p}_n + p_n^*)/2, p_n^*) h(p_n^*, p_0), \end{aligned}$$

and thus,

$$h^2((\hat{p}_n + p_n^*)/2, p_n^*) \lesssim \int \log \frac{\hat{p}_n + p_n^*}{2p_n^*} d(\mathbb{P}_n - P_0) + h((\hat{p}_n + p_n^*)/2, p_n^*) h(p_n^*, p_0).$$

If the second term dominates the preceding sum, we see that

$$h((\hat{p}_n + p_n^*)/2, p_n^*) \lesssim h(p_n^*, p_0),$$

so, as in (7.32), $h(\hat{p}_n, p_n^*) = O(h(p_n^*, p_0))$, and then also, by the triangle inequality,

$$h(\hat{p}_n, p_0) = O(h(p_n^*, p_0))$$

follows. Otherwise, we have

$$h^2\left(\frac{\hat{p}_n + p_n^*}{2}, p_n^*\right) \lesssim \int \log \frac{\hat{p}_n + p_n^*}{2p_n^*} d(\mathbb{P}_n - P_0),$$

which replaces Lemma 7.2.2 in the the proof of Theorem 7.2.1, which then applies here as well, with $p_n^* \in \mathcal{P}_n$ replacing p_0. ∎

7.2.2 The Information Geometry of the Likelihood Function

Many classical properties of maximum likelihood estimators $\hat{\theta}_n$ of regular parameters $\theta \in \Theta \subset \mathbb{R}^p$, such as asymptotic normality, are derived from the fact that the derivative of the log-likelihood function vanishes at $\hat{\theta}_n$, that is,

$$\frac{\partial}{\partial \theta} \ell_n(\theta)_{|\hat{\theta}_n} = 0. \tag{7.40}$$

This typically relies on the assumption that the true parameter θ_0 is interior to Θ so that by consistency $\hat{\theta}_n$ will eventually also be. In the infinite-dimensional setting, even if we can define an appropriate notion of derivative, this approach is usually not viable because $\hat{p}_n$ is, as we shall see, *never* an interior point in the parameter space, even when p_0 is.

We now investigate these matters in more detail in the setting where $\mathcal{P}$ consists of bounded probability densities. In this case, we can compute the Fréchet derivatives of the log-likelihood function on the vector space $L^\infty = L^\infty(\mathcal{X})$ of bounded functions equipped with the $\|\cdot\|_\infty$-norm. Recall that a real-valued function $L : U \to \mathbb{R}$ defined on an open subset U of a Banach space B is Fréchet differentiable at $f \in U$ if

$$\lim_{\|h\|_B \to 0} \frac{|L(f+h) - L(f) - DL(f)[h]|}{\|h\|_B} = 0, \tag{7.41}$$

for some linear continuous map $DL(f) : B \to \mathbb{R}$. If $g \in U$ is such that the line segment $(1-t)f + tg, t \in (0,1)$, joining f and g, lies in U (e.g., if U is convex), then the directional derivative of L at f in the direction g equals precisely

$$\lim_{t \to 0+} \frac{L(f + t(g-f)) - L(f))}{t} = DL(f)[g-f].$$

Higher Fréchet derivatives are defined in the usual way as derivatives of the mapping $f \mapsto DL(f)[h]$ for fixed $h \in B$.

The following proposition shows that the log-likelihood function ℓ_n is Fréchet differentiable on the open convex subset of L^∞ consisting of functions that are positive at the sample points. A similar result holds for ℓ if we restrict to functions that are bounded away from zero.

Proposition 7.2.4 *For any finite set of points $x_1, \ldots, x_n \in \mathcal{X}$, define*

$$\mathcal{U}(x_1, \ldots, x_n) = \left\{ f \in L^\infty : \min_{1 \le i \le n} f(x_i) > 0 \right\}$$

and

$$\mathcal{U} = \left\{ f \in L^\infty : \inf_{x \in \mathcal{X}} f(x) > 0 \right\}.$$

Then $\mathcal{U}(x_1, \ldots, x_n)$ and $\mathcal{U}$ are open subsets of $L^\infty(\mathcal{X})$. Let ℓ_n be the log-likelihood function from (7.20) based on $X_1, \ldots, X_n \sim^{i.i.d.} P_0$, and denote by P_n the empirical measure associated with the sample. Let ℓ be as in (7.21). For $\alpha \in \mathbb{N}$ and $f_1, \ldots, f_\alpha \in L^\infty$, the αth Fréchet derivatives of $\ell_n : \mathcal{U}(X_1, \ldots, X_n) \to \mathbb{R}$, $\ell : \mathcal{U} \to \mathbb{R}$ at a point $f \in \mathcal{U}(X_1, \ldots, X_n)$, $f \in \mathcal{U}$, respectively, are given by

$$D^\alpha \ell_n(f)[f_1, \ldots, f_\alpha] \equiv (-1)^{\alpha-1}(\alpha-1)P_n(f^{-\alpha} f_1 \cdots f_\alpha),$$
$$D^\alpha \ell(f)[f_1, \ldots, f_\alpha] \equiv (-1)^{\alpha-1}(\alpha-1)P_0(f^{-\alpha} f_1 \cdots f_\alpha).$$

Proof The set $\mathcal{U}(x_1, \ldots, x_n)$ equals $\cap_{i=1}^n \delta_{x_i}^{-1}((0,\infty))$ and hence is open since, by continuity of δ_x on L^∞, it is a finite intersection of open sets. Likewise, $\mathcal{U}$ is the preimage of $(0,\infty)$ under the continuous map $f \mapsto \inf_x f(x)$ and hence open. The derivatives of ℓ_n are easily computed using the chain rule for Fréchet differentiable functions, that δ_x is linear and continuous and hence Fréchet differentiable on L^∞ and that log is differentiable on $(0,\infty)$.

The derivatives of ℓ follow from the fact that ℓ_n is differentiable on $\mathcal{U} \subset \mathcal{U}(x_1,\dots,x_n)$ and from interchanging differentiation with respect to $f \in L^\infty$ and integration with respect to P_0, admissible in view of Exercise 7.2.4. ■

We deduce from the preceding proposition the intuitive fact that the limiting log-likelihood function has a derivative at the true point $p_0 > 0$ that is zero in all 'tangent space' directions h in

$$\mathcal{H} \equiv \left\{h : \int_{\mathcal{X}} h = 0\right\} \tag{7.42}$$

since

$$D\ell(p_0)[h] = \int_{\mathcal{X}} p_0^{-1} h dP_0 = \int_{\mathcal{X}} h = 0. \tag{7.43}$$

However, in the infinite-dimensional setting, the empirical counterpart of (7.43),

$$D\ell_n(\hat{p}_n)[h] = 0, \tag{7.44}$$

for $h \in \mathcal{H}$, and $\hat{p}_n$ the nonparametric maximum likelihood estimator is not true in general. The set $\mathcal{P}$ that the likelihood was maximised over will in typical nonparametric situations have empty interior in L^∞, and as discussed after (7.22), the maximiser $\hat{p}_n$ will lie at the boundary of $\mathcal{P}$ (see Proposition 7.2.9 for a concrete example). As a consequence, we cannot expect that $\hat{p}_n$ is a zero of $D\ell_n$.

In some situations there is a way around this problem: if the true value p_0 lies in the *interior* of $\mathcal{P}$ in the sense that local L^∞ perturbations of p_0 are contained in $\mathcal{P} \cap \mathcal{U}(X_1,\dots,X_n)$, then we can prove the following:

Lemma 7.2.5 *Let $\hat{p}_n$ be as in (7.23), and suppose that for some $\bar{g} \in L^\infty, \eta > 0$, the line segment joining $\hat{p}_n$ and $p_0 \pm \eta \bar{g}$ is contained in $\mathcal{P} \cap \mathcal{U}(X_1,\dots,X_n)$. Then*

$$|D\ell_n(\hat{p}_n)[\bar{g}]| \le (1/\eta)|D\ell_n(\hat{p}_n)(\hat{p}_n - p_0)|. \tag{7.45}$$

Proof Since $\hat{p}_n$ is a maximiser over $\mathcal{P}$, we deduce from differentiability of ℓ_n on $\mathcal{U}(X_1,\dots,X_n)$ that the derivative at $\hat{p}_n$ in the direction $p_0 + \eta\bar{g} \in \mathcal{P} \cap \mathcal{U}(X_1,\dots,X_n)$ necessarily has to be nonpositive; that is,

$$D\ell_n(\hat{p}_n)[p_0 + \eta\bar{g} - \hat{p}_n] = \lim_{t \to 0+} \frac{\ell_n(\hat{p}_n + t(p_0 + \eta\bar{g} - \hat{p}_n)) - \ell_n(\hat{p}_n)}{t} \le 0 \tag{7.46}$$

or, by linearity of $D\ell_n(\hat{p}_n)[\cdot]$,

$$D\ell_n(\hat{p}_n)[\eta\bar{g}] \le D\ell_n(\hat{p}_n)(\hat{p}_n - p_0). \tag{7.47}$$

Applying the same reasoning with $-\eta$, we see that

$$|D\ell_n(\hat{p}_n)[\eta\bar{g}]| \le D\ell_n(\hat{p}_n)(\hat{p}_n - p_0) = |D\ell_n(\hat{p}_n)(\hat{p}_n - p_0)|. \tag{7.48}$$

Divide by η to obtain the result. ■

The preceding lemma is interesting if we are able to show that

$$D\ell_n(\hat{p}_n)(\hat{p}_n - p_0) = o_{P_0^{\mathbb{N}}}(1/\sqrt{n}),$$

as then the same rate bound carries over to $D\ell_n(\hat{p}_n)[\bar{g}]$. This, in turn, can be used to mimic the finite-dimensional asymptotic normality proof of maximum likelihood estimators, which does not require (7.40) but only that the score is of smaller stochastic order of magnitude than $1/\sqrt{n}$. As a consequence, we will be able to obtain the asymptotic distribution of linear integral functionals of $\hat{p}_n$ and, more generally, for $\hat{P}_n$ the probability measure associated with $\hat{p}_n$, central limit theorems for $\sqrt{n}(\hat{P}_n - P)$ in 'empirical process–type' spaces $\ell^\infty(\mathcal{F})$ (comparable to results in Section 5.2.1 for linear estimators). To understand this better, we notice that Proposition 7.2.4 implies the following relationships: if we define the following projection of $g \in L^\infty$ onto $\mathcal{H}$,

$$\pi_0(g) \equiv (g - P_0 g)p_0 \in \mathcal{H}, \quad P_0(g) = \int_{\mathcal{X}} g dP_0, \tag{7.49}$$

and if we assume that $p_0 > 0$, then

$$\int_{\mathcal{X}} (\hat{p}_n - p_0) g d\mu = \int_{\mathcal{X}} p_0^{-2}(\hat{p}_n - p_0)(g - P_0 g)p_0 dP_0 = -D^2\ell(p_0)[\hat{p}_n - p_0, \pi_0(g)]$$

and

$$D\ell_n(p_0)(\pi_0(g)) = (P_n - P_0)g$$

so that the following is true:

Lemma 7.2.6 *Suppose that $p_0 > 0$. Let $\hat{p}_n$ be as in (7.23), and let $\hat{P}_n$ be the random probability measure induced by $\hat{p}_n$. For any $g \in L^\infty$ and P_n the empirical measure, we have*

$$|(\hat{P}_n - P_n)(g)| = \left| \int_{\mathcal{X}} gd(\hat{P}_n - P_n) \right| = |D\ell_n(p_0)[\pi_0(g)] + D^2\ell(p_0)[\hat{p}_n - p_0, \pi_0(g)]|. \tag{7.50}$$

Heuristically, the right-hand side equals, up to higher-order terms,

$$D\ell_n(\hat{p}_n)[\pi_0(g)] - D^2\ell_n(p_0)[\hat{p}_n - p_0, \pi_0(g)] + D^2\ell(p_0)[\hat{p}_n - p_0, \pi_0(g)]. \tag{7.51}$$

Control of (7.45) with choice $\bar{g} = \pi_0(g)$ at a rate $o_{P_0^{\mathbb{N}}}(1/\sqrt{n})$ combined with stochastic bounds on the second centred log-likelihood derivatives and convergence rates for $\hat{p}_n - p_0 \to 0$ thus give some hope that one may be able to prove that

$$(\hat{P}_n - P_0 - P_n + P_0)(g) = (\hat{P}_n - P_n)(g) = o_{P_0^{\mathbb{N}}}(1/\sqrt{n})$$

and that, thus, by the central limit theorem for $(P_n - P_0)g$,

$$\sqrt{n} \int_{\mathcal{X}} (\hat{p}_n - p_0) g d\mu \to^d N(0, P_0(g - P_0 g)^2)$$

as $n \to \infty$. We shall show how this can be made to work in a rigorous fashion in the two main examples we turn to now: the maximum likelihood estimator of a monotone density and of a density contained in a t-Sobolev ball for $t > 1/2$.

7.2.3 The Maximum Likelihood Estimator over a Sobolev Ball

In this section we investigate the likelihood principle in the prototypical nonparametric situation where the statistical model is a t-Sobolev ball of densities on $[0, 1]$. We define

Sobolev spaces H_2^m for $m \in \mathbb{N}$ as in Chapter 4 but shall work immediately with the equivalent wavelet definition, which makes sense for noninteger m as well: let

$$H^s \equiv H_2^s([0,1]) \equiv B_{22}^s([0,1])$$
$$= \left\{ f \in C([0,1]) : \|f\|_{H^s}^2 \equiv \sum_{l \geq -1} 2^{2ls} \sum_{k=0}^{2^l-1} |\langle f, \psi_{lk} \rangle|^2 < \infty \right\}, \quad s > 0, \tag{7.52}$$

where the $\{\psi_{lk}\}$ are a periodised Meyer-wavelet basis of $L^2([0,1])$ from Section 4.3.4, with the usual notational convention $\psi_{-1,k} \equiv 1$. We also could consider boundary-corrected wavelet bases on $[0,1]$ with only minor notational changes.

Existence and Basic Properties of the NPMLE

We observe i.i.d. random variables $X_1, \dots, X_n$ from law P_0 with Lebesgue density p_0 on $[0,1]$. For $t > 1/2$, we define the statistical model

$$\mathcal{P} \equiv \mathcal{P}(t,D) = \left\{ p \in C([0,1]),\ p \geq 0 \text{ on } [0,1],\ \int_0^1 p(x)dx = 1,\ \|p\|_{H^t} \leq D \right\}, \tag{7.53}$$

which equals a Sobolev ball of probability density functions. Note that $t > 1/2$ implies automatically (by (4.134)) that $p \in H^t$ defines a continuous and periodic function on $[0,1]$. Moreover, we tacitly assume $D > 1$ to ensure that $\mathcal{P}$ contains densities that are different from the constant one.

Recalling the likelihood function $\ell_n(p)$ from (7.19), a nonparametric maximum likelihood estimator is any element $\hat{p}_n \in \mathcal{P}$ such that

$$\ell_n(\hat{p}_n) = \sup_{p \in \mathcal{P}(t,D)} \ell_n(p). \tag{7.54}$$

The existence of $\hat{p}_n$ is ensured in the following proposition. It is shown in Exercise 7.2.5 that $\hat{p}_n$ is in fact unique.

Proposition 7.2.7 *Let $t > 1/2, D > 1$.*

(a) The set $\mathcal{P}(t,D)$ is a compact subset of the Banach space $C([0,1])$.
(b) The log-likelihood function ℓ_n is a continuous map from $\mathcal{P}(t,D)$ to $[-\infty,\infty)$ when $\mathcal{P}(t,D)$ is equipped with the topology of uniform convergence.
(c) There exists $\hat{p}_n \in \mathcal{P}(t,D)$ satisfying (7.54). The mapping $(X_1,\dots,X_n) \to \hat{p}_n$ can be taken to be Borel-measurable from $[0,1]^n \to C([0,1])$.

Proof (a) By Corollary 4.3.38, any fixed ball $\mathcal{B}$ of radius D in $H^t \subset B_{2\infty}^t, t > 1/2$, is a relatively compact subset of $C([0,1])$, and $\mathcal{B}$ is in fact compact because it is closed for uniform convergence: note first that $\mathcal{B}$ is the unit ball in the separable Hilbert space H^t, and hence $\mathcal{B}$ is sequentially compact for the weak topology of H^t (by the Banach-Alaoglu theorem). Since point evaluation $\delta_x \in C([0,1])^* \subset (H^t)^*$ for any $x \in [0,1], t > 1/2$, this implies that any sequence $f_m \in \mathcal{B}$ converges pointwise to $f \in \mathcal{B}$ along a subsequence. Hence, if f_m is any sequence in $\mathcal{B}$ that converges uniformly to some f^*, we infer from continuity and uniqueness of limits that $f^* = f$, so $\mathcal{B}$ is $\|\cdot\|_\infty$-closed.

Note next that the sets

$$\{f \in C([0,1]) : 0 \le f(x) < \infty\} = \cap_{x\in[0,1]}\delta_x^{-1}([0,\infty)), \quad \left\{f : \int_0^1 f(x)dx = 1\right\}$$

are both closed in $C([0,1])$ since point evaluation δ_x is continuous from $C([0,1])$ to $\mathbb{R}$ and since uniform convergence implies L^1-convergence on $[0,1]$, which implies convergence of integrals. Therefore, $\mathcal{P}$ is the intersection of a compact with two closed sets and hence is itself compact.

Finally, since $\mathcal{P}$ consists of nonnegative uniformly bounded functions, claim (b) is immediate because $\log 0 = -\infty$ continuously extends log to $[0,\infty)$. The first claim in (c) follows immediately from continuity of ℓ_n and compactness of $\mathcal{P}$. Measurability is discussed in Exercise 7.2.3. ■

The Sieved NPMLE

Let us consider next a sieved maximum likelihood estimator over a wavelet sieve

$$V_j = \text{span}(\psi_{lk} : -1 \le l \le j, k = 0,\ldots,2^l - 1),$$

that is, over the model

$$\mathcal{P}_j \equiv \mathcal{P}(t,D,j) = \{p \in \mathcal{P}(t,D), p \in V_j\}, \quad j < \infty. \tag{7.55}$$

The sieved MLE $\hat{p}_{n,j}$ is defined by

$$\ell_n(\hat{p}_{n,j}) = \sup_{p\in\mathcal{P}_j} \ell_n(p). \tag{7.56}$$

Since all that we have done is to intersect $\mathcal{P}(t,D)$ with a finite-dimensional linear space, it is evident that Proposition 7.2.7 carries over directly, and hence, there exists $\hat{p}_{n,j} \in \mathcal{P}(t,D,j)$ satisfying (7.56) whenever $t > 1/2$.

Uniform Consistency of the NPMLE

We now derive almost-sure uniform consistency of $\hat{p}_n$ under the assumption that $p_0 \in \mathcal{P}(t,D)$ is strictly positive on $[0,1]$. We can derive from it that $\hat{p}_n$ will eventually be the maximiser also over $\mathcal{P}$ restricted to densities that are bounded away from zero.

We give a direct consistency proof based on compactness and strong laws of large numbers only – the result also follows from the general rate of convergence theory applied to $\mathcal{P}(t,D)$ (see Proposition 7.2.10). We shall say that an event A_n happens eventually Pr almost surely if $\lim_m \Pr(\cap_{n\ge m} A_n) = 1$.

Proposition 7.2.8 *Assume that* $p_0 \in \mathcal{P}(t,D)$ *satisfies* $p_0 > 0$ *on* $[0,1]$, *and let* $\hat{p}_n$ *satisfy (7.54). Then, as* $n \to \infty$,

$$\|\hat{p}_n - p_0\|_\infty \to 0 \quad P_0^{\mathbb{N}}\ a.s.$$

In particular, for some $\zeta > 0$, *eventually* $P_0^{\mathbb{N}}$ *almost surely*

$$\hat{p}_n \in \mathcal{P}_\zeta(t,D) \equiv \mathcal{P}(t,D) \cap \{f \ge \zeta \text{ on } [0,1]\}.$$

Proof We note that by continuity, for some $0 < \xi < M < \infty$, we have $\xi \le p_0 \le M$ and hence, by the strong law of large numbers $P_0^{\mathbb{N}}$ a.s.,

$$\lim_n |\ell_n(p_0) - \ell(p_0)| = 0.$$

Moreover, for any constant $0 < \epsilon \le 1$ and $x \in [0,1]$, the mapping $p \mapsto \log(p(x) + \epsilon)$ is continuous from the compact metric space $(\mathcal{P}, \|\cdot\|_\infty)$ to $[\log \epsilon, \sup_{p\in\mathcal{P}} \|p\|_\infty)$ and, in particular, satisfies

$$\int_0^1 \sup_{p\in\mathcal{P}} |\log(p(x)+\epsilon)| dP_0(x) < \infty.$$

Hence, for every fixed $\epsilon > 0$, we have

$$\lim_n \sup_{p\in\mathcal{P}} |\ell_n(p+\epsilon) - \ell(p+\epsilon)| = 0 \ P_0^{\mathbb{N}} \ a.s. \tag{7.57}$$

by the strong law of large numbers in the separable Banach space $C[(\mathcal{P}, \|\cdot\|_\infty)]$ (Corollary 3.7.21) applied to the i.i.d. average $\{p \mapsto \frac{1}{n}\sum_{i=1}^n \log(p(X_i) + \epsilon)\}$ of continuous functions on $\mathcal{P}$.

In the rest of this proof we restrict ourselves to an event A of probability 1 on which the last two displayed limits hold and show that every subsequence $\hat{p}_{n'}$ of $\hat{p}_n$ has a further subsequence that converges to p_0 on this event. Take a decreasing sequence of positive real numbers $\epsilon_l \downarrow 0$, and note that by compactness of $\mathcal{P}$ there exists a subsequence $\hat{p}_{n''}$ of $\hat{p}_{n'}$ that converges uniformly to some $p^* \in \mathcal{P}$. By definition of $\hat{p}_{n''}$ and monotonicity of log, we see that

$$\begin{aligned}\ell_{n''}(p_0) &\le \ell_{n''}(\hat{p}_{n''}) \le \ell_{n''}(\hat{p}_{n''} + \epsilon_l) \\ &\le \ell(\hat{p}_{n''} + \epsilon_l) + \sup_{p\in\mathcal{P}} |\ell_n(p+\epsilon_l) - \ell(p+\epsilon_l)|.\end{aligned}$$

The first term on the right-hand side converges to $\ell(p^* + \epsilon_l)$ since $\ell(\cdot + \epsilon_l)$ is supnorm continuous on $\mathcal{P}$ (using the dominated convergence theorem), and hence, on the event A, we have, taking limits,

$$\ell(p_0) \le \ell(p^* + \epsilon_l). \tag{7.58}$$

The functions $\log(p^* + \epsilon_l)$ converge pointwise to $\log p^*$ and are bounded above by $\log(p^* + \epsilon_1)$, so, by monotone convergence, we deduce that $\ell(p_0) \le \ell(p^*)$. From (7.22), $p^* = p_0$ follows. ■

Geometric Interpretation of the Maximiser and H^t-Inconsistency of the MLE

While $\hat{p}_n$ is consistent in $\|\cdot\|_\infty$-loss, it is *not* consistent in the norm that defines the constraint of $\mathcal{P}(t,D)$. In fact, we now prove rigorously the intuitive fact that when maximising ℓ_n over $\mathcal{P}$ in (7.54), the MLE $\hat{p}_n$ always lies at the boundary of $\mathcal{P}(t,D)$ in the sense that it exhausts the Sobolev norm $\|\hat{p}_n\|_{H^t} = D$. Then, for any p_0 satisfying $\|p_0\|_{H^t} < D$ assuming consistency, $\|\hat{p}_n - p_0\|_{H^t} \to 0$ immediately leads to a contradiction. The fact that the MLE $\hat{p}_n$ necessarily lies on the D-sphere in H^t also implies, using the strict convexity of the Hilbert space H^t, the uniqueness of $\hat{p}_n$ (see Exercise 7.2.5).

Proposition 7.2.9 *Let $\hat{p}_n$ satisfy (7.54). Then we have for every $n \in \mathbb{N}$ that $\|\hat{p}_n\|_{H^t} = D$.*

Proof Suppose that the claim is not true; hence, $\|\hat{p}_n\|_{H^t} < D$. By continuity, there exists $z \in (0,1)$ different from the $X_1,\dots,X_n$ such that $\hat{p}_n(z) > 0$. Choose $\epsilon > 0$ small enough that $I = [z-2\epsilon, z+2\epsilon]$, $U = [X_1 - 2\epsilon, X_1 + 2\epsilon]$ and $\{X_j : X_j \neq X_1\}$ are all disjoint, and $\inf_{x\in I}\hat{p}_n(x) > 0$. There exists a compactly supported C^∞ function $\varphi : [0,1] \to [0,1]$ such that $\varphi = 1$ on $[X_1 - \epsilon, X_1 + \epsilon]$ and $\varphi = 0$ outside of U. Define $\bar{\varphi}(y) = \varphi(y + X_1 - z)$ whenever $y + X_1 - z \in [0,1]$ and equal to 0 otherwise. Note that $\bar{\varphi}$ is the translation of φ by $z - X_1$ and hence has disjoint support with φ. Then $g = \varphi - \bar{\varphi}$ integrates to 0, takes values in $[-1,1]$ and is contained in H^t because it is a C^∞ function with support interior to $[0,1]$. We thus can find $\eta > 0$ small enough such that

$$\tilde{p}_n = \hat{p}_n + \eta g \in \mathcal{P}.$$

Indeed, for η small, $\|\hat{p}_n + \eta g\|_{H^t} \leq D$, for $x \in [0,1]\setminus I$, we have $g(x) \geq 0$, so $\tilde{p}_n(x) \geq \hat{p}_n(x)$. For $x \in I$, we have by construction $\tilde{p}_n(x) \geq \hat{p}_n(x) - \eta \geq 0$. Now, since $\eta > 0$ and $g(X_1) = 1$, we see that

$$\tilde{p}_n(X_1) > \hat{p}_n(X_1),$$

but $\tilde{p}_n(X_j) = \hat{p}_n(X_j)$, for $j \neq 1$, so $\ell_n(\tilde{p}_n) > \ell_n(\hat{p}_n)$, a contradiction to $\hat{p}_n$ being the maximiser of ℓ_n. ■

Rates of Convergence

If $p_0 > 0$, we can directly apply Theorem 7.2.1 combined with the entropy estimates from Chapter 4 to obtain the following convergence rate estimates for $\hat{p}_n - p_0$ in Hellinger, L^2- and L^∞-distance:

Theorem 7.2.10 *Assume that $p_0 \in \mathcal{P}(t,D)$ for some $t > 1/2$ satisfies $p_0 > 0$ on $[0,1]$. Let $\hat{p}_n$ satisfy (7.54). Then we have, for some constants c, C depending only on $t, D, \inf_{x\in[0,1]} p_0(x)$, that*

$$P_0^{\mathbb{N}}\left(h(\hat{p}_n, p_0) > Cn^{-t/(2t+1)}\right) \leq c\exp(-n^{1/(2t+1)}/c), \tag{7.59}$$

$$P_0^{\mathbb{N}}\left(\|\hat{p}_n - p_0\|_2 > Cn^{-t/(2t+1)}\right) \leq c\exp(-n^{1/(2t+1)}/c), \tag{7.60}$$

and also

$$P_0^{\mathbb{N}}\left(\|\hat{p}_n - p_0\|_\infty > Cn^{-(t-1/2)/(2t+1)}\right) \leq c\exp(-n^{1/(2t+1)}/c). \tag{7.61}$$

Proof We note that for $\inf_{x\in[0,1]} |p_0(x)| \geq \zeta > 0$, the functions $(p + p_0)/2$ are bounded away from 0 by $\zeta/2$. Since $x \mapsto \sqrt{x}$ is Lipschitz on $[\zeta/2, \infty)$, the $L^2(\mu)$-bracketing metric entropy (for μ equal to Lebesgue measure on $[0,1]$) of the class of functions

$$\left\{\sqrt{\frac{p+p_0}{2}} : p \in \mathcal{P}_\zeta(t,D)\right\}$$

can be bounded by the supnorm metric entropy of a bounded subset of the space $H^t([0,1])$. By Corollary 4.3.38, this allows the choice

$$\delta^{(1-\frac{1}{2t})} \simeq \mathcal{J}(\delta) \geq C' \int_0^\delta (1/\varepsilon)^{1/2t} d\varepsilon,$$

for some $C' > 0$, so Theorem 7.2.1 gives (7.59). Since $\hat{p}_n, p_0$ are uniformly bounded, this rate is inherited by the L^2-distance

$$\|\hat{p}_n - p_0\|_2^2 \leq 2(\|\hat{p}_n\|_\infty + \|p_0\|_\infty)h^2(\hat{p}_n, p_0).$$

To obtain a rate for the supnorm distance, we can use results from Chapter 4 and interpolate: using $\|\hat{p}_n - p_0\|_{H^t} \leq D$, the uniform convergence of the wavelet series of elements of $H^t \subset B^{t-1/2}_{\infty\infty}$ for $t > 1/2$ and the Cauchy-Schwarz inequality, we see that

$$\begin{aligned}
\|\hat{p}_n - p_0\|_\infty &\lesssim \sum_l 2^{l/2} \max_k |\langle \hat{p}_n - p_0, \psi_{lk}\rangle| \\
&\lesssim 2^{j/2}\|\hat{p}_n - p_0\|_2 + \sum_{l>j} 2^{-l(t-1/2)} \\
&= O_{P_0^{\mathbb{N}}}\left(2^{j/2} n^{-t/(2t+1)} + 2^{-j(t-1/2)}\right),
\end{aligned}$$

which gives the desired rate by choosing $2^j \sim n^{1/(2t+1)}$. ■

Note that thus the MLE $\hat{p}_n$ achieves the minimax optimal rate of estimation in L^2-loss over the Sobolev ball $\mathcal{P}$ whenever $p_0 > 0$ on $[0,1]$; compare to Theorem 6.3.9 with $r = t, p = q = 2$. For the supnorm, the rate is not optimal, but the bound will still be useful later. It is an interesting open question whether the supnorm rate can be improved or not – see the notes at the end of this chapter for some discussion.

We can obtain a similar result for the sieved MLE from (7.56):

Proposition 7.2.11 *Assume that $p_0 \in \mathcal{P}(t,D)$ for some $t > 1/2$ satisfies $p_0 > 0$ on $[0,1]$. Let $\hat{p}_{n,j_n}$ be the sieved MLE from (7.56) with choice $2^{j_n} \sim n^{1/(2t+1)}$. Then we have*

$$h(\hat{p}_{n,j_n}, p_0) = O_P(n^{-t/(2t+1)}), \tag{7.62}$$

$$\|\hat{p}_{n,j_n} - p_0\|_2 = O_P(n^{-t/(2t+1)}) \tag{7.63}$$

and also

$$\|\hat{p}_{n,j_n} - p_0\|_\infty = O_P(n^{-(t-1/2)/(2t+1)}). \tag{7.64}$$

Proof See Exercise 7.2.6. ■

Stochastic Bounds for the Score Function

Taking notice of Lemma 7.2.5, we now investigate the behaviour of the likelihood derivative $D\ell_n(\hat{p}_n)$ in the direction of p_0.

Lemma 7.2.12 *Assume that $p_0 \in \mathcal{P}(t,D)$ for some $t > 1/2$ satisfies $p_0 > 0$ on $[0,1]$. Let $\hat{p}_n$ satisfy (7.54). Then*

$$|D\ell_n(\hat{p}_n)(\hat{p}_n - p_0)| = O_{P_0^{\mathbb{N}}}\left(n^{-1/2} n^{-(t-1/2)/(2t+1)}\right).$$

Proof By Theorem 7.2.10, we can restrict ourselves to the events

$$\{\|\hat{p}_n - p_0\|_2 \lesssim n^{-t/(2t+1)}\} \cap \{\|\hat{p}_n - p_0\|_\infty \lesssim n^{-(t-1/2)/(2t+1)}\},$$

and hence, for n large enough, further to events where $\hat{p}_n \geq \zeta$ for some $\zeta > 0$. On these events, both $\hat{p}_n$ and p_0 are contained in the set $\mathcal{U}(X_1, \ldots, X_n)$ from Proposition 7.2.4, which we can use to calculate the derivative $D\ell_n(\hat{p}_n)$. Moreover (Exercise 4.3.5), the functions $\hat{p}_n, \hat{p}_n^{-1}, p_0$ and pointwise products thereof vary in a fixed ball of $H^t, t > 1/2$, which has a small enough supnorm (and then also bracketing) metric entropy in view of Corollary 4.3.38. Using the moment inequality from Corollary 3.5.7 with $F = u = const, \sigma \simeq n^{-t/(2t+1)}$ and $H(x) = x^{1/t}$ (or Remark 3.5.14), (7.43) with $h = \hat{p}_n - p_0$ and again Theorem 7.2.10, we see that

$$\begin{aligned}|D\ell_n(\hat{p}_n)(\hat{p}_n - p_0)| &= |D\ell_n(\hat{p}_n)(\hat{p}_n - p_0) - D\ell(\hat{p}_n)(\hat{p}_n - p_0) + (D\ell(\hat{p}_n) - D\ell(p_0))(\hat{p}_n - p_0)| \\ &\leq \sup_{f:\|f\|_{H^t} \leq B, \|f\|_2 \lesssim n^{-t/(2t+1)}} |(P_n - P)(f)| + C\int_0^1 (\hat{p}_n - p_0)^2(x)dx \\ &= O_{P_0^{\mathbb{N}}}\left(\frac{1}{\sqrt{n}} n^{-(t-1/2)/(2t+1)} + n^{-2t/(2t+1)}\right),\end{aligned}$$

which implies the desired result. ■

This result now can be combined with Lemma 7.2.5: if p_0 is interior to $\mathcal{P}$ in the sense that

$$\inf_{x\in[0,1]} p_0(x) > 0, \|p_0\|_{H^t} < D, \tag{7.65}$$

we can construct a line segment around p_0 of the form $p_0 \pm \eta w$ where w is a fixed direction in $H^t \subset L^\infty$ that integrates to 0, and where $\eta \leq D - \|p_0\|_{H^t}$ is small enough. For general $g \in H^t$, we recall (7.49) and take projections $\pi_0(g) = (g - P_0 g)p_0$, which by the multiplication algebra property of $H^t, t > 1/2$ (Exercise 4.3.5), are again contained in H^t and integrate to 0. From Lemma 7.2.5 with $\bar{g} = \pi_0(g)$, which applies by convexity of $\mathcal{P}$, we thus obtain that for every $g \in H^t$ and p_0 satisfying (7.65) we have

$$|D\ell_n(\hat{p}_n)[\pi_0(g)]| \leq C(D - \|p_0\|_{H^t})^{-1} \|g\|_{H^t} \|p_0\|_{H^t} n^{-2t/(2t+1)} \tag{7.66}$$

for some fixed constant C independent of D, p_0, g.

Asymptotic Normality of the NPMLE

Lemma 7.2.13 *Let $\hat{p}_n$ satisfy (7.54), and assume that p_0 satisfies (7.65). Let $\hat{P}_n$ be the probability measure corresponding to $\hat{p}_n$, and let P_n be the empirical measure. Then, for any fixed ball $\mathcal{G}$ of $H^t, t > 1/2$, of radius B,*

$$\sup_{g\in\mathcal{G}} |(\hat{P}_n - P_n)(g)| = O_{P_0^{\mathbb{N}}}\left(Bn^{-1/2} n^{-(t-1/2)/(2t+1)}\right), \tag{7.67}$$

with constants independent of B.

Proof From Lemma 7.2.6 and Proposition 7.2.4, we have, using a pathwise mean value theorem with mean values $\bar{p}_n \in \mathcal{U} \cap \mathcal{P}$ on the line segment between $\hat{p}_n, p_0 \in \mathcal{U}$, and every $g \in \mathcal{G}$,

$$\begin{aligned}|(\hat{P}_n - P_n)(g)| &= |D\ell_n(p_0)[\pi_0(g)] + D^2\ell(p_0)[\hat{p}_n - p_0, \pi_0(g)]| \\ &= |D\ell_n(\hat{p}_n)[\pi_0(g)] - (D^2\ell_n(\bar{p}_n) - D^2\ell(p_0))[\hat{p}_n - p_0, \pi_0(g)]| \\ &\leq |D\ell_n(\hat{p}_n)(\pi_0(g))| + |(D^2\ell_n(\bar{p}_n) - D^2\ell(\bar{p}_n))[\hat{p}_n - p_0, \pi_0(g)]| \\ &\quad + |(D^2\ell(\bar{p}_n)) - D^2\ell(p_0))[\hat{p}_n - p_0, \pi_0(g)]|.\end{aligned}$$

The first term was bounded in (7.66). As in the proof of Lemma 7.2.12, the second term is bounded by

$$\sup_{f:\|f\|_{H^t}\le B, \|f\|_2 \lesssim n^{-t/(2t+1)}} |(P_n - P)(f)| = O_{P_0^{\mathbb{N}}}\left(n^{-1/2} n^{-(t-1/2)/(2t+1)}\right),$$

and the third term is bounded by a constant multiple of

$$\|g\|_\infty \int_0^1 (\hat{p}_n - p_0)^2 = O_{P_0^{\mathbb{N}}}(n^{-2t/(2t+1)})$$

in view of Theorem 7.2.10. ∎

The fact that this rate is faster than $1/\sqrt{n}$ allows us to push (7.67) further and, in particular, to treat classes $\mathcal{F}$ of functions that are independent of t. For $f \in \mathcal{F}$, we can decompose, for $\pi_{V_J}(f)$ the wavelet projection of $f \in \mathcal{F}$, onto the span V_J of wavelets up to resolution level J, that is,

$$\begin{aligned}\|\hat{P}_n - P_n\|_{\mathcal{F}} &\le \sup_{f\in\mathcal{F}} \left| \int_0^1 (\hat{p}_n - p_0)(f - \pi_{V_J}(f)) \right| \\ &\quad + \sup_{f\in\mathcal{F}} |(\hat{P}_n - P_n)(\pi_{V_J}(f))| + \sup_{f\in\mathcal{F}} |(P_n - P_0)(f - \pi_{V_J}(f))|. \end{aligned} \tag{7.68}$$

From this we can deduce the following theorem. Recall the definition of convergence of random probability measures in $\ell^\infty(\mathcal{F})$ and of the P-Brownian bridge process G_P from Section 3.7.

Theorem 7.2.14 *Let $\hat{p}_n$ satisfy (7.54), and assume that p_0 satisfies (7.65). Let $\hat{P}_n, P_0$ be the probability measures corresponding to $\hat{p}_n, p_0$, respectively, and let P_n be the empirical measure. For $s > 1/2$, let $\mathcal{F}$ be a bounded subset of the s-Sobolev space H^s or of the s-Hölder space C^s. Then*

$$\|\hat{P}_n - P_n\|_{\mathcal{F}} = \sup_{f\in\mathcal{F}} \left| \int_0^1 f d(\hat{P}_n - P_n) \right| = o_{P_0^{\mathbb{N}}}(1/\sqrt{n})$$

and, thus,

$$\sqrt{n}(\hat{P}_n - P_0) \to^d G_{P_0} \text{ in } \ell^\infty(\mathcal{F}).$$

In particular, for any $f \in H^s$ or $f \in C^s$, we have

$$\sqrt{n} \int_0^1 (\hat{p}_n(x) - p_0(x)) f(x) dx \to^d N(0, \|f - P_0 f\|^2_{L^2(P_0)}).$$

Proof Since $C^s \subset H^{s'}$ for any $s > s'$, it suffices to prove the case of $H^s, s > 1/2$ arbitrary. For the first estimate, assume that $s < t$ as otherwise the result is immediate from (7.67). Choose $2^{j_n} \sim n^{1/(2t+1)}$. Then we have, from the definition of the Sobolev norm,

$$\|\pi_{V_{j_n}}(f)\|_{H^t} \lesssim 2^{j_n(t-s)} \simeq n^{(t-s)/(2t+1)}$$

and, by Parseval's identity,

$$\|f - \pi_{V_{j_n}}(f)\|_2 \lesssim 2^{-j_n s} \lesssim n^{-s/(2t+1)}$$

and from uniform convergence of the wavelet series of $f \in C([0,1])$ also that

$$\|f - \pi_{V_{j_n}}(f)\|_\infty \lesssim n^{(-s+1/2)/(2t+1)}.$$

Combined with the fact that $\|\hat{p}_n - p_0\|_2 = O_{P_0^{\mathbb{N}}}(n^{-t/(2t+1)})$ from Theorem 7.2.10 and with (7.68), this gives the bound

$$\|\hat{P}_n - P_n\|_{\mathcal{F}} = O_{P_0^{\mathbb{N}}}\left(n^{(-t-s)/(2t+1)} + n^{-1/2}n^{(-t+1/2)/(2t+1)}n^{(t-s)/(2t+1)} + n^{-1/2}n^{(-s+1/2)/(2t+1)}\right),$$

with last term controlled as in the proof of Lemma 7.2.12. The overall bound is then $o_P(n^{-1/2})$ as soon as $s > 1/2$. The remaining claims follow from the fact that a ball in $H^s, s > 1/2$, is a uniform Donsker class (see Theorem 4.4.5). ■

We also can obtain the following result for the cumulative distribution function of P_0:

Theorem 7.2.15 *Under the assumptions of Theorem 7.2.14, let $\hat{F}_n = \int_0^t \hat{p}_n(x)dx$, $t \in [0,1]$, be the distribution function of the MLE, and let $F_0 = \int_0^t p_0(x)dx, t \in [0,1]$, be the true distribution function. Then*

$$\sqrt{n}\|\hat{F}_n - F_0\|_\infty = O_{P_0^{\mathbb{N}}}(1).$$

Proof In view of Corollary 3.7.39, we have

$$\|F_n - F_0\|_\infty = O_{P_0^{\mathbb{N}}}(1/\sqrt{n}),$$

where $F_n(t) = \int_0^t dP_n, 0 \le t \le 1$, is the empirical distribution function, and hence, it suffices to prove that $\|\hat{F}_n - F_n\|_\infty = O_{P_0^{\mathbb{N}}}(1/\sqrt{n})$. Since

$$f = 1_{[0,t]} \in B^1_{1\infty} \subset B^{1/2}_{2\infty}$$

from the theory in Section 4.3, we have

$$\|\pi_{V_j}(f)\|_{H^t} = \sum_{l \le j} 2^{lt} \sum_k |\langle f, \psi_{lk}\rangle|^2 \le C \sum_{l \le j} 2^{l(t-1/2)} \lesssim 2^{j(t-1/2)}$$

and

$$\|f - \pi_{V_j}(f)\|_2 = O(2^{-j/2}).$$

Take $2^{j_n} \sim n^{1/(2t+1)}$; then, from (7.68), Lemma 7.2.13 and Theorem 7.2.10, we deduce the bound

$$\begin{aligned}\|\hat{F}_n - F_n\|_\infty &= O_{P_0^{\mathbb{N}}}\left(n^{-t/(2t+1)}n^{-(1/2)/(2t+1)} + n^{-1/2}n^{(-t+1/2+t-1/2)/(2t+1)}\right) + o_{P_0^{\mathbb{N}}}(1/\sqrt{n}) \\ &= O_{P_0^{\mathbb{N}}}(n^{-1/2}),\end{aligned}$$

where we used, for the third term of the decomposition, that $\{\pi_{V_j}(1_{[0,t]}) : t \in [0,1]\}$ is a uniformly bounded VC-type class (it varies in a finite-dimensional linear space of functions) with variances σ bounded by

$$\|\pi_{V_j}(f) - f\|_{2,P_0} \le \|p_0\|_\infty \|\pi_{V_j}(f) - f\|_2 \to 0,$$

combined with the moment inequality Corollary 3.5.7. ■

These estimates are just marginally too weak to infer that $\sqrt{n}\|\hat{F}_n - F_n\|_\infty = o_{P_0^{\mathbb{N}}}(1)$. A similar gap appears when we study the MLE in generic directions ψ_{lk} for wavelets generating the Sobolev space. These satisfy $\|\psi_{lk}\|_{H^t} = 2^{lt}$, so for any wavelet $\psi_{lk}, l \geq -1, k = 0, \dots, 2^l - 1$,

$$|(\hat{P}_n - P_n)(\psi_{lk})| = O_{P_0^{\mathbb{N}}}\left(2^{lt} n^{-1/2} n^{-(t-1/2)/(2t+1)}\right). \tag{7.69}$$

It is an interesting open question whether these rates can be essentially improved or whether they are the correct rates for $\hat{p}_n$.

7.2.4 The Maximum Likelihood Estimator of a Monotone Density

Let $X_1, \dots, X_n$ be i.i.d. on $[0,1]$ with law P_0 and distribution function $F_0(x) = \int_0^x dP_0, x \in [0,1]$. Define the empirical measure $P_n = n^{-1}\sum_{i=1}^n \delta_{X_i}$ and the empirical cumulative distribution function $F_n(x) = \int_0^x dP_n, x \in [0,1]$. If P is known to have a monotone decreasing density p, then the associated maximum likelihood estimator $\hat{p}_n$ maximises the likelihood function $\ell_n(p)$ over

$$\mathcal{P} \equiv \mathcal{P}^{mon} = \left\{p : [0,1] \to [0,\infty), \int_0^1 p(x)dx = 1, p \text{ is nonincreasing}\right\};$$

that is,

$$\max_{p \in \mathcal{P}^{mon}} \ell_n(p) = \ell_n(\hat{p}_n). \tag{7.70}$$

It is easy to see that $\hat{p}_n$ is a left-continuous step function whose jumps can only occur at the observation points, or order statistics, $X_{(1)} < \cdots < X_{(n)}, X_{(0)} \equiv 0$. We can show, moreover, that the estimator has a simple geometric interpretation as the left derivative of the least concave majorant $\hat{F}_n$ of the empirical distribution function F_n (see Exercise 7.2.7 for details). The estimator $\hat{p}_n$ is also known as the *Grenander estimator*.

First Basic Properties and Rates of Convergence

We establish some first probabilistic properties of $\hat{p}_n$ that will be useful later: if p_0 is bounded away from 0, then so is $\hat{p}_n$ on the interval $[0, X_{(n)}]$, where $X_{(n)}$ is the last-order statistics (clearly, $\hat{p}_n(x) = 0$ for all $x > X_{(n)}$ in view of Exercise 7.2.7). Similarly, if p_0 is bounded above, then so is $\hat{p}_n$ with high probability.

Lemma 7.2.16 *(a) Suppose that the true density p_0 satisfies $\inf_{x\in[0,1]} p_0(x) > 0$. Then, for every $\epsilon > 0$, there exists $\xi > 0$ and a finite index $N(\epsilon)$ such that, for all $n \geq N(\epsilon)$,*

$$P_0^{\mathbb{N}}\left(\inf_{x\in[0,X_{(n)}]} \hat{p}_n(x) < \xi\right) = \Pr\left(\hat{p}_n(X_{(n)}) < \xi\right) < \epsilon$$

(b) Suppose that the true density p_0 satisfies $p_0(x) \leq K < \infty$ for all $x \in [0,1]$. Then, for every $\epsilon > 0$, there exists $0 < k < \infty$ such that, for all $n \in \mathbb{N}$,

$$P_0^{\mathbb{N}}\left(\sup_{x\in[0,1]} \hat{p}_n(x) > k\right) = \Pr\left(\hat{p}_n(0) > k\right) < \epsilon.$$

Proof (a) The first equality is obvious, since $\hat{p}_n$ is monotone decreasing. On each interval $(X_{(j-1)}, X_{(j)}]$, $\hat{p}_n$ is the slope of the least concave majorant of F_n (see Exercise 7.2.7). The least concave majorant touches $(X_{(n)}, 1)$ and at least one other order statistic (possibly $(X_{(0)}, 0)$), so

$$\{\hat{p}_n(X_{(n)}) < \xi)\} \subseteq \left\{X_{(n)} - X_{(n-j)} > j/(\xi n) \text{ for some } j = 1, \ldots, n\right\}.$$

Note next that since F_0 is strictly monotone, we have $X_i = F_0^{-1} F_0(X_i)$ and

$$F_0^{-1}F_0(X_{(n)}) - F_0^{-1}F_0(X_{(n-j)}) \leq \frac{1}{p_0(\eta)}\left(F_0(X_{(n)}) - F_0(X_{(n-j)})\right) \leq \zeta^{-1}\left(U_{(n)} - U_{(n-j)}\right),$$

where the $U_{(i)}$ are distributed as the order statistics of a sample of size n of a uniform random variable on $[0,1]$, and where $U_{(0)} = 0$ by convention. Hence, it suffices to bound

$$\Pr\left(U_{(n)} - U_{(n-j)} > \frac{\zeta j}{\xi n} \text{ for some } j = 1, \ldots, n\right). \tag{7.71}$$

By a standard computation involving order statistics (see Exercise 7.2.8), the joint distribution of $U_{(i)}$, $i = 1, \ldots, n$, is the same as the one of Z_i/Z_{n+1} where $Z_n = \sum_{l=1}^{n} W_l$, and where W_l are independent standard exponential random variables. Consequently, for $\delta > 0$, the probability in (7.71) is bounded by

$$\begin{aligned}
&\Pr\left(\frac{W_{n-j+1} + \ldots + W_n}{Z_{n+1}} > \frac{\zeta j}{\xi n} \text{ for some } j\right) \\
&= \Pr\left(\frac{n}{Z_{n+1}} \frac{W_{n-j+1} + \ldots + W_n}{n} > \frac{\zeta j}{\xi n} \text{ for some } j\right) \\
&\leq \Pr(n/Z_{n+1} > 1 + \delta) + \Pr\left(\frac{W_{n-j+1} + \ldots + W_n}{n} > \frac{\zeta j}{\xi n(1+\delta)} \text{ for some } j\right) \\
&= A + B.
\end{aligned}$$

To bound A, note that it is equal to

$$\Pr\left(\frac{1}{n+1}\sum_{l=1}^{n+1}(W_l - EW_l) < \frac{-\delta - (1+\delta)/n}{1+\delta}\frac{n}{n+1}\right),$$

which, since $\delta > 0$, is less than $\epsilon/2 > 0$ arbitrary, from some n onwards, by the law of large numbers. For the term B, we have, for ξ small enough and by Markov's inequality,

$$\begin{aligned}
&\Pr\left(W_{n-j+1} + \ldots + W_n > \frac{\zeta j}{\xi(1+\delta)} \text{ for some } j\right) \\
&\leq \sum_{j=1}^{n} \Pr\left(W_{n-j+1} + \ldots + W_n > \frac{\zeta j}{\xi(1+\delta)}\right) \\
&= \sum_{j=1}^{n} \Pr\left(\sum_{l=1}^{j}(W_{n-l+1} - EW_{n-l+1}) > \frac{\zeta j}{\xi(1+\delta)} - j\right) \\
&\leq \sum_{j=1}^{n} \frac{\xi^4 E(\sum_{l=1}^{j}(W_{n-l+1} - EW_{n-l+1}))^4}{j^4 C(\delta,\zeta)} \\
&\leq \xi^4 C'(\delta,\zeta)\sum_{j=1}^{n} j^{-2} \leq \xi^4 C''(\delta,\zeta) < \epsilon/2,
\end{aligned}$$

since, for $Y_l = W_{n-l+1} - EW_{n-l+1}$, by Hoffmann-Jørgensen's inequality (Theorem 3.1.22 with 'index set' T equal to a singleton),

$$\left\| \sum_{l=1}^{j} Y_l \right\|_{4,P} \leq K \left[\left\| \sum_{l=1}^{j} Y_l \right\|_{2,P} + \left\| \max_l Y_l \right\|_{4,P} \right] \leq K' \left(\sqrt{j} + j^{1/4} \right),$$

using the fact that $Var(Y_1) = 1$ and $EW_1^p = p$.
(b) From Exercise 7.2.7, we know that $\hat{p}_n$ is the left derivative of the least concave majorant of the empirical distribution F_n; hence

$$\|\hat{p}_n\|_\infty = \hat{p}_n(0) > M \iff F_n(t) > Mt, \quad \text{for some } t.$$

Since F_0 is concave and continuous, it maps $[0,1]$ onto $[0,1]$ and satisfies $F_0(t) \leq p_0(0)t \leq t\|p_0\|_\infty$, so we obtain

$$P_0^{\mathbb{N}}(\|\hat{p}_n\|_\infty > M) \leq P_0^{\mathbb{N}}\left(\sup_{t>0} \frac{F_n(t)}{F_0(t)} > M/\|p_0\|_\infty \right) = P_0^{\mathbb{N}}\left(\sup_{t\in[0,1]} \frac{F_n^U(t)}{t} > M/\|p_0\|_\infty \right),$$

where F_n^U is the empirical distribution function based on a sample of size n from the uniform distribution. The last probability can be made as small as desired for M large enough using Exercise 7.2.9. ■

We can now derive the rate of convergence of the maximum likelihood estimator of a monotone density. The rate corresponds to functions that are once differentiable in an L^1-sense, which is intuitively correct because a monotone decreasing function has a weak derivative that is a finite signed measure.

Theorem 7.2.17 *Suppose that $p_0 \in \mathcal{P}^{mon}$ and that p_0 is bounded. Let $\hat{p}_n$ satisfy (7.70). Then*

$$h(\hat{p}_n, p_0) = O_{P_0^{\mathbb{N}}}(n^{-1/3})$$

and also

$$\|\hat{p}_n - p_0\|_2 = O_{P_0^{\mathbb{N}}}(n^{-1/3}).$$

Proof In view of Lemma 7.2.16, part (b), we can restrict the set $\mathcal{P}$ over which $\hat{p}_n$ maximises the likelihood to $\mathcal{P}^{mon}$ intersected with a fixed $\|\cdot\|_\infty$-ball of radius k. The class

$$\left\{ \sqrt{(p+p_0)/2} : p \in \mathcal{P} \right\}$$

then consists of monotone decreasing functions that are uniformly bounded by a fixed constant. By Proposition 3.5.17, the $L^2(\mu)$-bracketing metric entropy of this class with respect to Lebesgue measure μ on $[0,1]$ is of order $(1/\varepsilon)$, and application of Theorem 7.2.1 gives the result in Hellinger distance. The result in L^2-distance follows as in Theorem 7.2.10 because $\hat{p}_n, p_0$ are uniformly bounded. ■

Admissible Directions and the Score Function

Similar to the situation with the NPMLE of a t-Sobolev density, the maximiser $\hat{p}_n$ is in some sense an object that lives on the boundary of $\mathcal{P}$ – it is piecewise constant with step discontinuities at the observation points, exhausting the possible 'roughness' of a monotone function. We can construct line segments in the parameter space through p_0, following the philosophy of Lemma 7.2.5. For instance, if we assume that $p_0 \geq \zeta > 0$ on $[0,1]$ and that its derivative p_0' exists, is bounded and is strictly negative, say, $|p_0'| \geq \xi > 0$ on $[0,1]$, then local perturbations of p_0 with ηh, where $h, h' \in L^\infty, \int h = 0$, will lie in $\mathcal{P}$: indeed, for η small enough, we then have

$$p_0 + \eta h \geq \zeta - \eta\|h\|_\infty > 0, \quad (p_0 + \eta h)' = p_0' + \eta h' \leq 0, \quad \int_0^1 (p_0 + \eta h) = 1. \tag{7.72}$$

For such p_0, $g, g' \in L^\infty$ and $\pi_0(g)$ as in (7.49), we thus obtain, from Lemma 7.2.5 with $\bar{g} = \pi_0(g)$, that on events of probability as close to 1 as desired and n large enough,

$$|D\ell_n(\hat{p}_n)[\pi_0(g)]| \leq d\|g\|_{C^1([0,1])}|D\ell_n(\hat{p}_n)(\hat{p}_n - p_0)|, \tag{7.73}$$

for some constant d that depends on ζ, ξ only. Note that $\|g\|_{C^1}$ also makes sense for not necessarily continuous g' and that the differential calculus from Proposition 7.2.4 applies because $\hat{p}_n, p_0$ as well as all points on the line segment $(1-t)\hat{p}_n + tp_0, t \in (0,1)$ lie in $\mathcal{U}(X_1, \ldots, X_n) \cap \mathcal{P}$ using Lemma 7.2.16, part (a).

We next need to derive stochastic bounds of the likelihood derivative at $\hat{p}_n$ in the direction of p_0.

Lemma 7.2.18 *Suppose that* p_0 *is bounded and satisfies* $\inf_{x\in[0,1]} p_0(x) > 0$. *For* $\hat{p}_n$ *satisfying (7.70), we have*

$$|D\ell_n(\hat{p}_n)(\hat{p}_n - p_0)| = O_{P_0^{\mathbb{N}}}(n^{-2/3}).$$

Proof By Lemma 7.2.16, we can restrict ourselves to an event where

$$0 < \xi \leq \inf_{x\in[0,X_{(n)}]} \hat{p}_n(x) \leq \sup_{x\in[0,1]} \hat{p}_n(x) \leq k < \infty$$

and, by Theorem 7.2.17, further to an event where

$$\|\hat{p}_n - p_0\|_{2,P_0} \leq \|p_0\|_\infty^{1/2}\|\hat{p}_n - p_0\|_2 \leq \|p_0\|_\infty^{1/2} M n^{-1/3}$$

for some finite constant M. We can write

$$\begin{aligned}
D\ell_n(\hat{p}_n)(\hat{p}_n - p_0) &= \frac{1}{n}\sum_{i=1}^n \frac{\hat{p}_n - p_0}{\hat{p}_n}(X_i)1_{[0,X_{(n)}]}(X_i) \\
&= \frac{1}{n}\sum_{i=1}^n \frac{\hat{p}_n - p_0}{\hat{p}_n}(X_i)1_{[0,X_{(n)}]}(X_i) - \int_0^1 \frac{\hat{p}_n - p_0}{\hat{p}_n}(x)1_{[0,X_{(n)}]}(x)p_0(x)dx \\
&\quad + \int_0^1 \frac{\hat{p}_n - p_0}{\hat{p}_n}(x)1_{[0,X_{(n)}]}(x)p_0(x)dx - \int_0^1 \frac{\hat{p}_n - p_0}{p_0}(x)1_{[0,X_{(n)}]}(x)p_0(x)dx \\
&\quad - \int_0^1 \frac{\hat{p}_n - p_0}{p_0}(x)1_{[X_{(n)},1]}(x)p_0(x)dx.
\end{aligned}$$

On the events from Lemma 7.2.16, the functions $\hat{p}_n^{-1}(\hat{p}_n - p_0)1_{[0,X_{(n)}]}$ are of bounded variation on $[0,1]$, with variation bounded by a fixed constant K that depends only on $k, \xi, \|p_0\|_\infty$. As a consequence, on these events, recalling Proposition 7.2.4 as well as (7.43), we have

$$
\begin{aligned}
&|D\ell_n(\hat{p}_n)(\hat{p}_n - p_0)| \\
&\lesssim \sup_{h:\|h\|_{BV}\le K, \|h\|_{2,P_0}\le\sigma} |(P_n - P_0)(h)| + \|\hat{p}_n - p_0\|_2^2 + \int_{X_{(n)}}^1 |\hat{p}_n - p_0| p_0^{-1} \qquad (7.74)\\
&= O_{P_0^{\mathbb{N}}}\left(n^{-1/2}n^{-1/6} + n^{-2/3} + \frac{\log n}{n}\right),
\end{aligned}
$$

where we have used the moment inequality Corollary 3.5.7 with

$$H = id, \sigma = (M/\zeta)\|p_0\|_\infty^{1/2} n^{-1/3}, \quad F = u = const,$$

combined with the uniform entropy bound Corollary 3.7.50 with $p = 1$, to control the supremum of the empirical process and Exercise 7.2.8 to deal with the last integral. ∎

From Lemma 7.2.5, we now deduce the following:

Proposition 7.2.19 *Suppose that $p_0 \in \mathcal{P}$ is differentiable and that p_0, p_0' are bounded and satisfy $p_0 \ge \zeta > 0, |p_0'| \ge \xi > 0$ on $[0,1]$. Let $g \in L^\infty$ have derivative $g' \in L^\infty$. Then*

$$|D\ell_n(\hat{p}_n)[\pi_0(g)]| = O_{P_0^{\mathbb{N}}}\left(\|g\|_{C^1} n^{-2/3}\right).$$

In particular, for $\hat{P}_n$ the probability measure with density $\hat{p}_n$, P_n the empirical measure and

$$\mathcal{B} = \{g : g, g' \in L^\infty, \|g\|_{C^1} \le B\},$$

we have

$$\sup_{g\in\mathcal{B}} |(\hat{P}_n - P_n)(g)| = O_{P_0^{\mathbb{N}}}(Bn^{-2/3}).$$

Proof We note that for such g, p_0 the function $\pi_0(g)$ from (7.49) has a bounded derivative and integrates to 0, so $p_0 + \eta\pi_0(g) \in \mathcal{P} \cap \mathcal{U}$ for η a small multiple of $\|(gp_0)'\|_{C^1}^{-1}$. The first claim of the proposition then follows from (7.73) and Lemma 7.2.18. To prove the second claim, we use Lemma 7.2.6, Proposition 7.2.4, $\hat{p}_n, p_0 \in \mathcal{U}(X_1, \dots, X_n)$ by Lemma 7.2.16 and a Taylor expansion up to second order to see that

$$
\begin{aligned}
|(\hat{P}_n - P_n)(g)| &= |D\ell_n(p_0)[\pi_0(g)] + D^2\ell(p_0)[\hat{p}_n - p_0, \pi_0(g)]| \\
&\le |D\ell_n(\hat{p}_n)[\pi_0(g)]| + |(D^2\ell_n(p_0) - D^2\ell(p_0))[\hat{p}_n - p_0, \pi_0(g)]| \\
&\quad + |(D^3\ell_n(\bar{p}_n)[\hat{p}_n - p_0, \hat{p}_n - p_0, \pi_0(g)]|,
\end{aligned}
$$

where $\bar{p}_n$ equals some mean values $p_n \in \mathcal{U}(X_1, \dots, X_n)$ between $\hat{p}_n$ and p_0. The first term is bounded using the first claim of this proposition, giving the bound $Bn^{-2/3}$ in probability. The second term is bounded similar as in (7.74) by

$$\sup_{h:\|h\|_{BV}\le K, \|h\|_{2,P_0}\le Mn^{-1/3}} |(P_n - P_0)(h)| = O_{P_0^{\mathbb{N}}}(\|g\|_{C^1} n^{-2/3}),$$

using also that the C^1-norm bounds the BV-norm so that $\pi_0(g)/\|g\|_{C^1}$ is contained in a fixed BV-ball. The third term can be recentred at a constant multiple of

$$\int_0^1 \frac{1}{\bar{p}_n^3(x)} (\hat{p}_n - p_0)^2(x) \pi_0(g)(x) 1_{[0, X_{(n)}]} = O_{P_0^{\mathbb{N}}}\left(\|\hat{p}_n - p_0\|_2^2\right) = O_{P_0^{\mathbb{N}}}(n^{-2/3}),$$

using again Lemma 7.2.16 to bound $\bar{p}_n$ from below on $[0,X_{(n)}]$. The centred process is now bounded again as in (7.74) using $\|\hat{p}_n - p_0\|_{BV} = O_{P_0^{\mathbb{N}}}(1)$ and noting that $\bar{p}_n$ as a convex combination of $\hat{p}_n, p_0$ has variation bounded by a fixed constant on $[0,X_{(n)}]$, so we can estimate this term by the supremum of the empirical process over a fixed BV-ball. ■

We can now approximate functions f in general classes $\mathcal{F}$ of functions to obtain, under the conditions of the preceding proposition, an asymptotic normality result for integral functionals of $\hat{p}_n - p_0$ that is uniform in $f \in \mathcal{F}$. We proceed as in (7.68): we can decompose, for $\pi_{V_J}(f)$ the wavelet projection of $f \in \mathcal{F}$ onto the span of wavelets up to resolution level J,

$$\|\hat{P}_n - P_n\|_{\mathcal{F}} \leq \sup_{f\in\mathcal{F}} \left| \int_0^1 (\hat{p}_n - p_0)(f - \pi_{V_J}(f)) \right| + \sup_{f\in\mathcal{F}} |(\hat{P}_n - P_n)(\pi_{V_J}(f))| + \sup_{f\in\mathcal{F}} |(P_n - P_0)(f - \pi_{V_J}(f))|. \tag{7.75}$$

For the following theorem, we recall that convergence in distribution in $\ell^\infty(\mathcal{F})$ of random probability measures towards the P_0-Brownian bridge G_{P_0} was defined as in Section 3.7.

Theorem 7.2.20 *Suppose that $p_0 \in \mathcal{P}$ is differentiable and that p_0, p_0' are bounded and satisfy $p_0 \geq \zeta > 0, |p_0'| \geq \xi > 0$ on $[0,1]$. Let $\mathcal{F}$ be a ball in the s-Hölder space C^s of order $s > 1/2$. Then*

$$\|\hat{P}_n - P_n\|_{\mathcal{F}} = o_{P_0^{\mathbb{N}}}(1/\sqrt{n})$$

as $n \to \infty$ and thus

$$\sqrt{n}(\hat{P}_n - P_0) \to^d G_{P_0} \text{ in } \ell^\infty(\mathcal{F}).$$

In particular, for any $f \in C^s$, we have

$$\sqrt{n}\int_0^1 (\hat{p}_n(x) - p_0(x)) f(x)dx \to^d N(0, \|f - P_0 f\|^2_{L^2(P_0)}).$$

Proof It is sufficient to prove the result for $1/2 < s < 1$. We recall from Chapter 4 that $C^s = B^s_{\infty\infty}$ for $s \notin \mathbb{N}$ and that the C^1-norm is bounded by the $B^1_{\infty 1}$-norm, so, for the wavelet partial sum $\pi_{V_j}(f)$ of $f \in C^s$, we have

$$\|\pi_{V_j}(f)\|_{C^1} \lesssim \sum_{l\leq j-1} 2^{3l/2} \max_k |\langle f, \psi_{lk}\rangle| \lesssim 2^{j(1-s)} \max_{l\leq j-1} 2^{l(s+1/2)} \sup_{l\leq j-1,k} |\langle f, \psi_{lk}\rangle| \leq 2^{j(1-s)} \|f\|_{B^s_{\infty\infty}}.$$

Moreover, by Parseval's identity,

$$\|\pi_{V_j}(f) - f\|_2 = O(2^{-js}).$$

Thus, taking $2^j \sim n^{1/3}$, we have

$$\sup_{f\in\mathcal{F}} |(\hat{P}_n - P_n)(\pi_{V_j}(f))| = O_{P_0^{\mathbb{N}}}(n^{-2/3} n^{(1-s)/3}) = o_{P_0^{\mathbb{N}}}(n^{-1/2})$$

since $s > 1/2$. Also, using Theorem 7.2.17 and the Cauchy-Schwarz inequality,

$$\left|\int_0^1 (\hat{p}_n - p_0)(f - \pi_{V_j}(f))\right| = O_{P_0^{\mathbb{N}}}(n^{-1/3}n^{-s/3}) = o_{P_0^{\mathbb{N}}}(1/\sqrt{n}),$$

and since the class $\{f - \pi_{V_j}(f)\}$ is contained in a fixed s-Hölder ball and has envelopes $\sup_f \|f - \pi V_j(f)\|_\infty \to 0$, the third term in (7.75) is also $o_{P_0^{\mathbb{N}}}(1/\sqrt{n})$, arguing as at the end of the proof of Theorem 7.2.14 (since the empirical process is tight and has a degenerate Gaussian limit). The remaining claims follow from the fact that $\mathcal{F}$ is a uniform Donsker class. ■

Exercises

7.2.1 Show that $h^2(p,p_0) \le Ch^2(\bar{p},p_0)$ for some universal constant C. *Hint*: Use

$$\frac{\bar{p}^{1/2} + p_0^{1/2}}{p^{1/2} + p_0^{1/2}} \le 2$$

for any density p.

7.2.2 Let $0 \le p_L \le p_U$, and consider $\bar{p}_L, \bar{p}_U$ as well as

$$g_L = \frac{1}{2}\log\frac{\bar{p}_L}{p_0}, \quad g_U = \frac{1}{2}\log\frac{\bar{p}_U}{p_0}$$

on the sets $\{p_0 > 0\}$. Show that $\rho_1(g^U - g^L) \le 2h(\bar{p}_l,\bar{p}_U)$. *Hint*: Use (7.33) and $g_U - g_L \ge 0$.

7.2.3 Let $(\mathcal{X},\mathcal{A})$ be a measurable space, Θ a compact metric space and $u : \mathcal{X} \times \Theta \to \mathbb{R}$ a function that is measurable in its first argument for every $\theta \in \Theta$ and continuous on Θ for every $x \in \mathcal{X}$. Show that there exists a Borel-measurable function $\hat{\theta} : (\mathcal{X},\mathcal{A}) \to \Theta$ such that $u(x,\hat{\theta}(x)) = \sup_{\theta\in\Theta} u(x,\theta)$. *Hint*: Reduce to $u(\hat{\theta}) = 0$, and realise Θ as a compact subset of $\mathbb{R}^{\mathbb{N}}$. Then show that the set of maximisers of u contains a largest element $\hat{\theta}$ for the lexicographic order. Then establish measurability of $\hat{\theta}$ by establishing measurability of each coordinate. Deduce Proposition 7.2.7, Part (c).

7.2.4 (Interchanging differentiation and integration in a Banach space setting.) Let V be an open subset of some Banach space E, and let $(S,\mathcal{A},\mu)$ be a measure space. Suppose that the function $f(v,s) : V \times S \to \mathbb{R}$ is contained in $L^1(S,\mathcal{A},\mu)$ for every $v \in V$. Assume that for every $v \in V$ and every $s \in S$ the Fréchet derivative $D_1 f(v,s)$ w.r.t. the first variable exists. Furthermore, assume that for every $s \in S$, the map $v \mapsto D_1 f(v,s)$ from V to E' is continuous. Suppose further that there exists a function $g \in L^1(S,\mathcal{A},\mu)$ such that $\|D_1 f(v,s)\|'_E \le g(s)$ for every $v \in V$ and $s \in S$. Then show that the function

$$\varphi : v \longmapsto \int_S f(v,s)d\mu(s)$$

from $V \subseteq E \to \mathbb{R}$ is Fréchet differentiable with derivative $D\varphi(v)(h) = \int_S D_1 f(v,s)(h)d\mu(s)$ for $h \in E$. Use this fact to justify the formula for $D^\alpha \ell(f)$ in Proposition 7.2.4.

7.2.5 Show that the MLE $\hat{p}_n$ over a Sobolev ball from (7.54) is unique. *Hint*: The function ℓ_n is concave on the convex set $\mathcal{P}$, so the set S of maximisers must be convex. By Proposition 7.2.9, we know that the maximiser $\hat{p}_n$ lies on the D-sphere of the space H^t, whose Hilbert norm is strictly convex, so S must be a singleton.

7.2.6 Prove Proposition 7.2.11. *Hint*: Combine the proof of Theorem 7.2.10 with Theorem 7.2.3. To construct a sequence in V_{j_n} such that (7.35) holds and

$$h(p_n^*,p_0) = O(\|p_n^* - p_0\|_2) = O(2^{-j_n t}) = O(n^{-t/(2t+1)}),$$

take $p_n^* = \pi_{V_j}(p_0)/\|\pi_{V_j}(p_0)\|_1$ and use $\|\pi_{V_j}(p_0) - p_0\|_\infty \to 0$ to show that $p_n^* \in \mathcal{P} \cap V_j$ for j large enough. Further, use $|\|\pi_{V_j}(p_0)\|_1 - \|p_0\|_1| \le \|\pi_{V_j}(p_0) - p_0\|_2$.

7.2.7 Show that the MLE $\hat{p}_n$ in (7.70) equals the left derivative $\hat{F}_n'$ of the least concave majorant $\hat{F}_n$ of the empirical distribution function F_n. (Note that $\hat{F}_n$ is the smallest concave function that satisfies $\hat{F}_n(x) \ge F_n(x)$ for all x.) Deduce that $\hat{p}_n$ is piecewise constant on $[0,1]$, left continuous, with jumps only at the observation points, and identically 0 on $(X_{(n)}, 1]$. *Hint*: First reduce to solutions that are of the form

$$p = \sum_i a_i 1_{[0, X_{(i)}]}, \quad a_i = \log(f_i/f_{i+1}), f_{n+1} = 1, \quad f_i \ge f_{i+1},$$

and deduce $\int \log f d\hat{F}_n \ge \int \log f dF_n$. Then show that $f = \hat{F}_n'$ gives equality in the last inequality, and use identifiability $\int \log \hat{p}_n d\hat{F}_n > \int \log f d\hat{F}_n, f \ne \hat{p}_n$, of the Kullback-Leibler distance (cf. 7.22) to obtain uniqueness.

7.2.8 Let $X_1, \dots, X_n$ be i.i.d. from distribution F on $[0,1]$ with density f, and let $X_{(1)} < \cdots < X_{(n)}$ be the corresponding order statistics. Show that the vector $(X_{(1)}, \dots, X_{(n)})$ then has density $n \prod_{i=1}^n f(x_i)$ on the set $x_1 < \cdots < x_n$ and that $X_{(i)}$ has density

$$nC_{i-1}^{n-1} F(x)^{i-1}(1 - F(x))^{n-i} f(x).$$

Deduce that for $\inf_{x \in [0,1]} f(x) \ge \zeta$ and $a_n \to 0$ such that $na_n \to \infty$, we have, for some constant $c = c(\zeta)$,

$$\Pr((1 - X_{(n)}) > a_n) = n \int_0^{1-a_n} F(x)^{n-1} f(x) dx \le (1 - ca_n)^n \to 0$$

as $n \to \infty$. For uniform random variables $F(t) = t$ on $[0,1]$, deduce that the $X_{(i)}$ have the same (joint) distribution as Z_i/Z_{n+1}, where Z_n is a sum of n independent Exp(1) random variables.

7.2.9 Let F_n^U be the empirical distribution function of a uniform $U(0,1)$ sample $U_1, \dots, U_n$ of size n. For $M \ge 1$, show that

$$\Pr(F_n^U(t) \le Mt \quad \forall t \in [0,1]) = 1 - 1/M.$$

Hint: The probability in question equals

$$n \int_{1/M}^1 \int_{(n-1)/nM}^{X_{(n)}} \cdots \int_{1/Mn}^{X_{(2)}} dx_{(1)} \cdots dx_{(n)},$$

and using the preceding exercise, the result follows from elementary integration.

7.2.10 Show that if a bounded monotone decreasing density p_0 has a jump discontinuity at $x_0 \in (0,1)$, then for $\hat{F}_n$ the distribution function of the MLE of a monotone density and F_n the empirical distribution function, we have

$$\hat{F}_n(x_0) - F_n(x_0) = o_P(1/\sqrt{n}).$$

Hint: Use Lemma 7.2.16, and note that (7.72) still remains true for $h = 1_{[0,x_0]}$ since $p_0' + \eta h'$ is then, for η small enough, the negative constant multiple of a Dirac measure. Then proceed as in the proof of Theorem 7.2.20. [See also Söhl (2015).]

7.3 Nonparametric Bayes Procedures

The Bayesian paradigm of statistical inference is in many respects fundamentally different from all the main ideas in this book because it does not view the unknown parameter f

indexing the distribution P_f of the sample as a point in a fixed parameter space $\mathcal{F}$ but rather itself as *random*. In this way, the Bayesian can accommodate subjective beliefs about the nature of f as encoded in a *prior distribution* Π on $\mathcal{F}$. For instance, this could model probabilities assigned to different hypotheses about a theory by a scientific community or, more generally, any kind of subjective beliefs about the nature of f. Given observations $X \sim P_f$, the simple 'updating' rule known as *Bayes' theorem* then computes the best subjective guess about f given X – known as the *posterior distribution* $\Pi(\cdot|X)$. For a subjective Bayesian, the story ends here. However, interesting questions arise if one analyses the posterior distribution under the frequentist sampling assumption that $X \sim P_{f_0}$ from a fixed f_0 in the support of the prior. This *frequentist Bayes approach* is nourished by the hope that the likelihood (i.e., the observations) may eventually dominate the choice of the prior, resulting in valid asymptotic frequentist inferences.

Despite the slight arbitrariness of the choice of the prior, in nonparametric models, the Bayesian approach can be very attractive from a methodological point of view: we have seen in the preceding section that likelihood-based methods need some complexity regularisation, such as constraining the model to a Sobolev ball or to the set of monotone densities. The posterior distribution gives another way to regularise the likelihood by averaging it out – or 'resampling' it – according to a prior distribution. At the same time, the Bayesian approach furnishes us with a posterior distribution that we can readily use for inference, such as for the construction of point estimators, tests and 'credible sets' (the Bayesian 'posterior' version of a confidence set). Belief in such inferences, however, needs to be founded on a thorough frequentist analysis. In this section we shall develop mathematical techniques that can be used to show that certain nonparametric Bayes methods give optimal frequentist procedures in some infinite-dimensional models.

Since the setting for Bayesian analysis involves conditional probabilities in infinite-dimensional models, let us first give a rigorous foundation for the notion of the posterior distribution. Consider a measurable space $(\mathcal{X},\mathcal{A})$ and a family $\{p_f : f \in \mathcal{F}\}$ of probability densities on $\mathcal{X}$ with respect to a common σ-finite dominating measure μ. Suppose further that $\mathcal{F}$ is equipped with a σ-field $\mathcal{B}$ for which the maps $(x, f) \mapsto p_f(x)$ are jointly measurable. If Π is a probability measure on $(\mathcal{F},\mathcal{B})$ – called the *prior distribution* – we obtain on $\mathcal{A} \otimes \mathcal{B}$ a product measure $\mu \otimes \Pi$, and define the probability space

$$(\mathcal{X} \times \mathcal{F}, \mathcal{A} \otimes \mathcal{B}, Q), \quad dQ(x, f) = p_f(x)d\mu(x)d\Pi(f), \tag{7.76}$$

the canonical setting for Bayesian analysis: by standard properties of conditional distributions (see Exercise 7.3.1), the distribution of the coordinate map X onto $\mathcal{X}$ conditional on f has density

$$X|f \sim \frac{p_f d\mu}{\int_{\mathcal{X}} p_f(x)d\mu(x)} = p_f d\mu, \tag{7.77}$$

and the distribution of the coordinate map f onto $\mathcal{F}$ conditional on X is

$$f|X \sim \frac{p_f(X)d\Pi(f)}{\int_{\mathcal{F}} p_f(X)d\Pi(f)} \tag{7.78}$$

in the sense that for any measurable set $B \in \mathcal{B}$, we have

$$\Pi(B|X) = \frac{\int_B p_f(X)d\Pi(f)}{\int_{\mathcal{F}} p_f(X)d\Pi(f)}, \tag{7.79}$$

the identity holding μ almost surely. The law of $f|X$ is known as the *posterior distribution*.

Let us illustrate what the preceding formulas give in the main sampling models considered in this book. Suppose, first, that we are given an i.i.d. sample $X_1, \dots, X_n$ on a measurable space $(\mathcal{X}, \mathcal{A})$ from a probability density p with respect to some dominating measure μ, and consider a prior distribution Π on a family $(\mathcal{P}, \mathcal{B})$ of such probability densities. The preceding formalism then applies directly with $p_f = p, \mathcal{F} = \mathcal{P}$. We find that the posterior distribution $\Pi(\cdot|X_1, \dots, X_n)$ of $p|X_1, \dots, X_n$ on $\mathcal{P}$ equals

$$d\Pi(p|X_1, \dots, X_n) \sim \frac{\prod_{i=1}^n p(X_i)d\Pi(p)}{\int_{\mathcal{P}} \prod_{i=1}^n p(X_i)d\Pi(p)}, \quad \mu^n \text{ a.s.} \tag{7.80}$$

When making the frequentist assumption that the X_i were drawn from a fixed density p_0, we will often write the preceding expression as

$$d\Pi(p|X_1, \dots, X_n) \sim \frac{\prod_{i=1}^n \frac{p}{p_0}(X_i)d\Pi(p)}{\int_{\mathcal{P}} \prod_{i=1}^n \frac{p}{p_0}(X_i)d\Pi(p)}. \tag{7.81}$$

For the Gaussian white noise model

$$dY(t) = f(t)dt + \frac{\sigma}{\sqrt{n}}dW(t), \quad t \in [0,1], \tag{7.82}$$

with $\sigma > 0, n \in \mathbb{N}$, we recall the results from Section 6.1.1: we take $(\mathcal{X}, \mathcal{A})$ equal to $C([0,1])$ equipped with its Borel σ-field and μ equal to P_0^Y, a scaled Wiener measure $(\sigma/\sqrt{n})P^W$. By Proposition 6.1.1 and the remark after it, the law P_f^Y of dY is, for every $f \in L^2$, absolutely continuous with respect to P_0^Y, with density

$$p_f(Y) = \exp\left\{\frac{n}{\sigma^2}\int_0^1 f dY - \frac{n}{2\sigma^2}\|f\|_2^2\right\}, \quad P_0^Y \text{ a.s.} \tag{7.83}$$

If Π is a prior distribution on the Borel sets $\mathcal{B}$ of $\mathcal{F} \subseteq L^2$, then we obtain from (7.78) that the posterior distribution $\Pi(\cdot|Y)$ of $f|Y$ on $\mathcal{F}$ is

$$d\Pi(f|Y) \sim \frac{p_f(Y)d\Pi(f)}{\int_{\mathcal{F}} p_f(Y)d\Pi(f)} = \frac{\exp\left\{\frac{n}{\sigma^2}\int_0^1 f dY - \frac{n}{2\sigma^2}\|f\|_2^2\right\}d\Pi(f)}{\int_{\mathcal{F}} \exp\left\{\frac{n}{\sigma^2}\int_0^1 f dY - \frac{n}{2\sigma^2}\|f\|_2^2\right\}d\Pi(f)}, \quad P_0^Y \text{ a.s.} \tag{7.84}$$

Under the frequentist assumption $dY \sim P_{f_0}^Y$ for some fixed $f_0 \in L^2$, we can rewrite the last expression as

$$d\Pi(f|Y) \sim \frac{\exp\left\{\frac{\sqrt{n}}{\sigma}\int_0^1 (f - f_0)dW - \frac{n}{2\sigma^2}\|f - f_0\|_2^2\right\}d\Pi(f)}{\int_{\mathcal{F}} \exp\left\{\frac{\sqrt{n}}{\sigma}\int_0^1 (f - f_0)dW - \frac{n}{2\sigma^2}\|f - f_0\|_2^2\right\}d\Pi(f)}, \quad P^W \text{ a.s.} \tag{7.85}$$

The posterior distribution in the white noise model has another representation using the isometry of L^2 with sequence space ℓ_2. The preceding arguments combined with Proposition 6.1.4 give for $f \in L^2$ and basis coefficients $(f_k = \langle f, e_k \rangle : k \in \mathbb{Z})$ the posterior distribution of $f|(Y_k : k \in \mathbb{Z})$ as

$$d\Pi(f|Y) \sim \frac{\exp\left\{\frac{n}{\sigma^2}\sum_k f_k Y_k - \frac{n}{2\sigma^2}\|f\|_{\ell_2}^2\right\} d\Pi(f)}{\int_{\mathcal{F}} \exp\left\{\frac{n}{\sigma^2}\sum_k f_k Y_k - \frac{n}{2\sigma^2}\|f\|_{\ell_2}^2\right\} d\Pi(f)}$$

$$= \frac{\exp\left\{\frac{\sqrt{n}}{\sigma}\sum_k (f_k - f_{0,k}) g_k - \frac{n}{2\sigma^2}\|f - f_{0,\cdot}\|_{\ell_2}^2\right\} d\Pi(f)}{\int_{\mathcal{F}} \exp\left\{\frac{\sqrt{n}}{\sigma}\sum_k (f_k - f_{0,k}) g_k - \frac{n}{2\sigma^2}\|f - f_{0,\cdot}\|_{\ell_2}^2\right\} d\Pi(f)} \tag{7.86}$$

a.s. under the infinite Gaussian product measure P^W of $(g_k : k \in \mathbb{Z}), g_k \sim^{i.i.d.} N(0,1)$, and where the last identity holds under the frequentist assumption $Y_k = f_{0,k} + (\sigma/\sqrt{n}) g_k$.

7.3.1 General Contraction Results for Posterior Distributions

We now give general conditions such that the posterior distribution $\Pi(\cdot|X)$ 'contracts' about the parameter f_0 in some metric d on $\mathcal{F}$ under the frequentist assumption that $X \sim P_{f_0}$. More precisely, for Π some prior distribution, we want to derive results of the kind

$$\Pi(f : d(f, f_0) > M\varepsilon_n | X) \to 0,$$

in P_{f_0} probability, for some fixed constant M, and where ε_n is known as the *rate of contraction* of $\Pi(\cdot|X)$ about f_0 in the metric d.

Contraction Results in the i.i.d. Sampling Model

We first consider the case of the i.i.d. sampling model with a general model $\mathcal{P}$ of densities on $(\mathcal{X}, \mathcal{A})$. The conditions involved in the following theorem require that the prior charges neighbourhoods of p_0 with a sufficient amount of probability, where neighbourhoods are in the 'correct topology' arising from the expected likelihood function in the sampling model. Moreover, if we want to derive a contraction rate ε_n in a metric d on $\mathcal{P}$, then a sufficiently large set in the support of the prior has to admit consistent tests with sufficiently good exponential error bounds for the nonparametric hypotheses

$$H_0 : p = p_0 \quad vs. \quad H_1 : \{p : d(p, p_0) > M\varepsilon_n\}$$

such as those studied in Chapter 6.2 and in Theorem 7.1.4. We write E_P for expectation under $P^{\mathbb{N}}$, where P has density $p \in \mathcal{P}$. We allow the prior Π to depend on n too in the following result.

Theorem 7.3.1 *Let $\mathcal{P}$ be a collection of probability densities on a measurable space $(\mathcal{X}, \mathcal{A})$ with respect to some σ-finite dominating measure μ, and let $\mathcal{B}$ be a σ-field over $\mathcal{P}$ such that the mappings $(x,p) \mapsto p(x)$ are jointly measurable. Let $\Pi = \Pi_n$ be a sequence of prior distributions on $\mathcal{B}$, suppose that $X_1, \dots, X_n$ are i.i.d. from density p_0 on $\mathcal{X}$, $dP_0 = p_0 d\mu$ and* *let* $\Pi(\cdot|X_1,\dots,X_n)$ *be the posterior distribution from (7.80). For* ε_n *a sequence of positive real numbers such that*

$$\varepsilon_n \to 0, \quad \sqrt{n}\varepsilon_n \to \infty,$$

and C, L *fixed constants, suppose that* Π *satisfies,*

$$\Pi\left(p \in \mathcal{P} : -E_{P_0}\log\frac{p}{p_0}(X) \le \varepsilon_n^2, E_{P_0}\left(\log\frac{p}{p_0}(X)\right)^2 \le \varepsilon_n^2\right) \ge e^{-Cn\varepsilon_n^2} \tag{7.87}$$

and that

$$\Pi(\mathcal{P}\setminus\mathcal{P}_n) \le Le^{-(C+4)n\varepsilon_n^2}, \tag{7.88}$$

for some sequence $\mathcal{P}_n \subset \mathcal{P}$ *for which we can find tests (indicator functions)* $\Psi_n \equiv \Psi(X_1,\dots,X_n)$ *such that, for every* $n \in \mathbb{N}, M > 0$ *large enough,*

$$E_{P_0}\Psi_n \to_{n\to\infty} 0, \qquad \sup_{p\in\mathcal{P}_n : d(p,p_0)\ge M\varepsilon_n} E_P(1-\Psi_n) \le Le^{-(C+4)n\varepsilon_n^2}. \tag{7.89}$$

Then the posterior $\Pi(\cdot|X_1,\dots,X_n)$ *contracts about* p_0 *at rate* ε_n *in the metric* d*; that is,*

$$\Pi(p : d(p,p_0) > M\varepsilon_n|X_1,\dots X_n) \to 0 \tag{7.90}$$

in $P_0^{\mathbb{N}}$ *probability as* $n \to \infty$.

Proof First,

$$E_{P_0}\left[\Pi\left(\{p \in \mathcal{P} : d(p,p_0) \ge M\varepsilon_n | X_1,\dots,X_n\}\right)\Psi_n\right] \le E_{P_0}\Psi_n \to 0$$

by assumption on the tests, so we only need to prove convergence in $P_0^{\mathbb{N}}$ probability to 0 of

$$\Pi(\{p \in \mathcal{P} : d(p,p_0) \ge M\varepsilon_n | X_1,\dots,X_n\})(1-\Psi_n)$$
$$= \frac{\int_{d(p,p_0)\ge M\varepsilon_n}\prod_{i=1}^n (p/p_0)(X_i)d\Pi(p)(1-\Psi_n)}{\int_{\mathcal{P}}\prod_{i=1}^n (p/p_0)(X_i)d\Pi(p)}.$$

Lemma 7.3.2 shows that for all $c > 0$ and probability measures ν with support in

$$B_n := \left\{p \in \mathcal{P} : -E_{P_0}\log\frac{p}{p_0}(X) \le \varepsilon_n^2,\ E_{P_0}\left(\log\frac{p}{p_0}(X)\right)^2 \le \varepsilon_n^2\right\},$$

we have

$$P_0^{\mathbb{N}}\left(\int\prod_{i=1}^n \frac{p}{p_0}(X_i)d\nu(p) \le e^{-(1+c)n\varepsilon_n^2}\right) \le \frac{1}{c^2 n\varepsilon_n^2}.$$

In particular, this result applied with $c = 1$ and ν equal to the normalised restriction of Π to B_n, together with condition (7.87) of the theorem, shows that if A_n is the event

$$A_n := \left\{\int_{B_n}\prod_{i=1}^n \frac{p}{p_0}(X_i)d\Pi(p) \ge \Pi(B_n)e^{-2n\varepsilon_n^2} \ge e^{-(2+C)n\varepsilon_n^2}\right\},$$

then $P_0^{\mathbb{N}}(A_n) \geq 1 - 1/n\varepsilon_n^2 \to 1$, and we can write, for every $\epsilon > 0$,

$$P_0^{\mathbb{N}}\left(\frac{\int_{d(p,p_0)\geq M\varepsilon_n}\prod_{i=1}^n (p/p_0)(X_i)d\Pi(p)(1-\Psi_n)}{\int_{\mathcal{P}}\prod_{i=1}^n (p/p_0)(X_i)d\Pi(p)} > \epsilon\right)$$
$$\leq P_0^{\mathbb{N}}(A_n^c) + P_0^{\mathbb{N}}\left(e^{(2+C)n\varepsilon_n^2}(1-\Psi_n)\int_{d(p,p_0)\geq M\varepsilon_n}\prod_{i=1}^n \frac{p}{p_0}(X_i)d\Pi(p) > \epsilon\right).$$

Now, using that

$$E_{P_0}\left[\prod_{i=1}^n \frac{p}{p_0}(X_i)\right] = \left[\int_{p_0>0} p\,d\mu\right]^n \leq 1, \quad E_{P_0}\left[\prod_{i=1}^n \frac{p}{p_0}(X_i)(1-\Psi_n)\right] \leq E_P(1-\Psi_n),$$

and that $0 \leq 1 - \Psi_n \leq 1$, we obtain

$$E_{P_0}\left[(1-\Psi_n)\int_{d(p,p_0)\geq M\varepsilon_n}\prod_{i=1}^n \frac{p}{p_0}(X_i)d\Pi(p)\right] \leq \Pi(\mathcal{P}\setminus\mathcal{P}_n) + \sup_{p\in\mathcal{P}_n:d(p,p_0)\geq M\varepsilon_n} E_P(1-\Psi_n).$$

Now the assumptions on $\mathcal{P}_n$ and on the tests combined with Markov's inequality give, for every $\epsilon > 0$,

$$P_0^{\mathbb{N}}\left((1-\Psi_n)\int_{d(p,p_0)\geq M\varepsilon_n}\prod_{i=1}^n \frac{p}{p_0}(X_i)d\Pi(p) > \frac{\epsilon}{e^{(2+C)n\varepsilon_n^2}}\right) \leq (2L/\epsilon)e^{-2n\varepsilon_n^2},$$

and the theorem follows by combining the preceding estimates, since $n\varepsilon_n^2 \to \infty$ as $n \to \infty$. ■

Lemma 7.3.2 *For every $\epsilon > 0$ and probability measure ν on the set*

$$B = \left\{p \in \mathcal{P} : -E_{P_0}\log\frac{p}{p_0}(X) \leq \epsilon^2, E_{P_0}\left(\log\frac{p}{p_0}(X)\right)^2 \leq \epsilon^2\right\},$$

we have, for every $c > 0$,

$$P_0^{\mathbb{N}}\left(\int_B \prod_{i=1}^n \frac{p}{p_0}(X_i)d\nu(p) \leq \exp\{-(1+c)n\epsilon^2\}\right) \leq \frac{1}{c^2 n\epsilon^2}.$$

Proof By Jensen's inequality,

$$\log\int\prod\frac{p}{p_0}(X_i)d\nu(p) \geq \sum_{i=1}^n \int \log\frac{p}{p_0}(X_i)d\nu(p),$$

and if $\sqrt{n}(P_n - P_0)$ is the empirical process, then the probability in question is bounded by

$$P_0^n\left(\sqrt{n}\int\int\log\frac{p}{p_0}d\nu(p)d(P_n-P_0) \leq -\sqrt{n}(1+c)\epsilon^2 - \sqrt{n}\int\int\log\frac{p}{p_0}(x)d\nu(p)dP_0(x)\right).$$

Now, by Fubini's theorem,

$$-\sqrt{n}\int\int\log\frac{p}{p_0}d\nu(p)dP_0(x) = \sqrt{n}\int -E_{P_0}\log\frac{p}{p_0}d\nu(p) \leq \sqrt{n}\epsilon^2,$$

and the last probability is thus further bounded, using the inequalities of Chebyshev and Jensen (again on ν), by

$$P_0^{\mathbb{N}}\left(\sqrt{n}\int\int\log\frac{p}{p_0}d\nu(f)d(P_n-P_0)\leq-\sqrt{n}c\epsilon^2\right)\leq\frac{\mathrm{Var}_{P_0}(\int\log(p/p_0)(X)d\nu(p))}{c^2n\epsilon^4}$$

$$\leq\frac{E_{P_0}\int(\log(p/p_0))^2d\nu(p)}{c^2n\epsilon^4}$$

$$\leq\frac{1}{c^2n\epsilon^2},$$

which completes the proof. ■

We can use the techniques from Section 7.1 to construct tests for alternatives that are separated in the Hellinger distance, which, in turn, give contraction rates of posterior distributions in the Hellinger distance h, under an entropy condition on a large support set $\mathcal{P}_n$ of $\mathcal{P}$.

Theorem 7.3.3 *In the setting of Theorem 7.3.1, assume for C,L fixed constants and ε_n s.t. $\varepsilon_n\to 0$, $\sqrt{n}\varepsilon_n\to\infty$, that Π satisfies,*

$$\Pi\left(p\in\mathcal{P}:-E_{P_0}\log(p/p_0)(X)\leq\varepsilon_n^2, E_{P_0}(\log(p/p_0)(X))^2\leq\varepsilon_n^2\right)\geq e^{-Cn\varepsilon_n^2} \tag{7.91}$$

and that

$$\Pi(\mathcal{P}\setminus\mathcal{P}_n)\leq Le^{-(C+4)n\varepsilon_n^2}, \tag{7.92}$$

for some sequence $\mathcal{P}_n\subset\mathcal{P}$ for which

$$\log N(\mathcal{P}_n,h,\varepsilon_n)\leq n\varepsilon_n^2. \tag{7.93}$$

Then the posterior $\Pi(\cdot|X_1,\ldots,X_n)$ contracts about p_0 at rate ε_n in Hellinger distance, and

$$\Pi(p:h(p,p_0)>M\varepsilon_n|X_1,\ldots X_n)\to 0 \tag{7.94}$$

in $P_0^{\mathbb{N}}$ probability as $n\to\infty$ for $M>0$ a large enough constant.

Proof Combine Theorems 7.3.1 and 7.1.4 with $\varepsilon_0=m\varepsilon_n$ for $m<M$ large enough constants and $N(\varepsilon)=N(\varepsilon_n)$ constant in ε to obtain suitable tests. ■

We can also use the testing tools from Chapter 6 instead of the Hellinger distance. This is illustrated in the white noise setting in the next subsection.

A General Contraction Theorem in Gaussian White Noise

An analogue of the results from the preceding subsection for the posterior (7.84) in the Gaussian white noise model is proved without difficulty, replacing Lemma 7.3.2 by the following result:

Lemma 7.3.4 *For every $\epsilon>0$ and probability measure ν on the set*

$$B=\left\{f\in\mathcal{F}:\|f-f_0\|_2^2\leq\epsilon^2/\sigma^2\right\},$$

we have, for every $c > 0$,

$$P^W\left(\int_B \exp\left\{\frac{\sqrt{n}}{\sigma}\int_0^1 (f-f_0)dW - \frac{n}{2\sigma^2}\|f-f_0\|_2^2\right\}d\nu(f) \le \exp\{-(1+c)n\epsilon^2/\sigma^2\}\right)$$
$$\le \frac{\sigma^2}{c^2 n\epsilon^2}.$$

Proof We set $\sigma = 1$ to expedite notation. By Jensen's inequality, the probability in question is hence less than or equal to

$$P^W\left(\int_B\left(\sqrt{n}\int_0^1 (f-f_0)dW - \frac{n}{2}\|f-f_0\|_2^2\right)d\nu(f) \le -(1+c)n\epsilon^2\right).$$

Using $\|f-f_0\|_2^2 \le \epsilon^2$ and that $\int_0^1 (f-f_0)dW \sim N(0, \|f-f_0\|_2^2)$ implies, again by Jensen,

$$\mathrm{Var}\left(\int_B\int_0^1 (f-f_0)dWd\nu(f)\right) \le \int_B E\left[\int_0^1 (f-f_0)dW\right]^2 d\nu(f) \le \|f-f_0\|_2^2,$$

so we can bound the last probability by

$$P^W\left(\int_B\int_0^1 (f-f_0)dWd\nu(f) \le -c\sqrt{n}\epsilon^2\right) \le \frac{\|f-f_0\|_2^2}{c^2 n\epsilon^4} \le \frac{1}{c^2 n\epsilon^2},$$

completing the proof. ∎

Given this lemma and (7.85), the proof of the following theorem is now a straightforward modification of the proof of Theorem 7.3.1 and is left as Exercise 7.3.2.

Theorem 7.3.5 *Let $\mathcal{F} \subset L^2$ be equipped with its Borel σ-field $\mathcal{B}$. Let $\Pi = \Pi_n$ be a sequence of prior distributions on $\mathcal{B}$, suppose that $dY \sim P^Y_{f0}$ is an observation in the white noise model (7.82) and let $\Pi(\cdot|Y)$ be the posterior distribution from (7.84). For ε_n a sequence of positive real numbers such that*

$$\varepsilon_n \to 0, \quad \sqrt{n}\varepsilon_n \to \infty$$

and C, L fixed constants, suppose that Π satisfies

$$\Pi\left(f \in \mathcal{F} : \|f - f_0\|_2 \le \varepsilon_n\right) \ge e^{-Cn\varepsilon_n^2} \tag{7.95}$$

and that

$$\Pi(\mathcal{F}\setminus\mathcal{F}_n) \le Le^{-(C+4)n\varepsilon_n^2}, \tag{7.96}$$

for some sequence of measurable sets $\mathcal{F}_n \subset \mathcal{F}$ for which we can find tests (indicator functions) $\Psi_n \equiv \Psi(Y)$ such that, for M a large enough constant,

$$E_{f0}\Psi_n \to_{n\to\infty} 0, \qquad \sup_{f\in\mathcal{F}_n : d(f,f_0)\ge M\varepsilon_n} E_f(1-\Psi_n) \le Le^{-(C+4)n\varepsilon_n^2}. \tag{7.97}$$

Then the posterior $\Pi(\cdot|Y)$ contracts about f_0 at rate ε_n in the metric d; that is,

$$\Pi(f : d(f, f_0) > M\varepsilon_n|Y) \to 0 \tag{7.98}$$

in P^Y_{f0} probability as $n \to \infty$.

To construct suitable tests, we can adapt the Hellinger testing theory to the white noise model, with h replaced by $\|\cdot\|_2$, the relevant information-theoretic distance, and proceed accordingly. Since L^2 is naturally compatible with approximation theory, we can instead use the results on nonparametric testing that were developed in Chapter 6, resulting in simple approximation-theoretic conditions on Π which can be easily verified for natural priors and which are at any rate closely related to an L^2-version of the entropy condition in Theorem 7.3.3. The following result is based on an approximation kernel K satisfying Conditions 5.1.1 with $A = [0,1]$:

Theorem 7.3.6 *Let $\mathcal{F} \subset L^2$ be equipped with its Borel σ-field $\mathcal{B}$. Let $\Pi = \Pi_n$ be a sequence of prior distributions on $\mathcal{B}$, suppose that $dY \sim P_{f_0}$ is an observation in white noise and let $\Pi(\cdot|Y)$ be the posterior distribution from (7.84). Let ε_n be a sequence of positive real numbers such that*

$$\varepsilon_n \to 0, \quad \sqrt{n}\varepsilon_n \to \infty,$$

and for C, L fixed constants, suppose that Π satisfies

$$\Pi(f \in \mathcal{F} : \|f - f_0\|_2 \le \varepsilon_n) \ge e^{-Cn\varepsilon_n^2}. \tag{7.99}$$

Further, let j_n be such that $2^{j_n} \sim n\varepsilon_n^2$, and suppose that for some sequence of measurable sets

$$\mathcal{F}_n \subset \{f \in L^2 : \|K_{j_n}(f) - f\|_2 \le \varepsilon_n\},$$

we have

$$\Pi(\mathcal{F}_n^c) \le Le^{-(C+4)n\varepsilon_n^2}. \tag{7.100}$$

Assume further that $\|K_{j_n}(f_0) - f_0\|_2 = O(\varepsilon_n)$. Then the posterior $\Pi(\cdot|Y)$ contracts about f_0 at rate ε_n in the L^2-distance; that is,

$$\Pi(f : \|f - f_0\|_2 > M\varepsilon_n|Y) \to 0 \tag{7.101}$$

in P_{f_0} probability as $n \to \infty$ for M large enough.

Proof We apply Theorem 7.3.5 with sieve set

$$\mathcal{F}_n = \{f \in L^2 : \|K_{j_n}(f) - f\|_2 \le \varepsilon_n\},$$

and all that is needed is to construct suitable tests verifying (7.97). Using

$$\|K_j(f - f_0) - f + f_0\|_2 \le 2\varepsilon_n \equiv B(j_n),$$

the existence of such tests follows from Corollary 6.2.15 with $H_0 = \{f_0\}$, $2^{j_n} \sim n\varepsilon_n^2$. ■

The approximation-theoretic approach to testing also works in the sampling setting, giving an alternative approach to Hellinger tests – we discuss some references in the notes at the end of this chapter.

7.3.2 Contraction Results with Gaussian Priors

It is time to put the general contraction theorems in the preceding section to a test for some natural prior choices. Whereas the theory indeed applies widely, some of the main mechanisms are best understood in the situation where the prior is Gaussian. We shall see that the use of very fine properties of Gaussian measures from Chapter 2, such as the isoperimetric theorem and small-ball asymptotics, allows us to verify the conditions in the preceding theorems to give minimax optimal contraction rates for nonparametric posterior distributions. We shall start with the conceptually simpler case of Gaussian white noise.

Contraction Rates with Integrated Brownian Motion Priors in the White Noise Model

Let us first illustrate how to apply Theorem 7.3.6 in the case where the prior on f is the random trajectory of a Brownian motion $W = (W(t) : t \in [0,1])$. Almost surely this process will have nondifferentiable trajectories that are 'almost' $1/2$-Hölder continuous. It also satisfies $W(0) = 0$, which is somewhat unnatural from a statistical point of view, and we thus release it at zero and consider $W^0 = g + W$, where $g \sim N(0,1)$ independent of W. A good test case is to assume that f_0 is contained in the Sobolev space $H_2^{1/2}([0,1])$, which intuitively means that the smoothness of f_0 is matched with the regularity of the trajectories of the prior. The posterior contraction rate we hope for over $H_2^{1/2} = B_{22}^{1/2}$ in light of the minimax results in Chapter 6 is $\varepsilon_n \simeq n^{-1/4}$.

Step 1. Small-ball estimate. We first need to guarantee that the prior charges an L^2-neighbourhood of f_0 with sufficient probability. The prior law of W^0 is a Gaussian Borel probability measure supported in L^2 with reproducing kernel Hilbert space (RKHS)

$$\mathbb{H} = H^1([0,1]) = B^1_{22}([0,1])$$

(see Proposition 2.6.24). By Proposition 2.6.19, we have

$$\Pi\left(f \in L^2 : \|f - f_0\|_2 \le \varepsilon_n\right) \ge \exp\left\{-\inf_{h \in \mathbb{H}: \|h-f_0\|_2 \le \varepsilon_n/2}\left[\frac{1}{2}\|h\|^2_{\mathbb{H}} - \log \Pr\left(\|W^0\|_2 < \varepsilon_n/2\right)\right]\right\}. \quad (7.102)$$

We see that the small-ball requirement (7.95) is governed by the small-ball probability of the prior at zero ($f_0 = 0$) and by the relative position of f_0 to the RKHS $\mathbb{H}$. Now, from Corollary 2.6.27 and $\|W^0\|_2 \le \|W^0\|_\infty$, we know, for some $c > 0$, that

$$-\log \Pr\left(\|W^0\|_2 < \varepsilon_n/2\right) \le c^{-1}\varepsilon_n^{-2} \simeq n\varepsilon_n^2$$

because $\varepsilon_n \simeq n^{-1/4}$. Moreover, we can approximate $f_0 \in B^{1/2}_{22}$ by its wavelet projection $\pi_{V_j}(f_0) \in \mathbb{H}$, with j large enough, $2^j \sim \sqrt{n}$, to give

$$\|\pi_{V_j}(f_0) - f_0\|_2 \lesssim \|f_0\|_{B^{1/2}_{22}} \sum_{l>j} 2^{-j/2} \le \varepsilon_n, \quad \|\pi_{V_j}(f_0)\|^2_{\mathbb{H}} \lesssim \|f_0\|^2_{B^{1/2}_{22}} 2^j \simeq \sqrt{n} \simeq n\varepsilon_n^2,$$

using (4.149), so indeed the probability in (7.102) is lower bounded by $e^{-Cn\varepsilon_n^2}$ for some constant C.

Step 2. Construction of approximating sets $\mathcal{F}_n$. For Φ the cumulative distribution function of a standard normal distribution (and $C = 1$, for instance), let

$$M_n = -2\Phi^{-1}(e^{-(C+4)n\varepsilon_n^2}),$$

and note that $a_n = \Pi(\|f\|_\infty \le \varepsilon_n) \ge e^{-Cn\varepsilon_n^2}$ from Corollary 2.6.27 (and by multiplying ε_n by a large enough constant). Consider the sets

$$\mathcal{F}_n = \{f = f_1 + f_2 : \|f_1\|_\infty \le \varepsilon_n, \|f_2\|_{\mathbb{H}} \le M_n\}, \tag{7.103}$$

to which the isoperimetric inequality Theorem 2.6.12 applies to give

$$\Pi(\mathcal{F} \setminus \mathcal{F}_n) \le 1 - \Phi(\Phi^{-1}(a_n) + M_n) \le 1 - \Phi(M_n/2) = \exp\{-(C+4)n\varepsilon_n^2\}. \tag{7.104}$$

Next, we control the approximation errors using a wavelet approximation $K_j(f)$, the wavelet characterisation of B^s_{22} spaces (4.149), and that $M_n \simeq \sqrt{n}\varepsilon_n$, to see

$$\|K_{j_n}(f_1 + f_2) - f_1 - f_2\|_2 \le C\|f_1\|_\infty + 2^{-j_n}\|f_2\|_{\mathbb{H}} \lesssim \varepsilon_n + (n\varepsilon_n^2)^{-1}\sqrt{n}\varepsilon_n \simeq \varepsilon_n^{-1}/\sqrt{n} \simeq \varepsilon_n$$

and

$$\|K_j(f_0) - f_0\|_2 \lesssim \|f_0\|_{B^{1/2}_{22}}(n\varepsilon_n^2)^{-1/2} = n^{-1/4},$$

completing the verification of the conditions of Theorem 7.3.6, so we conclude, for Π a Brownian motion released at zero, that

$$\Pi\left(f : \|f - f_0\|_2 > Mn^{-1/4}|Y\right) \to^{P^Y_{f_0}} 0 \tag{7.105}$$

as $n \to \infty$ whenever $f_0 \in B^{1/2}_{22}([0,1])$ for some large enough constant M.

To model smoother functions, we can take k primitives of Brownian motion, resulting in a function whose kth derivative is almost 1/2-Hölder continuous (see Section 2.6.3). Moreover, we wish to release these processes at zero and hence consider as prior Π the law of the process $(W^k(t) : t \in [0,1])$ from equation (2.82). The following theorem gives the minimax optimal contraction rate for the resulting posterior distributions for functions $f_0 \in B^{k+1/2}_{22}$, including the case of Brownian motion ($k = 0$) discussed in the preceding paragraphs:

Theorem 7.3.7 *Consider a prior Π on L^2 arising from the kth integrated Brownian motion $W^k, k \in \mathbb{N} \cup \{0\}$, released at zero, from (2.82). Suppose that we observe $dY \sim P^Y_{f_0}$ in the white noise model (7.82), where $f_0 \in B^\alpha_{22}$ for $\alpha = k + 1/2$. Then, for some large enough constant M,*

$$\Pi\left(f : \|f - f_0\|_2 > Mn^{-\alpha/(2\alpha+1)}|Y\right) \to^{P^Y_{f_0}} 0 \tag{7.106}$$

as $n \to \infty$.

Proof The proof follows the argument for standard Brownian motion combined with the results obtained in Section 2.6, where

$$\varepsilon_n = Mn^{-\alpha/(2\alpha+1)}.$$

Note that the RKHS $\mathbb{H} \equiv \mathbb{H}^k$ of W^k equals $H^{k+1}_2 = B^{k+1}_{22}$ in view of Proposition 2.6.24 and the results in Chapter 4 (more precisely, the version of (4.105) for spaces defined on $[0,1]$). We can use (7.102) combined with Corollary 2.6.30, $\|W\|_2 \le \|W\|_\infty$ and the estimate, for $2^j \simeq n^{1/(2\alpha+1)} \simeq n\varepsilon_n^2$,

$$\|\pi_{V_j}(f_0) - f_0\|_2 \lesssim \|f_0\|_{B^\alpha_{22}} \sum_{l>j} 2^{-j\alpha} \le \varepsilon_n, \quad \|\pi_{V_j}(f_0)\|^2_{\mathbb{H}} \lesssim \|f_0\|^2_{B^\alpha_{22}} 2^j \simeq n\varepsilon_n^2,$$

to obtain

$$\Pi\left(f \in L^2 : \|f - f_0\|_2 \le \varepsilon_n\right) \ge e^{-Cn\varepsilon_n^2}$$

for some constant C. As approximating set, we take $\mathcal{F}_n$ as in (7.103) with $\mathbb{H} = \mathbb{H}^k$ and the current choice of ε_n, so $\Pi(\mathcal{F} \setminus \mathcal{F}_n)$ is bounded just as in (7.104). The approximation errors are also bounded as earlier, that is,

$$\begin{aligned}\|K_{j_n}(f_1 + f_2) - f_1 + f_2\|_2 &\le C\|f_1\|_\infty + 2^{-j_n(\alpha+1/2)}\|f_2\|_{\mathbb{H}}\\ &\lesssim \varepsilon_n + (n\varepsilon_n^2)^{-\alpha-1/2}\sqrt{n}\varepsilon_n\\ &= \varepsilon_n(1 + n^{-(\alpha+1/2)/(2\alpha+1)}\sqrt{n}) = \varepsilon_n\end{aligned}$$

and

$$\|K_j(f_0) - f_0\|_2 \lesssim \|f_0\|_{B^\alpha_{22}}(n\varepsilon_n^2)^{-\alpha} \simeq \varepsilon_n,$$

completing the proof by application of Theorem 7.3.6. ■

This approach can be generalised to smoothness indices other than $k + 1/2, k \in \mathbb{N}$, by considering fractional Brownian motions instead of primitives of standard Brownian motion.

Contraction Rates with Gaussian Priors in Density Estimation

Consider next the situation where we observe $X_1, \dots, X_n$ i.i.d. from some density f on $[0,1]$ with respect to Lebesgue measure μ. A prior for f needs to accommodate a nonnegativity and integrability constraint, so we cannot use the trajectory of W^k directly as in the preceding subsection. However, we can consider priors of the form

$$p_w = \frac{e^w}{\int_0^1 e^w}, \qquad w \sim W^k. \tag{7.107}$$

The following auxiliary lemma relates the Hellinger distance as well as the information-theoretic quantities appearing in (7.87) of such densities to the uniform distance of the 'kernels' w:

Lemma 7.3.8 *For any measurable functions $v, w : \Leftarrow\mathcal{X}, \mathcal{A}) \to \mathbb{R}$ and p_v, p_w as in (7.107), we have*

(a) $h(p_v, p_w) \le \|v - w\|_\infty e^{\|v-w\|_\infty/2}$,
(b) $-\int \log(p_v/p_w)p_w \le \|v - w\|_\infty^2 e^{\|v-w\|_\infty}(1 + \|v - w\|_\infty)$, and
(c) $\int(\log(p/p_0))^2 p_0 \le \|v - w\|_\infty^2 e^{\|v-w\|_\infty}(1 + \|v - w\|_\infty)^2$.

Proof The proof is basic but somewhat technical, see Exercise 7.3.3. ■

The following result shows that the resulting posteriors contract in Hellinger distance h at the correct rates for Hölderian densities $p_0 \in B^\alpha_{\infty\infty}, \alpha = k + 1/2, k \in \mathbb{N}$.

Theorem 7.3.9 *Consider a prior Π on probability densities on $[0,1]$ arising from (7.107) with $W_k, k \in \mathbb{N} \cup \{0\}$, equal to a kth integrated Brownian motion released at zero (as in (2.82)). Suppose that we observe $X_1, \dots, X_n$ i.i.d. from density p_0 such that both p_0 and $w_0 =$*

$\log p_0 \in B^{\alpha}_{\infty\infty}([0,1])$ *for* $\alpha = k+1/2$. *Then, for some large enough constant* M,

$$\Pi\left(p : h(p,p_0) > Mn^{-\alpha/(2\alpha+1)}|X_1,\dots,X_n\right) \to^{P^{\mathbb{N}}_{f_0}} 0 \qquad (7.108)$$

as $n \to \infty$.

Proof We apply Theorem 7.3.3 with $\varepsilon_n = Mn^{-\alpha/(2\alpha+1)}$. By Lemma 7.3.8, we can lower bound the probability in (7.91) by

$$\Pr(\|W^k - w_0\|_\infty \leq \varepsilon_n).$$

Then, using Proposition 2.6.19 as in (7.102) with the L^∞-norm, replacing the L^2-norm, we deduce from Corollary 2.6.30 the desired lower bound for the centred small-ball probability. The approximation from the RKHS $\mathbb{H} = B^{k+1}_{22}$ also satisfies, for $2^j \simeq n^{1/(2\alpha+1)} \simeq n\varepsilon_n^2$,

$$\|\pi_{V_j}(p_0) - p_0\|_\infty \lesssim \|p_0\|_{B^{\alpha}_{\infty\infty}} \sum_{l>j} 2^{-j\alpha} \leq \varepsilon_n, \quad \|\pi_{V_j}(p_0)\|^2_{\mathbb{H}} \lesssim \|p_0\|^2_{B^{\alpha}_{\infty\infty}} 2^j \simeq n\varepsilon_n^2,$$

so (7.91) follows.

We next choose $\mathcal{P}_n = \{p_w : w \in \mathcal{F}_n\}$, where $\mathcal{F}_n$ is as in (7.103), with $\varepsilon_n = Mn^{-\alpha/(2\alpha+1)}$. As in (7.104), the Gaussian isoperimetric inequality verifies (7.92). In view of Corollary 4.3.38, the ε_n L^∞-metric entropy of a ball in H^k is of order $(M_n/\varepsilon_n)^{1/k}$, and this carries over to $\mathcal{F}_n$ because we are only adding an L^∞-ball of radius ε_n. By Lemma 7.3.8, this bound carries over to the Hellinger metric entropy of $\mathcal{P}_n = \{p_w : w \in \mathcal{F}_n\}$. Then, recalling $M_n \simeq \sqrt{n}\varepsilon_n$, we see that

$$\left(\frac{M_n}{\varepsilon_n}\right)^{1/(\alpha+1/2)} \lesssim n^{1/(2\alpha+1)} \leq n\varepsilon_n^2,$$

for M large enough, so Theorem 7.3.3 implies the conclusion. ∎

7.3.3 Product Priors in Gaussian Regression

Consider in this subsection the Gaussian white noise model

$$dY^{(n)}(t) = f(t)dt + \frac{1}{\sqrt{n}}dW(t), \quad t \in [0,1], \qquad (7.109)$$

from (7.82), where $f \in L^2 = L^2([0,1])$, and where we set $\sigma = 1$ for simplicity. As usual, we denote by P_f^Y the distribution of $dY = dY^{(n)}$ and by $E_f = E_f^Y$ the corresponding expectation operator.

Many common prior distributions Π on L^2 can be realised as a product measure on a suitable basis $\{e_k\}$ or, what is the same, are laws of a random function $\sum_k \phi_k \psi_k$, where the ϕ_k are independent real random variables. For instance, this is the case for any Gaussian prior on L^2 in view of Theorem 2.6.10, with ϕ_k i.i.d. $N(0,1)$. If the noise process $(g_k) = (\int \psi_k dW) = \mathbb{W}$ induces a Gaussian product measure $\mathcal{N} = \otimes_k N(0,1)$ on the ψ_k too, the posterior distribution $\Pi(\cdot\,|Y)$ is then also a product probability measure on the ψ_k. In the special case where the ϕ_k are themselves Gaussians, we can show that $\Pi(\cdot\,|Y)$ is also Gaussian – the so-called conjugate situation (see (7.115)).

In such product prior settings we can perform a direct coordinate-wise analysis of $\Pi(\cdot\,|Y)$ and obtain more precise results than through the general contraction theorems from earlier.

In particular, we will be able to obtain sharp results in loss functions that are not comparable to the L^2-distance, which appears to be difficult through the 'testing methods' from the preceding section (see also the notes at the end of this chapter).

We investigate this in what follows for product priors defined on general bases of L^2 that satisfy the following condition. Note that we use double-indexed 'wavelet notation' but that standard 'single-indexed' bases $\{e_k\}$ are admissible too in part(a).

Definition 7.3.10 Let $S \in \mathbb{N}$. By an S-regular basis $\{\psi_{lk} : l \in \mathcal{L}, k \in \mathcal{Z}_l\}$ of L^2 with index sets $\mathcal{L} \subset \mathbb{Z}, \mathcal{Z}_l \subset \mathbb{Z}$ and characteristic sequence a_l, we shall mean any of the following:

(a) $\psi_{lk} \equiv e_l$ is S-times differentiable with all derivatives in L^2, $|\mathcal{Z}_l| = 1$, $a_l = \max(2, |l|)$, and $\{e_l : l \in \mathcal{L}\}$ forms an ortho-normal basis of L^2.

(b) ψ_{lk} is S-times differentiable with all derivatives in L^2, $a_l = |\mathcal{Z}_l| = 2^l$, and $\{\psi_{lk} : l \in \mathcal{L}, k \in \mathcal{Z}_l\}$ forms an ortho-normal basis of L^2.

We consider priors of the form

$$f \sim \Pi, \ \Pi = \bigotimes_{l \in \mathcal{L}, k \in \mathcal{Z}_l} \Phi_{lk},$$

defined on the coordinates of the ortho-normal basis $\{\psi_{lk}\}$, where Φ_{lk} are probability distributions with Lebesgue density φ_{lk} on the real line. Further assume, for some fixed density φ on the real line, that

$$\varphi_{lk}(\cdot) = \frac{1}{\sigma_l}\varphi\left(\frac{\cdot}{\sigma_l}\right) \quad \forall k \in \mathcal{Z}_l, \quad \text{with } \sigma_l > 0.$$

To expedite notation, we shall write f_{lk} for $\langle f, \psi_{lk}\rangle$ when no confusion may arise, and we recall that the white noise model (7.109) can be expressed on the basis $\{\psi_{lk}\}$ as

$$Y_{lk} \equiv Y_{lk}^{(n)} = \int_0^1 \psi_{lk} dY^{(n)} = f_{lk} + \frac{1}{\sqrt{n}} g_{lk}, \tag{7.110}$$

and we shall use this notation repeatedly in this section. The posterior distribution $\Pi(\cdot|Y)$ given observations $Y = (Y_{lk})$ in the white noise model is then given in (7.86). When assuming $Y \sim P_{f_0}^Y$ from some fixed f_0, we have to require that $f_{0,lk}$ is in the interior of the support of φ_{lk} for all k, l. This is in some sense the analogue of the 'small-ball' conditions from the preceding section.

Let us summarise these hypotheses in the following assumption.

Condition 7.3.11 *(a) For $\{\psi_{lk}\}$ from Definition 7.3.10, consider product priors Π arising from the law of random series*

$$f(x) = \sum_{l \in \mathcal{L}} \sigma_l \sum_{k \in \mathcal{Z}_l} \phi_{lk} \psi_{lk}(x),$$

where the ϕ_{lk} are i.i.d. random variables from bounded density $\varphi : \mathbb{R} \to \mathbb{R}$, and where $\sum_{l,k} \sigma_l^2 < \infty$.

(b) Consider data Y generated from equation (7.109) *under a fixed function* $f_0 \in L^2$ *with coefficients* $\{f_{0,lk}\} = \{\langle f_0, \psi_{lk}\rangle\}$, *and suppose that*

(*P*1) *For a finite constant* $M > 0$,

$$\sup_{l\in\mathcal{L},k\in\mathcal{Z}_l} \frac{|f_{0,lk}|}{\sigma_l} \le M.$$

(*P*2) *For some* $\tau > M$ *and* $0 < c_\varphi \le C_\varphi < \infty$,

$$\varphi(x) \le C_\varphi \quad \forall x \in \mathbb{R}, \quad \varphi(x) \ge c_\varphi \quad \forall x \in (-\tau, \tau), \quad \int_{\mathbb{R}} x^2 \varphi(x) dx < \infty.$$

We allow for a rich variety of base priors φ, such as Gaussian, sub-Gaussian, Laplace, most Student laws or, more generally, any law with positive continuous density and finite second moment but also uniform priors with large enough support $(-\tau,\tau) \supset (-M,M)$. In view of the results in Sections 4.3.4 and 4.3.5, we can interpret Condition (P1) as a Hölder regularity condition on f_0 through the decay of the wavelet coefficients of f_0 relative to the regularity of the prior, modelled by the sequence (σ_l).

A Contraction Result for Marginal Posterior Second Moments

We first provide a result on the contraction of the marginal coordinates of the posterior distribution.

Theorem 7.3.12 *Consider observations dY in white noise (7.109), and for a prior and* f_0 *satisfying Condition 7.3.11, let* $\Pi(\cdot|Y)$ *be the resulting posterior distribution. Then we have for every fixed* l, k, *some constant* $0 < C < \infty$ *independent of* k, l, n *and every* $n \in \mathbb{N}$,

$$E^Y_{f_0} \int (f_{lk} - f_{0,lk})^2 d\Pi(f_{lk}|Y) \le C \min(\sigma_l^2, 1/n).$$

Proof We decompose the index set $\mathcal{L}$ into

$$\mathcal{J}_n := \{l \in \mathcal{L}, \sqrt{n}\sigma_l \ge S_0\}$$

and its complement $\mathcal{J}_n^c$, where S_0 is a fixed positive constant. Setting

$$B_{lk}(Y) := \int (f_{lk} - f_{0,lk})^2 d\Pi(f_{lk}|Y),$$

we shall show that

$$\sup_{l\in\mathcal{J}_n,k} E^Y_{f_0} B_{lk}(Y) \le C/n, \qquad \sup_{l\in\mathcal{J}_n^c,k} \sigma_l^{-2} E^Y_{f_0} B_{lk}(Y) \le C,$$

which implies the result. We write $E = E^Y_{f_0}$ throughout the proof to ease notation and let $\varepsilon_{lk} = \int \psi_{lk} dW$ be a sequence of i.i.d. $N(0,1)$ variables.

Using the independence structure of the prior and under $P^Y_{f_0}$, we have from (7.86) that

$$B_{lk}(Y) = \frac{\int (f_{lk} - f_{0,lk})^2 e^{-(n/2)(f_{lk}-f_{0,lk})^2 + \sqrt{n}\varepsilon_{lk}(f_{lk}-f_{0,lk})} \varphi_{lk}(f_{lk}) d f_{lk}}{\int e^{-\frac{n}{2}(f_{lk}-f_{0,lk})^2 + \sqrt{n}\varepsilon_{lk}(f_{lk}-f_{0,lk})} \varphi_{lk}(f_{lk}) d f_{lk}}$$

$$= \frac{1}{n} \frac{\int v^2 e^{-(v^2/2)+\varepsilon_{lk} v} \frac{1}{\sqrt{n}\sigma_l} \varphi\left(\frac{f_{0,lk}+n^{-1/2}v}{\sigma_l}\right) dv}{\int e^{-(v^2/2)+\varepsilon_{lk} v} \frac{1}{\sqrt{n}\sigma_l} \varphi\left(\frac{f_{0,lk}+n^{-1/2}v}{\sigma_l}\right) dv} =: \frac{1}{n} \frac{N_{lk}}{D_{lk}}(\varepsilon_{lk}).$$

Consider first the indices $l \in \mathcal{J}_n^c$. Restricting the integral to $[-\sqrt{n}\sigma_l, \sqrt{n}\sigma_l]$, we see that

$$D_{kl}(\varepsilon_{kl}) \geq \int_{-\sqrt{n}\sigma_l}^{\sqrt{n}\sigma_l} e^{-(v^2/2)+\varepsilon_{lk} v} \frac{1}{\sqrt{n}\sigma_l} \varphi\left(\frac{f_{0,lk}+n^{-1/2}v}{\sigma_l}\right) dv.$$

To simplify the notation, we suppose that $\tau > M+1$. The argument of the function φ in the preceding display stays in $[-M+1, M+1]$ under (P1). Under assumption (P2), this implies that the value of φ in the preceding expression is bounded from below by c_φ. Next, applying Jensen's inequality with the logarithm function with respect to $dv/(2\sqrt{n}\sigma_l)$ on $[-\sqrt{n}\sigma_l, \sqrt{n}\sigma_l]$, we obtain the lower bound

$$\begin{aligned} \log D_{kl}(\varepsilon_{kl}) &\geq \log(2c_\varphi) - \int_{-\sqrt{n}\sigma_l}^{\sqrt{n}\sigma_l} \frac{v^2}{2} \frac{dv}{2\sqrt{n}\sigma_l} + \varepsilon_{lk} \int_{-\sqrt{n}\sigma_l}^{\sqrt{n}\sigma_l} v \frac{dv}{2\sqrt{n}\sigma_l} \\ &= \log(2c_\varphi) - (\sqrt{n}\sigma_l)^2/6. \end{aligned}$$

Thus, $D_{kl}(\varepsilon_{kl}) \geq 2c_\varphi e^{-(\sqrt{n}\sigma_l)^2/6}$, which is bounded away from zero by a fixed constant for indices in $\mathcal{J}_n^c$.

To deal with the numerator, we have from Fubini's theorem as before and then changing variables back

$$\begin{aligned} EN_{lk} &= \int_{\mathbb{R}} v^2 e^{-(v^2/2)} E[e^{\varepsilon_{lk} v}] \frac{1}{\sqrt{n}\sigma_l} \varphi\left(\frac{f_{0,lk}+n^{-1/2}v}{\sigma_l}\right) dv \\ &= \int_{\mathbb{R}} \left(\sqrt{n}\sigma_l u - \sqrt{n}\sigma_l \frac{f_{0,lk}}{\sigma_l}\right)^2 \varphi(u) du \\ &\leq 2n\sigma_l^2 \left[\frac{f_{0,lk}^2}{\sigma_l^2} + \int_{-\infty}^{+\infty} u^2 \varphi(u) du\right]. \end{aligned}$$

Thus, using Condition 7.3.11, this term is bounded on $\mathcal{J}_n^c$ by a fixed constant times $n\sigma_l^2$, and as a consequence, there exists a fixed constant D independent of n, k, l such that $E(\sigma_l^{-2} B_{lk}(X)) \leq D$, for all k and all $l \in \mathcal{J}_n^c$.

Now, about the indices in $\mathcal{J}_n$, for such l, k, using (P1)–(P2), we can find $L_0 > 0$ depending only on S_0, M, τ such that, for any v in $(-L_0, L_0)$, $\varphi((f_{0,lk} + n^{-1/2}v)/\sigma_l) \geq c_\varphi$. Thus, the denominator $D_{lk}(\varepsilon_{lk})$ can be bounded from below by

$$D_{lk}(\varepsilon_{lk}) \geq c_\varphi \int_{-L_0}^{L_0} e^{-(v^2/2)+\varepsilon_{lk} v} \frac{1}{\sqrt{n}\sigma_l} dv.$$

Moreover, the numerator can be bounded above by

$$N_{lk}(\varepsilon_{lk}) \leq C_\varphi \int v^2 e^{-(v^2/2)+\varepsilon_{lk} v} \frac{1}{\sqrt{n}\sigma_l} dv,$$

Putting these two bounds together leads to

$$B_{lk}(\varepsilon_{lk}) \leq \frac{1}{n} \frac{C_\varphi}{c_\varphi} \frac{\int v^2 e^{-(v^2/2)+\varepsilon_{lk} v} dv}{\int_{-L_0}^{L_0} e^{-(v^2/2)+\varepsilon_{lk} v} dv}.$$

This last quantity has a distribution independent of l,k. Let us thus show that

$$Q(L_0) = E\left[\frac{\int v^2 e^{-(v-\varepsilon)^2/2} dv}{\int_{-L_0}^{L_0} e^{-(v-\varepsilon)^2/2} dv}\right]$$

is finite for every $L_0 > 0$, where $\varepsilon \sim N(0,1)$. In the numerator, we substitute $u = v - \varepsilon$. Using the inequality $(u+\varepsilon)^2 \le 2u^2 + 2\varepsilon^2$, the second moment of a standard normal variable appears, and this leads to the bound

$$Q(L_0) \le CE\left[\frac{1+\varepsilon^2}{\int_{-L_0}^{L_0} e^{-(v-\varepsilon)^2/2} dv}\right]$$

for some finite constant $C > 0$. Denote by g the density of a standard normal variable, by Φ its distribution function and by $\bar{\Phi} = 1 - \Phi$. It is enough to prove that the following quantity is finite

$$q(L_0) := \int_{-\infty}^{+\infty} \frac{(1+u^2)g(u)}{\bar{\Phi}(u-L_0) - \bar{\Phi}(u+L_0)} du = 2\int_0^{+\infty} \frac{(1+u^2)g(u)}{\bar{\Phi}(u-L_0) - \bar{\Phi}(u+L_0)} du,$$

since the integrand is an even function. Using the standard inequalities

$$\frac{1}{\sqrt{2\pi}} \frac{u^2}{1+u^2} \frac{1}{u} e^{-u^2/2} \le \bar{\Phi}(u) \le \frac{1}{\sqrt{2\pi}} \frac{1}{u} e^{-u^2/2}, \quad u \ge 1,$$

it follows that for any $\delta > 0$, we can find $M_\delta > 0$ such that,

$$(1-\delta)\frac{1}{u} e^{-u^2/2} \le \sqrt{2\pi}\,\bar{\Phi}(u) \le \frac{1}{u} e^{-u^2/2}, \quad u \ge M_\delta.$$

Set $A_\delta = 2L_0 \vee M_\delta$. Then, for $\delta < 1 - e^{-2L_0}$, we deduce

$$\begin{aligned} q(L_0) &\le 2\int_0^{A_\delta} \frac{(1+u^2)g(u)}{\bar{\Phi}(A_\delta - L_0) - \bar{\Phi}(A_\delta + L_0)} du \\ &\quad + 2\sqrt{2\pi} \int_{A_\delta}^{+\infty} (u-L_0)(1+u^2) \frac{e^{(u-L_0)^2/2} g(u)}{1-\delta-e^{-2L_0}} du \\ &\le C(A_\delta, L_0) + \frac{2e^{L_0^2/2}}{1-\delta-e^{-2L_0}} \int_{A_\delta}^{+\infty} u(1+u^2)e^{-L_0 u} du < \infty. \end{aligned}$$

We conclude that $\sup_{l\in\mathcal{J}_n,k} E_{f_0}^Y |B_{lk}(Y)| = O(1/n)$, completing the proof. ■

Contraction Rates in L^2-Norms

By summing the preceding bounds over all coordinates k,l and using Parseval's identity, we obtain an optimal L^2-contraction result for the posterior second moments (and by Markov's inequality also for the posterior distribution itself). In fact, the result immediately generalises to give contraction rates in general Sobolev norms too (see Exercise 7.3.5 and the proof of Theorem 7.3.19).

Corollary 7.3.13 *Set $\sigma_l = |l|^{-1\backslash 2-\gamma}$ or $\sigma_l = 2^{-(\gamma+1/2)l}$ depending on the chosen basis of type either (a) or (b) from Definition 7.3.10. Suppose that the conditions of Theorem 7.3.12 are satisfied. Then*

$$E_{f_0}^Y \int \|f - f_0\|_2^2 d\Pi(f|Y) = O\left(n^{-2\gamma/(2\gamma+1)}\right).$$

Moreover, denote by $\bar{f}_n := \bar{f}_n(Y) := \int f d\Pi(f|Y)$ *the posterior mean. Then we also have*

$$E_{f_0}^Y \|\bar{f}_n - f_0\|_2^2 = O\left(n^{-2\gamma/(2\gamma+1)}\right).$$

Proof For both types of bases, using Parseval's identity and Theorem 7.3.12, we have

$$\|f - f_0\|_2^2 \lesssim \sum_l |\mathcal{Z}_l|(\sigma_l^2 \wedge n^{-1}) = O(n^{-2\gamma\backslash 2\gamma+1}).$$

The second claim then follows from the Cauchy-Schwarz inequality

$$E_{f_0}^Y \|\bar{f}_n - f_0\|_2^2 = E_{f_0}^Y \sum_{l,k} \left[\int (f_{lk} - f_{0,lk}) d\Pi(f_{lk}|Y)\right]^2$$

$$\leq E_{f_0}^Y \sum_{l,k} \left[\int (f_{lk} - f_{0,lk})^2 d\Pi(f_{lk}|Y)\right]. \quad \blacksquare$$

The preceding choice of σ_l entails a regularity condition on f_0 through Condition (P1), namely, $\sup_k |f_{0,lk}| \leq M\sigma_l$. If $\sigma_l = 2^{-(\gamma+1/2)l}$ and we use a wavelet basis of $L^2([0,1])$ from Section 4.3.4 or 4.3.5 in Definition 7.3.10, then this amounts to a standard smoothness condition $f_0 \in C^\gamma([0,1])$, implying that the preceding rates are minimax optimal in view of the results in Section 6.3.

A Sub-Gaussian Bound on the Posterior Marginal Coordinates

For low frequencies l, we can refine Theorem 7.3.12 to a sub-Gaussian estimate on the posterior marginal coordinates. The following result foreshadows (and will be necessary to prove) the exact Gaussian asymptotics for the posterior distribution to be derived in the next section.

Proposition 7.3.14 *Under the conditions of Theorem 7.3.12, let* $l \in \mathbb{N}$ *be such that* $\sqrt{n}\sigma_l \geq S_0$ *for some fixed constant* $0 < S_0 < \infty$. *Then, for some fixed positive constant* $C = C(S_0)$ *independent of* l, k, n *and every* $t \in \mathbb{R}, n \in \mathbb{N}$, *we have*

$$E_{f_0}^Y E_\Pi \left(e^{t\sqrt{n}\langle f - Y, \psi_{lk}\rangle} | Y\right) \leq C e^{t^2/2}. \tag{7.111}$$

As a consequence, we also have, for some $0 < C' < \infty$ *and every* $n \in \mathbb{N}$,

$$E_{f_0}^Y E_\Pi \left[\max_{k \in \mathcal{Z}_l} |\langle f - f_0, \psi_{lk}\rangle| \, | Y\right] \leq C' \sqrt{\frac{\log(1 + |\mathcal{Z}_l|)}{n}}. \tag{7.112}$$

Proof We have from (7.86) that under $Y \sim P_{f_0}^Y$, with $\varepsilon_{lk} \sim^{i.i.d.} N(0,1)$ and substituting $v = \sqrt{n}(f_{lk} - f_{0,lk})$,

$$E\left(e^{t\sqrt{n}\langle f - Y, \psi_{lk}\rangle} | Y^{(n)}\right)$$

$$= \frac{\int \exp\left\{t\sqrt{n}(f_{lk} - f_{0,lk}) - t\varepsilon_{lk} + \sqrt{n}\varepsilon_{lk}(f_{lk} - f_{0,lk}) - \frac{n}{2}(f_{lk} - f_{0,lk})^2\right\} \varphi_{lk}(f_{lk}) d f_{lk}}{\int \exp\left\{\sqrt{n}\varepsilon_{lk}(f_{lk} - f_{0,lk}) - \frac{n}{2}(f_{lk} - f_{0,lk})^2\right\} \varphi_{lk}(f_{lk}) d f_{lk}}$$

$$= e^{-t\varepsilon_{lk}} \frac{\int \exp\{(t+\varepsilon_{lk})v - v^2/2\}\varphi\left(\frac{f_{0,lk}+v/\sqrt{n}}{\sigma_l}\right)dv}{\int \exp\{\varepsilon_{lk}v - v^2/2\}\varphi\left(\frac{f_{0,lk}+v/\sqrt{n}}{\sigma_l}\right)dv}.$$

Now, by Condition 7.3.11(b) and hypothesis on l, we have

$$\inf_{v\in[-L_0,L_0]} \varphi\left(\frac{f_{0,lk}+v/\sqrt{n}}{\sigma_l}\right) \geq c,$$

for some fixed positive constants L_0, c. Using that φ is also bounded, we see that the last expression in the preceding display is bounded by a constant multiple of

$$e^{-t\varepsilon_{lk}} \frac{\int e^{tv-(v-\varepsilon_{lk})^2/2}dv}{\int_{-L_0}^{L_0} e^{-(v-\varepsilon_{lk})^2/2}dv} = \frac{\int e^{tu-u^2/2}du}{\int_{-L_0}^{L_0} e^{-(v-\varepsilon_{lk})^2/2}dv} = \frac{\sqrt{2\pi}e^{t^2/2}}{\int_{-L_0}^{L_0} e^{-(v-\varepsilon_{lk})^2/2}dv}.$$

The expectation of the inverse of this integral is bounded by a fixed constant, arguing as at the end of the proof of Theorem 7.3.12, so the first bound of the proposition follows.

The second inequality of the proposition is obvious under Condition 7.3.10(a) as then $\mathcal{Z}_l$ is a singleton. In case (b), we have $|\mathcal{Z}_l| = 2^l$, so, by a standard bound for maxima of sub-Gaussian random variables (as in Section 2.3, using Lemma 2.3.4), the second bound follows too, noting that we can decompose

$$\langle f - f_0, \psi_{lk}\rangle = \langle f - dY, \psi_{lk}\rangle + n^{-1/2}\langle \psi_{lk}, dW\rangle,$$

the sum of two sub-Gaussian processes. ■

Contraction Rates in L^∞-Distance

Consider the case of a uniform wavelet prior $\Pi_{\gamma,B}$, where the ϕ_{lk} in Condition 7.3.11 are drawn from a uniform $U(-B,B)$ distribution on the interval $[-B,B], B>0$, based on basis functions ψ_{lk} from Definition 7.3.10. We choose

$$\sigma_l = 2^{-l(\gamma+\frac{1}{2})}, \quad \gamma > 0. \tag{7.113}$$

Recalling the results from Sections 4.3.4 and 4.3.5, we see that this models a function that lies in a Hölder-type space

$$C^\gamma([0,1]) = B^\gamma_{\infty\infty}([0,1]) = \left\{f : \|f\|_{C^\gamma} = \sup_{k,l} 2^{l(\gamma+1/2)}|\langle f, \psi_{lk}\rangle| < \infty\right\}$$

and has C^γ-norm no larger than B (Note the slight abuse of the C^γ notation in this subsection.). Such a prior hence draws a natural random function from a fixed Hölder ball. Assuming that the observations are generated from a fixed f_0 satisfying $\|f_0\|_{C^\gamma} < B$, we see that Condition 7.3.11 applies and obtain from Corollary 7.3.13 the L^2-contraction rate $n^{-\gamma/(2\gamma+1)}$ for the posterior distribution about f_0. We now refine this result and prove a minimax optimal contraction rate in the uniform norm over the given Hölder ball.

Proposition 7.3.15 *Consider data generated from equation* (7.109) *under a fixed function f_0 satisfying $\|f_0\|_{C^\gamma([0,1])} < B$. For the uniform wavelet prior $\Pi_{\gamma,B}$ from before (7.113) with* *resulting posterior distribution* $\Pi_{\gamma,B}(\cdot\,|Y)$*, we then have*

$$E^Y_{f_0} E_{\Pi_{\gamma,B}}[\|f - f_0\|_\infty |Y] \le M\left(\frac{\log n}{n}\right)^{\gamma/(2\gamma+1)},$$

for some constant $0 < M < \infty$.

Proof Choose j_n such that $2^{j_n} = (n/\log n)^{1/(2\gamma+1)}$, and note that $\sqrt{n}\sigma_l \ge \sqrt{\log n} \ge S_0$ for $l \le j_n$. We can estimate

$$\|f - f_0\|_\infty \lesssim \sum_{l \le j_n} \sqrt{\frac{2^l}{n}} \max_k \sqrt{n} |\langle f - f_0, \psi_{lk}\rangle| + \sum_{l > j_n} 2^{l/2} \max_k |\langle f - f_0, \psi_{lk}\rangle|.$$

Using (7.112), the first term has $E^Y_{f_0} \Pi_{\gamma,B}[\cdot|Y]$-expectation of order

$$\sum_{l \le j_n} \sqrt{\frac{2^l l}{n}} = O\left(\sqrt{\frac{2^{j_n} j_n}{n}}\right) = O\left(\frac{\log n}{n}\right)^{\gamma/(2\gamma+1)}.$$

Since prior and posterior and hence also $f - f_0$ concentrate on a fixed ball in $C^\gamma([0,1])$, the second term is less than or equal to a constant multiple of

$$\sum_{l > j_n} 2^{-l\gamma} = O(2^{-j_n\gamma}) = O\left(\frac{\log n}{n}\right)^{\gamma/(2\gamma+1)}$$

in view of the wavelet definition of the Hölder norm, completing the proof. ∎

For priors that have support in the whole Hölder space, the control of the high frequencies in the preceding theorem is not as simple. Under conditions on the exact tail of φ we can obtain supnorm rates too. Perhaps the most interesting case is where φ is a Gaussian distribution, where we can make an explicit conjugate analysis. For the general case, we give some references in the notes at the end of this chapter.

The Gaussian Conjugate Situation

Consider next the situation where the prior Π on L^2 is defined on a wavelet basis from Definition 7.3.10, part(b), as

$$f = \sum_l \sigma_l \sum_k g_{lk}\psi_{lk}, \quad \sum_{l,k} \sigma_l^2 < \infty, \tag{7.114}$$

where the $g_{lk} \sim^{i.i.d.} N(0,1)$, corresponding to a random wavelet series $f \in L^2$ as in the preceding subsection with the uniform random variables replaced by i.i.d. Gaussians.

Given an observation dY in (7.109), one shows, using the conjugacy of Gaussian random variables in each coordinate l,k (see Exercise 7.3.4), that the posterior distribution $\Pi(\cdot\,|Y)$ is also a Gaussian measure on L^2 which is given, conditional on $Y = (Y_{lk})$, by the law of the random wavelet series

$$f|Y = \sum_{l,k}\left[\frac{\sigma_l^2}{\sigma_l^2 + 1/n} y_{lk} + \left(\frac{\sigma_l^2}{n\sigma_l^2 + 1}\right)^{1/2} \bar{g}_{lk}\right]\psi_{lk} \tag{7.115}$$

$$= E_\Pi(f|Y) + \sum_{l,k}\left(\frac{\sigma_l^2}{n\sigma_l^2 + 1}\right)^{1/2} \psi_{lk}\bar{g}_{lk}, \tag{7.116}$$

where the $\bar{g}_{lk}$ variables are i.i.d. $N(0,1)$ independent of the $y_{lk} = \int_0^1 \psi_{lk} dY$. Explicit analysis of this Gaussian posterior distribution gives the following contraction result in the supremum norm:

Theorem 7.3.16 *Let Π be the law of the random function f from (7.114), where*

$$\sigma_l = 2^{-l(\gamma+1/2)} l^{-1/2}, \quad \gamma > 0,$$

and let $\Pi(\cdot|Y)$ be the posterior distribution given observations Y in (7.109). If $Y \sim P^Y_{f_0}$ for some fixed $f_0 \in C^\gamma([0,1])$, then there exist fixed constants $C, M_0 < \infty$ such that, for every $M_0 \le M < \infty$ and for all $n \in \mathbb{N}$,

$$E^Y_{f_0} \Pi \left(f : \|f - f_0\|_\infty > M \left(\frac{\log n}{n}\right)^{\gamma/(2\gamma+1)} \Big| Y \right) \le n^{-C^2(M-M_0)^2}. \tag{7.117}$$

Proof Let us write

$$\varepsilon_n = \left(\frac{\log n}{n}\right)^{\gamma/(2\gamma+1)}$$

throughout the proof. Under $P^Y_{f_0}$, we have

$$E^Y_{f_0} \Pi (\|f - f_0\|_\infty > M\varepsilon_n | Y)$$

$$= \Pr\left\{ \left\| \sum_{l,k} \left[\frac{-1/n}{\sigma_l^2 + 1/n} \langle f_0, \psi_{lk} \rangle + \frac{\sigma_l^2}{\sqrt{n}(\sigma_l^2 + 1/n)} g_{lk} + \left(\frac{\sigma_l^2}{n\sigma_l^2 + 1} \right)^{1/2} \bar{g}_{lk} \right] \psi_{lk} \right\|_\infty > M\varepsilon_n \right\} \tag{7.118}$$

$$= \Pr\left\{ \left\| E^Y_{f_0}(E_\Pi(f|Y) - f_0) + G \right\|_\infty > M\varepsilon_n \right\},$$

where G is the centred Gaussian process

$$G(t) = \sum_{l,k} \left[\frac{\sigma_l^2}{\sqrt{n}(\sigma_l^2 + 1/n)} g_{lk} + \left(\frac{\sigma_l^2}{n\sigma_l^2 + 1} \right)^{1/2} \bar{g}_{lk} \right] \psi_{lk}(t), \quad t \in [0,1].$$

We will apply Theorem 2.5.8 to the probability in (7.118), and for this we need to bound $\|E^Y_{f_0}(E_\Pi(f|Y) - f_0)\|_\infty$, $E\|G\|_\infty$ and $\|E(G^2(\cdot))\|_\infty$. Choose J_n such that $2^{J_n} \simeq (n/\log n)^{1/(2\gamma+1)}$.

First, since $f_0 \in C^\gamma([0,1])$ and $\left\|\sum_k |\psi_{lk}|\right\|_\infty \le C2^{l/2}$, we obtain

$$\begin{aligned}
\left\| E^Y_{f_0}(E_\Pi(f|Y) - f_0) \right\|_\infty &= \left\| \sum_{l,k} \frac{-1/n}{\sigma_l^2 + 1/n} \langle f_0, \psi_{l,k} \rangle \psi_{lk} \right\|_\infty \\
&\lesssim \sum_{l,k} |\psi_{lk}| \frac{2^{-l(\gamma+1/2)}}{n\sigma_l^2 + 1} \\
&\lesssim \left(\sum_{l \le J_n} \frac{2^{-l\gamma}}{n\sigma_l^2} + \sum_{l > J_n} 2^{-l\gamma} \right) \\
&\lesssim \left(\frac{\log n}{n}\right)^{\gamma/(2\gamma+1)}.
\end{aligned}$$

Second, to bound $E\|G\|_\infty$, recall from Section 2.3 that for any sequence of centred normal random variables Z_j,

$$E \max_{1\le j\le N} |Z_j| \le C\sqrt{\log N} \max_{j\le N} (EZ_j^2)^{1/2}, \tag{7.119}$$

where C is a universal constant. Therefore, using $\sigma_l^2 \lesssim n^{-1}$ for $l \ge J_n$ and $1/n \lesssim \sigma_l^2$ otherwise,

$$\begin{aligned}
E\|G\|_\infty &= E\left\| \sum_{l,k}\left[\frac{\sigma_l^2}{\sqrt{n}(\sigma_l^2+1/n)}g_{lk} + \left(\frac{\sigma_l^2}{n\sigma_l^2+1}\right)^{1/2} \bar{g}_{lk}\right]\psi_{lk}\right\|_\infty \\
&\lesssim \sum_l 2^{l/2} E\max_{k\le 2^l}|g_{lk}| \left(\frac{\sigma_l^4}{n(\sigma_l^2+1/n)^2} + \frac{\sigma_l^2}{n\sigma_l^2+1}\right)^{1/2} \\
&\lesssim \sum_l (l2^l)^{1/2}\left(\frac{\sigma_l^4}{n(\sigma_l^2+1/n)^2} + \frac{\sigma_l^2}{n\sigma_l^2+1}\right)^{1/2} \\
&\lesssim \left(\sum_{l\le J_n}\sqrt{\frac{2^l l}{n}} + \sum_{l>J_n}\sqrt{2^l l n\sigma_l^2} + \sum_{l>J_n}\sqrt{2^l l\sigma_l^2}\right) \\
&\lesssim \left(\sqrt{\frac{2^{J_n}J_n}{n}} + 2^{-J_n\gamma}\right) \le D\left(\frac{\log n}{n}\right)^{\gamma/(2\gamma+1)}.
\end{aligned} \tag{7.120}$$

Finally,

$$EG^2(t) = \sum_{l,k}\left(\frac{\sigma_l^4}{n(\sigma_l^2+1/n)^2} + \frac{\sigma_l^2}{n\sigma_l^2+1}\right)\psi_{lk}^2(t) \le C\left(\frac{2^{J_n}}{n} + 2^{-J_n(2\gamma+1)}\right) \le C_3\frac{2^{J_n}}{n}. \tag{7.121}$$

Summarising the preceding estimates and combining them with Theorem 2.5.8 give, for suitable constants $\bar{C}_1, \bar{C}_2$,

$$\begin{aligned}
&\Pr\left\{\left\|E_{f_0}^Y(E_\Pi(f|Y) - f_0) + G\right\|_\infty > M\varepsilon_n\right\} \\
&\le \Pr\left\{\|G\|_\infty - E\|G\|_\infty > M\varepsilon_n - \left\|E_{f_0}^Y(E_\Pi(f|Y) - f_0)\right\|_\infty - E\|G\|_\infty\right\} \\
&\le \Pr\left\{\|G\|_\infty - E\|G\|_\infty > (M - \bar{C}_1 - \bar{C}_2)\varepsilon_n\right\} \\
&\le \exp\left(-\frac{(M-\bar{C}_1-\bar{C}_2)^2\varepsilon_n^2}{C_3 2^{J_n}/n}\right).
\end{aligned} \tag{7.122}$$

Taking into account that $\varepsilon_n^2 \simeq 2^{J_n}J_n/n$ completes the proof. ∎

7.3.4 Nonparametric Bernstein–von Mises Theorems

A classical result in the theory of parametric statistical models is the Bernstein–von Mises (BvM) theorem: it states that the posterior is approximately distributed as a normal distribution, centred at the maximum likelihood (or, in fact, at any efficient) estimator and with covariance attaining the Cramér-Rao information bound. Remarkably, this is true under mild assumptions on the prior, effectively only requiring that the prior charges a neighbourhood of the true parameter point that generated the observations with positive probability.

A consequence is that posterior-based inference is asymptotically equivalent to standard frequentist inference procedures, including confidence sets and critical regions for tests. This provides a frequentist justification of the Bayesian approach to statistical inference that does not rely on any subjective belief in the prior distribution.

In this section we investigate the phenomena behind the Bernstein–von Mises theorem in the infinite-dimensional setting. The geometry of the space in which one can expect a Bernstein–von Mises theorem turns out to be of importance. In standard $\ell_2 \simeq L^2$-spaces, an analogue of the Bernstein–von Mises theorem can be shown not to hold true even in basic settings (see the notes at the end of this chapter). However, we shall show that for some other geometries that resemble topologies weaker than $\ell_2 \simeq L^2$, Bernstein–von Mises theorems hold true. The results we obtain are in some sense analogues of the asymptotic normality results for nonparametric likelihood estimators obtained in the first part of this chapter (Theorems 7.2.14 and 7.2.20). We shall concentrate on the situation of product priors in Gaussian models treated in the preceding section to lay out the main ideas. Some extensions and related results in sampling models are discussed in the notes at the end of this chapter.

The BvM Phenomenon for Finite-Dimensional Subspaces

We start with the easiest and, of course, in view of the classical theory from parametric models not at all surprising situation where one can expect a BvM theorem – the case of fixed finite-dimensional projection subspaces. Understanding the finite-dimensional situation is helpful to develop the main intuitions behind BvM-type results and at any rate will be needed as an ingredient of the proofs for the nonparametric settings considered later.

For Π any prior Borel probability distribution on L^2, the posterior distribution $\Pi_n \equiv \Pi(\cdot|Y)$ based on observing dY in white noise (7.109) defines a random probability measure on L^2. Let V be any of the finite-dimensional projection subspaces of L^2 spanned by the ψ_{lk} from Definition 7.3.10, equipped with the L^2-norm, and suppose that Π is a product measure on the coordinates $\{\psi_{lk}\}$. Let π_V denote the projection of any infinite vector $f = (f_{lk})$ onto V. For $z = (z_{lk})$, define the transformation

$$T_z \equiv T_{z,V} : f \mapsto \sqrt{n}\, \pi_V(f - z),$$

and consider the image measure $\Pi_n \circ T_z^{-1}$ of the posterior measure under T_z. The finite-dimensional space V carries a natural Lebesgue product measure on it.

Condition 7.3.17 *Suppose that Π is a product measure on the span of the $\{\psi_{lk}\}$ and that $\Pi \circ \pi_V^{-1}$ has a Lebesgue-density $d\Pi_V$ in a neighbourhood of $\pi_V(f_0)$ that is continuous and positive at $\pi_V(f_0)$. Suppose also that for every $\delta > 0$ there exists a fixed L^2-norm ball $C = C_\delta$ in V such that, for n large enough, $E^Y_{f_0}(\Pi(\cdot|Y) \circ T_{f_0}^{-1})(C^c) < \delta$.*

This condition requires that the projected prior have a continuous density at $\pi_V(f_0)$ and that the image of the posterior distribution under the finite-dimensional projection onto V concentrate on a $1/\sqrt{n}$-neighbourhood of the projection $\pi_V(f_0)$ of the true f_0 onto V.

For the main result of this section, denote by $\|\cdot\|_{TV}$ the total variation norm on the space of finite signed measures on V. We denote the observations dY and white noise dW as infinite vectors

$$\mathbb{Y} = (Y_{lk}) = \left(\int \psi_{lk} dY : k \in \mathcal{Z}_l, l \in \mathcal{L} \right), \quad \mathbb{W} = \left(\int \psi_{lk} dW : k \in \mathcal{Z}_l, l \in \mathcal{L} \right) \tag{7.123}$$

in the following theorem. We note that a draw from the shifted posterior random measure $\Pi(\cdot|Y) \circ T_{\mathbb{Y}}^{-1}$ is then simply $\sqrt{n}(\pi_V(f) - \pi_V(\mathbb{Y}))$, where f is drawn from the posterior. The following result says that this random variable is approximately a standard Gaussian measure $N(0,I)$ on V with diagonal covariance equal to the identity I and that this approximation holds, with high probability, in the strong sense of total variation distance. It is a version of the classical parametric Bernstein–von Mises theorem in a finite-dimensional Gaussian white noise model.

Theorem 7.3.18 *Consider $Y \sim P_{f_0}^Y$ generated in white noise* (7.109) *under a fixed function $f_0 \in L^2$. Assume Condition 7.3.17. Then we have, as $n \to \infty$,*

$$\|\Pi(\cdot|Y) \circ T_{\mathbb{Y}}^{-1} - N(0,I)\|_{TV} \to^{P_{f_0}^Y} 0.$$

Proof Under $P_{f_0}^Y$ we have $\mathbb{Y} = f_0 + n^{-1/2}\mathbb{W}$. Moreover, $W_V = \pi_V(\mathbb{W})$ is a standard Gaussian variable on V, and if $\tilde{\Pi}_{n,V} = \Pi_n \circ T_{f_0,V}^{-1}$, then it suffices to prove that $\|\tilde{\Pi}_{n,V} - N(W_V,I)\|_{TV}$ converges to zero in $P_{f_0}^Y$-probability. In the following, denote by λ the Lebesgue measure on V and by λ_C its restriction to a measurable subset C of V.

Define $\tilde{\Pi}_{n,V}^C$, the posterior distribution $\tilde{\Pi}_{n,V}$ based on the prior restricted to a measurable set C and renormalised; that is, for B a Borel subset of V and since Π is a product measure,

$$\tilde{\Pi}_{n,V}^C(B) = \frac{\int_B e^{-\|h\|^2/2 + \langle h, W_V \rangle} d\tilde{\Pi}_V^C(h)}{\int e^{-\|g\|^2/2 + \langle g, W_V \rangle} d\tilde{\Pi}_V^C(g)},$$

where $\tilde{\Pi}_V = \Pi \circ T_{f_0,V}^{-1}$, and where $\mu^C(B) = \mu(B \cap C)/\mu(C)$, for any probability measure μ. A simple computation shows that

$$E_{f_0}^Y \|\tilde{\Pi}_{n,V} - \tilde{\Pi}_{n,V}^C\|_{TV} \leq 2E_{f_0}^Y \tilde{\Pi}_{n,V}(C^c) < 2\delta,$$

using Condition 7.3.17 for the second inequality. Likewise, if $N^C(W_V,I)$ is the restricted and renormalised normal distribution, $\|N(W_V,I) - N^C(W_V,I)\|_{TV} < \delta$ almost surely, for every $\delta > 0$ and for $C = C_\delta$ a ball of large enough radius. It thus suffices to prove that

$$\|\tilde{\Pi}_{n,V}^C - N^C(W_V,I)\|_{TV} \to^{P_{f_0}^Y} 0.$$

The total variation distance $\|\tilde{\Pi}_{n,V}^C - N^C(W_V,I)\|_{TV}$ is bounded by twice

$$\int \left(1 - \frac{dN^C(W_V,I)(h)}{1_C e^{-\|h\|^2/2 + \langle h, W_V \rangle} d\tilde{\Pi}_V(h) / \int_C e^{-\|g\|^2/2 + \langle g, W_V \rangle} d\tilde{\Pi}_V(g)} \right)^+ d\tilde{\Pi}_{n,V}^C(h)$$

$$\leq \int \int \left(1 - \frac{e^{-\|g\|^2/2 + \langle g, W_V \rangle} d\tilde{\Pi}_V(g) dN^C(W_V,I)(h)}{e^{-\|h\|^2/2 + \langle h, W_V \rangle} d\tilde{\Pi}_V(h) dN^C(W_V,I)(g)} \right)^+ dN^C(W_V,I)(g) d\tilde{\Pi}_{n,V}^C(h)$$

$$\leq c \int \int \left(1 - \frac{d\tilde{\Pi}_V(g)}{d\tilde{\Pi}_V(h)} \right)^+ d\lambda_C(g) d\tilde{\Pi}_{n,V}^C(h),$$

where we used $(1-EY)^+ \le E(1-Y)^+$ in the first inequality, and where the constant $c \equiv c(W_V)$ in the preceding display is an upper bound for the density of $N^C(W_V, I)(g)$ with respect to λ_C. This constant is random but bounded in $P^Y_{f_0}$-probability since W_V is tight.

Now note that the preceding display is random through W_V only. Thus, considering convergence to zero under $P^Y_{f_0}$ amounts to considering convergence to zero under the marginal distribution $P^Y_{f_0,V}$ on the subspace V. Under $P^Y_{f_0,V}$, the variable W_V has law $N(0,I)$. We have to take the expectation of the display with respect to this law, which we denote by P_{W_V}. That is, dP_{W_V} has Lebesgue density proportional to $e^{-\|w\|^2/2}dw$ on V.

Define, for $c(V)$ a normalising constant,

$$dP^Y_C(w) = c(V)\left(\int e^{-\|k-w\|^2/2} d\tilde{\Pi}^C_V(k)\right) d\lambda(w) \tag{7.124}$$

$$= \left(\int e^{-\|k\|^2/2+\langle k,w\rangle} d\tilde{\Pi}^C_V(k)\right) dP_{W_V}(w),$$

a probability measure with respect to which dP_{W_V} is contiguous (see Exercise 7.3.6) so that it suffices to show convergence to zero under dP^Y_C instead of dP_{W_V}. The P^Y_C-expectation of the quantity in the preceding but one display equals the expectation of the integrand under

$$d\tilde{\Pi}^C_{n,V}(h) dP^Y_C(w) d\lambda_C(g) = c(V) e^{-\|h-w\|^2/2} dw\, d\tilde{\Pi}^C_V(h) d\lambda_C(g),$$

the latter identity following from Fubini's theorem and

$$\int_C e^{-(\|k\|^2/2)+\langle k,w\rangle} \frac{e^{-(\|h\|^2/2)+\langle h,w\rangle} d\tilde{\Pi}^C_V(h)}{\int e^{-(\|m\|^2/2)+\langle m,w\rangle} d\tilde{\Pi}^C_V(m)} d\tilde{\Pi}^C_V(k) e^{-(\|w\|^2/2)} dw = e^{-(\|h-w\|^2/2)} dw\, d\tilde{\Pi}^C_V(h).$$

We thus can obtain the bound, for n large enough and using that $d\Pi_V$ is continuous at and thus bounded near $\pi_V(f_0)$,

$$c' \int\int\int \left(1 - \frac{d\tilde{\Pi}_V(g)}{d\tilde{\Pi}_V(h)}\right)^+ e^{-\|h-w\|^2/2} dw\, d\lambda_C(g) d\lambda_C(h)$$

$$= c'' \int\int \left(1 - \frac{d\Pi_V(\pi_V(f_0)+g/\sqrt{n})}{d\Pi_V(\pi_V(f_0)+h/\sqrt{n})}\right)^+ d\lambda_C(g) d\lambda_C(h),$$

which converges to zero by dominated convergence and continuity of $d\Pi_V$ at $\pi_V(f_0)$. ■

Bernstein–von Mises Theorems in Negative-Order Sobolev Spaces

Recalling the results from Section 4.4.1, we now consider negative-order Sobolev spaces as a genuinely infinite-dimensional framework for BvM-type results. For basis functions ψ_{lk} from Definition 7.3.10, define

$$H_2^{s,\delta} \equiv \left\{ f : \|f\|^2_{H_2^{s,\delta}} := \sum_{l\in\mathcal{L}} \frac{a_l^{2s}}{(\log a_l)^{2\delta}} \sum_{k\in\mathcal{Z}_l} |\langle \psi_{lk}, f\rangle|^2 < \infty \right\}, \quad \delta \ge 0, s \in \mathbb{R},$$

which are Hilbert spaces satisfying the (compact) imbeddings $H_2^r \subset H_2^{r,\delta} \subset H_2^s$ for any real valued $s < r$. In particular, they contain L^2 for $s < 0$, and arguing as in Section 4.4.1, the white noise process $\mathbb{W} = (\int_0^1 \psi_{lk} dW)$ defines a tight Gaussian Borel probability measure $\mathcal{N}$ in

$$H(\delta) \equiv H_2^{-1/2,\delta},$$

for any $\delta > 1/2$, and for $f_0 \in L^2$, and then so does $\mathbb{Y} = (\int_0^1 \psi_{lk} dY)$. Similarly, any prior and posterior distribution on L^2 defines a Borel probability measure on $H(\delta)$ simply by the compact embedding $L^2 \subset H(\delta)$.

The following theorem shows that a Bernstein–von Mises theorem holds true in the space $H(\delta)$ for the product priors considered in the preceding section. Let β_S denote the bounded Lipschitz metric for weak convergence of probability measures in a metric space S (cf. Theorem 3.7.24).

Theorem 7.3.19 *Suppose that the prior Π and f_0 satisfy Condition 7.3.11 and that φ is continuous at $f_{0,lk}$ for all k,l. Let $\Pi(\cdot|Y)$ denote the posterior distribution from observing Y in white noise (7.82). Let $\tau : H(\delta) \to H(\delta)$ be the mapping $f \mapsto \sqrt{n}(f - \mathbb{Y})$, let $\Pi(\cdot|Y) \circ \tau^{-1}$ be the image of the posterior measure under τ (i.e., the law of $\sqrt{n}(f - \mathbb{Y})$) and let $\mathcal{N}$ be the Gaussian measure on $H(\delta)$, $\delta > 1/2$, which is the law of $\mathbb{W}$. Then*

$$\beta\left(\Pi(\cdot|Y) \circ \tau^{-1}, \mathcal{N}\right) \to^{P_{f_0}^Y} 0$$

as $n \to \infty$, where β is the BL metric for weak convergence in the space $H(\delta)$.

Proof A first observation is that from Theorem 7.3.12 we deduce for every $\delta' > 1/2$ and some $D > 0$ that

$$E_{f_0}^Y E_\Pi \left[\|f - f_0\|_{H(\delta')}^2 |Y\right] \le \sum_{l,k} a_l^{-1} (\log a_l)^{-2\delta'} E_{f_0}^Y E_\Pi \left[(f_{lk} - f_{0,lk})^2 |Y\right] \le D/n, \qquad (7.125)$$

which implies also, for V a fixed finite-dimensional space as in Condition 7.3.17, by continuity of the projection $\pi_V : H(\delta') \to V$, the bound

$$E_{f_0}^Y E_\Pi \left[\|f - f_0\|_V^2 |Y\right] \le D'/n, \qquad (7.126)$$

for some $D' > 0$. To prove the theorem, it is enough to show that for every $\varepsilon > 0$ there exists $N = N(\varepsilon)$ large enough such that, for all $n \ge N$,

$$P_{f_0}^Y \left(\beta(\Pi_n \circ \tau^{-1}, \mathcal{N}) > 4\varepsilon\right) < 4\varepsilon.$$

Fix $\varepsilon > 0$ and let V_J be the finite-dimensional subspace of L^2 spanned by $\{\psi_{lk} : k \in \mathcal{Z}_l, l \in \mathcal{L}, |l| \le J\}$, for an integer J. Writing $\tilde{\Pi}_n$ for $\Pi(\cdot|Y) \circ \tau^{-1}$, we see from the triangle inequality that

$$\beta(\tilde{\Pi}_n, \mathcal{N}) \le \beta(\tilde{\Pi}_n, \tilde{\Pi}_n \circ \pi_{V_J}^{-1}) + \beta(\tilde{\Pi}_n \circ \pi_{V_J}^{-1}, \mathcal{N} \circ \pi_{V_J}^{-1}) + \beta(\mathcal{N} \circ \pi_{V_J}^{-1}, \mathcal{N}).$$

The middle term converges to zero in $P_{f_0}^Y$-probability for every V_J, by Theorem 7.3.18, using that Condition 7.3.17 can be checked by (7.126) and hypothesis on φ, and since the total variation distance dominates β. Next,

$$\beta^2(\mathcal{N} \circ \pi_{V_J}^{-1}, \mathcal{N}) \le E\|\pi_{V_J}(\mathbb{W}) - \mathbb{W}\|_{H(\delta)}^2 = \sum_{l > J, k} \frac{a_l^{-1}}{(\log a_l)^{2\delta}} \to 0$$

as $J \to \infty$, so the last term in the preceding decomposition can be made as small as desired for J large.

Finally, we handle the first term in the preceding decomposition corresponding to approximate finite-dimensional concentration of the posterior measures. For $Q > 0$, consider the random subset D of $H(\delta')$ defined as

$$D = \{g : \|g + \mathbb{W}\|^2_{H(\delta')} \le Q\}.$$

Under $P^Y_{f_0}$, we have $\tilde{\Pi}_n(D) = \Pi(D_n|Y)$, where

$$D_n = \{f : \|f - f_0\|^2_{H(\delta')} \le Q/n\}.$$

Using (7.125) and Markov's inequality yields $P^Y_{f_0}(\tilde{\Pi}_n(D^c) > \varepsilon/4) \le \varepsilon$ for Q large enough.

If $X_n \sim \tilde{\Pi}_n$ (conditional on Y), then $\pi_{V_J}(X_n) \sim \tilde{\Pi}_n \circ \pi_{V_J}^{-1}$. For F any bounded function on $H(\delta)$ of Lipschitz norm less than 1,

$$\begin{aligned}\left|\int_{H(\delta)} F d\tilde{\Pi}_n - \int_{H(\delta)} F d(\tilde{\Pi}_n \circ \pi_{V_J}^{-1})\right| &= \left|E_{\tilde{\Pi}_n}\left[F(X_n) - F(\pi_{V_J}(X_n))\right]\right| \\ &\le E_{\tilde{\Pi}_n}\left[\|X_n - \pi_{V_J}(X_n)\|_{H(\delta)} 1_D(X_n)\right] + 2\tilde{\Pi}_n(D^c),\end{aligned}$$

where $E_{\tilde{\Pi}_n}$ denotes expectation under $\tilde{\Pi}_n$ (given dY). With $x_{lk} = \langle X_n, \psi_{lk}\rangle$,

$$\begin{aligned}E_{\tilde{\Pi}_n}\left[\|X_n - \pi_{V_J}(X_n)\|^2_{H(\delta)} 1_D(X_n)\right] &= E_{\tilde{\Pi}_n}\left[\sum_{l>J} a_l^{-1}(\log a_l)^{-2\delta} \sum_k |x_{lk}|^2 1_D(X_n)\right] \\ &= E_{\tilde{\Pi}_n}\left[\sum_{l>J} a_l^{-1}(\log a_l)^{2\delta'-2\delta-2\delta'} \sum_k |x_{lk}|^2 1_D(X_n)\right] \\ &\le (\log a_J)^{2\delta'-2\delta} E_{\tilde{\Pi}_n}\left[\|X_n\|^2_{H(\delta')} 1_D(X_n)\right] \le 2(\log a_J)^{2\delta'-2\delta}\left[Q + \|\mathbb{W}\|^2_{H(\delta')}\right].\end{aligned}$$

From the definition of β, we deduce that

$$\beta(\tilde{\Pi}_n, \tilde{\Pi}_n \circ \pi_{V_J}^{-1}) \le 2\tilde{\Pi}_n(D^c) + \sqrt{2}(\log a_J)^{\delta'-\delta}\sqrt{Q + \|\mathbb{W}\|^2_{H(\delta')}}.$$

Since $a_J \to \infty$ as $J \to \infty$, we conclude that $P^Y_{f_0}(\beta(\tilde{\Pi}_n, \tilde{\Pi}_n \circ \pi_{V_J}^{-1}) > \varepsilon) < 2\varepsilon$, for J large enough, combining the preceding deviation bound for $\tilde{\Pi}_n(D^c)$ and that $\|\mathbb{W}\|_{H(\delta')}$ is bounded in probability. This concludes the proof. ■

The proof of the preceding theorem gives in fact enough uniform integrability that convergence of moments (Bochner integrals)

$$\sqrt{n} E_\Pi[f - \mathbb{Y}|Y] \to^{P^Y_{f_0}} E\mathcal{N} \text{ in } H(\delta) \iff \|\bar{f}_n - \mathbb{Y}\|_{H(\delta)} = o_{P^Y_{f_0}}(1/\sqrt{n}) \qquad (7.127)$$

occurs; see Exercise 7.3.8 for details.

Bernstein–von Mises Theorems in Multiscale Spaces

We now show that a Bernstein–von Mises theorem holds also in the multiscale spaces $\mathcal{M}_0(w)$ from Section 5.2.2 if the coordinate densities φ of the product prior are sub-Gaussian. Any such product prior takes values in $\mathcal{M}_0(w)$, and from the results in

Section 5.2.2, it follows that the random variables $\mathbb{W} = (\int \psi_{lk} dW), \mathbb{Y} = (Y_{lk} = \int \psi_{lk} dY)$ define Gaussian Borel probability measures on $\mathcal{M}_0(w)$. Likewise, any probability measure on L^2 also defines a probability measure on $\mathcal{M}_0(w)$.

Theorem 7.3.20 *Consider a prior Π and f_0 on a wavelet basis $\{\psi_{lk}\}$ from Definition 7.3.10, part (b), that satisfy Condition 7.3.11 with $\sigma_l = 2^{-l(\alpha+1/2)}$, for some $\alpha > 0$, and where φ satisfies in addition that*

(a) $\varphi(x) \le Ce^{-a|x|^2}$ $\forall x \in \mathbb{R}$ for some finite positive constants a, C and
(b) φ is continuous at $\langle f_0, \psi_{lk}\rangle$ for all k, l.

Let $\Pi(\cdot\,|Y)$ denote the posterior distribution from observing Y in white noise (7.109). Let $\tau : \mathcal{M}_0(w) \to \mathcal{M}_0(w)$ be the mapping $f \mapsto \sqrt{n}(f - \mathbb{Y})$, let $\Pi(\cdot\,|Y) \circ \tau^{-1}$ be the image of the posterior measure under τ and let $\mathcal{N}$ be the Gaussian measure on $\mathcal{M}_0(w)$ which is the law of $\mathbb{W}$. Then, if w is admissible, we have

$$\beta\left(\Pi(\cdot\,|Y) \circ \tau^{-1}, \mathcal{N}\right) \to^{P^Y_{f_0}} 0,$$

as $n \to \infty$, where β is the BL metric for weak convergence in $\mathcal{M}_0(w)$.

Proof Main ideas and notation are as in the proof of Theorem 7.3.19, although stronger estimates on the marginal posterior coordinates are required to obtain a similar result in the multiscale space $\mathcal{M}_0(w)$. For π_{V_j} the projection operator onto V_j and $\tilde{\Pi}_n \equiv \Pi(\cdot\,|Y) \circ \tau^{-1}$, we decompose

$$\beta(\tilde{\Pi}_n, \mathcal{N}) \le \beta\left(\tilde{\Pi}_n, \tilde{\Pi}_n \circ \pi_{V_J}^{-1}\right) + \beta\left(\tilde{\Pi}_n \circ \pi_{V_J}^{-1}, \mathcal{N} \circ \pi_{V_J}^{-1}\right) + \beta\left(\mathcal{N}, \mathcal{N} \circ \pi_{V_J}^{-1}\right).$$

The second term converges to zero for every $J \in \mathbb{N}$ by Theorem 7.3.18, as in the proof of Theorem 7.3.19. The third term can be made as small as desired for admissible w and J large enough using

$$\beta(\mathcal{N} \circ \pi_{V_J}^{-1}, \mathcal{N}) \le E\|\pi_{V_J}(\mathbb{W}) - \mathbb{W}\|_{\mathcal{M}_0(w)} \le \sup_{l>J} \frac{\sqrt{l}}{w_l} E \sup_{k,l} \frac{|\mathbb{W}(\psi_{lk})|}{\sqrt{l}}$$

and since one shows, arguing as in Theorem 4.4.4b) and using also Theorem 2.1.20 a), that

$$E \sup_{k,l} |\mathbb{W}(\psi_{lk})|/\sqrt{l} < \infty. \tag{7.128}$$

Likewise, for the first term in the preceding decomposition, if $f \sim \Pi(\cdot\,|Y)$ conditional on Y, then it suffices to bound

$$\begin{aligned} &E^Y_{f_0} E_\Pi(\|\sqrt{n}(id - \pi_{V_J})(f - \mathbb{Y})\|_{\mathcal{M}(w)}|Y) \\ &\le \sqrt{n} \sup_{l>J} \frac{\sqrt{l}}{w_l} E^Y_{f_0} E_\Pi\left[\sup_{l>J} \frac{\max_k |\langle f - \mathbb{Y}, \psi_{lk}\rangle|}{\sqrt{l}} |Y\right]. \end{aligned} \tag{7.129}$$

The result thus follows for admissible w by choosing J large enough if we can bound the iterated expectation by a fixed constant divided by $1/\sqrt{n}$. To achieve the latter, let $j = j_n \in \mathbb{N}$ be such that

$$\sigma_j^{-1} = 2^{j(\alpha+1/2)} \simeq \sqrt{n}, \quad \sigma_l \lesssim \frac{1}{\sqrt{n}} \ \forall l > j,$$

and consider the decomposition in $\mathcal{M}_0(w)$, under $P^Y_{f_0}$,

$$\begin{aligned}\sqrt{n}(f-\mathbb{Y}) &= \sqrt{n}(\pi_{V_j}(f)-\pi_{V_j}(\mathbb{Y}))+\sqrt{n}(f-\pi_{V_j}(f))\\ &\quad+\sqrt{n}(\pi_{V_j}(f_0)-f_0)+(\pi_{V_j}(\mathbb{W})-\mathbb{W})\\ &= I+II+III+IV.\end{aligned}$$

We bound the multiscale norm from (7.129) for each of the terms I–IV separately and note that the term IV is bounded as in (7.128).

(III) This term is nonrandom, and we have, by definition of σ_l and Condition 7.3.11, for some constant $0<M<\infty$,

$$\sqrt{n}\sup_{l>j,k} l^{-1/2}|\langle f_0,\psi_{lk}\rangle| \lesssim M\sqrt{n}\sup_{l>j} l^{-1/2}\sigma_l \lesssim M/\sqrt{j}.$$

(II) For E the iterated expectation under $P^Y_{f_0}$ and $\Pi(\cdot\,|Y)$, we can bound

$$E\sup_{l>j,k} l^{-1/2}|\langle f,\psi_{lk}\rangle| \le \sum_{l>j} l^{-1/2}E\max_k|\langle f,\psi_{lk}\rangle|.$$

We are to bound the Laplace transform $E[e^{sf_{lk}}]$ for $s=t,-t$. Both cases are similar, so we focus on $s=t$,

$$E[e^{tf_{lk}}]=E\frac{\int e^{t(f_{0,lk}+(v/\sqrt{n}))}e^{-(v^2/2)+\varepsilon_{lk}v}\frac{1}{\sqrt{n}\sigma_l}\varphi\left(f_{0,lk}+(v/\sqrt{n}))\right)dv}{\int e^{-(v^2/2)+\varepsilon_{lk}v}\frac{1}{\sqrt{n}\sigma_l}\varphi\left((f_{0,lk}+(v/\sqrt{n}))\right)dv}=:E\frac{N_{lk}(t)}{D_{lk}}.$$

To bound the denominator D_{lk} from below, we apply the same technique as in the proof of Theorem 7.3.12. We first restrict the integral to $(-\sqrt{n}\sigma_l,\sqrt{n}\sigma_l)$ and notice that over this interval the argument of φ lies in a compact set; hence the function φ can be bounded below by a constant, using Condition 7.3.11. Next, we apply Jensen's inequality to obtain, for $l\ge j$, that $D_{lk}\ge e^{-C}$. To bound the numerator $N_{lk}(t)$, setting $w=f_{0,lk}+v/\sqrt{n}$ and using the subgaussianity of φ, we see that

$$\begin{aligned}EN_{lk} &\le \int e^{t\sigma_l w}E(e^{-\frac{n}{2}(w\sigma_l-f_{0,lk})^2+\varepsilon_{lk}\sqrt{n}(w\sigma l-f_{0,lk})})\varphi(w)dw\\ &\le \int e^{t\sigma_l w}\varphi(w)dw \le e^{d(\sigma_l t)^2},\end{aligned}$$

for some $d>0$. We conclude that, for some constant $D>0$ and all $t\in\mathbb{R}$,

$$E[e^{tf_{lk}}]\le De^{\sigma_l^2t^2/D},$$

so, from Lemmas 2.3.2 and 2.3.4, we deduce that

$$E\max_k|f_{lk}|\lesssim \sigma_l l^{1/2}.$$

This gives the overall bound

$$\sum_{l>j}2^{-l(1/2+\alpha)}\le 2^{-j(1/2+\alpha)}=O(1/\sqrt{n}).$$

(I) For the frequencies $l \le j_n$, we have from Proposition 7.3.14 the sub-Gaussian bound

$$E_{f_0}^Y E(e^{t\sqrt{n}(f_{lk}-Y_{lk})}|X) \le Ce^{t^2/2}. \tag{7.130}$$

Using the results from Section 2.3 and writing Pr for the law with expectation $E_{f_0}E(\cdot|Y)$, we have for all $v > 0$ and universal constants C, C' that

$$\Pr(\sqrt{n}|f_{lk} - X_{lk}| > v) \le C'e^{-Cv^2}.$$

We then bound, for M a fixed constant,

$$E_{f_0}^Y E\left(\sup_{l\le j} l^{-1/2} \max_k \sqrt{n}|f_{lk} - Y_{lk}||Y\right) \le M + \int_M^\infty \Pr\left(\sup_{l\le j,k} l^{-1/2} \max_k \sqrt{n}|f_{lk} - Y_{lk}| > u\right) du.$$

The tail integral can be further bounded as follows:

$$\sum_{l\le j,k} \int_M^\infty \Pr\left(\sqrt{n}|f_{lk} - Y_{lk}| > \sqrt{l}u\right) du \le \sum_{l\le j} 2^l \int_M^\infty e^{-Clu^2} du \lesssim \sum_{l\le j} 2^l e^{-CM^2 l} \le const,$$

for M large enough. This completes the proof. ■

Again, the proof of the preceding theorem gives in fact enough uniform integrability that convergence of moments (Bochner integrals)

$$\sqrt{n}E_\Pi[f - \mathbb{Y}|Y] \to^{P_{f_0}^Y} E\mathcal{N} \text{ in } \mathcal{M}_0(w) \iff \|\bar{f}_n - \mathbb{Y}\|_{\mathcal{M}_0(w)} = o_{P_{f_0}^Y}(1/\sqrt{n}) \tag{7.131}$$

occurs; see Exercise 7.3.8 for details.

Some Useful Facts about Weak Convergence in Probability of Posterior Measures

In Theorems 7.3.19 and 7.3.20 we have established weak convergence of the shifted and scaled random posterior measures $\tilde{\Pi}_n = \Pi(\cdot|Y) \circ \tau^{-1}$ towards the Gaussian measure $\mathcal{N}$ induced by a white noise $\mathbb{W}$ on $H(\delta)$ and $\mathcal{M}_0(w)$. Unlike in the classical finite-dimensional Bernstein–von Mises theorem (e.g., Theorem 7.3.18), however, we have not established convergence in total variation distance but only in the BL metric for weak convergence. In statistical applications, this can be a drawback since one often needs

$$\sup_{B\in\mathcal{B}} |\tilde{\Pi}_n(B) - \mathcal{N}(B)| \to^{P_{f_0}^Y} 0 \tag{7.132}$$

for sufficiently large classes of measurable sets $\mathcal{B}$. For instance, in the next subsection we will want to take $B = C_n$ a credible set of posterior measure $\tilde{\Pi}_n(C_n) = 1 - \alpha, 0 < \alpha < 1$, and the randomness of such C_n can be accommodated by taking a suitable supremum over measurable sets. Total variation convergence in $H(\delta), \mathcal{M}_0(w)$ would imply that (7.132) holds for *all* Borel sets of the respective space, which appears to be asking for too much in the infinite-dimensional setting. For instance, in the Gaussian conjugate situation (7.115), closeness of the posterior distribution to $\mathcal{N}$ in total variation distance would force these Gaussian measures to be eventually absolutely continuous to each other, which for $\sigma_l \to 0$ as $l \to \infty$ cannot be the case.

There is, however, still uniformity as in (7.132) for large classes $\mathcal{B}$ of sets. The idea is that weak convergence of measures implies uniformity in the family $\mathcal{B}$ of sets that have a uniformly regular boundary for the limiting measure. For a Borel subset A of a metric space (S,d), define

$$\partial_\epsilon A = \{x \in A : d(x,A) < \epsilon, d(x,A^c) < \epsilon\},$$

where as usual $d(x,A) = \inf_{y\in A} d(x,y)$. The proof of the following result consists of an application of standard arguments in weak convergence theory; see Exercise 7.3.7 for some hints.

Proposition 7.3.21 *Suppose that the probability measures μ_n on a separable metric space (S,d) converge weakly towards the probability measure μ. Let $\mathcal{B}$ be a family of measurable subsets of a metric space (S,d) which satisfies, as $n \to \infty$,*

$$\lim_{\epsilon\to 0} \sup_{B\in\mathcal{B}} \mu(\partial_\epsilon B) = 0. \tag{7.133}$$

Then

$$\sup_{B\in\mathcal{B}} |\mu_n(B) - \mu(B)| \to 0,$$

as $n \to \infty$.

From this proposition and Theorems 7.3.19 and 7.3.20 we can now deduce the following:

Corollary 7.3.22 *Let the conditions of Theorem 7.3.19 or Theorem 7.3.20 be satisfied, and denote by $\Pi(\cdot\,|Y)$ the posterior distribution from the corresponding theorem. Let $\mathcal{B}$ be a class of measurable subsets of $H(\delta), \mathcal{M}_0(w)$, respectively, which satisfies*

$$\lim_{\epsilon\to 0} \sup_{B\in\mathcal{B}} \mathcal{N}(\partial_\epsilon A) = 0.$$

Then, as $n \to \infty$,

$$\sup_{B\in\mathcal{B}} |\Pi(\cdot\,|Y)\circ\tau^{-1}(B) - \mathcal{N}(B)| \to 0 \tag{7.134}$$

in $P^Y_{f_0}$-probability.

Proof Suppose that the limit of $\Delta_n = \sup_{B\in\mathcal{B}} |\Pi(\cdot\,|Y)\circ\tau^{-1}(B) - \mathcal{N}(B)|$ is not zero; that is, along a subsequence of n and for some $\varepsilon_0 > 0$, we have

$$P^Y_{f_0}(\Delta_n \geq \varepsilon_0) > 0. \tag{7.135}$$

By either Theorem 7.3.19 or Theorem 7.3.20, we have for this subsequence

$$\beta(\Pi(\cdot\,|Y)\circ\tau^{-1}, \mathcal{N}) \to^{P^Y_{f_0}} 0,$$

which implies, by passing to a further subsequence if necessary, weak convergence of $\Pi(\cdot\,|Y)\circ\tau^{-1}$ towards $\mathcal{N}$ almost surely. Using Proposition 7.3.21, this implies that $\Delta_n \to 0$ almost surely along this subsequence, contradicting (7.135). ■

Balls in the spaces $H(\delta), \mathcal{M}_0(w)$ will be shown to be $\mathcal{N}$-uniformity classes, which is useful in applications to Bayesian credible/confidence sets, as we show in the next section. Further applications are discussed in the notes at the end of this chapter.

Using the same subsequence argument, one proves a continuous mapping theorem: suppose that F is a continuous mapping from either $H(\delta)$ or $\mathcal{M}_0(w)$ to a metric space

(S,d). Then, under the conditions of Theorems 7.3.19 and 7.3.20, respectively, we have, for $\tilde{\Pi}_n = \Pi(\cdot|Y)\circ\tau^{-1}$ the shifted posterior measures,

$$\beta_S(\tilde{\Pi}_n \circ F^{-1}, \mathcal{N}\circ F^{-1}) \to^{P^Y_{f_0}} 0, \tag{7.136}$$

as $n\to\infty$, where β_S is the BL metric for weak convergence in S.

Confident Bayesian Nonparametric Credible Sets

Let $B(0,t)$ be a ball of radius t in either $H(\delta)$ or $\mathcal{M}(w)$. We will show that the family $\mathcal{B} = \{B(0,t) : t\in[0,\infty)\}$ forms a uniformity class for weak convergence towards $\mathcal{N}$ in either of these spaces, and this implies the following result for posterior credible sets:

Theorem 7.3.23 *Let S equal $H(\delta)$ or $\mathcal{M}(w)$, for $\delta > 1/2$ or admissible w, respectively. Suppose that Π satisfies the conditions of Theorem 7.3.19 or Theorem 7.3.20, respectively. For $0<\alpha<1$, consider R_n such that*

$$C_n = \left\{f : \|f - T_n\|_S \le R_n/\sqrt{n}\right\},\ \Pi(C_n|Y) = 1-\alpha,$$

where either $T_n = \mathbb{Y}$ or $T_n = \bar{f}_n(Y)$ the posterior mean of $\Pi(\cdot|Y)$. Then the credible set C_n satisfies, as $n\to\infty$,

$$P^Y_{f_0}(f_0\in C_n)\to 1-\alpha,\quad R_n \to^{P^Y_{f_0}} const.$$

Proof By Exercise 2.4.4, the mapping

$$\Phi : t\mapsto \mathcal{N}(B(0,t)) = \mathcal{N}\circ(\|\cdot\|_S)^{-1}([0,t])$$

is uniformly continuous and increasing on $[0,\infty)$. In fact, the mapping is strictly increasing on $[0,\infty)$: using Theorem 2.4.5 and Corollary 2.6.18, it suffices to show that any shell $\{f : s < \|f\|_S < t\}, s<t$, contains an element of the RKHS L^2 of $\mathcal{N}$, which is obvious as L^2 is dense in S. Thus, Φ has a continuous inverse $\Phi^{-1} : [0,1)\to[0,\infty)$. Since Φ is uniformly continuous for every $\gamma>0$, there exists $\epsilon>0$ small enough that $|\Phi(t+\epsilon)-\Phi(t)|<\gamma$, for every $t\in[0,\infty)$. Now

$$\mathcal{N}(\partial_\epsilon B(0,t)) = \mathcal{N}(B(0,t+\epsilon)) - \mathcal{N}(B(0,t-\epsilon)) = |\Phi(t+\epsilon)-\Phi(t-\epsilon)| < 2\gamma,$$

for $\epsilon>0$ small enough, independently of t. We deduce that the balls $\{B(0,t)\}_{0\le t<\infty}$ form a $\mathcal{N}$-uniformity class, and from Corollary 7.3.22, we can thus conclude, with $T_n = \mathbb{Y}$, that

$$\sup_{0\le t<\infty}\left|\Pi(f : \|f-T_n\|_S \le t/\sqrt{n}|Y) - \mathcal{N}(B(0,t))\right| \to 0$$

in $P^Y_{f_0}$-probability, as $n\to\infty$. This combined with definition of C_n gives that

$$\mathcal{N}(B(0,R_n)) = \mathcal{N}(B(0,R_n)) - \Pi(f : \|f-T_n\|_S \le R_n/\sqrt{n}|Y) + 1-\alpha$$

converges to $1-\alpha$ as $n\to\infty$ in $P^Y_{f_0}$-probability, and thus, by the continuous mapping theorem,

$$R_n \to^{P^Y_{f_0}} \Phi^{-1}(1-\alpha), \tag{7.137}$$

as $n \to \infty$. Now, using this last convergence in probability,

$$\begin{aligned} P^Y_{f_0}(f_0 \in C_n) &= P^Y_{f_0}(f_0 \in B(\mathbb{Y}, R_n/\sqrt{n})) \\ &= P^Y_{f_0}(0 \in B(\mathbb{W}, R_n)) \\ &= P^Y_{f_0}(0 \in B(\mathbb{W}, \Phi^{-1}(1-\alpha))) + o(1) \\ &= \mathcal{N}(B(0, \Phi^{-1}(1-\alpha)) + o(1) \\ &= \Phi(\Phi^{-1}(1-\alpha)) + o(1) = 1 - \alpha + o(1), \end{aligned}$$

which completes the proof of the first claim. The second claim follows from the same arguments combined with convergence of moments (Exercise 7.3.8), which implies that

$$P^n_{f_0}(f_0 \in B(\bar{f}_n, R_n/\sqrt{n})) - P^n_{f_0}(f_0 \in B(\mathbb{Y}, R_n/\sqrt{n})) \to 0$$

in $P^Y_{f_0}$-probability, as $n \to \infty$. ∎

We can now proceed as in Section 6.4.2 to intersect the credible set with additional prior or posterior information. For instance, in the case of a uniform wavelet prior Π from Proposition 7.3.15, which also satisfies the conditions of the preceding theorem, we can naturally intersect C_n with the support of the posterior, which equals a ball in $C^\gamma([0,1])$.

Corollary 7.3.24 *Consider the* $1-\alpha$ *credible set*

$$\bar{C}_n = C_n \cap B_n, \quad B_n \equiv \{f : \|f\|_{C^\gamma} \le B\},$$

where C_n *is as in Theorem 7.3.23 with* $\mathcal{M}_0(w)$, *for the posterior* $\Pi_{\gamma,B}(\cdot\,|Y)$ *based on a uniform wavelet prior* $\Pi_{\gamma,B}$. *If* $Y \sim P^Y_{f_0}$ *for some fixed* f_0 *satisfying* $\|f_0\|_{C^\alpha([0,1])} < B$, *then*

$$P_{f_0}(f_0 \in \bar{C}_n) \to 1 - \alpha,$$

and the L^∞*-diameter of* C_n *satisfies*

$$|\bar{C}_n|_\infty = O_P\left((n/\log n)^{-\gamma/(2\gamma+1)} u_n\right),$$

where $u_n \to \infty$ *as slowly as desired.*

Proof Given Theorem 7.3.23, asymptotic coverage is immediate, and the $\mathcal{M}_0(w)$ diameter of $\bar{C}_n$ is of order $1/\sqrt{n}$. The rest of the proof is now the same as that of Proposition 6.4.9. ∎

Exercises

7.3.1 Prove (7.77) and (7.78).

7.3.2 Prove Theorem 7.3.5.

7.3.3 Prove Lemma 7.3.8 (See also van der Vaart and van Zanten (2008).)

7.3.4 For $f \in \mathbb{R}, \sigma^2 > 0$, let $Y|f \sim N(f, 1/n)$, and suppose that $f \sim N(0, \sigma^2)$. Show that

$$f|Y \sim N\left(\frac{\sigma^2}{\sigma^2 + 1/n} Y, \frac{\sigma^2}{n\sigma^2 + 1}\right).$$

Hint: Use (7.78).

7.3.5 Under the conditions of Corollary 7.3.13, prove that the contraction rate about f_0 in a $\|\cdot\|_{H^s_2}$ Sobolev norm, $0 < s < \gamma$, is $n^{-\gamma_s/(2\gamma+1)}$, where $\gamma_s = \gamma - s$.

7.3.6 Suppose that $P_C^Y(A_n) \to 0$ for some sequence of measurable sets, where P_C^Y is defined in (7.124). Then $P_{W_V}(A_n) \to 0$ or, in other words, P_{W_V} is contiguous with respect to P_C^n. *Hint*: Suppose that $P_C^Y(A_n) \to 0$, for a sequence of measurable sets A_n. This implies that

$$\int_{A_n} \left[\inf_{k \in C} e^{-\|k\|^2/2 + \langle k, w \rangle} \right] dP_{W_V}(w) \to 0.$$

Since C is compact, the infimum of the continuous function in the display is attained for some fixed γ in C. Thus,

$$\int_{A_n} e^{-\|\gamma\|^2/2 + \langle \gamma, w \rangle} e^{-\|w\|^2/2} d\lambda(w) = \int_{A_n} e^{-\|\gamma - w\|^2/2} d\lambda(w) \to 0.$$

Conclude by showing that $N(\gamma, I)$ and $N(0, I)$ are mutually contiguous (e.g., chapter 6 in van der Vaart (1998)).

7.3.7 Prove Proposition 7.3.21. *Hint*: Cover S by a countable partition of μ-continuity sets U_i of diameter less than δ; then $\mu_n(U_i) \to \mu(U_i)$, for all i by weak convergence. Moreover, for any U in $\mathcal{U}_\delta$, the σ-field generated by the U_i, we have

$$\sup_U |\mu_n(U) - \mu(U)| \le \sum_i |\mu_n(U_i) - \mu(U_i)| \to 0$$

by Scheffé's theorem. Deduce that we can always find a μ-uniformity class $\mathcal{V}_\delta$ such that for each $A \subset S$ there exist $V, W \in \mathcal{V}_\delta$ such that $W \subset A \subset V$ and $V \setminus W \subset \partial_\delta A$. From this observation, Proposition 7.3.21 follows easily. The result is due to Billingsley and Topsoe (1967).

7.3.8 Prove (7.127) and (7.131). *Hint*: Reduce to almost-sure weak convergence as in the proof of Corollary 7.3.22. Then, since second or exponential moments are bounded, we can use uniform integrability combined with weak convergence to deduce convergence of moments by standard arguments.

7.4 Notes

Section 7.1 The fundamental role of the Hellinger distance for estimating the distribution of a random sample was studied systematically by Le Cam (1973, 1986) and Birgé (1983, 1984). Theorem 7.1.2 is due to Birgé (1984), and the current proof is taken from Birgé (2012). The observation Theorem 7.1.4 is taken from Ghosal, Gosh and van der Vaart (2000). Recent developments in this area can be found in Birgé (2006, 2012) and Baraud (2011).

Section 7.2 The general convergence rate theory in the Hellinger distance for maximum likelihood estimators was developed in the papers by Birgé and Massart (1993) and van de Geer (1993), with important ideas dating back to Le Cam (1973). A version of Theorem 7.2.1 is due to Wong and Shen (1995), with important refinements in Birgé and Massart (1998) and van de Geer (2000). Our exposition partly follows the monograph by van de Geer (2000), where several further references and applications of the theory are given. Rates of convergence in stronger norms, such as L^∞, can be obtained by interpolation as in Theorem 7.2.10, but whether such rates are optimal is unclear. In the related Bayesian setting, optimal supremum norm convergence rates are given in Castillo (2014) – his ideas may be useful in answering this question also for MLEs.

The differential calculus of nonparametric likelihood derivatives and its connection to the asymptotic distribution of linear functionals of the NPMLE are taken from Nickl (2007), with some ideas implicit in the work of Wong and Severini (1991). The theory for the nonparametric MLE over a Sobolev ball was mostly developed in Nickl (2007), where Theorem 7.2.14 is proved and where also several applications to semiparametric functional estimation are discussed. Propositions 7.2.8 and 7.2.9 are taken from Gach and Pötscher (2011), where generalisations and applications of the results in Nickl (2007) to (simulation-based) robust statistical inference are given. The case of sieved MLEs over a Sobolev ball is treated in detail in Nickl (2009).

The maximum likelihood estimator of a monotone density was first derived in Grenander (1956), and hence is sometimes also called the Grenander estimator. While the global convergence theory in Hellinger distance for this estimator seems to require the empirical process techniques developed here, other aspects of the estimator can be analysed by more direct probabilistic tools: Prakasa Rao (1969) and Groeneboom (1985) obtain the exact pointwise limit distribution of the MLE of a monotone density, and this result is made uniform in certain subsets of $[0,1]$ in the more recent contribution Durot, Kulikov, Lopuhaä (2012). Kiefer and Wolfowitz (1976) showed under a strict curvature hypothesis on F that the distribution function $\hat{F}_n$ of the MLE of a monotone density satisfies

$$\|\hat{F}_n - F_n\|_\infty = o_P(n^{-1/2}), \quad \text{and hence } \sqrt{n}(\hat{F}_n - F_0) \to^d G_{P_0} \quad \text{in } \ell_\infty(\mathbb{R})$$

as $n \to \infty$, where F_n is the empirical distribution function. Balabdaoui and Wellner (2007) revisit this result. These results are similar in flavour (although formally different from) Theorem 7.2.20 which as such we do not have a reference for: some main ideas of the proof are implicit in Nickl (2007), and some generalisations can be found in Söhl (2015). Maximum likelihood estimators can be constructed for 'shape constraints' other than monotonicity, including (log-) concavity and convexity constraints, and in the regression setting, see Nemirovski, Polyak and Tsybakov (1985), Groeneboom, Jongbloed and Wellner (2001), Dümbgen and Rufibach (2009), Balabdaoui, Rufibach and Wellner (2009) and Doss and Wellner (2015) for some theory.

Section 7.3 A classical result on the frequentist consistency of Bayes procedures in general parameter spaces is Doob's (1949a) consistency theorem, which holds for almost all parameters under the prior. Further important references are Le Cam (1953), Freedman (1963) and Schwartz (1965), who focussed on consistency of Bayes procedures in weak metrics. Consistency in stronger metrics such as the Hellinger distance was studied in Barron, Schervish and Wasserman (1999). The general contraction theory in the Hellinger metric was developed in Ghosal, Gosh and van der Vaart (2000), Shen and Wasserman (2001), Ghosal and van der Vaart (2007) and van der Vaart and van Zanten (2008). Theorems 7.3.1 and 7.3.3, including in particular non-i.i.d. situations such as the one in Lemma 7.3.4, are due to Ghosal, Gosh and van der Vaart (2000). The approximation-theoretic approach that replaces Hellinger-type tests by general nonparametric tests from Chapter 6 was introduced in Giné and Nickl (2011), who focussed on the i.i.d. sampling setting – see also Ray (2013) for the white noise model case (including inverse problem settings). The elegant contraction theory for Gaussian process priors presented here is mostly due to van der Vaart and van Zanten (2008). Testing tools are not always approriate to obtain contraction rates, particularly not for some stronger loss functions such as supremum norm loss (see Hoffmann, Rousseau and Schmidt-Hieber (2015)). Semiparametric tools can give stronger results in specific situations; see Castillo (2014).

An explicit analysis of Gaussian product priors in the Gaussian white noise model has been undertaken in Zhao (2000) and, more recently, Giné and Nickl (2011). The relevant proof techniques are tied to the conjugate situation, and the nonconjugate analysis of the posterior in the general setting that is presented here is due to Castillo and Nickl (2013, 2014) and Castillo (2014). The proof of the finite-dimensional Bernstein–von Mises theorem based on contiguity arguments is due to Le Cam (1986), see also van der Vaart (1998). The nonparametric Bernstein–von Mises Theorems 7.3.19 and 7.3.20 and the resulting theory for confident credible sets was developed in Castillo and Nickl (2013, 2014), and these references contain several further applications as well as extensions to the more intricate i.i.d. sampling setting too. Extensions for adaptive priors of the results in Castillo and Nickl (2013, 2014) can be found in Ray (2014). It should be noted that several negative results for

Bernstein–von Mises theorems have been obtained earlier (see Cox (1993), Freedman (1999) and Leahu (2011)), but these are all relative to ℓ_2-type topologies. In this sense, the $H(\delta)$ and $\mathcal{M}(w)$ spaces can be considered the right choices for nonparamemtric Bernstein–von Mises results.

A large and important class of priors that we have not presented here is based on the Dirichlet process and variations of it; we refer to Ghosal (2010), Lijoi and Prünster (2010) and Teh and Jordan (2010) for an overview of this theory, which requires very different mathematical techniques than those presented in this book.

8

Adaptive Inference

A main motivation for the study of nonparametric models is that they do not impose potentially unrealistic finite-dimensional, or parametric, a priori restrictions. The minimax paradigm has revealed that the statistical performance of optimal nonparametric procedures depends heavily on structural properties of the parameter to be estimated and does not typically scale at the universal rate $1/\sqrt{n}$ encountered in classical parametric models. This dependence arises typically through the choice of *tuning parameters* which require choices of usually unknown aspects of the function f to be estimated, for instance, its smoothness r and the corresponding bound on the Besov norm $\|f\|_{B^r_{pq}}$. The question arises as to how fully automatic procedures that *do not* require the specification of such parameters can perform from a minimax point of view and whether procedures exist that 'adapt' to the unknown values of r, B. We shall show in this chapter that full adaptation is possible in many *testing* and *estimation* problems and that mild losses occur for some adaptive testing problems. In contrast, the theory of adaptive *confidence sets* – and, more generally, the problem of adaptive uncertainty quantification – is more intricate, and the price for adaptation can be severe unless some additional structural assumptions on the parameter space are imposed. We shall explicitly characterise the parameter regions in nonparametric models where this discrepancy between estimation and uncertainty quantification arises and reveal the underlying relationship to certain nonparametric hypothesis-testing problems.

The theory of adaptive inference in infinite-dimensional models reveals fundamental, and in this form previously unseen, information-theoretic differences between the three main pillars of statistics, that is, between estimation, testing and the construction of confidence sets. The insights drawn from the results in this chapter belong to the most intriguing statistical findings of the nonparametric theory, showcasing the genuine challenges of statistical inference in infinite dimensions. To meet this challenge, a class of 'self-similar' functions will be introduced, for which a unified theory of estimation, testing and confidence sets can be demonstrated to exist.

8.1 Adaptive Multiple-Testing Problems

In most situations encountered in Section 6.2, the construction of minimax optimal nonparametric test procedures depended strongly on regularity properties of the nonparametric model maintained – for instance, that f has $B^r_{p\infty}$-norm at most B. The crucial parameters r, B are usually not given in practice – it is thus desirable to construct an *adaptive test* of a null hypothesis H_0 which does not require knowledge of such parameters but still performs optimally for any given value of $r, B > 0$. We shall see that such adaptive tests exist for the signal-detection problem on $[0,1]$ and for the problem of testing for uniformity on $[0,1]$. When the alternative hypothesis H_1 is separated away from H_0 in L^2-distance, adaptivity comes at the expense of a marginal increase in the separation rate, which, as we will show, cannot be circumvented from a minimax point of view. However, we also show that in the case of separation in L^∞-distance, no price for adaptation has to be paid at all.

8.1.1 Adaptive Testing with L^2-Alternatives

Adaptive Minimax Signal Detection

Let us first turn to the signal-detection problem, where we wish to test

$$H_0\colon f = 0 \quad \text{vs.} \quad H_1 : \|f\|_2 \geq \rho_n$$

based on an observation

$$dY(t) = dY^{(n)}(t) = f(t)dt + \frac{\sigma}{\sqrt{n}}dW(t), \quad t \in [0,1], \tag{8.1}$$

in the Gaussian white noise model – we recall from Chapter 6 that dY has law P_f^Y on a suitable underlying sample space $\mathcal{Y}^n$, and we denote by E_f expectation under P_f^Y. The testing problem considered is thus about whether the observation has arisen from a pure Gaussian white noise or whether a sufficiently strong signal f has been present.

When the alternatives H_1 are further restricted to a ball of radius B in the Besov space $B^r_{2\infty}([0,1])$, then we have seen in Section 6.2 that the choice

$$j = j_n \in \mathbb{N}, \quad 2^{j_n} \simeq B^{2/(2r+1/2)} n^{1/(2r+1/2)}, \quad r, B > 0, \tag{8.2}$$

from before (6.36) ensures that the χ^2 test

$$\Psi_n(j) = 1\{|T_n| \geq \tau_n\}; \quad T_n = \|f_n(j)\|_2^2 - \sigma^2\frac{2^j}{n}, \quad \tau_n = \sigma^2 L\frac{2^{j/2}}{n}, \quad L > 0,$$

from Proposition 6.2.3 achieves the minimax separation rate

$$\rho_n = c' \max(1, B)^{(1/4r+1)} n^{-(r/2r+1/2)},$$

for given values of $r, B > 0$. The question arises whether a test of comparable statistical performance can be constructed that does *not* require knowledge of $r, B > 0$. We shall call any such test *adaptive* as it adapts to the unknown regularity parameters r, B of the alternative spaces H_1.

A starting point is to notice that the resolution levels j_n in (8.2) are at most of order $2^j \leq 2^{j_{\max}} \simeq n^2$, and we may thus reject H_0 as soon as *one* of the tests $\{\Psi_n(j), j \in [1, j_{\max}] \cap \mathbb{N}\}$ does. Controlling the type 1 error of this 'maximum test' will require enlarged critical values τ_n that accommodate the multiplicity of $j_{\max} \simeq \log n$ tests involved in the procedure. Enlarging the thresholds τ_n, in turn, has repercussions on the separation rates ρ_n that are required to control type 2 errors. Some analysis shows that the critical values need to be increased by a suitable power of $\log\log n$, and we shall show that the resulting increase in the separation rate is necessary from a minimax point of view for adaptation to unknown r, B.

Searching for a minimax test that adapts to the unknown smoothness $r > 0$ can be cast into the framework of Definition 6.2.1: We wish to control the maximum of type 1 and type 2 errors uniformly over the alternative space H_1 consisting of the union of all the alternatives $H_1(r,B), r,B > 0$, at separation rates $\rho_n(r,B)$ that are optimal in the sense that (6.20) holds for the given r. To focus on the main ideas, we restrict adaptation to $r \leq R$ in the following theorem, where R is arbitrary but fixed. For r-smooth functions with $r > R$, we thus obtain a performance of the adaptive test pertaining to the R-smooth case. The case of unbounded r also can be treated (see Exercise 8.1.1).

Theorem 8.1.1 *Let $R > 0$ be arbitrary. For real sequences $\rho \equiv (\rho_n(r,B) : n \in \mathbb{N}), B > 0, 0 < r \leq R$, consider testing*

$$H_0 : f = 0 \quad vs. \quad f \in H_1(\rho) = \bigcup_{0<r\leq R, B>0} H_1(r,B,\rho_n(r,B)),$$

based on observations $dY \sim P_f^Y$ in the white noise model (8.1), where

$$H_1(r,B,\rho_n(r,B)) = \left\{ f : \|f\|_{B^r_{2\infty}([0,1])} \leq B, \|f\|_2 \geq \rho_n(r,B) \right\}.$$

(a) For $n \in \mathbb{N}$ (and setting $\log\log n$ equal to an arbitrary positive constant for $n \leq e^e$), let

$$\rho^* \equiv \rho_n^*(r,B) = C \max(1,B)^{\frac{1}{4r+1}} n^{\frac{-r}{2r+1/2}} (\log\log n)^{\frac{r}{4r+1}}.$$

For every $\alpha > 0$, there exists a test $\Psi_n : \mathcal{Y}_n \to \{0,1\}$ such that, for every $n \in \mathbb{N}$ and $C > 0$ large enough,

$$E_0\Psi_n + \sup_{f \in H_1(\rho^*)} E_f(1-\Psi_n) \leq \alpha.$$

(b) Let $\rho_n(r,B)$ be any sequences such that $\rho_n(r,B) = o(\rho_n^*(r,B))$, for all r,B. Then

$$\liminf_n \inf_{\Psi_n} \left[E_0\Psi_n + \sup_{f \in H_1(\rho)} E_f(1-\Psi_n) \right] > 0,$$

where the infimum extends over all measurable functions $\Psi_n : \mathcal{Y}_n \to \{0,1\}$.

Proof We set $\sigma^2 = 1$ for notational simplicity, and let

$$\mathcal{J} = [1, j_{\max}] \cap \mathbb{N},$$

where $j_{\max} \equiv j_{\max,n}$ is a sequence of natural numbers such that $2^{j_{\max}} \simeq n^2$ for all $n \in \mathbb{N}$.

Part (a): Consider the test

$$\Psi_n = 1\{|T_n| \geq \tau_n\},$$

where

$$T_n = \max_{j \in \mathcal{J}} 2^{-j/2} \left| \|f_n(j)\|_2^2 - \frac{2^j}{n} \right|, \qquad \tau_n = \frac{L\sqrt{\log\log n}}{n}, \qquad L > 0,$$

and where $f_n(j)$ is as in (6.28) based on a R-regular wavelet basis of $L^2([0,1])$ from Section 4.3.5 (or Section 4.3.4 in the periodic case or on Haar wavelets; see Exercise 8.1.1).

Let us control the type 2 errors for f contained in any alternative $H_1(r,B,\rho_n(r,B))$: for j_n balancing

$$\frac{2^{j_n/2}\sqrt{\log\log n}}{n} \text{ and } B^2 2^{-2jr} \Rightarrow 2^{j_n} \simeq B^{2/(2r+1/2)} n^{1/(2r+1/2)} (\log\log n)^{1/(4r+1)},$$

we have

$$E_f(1-\Psi_n) \leq P_f^Y\left(\left|\|f_n(j_n)\|_2^2 - \frac{2^{j_n}}{n}\right| < L\frac{2^{j_n/2}\sqrt{\log\log n}}{n}\right) \to 0,$$

as $n \to \infty$, arguing as after (6.34) in the proof of Proposition 6.2.3, using that for C large enough depending only on L,

$$(\rho_n^*)^2 \geq 4L\frac{2^{j_n/2}\sqrt{\log\log n}}{n},$$

for this choice of j_n.

To control the type 1 error: under $f = 0$, we have, for L large enough depending only on α,

$$E_0\Psi_n = \Pr\left(\max_{j\in\mathcal{J}} 2^{-j/2}\left|\sum_{l\leq j-1}\sum_k (g_{lk}^2 - 1)\right| \geq L\sqrt{\log\log n}\right) \leq \alpha$$

by Exercise 3.1.11, noting that $\sum_{l\leq j}\sum_k (g_{lk}^2 - 1)$ is a centred sum of 2^j-many i.i.d. squared $N(0,1)$ variables. (An alternative to using that exercise is to notice that $\mathcal{J}$ consists of at most approximately $\log n$ terms and to apply the 'diagonal' case of Theorem 3.1.9 to deduce, for some universal constant $D > 0$ and n large enough, the bound

$$\sum_{j=1}^{j_{\max}} \frac{1}{D}\exp\{-DL^2\log\log n\} \lesssim \frac{\log n}{(\log n)^{DL^2}} \tag{8.3}$$

for the last probability. Choosing L large enough that this quantity converges to zero as $n \to \infty$.)

Part (b): Assume without loss of generality that $R > 2$, fix $B = 1$ and take a dissection of $[1,2]$ into $|\mathcal{S}_n| \approx \log n$ smoothness levels $\mathcal{S}_n = \{s_i\}_{i=1}^{|\mathcal{S}_n|}$, with corresponding distinct resolution levels $j_s \in \mathcal{J}$, such that

$$2^{j_s} = n^{1/(2s+1/2)}(\log\log n)^{-1/(4s+1)}, \quad s \in \mathcal{S}_n.$$

Pick σ at random from $\mathcal{S}_n$ with equal probability $1/|\mathcal{S}_n|$, and define the functions

$$f_m = \epsilon 2^{-j_\sigma(\sigma+1/2)} \sum_{k\in\mathcal{Z}_{j_\sigma}} \beta_{mk}\psi_{jk},$$

as in the proof of part (c) of Theorem 6.2.11 but now with r chosen at random through σ. By hypothesis on $\rho(r,1)$ for n large enough, these functions are in $H_1(s,1,\rho_n(s,1))$ for some $s \in [1,2]$. Part (b) of the theorem thus will follow from (6.23), where

$$Z = \frac{1}{|\mathcal{S}_n|}\sum_{\sigma\in\mathcal{S}_n} Z_\sigma,$$

and Z_σ is the average likelihood ratio from the proof of Theorem 6.2.11, part (c), for fixed $\sigma = r$. By independence of the $\{g_{jsk}, s \in \mathcal{S}_n\}$ and since $E_0 Z_\sigma = 1$, we have

$$E_0(Z-1)^2 = \frac{1}{|\mathcal{S}_n|^2} \sum_{\sigma \in \mathcal{S}_n} E_0(Z_\sigma - 1)^2.$$

Now we bound $E_0 Z_\sigma^2$ as in the proof of Theorem 6.2.11, part(c), but with

$$\gamma_n' \equiv \epsilon^2 n 2^{-j_\sigma(2\sigma+1)} \equiv 2^{-j_\sigma/2}\gamma_n \simeq 2^{-j_\sigma/2}\epsilon^2\sqrt{\log\log n}$$

and using

$$\exp\{D_0 \gamma_n'^2/2\} \lesssim (\log n)^{(D_0/2)\epsilon^2}, \quad D_0 > 0,$$

to see that, for $\epsilon > 0$ small and n large enough,

$$E_0(Z-1)^2 \lesssim \frac{(\log n)^{D_0\epsilon^2}}{|\mathcal{S}_n|} \to 0, \tag{8.4}$$

completing the proof. ■

Intuitively speaking, the penalty of order a power of $\log\log n$ arises from the fact that adapting to the unknown smoothness r is, for fixed sample size n, equivalent to an alternative space that contains approximately $\log n$ 'independent copies' of the original testing problems. Since sub-Gaussian bounds are available for each test, we can control maxima by a penalty of the order of an iterated logarithm.

Adaptive Tests of Uniformity on $[0,1]$

We now turn to the i.i.d. sampling analogue of the signal-detection problem; that is, we consider the adaptive version of the minimax test for uniformity on $[0,1]$ from Theorem 6.2.9. The situation is quite similar to the preceding and only needs adaptation of the probabilistic tools from the proof of Theorem 8.1.1.

Theorem 8.1.2 *Let $R > 0$ be arbitrary. For real sequences $\rho \equiv (\rho_n(r,B) : n \in \mathbb{N}), B > 0, 0 < r \le R$, consider testing*

$$H_0 : f = f_0 \equiv 1 \quad vs. \quad f \in H_1(\rho) = \bigcup_{0<r\le R, B>0} H_1(r,B,\rho_n(r,B)),$$

based on observations $X_1,\dots,X_n \sim^{i.i.d.} f$ on $[0,1]$, where

$$H_1(r,B,\rho_n(r,B)) = \left\{\max(\|f\|_\infty, \|f\|_{B^r_{2\infty}([0,1])}) \le B, \|f-1\|_2 \ge \rho_n(r,B)\right\}.$$

(a) For $n \in \mathbb{N}$ (and setting $\log\log n$ equal to an arbitrary positive constant for $n \le e^e$), let

$$\rho^* \equiv \rho_n^*(r,B) = C\max(1,B)^{\frac{1}{4r+1}} n^{\frac{-r}{2r+\frac{1}{2}}} (\log\log n)^{\frac{r}{4r+1}}.$$

For every $\alpha > 0$, there exists a test $\Psi_n : [0,1]^n \to \{0,1\}$ such that, for every $n \in \mathbb{N}$ and $C > 0$ large enough,

$$E_{f_0}\Psi_n + \sup_{f \in H_1(\rho^*)} E_f(1-\Psi_n) \le \alpha.$$

(b) Let $\rho_n(r,B)$ be any sequences such that $\rho_n(r,B)=o(\rho_n^*(r,B))$ for all r,B. Then

$$\liminf_n \inf_{\Psi_n}\left[E_{f_0}\Psi_n + \sup_{f\in H_1(\rho)} E_f(1-\Psi_n)\right] > 0,$$

where the infimum extends over all measurable functions $\Psi_n : [0,1]^n \to \{0,1\}$.

Proof We use the notation of the proof of Theorem 8.1.1 and sketch the necessary adaptations of it relevant for the sampling situation.

Part (a): We consider a slight modification of the test statistic used in Proposition 6.2.5, namely,

$$\Psi_n = 1\{|T_n| \geq \tau_n\},$$

where $\tau_n = L\sqrt{\log\log n}/n$, and

$$T_n = \max_{j\in\mathcal{J}} 2^{-j/2}\left|\frac{2}{n(n-1)}\sum_{i<i'}\sum_{l\leq j-1}\sum_k (\psi_{lk}(X_i)-\langle 1,\psi_{lk}\rangle)(\psi_{lk}(X_{i'})-\langle 1,\psi_{lk}\rangle)\right|,$$

where the $\{\psi_{lk}\}$ form an R-regular wavelet basis of $L^2([0,1])$ as in Section 4.3.5 (or Section 4.3.4 for the periodic case). The control of type 2 errors of this test is obtained as in the proof of Theorem 8.1.1 after direct adaptations of the arguments of Proposition 6.2.5. Likewise, type 1 errors can be bounded as in (8.3), where Theorem 3.1.9 is replaced by Theorem 3.4.8, applied as in the proof of Theorem 6.2.17, with $f = f_0, d_n = L\sqrt{\log\log n}$ (cf. also (6.81)).

To prove part (b), we take functions

$$f_m = 1 + \epsilon 2^{-j_\sigma(\sigma+1/2)}\sum_{k\in\mathcal{Z}_{j_\sigma}} \beta_{mk}\psi_{jk} \in H_1,$$

as in the proof of Theorem 6.2.9, but with σ random as in the proof of Theorem 8.1.1. Proceeding as in that proof, the standard inequality (6.23) and the proof of Theorem 6.2.9 then give a bound similar to (8.4) and hence the result. ∎

8.1.2 Adaptive Plug-in Tests for L^∞-Alternatives

We now show that the penalty $(\log\log n)^{1/(4r+1)}$ that occurred in the preceding two theorems is specific to L^2-separation – and that it does not occur when the alternative hypothesis is separated in L^∞-distance. Full adaptation is thus possible in the setting of Theorem 6.2.7 at no cost in the separation rates.

Theorem 8.1.3 *Let R be arbitrary. For*

$$\rho \equiv \rho_n(r,B) = CB^{\frac{1}{2r+1}}\left(\frac{\log n}{n}\right)^{\frac{r}{2r+1}}, \quad 0<r\leq R, B>0,$$

consider testing

$$H_0: f=0 \quad vs. \quad f\in H_1(\rho) = \bigcup_{0<r\leq R, B>0} H_1(r,B,\rho_n(r,B)),$$

based on observations $dY \sim P_f^Y$ in the white noise model (8.1), where

$$H_1(r,B,\rho_n(r,B)) = \left\{\|f\|_{B^r_{\infty\infty}([0,1])} \le B, \|f\|_\infty \ge \rho_n(r,B)\right\}.$$

For every $\alpha > 0$, there exists a test $\Psi_n : \mathcal{Y}_n \to \{0,1\}$ such that, for every $n \in \mathbb{N}$ and $C > 0$ large enough,

$$E_0\Psi_n + \sup_{f \in H_1(\rho^*)} E_f(1-\Psi_n) \le \alpha.$$

Proof We take linear wavelet estimators

$$f_n = f_n(j) = \sum_{l \le j}\sum_k \left(\int_0^1 \psi_{lk}(t)dY(t)dt\right)\psi_{lk}$$

based on an R-regular wavelet basis of $L^2([0,1])$ as in Proposition 5.1.12, where j varies in the discrete grid

$$\mathcal{J} = [1, j_{\max}] \cap \mathbb{N}, \quad 2^{j_{\max}} \sim n.$$

Consider the test

$$\Psi_n = 1\left\{\max_{j \in \mathcal{J}} \sqrt{\frac{1}{2^j j}}\|f_n(j)\|_\infty \ge \frac{L}{\sqrt{n}}\right\}.$$

If $f \in H_1(\rho)$, then $f \in H_1(r,B,\rho_n(r,B))$ for some r,B, and if $j_n \in \mathcal{J}$ is such that

$$2^{j_n} \simeq B^{2/(2r+1)}(n/\log n)^{1/(2r+1))},$$

then, for C and in turn L' large enough, we have from Proposition 5.1.7 and Markov's inequality and since $\|Ef_n(j_n) - f\|_\infty \le cB2^{-j_n r}$ that

$$\begin{aligned}
E_f(1-\Psi_n) &\le P_f^Y\left(\|f_n(j_n)\|_\infty < L\sqrt{\frac{2^{j_n}j_n}{n}}\right) \\
&\le P_f^Y\left(\|f_n(j_n) - f\|_\infty > \|f\|_\infty - L\sqrt{\frac{2^{j_n}j_n}{n}}\right) \\
&\le P_f^Y\left(\|f_n(j_n) - Ef_n(j_n)\|_\infty > L'\sqrt{\frac{2^{j_n}j_n}{n}}\right) \le \alpha/2,
\end{aligned}$$

as $n \to \infty$. For type 1 errors, we have, for L large enough,

$$\begin{aligned}
E_0\Psi_n &= P_0^Y\left(\max_{j\in\mathcal{J}}\sqrt{\frac{1}{2^j j}}\|f_n(j) - Ef_n(j)\|_\infty \ge L/\sqrt{n}\right) \\
&\le \sum_{j \in \mathcal{J}} P_0^Y\left(\|f_n(j) - Ef_n(j)\|_\infty \ge L\sqrt{\frac{2^j j}{n}}\right) \\
&\le c\sum_{j\in\mathcal{J}} e^{-L^2 j/c} \le \alpha/2
\end{aligned}$$

as $n \to \infty$, using Proposition 5.1.12. ■

A sampling analogue of the preceding theorem can be proved by standard adaptation of the last proof (see Exercise 8.1.2).

Exercises

8.1.1 Prove versions of Theorems 8.1.1 and 8.1.2, where $0 < r < \infty$ is unrestricted in the alternative hypothesis. *Hint*: Use test statistics based on Haar wavelets and Remark 6.2.4.

8.1.2 Formulate and prove a sampling analogue of Theorem 8.1.3. *Hint*: Proceed as in the proof of that theorem, but use Theorems 5.1.5 and 5.1.13 in place of Theorem 5.1.2 and Proposition 5.1.12.

8.2 Adaptive Estimation

We now turn to the crucial question of whether minimax optimal estimation of the functional objects f from an observation in white noise or the i.i.d. sampling model is possible in a fully automatic way. That is, we are searching for fully data-driven algorithms that recover the minimax estimation rates from Chapter 6 for the unknown parameter f contained in a suitable scale of functional smoothness classes. We shall show that, remarkably, for L^2- and L^∞-risk, full adaptive estimation is possible at no cost in the minimax rate of convergence. There are several methods that lead to adaptation, and we discuss two key ones: *Lepski's method* and *wavelet thresholding*. Other methods exist, and all are of a related nature, as discussed in the notes at the end of this chapter.

8.2.1 Adaptive Estimation in L^2

For both observational models considered here (white noise and i.i.d. sampling), we have seen in Proposition 5.1.7 that suitable choice of the estimation method and resolution level j_n (or related bandwidth h_n) produced estimators $f_n(j_n)$ for which the risk bound

$$\sup_{f:\|f\|_{B^s_{2\infty}} \le B} E_f \| f_n(j_n) - f \|_2 \le C \max(1,B)^{1/(2s+1)} n^{-s/(2s+1)} \tag{8.5}$$

could be established for any fixed $s, B > 0$ and some constant $C > 0$. In Theorems 6.3.8 and 6.3.9, we proved that such a bound cannot be improved in the sense that $n^{-s/(2s+1)}$ is the minimax rate of convergence in L^2-risk over any ball of functions/densities in the Besov spaces $B^s_{2\infty}$, when s is given. The estimator $f_n(j_n)$ depended, through the choice of j_n, on the values s, B, which are typically unknown to the statistician. We now show that knowledge of s, B is not necessary and that a single estimator $\hat{f}_n$ exists that achieves the performance from (8.5) for any value $s, B > 0$. We say that $\hat{f}_n$ adapts to the unknown values of s and B – from the point of view of minimax L^2-rates of convergence, there is thus no loss of information incurred from not knowing s, B.

Lepski's Method

We recall that j_n in Proposition 5.1.7 was chosen to balance the bounds obtained for bias and variance of the unbiased estimator $f_n(j)$ of the projection $K_j(f)$:

$$E_f \| f_n(j) - E_f f_n(j) \|_2^2 \le L^2 \frac{2^j}{n} \quad \text{and} \quad \|K_j(f) - f\|_2^2 \le c^2 \|f\|^2_{B^s_{2\infty}} 2^{-2js}.$$

The dependence on s, B is thus entirely induced by the nonstochastic approximation error of f by $K_j(f)$ and, roughly speaking, j is chosen to balance the antagonistic terms

$$\frac{2^j}{n} \approx B^2 2^{-2js}$$

to provide a minimax optimal procedure over a given ball $\{f : \|f\|_{B^s_{2\infty}} \leq B\}$.

A first attempt at adaptation, which has come to be known as *Lepski's method* (see the notes at the end of this chapter for historical remarks), works for many such bias-variance tradeoff situations and, in the setting of L^2-risk, is based on considering statistics

$$T_n(j,l) = \|f_n(j) - f_n(l)\|_2^2,$$

where the parameters $j < l$ vary in a finite grid $\mathcal{J}$ bounded between 'minimal' and 'maximal' resolution levels $j_{\min}, j_{\max}$, respectively. If we use wavelet estimators, we can naturally restrict to all $j \in [j_{\min}, j_{\max}]$ and take $j_{\min} = 1$ (although sometimes it will be seen that letting $j_{\min}$ diverge with n is natural). For kernel estimators, we can discretise the set of possible bandwidths h by the dyadic conversion $h = 2^{-j}, j \in \mathcal{J}$, or construct a grid of a similar nature directly. We can write $T_n(j,l)$ as

$$\|f_n(j) - f_n(l)\|_2^2 = \|f_n(j) - E_f f_n(j) - (f_n(l) - E_f f_n(l)) + K_j(f) - K_l(f)\|_2^2.$$

In the wavelet case and for observations in Gaussian white noise, this decomposes (up to a typically negligible cross-term) into

$$\sum_{\ell=j}^{l-1} \sum_k (y_{\ell k} - E_f y_{\ell k})^2 \quad \text{and} \quad \sum_{\ell=j}^{l-1} f_{\ell k}^2$$

and thus into the variance and bias term restricted to the window $[j,l]$. As there are 2^l wavelets at level l, the first term has stochastic order $2^l/n$, and hence, when $T_n(j,l)$ is of that order, it indicates that the bias is comparably small. However, if $T_n(j,l)$ is significantly larger than $2^l/n$, this indicates that the second summand in the preceding display must contribute, and hence the presence of too large approximation error, so increasing j will be necessary to obtain a minimax performance. Intuitively speaking, we should choose j as the smallest resolution level j for which $T_n(j,l), l > j$, does not significantly exceed $2^l/n$, for all $l > j$, suggesting that j is the most parsimonious 'model dimension' for which the approximation error does not dominate the stochastic error.

We detail the preceding ideas now in the setting of the Gaussian white noise model (8.1) and in the setting of periodic signals, $f \in B^{s,per}_{2\infty}([0,1])$. As shown in Section 4.3.4, this space can be defined with a periodised translation operator or, equivalently, via wavelets. Throughout we denote $B^{s,per}_{2\infty}([0,1])$ from (4.130) by $B^s_{2\infty}$ in this subsection, with wavelet norm based on periodised Meyer wavelets (cf. Remark 4.3.25 and Theorem 4.2.9). Restriction to periodicity is by no means necessary but is convenient to lay out the main ideas. Using the periodised wavelet basis

$$\{\psi_{lk}\} \equiv \{\psi_{-10} = 1, \psi_{lk} : l \geq 0, k = 0, \ldots, 2^l - 1\}$$

of $L^2([0,1])$, the observations dY can be mapped into sequence space as

$$y_{lk} = \int_0^1 \psi_{lk}(t)dY(t)dt = \langle f, \psi_{lk}\rangle + \frac{1}{\sqrt{n}}\int_0^1 \psi_{lk}(t)dW(t)$$
$$\equiv f_{lk} + \frac{1}{\sqrt{n}} g_{lk}, \quad g_{lk} \sim^{i.i.d.} N(0,1). \tag{8.6}$$

For the resulting linear wavelet estimator

$$f_n(j) = \sum_{l\leq j-1}\sum_k y_{lk}\psi_{lk}, \quad E_f f_n(j) = \sum_{l\leq j-1}\sum_k f_{lk}\psi_{lk}, \tag{8.7}$$

we then have, with $c(s) = 1/(1-2^{-2s})$ and for any $f \in B^s_{2\infty}, s > 0$ (as the Meyer-basis is S-regular for every value of $S \in \mathbb{N}$), that

$$\|E_f f_n(j) - f\|_2^2 = \|K_j(f) - f\|_2^2 \leq c(s)\|f\|^2_{B^s_{2\infty}} 2^{-2js} \equiv B(j, f), \tag{8.8}$$

as well as

$$E_f\|f_n(j) - E_f f_n(j)\|_2^2 = \frac{1}{n} E_f \sum_{l\leq j-1}\sum_k g^2_{lk} = \frac{2^j}{n}. \tag{8.9}$$

Take $j_{\max} \in \mathbb{N}$ such that $n \simeq 2^{j_{\max}} \leq n$, and define a discrete grid $\mathcal{J}$ of resolution levels

$$\mathcal{J} = \{j \in \mathbb{N} : j \in [1, j_{\max}]\} \tag{8.10}$$

which has approximately $\log n$ elements. For $f \in B^s_{2\infty}$, we define

$$j^* = j_n^*(f) = \min\left\{j \in \mathcal{J} : B(j, f) \leq \frac{2^j}{n}\right\}, \tag{8.11}$$

which, by monotonicity, balances bias and variance terms in the sense that

$$B(j, f) = c(s)2^{-2js}\|f\|^2_{B^s_{2\infty}} \leq \frac{2^j}{n}, \quad \forall j \geq j_n^*,$$
$$B(j, f) = c(s)2^{-2js}\|f\|^2_{B^s_{2\infty}} > \frac{2^j}{n}, \quad \forall j < j_n^*.$$

In particular, we see that j_n^* is such that

$$2^{j_n^*} \simeq \|f\|^{2/(2s+1)}_{B^s_{2\infty}} n^{1/(2s+1)},$$

hence theoretically producing an estimator $f_n(j_n^*)$ that is minimax optimal for f contained in a ball of $B^s_{2\infty}$. Now we estimate (8.11) from the observations, and define

$$\bar{j}_n = \min\left\{j \in \mathcal{J} : \|f_n(j) - f_n(l)\|_2^2 \leq \tau\frac{2^l}{n} \ \forall l > j, l \in \mathcal{J}\right\}, \tag{8.12}$$

where τ is a constant to be chosen. In case the set defining the preceding minimum is empty, we define by convention $\bar{j}_n = j_{\max}$. The following lemma shows that, for

n large, this procedure selects a resolution level that exceeds j_n^* only with very small probability:

Lemma 8.2.1 *Assume that $f \in B^s_{2\infty}$ for some $s > 0$, and let j_n^* be as in (8.11). Let $\bar{j}_n$ be as in (8.12), based on observations $dY \sim P_f^Y$ generated from (8.1). For some constant $C > 0$ that depends only on σ, every $\tau \geq 4$, every $j > j_n^*$ and every $n \in \mathbb{N}$, we have*

$$P_f^Y(\bar{j}_n = j) \leq \frac{1}{C} \exp\{-C\tau 2^j\}$$

and

$$P_f^Y(\bar{j}_n > j_n^*) \leq \frac{1}{C} \exp\{-C\tau 2^{j_n^*}\}.$$

Proof We set the variance $\sigma = 1$ without loss of generality. The case where $j_n^* = j_{\max}$ is obvious. Pick any $j \in \mathcal{J}, j > j_n^*$, and denote by $j^- = j - 1 \geq j_n^*$ the previous element in the grid $\mathcal{J}$. If $\bar{j}_n = j$, then one of the tests in (8.12) with j^- must have exceeded the threshold, and hence, by a union bound,

$$P_f^Y(\bar{j}_n = j) \leq \sum_{l \in \mathcal{J}: l \geq j} P_f^Y\left(\| f_n(j^-) - f_n(l) \|_2^2 > \tau \frac{2^l}{n} \right). \tag{8.13}$$

By Parseval's identity,

$$\| f_n(j^-) - f_n(l) \|_2^2 = \sum_{\ell=j^-}^{l-1} \sum_k (y_{\ell k} - E_f y_{\ell k})^2 + \sum_{\ell=j^-}^{l-1} \sum_k f_{\ell k}^2 + \frac{2}{\sqrt{n}} \sum_{\ell=j^-}^{l-1} \sum_k g_{\ell k} f_{\ell k}.$$

Since $f \in B^s_{2\infty}$ (and recalling (4.131)), we have, from (8.8), $l \geq j^- \geq j_n^*$ and the definition of j_n^*, that

$$\sum_{\ell=j^-}^{l-1} \sum_k f_{\ell k}^2 \leq c(s) \| f \|_{B^s_{2\infty}}^2 2^{-2j^- s} \leq B(j_n^*, f) \leq \frac{2^{j_n^*}}{n} \leq \frac{2^l}{n}. \tag{8.14}$$

Consequently, each probability in (8.13) is bounded from above by the sum of

$$P_f^Y \left\{ \frac{1}{n} \sum_{\ell=j^-}^{l-1} \sum_k g_{\ell k}^2 > \frac{\tau - 1}{2} \frac{2^l}{n} \right\} \leq P_f^Y \left\{ \sum_{\ell=j^-}^{l-1} \sum_k (g_{\ell k}^2 - 1) > \frac{\tau - 3}{2} 2^l \right\} \tag{8.15}$$

and

$$P_f^Y \left\{ \left| \frac{2}{\sqrt{n}} \sum_{\ell=j^-}^{l-1} \sum_k g_{\ell k} f_{\ell k} \right| > \frac{\tau - 1}{2} \frac{2^l}{n} \right\} = \Pr \left\{ |Z(j^-)| > \frac{\tau - 1}{4} \frac{2^l}{\sqrt{n}} \right\}, \tag{8.16}$$

where $Z(j^-)$ is a Gaussian random variable with variance $\sum_{\ell=j^-}^{l-1} \sum_k f_{lk}^2 \leq B(j_n^*, f)$.

Each probability in (8.15) is, by Theorem 3.1.9, less than or equal to a constant multiple of

$$\exp\left\{ -D \frac{(\tau - 3)^2 2^{2l}}{(2^l + (\tau - 3) 2^l)} \right\} \leq e^{-c\tau 2^l}$$

for universal constants c, D, and by (8.14) and a standard Gaussian tail bound, the probability in (8.16) is bounded by a constant multiple of

$$\exp\left\{-D'\frac{(\tau-1)^2 2^{2l}}{nB(j_n^*, f)}\right\} \le e^{-c'\tau 2^l},$$

for some constants $D', c' > 0$. We thus obtain the overall bounds

$$P_f^Y(\bar{j}_n = j) \le \sum_{l=j}^{j_{\max}} (1/c)e^{-c\tau 2^l}, \quad j > j_n^*, \quad P_f^Y(\bar{j}_n > j_n^*) \le \sum_{j > j_n^*} \sum_{l=j}^{j_{\max}} (1/c)e^{-c\tau 2^l}.$$

The result follows by summing these series. ■

We cannot control in general the probability that $\bar{j}_n < j_n^*$ in a similar way (unless we make additional assumptions on f; see Lemma 8.3.17), but for adaptive estimation of f by $\hat{f}_n = f_n(\bar{j}_n)$, this is immaterial, as the proof of the following theorem shows.

Theorem 8.2.2 *The estimator $\hat{f}_n = f_n(\bar{j}_n)$ with f_n as in (8.7), $\bar{j}_n$ as in (8.12) and any choice $\tau > 4$, based on observations in a Gaussian white noise model (8.1), satisfies, for all $s, B > 0$ fixed and every $n \in \mathbb{N}$,*

$$\sup_{f : \|f\|_{B^s_{2\infty}} \le B} E_f \|\hat{f}_n - f\|_2 \le D \max(1, B)^{1/(2s+1)} n^{-s/(2s+1)},$$

where D is a constant that depends only on s, σ, τ.

Proof Consider the cases $\{\bar{j}_n \le j_n^*\}$ and $\{\bar{j}_n > j_n^*\}$ separately. First, by the definition of $\bar{j}_n, j_n^*$ and (8.8), (8.9), for some constant $D > 0$,

$$\begin{aligned} E_f \|f_n(\bar{j}_n) - f\|_2^2 I_{\{\bar{j}_n \le j_n^*\}} &\lesssim E_f \left(\|f_n(\bar{j}_n) - f_n(j_n^*)\|_2^2 + \|f_n(j_n^*) - f\|_2^2\right) I_{\{\bar{j}_n \le j_n^*\}} \\ &\le \tau \frac{2^{j_n^*}}{n} + E_f \|f_n(j_n^*) - E_f f_n(j_n^*)\|_2^2 + \|K_{j_n^*}(f) - f\|_2^2 \qquad (8.17) \\ &\le (D/2)^2 \max(1, B)^{2/(2s+1)} n^{-2s/2s+1}. \end{aligned}$$

On the event $\{\bar{j}_n > j_n^*\}$, we can use (8.9), the Cauchy-Schwarz inequality and Lemma 8.2.1 to see that

$$\begin{aligned} E_f \|f_n(\bar{j}_n) - f\|_2 I_{\{\bar{j}_n > j_n^*\}} &\le \sum_{j \in \mathcal{J} : j > j_n^*} \left(E_f \|f_n(j) - f\|_2^2\right)^{1/2} \left(E_f I_{\{\bar{j}_n = j\}}\right)^{1/2} \\ &\le \sum_{j \in \mathcal{J} : j > j_n^*} \left[\sqrt{\frac{2^j}{n}} + \sqrt{B(j, f)}\right] \cdot \sqrt{P_f^Y(\bar{j}_n = j)} \\ &\lesssim \sum_{j \in \mathcal{J} : j > j_n^*} \sqrt{P_f^Y(\bar{j}_n = j)} \\ &\le \sqrt{j_{\max} P_f^Y(\bar{j}_n > j_n^*)} = o(n^{-s/2s+1}), \end{aligned}$$

completing the proof. ■

The preceding theorem holds in the i.i.d. sampling model just as well: if $X_1,\dots,X_n$ are i.i.d. from bounded density $f:[0,1]\to[0,\infty)$, we take in place of (8.7) the density estimators

$$f_n(j)=\sum_{l\le j-1}\sum_k y_{lk}\psi_{lk},\quad \text{where now}\quad y_{lk}=\int_0^1\psi_{lk}(t)dP_n(t),\quad P_n=\frac{1}{n}\sum_{i=1}^n\delta_{X_i}. \tag{8.18}$$

The procedure defining $\hat f_n=f_n(\bar j_n)$ can be adapted directly, but the thresholds need to be modified slightly because the stochastic error $f_n(j)-E_f f_n(j)$ depends now, unlike in the white noise case, also on the unknown density f. If f is bounded by a constant U, we can take

$$\tau=\tilde\tau U,\quad U\ge 1, \tag{8.19}$$

in (8.12). Clearly, U will usually not be available in practice, but we can replace it by $\|f_n(\tilde j_n)\|_\infty$, with $\tilde j_n\in\mathbb{N}$ chosen such that $2^{\tilde j_n}\sim(n/\log^2 n)$ – the proof of the following theorem goes through with this random choice of τ_n too (see Exercise 8.2.1), and hence we restrict to U known for simplicity. The choice of the constant $\tilde\tau$ is discussed in the notes at the end of this chapter. We also note that for probability densities, we always have $\int f=1$ and hence $\|f\|_{B^s_{2\infty}}\ge 1$ at least, so the restriction to $B\ge 1$ is natural in the following theorem:

Theorem 8.2.3 *Consider the estimator $\hat f_n=f_n(\bar j_n)$ with f_n as in (8.18), based on i.i.d. observations $X_1,\dots,X_n$ from density f on $[0,1]$. Let $\bar j_n$ as in (8.12) with τ as in (8.19) and $\tilde\tau$ a large enough universal constant. Then, for all $s>0,B\ge 1,U\ge 1$ and every $n\in\mathbb{N}$,*

$$\sup_{f:\|f\|_{B^s_{2\infty}}\le B,\|f\|_\infty\le U} E_f\|\hat f_n-f\|_2\le DB^{1/(2s+1)}n^{-s/(2s+1)},$$

where D is a constant that depends only on $s,U,\tilde\tau$.

Proof Given the following lemma the proof of the theorem is the same as that of Theorem 8.2.2, noting that (8.9) also holds (up to multiplicative constants) for the density estimator $f_n(j)$ by Theorem 5.1.5. ■

Lemma 8.2.4 *Let $X_1,\dots,X_n$ be drawn i.i.d. on $[0,1]$ from density $f\in B^s_{2\infty}, s>0$, satisfying $\|f\|_\infty\le U$. Let j_n^* be as in (8.11). Then, for every $\tilde\tau$ large enough, every $j>j_n^*$ and every $n\in\mathbb{N}$, there exists a universal constant $C>0$ such that we have*

$$P_f^{\mathbb{N}}(\bar j_n=j)\le\frac{1}{C}\exp\{-C\tilde\tau\min(2^j,\sqrt{n})\}$$

and

$$P_f^{\mathbb{N}}(\bar j_n>j^*)\le\frac{1}{C}\exp\{-C\tilde\tau\min(2^{j^*},\sqrt{n})\}.$$

Proof The proof is similar to that of Lemma 8.2.1: using the triangle inequality and Parseval's identity,

$$\|f_n(l)-f_n(j^-)\|_2\le\|f_n(l)-E_f f_n(l)-(f_n(j^-)-E_f f_n(j^-))\|_2+\sqrt{\sum_{\ell=j^-}^{l-1} f_{lk}^2},$$

which combined with (8.14) allows us to bound $P_f^{\mathbb{N}}(\bar{j}_n = j), j > j_n^*$, by

$$\sum_{l \in \mathcal{J}, l \geq j} P_f^{\mathbb{N}}\left(\|f_n(l) - E_f f_n(l) - (f_n(j^-) - E_f f_n(j^-))\|_2 > (\sqrt{\tilde{\tau} U} - 1)\sqrt{\frac{2^l}{n}} \right)$$

$$\leq \sum_{l \in \mathcal{J}, l \geq j} P_f^{\mathbb{N}}\left(n\|f_n(l) - E_f f_n(l)\|_2 > \sqrt{\tilde{\tau} n 2^{l-1} \|f\|_\infty} \right)$$

$$+ \sum_{l \in \mathcal{J}, l \geq j} P_f^{\mathbb{N}}\left(n\|f_n(j^-) - E_f f_n(j^-)\|_2 > \sqrt{\tilde{\tau} n 2^{l-1} \|f\|_\infty} \right),$$

for $\tilde{\tau}$ large enough. The second inequality in Theorem 5.1.13 with $p = 2$, $x = c\tilde{\tau}\min(2^l, \sqrt{n})$, $\tilde{\tau}$ large enough and suitable $c > 0$ (noting that $2^j \lesssim n$ for $j \in \mathcal{J}$ and that the L^p-norms of Φ in that Theorem can be taken to be bounded by a universal constant for the fixed periodised Meyer wavelet basis) implies a bound of order $e^{-c'\tilde{\tau}\min(2^l, \sqrt{n})}$ for each of the summands in the last term. The remainder of the proof of the lemma again proceeds as in Lemma 8.2.1. ■

8.2.2 Adaptive Estimation in L^∞

We now replace L^2-risk by the (on $[0,1]$ strictly dominant) L^∞-risk and investigate the problem of adaptation, where quite naturally the spaces $B^s_{2\infty}$ from the preceding subsection are replaced by the Hölder-Besov spaces $B^s_{\infty\infty}$. Both in Gaussian white noise and in the i.i.d. sampling model, we have seen in Proposition 5.1.7 that linear estimators $f_n(j_n)$ can achieve the risk bound

$$\sup_{f : \|f\|_{B^s_{\infty\infty}} \leq B} E_f \|f_n(j_n) - f\|_\infty \leq C \max(1, B)^{1/2s+1} \left(\frac{\log n}{n}\right)^{s/(2s+1)}, \tag{8.20}$$

for any fixed $s, B > 0$, some constant $C > 0$ and every $n \geq 2, n \in \mathbb{N}$. This is minimax optimal over balls in $B^s_{\infty\infty}$ in view of Theorems 6.3.5 and 6.3.7.

Just as earlier, the construction of this estimator depends on knowledge of the value s, B, and the question arises as to whether such knowledge can be circumvented by an adaptive estimator. The answer is, as in the L^2-setting, affirmative, and Lepski's method can be used to this effect (another method will be discussed later). Let us quickly describe the necessary modifications in the setting of Gaussian white noise with $\sigma = 1$ – the sampling model case is treated in Exercise 8.2.2. We consider as earlier in (8.7) the wavelet estimator $f_n(j)$ based on periodised Meyer wavelets. If $f \in B^s_{\infty\infty}$, for any $s > 0$, then, by the results in Section 4.3.4,

$$\|E_f f_n(j) - f\|_\infty = \|K_j(f) - f\|_\infty \leq c(s)\|f\|_{B^s_{\infty\infty}} 2^{-js} \equiv B(j, f), \tag{8.21}$$

as well as, for some universal constant $c_0 > 0$,

$$E_f \|f_n(j) - E_f f_n(j)\|_\infty \leq c_0 \sqrt{\frac{2^j j}{n}}, \tag{8.22}$$

from Theorem 5.1.5.

Take integers $1 \leq j_{\max} \in \mathbb{N}$ such that $(n/\log n) \simeq 2^{j_{\max}} \leq (n/\log n)$ and

$$\mathcal{J} = \{j \in \mathbb{N} : j \in [1, j_{\max}]\}. \tag{8.23}$$

For $f \in B^s_{\infty\infty}$, $B(j,f)$ as in (8.21), we define

$$j^* = j_n^*(f) = \min\left\{j \in \mathcal{J} : B(j,f) \le c_0\sqrt{\frac{2^j j}{n}}\right\}, \tag{8.24}$$

which is such that

$$2^{j_n^*} \simeq \|f\|_{B^s_{\infty\infty}}^{2/(2s+1)} (n/\log n)^{1/(2s+1)},$$

hence theoretically producing an estimator $f_n(j_n^*)$ that is minimax optimal for estimating f contained in a ball of $B^s_{\infty\infty}$ in supnorm loss. Following (8.12), the estimated resolution level is

$$\bar{j}_n = \min\left\{j \in \mathcal{J} : \|f_n(j) - f_n(l)\|_\infty \le \tau\sqrt{\frac{2^l l}{n}} \quad \forall l > j, l \in \mathcal{J}\right\}, \tag{8.25}$$

where τ is a threshold constant. In case the set defining the preceding minimum is empty, we simply set $\bar{j}_n = j_{\max}$.

Theorem 8.2.5 *The estimator $\hat{f}_n = f_n(\bar{j}_n)$, with f_n as in (8.7) and $\bar{j}_n$ as in (8.25) based on observations in a Gaussian white noise model (8.1), satisfies, for τ large enough depending only on an upper bound $S > s$, all $0 < s < S, B > 0$ and every $n \in \mathbb{N}$,*

$$\sup_{f : \|f\|_{B^s_{\infty\infty}} \le B} E_f \|\hat{f}_n - f\|_\infty \le D \max(1,B)^{1/(2s+1)} (n/\log n)^{-s/(2s+1)},$$

where D is a constant that depends only on s, σ, τ.

Proof Given the proof of Theorem 8.2.2, all that is needed is an analogue of Lemma 8.2.1, given by the following result, which, due to the L^∞-structure, has a slightly 'worse' (but still good enough) exponential tail in j or j^*:

Lemma 8.2.6 *Assume that $f \in B^s_{\infty\infty}$ for some $s > 0$, and let j_n^* be as in (8.24). For some constant $C > 0$ that depends only on σ, every τ large enough, $j > j_n^*$ and every $n \in \mathbb{N}$, we have*

$$P_f^Y(\bar{j}_n = j) \le \frac{1}{C}\exp\{-C\tau j\}$$

and

$$P_f^Y(\bar{j}_n > j^*) \le \frac{1}{C}\exp\{-C\tau j^*\}.$$

Proof With j, j^*, j^- as before (8.13), we have

$$P_f^Y(\bar{j}_n = j) \le \sum_{l \in \mathcal{J} : l \ge j} P_f^Y\left(\left\|f_n(j^-) - f_n(l)\right\|_\infty > \tau\sqrt{\frac{2^l l}{n}}\right). \tag{8.26}$$

By the triangle inequality,

$$\left\|f_n(j^-) - f_n(l)\right\|_\infty \le \|f_n(j^-) - E_f f_n(j^-) - (f_n(l) - E_f f_n(l))\|_\infty + \|K_l(f) - K_{j^-}(f)\|_\infty.$$

Since $f \in B^s_{\infty\infty}$ (and recalling (4.131)), we have, from (8.26), $l \geq j^- \geq j^*$ and the definition of j^*, that

$$\|K_l(f) - K_{j^-}(f)\|_\infty \leq c(s)\|f\|_{B^s_{2\infty}} 2^{-j^- s} \leq B(j^*, f) \leq \sqrt{\frac{2^{j^*} j^*}{n}} \leq \sqrt{\frac{2^l l}{n}}.$$

Consequently, each probability in (8.26) is bounded from above by

$$P_f^Y\left\{\|f_n(j^-) - E_f f_n(j^-) - (f_n(l) - E_f f_n(l))\|_\infty > \frac{\tau}{2}\sqrt{\frac{2^l l}{n}}\right\}.$$

For τ large enough, we can apply Proposition 5.1.12 with $p = \infty$, $x = c\tau l$ to the last probability and obtain the overall bound

$$P_f^Y(\bar{j}_n > j_n^*) \leq \sum_{j > j^*} \sum_{l=j}^{j_{\max}} \sum_{\ell=j}^{l} (1/c) e^{-c\tau l}$$

The result follows from summing the preceding series. ■

The rest of the proof is as in Theorem 8.2.2, noting that, for τ large enough, we can make $\Pr(\hat{j}_n = j), j > j_n^*$, asymptotically offset any polynomial growth in n. ■

Wavelet Thresholding

An alternative to Lepski's method to construct an adaptive estimate is *wavelet thresholding*, particularly suited to the situation of L^∞-adaptation. Let us discuss the main ideas again first with observations dY in the Gaussian white noise model (8.1). Instead of attempting to select the correct resolution level j^* at which the linear wavelet estimator (8.7) should be truncated, we take the full estimated wavelet series $f_n(j_{\max})$, where

$$j_{\max} \in \mathbb{N}, \quad 2^{j_{\max}} \simeq n.$$

In the expansion

$$f_n(j_{\max}) = \sum_{l \leq j_{\max}-1} \sum_k Y_{lk}\psi_{lk}, \quad Y_{lk} = \int_0^1 \psi_{lk}(t)dY(t),$$

however, we only retain the estimated coefficients Y_{lk} whose absolute values exceed a certain threshold. The idea is that we should not keep the coefficients whose estimated values suggest that $f_{lk} = \langle f, \psi_{lk}\rangle$ is actually zero. This leads to the *thresholded wavelet estimator*

$$f_n^T(x) = \sum_{l \leq j_{\max}-1} \sum_k Y_{lk} 1\{|Y_{lk}| > \tau_n\}\psi_{lk}(x), \quad x \in [0,1], \quad \tau_n \equiv \tau\sqrt{\frac{\log n}{n}}, \tag{8.27}$$

where τ is a thresholding constant to be specified, and where the $\{\psi_{lk}\}$ constitute a periodised Meyer wavelet basis of $L^2([0,1])$ (general S-regular wavelet bases can be used as well if we restrict adaptation to unknown smoothness less than S). We note that just like a Lepski-type estimator $f_n(\bar{j}_n)$ from above, f_n^T is nonlinear in the observations dY.

The following theorem implies that this estimator is fully rate adaptive, just as the estimator from Theorem 8.2.5, and provides a perhaps conceptually simpler solution of the adaptation problem. Again, Besov spaces are to be understood as periodic spaces in what follows.

Theorem 8.2.7 *Consider observations in the Gaussian white noise model (8.1). The estimator f_n^T from (8.27) with τ a large enough constant satisfies, for all $s, B > 0$ fixed and every $n \geq 2$,*

$$\sup_{f:\|f\|_{B^s_{\infty\infty}} \leq B} E_f \|f_n^T - f\|_\infty \leq D \max(1,B)^{1/(2s+1)} \left(\frac{\log n}{n}\right)^{s/(2s+1)},$$

where D is a constant that depends only on s, σ, τ.

Proof We will prove the result for unspecified constants to simplify the exposition. We will write f_{lk} for $\langle f, \psi_{lk}\rangle$ and use repeatedly that, for constants c, C, C' and $g_{lk} \sim N(0,1)$,

$$E_f \max_k |Y_{lk} - f_{lk}| = \frac{1}{\sqrt{n}} E \max_k |g_{lk}| \leq c\sqrt{\frac{l}{n}},$$
$$\left\|\sum_k |\psi_{lk}|\right\|_\infty \leq C 2^{l/2}$$

and

$$\|f\|_{B^s_{\infty\infty}} \leq B \Rightarrow \max_k |f_{lk}| \leq C' B 2^{-l(s+1/2)} \quad \forall l, \tag{8.28}$$

in view of the results in Sections 2.3 and 4.3.4.

We first have, for all $x \in [0,1]$,

$$|f_n^T(x) - f(x)| \leq C \sum_{l \geq j_{\max}} 2^{l/2} |f_{lk}| + \left| \sum_{l \leq j_{\max}-1} \sum_k (Y_{lk} 1\{|Y_{lk}| > \tau_n\} - f_{lk})\psi_{lk}(x) \right|.$$

By the preceding decay estimate (8.28) on the $|f_{lk}|$, and since $s > s/(2s+1)$, the first term is seen to be of order

$$2^{-j_{\max} s} \simeq n^{-s} = o\left(\frac{\log n}{n}\right)^{s/(2s+1)}.$$

We decompose the second term as

$$\begin{aligned}
&\sum_{l \leq j_{\max}-1} \sum_k (Y_{lk} - f_{lk})\psi_{lk} (1\{|Y_{lk}| > \tau_n, |f_{lk}| > \tau_n/2\} + 1\{|Y_{lk}| > \tau_n, |f_{lk}| \leq \tau_n/2\}) \\
&\quad - \sum_{l \leq j_{\max}-1} \sum_k f_{lk}\psi_{lk} (1\{|Y_{lk}| \leq \tau_n, |f_{lk}| > 2\tau_n\} + 1\{|Y_{lk}| \leq \tau_n, |f_{lk}| \leq 2\tau_n\}) \\
&= I + II + III + IV
\end{aligned}$$

and treat these four terms separately.

About term (I): Let $j_1(s) \leq j_{\max}$ be such that

$$2^{j_1(s)} \simeq \|f\|_{B^s_{\infty\infty}}^{2/(2s+1)} (n/\log n)^{1/(2s+1)}.$$

Then, by the estimates at the beginning of the proof,

$$
\begin{aligned}
E_f & \left\| \sum_{l \le j_1(s)-1} \sum_k (Y_{lk} - f_{lk}) \psi_{lk} I_{[Y_{lk}|>\tau_n, |f_{lk}|>\tau_n/2]} \right\|_\infty \\
& \le \sum_{l \le j_1(s)-1} \left\| \sum_k |\psi_{lk}| \right\|_\infty E_f \max_k |Y_{lk} - f_{lk}| \\
& \le cC \sum_{l \le j_1(s)-1} \sqrt{\frac{2^l l}{n}} = O\left(\left(\frac{\log n}{n} \right)^{s/(2s+1)} \right).
\end{aligned}
$$

For the second part of (I), using the definition of τ_n, we have

$$
\begin{aligned}
E_f & \left\| \sum_{l=j_1(s)}^{j_{\max}-1} \sum_k (Y_{lk} - f_{lk}) \psi_{lk} I_{[|Y_{lk}|>\tau_n, |f_{lk}|>\tau_n/2]} \right\|_\infty \\
& \le \sum_{l=j_1(s)}^{j_{\max}-1} E_f \max_k |Y_{lk} - f_{lk}| \frac{2}{\tau} \sqrt{\frac{n}{\log n}} \max_k |f_{lk}| \left\| \sum_k |\psi_{lk}| \right\|_\infty \\
& \le C \sum_{l=j_1(s)}^{j_{\max}-1} 2^{-ls} = O\left(\left(\frac{\log n}{n} \right)^{s/(2s+1)} \right).
\end{aligned}
$$

For term (II), we have, using the Cauchy-Schwarz inequality and

$$
\{|Y_{lk}| > \tau_n, |f_{lk}| \le \tau_n/2\} \subset \{|Y_{lk} - f_{lk}| > \tau_n/2\}, \tag{8.29}
$$

for τ large enough,

$$
\begin{aligned}
E_f & \left\| \sum_{l \le j_{\max}-1} \sum_k (Y_{lk} - f_{lk}) \psi_{lk} I_{[|Y_{lk}|>\tau_n, |f_{lk}|\le\tau_n/2]} \right\|_\infty \\
& \le \sum_{l \le j_{\max}-1} \sum_k \|\psi_{lk}\|_\infty \sqrt{E_f(|Y_{lk} - f_{lk}|)^2} \sqrt{P_f^Y(|Y_{lk} - f_{lk}| > \tau_n/2)} \\
& \lesssim \sum_{l \le j_{\max}-1} 2^{3l/2} n^{-1/2} e^{-\tau^2 \log n/16} = o\left(\left(\frac{\log n}{n} \right)^{s/(2s+1)} \right)
\end{aligned}
$$

since, using the standard Gaussian tail inequality,

$$
P_f^Y(|Y_{lk} - f_{lk}| > \tau_n/2) = \Pr(|g_{lk}| > \tau\sqrt{\log n}/2) \le 2\exp\left\{-\tau^2 \log n/8\right\}.
$$

For term (*III*), using the same Gaussian tail estimate and an inclusion similar to (8.29),

$$E_f \left\| \sum_{l \le j_{\max}-1} \sum_k f_{lk} \psi_{lk} I\{|Y_{lk}| \le \tau_n, |f_{lk}| > 2\tau_n\} \right\|_\infty$$
$$\le \sum_{l \le j_{\max}-1} \sum_k \|\psi_{lk}\|_\infty |f_{lk}| P_f^Y(|Y_{lk} - f_{lk}| > \tau_n)$$
$$\le D' \sum_{l \le j_{\max}-1} 2^l 2^{-ls} e^{-\tau^2 \log n/2} = o\left(\left(\frac{\log n}{n}\right)^{s/(2s+1)}\right),$$

for τ large enough.

Finally, for term (*IV*), we have

$$\left\| \sum_{l \le j_{\max}-1} \sum_k f_{lk} \psi_{lk} I_{[|Y_{lk}| \le \tau_n, |f_{lk}| \le 2\tau_n]} \right\|_\infty \le \sum_{l \le j_{\max}-1} \left\| \sum_k |\psi_{lk}| \right\|_\infty \max_k |f_{lk}| I_{[|f_{lk}| \le 2\tau_n]}$$
$$\le CC' \sum_{l \le j_{\max}-1} \min(2^{l/2}\tau_n, 2^{-ls}).$$

If $j_1(s)$ is as above, then we can estimate the last quantity by

$$c\sqrt{\frac{\log n}{n}} \sum_{l \le j_1(s)-1} 2^{l/2} + c \sum_{j_1(s) \le l \le j_{\max}-1} 2^{-ls} = O\left(\left(\frac{\log n}{n}\right)^{s/(2s+1)}\right), \tag{8.30}$$

completing the proof. ■

A version of the preceding theorem in the i.i.d. sampling model on $[0,1]$ can be proved by replacing Gaussian tail inequalities for centred wavelet coefficients Y_{lk} by Bernstein's inequality (i.e., by Theorem 3.1.7), using that the ψ_{lk} are all uniformly bounded. When the density f does not have compact support, the situation is somewhat more difficult because the sums over k at each resolution level l are then not finite any longer – Bernstein's inequality then has to be replaced by concentration inequalities for suprema of empirical processes, as will be done in the proof of the next theorem.

For i.i.d. observations $X_1, \dots, X_n$ from density f on the real line, the thresholding density estimator is

$$f_n^T(x) = \sum_{l \le J_{\max}-1} \sum_{k \in \mathbb{Z}} Y_{lk} 1\{|Y_{lk}| > \tau_n\} \psi_{lk}(x), \quad x \in \mathbb{R}, \tag{8.31}$$

with

$$Y_{lk} = \frac{1}{n} \sum_{i=1}^n \psi_{lk}(X_i), \quad J_{\max} \in \mathbb{N}, \quad 2^{J_{\max}} \simeq \frac{n}{\log n}, \quad \tau_n \equiv \tau\sqrt{U}\sqrt{\frac{\log n}{n}},$$

where U is a bound on $\|f\|_\infty$. If no such bound U is available, it can be replaced by an estimate (see Exercise 8.2.1).

If we use compactly supported (e.g., Daubechies) wavelets in the construction of the preceding estimator, then the sums over k in the definition of f_n^T are all finite, and computation of f_n^T is straightforward. This comes at the expense of adapting only up to smoothness S in the following theorem, where S is the regularity of the wavelet basis. If we choose an infinitely regular wavelet basis such as the Meyer basis, then, as in the preceding results, adaptation holds for all $s > 0$, but the sums over k are infinite (which, in practice, is no problem either because the localisation of the Meyer wavelet implies that only finitely many Y_{lk} are nonzero within machine precision). We note again that f being a density, the assumption $B \geq 1$ is natural in the following theorem:

Theorem 8.2.8 *Consider i.i.d. observations $X_1,\dots,X_n$ from density $f:\mathbb{R}\to[0,\infty)$. The estimator f_n^T from (8.31), with τ a large enough constant and based on S-regular wavelets, satisfies, for all $B \geq 1$, $0 < s < S$ and every $n \geq 2$,*

$$\sup_{f:\|f\|_{B^s_{\infty\infty}(\mathbb{R})}\leq B} E_f\|f_n^T - f\|_\infty \leq DB^{1/(2s+1)}(n/\log n)^{-s/(2s+1)},$$

where D is a constant that depends only on s.

Proof The proof is similar to that of Theorem 8.2.7, with some modifications pertaining to the possibly unbounded support of f. Particularly, we will use repeatedly that

$$E_f \sup_{k\in\mathbb{Z}}|Y_{lk} - f_{lk}| \leq \sqrt{E_f\left(\sup_{k\in\mathbb{Z}}|Y_{lk} - f_{lk}|\right)^2} \leq c\sqrt{l/n}, \tag{8.32}$$

for any bounded density f (see Proposition 5.1.8 and Remark 5.1.16) that $\left\|\sum_{k\in\mathbb{Z}}|\psi_{lk}|\right\|_\infty \lesssim 2^{l/2}$ and that (8.28) holds with suprema over $k\in\mathbb{Z}$ as well when $f\in B^s_{\infty\infty}(\mathbb{R})$. As in the preceding proof, we have

$$E_f\|f_n^T - f\|_\infty \leq E_f\left\|\sum_{l\leq J_{\max}-1}\sum_k (Y_{lk}1_{|Y_{lk}|>\tau_n} - f_{lk})\psi_{lk}\right\|_\infty + O(2^{-J_{\max}s}),$$

the second 'bias' term is negligible, and the quantity inside the expectation of the supremum of the second term can be decomposed as

$$\sum_{l\leq J_{\max}-1}\sum_k (Y_{lk} - f_{lk})\psi_{lk}\left(1_{|Y_{lk}|>\tau_n,|f_{lk}|>\tau_n/2} + 1_{|Y_{lk}|>\tau_n,|f_{lk}|\leq\tau_n/2}\right)$$
$$- \sum_{l\leq J_{\max}-1}\sum_k f_{lk}\psi_{lk}\left(1_{|Y_{lk}|\leq\tau_n,|f_{lk}|>2\tau_n} + 1_{|Y_{lk}|\leq\tau_n,|f_{lk}|\leq 2\tau_n}\right) = I + II + III + IV.$$

We first treat the large deviation terms (*II*) and (*III*). For (*II*), using (8.32) and the Cauchy-Schwarz's inequality, we have

$$E_f\sup_{x\in\mathbb{R}}\left|\sum_{l\leq J_{\max}-1}\sum_k (Y_{lk} - f_{lk})\psi_{lk}(x)1_{|Y_{lk}|>\tau_n,|f_{lk}|\leq\tau_n/2}\right| \tag{8.33}$$
$$\leq E_f\left[\sum_{l\leq J_{\max}-1}\sup_k|Y_{lk} - f_{lk}|\sup_k 1_{|Y_{lk}|>\tau_n,|f_{lk}|\leq\tau_n/2}\left\|\sum_k|\psi_{lk}|\right\|_\infty\right]$$
$$\leq \sum_{l\leq J_{\max}-1} 2^{l/2}c\left[E_f\sup_k|Y_{lk} - f_{lk}|^2\right]^{1/2}\left[E_f\sup_k 1_{|Y_{lk}|>\tau_n,|f_{lk}|\leq\tau_n/2}\right]^{1/2}.$$

We have, using Theorem 5.1.13 with $p=\infty$ and (5.23) and (5.24), choosing τ large enough and using that $(2^l l/n)^{1/2}$ is bounded by a fixed constant independent of $l \le J_{\max}$,

$$
\begin{aligned}
E_f \sup_k 1_{|Y_{lk}|>\tau_n, |f_{lk}|\le \tau_n/2} &\le E_f \sup_k 1_{|Y_{lk}-f_{lk}|>\tau_n/2} \le E_f 1_{\sup_k |Y_{lk}-f_{lk}|>\tau_n/2} \qquad (8.34)\\
&\le P_f^Y\left(\sup_k |Y_{lk}-f_{lk}| > \tau \|f\|_\infty^{1/2}\sqrt{\frac{\log n}{n}}\right) \le K e^{-\tau^2 \log n/K},
\end{aligned}
$$

for some fixed constant $K>0$, so *(II)* is negligible by choosing τ large enough.

For term *(III)*, using (8.34) as well as $\sum_k |f_{lk}| \lesssim 2^{l/2}$ for any density $f \in L^1 \subset B^0_{1\infty}(\mathbb{R})$, we have, for τ large enough,

$$
\begin{aligned}
&E_f \sup_{x\in\mathbb{R}} \left| \sum_{l\le J_{\max}-1} \sum_k f_{lk}\psi_{lk}(x) 1_{|Y_{lk}|\le \tau_n, |f_{lk}|>2\tau_n} \right| \\
&\le \sum_{l\le J_{\max}-1} 2^{l/2}\|\psi\|_\infty \sum_k |f_{lk}| P_f^Y(|Y_{lk}|\le \tau_n, |f_{lk}|>2\tau_n) \\
&\le C''' e^{-\tau^2\log n/K} \sum_{l\le J_{\max}-1} 2^l = o(n^{-1/2}).
\end{aligned}
$$

We now bound term *(I)*. Let $j_1(s)$ be as in the proof of Theorem 8.2.7. Then, by (8.32),

$$
\begin{aligned}
&E_f \sup_{x\in\mathbb{R}} \left| \sum_{l\le j_1(s)-1} \sum_k (Y_{lk}-f_{lk})\psi_{lk}(x) 1_{|Y_{lk}|>\tau_n, |f_{lk}|>\tau_n/2} \right| \\
&\le c \sum_{l\le j_1(s)-1} E_f \sup_k |Y_{lk}-f_{lk}| 2^{l/2} \le D \sum_{l\le j_1(s)-1} \sqrt{\frac{2^l \log n}{n}} \\
&\le D'' G\left(\frac{\log n}{n}\right)^{\frac{s}{2s+1}},
\end{aligned}
$$

where $D''>0$ is some constant. For the second part, we use the definition of τ, $f\in B^s_{\infty\infty}$ and again (8.32) to obtain

$$
\begin{aligned}
&E_f \sup_{x\in\mathbb{R}} \left| \sum_{l=j_1(s)}^{J_{\max}-1} \sum_k (Y_{lk}-f_{lk})\psi_{lk}(x) 1_{|Y_{lk}|>\tau_n, |f_{lk}|>\tau_n/2} \right| \\
&\le c \sum_{l=j_1(s)}^{J_{\max}-1} E_f \sup_k |Y_{lk}-f_{lk}| \frac{2}{\tau}\sqrt{\frac{n}{\log n}} \sup_k |f_{lk}| 2^{l/2} \\
&\le D''' \sum_{l=j_1(s)}^{J_{\max}-1} 2^{-ls} \le D'''' \left(\frac{\log n}{n}\right)^{\frac{s}{2s+1}},
\end{aligned}
$$

where D'''' is some constant.

To complete the proof, we control term (*IV*): by (8.32), we see that

$$\sup_{x\in\mathbb{R}}\left|\sum_{l\le J_{\max}-1}\sum_k f_{lk}\psi_{lk}(x)1_{|Y_{lk}|\le\tau_n,|f_{lk}|\le 2\tau_n}\right| \le c\sum_{l\le J_{\max}-1}\sup_k 2^{l/2}|f_{lk}|1_{|f_{lk}|\le 2\tau_n}$$
$$\le c'\sum_{l\le J_{\max}-1}\min\left(2^{l/2}\sqrt{\frac{\log n}{n}},2^{-ls}\right),$$

which is controlled as in (8.30), completing the proof. ■

Exercises

8.2.1 Show that Theorems 8.2.3 and 8.2.8 remain true if U is replaced by $\|f_n(\tilde{j}_n)\|_\infty$, where $\tilde{j}_n$ is such that $2^{\tilde{j}_n} = (n/\log^2 n)$. *Hint*: Use a concentration inequality for $\|f_n(\tilde{j}_n) - Ef_n(\tilde{j}_n)\|_\infty$ and a bias bound for uniformly continuous f to infer concentration of $\|f_n(\tilde{j}_n)\|_\infty$ about the true value $\|f\|_\infty$ for n large enough.

8.2.2 Prove a version of Theorem 8.2.5 in the i.i.d. sampling setting.

8.3 Adaptive Confidence Sets

In the preceding section we saw that adaptive estimation is possible at no loss of precision in terms of minimax rates of convergence in the L^2- and L^∞-risk. In this section we investigate whether confidence sets in the sense of Section 6.4 exist that resemble these adaptive risk bounds. From a statistical point of view, this question is of central importance – a positive answer would mean that we can take advantage of the construction of adaptive estimators in the preceding section for purposes of uncertainty quantification and inference. Perhaps somewhat surprisingly, however, and in contrast to the situation of classical parametric statistics, the construction of confidence sets for adaptive estimators is a far from obvious task. The challenges are not just technical but fundamental, and we shall reveal that whether adaptive confidence sets exist or not depends on deeper information-theoretic properties of the statistical problem at hand. Moreover, a qualitative difference between L^2- and L^∞-theory exists, and geometric considerations will be seen to play a key role in our results. The reason behind is related to the fact that 'adaptive estimation of a function' and 'estimation of the accuracy of adaptive estimation' are statistically quite distinct problems in nature: the former just requires the existence of a method that is adaptive optimal, whereas the latter implicitly requires the estimation of aspects of the unknown function, such as its smoothness, which will be seen to be a fundamentally more difficult problem.

We shall first develop the theory in a simple two-class adaptation problem, where the main mechanisms can be explained fairly easily by establishing a relationship between adaptive confidence sets and certain composite testing problems from Chapter 6.2.4. These mechanisms are shown to form a general statistical principle of 'adaptive inference' and are not specific to function estimation problems. In Sections 8.3.2 and 8.3.3 we move on to the more concrete problem of constructing confidence sets for common adaptive estimators over a continuous scale of Hölder or Besov balls. We will see that full adaptive inference is possible for certain subclasses of Sobolev and Hölder balls characterised by a 'self-similarity' property of the wavelet expansion of the function on which we want to make inference. These subclasses will be shown to be generic in several ways – in particular, they provide necessary and sufficient conditions for the possibility of inference in general adaptation problems.

We shall develop the theory in this section entirely in the Gaussian white noise model (8.1). The results for the sampling model are the same up to fairly obvious modifications given the material already developed in this book.

8.3.1 Confidence Sets in Two-Class Adaptation Problems

We first turn our attention to the problem of confidence sets that adapt to only two fixed smoothness degrees and with a known bound on the Besov norm on the functions involved. Whereas such a 'toy' situation is perhaps not practically relevant, it highlights the subtleties that we have to expect in the general case and also provides some first lower bounds that disprove the existence of adaptive confidence sets even in simple situations. The results will strongly depend on the geometry through the choice of either the L^∞- or L^2-risk, and we consider the least favourable L^∞ case first.

Adaptive Confidence Bands for Two Nested Hölder Balls

Consider observations $dY \sim P_f^Y$ in the Gaussian white noise model (8.1), where we know that f is contained in a fixed Hölder ball

$$\Sigma(r) \equiv \Sigma(r,B) = \{f : [0,1] \to \mathbb{R}, \|f\|_{B^r_{\infty\infty}} \leq B\} \tag{8.35}$$

of radius $B > 0$ and smoothness level $r > 0$. We take here the Hölder-Besov norm $\|\cdot\|_{B^r_{\infty\infty}}$ arising from a wavelet basis of $L^2([0,1])$, such as the ones from Section 4.3.4 or Section 4.3.5.

We will assume for now that the radius B is known (and hence suppress the dependence in the notation), and consider adaptation to the smoothness parameter only. The adaptation hypothesis is that f is possibly much more regular, say, contained in $\Sigma(s)$ for some $s > r$, but we do not know whether this is the case or not. An adaptive estimator exists – for instance, if we take the wavelet thresholding estimator from Theorem 8.2.7 (which does not even require knowledge of B), then

$$\sup_{f\in\Sigma(t)} E_f\|f_n^T - f\|_\infty \lesssim \left(\frac{\log n}{n}\right)^{t/(2t+1)} \quad \forall\, t \in \{s,r\}.$$

In other words, the estimator f_n^T picks up the minimax L^∞-risk over $\Sigma(s)$ whenever f indeed is smooth enough and otherwise attains the optimal convergence rate over the maximal model $\Sigma(r)$. Following the ideas laid out in Section 6.4, and given significance levels $0 < \alpha, \alpha' < 1$, we are interested in a corresponding adaptive confidence set, that is, in a random subset C_n of L^∞ that has at least asymptotically honest coverage over the full model $\Sigma(r) \supset \Sigma(s)$, that is,

$$\liminf_n \inf_{f\in\Sigma(r)} P_f^Y(f \in C_n) \geq 1-\alpha, \tag{8.36}$$

and whose L^∞-diameter $|C_n|_\infty$ shrinks at the minimax optimal rate (recalling Definition 6.4.2 and Theorem 6.3.5) in the sense that

$$\sup_{f\in\Sigma(t)} P_f^Y\left(|C_n|_\infty > L\left(\frac{\log n}{n}\right)^{t/(2t+1)}\right) \leq \alpha', \quad \text{for all } t \in \{s,r\} \text{ and some } L > 0. \tag{8.37}$$

This inequality is usually only required (and, since L is not specified, only interesting) for n large enough. Because we measure the diameter of C_n in L^∞-distance we can always replace C_n by the smallest L^∞-ball that contains C_n without changing this problem. In this way, we can think of C_n as a 'confidence band' that gives pointwise control of the estimation error at all points $x \in [0,1]$ simultaneously. The significance levels α, α' are to be chosen by the statistician – the number $1-\alpha$ has the traditional interpretation as a coverage probability. The number α' specifies the exceptional probability for which C_n may not be adaptive – typically we will require (8.37) to hold for every $\alpha' > 0$ if $L = L(\alpha')$ is chosen large enough. Instead of introducing α', we could have insisted on the stronger requirement of a bound for $E_f|C_n|_\infty$ in (8.37) – as the following results are particularly interesting from the point of view of lower bounds, however, we prefer the weaker formulation involving α'.

Our more refined results later will in particular imply the following negative result:

Theorem 8.3.1 *Consider observations $dY \sim P_f^Y$ in (8.1). A confidence set $C_n \equiv C(dY, \alpha, \alpha', B)$ satisfying both (8.36) and (8.37) does not exist. In fact, any confidence set C_n satisfying (8.36) cannot also satisfy*

$$\sup_{f \in \Sigma(s)} P_f^Y(|C_n|_\infty > r_n) \leq \alpha',$$

for $s > r$, every n large enough and every $\alpha' > 0$ at any rate

$$r_n = o\left(\left(\frac{\log n}{n}\right)^{r/(2r+1)}\right).$$

In words, this theorem shows that adaptive confidence bands do not exist over the whole of $\Sigma(r)$. Moreover, this is not just a problem of possibly paying a mild penalty for adaptation (as in Section 8.1.1, for instance): if we want an adaptive confidence set over all of $\Sigma(r)$, then the price for adaptation is maximal in the sense that only the minimax rate of the maximal model can be attained over the submodel $\Sigma(s)$. Note that this is not a shortcoming of a particular procedure but an information-theoretic lower bound on *any* procedure.

One may construe the preceding theorem as saying that adaptive minimax confidence bands simply do not exist over Hölder balls $\Sigma(r)$. Not being able to quantify the uncertainty in an adaptive estimator poses serious doubts as to whether adaptive procedures are useful in statistics. Before drawing such strong conclusions, let us try to analyse the situation more closely: for $\rho \geq 0$, let us introduce sets

$$\tilde{\Sigma}(r,\rho) = \left\{ f \in \Sigma(r) : \inf_{g \in \Sigma(s)} \|f - g\|_\infty \geq \rho \right\} \tag{8.38}$$

which consist of the elements of the full model $\Sigma(r)$ that are separated away from the adaptation hypothesis $\Sigma(s)$ by uniform distance $\|f-g\|_\infty$ at least ρ. Clearly, $\tilde{\Sigma}(r,0) = \Sigma(r)$, but otherwise (the following proofs imply that) we are removing elements from $\Sigma(r) \setminus \Sigma(s)$. Instead of requiring coverage over all of $\Sigma(r)$ as in (8.36), we shall now only require, for some sequence $(\rho_n : n \in \mathbb{N})$, the weaker coverage inequality

$$\liminf_n \inf_{f \in \Sigma(s) \cup \tilde{\Sigma}(r,\rho_n)} P_f^Y(f \in C_n) \geq 1 - \alpha, \tag{8.39}$$

as well as the following adaptation properties: for some fixed $\alpha' > 0$, constant $L > 0$, sequence r_n to be chosen, we have, for all n large enough, that

$$\sup_{f \in \Sigma(s)} P_f^Y(|C_n|_\infty > Lr_n) \le \alpha' \tag{8.40}$$

and

$$\sup_{f \in \tilde{\Sigma}(r,\rho_n)} P_f^Y\left(|C_n|_\infty > L\left(\frac{\log n}{n}\right)^{r/(2r+1)}\right) \le \alpha'. \tag{8.41}$$

When $\rho = 0$ and $r_n = (n/\log n)^{-s/(2s+1)}$, this just reproduces the situation of Theorem 8.3.1, but allowing for flexible choices of ρ_n, r_n will reveal some interesting features of the problem at hand. Let us write

$$r_n(t) = \left(\frac{\log n}{n}\right)^{t/(2t+1)}$$

in what follows to expedite notation.

Theorem 8.3.2 *Consider observations $dY \sim P_f^Y$ in the Gaussian white noise model (8.1).*

(a) Suppose that for every $\alpha, \alpha' > 0$, any sequence $r_n = o(r_n(r))$ and some sequence $(\rho_n : n \in \mathbb{N})$, a confidence set $C_n = C(dY, \alpha, \alpha')$ satisfies (8.39), (8.40) and (8.41). Then, necessarily,

$$\liminf_n \frac{\rho_n}{r_n(r)} > 0. \tag{8.42}$$

(b) If for some large enough constant $c > 0$ and all n large enough

$$\rho_n \ge c r_n(r), \tag{8.43}$$

then for every $\alpha, \alpha' > 0$ and such ρ_n there exists a confidence set $C_n = C(dY, \alpha, \alpha', B)$ satisfying (8.39), (8.40) and (8.41) with $r_n = (n/\log n)^{-s/(2s+1)}$.

Remark 8.3.3 Part (a) implies, in particular, by way of contradiction, Theorem 8.3.1.

Proof Part (a): We will argue by contradiction and assume that a confidence set exists for ρ_n such that the limit inferior in (8.42) is zero. By passing to a subsequence, we can assume without loss of generality that the limit inferior is a limit, that is,

$$\rho_n = o(r_n(r)).$$

As in the proof of Theorem 6.2.11, part (b), we take distinct functions

$$\{f_m = \epsilon 2^{-j(r+1/2)} \psi_{jm} : m = 1, \dots, M\}, \quad M \simeq 2^j,$$

which all satisfy $\|f_m\|_{B^r_{\infty\infty}} = \epsilon < B$ for ϵ small enough. Then, for any $g \in \Sigma(s)$, by definition of the $B^s_{\infty\infty}$-norm and since $|\langle h, \psi_{lk}\rangle| \le \|\psi_{lk}\|_1 \|h\|_\infty \le 2^{-l/2}\|h\|_\infty$,

$$\begin{aligned}\|f_m - g\|_\infty &\ge \sup_{l,k} 2^{l/2} |\langle f_m, \psi_{lk}\rangle - \langle g, \psi_{lk}\rangle| \\ &\ge \epsilon 2^{-jr} - B 2^{-js} \ge (\epsilon/2) 2^{-jr},\end{aligned}$$

for j large enough, depending only on r, s, B, ϵ. Now let $(\rho_n' : n \in \mathbb{N})$ be such that

$$\max(r_n, \rho_n) \ll \rho_n' \ll r_n(r), \quad n \in \mathbb{N};$$

such a sequence exists by the hypotheses on r_n, ρ_n. If j_n^* is such that $2^{-j_n^* r} \simeq r_n(r)$, then we can find $j_n > j_n^*$ such that $(\epsilon/2)2^{-j_n r} \geq \rho_n'$, and hence, for such $j = j_n$, all the f_m are contained in $\tilde{\Sigma}(r, \rho_n)$.

Suppose now that C_n is a confidence band that is adaptive and honest over $\Sigma(s) \cup \tilde{\Sigma}(r, \rho_n)$, and consider testing

$$H_0 : f = 0 \quad \text{against} \quad H_1 : f \in \{f_1, \dots, f_M\} = \mathcal{M}.$$

Define a test Ψ_n as follows: if $C_n \cap \mathcal{M} = \emptyset$, then $\Psi_n = 0$, but as soon as one f_m is contained in C_n, then $\Psi_n = 1$. We control the type 1 and type 2 error probabilities of this test. Using (8.39) and (8.40) and noting that $r_n = o(\rho_n')$, we deduce, for n large enough,

$$\begin{aligned}
P_0^Y(\Psi_n \neq 0) &= P_0^Y(f_m \in C_n \text{ for some } m) \\
&= P_0^Y(f_m, 0 \in C_n \text{ for some } m) + P_0^Y(f_m \in C_n \text{ for some } m, 0 \notin C_n) \\
&\leq P_0^Y(\|f_m - 0\|_\infty \leq |C_n| \text{ for some } m) + \alpha + o(1) \\
&\leq P_0^Y(|C_n| \geq \rho_n') + \alpha + o(1) \leq \alpha' + \alpha + o(1).
\end{aligned}$$

Under any alternative $f_m \in \tilde{\Sigma}(r, \rho_n')$ and invoking honesty of C_n, we have

$$P_{f_m}^Y(\Psi_n = 0) \leq P_{f_m}^Y(f_m \notin C_n) \leq \alpha + o(1),$$

so, summarizing, we have

$$\limsup_n \left(E_0 \Psi_n + \sup_{f \in \mathcal{M}} E_f(1 - \Psi_n) \right) \leq 2\alpha + \alpha'.$$

For α, α' small enough, this contradicts the testing lower bound from Theorem 6.2.11, part (b), which implies that

$$\liminf_n \left(E_{f_0} \Psi_n + \sup_{f \in \mathcal{M}} E_f(1 - \Psi_n) \right) > 0. \tag{8.44}$$

Part (b): Let α, α' be given. For every $\beta > 0$, we can use Proposition 6.2.13 (noting that (6.52) holds with the choice $r_n = r_n(r)$ for the uniform norm metric d in view of Theorem 6.3.5) to construct a test Ψ_n of

$$H_0 : f \in \Sigma(s) \quad vs. \quad H_1 : f \in \tilde{\Sigma}(r, \rho_n) \quad \rho_n = c r_n(r), \quad c > 0,$$

with type 1 and type 2 errors bounded by β. If H_0 is accepted, we take as confidence band $C_n = [f_n^T \pm L r_n(s)]$ and otherwise $C_n = [f_n^T \pm L r_n(r)]$, where f_n^T is the adaptive estimator from Theorem 8.2.7, and where L is a large enough constant. We then have

$$\inf_{f \in \Sigma(s)} P_f^Y(f \in C_n) \geq 1 - \sup_{f \in \Sigma(s)} P_f^Y\left(\|f_n^T - f\|_\infty > L r_n(s)\right) \geq 1 - \alpha,$$

for L large enough by adaptivity of f_n^T (Theorem 8.2.7 and Markov's inequality). When $f \in \tilde{\Sigma}(r, \rho_n)$,

$$\inf_{f \in \tilde{\Sigma}(r, \rho_n)} P_f^Y(f \in C_n) \geq 1 - \frac{\sup_{f \in \tilde{\Sigma}(r, \rho_n)} E_f \|f_n^T - f\|_\infty}{L r_n(r)} - \sup_{f \in \tilde{\Sigma}(r, \rho_n)} P_f^Y(\Psi_n = 0),$$

and, as earlier, the first term subtracted can be made smaller than $\alpha/2$ for L large enough. The second term is also less than any $\alpha/2$ for c large enough and level β small enough.

Moreover, this confidence band is adaptive: When $f \in \tilde{\Sigma}(r, \rho_n)$, there is nothing to prove, and when $f \in \Sigma(s)$, the confidence band has diameter $Lr_n(s)$ unless $\Psi_n = 1$ has occurred. The probability of that exceptional event is again controlled at any level α' by taking the test Ψ_n to have level β small enough, completing the proof. ■

Note that part (a) actually holds for any α, α' such that $2\alpha + \alpha' < 1$ because the lower bound in (8.44) can be taken to be 1, as in (6.49).

Another, more statistically intuitive way of formulating the preceding theorem for the specific choice $r_n = (n/\log n)^{-s/(2s+1)}$ is the following:

Theorem 8.3.4 *An honest and adaptive (with $r_n = r_n(s)$) L^∞-confidence set over $\Sigma(s) \cup \tilde{\Sigma}(r, \rho_n)$ exists if and only if ρ_n exceeds, up to a multiplicative constant, the minimax rate of testing between the hypotheses*

$$H_0 : f \in \Sigma(s) \quad vs. \quad H_1 : f \in \tilde{\Sigma}(r, \rho_n).$$

This 'testing' interpretation will be investigated further later in a general decision-theoretic framework (Proposition 8.3.6). This result shows that adaptive confidence bands exist in the two-class model precisely whenever the level of smoothness $t \in \{s, r\}$ can be consistently tested from the observations. This is in strict contrast to the existence of adaptive estimators in Theorem 8.2.7 for the whole scale of Hölder balls $\Sigma(s), s > r$. Adaptive L^∞-confidence sets do not exist for parameters that are too close to the adaptation hypothesis $\Sigma(s)$, and the separation rate required is dictated by the minimax rate of the associated composite testing problem.

Adaptive L^2-Confidence Balls and Unbiased Risk Estimation

Let us next investigate the situation where the L^∞-risk is replaced by the weaker L^2-risk or, what is the same, when the performance of the confidence set C_n is measured by its L^2-diameter $|C_n|_2$. The situation is qualitatively different from the L^∞ case considered in the preceding section.

Consider again an observation $dY \sim P_f^Y$ in the Gaussian white noise model (8.1), where we assume now that f is contained in a Besov ball

$$\mathcal{S}(r) \equiv \mathcal{S}(r, B) = \{f : [0,1] \to \mathbb{R}, \|f\|_{B^r_{2\infty}} \le B\} \tag{8.45}$$

of radius B and smoothness level $r > 0$. We again take the Besov norm generated by a wavelet basis of $L^2([0,1])$ from Sections 4.3.4 and 4.3.5. All that follows also works for standard bases $\{e_k : k \in \mathbb{Z}\}$ of L^2 if one replaces $\mathcal{S}(r)$ by a Sobolev ball (similar to the proof of Theorem 8.3.16).

We can split the sample into two 'halves' as in (6.10) and compute the adaptive estimator $\hat{f}_n = f_n(\bar{j}_n)$ from Theorem 8.2.2 based on the first subsample. (In fact, any adaptive estimator could be used.) Based on the second subsample, we can then construct the confidence set C_n from (6.139), where we choose $j = j_n$ such that $2^j \simeq n^{1/(2r+1/2)}$ and where $B(j) = \bar{B}2^{-2jr}$ with $\bar{B}$ a large enough constant depending only on B. As will be shown in the next theorem, the expected diameter $E|C_n|_2$ is then of order

$$\max\left(n^{-r/(2r+1/2)}, n^{-s/(2s+1)}\right) \quad \text{which is} \quad O(n^{-s/(2s+1)}), \quad \text{if } s \le 2r; \tag{8.46}$$

hence, such C_n provides adaptation in the smoothness window $s \in [r, 2r]$, highlighting a remarkable difference to the L^∞ situation from the preceding subsection. However, adaptation only to the window $[r, 2r]$ is rather limited, and the question arises as to whether one can adapt to $s > 2r$ too. The answer to this question is negative, as will be shown by using the separation approach from the preceding subsection. However, in the L^2 setting, the separation rates are different, pertaining to the fact that the minimax testing rates are sensitive to whether separation occurs in L^2 or L^∞. Let us rigorously collect all these findings now in a theorem.

For $\rho \geq 0$, we again consider separated sets

$$\tilde{\mathcal{S}}(r, \rho) = \left\{ f \in \mathcal{S}(r) : \inf_{g \in \mathcal{S}(s)} \|f - g\|_2 \geq \rho \right\}, \tag{8.47}$$

where the separation distance is now the L^2-metric. For fixed $0 < \alpha < 1$ and some sequence $(\rho_n : n \in \mathbb{N})$, we require asymptotic coverage

$$\liminf_n \inf_{f \in \mathcal{S}(s) \cup \tilde{\mathcal{S}}(r, \rho_n)} P_f^Y(f \in C_n) \geq 1 - \alpha \tag{8.48}$$

and the following adaptation properties: for some fixed $\alpha' > 0$, constant $L > 0$ and sequence r_n to be chosen, we have

$$\sup_{f \in \mathcal{S}(s)} P_f^Y(|C_n|_2 > L r_n) \leq \alpha' \tag{8.49}$$

and

$$\sup_{f \in \tilde{\mathcal{S}}(r, \rho_n)} P_f^Y\left(|C_n|_2 > L n^{-r/(2r+1)}\right) \leq \alpha', \tag{8.50}$$

for all n large enough.

Theorem 8.3.5 *Consider observations dY in the Gaussian white noise model (8.1).*

(a) Suppose that for every $\alpha, \alpha' > 0$, any sequence $r_n = o(n^{-r/(2r+1/2)})$ (so, in particular, with $r_n = n^{-s/(2s+1)}$, for $s > 2r$,) and some sequence $(\rho_n : n \in \mathbb{N})$, a confidence set $C_n = C(dY, \alpha, \alpha')$ satisfies (8.48), (8.49) and (8.50). Then, necessarily,

$$\liminf_n \frac{\rho_n}{n^{-r/(2r+1/2)}} > 0. \tag{8.51}$$

(b) For every $\alpha, \alpha' > 0$, a confidence set $C_n = C(dY, \alpha, \alpha', B)$ satisfying (8.48), (8.49) and (8.50) with $r_n = n^{-s/(2s+1)}$ exists if one of the following conditions is satisfied:

(i) $s \leq 2r$ and $\rho_n = 0$ $\forall n$, or

(ii) $s > 2r$ and, for some large enough constant $c > 0$ and every $n \in \mathbb{N}$ large enough,

$$\rho_n \geq c n^{-r/(2r+1/2)}. \tag{8.52}$$

Proof We first prove part (b)(i) and invoke Theorem 6.4.8, as discussed in the paragraph containing (8.46), with $\mathcal{F} = \mathcal{S}(r)$, $\tilde{f}_n = \hat{f}_n$, where $\hat{f}_n$ is the adaptive estimator from Theorem 8.2.2, and j_n such that $2^{-j_n r} \simeq n^{-r/(2r+1/2)}$. Applying Lemma 8.2.1 with r in place of s, and since $j_n > j_n^*$ holds for n large enough, we see that

$$\sup_{f \in \mathcal{S}(r)} \|K_j(f - \hat{f}_n) - (f - \hat{f}_n)\|_2^2 = O_P(2^{-2j_n r}) = O_P(n^{-2r/(2r+1/2)}),$$

verifying the first hypothesis of Theorem 6.4.8. For $f \in \mathcal{S}(s)$, we can further uniformly bound $E\tau_n(f)$ by a term of order

$$\frac{1}{\sqrt{n}} E\|\hat{f}_n - f\|_2 + \frac{2^{j_n/2}}{n} = O\left(\frac{1}{\sqrt{n}} n^{-s/(2s+1)} + n^{-2r/(2r+1/2)}\right) = O(n^{-2s/(2s+1)})$$

in view of Theorem 8.2.2, and since $s \leq 2r$, implying that the L^2-diameter of C_n is indeed of the required adaptive order. Coverage of C_n follows from Theorem 6.4.8, completing the proof of this part of the theorem.

Part (b)(ii) is proved in a similar fashion as Theorem 8.3.2, part (b), where Proposition 6.2.13 is replaced by Theorem 6.2.20 to first construct a level α test for the testing problem

$$H_0 : f \in \mathcal{S}(s) \quad vs. \quad H_1 : f \in \tilde{\mathcal{S}}(r, \rho_n),$$

and where f_n^T is replaced by $\hat{f}_n$ from Theorem 8.2.2. The result also follows from the general Proposition 8.3.7.

Finally, to prove part (a), we apply the same testing reduction as in the proof of Theorem 8.3.2, but since the role of Theorem 6.2.11, part (b), has to be replaced by Theorem 6.2.11, part (c), there, the difference in the minimax testing rates explains the difference in the required separation rates. Again, the result also follows from the general considerations in Proposition 8.3.6 (combined with Theorem 6.2.20). ∎

Note that the bounds in parts (a) and (b)(ii) complement each other and imply that for $s > 2r$, the separation rate $n^{-r/(2r+1/2)}$ is necessary and sufficient for the existence of adaptive confidence sets.

A Decision-Theoretic Perspective on Adaptive Minimax Confidence Sets

Having seen that the geometry of the inference problem affects the nature of the problem of existence of adaptive confidence sets, we wish to show in this section that this mechanism is in fact a general decision-theoretic principle and is not particular to the statistical function estimation problems considered here. We restrict again to the two-class problem, where the exposition of the main ideas is clearest.

Suppose that we are given observations $(X^{(n)} : n \in \mathbb{N})$ taking values in some measurable space $(\mathcal{X}^{(n)}, \mathcal{A}^{(n)})$ with distribution P_f, where f is indexed by some parameter space Σ. We suppose that Σ is endowed with a metric d, and we denote by $r_n(\Sigma)$ the minimax rate of estimation on this space, that is

$$\inf_{T_n} \sup_{f \in \Sigma} E_f d(T_n, f) \sim r_n(\Sigma), \quad r_n(\Sigma) \to_{n\to\infty} 0, \tag{8.53}$$

the infimum extending over all estimators $T_n : (\mathcal{X}^{(n)}, \mathcal{A}^{(n)}) \to \Sigma$.

We consider an arbitrary subset $\Sigma_0 \subseteq \Sigma$ for which the minimax rate of estimation is $r_n(\Sigma_0)$. If $r_n(\Sigma_0) = o(r_n(\Sigma))$ as $n \to \infty$, it is sensible to speak of Σ_0 as an adaptation hypothesis – we may wish to construct an estimator T_n that attains this rate when $f \in \Sigma_0$ while still performing optimally when $f \in \Sigma \setminus \Sigma_0$. We have seen in this chapter that such estimators exist in a variety of situations; for instance, Σ could be a Sobolev or Hölder ball in $B^r_{p\infty}$ and Σ_0 a ball in the same scale of spaces but of higher smoothness $s > r$.

We are interested not simply in the construction of an adaptive estimator but also in an adaptive confidence set C_n for f, that is, a random subset $C_n = C(X^{(n)})$ of Σ based on the observations only such that C_n contains f with prescribed probability and such that the random diameter

$$|C_n| = \sup_{f,g \in C_n} d(f,g)$$

of C_n for the d-metric reflects the rates of adaptive estimation $r_n(\Sigma_0), r_n(\Sigma)$, depending on whether $f \in \Sigma_0$ or not. Note that, unlike an adaptive estimator, any such confidence set C_n provides an estimate $|C_n|$ of the minimax accuracy r_n of estimation. We shall see that this is in general a harder problem than adaptive estimation, whose solution depends on the 'information geometry' of the triplet (Σ, Σ_0, d).

To develop this idea, we *separate* Σ_0 from Σ: Setting $d(f,\Sigma_0) = \inf_{v \in \Sigma_0} d(f,v)$, we define

$$\tilde{\Sigma}(\rho) = \{f \in \Sigma : d(f,\Sigma_0) \geq \rho\} \tag{8.54}$$

and study the composite testing problem

$$H_0\colon f \in \Sigma_0 \quad vs. \quad H_1 : f \in \tilde{\Sigma}(\rho). \tag{8.55}$$

A test Ψ_n for this problem is a measurable function $\Psi_n = \Psi(X^{(n)})$ taking values in $\{0,1\}$. Following Chapter 6, a sequence $(\rho_n^* : n \in \mathbb{N})$ is called the *minimax rate of testing* for this problem if the following two requirements are satisfied:

(a) For every β, there exists a constant $L = L(\beta)$ and a test Ψ_n such that

$$\sup_{f \in \Sigma_0} E_f \Psi_n + \sup_{f \in \tilde{\Sigma}(L\rho_n^*)} E_f(1-\Psi_n) \leq \beta. \tag{8.56}$$

(b) For some $\beta' > 0$ and any $\rho_n' = o(\rho_n^*)$, we have

$$\liminf_n \inf_{\Psi} \left[\sup_{f \in \Sigma_0} E_f \Psi_n + \sup_{f \in \tilde{\Sigma}(\rho_n')} E_f(1-\Psi_n) \right] \geq \beta'. \tag{8.57}$$

We shall consider confidence sets that are adaptive and honest for the model

$$\mathcal{P}(\rho_n) = \Sigma_0 \cup \tilde{\Sigma}(\rho_n)$$

for a suitable sequence $\rho_n \geq 0$. Note that $\rho_n = 0$ is admissible in principle. Formally, we require C_n to satisfy, for some sequence ρ_n, some constant $0 < K < \infty$ and every $0 < \alpha$, $\alpha' < \beta'/3$, the following three requirements:

$$\liminf_n \inf_{f \in \mathcal{P}(\rho_n)} P_f(f \in C_n) \geq 1 - \alpha, \tag{8.58}$$

$$\limsup_n \sup_{f \in \Sigma_0} P_f(|C_n| \geq K r_n(\Sigma_0)) \leq \alpha', \tag{8.59}$$

$$\limsup_n \sup_{f \in \tilde{\Sigma}(\rho_n)} P_f(|C_n| \geq K r_n(\Sigma)) \leq \alpha'. \tag{8.60}$$

The first result is the following lower bound, which says that if the rate of adaptive estimation in Σ_0 is faster than the rate ρ_n^* of testing in (8.55), then an adaptive confidence set over $\mathcal{P}(\rho_n')$ cannnot exist for any $\rho_n' = o(\rho_n^*)$, particularly not for all of Σ as soon as $\liminf_n \rho_n^* > 0$.

Note that the following proof applies for any (not necessarily minimax estimation) rate $r_n(\Sigma_0) = o(\rho_n^*)$.

Proposition 8.3.6 *Let ρ_n^* be the minimax rate of testing for the problem (8.55), and let $r_n(\Sigma_0)$ be the minimax rate of estimation over Σ_0 in d-risk. Suppose that*

$$r_n(\Sigma_0) = o(\rho_n^*). \tag{8.61}$$

Then an adaptive and honest confidence set C_n that satisfies (8.58), (8.59) and (8.60) for any α, α' such that $0 < 2\alpha + \alpha' < \beta'$ and any $\rho_n = o(\rho_n^*)$ does not exist.

Proof Suppose that such C_n exist. For any sequence $\rho_n' \geq \rho_n$, the properties (8.58), (8.59) and (8.60) with ρ_n replaced by ρ_n' hold as well. We can choose such ρ_n' in a way that

$$r_n(\Sigma_0) = o(\rho_n'), \quad \rho_n' = o(\rho_n^*).$$

We consider testing (8.55) for $\rho = \rho_n'$, and construct

$$\Psi_n = 1\{C_n \cap \tilde{\Sigma}(\rho_n') \neq \emptyset\},$$

so we reject H_0 as soon as C_n contains any of the alternatives. Then, using coverage (8.58) and also (8.59) combined with $r_n(\Sigma_0) = o(\rho_n')$,

$$\begin{aligned}\sup_{f\in\Sigma_0} E_f \Psi_n &= \sup_{f\in\Sigma_0} P_f(C_n \cap \tilde{\Sigma}(\rho_n') \neq \emptyset) \\ &\leq \sup_{f\in\Sigma_0} P_f(f \in C_n, C_n \cap \tilde{\Sigma}(\rho_n') \neq \emptyset) + \alpha + o(1) \\ &\leq \sup_{f\in\Sigma_0} P_f(|C_n| \geq \rho_n') + \alpha + o(1) \leq \alpha' + \alpha + o(1).\end{aligned}$$

Likewise, using again (8.58),

$$\sup_{f\in\tilde{\Sigma}(\rho_n')} E_f(1-\Psi_n) = \sup_{f\in\tilde{\Sigma}(\rho_n')} P_f(C_n \cap H_1 = \emptyset) \leq \sup_{f\in\tilde{\Sigma}(\rho_n')} P_f(f \notin C_n) \leq \alpha + o(1).$$

Summarising, using the bounds on α, α', this test verifies that

$$\limsup_n \left[\sup_{f\in\Sigma_0} E_f \Psi_n + \sup_{f\in\tilde{\Sigma}(\rho_n')} E_f(1-\Psi_n) \right] < 2\alpha + \alpha' < \beta',$$

which contradicts (8.57) and completes the proof. ■

However, as soon as the testing problem (8.55) can be solved, then adaptive confidence sets exist as soon as adaptive estimators do, as the following proposition shows. One can also show in general that when $\rho_n^* = O(r_n(\Sigma_0))$, then adaptive honest confidence sets exist without the necessity of separation in Proposition 8.3.6 (see the notes at the end of this chapter).

Proposition 8.3.7 *Suppose that there exists an estimator $\hat{f}_n = \hat{f}(X^{(n)})$ that is adaptive, that is, such that $\forall \varepsilon > 0$ there exists $L' = L'(\varepsilon)$ such that*

$$\sup_{f\in\Sigma_0} P_f(d(\hat{f}_n, f) > L' r_n(\Sigma_0)) < \varepsilon \quad \textit{and} \quad \sup_{f\in\Sigma} P_f(d(\hat{f}_n, f) > L' r_n(\Sigma)) < \varepsilon.$$

Suppose, moreover, that for every $\beta' > 0$ there exists a test Ψ_n satisfying (8.56). Then, for every $\alpha, \alpha' > 0$, there exists $L = L(\alpha, \alpha')$ and a confidence set C_n satisfying (8.58), (8.59) and (8.60) for $\rho_n = L\rho_n^*$.

Proof Define, for $0 < M < \infty$ to be chosen,

$$C_n = \begin{cases} \{f \in \Sigma : d(f, \hat{f}_n) \le M r_n(\Sigma_0)\}, & \text{if } \Psi_n = 0, \\ \{f \in \Sigma : d(f, \hat{f}_n) \le M r_n(\Sigma)\}, & \text{if } \Psi_n = 1. \end{cases}$$

To establish coverage: for $f \in \Sigma_0$, we have from adaptivity of $\hat{f}_n$ that

$$\begin{aligned} \inf_{f \in \Sigma_0} P_f(f \in C_n) &\ge 1 - \sup_{f \in \Sigma_0} P_f\left(d(\hat{f}_n, f) > M r_n(\Sigma_0)\right) \\ &\ge 1 - \alpha, \end{aligned}$$

for $M \ge L'(\alpha)$ large enough. When $f \in \tilde{\Sigma}(\rho_n)$,

$$\inf_{f \in \tilde{\Sigma}(\rho_n)} P_f(f \in C_n) \ge 1 - \sup_{f \in \Sigma_0} P_f\left(d(\hat{f}_n, f) > M r_n(\Sigma)\right) - \sup_{f \in \tilde{\Sigma}(\rho_n)} E_f(1 - \Psi_n) \ge 1 - \alpha,$$

using again the adaptivity of $\hat{f}_n$ with $M \ge L'(\alpha/2)$ and (8.56) for $L = L(\alpha/2)$ large enough. This proves that C_n satisfies (8.58). For adaptivity: by the definition of C_n, we always have (8.60) with $K \ge 2M$. If $f \in \Sigma_0$, then, again with $K \ge 2M$,

$$P_f\{|C_n| > L r_n(\Sigma_0)\} \le P_f\{\Psi_n = 1\} \le \alpha',$$

for $L = L(\alpha')$ large enough, completing the proof. ■

8.3.2 Confidence Sets for Adaptive Estimators I

The results from the preceding subsection reveal some of the intrinsic difficulties associated with the theory of adaptive confidence sets. There is still need to provide uncertainty quantification for adaptive estimators, and the goal we shall set ourselves here is to find maximal parameter spaces for which honest inference with adaptive estimators is possible. *That is, we want confidence sets that reflect the actual accuracy of adaptive estimation and that are valid for as many points in the parameter space as possible.* The theory is made more difficult by the fact that in most nonparametric adaptation problems the target we want to adapt to (smoothness and perhaps also the Hölder or Sobolev norm of the function to be estimated) is not a discrete but a *continuous* parameter, and the two-class theory from the preceding section does not obviously generalise to the continuous case. We shall investigate this problem in this (and the subsequent) subsection.

Risk Estimation and Canonical Discretisations

Let $dY \sim P_f^Y$ be an observation in the white noise model (8.1). Consider the situation where we know a priori that f belongs to a Sobolev-Besov ball

$$\mathcal{S}(r, B_0) = \{f : [0,1] \to \mathbb{R}, \|f\|_{B^r_{2\infty}} \le B_0\}, \quad r > 0, B_0 > 0,$$

and we are interested in adapting to the unknown smoothness $s \in [r, R]$ for some fixed but arbitrary $R > r$ and to norm $\|f\|_{B^s_{2\infty}} \le B, B \le B_0$.

The idea is to split the sample as in (6.10) to compute one's favourite adaptive estimator $\hat{f}_n$ based on the first subsample and to estimate its resulting risk $E_f\|\hat{f}_n - f\|_2^2$ based on the second subsample. Indeed, if we consider a 'canonical discretisation' of the adaptation window $[r,R]$ equal to

$$\mathcal{R} = \{r, 2r, 4r, 8r, \dots, 2^{N-1}r\} = \{s_i\}_{i=0}^{N-1}, \quad \text{where } N \text{ is such that } s_N \equiv 2^N r > R,$$

then in view of the proof of Theorem 8.3.5, we know that for every fixed window $[s_i, s_{i+1}]$, we can estimate $E_f\|\hat{f}_n - f\|_2^2$ at a precision that is compatible with adaptation, and hence, adaptive confidence sets exist for $f \in \Sigma(s_i, B_0)$ in that window. The idea is now to try to estimate the 'true' $s = s(f) \in \mathcal{R}$ and then to use the procedure from Theorem 8.3.5 for the estimated value of s. We call the discretisation $\mathcal{R}$ *canonical* because it reflects the fact, specific to L^2-theory, that the unknown smoothness s of the function f needs to be estimated only at accuracy within a constant to construct adaptive confidence sets. We note in advance that such a construction does not work for L^∞-confidence bands, where the unknown smoothness needs to be estimated consistently (as will be discussed later).

Using the notation $\|f - \mathcal{T}\|_2 = \inf_{g \in \mathcal{T}} \|f - g\|_2$ for any set of functions $\mathcal{T}$ and defining

$$\rho_n(t) = n^{-t/(2t+1/2)}, \quad t \in \mathcal{R}, \tag{8.62}$$

we define the model

$$\mathcal{F}_n = \mathcal{S}(s_N, B_0) \cup \left(\bigcup_{s \in \mathcal{R}} \{f \in \mathcal{S}(s, B_0) : \|f - \mathcal{S}(t, B_0)\|_2 \geq L\rho_n(s) \ \forall t > s, t \in \mathcal{R}\} \right), \tag{8.63}$$

where L is a large enough constant to be chosen. We will construct a confidence set that adapts to smoothness $s \in [r,R]$ in L^2-diameter and is honest over $\mathcal{F}_n$ if r and B_0 are known. This confidence set can in fact be taken to be centred at an arbitrary adaptive estimator $\hat{f}_n$ whose performance is estimated by a statistic constructed in the proof of the following theorem.

Theorem 8.3.8 *Let $dY \sim P_f^Y$ be an observation in the Gaussian white noise model (8.1), and let $0 < r < R < \infty$. For every $\alpha, \alpha' > 0$, there exists L large enough and a confidence set $C_n = C(\alpha, \alpha', B_0, r, R, dY)$ that is honest over $\mathcal{F}_n$ from (8.63)*

$$\liminf_n \inf_{f \in \mathcal{F}_n} P_f^Y(f \in C_n) \geq 1 - \alpha \tag{8.64}$$

and that is adaptive in the sense that, for some fixed constant $L' > 0$,

$$\limsup_n \sup_{f \in \mathcal{S}(s,B) \cap \mathcal{F}_n} P_f^Y(|C_n|_2 \geq L' B^{1/(2s+1)} n^{-s/(2s+1)}) \leq \alpha' \quad \forall s \in [r,R], \quad B \in [1, B_0]. \tag{8.65}$$

Remark 8.3.9 We remark that adaptation in (8.65) holds both with respect to smoothness $s \in [r,R]$ and radius $B \in [1, B_0]$. Also note that the intersection with $\mathcal{F}_n$, in (8.65) is in fact not necessary, but since coverage cannot be guaranteed for elements not contained in $\mathcal{F}_n$, this is only of theoretical interest.

Proof Writing $\mathcal{S}(s) \equiv \mathcal{S}(s, B_0)$ throughout this proof, we begin with a test

$$H_0 : \mathcal{S}(2r) \quad vs. \quad H_1 : \tilde{\mathcal{S}}(r, \rho_n(r)) \equiv \{f \in \mathcal{S}(r) : \|f - H_0\|_2 \geq \rho_n(r)\}.$$

That is, we test whether the signal f that has generated the observations dY is possibly $2r$ smooth or not. If the test rejects, we set $\hat{s}_n = r$; if it accepts, we continue to test whether f is perhaps $4r$ smooth or not, and so on; that is, we define $\hat{s}_n$ to be the first element of $\mathcal{R}$ in this upwards procedure for which the test between

$$H_0 : f \in \mathcal{S}(s_i) \quad \text{and} \quad H_1 : f \in \mathcal{S}(s_{i-1}), \|f - H_0\|_2 \geq \rho_n(s_{i-1})$$

rejects the null hypothesis. For each of these composite testing problems, the minimax rate of testing was seen to equal

$$\rho_n(s_{i-1}) = n^{-(s_{i-1})/(2s_{i-1}+1/2)}, \quad i = 1, \ldots, N-1,$$

in Section 6.2.4, and we can use the test function $\Psi_n(i)$ constructed in Theorem 6.2.20 for each of these individual testing problems. (Alternatively, if $r > 1/4$, we can use the test from Corollary 6.2.16, noting that (6.70) is then verified with $s = s_i > 1/2$ for $i > 1$ in view of Theorem 4.3.36). Since there is only a finite number $N - 1$ of tests that are independent of sample size n, we can apply a trivial multiple-testing correction (by tuning each test to have level α/N instead of α). For each $f \in \mathcal{F}_n$, we either have $f \in \mathcal{S}(s_{N-1})$, in which case we set $s(f) = s_{N-1}$, or otherwise $f \in \tilde{\mathcal{S}}_{s(f)}$ for some unique $s(f)$. Using the minimax property of all the tests $\Psi_n(i)$, we deduce that, for $L = L(\alpha/(2N))$ and n large enough, we have

$$\sup_{f \in \mathcal{F}_n} P_f^Y(\hat{s}_n \neq s(f)) \leq \alpha/2. \tag{8.66}$$

Hence, the confidence set C_n constructed in the proof of Theorem 8.3.5 with r there replaced by $\hat{s}_n$ and tuned to have coverage $\alpha/2$ has the desired property (using the proof there on the event $\hat{s}_n = s(f)$). ■

We note that this construction is optimal in the sense that the separation sequences $\rho_n(t)$ in (8.62) cannot be taken any faster because otherwise we would arrive at a contradiction with Theorem 8.3.5, part (a). In particular, the set removed from $\mathcal{S}(r, B_0)$ 'disappears' in the large sample limit, and any f will eventually be contained in $\mathcal{F}_n$ if n is only large enough. The fact that the set $\mathcal{F}_n$ grows dense in the full model $\mathcal{S}(r, B_0)$ implies, in particular, the following 'misleading' corollary:

Corollary 8.3.10 *For every α, α', there exist confidence sets $C_n = C(\alpha, \alpha', r, R, B_0, dY)$ such that*

$$\liminf_n P_f^Y(f \in C_n) \geq 1 - \alpha \quad \forall f \in \mathcal{S}(r, B_0)$$

and such that

$$|C_n|_2 = O_{P_f^Y}(n^{-s/(2s+1)}) \quad \forall f \in \mathcal{S}(s, B_0), \quad s \in [r, R].$$

This result seemingly suggests that adaptive confidence sets exist for the whole model $\mathcal{S}(r, B_0)$. However, the result is pointwise in f, and the index n from when onwards coverage holds depends on f, so this is a bad use of asymptotics. We need to insist on honest (uniform in f) coverage to reveal the additional complexity behind the existence of adaptive confidence sets.

Next to optimality of the separation sequences $\rho_n(t)$, we may ask whether the restriction to known upper bounds on r, B_0 in Theorem 8.3.8 and Corollary 8.3.10 could be removed.

We shall now turn to proving a general lower bound that shows in particular that Corollary 8.3.10 (and a fortiori Theorem 8.3.8) cannot hold true if no bound on B_0 is required.

Impossibility of Confidence Sets without Qualitative Constraints

Theorem 8.3.8 (and Corollary 8.3.10) shows that the 'pathological' set removed from consideration to construct L^2-adaptive confidence sets over a fixed Sobolev ball vanishes as $n \to \infty$. This is only true thanks to the fact that we have a priori restricted to a fixed ball in $B^r_{2\infty}$, which in itself means that some functions $f \in B^r_{2\infty}$ are permanently excluded from consideration. We may ask whether such a qualitative restriction (a bound on the radius) is indeed necessary. We show now that this is the case, even if one is only after 'pointwise in f' confidence sets. This implies that full adaptive inference in the space $B^r_{2\infty}$ is impossible when $R > 2r$. It shows, moreover, that one has to make qualitative assumptions of some kind on the parameter space in question if one wants adaptive confidence statements.

Theorem 8.3.11 *Consider observations $dY \sim P^Y_f$ in the Gaussian white noise model (8.1), and let $0 < \alpha < 1/2$. A random subset $C_n = C(\alpha, dY)$ of L^2 cannot simultaneously satisfy*

$$\liminf_n P^Y_f(f \in C_n) \geq 1 - \alpha \;\; \forall f \in B^r_{2\infty} \tag{8.67}$$

and

$$|C_n|_2 = o_P(r_n) \tag{8.68}$$

at any rate

$$r_n = o(n^{-r/(2r+1/2)}).$$

Remark 8.3.12 When $R > 2r$, we have $n^{-s/(2s+1)} = o(n^{-r/(2r+1/2)})$, implying the desired impossibility result. For $R \leq 2r$, no restrictions are necessary for adaptation in view of Theorems 8.3.5, part (b)(i).

Proof Let $\{\psi_{lk}\}$ denote a wavelet basis of $L^2([0,1])$ that generates all the norms of the spaces $B^s_{2\infty}([0,1]), s \in [r,R]$, as in Section 4.3.4 or Section 4.3.5. For each l, consider disjoint index sets $\mathcal{K}^{(1)}_l, \mathcal{K}^{(2)}_l$ such that $|\mathcal{K}^{(2)}_l| \geq c2^l$ for some $c > 0$. Fix $s' > 2r$, and take $f_0 \in B^{s'}_{2\infty}$ any function for which the coefficients $\langle \psi_{lk}, f_0 \rangle$ are zero for all $l \in \mathbb{N}, k \in \mathcal{K}^{(2)}_l$. (For the proof of this theorem, we could simply take $f_0 = 0, \mathcal{K}^{(1)}_l = \emptyset$ and $\mathcal{K}^{(2)}_l = \{0,1,\dots,2^l - 1\}$, but other choices will be of interest later, so we give this proof in this slightly more general setting.) For $m \in \mathbb{N}$ and some coefficients $\beta_{j_ik} = \pm 1$ to be chosen later, we will define functions

$$f_m = f_0 + \sum_{i=1}^{m} \sum_{k \in \mathcal{K}^{(2)}_{j_i}} 2^{-j_i(r+1/2)} \beta_{j_ik} \psi_{j_ik}, \quad m \in \mathbb{N}. \tag{8.69}$$

The monotone increasing sequence j_m is chosen inductively as follows: set $\delta = (1 - 2\alpha)/5$. We have already defined f_0 and set $n_0 = 1$. Given $f_{m-1} \in B^r_{2\infty}$, we can use the hypotheses of the theorem to find $n_m > n_{m-1}$ depending only on f_{m-1} such that both

$$P^Y_{f_{m-1}}(f_{m-1} \notin C_{n_m}) \leq \alpha + \delta \quad \text{and} \tag{8.70}$$

$$P^Y_{f_{m-1}}(|C_{n_m}|_2 \geq r_{n_m}) \leq \delta \tag{8.71}$$

hold true. We then choose $j_m, m \in \mathbb{N}$, through

$$n_m \simeq C\, 2^{j_m(2r+1/2)},$$

where $C > 0$ is a small enough constant chosen later depending only on δ, and require in addition that $j_m/j_{m-1} \geq 1 + (1/2r)$. (The last inequality can always be achieved by choosing n_m sufficiently large in each step.)

To choose the $\beta_{j_i k}$ in each step, consider functions

$$f_{m,\beta} = f_{m-1} + g_\beta,\ g_\beta = 2^{-j_m(r+1/2)} \sum_{k \in \mathcal{K}^{(2)}_{j_m}} \beta_{j_m k} \psi_{j_m k},$$

where β is a point in the discrete hypercube $\{-1,1\}^{\mathcal{K}^{(2)}_{j_m}}$. For

$$\gamma'_{n_m} = n_m 2^{-(2r+1)j_m}$$

and $g_k \sim N(0,1)$, we see from Proposition 6.1.1 that the likelihood ratio between observations from $f_{m,\beta}$ and from f_{m-1} (and at noise level $1/n_m$), as well as its averaged version are given by

$$Z_\beta = \frac{dP^Y_{f_{m,\beta}}}{dP^Y_{f_{m-1}}} = \prod_{k \in \mathcal{K}^{(2)}_{j_m}} \exp\{\beta_{j_m k} \sqrt{\gamma'_{n_m}} g_k - \gamma'_{n_m}/2\}, \quad \text{and} \quad Z = \frac{1}{2^{|\mathcal{K}^{(2)}_{j_m}|}} \sum_\beta Z_\beta,$$

respectively. Thus, we have

$$\begin{aligned}
E_{f_{m-1}}(Z^2) &= E_{f_{m-1}} \left(2^{-|\mathcal{K}^{(2)}_{j_m}|} \sum_\beta \prod_{k \in \mathcal{K}^{(2)}_{j_m}} \exp\{\beta_k \sqrt{\gamma'_{n_m}} g_k - \gamma'_{n_m}/2\} \right)^2 \\
&= 2^{-2|\mathcal{K}^{(2)}_{j_m}|} \sum_{\beta,\beta'} E_0 \left[\prod_{k \in \mathcal{K}^{(2)}_{j_m}} \exp\{(\beta_k + \beta'_k)\sqrt{\gamma'_{n_m}} g_k - \gamma'_{n_m}\} \frac{dP^Y_{f_{m-1}}}{dP^Y_0} \right] \\
&= 2^{-2|\mathcal{K}^{(2)}_{j_m}|} \sum_{\beta,\beta'} \prod_{k \in \mathcal{K}^{(2)}_{j_m}} \exp\{(\beta_k + \beta'_k)^2 \gamma'_{n_m}/2 - \gamma'_{n_m}\} \\
&= 2^{-2|\mathcal{K}^{(2)}_{j_m}|} \sum_{\beta,\beta'} \exp\left\{ \gamma'_{n_m} \sum_{k \in \mathcal{K}^{(2)}_{j_m}} \beta_k \beta'_k \right\} \\
&= E(\exp[\gamma'_{n_m} Y_{j_m}])
\end{aligned}$$

using independence of the $\{g_k : k \in \mathcal{K}^{(2)}_{j_m}\}$ with $dP^Y_{f_{m-1}}/dP^Y_0$, the identities

$$E_0[dP^Y_{f_{m-1}}/dP^Y_0] = 1, \quad Ee^{ug_k} = e^{u^2/2},$$

and where $Y_{j_m} = \sum_{k \in \mathcal{K}^{(2)}_{j_m}} R_k, R_k \sim^{i.i.d.} \pm 1$, is a Rademacher average. We conclude as in the last two displays of the proof of Theorem 6.2.9 that, for C small enough depending only on δ,

$$E_{f_{m-1}}(Z-1)^2 \leq \delta^2.$$

As a consequence, if we consider the statistic

$$T_{n_m} = 1\left\{\exists f \in C_{n_m}, \|f - f_{m-1}\|_2 \geq r_{n_m}\right\}, \tag{8.72}$$

then, by the usual testing bound (as in (6.21)) and the Cauchy-Schwarz inequality,

$$P^Y_{f_{m-1}}(T_{n_m} = 1) + \max_\beta P^Y_{f_\beta}(T_{n_m} = 0) \geq 1 + E_{f_{m-1}}[(Z-1)1\{T_{n_m} = 0\}] \geq 1 - \delta. \tag{8.73}$$

Now we set $f_m = f_{m,\beta}$ for β maximising the left-hand side of the preceding inequality. By definition of j_m, we have

$$\|f_\infty - f_m\|_2^2 \lesssim \sum_{i=m+1}^{\infty} 2^{-2j_i r} \lesssim 2^{-2j_{m+1} r} \lesssim 2^{-j_m(2r+1)}; \tag{8.74}$$

in particular, the sequence f_m is uniformly Cauchy and converges to a limiting function f_∞ that is contained in $B^r_{2\infty}$ (since its wavelet coefficients are all equal to either zero, $\langle f_\infty, \psi_{lk}\rangle = \epsilon 2^{-l(2r+1)}(\pm 1)$ or $\langle f_0, \psi_{lk}\rangle$). Considering likelihood ratios (and writing $g_k = g_{ik}$ in slight abuse of notation),

$$Z' = \frac{dP^Y_{f_\infty}}{dP^Y_{f_m}} = \prod_{i=m+1}^{\infty} \prod_{k \in \mathcal{K}^{(2)}_{j_i}} \exp\{\beta_k \sqrt{n_m 2^{-(2r+1)j_i}} g_k - n_m 2^{-(2r+1)j_i}/2\},$$

we show from a similar computation as earlier and using (8.74) that

$$E_{f_m}[(Z')^2] = \exp\{n_m \|f_\infty - f_m\|_2^2\} \leq \exp\{D 2^{-j_m/2}\} \leq 1 + \delta^2,$$

for m large enough and some fixed constant $D > 0$. We obtain, using also (8.73) evaluated at the maximiser in β,

$$\begin{aligned}
&P^Y_{f_{m-1}}(T_{n_m} = 1) + P^Y_{f_\infty}(T_{n_m} = 0) \\
&= P^Y_{f_{m-1}}(T_{n_m} = 1) + P^Y_{f_m}(T_{n_m} = 0) + P^Y_{f_\infty}(T_{n_m} = 0) - P^Y_{f_m}(T_{n_m} = 0) \\
&\geq 1 - \delta + E_{f_m}[(Z' - 1)1\{T_{n_m} = 0\}] \\
&\geq 1 - 2\delta.
\end{aligned}$$

Now, if C_n is a confidence set as in the theorem, then we have, from (8.70) and along the chosen subsequence n_m of n,

$$P^Y_{f_{m-1}}(T_{n_m} = 1) \leq P^Y_{f_{m-1}}(f_{m-1} \notin C_{n_m}) + P^Y_{f_{m-1}}(|C_{n_m}| \geq r_{n_m}) \leq \alpha + 2\delta, \tag{8.75}$$

which combined with the preceding display gives

$$P^Y_{f_\infty}(T_{n_m} = 0) \geq 1 - 2\delta - P^Y_{f_{m-1}}(T_{n_m} = 1) \geq 1 - \alpha - 4\delta.$$

Moreover, for m large enough and positive constants d, D',

$$\|f_\infty - f_{m-1}\|_2^2 \geq \sum_{k \in \mathcal{K}^{(2)}_{j_m}} 2^{-(2r+1)j_m} \geq d 2^{-2j_m r} \geq D' n_m^{-2r/(2r+1/2)} \gg r_{n_m}^2 \tag{8.76}$$

by hypothesis on r_n. Consequently, on the event $f_\infty \in C_{n_m}$ and by (8.76), the set C_{n_m} contains an element f_∞ that is at distance from f_{m-1} by more than r_{n_m}, and we conclude overall that

$$P^Y_{f_\infty}(f_\infty \in C_{n_m}) \leq P^Y_{f_\infty}(T_{n_m} = 1) \leq \alpha + 4\delta = 1 - \alpha - \delta < 1 - \alpha,$$

giving a contradiction to (8.67) and completing the proof. ■

8.3.3 Confidence Sets for Adaptive Estimators II: Self-Similar Functions

Theorem 8.3.11 exhibits limitations for adaptive confidence sets and shows that if we are not willing to make some a priori restrictions on the functional parameter f, then the adaptation window $[r, 2r]$ from Theorem 8.3.5 cannot be improved on in the L^2 setting. In Theorem 8.3.8, a priori upper bounds B_0, r on the parameters B, s were assumed in the inequality $\|f\|_{B^s_{2\infty}} \le B$, so a testing suite could be used to estimate the 'smoothness window' that f belongs to. In the L^∞ setting, no such adaptation window exists at all (Theorem 8.3.2), so this approach cannot be used. Moreover, when adaptation for full ranges of smoothness parameters s, B is desired, the 'testing and separation' approach from earlier cannot be directly implemented as it relies on knowledge of B_0.

The question arises as to whether a qualitative assumption exists other than a bound on B_0 under which adaptive inference is possible over a full range of parameters. We study such a condition in this subsection. It models 'typical' elements of $B^s_{p\infty}$ that are 'identifiably s-smooth' in some sense. These conditions will be coined *self-similarity* conditions for reasons to be discussed later. How much 'identifiability' is needed will be seen to depend on the information-theoretic structure of the problem. The conditions we will introduce will be shown in Section 8.3.4 to give a natural model for s-regular functions, but more importantly perhaps, they give adaptive confidence sets for standard adaptive estimators (such as those constructed from Lepski's method) without requiring any knowledge of unknown parameters such as B_0.

Self-Similarity for Hölder Balls and Estimating the Hölder Exponent

We start by introducing the concept of self-similarity in the L^∞ setting, where adaptation is sought after in supremum-norm loss. In this case, the natural smoothness classes are Hölder-Besov balls

$$\Sigma_\infty(s, B) \equiv \{f : \|f\|_{B^s_{\infty\infty}} \le B\}.$$

For such f, we know that the approximation errors of wavelet projections (or convolution kernels) scale as

$$\|K_j(f) - f\|_\infty \le c\|f\|_{B^s_{\infty\infty}} 2^{-js}. \tag{8.77}$$

Note that such a bound does not identify the smoothness of f – no statement is made about the *maximal* value of s for which $f \in B^s_{\infty\infty}$ holds true. In fact, the Hölder exponent

$$s(f) = \sup\{s : f \in B^s_{\infty\infty}\}$$

need not be attained for a given function, and hence, we cannot in general define a unique s that describes the 'true smoothness' of f.

One way around this problem is to assume, in addition, that $f \in B^s_{\infty\infty}$ also satisfies a *lower bound* matching (8.77); that is, for some $\epsilon > 0$ and some $J_0 \in \mathbb{N}$,

$$\|K_j(f) - f\|_\infty \ge \epsilon\|f\|_{B^s_{\infty\infty}} 2^{-js} \quad \forall j \ge J_0. \tag{8.78}$$

Intuitively speaking, (8.77) and (8.78) require the approximation errors (=bias) to behave similarly across all scales $j \ge J_0$, uniquely identifying the smoothness $s(f)$ of f.

In basic examples, the requirement (8.78) can be quite reasonable: for instance, when the approximation $K_j(f) = (2^j K(\cdot/2^j)) * f$ is based on a convolution kernel $K = 1_{[-1/2,1/2]}$, we have

$$\|K_j(f) - f\|_\infty = \sup_{x \in \mathbb{R}} \left| \int_{-1/2}^{1/2} (f(x - u2^{-j}) - f(x)) du \right|. \tag{8.79}$$

Suppose now that f is infinitely differentiable except at x_0, where f behaves locally as $|x - x_0|$ so that $f \in B^1_{\infty\infty}$ but $f \notin B^{1+\gamma}_{\infty\infty}$, for any $\gamma > 0$; hence, the Hölder exponent is $s(f) = 1$. The integrand in (8.79), for $x = x_0$, equals $2^{-j}|u|$, so $\|K_j(f) - f\|_\infty \geq \epsilon 2^{-j}$ indeed follows.

We shall show in the next subsection that (8.78) holds for 'representative' or 'typical' elements of $B^s_{\infty\infty}$ and hence may be considered a reasonable modelling assumption. For the moment, however, let us demonstrate why (8.78) is a condition useful in statistical analysis of adaptive inference procedures: it allows us to complement Lemma 8.2.6 by a corresponding result for small values of $j < j^*$.

Lemma 8.3.13 *Suppose that* $f \in B^s_{\infty\infty}$ *satisfies (8.78) for some* $\epsilon, s > 0, J_0 \in \mathbb{N}$. *Let* $j^* = j^*_n(f), \bar{j}_n$ *be as in (8.24) and (8.25), respectively, where* $f_n(j)$ *is as in (8.7) based on observations* dY *in Gaussian white noise (8.1) and with a general grid* $\mathcal{J} = \{[j_{\min}, j_{\max}] \cap \mathbb{N}\}$ *for some* $J_0 \leq j_{\min} < j_{\max}$. *Then there exists* $m \in \mathbb{N}$ *depending only on* ϵ, τ, s *such that, for every* $j < j^*_n(f) - m, j \in \mathcal{J}$, *and some universal constant* c', *we have*

$$P^Y_f(\bar{j}_n = j) \leq (1/c') \exp\{-c'\epsilon^2 2^{2ms} j\}. \tag{8.80}$$

In particular,

$$P^Y_f(\bar{j}_n < j^*_n(f) - m) \leq (1/c') \exp\{-c'\epsilon^2 2^{2ms} j_{\min}\}. \tag{8.81}$$

Remark 8.3.14 When adapting to s in a fixed window $[s_{\min}, s_{\max}]$, it is natural to let $j_{\min} \to \infty$ with n, so the requirement $j_{\min} \geq J_0$ is met for n large enough. A version of this lemma can be proved for the choice $j_{\min} = 1$ as well by slightly modifying the thresholds in (8.25): we replace $\sqrt{(2^l l)/n}$ by $\sqrt{(2^l \max(l, \log n))/n}$, in which case $j, j_{\min}$ in (8.80) and (8.81) can be replaced by $\log n$, as inspection on the following proof shows:

Proof Fix $j \in \mathcal{J}, j < j^* - m$, where $m \in \mathbb{N}$, and observe that

$$P^Y_f(\hat{j}_n = j) \leq P^Y_f\left(\|f_n(j) - f_n(j^*)\|_\infty \leq \tau \sqrt{\frac{2^{j^*} j^*}{n}} \right). \tag{8.82}$$

Now, using the triangle inequality, we deduce

$$\begin{aligned} \|f_n(j) - f_n(j^*)\|_\infty \geq \|K_j(f) - f\|_\infty - \|K_{j^*}(f) - f\|_\infty - \|f_n(j) \\ - Ef_n(j) - f_n(j^*) + Ef_n(j^*)\|_\infty \end{aligned}$$

so that, using Conditions (8.77) and (8.78) and the definition of j^*, the probability in (8.82) is bounded by

$$P^Y_f\left(\|f_n(j) - Ef_n(j) - f_n(j^*) + Ef_n(j^*)\|_\infty \geq (\epsilon 2^{ms} c''(s) - \tau) \sqrt{\frac{2^{j^*} j^*}{n}} \right),$$

for some constant $c''(s) > 0$. We can now choose $m > 2$ sufficiently large but finite and only depending on ϵ, s, τ so that, using Theorem 5.1.12 with $p = \infty$, the preceding probability can be bounded by the right-hand side of (8.80). The bound (8.81) follows from summing the first bound over $j_{\min} \leq j \leq j^* - m$. ■

This result, combined with Lemma 8.2.6, implies that the Lepski-type estimate $\bar{j}_n$ approximates the optimal choice $j_n^*(f)$ within a fixed constant m: with high probability,

$$\bar{j}_n \in [j_n^*(f), j_n^*(f) - m].$$

Asymptotically, this means that $\bar{j}_n$ concentrates on a finite number of choices, and hence, the theory of confidence sets for linear estimators from Section 6.4 can be carried over to the fully data-driven 'nonlinear' estimators $f_n(\bar{j}_n)$. As a consequence, adaptive confidence bands can be constructed for f that are honest over classes of self-similar functions. In the easiest case, we split the sample into two (cf. (6.10)) and use half the sample for construction of $\bar{j}_n$ and the other half for $f_n(j)$. See Exercise 8.3.1 for some details and hints. Instead of pursuing this direction, we discuss now an alternative way to use the preceding result to construct an adaptive confidence band for f which is related to direct estimation of the Hölder exponent.

The starting point is to notice that $\bar{j}_n$ can be turned into a consistent estimate of the unknown smoothness $s(f)$ of a self-similar function f (satisfying (8.77) and (8.78)). From the discussion after (8.24) we see that

$$j^* = \frac{2}{2s+1}\left(D_n + \log_2 \|f\|_{B^s_{\infty\infty}}\right) + \frac{1}{2s+1}\log_2 N, \quad N \equiv \frac{n}{\log n},$$

where $D_n = O(1)$ is a sequence of universal constants. Hence,

$$s(f) = \frac{D_n + \log_2 \|f\|_{B^s_{\infty\infty}}}{j^*} + \frac{\log_2 N}{2j^*} - \frac{1}{2}. \tag{8.83}$$

We can then define the estimate

$$\hat{s} = \frac{\log_2 N}{2\bar{j}_n} - \frac{1}{2}, \tag{8.84}$$

which has accuracy, in view of Lemmas 8.2.6 and 8.3.13 and by definition of j_n^*,

$$|\hat{s} - s(f)| = \left|\frac{(j^* - \bar{j}_n)\log_2 N}{2\bar{j}_n j_n^*}\right| + O\left(\frac{1}{j_n^*}\right) = O_{P_f^Y}\left(\frac{m}{\log n}\right). \tag{8.85}$$

We conclude that under self-similarity, the Hölder exponent $s(f)$ of f can be estimated consistently, in fact, with $1/\log n$ rate.

Splitting the sample into two parts dY, dY' as in (6.10), we can use one half to estimate $\hat{s}$ and the other half to construct a multiscale confidence band C_n from (6.142), where r is selected to equal $\hat{s}$ with a small 'undersmoothing correction'. To fix ideas, let us consider adaptation to self-similar functions

$$f \in \tilde{\Sigma}(s,\epsilon) \equiv \Sigma_\infty(s,B) \cap \{f \text{ satisfies } (8.78)\}, \tag{8.86}$$

where $s \in [s_{\min}, s_{\max}], 0 < s_{\min} < s_{\max} < \infty$, and where $\epsilon > 0, J_0 \in \mathbb{N}$ are arbitrary but fixed.

Theorem 8.3.15 *Let dY be observations in the Gaussian white noise model (8.1), split into two halves dY', dY''. Let u_n, v_n be any sequences such that $u_n, v_n \to \infty, v_n = o(\log_2 n)$. For $\hat{s} = \hat{s}(dY'')$ as in (8.84) with $j_{\min}$ such that $2^{j_{\min}} \simeq n^{1/(2s'+1)}, s' > s_{\max}$, define $\tilde{s} = \hat{s} - (v_n/\log_2 N)$.*

Consider the multiscale confidence band $C_n = C(\alpha, dY')$ from (6.142), where $r = \tilde{s}$. Then

$$\sup_{f \in \cup_{s \in [s_{\min}, s_{\max}]} \tilde{\Sigma}(s,\epsilon)} |P_f^Y(f \in C_n) - (1-\alpha)| \to 0, \tag{8.87}$$

as $n \to \infty$, and whenever $f \in \tilde{\Sigma}(s,\epsilon)$ for some $s \in [s_{\min}, s_{\max}]$, the L^∞-diameter of C_n is of order

$$|C_n|_\infty = O_P\left(B^{1/(2s+1)}\left(\frac{\log n}{n}\right)^{s/(2s+1)} v_n\right), \tag{8.88}$$

where $v_n = \max(u_n, 2^{O(v_n)}) \to \infty$ can be taken to grow as slowly as desired.

Proof The modified estimate $\tilde{s}$ still satisfies, as in (8.85), that

$$|\tilde{s} - s(f)| = O_P\left(\frac{m}{\log n} + \frac{v_n}{\log N}\right) \to 0,$$

as $n \to \infty$, and also, for n large enough depending only on $B, s_{\min}, s_{\max}$, by Lemma 8.3.13,

$$\begin{aligned} P_f^Y(\tilde{s}_n > s(f)) &= P_f^Y\left(\frac{(\log_2 N)(j^* - \bar{j}_n)}{2j^*\bar{j}_n} > \frac{v_n}{\log_2 N} - \frac{D_n + \log_2 \|f\|_{B^s_{\infty\infty}}}{j^*}\right) \\ &\le P_f^Y\left(\frac{C(m, s_{\min}, s_{\max})}{\log_2 N} > \frac{v_n}{2\log_2 N}\right) + o(1) \to 0. \end{aligned}$$

The proof is now similar to that for Theorem 6.4.9: since the events $\{\tilde{s}_n \le s(f)\}$ have probability approaching 1 as $n \to \infty$, we have

$$\sup_{f \in \tilde{\Sigma}(s,\epsilon)} P_f^Y(\|f\|_{B^{\tilde{s}}_{\infty\infty}} \le u_n) \to 1,$$

and thus (8.87) follows from (6.141). The diameter bound (8.88) follows as in the last two displays in the proof of Theorem 6.4.9, with choice of j_n such that $2^{j_n} \simeq B^{2/(2s+1)} N^{1/(2s+1)}$, and using for the high frequencies that, again on the events $\{\tilde{s}_n \le s(f)\}$,

$$\|f\|_{B^{\tilde{s}_n}_{\infty\infty}} 2^{-j_n \tilde{s}_n} \le B 2^{-j_n s} 2^{-j_n(\tilde{s}_n - s)} = O_P\left(2^{-j_n s} 2^{O(v_n)}\right)$$

is of the desired order. ■

Self-Similarity for Sobolev Balls

We continue by investigating the concept of self-similarity in the setting of L^2-confidence sets, that is, when the diameter $|C_n|$ is measured in the weaker L^2-norm instead of in L^∞. As is perhaps expected, the theory here is somewhat more subtle: optimal results hold under comparably weak self-similarity conditions, but proving this requires some additional effort.

Since wavelet bases are not of particular importance in the L^2 setting, we cast our results in the general sequence space setting under the isometry of L^2 via an arbitrary ortho-normal basis $\{e_k : k \in \mathbb{N}\}$ of L^2. (A result for wavelet bases can be obtained by enumerating the wavelet functions in lexicographic order.) Thus, consider observations $Y = (y_k : k \in \mathbb{N})$ in the Gaussian sequence space model

$$y_k = f_k + \frac{1}{\sqrt{n}} g_k, \quad g_k \overset{i.i.d.}{\sim} N(0,1), \quad k \in \mathbb{N}, \tag{8.89}$$

write $\Pr_f$ for the law of $(y_k : k \in \mathbb{N})$ and recall that E_f denotes expectation under the law $\Pr_f$. Let us assume that the unknown sequence of interest $f = (f_k) \in \ell_2$ belongs to a Sobolev ball, that is, an ellipsoid in ℓ_2 of the form

$$S^s(B) = \{f \in \ell_2 : \|f\|_{s,2} \le B\}, \quad s > 0,\ B > 0,$$

where the Sobolev norm is given by

$$\|f\|_{s,2}^2 = \sum_{k=1}^{\infty} f_k^2 k^{2s}.$$

We will consider adaptation to smoothness degrees s in any fixed window $[s_{\min}, s_{\max}]$ and to the radius $B \in [b, \infty)$. Here $0 < s_{\min} < s_{\max} < \infty$ are fixed and known parameters, whereas $b > 0$ is a (not necessarily known) lower bound for B.

For $s \in [s_{\min}, s_{\max}]$, self-similarity function $\varepsilon : [s_{\min}, s_{\max}] \to (0,1]$, $J_0 \in \mathbb{N}$, $0 < b < B < \infty$, and constant $c(s) = 16 \times 2^{2s+1}$ define self-similar classes

$$\begin{aligned} S^s_{\varepsilon(s)} &\equiv S^s_{\varepsilon(s)}(b, B, J_0) \qquad (8.90)\\ &\equiv \left\{ f \in \ell_2 : \|f\|_{s,2} \in [b,B] : \sum_{k=2^{J(1-\varepsilon(s))}}^{2^J} f_k^2 \geq c(s) \|f\|_{s,2}^2 2^{-2Js} \; \forall J \in \mathbb{N}, J \geq J_0 \right\}, \end{aligned}$$

where the notation $\sum_{k=a}^{b} c_k$ for $a, b \in \mathbb{R}$ stands for $\sum_{k=\lceil a \rceil}^{\lfloor b \rfloor} c_k$ throughout the remainder of this subsection. Note that $\|f\|_{s,2} < \infty$ implies, for all $J \in \mathbb{N}$,

$$\sum_{k \geq 2^{J(1-\varepsilon(s))}} f_k^2 \leq \|f\|_{s,2}^2 2^{-2J(1-\varepsilon(s))s} = \|f\|_{s,2}^2 2^{-2Js} \times 2^{2J\varepsilon(s)s},$$

and for self-similar functions, this upper bound needs to be matched by a lower bound, accrued repeatedly over coefficient windows $k \in [2^{J(1-\varepsilon(s))}, 2^J], J \geq J_0$, that is not off by more than a factor of $2^{2J\varepsilon(s)s}/c(s)$. The condition is thus comparable to (8.78) in the L^∞ setting, although it is in a certain sense weaker: the proofs will reveal that the Sobolev exponent of f is only *approximately* identified, even for large scales $J \geq J_0$.

If condition (8.90) holds for some $\varepsilon(s) > 0$, then it also holds for $c(s) = 16 \times 2^{2s+1}$ replaced by an arbitrary small positive constant and any $\varepsilon'(s) > \varepsilon(s)$ (for J_0 chosen sufficiently large). In this sense, the particular value of $c(s)$ is somewhat arbitrary and chosen here only for convenience.

Larger values of $\varepsilon(s)$ correspond to weaker assumptions on f: indeed, increasing the value of $\varepsilon(s)$ makes it easier for a function to satisfy the self-similarity condition as the lower bound is allowed to accrue over a larger window of 'candidate' coefficients and since the 'tolerance factor' $2^{2J\varepsilon(s)s}$ in the lower bound increases. In contrast, smaller values of $\varepsilon(s)$ require a strong enough signal in blocks of comparably small size.

A heuristic summary of what we shall prove is that signal-strength conditions enforced through the self-similarity function $\varepsilon(s)$ allow for the construction of honest adaptive confidence balls over the parameter space

$$\bigcup_{s_{\min} \leq s \leq s_{\max}} S^s_{\varepsilon(s)},$$

for arbitrary values of B. We will effectively show that

$$\varepsilon(s) < \frac{1}{2} \; \forall s$$

is a necessary condition for the construction of such adaptive confidence sets (when $s_{\max} > 2s_{\min}$), whereas a sufficient condition is

$$\varepsilon(s) < \frac{s}{2s + 1/2} \ \forall s.$$

As $s \to \infty$, we have $s/(2s + 1/2) \to 1/2$, showing that the necessary condition cannot be improved on.

To formulate our main results, let us introduce the notation

$$\mathbb{S}(\varepsilon) = \mathbb{S}(\varepsilon, b, B, J_0) \equiv \cup_{s \in [s_{\min}, s_{\max}]} S^s_{\varepsilon(s)}(b, B, J_0) \tag{8.91}$$

for the collection of self-similar functions with regularity ranging between $[s_{\min}, s_{\max}]$ and function $\varepsilon : [s_{\min}, s_{\max}] \mapsto (0, 1)$.

A Sharp Adaptive Confidence Ball

In this subsection we give an algorithm which provides asymptotically honest and adaptive confidence sets over the collection $\mathbb{S}(\varepsilon)$ of self-similar functions whenever the function $\varepsilon(\cdot)$ satisfies

$$\sup_{s \in [s_{\min}, s_{\max}]} \varepsilon(s) \frac{2s + 1/2}{s} \le m < 1, \tag{8.92}$$

for some known parameter $0 < m < 1$ that is fixed in advance. As a first step, we split the 'sample' into two parts that we denote, in slight abuse of notation, by $y = (y_k)$ and $y' = (y'_k)$ (with Gaussian noise g_k and g'_k drawn from $N(0, 2)$, respectively, as in (6.10)), inflating the variance of the noise by 2. We denote the laws by $\Pr_1$ and $\Pr_2$ and the expectations by E_1 and E_2, respectively. Furthermore, we denote by $\Pr$ or $\Pr_f$ and E or E_f the joint distribution and corresponding expected value, respectively.

Using the first sample y, we denote by $f_n(j)$ the linear estimator with 'resolution level' $j \in \mathbb{N}$

$$f_n(j) \equiv (y_k)_{1 \le k \le 2^j}, \quad E_1 f_n(j) = (f_k)_{1 \le k \le 2^j} = K_j(f), \tag{8.93}$$

where K_j denotes the projection operator onto the first 2^j coordinates. Let us consider minimal and maximal truncation levels $j_{\min} = \underline{\sigma} \log_2 n$, $j_{\max} = \overline{\sigma} \log_2 n$ – for concreteness we take $\underline{\sigma} = 1/(2s' + 1)$ for arbitrary $s' > s_{\max}$ and $\overline{\sigma} = 1$, but other choices are possible. We define a discrete grid $\mathcal{J}$ of resolution levels

$$\mathcal{J} = \{j \in \mathbb{N} : j \in [j_{\min}, j_{\max}]\}$$

that has approximately $\log_2 n$ elements. Using Lepski's method, we define a first estimator by

$$\bar{j}_n \equiv \min\left\{ j \in \mathcal{J} : \| f_n(j) - f_n(l) \|_2^2 \le 4 \times \frac{2^{l+1}}{n} \ \forall l > j, l \in \mathcal{J} \right\}. \tag{8.94}$$

While $\bar{j}_n$ is useful for adaptive estimation via $f_n(\bar{j}_n)$ (as in Theorem 8.2.2), for adaptive confidence sets we shall need to systematically increase $\bar{j}_n$ by a certain amount – approximately by a factor of 2. To achieve this, let us choose parameters $0 < \kappa_1 < 1$ and $0 < \kappa_2 < 1$ that satisfy

$$m < \frac{2s_{\min} + 1/2}{s_{\min} + (s_{\min} + 1/2)/\kappa_1} < 1 \quad \text{and} \quad 0 < \frac{1 + \kappa_1}{2\kappa_2} < \kappa_2 < 1. \tag{8.95}$$

Intuitively, given $\delta > 0$, we can choose m, κ_1, κ_2 such that all lie in $(1-\delta, 1)$ – the reader thus may think of the κ_i as constants that are arbitrarily close to 1. Next, an undersmoothed estimate $\hat{J}_n > \bar{j}_n$ is defined as

$$\hat{J}_n = \lceil J_n \rceil, \quad \text{where} \quad \frac{1}{J_n} \equiv \frac{1}{2\kappa_2} \frac{1}{\bar{j}_n} - \frac{1-\kappa_2}{2\kappa_2} \frac{1}{\log_2 n}. \tag{8.96}$$

With $\hat{J}_n$ in hand, we use again the sample y to construct any standard adaptive estimator $\hat{f}_n$ of f in ℓ_2 loss; for concreteness, let us take $\hat{f}_n = f_n(\bar{j}_n)$ (for which an analogue of Theorem 8.2.2 is easily proved using Lemma 8.3.17, part (a); other adaptive estimators could be used as well). Adapting ideas from Theorem 6.4.8 we then use the second subsample y' to estimate the squared ℓ_2 risk of $\hat{f}_n$: the statistic

$$U_n(\hat{f}_n) = \sum_{k \le 2^{\hat{J}_n}} (y'_k - \hat{f}_{n,k})^2 - \frac{2^{\hat{J}_n+1}}{n}$$

has expectation (conditional on the first subsample)

$$E_2 U_n(\hat{f}_n) = \sum_{k \le 2^{\hat{J}_n}} (f_k - \hat{f}_{n,k})^2 = \|K_{\hat{J}_n}(f - \hat{f}_n)\|_2^2. \tag{8.97}$$

Our ℓ_2-confidence ball is defined as

$$C_n = \left\{ f : \|f - \hat{f}_n\|_2^2 \le U_n(\hat{f}_n) + \sqrt{8}\gamma_\alpha \frac{2^{\hat{J}_n/2}}{n} \right\}, \tag{8.98}$$

where γ_α denotes the $1-\alpha$ quantile of the standard normal $N(0,1)$ random variable, $0 < \alpha < 1$. We note that we do not require knowledge of any self-similarity or radius parameters in the construction; we only used the knowledge of $s_{\max}$ in the construction of the discrete grid $\mathcal{J}$ and the parameters m and $s_{\min}$ in the choice of κ_2.

Theorem 8.3.16 *For any $0 < b < B < \infty, J_0 \in \mathbb{N}$, and self-similarity function ε satisfying* (8.92)*, the confidence set C_n defined in* (8.98) *has exact honest asymptotic coverage $1-\alpha$ over the collection of self-similar functions $\mathbb{S}(\varepsilon)$; that is,*

$$\sup_{f \in \mathbb{S}(\varepsilon, b, B, J_0)} \left| \Pr_f(f \in C_n) - (1-\alpha) \right| \to 0,$$

as $n \to \infty$. Furthermore, the ℓ_2-diameter $|C_n|_2$ of the confidence set is rate adaptive: for every $s \in [s_{\min}, s_{\max}], B > b, J_0 \in \mathbb{N}$ and $\delta > 0$, there exists $C(s,\delta) > 0$ such that

$$\limsup_{n \to \infty} \sup_{f \in S^s_{\varepsilon(s)}(b, B, J_0)} \Pr_f(|C_n|_2 \ge C(s,\delta) B^{1/(2s+1)} n^{-s/(2s+1)}) \le \delta.$$

Proof As a first step in the proof, we investigate the estimator of the optimal resolution level $\bar{j}_n$ balancing out the bias and variance terms in the estimation. The linear estimator $f_n(j)$ defined in (8.93) has bias and variance such that

$$\|E_1 f_n(j) - f\|_2^2 \le \|f\|_{s,2}^2 2^{-2js} \equiv B(j, f) \tag{8.99}$$

and, recalling that by sample splitting the g_k are i.i.d. $N(0,2)$,

$$E_1 \| f_n(j) - E_1 f_n(j) \|_2^2 = \frac{1}{n} E_1 \sum_{k=1}^{2^j} g_k^2 = \frac{2^{j+1}}{n}. \tag{8.100}$$

Our goal is to find an estimator which balances out these two terms. For this we use Lepski's method in (8.94). For $f \in S^s(B)$, we define

$$j_n^* = j_n^*(f) = \min\{j \in \mathcal{J} : B(j,f) \leq 2^{j+1}/n\}, \tag{8.101}$$

which implies, by monotonicity, that

$$B(j,f) = 2^{-2js} \| f \|_{s,2}^2 \leq \frac{2^{j+1}}{n}, \quad \forall j \geq j_n^*, \quad j \in \mathcal{J}, \tag{8.102}$$

$$B(j,f) = 2^{-2js} \| f \|_{s,2}^2 > \frac{2^{j+1}}{n}, \quad \forall j < j_n^*, \quad j \in \mathcal{J}.$$

We note that, for n large enough (depending only on b and B, and recalling the definition of $\underline{\sigma}$), the inequalities $j_n^* < \lfloor \log_2 n \rfloor$ and $j_n^* > \lceil (\log_2 n)/(2s'+1) \rceil$ hold; hence, we also have

$$2^{2s+1} 2^{-2j_n^* s} \| f \|_{s,2}^2 \geq \frac{2^{j_n^*+1}}{n}. \tag{8.103}$$

Therefore, we can represent j_n^* and the given value of s as

$$j_n^* = \frac{\log_2 n + 2(\log_2 (\| f \|_{s,2}) + c_n)}{2s+1} \tag{8.104}$$

and

$$s = \frac{\log_2 n}{2j_n^*} + \frac{\log_2 (\| f \|_{s,2}) + c_n}{j_n^*} - \frac{1}{2}, \tag{8.105}$$

respectively, where $c_n \in [-1/2, s_{\max}]$.

The following lemma shows that $\bar{j}_n$ is a good estimator for the optimal resolution level j_n^* in the sense that with probability approaching 1, it lies between $(1 - \varepsilon(s))j_n^*$ and j_n^* whenever f is a self-similar function in the sense of (8.90).

Lemma 8.3.17 *Assume that $f \in S^s(B)$ for some $s \in [s_{\min}, s_{\max}]$ and any $B > 0$.*

(a) We have, for all $n \in \mathbb{N}$ and some universal constant $C > 0$,

$$\Pr\nolimits_1(\bar{j}_n \geq j_n^*) \leq C \exp\{-2^{j_n^*}/8\}.$$

(b) Furthermore, if the self-similarity condition (8.90) *holds, we also have, for all $n \in \mathbb{N}$ such that $j_n^* \geq J_0$, that*

$$\Pr\nolimits_1 \left(\bar{j}_n < j_n^*(1 - \varepsilon(s)) \right) \leq j_n^* \exp\{-(9/8) 2^{j_n^*}\}.$$

Proof Part (a) is effectively the same as Lemma 8.2.1 and hence left to the reader. For part (b), fix $j \in \mathcal{J}$ such that $j < j_n^*(1-\varepsilon)$, where $\varepsilon = \varepsilon(s)$. Then, by definition of $\bar{j}_n$,

$$\Pr\nolimits_1(\bar{j}_n = j) \leq \Pr\nolimits_1 \left(\| f_n(j) - f_n(j_n^*) \|_2 \leq 2\sqrt{2^{j_n^*+1}/n} \right). \tag{8.106}$$

Now, using the triangle inequality,

$$\begin{aligned}\|f_n(j) - f_n(j_n^*)\|_2 &= \|f_n(j) - f_n(j_n^*) - E_1(f_n(j) - f_n(j_n^*)) + E_1(f_n(j) - f_n(j_n^*))\|_2\\ &\geq \|E_1(f_n(j) - f_n(j_n^*))\|_2 - \|f_n(j) - f_n(j_n^*) - E_1(f_n(j) - f_n(j_n^*))\|_2\\ &= \sqrt{\sum_{k=2^j+1}^{2^{j_n^*}} f_k^2} - \frac{1}{\sqrt{n}}\sqrt{\sum_{k=2^j+1}^{2^{j_n^*}} g_k^2}.\end{aligned}$$

Since $j < j_n^*(1-\varepsilon)$, we have, from the definition of self-similarity (8.90) and (8.103), that

$$\sqrt{\sum_{k=2^j+1}^{2^{j_n^*}} f_k^2} \geq \sqrt{\sum_{k=2^{j_n^*(1-\varepsilon)}}^{2^{j_n^*}} f_k^2 \geq 4 \times 2^{s+1/2}\|f\|_{s,2}2^{-j_n^* s}} \geq 4 \times \sqrt{\frac{2^{j_n^*+1}}{n}},$$

so the probability on the right-hand side of (8.106) is less than or equal to

$$\begin{aligned}&\Pr_1\left(\frac{1}{\sqrt{n}}\sqrt{\sum_{k=2^j+1}^{2^{j_n^*}} g_k^2} \geq \sqrt{\sum_{k=2^j+1}^{2^{j_n^*}} f_k^2} - 2\sqrt{\frac{2^{j_n^*+1}}{n}}\right) \leq \Pr_1\left(\sum_{k=1}^{2^{j_n^*}} g_k^2 > (4-2)^2 2^{j_n^*+1}\right)\\ &= \Pr_1\left(\sum_{k=1}^{2^{j_n^*}} ((g_k/\sqrt{2})^2 - 1) > 3 \times 2^{j_n^*}\right).\end{aligned}$$

This probability is bounded by $\exp\{-(9/8)2^{j_n^*}\}$ using (3.29). The overall result follows by summing the preceding bound in $j < (1-\varepsilon)j_n^* < j_n^*, j \in \mathcal{J}$. ■

We note that by definition $j_n^* \geq \log n/(2s'+1) \to \infty$ and, hence for n large enough, $j_n^* \geq J_0$ holds uniformly over $f \in \mathbb{S}(\varepsilon, b, B, J_0)$.

As a next step we examine the new (undersmoothed) estimator of the resolution level $\hat{J}_n$. Assuming that $f \in S^s_{\varepsilon(s)}(b, B, J_0)$, the estimate $\bar{j}_n$ of j_n^* can be converted into an estimate of s. We note that a given f does not necessarily belong to a unique self-similar class $S^s_{\varepsilon(s)}(b, B, J_0)$, but the following results hold for any class to which f belongs. We estimate s simply by

$$\bar{s}_n = \frac{\log_2 n}{2\bar{j}_n} - \frac{1}{2},$$

ignoring lower-order terms in (8.105). We then have from (8.105) that

$$\begin{aligned}\bar{s}_n - s &= \frac{\log_2 n}{2\bar{j}_n} - \frac{1}{2} - \frac{\log_2 n}{2j_n^*} - \frac{\log_2(\|f\|_{s,2}) + c_n}{j_n^*} + \frac{1}{2}\\ &= \frac{\log_2 n}{2}\left(\frac{j_n^* - \bar{j}_n}{j_n^* \bar{j}_n}\right) - \frac{\log_2(\|f\|_{s,2}) + c_n}{j_n^*}.\end{aligned}$$

Now choose a constant $\kappa_3 \in (\kappa_2, 1)$ such that

$$0 < \frac{1+\kappa_1}{2\kappa_2} < \kappa_2 < \kappa_3 < 1,$$

which is possible, recalling (8.95). From Lemma 8.3.17, part (a), we have

$$\Pr_f(\bar{j}_n - j_n^* < 0) \to 1$$

uniformly over $f \in \cup_{s\in[s_{\min},s_{\max}]} S^s(B)$; hence, from the inequality $j_n^* \geq (\log_2 n)/(2s'+1)$, we have, for some constant $C = C(B,s'), B \geq \|f\|_{s,2}$,

$$\begin{aligned}
\Pr_f(\bar{s}_n \leq \kappa_3 s) &= \Pr_f(\bar{s}_n - s \leq (\kappa_3 - 1)s) \\
&\leq \Pr_f\left(\frac{\log_2 n}{2}\left(\frac{\bar{j}_n - j_n^*}{j_n^*\bar{j}_n}\right) + \frac{\log_2(\|f\|_{s,2}) + c_n}{j_n^*} \geq (1-\kappa_3)s_{\min}\right) \\
&\leq \Pr_f\left(C/\log_2 n > (1-\kappa_3)s_{\min}\right) + o(1) \to 0,
\end{aligned}$$

as $n \to \infty$. However, we also have from Lemma 8.3.17, (8.104) and $0 < \varepsilon(s) \leq 1$ that

$$\begin{aligned}
&\Pr_f(\bar{s}_n \geq (1+\kappa_1)s) = \Pr_f(\bar{s}_n - s \geq \kappa_1 s) \\
&\leq \Pr_f\left(\frac{\log_2 n}{2}\left(\frac{j_n^* - \bar{j}_n}{j_n^*\bar{j}_n}\right) - \frac{\log_2(\|f\|_{s,2}) + c_n}{j_n^*} \geq \kappa_1 s\right) \\
&\leq \Pr_f\left(\frac{j_n^* - \bar{j}_n}{j_n^*\bar{j}_n} \geq \frac{2\kappa_1 s}{\log_2 n} + \frac{2(\log_2(\|f\|_{s,2}) + c_n)}{j_n^* \log_2 n}\right) \\
&\leq \Pr_f\left(\varepsilon(s) j_n^* > \frac{2\kappa_1 s(1-\varepsilon(s))(j_n^*)^2}{\log_2 n} + \frac{2(1-\varepsilon(s))j_n^*(\log_2(\|f\|_{s,2}) + c_n)}{\log_2 n}\right) + o(1) \\
&= \Pr_f\left(\varepsilon(s) > \frac{2\kappa_1 s(1-\varepsilon(s))}{2s+1} + \frac{2\kappa_1 s(1-\varepsilon(s))}{\log_2 n}\right. \\
&\quad \left.\times \frac{2(\log_2(\|f\|_{s,2}) + c_n)}{2s+1} + \frac{2(1-\varepsilon(s))(\log_2(\|f\|_{s,2}) + c_n)}{\log_2 n}\right) + o(1) \\
&\leq \Pr_f\left(\varepsilon(s) > \frac{2\kappa_1 s(1-\varepsilon(s))}{2s+1} + \frac{2(\kappa_1+1)s+1}{2s+1} \times \frac{2\log_2(b/2) \wedge 0}{\log_2 n}\right) + o(1) \\
&= \Pr_f\left(\varepsilon(s) > \frac{\kappa_1 s}{(1+\kappa_1)s + 1/2} + \frac{2\log_2(b/2) \wedge 0}{\log_2 n}\right) + o(1).
\end{aligned}$$

The probability on the right-hand side tends to zero for n large enough (depending only on b) because

$$\varepsilon(s) \leq m\frac{s}{2s+1/2} < \frac{\kappa_1(2s_{\min}+1/2)}{(1+\kappa_1)s_{\min}+1/2} \times \frac{s}{2s+1/2} \leq \frac{\kappa_1 s}{(1+\kappa_1)s+1/2}$$

by definition of κ_1 given in (8.95) and the monotone increasing property of the function $g(s) = (2s+1/2)/[(1+\kappa_1)s+1/2]$. Therefore, we see that on an event of probability approaching 1, we have

$$\bar{s}_n \in (\kappa_3 s, (1+\kappa_1)s), \tag{8.107}$$

and, hence, if we define

$$\hat{s}_n = \bar{s}_n/(2\kappa_2),$$

we see that

$$\Pr_f\left(\hat{s}_n \in \left(\frac{\kappa_3}{2\kappa_2}s, \frac{1+\kappa_1}{2\kappa_2}s\right)\right) \to 1, \tag{8.108}$$

as $n \to \infty$. By choice of the κ_i, we see that $\hat{s}_n$ systematically underestimates the smoothness s and is contained in a closed subinterval of $(s/2, s)$ with probability approaching 1. The resolution level J corresponding to $\hat{s}_n$ is $\hat{J}_n$: easy algebraic manipulations imply that

$$2n^{1/(2\hat{s}_n+1/2)} > 2^{\hat{J}_n} \geq n^{1/(2\hat{s}_n+1/2)} \tag{8.109}$$

(where $\hat{J}_n$ was defined in (8.96)). Furthermore, we note that from (8.96) and since $\bar{j}_n \in \mathcal{J}$, we have

$$\hat{J}_n \in \left[\frac{2\kappa_2}{2s'+\kappa_2}\log_2 n, \lceil 2\log_2 n\rceil\right]. \tag{8.110}$$

Next, we turn our attention to an analysis of the confidence set C_n given in (8.98). First of all, note that

$$\begin{aligned} U_n(\hat{f}_n) - E_2 U_n(\hat{f}_n) &= \frac{1}{n}\sum_{k\leq 2^{\hat{J}_n}} ((g_k')^2 - 2) + \frac{2}{\sqrt{n}}\sum_{k\leq 2^{\hat{J}_n}} (f_k - \hat{f}_{n,k})g_k' \\ &\equiv -A_n - B_n. \end{aligned} \tag{8.111}$$

We deal with the two random sums A_n and B_n on the right-hand side separately. First, we show that $B_n = O_{\Pr_f}(n^{-(2s+1/2)/(2s+1)})$. Note that conditionally on the first sample, the random variable B_n has Gaussian distribution with mean zero and variance

$$\frac{8}{n}\sum_{k\leq \hat{J}_n}(f_k - \hat{f}_{n,k})^2 \leq \frac{8}{n}\|f - \hat{f}_n\|_2^2.$$

Furthermore, note that $\|f - \hat{f}_n\|_2^2 = O_{\Pr_1}(n^{-(2s)/(2s+1)})$ by adaptivity of the estimator $\hat{f}_n$. Hence, we can conclude, by independence of the samples y and y', that for every $\delta > 0$ there exists a large enough constant K such that $B_n \geq Kn^{-(2s+1/2)/(2s+1)}$ with $\Pr_f$ probability less than δ.

It remains to deal with A_n. In view of sample splitting, the centred variables $(2-(g_k')^2)$ are independent of $\hat{J}_n$ and have variance $\sigma^2 = 8$ and finite skewness $\rho > 0$. From the law of total probability, (8.107), (8.110) and Berry-Esseen's theorem (see Exercise 8.3.5), we deduce that

$$\begin{aligned} &\left|\Pr_f\left(A_n \leq \frac{\sigma\gamma_\alpha 2^{\hat{J}_n/2}}{n}\right) - (1-\alpha)\right| \\ &\leq \sum_{j=2\kappa_2\log_2 n/(2s'+\kappa_2)}^{\lceil 2\log_2 n\rceil} \left|\Pr_2\left(\frac{1}{\sigma 2^{j/2}}\sum_{k=1}^{2^j}(2-(g_k')^2) \leq \gamma_\alpha\right) - (1-\alpha)\right| \Pr_1(\hat{J}_n = j) \\ &\leq (3\rho/\sigma^3)2^{-\kappa_2\log_2 n/(2s'+\kappa_2)} = o(1). \end{aligned} \tag{8.112}$$

Next, note that in view of $f \in S^s(B)$, and since for $\hat{f}_n = f_n(\bar{j}_n)$ we have $K_{\hat{J}_n}(\hat{f}_n) = \hat{f}_n$ as $\hat{J}_n > \bar{j}_n$, the bias satisfies

$$\|K_{\hat{J}_n}(f - \hat{f}_n) - (f - \hat{f}_n)\|_2^2 = O(2^{-2\hat{J}_n s}) = o(2^{-2\hat{J}_n \hat{s}_n}) \tag{8.113}$$

using also

$$s > [(\kappa_1 + 1)/(2\kappa_2)]s > \hat{s}_n.$$

Furthermore, from (8.109), we have $2^{2\hat{J}_n \hat{s}_n} \geq n2^{-\hat{J}_n/2}$. Then, by using Pythagoras' theorem, (8.113) and (8.111), we deduce

$$\begin{aligned} \|f - \hat{f}_n\|_2^2 &= \|K_{\hat{J}_n}(f - \hat{f}_n)\|_2^2 + \|K_{\hat{J}_n}(f - \hat{f}_n) - (f - \hat{f}_n)\|_2^2 \\ &= E_2 U_n(\hat{f}_n) + o(2^{\hat{J}_n/2}/n) \\ &= U_n(\hat{f}_n) + A_n + B_n + o(2^{\hat{J}_n/2}/n). \end{aligned} \tag{8.114}$$

Following from (8.108) and (8.109), we obtain that (uniformly over $\mathbb{S}(\varepsilon, b, B, J_0)$) with $\Pr_f$ probability tending to 1,

$$2^{\hat{J}_n/2}/n \gtrsim n^{-(s(1+\kappa_1)/\kappa_2)/(s(1+\kappa_1)/(\kappa_2)+1/2)} \gg n^{-2s/(2s+1/2)}.$$

Furthermore, by $B_n = O_{\Pr_f}(n^{-(2s+1/2)/(2s+1)})$ and $n^{-(2s+1/2)/(2s+1)} \ll n^{-2s/(2s+1/2)}$, we see that the right-hand side of (8.114) can be rewritten as

$$U_n(\hat{f}_n) + A_n + o(2^{\hat{J}_n/2}/n). \tag{8.115}$$

Therefore, following from (8.114), (8.115) and (8.112), we deduce that the confidence set C_n given in (8.98) has exact asymptotic coverage $1 - \alpha$

$$\begin{aligned} \Pr_f(f \in C_n) &= \Pr_f\left(\|f - \hat{f}_n\|_2^2 \leq U_n(\hat{f}_n) + 2\gamma_\alpha \frac{2^{\hat{J}_n/2}}{n}\right) \\ &= \Pr_f\left(A_n \leq \left(2\gamma_\alpha + o(1)\right)\frac{2^{\hat{J}_n/2}}{n}\right) \to 1 - \alpha. \end{aligned}$$

Finally, we show that the radius of the confidence set is rate adaptive. First, we note that

$$2^{\hat{J}_n/4}/\sqrt{n} \leq 2^{1/4} n^{-\hat{s}_n/(2\hat{s}_n+1/2)} = o_{\Pr_1}(n^{-s/(2s+1)}),$$

by $\hat{s}_n > s\kappa_3/(2\kappa_2) > s/2$ and (8.109). Then, following from (8.97) and adaptivity of $\hat{f}_n$ (as in Theorem 8.2.3), we conclude that

$$E_f U_n(\hat{f}_n) = E_1 \|K_{\hat{J}_n}(f - \hat{f}_n)\|_2^2 \leq E_1 \|f - \hat{f}_n\|_2^2 \leq K(s) B^{1/(1+2s)} n^{-s/(1+2s)},$$

so the second claim of Theorem 8.3.16 follows from Markov's inequality. ■

Minimality of Self-Similarity Conditions

The main results for adaptive confidence sets presented so far in this subsection (Theorems 8.3.15 and 8.3.16) show that assuming self-similarity is an alternative to requiring an a priori bound on the Sobolev norm as in Theorem 8.3.8. From Theorem 8.3.11, we already know that certain additional conditions will be necessary if the a priori norm bound is dropped, but once self-similarity conditions are introduced, we can reasonably ask for the weakest possible ones.

In the ℓ_2 setting, the proof of Theorem 8.3.11 can be adapted to imply the following result – similar lower bounds in the L^∞ setting are discussed in the notes at the end of this chapter.

Theorem 8.3.18 *Fix $\alpha \in (0,1/2)$, $0<\varepsilon(\cdot)\equiv\varepsilon<1$, $0<r'<r<r'/(1-\varepsilon)$, and let $s \in (r, r'/(1-\varepsilon))$ be arbitrary. Then there does not exist a confidence set C_n in ℓ_2 which satisfies, for every $0<b<B, J_0 \in \mathbb{N}$,*

$$\liminf_{n\to\infty} \inf_{f\in S_\varepsilon^{r'}(b,B,J_0)\cup S_\varepsilon^{s}(b,B,J_0)} \Pr_f(f\in C_n)\geq 1-\alpha, \tag{8.116}$$

$$\sup_{f\in S_\varepsilon^{s}(b,B,J_0)} \Pr_f(|C_n|>r_n)\to_{n\to\infty} 0, \tag{8.117}$$

for any sequence $r_n = o(n^{-r/(2r+1/2)})$.

Proof We follow the proof of Theorem 8.3.11 and mimic a wavelet basis: partition $\mathbb{N}$ into sets of the form $Z_i^0 = \{2^i, 2^i+1, \dots, 2^i+2^{i-1}-1\}$ and $Z_i^1 = \{2^i+2^{i-1}, 2^i+2^{i-1}+1, \dots, 2^{i+1}-1\}$. Let us choose a parameter $s'>s$ satisfying $r'>s'(1-\varepsilon)>s(1-\varepsilon)$ and define self-similar sequences $f_m = (f_{m,k})$, for $m\in\mathbb{N}$,

$$f_{m,k} = \begin{cases} 2^{-(s'+1/2)l}, & \text{for } l\in\mathbb{N}\cup\{0\} \text{ and } k\in Z_l^0,\\ 2^{-(r+1/2)j_i}\beta_{j_i,k}, & \text{for } i\leq m \text{ and } k\in Z_{j_i}^1,\\ 0, & \text{else,}\end{cases}$$

for some monotone increasing sequence j_i and coefficients $\beta_{j_i,k}=\pm 1$ to be defined later. First, we show that independent of the choice of the sequence j_i and of the coefficients $\beta_{j_i,k}=\pm1$, the signals f_m and $f_\infty = \ell_2 - \lim_m f_m$ satisfy the self-similarity condition.

Using the definition of f_m and the monotone decreasing property of the function $f(x) = x^{-1-2(s'-s)}$, we can see that

$$\begin{aligned}\|f_m\|_{s,2}^2 &= \sum_{k=1}^{\infty} f_{m,k}^2 k^{2s} \leq 2^{2s'+1}\sum_{k=1}^{\infty} k^{-1-2(s'-s)} + 2^{2s}\sum_{i=1}^{m}\sum_{k\in Z_{j_i}^1} 2^{j_i(2s-2r-1)}\\ &\leq 2^{2s'+1}\Big(1+\int_1^\infty x^{-1-2(s'-s)}dx\Big) + 2^{2s-1}\sum_{i=1}^{m} 2^{j_i(2s-2r)}\\ &\leq 2^{2s'+1}\Big(1+\frac{1}{2(s'-s)}\Big) + 2^{2s-1}\frac{2^{j_m(2s-2r)}}{1-2^{-(2s-2r)}} \equiv B(s,s',r,j_m),\end{aligned} \tag{8.118}$$

for some constant $B(s,s',r,j_m)$ depending only on s, s', r and j_m. Furthermore, for $J\geq J_0$,

$$\sum_{k=2^{(1-\varepsilon)J}}^{2^J} f_{m,k}^2 \geq \sum_{k\in Z^0_{\lceil(1-\varepsilon)J\rceil}} f_{m,k}^2 = 2^{-(2s'+1)\lceil(1-\varepsilon)J\rceil}\times 2^{\lceil(1-\varepsilon)J\rceil-1} = 2^{-2s'\lceil(1-\varepsilon)J\rceil}/2. \tag{8.119}$$

Then, in view of the upper bound on the norm (8.118) and the inequalities $s'(1-\varepsilon)<r'<s$, the right-hand side of (8.119) is further bounded from below by

$$2^{-2r'J}/2 \geq 16\times 2^{2s+1}B(s,s',r,j_m)2^{-2sJ}\geq 16\times 2^{2s+1}\|f_m\|_{s,2}^2 2^{-2sJ},$$

for $J>J_0$ (where J_0 depends on s,s',r,r',ε and j_m). (The reader should note that the dependence of J_0 on j_m is harmless because j_m will remain independent of n.) Finally, the lower bound on the Sobolev norm can be obtained via

$$\|f_m\|_{s,2}^2 \geq \sum_{k\in Z_1^0} f_{m,k}^2 k^{2s} = 2^{-1-2(s'-s)} > 2^{-1-2(s'-r')} \equiv b^2. \tag{8.120}$$

Next, we show that f_∞ is r' self-similar. First, we note that the existence of f_∞ follows from the Cauchy property of the sequence (f_m) in ℓ_2. Furthermore, by definition, we have that $f_{\infty,k} = f_{m,k}$, for all $k \le 2^{j_m}, m \in \mathbb{N}$. Therefore, similar to (8.120) and (8.118), the signal f_∞ satisfies $\| f_\infty \|_{r',2} \ge b$, and

$$\| f_\infty \|_{r',2}^2 = \sum_{k=1}^{\infty} f_{\infty,k}^2 k^{2r'} \le 2^{2r'+1} \sum_{k=1}^{\infty} k^{-1-2(r-r')}$$
$$\le 2^{2r'+1}\left(1 + \frac{1}{2(r'-r)}\right) \equiv B(r,r'); \tag{8.121}$$

hence, it belongs to the Sobolev ball $S^{r'}(B)$ with radius $B = B(r,r')$ depending only on r and r'. Then, similar to (8.119), we deduce from (8.121) and the inequality $(1-\varepsilon)s' < r'$ that

$$\sum_{k=2^{(1-\varepsilon)J}}^{2^J} f_{\infty,k}^2 \ge 2^{-2s'\lceil (1-\varepsilon)J \rceil}/2 \ge 16 \times 2^{2r'+1} B(r,r') 2^{-2r'J}$$
$$\ge 16 \times 2^{2r'+1} \| f_\infty \|_{r',2}^2 2^{-2r'J},$$

for $J > J_0$ (where J_0 depends only on ϵ, r, s' and r').

The proof now proceeds similarly as in Theorem 8.3.11, with f_0, f_m, f_∞ there replaced by the current choices, which by the preceding arguments lie within the class of self-similar functions. The details are hinted in Exercise 8.3.2. ∎

Corollary 8.3.19 *Assume that $s_{\max} > 2s_{\min}$ and $\varepsilon(\cdot) \equiv \varepsilon > 1/2$. Then there does not exist a confidence set C_n in ℓ_2 which satisfies, for every $0 < b < B, J_0 \in \mathbb{N}$,*

$$\liminf_{n\to\infty} \inf_{f \in \cup_{s\in[s_{\min},s_{\max}]} S_\varepsilon^s(b,B,J_0)} \Pr_f(f \in C_n) \ge 1 - \alpha, \tag{8.122}$$

and, for all $s \in [s_{\min}, s_{\max}]$, $\delta > 0$ and some constant $K > 0$ depending on δ,

$$\limsup_{n\to\infty} \sup_{f \in S_\varepsilon^s(b,B,J_0)} \Pr_f(|C_n| > K n^{-s/(1+2s)}) \le \delta. \tag{8.123}$$

Proof Assume that there exists an honest confidence set C_n satisfying (8.122) and (8.123). Then take any $s \in (2s_{\min}, s_{\max})$, and choose the parameters r, r' such that they satisfy $s/2 > r > r' > \max\{(1-\varepsilon)s, s_{\min}\}$. Following from Theorem 8.3.18, if assertion (8.122) holds, then (8.117) cannot be true; that is, the size of the confidence set for any $f \in S_\varepsilon^s(b,B,J_0)$ cannot be of a smaller order than $n^{-r/(2r+1/2)}$. However, since $r < s/2$, we have $n^{-s/(2s+1)} \ll n^{-r/(2r+1/2)}$. Hence, the size of the honest confidence set has to be of a polynomially larger order than $n^{-s/(2s+1)}$, which contradicts (8.123). ∎

8.3.4 Some Theory for Self-Similar Functions

We have argued in this chapter that adaptive testing and estimation are possible without any, or at least without any substantial, price to pay for the statistician. For *inference* procedures such as confidence sets, however, the theory is more subtle, and adaptation is not possible over the entire parameter space. Classes of self-similar functions were shown to constitute statistical models for which a unified theory of estimation, testing and confidence sets is possible, serving perhaps as a paradigm for an 'honest' nonparametric statistical model. In this section we try to shed some more light on the classes of self-similar functions introduced in (8.86) and (8.90): we shall show, from three different perspectives, that the essential features and complexity of a nonparametric model for Sobolev or Hölder functions are revealed already by the 'restricted' self-similar classes of Sobolev or Hölder functions.

More precisely, we will show that within a given Hölder or Sobolev ball of functions,

(a) The information-theoretic complexity of a model is not decreased by introducing a self-similarity constraint;
(b) The non-self-similar functions are *nowhere dense* (and hence topologically negligible); and
(c) Natural nonparametric prior probability distributions draw self-similar functions *almost surely*.

Minimax Exhaustion for Self-Similar Functions

We start by showing that the minimax rates of convergence over a given Sobolev of Hölder ball are not changed after adding a self-similarity constraint. Note that the reference rates of convergence (without self-similarity) were derived in Theorems 6.3.5 and 6.3.8.

Theorem 8.3.20 *Let $r > 0, \varepsilon > 0$, be arbitrary. For the self-similar classes $\tilde{\Sigma}(r,\varepsilon)$ and S^r_ε from (8.86) and (8.90), respectively, the minimax convergence rates of estimation in L^∞ and ℓ_2-risk (in the sense of Definition 6.3.1) are of the order*

$$\left(\frac{\log n}{n}\right)^{r/(2r+1)}, \qquad n^{-r/(2r+1)},$$

respectively.

Proof Only the lower bound needs to be proved. The proof is similar to Theorems 6.3.5 and 6.3.8 after adding a self-similar base function f_0 to the alternatives f_m appearing in these proofs. Let us give some details for the ℓ_2 case; the L^∞ case is left as Exercise 8.3.3.

Take $s > r$ such that $r > (1-\varepsilon)s$, and using the notations of Theorem 8.3.18, let $Z_i^0 = \{2^i, 2^i+1, \dots, 2^i+2^{i-1}-1\}$ and $Z_i^1 = \{2^i+2^{i-1}, 2^i+2^{i-1}+1, \dots, 2^{i+1}-1\}$. Then we define $f_0, f_{m,j} \in \ell_2$ as

$$f_{0,k} = \begin{cases} K_1 2^{-(s+1/2)l}, & \text{for } l \in \mathbb{N} \text{ and } k \in Z_l^0, \\ 0, & \text{else,} \end{cases}$$

and

$$f_{m,j,k} = \begin{cases} K_1 2^{-(s+1/2)l}, & \text{for } l \in \mathbb{N} \text{ and } k \in Z_l^0, \\ \delta\beta_{m,j,k} 2^{-(r+1/2)j}, & \text{for } k \in Z_j^1, \\ 0, & \text{else,} \end{cases}$$

for some coefficients $\beta_{m,j,k} \in \{-1, 1\}$ and $K_1, \delta > 0$ to be defined later. Next, we show that all the preceding sequences f_0 and $f_{m,j}$ are self-similar.

First of all, we show that their $\|\cdot\|_{r,2}$-norm is bounded from below by b. By definition, we have

$$\|f_{m,j}\|_{r,2}^2 \geq \|f_0\|_{r,2}^2 = K_1^2 \sum_{l\in\mathbb{N}} \sum_{k\in Z_l^0} 2^{-(1+2s)l} k^{2r},$$

where the right-hand side depends only on the choice of s and r. We choose K_1 such that the right-hand side of this display is equal to b^2.

As a next step, we verify that f_0 and $f_{m,j}$ are in $S^r(B)$:

$$\begin{aligned}\|f_0\|_{r,2}^2 \leq \|f_{m,j}\|_{r,2}^2 &= \sum_{k=1}^{\infty} f_{m,j,k}^2 k^{2r} \\ &\lesssim \sum_{l} \sum_{k\in Z_l^0} 2^{-(1+2s)l} k^{2r} + 2^{2r}\delta^2 \sum_{k\in Z_j^1} \beta_{m,j,k}^2 2^{-j} \\ &\lesssim b^2 + \delta^2 2^{2r-1}.\end{aligned}$$

It is easy to see that for a small enough choice of the parameter $\delta > 0$, the right-hand side is bounded above by B^2 (the choice $\delta^2 < (B^2 - b^2)2^{1-2r}$ is sufficiently good); hence, both f_0 and $f_{m,j}$ belong to the Sobolev ball $S^r(B)$. Then we show that f_0 satisfies the lower bound (8.90) as well. Similar to the proof of Theorem 8.3.18, we have from $s(1-\varepsilon) < r$ that

$$\begin{aligned}\sum_{k=2^{(1-\varepsilon)J}}^{2^J} f_{0,k}^2 \geq \sum_{k\in Z_{\lceil(1-\varepsilon)J\rceil}^0} f_{0,k}^2 &= K_1^2 2^{-2s\lceil(1-\varepsilon)J\rceil}/2 \\ &\geq 16\times 2^{1+2r} B^2 2^{-2rJ} \geq 16\times 2^{1+2r} \|f_0\|_{r,2}^2 2^{-2rJ},\end{aligned}$$

for $J > J_0$ (where the parameter J_0 depends only on r, s, B and ε). The self-similarity of the functions $f_{m,j}$ follows exactly the same way.

Next, we define the sequences f_m ($m \in \mathcal{M}$) with the help of the sequences $f_{m,j}$ such that the ℓ_2-distance between them is sufficiently large. It is easy to see that

$$\|f_{m,j,k} - f_{m',j,k}\|_2^2 = 2^{-j(2r+1)}\delta^2 \sum_{k\in Z_j^1} (\beta_{m,j,k} - \beta_{m',j,k})^2.$$

Then, by the Varshamov-Gilbert bound, Example 3.1.4, there exists a subset $\mathcal{M} \subset \{-1,1\}^{|Z_j^1|}$ with cardinality $M = 3^{c'2^j}, c' > 0$, such that

$$\sum_{k\in Z_j^1} (\beta_{m,j,k} - \beta_{m',j,k})^2 \geq 2^j/8,$$

for any $m \neq m'$. Therefore,

$$\|f_{m,j} - f_{m',j}\|_2^2 \geq (\delta^2/8)2^{-2jr}.$$

Then, choosing $j = j_n$ such that $2^{j_n} = n^{1/(1+2r)}$, the $f_m \equiv f_{m,j_n}$ sequences are $c(\delta)n^{-r/(2r+1)}$ separated and are satisfying the self-similarity condition.

As in (6.16), the KL divergence is bounded by

$$K(P^Y_{f_0}, P^Y_{f_m}) = \frac{n}{2}\|f_m - f_0\|_2^2 = \frac{n}{2} 2^{-j(2r+1)}\delta^2 \sum_{k\in\mathcal{Z}^1_{jn}} \beta^2_{m,j_n,k} = \frac{2^j\delta^2}{4} \lesssim \delta^2 \log M.$$

Therefore, we can conclude the proof by applying Theorem 6.3.2 with δ small enough. ■

Topological Genericity and Self-Similarity

The self-similarity constraints (8.78) and (8.90) pose an additional restriction on the elements of a fixed Hölder ball in $B^s_{\infty\infty}$ or on a Sobolev ball S^s. It is thus natural to investigate whether the exceptional set that was removed is 'small' in some sense. In the absence of a volume measure on these infinite-dimensional sets, we can resort to topological quantifications, and a natural topology to consider is the 'norm' topology which makes the Hölder of Sobolev ball a 'ball' (open set). Notably, in the sense of Baire categories for this topology, the exceptional sets are *nowhere dense* and, hence, in this sense negligible.

We shall now make these statements rigorous. It is somewhat easier to make topological statements in the whole normed space (without the norm restriction $\|f\| \le B$). The proofs carry over to the trace topology on balls as well, but to present the main ideas, we only study self-similar functions as subsets of the full Hölder or Sobolev spaces. In this case, the presence of the norms $\|f\|_{B^s_{\infty\infty}}, \|f\|_{s,2}$ in (8.78) and (8.90) is immaterial (as inspection of the proofs shows).

For the next proposition, recall condition (8.78) with the norm in the lower bound removed (as just discussed). The proof of the following result shows that for wavelet projection kernels, the exceptional set of non-self-similar functions is contained in the complement of an open and dense subset of the Hölder space $B^s_{\infty\infty}$.

Proposition 8.3.21 *For $f \in B^s_{\infty\infty}$, let $K_j(f), j \in \mathbb{N}$, be the projection onto a wavelet basis generating $B^s_{\infty\infty}$. Then the set of functions*

$$\mathcal{N}_s = \left\{ f \in B^s_{\infty\infty} : \text{there do not exist } \epsilon > 0 \text{ and } J_0 \in \mathbb{N} \text{ s.t. } \|K_j(f) - f\|_\infty \ge \epsilon 2^{-js}\ \forall j \ge J_0 \right\}$$

is nowhere dense in (the norm topology of) the Banach space $B^s_{\infty\infty}$.

Proof Write $\beta_{lk}(g)$ for the wavelet coefficients $\langle g, \psi_{lk}\rangle$ of g. Since

$$|\beta_{lk}(g)| = \left| 2^{l/2} \int \psi(2^l x - k) g(x) dx \right| \le 2^{-l/2}\|\psi\|_1 \|g\|_\infty,$$

for every l, k and every bounded function g, we have, for $g = K_j(f) - f$, whose wavelet coefficients are 0 for $l < j$, that

$$\|K_j(f) - f\|_\infty \ge \|\psi\|_1^{-1} \sup_{l\ge j,k} |2^{l/2}\beta_{lk}(f)|. \tag{8.124}$$

For k arbitrary, take

$$E_m(k) = \{f \in B^s_{\infty\infty} : |\beta_{lk}(f)| \ge 2^{-l(s+1/2)}2^{-m} \text{ for every } l \in \mathbb{N}\},$$

and with wavelet norm $\|g\|_{B^s_{\infty\infty}} = \sup_{l,k} 2^{l(s+1/2)}|\beta_{lk}(g)|$, define the neighbourhoods

$$A_m(k) = \{h \in B^s_{\infty\infty} : \|h - f\|_{B^s_{\infty\infty}} < 2^{-m-1} \text{ for some } f \in E_m(k)\}$$

so that, for every $h \in A_m(k)$ and every l, $|\beta_{lk}(h)| \geq 2^{-l(s+1/2)}2^{-m-1}$. Consequently, using (8.124), we have

$$\|K_j(h) - h\|_\infty \geq \|\psi\|_1^{-1} 2^{-m-1} 2^{-js}, \tag{8.125}$$

for every nonnegative integer j and every $h \in A_m(k)$. Define now

$$A = \cup_{m \geq 0, k} A_m(k),$$

all of whose elements satisfy the desired lower bound for some m, and therefore, $A \subset \mathcal{N}_s^c$ (with the complement taken in the ambient Banach space $B^s_{\infty\infty}$).

The set A is clearly open, and it is also dense in $B^s_{\infty\infty}$: let $g \in B^s_{\infty\infty}$ be arbitrary, and define the function g_m by its wavelet coefficients $\beta_{lk}(g_m)$ equal to $\beta_{lk}(g)$ when $|\beta_{lk}(g)| > 2^{-l(s+1/2)}2^{-m}$ and equal to $2^{-l(s+1/2)}2^{-m}$ otherwise. Clearly, $g_m \in A$, for every m, and for $\varepsilon > 0$ arbitrary, we can choose m large enough such that

$$\|g - g_m\|_{B^s_{\infty\infty}} = \sup_{l,k} 2^{l(s+1/2)} \left|\beta_{lk}(g) - 2^{-l(s+1/2)}2^{-m}\right| 1_{|\beta_{lk}(g)| \leq 2^{-l(s+1/2)}2^{-m}} \leq 2^{-m+1} < \epsilon.$$

This proves that $\mathcal{N}_s$ is contained in the complement of an open and dense set and hence itself must be nowhere dense. ■

The driving force in this result is a generic lower bound on the wavelet coefficients

$$\max_k |\langle f, \psi_{lk} \rangle| \geq \varepsilon 2^{-l(s+1/2)},$$

for all l (large enough) of 'typical' functions $f \in B^s_{\infty\infty}$. This can be compared to the ℓ_2 self-similarity assumption defined in (8.90). This condition, transposed into double-indexed wavelet notation (by numbering the wavelet basis functions in lexicographic order), requires the 'energy packets' $\sum_k \langle f, \psi_{lk} \rangle^2$ accruing over windows of resolution levels l to be large enough; more precisely, for some $\varepsilon > 0$,

$$\sum_{l=J(1-\varepsilon)}^{J} \sum_k f_{lk}^2 \geq c(s) 2^{-2Js} \quad \forall J \geq J_0, \quad f_{lk} = \langle f, \psi_{lk} \rangle. \tag{8.126}$$

In view of the adaptive estimation result (Theorem 8.2.2), the appropriate maximal topology to study such ℓ_2-type self-similarity conditions in the wavelet setting is the Besov space $B^s_{2\infty}$. The result paralleling Proposition 8.3.21 is then the following:

Proposition 8.3.22 *The set*

$$\mathcal{M}_s = \left\{ f \in B^s_{2\infty} : \textit{there do not exist } \varepsilon > 0 \textit{ and } J_0 \in \mathbb{N} \textit{ s.t. (8.126) holds} \right\}$$

is nowhere dense in (the norm topology) of the Banach space $B^s_{2\infty}$.

Proof Consider the set of sequences

$$E_m = \left\{ f \in B^s_{2\infty} : \sum_k f_{lk}^2 \geq 2^{-2ls} 2^{-2m} \; \forall l \in \mathbb{N} \right\}, \quad m \in \mathbb{N},$$

and define, for $\delta > 0$ to be chosen later, the open sets

$$A_m = \{ h \in B^s_{2\infty} : \|f - h\|_{B^s_{2\infty}} < \delta 2^{-m}, f \in E_m \},$$

where we recall the definition of the norm $\|g\|_{B^s_{2\infty}} = \sup_l 2^{ls}\sqrt{\sum_k g_{lk}^2}$. For $h \in A_m$ and $\varepsilon > 0$, we have, by the triangle inequality $\sqrt{\sum_l a_l} \le \sum_l \sqrt{a_l}$, for $\delta > 0$ small enough and J large enough, that

$$\begin{aligned}\sqrt{\sum_{l=J(1-\varepsilon)}^{J}\sum_k h_{lk}^2} &\ge \sqrt{\sum_{l=J(1-\varepsilon)}^{J}\sum_k f_{lk}^2} - \sum_{l=J(1-\varepsilon)}^{J}\sqrt{\sum_k (f_{lk}-h_{lk})^2} \\ &\ge 2^{-m}\sqrt{\sum_{l=J(1-\varepsilon)}^{J} 2^{-2ls}} - \sum_{l=J(1-\varepsilon)}^{J} 2^{-ls}\|f-h\|_{B^s_{2\infty}} \\ &\ge c'(s)2^{-m-1}2^{-J(1-\varepsilon)s} \ge \sqrt{c(s)}2^{-Js}.\end{aligned}$$

Conclude that the elements of A_m are all self-similar in the sense of (8.126) and that the union $A = \bigcup_{m\in\mathbb{N}} A_m$ is open in $B^s_{2\infty}$. To proof is completed as in Proposition 8.3.21 by showing that A is also dense in $B^s_{2\infty}$: we approximate $g \in B^s_{2\infty}$ arbitrary by g_m defined via wavelet coefficients $g_{m,lk} = g_{lk}$ whenever $\sum_k g_{lk}^2 > 2^{-2ls}2^{-2m}$ at level l and equal to $g_{m,lk} = 2^{-l(s+1/2)}2^{-m}$ otherwise. Then

$$\|g - g_m\|_{B^s_{2\infty}} = \sup_l 2^{ls}\sqrt{\sum_k (g_{lk} - 2^{-ls}2^{-m})^2 1_{\sum_k g_{lk}^2 \le 2^{-2ls}2^{-2m}}} \le 2^{-m+1}$$

can be made as small as desired for m large enough. ■

A Bayesian Perspective on Self-Similarity

We finally take a Bayesian perspective and show that natural Bayesian nonparametric priors for Hölder or Sobolev functions charge self-similar functions with probability 1. Take a periodised wavelet basis $\{\psi_{lk}\}$ of $L^2([0,1])$. The wavelet characterisation of the Hölder-Besov space $B^s_{\infty\infty}$ motivates us to distribute the basis functions ψ_{lk} randomly on a fixed ball of radius B in $B^s_{\infty\infty}$ as follows: take u_{lk} i.i.d. uniform random variables on $[-B, B]$, and define the random wavelet series

$$U_s(x) = \sum_l \sum_k 2^{-l(s+1/2)} u_{lk}\psi_{lk}(x), \qquad (8.127)$$

for which we have

$$\|U_s\|_{B^s_{\infty\infty}} \le \sup_{k,l}|u_{lk}| \le B \ \ a.s.,$$

so its law is a natural prior on an s-Hölder ball.

Proposition 8.3.23 *Let $\epsilon > 0, s > 0, j \in \mathbb{N}$. Then*

$$\Pr\left\{\|K_j(U_s) - U_s\|_\infty < \epsilon B 2^{-js}\right\} \le e^{-\log(1/\epsilon)2^j};$$

in particular, the set $\mathcal{N}_s$ from Proposition 8.3.21 has probability 0 under the law of U_s.

Proof We have

$$\|K_j(U_s) - U_s\|_\infty \ge \|\psi\|_1^{-1}\sup_{l\ge j,k} 2^{l/2}|\langle\psi_{lk}, U_s\rangle| \ge \|\psi\|_1^{-1}2^{-js}\max_{k=1,\ldots,2^j}|u_{jk}|.$$

The variables u_{jk}/B are i.i.d. $U(-1,1)$, and so the U_k, $U_k := |u_{jk}/B|$, are i.i.d. $U(0,1)$ with maximum equal to the largest order statistic $U_{(2^j)}$. Since $\|\psi\|_1 \le \|\psi\|_2 \le 1$, we deduce

$$\Pr\left(\|K_j(U_s) - U_s\|_\infty < \epsilon B 2^{-js}\right) \le \Pr\left(U_{(2^j)} < \epsilon\right) = \epsilon^{2^j}$$

to complete the proof of the first claim. For the second claim, suppose that the set $\mathcal{N}_s$ has positive probability, say, $p_0 > 0$. Then, choosing ϵ small enough such that $\epsilon^{2^j} < p_0$, we have a contradiction, completing the proof. ■

A similar result can be proved for self-similarity conditions in the ℓ_2 setting, and for sake of exposition, let us again only analyse the wavelet version of condition (8.90) given in (8.126). We now consider Gaussian product priors

$$\Pi_s = \bigotimes_{k,l} \gamma_{k,l}, \quad \gamma_{k,l} \sim N(0, 2^{-2l(s+1/2)}), \quad s > 0.$$

Just as in the proof of Theorem 4.4.3, we show that a signal $p = (p_k : k \in \mathbb{N})$ drawn from Π_s lies in $B^s_{2\infty}$ almost surely. Moreover, any such draw concentrates almost surely on the self-similar elements of the space $B^s_{2\infty}$, as the following proposition shows.

Proposition 8.3.24 *For $s > 0$, consider the Gaussian probability measure Π_s on $B^s_{2\infty}$, and let $\mathcal{M}_s$ be as in Proposition 8.3.22. Then*

$$\Pi_s(\mathcal{M}_s) = 0.$$

Proof We realise that $p \sim \Pi_s$ as $(p_{lk}) = (2^{-l(s+1/2)} g_{lk})$ for i.i.d. standard normals g_{lk}. For $\varepsilon > 0$ fixed, the probability $\Pr(A_J)$ that p does not satisfy (8.126) at scale $J \in \mathbb{N}$ equals

$$\Pr(A_J) \equiv \Pr\left(\sum_{l=J(1-\varepsilon)}^{J} \sum_k 2^{-l(2s+1)} g_{lk}^2 < c(s) 2^{-2Js}\right) \tag{8.128}$$

$$= \Pr\left(\sum_{l=J(1-\varepsilon)}^{J} \sum_k 2^{-l(2s+1)} (g_{lk}^2 - 1) < c(s) 2^{-2Js} - c'(s) 2^{-2Js(1-\varepsilon)}\right) \tag{8.129}$$

$$\le \Pr\left(\left|\sum_{l=J(1-\varepsilon)}^{J} \sum_k 2^{-l(2s+1)} (g_{lk}^2 - 1)\right| > c'' 2^{-2Js(1-\varepsilon)}\right), \tag{8.130}$$

for $J \ge J_0$ large enough depending on s, ε, and some constants $c'(s), c'' > 0$. Using independence in the variance bound

$$\mathrm{Var}\left(\sum_{l=J(1-\varepsilon)}^{J} \sum_k 2^{-l(2s+1)} (g_{lk}^2 - 1)\right) \lesssim \sum_{l=J(1-\varepsilon)}^{J} 2^{-l(4s+1)} \lesssim 2^{-J(4s+1)(1-\varepsilon)}$$

and Chebyshev's inequality, the preceding probabilities satisfy $\sum_{J \ge J_0} \Pr(A_J) < \infty$. By the Borel-Cantelli lemma, we conclude that $\Pr(A_J \text{ i.o.}) = 0$ and hence that (8.126) holds almost surely. ■

Exercises

8.3.1 Consider the Haar wavelet estimator $f_n(j)$ from (6.120), and evaluate it at the bandwidth $j = \bar{j}_n + u_n$, where $\bar{j}_n$ is as in Lemma 8.3.13, computed from a sample of a Gaussian white noise independent of the sample used for the construction of $f_n(j)$ (e.g., by sample splittling). Show that (6.121) holds true uniformly in self-similar classes of functions, and deduce a confidence band as in Theorem 8.3.15 (see Giné and Nickl (2010)). *Hint*: For suitable random variables $Z_n(j)$, use Lemma 8.3.13 and independence to show that

$$P_f^Y(Z_n(\bar{j}_n) \leq t) = \sum_{j\in[j^*-m, j^*]} P_f^Y(Z_n(j) \leq t) P_f^Y(\bar{j}_n = j) + o(1),$$

and deduce the limit distribution from Theorem 2.7.1. For i.i.d. data, we can construct similar results by appealing to Propositions 5.1.22 and 5.1.23.

8.3.2 Complete the proof of Theorem 8.3.18. *Hint*: Proceed as in Theorem 8.3.11, with the function f_0 representing the first half of the functions f_m in Theorem 8.3.18 (see also Nickl and Szabo (2014)).

8.3.3 Prove the L^∞ case of Theorem 8.3.20.

8.3.4 Prove a version of Proposition 8.3.24 for uniform wavelet priors as in (8.127).

8.3.5 (Berry-Esseen bound.) For $X_1, \dots, X_n$ i.i.d. random variables with $EX_i = 0, EX_i^2 = 1, E|X_i|^3 = \rho < \infty$ and Φ the standard normal c.d.f., prove that

$$\left| \Pr\left(\sum_{i=1}^n X_i \leq t\sqrt{n} \right) - \Phi(t) \right| \lesssim \frac{C\rho}{\sqrt{n}},$$

where $C > 0$ is a universal constant. *Hint*: Adapt the proof of Lemma 3.7.45 or see Durrett (1996).

8.4 Notes

Section 8.1 Theorem 8.1.1 and the key ideas that derive from it are due to Spokoiny (1996). Related approaches are studied, for instance, in Baraud (2002) and Baraud, Huet and Laurent (2003). Results comparable to Theorem 8.1.2 in the setting of sampling models (in fact, in the more general deconvolution model) are obtained in Butucea, Matias and Pouet (2009). A multi-scale approach to certain adaptive testing problems is studied in Dümbgen and Spokoiny (2001). Exact minimax constants for adaptive tests are studied in Lepski and Tsybakov (2000). The monograph by Ingster and Suslina (2003) contains various materials on adaptive testing problems, and multiple testing corrections such as the ones encountered here are a large subject in statistics on its own, which we cannot survey here.

Section 8.2 Adaptive function estimation is by now a classical and well-studied topic. This section only contains some main ideas, and the following references are by no means complete.

Early key contributions to adaptive estimation from a minimax point of view are Efromovich and Pinsker (1984), Golubev (1987), Lepski (1990), Golubev and Nussbaum (1992) – see also Tsybakov (1998), Cavalier and Tsybakov (2001) and Tsybakov (2003), including also a discussion of the relationships to Stein's phenomenon and exact adaptation to Pinsker's minimax constant. The general purpose adaptation principle known as Lepski's method was introduced in Lepski (1990), see also Lepski, Mammen and Spokoiny (1997). Lepski also noted that in some situations a penalty for adaptive estimation can occur (for instance, when estimating a linear functional adaptively, such as in pointwise loss – see Tsybakov (1998) and Cai and Low (2005)). Wavelet thresholding ideas were introduced in the influential papers Donoho and Johnstone (1995) as well as Donoho, Johnstone, Kerkyacharian and Picard (1995, 1996). Another related approach to adaptation (that works best in the case of L^2-loss) is based on model selection, see Barron, Birgé and Massart (1999), Birgé and Massart (2001). More sophisticated versions of Lepski's method that are compatible also with ('anisotropic') multivariate situations have recently been suggested in Goldenshluger and Lepski (2011, 2014), and another popular method of adaptation is known as *aggregation*, see, for instance Tsybakov (2003), Bunea, Tsybakov and Wegkamp (2007) and Juditsky, Rigollet and Tsybakov (2008).

Most of the original adaptive estimation results were in the Gaussian white noise model and results in the i.i.d. sampling model are, with a few exceptions, more recent, e.g., Efromovich (1985, 2008), Golubev (1992), Giné and Nickl (2009, 2009a), Goldenshluger and Lepski (2011, 2014) and Lepski (2013). The global thresholding result Theorem 8.2.8 is due to Lounici and Nickl (2011). (Although an essential precursor, with some unnecessary moment conditions, was proved in Giné and Nickl (2009).)

A critical problem not addressed here in much detail is the choice of the thresholding constant τ. Resampling and symmetrisation approaches have been suggested, see for instance Giné and Nickl (2010a). Another approach is based on the idea of 'minimal penalties' in model selection, see Birgé and Massart (2007). In practice some kind of numerical calibration or cross-validation procedure can be advocated as well.

Section 8.3 That adaptive estimators do not automatically translate into adaptive confidence sets was noticed in the paper by Low (1997) in the setting of density estimation – he essentially proved Theorem 8.3.1 (for pointwise loss instead of uniform loss) – in fact, he showed the stronger result that the worst-case diameter can occur at any given function $f \in \Sigma(s)$. See also Cai and Low (2004), where these findings are cast into a general decision-theoretic framework, and Genovese and Wasserman (2008). The main observation behind Theorem 8.3.5, part (b)(i), is due, independently, to Juditsky and Lambert-Lacroix (2003), Cai and Low (2006) and Robins and van der Vaart (2006), with important ideas already implicit in Li (1988), Lepski (1999), Beran and Dümbgen (2003), Hoffmann and Lepski (2002) and Baraud (2004). As discussed in Robins and van der Vaart (2006), the connection to nonparametric testing problems is lurking in the background of many of these results – the explicit equivalence of adaptive confidence sets with smoothness testing problems as highlighted by Theorem 8.3.2 is from Hoffmann and Nickl (2011) and is further investigated in Bull and Nickl (2013) in the L^2 setting and in Carpentier (2013) in the L^p setting. The mechanism behind applies more generally as discussed in the subsection surrounding Proposition 8.3.6 – see Nickl and van de Geer (2013) for an application of these ideas to high-dimensional sparse regression. Moreover, the proof of theorem 3.5 in Carpentier (2013) can be adapted to show that a converse to Proposition 8.3.6 holds true in the sense that when $\rho_n^* = O(r_n(\Sigma_0))$, then adaptive and honest confidence sets can be constructed without any removal of parameters.

Honest inference for continuous smoothness parameters and unbounded radius is a more challenging task. The results from Section 8.3.2 are due to Bull and Nickl (2013) in the i.i.d. sampling model, and the respective positive results are inspired by Robins and van der Vaart (2006). In particular, the lower bound theorem 8.3.11 from Bull and Nickl (2013) (somewhat refined in Nickl and Szabó (2014)) shows how the nonparametric testing connection can be exploited to give rather strong negative results even for 'pointwise in f' confidence sets. Some further interesting positive results in this direction are found in Cai, Low and Ma (2014).

The approach to adaptive inference via self-similar functions was developed in Giné and Nickl (2010) in the L^∞ setting, with ideas going back to Picard and Tribouley (2000). The ℓ_2-theory for self-similar functions as presented here is due to Nickl and Szabò (2014). Minimal self-similarity assumptions in the L^∞ setting (paralleling Theorem 8.3.18) are studied in Bull (2012), who also gives a construction of a confidence band that adapts to possibly unbounded Hölder norm in the L^∞ setting if self-similarity parameters are known.

The theory about the genericity of self-similar functions was developed in Giné and Nickl (2010) and Hoffmann and Nickl (2011) in the i.i.d. sampling model, where Propositions 8.3.21 and 8.3.23 were proved, with some ideas going back to conjectures in mathematical physics and multifractal analysis (Jaffard (2000)). Further recent references that employ notions of self-similarity are Chernozhukov, Chetverikov and Kato (2014a) and Szabò, van der Vaart and van Zanten (2015). Another possibility to construct adaptive confidence sets for certain ranges of smoothness levels – not discussed in this book – is to assume shape constraints (such as monotonicity) of the function involved; see Dümbgen (2004) and Cai, Low and Xia (2013).

References

Adamczak, R. Moment inequalities for U-statistics. *Ann. Probab.* **34** (2006), 2288–314.

Adamczak, R. A tail inequality for suprema of unbounded empirical processes with applications to Markov chains. *Electronic J. Probab.* **34** (2008), 1000–34.

Akaike, H. Approximation to the density function. *Ann. Stat. Math. Tokyo* **6** (1954), 127–32.

Albin, J. M. P., and Choi, H. A new proof of an old result by Pickands. *Electronic Comm. Probab.* **15** (2010), 339–45.

Alexander, K. Probability inequalities for empirical processes and a law of the iterated logarithm. *Ann. Probab.* **12** (1984), 1041–67

Alexander, K. S. The central limit theorem for empirical processes on Vapnik-Červonenkis classes. *Ann. Probab.* **15** (1987), 178–203.

Andersen, N. T. *The Calculus of Non-measurable Functions and Sets.* (Various Publications Series No. 36). Aarhus Universitet, Matematisk Institut, Aarhus, Denmark, 1985.

Andersen, N. T., and Dobrić, V. The central limit theorem for stochastic processes. *Ann. Probab.* **15** (1987), 164–7.

Andersen, N. T., Giné, E., and Zinn, J. The central limit theorem and the law of iterated logarithm for empirical processes under local conditions. *Probab. Theory Related Fields* **77** (1988), 271–305.

Anderson, T. W. The integral of a symmetric unimodal function over a symmetric convex set and some probability inequalities. *Proc. Am. Math. Soc.* **6** (1955), 170–6.

Arcones, M. A., and Giné, E. Limit theorems for U-processes. *Ann. Probab.* **21** (1993), 1494–542.

Aronszajn, N. Theory of reproducing kernels. *Trans. Am. Math. Soc.* **68** (1950), 337–404.

Baernstein II, A., and Taylor, B. A. Spherical rearrangements, subharmonic functions, and *-functions in n-space. *Duke Math. J.* **43** (1976), 245–68.

Balabdaoui, F., Rufiback, K., and Wellner, J. Limit distribution theory for maximum likelihood estimation of a log-concave density. *Ann. Stat.* **37** (2009), 1299–331.

Balabdaoui, F., and Wellner, J. A Kiefer-Wolfowitz theorem for convex densities. *IMS Lectures Notes* **55** (2007), 1–31.

Baldi, P., and Roynette, B. Some exact equivalents for Brownian motion in Hölder norms. *Probab. Theory Rel. Fields* **93** (1992), 457–84.

Ball, K. Isometric problems in ℓ_p and sections of convex sets. Ph.D. thesis, University of Cambridge, 1986.

Ball, K., and Pajor, A. The entropy of convex bodies with 'few' extreme points. *London Mathematical Society Lecture Note Series* **158** (1990), 25–32 [Geometry of Banach spaces, Proceedings of a Conference, Strobl, Austria, 1999. P. F. X. Müller and W. Schachermayer, eds.].

Baraud, Y. Non-asymptotic minimax rates of testing in signal detection. *Bernoulli* **8** (2002), 577–606.

Baraud, Y. Confidence balls in Gaussian regression. *Ann. Stat.* **32** (2004), 528–51.

Baraud, Y. Estimator selection with respect to Hellinger-type risks. *Probab. Theory Relat. Fields* **151** (2011), 353–401.

Baraud, Y., Huet, S., and Laurent, B. Adaptive tests of linear hypotheses by model selection. *Ann. Stat.* **31** (2003), 225–51.

Baraud, Y., Huet, S., and Laurent, B. Testing convex hypotheses on the mean of a Gaussian vector: application to testing qualitative hypotheses on a regression function. *Ann. Stat.* **33** (2005), 214–57.

Barron, A., Birgé, L., and Massart, P. Risk bounds for model selection via penalization. *Probab. Theory Relat. Fields* **113** (1999), 301–413.

Barron, A., Schervish, M.J., and Wasserman, L. The consistency of posterior distributions in nonparametric problems. *Ann. Stat.* **27** (1999), 536–61.

Bartlett, P., Boucheron, S., and Lugosi, G. Model selection and error estimation. *Mach. Learn.* **48** (2002), 85–113.

Bass, R. F. Law of the iterated logarithm for set-indexed partial sum processes with finite variance. *Z. Wahrsch. Verw. Gebiete* **70** (1985), 591–608.

Beckner, W. Inequalities in Fourier analysis. *Ann. Math.* **102** (1975), 159–82.

Bednorz, W., and Latała, R. On the boundedness of Bernoulli processes. *Ann. Math.* **180** (2014), 1167–203.

Beers, G. *Topologies on Closed and Closed Convex Sets.* Klüwer, Dordrecht, 1993.

Beirlant, J., and Mason, D. M. On the asymptotic normality of L_p-norms of empirical functionals. *Math. Methods Stat.* **4** (1995), 1–19.

Bennett, G. Probability inequalities for the sum of independent random variables. *J. Am. Stat. Assoc.* **57** (1962), 33–45.

Benyamini, Y. Two point symmetrization, the isoperimetric inequality on the sphere and some applications. *Longhorn Notes, Texas Functional Analysis Seminar 1983–1984.* University of Texas, Austin, 1984, pp. 53–76.

Beran, R., and Dümbgen, L. Modulation of estimators and confidence sets. *Ann. Stat.* **26** (1998), 1826–56.

Bergh, J., and Löfström, J. *Interpolation Spaces.* Springer, Berlin, 1976.

Berman, S. M. Limit theorems for the maximum term in stationary sequences. *Ann. Math. Stat.* **35** (1964), 502–16.

Berman, S. M. Excursions above high levels for stationary Gaussian processes. *Pacific J. Math.* **36** (1971), 63–79.

Berman, S. M. Asymptotic independence of the number of high and low level crossings of stationary Gaussian processes. *Ann. Math. Stat.* **42** (1971a), 927–45.

Besov, O. V. On a family of function spaces: Embedding theorems and applications. *Dokl. Akad. Nauk SSSR* **126** (1959), 1163–65 (in Russian).

Besov, O. V. On a family of function spaces in connection with embeddings and extensions. *Trudy Mat. Inst. Steklov* **60** (1961), 42–81 (in Russian).

Bickel, P. and Ritov, Y. Estimating integrated squared density derivatives: sharp best order of convergence estimates. *Sankhya* **50** (1988), 381–93.

Bickel, P. J., and Ritov, Y. Nonparametric estimators that can be 'plugged-in'. *Ann. Stat.* **31** (2003), 1033–53.

Bickel, P. J., and Rosenblatt, M. On some global measures of the deviations of density function estimates. *Ann. Stat.* **1** (1973), 1071–95; correction, *ibid.* **3**, p. 1370.

Billingsley, P., and Topsoe, F. Uniformity in weak convergence. *Z. Wahrsch. Verw. Gebiete* **7** (1967), 1–16.

Birgé, L. Approximation dans les espace métriques et théorie de l'estimation. *Z. für Wahrscheinlichkeitstheorie und Verw. Geb.* **65** (1983), 181–238.

Birgé, L. Sur un théorème de minimax et son application aux tests. *Probab. Math. Stat.* **3** (1984), 259–82.

Birgé, L. Model selection via testing: an alternative to (penalized) maximum likelihood estimators. *Ann. Inst. H. Poincaré B* **20** (2006), 201–23.

Birgé, L. Robust tests for model selection. *IMS Collections* **9** (2012), 47–64.

Birgé, L., and Massart, P. Rates of convergence for minimum contrast estimators. *Probab. Theory Rel. Fields* **97** (1993), 113–50.

Birgé, L., and Massart, P. Estimation of integral functionals of a density. *Ann. Stat.* **23** (1995), 11–29.

Birgé, L., and Massart, P. Minimum contrast estimators on sieves: exponential bounds and rates of convergence. *Bernoulli* **4** (1998), 329–75.

Birgé, L., and Massart, P. Gaussian model selection. *J. Eur. Math. Soc.* **3** (2001), 203–68.

Birgé, L., and Massart, P. Minimal penalties for Gaussian model selection. *Probab. Theory Related Fields* **138** (2007), 33–73.

Birman, M. S., and Solomjak, M. Z. Piecewise polynomial approximations of functions of the classes W_p^α. *Math. Sb.* **73** (1967), 331–55 (in Russian).
Blum, J. R. On the convergence of empirical distribution functions. *Ann. Math. Stat.* **26** (1955), 527–9.
Bobkov, S. G., and Ledoux, M. From Brunn-Minkowski to Brascamp-Lieb and to logarithmic Sobolev inequalities. *Geom. and Funct. Analysis* **10** (2000), 1028–52
Bogachev, V. L. *Gaussian Measures*. American Mathematical Society, Providence, RI, 1998.
Bonami, A. Étude des coefficients de Fourier des fonctions de $L^p(G)$. *Ann. Inst. Fourier* **20** (1970), 335–402.
Borell, C. The Brunn-Minkowski inequalty in Gauss space. *Invent. Math.* **30** (1975), 207–16.
Borell, C. Gaussian Radon measures on locally convex spaces. *Math. Scand.* **38** (1976), 265–84.
Borell, C. On the integrability of Banach space valued Walsh Polynomials. *Seminaire de Probabilités XIII, Lect. Notes in Math.* **721** (1979), 1–3.
Borisov, I. S. Some limit theorems for empirical distributions. *Abstracts of Reports, Third Vilnius Conf. Probab. Theory Math. Stat.* **1** (1981), 71–2.
Borovkov, A., and Mogulskii, A. On probabilities of small deviations for stochastic processes. *Siberian Adv. Math.* **1** (1991), 39–63.
Boucheron, S., Bousquet, O., Lugosi, G., and Massart, P. Moment inequalities for functions of independent random variables. *Ann. Prob.* **33** (2005), 514–60.
Boucheron, S., Lugosi, G., and Massart, P. A sharp concentration inequality with applications. *Random Structures and Algorithms* **16** (2000), 277–92.
Boucheron, S., Lugosi, G., and Massart, P. *Concentration Inequalities: A Nonasymptotic Theory of Independence.* Oxford University Press, Oxford, UK, 2013.
Bourdaud, G., de Cristoforis, M. L., and Sickel, W. Superposition operators and functions of bounded p-variation. *Rev. Mat. Iberoam.* **22** (2006), 455–87.
Bousquet, O. Concentration inequalities for sub-additive functions using the entropy method. In *Stochastic Inequalities and Applications*, E. Giné, C. Houdré, and D. Nualart, eds. Birkhäuser, Boston, 2003, pp. 180–212.
Bretagnolle, J., and Huber, C. Estimation des densités: risque minimax. *Z. für Wahrscheinlichkeitstheorie und verw. Geb.* **47** (1979), 199–37.
Brown, L. D., Cai, T. T., Low, M., and Zhang, C. H. Asymptotic equivalence theory for nonparametric regression with random design. *Ann. of Stat.* **30** (2002), 688–707.
Brown, L. D., and Low, M. Asymptotic equivalence of nonparametric regression and white noise. *Ann. Stat.* **24** (1996), 2384–98.
Brown, L. D., and Zhang, C. H. Asymptotic nonequivalence of nonparametric experiments when the smoothness index is $1/2$. *Ann. Stat.* **26** (1998), 279–87.
Bull, A. D. Honest adaptive confidence bands and self-similar functions. *Elect. J. Stat.* **6** (2012), 1490–516.
Bull, A. D. A Smirnov-Bickel-Rosenblatt theorem for compactly-supported wavelets. *Construct. Approx.* **37** (2013), 295–309.
Bull, A. D., and Nickl, R. Adaptive confidence sets in L^2. *Probability Theory and Related Fields* **156** (2013), 889–919.
Bunea, F., Tsybakov, A. B., and Wegkamp, M. Aggregation for Gaussian regression. *Ann. Stat.* **35** (2007), 1674–97.
Butucea, C., Matias, C., and Pouet, C. Adaptive goodness-of-fit testing from indirect observations. *Ann. Inst. Henri Poincaré Probab. Stat.* **45** (2009), 352–72.
Butucea, C., and Tsybakov, A. B. Sharp optimality in density deconvolution with dominating bias, I (Russian summary). *Teor. Veroyatn. Primen.* **52** (2007), 111–28; trans. in *Theory Probab. Appl.* **52** (2008a), 24–39.
Butucea, C., and Tsybakov, A. B. Sharp optimality in density deconvolution with dominating bias, II (Russian summary). *Teor. Veroyatn. Primen.* **52** (2007), 336–49; trans. in *Theory Probab. Appl.* **52** (2008b), 237–49.
Cai, T. T., and Low, M. G. An adaptation theory for nonparametric confidence intervals. *Ann. Stat.* **32** (2004), 1805–40.
Cai, T. T., and Low, M. G. On adaptive estimation of linear functionals, *Ann. Stat.* **33** (2005), 2311–43.
Cai, T. T., and Low, M. Nonquadratic estimators of a quadratic functional. *Ann. Stat.* **33** (2005), 2930–56.

Cai, T. T., and Low, M. G. Adaptive confidence balls. *Ann. Stat.* **34** (2006), 202–28.

Cai, T. T., Low, M. G., and Ma, Z. Adaptive confidence bands for nonparametric regression functions. *J. Amer. Stat. Assoc.* **109** (2014), 1054–70.

Cai, T. T., Low, M. G., and Xia, Y. Adaptive confidence intervals for regression functions under shape constraints. *Ann. Stat.* **41** (2013), 722–50.

Cameron, R. H., and Martin, W. T. Transformations of Wiener integrals under translations. *Ann. Math.* **45** (1944), 386–96.

Cantelli, F. P. Sulla determinazione empirica delle leggi di probabilitá. *Giorn. Ist. Ital. Attuari* **4** (1933), 421–4.

Carl, B. Metric entropy of convex hulls in Hilbert spaces. *Bull. London Math. Soc.* **29** (1997), 452–8.

Carpentier, A. Honest and adaptive confidence sets in L^p. *Elect. J. Stat.* **7** (2013), 2875–2923.

Carpentier, A. Testing the regularity of a smooth signal. *Bernoulli*, **21** (2015), 465–88.

Castillo, I. On Bayesian supremum norm contraction rates, *Ann. Stat.* **42** (2014), 2058–91.

Castillo, I., and Nickl, R. Nonparametric Bernstein-von Mises theorems in Gaussian white noise. *Ann. Stat.* **41** (2013), 1999–2028.

Castillo, I., and Nickl, R. On the Bernstein–von Mises phenomenon for nonparametric Bayes procedures. *Ann. Stat.* **42** (2014), 1941–69.

Cavalier, L. Nonparametric statistical inverse problems. *Inverse Problems* **24** (2008).

Cavalier, L., and Tsybakov, A. B. Penalized blockwise Stein's method, monotone oracles and sharp adaptive estimation. *Math. Meth. Stat.* **10** (2001), 247–82.

Čencov, N. N. A bound for an unknown distribution density in terms of the observations. *Dokl. Akad. Nauk. SSSR* **147** (1962), 45–8.

Cencov, N. N. (1972) *Statistical Decision Rules and Optimal Inference*. Nauka, Moscow. English translation in Translations of Mathematical Monographs, 53, AMS, Providence, RI, 1982.

Chernozhukov, V., Chetverikov, D., and Kato, K. Gaussian approximation of suprema of empirical processes. *Ann. Stat.* **42** (2014), 1564–97.

Chernozhukov, V., Chetverikov, D., and Kato, K. Anti-concentration and honest adaptive confidence bands. *Ann. Stat.* **42** (2014a), 1787–818.

Chevet, S. Mesures de Radon sur $\mathbb{R}^n$ et mesures cylindriques. *Ann. Fac. Sci. Univ. Clermont* **43** (1970), 91–158.

Ciesielski, Z., Kerkyacharian, G., and Roynette, B. Quelques espaces fonctionnels associés à des processus Gaussiens. *Studia Math.* **107** (1993), 171–204.

Claeskens, G., and van Keilegom, I. Bootstrap confidence bands for regression curves and their derivatives. *Ann. Stat.* **31** (2003), 1852–84.

Cohen, A., Daubechies, I., and Vial, P. Wavelets on the interval and fast wavelet transforms. *Appl. Comput. Harmon. Anal.* **1** (1993), 54–84.

Coulhon, T., Kerkyacharian, G., and Petrushev, P. Heat kernel generated frames in the setting of Dirichlet spaces. *J. Fourier Anal. Appl.* **18** (2012), 995–1066.

Cox, D. D. An analysis of Bayesian inference for nonparametric regression. *Ann. Stat.* **21** (1993), 903–23.

Csörgö, M., and Horváth, L. Central limit theorems for L_p-norms of density estimators. *Z. Wahrscheinlichkeitstheorie Verw. Geb.* **80** (1988), 269–91.

Dalalyan, A., and Reiß, M. Asymptotic statistical equivalence for scalar ergodic diffusions. *Probability Theory and Related Fields*, **134** (2006), 248–82.

Daubechies, I. Orthonormal bases of compactly supported wavelets. *Comm. Pure Appl. Math.* **41** (1988), 909–96.

Daubechies, I. *Ten Lectures on Wavelets.* Society for Industrial and Applied Mathematics, Philadelphia, 1992.

Davies, E. B., and Simon, B. Ultracontractivity and the heat kernel for Schrödinger operators and Dirichlet Laplacians. *J. Funct. Anal.* **59** (1984), 335–95.

Davies, P. L., and Kovac, A. Local extremes, runs, strings and multiresolution. *Ann. Stat.* **29** (2001), 1–65.

Davies, P. L., Kovac, A., and Meise, M. Nonparametric regression, confidence regions and regularization. *Ann. Stat.* **37** (2009), 2597–625.

de Acosta, A. Small deviations in the functional central limit theorem with applications to functional laws of the iterated logarithm. *Ann. Prob.* **11** (1983), 78–101.

DeHardt, J. Generalizations of the Glivenko-Cantelli theorem. *Ann. Math. Stat.* **42** (1971), 2050–5.

Deheuvels, P. Uniform limit laws for kernel density estimators on possibly unbounded intervals. In *Recent Advances in Reliability Theory: Methodology, Practice and Inference*, N. Limnios and M. Nikulin, eds., Birkhäuser, Boston, 2000, pp. 477–92.

de la Peña, V., and Giné, E. *Decoupling, from Dependence to Independence.* Springer, Berlin, 1999.

DeVore, R., and Lorentz, G. G. *Constructive Approximation.* Springer, Berlin, 1993.

Devroye, L. *A Course in Density Estimation.* Birkhäuser, Boston, 1987.

Devroye, L. Exponential inequalities in nonparametric estimation. In *Nonparametric Functional Estimation and Related Topics*, G. Roussas, ed. NATO ASI Series. Kluwer, Dordrecht, 1991, pp. 31–44.

Devroye, L. A note on the usefulness of super kernels in density estimation. *Ann. Stat.* **20** (1992), 2037–56.

Devroye, L., and Lugosi, G. *Combinatorial Methods in Density Estimation*. Springer, New York, 2001.

Donoho, D. L., and Johnstone, I. M. Adapting to unknown smoothness via wavelet shrinkage, *J. Am. Stat. Assoc.* **90** (1995), 1200–24.

Donoho, D. L., and Johnstone, I. M. Minimax estimation via wavelet shrinkage. *Ann. Stat.* **26** (1998), 879–921.

Donoho, D. L., Johnstone, I. M., Kerkyacharian, G., and Picard, D. Wavelet shrinkage: asymptopia? *J. R. Stat. Soc. B* **57** (1995), 301–69.

Donoho, D. L., Johnstone, I. M., Kerkyacharian, G., and Picard, D. Density estimation by wavelet thresholding, *Ann. Stat.* **24** (1996), 508–39.

Donoho, D., and Nussbaum, M. Minimax quadratic estimation of a quadratic functional. *J. Complexity* **6** (1990), 290–323.

Donsker, M. D. Justification and extension of Doob's heuristic approach to the Komogorov-Smirnov theorems. *Ann. Math. Stat.* **23** (1952), 277–81.

Doob, J. L. Heuristic approach to the Kolmogorov-Smirnov theorems. *Ann. Math. Stat.* **20** (1949), 393–403.

Doob, J. L. Application of the theory of martingales. In *Le Calcul des Probabilités et ses Applications* (1949a), pp. 23–7. Colloques Internationaux du Centre National de la Recherche Scientifique, no. 13, *Centre National de la Recherche Scientifique, Paris.*

Doss, C., and Wellner, J. Global rates of convergence of the MLEs of logconcave and s-concave densities, arxiv preprint (2015).

Dudley, R. M. Weak convergences of probabilities on nonseparable metric spaces and empirical measures on Euclidean spaces. *Illinois J. Math.* **10** (1966), 109–26.

Dudley, R. M. The sizes of compact subsets of Hilbert space and continuity of Gaussian processes. *J. Funct. Anal.* **1** (1967), 290–330.

Dudley, R. M. Measures on non-separable metric spaces. *Illinois J. Math.* **11** (1967a), 449–53.

Dudley, R. M. Sample functions of the Gaussian process. *Ann. Probab.* **1** (1973), 66–103.

Dudley, R. M. Central limit theorems for empirical processes. *Ann. Probab.* **6** (1978), 899–929.

Dudley, R. M. A course on empirical processes. *Lecture Notes in Math.* **1097** (1984) 1–142.

Dudley, R. M. An extended Wichura theorem, definitions of Donsker class, and weighted empirical distributions, (Probability in Banach spaces, V). *Lecture Notes in Math.* **1153** (1985), 141–78.

Dudley, R. M. Universal Donsker classes and metric entropy. *Ann. Probab.* **15** (1987), 1306–26.

Dudley, R. M. Nonlinear functionals of empirical measures and the bootstrap (Probability in Banach spaces, 7). *Progr. Probab.* **21** (1990), 63–82.

Dudley, R. M. Fréchet differentiability, p-variation and uniform Donsker classes. *Ann. Probab.* **20** (1992), 1968–82.

Dudley, R. M. *Uniform Central Limit Theorems* (1st ed. 1999), 2nd ed. Cambridge University Press, 2014.

Dudley, R. M. *Real Analysis and Probability*, 2nd ed. Cambridge University Press, 2002.

Dudley, R. M., Giné, E., and Zinn, J. Uniform and universal Glivenko-Cantelli classes. *J. Theoret. Probab.* **4** (1991), 485–510.

Dudley, R. M., and Kanter, M. Zero-one laws for stable measures. *Proc. Am. Math. Soc.* **45** (1974), 245–52.

Dudley, R. M., and Norvaisa, R. *Concrete Functional Calculus.* Springer, New York, 2011.

Dudley R. M., and Philipp, W. Invariance principles for sums of Banach space valued random elements and empirical processes. *Z. Wahrsch. Verw. Gebiete* **62** (1983), 509–52.

Dümbgen, L. Optimal confidence bands for shape-restricted curves. *Bernoulli* **9** (2003), 423–49.

Dümbgen, L., and Rufibach, K. Maximum likelihood estimation of a log-concave density and its distribution function: basic properties and uniform consistency. *Bernoulli* **15** (2009), 40–68.

Dümbgen, L., and Spokoiny, V. Multiscale testing of qualitative hypotheses. *Ann. Stat.* **29** (2001), 124–52.

Durot, C., Kulikov, V. N., and Lopuhaä, H. P. The limit distribution of the L_∞-error of Grenander-type estimators. *Ann. Stat.* **40** (2012), 1578–608.

Durrett, R. *Probability: Theory and Examples.* Duxbury, Pacific Grove, CA, 1996.

Durst, M., and Dudley, R. M. Empirical processes, Vapnik-Červonenkis classes and Poisson processes. *Probab. Math. Stat.* **1** (1981), 109–15.

Dvoretzky, A., Kiefer, J., and Wolfowitz, J. Asymptotic minimax character of a sample distribution function and of the classical multinomial estimator. *Ann. Math. Stat.* **33** (1956), 642–69.

Edmunds, D. E., and Triebel, H. *Function Spaces, Entropy Numbers, Differential Operators.* Cambridge University Press, 1996.

Efromovich, S. Y. Nonparametric estimation of a density of unknown smoothness. *Theory Probability and Applications* **30** (1985), 524–34.

Efromovich, S. Y. Adaptive estimation of and oracle inequalities for probability densities and characteristic functions. *Ann. Stat.* **36** (2008), 1127–55.

Efromovich, S. Y., and Pinsker, M. S. Learning algorithm for nonparametric filtering, *Automation and Remote Control* **11** (1984), 1434–40.

Eggermont, P. P. B., and LaRiccia, V. N. *Maximum Penalized Likelihood Estimation, Vol. 1; Density Estimation.* Springer, New York, 2000.

Einmahl, U., and Mason, D. M. Some universal results on the behavior of increments of partial sums. *Ann. Probab.* **24** (1996), 1388–407.

Einmahl, U., and Mason, D. M. An empirical process approach to the uniform consistency of kernel-type function estimators. *J. Theoret. Probab.* **13** (2000), 1–37.

Einmahl, U., and Mason, D. Uniform in bandwidth consistency of kernel type function estimators. *Ann. Stat.* **33** (2005), 1380–403.

Ermakov, M. S. Minimax detection of a signal in a white Gaussian noise, *Theory Probab. Appl.* **35** (1990), 667–79.

Fan, J. Global behavior of deconvolution kernel estimates. *Stat. Sinica* **1** (1991), 541–51.

Fan, J. Adaptively local one-dimensional subproblems with application to a deconvolution problem. *Ann. Stat.* **21** (1993), 600–10.

Feller, W. *An Introduction to Probability Theory and Its Applications,* Vol. 1. Wiley, New York, 1968.

Fernique, X. Une demonstration simple du théorème de R. M. Dudley et M. Kanter sur les lois zéro-un pour les mésures stables. *Lecture Notes in Math.* **381** (1974), 78–9.

Fernique, X. Regularité des trajectoires des fonctions aléatoires gaussiennes. *Lecture Notes in Math.* **480** (1975), 1–96.

Fernique, X. *Fonctions Aléatoires Gaussiennes, Vecteurs Aléatoires Gaussiens.* Les Publications CRM, Montreal, 1997.

Figiel, T., Lindenstrauss, J., and Milman, V. D. The dimension of almost spherical sections of convex bodies. *Acta Math.* **139** (1977), 53–94.

Fisher, R. A. On the mathematical foundations of theoretical statistics. *Philos. Trans. R. Soc. A.* **222** (1922), 309–68.

Fisher, R. A. Theory of statistical estimation. *Proc. Cambridge Philos. Soc.* **22** (1925a), 700–25.

Fisher, R.A. *Statistical Methods for Research Workers.* Oliver and Body, Edinburgh, 1925b.

Folland, G. B. *Real Analysis: Modern Techniques and Their Applications.* Wiley, New York, 1999.

Frazier, M., Jawerth, B., and Weiss, G. *Littlewood-Paley Theory and the Study of Function Spaces.* AMS monographs, Providence, RI, 1991.

Freedman, D. A. On the asymptotic behavior of Bayes' estimates in the discrete case. *Ann. Math. Stat.* **34** (1963), 1386–403.

Freedman, D. A. On the Bernstein-von Mises theorem with infinite-dimensional parameters. *Ann. Stat.* **27** (1999), 1119–40.
Fromont M., and Laurent, B. Adaptive goodness-of-fit tests in a density model. *Ann. Stat.* **34** (2006), 680–720.
Gach, F., and Pötscher, B. M. Nonparametric maximum likelihood density estimation and simulation-based minimum distance estimators. *Math. Methods Stat.* **20** (2011), 288–326.
Gänssler, P., and Stute, W. Empirical processes: a survey of results for independent and identically distributed random variables. *Ann. Probab.* **7** (1979), 193–243.
Gayraud, G., and Pouet, C. Adaptive minimax testing in the discrete regression scheme. *Probab. Theory Related Fields* **133** (2005), 531–58.
Gardner, R. J. The Brunn-Minkowski inequality. *Bull AMS* **39** (2002), 355–405.
Gauß, C. F. *Theoria Motus Corporum Coelestium.* Perthes, Hamburg, 1809; reprint of the original at Cambridge University Press, 2011.
Gelfand, I. M., and Vilenkin, N. Y. *Les distributions*, Tome 4: *Applications de l'analyse harmonique.* Dunod, Paris, 1967 (translated from Russian).
Geller, D., and Pesenson, I. Z. Band-limited localized parseval frames and Besov spaces on compact homogeneous manifolds. *J. Geom. Anal.* **21** (2011), 334–71.
Genovese, C., and Wasserman, L. Adaptive confidence bands. *Ann. Stat.* **36** (2008), 875–905.
Ghosal, S. The Dirichlet process, related priors and posterior asymptotics. In *Bayesian Nonparametrics,* Cambridge University Press, 2010, pp. 35–79.
Ghosal, S., Ghosh, J. K., and van der Vaart, A. W. Convergence rates for posterior distributions. *Ann. Stat.* **28** (2000), 500–31.
Ghosal, S., and van der Vaart, A. W. Convergence rates of posterior distributions for non-i.i.d. observations. *Ann. Stat.* **35** (2007), 192–223.
Giné, E. Invariant tests for uniformity on compact Riemannian manifolds based on Sobolev norms. *Ann. Stat.* **3** (1975), 1243–66.
Giné, E. Empirical processes and applications: an overview. *Bernoulli* **2** (1996), 1–28.
Giné, E., and Guillou, A. On consistency of kernel density estimators for randomly censored data: rates holding uniformly over adaptive intervals *Ann. Inst. H. Poincaré Probab. Stat.* **37** (2001), 503–22.
Giné, E., and Guillou, A. Rates of strong uniform consistency for multivariate kernel density estimators. *Ann. Inst. H. Poincaré Prob. Stat.* **38** (2002), 907–21.
Giné., E., Güntürk C. S., and Madych, W. R. On the periodized square of L_2 cardinal splines. *Exp. Math.* **20** (2011), 177–88.
Giné, E., and Koltchinskii, V. Concentration inequalities and asymptotic results for ratio type empirical processes. *Ann. Probab.* **34** (2006), 1143–216.
Giné, E., Koltchinskii, V., and Sakhanenko, L. Kernel density estimators: convergence in distribution for weighted sup-norms. *Probab. Theory Related Fields* **130** (2004), 167–98.
Giné, E., Koltchinskii, V., and Wellner, J. Ratio limit theorems for empirical processes. In *Stochastic Inequalities and Applications*, Vol. 56. Birkhäuser, Basel, 2003, pp. 249–78.
Giné, E., Koltchinskii, V., and Zinn, J. Weighted uniform consistency of kernel density estimators. *Ann. Probab.* **32** (2004), 2570–605.
Giné, E., Latała, R., and Zinn, J. Exponential and moment inequalities for U-statistics. In *High Dimensional Probability*, Vol. 2. Birkhäuser, Boston, 2000, pp. 13–38.
Giné, E., and Madych, W. On wavelet projection kernels and the integrated squared error in density estimation. *Probab. Lett.* **91** (2014), 32–40.
Giné, E., and Mason, D. M. On the LIL for self-normalized sums of i.i.d. random variables. *J. Theoret. Probab.* **11** (1998), 351–70.
Giné, E., and Mason, D. M. On local U-statistic processes and the estimation of densities of functions of several sample variables. *Ann. Stat.* **35** (2007), 1105–45.
Giné, E., Mason D., and Zaitsev, A. The L_1-norm density estimator process. *Ann. Probab.* **31** (2003), 719–68.
Giné, E., and Nickl, R. Uniform central limit theorems for kernel density estimators. *Probab. Theory Relat. Fields* **141** (2008), 333–87.

Giné, E., and Nickl, R. A simple adaptive estimator of the integrated square of a density. *Bernoulli* **14** (2008a), 47–61.

Giné, E., and Nickl, R. Uniform limit theorems for wavelet density estimators. *Ann. Probab.* **37** (2009), 1605–46.

Giné, E., and Nickl, R. An exponential inequality for the distribution function of the kernel density estimator, with applications to adaptive estimation. *Probab. Theory Related Fields* **143** (2009a), 569–96.

Giné, E., and Nickl, R. Confidence bands in density estimation. *Ann. Stat.* **38** (2010), 1122–70.

Giné, E., and Nickl, R. Adaptive estimation of a distribution function and its density in sup-norm loss by wavelet and spline projections. *Bernoulli* **16** (2010a), 1137–63.

Giné, E., and Nickl, R. Rates of contraction for posterior distributions in L^r-metrics, $1 \leq r \leq \infty$. *Ann. Stat.* **39** (2011), 2883–911.

Giné, E., and Zinn, J. Central limit theorems and weak laws of large numbers in certain Banach spaces. *Z. Warscheinlichkeitstheorie Verw. Geb.* **62** (1983), 323–54.

Giné, E., and Zinn, J. Some limit theorems for empirical processes. *Ann. Probab.* **12** (1984), 929–89.

Giné, E., and Zinn, J. Lectures on the central limit theorem for empirical processes (probability and Banach spaces). *Lecture Notes in Math.* **1221** (1986), 50–113.

Giné, E., and Zinn, J. Empirical processes indexed by Lipschitz functions. *Ann. Probab.* **14** (1986a), 1329–38.

Giné, E., and Zinn, J. Bootstrapping general empirical measures. *Ann. Probab.* **18** (1990), 851–69.

Giné, E., and Zinn, J. Gaussian characterization of uniform Donsker classes of functions. *Ann. Probab.* **19** (1991), 758–82.

Glivenko, V. I. Sulla determiniazione empirica delle leggi di probabilitá. *Giorn. Ist. Ital. Attuari* **4** (1933), 92–9.

Goldenshluger, A., and Lepski, O. V. Bandwidth selection in kernel density estimation: oracle inequalities and adaptive minimax optimality. *Ann. Stat.* **39** (2011), 1608–32.

Goldenshluger, A., and Lepski, O. V. On adaptive minimax density estimation on R^d. *Probab. Theory Relat. Fields* **159** (2014), 479–543.

Golubev, G. K. Adaptive asymptotically minimax estimates of smooth signals, *Prob. Inf. Transm.* **23** (1987), 57–67.

Golubev, G. K. Nonparametric estimation of smooth densities of a distribution in L2. *Problems of Information Transmission*, **28** (1992), 44–54.

Golubev, G. K., and Nussbaum, M. Adaptive spline estimates in a nonparametric regression model. *Theory Prob. Appl.* **37** (1992), 521–9.

Golubev, G. K., Nussbaum, M., and Zhou, H. H. Asymptotic equivalence of spectral density estimation and Gaussian white noise. *Ann. Stat.* **38** (2010), 181–214.

Grama, I., and Nussbaum, M. Asymptotic equivalence for nonparametric regression. *Math. Methods Stat.* **11** (2002), 1–36.

Grenander, U. On the theory of mortality measurement. *Scand. Actuarial J.* **2** (1956), 125–53.

Groeneboom, P. Estimating a monotone density. In *Proceedings of the Berkeley Conference in Honor of Jerzy Neyman and Jack Kiefer,* Vol. II. Berkeley, CA, 1985, pp. 539–55.

Groeneboom, P., Jongbloed, G., and Wellner, J. A. Estimation of a convex function: characterizations and asymptotic theory. *Ann. Stat.* **29** (2001), 1653–98.

Gross, L. Abstract Wiener spaces. In *Proceedings of the Fifth Berkeley Symposium on Mathematics Statistics, and Probability*, Vol. II: *Contributions to Probability Theory, Part 1*. University of California Press, Berkeley, 1967, pp. 31–42.

Gross, L. Logarithmic Sobolev inequalities. *Am. J. Math.* **97** (1975), 1061–83.

Haagerup, U. The best constants in the Khintchine inequality. *Studia Math.* **70** (1982), 231–83.

Haar, A. Zur Theorie der orthogonalen Funktionensysteme. *Math. Ann.* **69** (1910), 331–71.

Hall, P. The rate of convergence of normal extremes. *J. Appl. Probab.* **16** (1979), 433–9.

Hall, P. Central limit theorem for integrated square error of multivariate nonparametric density estimators. *J. Multivariate Analysis* **14** (1984), 1–16.

Hall, P. Effect of bias estimation on coverage accuracy of bootstrap confidence intervals for a probability density. *Ann. Stat.* **20** (1992), 675–94.

Hanson, D. L., and Wright, F. T. A bound on tail probabilities for quadratic forms in independent random variables. *Ann. Math. Stat.* **42** (1971), 1079–83.

Härdle, W., Kerkyacharian, G., Picard, D., and Tsybakov, A. *Wavelets, Approximation and Statistical Applications* (Springer Lecture Notes in Statistics 129), Springer, New York, 1998.

Hardy, G. H., Littlewood, J. E., and Pólya, G. *Inequalities.* Cambridge University Press, 1967.

Haussler, D. Sphere packing numbers for subsets of the Boolean n-cube with bounded Vapnik-Červonenkis dimension. *J. Comb. Theory* **69** (1995), 217–32.

Hoeffding, V. Probability inequalities for sums of bounded random variables. *J. Am. Stat. Assoc.* **58** (1963), 13–30.

Hoffmann, M., and Lepski, O. V. Random rates in anisotropic regression. *Ann. Stat.* **30** (2002), 325–96 (with discussion).

Hoffmann, M., and Nickl, R. On adaptive inference and confidence bands, *Ann. Stat.* **39** (2011), 2383–409.

Hoffmann, M., Rousseau, J., and Schmidt-Hieber, J. On adaptive posterior concentration rates. *Ann. Stat.* (2015) (in press).

Hoffmann-Jørgensen, J. Sums of independent Banach space valued random variables. *Studia Math.* **52** (1974), 159–86.

Hoffmann-Jørgensen, J. The law of large numbers for non-measurable and non-separable random elements. *Asterisque* **131** (1985), 299–356.

Hoffmann-Jørgensen, J. *Stochastic Processes on Polish Spaces* (Various Publications Series **39**). Aarhus Universitet, Mathematisk Institute, 1991, Aarhus, Denmark.

Hoffmann-Jørgensen, J., Shepp, L. A., and Dudley, R. M. On the lower tail of Gaussian seminorms. *Ann. Probab.* **7** (1979), 319–42.

Houdré, C., and Reynaud-Bouret, P. Exponential inequalities with constants for U-statistics of order two. In *Stochastic Inequalities and Applications* (Progress in Probability 56). Birkhäuser, Boston, 2003, pp. 55–69.

Ibragimov, I. A., and Khasminskii, R. Z. On the estimation of an infinite-dimensional parameter in Gaussian white noise. *Sov. Math. Dokl.* **18** (1977), 1307–8.

Ibragimov, I. A., and Khasminskii, R. Z. On the nonparametric estimation of functionals. In *Symposium in Asymptotic Statistics*, J. Kozesnik, ed. Reidel, Dordrecht, 1978, pp. 42–52.

Ibragimov, I. A., and Khasminskii, R. Z. *Statistical Estimation. Asymptotic Theory.* Springer, New York, 1981.

Ibragimov, I. A., and Khasminskii, R. Z. Bounds for the risks of nonparametric regression estimates. *Theory of Probability and its Applications*, **27** (1982), 84–99.

Ibragimov, I. A., Nemirovskii, A. S., and Khasminskii, R. Z. Some problems of nonparametric estimation in Gaussian white noise. *Theory of Probability and its Applications*, **31** (1987), 391–406.

Ingster, Y. I. On the minimax nonparametric detection of a signal with Gaussian white noise. *Prob. Inform. Transm.* **28** (1982), 61–73 (in Russia).

Ingster, Y. I. Minimax testing of nonparametric hypotheses on a distribution density in the L^p-metrics, *Theory Probab. Appl.* **31** (1986), 333–7.

Ingster, Y. I. Asymptotically minimax hypothesis testing for nonparametric alternatives, Part I, II, and III. *Math. Methods Stat.* **2** (1993), 85–114, 171–89, 249–68.

Ingster, Y. I., and Suslina, I. A. *Nonparametric Goodness-of-Fit Testing under Gaussian Models* (Lecture Notes in Statistics). Springer, New York, 2003.

Jaffard, S. On the Frisch-Parisi conjecture. *J. Math. Pures Appl.* **79** (2000), 525–52.

Jain, N. C., and Marcus, M. B. Central limit theorems for $C(S)$-valued random variables. *J. Funct. Anal.* **19** (1975), 216–31.

Johnson, W. B., Schechtman, G., and Zinn, J. Best constants in moment inequalities for linear combinations of independent and exchangeable random variables. *Ann. Probab.* **13** (1985), 234–53.

Juditsky, A., and Lambert-Lacroix, S. Nonparametric confidence set estimation. *Math. Methods Stat.* **12** (2003), 410–28.

Juditsky, A., Rigollet, P., and Tsybakov, A. B. Learning by mirror averaging, *Ann. Stat.* **36** (2008), 2183–206.

Kahane, J. P. Sur les sommes vectorielles $\sum \pm u_n$. *Comptes Rendus Acad. Sci. Paris* **259** (1964), 2577–80.

Kahane, J. P. *Some Random Series of Functions*. D. C. Heath, Lexington, MA, 1968.

Kakutani, S. On equivalence of infinite product measures. *Ann. Math.* **49** (1948), 214–24.

Kallianpur, G. Abstract Wiener processes and their reproducing kernel Hilbert spaces. *Zeits. Wahrsch. Verb. Geb.* **17** (1971), 113–23.

Karhunen, K. Über lineare Methoden in der Wahrscheinlichkeitsrechnung. *Ann. Acad. Sci. Fennicae. Ser. A. I. Math.-Phys.* **37** (1947), 1–79.

Katz, M. F., and Rootzén, H. On the rate of convergence for extremes of mean square differentiable stationary normal processes. *J. Appl. Probab.* **34** (1997), 908–23.

Kerkyacharian, G., Nickl, R., and Picard, D. Concentration inequalities and confidence bands for needlet density estimators on compact homogeneous manifolds. *Probab. Theory Related Fields* **153** (2012), 363–404.

Kerkyacharian, G., and Picard, D. Density estimation in Besov spaces. *Stat. Probab. Letters* **13** (1992), 15–24.

Khatri, C. G. On certain inequalities for normal distributions and their applications to simultaneous confidence bounds. *Ann. Math. Stat.* **38** (1967), 1853–67.

Khinchin, A. Uber dyadische Brüche. *Math. Zeits.* **18** (1923), 109–16.

Kiefer, J., and Wolfowitz, J. Asymptotically minimax estimation of concave and convex distribution functions. *Z. Wahrscheinlichkeitstheorie und Verw. Gebiete* **34** (1976), 73–85.

Klein, T. Une inégalité de concentration à gauche pour les processus empiriques. *C. R. Math. Acad. Sci. Paris* **334** (2002), 501–4.

Klein, T., and Rio, E. Concentration around the mean for maxima of empirical processes. *Ann. Probab.* **33** (2005), 1060–77.

Klemelä, J. Sharp adaptive estimation of quadratic functionals. *Probab. Theory Relat. Fields* **134** (2006), 539–64.

Kolmogorov, A. N. *Grundbegriffe der Wahrscheinlichkeitstheorie.* Springer, Berlin, 1933.

Kolmogorov, A. N. Sulla determiniazione empirica di una legge di distribuzione. *Giorn. Ist. Ital. Attuari* **4** (1933a), 83–91.

Kolmogorov, A. N., and Tikhomirov, V. M. The ε-entropy and ε-capacity of sets in functional spaces. *Am. Math. Soc. Trans* **17** (1961), 277–364.

Koltchinskii, V. I. On the central limit theorem for empirical measures. *Theor. Probab. Math. Stat.* **24** (1981), 63–75.

Koltchinskii, V. Rademacher penalties and structural risk minimization. *IEEE Trans. Inform. Theory* **47** (2001), 1902–14.

Koltchinskii, V. Local Rademacher complexities and oracle inequalities in risk minimization. *Ann. Stat.* **34** (2006), 2593–656.

Konstant, D. G., and Piterbarg, V. I. Extreme values of the cyclostationary Gaussian processes. *J. Appl. Probab.* **30** (1993), 82–97.

Korostelev, A. P., and Tsybakov, A. B. *Minimax Theory of Image Reconstruction* (Lecture Notes in Statistics 82). Springer, New York, 1993.

Kuelbs, J., and Li, W. Metric entropy and the small ball problem for Gaussian measures. *J. Funct. Anal.* **116** (1993), 133–57.

Kuelbs, J., Li, W., and Linde, W. The Gaussian measure of shifted balls. *Probab. Theory Related Fields* **98** (1994), 143–62.

Kullback, S. A lower bound for discrimination information in terms of variation. *IEEE Trans. Inform. Theory* **13** (1967), 126–7.

Kwapień, S., and Woyczynski, W. *Random Series and Stochastic Integrals: Single and Multiple*. Birkhäuser, Boston, 1992.

Latała, R. Estimation of moments of sums of independent random variables. *Ann. Probab.* **25** (1997), 1502–13.

Latała, R. Estimates of moments and tails of Gaussian chaoses. *Ann. Probab.* **34** (2006), 2315–31.

Latała, R., and Oleskiewicz, K. On the best constant in the Khinchin-Kahane inequality. *Studia Math.* **109** (1994), 101–4.

Laurent, B. Efficient estimation of integral functionals of a density. *Ann. Stat.* **24** (1996), 659–81.

Leadbetter, M. R., Lindgren, G., and Rootzén, H. *Extremes and Related Properties of Random Sequences and Processes*. Springer, New York, 1983.

Leahu, H. On the Bernstein–von Mises phenomenon in the Gaussian white noise model. *Elect. J. Stat.* **5** (2011), 474–4.

Le Cam, L. M. On some asymptotic properties of maximum likelihood estimates and related Bayes' estimates. *Univ. California Publ. Stat.* **1** (1953), 277–329.

Le Cam, L. M. Convergence of estimates under dimensionality restrictions. *Ann. Stat.* **1** (1973), 38–53.

Le Cam, L. *Asymptotic Methods in Statistical Decision Theory*. Springer, New York, 1986.

Le Cam, L., and Yang, G. L. *Asymptotics in Statistics: Some Basic Concepts*. Springer, New York, 1990.

Ledoux, M. Isoperimetry and Gaussian analysis. *Lecture Notes in Math.* **1648** (1996), 165–294.

Ledoux, M. On Talagrand's deviation inequalities for product measures. *ESAIM: Probab. Stat.* **1** (1997), 63–87.

Ledoux, M. *The Concentration of Measure Phenomenon*. American Math. Soc., Providence, RI, 2001.

Ledoux, M., and Talagrand, M. Conditions d'integrabilité pour les multiplicateurs dans le TLC banachique. *Ann. Probab.* **14** (1986), 916–21.

Ledoux, M., and Talagrand, M. Comparison theorems, random geometry and some limit theorems for empirical processes. *Ann. Probab.* **17** (1989), 596–631.

Ledoux, M., and Talagrand, M. *Probability in Banach Spaces*. Springer-Verlag, Berlin, 1991.

Leindler, L. On a certain converse of Hölder's inequality. *Acta Sci. Math.* **34** (1973), 335–43.

Lepski, O. V. On a problem of adaptive estimation in Gaussian white noise, *Theory Prob. Appl.* **35** (1990), 454–66.

Lepski, O. V. How to improve the accuracy of estimation. *Math. Meth. Stat.* **8** (1999), 441–86.

Lepski, O. V. Multivariate density estimation under sup-norm loss: oracle approach, adaptation and independence structure. *Ann. Stat.* **41** (2013), 1005–34.

Lepski, O. V., Mammen, E., and Spokoiny, V. Optimal spatial adaptation to inhomogeneous smoothness: an approach based on kernel estimators with variable bandwidth selectors. *Ann. Stat.* **25** (1997), 929–47.

Lepski, O. V., and Tsybakov, A.B. Asymptotically exact nonparametric hypothesis testing in sup-norm and at a fixed point. *Probab. Theory Related Fields* **117** (2000), 17–48.

Levit, B. Ya. Asymptotically efficient estimation of nonlinear functionals. *Prob. Inform. Transm.* **14** (1978), 204–9.

Lévy, P. *Problèmes concrets d'analyse fonctionelle*. Gauthier-Villars, Paris, 1951.

Li, K. C. Honest confidence regions for nonparametric regression, *Ann. Stat.* **17** (1989), 1001–8.

Li, W., and Linde, W. Approximation, metric entropy and small ball estimates for Gaussian measures. *Ann. Probab.* **27** (1999), 1556–78.

Li, W. V., and Shao, Q.-M. Gaussian processes: inequalities, small ball probabilities and applications. In *Stochastic Processes: Theory and Methods* (Handbook of Statistics), Vol. 19, C. R. Rao and D. Shanbhag, eds. Elsevier, New York, 2001, pp. 533–98.

Lijoi, A., and Prünster, I. Models beyond the Dirichlet process. In *Bayesian Nonparametrics. 35–79*, (Camb. Ser. Stat. Probab. Math.). Cambridge University Press, 2010, pp. 35–79.

Littlewood, J. E. On bounded bilinear forms in an infinite number of variables. *Q. J. Math., Oxford Ser.* **1** (1930), 164–74.

Littlewood, J. E., and Paley, R. E. A. C. Theorems on Fourier series and power series I, II. *J. Lond. Math. Soc.* **6** (1931), 230–3 (I); *Proc. Lond. Math. Soc.* **42** (1936), 52–89 (II).

Loève, M. *Probability Theory* (Graduate Texts in Mathematics 46), Vol. II, 4th ed. Springer-Verlag, Berlin, 1978.

Lorentz, G. G., Golitscheck, M. V., and Makovoz, Y. *Constructive Approximation: Advanced Problems*. Springer, Berlin, 1996.

Lounici, K., and Nickl, R. Global uniform risk bounds for wavelet deconvolution estimators. *Ann. Stat.* **39** (2011) 201–31.

Love, E. R., and Young, L. C. Sur une classe de fonctionelles linéaires. *Fund. Math.* **28** (1937), 243–57.

Low, M. G. On nonparametric confidence intervals, *Ann. Stat.* **25** (1997), 2547–54.

Mallat, S. Multiresolution approximation and wavelet orthonormal bases of $L^2(\mathbb{R})$. *Trans. Am. Math. Soc.* **315** (1989), 69–87.

Marcus, D. J. Relationships between Donsker classes and Sobolev spaces, *Z. Wahrscheinlichkeitstheorie verw. Gebiete* **69** (1985), 323–30.

Marcus, M. B., and Shepp, L. A. Sample behavior of Gaussian processes. In *Proceedings of the Sixth Berkeley Symposium on Mathematical Statistics and Probability (Univ. California, Berkeley, Calif., 1970–1971)*, Vol. II: *Probability Theory*. University California Press, Berkeley, 1972, pp. 423–41.

Massart, P. The tight constant in the Dvoretzky-Kiefer-Wolfowitz inequality. *Ann. Probab.* **18** (1990), 1269–83.

Massart, P. About the constants in Talagrand's concentration inequalities for empirical processes. *Ann. Probab.* **28** (2000), 863–84.

Massart, P. Some applications of concentration inequalities in statistics. *Ann. Fac. Sci. Toulouse Math.* **9** (2000), 245–303.

Massart, P. *Concentration Inequalities and Model Selection* (Lectures from the 33rd Summer School on Probability Theory Held in Saint-Flour, July 6–23, 2003; Lecture Notes in Mathematics 1896). Springer, Berlin, 2007.

Meister, A. *Deconvolution Problems in Nonparametric Statistics* (Lecture Notes in Statistics 193). Springer, Berlin, 2009.

Maurer, A. Thermodynamics and concentration. *Bernoulli* **18** (2012), 434–54.

Mattila, P. *Geometry of Sets and Measures in Euclidean Spaces*. Cambridge University Press, 1995.

McDiarmid, C. On the method of bounded differences. In *Surveys in Combinatorics*. Cambridge University Press, 1989, pp. 148–88.

McDiarmid, C. Concentration. In *Probabilistic Methods for Algorithmic Discrete Mathematics* (Algorithms Combin, 16). Springer, Berlin, 1998, pp. 195–248.

Meyer, Y. *Wavelets and Operators*. Cambridge University Press, 1992.

Meyer, Y., Sellan, F., and Taqqu, M. S. Wavelets, generalized white noise and fractional integration: the synthesis of fractional Brownian motion. *J. Fourier Anal. Appl.* **5** (1999), 465–94.

Montgomery-Smith, S. Comparison of sums of independent identically distributed random variables. *Prob. Math. Stat.* **14** (1994), 281–5.

Mourier, E. Lois de grandes nombres et théorie ergodique. *C. R. Acad. Sci. Paris* **232** (1951), 923–5.

Nadaraya, E. A. On estimating regression. *Theory Probab. Appl.*, **9** (1964), 141–2.

Nemirovskii, A. S. Nonparametric estimation of smooth regression functions. *Soviet J. of Computer and Systems Sciences*, **23** (1985), 1–11.

Nemirovskii, A. S. On necessary conditions for the efficient estimation of functionals of a nonparametric signal which is observed in white noise. *Theory of Probability and its Applications*, 35 (1990), 94–103.

Nemirovski, A. Topics in Non-parametric Statistics. Ecole d'Ete de Probabilités de Saint-Flour XXVIII – 1998. *Lecture Notes in Mathematics, v. 1738*. Springer, New York (2000).

Nemirovskii A. S., Polyak B. T., and Tsybakov, A. B. Rate of convergence of nonparametric estimators of maximum-likelihood type. *Problems of Information Transmission*, **21** (1985), 258–72.

Nickl, R. Empirical and Gaussian processes on Besov classes. In *High Dimensional Probability IV* (IMS Lecture Notes 51), E. Giné, V. Koltchinskii, W. Li, and J. Zinn, eds. Springer, Berlin, 2006, pp. 185–95.

Nickl, R. Donsker-type theorem for nonparametric maximum likelihood estimators. *Probab. Theory Related Fields* **138** (2007), 411–49; erratum (2008) *ibid.*

Nickl, R. Uniform central limit theorems for sieved maximum likelihood and trigonometric series estimators on the unit circle. In *High Dimensional Probability V*. Birkhaeuser, Boston, 2009, pp. 338–56.

Nickl, R., and Pötscher, B. M. Bracketing metric entropy rates and empirical central limit theorems for function classes of Besov- and Sobolev-type. *J. Theoret. Probab.* **20** (2007), 177–99.

Nickl, R., and Reiß, M. A Donsker theorem for Lévy measures. *J. Funct. Anal.* **263** (2012), 3306–32.

Nickl, R., Reiß, M., Söhl, J., and Trabs, M. High-frequency Donsker theorems for Lévy measures. *Probab. Theory Related Fields* (2015).

Nickl, R., and Szabó, B. A sharp adaptive confidence ball for self-similar functions. *Stoch. Proc. Appl.*, in press (2015).

Nickl, R., and van de Geer, S. Confidence sets in sparse regression, *Ann. Stat.* **41** (2013), 2852–76.

Nolan, D., and Pollard, D. U-processes: rates of convergence. *Ann. Stat.* **15** (1987), 780–99.

Nussbaum, M. Asymptotic equivalence of density estimation and Gaussian white noise. *Ann. Stat.* **24** (1996), 2399–430.

Ossiander, M. A central limit theorem under metric entropy with L_2 bracketing. *Ann. Probab.* **15** (1987), 897–919.

Ottaviani, G. Sulla teoria astratta del calcolo delle probabilità proposita dal Cantelli. *Giorn. Ist. Ital. Attuari* **10** (1939), 10–40.

Oxtoby, J. C., and Ulam, S. On the existence of a measure invariant under a transformation. *Ann. Math.* **40** (1939), 560–6.

Paley, R. E. A. C., and Zygmund, A. On some series of functions, part 1. *Proc. Cambridge Philos. Soc.* **28** (1930), 266–72.

Paley, R. E. A. C., and Zygmund, A. A note on analytic functions on the unit circle. *Proc. Cambridge Philos. Soc.* **28** (1932), 266–72.

Parzen, E. On estimation of a probability density function and mode. *Ann. Math. Stat.* **33** (1962), 1065–76.

Pearson, K. On the criterion that a given system of deviations from the probable in the case of a correlated system of variables is such that it can be reasonably supposed to have arisen from random sampling. *Philos. Mag. Series 5* **50** (1900), 157–75.

Peetre, J. *New Thoughts on Besov Spaces* (Duke University Math. Series). Duke University Press, Durham, NC, 1976.

Pensky, M., and Vidakovic, B. Adaptive wavelet estimator for nonparametric density deconvolution. *Ann. Stat.* **27** (1999), 2033–53.

Picard, D., and Tribouley, K. Adaptive confidence interval for pointwise curve estimation. *Ann. Stat.* **28** (2000), 298–335.

Pickands, J., III. Asymptotic properties of the maximum in a stationary Gaussian process. *Trans. Am. Math. Soc.* **145** (1969), 75–86.

Pinelis, I. Optimum bounds for the distributions of martingales in Banach spaces. *Ann. Probab.* **22** (1994), 1679–706.

Pinsker, M. S. *Information and Information Stability of Random Variables and Processes.* Holden-Day, San Francisco, 1964.

Pisier, G. Remarques sur un résultat non publié de B. Maurey. In *Séminaire d'Analyse Fonctionelle 1980–1981*, Vols. 1–12. École Polytechnique, Palaiseau, France.

Pisier, G. Some applications of the metric entropy condition to harmonic analysis. *Lecture Notes in Math.* Vol. **995**. Springer, Berlin, pp. 123–54.

Pisier, G. Probabilistic methods in the geometry of Banach spaces. *Lecture Notes in Math.*, Vol. **1206**. Springer, Berlin, 1986, 105–136.

Pisier, G. *The Volume of Convex Bodies and Banach Space Geometry*. Cambridge University Press, 1989.

Piterbarg, V. I. *Asymptotic Methods in the Theory of Gaussian Processes and Fields* (Translations of Math. Monographs 148). AMS, Providence, RI, 1996.

Piterbarg, V. I., and Seleznjev, O. Linear interpolation of random processes and extremes of a sequence of Gaussian non-stationary processes. Technical report, Center for Stochastic Processes, North Carolina University; Chapel Hill, NC, 1994, p. 446.

Plackett, R. L. The discovery of the method of least squares. *Biometrika* **59** (1972), 239–51.

Pollard, D. Limit theorems for empirical processes. *Z. Wahrsch. Verw. Gebiete* **57** (1981), 181–95.

Pollard, D. A central limit theorem for empirical processes. *J. Austral. Math. Soc., Ser. A* **33** (1982), 235–48.

Pouet, C. Test asymptotiquement minimax pour une hypothèse nulle composite dans le modèle de densité. *C.R. Math. Acad. Sci. Paris* **334** (2002), 913–16.

Prakasa Rao, B. L. S. Estimation of a unimodal density. *Sankhya Ser. A* **31** (1969), 23–36.

Prékopa, A. On logarithmically concave measures and functions. *Acta Sci. Math.* **33** (1972), 217–23.

Radulović, D., and Wegkamp, M. Uniform central limit theorems for pregaussian classes of functions. In *High Dimensional Probability V: The Luminy Volume* (Inst. Math. Stat. Collect. 5). IMS, Beachwood, OH, 2009, pp. 84–102.

Ray, K. Random Fourier series with applications to statistics. *Part III Essay in Mathematics*, University of Cambridge, 2010.

Ray, K. Bayesian inverse problems with non-conjugate priors, *Elect. J. Stat.* **7** (2013), 2516–49.

Ray, K. Bernstein–von Mises theorems for adaptive Bayesian nonparametric procedures, Preprint (2014), *arxiv 1407.3397*.

Reiß, M. Asymptotic equivalence for nonparametric regression with multivariate and random design. *Ann. Stat.* **36** (2008), 1957–82.

Rhee, WanSoo T. Central limit theorem and increment conditions. *Stat. Probab. Lett.* **4** (1986), 191–5.

Rio, E. Local invariance principles and their application to density estimation. *Probab. Theory Related Fields* **98** (1994), 21–45.

Rio, E. Inégalités de concentration pour les processus empiriques de classes de parties. *Probab. Theory Related. Fields* **119** (2001), 163–75.

Rio, E. Une inégalité de Bennett pour les maxima de processus empiriques. *Ann. I. H. Poincaré Prob.* **38** (2002), 1053–7.

Rio, E. Inegalités exponentielles et inegalités de concentration. Lectures at University Bordeaux Sud-Ouest, 2009.

Rio, E. Sur la function de taux dans les inegalités de Talagrand pour les processus empiriques. *C. R. Acad. Sci. Paris Ser. I* **350** (2012), 303–5.

Robins, J., and van der Vaart, A. W. Adaptive nonparametric confidence sets. *Ann. Stat.* **34** (2006), 229–53.

Rootzén, H. The rate of convergence of extremes of stationary normal sequences. *Adv. Appl. Probab.* **15** (1983), 54–80.

Rosenblatt, M. Remarks on some nonparametric estimates of a density function. *Ann. Math. Stat.* **27** (1956), 832–7.

Salem, R., and Zygmund, A. Some properties of trigonometric series whose terms have random signs. *Acta Math.* **91** (1954), 245–301.

Samson, P.-M. Concentration of measure inequalities for Markov chains and ϕ-mixing processes. *Ann. Probab.* **28** (2000), 416–61.

Sauer, N. On the density of families of sets. *J. Comb. Theory* **13** (1972), 145–7.

Schmidt, E. Die Brunn-Minkowskische Ungleichung und ihr Spiegelbild sowie die isoperimetrische Eigenschaft der Kugel in der euklidischen und nichteuklidischen Geometrie. *Math. Nachr.* **1** (1948), 81–157.

Schwartz, L. On Bayes procedures. *Z. Wahrscheinlichkeitstheorie und Verw. Gebiete* **4** (1965), 10–26.

Shannon, C. E. Communication in the presence of noise, *Proc. Institute of Radio Engineers* **37** (1949), 10–21.

Sheehy, A., and Wellner, J. Uniform Donsker classes of functions. *Ann. Probab.* **20** (1992), 1983–2030.

Shelah, S. A combinatorial problem: stability and order for models and theories in infinitary languages. *Pacific J. Math.* **41** (1992), 247–61.

Shen, X., and Wasserman, L. Rates of convergence of posterior distributions. *Ann. Stat.* **29** (2001), 687–714.

Sidak, Z. Rectangular confidence regions for the means of multivariate normal distributions. *J. Am. Stat. Assoc.* **62** (1967), 626–33.

Sidak, Z. On multivariate normal probabilities of rectangles: their dependence on correlations. *Ann. Math. Stat.* **39** (1968), 1425–34.

Slepian, D. The one sided barrier problem for Gaussian noise. *Bell Systems Tech. J.* **41** (1962), 463–501.

Smirnov, N. V. Estimation of the deviation between empirical distribution curves of two independent samples. *Bull. Univ. Moscow* **2** (1939), 3–14 (in Russian).

Smirnov, N. V. On the construction of confidence regions for the density of distribution of random variables. *Dokl. Akad. Nauk SSSR* **74**, (1950), 184–91 (in Russian).

Sobolev, S. L. The Cauchy problem in a functional space. *Dokl. Akad. Nauk SSSR* **3** (1935), 291–4 (in Russian).

Sobolev, S. L. On a theorem of functional analysis. *Mat. Sb.* **4** (1938), 471–97 (in Russian).

Söhl, J. Uniform central limit theorems for the Grenander estimator. *Elect. J. Stat.* **9** (2015) 1404–23.

Spokoiny, V. Adaptive hypothesis testing using wavelets. *Ann. Stat.* **24** (1996), 2477–98.

Stefanski, L. A., and Carroll, R. J. Deconvoluting kernel density estimators. *Stat. 21* (1990), 169–84.

Stigler, S. M. Gauss and the invention of least squares. *Ann. Stat.* **9** (1981), 465–74.

Stolz, W. Une méthode elementaire pour l'evaluation de petities boules browniennes. *C. R. Acad. Sci. Paris* **316** (1994), 1217–20.

Stone, C. J. Optimal rates of convergence for nonparametric estimators, *Ann. Stat.* **8** (1980), 1348–60.

Stone, C. J. Optimal global rates of convergence for nonparametric regression, *Ann. Stat.* **10** (1982), 1040–53.

Strassen, V., and Dudley, R. M. The central limit theorem and epsilon-entropy. In *Probability and Information Theory* (Lecture Notes in Math. **1247**). Springer, Berlin, 1969, pp. 224–31.

Strobl, F. On the reversed submartingale property of empirical discrepancies in arbitrary sample spaces. *J. Theoret. Probab.* **8** (1995), 825–31.

Stute, W. The oscillation behavior of empirical processes. *Ann. Probab.* **10** (1982), 86–107.

Stute, W. The oscillation behavior of empirical processes: the multivariate case. *Ann. Probab.* **12** (1984), 361–79.

Sudakov, V. N. Gaussian measures, Cauchy measures and ε-entropy. *Soviet Math. Dokl.* **10** (1969), 310–13.

Sudakov, V. N. A remark on the criterion of continuity of Gaussian sample functions. *Lecture Notes in Math.* **330** (1973), 444–54.

Sudakov, V. N., and Tsirelson, B. S. Extremal properties of half-spaces for spherically invariant measures. *Zap. Naucn. Sem. Leningrad. Otdel. Mat. Inst. Steklov. (LOMI)* **41** (1974), 14–24.

Szabó, B., van der Vaart, A. W., and van Zanten, J. H. Frequentist coverage of adaptive nonparametric Bayesian credible sets (with discussion). *Ann. Stat.* **43** (2015), 1391–428.

Szarek, S. J. On the best constants in the Khinchin inequality. *Studia Math.* **58** (1976), 197–208.

Talagrand, M. Donsker classes and random geometry. *Ann. Probab.* **15** (1987), 1327–38.

Talagrand, M. Regularity of Gaussian processes. *Acta Math.* **159** (1987a), 99–149.

Talagrand, M. Donsker classes of sets. *Probab. Theory Related Fields* **78** (1988), 169–91.

Talagrand, M. An isoperimetric theorem on the cube and the Kintchine-Kahane inequalities. *Proc. Am. Math. Soc.* **104** (1988a), 905–9.

Talagrand, M. Isoperimetry and integrability of the sum of independent Banach space valued random variables. *Ann. Probab.* **17** (1989), 1546–70.

Talagrand, M. Sharper bounds for Gaussian and empirical processes. *Ann. Probab.* **22** (1994), 28–76.

Talagrand, M. The supremum of some canonical processes. *Am. J. Math.* **116** (1994a), 283–325.

Talagrand, M. Concentration of measure and isoperimetric inequalities in product spaces. *Inst. Hautes Études Sci. Publ. Math.* **81** (1995), 73–205.

Talagrand, M. New concentration inequalities in product spaces. *Invent. Math.* **126** (1996), 505–63.

Talagrand, M. *The Generic Chaining*. Springer, Berlin, 2005.

Teh, Y. W., and Jordan, M. I. Hierarchical Bayesian nonparametric models with applications. In *Bayesian Nonparametrics*, (Camb. Ser. Stat. Probab. Math.) Cambridge University Press, 2010, pp. 35–79.

Triebel, H. *Theory of Function Spaces*. Birkhäuser, Basel, 1983.

Tsybakov, A. B. Pointwise and sup-norm sharp adaptive estimation of functions on the Sobolev classes. *Annals of Statistics*, **26** (1998), 2420–69.

Tsybakov, A. B. Optimal rates of aggregation. Computational Learning Theory and Kernel Machines. Proc. 16th Annual Conference on Learning Theory (COLT), B. Scholkopf and M. Warmuth, eds. *Lecture Notes in Artificial Intelligence*, v.2777. Springer, Heidelberg, 303–13 (2003).

Tsybakov, A. B. *Introduction to Nonparametric Estimation, Springer Series in Statistics*, Springer, New York (2009).

van de Geer, S. The entropy bound for monotone functions. Report TW 91–10, University of Leiden, 1991.

van de Geer, S. Hellinger-consistency of certain nonparametric maximum likelihood estimators. *Ann. Stat.* **21** (1993), 14–44.

van de Geer, S. The method of sieves and minimum contrast estimators, *Math. Methods Stat.* **4** (1995), 20–28.

van de Geer, S. *Empirical Processes in M-Estimation.* Cambridge University Press, 2000.

van der Vaart, A. W. Weak convergence of smoothed empirical processes. *Scand. J. Stat.* 21 (1994), 501–4.

van der Vaart, A. W. *Asymptotic Statistics.* Cambridge University Press, 1998.

van der Vaart, A. W., and van Zanten, J. H. Rates of contraction of posterior distributions based on Gaussian process priors. *Ann. Stat.* **36** (2008), 1435–63.

van der Vaart, A. W., and van Zanten, J. H. Reproducing kernel Hilbert spaces of Gaussian priors. *IMS Collections: Pushing the Limits of Contemporary Statistics: Contributions in honor of Jayantha K. Ghosh* **3** (2008a), 200–22.

van der Vaart, A. W., and Wellner, J. *Weak convergence and empirical processes. With Applications to Statistics.* Springer, New York, 1996.

van der Vaart, A. W., and Wellner, J. A local maximal inequality under uniform entropy. *Electronic J. Stat.* **5** (2011), 192–203.

Vapnik, V. N., and Červonenkis, A. Ya. On the uniform convergence of frequencies of occurrence of events to their probabilities. *Probab. Theory Appl.* **26** (1968), 264–80.

Vapnik, V. N., and Červonenkis, A. Ya. Necessary and sufficient conditions for the uniform convergence of means to their expectations. *Theory Probab. Appl.* **16** (1971), 264–80.

Vapnik, V. N., and Červonenkis, A. Ya. *Theory of Pattern Recognition: Statistical Problems on Learning.* Nauka, Moscow, 1974 (in Russian).

Watson, G. S. Smooth regression analysis. *Sankhya, Ser. A* **26** (1964), 359–72.

Wong, W. H., and Severini, T. A. On maximum likelihood estimation in infinite dimensional parameter spaces. *Ann. Stat.* **19** (1991), 603–32.

Wong, W. H., and Shen, X. Probability inequalities for likelihood ratios and convergence rates of sieve MLEs. *Ann. Stat.* **23** (1995), 33–362.

Yukich, J. Weak convergence of smoothed empirical processes. *Scand. J. Stat.* **19** (1992), 271–9.

Zhao, L. H. Bayesian aspects of some nonparametric problems. *Ann. Stat.* **28** 532–52.

Zygmund, A. Smooth functions. *Duke Math. J.* **12** (1945), 47–76.

Zygmund, A. *Trigonometric Series* Vols. I and II, 3rd edn. Cambridge University Press, 2002.

Author Index

Index